CONTROL SYSTEMS DESIGN 2003
(CSD '03)

A Proceedings volume from the 2nd IFAC Conference,
Bratislava, Slovak Republic, 7 – 10 September 2003

Edited by

Š. KOZÁK and M. HUBA

Department of Automatic Control Systems,
Faculty of Electrical Engineering and Information Technology,
Bratislava, Slovak Republic

Published for the

INTERNATIONAL FEDERATION OF AUTOMATIC CONTROL

by

ELSEVIER LTD

ELSEVIER Ltd
The Boulevard, Langford Lane
Kidlington, Oxford OX5 1GB, UK

Elsevier Internet Homepage
http://www.elsevier.com

Consult the Elsevier Homepage for full catalogue information on all books, journals and electronic products and services.

IFAC Publications Internet Homepage
http://www.elsevier.com/locate/ifac

Consult the IFAC Publications Homepage for full details on the preparation of IFAC meeting papers, published/forthcoming IFAC books, and information about the IFAC Journals and affiliated journals.

First edition 2004

Library of Congress Cataloging in Publication Data

A catalogue record for this book is available from the Library of Congress

British Library Cataloguing in Publication Data

A catalogue record for this book is available from the British Library

ISBN 0-08-044175 0
ISSN 1474-6670

Transferred to digital print, 2008
Printed and bound by CPI Antony Rowe, Eastbourne

To Contact the Publisher

Elsevier welcomes enquiries concerning publishing proposals: books, journal special issues, conference proceedings, etc. All formats and media can be considered. Should you have a publishing proposal you wish to discuss, please contact, without obligation, the publisher responsible for Elsevier's industrial and control engineering publishing programme:

Christopher Greenwell
Publishing Editor
Elsevier Ltd
The Boulevard, Langford Lane
Kidlington, Oxford
OX5 1GB, UK

Phone: +44 1865 843230
Fax: +44 1865 843920
E.mail: c.greenwell@elsevier.com

General enquiries, including placing orders, should be directed to Elsevier's Regional Sales Offices – please access the Elsevier homepage for full contact details (homepage details at the top of this page).

2nd IFAC CONFERENCE ON CONTROL SYSTEMS DESIGN 2003

Sponsored by
International Federation of Automatic Control (IFAC)
Technical Committee on Control Design

Organized by
Slovak Society for Cybernetics and Informatics
Faculty of Electrical Engineering and Information Technology

PREFACE

The 2nd IFAC Conference "Control Systems Design" (CSD'03) held on September 7 – 10, 2003 in Bratislava, Slovak Republic, was organized jointly by the Slovak Society for Cybernetics and Informatics which is the Slovak IFAC NMO, and the Faculty of Electrical Engineering and Information Technology, and sponsored by the IFAC Technical Committee "Control Design".

Being actually the fourth in the series of the IFAC technical events on Control Systems Design organized in the Slovak Republic (preceded by the two IFAC Workshops "New Trends in Design of Control Systems" held in 1994, 1997 in Smolenice and the 1st IFAC Conference "Control Systems Design" held in 2000 in Bratislava) the Conference outlined present developments and novel directions in Control Engineering as well as their various control applications.

The aim of the Conference was to provide a forum for professionals, theorists, engineers, researchers and students involved in the broad field of Control Engineering allowing to integrate and exchange their ideas, knowledge, experience and results in various fields of control system design and to explore present theoretical developments as well as their application to engineering problems.

The broad and interdisciplinary scope of the conference covered the following topics:
PID Control, Model Predictive Control, Adaptive and Selftuning Control, Robust Control, Intelligent Control (including Neural, Fuzzy and Genetic Control), DEDS Control and *Control Applications*.

Attendees from 24 countries presented 91 papers in total. The papers were organized in 1 Plenary Session, 6 regular sessions, and 1 Invited Session.

We believe that within the technical program and the rich social program the Conference provided to its participants many opportunities to discuss topical issues as well as to establish fruitful professional and personal contacts; according to feedback received from its attendees, the 2nd IFAC Conference "Control Systems Design" succeeded in doing so and we hope to repeat it in three years again.

Finally, we would like to express our gratitude and appreciation to the IFAC, especially to its TC "Control Design" and to the IFAC Secretariat for the permanent support and assistance during the preparation and organization of the Conference.

Štefan Kozák
Chairman IPC

Mikuláš Huba
Chairman NOC

CONTENTS

PLENARY PAPERS

PID CONTROL SYSTEM DESIGN

INTELLIGENT CONTROL SYSTEM DESIGN

DEDS CONTROL SYSTEM DESIGN

CONTROL SYSTEM DESIGN I

CONTROL SYSTEM DESIGN II

ROBUST CONTROL SYSTEM DESIGN

CONTROL SYSTEM APPLICATIONS

INVITED PAPERS

Advances in Model Predictive Control

STABILIZING MODEL PREDICTIVE CONTROL OF NONLINEAR CONTINUOUS TIME SYSTEMS

L. Magni and R. Scattolini *,[1]

* *Dipartimento di Informatica e Sistemistica, Università di Pavia, via Ferrata 1, 27100 Pavia, Italy*
e-mail: {lalo.magni, riccardo.scattolini}@unipv.it
WEB: http://sisdin.unipv.it/lab/

Abstract: This paper surveys some of the main design strategies of nonlinear MPC (Model Predictive Control). The system under control, the performance index to be minimized and the state and control constraints to be fulfilled are defined in the continuous time. The considered algorithms are analyzed and compared in terms of stability, performance and implementation issues. In particular, it is shown that the solution of the optimization problem underlying the MPC formulation calls for (a) a suitable parametrization of the control variable, (b) the use of a suitable discretization of time, that is of a "sampled" control law and, (c) the numerical integration of the system over the considered prediction horizon. In turn, these implementation aspects are such that many theoretical results concerning stability have to be critically evaluated. In order to cope with these problems, two different methods guaranteeing stability are presented. One of them is used to global stabilize a pendulum. *Copyright © 2003 IFAC*

Keywords: Model Predictive Control, Nonlinear Continuous time systems, Sampled data systems

1. INTRODUCTION

The extraordinary industrial success of Model Predictive Control (MPC) techniques based on linear plant models, see e.g. the survey paper of (Qin and Badgwell, 1996), motivates the development of MPC algorithms for nonlinear systems. Nowadays there are many theoretical results, see (Mayne *et al.*, 2000) and (Magni, 2003), as well as industrial applications, see (Qin and Badgwell, 2000), which witness that MPC for nonlinear systems is going to have a diffusion and popularity similar to the one achieved by MPC algorithms for linear systems.

MPC methods for nonlinear systems are developed by assuming that the plant under control is either described by a continuous time model, see (Mayne and Michalska, 1990), (Michalska and Mayne, 1993), (Chen and Allgöwer, 1998), (Magni and Sepulchre, 1997), (Jadbabaie and Hauser, 2001), (Jadbabaie *et al.*, 2001), or by a discrete time one, see (Keerthi and Gilbert, 1988), (De Nicolao *et al.*, 1998), (Magni *et al.*, 2001). A continuous time representation is much more natural, since the plant model is usually derived by resorting to first principles equations, but it results in a more difficult development of the MPC control law, which in principle calls for the solution of a functional optimization problem. As a matter of fact, the performance index to be minimized is defined in a continuous time setting and the overall optimization procedure is assumed to be continuously repeated after any vanishingly small sampling time, which often turns out to be a computationally intractable task. On the contrary, MPC algorithms based on a discrete time system representation are computationally simpler, but require the discretization of the model equations, so that they rely from the very beginning on an approximate system representation. Moreover,

[1] The authors acknowledge the partial financial support by $MURST$ Project "New techniques for the identification and adaptive control of industrial systems"

the performance index to be minimized as well as the state constraints only consider the system behavior in the sampling instants, so ignoring the intersample behavior, which in some cases could be significant in the evaluation of the control performance.

In this paper, the hybrid nature of sampled data control systems is fully considered. The plant under control, the state and control constraints and the performance index to be minimized are described in continuous time, while the manipulated signals are allowed to change at fixed and uniformly distributed sampling times. It is shown that a proper choice of the terminal penalty and terminal inequality constraint is necessary in order to guarantee closed stability. Among the other possibilities, the stabilizing methods proposed in (Keerthi and Gilbert, 1988), (Mayne and Michalska, 1990), (Michalska and Mayne, 1993), (Chen and Allgöwer, 1998), (De Nicolao *et al.*, 1998), (Magni *et al.*, 2001) are presented and compared in term of performance, enlargement properties and computational issues. In the second part of the paper, the effects of two approximation which must be introduced in the numerical solution of the optimization problem are considered: namely the parametrization of the input profile and the numerical integration of the system over the prediction horizon. Irrespective of the adopted algorithm, the parametrization of the input profile is required to limit the number of the optimization variables. In turn it forces the use of a congruent auxiliary control law, that is of a control law producing a signal compatible with the adopted parametrization. As for the numerical integration over the future prediction horizon, it must be performed at any sampling time for the on line solution of the optimization problem. However, in so doing the state constraints, originally posed in the continuous time, can only be checked at the integration time instants. Although the integration step can be definitely smaller than the sampling time, this means that a-priori there is not any guarantee that these state constraints are fulfilled everywhere. When these approximations are considered, the MPC algorithms previously considered do not guarantee stability and constraints satisfaction, therefore two different non trivial schemes are suggested in order to recover closed-loop stability (Magni *et al.*, 2002), (Magni and Scattolini, 2002). One of them is used in the final section of the paper to globally stabilize a pendulum as well as to improve the performance and the stability region of the nonlinear Energy-control proposed in (Åström and Furuta, 2000).

2. PROBLEM STATEMENT AND PRELIMINARY RESULTS

Consider a plant P described by the nonlinear continuous-time dynamic system

$$\dot{x}(t) = f(x(t), u(t)),\ t \geq 0,\ x(0) = x_0 \quad (1)$$

where $x \in R^n$ is the state, $u \in R^m$ is the input, $f(0,0) = 0$ and $f(\cdot,\cdot)$ is a C^1 function of its arguments. The state and control variables are restricted to fulfill the following constraints

$$x(t) \in X, \quad u(t) \in U, \quad t \geq 0 \quad (2)$$

where X and U are compact subsets of R^n and R^m, respectively, both containing the origin as an interior point. The solution of (1) from the initial time $\bar{t}$ and initial state $x(\bar{t})$ for a control signal $u(\cdot)$ is denoted by $\varphi(t, \bar{t}, x(\bar{t}), u(\cdot))$.

Define by T_s a suitable sampling period and let $t_k = kT_s$, k nonnegative integer, be the sampling instants; the goal is to determine a "sampled" feedback control law

$$u(t) \equiv \kappa(t, x(t_k)),\ \kappa(t,0) = 0,\ t \in [t_k, t_{k+1}) \quad (3)$$

which asymptotically stabilizes the origin of the associated closed-loop system.

The description of the hold mechanism implicit in (3) calls for a state augmentation. Letting $x_c := [x' \ x_1']' \in R^{2n}$, the closed loop system (1)-(3) is

$$\dot{x}_c(t) = \begin{bmatrix} f(x(t), \kappa(t, x_1(t))) \\ 0_{n,1} \end{bmatrix},\ t \in [t_k, t_{k+1}) \quad (4)$$
$$x_c(t_k) = \begin{bmatrix} x(t_k^-) \\ x(t_k^-) \end{bmatrix}$$

and its solution from the initial time $\bar{t}$ and initial state $x_c(\bar{t})$ is denoted by

$$\varphi_c(t, \bar{t}, x_c(\bar{t})) = \begin{bmatrix} \varphi_c^x(t, \bar{t}, x_c(\bar{t})) \\ \varphi_c^{x_1}(t, \bar{t}, x_c(\bar{t})) \end{bmatrix}$$
$$\varphi_c^x \in R^n,\ \varphi_c^{x_1} \in R^n$$

With reference to the closed-loop system (4), define the following sets.

Definition 1. The *sampled output admissible set* associated to (4) is a set $X_k^c(\kappa) \in R^n$ such that for all $x \in X_k^c(\kappa)$, $\varphi_c^x(t_{k+1}, t_k, [x' \ x']') \in X_k^c(\kappa)$, $\varphi_c^x(t, t_k, [x' \ x']') \in X$, $\kappa(t, \varphi_c^{x_1}(\tau, t, x_c)) \in U$, $t \in [t_k, t_{k+1})$, $\lim_{t \to \infty} \|\varphi_c^x(t, \bar{t}, x_c(\bar{t})))\| = 0$. In other words, $X_k^c(\kappa)$ is a state invariant set, associated to the closed loop system (4) and defined at the sampling instants t_k, such that (i) the state and control constraints (2) are satisfied in all the future continuous time instants, (ii) the regulation problem is asymptotically solved.

Definition 2. The *output admissible set* associated to (4) is a set $X^c(t, \kappa) \in R^{2n}$ such that for all $x_c \in X^c(t, \kappa)$, $\varphi_c^x(t_k, t, x_c) \in X_k^c(\kappa)$, where t_k is the closest sampling time in the future, $\varphi_c^x(\tau, t, x_c) \in X$, $\kappa(t, \varphi_c^{x_1}(\tau, t, x_c)) \in U$, $\tau \in [t, t_k)$. In other words, $X^c(t, \kappa)$ is a set, defined at any continuous time instant t, of states of the closed-loop system (4) such that (i) the state of (1) at the closest sampling time in the future belongs to $X_k^c(\kappa)$ and (ii) the state and control constraints (2) are satisfied in all the future continuous time instants.

The regulation problem can now be formally stated as the problem of finding a sampled control law (3) such that its output admissible set is non empty. Such a control law will be called *feasible* hereafter. Besides, one can also wish to find the control law (3) with the largest output admissible set X^c and which minimizes the IH (Infinite-Horizon) cost function

$$J_{IH}(x(t_k), u(\cdot)) = \int_{t_k}^{\infty} \left\{ \|x(\tau)\|_Q^2 + \|u(\tau)\|_R^2 \right\} d\tau \tag{5}$$

subject to (1) and (2). In (5) Q and R are positive definite weighting matrices. Let X^{IH} be the set of states x such that the IH problem is solvable, $u^{IH}(\cdot)$ the optimal solution and $\kappa^{IH}(t, x(t_k)) = u^{IH}(t)$, $t \in [t_k, t_{k+1})$ the optimal IH control law, so X^{IH} is an output admissible set for (4) with $\kappa(\cdot,\cdot) = \kappa^{IH}(\cdot,\cdot)$. In general, the IH nonlinear optimal control problem is computationally intractable since minimization must be performed with respect to functions. Nevertheless, it constitutes a touchstone for suboptimal approaches yielding to nonlinear control laws $\kappa(t, x(t_k))$ with the following properties:

(i) *performance/complexity trade-off:* by suitably tuning the design parameters of the control synthesis algorithm, it should be possible to obtain a fair compromise between an arbitrarily good approximation of the optimal (but computationally intractable) IH controller κ^{IH} and a computationally cheap feasible control law;

(ii) *enlargement property*: the output admissible set associated with $\kappa(\cdot,\cdot)$ should be larger than the output admissible set of a possible already known feasible control law.

An effective strategy to design suboptimal controllers based on the MPC strategy will be presented in the following Section.

3. MODEL PREDICTIVE CONTROL

In order to introduce the MPC algorithm, a finite-horizon optimization problem is first defined. Let $u_{t_1,t_2} : [t_1, t_2] \rightarrow R^m$ be a finite time control signal.

Finite Horizon Optimal Control Problem ($FHOCP^1$). Given X_f, a compact subset of R^n containing the origin, a sampling time T_s, a prediction horizon N_p, two positive definite matrices Q and R, a penalty function $V_f(\cdot) : R^n \rightarrow R$, at every sampling time instant t_k, minimize, with respect to $u_{t_k,t_{k+N_p}}$, the performance index

$$\begin{aligned} &J_{FH}(x_{t_k}, u_{t_k,t_{k+N_p}}, N_p) \\ &= \int_{t_k}^{t_{k+N_p}} \left\{ \|x(\tau)\|_Q^2 + \|u(\tau)\|_R^2 \right\} d\tau \\ &+ V_f(\varphi(t_{k+N_p}, t_k, x(t_k), u_{t_k,t_{k+N_p}})) \end{aligned} \tag{6}$$

The minimization of (6) must be performed under the following constraints:

(i) the state dynamics (1) with $x(t_k) = x_{t_k}$;
(ii) the constraints (2), $t \in [t_k, t_{k+N_p})$;
(iii) the terminal state constraint $x(t_{k+N_p}) \in X_f$. ■

According to the *Receding Horizon* approach, the state-feedback MPC control law is derived by solving $FHOCP^1$ at every sampling time instant t_k, and applying the control signal $u(t) = u^o_{t_k,t_{k+1}}$, $t \in [t_k, t_{k+1})$ where $u^o_{t_k,t_{k+1}}$ is the first part of the optimal signal $u^o_{t_k,t_{k+N_p}}$. In so doing, one implicitly defines the sampled state-feedback control law

$$u(t) = \kappa^{RH}(t, x(t_k)) \quad , \quad t \in [t_k, t_{k+1}) \tag{7}$$

In order to establish the properties of the control law (7), first let

$$\varphi^{RH}(t, \bar{t}, x_c(\bar{t})) = \begin{bmatrix} \varphi_x^{RH}(t, \bar{t}, x_c(\bar{t})) \\ \varphi_{x_1}^{RH}(t, \bar{t}, x_c(\bar{t})) \end{bmatrix}$$

$$\varphi_x^{RH} \in R^n, \quad \varphi_{x_1}^{RH} \in R^n$$

be the solution of (4) with $\kappa(\cdot,\cdot) = \kappa^{RH}(\cdot,\cdot)$. Then, define the following sets.

Definition 3. Let $X_k^0(N_p) \in R^n$ be the set of states x_{t_k} of system (1) at the sampling times t_k such that there exists a feasible control sequence $u_{t_k,t_{k+N_p}}$ for $FHOCP^1$.

Definition 4. Let $X^0(t, N_p) \in R^{2n}$ be the set of states x_c such that for all $x_c(t) \in X^0(t, N_p)$, $\varphi_x^{RH}(t_k, t, x_c) \in X_k^0(N_p)$, where t_k is the closest sampling time in the future, $\varphi_x^{RH}(\tau, t, x_c) \in X$, $\kappa^{RH}(\tau, \varphi_{x_1}^{RH}(\tau, t, x_c)) \in U$, $\tau \in [t, t_k)$.

In order to guarantee the stability of the MPC closed-loop system, the terminal set X_f and the terminal cost function V_f introduced in the $FHOCP^1$ must be properly chosen.

Assumption 1. There exist an auxiliary control law $\kappa_f(x)$, a terminal set X_f and a terminal penalty V_f such that, letting $\varphi_f(t, \bar{t}, x(\bar{t}))$ the solution of the closed loop system

$$\dot{x}(t) = f(x(t), \kappa_f(x(t))), \tag{8}$$

from the initial time $\bar{t}$ and initial state $x(\bar{t})$, the following conditions hold:

- $X_f \subset X$, X_f closed, $0 \in X_f$;

- $\kappa_f(x) \in U, \forall x \in X_f$;
- X_f is positively invariant for (8);
- $V_f(\cdot) : R^n \to R$ is such that $\forall x(t_k) \in X_f$

$$V_f(\varphi_f(t_{k+1}, t_k, x(t_k))) - V_f(x(t_k)) \quad (9)$$
$$\leq - \int_{t_k}^{t_{k+1}} \Big\{ \|\varphi_f(\tau, t_k, x(t_k))\|_Q^2$$
$$+ \|\kappa_f(\varphi_f(\tau, t_k, x(t_k)))\|_R^2 \Big\} d\tau$$

■

Note that in Assumption 1, at this stage the auxiliary control law $\kappa_f(x)$ is not required to be a "sampled" control law because, as it will be clarified below in the description of some well known MPC algorithms, it is never applied to the systems but it is only used in simulation in order to obtain the terminal set and the terminal penalty.

Theorem 2. Given an auxiliary control law κ_f, a terminal set X_f and a terminal penalty V_f satisfying Assumption 1:

(i) the origin is an asymptotically stable equilibrium point for the closed-loop system formed by (1) and (7) with output admissible set $X^0(t, N_p)$;
(ii) $X_k^0(N_p + 1) \supseteq X_k^0(N_p), \forall N_p$;
(iii) $X_k^0(N_p) \supseteq X_f, \forall N_p$;
(iv) there exist a finite $\bar{N}_p$ such that $X_k^0(\bar{N}_p) \supseteq \bar{X}_f$, where $\bar{X}_f \subset X$ is any closed positive invariant set for (8) such that $\kappa_f(x) \in U, \forall x \in \bar{X}_f$.

Proof of Theorem 2: In view of Definitions 1-4, if $X_k^0(N_p)$ is a sampled output admissible set of (4) with $\kappa(\cdot,\cdot) = \kappa^{RH}(\cdot,\cdot)$ then $X^0(t, N_p)$ is an output admissible set of (4) with $\kappa(\cdot,\cdot) = \kappa^{RH}(\cdot,\cdot)$. Moreover by Assumption 1 it follows that X_f is non-empty, then also $X^0(t, N_p)$ is non-empty. Let now show that $X_k^0(N_p)$ is a sampled output admissible set for system (1) and (7). In fact letting $x(t_k) := x_{t_k} \in X_k^0(N_p)$ and the associated solution $u^o_{t_k,t_{k+N_p}}$ of the $FHOCP^1$ at time t_k, a feasible solution at time t_{k+1} for the $FHOCP^1$ is

$$\tilde{u}_{t_{k+1},t_{k+N_p+1}} \quad (10)$$
$$:= \begin{cases} u^o_{t_{k+1},t_{k+N_p}} & t \in [t_{k+1}, t_{k+N_p}) \\ \kappa_f(\varphi_f(t, t_{k+N_p}, \xi^{N_p}) & t \in [t_{k+N_p}, t_{k+N_p+1}) \end{cases}$$

with $\xi^{N_p} := \varphi(t_{k+N_p}, t_k, x_{t_k}, u^o_{t_k,t_{k+N_p}})$. Then, by definition, $\xi := \varphi_x^{RH}(t_{k+1}, t_k, [x'_{t_k}\ x'_{t_k}]') \in X_k^0(N_p)$ and, in view of constraints (ii) of the $FHOCP^1$, (2) are satisfied along the trajectory of (4) with $\kappa(\cdot,\cdot) = \kappa^{RH}(\cdot,\cdot)$.

Let now show that the origin is an asymptotically stable equilibrium point for the closed-loop system (1), (7). To this end define

$$V(x_c(t), t) := J_{FH}(x(t), u^o_{t_k,t_{k+N_p}}, N_p)$$

if $t = t_k$ and

$$V(x_c(t), t)$$
$$:= \int_t^{t_{k+1}} \Big\{ \left\|\varphi_x^{RH}(\tau, t_k, x_c(t_k))\right\|_Q^2 +$$
$$+ \left\|\kappa^{RH}(\tau, \varphi_{x_1}^{RH}(\tau, t_k, x_c(t_k)))\right\|_R^2 \Big\} d\tau$$
$$+ J_{FH}(\varphi_x^{RH}(t_{k+1}, t, x_c(t)), u^o_{t_{k+1},t_{k+N_p}}, N_p - 1)$$

if $t \in (t_k, t_{k+1})$. Note that $V(x_c(t_k), t_k)$ is bounded $\forall x \in X_k^o(N_p)$. Moreover

- $\forall t \in [t_k, t_{k+1})$

$$V(\varphi^{RH}(t, t_k, x_c(t_k)), t)$$
$$= V(x_c(t_k), t_k) - \int_{t_k}^{t} \Big\{ \left\|\varphi_x^{RH}(\tau, t_k, x_c(t_k))\right\|_Q^2$$
$$+ \left\|\kappa^{RH}(\tau, \varphi_{x_1}^{RH}(\tau, t_k, x_c(t_k)))\right\|_R^2 \Big\} d\tau \quad (11)$$

- At time $t = t_{k+1}$, $\tilde{u}_{t_{k+1},t_{k+N_p+1}}$ given by (10) is a (sub-optimal) feasible solution for the new $FHOCP^1$ so that

$$V(\varphi^{RH}(t_{k+1}, t_k, x_c(t_k)), t_{k+1})$$
$$\leq J_{FH}(\varphi_x^{RH}(t_{k+1}, t_k, x_c(t_k)), \tilde{u}_{t_{k+1},t_{k+N_p+1}} N_p)$$
$$= V(\varphi^{RH}(t_{k+1}^-, t_k, x_c(t_k)), t_{k+1}^-)$$
$$+ \int_{t_{k+N_p}}^{t_{k+N_p+1}} \Big\{ \left\|\varphi_c^x(\tau, t_{k+N_p}, \xi^{N_p})\right\|_Q^2$$
$$+ \left\|\kappa_f(\tau, \varphi_f(\tau, t_{k+N_p}, \xi^{N_p}))\right\|_R^2 \Big\} d\tau$$
$$+ V_f(\varphi_f(t_{k+N_p+1}, t_{k+N_p}, \xi^{N_p})) - V_f(\xi^{N_p})$$
$$\leq V(\varphi^{RH}(t_{k+1}^-, t_k, x_c(t_k)), t_{k+1}^-) \quad (12)$$

In conclusion using (11) for $t \in [t_{k+i}, t_{k+i+1})$, $i \geq 0$, and (12) for $t = t_{k+i+1}$, $i \geq 0$, $\forall t \geq t_k$

$$V(\varphi^{RH}(t, t_k, x_c(t_k)), t)$$
$$+ \int_{t_k}^{t} \Big\{ \left\|\varphi_x^{RH}(\tau, t_k, x_c(t_k))\right\|_Q^2$$
$$+ \left\|\kappa^{RH}(\tau, \varphi_{x_1}^{RH}(\tau, t_k, x_c(t_k)))\right\|_R^2 \Big\} d\tau$$
$$\leq V(x_c(t_k), t_k)$$

and, since Q and R are positive definite matrices, if $x_c(t_k) \in X^0(t_k, N_p)$, both $V(\varphi^{RH}(t, t_k, x_c(t_k)), t)$ and

$$\int_{t_k}^{t} \Big\{ \left\|\varphi_x^{RH}(\tau, t_k, x_c(t_k))\right\|_Q^2 +$$
$$+ \left\|\kappa^{RH}(\tau, \varphi_{x_1}^{RH}(\tau, t_k, x_c(t_k)))\right\|_R^2 \Big\} d\tau$$

are bounded. These facts prove that

$$lim_{t\to\infty}\varphi_x^{RH}(t, t_k, x_c(t_k))) = 0$$

(Michalska and Vinter, 1994).

The proof of $(ii) - (iv)$ can be derived as in Theorem 6 in (Magni *et al.*, 2001).

4. STABILIZING MPC CONTROL ALGORITHM

Many stabilizing MPC algorithms can be obtained depending on the choices made to satisfy Assumption 1. Here, the main algorithms proposed in the literature are briefly described.

4.1 *Terminal equality constraint (EC)*

The first algorithm presented in literature is characterized by a terminal equality constraint $x(t_{k+N_p}) = 0$ so that $X_f = \{0\}$ (Keerthi and Gilbert, 1988), (Mayne and Michalska, 1990). The auxiliary control law and the terminal penalty are defined only in the origin so that the following trivial functions can be chosen: $V_f(x) \equiv 0$ and $\kappa_f(x) \equiv 0$. The fulfillment of Assumption 1 is easily checked. In fact $X_f = \{0\} \in X$, $\kappa_f(0) = 0 \in U$, $f(0, \kappa_f(0)) = 0$ so that X_f is positively invariant and $V_f(0) - V_f(0) \leq 0$.

Performance/complexity trade-off If a short optimization horizon N_p is used, the terminal constraint forces an excessive control effort, while increasing the optimization horizon increases the computational load.

Enlargement property The output admissible set coincides with the (constrained) controllability region $X^{con}(N_p)$ (Gilbert and Tan, 1991). Note that $X^{con}(N_p)$ may be "small". In particular, there is no guarantee that $X^c(N_p)$ is larger than the output admissible set guaranteed by the trivial control law $\kappa(x) \equiv 0$.

4.2 *Quadratic terminal penalty (QP)*

A second well known method was presented in (Chen and Allgöwer, 1998) where a quadratic terminal penalty and a linear auxiliary control law are considered. More precisely, assume that system (1) is linearizable and denote the linearized matrices with

$$A = \left.\frac{\partial f}{\partial x}\right|_{x=0,u=0}, B = \left.\frac{\partial f}{\partial u}\right|_{x=0,u=0}.$$

The auxiliary control law is given by $\kappa_f(x) = Kx$ where K is such that $A_{cl} = A + BK$ is Hurwitz. The terminal penalty is a quadratic function $V_f(x) = x'Px$ where P is the solution of the following Lyapunov function

$$(A_{cl} + \kappa_\varepsilon I)'P + P(A_{cl} + \kappa_\varepsilon I) = \bar{Q} \quad (13)$$

where $\bar{Q} = Q + K'RK$ and the scalar $\kappa_\varepsilon \in [0, \infty)$ satisfies

$$\kappa_\varepsilon < -\lambda_{\max}(A_{cl}).$$

The terminal region is defined as a level set of the terminal penalty.

$$X_f := \{x \in R^n | x'Px \leq \alpha\} \subset X \quad (14)$$

such that:

(i) $Kx \in U$, for all $x \in X_f$;
(ii) X_f is a positively invariant for the closed-loop systems with $u = Kx$;
(iii) $\forall x \in X_f$

$$\frac{d}{dt}x'Px \leq -x'(Q + K'RK)x$$

subject to the closed-loop dynamic with $u = Kx$.

Satisfaction of Assumption 1 is easily checked. In fact $X_f \subset X$, X_f closed, $0 \in X_f$ in view of (14); $\kappa_f(x) \in U, \forall x \in X_f$ from (i); X_f is positively invariant for the closed-loop system (8) from (ii); finally $\forall x(t_k) \in X_f$

$$\begin{aligned} & V_f(\varphi_f(t_{k+1}, t_k, x(t_k))) - V_f(x(t_k)) \\ &= \int_{t_k}^{t_{k+1}} \|\varphi_f(\tau, t_k, x(t_k))\|_P^2 \, d\tau \\ &\leq - \int_{t_k}^{t_{k+1}} \|\varphi_f(\tau, t_k, x(t_k))\|_{Q+K'RK}^2 \, d\tau \\ &= - \int_{t_k}^{t_{k+1}} \left\{ \|\varphi_f(\tau, t_k, x(t_k))\|_Q^2 \right. \\ & \left. + \|\kappa_f(\varphi_f(\tau, t_k, x(t_k)))\|_R^2 \right\} d\tau \end{aligned}$$

Performance/complexity trade-off In view of the constant κ_ε introduced in the Lyapunov equation (13), the infinite horizon optimal performance are not recovered even if the auxiliary control law is locally optimal and $\|x\| \to 0$. The infinite horizon optimal control law can be reached only at the cost of a "long" prediction (optimization) horizon N_p.

Enlargement property For any horizon N_p the output admissible set includes the terminal set X_f, but a sufficient long optimization horizon N_p occurs in order to enlarge the maximum output admissible set of the auxiliary control law.

4.3 *Infinite-horizon closed-loop costing (CL)*

A third method was presented in (De Nicolao *et al.*, 1998) where the infinite horizon cost associated with a generally nonlinear auxiliary control law is used as the terminal penalty. More precisely, assume that an auxiliary locally stabilizing control law is given by $\kappa_f(x)$. The terminal penalty is given by

$$V_f(x(t_k)) = \int_{t_k}^{\infty} \left\{ \|\varphi_f(\tau, t_k, x(t_k))\|_Q^2 + \|\kappa_f(\varphi_f(\tau, t_k, x(t_k)))\|_R^2 \right\} d\tau$$

The terminal region is implicitly defined as

$$X_f := \{\bar{x} \in R^n | \varphi_f(t, t_k, x(t_k)) \in X, \quad \kappa_f(\varphi_f(t, t_k, x(t_k))) \in U, t > \bar{t}, \quad V_f(\bar{x}) \text{ bounded}\} \subset X \tag{15}$$

The fulfillment of Assumption 1 is easily checked. In fact $X_f \subset X$, X_f closed, $0 \in X_f$, $\kappa_f(x) \in U$, $\forall x \in X_f$ in view of (15); X_f is positively invariant for (8) because, from the definition of X_f, it follows that if $\bar{x} \in X_f$, $\varphi_f(t, t_k, x(t_k) \in X_f, t > \bar{t}$; finally $\forall x(t_k) \in X_f$

$$V_f(x(t_{k+1})) - V_f(x(t_k)) = - \int_{t_k}^{t_{k+1}} \left\{ \|\varphi_f(\tau, t_k, x(t_k))\|_Q^2 + \|\kappa_f(\varphi_f(\tau, t_k, x(t_k)))\|_R^2 \right\} d\tau$$

Performance/complexity trade-off If the auxiliary control law is locally optimal, then also the MPC control law is locally optimal. If the auxiliary control law is the solution of the unconstrained nonlinear infinite horizon control problem, then the MPC control law is the solution of the constrained infinite horizon optimal control problem.

Enlargement property The output admissible set guaranteed by this control scheme is larger than the maximum output admissible set of the auxiliary control law for any optimization horizon N_p. In fact X_f is equal to the maximum output admissible set of the auxiliary control law. Note that in the QP algorithm the terminal region X_f must be explicitly computed off-line, while in this one the explicit computation of X_f is not needed. This difference is crucial in view of the difficulty to compute an output admissible set for a nonlinear system with constraints, so that only very conservative approximations can be obtained.

4.4 *Infinite-horizon closed-loop costing with control and prediction horizons (CL-2H)*

The CL algorithm achieves better properties with respect to EC and QP both for performance and for the enlargement property with a lower computational burden. Its drawback is that the terminal penalty cannot be computed exactly because the integration of the closed-loop system (8) should be performed for an infinite time. However, since the auxiliary control law is stabilizing, in practice it is possible to compute the cost function until the state is sufficiently close to the origin. In (Magni *et al.*, 2001) this problem is analyzed and a rule to compute a finite integration horizon that preserves closed-loop stability is given. Otherwise, it is necessary to guarantee that the state is in the output admissible set of the auxiliary control law. Consequently, an output admissible set for the auxiliary control law must be computed off line as in the QP algorithm. In order to recover the properties of the CL algorithm without increasing the computational burden, the use of a control (optimization) horizon shorter than the prediction horizon was proposed in (Magni *et al.*, 2001). In particular the terminal penalty and the terminal inequality constraint are imposed at the end of the prediction horizon, while the optimization is performed only with respect to a shorter horizon, the so called control horizon. The control signal from the end of the control horizon to the end of the prediction horizon is given by the auxiliary control law. Accordingly, the optimization problem is changed in the following way.

Finite Horizon Optimal Control Problem ($FHOCP^2$). Given X_f, a compact subset of R^n containing the origin, the sampling time T_s, the control horizon N_c, the prediction horizon N_p, an auxiliary control law $u = \kappa(x)$, two positive definite matrices Q and R, a penalty function $V_f(\cdot) : R^n \to R$, at every sampling time instant t_k, minimize, with respect to $u_{t_k, t_{k+N_c}}$, the performance index

$$J_{FH}(x_{t_k}, u_{t_k, t_{k+N_c}}, N_c, N_p) = \int_{t_k}^{t_{k+N_p}} \left\{ \|x(\tau)\|_Q^2 + \|u(\tau)\|_R^2 \right\} d\tau + V_f(\varphi(t_{k+N_p}, t_k, x(t_k), u_{t_k, t_{k+N_p}}))$$

The minimization of (6) must be performed under the following constraints:

(i) the state dynamics (1) with $x(t_k) = x_{t_k}$;

(ii) $$u(t) := \begin{cases} u_{t_k, t_{k+N_c}} & t \in [t_k, t_{k+N_c}) \\ \kappa_f(\varphi_f(t, t_{k+N_c}, \xi^{N_c})) & t \in [t_{k+N_c}, t_{k+N_p}) \end{cases}$$
where $\xi^{N_c} := \varphi(t_{k+N_c}, t_k, x_{t_k}, u_{t_k, t_{k+N_c}})$.

(iii) the constraints (2), with $u(t) = \kappa(x)$, $t \in [t_k, t_{k+N_p})$;

(iv) the terminal state constraint $x(t_{k+N_p}) \in X_f$. ■

Performance/complexity trade-off If the auxiliary control law is locally optimal, then also the MPC control law obtained with $N_p \to \infty$, is locally optimal. If the auxiliary control law is the solution of the unconstrained nonlinear infinite horizon control problem, then the MPC control law obtained with $N_p \to \infty$, is the solution of the constrained infinite horizon optimal control problem.

Enlargement property The output admissible set guaranteed by this control scheme is larger than the maximum output admissible set of the auxiliary control law for any optimization horizon N_c and for a sufficiently large prediction horizon N_p. Note that the computational burden is mostly related the length of the control horizon.

5. IMPLEMENTATION AND NOMINAL STABILITY

In view of Theorem 2, the EC, QP, CL and $CL-2H$ MPC algorithms surveyed in the previous Section guarantee nominal stability provided that the underlying optimization problem is efficiently solved on line. As a matter of fact, this is not possible in practice for a couple of reasons, namely the requirement to fulfill the state and control constraints at any continuous time instant and the necessity to use a suitable parametrization of the input signal. As for the first issue, it is apparent that the numerical integration of the system over the future prediction horizon is such that the state (and control) constraints can be verified only at the integration time instants. However, in the original problem formulation they are required to be fulfilled at any continuous time t. The second problem is even more intriguing and can be explained as follows: the on line optimization problem underlying the MPC algorithms can be completed in the sampling period only with respect to a finite number of parameters, instead of functions, so that the input profile must be parametrized. For example, a typical procedure consists in assuming that the input signal is held constant between two successive sampling instants. In turn, this means that also the adopted auxiliary control law must produce a congruent signal, that is a signal that satisfies the adopted parametrization, otherwise the signal $\tilde{u}_{t_{k+1},t_{k+N_p+1}}$ computed at t_k and given in (10) would not be a feasible solution for the (parametrized) $FHOCP$ at time t_{k+1}. With respect to this problem, note that the EC method (where the auxiliary control law is obtained by setting equal to zero the control variable) cannot be solved in a finite number of iterations in view of the zero terminal equality constraints, while in all the other MPC algorithms the stability proof relies on an auxiliary control law providing a general continuous time control signal, a priori not congruent with the adopted parametrization.

In the following, the two implementation issues above discussed are analyzed in a reverse order. Specifically, two algorithms are first presented guaranteeing closed-loop stability when a piece-wise constant control parametrization is used. Then, the problem of the fulfillment of (2) is solved by specifying the (more restrictive) constraints to be imposed at any integration time during the on line optimization.

5.1 *The piece-wise constant MPC control law*

In (Magni and Scattolini, 2002) a piece-wise constant signal parametrization has been considered in the numerical solution of the optimization problem. Specifically, it has been suggested to use a time invariant auxiliary control law

$$u(t) \equiv \kappa(x(t_k)) \quad , t \in [t_k, t_{k+1}) \,, \tag{16}$$

with $\kappa(0) = 0$, satisfying the following assumption.

Assumption 3. The feasible control law (16) is a C^1 function with Lipschitz constant L_κ.

For control law (16), an associated sampled output admissible set can be computed as follows. First, define the linearization of system (1) at the origin

$$\dot{x}(t) = Ax(t) + Bu(t), \tag{17}$$

Then introduce the discretization of (17) given by

$$x(t_{k+1}) = A_D x(t_k) + B_D u(t_k), \; x(0) = x_0 \tag{18}$$

with

$$A_D := e^{AT_s}, \quad B_D := \int_0^{T_s} e^{A\eta} B d\eta$$

Finally let

$$K = \left. \frac{\partial \kappa(x)}{\partial x} \right|_0$$

In view of the feasibility of (16), it is then easy to show that the closed-loop matrix $A_D^{cl} := A_D + B_D K$ of the linearized discrete-time system (18) is Hurwitz and the following result holds.

Lemma 4. Let $\kappa(x)$ be a feasible control law, suppose that Assumption 3 is satisfied and consider a positive definite matrix $\tilde{Q}$ and two real positive scalars γ and γ_2 such that $\gamma < \lambda_{\min}(\tilde{Q})$. Define by Π the unique symmetric positive definite solution of the following Lyapunov equation:

$$A_D^{cl\prime} \Pi A_D^{cl} - \Pi + \bar{Q} = 0 \tag{19}$$

where

$$\bar{Q} = \int_0^{T_s} A_c^{ZOH}(\eta)' \tilde{Q} A_c^{ZOH}(\eta) d\eta + \gamma_2 I_n$$

and

$$A_c^{ZOH}(t) := e^{At} + \left(\int_0^t e^{A(t-\tau)} d\tau \right) BK.$$

Then, there exist two constants $T_s \in (0, \infty)$ and $c \in (0, \infty)$ specifying a neighborhood $\Omega_c(\kappa, T_s)$ of the origin of the form

$$\Omega_c(\kappa, T_s) = \left\{ x \in \Re^n \mid \|x\|_\Pi^2 \le c \right\} \quad (20)$$

such that $\forall x \in \Omega_c(\kappa, T_s)$:

(i) $\varphi_c^x(t, t_k, [x' \; x']') \in X, t \in [t_k, t_{k+1}), \kappa(x) \in U$;
(ii)

$$\begin{aligned} & \|\varphi_c^x(t_{k+1}, t_k, [x'x']')\|_\Pi^2 - \|x\|_\Pi^2 \\ & \le -\gamma \int_{t_k}^{t_{k+1}} \|\varphi_c^x(\eta, t_k, [x'x']')\|^2 \, d\eta \\ & -\gamma_2 \|x\|^2 \end{aligned} \quad (21)$$

The Lemma states that, in view of (i) and (ii), $\Omega_c(\kappa, T_s)$ is a sampled output admissible set for (4); moreover, from (ii) $V_L(x) = x'\Pi x$ is a positive definite function decreasing in the sampling times along the trajectory of (4).

Remark 1. An obvious way to determine a feasible sampled control law is to choose a suitable T_s, to consider the linearization of (1) around the origin and the sampled linear model described by (18) and to synthesize with any standard linear control synthesis technique, a linear control law

$$u(t) = Kx(t_k) \quad , t \in [t_k, t_{k+1}) \quad (22)$$

such that $A_D + B_D K$ is Hurwitz.

Assume to know the *auxiliary control law* $u = \kappa(x)$, together with the associated sampled output admissible set and the Lyapunov function both given in Lemma 4. It is now shown how MPC allows one to stabilize the closed-loop system, to extend the output admissible set of κ and to improve the control performance by minimizing a cost function suitably chosen by the designer.

To this end, given a control sequence

$$\bar{u}_{1,N_c}(t_k) := [u_{1_{t_k}}, u_{2_{t_k}}, ..., u_{N_{ct_k}}]$$

with $N_c \ge 1$, define the *Finite Horizon* piece-wise constant control signal

$$u_{t_k}^{FH}(t) = \begin{cases} u_{j_{t_k}} \\ \quad t \in [t_{k+j-1}, t_{k+j}), \\ \quad j = 1, ..., N_c \\ \kappa(\varphi(t_{k+j-1}, t_k, x(t_k), u_{t_k}^{FH}(\cdot))) \\ \quad t \in [t_{k+j-1}, t_{k+j}), \\ \quad j = N_c + 1, ..., N_p \end{cases} \quad (23)$$

where $N_p \ge N_c$. Moreover denote by $\bar{u}_{t_k}^{FH}(t_{fin}, t_{in})$ the signal $u_{t_k}^{FH}(t)$ in the interval $t \in [t_{in}, t_{fin})$.

For system (1) the MPC control problem here considered is based on the solution of the following

Finite Horizon Optimal Control Problem[3] (FHOCP[3]). Given the sampling time T_s, the control horizon N_c, the prediction horizon N_p, $N_c \le N_p$, two positive definite matrices Q and R, a feasible auxiliary control law $\kappa(x)$, the matrix Π and the region $\Omega_c(\kappa, T_s)$ given in Lemma 4 with $\gamma > \lambda_{\max}(Q)$ and $\gamma_2 > T_s \lambda_{\max}(R) L_\kappa$, at every sampling time instant t_k, minimize, with respect to $\bar{u}_{1,N_c}(t_k)$, the performance index

$$\begin{aligned} & J_{FH}(x_{t_k}, \bar{u}_{1,N_c}(t_k), N_c, N_p) \\ & = \int_{t_k}^{t_{k+N_p}} \left\{ \|x(\tau)\|_Q^2 + \|u(\tau)\|_R^2 \right\} d\tau \\ & + V_f(\varphi(t_{k+N_p}, t_k, x(t_k), \bar{u}_{t_k}^{FH}(t_{k+N_p}, t_k))) \end{aligned} \quad (24)$$

where the terminal penalty V_f is selected as

$$V_f(x) = \|x\|_\Pi^2$$

The minimization of (24) must be performed under the following constraints:

(i) the state dynamics (1) with $x(t_k) = x_{t_k}$;
(ii) the constraints (2), $t \in [t_k, t_{k+N_p})$ with u given by (23);
(iii) the terminal state constraint $x(t_{k+N_p}) \in \Omega_c(\kappa, T_s)$. ■

According to the *Receding Horizon* approach, the state-feedback MPC control law is derived by solving $FHOCP^3$ at every sampling time instant t_k, and applying the constant control signal $u(t) = u_{1_{t_k}}^o$, $t \in [t_k, t_{k+1})$ where $u_{1_{t_k}}^o$ is the first column of the optimal sequence $\bar{u}_{1,N_c}^o(t_k)$. In so doing, one implicitly defines the sampled state-feedback control law

$$u(t) = \kappa^{RH}(x(t_k)) \quad , \quad t \in [t_k, t_{k+1}) \quad (25)$$

Remarkably, the algorithm proposed here satisfies all the assumptions of Theorem 2, so that closed-loop stability can be guaranteed.

5.2 *Prestabilized MPC control scheme*

In many cases, a stabilizing continuous time control law $\kappa(x(t))$ is already known and applied to the plant. For this reason, the scheme already proposed in (Magni *et al.*, 2002) is presented here. It can be used to improve the performance provided by $\kappa(x(t))$ with a reduced computational effort.

Given the control law $\kappa(x(t))$, the problem is to determine with the MPC approach an additive feedback control signal $v(t)$, such that the overall resulting control law

$$u(t) = \kappa(x(t)) + v(t) \quad (26)$$

enlarges the stability region of $\kappa(x(t))$ and enhances the overall control performance with the fulfillment of the constraints (2).

The closed-loop system (1), (26) is described for $t \geq \bar{t}$ by

$$\dot{x}(t) = f(x(t), \kappa(x(t)) + v(t)) \qquad (27)$$

with $x(\bar{t}) = \bar{x}$. Hence, for system (27) the MPC problem can be formally stated as follows: consider the control sequence

$$\bar{v}_{1,N_c}(t_k) := [v_{1_{t_k}}, v_{2_{t_k}}, ..., v_{N_{ct_k}}]$$

with $N_c \geq 1$, for any $t \geq t_k$ define the associated piece-wise constant control signal

$$v(t) = \begin{cases} v_{j_{t_k}} & t \in [t_{k+j-1}, t_{k+j}),\ j = 1, ..., N_c \\ 0 & t \geq t_k + N_c T_s \end{cases} \qquad (28)$$

and consider the following

Finite Horizon Optimal Control Problem (FHOCP[3]). Given the positive integers N_c and N_p, $N_c \leq N_p$ at every "sampling time" instant t_k, minimize, with respect to $\bar{v}_{1,N_c}(t_k)$, the performance index

$$J_{FH}(x_{t_k}, \bar{v}_{1,N_c}(t_k), N_c, N_p) \qquad (29)$$
$$= \int_{t_k}^{t_k+N_pT_s} \left\{ \|x(\tau)\|_Q + \|v(\tau)\|_R \right\} d\tau$$
$$+ V_f(x(t_k + N_p T_s))$$

As for the terminal penalty V_f, it is here selected such that Assumption 1 is satisfied with $\kappa_f(x) = 0$.

The minimization of (29) must be performed under the following constraints:

(i) the state dynamics (27) with $x(t_k) = x_{t_k}$;
(ii) the constraints (2), $t \in [t_k, t_k + N_p T_s)$ with u given by (26);
(iii) $v(t)$ given by (28);
(iv) the terminal state constraint $x(t_k + N_p T_s) \in X_f$, where X_f is a set satisfying Assumption 1 with $\kappa_f(x) \equiv 0$. ■

According to the well known *Receding Horizon* approach, the state-feedback MPC control law is derived by solving the $FHOCP^3$ at every sampling time instant t_k, and applying the constant control signal $u(t) = \kappa(x) + v^o_{1_{t_k}}$, $t \in [t_k, t_{k+1})$ where $v^o_{1_{t_k}}$ is the first column of the optimal sequence $\bar{v}^o_{1,N_c}(t_k)$. In so doing, one implicitly defines the discontinuous (with respect to time) state-feedback control law

$$v(t) = \kappa(x) + \kappa^{RH}(x_k), \quad t \in [t_k, t_{k+1}) \qquad (30)$$

Again, all the assumptions of Theorem 2 are satisfied by the algorithm, so that closed-loop stability is achieved.

5.3 *Continuous time state space constraint fulfillment*

Recall that the second issue related to the practical implementation of MPC algorithms for continuous time systems was related to the fulfillment of the continuous time constraints (2) at any time instant t. This problem can be easily solved by forcing in the on line optimization a finite number of suitable constraints on the state variable at any integration time step. In particular, letting δ be the (maximum) integration step used in the optimization phase to simulate the plant (1) with the control signal (23) and defining by $B^\nu_r = \{x \in R^n : \|x\|_\nu \leq r\}, \nu > 0$, the following result holds

Theorem 5. Let

$$M := \left\| \max_{x \in B^\nu_g, u \in U} f(x, u) \right\|_\nu$$

if (a) $0 < \delta < \frac{g}{M}, g > 0$, (b) $x(\bar{t}) \in B^\nu_{\bar{g}}, \bar{g} = g - \delta M$, then $\varphi(\bar{t} + t, \bar{t}, x(\bar{t}), \bar{u}) \in B^\nu_g, \forall t \in [0, \delta), \bar{u} \in U$.

From this result it is clear that one can choose the maximum integration step δ and a more conservative discrete-time state constraint (defined by $\bar{g}$) so as to guarantee continuous-time state constraint satisfaction. More precisely, given ν and c such that $B^\nu_c \subseteq X$, two nonnegative integers n_s and n_δ, a constant integration step $\delta = T_s/n_s$, condition (ii) in the $FHOCP$ can be replaced by

$$\|\varphi(t_k + n_\delta \delta, t_k, x(t_k), u^{FH}_{t_k}(\cdot))\|_\nu \leq \bar{g},$$
$$n_\delta = 1, 2, \ldots, < N_p n_s,$$
$$u(t_{k+j}) \in U,\ j = 0, \ldots, N_p$$

The use of a more conservative constraints set has already been proposed for linear systems in (Bernardi *et al.*, 2001). Notably the conservatism introduced is substantially less than the one of the discrete-time MPC, because the maximum integration time δ can be chosen much smaller than the sampling time T_s.

6. GLOBAL STABILIZATION OF A PENDULUM

In this section the global stabilization of a pendulum is solved using as pre-stabilizing control law the nonlinear Energy-control proposed in (Åström and Furuta, 2000). The MPC control law, according to the scheme described in Section 5.2, is used to improve performance and to achieve global stability. The equation of motion of a pendulum, written in normalized variables (Åström and Furuta, 2000), is

$$\ddot{\theta}(t) - \sin\theta(t) + u(t)\cos\theta(t) = 0, \qquad (31)$$

where θ is the angle between the vertical and the pendulum, assumed to be positive in the clockwise direction, and u is the normalized acceleration, positive if directed as the positive real axis. The system has two state variables, the angle θ and its rate of change $\dot{\theta}$ (i.e. $x = [\theta\ \dot{\theta}]'$), defined taking θ modulo 2π, with two equilibria, i.e. $u = 0$, $\theta = 0$, $\dot{\theta} = 0$, and $u = 0$, $\theta = \pi$, $\dot{\theta} = 0$. Moreover, it is assumed that $|u| \leq n$.

The normalized total energy of the uncontrolled system ($u = 0$) is

$$E_n(t) = \frac{1}{2}\dot{\theta}^2(t) + \cos\theta(t) - 1$$

Consider now the energy control law

$$u(t) = sat_n(k_u E_n(t) sign(\dot{\theta}(t)\cos\theta(t))) \quad (32)$$

where sat_n is a linear function which saturates at n. In (Åström and Furuta, 2000) it is shown that the control law (32) is able to bring the pendulum at the upright position provided that its initial condition does not coincide with the download stationary position (in fact, with $\theta = \pi$, $\dot{\theta} = 0$, (32) gives $u = 0$ so that the pendulum remains in the download equilibrium). However, the upright equilibrium is an unstable saddle point. For this reason, when the system approaches the origin of the state space, a different strategy is used to locally stabilize the system. In the reported simulations, a linear control law computed with the LQ method applied to the linearized system has been used. This switching strategy, synthetically called in the sequel again "energy control", is described by the control law

$$\kappa(x) \quad (33)$$
$$= \begin{cases} sat_n(k_u E_n sign(\dot{\theta}\cos\theta)) & if\ x_{2\pi} \notin \Omega(K) \\ -Kx'_{2\pi} & if\ x_{2\pi} \in \Omega(K) \end{cases}$$

where $x_{2\pi} := [mod_{2\pi}(\theta)\ \dot{\theta}]$, K is the gain of the locally stabilizing LQ control law and $\Omega(K)$ is an associated output admissible set.

The $NMPC$ control algorithm described in Section 5.2 has been applied to the closed-loop system (31), (33), with the aim of enhancing the performance provided by (33) in terms of the energy required to swing up the pendulum and of the time required to reach the upright position.

For this reason, the stage-cost of the $FHOCP$ is given by

$$\psi(x, u) = \phi(\theta)E_n^2 + (1 - \phi(\theta))V_n \quad (34)$$

where

$$V_n = k_{v1}\frac{1}{2}\sin^2\left(\frac{\theta}{2}\right) + \frac{1}{2}\dot{\theta}^2 \quad (35)$$

and

$$\phi(\theta) = \frac{\beta\tan^2\left(\frac{\theta}{2}\right)}{1 + \beta\tan^2\left(\frac{\theta}{2}\right)} \quad (36)$$

The function V_n given by (35) penalizes the state deviation from the origin, while $\phi(\theta)$ allows to balance the need to reduce the total energy applied and to bring the state to zero. The dependence of $\phi(\theta)$ from the parameter β is shown in Fig. 1.

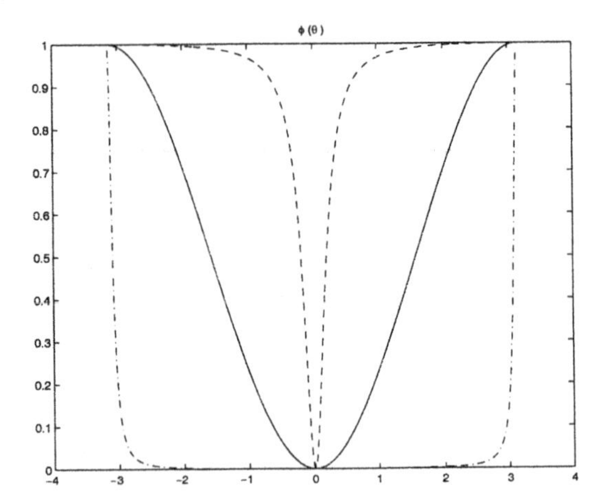

Fig. 1. $\phi(\theta)$ with $\beta = 0.01$ (dash-dot line), $\beta = 1$ (continuous line), $\beta = 100$ (dashed line)

In the following simulation examples the saturation limit is $n = 0.29$, the $FHOCP$ is solved every $T_s = 0.1$ sec and the following parameters are used to synthesize the $NMPC$ control law.

- Auxiliary control law (33): $k_u = 100$, K is the LQ control gain with state penalty matrix $Q = diag\{2.5, 1\}$, and control penalty matrix $R = 1$; $\Omega(K)$ is given by

  $$\Omega(K) = \{x \in \Re^n \mid x'_{2\pi}Px_{2\pi} \leq c\}$$

  where P is the solution of the Riccati equation for the solution of the LQ control problem and $c = 0.001$.
- $FHOCP$: $N_p = 2500$, $k_{v1} = 10$, $\Omega_c(\kappa) = \Omega(K)$, $\tilde{Q} = Q + K'RK$, $\gamma = \lambda_{\min}(\tilde{Q})/20$.

For different choices of the tuning parameters, and starting from the initial condition $[\pi, 0]$, the results summarized in the Table have been obtained. In the table, J_{IH} is the infinite horizon performance index with stage cost (34) and % is the variation with respect to the performance provided by the "energy control" strategy. Note that for $N_c = 0$, when the "energy control" strategy is used, a numerical error is sufficient to move the pendulum in the output admissible set guaranteed by the energy control strategy. On the contrary, with $N_c \geq 1$, the MPC control law guarantees the global stabilization of the inverted pendulum. Moreover note that the best improvement is obtained with a low value of β because in this case the energy is not considered in the cost function. In Fig. 2-4 the movement of the angle position, of the velocity and of the control signal are reported for the control strategies with $\beta = 0$ and for different control horizons N_c.

$\beta = 0$				
N_c	0	2	4	8
J_{IH}	73.9	70.0	68.7	67.0
%	0	-5.3	-7.0	-9.4
$\beta = 0.01$				
N_c	0	2	4	8
J_{IH}	64.7	60.6	61.0	58.7
%	0	-6.3	-5.7	-9.2
$\beta = 1$				
N_c	0	2	4	8
J_{IH}	37.0	35.9	35.5	35.5
%	0	-3.2	-4.2	-4.2
$\beta = 100$				
N_c	0	2	4	8
J_{IH}	26.4	26.2	25.9	25.7
%	0	-0.6	-1.8	-2.3

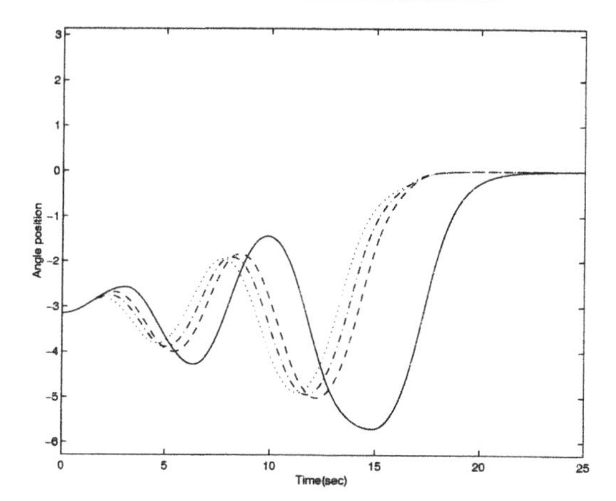

Fig. 2. Angle position movement with "energy control" (continuous line), MPC with $\beta = 0$ and $N_c = 2$ (dashdot line), $N_c = 4$ (dashed line) and $N_c = 8$ (dotted line)

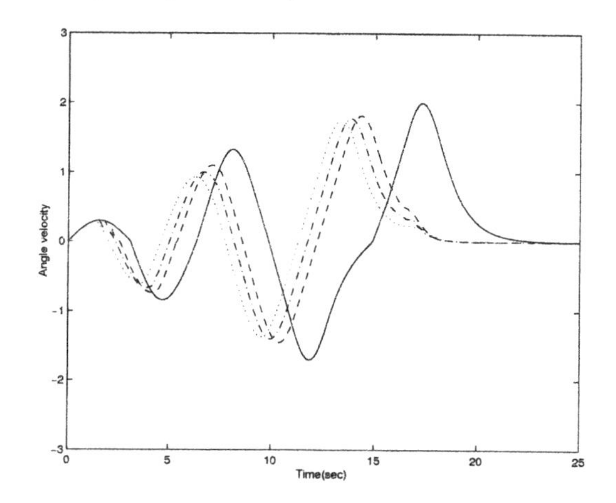

Fig. 3. Angle velocity movement with "energy control" (continuous line), MPC with $\beta = 0$ and $N_c = 2$ (dashdot line), $N_c = 4$ (dashed line) and $N_c = 8$ (dotted line)

7. CONCLUSIONS

In this paper, the main algorithms for the stabilization of continuous time nonlinear systems with the MPC approach have been critically analyzed. In particular, emphasis has been given on the necessity to resort to a "sampled" implementation of the methods and on all the theoretical issues related to the implementation phase. It has been shown how the need to use a suitable parametrization of the control signal forces the selection of a coherent auxiliary control law, usually required to compute the terminal penalty and the terminal state constraint providing closed-loop stability. Also the problem to guarantee the satisfaction of the state constraints has been addressed and solved by reformulating these constraints in the integration times. Many problems are still open and widely studied in MPC for nonlinear systems, among them we recall the development of stabilizing output feedback solutions, for which some significant results have already been suggested in (Magni *et al.*, 1998), (Findeisen *et al.*, 2003), or the analysis of the algorithms when disturbances act on the system.

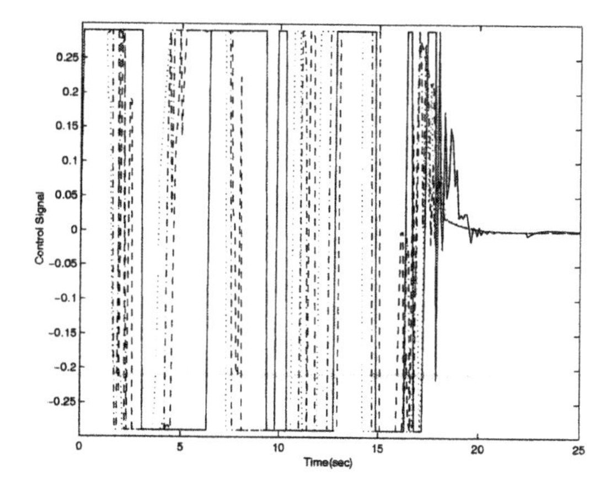

Fig. 4. Control signal with "energy control" (continuous line), MPC with $\beta = 0$ and $N_c = 2$ (dashdot line), $N_c = 4$ (dashed line) and $N_c = 8$ (dotted line)

8. REFERENCES

Åström, K. J. and K. Furuta (2000). Swinging up a pendulum by energy control. *Automatica* **36**, 287–295.

Bernardi, L., E. De Santis, M. D. Di Benedetto and G. Pola (2001). Controlled safe sets for continuous time linear systems. In: *Proceedings of European Control Conference - ECC'01 (J.L. Martins de Carvalho, F. A. C. C. Fontes, and M. D. R. De Pinho, Editors), Porto, Portugal, Pp. 803-808.*

Chen, H. and F. Allgöwer (1998). A quasi-infinite horizon nonlinear model predictive control scheme with guaranteed stability. *Automatica* **34**, 1205–1217.

De Nicolao, G., L. Magni and R. Scattolini (1998). Stabilizing receding-horizon control of nonlinear time-varying systems. *IEEE Trans. on Automatic Control* **AC-43**, 1030–1036.

Findeisen, R., L. Imsland, F. Allgöwer and B. A. Foss (2003). Output feedback stabilization of constrained systems with nonlinear predictive control. *International Journal of Robust and Nonlinear Control* **13**, 211–228.

Gilbert, E. G. and K. T. Tan (1991). Linear systems with state and control constraints: the theory and application of maximal output admissible sets. *IEEE Transaction on Automatic Control* **AC-36**, 1008–1020.

Jadbabaie, A. and J. Hauser (2001). Unconstrained receding-horizon control of nonlinear systems. *IEEE Transactions on Automatic Control* **5**, 776–783.

Jadbabaie, A., J. Primbs and J. Hauser (2001). Unconstrained receding horizon control with no terminal cost. In: *Proceedings of the American Control Conference, Arlington, VA June 25-27.*

Keerthi, S. S. and E. G. Gilbert (1988). Optimal, infinite-horizon feedback laws for a general class of constrained discrete-time systems. *J. Optimiz. Th. Appl.* **57**, 265–293.

Magni, L. (2003). Editorial of the special issue on control of nonlinear systems with model predictive control. *International Journal of Robust and Nonlinear Control.*

Magni, L. and R. Scattolini (2002). State-feedback MPC with piecewise constant control for continuous-time systems. In: *IEEE Conference on Decision and Control, Las Vegas, Nevada, USA December 10 - 13.*

Magni, L. and R. Sepulchre (1997). Stability margins of nonlinear receding horizon control via inverse optimality. *System & Control Letters* **32**, 241–245.

Magni, L., G. De Nicolao and R. Scattolini (1998). Output feedback receding-horizon control of discrete-time nonlinear systems. In: *IFAC NOLCOS '98*. Enschede, The Netherlands.

Magni, L., G. De Nicolao, L. Magnani and R. Scattolini (2001). A stabilizing model-based predictive control for nonlinear systems. *Automatica* **37**, 1351–1362.

Magni, L., R. Scattolini and K. J. Åström (2002). Global stabilization of the inverted pendulum using model predictive control. In: *15th IFAC World Congress, Barcelona, Spain.*

Mayne, D. Q. and H. Michalska (1990). Receding horizon control of nonlinear systems. *IEEE Trans. on Automatic Control* **35**, 814–824.

Mayne, D. Q., J. B. Rawlings, C. V. Rao and P. O. M. Scokaert (2000). Constrained model predictive control: Stability and optimality. *Automatica* **36**, 789–814.

Michalska, H. and D. Q. Mayne (1993). Robust receding horizon control of constrained nonlinear systems. *IEEE Trans on Automatic Control* **38**, 1623–1633.

Michalska, H. and R. B. Vinter (1994). Nonlinear stabilization using discontinuous moving-horizon control. *IMA J. Math Control Inform.* **11**, 321–340.

Qin, S. J. and T. A. Badgwell (1996). An overview of industrial model predictive control technology. In: *Fifth International Conference on Chemical Process Control* (J. C. Kantor, C. E. Garcia and B. Carnahan, Eds.). pp. 232–256.

Qin, S. J. and T. A. Badgwell (2000). An overview of nonlinear model predictive control applications. In: *Nonlinear Model Predictive Control* (F. Allgower and A. Zheng, Eds.). pp. 369–392. Birkhauser Berlin.

www.elsevier.com/locate/ifac

A PARAMETERIZATION OF EXPONENTIALLY STABILIZING CONTROLLERS FOR LINEAR MECHANICAL SYSTEMS SUBJECT TO NON-SMOOTH IMPACTS

S. Galeani, L. Menini and A. Tornambè [*,1]

[*] *Dip. Informatica, Sistemi e Produzione*
Università di Roma Tor Vergata
Via del Politecnico, 1
00133, Roma, Italy
Email: [galeani,menini,tornambe]@disp.uniroma2.it

Abstract: The goal of this paper is to propose a family of output feedback compensators (whose state is subject to jumps at the impact times) that stabilize a given n degrees of freedom mechanical system, with linear continuous-time dynamics, subject to non-smooth impacts. An example is used in order to show how the available degrees of freedom can be used to satisfy important requirements, apart of asymptotic stability. *Copyright © 2003 IFAC*

Keywords: Non-smooth impacts, stabilization, Liapunov method.

1. INTRODUCTION

Mechanical systems subject to impacts occur frequently in every day's life as well as in the industry. Relevant examples arise in robotics, where the impacts can be either a part of the robot's task (hopping (Koditschek and Bühler, 1991), walking (Hurmuzlu, 1993; Hurmuzlu *et al.*, n.d.) or juggling (Zumel and Erdmann, 1994; Swanson *et al.*, 1995) robots, tasks like hammering (Izumi and Hitaka, 1997) or inserting a peg into a hole (Prokop and Pfeiffer, 1998)) or, more frequently, undesired collisions, *e.g.* in space robotics (Nenchev and Yoshida, 1999; Wee and Walker, 1993) or during the so-called "transition phases" between uncostrained motion and a subtask to be executed in permanent contact (Mills and Lokhorst, 1993; Marth *et al.*, 1994).

In all these cases, a suitable model of the impact phenomena is needed; unfortunately the problem of modeling mechanical systems subject to impacts is far from being solved in its generality, especially when multiple impacts or the effects of friction have to be taken into account. In (Tornambè, 1999) two different methods are used to model dynamic systems subject to inequality constraints: the Valentine variables method (which leads to the concept of "non-smooth" impacts) and the penalizing functions method. The efficacy of simple PD control laws for obtaining global asymptotic stability is theoretically proven, for both classes of models, and experimentally tested on a single-link flexible robot arm. Interesting results concerned with the modeling of impacting systems can be found also in (Zheng and Hemami, 1985; Hurmuzlu and Marghitu, 1994; Hurmuzlu and Chang, 1992).

It is important to stress that in most of the existing work on the control of impacting systems

[1] Supported by ASI (I/R/125/02) and MIUR (COFIN:Matrics)

the impacts are considered "smooth", *i.e.* some deformation of the bodies involved is assumed. Here, on the contrary, we consider "non-smooth" impacts, in view of the consideration that, in many cases, the elastic constants which can be used to describe the flexibilities of the impacting bodies are very high, so that it is extremely burdensome to simulate the system's behaviour; for the same reason, the possibility of taking into account the deformations arising during the impacts in the control system design often relies on the availability of very powerful devices for the controller implementation.

In this paper, a family of compensators is proposed for a class of linear mechanical systems subject to a linear inequality constraint, in order to stabilize the origin, which corresponds to a condition of contact, *i.e.*, a configuration in which the constraint is satisfied with the equality sign. Although the mechanical systems considered are linear, when only their unconstrained motion is described, the presence of a linear inequality constraint (and the consequent impulses generated to prevent the possible violation of the constraint at the impact times) renders the system under study nonlinear. The mechanical systems considered have n degrees of freedom and are fully actuated; only the generalized position coordinates are assumed to be measured, whereas the generalized velocities are not available for measure (this is congruent with the practical impossibility of accurately measuring the velocities in neighborhoods of the impact times). The proposed family of compensators is a modification of the Youla-Kučera parameterization of all stabilizing linear compensators that could be derived for the mechanical system under consideration in absence of the inequality constraint; such a modification (which renders nonlinear also the controllers) is just introduced for taking into account the jumps in the generalized velocities at the impact times. Each one of the compensators belonging to the proposed class guarantees the exponential stability of the origin, and the BIBS (Bounded-Input Bounded-State) stability for a large class of input functions (the latter property is particularly important for impacting mechanical systems).

The work presented here is based on preliminary results reported in (Menini and Tornambè, 2001*b*; Menini and Tornambè, 2001*d*; Menini and Tornambè, 2001*a*; Menini and Tornambè, 2001*c*).

Notations. In the following, $\mathbb{R}$ will denote the set of real numbers, $\mathbb{R}^+$ the set of non-negative real numbers, $\mathbb{Z}^+$ the set of non-negative integers, $\mathbb{N}$ the set of positive integers, $\mathbb{R}^\nu$ the set of vectors of dimension ν and $\mathbb{R}^{\nu\times\nu}$ the set of real matrices of dimensions $\nu \times \nu$. If $\mathbf{g}$ is a vector, with g_i we denote its i-th entry. Moreover, for a given $t_0 \in \mathbb{R}$, $\overline{\mathcal{C}}^0([t_0, +\infty))$ will denote the set of all vector functions $\mathbf{g}(\cdot)$ having entries $g_h(\cdot) : [t_0, +\infty) \to \mathbb{R}$ that are bounded on every compact subset of $[t_0, +\infty)$. For the sake of brevity, the shorthand notations $\mathbf{g}(\tau^-)$ and $\mathbf{g}(\tau^+)$ will be used in place of $\lim_{t\to\tau^-} \mathbf{g}(t)$ and $\lim_{t\to\tau^+} \mathbf{g}(t)$, respectively, for any vector function $\mathbf{g}(t)$. At each time $t = \tau$ at which at least one entry of $\mathbf{g}(t)$ has a corner point, symbol $\dot{\mathbf{g}}(\tau)$ has to be understood as $\dot{\mathbf{g}}(\tau^-)$ and $\dot{\mathbf{g}}(\tau^+)$. For any matrix $\mathbf{F}$, $\mathbf{F}_{h,:}$ will denote the h-th row of $\mathbf{F}$, $\overline{\sigma}(\mathbf{F})$ and $\underline{\sigma}(\mathbf{F})$ will denote the minimum and the maximum singular values of $\mathbf{F}$, respectively, whereas, if $\mathbf{F}$ is square, $\lambda_m(\mathbf{F})$ and $\lambda_M(\mathbf{F})$ will denote the smallest and the largest eigenvalue of $\mathbf{F}$, respectively. Finally, $\mathbf{I}_\nu$ will denote the $\nu \times \nu$ identity matrix and the symbol $\|\cdot\|$ will denote the Euclidean norm of the vector at argument or the corresponding induced norm if the argument is a matrix. In view of this choice, for any matrix $\mathbf{F}$, we have $\|\mathbf{F}\| = \overline{\sigma}(\mathbf{F})$. □

2. THE CLASS OF CONSIDERED MECHANICAL SYSTEMS

In this paper, we consider n-degrees of freedom fully actuated linear mechanical systems whose unconstrained motion (*i.e.*, in absence of impacts and contacts) can be described by:

$$\mathbf{T}\,\ddot{\mathbf{q}}(t) + \mathbf{U}_2\,\mathbf{q}(t) + \mathbf{u}_1(t) = \boldsymbol{\tau}(t), \quad \forall t \geq t_0, \tag{1}$$

where $t_0 \in \mathbb{R}$ is the initial time, $\mathbf{q}(t) \in \mathbb{R}^n$ is the vector of the Lagrangian coordinates, uniquely describing the configuration of the system at time t, $\mathbf{T} \in \mathbb{R}^{n\times n}$ is the symmetric and positive definite inertia matrix, $\mathbf{U}_2 \in \mathbb{R}^{n\times n}$ is symmetric, $\mathbf{u}_1(t)$ is the vector of the (possibly equal to 0, or time-varying) known external forces acting on the system (it can include also potential forces other than $\mathbf{U}_2\,\mathbf{q}(t)$, *e.g.*, the gravity force) and $\boldsymbol{\tau}(t) \in \mathbb{R}^n$ is the vector of the control forces. Assume that $\mathbf{u}_1(\cdot)$, $\boldsymbol{\tau}(\cdot) \in \overline{\mathcal{C}}^0([t_0, +\infty))$. Notice that it is assumed that there is no friction of any kind acting on the system: under standard assumptions, it is easy to extend the theory presented for when the system is subject to dissipative forces proportional to the generalized velocities, whereas nonlinear dependences (as in the case of "stiction") are to be omitted because they destroy the linearity of (1). The description of the mechanical system under study is completed by considering the following linear inequality constraint:

$$\mathbf{J}\,\mathbf{q}(t) \leq 0, \quad \forall t \geq t_0, \tag{2}$$

where $\mathbf{J}^T \in \mathbb{R}^n$, $\mathbf{J} \neq \mathbf{0}$. A time $\hat{t} \in \mathbb{R}$, $\hat{t} > t_0$, is an *impact time* if $\mathbf{J}\,\mathbf{q}(\hat{t}) = 0$ and $\mathbf{J}\,\dot{\mathbf{q}}(\hat{t}^-) > 0$.

However, in the following, especially when only the position $\mathbf{q}(t)$ will be assumed to be measured, it will be difficult to distinguish (in practice) an impact time from a *degenerate impact time*, *i.e.*, a time $\hat{t} \in \mathbb{R}$, $\hat{t} > t_0$, such that $\mathbf{J}\,\mathbf{q}(\hat{t}) = 0$, $\mathbf{J}\,\dot{\mathbf{q}}(\hat{t}^-) = 0$ and there exists $\hat{\varepsilon} \in \mathbb{R}$, $\hat{\varepsilon} > 0$, such that $\mathbf{J}\,\mathbf{q}(\hat{t} - \varepsilon) < 0$ for all $\varepsilon \in (0,\, \hat{\varepsilon})$. Hence, for the sake of clarity, from now on we will denote the impact times (degenerate or not) with the notation t_i, $i \in \mathbb{N}$, ordering them so that $t_{i+1} > t_i$, for all $i \in \mathbb{N}$.

We consider the case of a single inequality constraint for simplicity, as it is easy to extend the theory here reported to the case in which more inequality constraints are present, but with more cumbersome notations. In such a case, the only further needed assumption is that no multiple impacts occur, since, to the best of the authors' knowledge, no satisfactory theory for modeling multiple impacts exists (*i.e.*, by excluding special cases, it is not possible to compute in a unique manner the post-impact velocities after a multiple impact).

Remark 1. The previously introduced notation about the impact times actually implies some loss of generality. As a matter of fact, it is not difficult to create examples in which the impact times have one finite accumulation point, followed by other impacts, or even more than one (and, possibly, an infinite number of) finite accumulation points (this occurs especially for inelastic impacts, *i.e.*, with coefficient of restitution $e < 1$). In such a case, assuming that there are no "backward" accumulation points of the impact times, let T_j be the j-th accumulation point of the impact times (with $T_{j+1} > T_j$); under the assumption that there is no finite accumulation point of the times T_j, let $T_0 = t_0$ and let $t_{j,i}$ be the i-th impact time after the $(j-1)$-th accumulation point of the impact times (with $t_{j,\,i+1} > t_{j,\,i}$). For the sake of simplicity, in this paper it will be always assumed that there is no finite accumulation point of the finite accumulation points of the impact times and also that all the finite accumulation points of the impact times are "forward" accumulation points, *i.e.*, that $T_j = \lim_{i \to +\infty} t_{j,\,i}$. Moreover, with a slight abuse of notation, the more complicate notation $t_{j,\,i}$ introduced in this remark will be used only in those assumptions and parts of the proofs in which it is actually needed, whereas the proposed results will be stated as if there was no finite accumulation point of the impact times, or at most one finite accumulation point T_1 but without further impacts in the interval $(T_1,\, +\infty)$, since these two cases can be dealt with by means of the simpler notation t_i for the impact times. Nevertheless, the proposed results will be valid in general, with the exception of the cases when there are finite accumulation points of the finite accumulation points of the impact times, or when there are "backwards" accumulation points, which cases are omitted. Finally, the notation t_i, $i \in \mathbb{N}$, or $i \in \mathbb{Z}^+$, will not necessarily imply that the impacts are infinite in number, and the notation $(t_i,\, t_{i+1})$ has to be understood as $(t_i,\, +\infty)$ if there are no impact times greater than t_i. □

For $t \in (t_i,\, t_{i+1})$, $i \in \mathbb{Z}^+$, the system is described by the following differential equation:

$$\mathbf{T}\,\ddot{\mathbf{q}}(t) + \mathbf{U}_2\,\mathbf{q}(t) + \mathbf{u}_1(t) = \boldsymbol{\tau}(t) + \mathbf{R}(t), \qquad (3)$$

where $\mathbf{R}(t)$ is the reaction force due to possible contacts, which can be different from zero only in the intervals of time (which necessarily start at degenerate impact times) during which there is a permanent contact. In particular, for $t \in (t_i,\, t_{i+1})$, $i \in \mathbb{Z}^+$, we have

$$\mathbf{R}(t) = \begin{cases} \mathbf{0}, & \text{if } (\mathbf{J}\,\mathbf{q}(t) < 0) \text{ or } \\ & (\mathbf{J}\,\mathbf{q}(t) = 0 \text{ and } c(t) \leq 0), \\ -\dfrac{c(t)}{\mathbf{J}\,\mathbf{T}^{-1}\,\mathbf{J}^T}\,\mathbf{J}^T, & \text{if } \mathbf{J}\,\mathbf{q}(t) = 0 \text{ and } \\ & c(t) > 0, \end{cases} \qquad (4)$$

where

$$c(t) := \mathbf{J}\,\mathbf{T}^{-1}\,(\boldsymbol{\tau}(t) - \mathbf{U}_2\,\mathbf{q}(t) - \mathbf{u}_1(t)). \qquad (5)$$

When considering also the impact times $t = t_i$, $\mathbf{R}(t)$ includes also the impulsive terms, which guarantee that the inequality constraint $\mathbf{J}\,\mathbf{q}(t) \leq 0$ holds for the times t immediately after t_i; however, it is stressed that, for $t \in (t_i,\, t_{i+1})$, $i \in \mathbb{Z}^+$, if $\mathbf{u}_1(t)$ and $\boldsymbol{\tau}(t)$ are bounded, then $\mathbf{R}(t)$ is also bounded. In order to compute the post-impact velocity vector $\mathbf{q}(t_i^+)$ as a function of the pre-impact velocity vector $\mathbf{q}(t_i^-)$, we use the approach described in (Menini and Tornambè, 1999), which is completely equivalent to the *kinetic metric approach* (Brogliato, 1996, Ch. 6). Let $\mathbf{W} \in \mathbb{R}^{n \times n}$ be a nonsingular matrix that simultaneously diagonalizes $\mathbf{T}$ and $\mathbf{J}^T\,\mathbf{J}$, according to:

$$\mathbf{W}^T\,\mathbf{T}\,\mathbf{W} = \mathbf{I}, \qquad (6a)$$

$$\mathbf{W}^T\,\mathbf{J}^T\,\mathbf{J}\,\mathbf{W} = \mathrm{diag}(\mathbf{J}\,\mathbf{T}^{-1}\,\mathbf{J}^T,\, \underbrace{0, ..., 0}_{(n-1)\,\text{times}}\,). \qquad (6b)$$

The following lemma, whose proof is rather standard and therefore is omitted here, states the existence of a matrix $\mathbf{W}$ that satisfies (6) and gives formulae for its computation.

Lemma 1. Let $\mathbf{T} \in \mathbb{R}^{n \times n}$ be symmetric and positive definite, and let $\mathbf{J}^T \in \mathbb{R}^n$, $\mathbf{J} \neq \mathbf{0}$. Then, there exists a nonsingular matrix $\mathbf{W} \in \mathbb{R}^{n \times n}$ that satisfies (6), which can be computed as $\mathbf{W} = \mathbf{H}^{-T}\,[\mathbf{w}_1 \ldots \mathbf{w}_n]$, where $\mathbf{H} \in \mathbb{R}^{n \times n}$ is such that $\mathbf{T} = \mathbf{H}\mathbf{H}^T$, $\mathbf{w}_1 := \dfrac{\mathbf{H}^{-1}\mathbf{J}^T}{\|\mathbf{H}^{-1}\mathbf{J}^T\|}$, and $\{\mathbf{w}_1, \mathbf{w}_2, ..., \mathbf{w}_n\}$ is an orthonormal basis for $\mathbb{R}^n$.

Now, let $e \in [0,\ 1]$ be the coefficient of restitution characterizing the impacts. Let $\mathbf{A}(e) \in \mathbb{R}^{n\times n}$ be defined as $\mathbf{A}(e) := \mathrm{diag}(-e, 1, ..., 1)$ and $\mathbf{Z}(e) := \mathbf{W}\,\mathbf{A}(e)\,\mathbf{W}^{-1}$. The rule for computing the post-impact velocity vector is the following:

$$\dot{\mathbf{q}}(t_i^+) = \mathbf{Z}(e)\,\dot{\mathbf{q}}(t_i^-), \quad i \in \mathbb{N}, \tag{7}$$

whereas the position $\mathbf{q}(t)$ is a continuous function of time, *i.e.*: $\mathbf{q}(t_i^+) = \mathbf{q}(t_i^-)$, $i \in \mathbb{N}$. The mechanical system under consideration is therefore constituted by equation

$$\mathbf{T}\,\ddot{\mathbf{q}}(t) + \mathbf{U}_2\,\mathbf{q}(t) + \mathbf{u}_1(t) = \boldsymbol{\tau}(t) + \mathbf{R}(t), \quad t \in (t_i,\ t_{i+1}),\ i \in \mathbb{Z}^+, \tag{8}$$

together with (4) and (7), to be solved starting from the initial condition $[\mathbf{q}^T(t_0)\ \dot{\mathbf{q}}^T(t_0)]^T \in \mathcal{A}$. Notice that the presence of the inequality constraint (2) (which is translated into the rule (7) and the possible presence of a non null $\mathbf{R}(t)$) renders the dynamical system under study nonlinear, although (1) is linear.

Rule (7) can be used to justify the choice of the following set of admissible initial conditions:

$$\mathcal{A} := \{[\mathbf{q}^T\ \mathbf{v}^T]^T \in \mathbb{R}^{2\,n} : (\mathbf{J}\,\mathbf{q} \le 0) \text{ and } (\mathbf{J}\,\mathbf{v} \le 0 \text{ if } \mathbf{J}\,\mathbf{q} = 0)\}.$$

Remark 2. Note that there is no loss of generality in assuming $[\mathbf{q}^T(t_0)\ \dot{\mathbf{q}}^T(t_0)]^T \in \mathcal{A}$, as if $\mathbf{J}\,\mathbf{q}(t_0) = 0$ and $\mathbf{J}\,\dot{\mathbf{q}}(t_0) > 0$, then t_0 is an impact time and it is sufficient to take as initial condition the post-impact state $[\mathbf{q}^T(t_0)\ \ \mathbf{Z}(e)\,\dot{\mathbf{q}}^T(t_0)]^T$, which belongs to $\mathcal{A}$. □

3. THE FAMILY OF PROPOSED COMPENSATORS

In this section, a family of feedback compensators from the measured output $\mathbf{y}(t) = \mathbf{q}(t)$ is introduced for the mechanical system subject to non-smooth impacts described by (8) and (7), in order to globally stabilize the origin of the state space. The structure of such a family is closely related to the well-known Youla-Kučera parameterization of all stabilizing controllers for linear time-invariant systems (we followed the approach in (Colaneri *et al.*, 1997), see also (Kučera, 1974; Youla *et al.*, 1976; Francis, 1987; Maciejowski, 1989)). As a matter of fact, the compensators proposed here are described by:

$$\begin{aligned}
\boldsymbol{\tau}(t) &= \mathbf{U}_2\,\mathbf{y}(t) + \mathbf{u}_1(t) - \mathbf{K}_p\,\hat{\mathbf{q}}(t) - \\
&\qquad \mathbf{K}_v\,\hat{\mathbf{v}}(t) + \mathbf{y}_Q(t), \quad t \in (t_i,\ t_{i+1}),\ i \in \mathbb{Z}^+, \\
\dot{\hat{\mathbf{q}}}(t) &= \hat{\mathbf{v}}(t) + \mathbf{K}_1\,(\mathbf{y}(t) - \hat{\mathbf{q}}(t) - \mathbf{r}(t)), \\
&\qquad t \in (t_i,\ t_{i+1}),\ i \in \mathbb{Z}^+, \\
\mathbf{T}\,\dot{\hat{\mathbf{v}}}(t) &= -\mathbf{K}_p\,\hat{\mathbf{q}}(t) - \mathbf{K}_v\,\hat{\mathbf{v}}(t) + \mathbf{y}_Q(t) + \mathbf{R}(t) + \\
&\qquad \mathbf{K}_2\,(\mathbf{y}(t) - \hat{\mathbf{q}}(t) - \mathbf{r}(t)), \\
&\qquad t \in (t_i,\ t_{i+1}),\ i \in \mathbb{Z}^+, \\
\dot{\mathbf{x}}_Q(t) &= \mathbf{A}_Q\,\mathbf{x}_Q(t) + \mathbf{B}_Q\,(\mathbf{y}(t) - \hat{\mathbf{q}}(t) - \mathbf{r}(t)), \\
&\qquad t \in (t_i,\ t_{i+1}),\ i \in \mathbb{Z}^+, \\
\mathbf{y}_Q(t) &= \mathbf{C}_Q\,\mathbf{x}_Q(t) + \mathbf{D}_Q\,(\mathbf{y}(t) - \hat{\mathbf{q}}(t) - \mathbf{r}(t)), \\
&\qquad t \in (t_i,\ t_{i+1}),\ i \in \mathbb{Z}^+, \\
\hat{\mathbf{q}}(t_i^+) &= \mathbf{Z}(e)\,\hat{\mathbf{q}}(t_i^+), \quad i \in \mathbb{N}, \\
\hat{\mathbf{v}}(t_i^+) &= \mathbf{Z}(e)\,\hat{\mathbf{v}}(t_i^+), \quad i \in \mathbb{N}, \\
\mathbf{x}_Q(t_i^+) &= \mathbf{x}_Q(t_i^-), \quad i \in \mathbb{N},
\end{aligned}$$

where $\mathbf{y}_Q(t) \in \mathbb{R}^n$, $\mathbf{x}_Q(t) \in \mathbb{R}^m$, $m \in \mathbb{Z}^+$, $\mathbf{A}_Q \in \mathbb{R}^{m\times m}$ has all the eigenvalues with negative real part, $\mathbf{B}_Q$, $\mathbf{C}_Q$ and $\mathbf{D}_Q$ are real matrices of suitable dimensions, such that the pairs $(\mathbf{A}_Q,\ \mathbf{B}_Q)$, and $(\mathbf{C}_Q,\ \mathbf{A}_Q)$ are reachable and observable, respectively, $\mathbf{R}(t)$ is given by (4),

$$\begin{aligned}
\mathbf{K}_p &:= \mathbf{W}^{-T}\,\overline{\mathbf{K}}_p\,\mathbf{W}^{-1}, \quad \mathbf{K}_v := \mathbf{W}^{-T}\,\overline{\mathbf{K}}_v\,\mathbf{W}^{-1}, \\
\overline{\mathbf{K}}_p &:= \mathrm{diag}\,(k_{p,1}, \ldots, k_{p,n}), \\
\overline{\mathbf{K}}_v &:= \mathrm{diag}\,(k_{v,1}, \ldots, k_{v,n}),
\end{aligned}$$

$k_{p,h}$, $k_{v,h} > 0$, for all $h = 1, \ldots, n$, and $\mathbf{r}(t) \in \overline{\mathcal{C}}^0([t_0,\ +\infty))$ is the new control signal, whereas

$$\mathbf{K}_1 := \mathbf{W}\,\mathrm{diag}(k_{1,1}, ..., k_{1,n})\,\mathbf{W}^{-1}, \tag{9a}$$
$$\mathbf{K}_2 := \mathbf{W}^{-T}\,\mathrm{diag}(k_{2,1}, ..., k_{2,n})\,\mathbf{W}^{-1}, \tag{9b}$$

with $k_{h,j} \in \mathbb{R}$, $k_{h,j} > 0$, $\forall h \in \{1, 2\}$, $\forall j \in \{1, ..., n\}$.

It can be seen that if the subsystem having as state vector $\mathbf{x}_Q$ is absent, *i.e.*, if $\mathbf{y}_Q(t) = \mathbf{0}$, for all $t \ge t_0$, the proposed compensator is a slight modification of the compensator proposed in (Menini and Tornambè, 2001*a*), whereas, by varying m and the matrices $\mathbf{A}_Q$, $\mathbf{B}_Q$, $\mathbf{C}_Q$ and $\mathbf{D}_Q$, apart from the jumps at the impact times, if any, we obtain the family of all stabilizing compensators for the unconstrained system. The compensators described above are nonlinear systems (in view of the jumps imposed to some state variables at the times t_i, which are actually functionals of the input signal $\mathbf{y}(t)$). However, by limiting our attention to an interval $(t_i,\ t_{i+1})$, and assuming that t_i is a non-degenerate impact time (so that $\mathbf{R}(t) = \mathbf{0}$ for all $t \in (t_i,\ t_{i+1})$), the control input can be written as $\boldsymbol{\tau}(t) = \mathbf{U}_2\,\mathbf{y}(t) + \mathbf{u}_1(t) + \overline{\boldsymbol{\tau}}(t)$, where the signal $\overline{\boldsymbol{\tau}}(t)$ is the output of a linear system having as input $\mathbf{r}(t) - \mathbf{y}(t)$ and suitable initial conditions at time t_i^+. The transfer matrix of such system can be computed as

$$\mathbf{K}(s) = -\Big(\mathbf{Y}(s) + \mathbf{M}(s)\,\mathbf{Q}(s)\Big)\Big(\mathbf{X}(s) + \mathbf{N}(s)\,\mathbf{Q}(s)\Big)^{-1},$$

where

$$\mathbf{Q}(s) = \mathbf{D}_Q + \mathbf{C}_Q\,(s\,\mathbf{I}_n - \mathbf{A}_Q)\,\mathbf{B}_Q, \quad (10)$$

$$\mathbf{M}(s) = \left(\mathbf{I}_n + \frac{1}{s}\mathbf{K}_v\,\mathbf{T}^{-1} + \frac{1}{s^2}\,\mathbf{K}_p\,\mathbf{T}^{-1}\right)^{-1}, \quad (11)$$

$$\mathbf{N}(s) = \boldsymbol{\Delta}^{-1}(s), \quad (12)$$

$$\mathbf{X}(s) = \mathbf{I}_n + \frac{1}{s}\,\boldsymbol{\Delta}^{-1}(s)\cdot \Big((s\,\mathbf{I}_n + \mathbf{T}^{-1}\,\mathbf{K}_v)\,\mathbf{K}_1 + \mathbf{T}^{-1}\,\mathbf{K}_2\Big), \quad (13)$$

$$\mathbf{Y}(s) = -\frac{1}{s}\,\mathbf{K}_p\,\boldsymbol{\Delta}^{-1}(s)\cdot \left((s\,\mathbf{I}_n + \mathbf{T}^{-1}\,\mathbf{K}_v)\,\mathbf{K}_1 + \frac{1}{s}\,\mathbf{T}^{-1}\,\mathbf{K}_2\right) + \mathbf{K}_p\,\boldsymbol{\Delta}^{-1}(s)\left(\frac{1}{s}\,\mathbf{T}^{-1}\,\mathbf{K}_p\,\mathbf{K}_1 - \mathbf{T}^{-1}\,\mathbf{K}_2\right), \quad (14)$$

with $\boldsymbol{\Delta}(s) = s\,\mathbf{I}_n + \mathbf{T}^{-1}\,\mathbf{K}_v + \frac{1}{s}\,\mathbf{T}^{-1}\,\mathbf{K}_p$.

The closed-loop system can be rewritten as follows, using the variables $\mathbf{v} = \dot{\mathbf{q}}$, $\tilde{\mathbf{q}} = \mathbf{q} - \hat{\mathbf{q}}$ and $\tilde{\mathbf{v}} = \mathbf{v} - \hat{\mathbf{v}}$:

$$\dot{\mathbf{q}}(t) = \mathbf{v}(t), \quad t \in (t_i,\, t_{i+1}),\, i \in \mathbb{Z}^+, \quad (15a)$$

$$\mathbf{T}\,\dot{\mathbf{v}}(t) = -\mathbf{K}_p\,(\mathbf{q}(t) - \tilde{\mathbf{q}}(t)) + \mathbf{D}_Q\,\tilde{\mathbf{q}}(t) - \mathbf{K}_v\,(\mathbf{v}(t) - \tilde{\mathbf{v}}(t)) + \mathbf{C}_Q\,\mathbf{x}_Q(t) - \mathbf{D}_Q\,\mathbf{r}(t) + \mathbf{R}(t), \quad t \in (t_i,\, t_{i+1}),\, i \in \mathbb{Z}^+, \quad (15b)$$

$$\dot{\tilde{\mathbf{q}}}(t) = \tilde{\mathbf{v}}(t) - \mathbf{K}_1\,\tilde{\mathbf{q}}(t) + \mathbf{K}_1\,\mathbf{r}(t), \quad t \in (t_i,\, t_{i+1}),\, i \in \mathbb{Z}^+, \quad (15c)$$

$$\mathbf{T}\,\dot{\tilde{\mathbf{v}}}(t) = -\mathbf{K}_2\,\tilde{\mathbf{q}}(t) + \mathbf{K}_2\,\mathbf{r}(t), \quad t \in (t_i,\, t_{i+1}),\, i \in \mathbb{Z}^+, \quad (15d)$$

$$\dot{\mathbf{x}}_Q(t) = \mathbf{A}_Q\,\mathbf{x}_Q(t) + \mathbf{B}_Q\,\tilde{\mathbf{q}}(t) - \mathbf{B}_Q\,\mathbf{r}(t), \quad t \in (t_i,\, t_{i+1}),\, i \in \mathbb{Z}^+, \quad (15e)$$

$$\mathbf{v}(t_i^+) = \mathbf{Z}(e)\,\mathbf{v}(t_i^-), \quad i \in \mathbb{N}, \quad (15f)$$

$$\tilde{\mathbf{q}}(t_i^+) = \mathbf{Z}(e)\,\tilde{\mathbf{q}}(t_i^-), \quad i \in \mathbb{N}, \quad (15g)$$

$$\tilde{\mathbf{v}}(t_i^+) = \mathbf{Z}(e)\,\tilde{\mathbf{v}}(t_i^-), \quad i \in \mathbb{N}. \quad (15h)$$

Denote by

$$\mathbf{x}_{cc}(t) = \left[\, \mathbf{q}^T(t)\ \mathbf{v}^T(t)\ \tilde{\mathbf{q}}^T(t)\ \tilde{\mathbf{v}}^T(t)\ \mathbf{x}_Q^T \,\right]^T$$

the state of the system (15).

Assumption 1. Let $\mathbf{x}_{cc}(t_0) \in \mathcal{A} \times \mathbb{R}^{2n+m}$ and $\mathbf{r}(\cdot) \in \overline{\mathcal{C}}^0([t_0,\ +\infty))$ be such that the solution of the closed-loop system (15) has no finite accumulation points of the impact times.

Theorem 1. Let $e \in [0,\,1]$. If $k_{p,j} > 0$, $k_{v,j} > 0$, $k_{h,j} > 0$, $\forall h \in \{1,\,2\}$, $\forall j \in \{1,\,...,\,n\}$, and all the eigenvalues of $\mathbf{A}_Q$ have negative real part, then

(i) the equilibrium $\mathbf{x}_{cc} = \mathbf{0}$ of system (15) is a globally exponentially stable equilibrium with respect to all the initial states satisfying Assumption 1 together with $\mathbf{r}(\cdot) = \mathbf{0}$, i.e., there exist $\alpha > 0$ and $\beta > 0$ such that, if $\mathbf{x}_{cc}(t_0)$ and $\mathbf{r}(\cdot) = \mathbf{0}$ satisfy Assumption 1, then the following relation holds for $\mathbf{r}(\cdot) = \mathbf{0}$:

$$\|\mathbf{x}_{cc}(t)\| \leq \alpha \exp\left(-\beta(t - t_0)\right) \|\mathbf{x}_{cc}(t_0)\|, \quad \forall t \geq t_0.\forall t \geq t_0; \quad (16)$$

(ii) the closed-loop system (15) is globally bounded-input bounded-state (shortly, BIBS) stable for all the initial states and inputs satisfying Assumption 1, i.e., there exist two functions $\gamma_x(\cdot,\,\cdot) : \mathbb{R}^+ \times \mathbb{R}^+ \to \mathbb{R}^+$ and $\gamma_r(\cdot) : \mathbb{R}^+ \to \mathbb{R}^+$, both strictly increasing with respect to the first argument, $\gamma_x(s,\,\cdot)$ decreasing with $\lim_{z \to +\infty} \gamma_x(s,\,z) = 0$ for every $s > 0$, and such that $\gamma_x(0,\,z) = 0$, for all $z \in \mathbb{R}^+$ and $\gamma_r(0) = 0$, such that, if $\mathbf{x}_{cc}(t_0)$ and $\mathbf{r}(\cdot)$ satisfy Assumption 1 and $\|\mathbf{r}(t)\| \leq r_M$, $r_M \in \mathbb{R}^+$, for all $t \geq t_0$, then the solution $\mathbf{x}_{cc}(\cdot)$ of system (15) satisfies:

$$\|\mathbf{x}_{cc}(t)\| \leq \gamma_x(\|\mathbf{x}_{cc}(t_0)\|,\, t - t_0) + \gamma_r(r_M), \quad \forall t \geq t_0. \quad (17)$$

Remark 3. One of the nicest properties of the Youla-Kučera parameterization for linear time-invariant systems is that it allows to represent by means of the parameter $\mathbf{Q}(s)$ the class of all linear stabilizing compensators for the given system. Theorem 1 states a weaker property of the family of compensators proposed here for linear mechanical systems subject to impacts, since it only states that each compensator in the family is a stabilizing one, but it does not exclude the existence of other stabilizing compensators. This is actually due to the presence of the inequality constraint, which renders the system nonlinear, and (in some situations) can exert an additional "stabilizing" action. As an example, in the case of $e = 1$ (no dissipation occurs at the impact times), consider the constrained one-degree of freedom system described by $\ddot{q} = \tau$, $q \leq 0$, and the control law $\tau = 9.81 - k_v\,\dot{q}$, with $k_v > 0$. It can be easily seen that the origin $q = 0$, $\dot{q} = 0$ is an asymptotically stable equilibrium point for the closed-loop system in presence of the inequality constraint (it is the model of a mass subject to viscous friction, bouncing on the ground under the action of the gravity force), whereas it is not (it is not even an equilibrium) in absence of the constraint. □

Proof of Theorem 1. By rewriting system (15) in the variables $\mathbf{z} = \mathbf{W}^{-1}\mathbf{q}$, $\mathbf{v}_z = \mathbf{W}^{-1}\mathbf{v}$, $\tilde{\mathbf{z}} = \mathbf{W}^{-1}\tilde{\mathbf{q}}$ and $\tilde{\mathbf{v}}_z = \mathbf{W}^{-1}\tilde{\mathbf{v}}$, where $\tilde{\mathbf{q}}(t) := \mathbf{q}(t) - \hat{\mathbf{q}}(t)$ and $\tilde{\mathbf{v}}(t) := \dot{\mathbf{q}}(t) - \hat{\mathbf{v}}(t)$ we obtain:

$$\dot{\mathbf{z}}(t) = \mathbf{v}_z(t), \quad t \in (t_i, t_{i+1}), i \in \mathbb{Z}^+, \tag{18a}$$

$$\begin{aligned}\dot{\mathbf{v}}_z(t) = &-\overline{\mathbf{K}}_p(\mathbf{z}(t) - \tilde{\mathbf{z}}(t)) + \overline{\overline{\mathbf{D}}}_Q\,\tilde{\mathbf{z}}(t) - \\ &\overline{\mathbf{K}}_v(\mathbf{v}_z(t) - \tilde{\mathbf{v}}_z(t)) + \overline{\mathbf{C}}_Q\,\mathbf{x}_Q(t) - \\ &\overline{\mathbf{D}}_Q\,\mathbf{r}(t) + \mathbf{R}_z(t), \\ &t \in (t_i, t_{i+1}), i \in \mathbb{Z}^+,\end{aligned} \tag{18b}$$

$$\dot{\tilde{\mathbf{z}}}(t) = \tilde{\mathbf{v}}_z(t) - \operatorname{diag}(k_{1,1}, \dots, k_{1,n})\,\tilde{\mathbf{z}}(t) + \mathbf{W}^{-1}\mathbf{K}_1\mathbf{r}(t), \quad t \in (t_i, t_{i+1}), i \in \mathbb{Z}^+, \tag{18c}$$

$$\dot{\tilde{\mathbf{v}}}_z(t) = -\operatorname{diag}(k_{1,1}, \dots, k_{1,n})\,\tilde{\mathbf{z}}(t) + \mathbf{W}^T\mathbf{K}_2\mathbf{r}(t), \quad t \in (t_i, t_{i+1}), i \in \mathbb{Z}^+, \tag{18d}$$

$$\dot{\mathbf{x}}_Q(t) = \mathbf{A}_Q\,\mathbf{x}_Q(t) + \overline{\mathbf{B}}_Q\,\tilde{\mathbf{z}}(t) - \mathbf{B}_Q\,\mathbf{r}(t), \quad t \in (t_i, t_{i+1}), i \in \mathbb{Z}^+, \tag{18e}$$

$$\mathbf{v}_z(t_i^+) = \mathbf{A}(e)\,\mathbf{v}_z(t_i^-), \quad i \in \mathbb{N}, \tag{18f}$$

$$\tilde{\mathbf{z}}(t_i^+) = \mathbf{A}(e)\,\tilde{\mathbf{z}}(t_i^-), \quad i \in \mathbb{N}, \tag{18g}$$

$$\tilde{\mathbf{v}}_z(t_i^+) = \mathbf{A}(e)\,\tilde{\mathbf{v}}_z(t_i^-), \quad i \in \mathbb{N}. \tag{18h}$$

where $\overline{\mathbf{B}}_Q = \mathbf{B}_Q\mathbf{W}$, $\overline{\mathbf{C}}_Q = \mathbf{W}^T\mathbf{C}_Q$, $\overline{\mathbf{D}}_Q = \mathbf{W}^T\mathbf{D}_Q$, $\overline{\overline{\mathbf{D}}}_Q = \mathbf{W}^T\mathbf{D}_Q\mathbf{W}$ and $\mathbf{R}_z(t) := \mathbf{W}^T\mathbf{R}(t)$, $t \in (t_i, t_{i+1})$, $i \in \mathbb{Z}^+$. By reordering the state variables in the new state vector $\mathbf{x}_{zc} = \begin{bmatrix}\mathbf{x}_z^T & \tilde{\mathbf{x}}_o^T & \mathbf{x}_Q^T\end{bmatrix}^T$, where

$$\mathbf{x}_z := [z_1 \quad v_{z,1} \dots z_n \quad v_{z,n}]^T$$
$$\tilde{\mathbf{x}}_o := \begin{bmatrix}\tilde{z}_1\ \tilde{v}_{z,1}\ \dots\ \tilde{z}_n\ \tilde{v}_{z,n}\end{bmatrix}^T$$

and letting

$$S_m := \min\left\{\frac{1}{\overline{\sigma}(\mathbf{W})}, 1\right\},$$
$$S_M := \max\left\{\frac{1}{\underline{\sigma}(\mathbf{W})}, 1\right\},$$

the following inequalities can be easily derived:

$$S_m\,\|\mathbf{x}_{cc}\| \le \|\mathbf{x}_{zc}\| \le S_M\,\|\mathbf{x}_{cc}\|. \tag{19}$$

Notice also that, in view of the special structure of $\mathbf{R}(t)$, given by (4), only the first component of the vector $\mathbf{R}_z(t)$ can be different from zero, *i.e.*, we have

$$\mathbf{R}_z(t) = \begin{bmatrix}\lambda(t)\\ 0\\ \vdots\\ 0\end{bmatrix}, \quad t \in (t_i, t_{i+1}), i \in \mathbb{Z}^+, \tag{20}$$

where, recalling (5), we have

$$\begin{aligned}\lambda(t) = &\,k_{p,1}(z_1(t) - \tilde{z}_1(t)) + \\ &k_{v,1}(v_{z,1}(t) - \tilde{v}_{z,1}(t)) - \\ &\overline{\mathbf{C}}_{Q\,1,:}\,\mathbf{x}_Q(t) - \overline{\overline{\mathbf{D}}}_{Q\,1,:}\,\tilde{\mathbf{z}}(t) + \overline{\mathbf{D}}_{Q\,1,:}\,\mathbf{r}(t).\end{aligned}$$

Due to the fact that, if $\mathbf{R}_z(t) \neq 0$ then $z_1(t) = 0$ and $v_{z,1}(t) = 0$, it is easy to see that equations (18a)-(18e) can be rewritten in compact form in terms of the state vector $\mathbf{x}_{zc}$, as follows:

$$\dot{\mathbf{x}}_{zc}(t) = \begin{cases}\mathbf{A}_{zc}\,\mathbf{x}_{zc}(t) + \mathbf{B}_{zc}\,\mathbf{r}(t), & \text{if } \mathbf{R}_z(t) = \mathbf{0},\\ \overline{\mathbf{A}}_{zc}\,\mathbf{x}_{zc}(t) + \overline{\mathbf{B}}_{zc}\,\mathbf{r}(t), & \text{if } \mathbf{R}_z(t) \neq \mathbf{0},\end{cases}$$

where the only difference between the matrices $\overline{\mathbf{A}}_{zc}$ and $\mathbf{A}_{zc}$ (between the matrices $\overline{\mathbf{B}}_{zc}$ and $\mathbf{B}_{zc}$) is that the components in positions $(2, 2n+1)$, ..., $(2, 4n+m)$ of matrix $\overline{\mathbf{A}}_{zc}$ are zero (the second row of matrix $\overline{\mathbf{B}}_{zc}$ is zero), whereas the corresponding components of $\mathbf{A}_{zc}$ (the second row of $\mathbf{B}_{zc}$) can be different from zero, in general. Let

$$\mathbf{A}_z := \operatorname{blockdiag}\left(\begin{bmatrix}0 & 1\\ -k_{p,1} & -k_{v,1}\end{bmatrix}, \dots, \begin{bmatrix}0 & 1\\ -k_{p,n} & -k_{v,n}\end{bmatrix}\right),$$

$$\mathbf{B}_z := \operatorname{blockdiag}\Big(\underbrace{\begin{bmatrix}0\\1\end{bmatrix}, \dots, \begin{bmatrix}0\\1\end{bmatrix}}_{n \text{ times}}\Big),$$

and

$$\mathbf{A}_o := \operatorname{blockdiag}\left(\begin{bmatrix}-k_{1,1} & 1\\ -k_{2,1} & 0\end{bmatrix}, \dots, \begin{bmatrix}-k_{1,n} & 1\\ -k_{2,n} & 0\end{bmatrix}\right).$$

Consider the quadratic function $V_{zc} = \mathbf{x}_{zc}^T\,\mathbf{P}_{zc}\,\mathbf{x}_{zc}$, where $\mathbf{P}_{zc} = \operatorname{blockdiag}(\mathbf{P}_z, \varepsilon\,\mathbf{P}_o, \eta\,\mathbf{P}_Q)$, with $\mathbf{P}_z$, $\mathbf{P}_o$ and $\mathbf{P}_Q$ are, respectively, the solutions of the following Liapunov equations

$$\mathbf{P}_z\,\mathbf{A}_z + \mathbf{A}_z^T\,\mathbf{P}_z = -\mathbf{I}_{2n},$$
$$\mathbf{P}_o\,\mathbf{A}_o + \mathbf{A}_o^T\,\mathbf{P}_o = -\mathbf{I}_{2n},$$
$$\mathbf{P}_Q\,\mathbf{A}_Q + \mathbf{A}_Q^T\,\mathbf{P}_Q = -\mathbf{I}_m,$$

and ε and η are positive parameters, whose values will be fixed later. By computing the matrices

$$\mathbf{Q}_{zc} = -\left(\mathbf{P}_{zc}\,\mathbf{A}_{zc} + \mathbf{A}_{zc}^T\,\mathbf{P}_{zc}\right),$$
$$\overline{\mathbf{Q}}_{zc} = -\left(\mathbf{P}_{zc}\,\overline{\mathbf{A}}_{zc} + \overline{\mathbf{A}}_{zc}^T\,\mathbf{P}_{zc}\right),$$

in view of the block diagonal form of $\mathbf{P}_{zc}$, it can be seen that

$$\mathbf{Q}_{zc} = \begin{bmatrix}\mathbf{I}_{2n} & \mathbf{Q}_1 & \mathbf{Q}_2\\ \mathbf{Q}_1^T & \varepsilon\,\mathbf{I}_{2n} & \eta\,\mathbf{Q}_3\\ \mathbf{Q}_2^T & \eta\,\mathbf{Q}_3^T & \eta\,\mathbf{I}_m\end{bmatrix},$$

$$\overline{\mathbf{Q}}_{zc} = \begin{bmatrix}\mathbf{I}_{2n} & \overline{\mathbf{Q}}_1 & \overline{\mathbf{Q}}_2\\ \overline{\mathbf{Q}}_1^T & \varepsilon\,\mathbf{I}_{2n} & \eta\,\mathbf{Q}_3\\ \overline{\mathbf{Q}}_2^T & \eta\,\mathbf{Q}_3^T & \eta\,\mathbf{I}_m\end{bmatrix},$$

where $\overline{\mathbf{Q}}_1$ and $\overline{\mathbf{Q}}_2$ are equal to $\mathbf{Q}_1$ and $\mathbf{Q}_2$, respectively, apart from the first two rows of $\overline{\mathbf{Q}}_1$ and $\overline{\mathbf{Q}}_2$, which are always null, whereas the first two rows of $\mathbf{Q}_1$ and $\mathbf{Q}_2$ need not be null. Then, letting $\chi_h = \|\mathbf{Q}_h\|$, $h = 1, 2, 3$, if $\mathbf{r}(\cdot) = \mathbf{0}$, we have

$$\dot{V}_{zc} \leq -\left[\, \|\mathbf{x}_z\| \; \|\tilde{\mathbf{x}}_o\| \; \|\mathbf{x}_Q\| \,\right] \widehat{\mathbf{Q}}_{zc} \begin{bmatrix} \|\mathbf{x}_z\| \\ \|\tilde{\mathbf{x}}_o\| \\ \|\mathbf{x}_Q\| \end{bmatrix}, \; \forall t \geq t_0,$$

where

$$\widehat{\mathbf{Q}}_{zc} = \begin{bmatrix} 1 & -\chi_1 & -\chi_2 \\ -\chi_1 & \varepsilon & -\eta\,\chi_3 \\ -\chi_2 & -\eta\,\chi_3 & \eta \end{bmatrix}.$$

By computing the three leading principal minors of $\widehat{\mathbf{Q}}_{zc}$:

$$\{1,\, \varepsilon - \chi_1^2,\; \varepsilon\,\eta - \eta^2\,\chi_3^2 - \eta\,\chi_1^2 - 2\,\eta\,\chi_1\,\chi_2\,\chi_3 - \varepsilon\,\chi_2^2\},$$

it can be seen that the choices $\eta = 2\,\chi_2^2$ (the fact that the pair $(\mathbf{C}_Q,\, \mathbf{A}_Q)$ is observable implies that $\mathbf{C}_Q$ cannot be zero, and this implies that $\chi_2 > 0$) and $\varepsilon > \varepsilon^*$, with $\varepsilon^* := 4\,\chi_3^2\,\chi_2^2 + 2\,\chi_1^2 + 4\,\chi_1\,\chi_2\,\chi_3$, render matrix $\widehat{\mathbf{Q}}_{zc}$ (and therefore both matrices $\mathbf{Q}_{zc}$ and $\overline{\mathbf{Q}}_{zc}$) positive definite. Whence, with such choices, V_{zc} is a Liapunov function for system (18). Let $\xi_c := \min\{\lambda_m(\mathbf{Q}_{zc}),\, \lambda_m(\overline{\mathbf{Q}}_{zc})\}$.

By computations wholly similar to the ones used in (Menini and Tornambè, 2001*b*), we can write

$$\|\mathbf{x}_{cc}(t)\| \leq \frac{S_M}{S_m}\sqrt{\frac{\lambda_M(\mathbf{P}_{zc})}{\lambda_m(\mathbf{P}_{zc})}}\,\|\mathbf{x}_{cc}(t_0)\| \cdot \exp\left(-\frac{\xi_c}{2\,\lambda_M(\mathbf{P}_{zc})}(t-t_0)\right),$$

which is relation (16) with $\alpha := \frac{S_M}{S_m}\sqrt{\frac{\lambda_M(\mathbf{P}_{zc})}{\lambda_m(\mathbf{P}_{zc})}}$ and $\beta := \frac{\xi_c}{2\,\lambda_M(\mathbf{P}_{zc})}$. As for assertion (ii), again using computations wholly similar to the ones in (Menini and Tornambè, 2001*b*), if $\|\mathbf{r}(t)\| < r_M$, $\forall t \geq t_0$, we can write

$$\|\mathbf{x}_{zc}(t)\| \leq \sqrt{\frac{\lambda_M(\mathbf{P}_{zc})}{\lambda_m(\mathbf{P}_{zc})}}\,\|\mathbf{x}_{zc}(t_0)\| \cdot \exp\left(-\frac{\xi_c}{2\,\lambda_M(\mathbf{P}_{zc})}(t-t_0)\right) + \frac{2\,\lambda_M^2(\mathbf{P}_{zc})}{\lambda_m(\mathbf{P}_{zc})\,\xi_c}\,\|\mathbf{B}_{zc}\|\,\|\mathbf{W}\|\,r_M,$$

from which, by (19), we derive that relation (17) holds with

$$\gamma_x(s,z) := \frac{S_M}{S_m}\sqrt{\frac{\lambda_M(\mathbf{P}_{zc})}{\lambda_m(\mathbf{P}_{zc})}}\exp\left(-\frac{\xi_c}{2\,\lambda_M(\mathbf{P}_{zc})}\,z\right) s,$$

$$\gamma_r(s) := \frac{2\,\overline{\sigma}(\mathbf{W})\,\lambda_M^2(\mathbf{P}_{zo})\,\|\mathbf{B}_{zc}\|}{S_m\,\lambda_m(\mathbf{P}_{zc})\,\xi_c}\,s. \qquad \blacksquare$$

The example described in next section shows how the parameterization of a family of stabilizing compensators can be used to deal with control problems involving different requirements in addition to asymptotic and BIBS stability.

4. AN EXAMPLE OF APPLICATION

Consider a dimensionless body having mass M that moves (without friction) on a vertical plane, subject to the gravity acceleration, of magnitude g. Letting $q_1(t)$ and $q_2(t)$ denote the horizontal and vertical position coordinates of the body at time t, respectively, and assuming that it is possible to exert an horizontal and a vertical force on the body, which are denoted by $\tau_1(t)$ and $\tau_2(t)$, the equations of motion for the uncostrained body are:

$$\ddot{q}_1(t) = \tau_1(t)/M,$$
$$\ddot{q}_2(t) = \tau_2(t)/M - g.$$

Let the body be constrained to move on the upper half plane delimited by $q_2(t) \geq q_1(t)$, so that, in our setting, $\mathbf{J} = [1 \;\; -1]$, and let $e = 1$; it can be computed that, at the impact times, we have $\dot{q}_1(t_i^+) = \dot{q}_2(t_i^-)$ and $\dot{q}_2(t_i^+) = \dot{q}_1(t_i^-)$. Let only the position coordinates $\mathbf{q}(t) := [q_1(t)\; q_2(t)]^T$ be available for measurement. We assume, further, that the mass of the body is not exactly known, and only its nominal value M_0 is available for control design. It is clear that the goal of regulating the position of the body to the origin $\mathbf{q} = \mathbf{0}$ cannot be obtained robustly with respect to the mass variation by means of the PD compensator from the observed state proposed in (Menini and Tornambè, 2001*a*) (which is practically the one proposed here with the "Q" subcompensator set equal to zero), since the compensation of the gravity acceleration is not exact if $M \neq M_0$. A first set of simulations has been carried out to confirm this fact, by using the output feedback compensator proposed in (Menini and Tornambè, 2001*a*), with gains $k_{p,1} = k_{p,2} = 3$, $k_{v,1} = k_{v,2} = 4$, $k_{1,1} = k_{1,2} = 15$ and $k_{2,1} = k_{2,2} = 50$. In Figure 1, the results of three of such simulations, corresponding to $M = M_0 = 1$ kg (the nominal case, represented by the bold lines), $M = M_1 = 0.95$ kg (represented by the dashed lines) and to $M = M_2 = 1.1$ kg (represented by the continuous lines), respectively, are reported. In the plots (b) and (c), horizontal lines represent the equilibrium positions of the constrained closed-loop system. In both the perturbed cases, since the gravity force is not compensated exactly, such equilibrium positions are different from the origin. Notice that, for $M = M_1$ (dashed lines), such an equilibrium coincides with the one of the uncostrained system (it is not a contact configuration), whereas, for $M = M_2$, the equilibrium of the uncostrained system does not belong to the admissible region,

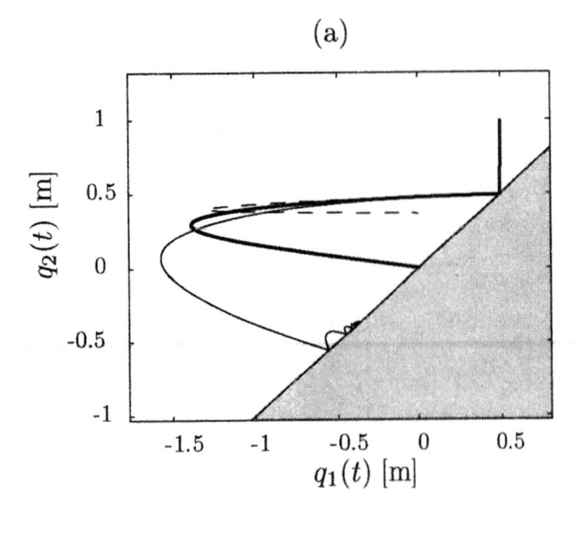

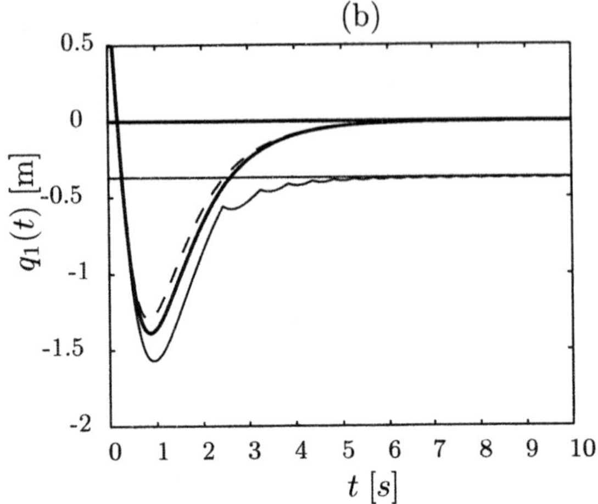

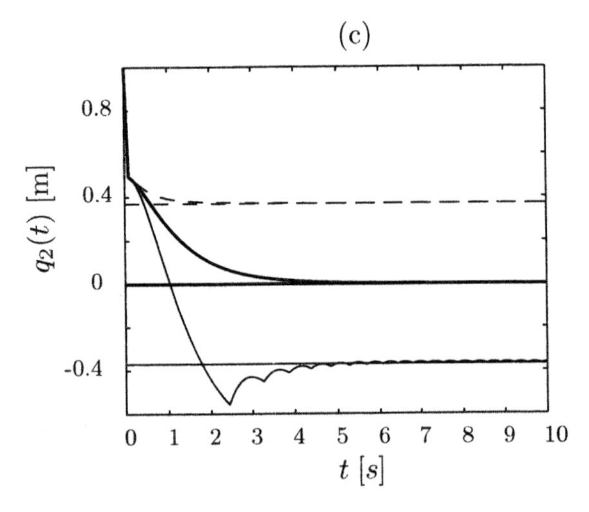

Fig. 1. Results with the PD compensator from the observed state proposed in (i.e., without "Q" subcompensator). The bold lines represent the nominal case $M = 1$ kg, the dashed lines the case of $M = 0.95$ kg, whereas the continuous lines the case of $M = 1.1$ kg.

thus implying that the system experiences an infinite number of impacts approaching the only equilibrium of the constrained system.

In order to solve the mentioned robust regulation problem, the degrees of freedom for control design provided by the parameterization proposed in this section can be used. Observe, first, that the system to be controlled is constituted, if the constraint is neglected, by two SISO decoupled subsystems, having transfer function $q_h(s)/\tau_h(s) = 1/\left(M\,s^2\right)$, $h = 1, 2$, which interact only at the impact times. Therefore, it is very easy to design, for each of such subsystems, a linear time invariant compensator which solves the robust regulation problem, just by modelling the wrong compensation of the gravity acceleration as an additional constant input disturbance, to be asymptotically rejected. To this end, two identical linear time invariant compensators having transfer functions $\tau_h(s)/(q_h(s) - r_h(s)) = C_h(s) = (b_2\,s^2 + b_1\,s + b_0)/(s^2 + a_1\,s)$, $h = 1, 2$, have been designed, by placing the closed-loop poles of the nominal system at the same locations obtained by the PD compensator mentioned above. Then, by using the formulae given in (Colaneri *et al.*, 1997, Sec. 3.4), such compensators have been transformed into the connection of the PD compensator used above with the "Q" subsystem, which turns out to be a suitable static gain. As a final step, the jumps (15g) and (15h) have been incorporated in the compensator. The results of the three simulations of the closed-loop systems corresponding to $M = M_0$, $M = M_1$ and $M = M_2$ (with the same initial conditions of the simulations in Figure 1) are reported in Figure 2. It can be appreciated that regulation is obtained in all three cases. In addition, since two identical compensators have been used for the "horizontal" and "vertical" subsystems, asymptotic rejection of constant input disturbances is obtained with such a compensator.

5. CONCLUDING REMARKS

The strongest assumption made in this paper is the linearity of the mechanical system when undisturbed by the impacts. This assumption allows one to write in closed form a family of compensators guaranteeing for the closed-loop system both the Liapunov stability and the BIBS stability; if the unconstrained system is nonlinear, then the compensators proposed can be applied only locally. It seems difficult to overcome the mentioned assumption, since, also when the mechanical system is unconstrained, it is hard (if not impossible) to design the family of all globally stabilizing controllers.

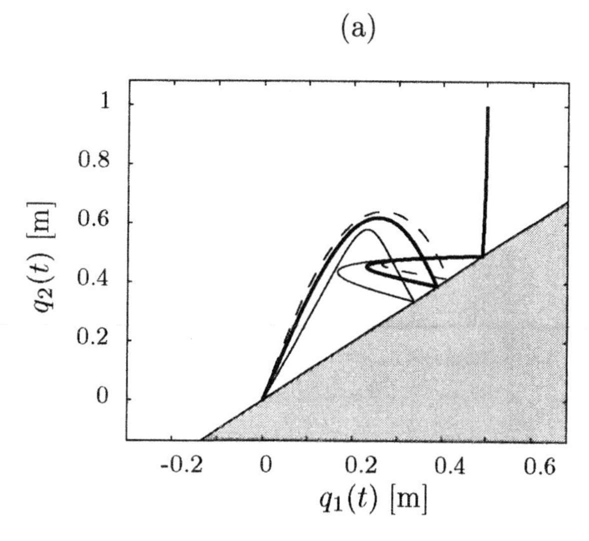

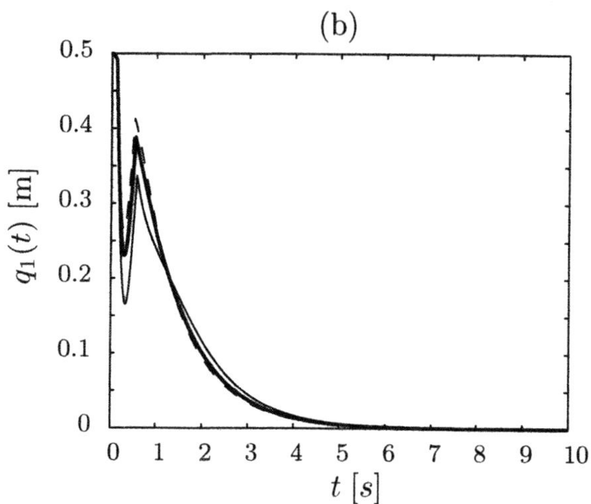

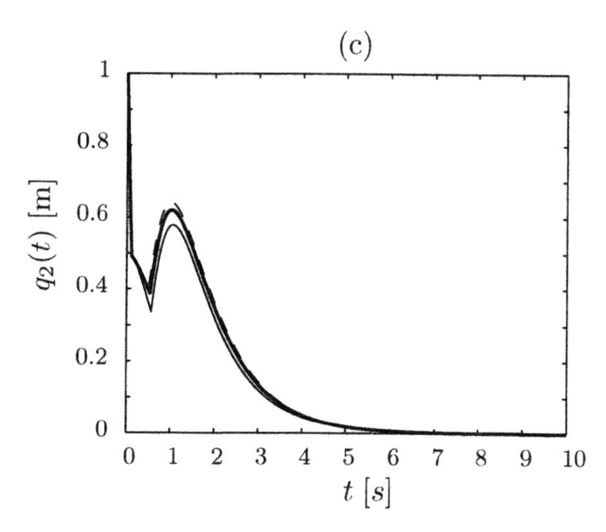

Fig. 2. Results with the compensator designed for robust regulation. As in Fig. 1, the bold lines represent the nominal case $M = 1$ kg, the dashed lines the case of $M = 0.95$ kg, whereas the continuous lines the case of $M = 1.1$ kg.

Future work will be devoted to study choices of the free parameter $\mathbf{Q}(s)$ so that we can guarantee additional properties for the closed-loop system, such as parametric or non-parametric robustness, asymptotic or practical trajectory tracking (possibly, involving a large number of or an infinite number of impacts), asymptotic or practical disturbance rejection (possibly guaranteeing a condition of permanent contact also in the presence of disturbances).

REFERENCES

Brogliato, B. (1996). *Nonsmooth impact mechanics.* Springer-Verlag. London.

Colaneri, P., J. C. Geromel and A. Locatelli (1997). *Control theory and design: an RH_2 and RH_∞ viewpoint.* Academic Press.

Francis, B. A. (1987). *A course in H_∞ control theory.* Springer-Verlag.

Hurmuzlu, Y. (1993). Dynamics of bipedal gait: Part I and part II – stability analysis of a planar five-link biped. *J. of Applied Mechanics* **60**, 331–343.

Hurmuzlu, Y. and D. B. Marghitu (1994). Rigid body collision of planar kinematic chains with multiple contact points. *Int. J. of Robotics Research* **13**(1), 82–92.

Hurmuzlu, Y. and T. H. Chang (1992). Rigid body collisions of a special class of planar kinematic chain. *IEEE Trans. Systems, Man and Cybernetics* **22**(5), 964–971.

Hurmuzlu, Y., F. Génot and B. Brogliato (n.d.). Dynamics and control of bipedal locomotion systems - a tutorial survey. Submitted.

Izumi, T. and Y. Hitaka (1997). Hitting from any direction in 3-D space by a robot with a flexible link hammer. *IEEE Trans. on Robotics and Autom.* **13**(2), 296–301.

Koditschek, D. E. and M. Bühler (1991). Analysis of a simplified hopping robot. *Int. J. of Robotics Research* **10**(6), 586–605.

Kučera, V. (1974). Algebraic theory of discrete optimal control for multivariable systems. *Kybernetika* **10–12**, 1–240.

Maciejowski, J. M. (1989). *Multivariable feedback design.* Addison Wesley.

Marth, G. T., T.J. Tarn and A. K. Bejczy (1994). An event based approach to impact control: Theory and experiments. In: *Proc. IEEE Conf. Robotics Autom.*. San Diego, CA.

Menini, L. and A. Tornambè (1999). The method of penalizing functions for elastic/anelastic impacts. In: *European Control Conference.* Karlsruhe (Germany).

Menini, L. and A. Tornambè (2001*a*). Dynamic position feedback stabilization of multi-degrees-of-freedom linear mechanical systems subject to non-smooth impacts.

IEE Proc. Control Theory and Applications **148**(2), 147–155.

Menini, L. and A. Tornambè (2001*b*). Exponential and BIBS stabilization of one-degree-of-freedom mechanical systems subject to single non-smooth impacts. *IEE Proc. Control Theory and Applications* **148**(2), 147–155.

Menini, L. and A. Tornambè (2001*c*). Modification of the youla-kucera parameterization for linear mechanical systems subject to non-smooth impacts. In: *40th IEEE Conference on Decision and Control*. Orlando (FL, USA).

Menini, L. and A. Tornambè (2001*d*). Velocity observers for linear mechanical systems subject to single non-smooth impacts. *Systems & Control Letters* **43**, 193–202.

Mills, J.K. and D. M. Lokhorst (1993). Stability and control of robotic manipulators during contact/noncontact task transition. *IEEE Trans. Robotics and Automation* **9**(3), 335–345.

Nenchev, D.N. and K. Yoshida (1999). Impact analysis and post-impact motion control issues of a free-floating space robot subject to a force impulse. *IEEE Transactions on Robotics and Automation* **15**(3), 548 – 557.

Prokop, G. and F. Pfeiffer (1998). Synthesis of robot dynamic behavior for environmental interaction. *IEEE Trans. on Robotics and Autom.* **14**(5), 718 – 731.

Swanson, P. J., R. R. Berridge and D. E. Koditschek (1995). Global asymptotic stability of a passive juggler: A parts feeding strategy. In: *Proc. IEEE Conf. Robotics and Automation*. pp. 1983–1988.

Tornambè, A. (1999). Modeling and control of impact in mechanical systems: theory and experimental results. *IEEE Trans. Autom. Control* **44**(2), 294 – 309.

Wee, Liang-Boon and M.W. Walker (1993). On the dynamics of contact between space robots and configuration control for impact minimization. *IEEE Trans. on Robotics and Autom.* **9**(5), 581 – 591.

Youla, D. C., H. A. Jabr and J. J. Bongiorno (1976). Modern Wiener-Hopf design of optimal controllers, part II: the multivariable case. *IEEE Trans. on Automatic Control* **21**, 319–338.

Zheng, Y. F. and H. Hemami (1985). Mathematical modelling of a robot collision with its environment. *J. of Robotics Systems* **2**(3), 289–307.

Zumel, N. B. and M. A. Erdmann (1994). Balancing of a planar bouncing object. In: *Proc. IEEE Conf. Robotics and Automation*. pp. 2949–2954.

CRITERIA FOR DESIGN OF PID CONTROLLERS

Ari Ingimundarson **Tore Hägglund and Karl Johan Åström**

ESAII, Edifici TR11
Rambla de Sant Nebridi s/n,
08222 Terrassa, Spain
ari.ingimundarson@upc.es
Fax: +34 93 739 8628

Department of Automatic Control
Lund Institute of Technology
Box 118, SE-221 00 Lund, Sweden
{tore|kja}@control.lth.se
Fax: +46 46 13 81 18

Abstract This paper treats robustness constraints and performance criteria for design of PID controllers. It is shown that robustness constraints are essential, and that controller design based on maximization of performance criteria without robustness constraints give poor controllers. It is also shown that the robustness constraints have a significant influence on the performance, and that the performance often can be increased significantly by reducing the robustness constraints. Some comments relating to industrial practice are offered in this context. *Copyright © 2003 IFAC*

Keywords PID control, robust control, design criteria

1. INTRODUCTION

In spite of all the advances in control over the past 50 years the PID controller is still the most common controller, see (Åström and Hägglund, 2001). A survey of more than eleven thousand controllers in the refining, chemicals, and pulp and paper industries showed that 97% of regulatory controllers had the PID structure, see (Desbourough and Miller, 2002). Even when more sophisticated control laws are used, it is common practice to have an hierarchical structure with PID control at the lowest level, see (Qin and Badgwell, 1997).

Because of the wide spread use of PID control it is highly desirable to have efficient manual and automatic methods of tuning the controllers. A good insight into PID tuning is also useful in developing more schemes for automatic tuning and loop assessment.

In this paper, the compromise between robustness and performance is discussed. Some questions that will be answered are: What are the important aspects of the criteria for robustness and performance? For design of PID controllers can we neglect the robustness and simply focus on the performance? What are the essential differences between the different robustness criteria?

2. PRELIMINARIES

Figure 1 shows a simple feedback loop, with a process with transfer function $G(s)$ and a controller with two degrees of freedom. Three external signals act on the control loop, namely set point y_{sp}, load disturbance l and measurement noise n. The load disturbance drives the process variables away from their desired values and the measurement noise corrupts the information obtained from the sensors.

Therefore, the design should express requirements on set point, load disturbance and measurement noise response as well as robustness to model uncertainties.

Since the controller has two degrees of freedom, it is possible to separate the requirements. The response to measurement noise can be treated by designing a low-pass filter for the measurement signal. The desired set-point response can be obtained by feed forward. It is also possible to feed the set point through filters or ramping modules.

Load disturbances are often the most common disturbances in process control. See (Shinskey, 1996). Most design methods should therefore focus on load disturbances, and try to find a suitable compromise between demands on performance at load disturbances and robustness. It is a great advantage if this compromise can be decided by the user by a tuning parameter. There

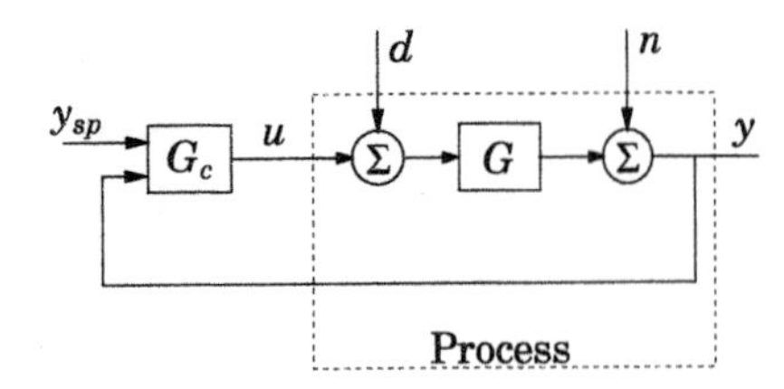

Fig. 1 Block diagram of a system with a process and a controller having two degrees of freedom.

are two types of tuning parameters, those where the robustness is specified, and those where the performance is specified.

3. PERFORMANCE AND ROBUSTNESS

Regulation performance is often of primary importance since most PID controllers operate as regulators. Regulation performance is often expressed in terms of the control error obtained for certain disturbances. A disturbance is typically applied at the process input. Typical criteria are to minimize a loss function of the form

$$I = \int_0^\infty t^n |e(t)|^m dt \tag{1}$$

where the error is defined as $e = y_{sp} - y$. Common cases are IAE ($n = 0$, $m = 1$), ISE ($n = 0$, $m = 2$), or ITSE ($n = 1$, $m = 2$). Quadratic criteria are particularly popular since they admit analytical solutions. Criteria which put penalty on the control signal have also been used.

A related error is integrated error (IE) defined as

$$IE = \int_0^\infty e(t)dt \tag{2}$$

The criteria IE and IAE are the same if the error does not change sign. Notice, however, that IE can be very small even if the error is not. For IE to be relevant it is necessary to add conditions that ensure that the error is not too oscillatory.

One reason for using IE is that its value is directly related to the integral gain k_i of the PID controller, see (Åström and Hägglund, 2001). For a load disturbance in the form of unit step, $IE = 1/k_i$.

The criteria described above are used to *optimize* performance. Another approach is to *specify* a desired performance. This is used e.g. in Lambda-tuning, where the desired time constant of the closed-loop system is specified.

3.1 *Robustness Criteria*

Robustness is an important consideration in control design. In this section we will discuss some robustness criteria and we will also derive the constraints robustness imposes on the parameters of a PID controller.

There are many different criteria for robustness. Many of them can be expressed as restrictions on the Nyquist curve of the loop transfer function. The constraint that sensitivity function $S(i\omega)$ is less than a given value M_s implies that the loop transfer function should be outside a circle with radius $1/M_s$ and center at -1. The constraint that complementary sensitivity $T(i\omega)$ is less than a given value M_t also implies that the loop transfer function is outside a circle. In (Åström *et al.*, 1998) it is shown that a constraint on both M_s and M_t can be replaced by a slightly more conservative constraint M, assuming that $M_s = M_t = M$. A given value of M implies that the loop transfer function is outside a circle with center C and radius R where

$$C = \frac{2M^2 - 2M + 1}{2M(M-1)}$$
$$R = \frac{2M - 1}{2M(M-1)}$$

By choosing such a constraint one can capture robustness by one parameter only.

Good insight into the robustness problem has been provided by the $\mathcal{H}_\infty$ theory, see (Panagopoulos and Åström, 2000). According to this theory it follows that the closed loop system is robust to perturbations if the $\mathcal{H}_\infty$-norm of the transfer function

$$H = \frac{1}{1 + GG_c}\begin{pmatrix} 1 & G \\ G_c & GG_c \end{pmatrix}$$

is sufficiently small. H is the transfer function from n and l to y and u. In (Panagopoulos and Åström, 2000) it was shown that with a scaling to which the loop transfer function is invariant, the norm is equivalent to:

$$\begin{aligned} \gamma &= \sup_\omega \frac{1 + |G(i\omega)G_c(i\omega)|}{|1 + G(i\omega)G_c(i\omega)|} \\ &= \sup_\omega (|S(i\omega)| + |T(i\omega)|) \end{aligned} \tag{3}$$

There it is also shown that the norm is always less than $\gamma_0(M) = \sqrt{4M^2 - 4M + 2}$.

The different robustness criteria imply that the Nyquist curve of the loop transfer function should avoid certain regions. Figure 2 shows these regions for constraints defined by M_s, M_t, M and γ.

3.2 *The Robustness Region*

Requirements on robustness imposes constraints on the admissible parameters of a

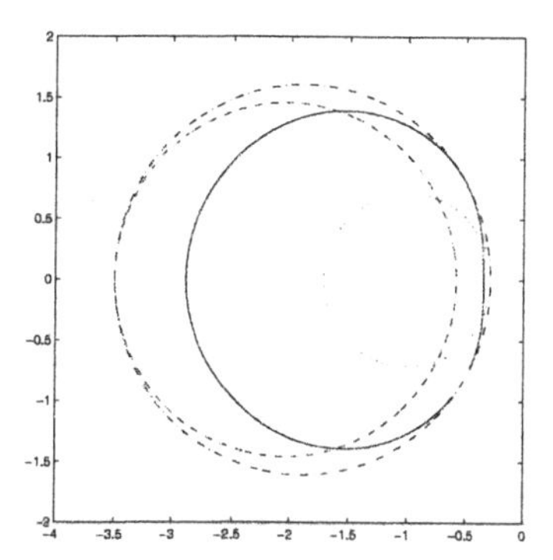

Fig. 2 Geometric illustration of the robustness criteria. The figure shows the boundaries of the regions in the Nyquist plot that must be avoided to have $M_s \leq 1.4$ (dotted), $M_t \leq 1.4$ (dashed), $M \leq 1.4$ (dash-dotted) and $\gamma_0(1.4) = 2.06$.

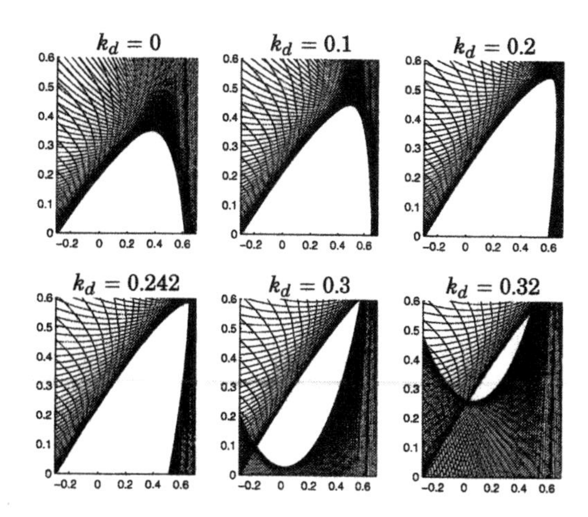

Fig. 3 Robustness regions for constant derivative gain k_d. The figure shows k_i versus k with fixed k_d, for PID control of the process $P(s) = e^{-s}/(s+1)$. The robustness region is computed for $M = 1.4$. Notice the sharp corners of the region for large k_d ("the derivative cliff").

PID controller. Consider a process with transfer function

$$G(i\omega) = a(\omega) + ib(\omega) = r(\omega)e^{i\varphi(\omega)} \tag{4}$$

and the PID controller

$$G_c(s) = k + \frac{k_i}{s} + k_d s$$

The condition that the loop transfer function is outside a circle with center at $-C$ and radius R can be expressed as

$$\begin{aligned} 1 \;\leq\; & \left(\frac{r}{\omega R}\right)^2 \left(k_i + \frac{b(\omega)\omega C}{R^2} - \omega^2 k_d\right)^2 \\ + \; & \left(\frac{r}{R}\right)^2 \left(k + \frac{a(\omega)C}{r^2}\right)^2 \end{aligned} \tag{5}$$

see (Panagopoulos *et al.*, 2002). For fixed ω and k_d this equation represents the exterior of an ellipse in the k-k_i plane. For fixed k_d the projection of the constraint is thus the envelope of the ellipses generated by varying ω. The complete constraint surface is obtained by sweeping over k_d.

4. SOME OBSERVATIONS

4.1 *The Derivative Cliff*

Figure 3 shows integral gain k_i as a function of k for different constant values of k_d for the process

$$P(s) = \frac{e^{-s}}{s+1} \tag{6}$$

Consider first the case of PI control, i.e. $k_d = 0$. Integral gain k_i is a nice smooth function which has a maximum $k_i = 0.33$. The value of k_i can be increased by introducing derivative action. As k_d is increased, the maximum of $k_i = 0.6$ does, however, occur at an edge. Integral gain decreases as k_d is increased further. The robustness region shrinks and disappears for k_d slightly larger than 0.33. The largest value of k_i is obtained for $k_d = 0.3$. When the maximum is at an edge, the performance is very sensitive to variations in the controller parameters. Figure 3 clearly illustrates the difficulties with derivative action. If the sensitivity region is represented by a volume in the k_i, k, k_d plane it follows that the volume has a sharp-pointed edge. This is the reason why the robustness region can be called the derivative cliff, see (Wallén *et al.*, 2002). The robustness regions are inside the stabilizing region which also have edges, see (Datta *et al.*, 2000).

The situation illustrated in Figure 3 is not an exceptional case, but it will occur in most cases. This is clearly not a desirable situation.

Figure 4 shows the robustness region for the process

$$P(s) = \frac{e^{-s}}{10s+1} \tag{7}$$

The figure is similar to Figure 3 but the integral gain can be increased more because the process is lag dominated.

When there is an edge in the envelope, the Nyquist curve of the loop transfer function touches the robustness circle in two points. In such cases the Nyquist curve has a cusp as discussed in (Panagopoulos *et al.*, 2002). This implies that the controller has excessive lead and that the responses are oscillatory. Additional conditions are therefore required to avoid these difficulties. In (Panagopoulos *et al.*, 2002) it is suggested that the loop transfer function should have a decreasing phase curve and that the curvature of the loop transfer should be negative in the neighborhood of the M-circle. These

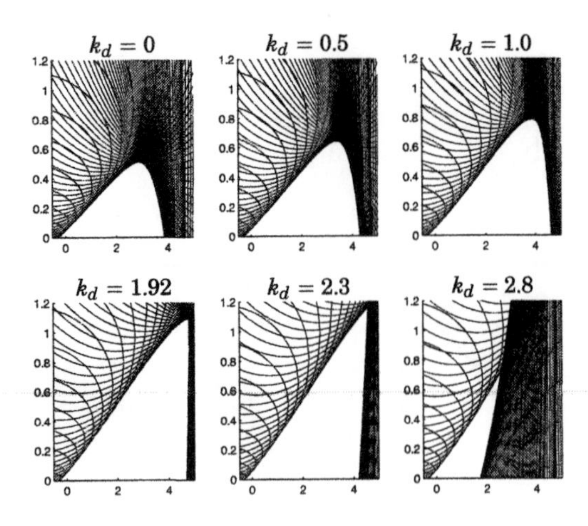

Fig. 4 Robustness region for constant derivative gain k_d. The figure shows k_i versus k with fixed k_d, for PID control of the process $P(s) = e^{-s}/(10s + 1)$. The robustness region is computed for $M = 1.4$. Notice the sharp corners of the region for large k_d ("the derivative cliff").

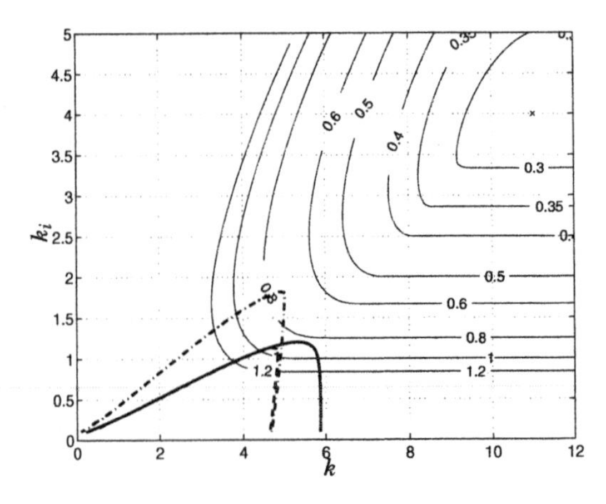

Fig. 5 Robustness region for constant derivative gain $k_d = 2.2$. The figure shows k_i versus k with fixed k_d, for PID control of the process $P(s) = e^{-s}/(1 + 10s)$ and the robustness regions computed for $\gamma = 2.06$ (solid), $M = 1.4$ (dashed), and $M_s = 1.4$ (dash-dotted). Level curves for IAE are also shown in the figure. The M region is approximately the intersection between the M_s region and γ region.

constraints give a nice shape of the loop gain but can lead to unnecessarily conservative designs. Another drawback is that computation of the curvature involves two differentiations of the frequency response of the process. This is particularly serious when the transfer function is obtained experimentally, see (Wallén, 2000). In (Wallén *et al.*, 2002) excessive derivative action is avoided by constraining the ratio to $T_i/T_d = 4$. It is shown that this gives reasonable controllers for processes whose dynamics is balanced, but some performance is sacrificed.

In (Åström and Hägglund, 2003) another way to obtain large integral gain and avoid excessive derivative action is proposed. It is suggested to use the largest value of the derivative gain such that the slope $\partial k_i/\partial k$ is nonnegative. This means that derivative gain should be chosen so that the curve $k_i(k)$ has a horizontal tangent. In Figure 3 this corresponds to the case $k_d = 0.242$ and in Figure 3 it corresponds to $k_d = 1.92$. Evaluations of a large number of cases have shown that the design method gives controllers with good properties. It is also straight forward to determine controller parameters for this constraint, see (Åström and Hägglund, 2003).

4.2 *Design Choices*

A good controller must satisfy some robustness requirement. The controller should also give a closed-loop system with a good performance. Since there are many ways to express the robustness and many performance criteria there are many different ways to formulate the design problem. In this section we will present results that give insight into the different choices. In particular we will try to find the issues that are of prime importance for control.

We will start by investigating a lag dominated process with the transfer function

$$G(s) = \frac{e^{-s}}{1 + 10s}$$

A large number of controllers have been investigated by gridding the controller parameters and evaluating the closed-loop system obtained by computing the performance criteria IAE and IE when a step was applied to the plant input. In addition, the robustness constraints M_s, M, and γ were calculated for each controller. The results are presented in Figure 5. This figure shows how the criteria on robustness and performance depend on the controller parameters. In particular the figure shows the dependence on k and k_i for fixed k_d.

Figure 5 also shows the robustness constraints $M_s = 1.4$, $M = 1.4$, and $\gamma = 2.06$, where the values are matched as described in the previous section. It means that M encloses both M_s and γ, see Figure 2. Figure 5 shows that M is the most restrictive constraint, and that the constraints on M and γ are close. The γ constraint admits slightly higher controller gain since it deviates from the circular shape, see Figure 2. Figure 6 shows the Nyquist curves of the loop transfer functions obtained when minimizing IAE with respect to robustness constraints on M_s, M, and γ, respectively.

A comparison of the curves of constant robustness with the level curves for IAE in Figure 5 shows that the achievable performance is limited by the robustness constraints. The figure

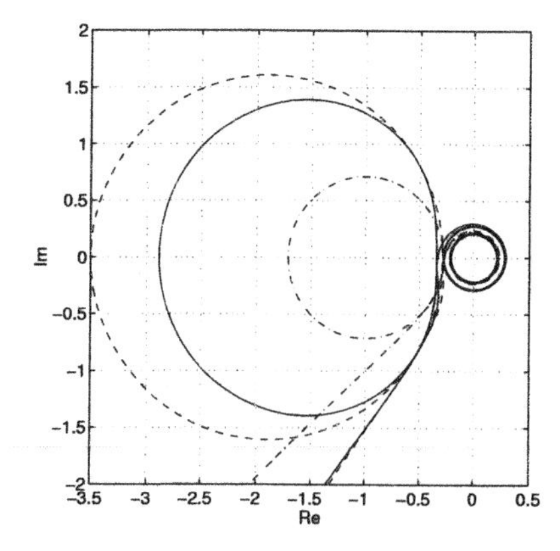

Fig. 6 Nyquist curves of the loop transfer functions obtained when minimizing IAE with respect to robustness constraints on M_s, M, and γ, respectively

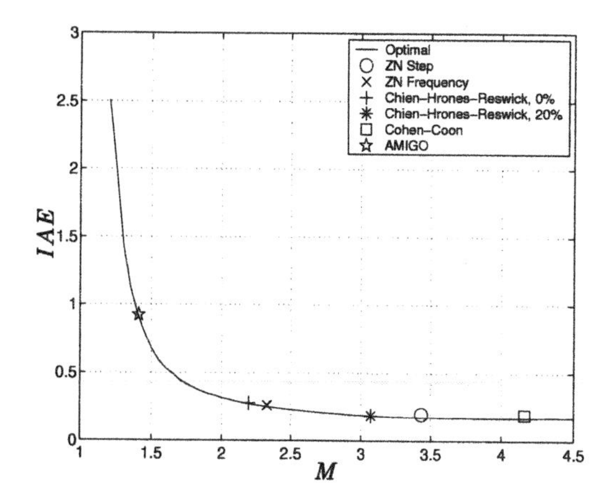

Fig. 8 Optimal IAE as a function of M with some common PID tuning rules. The process is $P(s) = e^{-s}/(10s+1)$.

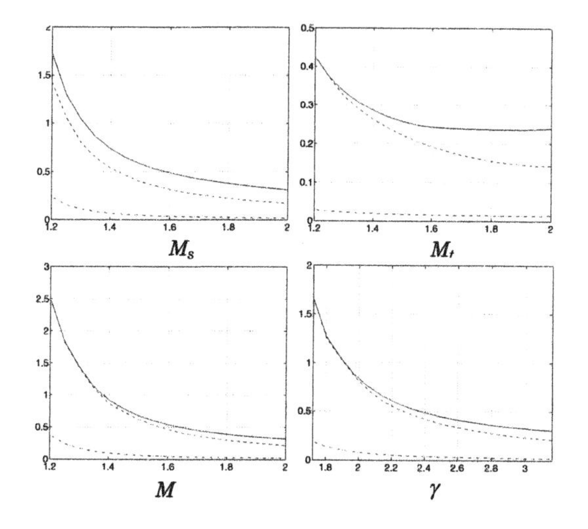

Fig. 7 Optimal IAE (solid), IE (dashed) and ISE (dash-dotted) values obtained for different values of the robustness constraints M_s, M_t, M, and γ, respectively.

also shows that for maximization of IAE and IE leads to similar parameter values because of the knee in the robustness constraint.

The figure shows clearly that it is not reasonable to neglect the robustness constraints. Maximization of IAE without consideration of robustness gives $IAE = 0.25$ for $k = 11.0$ and $k_i = 4.0$. These controller parameters do, however, give very poor robustness: $M_s = 3.0$, $M_t = 2.4$, $M = 3.2$ and $\gamma = 5.3$. The optimal over all k_d values on the other hand resulted in an optimal $IAE = 0.17$, with $k_d = 6.4$, $M_s = 3.9$, $M_t = 3.0$, $M = 4.0$ and $\gamma = 6.9$. This situation is representative for many different processes and other criteria such as IE, ISE, ISTE etc. We can thus conclude that it is essential to consider the robustness constraints.

Figure 7 illustrates the compromise between robustness and performance. The figure shows optimal IAE, IE and ISE values obtained for different values of the robustness constraints M_s, M_t, M, and γ, respectively. Note that ISE has different unit from IAE and IE so the numerical values should not be compared.

Figure 7 shows that most of the robustness constraints have a significant influence on the performance. An exception is M_t. Reasonable values of the robustness constraints must be used to obtain good control, but in process control it is common practice to increase the constraints further, typically by detuning the controller, to ensure good control when dynamics change due to nonlinearities in the plant. Figure 7 shows that the prize in terms of losses in performance may be high. In many cases it is a better solution to relax the robustness constraints, and compensate for changes in dynamics using gain scheduling or adaptation.

The figure also shows that the IAE and IE values coincide for large robustness constraints in terms of M and γ. In the case of M_s, there is always a difference between IE and IAE indicating that the optimal controller is not strictly damped even for low M_s values. This shows the necessity to include extra constraints in the design problem as discussed in Section 3.1.

Audits conducted in the process industry have shown that problems due to controller tuning are frequent. The problems appear as oscillations due to tight tuning or sluggishness due to loose tuning, see (Bialkowski, 1993). Systematic tuning of PID controllers in the process industry is usually done using a tuning rule. This applies even if the controller is automatically tuned. In Figures 8 and 9 the IAE and M are shown for some common tuning rules. It should be noted that many of these rules do not aim specifically for good IAE performance. For the process $e^{-s}/(10s+1)$, all of the rules yield controllers with performance close to the optimal one. However, the robustness varies considerably for the

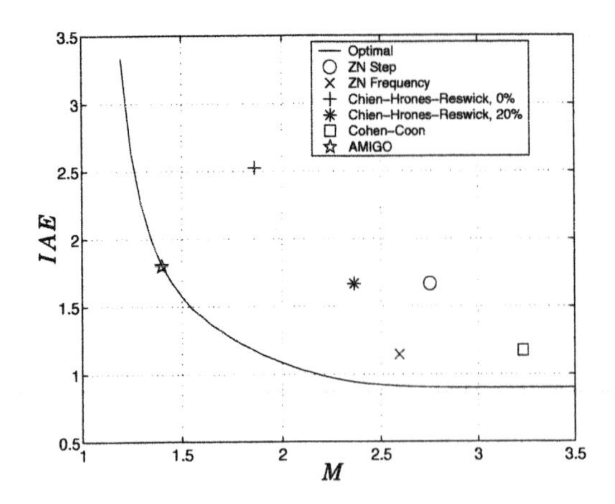

Fig. 9 Optimal IAE as a function of M with some common PID tuning rules. The process is $P(s) = e^{-s}/(s+1)$.

rules considered. Many of them have poor robustness. For the process $e^{-s}/(s+1)$ the robustness is slightly better but the performance differs from the optimal one considerably. Use of these rules without careful consideration to robustness is likely to result in problems later on when process dynamics change.

5. CONCLUSIONS

This paper has discussed robustness constraints and performance criteria for design of PID controllers. One conclusion is that robustness constraints are essential. Maximizing performance criteria without robustness constraints give very poor controllers.

Another conclusion is that the robustness constraints have a significant influence on the performance. The performance can often be increased significantly if the robustness constraints are relaxed, e.g. by improving the model certainty using gain scheduling or adaptation.

Optimization of IE and IAE give similar controllers for most of the robustness constraints. It was concluded that M and γ are more suitable robustness measures since it is reasonable to obtain an over-damped system for large robustness values. The importance of additional constraints on the designs to step away from the "derivative cliff" was demonstrated as well.

For M and γ, IAE and IE are equal for the optimal controllers when robustness is large. The IAE and IE values differ when the robustness is specified in terms of M_s. It is desirable that a design for large robustness results in an over-damped system. In light of this it seems that M or γ are preferable over M_s. It was shown that the use of common tuning rules might result in a controller with good performance but poor robustness.

6. ACKNOWLEDGMENTS

The authors acknowledge the support received by the Research Commission of the "Generalitat de Catalunya" (group SAC ref.2001/SGR/00236).

7. REFERENCES

Åström, K. J. and T. Hägglund (2001): "The future of PID control." *Control Engineering Practice*, **9**, pp. 1163–1175.

Åström, K. J. and T. Hägglund (2003): "Revisiting the Ziegler-Nichols step response method for PID control." *Submitted to Journal of Process Control.*

Åström, K. J., H. Panagopoulos, and T. Hägglund (1998): "Design of PI controllers based on non-convex optimization." *Automatica*, **34:5**, pp. 585–601.

Bialkowski, W. (1993): "Dreams versus reality: a view from both sides of the gap." *Pulp and Paper Canada*, **94**.

Datta, A., M. Ho, and S. Bhattacharyya (2000): *Structure and synthesis of PID controllers.* Springer-Verlag.

Desbourough, L. and R. Miller (2002): "Increasing customer value of industrial control performance monitoring - Honeywell's experience." In *Sixth International Conference on Chemical Process Control.* AIChE Symposium Series Number 326 (Volume 98).

Panagopoulos, H. and K. J. Åström (2000): "PID control design and H_∞ loop shaping." *Int. J. Robust Nonlinear Control*, **10**, pp. 1249–1261.

Panagopoulos, H., K. J. Åström, and T. Hägglund (2002): "Design of PID controllers based on constrained optimisation." *IEE Proc. Control Theory Appl.*, **149:1**, pp. 32–40.

Qin, S. J. and T. A. Badgwell (1997): *An overview of industrial model predictive control technology*, pp. 232–256. CACHE, AIChE.

Shinskey, F. G. (1996): *Process-Control Systems. Application, Design, and Tuning*, 4 edition. McGraw-Hill, New York.

Wallén, A. (2000): *Tools for Autonomous Process Control.* PhD thesis ISRN LUTFD2/TFRT--1058--SE, Department of Automatic Control, Lund Institute of Technology, Sweden.

Wallén, A., K. J. Åström, and T. Hägglund (2002): "Loop-shaping design of PID controllers with constant Ti/Td ratio." *Asian Journal of Control*, **4:4**, pp. 403–409.

A NEW WINDUP PREVENTION SCHEME FOR STABLE AND UNSTABLE SYSTEMS

P. Hippe

Lehrstuhl für Regelungstechnik, Universität Erlangen-Nürnberg
Cauerstraße 7, D-91058 Erlangen, Germany
Fax: 49-9131-8528715
†*email:* P.Hippe@rzmail.uni-erlangen.de, *phone: 49-9131-8528592*

Abstract: Most of the existing windup prevention schemes can only be applied in the presence of stable plants. In this paper a new windup prevention scheme for stable and unstable systems is presented, allowing to handle arbitrary reference inputs in the presence of arbitrary disturbances with known maximum amplitudes. This becomes possible by introducing a nonlinear model based reference shaping filter in conjunction with a disturbance rejecting controller. *Copyright © 2003 IFAC*

Keywords: Constrained systems, Windup prevention, Unstable systems

1 INTRODUCTION

Considered are linear time invariant systems with linear time invariant controllers, where input saturation is the only nonlinearity in the loop. The undesired effects of input saturation can be separated into two different phenomena, namely the *controller windup* and the *plant windup*. Controller windup is the usually considered form of windup, also called *integral windup*. It can happen whenever the controller contains unstable modes. The more popular approaches to controller windup prevention are the Conditioning Technique (Hanus, 1987), the Generalized Anti Windup Control (GAWC) (Aström & Wittenmark, 1984), and the Observer Technique (Hippe & Wurmthaler, 1999). For a generalized treatment of these methods see e.g. Kothare et al. (1994).

But also in the absence of controller windup, the closed loop is a nonlinear system that can become unstable. This effect is due to what Hippe & Wurmthaler (1999) called *plant windup*. Whereas controller windup can be removed by structural methods when realizing the controller, one usually needs additional controller dynamics to prevent plant windup. Methods for the prevention of plant windup (called "short sightedness" of the conditioning technique in Rönnbäck et al. (1991)), are the *filtered setpoint* in Rönnbäck et al. (1991) or the *additional network* in Hippe & Wurmthaler (1999). A method that removes both types of windup at the same time was presented by Teel & Kapoor (1997).

All the above methods for plant windup prevention only work for stable plants. There have been attempts to use the Teel & Kapoor (1997) approach to prevent windup for exponentially unstable systems (see Teel, 1999; Hippe & Deutscher, 2001). The plant windup prevention schemes by Teel & Kapoor (1997) and Hippe & Wurmthaler (1999) only become active after the input signal has hit the saturation limit for the first time. Any such scheme applied to unstable systems can be destabilized by disturbances.

There are papers on constrained control of unstable systems with proven stability. In Kapasouris & Stein (1990) a reference "governor" is presented which, via a time varying rate operator, shapes the reference input such that the control input does not saturate. The computations for constructing the time-varying rate are, however, quite complicated. In Seron et al. (1995) the set point tracking problem was investigated using an internal model control configuration, and rules of thumb were presented giving bounds on amplitude and frequency for sinusoidal references ensuring stable saturated operation of the nonlinear reference filter. In Turner et al. (2000) the tracking of constant reference signals is considered and regions of initial conditions are computed for which stability is guaranteed.

Disturbance rejection for unstable systems with input saturation was also investigated by various authors. The rejection of input additive disturbances was discussed for state feedback control e.g. in Saberi et al. (1996). Teel (1999) considers the windup problem for exponentially unstable systems in the presence of reference and disturbance inputs. The disturbances, however, are restricted not to act on the exponentially unstable modes. The joint problem of tracking a certain reference level in the presence of (constant) disturbances was investigated in Tabouriech et al. (2000).

In this contribution, the control problem for stable and unstable linear systems with saturating input nonlinearities is solved in the presence of arbitrarily time varying disturbances of known maximum amplitudes. After repeating basic facts on the relation between time and frequency domain representations of linear controllers (incorporating signal models for disturbance rejection) and on the controller and plant windup prevention methods in Section 2, Section 3 presents the new windup prevention scheme for stable and unstable systems. It uses a nonlinear model based reference shaping filter such that there remains a input signal margin for disturbance rejection. In Section 4 the design of the reference shaping filter is discussed, where two different structures for stable and unstable systems exist. An example and a short summary conclude the paper.

2 PRELIMINARIES

Considered are completely controllable and observable linear, time invariant SIMO systems of nth order with one input $u_s(t)$ and m outputs y(t) with state equations

$$\begin{aligned}\dot{x}(t) &= Ax(t) + bu_s(t) + b_d d(t)\\ y(t) &= Cx(t) + C_d d(t)\end{aligned} \quad (2.1)$$

There is a disturbance input d(t) and for the remainder of this section, it is assumed that the system is stable with its eigenvalues having negative real parts; but simple eigenvalues on the imaginary axis are also allowed. The output $y_1(t)$ is supposed to be the controlled variable. In the frequency domain these systems are characterized by

$$y(s) = G(s)u_s(s) + G_d(s)d(s) \quad (2.2)$$

where

$$G(s) = C(sI - A)^{-1}b = N(s)D^{-1}(s) \quad (2.3)$$

and N(s) a vector of polynomials $N_i(s)$, i = 1,2,...,m.

A linear time invariant controller of order n_C has the transfer behaviour

$$u_C(s) = -G_{CY}(s)y(s) + G_{CR}(s)r(s) \quad (2.4)$$

with

$$G_{CY}(s) = D_C^{-1}(s)N_Y^T(s) \quad (2.5)$$

i.e. the row vector $N_Y^T(s)$ contains the polynomials $N_{Yi}(s)$, i = 1,2,...,m. The reference input r(t) for the controlled variable $y_1(t)$ is injected via

$$G_{CR}(s) = D_C^{-1}(s)N_R(s) \quad (2.6)$$

such that the steady state error $y_1(\infty) - r(\infty)$ vanishes. The nominal loop, characterized by $u_s = u_C$, is supposed to be internally stable and to give desired disturbance rejection. There is a saturation nonlinearity

$$u_s(t) = \mathcal{N}(u_C(t)) \quad (2.7)$$

at the plant input, i.e. $u_s(t) = u_C(t)$ when $|u_C(t)| < u_{sat}$ and $u_s(t) = \text{sign}(u_C(t))u_{sat}$ when $|u_C(t)| \geq u_{sat}$.

The undesired effects of input saturation, the *windup*, can best be demonstrated in an observer based state control configuration. Suppose a state feedback

$$u_C(t) = -k^T x(t) + Lr(t) \quad (2.8)$$

with L such that $y_1(\infty) = r(\infty)$ has been designed to achieve a desired closed loop behaviour and denote the corresponding characteristic polynomial by

$$\tilde{D}(s) = \det(sI - A + bk^T) \quad (2.9)$$

Now use an observer of order n_O with state equations

$$\dot{z}(t) = Fz(t) + Dy(t) + Tbu_C(t) \quad (2.10)$$

yielding a state estimate

$$\hat{x}(t) = \Psi y(t) + \Theta z(t) \quad (2.11)$$

Substituting $\hat{x}(t)$ in (2.8) the controller's transfer behaviour is given by

$$G_{CY}(s) = k^T\Theta(sI - F + Tbk^T\Theta)^{-1}(D - Tbk^T\Psi) + k^T\Psi \quad (2.12)$$

and

$$G_{CR}(s) = k^T\Theta(sI - F + Tbk^T\Theta)^{-1}TbL + L = D_C^{-1}(s)\det(sI - F)L \quad (2.13)$$

If one or more roots of $D_C(s) = \det(sI - F + Tbk^T\Theta)$ are located on the imaginary axis or in the right half s-plane, distinctive windup effects can occur if input saturation becomes active. This type of windup is related to the controller dynamics and it is therefore a *controller windup* (Hippe & Wurmthaler, 1999). This effect was first observed in classical controllers with integral action. The integral part is "winding up" and hence this effect is also called *integral windup*.

When feeding $Tbu_s(t)$ instead of $Tbu_C(t)$ into the observer (2.10), the unstable controller denominator $D_C(s)$ is replaced by

$$\Delta(s) = \det(sI - F) \quad (2.14)$$

during saturation, and this is a hurwitz polynomial. Consequently *controller windup* no longer occurs. This windup prevention technique is called the *Observer Technique* (Hippe & Wurmthaler, 1999). Fig. 1 shows the frequency domain representation of it.

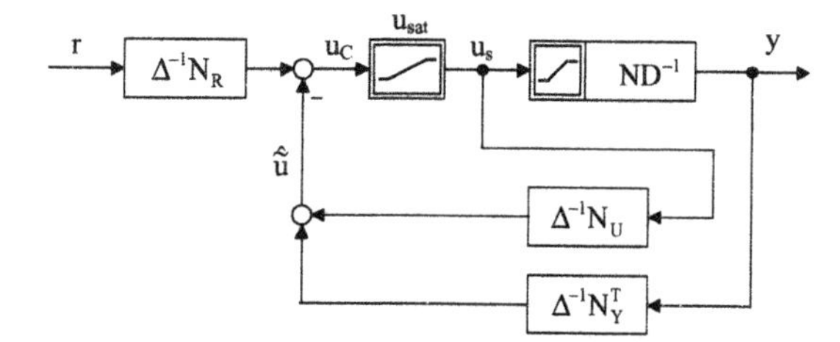

Fig. 1. Observer Technique for windup prevention

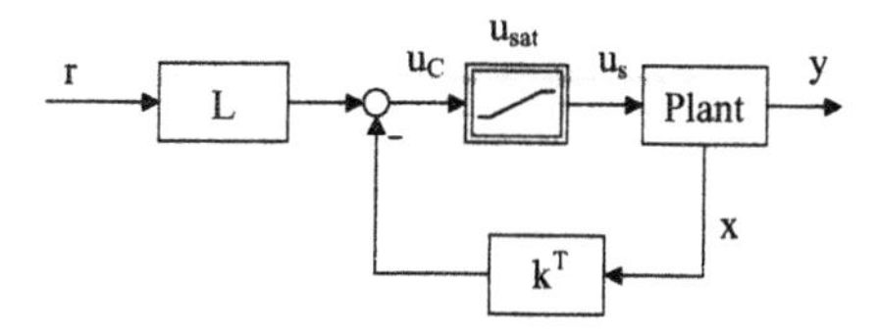

Fig. 2. Constant state feedback with input saturation

If the polynomial $N_U(s)$ in Fig. 1 has the form

$$N_U(s) = D_C(s) - \Delta(s) \tag{2.15}$$

the linear closed loop in Fig. 1 is identical with the nominal one (described by G(s), $G_{CY}(s)$ and $G_{CR}(s) = D_C^{-1}(s)\Delta(s)L$, i.e. $N_R(s) = \Delta(s)L$). Reference inputs do not cause observation errors, and due to the observer technique, input saturation does not cause observation errors either. Consequently the reference behaviour of the loop in Fig.1 is exactly the same as that of the constant state feedback control in Fig. 2.

Since there are no controller states in the loop of Fig. 2, controller windup cannot occur. The loop in Fig. 2 remains of course nonlinear, and it is well known from the describing function method, that if the Nyquist plot of

$$F_L(s) = k^T(sI - A)^{-1}b \tag{2.16}$$

intersects the negative real axis left of –1, limit cycles can appear. Using the well known relationship (Kailath, 1980)

$$1 + k^T(sI - A)^{-1}b = \tilde{D}(s)D^{-1}(s) \tag{2.17}$$

the transfer behaviour (2.16) can be written in frequency domain quantities as

$$F_L(s) = \frac{\tilde{D}(s)}{D(s)} - 1 \tag{2.18}$$

Consequently, an intersection of the negative real axis is equivalent to the phase $\Phi\left(\frac{\tilde{D}(j\omega)}{D(j\omega)}\right)$ crossing $\pm 180°$. But also for phase angles $|\Phi(\cdot)| < 180°$ there may be the danger of an oscillatory behaviour due to input saturation. Since this is definitely related to the dynamics of the controlled plant, it is called *plant windup* (Hippe & Wurmthaler, 1999).

In order to get a strong proof of stability for the loop in Fig. 2, the phase Φ should not surpass $\pm 90°$ (Circle criterion). It has been demonstrated in Hippe & Wurmthaler (1999), that this is unnecessarily restrictive. Instead a *phase design aid* has been presented, relaxing the 90° limit while assuring that input saturation does not cause an oscillating behaviour of the closed loop transients.

If the closed loop dynamics, characterized by the polynomial $\tilde{D}(s)$, are such that $\Phi\left(\frac{\tilde{D}(j\omega)}{D(j\omega)}\right)$ indicates the danger of plant windup, it can be prevented by an additional dynamic network (ADN) (Hippe & Wurmthaler, 1999). This ADN results when replacing $u_C(t)$ by $u_C(t) - \eta(t)$ in the Figs. 1 and 2, and η is given by

$$\eta(s) = \frac{\tilde{\Omega}(s) - \Omega(s)}{\Omega(s)}\left[u_C(s) - u_s(s)\right] \tag{2.19}$$

Now the hurwitz polynomials $\Omega(s)$ and $\tilde{\Omega}(s)$ of equal degree can be chosen such that the modified linear part $u_C(s) = -F_L(s)u_s(s)$ with

$$F_L(s) = \frac{\tilde{D}(s)\Omega(s)}{D(s)\tilde{\Omega}(s)} - 1 \tag{2.20}$$

has a phase $\Phi\left(\frac{\tilde{D}(j\omega)\Omega(j\omega)}{D(j\omega)\tilde{\Omega}(j\omega)}\right)$ staying within the allowed region (see above).

To cover more general controllers, the incorporation of signal models of order n_d with characteristic polynomial $N_D(s)$ is now considered. The signal forms of the disturbances are characterized by $N_D(s)$ ($N_D(s) = s$ step like, $N_D(s) = s^2 + \omega_0^2$ sinusoidal, a.s.o.). There are two different approaches for the accommodation of disturbances, the *disturbance observer* by Johnson (1971) and the controlled *disturbance model* by Davison (1975). Since Davison's approach is robust to model uncertainties and to changing disturbance input locations, it is considered here.

The frequency domain design of such disturbance rejecting controllers is as follows. Choose

$$\begin{aligned} D_C(s) &= \overline{D}_C(s)N_D(s); \\ N_{Yi}(s) &= \overline{N}_{Yi}(s)N_D(s), i = 2,3,\ldots,m \end{aligned} \tag{2.21}$$

(recall that y_1 is the controlled variable) and suppose the characteristic polynomials $\tilde{D}(s)$ (degree n, for the controlled plant), $\Delta_O(s)$ (degree n_O, for the state observer), and $\Delta_d(s)$ (degree n_d, for the controlled disturbance model) are known. Then defining the observer polynomial

$$\Delta(s) = \Delta_O(s)\Delta_d(s) \tag{2.22}$$

of degree $n_C = n_O + n_d$, the unknown polynomials $\overline{D}_C(s), N_{Y1}(s), \overline{N}_{Yi}(s), i = 2,3,\ldots,m$ can be obtained from a comparison of polynomial coefficients in the closed loop characteristic polynomial

$$N_Y^T(s)N(s) + D_C(s)D(s) = \Delta(s)\tilde{D}(s) \tag{2.23}$$

Now computing $N_U(s)$ from (2.15), the closed loop can be represented in the form of Fig. 1, and thus one has $\hat{u}(t) = k^T x(t)$ for reference inputs.

3 THE NEW WINDUP PREVENTION SCHEME

Assume a nominal controller has been designed (with or without disturbance model) to obtain a desired disturbance rejection, and that it has been realized in the observer structure, so that controller windup does not occur. Different from Section 2 the plant is now allowed to be exponentially unstable. Fig. 3 shows the new scheme.

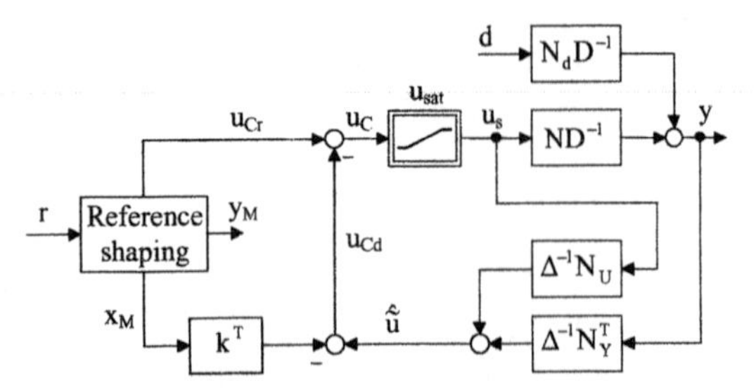

Fig. 3. The new windup prevention scheme

A nonlinear model based reference shaping filter (see Section 4) has been added. The maximum amplitude of the output $u_{Cr}(t)$ of the filter is less than u_{sat} for arbitrary reference inputs r(t). If the plant model is correct and no disturbances act on the system ($d(t) \equiv 0$), one has

$$\hat{u}(t) = k^T x(t) = k^T x_M(t) \tag{3.1}$$

Consequently the signal $u_{Cd}(t)$ vanishes for reference inputs. If, however, the states x(t) of the system and $x_M(t)$ of the model differ (this can happen due to disturbances or due to the unstable plant modes) the nominal controller gives the desired disturbance rejection and stabilizes the plant.

In the case of exponentially unstable plants, the maximum amplitude u_{Cdlim} of $u_{Cd}(t)$ required for disturbance rejection is supposed to be

$$u_{Cd\,lim} = \beta u_{sat} \quad \text{with } \beta < 1 \tag{3.2}$$

This is a necessary condition for a stable closed loop. If disturbances drove the plant input signal beyond the saturation limit, the closed loop would be open. Thus the exponentially unstable modes could develop freely, entailing the danger that the system can no longer be stabilized by the limited input.

If the maximum amplitude d_{lim} of an arbitrary disturbance d(t) is known ($r(t) \equiv 0$ and u_{sat} not active), the resulting maximum controller output signal amplitude u_{Cdlim} can be computed from the L_1-norm of the transfer function

$$u_{Cd}(s) = \frac{N_Y^T(s)N_d(s)}{\Delta(s)\tilde{D}(s)} d(s) \tag{3.3}$$

Given a linear time invariant system with transfer function G(s) and a corresponding weighting function h(t). Then, when an arbitrary input signal u(t) with known maximum amplitude u_{lim} is applied to the system, the maximum output amplitude y_{lim} of this system is bounded by

$$y_{lim} \le u_{lim} \int_0^\infty |h(\tau)| d\tau = \|G(s)\|_1 u_{lim} \tag{3.4}$$

where $\|G(s)\|_1$ is the L_1-norm of G(s) (Dahleh & Pearson, 1987, Ohta et al., 1992). With this result, the L_1-norm of the transfer function (3.3) can be computed and given d_{lim}, β in (3.2) is known.

When confining the input signal $u_{Cr}(t)$ to the range

$$-(1-\beta)u_{sat} \le u_{Cr}(t) \le (1-\beta)u_{sat} \tag{3.5}$$

plant input saturation never becomes active for arbitrary disturbances d(t) with maximum amplitude d_{lim} and for arbitrary reference signals r(t). Consequently the loop with exponentially unstable plant remains stable in spite of the saturating input nonlinearity.

In the case of stable plants, input saturation must not be avoided under all circumstances. One can either allow the maximum amplitude of $u_{Cr}(t)$ to be close to u_{sat}, or one can reduce this amplitude to an extent, where only worst case disturbances cause input saturation. If the plant input signal saturates, plant windup could be triggered if the phase $\Phi\left(\frac{\tilde{D}(j\omega)}{D(j\omega)}\right)$ was outside the allowed region. If so, an ADN must be added to the scheme in Fig. 3.

4 THE REFERENCE SHAPING FILTER

4.1 The case of exponentially unstable systems

For the following it is assumed that the system does not have eigenvalues at s = 0 (an assumption that will be dropped later). Suppose there is a model $G_M(s)$ for the plant G(s) and that it is correct, i.e. $G_M(s) = G(s)$, and that there is a state space realization (A_M, b_M, C_M) for it. Now consider the scheme in Fig. 4, which is based on an idea of Bühler (2000).

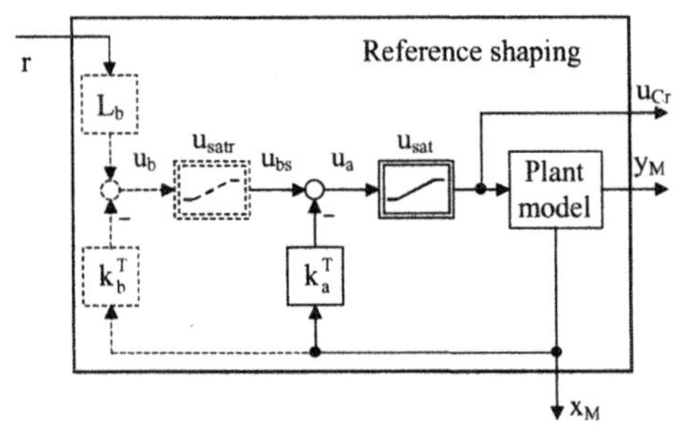

Fig. 4. Reference shaping filter for unstable systems

There is a stabilizing (inner) state feedback

$$u_a(t) = -k_a^T x_M(t) + u_{bs}(t) \tag{4.1}$$

for the exponentially unstable system model, and in view of the input saturation u_{sat} the roots of

$$\tilde{D}_a(s) = \det(sI - A_M + b_M k_a^T) \tag{4.2}$$

are the stable eigenvalues of A_M and the "stabilized" unstable eigenvalues of A_M having moderate negative real parts.

If u_{sat} is not active, one has (the parts in broken lines removed from Fig. 4 for the moment)

$$[1+k_a^T(sI-A_M)^{-1}b_M]u_a(s)=u_{bs}(s) \qquad (4.3)$$

and with the relation (2.17) one obtains

$$\frac{u_a(s)}{u_{bs}(s)}=\frac{D_M(s)}{\tilde{D}_a(s)}=G_{IL}(s) \qquad (4.4)$$

Let the L_1-norm of $G_{IL}(s)$ be $\alpha_{IL} > 0$. Then when confining the signal $u_{bs}(t)$ to the region

$$-u_{satr} \le u_{bs}(t) \le u_{satr} \qquad (4.5)$$

with

$$u_{satr}=\frac{u_{sat}}{\alpha_{IL}}(1-\beta) \qquad (4.6)$$

the amplitude of $u_{Cr}(t) = u_a(t)$ never becomes bigger than $u_{sat}(1 - \beta)$, and this is true for arbitrary input signals $u_b(t)$! The amplitude y_{1safe} of the controlled variable $y_1(t)$ achievable with this scheme is given by

$$y_{1safe}=\frac{u_{sat}}{\alpha_{IL}}\frac{N_{M1}(0)}{\tilde{D}_a(0)}(1-\beta) \qquad (4.7)$$

Since the nonlinearity u_{sat} never becomes active, it can be removed from the scheme in Fig. 4. Consequently, the inner system is a stable linear system with input nonlinearity u_{satr}. Adding another loop (broken lines in Fig. 4) and assigning the roots of

$$\tilde{D}_b(s)=\det(sI-A_M+b_M(k_a^T+k_b^T)) \qquad (4.8)$$

such that the phase $\Phi\left(\frac{\tilde{D}_b(j\omega)}{\tilde{D}_a(j\omega)}\right)$ stays within the ±90° region, the loop shown in Fig. 4 is globally asymptotically stable for arbitrary reference inputs r(t). With $L_b=\frac{\tilde{D}_b(0)}{N_{M1}(0)}$ one assures vanishing steady state errors $y_{1M}(\infty) - r(\infty)$.

4.2 The case of stable systems

For stable systems (again no eigenvalues at s = 0 assumed for the moment) the inner stabilizing loop of the filter in Fig. 4 can be omitted. Consequently, it now has the form shown in Fig. 5 (full lines).

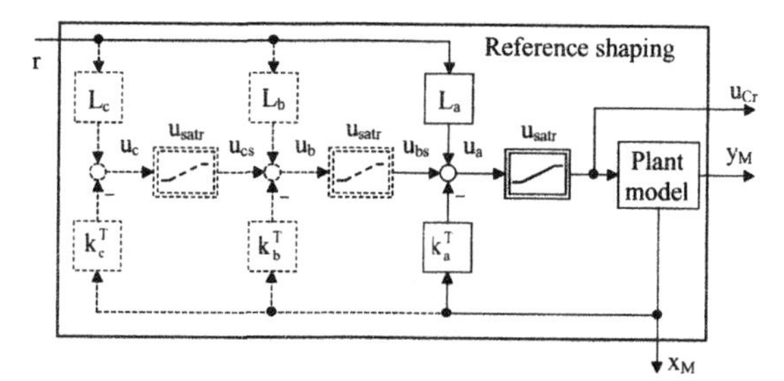

Fig. 5. Reference shaping filter for stable systems

With $u_{satr} = u_{sat}$ the total input amplitude range could be used for reference tracking. However, in view of an ADN probably needed in the scheme of Fig. 3, u_{satr} should at least be reduced by 1% in relation to u_{sat}. If a bigger margin βu_{sat} has to be reserved for disturbance rejection, choose $u_{satr} = (1 - \beta)u_{sat}$.

With $u_a(t)=-k_a^Tx_M(t)+L_ar(t)$ such that the phase $\Phi\left(\frac{\tilde{D}_a(j\omega)}{D_M(j\omega)}\right)$ stays within the allowed range of the circle criterion, one has a reference shaping filter with proven stability. However, one can add one, two (as indicated in broken lines in Fig. 5) or more nested loops to the reference shaping filter. The polynomials $\tilde{D}_b(s)=\det(sI-A_M+b_M(k_a^T+k_b^T))$, $\tilde{D}_c(s)=\det(sI-A_M+b_M(k_a^T+k_b^T+k_c^T))$, ... should be such that the phases $\Phi\left(\frac{\tilde{D}_b(j\omega)}{\tilde{D}_a(j\omega)}\right)$, $\Phi\left(\frac{\tilde{D}_c(j\omega)}{\tilde{D}_b(j\omega)}\right)$, ... all stay within the allowed region. This, however, only guarantees closed loop stability if the saturations and desaturations take place in the correct order, namely saturation starting in the outer cascade first and in the inner last, and desaturation vice versa. This can be assured by assigning well damped dynamics to the cascaded loops and by choosing the reference injections L_a, L_b, L_c, ... such that

(i) the steady state error $y_{1M}(\infty) - r(\infty)$ vanishes

(ii) all stationary values of $u_a(\infty)$, $u_b(\infty)$, $u_c(\infty)$, ... coincide.

There is no proof of stability for the additional cascaded loops, but under the above assumptions they work fine and allow to speed the achievable reference transients especially for small input amplitudes.

Remark 4.1. If the system contains eigenvalues at s = 0, there are two possible ways to design the reference tracking filter. If the desired output signal amplitudes are confined to a limited region $-y_{1max} < y_1 < y_{1max}$, all eigenvalues of the inner loop (k_a^T) could be given negative real parts. If, however, one wants to exploit the potential of the system to reach unlimited output amplitudes in spite of input saturation, the inner loop should contain an eigenvalue at s = 0. ./.

5 AN EXAMPLE

Consider a system of seventh order with four exponentially unstable modes and transfer function

$$G(s)=\frac{N(s)}{D(s)}=\frac{96}{(s-0.5)(s^2-2s+2)(s-3)(s^2+4s+8)(s+4)}$$

The saturation limit is $u_{sat} = 1$. For disturbance rejection a state feedback control has been designed s.t.

$$\tilde{D}(s)=\det(sI-A+bk^T)=(s+2.8)^4(s^2+4s+8)(s+4)$$

and suppose it is known that the attenuation of the worst case disturbances requires 50% of the available input amplitude, i.e. β = 0.5. Consequently, 50% of the available input signal amplitude (here 0.5) can safely be used to apply reference signals.

In the reference shaping filter the inner loop is s. t.

$$\tilde{D}_a(s)=\det(sI-A+bk_a^T)=(s+0.25)^4(s^2+4s+8)(s+4).$$

This gives an L_1-norm $\alpha_{IL} = 795$ of (4.4). Consequently when assigning (see (4.6)) $u_{satr} = 6.25e\text{-}04$, arbitrary signals $u_b(t)$ (see Fig. 4) do not cause signals $u_a(t) = u_{Cr}(t)$ surpassing an amplitude 0.5, so that the plant input signal $u_C(t)$ (see Fig. 3) never hits the saturation limit $u_{sat} = 1$. However, the maximum output amplitude resulting from reference inputs is now limited by $y_{1safe} = 0.483$ (see (4.7)). Adding a state feedback $u_b(t) = -k_b^T x_M(t) + L_b r(t)$ such that $\tilde{D}_b(s) = (s+0.56)^4(s^2+4s+8)(s+4)$ (see (4.8)), the phase $\Phi\left(\frac{\tilde{D}_b(j\omega)}{\tilde{D}_a(j\omega)}\right)$ does not surpass 90° and consequently the reference shaping filter is asymptotically stable. Figure 6 shows reference step responses of the closed loop according to Fig. 3 having amplitudes between 0 and 0.48.

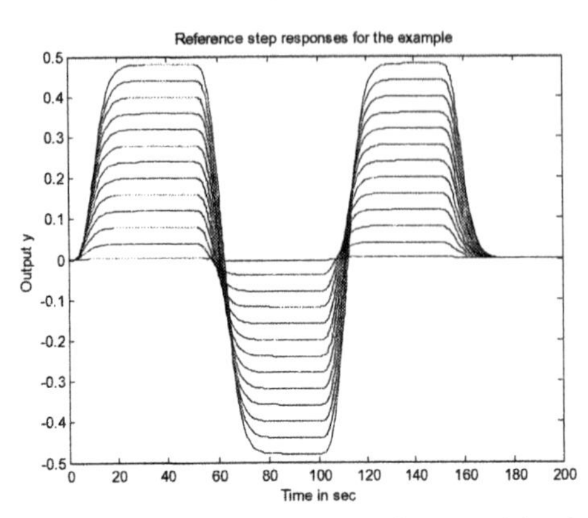

Fig. 6. Reference step responses for unstable plant

6 SUMMARY

Presented is a novel windup prevention scheme for stable and unstable systems, guaranteeing a stable closed loop behaviour for completely arbitrary reference inputs in the presence of arbitrary disturbance inputs with known maximum amplitudes. This is achieved by using one part of the control scheme for reference tracking and the other for disturbance rejection. The reference tracking uses an appropriate nonlinear reference shaping filter and the nominal controller is only active for disturbance rejection. Stability for unstable plants is guaranteed by splitting the limited control input amplitudes into one part necessary for disturbance rejection and the remaining part for reference tracking. Thus input saturation never becomes active and the stability of the nonlinear loop is consequently guaranteed. For stable systems, input saturation can be allowed, at least to some extent, and the possibly existing danger of plant windup can be avoided by introducing an additional dynamic network.

REFERENCES

Aström, K. J., and Wittenmark, B. (1984). *Computer Controlled Systems*: Theory and Design. Prentice Hall, Englewood Cliffs, NJ.

Bühler, H.-R. (2000). Regelkreise mit Begrenzungen. Fortschritt-Berichte VDI, Nr. 828.

Dahleh, M. A. und Pearson, J. B.: L_1-optimal compensators for continuous-time systems. *IEEE Transactions Automatic Control* 32 (1987), S. 889-895.

Davison, E. J. (1975). Robust control of a general servomechanism problem: The servo-compensator. *Automatica*, 11, 461-471.

Hanus, R., Kinnaert, M., & Henrotte, J. L. (1987). Conditioning technique, a general antiwindup and bumpless transfer method. *Automatica*, 23, 729-739.

Hippe, P., & Wurmthaler, Ch. (1999). Systematic closed loop design in the presence of input saturation. *Automatica*, 35, 689-695.

Hippe, P., & Deutscher, J. (2001). A windup prevention scheme for exponentially unstable plants. Proceedings of the sixth European Control Conference, Porto, 2001.

Johnson, C. D.: Accommodation of external disturbances in linear regulator and servomechanism problems. *IEEE Transactions Automatic Control* 16 (1971), S. 635-644.

Kailath, T. (1980). *Linear Systems*. Prentice Hall. Englewood Cliffs, N.J..

Kapasouris, M. A., & Stein, G. (1990). Design of feedback control systems for unstable plants with saturating actuators. Proceedings of the IFAC Symposium on Nonlinear Control System Design.

Kothare, M. V., Campo, P. J., Morari, M., & Nett, C. N. (1994). A unified framework for the study of anti-windup designs. *Automatica*, 30, 1869-1883.

Ohta, Y., Maeda, H., & Kodama, S. (1992). Rational approximation of L_1 optimal controllers for SISO systems. *IEEE Transactions Automatic Control*, 37, 1683-1691.

Rönnbäck, S., Walgama, K. S., & Sternby, J. (1991). An extension to the generalized anti-windup compensator. 13th IMACS World Congress on Scientific Computation, Dublin, Ireland.

Saberi, A., Lin, Z., & Teel, A. R. (1996). Control of linear systems with saturating actuators. *IEEE Transactions Automatic Control*, 41, 368-378.

Seron, M. M., Goodwin, G. C., & Graebe, S. F. (1995). Control system design issues for unstable linear systems with saturated inputs. *IEE Proc. Control Theory and Applications*, 142, 335-344.

Tabouriech, S., Pittet, C., & Burgat, C. (2000). Output tracking problem for systems with input saturations via nonlinear integrating actions. *Intern. Journal Robust and Nonlinear Control*, 10, 489-512.

Teel, A. R., & Kapoor, N. (1997). The L_2 anti-windup problem: its definition and solution. Proceedings of the Fourth European Control Conference, Brussels, 1997.

Teel, A. R. (1999). Anti-windup for exponentially unstable systems. *Intern. Journal Robust and Nonlinear Control*, 9, 701-716.

Turner, M. C., Postlethwaite, I., & Walker, D. J. (2000). Non-linear tracking control for multivariable constrained input linear systems. *Int. J. Control*, 73, 1160-1172.

www.elsevier.com/locate/ifac

THE PIL (NOT PID?) CONTROLLER AS EIGENVALUE ASSIGNER AND OPTIMUM STABILISER

Annraoi de Paor

*Department of Electronic and Electrical Engineering,
National University of Ireland,
Belfield, Dublin 4, Ireland.
Email: annraoi.depaor@ucd.ie*

Abstract: For reasons of saturation avoidance and limitation of high frequency noise, the classical PID controller is usually implemented with a lowpass filter on the derivative channel. As pointed out by the author some years ago, this has profound implications for the stability of digital control systems based on discretised analog designs. However, it then ceases to be a PID controller and might be better described as PIL, where L normally stands for phase "lead," but might in exceptional circumstances be phase "lag." Such controllers are investigated as eigenvalue assigners for an inherently unstable magnetic levitation (maglev) system and for a first order lag with pure time delay, where the time delay is approximated by a first order Padé approximation for design purposes and then simulated exactly for performance evaluation. Explicit controller designs are derived with all closed loop eigenvalues placed at the same location. It is demonstrated, using root locus arguments, that this gives an optimally stable system in the sense that as any single controller or process parameter is varied with respect to its nominal value, the rightmost eigenvalue, on passage through the nominal parameter setting, is as deep in the left half plane as possible. *Copyright © 2003 IFAC*

Keywords: Three-term control, eigenvalue assignment, optimal control, root locus, time delay, approximate analysis.

1 THE PIL CONTROLLER

A very common form of the PID controller transfer function is

$$C(s) = K[1+1/(sT_i) + sT_d] \qquad (1),$$

where K, T_i, and T_d are, respectively, the proportional gain, integral action time and derivative action time. If the terms are combined into a single rational function, it is readily seen to be *improper*—a second order numerator over a first order denominator. It was shown by the author some years ago (de Paor, 1995), using the famous double integrator process as a test case (Åstrom and Wittenmark, 1990) that this feature leads to failure of many common discretisation algorithms to produce an asymptotically stable translation of an apparently well behaved analog design. The difficulty is removed by incorporating a lowpass filter on the differentiator, as proposed over the years by many authors (e.g., Åstrom and Wittenmark, 1990) for reasons of saturation avoidance and limitation of high frequency noise. When the lowpass filter is applied, the controller transfer function is modified to

$$C(s) = K[1 + 1/(sT_i) + sT_d/(1+s\tau)] \qquad (2),$$

where τ is the time constant of the lowpass filter. The condensed controller rational transfer function is now *balanced*, i.e., it has equal numerator and denominator degrees. All

the stability difficulties associated with discretisation disappear (de Paor, 1995).
The nature of the controller has now changed, as can be appreciated by rearranging eqn.(2) as

$$C(s) = K[1+T_d/\tau].[1 + 1/(\check{T}_i s) - \alpha/(1 + s\tau)] \quad (3),$$

where

$$\check{T}_i = T_i[1 + T_d/\tau] \quad (4)$$

and

$$\alpha = T_d/[T_d + \tau] \quad (5).$$

The lowpass filter can be implemented as an integrator with integral action time τ, i.e., $1/(s\tau)$, in a unity negative feedback loop, so that there is no longer any differentiator involved in eqn. (3). In what was formerly the derivative channel, the negative gain $-\alpha$ contributes a phase lead of π and the lowpass filter contributes, at angular frequency ω, a phase lag of $\tan^{-1}(\omega\tau)$ radians, giving an overall phase lead of $\pi - \tan^{-1}(\omega\tau)$. Hence the name PIL controller. When using any of the multitude of PID tuning rules (O'Dwyer, 2003), α is positive and thus "L" denotes "lead." However, it is conceiveable that, in some problems not yet tackled, the eigenvalue assigning controllers to be discussed below could give α negative, and "L" would then denote "lag."

Controllers of the structure in eqn. (3) are now discussed, derived not from the PID controller, but from eigenvalue assigning controllers for second order processes. These, with integral action incorporated for static rejection of a disturbance applied at the process input, have the general transfer function (Cogan and de Paor, 2000)

$$C(s) = k_2[s^2 + f_1 s + f_0]/[s(s + h)] \quad (6).$$

When controlling an approximated or exact second order process with proper rational transfer function

$$G(s) = k_1[s + b_0]/[s^2 + a_1 s + a_0] \quad (7),$$

the characteristic polynomial

$$P(s) = s[s^2 + a_1 s + a_0][s + h] + k_1 k_2[s + b_0][s^2 + f_1 s + f_0] \quad (8)$$

contains just the right number of free parameters, h, k_2, f_1 and f_0 to permit arbitrary assignment of the four system eigenvalues, restricted only by complex pairing.

2 CONTROL OF AN UNSTABLE MAGNETIC LEVITATION SUSPENSION SYSTEM

The linearised equation of motion of a magnetically levitated body (Franklin et al., 2002; Power and Simpson, 1978), with deviation of electromagnet current from a reference value as input, and deviation of displacement from the corresponding unstable equilibrium position as output, has the form

$$G(s) = k_1/[(s + a)(s - a)], \text{ with } a>0. \quad (9).$$

The characteristic polynomial, to which four equal roots are assigned at s = -b, with the controller defined by eqn.(6), becomes

$$(s + b)^4 = s(s + a)(s - a)(s + h) + k_1 k_2(s^2 + f_1 s + f_0) \quad (10).$$

Expanding the left hand polynomial and dividing it by the *known* terms $s(s + a)(s - a) = s^3 - a^2 s$, leads immediately to the result

$$h = 4b \quad (11),$$

$$k_2(s^2 + f_1 s + f_0) = [(6b^2 + a^2)s^2 + 4b(a^2 + b^2)s + b^4]/k_1 \quad (12),$$

thus furnishing the controller design.

As a simple illustration, for $k_1 = 1$, $a = 1$, $b = 2$, the controller transfer function is

$$C(s) = 25[s^2 + 1.6s + 0.64]/[s(s + 8)] \quad (13).$$

By partial fraction expansion and slight rearrangement, this may be written as

$$C(s) = 25[1 + 1/(12.5s) - 0.81/(1 + 0.125s)] \quad (14),$$

which is in the PIL form of eqn. (3).

Fig. (1) shows process output in response to a step disturbance input (bottom curve), to a step reference (top curve) and to a step reference applied through a first lowpass filter of experimentally determined time constant 1.65 seconds (middle curve). Control signals in the first and third of these situations are shown on Fig. (2). The lower curve corresponds to a disturbance input, the upper to the filtered step reference.

As pointed out by Cogan and de Paor (2000), placing all eigenvalues together at the location $s = -b$ ($b = 2$ here) is a very simple idea but has a very profound consequence: the system becomes optimally stable in the sense that if any one parameter from either process or controller is varied, all other parameters being held at their nominal values, then, as the parameter in question passes through its nominal value, the rightmost eigenvalue passes through a location as deep in the left half plane as possible. This property is most readily demonstrated by root locus plots. For illustration, relevant portions of root loci are shown in Figs. (3), (4) and (5), for variation of k_2, a and f_1, respectively. Fig. (3) is an illustration of a root locus made up of two famous curves, the circle and strophoid (de Paor, 2000).

3 CONTROL OF A FIRST ORDER LAG WITH TIME DELAY

PIL control is now applied to a process with transfer function

$$G(s) = [K_p/(1 + sT)].\exp(-sL) \quad (15).$$

This is a classic problem, going back to the "process reaction curve" method of Ziegler and Nichols in 1942 and that of Chien, Hrones and Resnick in 1952 (Stefani et al., 2002). It is brought within the scope of the PIL controller as eigenvalue assigner by replacing the time delay transfer function by a first order Padé approximant

$$\exp(-sL) \approx [1 - sL/2]/[1 + sL/2] \quad (16).$$

With this approximation, the process transfer function becomes

$$G(s) = -K_p/T.(s - 2/L)/[(s + 1/T)(s + 2/L)] \quad (17).$$

A controller in the form of eqn.(6) is designed and its efficacy verified by simulation, using an exact process delay representation. In general, good results would be expected for L/T sufficiently less than 1. In this example, L/T = 0.25. More detailed recent treatments of the control of processes with time delay are given in Visioli (2001) and de Paor (2002).

Subject to eqn.(6) and setting the characteristic polynomial to $P(s) = (s + b)^4$, the polynomial equation to be solved is

$$(s + b)^4 = s(s + 1/T)(s + 2/L)(s + h) - [K_pk_2/T].(s - 2/L)(s^2 + f_1s + f_0) \quad (18).$$

Dividing through by the known polynomials gives

$$(s + b)^4/[s(s + 1/T)(s + 2/L)(s - 2/L)] = (s + h)/(s - 2/L) - [K_pk_2/T].(s^2 + f_1s + f_0)/[s(s + 1/T)(s + 2/L)] \quad (19).$$

The first term on the right hand side may be rewritten as

$$(s + h)/(s - 2/L) = \{1 + (h + 2/L)/(s - 2/L)\} \quad (20).$$

A partial fraction expansion of the expression on the left hand side of eqn. (19), results in the form

$$(s + b)^4/[(s - 2/L)s(s + 1/T)(s + 2/L)] = \{1 + R_1/(s - 2/L)\} + \{R_2/s + R_3/(s + 1/T) + R_4/(s + 2/L)\} \quad (21).$$

The residues R_i follow from Heaviside's rule:

$$R_1 = [T(2 + Lb)^4]/[8L(2T + L)]$$

$$R_2 = [-b^4TL^2]/4$$

$$R_3 = [L^2(bT - 1)^4]/[T(4T^2 - L^2)]$$

$$R_4 = [-T(2 - Lb)^4]/[8L(2T - L)] \quad (22).$$

Now, equating eqns.(21) and (20) with (19) gives

$$h = R_1 - 2/L$$

$$k_2 = [-T/K_p].[R_2 + R_3 + R_4]$$

$$f_1 = [-T/(K_pk_2)].[R_2(L+ 2T)/(TL) + 2R_3/L + R_4/T]$$

$$f_0 = [-T/(K_pk_2)].[2R_2/(LT)] \quad (23).$$

As an illustration, $K_p = 3$, $T = 2$, $L = 0.5$, $b=1$, giving

$$C(s) = 0.5601852(s^2 + 1.06198343s + .297520653)/[s(s + .340278)] \quad (24).$$

Fig.(6) shows the process output, obtained using an exact time delay, implemented with the Swedish program SIMNON, to a step disturbance (leftmost curve) and a step reference applied directly and through a first

order lowpass filter with time constant 1.8 seconds. Fig. (7) shows control signals in response to the disturbance and filtered reference inputs, with the process time delay being evident in the disturbance response.
In order to illustrate the optimum stability property, Fig. (8) shows the root locus of the characteristic polynomial for the Padé Delay approximated system, for variation of k_2. This is also a strophoid and circle root locus (de Paor, 2000).

4 DISCUSSION

The classical PID controller when fitted with a lowpass filter in the derivative channel, loses its differentiator and might be better described as a PIL controller, where L generally denotes phase lead. This controller is balanced—numerator has the same degree as denominator—and has very favourable properties when subjected to discretisation, in the transfer of analog to digital designs. It is also identical in form with cascade eigenvalue assigning controllers for second order processes with proper rational transfer functions. Two examples are given to illustrate the performance of these controllers. The first is a specific system—magnetic levitation suspension, such as is now in use in some trains throughout the world. The second example is generic—a first order lag process with time delay approximated at the design stage by a first order Padé Delay. Efficacy of the designed controller is verified by simulation using an exact delay representation. The optimum stability properties of systems designed by placing all eigenvalues at the same location are illustrated by root locus diagrams. A partial fraction expansion technique has been illustrated for solving a polynomial equation involved in controller design. It is hoped that the paper will be of interest in drawing together a lot of elementary ideas and applying them to practical designs.

REFERENCES

Åstrom, K.J., and B. Wittenmark (1990). *Computer-controlled systems: theory and design*, Prentice-Hall, New Jersey. (2nd edition).

Cogan, B., and A.M. de Paor (2000). Optimum stability and minimum complexity as desiderata in feedback control system design, *Control Systems Design (IFAC Conference), Bratislava, Slovakia, June 18-20*, 51-53.

de Paor, A.M. (2002). On the control of integrating and unstable processes with time delay, *Journal of Electrical Engineering (Bratislava)*, **53**, No. 5-6, 1-7.

de Paor, A.M. (2000). The root locus method: famous curves, control designs and non-control applications, *Int. J. Elec. Eng. Education*, **37**, 344-356.

de Paor, A.M. (1995). Control of the double integrator process: a cautionary tale in discretization, *Int. J. Elec. Eng. Education*, **32**, 307-318.

Franklin, G., J.D. Powell and A. Emami-Naeini (2002). *Feedback control of dynamic systems*, Addison-Wesley, Reading, Massachusetts. (4th edition).

O'Dwyer, A. (2003). *Handbook of PI and PID controller rules*, Imperial College Press, London.

Power, H.M. and R.J. Simpson (1978). *Introduction to dynamics and control*, Mc Graw-Hill, Maidenhead, UK.

Stefani, R.T., B. Shahian, C.J. Savant and G.H. Hostetter (2002). *Design of feedback control systems*, Oxford University Press, Oxford and New York. (4th edition).

Visioli, A. (2001). Optimal tuning of PID controllers for integral and unstable processes, *IEE Proceedings-Control theory and applications*, **142**, 180-184.

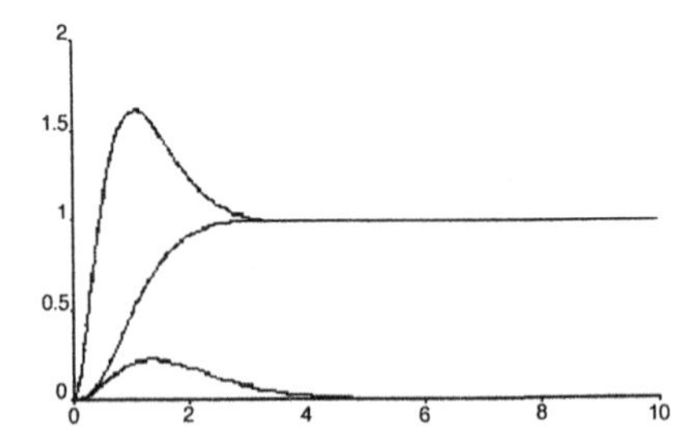

Fig. 1 Process outputs of maglev system

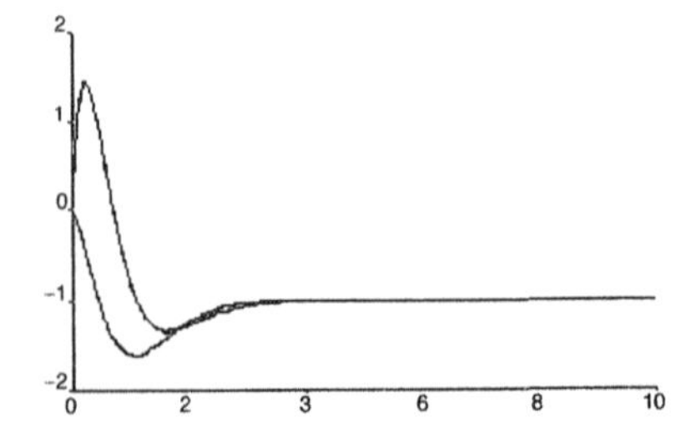

Fig. 2 Control signals of maglev system

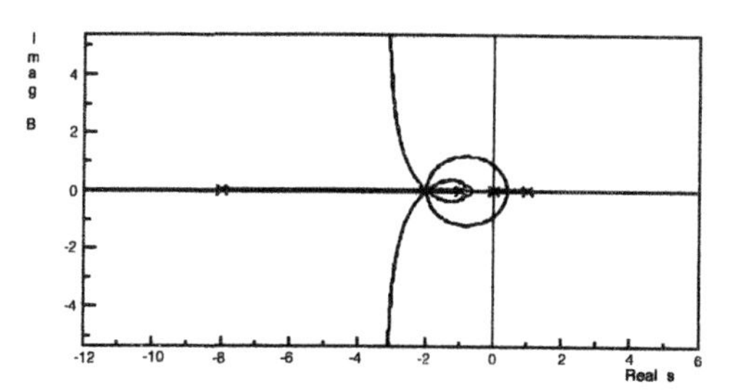

Fig.3 Maglev root locus with k_2 as parameter

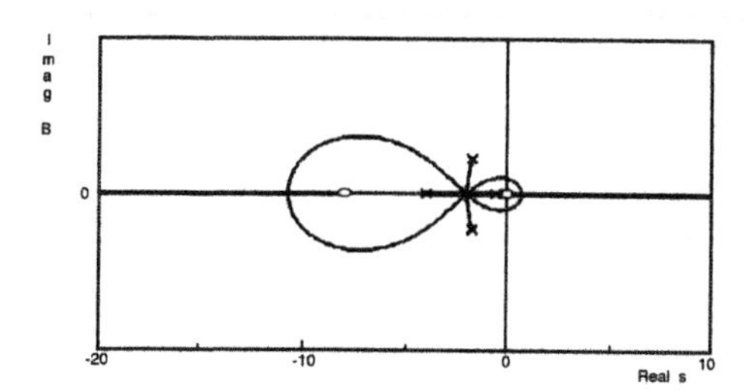

Fig. 4 Maglev root locus with a as parameter

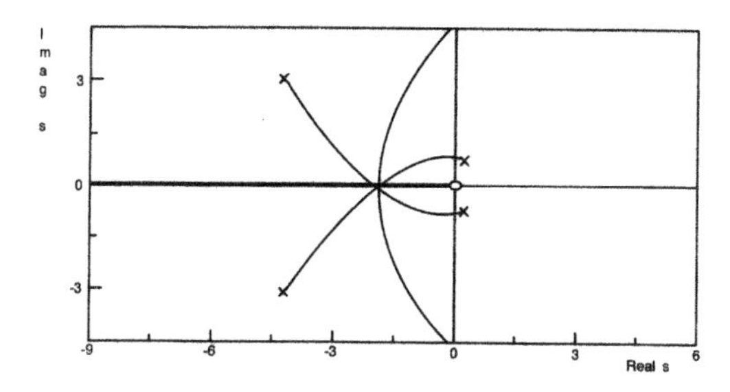

Fig. 5 Maglev root locus with f_1 as parameter

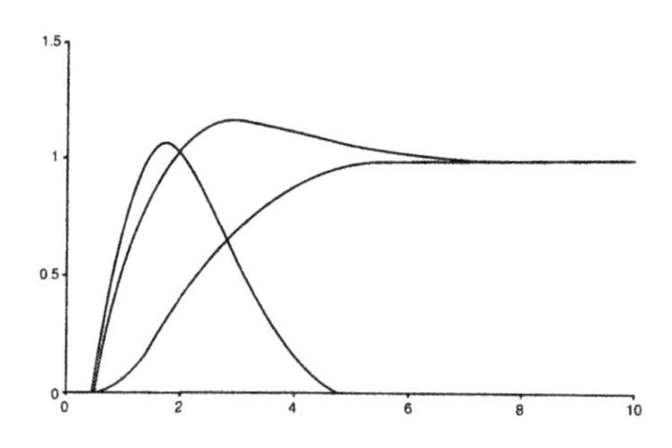

Fig. 6 Process outputs for delayed system

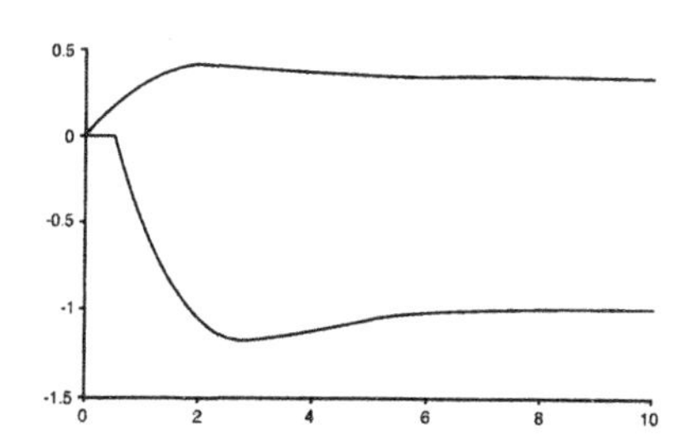

Fig. 7 Control signals for delayed system

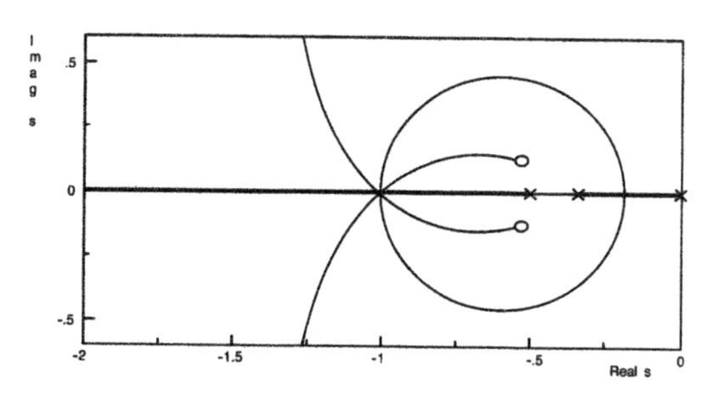

Fig. 8 Root locus of Padé delayed system with k_2 as parameter

www.elsevier.com/locate/ifac

IMPROVING TRACKING PERFORMANCE ON DISTURBANCE-REJECTION CONTROLLERS

Damir Vrančić, Stanko Strmčnik and Mikulas Huba[*]

J. Stefan Institute, Department of Systems and Control,
Jamova 39, SI-1000 Ljubljana, Slovenia
FAX: +386-1-4257-009, e-mail: damir.vrancic@ijs.si
[*] *Department of Automation and Control, Faculty of Electrical Engineering and Information Technology, Slovak Technical University, SK-812 19 Bratislava, Slovakia*

Abstract: Disturbance rejection performance is usually the most important aspect in designing controller in the process and the chemical industries. On the other hand, such tuned control loop may exhibit large overshoots on reference tracking. The only way to decrease overshoot is to use two-degrees-of-freedom (2-DOF) controllers. Unfortunately, reducing the overshoot usually results in slower tracking response when using some of the most common 2-DOF controller structures. This paper suggests using additional first-order filter on the set-point of the PI controller in order to improve tracking performance. The calculation of compensator's parameters is straightforward and can be easily performed in practice. Several examples confirmed the efficiency of the proposed approach. *Copyright © 2003 IFAC*

Keywords: PID control, Magnitude Optimum, Controller Tuning, Disturbance Rejection, Tracking

1. INTRODUCTION

PID controllers are the most widely used controllers in the process industry. It has been acknowledged that more than 95% of the control loops used in the process control are of the PID type, of which most are the PI type (Åström and Hägglund, 1995).

Tuning of PID controllers has been attracting interest for almost six decades. Numerous tuning methods have been proposed so far. Some of them are concentrating on improving tracking, and some on disturbance rejection performance of the closed-loop.

In chemical and process industries, disturbance rejection performance is often more important (Åström et al., 1995; Åström et al., 1998; Vrančić and Strmčnik, 1999), since usually the set-point is fixed. However, if both, tracking and disturbance rejection performances are important (e.g. in batch processes), the two-degrees-of-freedom (2-DOF) controller should be used (Åström and Hägglund, 1995; Åström et al., 1998, Huba et al., 1997, Taguchi and Araki, 2000).

The most popular realisation of 2-DOF PI controller is the so-called set-point weighting PI controller (Hang et al., 1991). This type of controller significantly reduces overshoots in reference tracking at the cost of slower tracking speed.

In this paper it will be shown that tracking performance of the PI controller can be significantly improved by the introduction of simple filters on the set-point signal.

2. PI CONTROLLERS

Fig. 1 shows one-degree-of-freedom (1-DOF) PI controller (G_C) and the process (G_P) in the closed-loop configuration.

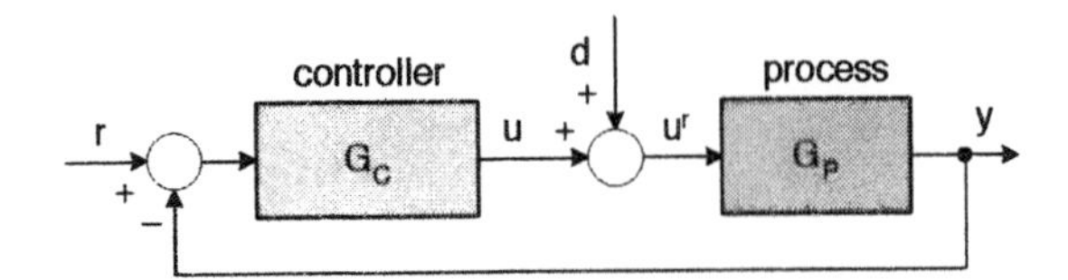

Fig. 1. The closed-loop configuration using 1-DOF controller

Case 1

Let us assume the following process transfer function:

$$G_P = \frac{1}{(1+s)(1+9s)} \tag{1}$$

and the following PI controller:

$$G_{C1} = K_{P1} + \frac{K_{I1}}{s}, \tag{2}$$

where parameters K_{P1} and K_{I1} are calculated by using the magnitude optimum tuning (MO) method (Åström and Hägglund, 1995; Hanus, 1975; Kessler, 1955; Vrančić et al., 1999; Vrančić et al., 2001):

$$K_{P1} = 4.56,\ K_{I1} = 0.506 \tag{3}$$

Fig. 2. shows tracking (see upper figure; set-point r changed from 0 to 1) and disturbance rejection performance (see lower figure; disturbance d changed from 0 to 1). It can be seen that tracking performance is good, while disturbance rejection is somehow sluggish, taking into account visible long-tail.

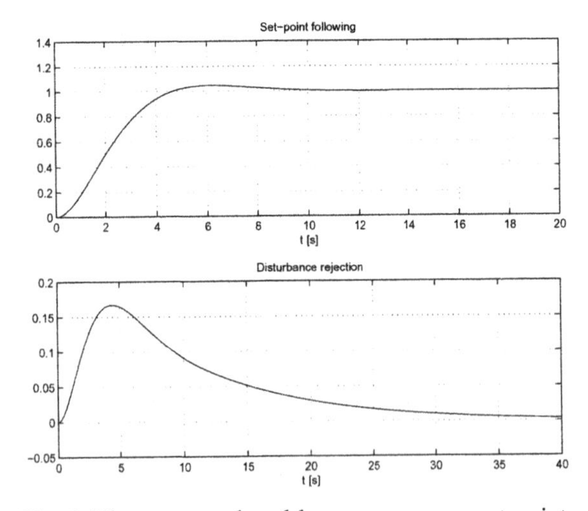

Fig. 2. The process closed-loop response on set-point change (upper figure) and disturbance (lower figure).

Case 2

In order to improve disturbance rejection performance, the controller parameters are calculated according to the modified magnitude optimum (MMO) tuning method (Vrančić and Strmčnik, 1999):

$$K_{P2} = 4.56,\ K_{I2} = 1.54\,. \tag{4}$$

Tracking and control performance are shown in Fig. 3. Disturbance rejection performance is now significantly improved. On the other hand, significant overshoot exists in reference tracking.

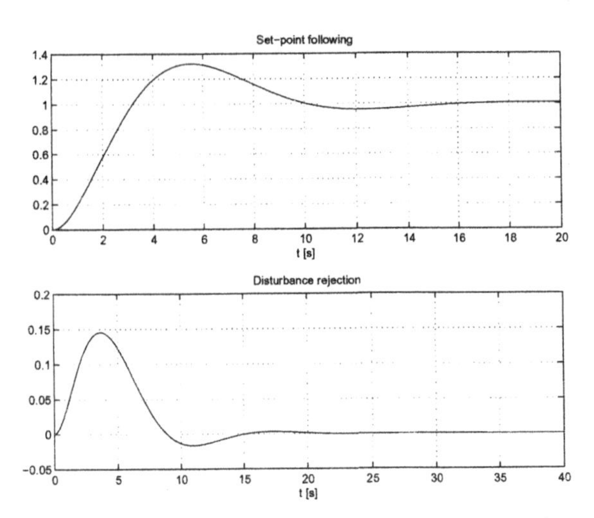

Fig. 3. The process closed-loop response on set-point change (upper figure) and disturbance (lower figure).

Case 3

In order to decrease the overshoot in reference tracking, 2-DOF controller should be applied. One of the most common 2-DOF PI controller realisation is the so-called "set-point weighting" PI controller (Åström and Hägglund, 1995; Hang et al., 1991). The controller transfer function is the following:

$$U = \left(K_P b + \frac{K_I}{s}\right)R - \left(K_P + \frac{K_I}{s}\right)Y\,, \tag{5}$$

where U, R and Y are Laplace transforms of controller output, set-point and process output, while b is the "set-point weighting" factor. By choosing b=0 and keeping the same controller parameters as in case 2, the overshoot on reference tracking can be significantly reduced, while remaining the same disturbance rejection performance, as shown in Fig. 4.

However, tracking speed is now significantly reduced when compared to results from Cases 1 and 2. The ideal response would be combination of tracking performance of case 1 and disturbance rejection performance of case 2 (or 3).

In the next section, the modified controller structure, which tries to accomplish the aforementioned task, will be described.

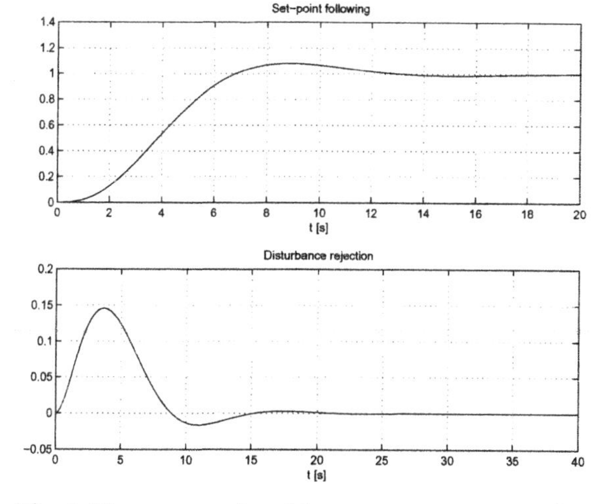

Fig. 4. The process closed-loop response on set-point change (upper figure) and disturbance (lower figure).

3. SET-POINT FILTERS

The ideal closed-loop transfer functions would be the following (see Fig. 1):

$$\frac{Y}{R}=\frac{G_{C1}G_P}{1+G_{C1}G_P}, \tag{6}$$

$$\frac{D}{R}=\frac{G_P}{1+G_{C2}G_P}, \tag{7}$$

where G_{C1} denotes controller tuned for reference following (e.g. controller (3)) and G_{C2} for disturbance rejection (e.g. controller (4)). In general, this cannot be realised by using 1-DOF controller.

By using 2-DOF controller, shown in Fig. 5, the closed-loop transfer functions become:

$$\frac{Y}{R}=\frac{G_R G_P}{1+G_Y G_P}, \tag{8}$$

$$\frac{D}{R}=\frac{G_P}{1+G_Y G_P}. \tag{9}$$

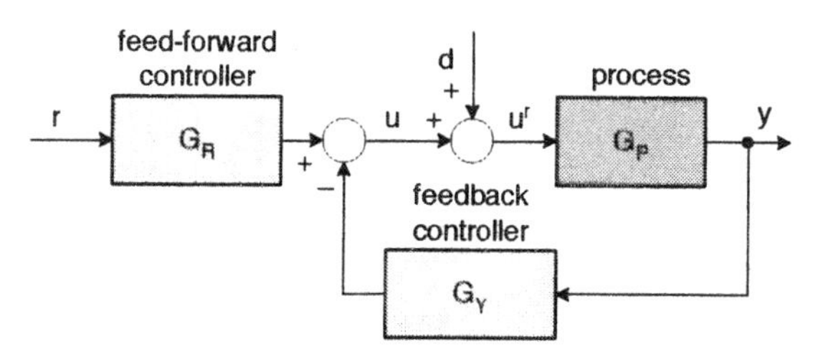

Fig. 5. The closed-loop configuration using 2-DOF controller

When comparing expressions (6) and (7) to (8) and (9), it follows that:

$$G_Y=G_{C2}, \tag{10}$$

$$G_R=G_{C1}\frac{1+G_{C2}G_P}{1+G_{C1}G_P}, \tag{11}$$

When applying the PI controller transfer function (2) instead of G_{C1} and G_{C2} (by using index 2 instead of 1), and developing the process into the following infinite series[1]:

$$G_P=G_0+G_1 s+G_2 s^2+\cdots, \tag{12}$$

the expression for G_R becomes:

$$G_R=\frac{f_0+f_1 s+f_2 s^2+\cdots}{e_1 s+e_2 s^2+\cdots}, \tag{13}$$

where

$$\begin{aligned}
f_0 &= K_{I1}K_{I2}G_0\\
f_1 &= K_{I1}(1+K_{P2}G_0+K_{I2}G_1)+K_{P1}K_{I2}G_0\\
f_2 &= K_{I1}(K_{I2}G_2+K_{P2}G_1)+\\
&\quad +K_{P1}(1+K_{P2}G_0+K_{I2}G_1)\\
&\ \vdots\\
e_1 &= K_{I1}G_0\\
e_2 &= 1+K_{P1}G_0+K_{I1}G_1\\
&\ \vdots
\end{aligned} \tag{14}$$

Note that process is represented by expression (12), which is not related to any conventional process model. As will be shown later, this kind of representation by the infinite series will result in simple calculation of controller parameters.

In general, due to the infinite order of expression (12), transfer function G_R is of an infinite order. In practice, the order should be appropriately decreased. Higher order means better approximation of expression (13), while lower order results in much simpler realisation of the controller. In our case, the second-order transfer function G_R has been chosen (f_i=0, e_i=0; i>2). In this case the realisation becomes relatively simple, resulting in the first-order set-point compensator, as shown in Fig. 6.

Parameters T_{PF}, T_{ZF} and b_F can be calculated from (14):

$$T_{PF}=\frac{1+K_{P1}G_0+K_{I1}G_1}{K_{I1}G_0}, \tag{15}$$

$$T_{ZF}=\frac{\begin{bmatrix}K_{I1}(K_{P2}G_1+K_{I2}G_2)+\\ +K_{P1}(1+K_{P2}G_0+K_{I2}G_1)\end{bmatrix}}{K_{I1}(1+K_{P2}G_0+K_{I2}G_1)+K_{P1}K_{I2}G_0}, \tag{16}$$

[1] The process time delay can be developed into infinite Taylor (or Pade) series (see Vrančić et al., 1999).

$$b_F = \frac{K_{I1}(1 + K_{P2}G_0 + K_{I2}G_1) + K_{P1}K_{I2}G_0}{K_{P2}K_{I1}G_0}, \tag{17}$$

where K_{P1} and K_{I1} denote parameters of controller G_{C1} and K_{P2} and K_{I2} denote parameters of controller G_{C2}.

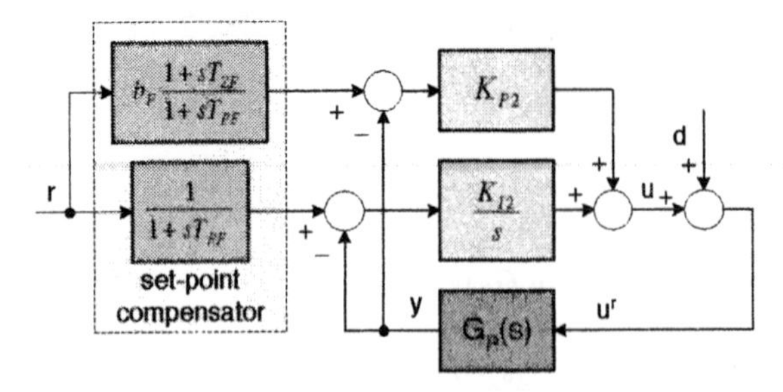

Fig. 6. 2-DOF controller with set-point compensator

Derivations above are based on representation of the process model by the infinite series (12). Calculation of parameters G_{Pi} (i=0,1,2,...) is relatively simple. If the process can be expressed by the following transfer function:

$$G_P = K_{PR}\frac{1 + b_1 s + b_2 s^2 + \cdots}{1 + a_1 s + a_2 s^2 + \cdots}e^{-sT_{del}}, \tag{18}$$

then parameters G_{Pi} in expression (12) can be calculated as follows (Preuss, 1991; Rake, 1987):

$$G_{Pi} = (-1)^i A_i, \tag{19}$$

where A_i are so-called characteristic areas:

$$\begin{aligned}
A_0 &= K_{PR} \\
A_1 &= K_{PR}(a_1 - b_1 + T_{del}) \\
A_2 &= K_{PR}\left(b_2 - a_2 - T_{del}b_1 + \frac{T_{del}^2}{2}\right) + A_1 a_1 \\
A_3 &= K_{PR}\left(a_3 - b_3 + T_{del}b_2 - \frac{T_{del}^2 b_1}{2} + \frac{T_{del}^3}{3!}\right) + \\
&\quad + A_2 a_1 - A_1 a_2 \\
&\vdots
\end{aligned} \tag{20}$$

Note that characteristic areas can also be calculated directly from the measurement of the process steady-state change by using simple summations (Vrančić et al., 1999; 2001). Therefore, explicit parameters of the process transfer function are not required in practice. Such simple calculation is achieved due to representation of the process model by the infinite series (12).

Case 4

Let us now calculate parameters of controller G_R for the process G_P (1), controller G_{C1} (3), and controller G_{C2} (4). First, characteristic areas are calculated from the process transfer function by using expressions (20): A_0=1, A_1=10, A_2=91, and A_3=820. Then, parameters T_{PF}, T_{ZF} and b_F are calculated from expressions (15-17):

$$T_{PF} = 0.9881,\ T_{ZF} = 1.4519,\ b_F = 0.8856. \tag{21}$$

As shown in Fig. 7, the reference tracking is now significantly improved when compared to cases 2 or 3 (Figs 3 or 4), while retaining the same disturbance rejection performance. As expected, tracking performance is close to one obtained in case 1.

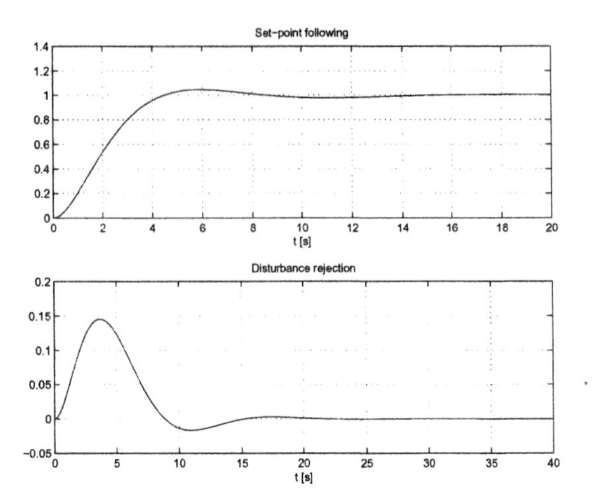

Fig. 7. The process closed-loop response on set-point change (upper figure) and disturbance (lower figure).

4. ILLUSTRATIVE EXAMPLES

The following process models (the first three examples in Åström et al., 1998) will be used in simulations:

$$\begin{aligned}
G_{P1} &= \frac{1}{(1+s)^3} \\
G_{P2} &= \frac{1}{(1+s)(1+0.2s)(1+0.04s)(1+0.008s)} \\
G_{P3} &= \frac{e^{-15s}}{(1+s)^3}
\end{aligned} \tag{22}$$

Controller parameters for set-point tracking are calculated by using the MO tuning method (Vrančić et al., 1999). First, the characteristic areas are calculated from (20):

Table 1. Characteristic areas

Process	A_0	A_1	A_2	A_3
G_{P1}	1	3	6	10
G_{P2}	1	1.248	1.30	1.31
G_{P3}	1	18	163.5	1000

Then, the PI controller parameters are calculated by using the following formulae (Vrančić et al., 1999):

$$K_P = \frac{A_3}{2(A_1A_2 - A_0A_3)}$$
$$K_I = \frac{A_2}{2(A_1A_2 - A_0A_3)} \quad (23)$$

The tracking parameters are given in table 2:

Table 2. Set-point following controller parameters

Process	K_{P1}	K_{I1}
G_{P1}	0.625	0.375
G_{P2}	2.1	2.083
G_{P3}	0.257	0.0421

Disturbance rejection parameters were kept the same as given in Åström et al. (1998), by choosing maximum sensitivity M_S=1.6 (see table 3).

Table 3. Disturbance rejection controller parameters

Process	K_{P2}	K_{I2}	b
G_{P1}	0.862	0.461	0.93
G_{P2}	2.74	4.078	0.75
G_{P3}	0.208	0.0354	1.00

Controller parameters K_P and K_I have been calculated by means of optimisation in frequency-domain, in order to achieve maximum K_I at chosen maximum sensitivity M_S. Parameter b is the already mentioned "set-point weighting" factor in expression (5). It has been calculated so as to sufficiently decrease the overshoot on set-point change (Åström et al., 1998).

Parameters of the set-point compensator (T_{PF}, T_{ZF}, and b_F) can be calculated from expressions (15-17), by using controller parameters in Tables 2 and 3:

Table 4. Set-point compensator parameters

Process	T_{PF}	T_{ZF}	b_F
G_{P1}	1.333	0.784	1.45
G_{P2}	0.24	0.188	1.01
G_{P3}	11.89	7.04	3.78

The closed-loop responses on set-point change are shown in Figs. 8 (process outputs) and 9 (process inputs), where solid lines represent the proposed method and broken lines show the responses originally obtained by (Åström et al., 1998), where the controller is given by expression (5). It can be seen that the proposed method results in somehow better tracking performance.

Note that the calculation of the set-point compensator parameters can be entirely based on simple measurement of the process steady-state change (by measuring characteristic areas) an does not require explicit parameters of the process transfer function.

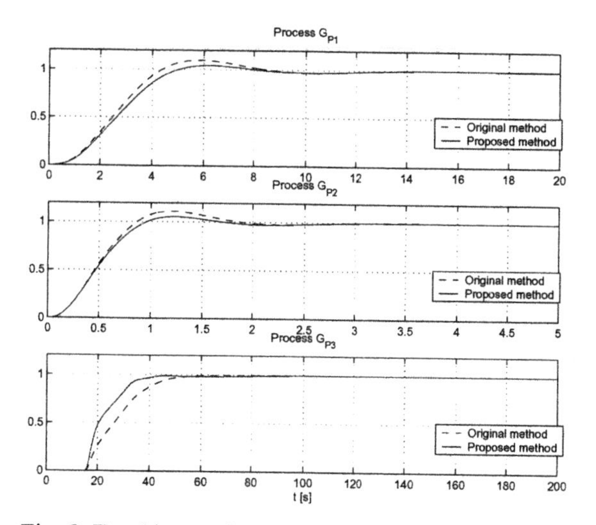

Fig. 8. Tracking performance for all three process models by using the original method (Åström et al., 1998) and the proposed method.

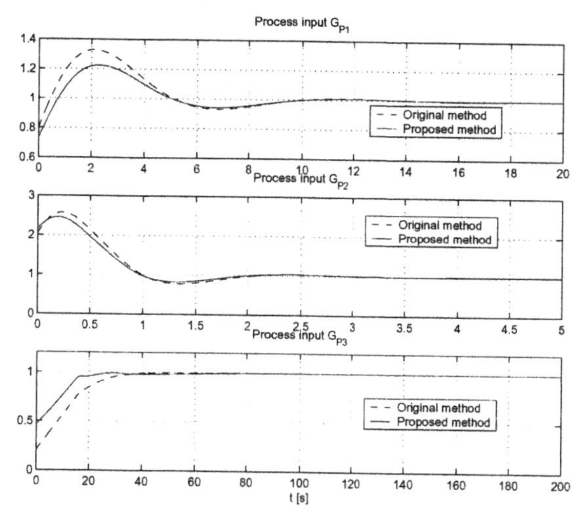

Fig. 9. Process inputs of all three process models by using the original method (Åström et al., 1998) and the proposed method.

5. CONCLUSIONS

The proposed set-point compensator may improve the tracking performance of the PI controllers, which parameters (K_P and K_I) are already selected (e.g. for improving disturbance rejection performance). The structure of the compensator is simple, since it consists of two first-order filters. The tracking performance depends on the chosen parameters of the "tracking PI controller". Parameters can be calculated by using any available tuning method. The calculation of the compensator's parameters is relatively simple, since it requires only characteristic areas. These areas can be calculated directly from the process transfer function or in time-domain by measuring the process steady-state change and performing simple summations (Vrančić et al., 1999).

The efficiency of the proposed set-point compensator is proven on four different process transfer functions. The results of experiments showed improved tracking performance when compared to the most common 2-DOF PI controller (by using set-point weighting). Moreover, the introduction of the compensator has no influence on the closed-loop stability, since it is not placed in the feedback path.

However, it should be noted that the proposed low-order structure of the compensator is only approximation of the function G_R (13). Therefore, the tracking response will only be similar to the one chosen by parameters of set-point tracking controller G_{C1} (2). The accuracy of the response might be improved by increasing the compensator's order but at the cost of more complex realisation and calculation (with more demanding accuracy of the identified process model or characteristic areas).

6. ACKNOWLEDGEMENT

The support of the Multi-Agent Control Research Training Network, supported by EC TMR grant HPRN-CT-1999-00107, and the Ministry of Education, Science and Sports of the Republic of Slovenia are gratefully acknowledged.

Reviewers' comments are gratefully acknowledged.

7. REFERENCES

Åström, K. J., and T. Hägglund (1995). PID Controllers: Theory, Design, and Tuning. *Instrument Society of America*, 2nd edition.

Åström, K. J., H. Panagopoulos, and T. Hägglund (1998). Design of PI Controllers based on Non-Convex Optimization. *Automatica*, **34**, 5, pp. 585-601.

Hang, C. C., K. J. Åström, and W. K. Ho (1991). Refinements of the Ziegler-Nichols tuning formula. IEE Proceedings, D, **138** (2), pp. 111-118.

Hanus, R. (1975). Determination of controllers parameters in the frequency domain. Journal A, **XVI** (3).

Huba, M., P. Kul'ha, P. Bisták, and Z. Skachová (1997). Dynamical Classes of PI Controllers for the 1st Order Loops. *Proceedings of the IFAC workshop "New Trends in Design of Control Systems"*, Oxford: Pergamon, 1998., pp. 189-194.

Kessler, C. (1955). Über die Vorausberechnung optimal abgestimmter Regelkreise Teil III. Die optimale Einstellung des Reglers nach dem Betragsoptimum. Regelungstechnik, Jahrg. **3**, pp. 40-49.

Preuss, H. P. (1991). Prozessmodellfreier PID-regler-Entwurf nach dem Betragsoptimum. *Automatisierungstechnik*, **39** (1), pp. 15-22.

Rake, H. (1987). Identification: Transient- and Frequency-Response Methods. Systems & Control Encyclopaedia; Theory, Technology, Applications, Madan G Singh, ed., *Pergamon Press*, pp. 2320-2325.

Taguchi, H., and M. Araki (2000). Two-degree-of-freedom PID controllers – Their functions and optimal tuning. *Pre-prints of the IFAC Workshop on Digital control (PID' 00)*, Terrassa, 5-7. April 2000, pp. 95-100.

Vrančić, D., and S. Strmčnik (1999). Achieving Optimal Disturbance Rejection by Using the Magnitude Optimum Method. *Pre-prints of the CSCC'99 Conference*, Athens, pp. 3401-3406.

Vrančić, D., Y. Peng, and S. Strmčnik (1999). A new PID controller tuning method based on multiple integrations. *Control Engineering Practice*, **VII**, 5, pp. 623-633.

Vrančić, D., S. Strmčnik, and Đ. Juričić (2001). A magnitude optimum multiple integration tuning method for filtered PID controller. *Automatica*, **37**, pp. 1473-1479.

www.elsevier.com/locate/ifac

SIMPLE ANALYTIC RULES FOR BALANCED TUNING OF PI CONTROLLERS

P. Klán *,[1], **R. Gorez** **

* *Institute of Computer Science, Pod vodárenskou věží 2, 182 07 Prague 8, Czech Republic*
** *Centre for Systems Engineering and Applied Mechanics Univ. Louvain, Avenue Georges Lemaitre 4, B-1348 Louvain-la-Neuve, Belgium*

Abstract: New explicit design relations are proposed for setting the parameters of PI controllers, from the knowledge of three parameters: gain, average residence time, and normalized dead time, characterizing the process to be controlled. These parameters can be obtained from the process step response or from the closed-loop response with any stabilizing PI controller. Simulation results for different types of process transfer function, including non minimum-phase processes or processes with dead time, show well damped responses with smooth control actions minimizing the total variation of the controller input. *Copyright © 2003 IFAC*

Keywords: PI Control; Tuning methods; Balanced tuning; New design relations

1. INTRODUCTION

PI control provides adequate performance in a vast majority of applications (Shinskey 1995). Nevertheless, despite continual advances in control theory (Åström and Hägglund 1995) there remains some interest for research on PI control. The Web of Science refers about more than 400 papers which were published on this topics since 1980. More than a hundred of rules for tuning PI controllers can be found in the literature; see e.g. (Åström and Hägglund 1995, O'Dwyer 2000, Hang *et al.* 2002, Shinskey 1995, Seborg *et al.* 1989, Taguchi and Araki 2000).

The first rules for tuning PI and PID controllers were published in (Ziegler and Nichols 1942); nowadays they are given in most textbooks on process control, see e.g. (Åström and Hägglund 1995, ?). These rules are based on the ultimate gain and ultimate period of the process to be controlled: "the ultimate gain K_{cu}, is the largest value of the controller gain that results in closed-loop stability when a proportional-only controller is used; the period of the resulting sustained oscillation is referred to as the ultimate period, P_u" (Seborg *et al.* 1989). With the original Ziegler-Nichols settings, the controller gain K_c should be equal to $K_{cu}/2$ for a proportional-only controller and slightly reduced ($0.45K_{cu}$) for a PI controller whose integral time constant is set to $T_i = P_u/2$. These settings provide a 1/4 decay ratio, that is to say a 50% overshoot for set-point changes, but "such responses are often judged too oscillatory by plant operating personnel" (Seborg *et al.* 1989). In fact, the controller is designed to be rather aggressive. It means that it is based on mostly proportional-only control, and the integral action is introduced a posteriori.

Other approaches to controller design are possible, that emphasize the role of the integral

[1] Corresponding author. Tel.: +420-266053850; fax.: +420-286585789; e-mail: pklan@cs.cas.cz. This research is supported by the Grant agency of the Czech rep. via project 102/03/0625.

term (Liu and Daley 2001, Vrancic *et al.* 1999) or try to balance the two control actions (Klán and Gorez 2000). Most industrial processes are self-regulating, so that they can be controlled by integral-only controllers. However, such controllers are seldom used; the response to changes in the set point or to disturbances in the measurement signal is to slow, as a consequence of little control action until the error signal has persisted for some time. Therefore, some proportional control action is desirable to have immediate corrective action as soon as an error is detected.

However, when applying some selected rule, control operators often have difficulties to predict the resulting controller functioning, so that tuning a PI controller requires some subsequent investigations. PI controller tuning therefore becomes rather subjective than optimal and operators a priori ask the following key questions: will the PI controller tuning produce satisfactory closed loop responses, and will it fulfill the required performance specifications? In some cases a partial answer on these questions is known; Ziegler-Nichols tuning, for example, often leads to a rather oscillatory response to set point changes (Liu and Daley 2001). However, generally there is no such easy answer.

PI tuning relations guarantying some predetermined property can be used as a possible solution to this "information embargo". For example, PI controllers can be tuned with a view to some predetermined maximum sensitivity of the closed loop, and tuning relations for two or three values of the latter can be found in (Åström and Hägglund 1995) and in (Gorez 2003). Here, another property is used for the design of PI controllers; it is the "balanced tuning" of the controller which involves the balance of the contributions of the two terms of the controller, with a clear intuitive interpretation. In other words, balanced tuning means that the proportional action and the integral action of the controller are equal in the average. Balanced tuning was proposed in (Klán and Gorez 2000), and from simulation results obtained for a batch of typical processes given in (Åström and Hägglund 1995), it provides robust control with smooth responses having no or negligibly small overshoot.

The balance of the two control actions can be performed with a view to quiet process operation, which is generally desired by process engineers and plant operators. Therefore, the variations of the control variable during the transients following a disturbance or a change in the set point should be kept as small as possible. For that purpose, in modern control system design methods such as linear-quadratic optimal control or model-based predictive control, a term proportional to the squared time–derivative (or time–increment) of the control variable is introduced in the performance index to be minimized.

Here, balanced tuning is derived for first order processes characterized by three parameters: the static gain K, a global time constant T, and an apparent dead time L. This is the most common process model used in papers on PID controller tuning. As pointed in (Åström and Hägglund 1995), the difficulty of controlling a given process can be characterized by the normalized dead time of the latter, that is to say the ratio $\tau = L/(L + T)$. Using these parameters for characterizing the process to be controlled simple analytical tuning rules for balanced tuning of a PI controller can be proposed.

The paper is organized as follows. Section 2 introduces the balance between P and I control actions. Section 3 proposes simple analytical rules for balanced control. Section 4 presents and compares some simulation results obtained with these design rules and that proposed by Ziegler–Nichols (Ziegler and Nichols 1942), Cohen–Coon (Cohen and Coon 1953) and Åström–Hägglund (Åström and Hägglund 1995). Section 5 establishes some link between balanced control and the minimization of the total variation of the controller input. Conclusions are given in Section 6.

2. CHARACTERIZATION OF BALANCED PI CONTROL

The PI control law is usually expressed as

$$u(t) = K\,[e(t) + \frac{1}{T_I}\int_0^t e(\tau)\,d\tau], \tag{1}$$

where u is the controller output variable, e denotes the error signal resulting from the difference between the controller set point and the process output, and K and T_I are the proportional gain and the integral time constant of the controller, respectively. The control law can also be expressed in the so-called velocity form (Åström and Hägglund 1995)

$$\dot{u}(t) = K\,[\dot{e}(t) + \frac{1}{T_I}e(t)]. \tag{2}$$

Clearly the well-known performance index (Seborg *et al.* 1989)

$$\text{ITAE} = \int_0^\infty t\,|e(t)|\,dt, \tag{3}$$

which is often used for obtaining well-damped closed-loop responses, can be related to the second term of (2); thus, it can be used to characterize the integral term of the PI controller. A similar performance index was introduced in (Klán and Gorez 2000) to characterize the proportional control action as follows:

$$\mathrm{ITAD} = T_I \int_0^\infty t\,|\dot{e}(t)|\,dt, \qquad (4)$$

with the acronym ITAD being used to mean the integral of time × the absolute value of the time derivative of the error signal. When multiplied by the common factor K/T_I the two performance indices (4) and (3) can be viewed as measures of the energy used in the P and I control actions. In (Klán and Gorez 2000) it is shown that *balanced tuning* where the ITAE performance index is minimized under the constraint of equal values of ITAE and ITAD provides well-damped closed-loop responses.

For illustrating the properties of balanced tuning, Figure 1 shows the closed-loop responses obtained with PI control of a given process and different settings of the controller parameters; the values of the latter and that of the resulting ITAE and ITAD indices are given in Table 1. Note that the normalized dead time of this process is $\tau = 0.54$, which means that the process is in the mid-range of the difficulty of controlling. It turns out that Ziegler-Nichols and Cohen-Coon controller tunings provide fast poorly damped responses. Clearly, these controllers have a high gain resulting into an "aggressive" control action, as indicated by the value of the ITAD index which is much higher than that of the ITAE index. At the opposite, the Åström–Hägglund design relations for a maximum sensitivity $M_s = 1.4$ lead to a slow overdamped response; this controller tuning is clearly not aggressive, with a small controller gain and a value of ITAD much lower than that of ITAE. As for balanced tuning of the controller, with ITAD ≈ ITAE, it provides a smooth response with negligible overshoot and the lowest settling time.

Table 1. PI control of the process $G(s) = e^{-s}/(s+1)^2$.

Controller settings	T_I	K	ITAE	ITAD
Ziegler–Nichols	3.84	1.22	13.50	32.70
Cohen–Coon	1.67	0.96	14.70	21.10
Balanced tuning	1.76	0.55	6.55	6.68
Åström–Hägglund	1.10	0.21	17.60	5.85

In order to establish advances of balanced tuning, simulation runs have been performed on the test batch of 17 processes proposed in (Åström and Hägglund 1995), with the process normalized dead time ranging from 0.2 to 0.8. Various design rules have been used for tuning the PI controller and the value of ITAE performance index has been

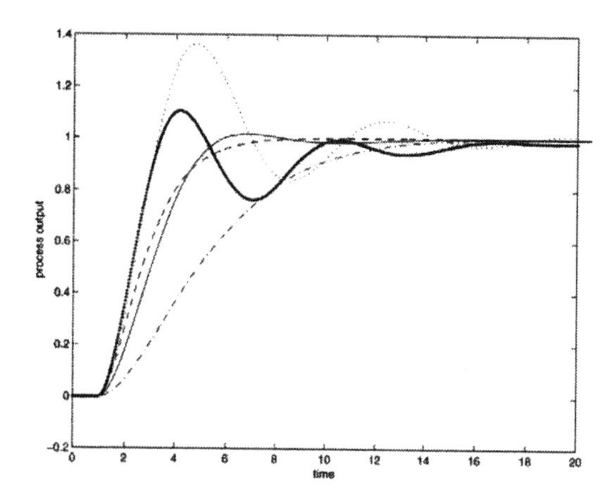

Fig. 1. Step response of the process $G(s) = e^{-s}/(s+1)^2$ (broken line), and closed-loop responses of this process with PI control and different controller settings: Ziegler–Nichols (thick line), Cohen–Coon (dotted line), Åström–Hägglund (dash–dotted line), Balanced tuning (full line).

recorded for each case. The average of the ITAE values obtained with the set of tuning rules for the 17 processes is given in Table 2. It can be observed that the design relations proposed by Ziegler-Nichols, Cohen-Coon and Åström–Hägglund lead to values of the ITAE index which are almost twice as big as that provided by balanced tuning with minimization of the ITAE index. Besides, the closed-loop response of the process is very close to its natural step response (Klán and Gorez 2000). One can say that it is relatively easy to achieve a closed-loop step response faster than the process natural response for processes characterized by $\tau \leq 0.5$, but this becomes more and more difficult as the process normalized dead time τ is increasing beyond $\tau = 0.5$. Balanced tuning with the a priori effect of keeping the natural step response may achieve some compromise whereas the effort for speeding up the transient response may result into some technical or economical problems.

Table 2. Average value of the ITAE performance index for different settings of PI controller parameters.

Tuning method	*ITAE*
Ziegler–Nichols	168.8
Åström–Hägglund ($M_s = 1.4$)	64.8
Cohen–Coon	42.6
Balanced tuning	29.8
Min ITAE for responses without oveshoot	25.2
Global min ITAE	20.0

3. DESIGN RULES FOR BALANCED PI CONTROL

Balanced tuning provides a response close to the natural dynamics of the process to be controlled

(Klán and Gorez 2000). For the three-parameter model

$$G(s) = \frac{K_P}{Ts+1} e^{-sL} \tag{5}$$

the control error after a step in the controller set point is

$$e(t) = K_P e^{-(t-L)/T} \tag{6}$$

for $t \geq L$, with $e(t) = K_P$ for $0 \leq t < L$. The balance condition ITAE = ITAD for PI control then leads to an explicit design relation for setting the controller integral time constant as

$$T_I = \frac{T_{ar}^2 + T^2}{2T_{ar}} = T_{ar} \frac{1 + (1-\tau)^2}{2}, \tag{7}$$

where $T_{ar} = L + T$ and $\tau = L/T_{ar}$ are the average residence time and the normalized dead time of (5). In particular, if the controlled process reduces to a simple time lag ($L = 0$) the integral time constant of the controller will cancel that of the process, while in the case of pure time delay ($T = 0$) the relation (7) leads to $T_I = L/2$.

Another condition is needed to obtain a relation for the controller gain K. Since balanced tuning provides closed-loop step responses close to the process step response, this condition can be formulated as follows: keep the closed-loop average residence time equal to that of the process. Then, the controller gain K will be selected to ensure this condition. For any stable transfer function $F(s)$, the average residence time is given by $T_{ar} = -F'(0)/F(0)$ (Åström and Hägglund 1995). Then, for the closed-loop transfer function relating the process output to the controller set point

$$\frac{K_P K e^{-sL}(T_I s + 1)}{T_I s(Ts+1) + K_P K e^{-sL}(T_I s + 1)}, \tag{8}$$

the average residence time is given by T_I/K_l, where $K_l = KK_P$ is the loop gain. The previous condition on the equality of the closed-loop and open-loop average residence times reduces to $T_I/K_l = T + L$, and it provides the following explicit design relation for setting the controller gain:

$$K = \frac{1}{K_P} \frac{T_I}{T+L} = \frac{1}{K_P} \frac{1+(1-\tau)^2}{2}. \tag{9}$$

Again, in the particular cases $L = 0$ and $T = 0$, the controller gain will be equal to $1/K_P$ and $0.5/K_P$, respectively. As for the mid-range case $\tau = 0.5$, that is to say equal apparent dead time and time lag, (7) and (9) yield $T_I = 1.25L$ and $K = 0.625/K_P$.

Since $1 + (1-\tau)^2)/2 \approx 1 - \tau$ for $\tau \ll 1$, the comparison of the new rules (7) and (9) with that based on the IMC design (Åström and Hägglund 1995, Seborg *et al.* 1989, Skogestad 2003) shows that balanced tuning is equivalent to the IMC design for the model (5) and $\tau \ll 1$.

4. SIMULATION RESULTS

Simple analytical design relations derived in the previous section have been applied to the four following typical processes whose normalized dead time values cover the full range of the control difficulty:

$$\frac{1}{(3s+1)(0.7s+1)(0.16s+1)(0.04s+1)} \qquad \tau = 0.2$$

$$\frac{e^{-0.8s}}{(1.6s+1)^2} \qquad \tau = 0.4$$

$$\frac{1-s}{(s+1)^3} \qquad \tau = 0.6$$

$$\frac{1}{(0.13s+1)^{30}} \qquad \tau = 0.8$$

Simulation results are presented in Fig. 2. Responses obtained with the controller parameters determined via the relations (7) and (9) are shown by full lines plots; other settings were used for comparison.

From these results it turns out that a controller design based on the three following parameters to characterize the controlled process: static gain, normalized dead time and apparent dead time, provides a relatively fast and smooth reset, for all the process considered here. From the experiments it turns out that the maximum loop sensitivity provided by balanced tuning determined via (7) and (9) is between 1.2 and 2.2 for most processes; it is slightly higher than 2 only for processes with high normalized dead time (> 0.8). Such values guarantee the robust stability of the control loop for all the processes considered in the test batch. Furthermore, based on inequalities in (Åström and Hägglund 1995), the gain and phase margins are better than 2 and 30 degrees, respectively. These values also provide a good disturbance rejection.

5. LINK BETWEEN BALANCED TUNING AND CONTROLLER RESPONSE

For self-regulating processes, the control variable would remain constant over the transients consecutive to a step in the set point if the two following conditions are met:

(1) Assuming that the process was in steady-state operation and a step in the set point is

applied at $t = 0$, the initial and final values of the control variable are equal:

$$u(0) = u(\infty),$$

where 0 actually means 0_+, the initial value being considered immediately after the step has been applied.

(2) At any time following the initial step the time derivative of control variable is equal to zero:

$$\dot{u}(t) = 0 \qquad (10)$$

for all $t > 0$. For a PI controller given by (2) it leads to the constraint

$$T_I \dot{e}(t) = -e(t). \qquad (11)$$

Except in the simplest case of a first-order process ($\tau = 0$) the above constraint cannot be satisfied at any time, but it can be fulfilled in the average by integrating and balancing both sides of (11) via the condition ITAE=ITAD. Since the controller output should be at least approximately kept at a constant value equal to the required steady-state value, many plant operators would appreciate this feature.

Moreover, balanced control substantially reduces the variations of the controller output. The latter can be appreciated through the integral $\int_0^\infty |\dot{u}(t)|dt$, or with a proper time discretisation (Skogestad 2003), through

$$TV = \sum_{i=1}^{\infty} |u_{i+1} - u_i|, \qquad (12)$$

where TV means the total variation of the controller output u. This provides a good measure of the *smoothness* of the actuating variable, and for a satisfying process control TV should be as small as possible.

The basic condition (10) for balanced tuning provides a substantial reduction of the performance index (12), as confirmed by the experiments with different controller settings for the test batch processes considered here (e.g. more than 100 times lower in average compared with Ziegler–Nichols tuning).

6. CONCLUSIONS

The concept of balanced PI control has been resumed and it has provided simple analytical rules allowing the designer of the control system to set the parameters of a PI controller in an easy and intuitive way. The nice features of balanced control have been illustrated by simulation results, which also establish some link with the minimization of the total variance of controller output provided by balanced tuning. These results show that the design rules proposed here are appropriate at least for self-regulating processes with normalized dead time lower than 0.8.

REFERENCES

Åström, K.J. and T. Hägglund (1995). *PID controllers: Theory, Design and Tuning.* Instrument Society of America, Research Triangle Park. NC, U.S.A.

Cohen, K.H. and G.A. Coon (1953). Theoretical consideration of retarded control. *Trans. ASME* **75**, 827–834.

Gorez, R. (2003). New design relations for 2-DOF PID-like control systems. *Automatica* **39**, 901–908.

Hang, C.C., K.J. Åström and Q.G. Wang (2002). Relay feedback auto-tuning of process controllers. A tutorial review. *Journal of Process Control* **12**, 143–162.

Klán, P. and R. Gorez (2000). Balanced tuning of PI controllers. *European Journal of Control* **6**, 541–550.

Liu, G.P. and S. Daley (2001). Optimal tuning PID control for industrial systems. *Control Engineering Practice* **9**, 1185–1194.

O'Dwyer, A. (2000). A summary of PI and PID controller tuning rules for processes with time delay. Part I, Part II. In: *Preprints of IFAC Workshop on Digital Control PID'00: Past, Present and Future of PID Control.* pp. 175–180,242–247. Terrassa, Spain.

Seborg, D., T. Edgar and T. Mellichamp (1989). *Process Dynamics and Control.* John Wiley & Sons.

Shinskey, F.G. (1995). *Process Control Systems: Application, Design and Tuning.* Mc Graw–Hill.

Skogestad, S. (2003). Simple analytic rules for model reduction and PID controller tuning. *Journal of Process Control* **13**, 291–309.

Taguchi, H. and M. Araki (2000). Two-degree-of-freedom PID controllers. Their functions and optimal tuning. In: *Preprints of IFAC Workshop on Digital Control PID'00: Past, Present and Future of PID Control.* pp. 95–100. Terrassa, Spain.

Vrancic, D., Y. Peng and S. Strmčnik (1999). A new PID controller tuning method based on multiple integrations. *Control Engineering Practice* **7**, 623–633.

Ziegler, J.G. and N.B. Nichols (1942). Optimum settings for automatic controllers. *Trans. ASME* **64**, 759–768.

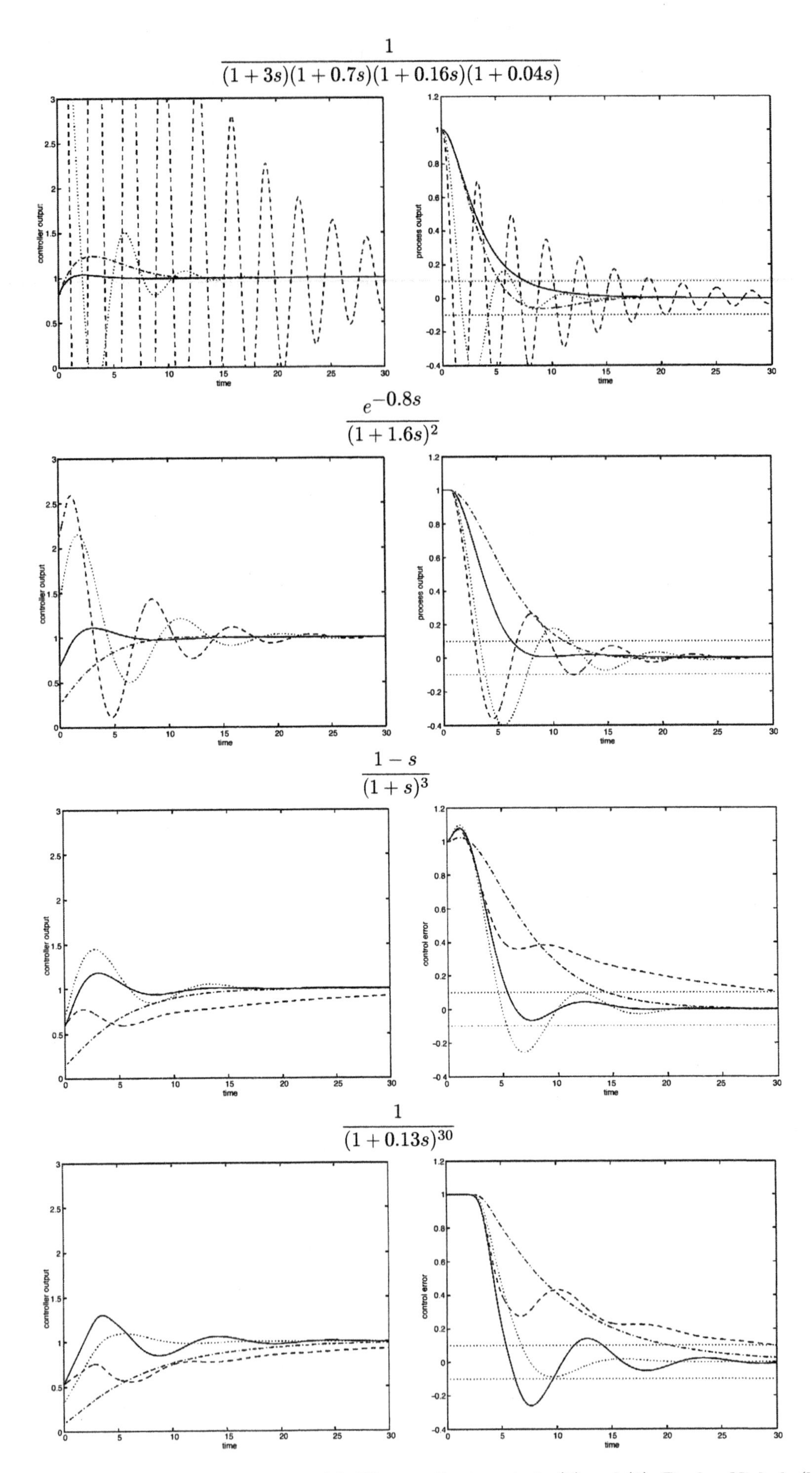

Fig. 2. Closed-loop step responses with PI controller tuned via (7) and (9): Ziegler–Nichols (broken line), Cohen–Coon (dotted line), Åström–Hägglund (dash- dotted line), Balanced tuning (full line).

GAIN SCHEDULED PI LEVEL CONTROL OF A TANK WITH VARIABLE CROSS SECTION

Mikuláš Huba

Slovak University of Technology in Bratislava,
Faculty of Electrical Eng. & Info. Technology, Ilkovičova 3,
812 19 Bratislava, Slovakia, huba @elf.stuba.sk

Abstract: This paper deals with the PI control of nonlinear 1st order plants. Different realizations of the parallel gain scheduled controllers are compared with the windupless ones introduced by Huba et al. (1998) and by the extended exact linearization based on the reconstruction and compensation of input disturbances via the inverse plant dynamics. The achieved results show strong necessity for rethinking the foundations of the most spread type of controllers. *Copyright © 2003 IFAC*

Keywords: Nonlinear control; PI control; gain scheduling; exact linearization; saturation.

1. INTRODUCTION

According to different surveys, the PI controller represent dominant majority of control applications. Aström and Hägglund (1995) report in their book about PID-control following statistics: „In process control, more than 95% of the control loops are of PID type, most loops are actually PI control." They also cite an audit of paper mills made by Bialkowski that: „A typical mill has more than 2000 control loops and that 97% use PI control." So, taking into account its simplicity, importance and "age" approaching slowly whole century, one could expect an absolute transparency in its properties, setting and use. However, reality shows to be far from this expectation. Already under restriction on linear control one can find decades, or even hundreds of optimal settings (see e.g. overview made by O'Dwyer, 2000). Furthermore, practically at each conference devoted to the control design, it is possible to see new ones!

It is not to wonder that in nonlinear control loops the situation becomes yet more obscure. The first problems are usually connected with the windup problem. There exist many solutions (see e.g. Kothare et al., 1994). Since they often lead to a better performance than guaranteed by linear controller, the proclaimed aimed of the anti-windup circuits: "to suppress effect of the control signal saturation" and "to recover the linear behavior", looks inadequate and indicates some vagueness in the linear design!

It is well known that the nonlinear plant character leads to a state dependent performance. This can be at least partially eliminated by different gain scheduling methods based on the plant linearization above fixed operating points, which are later relaxed to move (see e.g. Khalil, 1996).

In this paper it will be stressed that the windup effect is caused by unsuitable structure of the parallel PI-controller that introduces additional zero into the closed loop transfer function. In the case of nonlinear plant, this zero is state dependent and cannot be fully eliminated by usual scheduling methods.

Properties of the classical PI-controller are compared with the new windupless solutions. These show better properties already for linear plant. Soundness of new solutions is yet more exhibited by applications in nonlinear control.

The paper is organized as follows: Chapter 2 introduces new windupless PI_1 controller. Chapter 3 states the nonlinear problem and derives the plant linearization above a fixed operating point. In Chapter 4 the design of the parallel PI-controller based on the linearization above a fixed operating point is shown. Chapter 5 evaluates properties of different realizations of the gain scheduled parallel PI-controllers. Chapter 6 is devoted to the evaluation of the gain scheduled windupless controller. In Chapter 7, nonlinear windupless modification of the PI controller are proposed based on extension of the linearization via feedback by a disturbance-reconstruction based I-action

2. DYNAMICAL CLASSES OF PI CONTROL

PI-control is devoted to the control of loops with dominant 1st order dynamics. When speaking about 1st order loop dynamics (see Huba et al. 1998), one should distinguish 2 basic loop configurations (Fig.1) characteristic by two different dynamical classes of the PI-control.

Huba et al. (1998) showed that the linear PI-controller represents the optimal solution just for the feedback-dominated loops. This will be denoted as PI_0-controller according to no peak in the control signal response corresponding to a step change of the reference signal.

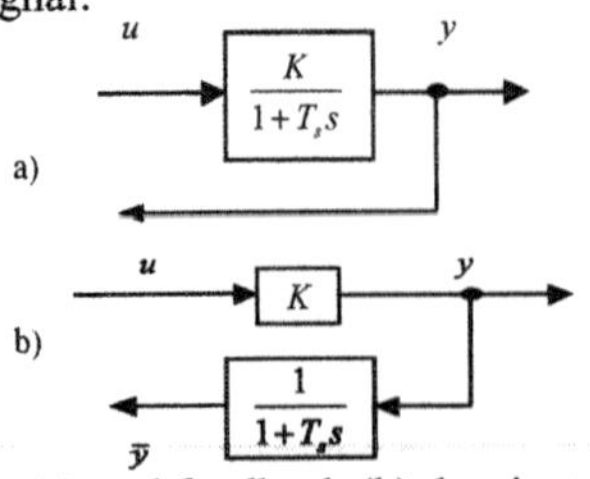

Fig.1. Plant (a) and feedback (b) dominated 1st order loop.

For the plant dominant dynamics placed in the feedforward path, new optimal solutions (Fig.2) can be derived denoted here as the PI_1-controller. It can be interpreted as the P-controller extended by static feedforward and by inverse dynamics based disturbance reconstruction and compensation.

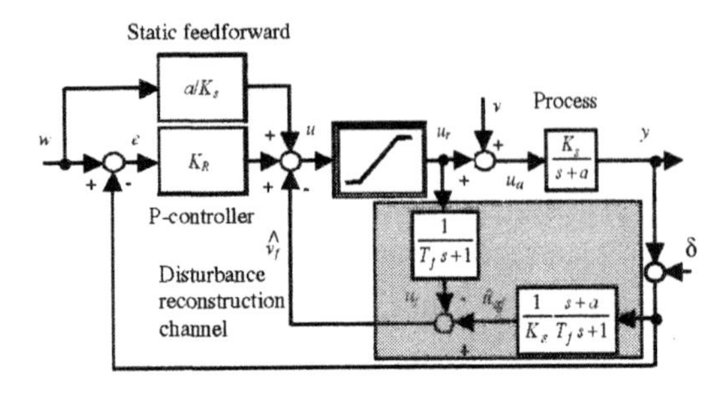

Fig.2. PI_1-controller: P-controller extended by static feedforward and inverse dynamics based disturbance reconstruction and compensation.

The equivalent loop (Fig.3) is typical by the 1st order lag in the feedback around the saturation. Fractions of this solution can be found in different anti-windup schemes (see e.g. Aström and Hägglund, 1995; p.91, Fig.38A). The control response corresponding to a reference signal step change is typical by one interval at the saturation limit that can occur at sufficiently large deviations.

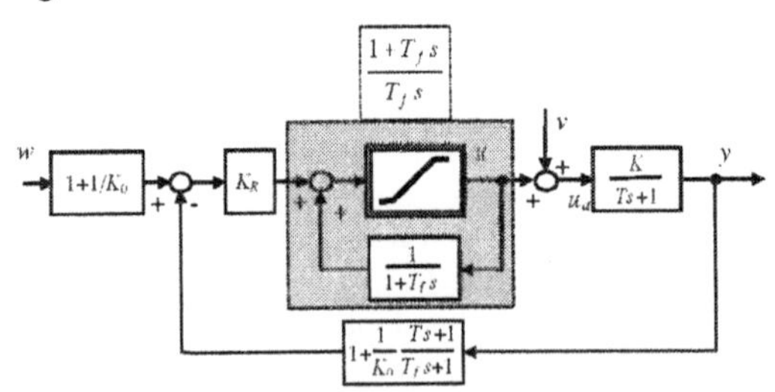

Fig.3. Equivalent scheme of PI_1-controller. $K = K_s / a$; $T = 1/a$; $K_0 = K_R K$;

The controller gain corresponding to the chosen closed loop pole $\gamma < 0$ can be expressed as

$$K_R = -(\gamma + a)/K_s \tag{1}$$

Loop stability is guaranteed for

$$K_R K_s > -a \tag{2}$$

It is important to note that if the closed loop pole chosen for control is "slower" than the process pole, then

$$K_R K_s < 0 \tag{3}$$

Naturally, one can expect that the optimal solution will be given by the natural distribution of the loop dynamics. However, it is to note that:

- By the controller tuning based just on measurement of the process output, one has no information to decide about the loop distribution!
- A step $\Delta\delta$ in the measurement noise produces a "kick" in the control signal

$$\Delta u_1 = \left[K_R + \frac{1}{K_s T_f} \right] \Delta\delta \; ; \; K_s = \frac{K}{T_s} \tag{4}$$

It can be decreased by putting $K_R = 0$, what corresponds to the PI_0-controller! So, in a noisy industrial environment one can intentionally choose controller corresponding to a lower dynamical class as naturally given by the loop configuration!

3. STATEMENT OF THE PROBLEM

Let us consider fluid level control in a tank with variable cross section ($c_v = 2; \; r_0 = 0; \; k = 1$, Fig.4)

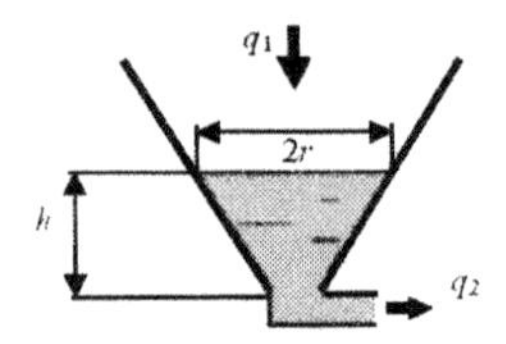

Fig.4. Fluid level control in a tank with variable cross section $S(h) = \pi r^2 = \pi (r_0 + kh)^2$

After introducing the denotation $y=h$, the process can be described as

$$\dot{y} = \frac{1}{S(y)} \left(u - c_v \sqrt{y} \right) = f(y,u) \; ; \qquad y \in \langle 0, H \rangle \; ; \; u \in \langle 0, U \rangle \tag{5}$$

Possible steady states characterized by $\dot{y} = 0$, when

$$y_0 = \alpha \qquad u_0 = c_v \sqrt{\alpha} \tag{6}$$

will be used for parametrization of linear models. By linearizing process above such equilibrium points one gets

$$\dot{y} = f(y_0, u_0) - a(\alpha)\Delta y + b(\alpha)\Delta u \tag{7}$$

whereby

$$a(\alpha) = -\left[\frac{\partial f}{\partial y} \right]_{y=\alpha, u=c_v\sqrt{\alpha}} = \frac{c_v}{2S(\alpha)\sqrt{\alpha}} = \frac{c_v}{2\pi (r_0 + k\alpha)^2 \sqrt{\alpha}} \tag{8a}$$

$$b(\alpha) = \left[\frac{\partial f}{\partial u} \right]_{y=\alpha, u=c_v\sqrt{\alpha}} = \frac{1}{S(\alpha)} = \frac{1}{\pi (r_0 + k\alpha)^2} \tag{8b}$$

$$f(y_0, u_0) = 0 \qquad \Delta y = y - \alpha \; ; \; \Delta w = w - \alpha \qquad \Delta\dot{y} = \dot{y} \; ; \qquad \Delta u = u - c_v \sqrt{\alpha} \tag{9}$$

Such system can be represented by the transfer function

$$S_\alpha(s)=\frac{\Delta Y(s)}{\Delta U(s)}=\frac{K(\alpha)}{T(\alpha)s+1}$$

$$K(\alpha)=\frac{b(\alpha)}{a(\alpha)}=\frac{2\sqrt{\alpha}}{c_v}\ ;\ T(\alpha)=\frac{1}{a(\alpha)}=\frac{2S(\alpha)\sqrt{\alpha}}{c_v} \qquad (10)$$

$$K_s(\alpha)=\frac{K(\alpha)}{T(\alpha)}=\frac{1}{S(\alpha)}$$

4. PARALLEL PI-CONTROLLER

For system linearized above fixed operating point $\alpha = const$, one possible realization of the PI-controller (Fig.5a) is given by equations

$$\Delta u = \lfloor K_p(\alpha)e + K_I(\alpha)\sigma \rfloor$$

$$\dot{\sigma} = e = w - y = \Delta w - \Delta y \qquad (11)$$

$$\Delta w = w - \alpha$$

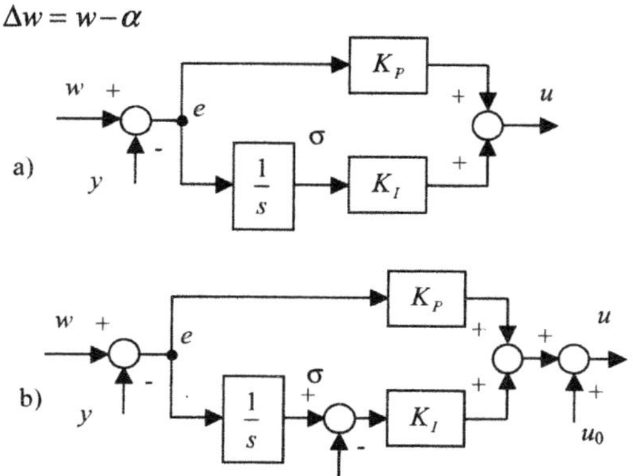

Fig.5. Two realizations of the parallel PI controller.

For chosen closed loop poles γ_1, γ_2 (see e.g. Khalil, 1996) one gets

$$K_p(\alpha)=\frac{-a(\alpha)-\gamma_1-\gamma_2}{b(\alpha)}=-\overbrace{\frac{c_v}{2\sqrt{\alpha}}}^{1/K(\alpha)}-[\gamma_1+\gamma_2]S(\alpha) \quad (12)$$

$$K_I(\alpha)=\frac{\gamma_1\gamma_2}{b(\alpha)}=\gamma_1\gamma_2 S(\alpha)$$

The closed loop is described by

$$F_w(s)=\frac{(K_p(\alpha)s+K_I(\alpha))b(\alpha)}{s^2+s[b(\alpha)K_p(\alpha)+a(\alpha)]+b(\alpha)K_I(\alpha)} \qquad (13)$$

Transients are typically state-dependent (Fig.6).

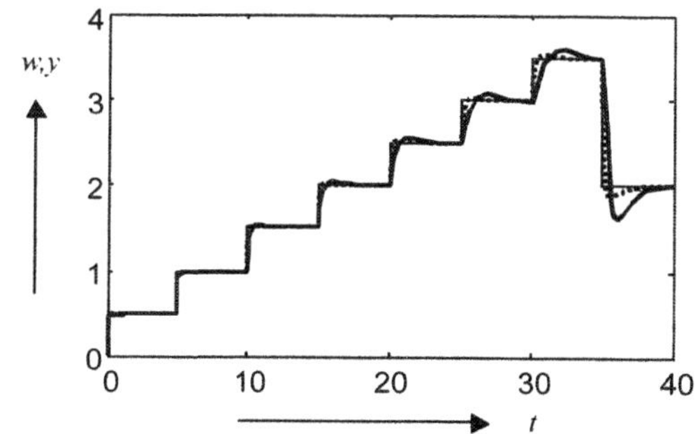

Fig.6. Parallel PI-controller: Transients corresponding to linearization $\alpha = H/2 = 2$ (full line) and $\alpha = H = 4$ (dotted). $y_0 = 0.01$; $\gamma_{1,2} = -2$.

This state dependant character is yet more exhibited by the control saturation (Fig.7).

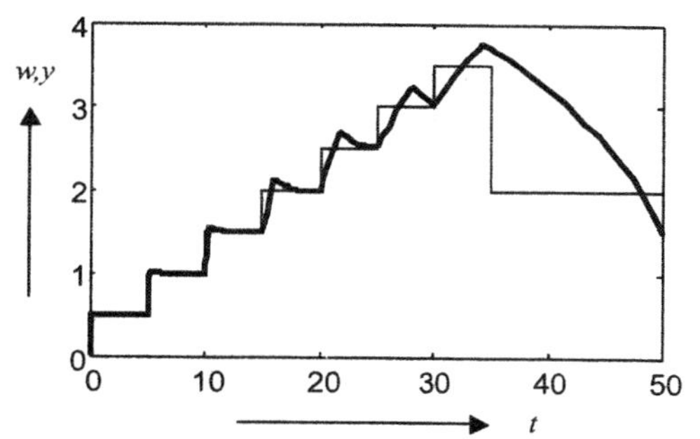

Fig.7. Parallel PI-controller: Transients corresponding to $\alpha = H = 4$ and constrained control signal $u \in \langle 0,10 \rangle$; $y_0 = 0.01$; $\gamma_{1,2} = -2$;

5. GAIN SCHEDULED PI CONTROLLER

Fig.8 demonstrates that by relaxing the operating point α frozen in the process of the linearization it is possible to decrease the state-dependent character of transients.

In principle, there exist several possibilities to choose appropriate scheduling variables. Intuition leads us to schedule the controller parameters by the process output y, or by the reference signal w. In the case of the controller realization according to Fig.5a the input scheduling $\alpha = w$ preserves the closed loop poles. However, the transfer function achieved by combination of the transfer function of the linearized process and of the controller is not equal (due to different zero) to the transfer function of the closed loop linearization with the parametrized nonlinear controller!

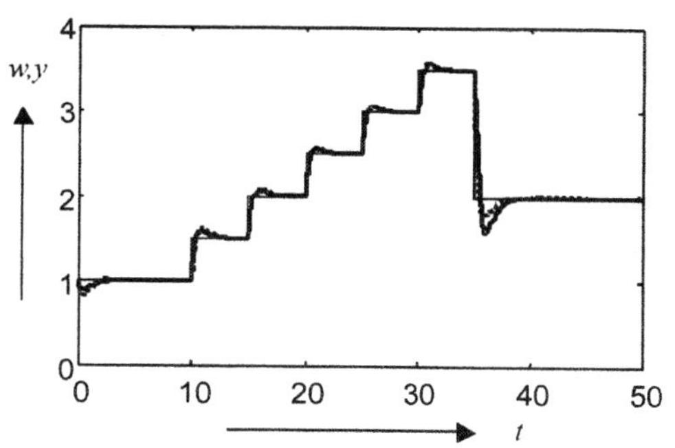

Fig.8. Parallel PI-controller: Transients corresponding to $\alpha = w$ (full line) and $\alpha = y$ (dotted). $y_0 = 1$; $\gamma_{1,2} = -2$; no control constraints

The small overshoot is typical result of the zero introduced into the loop by the parallel I-action.

It is to note that the output scheduling with $\alpha = y$, changes the closed loop poles (poles of the linearization of the closed loop system with the nonlinear parametrized PI controller). A direct conclusion of this fact makes itself felt by instability occurring for relatively small initial conditions. So, in Fig.8 the initial conditions have been limited to $y_0 = 1$, while in Fig.5 it was possible to start already with $y_0 = 0.01$. Besides of this, the differences in the loop dynamics caused by the particular choice of the scheduling variable are to see especially for larger changes.

In the gain scheduling research, a high attention has been given to the local linear equivalence conditions (Lawrence and Rugh 1995; Kaminer et al. 1995). They require: "At any operating point, the linearization of the feedback system consisting of the gain-scheduled controller and the nonlinear plant exhibits the same internal and input-output properties as the feedback connection of the linearized plant and the corresponding linear controller."

Following this property, Khalil (1996) cites two solutions for implementation of the PI-controller:

The 1st one (Fig.5b) extends the PI-controller by state dependent bias signals $\sigma_0 = \sigma_0(\alpha)$ and $v_0 = u_0(\alpha)$, which should satisfy the condition

$$K_I(\alpha)\frac{\partial \sigma_0(\alpha)}{\partial \alpha} = -\frac{\partial u_0}{\partial \alpha} \qquad (14)$$

For the given system one gets

$$\frac{\partial \sigma_0}{\partial \alpha} = -\frac{1}{2}\frac{c_c}{\gamma_1\gamma_2\pi\sqrt{\alpha}(r_0 + k\alpha)^2} \qquad (15)$$

Solving this equation for σ_0 one gets

$$\sigma_0 = -\frac{c_v}{2\gamma_1\gamma_2\pi r_0}\left[\frac{\sqrt{\alpha}}{r_0 + k\alpha} + \frac{arctg\left(\sqrt{\frac{k\alpha}{r_0}}\right)}{\sqrt{kr_0}}\right]; \; r_0 \neq 0 \qquad (16)$$

$$\sigma_0 = \frac{c_v}{3\gamma_1\gamma_2\pi k^2\alpha\sqrt{\alpha}}; \; r_0 = 0$$

To avoid problems caused by the control signal saturation, the different realizations are compared for relatively small reference signal values, at first. But, also under these conditions, the transient responses in Fig.9 achieved for $\alpha = w$ do not show any improvement of the control quality. For $\alpha = y$ the transients are yet worse – system is unstable for a broad range of initial states!

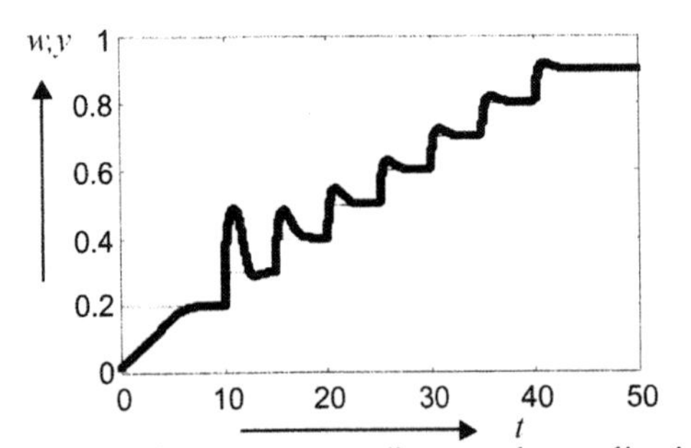

Fig.9. Transients corresponding to the realization according to Fig.5b for $\alpha = w$; $\gamma_{1,2} = -2$;

The 2nd solution proposed by Kaminer et al. (1995) is based on positioning the integrator at the controller output (Fig.10).

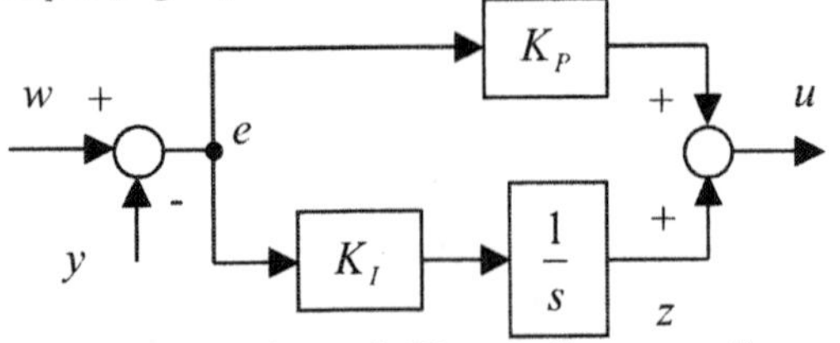

Fig.10. Realization of PI-controller according to Kaminer et al. (1995)

Achieved transient responses with $\alpha = w$ are stable just for $w > 0.4$. The dependence of the control performance on state was decreased (Fig.11), but not fully eliminated. This can be seen for larger reference values, when the control signal saturation occurs. After larger reference signal steps (at the end of the responses), instability occurs.

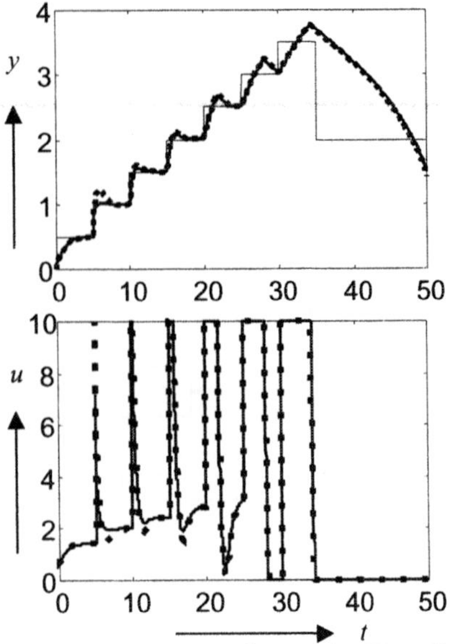

Fig.11. Transients corresponding to the PI realization according to Fig.10 (full line) and to Fig.5a (dotted); $\alpha = w$; $\gamma_{1,2} = -2$; $y_0 = 0.01$; $u \in \langle 0, 10 \rangle$;

Of course, the transient responses can be reasonably improved by introducing different anti-windup measures. However, to be rigorous, it is then not possible to say that the function of the anti-windup measures is to recover the linear behavior, because the not satisfactory control has its roots in the not appropriate linear structure!

6. WINDUPLESS PI_1 CONTROLLER

For computing Δu, the windupless PI_1 will be implemented according to Fig.2.

The controller will be based on the generalized scheduling variable

$$\alpha = mw + (1-m)y \; ; \; m \in \langle 0,1 \rangle \qquad (17)$$

The P- and I-channel outputs will be given (according to (9)) as

$$u_p = u_0 + \Delta u =$$
$$= c_v\sqrt{\alpha} + K_R(\alpha)(w-y) + \frac{1}{K(\alpha)}(w-\alpha) \qquad (18a)$$
$$\hat{v} = \frac{1}{K}\frac{1+T(\alpha)s}{1+T_f s}\Delta y - \frac{1}{1+T_f s}\Delta u$$

or (for $m \to 1$) also by

$$u_p = K_R(\alpha)(w-y) + \frac{1}{K(\alpha)}w$$
$$\hat{v} = \frac{1}{K}\frac{1+T(\alpha)s}{1+T_f s}y - \frac{1}{1+T_f s}u \qquad (18b)$$

The controller shows reasonably higher homogeneity of reference following (Fig.12) and also good disturbance rejection. But, also here it is possible to see some stability problems occurring for low-level values.

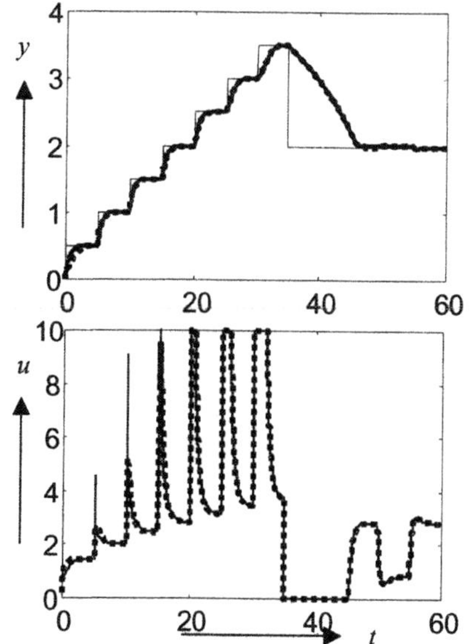

Fig.12. Transients corresponding to the gain scheduled windupless PI$_1$-controller realized according to Fig.2 (dotted) with $T_s = 1/a(\alpha)$, $K_s = K(\alpha)a(\alpha)$, $\alpha = w$, and transient responses corresponding to the extended exact linearization (full line); For t=50 input disturbance $v = 2$ occurred and for t=55 the disturbance $v = -2$; $\gamma = -2$, $T_f = 0.5$; $T_d = 0.01$; $y_0 = 0.01$; $u \in \langle 0, 10 \rangle$

The stability problems occurring for small output values arise in region, where the plant linearization has faster dynamics as that one prescribed by the closed loop poles. Here, the controller proportional gain has negative values, which are close to stability boundary. So, already small deviations of the supposed dynamics caused by process nonlinearity lead to instability. The "braking" function of the controller is to see during the 1st stage of the control signal in Fig.12. Stability can be retrieved by limiting the proportional gain to positive values, by limiting the reference signal dynamics (ramp reference changes), and or by choosing faster poles. Fig.13 illustrates the last possibility. Here, the range of negative values of the controller gain coincides with the range of stability problems in controlling the nonlinear process. It is also to see that choosing faster poles can enlarge the range of positive values of the gain.

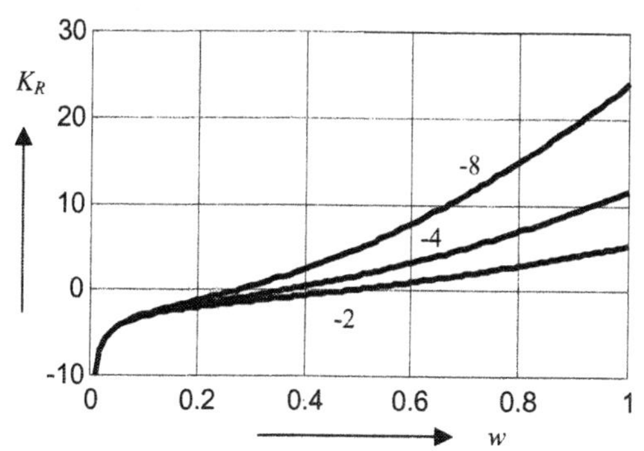

Fig.13. Proportional gain K_R of the windupless PI$_1$ controller as the function of the reference variable w for different poles γ.

Under constrained control, it is not possible to achieved regular (linear) input-output behavior in the full range of operation. But, the responses remain close to the fastest possible monotonous ones.

7. EXTENDED EXACT LINEARIZATION

Due to its properties, the exact linearization via feedback can be considered as a performance standard for evaluation of other linearization based controllers.

Although some authors speak about special kind of linearization, which, in contrast to the linearization above constant operating point, guarantees global linearization properties, some others interpret it as a special type of gain scheduling control (see e.g. Åström and Wittenmark (1995). Huba and Pauer (1991) showed that it can also be considered as the Taylor linearization above nonequilibrium point – above the actual state $\mathbf{x}_0 = \mathbf{x}$, when all 1st order terms vanish and the zero-term of the Taylor expansion can be compensated by the control signal.

In our example with the closed loop pole $\gamma < 0$, the controller equation is given as

$$u_p = S(y)K_R(w - y) + c_v\sqrt{y}\ ;\ K_R = -\gamma \qquad (19)$$

Under **Extended Exact Linearization (EEL)** controller it is to understand the exact linearization extended by the inverse dynamics based disturbance reconstruction and compensation (I-action) according to

$$u_I = -\frac{1}{1+T_f s}\left[S(y)\dot{y} + c_v\sqrt{y} - u_r\right] \qquad (20)$$

The total controller output is computed according to

$$u_r = sat\{u_p + u_I\} = \begin{cases} U_{\max};\ u_p + u_I > u_{\max} \\ u_p + u_I \in \langle U_{\min}, U_{\max}\rangle \\ U_{\min};\ u_p + u_I < U_{\min} \end{cases} \qquad (21)$$

In the non-saturated region, the EEL-control enables "ideal" nonlinearity compensation (Fig.12) both in the reference following, as well as in disturbance rejection. Under control saturation, the controller is no more able to guarantee linear input-output properties.

The responses are close to those achieved with the gain scheduled windupless PI$_1$-controller in Fig.12. It was shown by (Huba, 2003a) that the windupless PI$_1$-controller is for $\alpha = y$ even identical with the EEL-control!

The disturbance rejection initiated by the disturbance steps $\Delta v = \pm 2$ produced at the time instants t=50 and t=55 is dominated by the time constant T_f.

While in computing the output derivative

$$\dot{y} \approx \frac{s}{1+T_d s} y \qquad (22)$$

it is recommended to work with as small T_d as possible, an appropriate noise attenuation can be achieved by choosing the time constant T_f, which determines the dynamics of the disturbance reconstruction and compensation. In this loop, a step of the noise signal $\Delta\delta$ will produce a kick in the control signal with contributions of the exact linearization channel of the controller

$$\Delta u_p = \left[S(y)\gamma + \frac{c_v}{2\sqrt{y}} \right] \Delta\delta \qquad (23)$$

For $T_d << T_f$, the contribution of the disturbance reconstruction & compensation channel can be roughly estimated as

$$\Delta u_I = -\left[\frac{S(y)}{T_f} + \frac{c_v}{2T_f\sqrt{y}} \right] \Delta\delta \qquad (24)$$

The estimates of the noise influence coincides well with the simulation results (Fig.14).

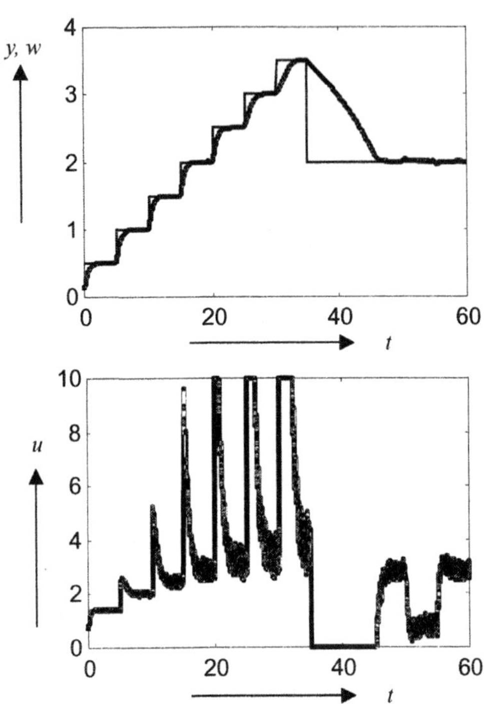

Fig.14. Constrained EEL-control with measurement noise $|\delta| < 0.01$; $u \in \langle 0,10 \rangle$; $\gamma = -2; T_f = 0.5$; $T_d = 0.01$; For t=50 input disturbance $v = 2$ occurred and for t=55 the disturbance $v = -2$

8. CONCLUSIONS

When evaluating results of the carried out analysis, one can agree with comments made by Leith and Leithead (1998) that the local linear equivalence conditions have just limited use in designing well behaving gain-scheduled controllers.

In our particular example, much more depends on the type of the controller used. Since the parallel parametrized PI-controller introduces a state-dependent zero into the loop linearization, which then dominates also in loops with different nonlinear realizations of the controller, it does not enable a generally valid optimization. On the other hand, the new windupless solutions not only explain some already known structures of the anti-windup circuitry. In this case the solution based on gain scheduling is very close, or even identical to the exact linearization extended by the disturbance reconstruction and compensation via the inverse dynamics. Both solutions guarantee behavior, which can be asymptotically close to the optimal one. Practical aspects of the application of these new structures are treated in Huba (2003b).

ACKNOWLEDGEMENTS

This work has been partially supported by the Slovak Scientific Grant Agency, Grant No. KG-57-HUBA-Sk1

REFERENCES

Aström, K. J., T.Hägglund (1995). *PID controllers: Theory, design, and tuning* – 2nd ed., ISA, Research Triangle Park, NC

Bialkowski, W.L. (1993) Dreams versus reality: A view from both sides of the gap. *Pulp and paper*, Canada, 94:11.

Huba,M. and A.Pauer (1991). Linearisierungsverfahren zur nichtlinearen Reglersynthese. *at* 39, 35-36.

Huba, M., Kuľha, P.,Bisták, P., Skachová, Z. (1998). Two Dynamical Classes of PI Controllers for the 1st Order Loops. *In: New Trends in Design of Control Systems Smolenice 1997*, Pergamon Press Oxford, 189-194.

Huba, M., Bisták, P. (1999). Dynamic Classes in the PID Control. In: *Proc. 1999 American Control*

Huba, M. (2003a). *Nelineárne systémy (Nonlinear Control Systems.)* STU Bratislava. *Conference.* San Diego: AACC.

Huba, M. (2003b). *Syntéza systémov s obmedzeniami (Synthesis of Constrained Systems).* STU Bratislava.

Kaminer, I., Pascoal, A.M., Khargonekar, P.P. and E.E. Coleman (1995). A Velocity Algorithm for the Implementation of Gain-scheduled Controllers. *Automatica* 31, 8, 1185-1191.

Khalil, H.K. (1996). *Nonlinear Systems*, 2nd Ed. Prentice Hall Int. London, pp. 494-498

Kothare,M.V., Campo,P.J., Morari,M. and C.V.Nett (1994). A Unified Framework for the Study of Anti-windup Designs. *Automatica*, Vol 30, 1869-1883.

Lawrence, D.A. and W.J. Rugh (1995). Gain Scheduling Dynamic Linear Controllers for a Nonlinear Plant. *Automatica* 31, No. 3, 381-390.

Leith, D.J. and W.E.Leithead (1998). Comments on „Gain Scheduling Dynamic Linear Controllers for a Nonlinear Plant". *Automatica* 34, No.8, 1041-1043.

O'Dwyer, A.(2000). PI and PID controller tuning rules for time delay processes: a summary. *Technical Report AOD-00-01*, Edition 1. School of Control Systems and Electrical Engineering, Dublin Institute of Technology, Ireland

www.elsevier.com/locate/ifac

ANALYTICAL DIGITAL AND ANALOG CONTROLLER TUNING METHOD FOR PROPORTIONAL NON-OSCILLATORY PLANTS WITH TIME DELAY

Miluše Vítečková[*], Antonín Víteček[**]

Department of Control Systems and Instrumentation, Faculty of Mechanical Engineering Technical University of Ostrava, 17. listopadu, 708 33 Ostrava-Poruba, Czech Republic
[*]Fax: 00420/69/6916490, E-mail: miluse.viteckova@vsb.cz
[**]Fax: 00420/69/6916490, E-mail: antonin.vitecek@vsb.cz

Abstract: The paper deals with the analytical digital and analog controller tuning method for the non-oscillatory proportional plants with time delays. The described method is based on the real dominant multiple pole and the D-transform. The resultant control process is closed to the marginal non-oscillatory course. The application is shown on the examples. *Copyright © 2003 IFAC*

Key words: time delay, controller tuning, PID controller, dominant pole

1. INTRODUCTION

The article deals with a unified approach to the conventional digital and analog controller tuning by the analytical method for the non-oscillatory proportional plants with time delays, whose L-transfer function has the form

$$G_P(s) = \frac{k_1}{T_1 s + 1} e^{-T_{d1} s} \tag{1}$$

where k_1 is the plant gain, T_1 - the plant time constant, T_{d1} - the plant time delay, s - the complex variable in the L-transform.

2. D-TRANSFORM

It is appropriate to use the D-transform which allowed to get computational formulas for adjustable parameters by the unified analytic way for digital controllers as well as for analog ones. This approach is based on the delta models and behaviour of a relative forward difference

$$\delta x(kT) = \frac{x[(k+1)T] - x(kT)}{T} \tag{2}$$

for what holds

$$\lim_{T \to 0} \delta x(kT) = \dot{x}(kT) \tag{3}$$

where δ is the relative forward difference operator (delta operator), T – the sampling period, k – the relative discrete time ($k = 0, 1, 2, \ldots$).

The D-transform is defined by the relations (Middleton and Goodwin, 1990; Feuer and Goodwin, 1996; Goodwin, Graebe and Salgado, 2001; Vítečková, et al., 2002)

$$X(\gamma) = D\{x(kT)\} = T \sum_{k=0}^{\infty} x(kT)(1 + \gamma T)^{-k} \tag{4a}$$

$$x(kT) = D^{-1}\{X(\gamma)\} = \frac{1}{2\pi j} \oint_C X(\gamma)(1 - \gamma T)^{k-1} d\gamma \tag{4b}$$

where x is the original, X – the transform, D and D^{-1} - the direct and inverse D-transform operators, γ – the complex variable in the D-transform.

The closed integration path C lays within convergence domain of the transform $X(\gamma)$ and contains its all singular points. The conditions for the original are the same like for the Z-transform. More detailed information can be found for example in publications (Graebe and Salgado, 2001; Vítečková, et al., 2002).

The simple transfer relations among complex variables, transforms and transfer functions hold:

complex variables

$$s = \lim_{T \to 0} \gamma \tag{5a}$$

$$z = \gamma T + 1 \tag{5b}$$

transforms of variables

$$X(s) = \lim_{T \to 0} X(\gamma) \tag{6a}$$

$$X(z) = \frac{1}{T} X(\gamma) \Big|_{\gamma = \frac{z-1}{T}} \tag{6b}$$

transfer functions

$$G(s) = \lim_{T \to 0} G(\gamma) \tag{7a}$$

$$G(z) = G(\gamma) \Big|_{\gamma = \frac{z-1}{T}} \tag{7b}$$

where z is the complex variable in the Z-transform.

In the case when the D/A converter in a control system with a digital controller (which corresponds to a zero-order holder) is used than the transfer function of the controlled plant is given by the formula (Midleton and Goodwin, 1990; Feuer and Goodwin, 1996; Goodwin, Graebe and Salgado, 2001; Vítečková, 2001a, b; Vítečková, 2002)

$$G_P(\gamma) = \frac{\gamma}{\gamma T + 1} D\left\{ L^{-1}\left\{ \frac{G_P(s)}{s} \right\} \Big|_{t=kT} \right\} \tag{8}$$

The extensive tables of the D-transforms are worked up, see for example (Vítečková, 2002). They substantially do easy their use.

For the L-transfer function of the plant (1) the corresponding D-transfer function

$$G_P(\gamma) = \frac{ak_1}{\gamma T + a} (\gamma T + 1)^{-d}, \quad d = \frac{T_{d1}}{T}, \quad a = 1 - e^{-\frac{T}{T_1}} \tag{9}$$

is obtained, where d is the relative time delay (for simplicity only integer number is considered).

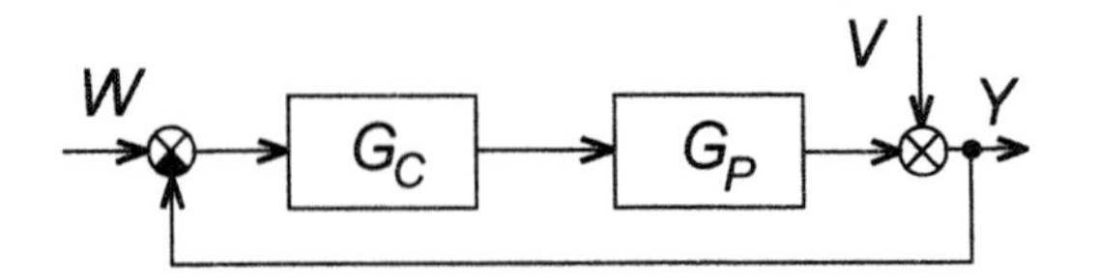

Fig. 1. Control system

It is supposed that the conventional controller with the transfer function G_C (Tab. 1) is used, which ensures the non-oscillatory control process with the zero steady error for the step change of the desired variable and the disturbance variable which are caused at the output of the plant (see Fig. 1), where W, V and Y are the corresponding transforms of the desired variable w, the disturbance variable v and the controlled variable y.
Forasmuch as the plant (1) is proportional, the conventional controllers with the integral action are only considered in accordance with Tab. 1, where k_P is the controller gain, T_I - the integral time, T_D - the derivative time.

3. DOMINANT MULTIPLE POLE METHOD

The dominant multiple pole method is based on the assumption that the dominant pole of the control system is multiple and real, which ensures the stable non-oscillatory control process closed to the marginal process. Simultaneously, it is supposed that the influence of the non-dominant zeros and poles can be neglected. The multiplicity is given by the number of the controller adjustable parameters which is increased to 1 (Vítečková, 2001a, b).

The characteristic polynomial of the control system $N(\gamma)$ can be determined on the basis of the open-loop transfer function (see Fig. 1)

$$G_o(\gamma) = G_C(\gamma) G_P(\gamma) = \frac{M_o(\gamma)}{N_o(\gamma)}$$

$$N(\gamma) = N_o(\gamma) + M_o(\gamma)$$

The multiple dominant pole is given by the solution of the equation system

$$\frac{\mathrm{d}^i N(\gamma)}{\mathrm{d}^i \gamma} = 0, \quad i = 0, 1, \ldots, m \tag{10}$$

where m is the number of the controller adjustable parameters.

The procedure of the determination of the controller parameters will be shown on the example of the I controller.

Table 1 Controller transfer functions

m	Type	$G_C(\gamma)$	$G_C(s)$	$G_C(z)$
1	I	$\frac{\gamma T+1}{T_I\gamma}$	$\frac{1}{T_I s}$	$\frac{T}{T_I}\frac{z}{z-1}$
2	PI	$k_P\left(1+\frac{\gamma T+1}{T_I\gamma}\right)$	$k_P\left(1+\frac{1}{T_I s}\right)$	$k_P\left(1+\frac{T}{T_I}\frac{z}{z-1}\right)$
3	PID	$k_P\left(1+\frac{\gamma T+1}{T_I\gamma}+\frac{T_D\gamma}{\gamma T+1}\right)$	$k_P\left(1+\frac{1}{T_I s}+T_D s\right)$	$k_P\left(1+\frac{T}{T_I}\frac{z}{z-1}+\frac{T_D}{T}\frac{z-1}{z}\right)$

The open-loop transfer function and the characteristic polynomial have forms

$$G_o(\gamma)=G_C(\gamma)G_P(\gamma)=\frac{ak_1}{T_I\gamma(\gamma T+a)}(\gamma T+1)^{-(d-1)}$$

$$N(\gamma)=T_I\gamma(\gamma T+a)(\gamma T+1)^{d-1}+ak_1$$

The I controller has one adjustable parameter T_I, therefore m = 1 and in accordance with (10) the two equations were obtained

$$N(\gamma)=0\Rightarrow T_I\gamma(\gamma T+a)(\gamma T+1)^{d-1}+ak_1=0$$

$$\frac{\mathrm{d}N(\gamma)}{\mathrm{d}\gamma}=0\Rightarrow(d+1)\gamma^2T^2+(ad+2)\gamma T+a=0$$

On the basis of the second equation the dominant double pole (11) in the γ-domain was obtained. Before radical the sign plus for dominance of the double pole reason must be considered. It is easy to verify that the dominant double pole (11) is stable.

From the first equation the relation for the integral time (12) was obtained. After substitution of the dominant pole value (11) in the relation (12) the integral time value for the digital I controller can be obtained.

By the limit transition for $T\to 0$ the corresponding formulas (13) (the dominant double pole in the s-domain) and (14) for the analog I controller are got.

In the same way the formulas (15) – (20) and (22) – (29) were obtained. The resultant values are marked by a star *.

I Controller (digital)

$$\gamma_2^*=-\frac{ad+2}{2(d+1)T}+\sqrt{\frac{a^2d^2-4a+4}{4(d+1)^2T^2}}<0 \qquad (11)$$

$$T_I^*=T_I^*(\gamma_2^*)=-\frac{ak_1}{\gamma_2^*(\gamma_2^*T+a)(\gamma_2^*T+1)^{d-1}} \qquad (12)$$

I Controller (analog)

$$s_2^*=\lim_{T\to 0}\gamma_2^*=-\frac{1}{T_{d1}}-\frac{1}{2T_1}+\sqrt{\frac{1}{T_{d1}^2}+\frac{1}{4T_1^2}}<0 \qquad (13)$$

$$T_I^*=T_I^*(s_2^*)=\lim_{T\to 0}T_I^*(\gamma_2^*)=-\frac{k_1}{s_2^*(T_1s_2^*+1)e^{T_{d1}s_2^*}} \qquad (14)$$

PI Controller (digital)

$$\gamma_3^*=-\frac{ad+4}{2(d+2)T}+\sqrt{\frac{(a^2d^2+a^2d-8a+8)d}{4(d+1)(d+2)^2T^2}}<0 \qquad (15)$$

$$k_P^*=k_P^*(\gamma_3^*)=$$
$$=-\frac{1}{ak_1}\left[T^2(d+1)\gamma_3^{*2}+T(ad+2)\gamma_3^*+a\right](\gamma_3^*T+1)^d \qquad (16)$$

$$T_I^*=T_I^*(\gamma_3^*)=$$
$$=-\frac{\left[T^2(d+1)\gamma_3^{*2}+T(ad+2)\gamma_3^*+a\right](\gamma_3^*T+1)}{\gamma_3^{*2}T\left[T(d+1)\gamma_3^*+ad+1\right]} \qquad (17)$$

PI Controller (analog)

$$s_3^*=\lim_{T\to 0}\gamma_3^*=-\frac{2}{T_{d1}}-\frac{1}{2T_1}+\sqrt{\frac{2}{T_{d1}^2}+\frac{1}{4T_1^2}}<0 \qquad (18)$$

$$k_P^*=k_P^*(s_3^*)=\lim_{T\to 0}k_P^*(\gamma_3^*)=$$
$$=-\frac{1}{k_1}\left[T_{d1}T_1s_3^{*2}+(2T_1+T_{d1})s_3^*+1\right]e^{T_{d1}s_3^*} \qquad (19)$$

$$T_I^*=T_I^*(s_3^*)=\lim_{T\to 0}T_I^*(\gamma_3^*)=$$
$$=-\frac{T_{d1}T_1s_3^{*2}+(2T_1+T_{d1})s_3^*+1}{s_3^{*2}(T_{d1}T_1s_3^*+T_1+T_{d1})} \qquad (20)$$

The stable zero in the numerator of the control system transfer function has very positive influence on the rise time. This stable zero causes for $T_{d1}<T_1$ overshoot, which is practically zero for $T_{d1}>0.7T_1$. For PI digital controller the overshoot depends on the sampling period T too. For bigger value of T the overshoot arises at the less values of the ratio T_{d1}/T_1.

For $0.05T_1 < T_{d1} < 0.8T_1$ the formula

$$T_I^* = T_1 - 0.5T \tag{21}$$

for the integral time can be used and the control gain k_P^* keep on the calculated value. In this case the overshoot will be under 2%. In every case in rise the unmeant overshoot the control process can be fine-tuned by increasing the value of the integral time T_I (not by decreasing the controller gain k_P).

PID Controller (digital)

$$\gamma_4^* = -\frac{ad+6}{2(d+3)T} + \sqrt{\frac{a^2d^3 + 2a^2d^2 + 12(1-a)d}{4(d+2)(d+3)^2T^2}} < 0 \tag{22}$$

$$\begin{aligned} k_P^* = k_P^*(\gamma_4^*) = \frac{1}{ak_1}\{ & T^4\gamma_4^{*4}(d+1)(d+2) + T^3\gamma_4^{*3}[(a+1)d^2 + (a+6)d + 5] + T^2\gamma_4^{*2}[ad^2 + (2a+3)d + 3] + \\ & + aT\gamma_4^*(d-1) - a\}(\gamma_4^*T + 1)^{d-1} \end{aligned} \tag{23}$$

$$\begin{aligned} T_I^* = T_I^*(\gamma_4^*) = -2\{ & T^4\gamma_4^{*4}(d+1)(d+2) + T^3\gamma_4^{*3}[(a+1)d^2 + (a+6)d + 5] + T^2\gamma_4^{*2}[ad^2 + (2a+3)d + 3] + \\ & + aT\gamma_4^*(d-1) - a\} / \{T^2\gamma_4^{*3}(d+1)[T(d+2)\gamma_4^* + ad + 2]\} \end{aligned} \tag{24}$$

$$\begin{aligned} T_D^* = T_D^*(\gamma_4^*) = -\frac{T}{2}\{ & T^4\gamma_4^{*4}(d+1)(d+2) + T^3\gamma_4^{*3}[(a+2)d^2 + (a+10)d + 8] + T^2\gamma_4^{*2}[(2a+1)d^2 + (4a+11)d + 12] + \\ & + T\gamma_4^*[ad^2 + (5a+4)d + 8] + 2(ad+1)\} / \{T^4\gamma_4^{*4}(d+1)(d+2) + T^3\gamma_4^{*3}[(a+1)d^2 + (a+6)d + 5] + \\ & + T^2\gamma_4^{*2}[ad^2 + (2a+3)d + 3] + aT\gamma_4^*(d-1) - a\} \end{aligned} \tag{25}$$

PID Controller (analog)

$$s_4^* = \lim_{T \to 0} \gamma_4^* = -\frac{3}{T_{d1}} - \frac{1}{2T_1} + \sqrt{\frac{3}{T_{d1}^2} + \frac{1}{4T_1^2}} < 0 \tag{26}$$

$$k_P^* = k_P^*(s_4^*) = \lim_{T \to 0} k_P^*(\gamma_4^*) = \frac{1}{k_1}\left[T_{d1}^2T_1s_4^{*3} + (3T_{d1}T_1 + T_{d1}^2)s_3^{*2} + T_{d1}s_4^* - 1\right]e^{T_{d1}s_4^*} \tag{27}$$

$$T_I^* = T_I^*(s_4^*) = \lim_{T \to 0} T_I^*(\gamma_4^*) = -2\frac{T_{d1}^2T_1s_4^{*3} + (3T_{d1}T_1 + T_{d1}^2)s_4^{*2} + T_{d1}s_4^* - 1}{s_4^{*3}(T_{d1}^2T_1s_4^* + 2T_{d1}T_1 + T_{d1}^2)} \tag{28}$$

$$T_D^* = T_D^*(s_4^*) = \lim_{T \to 0} T_D^*(\gamma_4^*) = -\frac{1}{2}\frac{T_{d1}^2T_1s_4^{*2} + (4T_{d1}T_1 + T_{d1}^2)s_4^* + 2T_{d1} + 2T_1}{T_{d1}^2T_1s_4^{*3} + (3T_{d1}T_1 + T_{d1}^2)s_4^{*2} + T_{d1}s_4^* - 1} \tag{29}$$

For the PID digital and analog controllers the overshoot do not rising for $T_{d1} \geq 2T_1$. The overshoot is practically very small for $T_{d1} > 1.4T_1$. For bigger value of T the overshoot arises at the less values of the ratio T_{d1}/T_1. For $0.05T_1 < T_{d1} < 1.6T_1$ the formula

$$T_I^* = 1.2T_1 - T \tag{30}$$

for the integral time can be used and the control gain k_P^* keep on the calculated value. In this case the overshoot will be under 2%. In every case in rise the unmeant overshoot the control process can be fine-tuned by increasing the value of the integral time T_I (not by decreasing the controller gain k_P).

4. EXAMPLES

Example 1

For the control system with the controlled plant with the transfer function

$$G_P(s) = \frac{1}{4s+1}e^{-10s}$$

it is necessary to determine the values of the I, PI and PID digital and analog controller adjustable parameters by the dominant multiple pole tuning method.

Solution:

The plant parameters are: $k_1 = 1$, $T_1 = 4\,\text{s}$, $T_{d1} = 10\,\text{s}$.

For the digital controller the sampling period T = 1s is chosen. Then the relative discrete time delay is $d = \frac{T_{d1}}{T} = 10$ and constant $a = 1 - e^{-\frac{T}{T_1}} \doteq 0.2212$.

I Controller (digital) [relations (11) and (12)]

$\gamma_2^* \doteq -0.0628;\quad T_I^* \doteq 39.864$

I Controller (analog) [relations (13) and (14)]

$s_2^* \doteq -0.0649;\quad T_I^* \doteq 39.824$

PI Controller (digital) [relations (15) ÷ (17)]

$\gamma_3^* \doteq -0.1235;\quad k_P^* \doteq 0.159;\quad T_I^* \doteq 4.068$

PI Controller (analog) [relations (18) ÷ (20)]

$s_3^* \doteq -0.1363;\quad k_P^* \doteq 0.182;\quad T_I^* \doteq 4.473$

PID Controller (digital) [relations (22) ÷ (25)]

$\gamma_4^* \doteq -0.1789;\quad k_P^* \doteq 0.291;\quad T_I^* \doteq 5.835;$

$T_D^* \doteq 1.400$

PID Controller (analog) [relations (26) ÷ (29)]

$s_4^* \doteq -0.2114;\quad k_P^* \doteq 0.355;\quad T_I^* \doteq 6.519;$

$T_D^* \doteq 1.546$

The courses of the step responses are shown in Fig. 2.

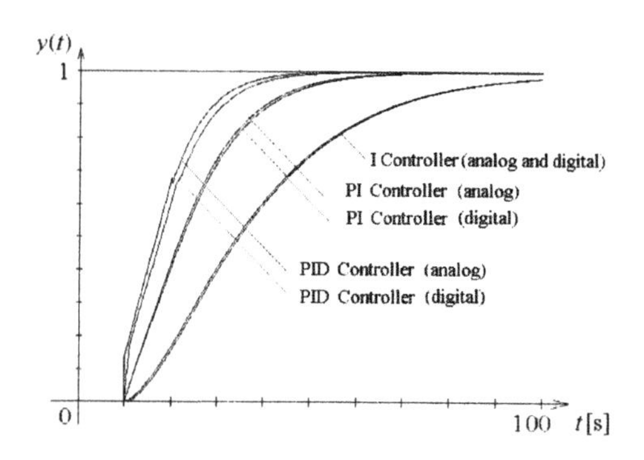

Fig. 2 The step responses of the control system from example 1

Example 2

The analog PI controller must be tuned by the multiple dominant pole method for the plant with the transfer function

$$G_P(s) = \frac{2}{6s+1}e^{-2s}$$

so that the control process will be without overshoot.

Solution:

The plant parameters are: $k_1 = 2$, $T_1 = 6\text{s}$, $T_{d1} = 2\text{s}$.

On the basis of the relations (18) – (20) the values

$s_3^* \doteq -0.3713;\quad k_P^* \doteq 0.605;\quad T_I^* \doteq 5.206$

were obtained.

The computed step response (a) is shown in Fig. 3. If the formula (21) for $T = 0$ is used than $T_I^* = T_1 = 6\,\text{s}$ and overshoot is very small (b). The step response without overshoot got by decreasing the controller gain k_P (c) is shown in Fig. 3 too.

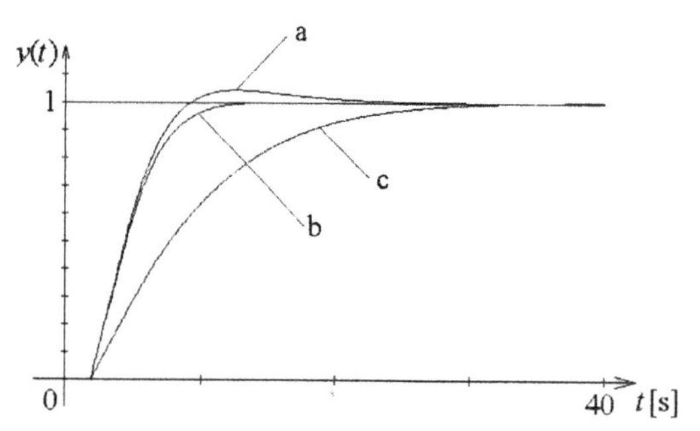

Fig. 3 The step responses of the control system from example 2

5. CONCLUSIONS

The analytical dominant multiple pole method for conventional digital and analog controller tuning for the non-oscillatory proportional plants with time delays gives surprisingly good results. The control process can be regarded as marginal. The use of the D-transform allowed to get the computational formulas by the analytic way for digital controllers as well as for analog ones by a unified approach. In the case when a faster response is required, it is possible to correct the control process by the decreasing of the controller integral time.

The paper was supported by the grant project MSM 272300012.

REFERENCES

Farana, R. (1996). *Universal Simulation Program SIPRO 3.4* (in Czech). Technical University of Ostrava, Ostrava.

Filasová, A. (1999). Robust Decentralized Regulator for System with Unknown Uncertainty Boundaries. In *Process Control '99: Proceedings of the 12th Conference, Vol. 2.*, Tatranské Matliare, STU Bratislava, pp. 86-90.

Feuer, A. and G. C. Goodwin (1996). *Sampling in Digital Signal Processing and Control.* Birkhäuser, Boston.

Goodwin, G. C., S. F. Graebe and M. E. Salgado (2001). *Control System Design.* Prentice-Hall, New Jersey.

Hudzovič, P. and A. Kozáková (2001). A Contribution to the Synthesis of PI Controllers. In *Abstract Proceedings of Conference "Cybernetics and Informatics"* .April 5–6. STU Bratislava, Piešťany, pp. 31-34.

Krokavec, D. and A. Filasová (2002) *Optimal Stochastic Systems.* 2nd Edition. Elfa, Bratislava.

Middleton, R. H. and G. C. Goodwin (1990). *Digital Control and Estimation. A Unified Approach.* Prentice-Hall, Englewood Cliffs.

Vítečková, M. (2001a). Use of delta models for Analog and Digital Controllers Tuning (in Czech). In *Proceedings of XXVI. Seminar ASR'2001 "Instruments and Control".* Ostrava, April 26-27. Paper 75, pp. 1-10.

Vítečková, M. (2001b). Use of D-transform in Dominant Multiple Poles Method for Controller Tuning (in Czech). In *Proceedings of the Conference on Information Engineering and Process Control.* Praha : ČVUT Praha, September 4, pp. 65-66.

Vítečková, M. et al. (2002). *Control System Synthesis Methods Based on Delta Models* (in Czech). Technical Report of Grant 102/00/0186. VŠB-Technical University of Ostrava, Ostrava.

MINIMUM TIME PD-CONTROLLER DESIGN

Mikuláš Huba and Pavol Bisták

Slovak University of Technology in Bratislava, Faculty of Electrical Eng. and Information Technology, Ilkovičova 3, 812 19 Bratislava, Slovakia, huba, bistak@elf.stuba.sk

Abstract: This work deals with the minimum time sampled-data control algorithms for linear 2nd order systems. Based on the general theory (Desoer and Wing, 1961) and its modification for the non-symmetrical amplitude constraints and the fully digital implementation (Huba, 1992), it is focused to the control processes having more than one sign variation of the optimal control. A till now unknown non-uniqueness of the optimal control arising at specific sampling periods in systems with complex poles is analyzed and possibilities for further simplification of the minimum time (pole assignment) control are shown. *Copyright © 2003 IFAC*

Keywords: Control system design; minimum time control; optimal control; saturation.

1. INTRODUCTION

In the sampled data control design, one has to conclude that the present standard textbooks lack on principles, which would reflect the importance of the control signal constraints in combination with the discrete time essence of control. These can neither be replaced by the principles of rely (continuous-time) minimum time control, or by those of the linear control extended by some anti-windup measures, which are still not able to deal effectively with the situation, when the control signal hits both the upper and the lower saturation limit (Rönnbäck, 1996). The impact of this situation is evident e.g. by inflation of different "optimal" PID tunings appearing at practically each control conference, or in conclusions that the PD controller cannot be optimally tuned (Aström and Hägglund, 1995). The early works on the minimum time sampled data systems (Desoer and Wing, 1961), mentioned yet in some textbooks in 70s (Kuo, 1970) have been practically forgotten.

What is suggested to improve situation in this field? It is well known that the already mentioned minimum time control is mostly not directly applicable in practice. Therefore, many alternative solutions have been proposed and the constrained systems became again to be a popular topic of research. One of the modifications denoted as the minimum time pole assignment control is based on putting additional constraints on the rate of the state and of the control signal changes of the minimum time controller (Huba et al., 1999, Huba and Bisták, 1999). In this way it is possible to decrease the system sensitivity to the measurement noise, unmodelled dynamics, parameter fluctuations, etc. However, in controlling oscillatory systems, the new concept is not applicable for arbitrary initial states (for transient responses with more than one sign change of the control signal). Furthermore, the minimum time control algorithms are simpler and so more appropriate for the 1st phase of education. Their treatment can effectively show, why the optimal control cannot be sufficiently approximated by the linear PD control (Huba, 1992).

The aim of this paper is to show that in the area, where the minimum time pole assignment control is not directly applicable, the pure minimum time control is not unique and relatively not sensitive to the actual choice of the control signal. This approves use of simplified control algorithms in this area. Simultaneously, simplified interpretation of optimal control valid rigorously just for a limited range of initial states is formulated, which is easier to implement and also to understand.

In chapters 2-8, basic properties of minimum time control systems are summarized based on previous works by Desoer and Wing (1961) and Huba (1992). Chapters 8 and 9 show properties of minimum time control occurring for special sampling periods, which explain possibilities of the controller simplification.

2. STATEMENT OF THE PROBLEM

The paper considers sampled data control of a continuous linear system ($\mathbf{A}_c$, $\mathbf{b}_c$) described as

$$\mathbf{x}_{k+1} = \mathbf{A}\mathbf{x}_k + \mathbf{b}u_{k+1}$$

$$\mathbf{A} = \mathbf{A}(T) = e^{-\mathbf{A}_c T} \; ; \; \mathbf{b} = \mathbf{b}(T) = \int_0^T e^{\mathbf{A}_c \tau}\mathbf{b}_c d\tau \qquad (1)$$

whereby $\mathbf{x}_k$ is the phase vector at time $t=kT$ and

$$u_{k+1} \in \langle U_1, U_2 \rangle ; \; U_1 < 0 \; ; \; U_2 > 0 \qquad (2)$$

is the control over the interval $kT \le t < (k+1)T$. The minimum time regulator problem may be stated as follows: given any admissible initial state $\mathbf{x}_0$, it is required to bring the system to the demanded state $\mathbf{w}=\mathbf{0}$ with a control constrained by (2).

Applying the control sequence $\{u_i\}_1^N ; \; u_i \in \langle U_1, U_2 \rangle$ for a sampling period T one gets series of states

$$\begin{aligned} \mathbf{x}_1 &= \mathbf{A}\mathbf{x}_0 + \mathbf{b}u_0 \\ \mathbf{x}_2 &= \mathbf{A}\mathbf{x}_1 + \mathbf{b}u_1 = \mathbf{A}^2\mathbf{x}_0 + \mathbf{A}\mathbf{b}u_0 + \mathbf{b}u_1 \\ &\vdots \\ \mathbf{x}_N &= \mathbf{A}^N\mathbf{x}_0 + \mathbf{A}^{N-1}\mathbf{b}u_0 + \ldots + \mathbf{A}\mathbf{b}u_{N-2} + \mathbf{b}u_{N-1} \end{aligned} \qquad (3)$$

If after that $\mathbf{x}_N=\mathbf{0}$, Eqs. (3) can be rearranged by introducing vectors

$$\mathbf{v}_k = -\mathbf{A}^{-k}\mathbf{b} = -\mathbf{A}(-kT)\mathbf{b}(T) \; ; \; k = 1,2,.... \qquad (4)$$

into the canonical representation of the initial states

$$\mathbf{x}_0 = \mathbf{v}_1 u_1 + \mathbf{v}_2 u_2 + \cdots + \mathbf{v}_N u_N \qquad (5)$$

3. THE SETS $\mathbf{R}_N$ AND $\mathbf{I}_n$; BOUNDARY STATES

$\mathbf{R}_N$ is the set of all initial states that can be brought to the origin by the control (2) in N sampling period or less:

$$R_N = \left\{ \mathbf{x}_0 \mid \mathbf{x}_0 = \sum_{i=1}^{N} \mathbf{v}_i u_i \; ; \; u_i \in \langle U_1 ; U_2 \rangle ; i = 1,2,\ldots,N \right\} \qquad (6)$$

Desoer and Wing have shown that $\mathbf{R}_N$ are convex: $\mathbf{R}_1$ is the line segment with vertices

$$V_1^1 = \mathbf{v}_1 U_1 \; ; \; V_1^2 = \mathbf{v}_1 U_2 \qquad (7)$$

$\mathbf{R}_2$ is the parallelogram whose edges are parallel to $\mathbf{v}_1$ and $\mathbf{v}_2$. Its vertices (see also Huba, 1992) are:

$$\begin{aligned} V_2^{11} &= \mathbf{v}_1 U_1 + \mathbf{v}_2 U_1 \; ; \; V_2^{12} = \mathbf{v}_1 U_1 + \mathbf{v}_2 U_2 \\ V_2^{21} &= \mathbf{v}_1 U_2 + \mathbf{v}_2 U_1 \; ; \; V_2^{22} = \mathbf{v}_1 U_2 + \mathbf{v}_2 U_2 \end{aligned} \qquad (8)$$

For $N>2$, $\mathbf{R}_N$ is obtained inductively. Suppose $\mathbf{R}_N$ is known, let us construct $\mathbf{R}_{N+1}$. $\mathbf{R}_N$ being a convex polygon is completely described by its vertices. Let us classify these vertices in two classes: let δ be an arbitrarily small positive number. The class P_N^2 contains all the vertices V_N of $\mathbf{R}_N$, such that $V_N + \delta\mathbf{v}_{N+1} \notin \mathbf{R}_N$. P_N^1 contains all vertices V_N of $\mathbf{R}_N$, such that $V_N - \delta\mathbf{v}_{N+1} \notin \mathbf{R}_N$.

These two classes usually have common vertices characterized by the maximum distance from a line going through the origin parallel to $\mathbf{v}_{N+1}$. The vertices of $\mathbf{R}_{N+1}$ are obtained from those of $\mathbf{R}_N$ as follows:

- to each vertex V_N of P_N^1 corresponds the vertex of $\mathbf{R}_{N+1}$: $V_N + \mathbf{v}_{N+1}U_1$.
- to each vertex V_N of P_N^2 corresponds the vertex of $\mathbf{R}_{N+1}$: $V_N + \mathbf{v}_{N+1}U_2$.

The boundary of $\mathbf{R}_{N+1}$ is obtained from that of $\mathbf{R}_N$ by adding vectors $\mathbf{v}_{k+1}U_j$ in an outward direction. In all linear 2nd order systems the origin is surrounded by sets $\mathbf{R}_N$ with $2N$ vertices of the form

$$V_{k,N-k}^j = U_j \sum_{i=1}^{k} \mathbf{v}_i + U_{3-j} \sum_{i=k+1}^{N} \mathbf{v}_i \qquad (9)$$

corresponding to sequences where: 1) each u_i is maximum in absolute value (i.e. equal to U_1 or U_2, respectively); 2) the sequence consists of 2 subsequences of u_i's that have the same sign and each subsequence has a sign opposite to that of the another one, k being the number of steps of the 1st subsequence with $u_i=U_j$, N-k the number of steps with the opposite limit value U_{3-j}.

In systems with complex poles there are also possible sequences consisting of more than 2 subsequences (with more than 1 sign variation). A vertex with 2 sign variations can then be expressed as

$$V_{k,l,N-k-l}^j = U_j \sum_{i=1}^{k} \mathbf{v}_i + U_{3-j} \sum_{i=k+1}^{k+l} \mathbf{v}_i + U_j \sum_{i=k+l+1}^{N} \mathbf{v}_i \qquad (10)$$

with k and l denoting the number of steps of the 1st or of the 2nd subsequence, respectively. It is yet convenient to partition the set off all initial states into sets $\mathbf{I}_n$ in terms of the number of trains of pulses with a constant sign in the optimal control sequence.

From the construction of $\mathbf{R}_N$ it is obvious that all its edges are parallel to one of the vectors. Points of an edge of $\mathbf{I}_2$ parallel to $\mathbf{v}_m$ can then be expressed as

$$\begin{aligned} B_N^j &= \mathbf{v}_1 U_j + \cdots + \mathbf{v}_{m-1} U_j + \mathbf{v}_m u_m + \mathbf{v}_{m+1} U_{3-j} + \cdots + \mathbf{v}_N U_{3-j} \\ u_m &\in \langle U_1, U_2 \rangle , \; m \ne N ; \; u_m \in \langle 0, U_{3-j} \rangle , \; m = N \end{aligned} \qquad (11)$$

In $\mathbf{I}_3$ the boundary states can be expressed as

$$\begin{aligned} B_N^j &= U_j \sum_{i=1}^{m-1} \mathbf{v}_i + \mathbf{v}_m u_m + U_{3-j} \sum_{i=m+1}^{m+l} \mathbf{v}_i + U_j \sum_{i=m+l+1}^{N} \mathbf{v}_i \\ u_m &\in \langle U_1, U_2 \rangle \end{aligned} \qquad (12)$$

$$\begin{aligned} B_N^j &= U_j \sum_{i=1}^{k} \mathbf{v}_i + U_{3-j} \sum_{i=k+1}^{m-1} \mathbf{v}_i + \mathbf{v}_m u_m + U_j \sum_{i=m+1}^{N} \mathbf{v}_i \\ u_m &\in \langle U_1, U_2 \rangle , \; m \ne N ; \; u_m \in \langle 0, U_j \rangle , \; m = N \end{aligned} \qquad (13)$$

4. UNIQUENESS PROPERTIES

Desoer and Wing (1961) have shown that if $\mathbf{x}_0 = \mathbf{v}_1 u_1 \in \mathbf{R}_1$, one and only one value of $\mathbf{x}_0$ corresponds to each value of u_1 and the canonical representation is unique. If $\mathbf{x}_0 = \mathbf{v}_1 u_1 + \mathbf{v}_2 u_2 \in \mathbf{R}_2$, two rows of this vector equation uniquely determine the values u_1 and u_2 under the condition that the vectors $\mathbf{v}_1$ and $\mathbf{v}_2$ are not parallel, which is equivalent to the controllability condition $\det[\mathbf{b} \;\; \mathbf{Ab}] \ne 0$.

The expression of the vertices and the points B_N on the boundary of $\mathbf{R}_N$ is also unique. The canonical representation (6) is, however, not unique in interior states of $\mathbf{R}_N$, $N>2$. For these points they have proposed a recursive construction of canonical representation, which has seemed to them to be the simplest one: An interior point of $\mathbf{R}_N$ can be obtained from a boundary point of $\mathbf{R}_{N-1}$ by adding the vector $\mathbf{v}_N u_N$ in an outward direction

$$\mathbf{x}_0 = B^j_{N-1} + \mathbf{v}_N u_N \qquad (14)$$

From (11-14) it follows that an optimal control sequence can involve, besides the trains of saturated u_i's, two non-saturated steps: the first occurring at the sign change and the second in the last control step. However, Desoer and Wing have not observed another significant feature of the construction (14): It gives $|u_N| = \min$! From (3) it immediately follows that the control error at the last-but-one sampling instant takes minimal possible value:

$$\mathbf{x}_{N-1} = -\mathbf{A}(-T)\mathbf{b}(T) = \mathbf{v}_1 u_N \qquad (15)$$

The output variable reaches an ε-neighborhood of the origin in the minimum time!

5. PROPORTIONAL BAND

In an usual application it is not necessary to produce the optimal control sequence all at once at time t=0. Instead, it is more advantageous to compute only the control signal u_1 for the next control interval and to repeat the computation at each new sampling instant. This does note change the system behavior and at the same time simplifies the computer operation and enables registration of unexpected disturbances.

Let us consider those $\mathbf{x}_0 \in \mathbf{I}_2$ for which $u_1 \in \langle U_1, U_2 \rangle$. From (11) and (14) for m=1 it follows

$$\mathbf{x}_0 = \mathbf{v}_1 u_1 + U_j \sum_{i=2}^{N-1} \mathbf{v}_i + \mathbf{v}_N u_N$$
$$u_1 \in \langle U_1, U_2 \rangle; \; u_N \in (0, U_j \rangle; \; j = 1,2 \qquad (16)$$

Equivalently, for $\mathbf{x}_0 \in \mathbf{I}_3$ from (12) and (14)

$$\mathbf{x}_0 = \mathbf{v}_1 u_1 + U_j \sum_{i=2}^{l+1} \mathbf{v}_i + U_{3-j} \sum_{i=l+2}^{N-1} \mathbf{v}_i + \mathbf{v}_N u_N$$
$$u_1 \in \langle U_1, U_2 \rangle; \, u_N \in (0, U_{3-j} \rangle; \; j = 1,2 \qquad (17)$$

The set of all points with $u_1 \in \langle U_1, U_2 \rangle$ is denoted as the **Proportional Band ($\mathbf{P_b}$)**. By the second term in (15) and (16) denoted as $\mathbf{z}^j_N$, whereby in $\mathbf{I}_2$

$$\mathbf{z}^j_N = U_j \sum_{i=2}^{N-1} \mathbf{v}_i + \mathbf{v}_N u_N \; ; \; u_N \in (0, U_j \rangle; \; j = 1,2 \; (18)$$

the **Critical** or **Zero Curve (ZC)** is defined, which consists of points at which u_1=0 (dividing the phase plane according to the sign of u_1). In $\mathbf{I}_2$ this polygonal curve is obtained by joining vertices

$$Z^j_N = \begin{bmatrix} z^j_N \\ \dot{z}^j_N \end{bmatrix} = U_j \sum_{i=2}^{N} \mathbf{v}_i, \; N > 1; \; Z^j_0 = \mathbf{0}; \; j = 1,2 \qquad (19)$$

ZC of $\mathrm{P_b}$ in $\mathbf{I}_3$ has interior points

$$\mathbf{z}^j_N = U_j \sum_{i=2}^{l+1} \mathbf{v}_i + U_{3-j} \sum_{i=l+2}^{N-1} \mathbf{v}_i + \mathbf{v}_N u_N \; ; u_N \in (0, U_{3-j} \rangle; \; j = 1,2 \quad (20)$$

and vertices

$$Z^j_N = U_j \sum_{i=2}^{l+1} \mathbf{v}_i + U_{3-j} \sum_{i=l+2}^{N} \mathbf{v}_i \, ; \; j = 1,2 \qquad (21)$$

For u_1=U_j in (16) and (17) a new polygonal curve is defined, referred originally as the **Switching Curve** (SC) due to its relation to the switching curve of the relay minimum time systems. In $\mathbf{I}_2$ it has vertices

$$X^j_N = V^j_{N,0} = Z^j_N + \mathbf{v}_1 U_j = U_j \sum_{i=1}^{N} \mathbf{v}_i \; ; \; j = 1, \, 2 \quad (22)$$

In $\mathbf{I}_2$ SC represents an invariant set, along which the system is braked to the origin, it could also be denoted as the **Reference Braking Curve (RBC)**. Since the last property does not hold in $\mathbf{I}_n$ n>2, in such situations it will be denoted as the **Full-Braking-Boundary (FBB)**. In $\mathbf{I}_3$

$$X^j_N = Z^j_N + \mathbf{v}_1 U_j = U_j \sum_{i=1}^{l+1} \mathbf{v}_i + U_{3-j} \sum_{i=l+2}^{N} \mathbf{v}_i \, ; \; j = 1,2 \qquad (23)$$

Putting u_1=U_{3-j} in (16) and (17), vertices of the **Full-Acceleration-Boundary (FAB)** are defined

$$Y^j_N = V^{3-j}_{1,N-1} = Z^j_N + \mathbf{v}_1 U_{3-j} = X^j_N + (U_{3-j} - U_j)\mathbf{v}_1 \; ; \; j = 1, 2 \qquad (24)$$

Both these polygonal curves represent boundaries of $\mathrm{P_b}$, in which the control signal is not saturated and therefore has to be computed. Outside of $\mathrm{P_b}$ u_1=U_1 or U_2, which depends only on the representative point position with respect to the ZC.

The proportional zone is divided into two parts by a line parallel to $\mathbf{v}_1$ and crossing the origin

$$\mathbf{x} = q\mathbf{v}_1, \; q \in (-\infty, \infty) \qquad (25)$$

By eliminating q the line equation can be written as

$$p = 0 \; ; \quad p = y - \dot{y}\frac{v_1}{\dot{v}_1} \qquad (26)$$

To the left of p, j=1 will be substituted into (17-24) and to the right j=2. In order to avoid the ambiguous situation p=0, "j" can be determined according to

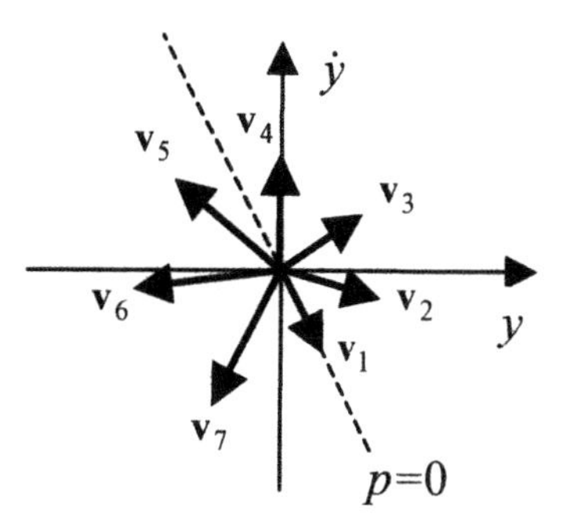

Fig.1. Vektors $\mathbf{v}_k$ of the oscillating plant (a_0= 0.82; a_1= 0.2; T=1).

$$\text{if } p < 0 \text{ then } j\text{=1 else } j\text{=2} \qquad (27)$$

Now the question is, how long $\mathrm{P_b}$ can be constructed according to the formulas for $\mathbf{I}_2$ and when one has to use formulas derived for $\mathbf{I}_n$, n>2. Let the characteristic polynomial of the original system with complex poles be $A(s) = s^2 + a_1 s + a_0$. Observe that increasing k the length of vectors $\mathbf{v}_k$ (4) is increasing in stable systems (Fig.1) and decreasing in unstable systems. Simultaneously, $\mathbf{v}_k$ also rotate in each step by an angle $\Delta\varphi$. If the formula (19) is used for $N = \mathrm{int}(\pi / \Delta\varphi) + 1$, int=integer part, the new point belongs to $\mathbf{R}_{N-1}$, $\mathbf{v}_N U_j$ points from Z^j_{N-1} into $\mathbf{R}_{N-1}$ (the angle between $\mathbf{v}_1$ and $\mathbf{v}_N$ is greater then π). Therefore, the new vertex Z^j_N of ZC can only be constructed from the previous one by adding the vector $\mathbf{v}_N U_{3-j}$ pointing in the outwards direction. That holds for all $\mathbf{v}_N$ having tips to the left of p: The new vertex should be constructed using recursion

$$Z^j_N = Z^j_{N-1} + \mathbf{v}_N U_i \qquad (28)$$

whereby

$i=j$ ak $p(\mathbf{v}_N) > 0$ (29)
$i=3-j$ ak $p(\mathbf{v}_N) < 0$

Equivalently, points $\mathbf{z}_N \in Z_{N-1}^j Z_N^j$ are expressed as

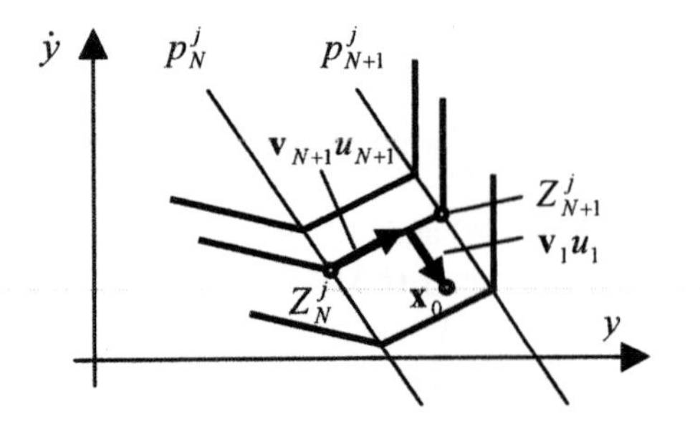

Fig.2. Computation of the control signal u_1.

$$\mathbf{z}_N^j = Z_{N-1}^j + \mathbf{v}_N u_N \; ; \; u_N \in (0, U_i) \tag{30}$$

6. CONTROL ALGORITHM

From modified expressions (16-20)

$$x_0 = \mathbf{v}_1 u_1 + \mathbf{z} = \mathbf{v}_1 u_1 + Z_N^j + \mathbf{v}_{N+1} u_{N+1}$$
$$N = 2,3,...; \; j = 1,2 \; ; \tag{31}$$
$$u_1 \in < U_1, U_2 > \; ; \; u_{N+1} \in (0, U_i >$$

whereby the value U_i is given by (29), it is obvious that the value u_1 can be determined as the distance of the initial state $\mathbf{x}_0$ to ZC in the direction of $\mathbf{v}_1$.

The line crossing $\mathbf{x}_0$ in direction $\mathbf{v}_1$ (Fig.2) is given by the equation $\mathbf{x} = \mathbf{z} + \mathbf{v}_1 u_1 \; ; \; u_1 \in (-\infty, \infty)$ whereby $\mathbf{z} = Z_N^j + \mathbf{v}_{N+1} u_{N+1}$. Eliminating $\mathbf{z}$ and u_N yields

$$u_1 = (\dot{v}_{N+1} y - v_{N+1} \dot{y} + \dot{z}_N v_{N+1} - z_N \dot{v}_{N+1}) / (\dot{v}_{N+1} v_1 - v_{N+1} \dot{v}_1) = r_0(N) y + r_1(N) \dot{y} + r_c(N) \tag{32}$$

7. STRIPS S_N^j; DETERMINATION OF N

After limiting the calculated value (32) according to (2), the control algorithms can be used in all points of the strip $S_N{}^j$ determined by parallel lines p_{N+1}^j and p_N^j crossing Z_{N+1}^j and Z_N^j in direction of $\mathbf{v}_1$. p_N^j is described as $\mathbf{x} = Z_N^j + q\mathbf{r}_1 \; ; \; q \in (-\infty, \infty)$, or after eliminating the parameter q by the equation $p_N^j(\mathbf{x}) = 0$;

$$p_N^j(\mathbf{x}) = y - \frac{v_1(\dot{y} - \dot{z}_N^j)}{\dot{v}_1} - z_N^j = p + c_N^j \tag{33}$$

The value of the parameter N in (32) is determined for the point $\mathbf{x}_0$ to be between lines p_N^j and p_{N+1}^j ($\mathbf{x}_0 \in S_N^j$), which can be realized by an iterative procedure, until the inequalities

$$(p.p_{N+1}^j < 0) \; AND \; (p.p_N^j > 0) \tag{34}$$

are fulfilled (Huba, Sovišová and Spurná, 1987).

8. BEHAVIOUR IN $\mathbf{I}_2$

Considering an initial point $\mathbf{x}_0 \in P_b \subset \mathbf{I}_2$ (16) and the applied optimal control u_1, the representative point takes in the next sampling instant value

$$\mathbf{x}_1 = \mathbf{A}\left[\mathbf{v}_1 u_1 + U_j \sum_{i=2}^{N-1} \mathbf{v}_i + \mathbf{v}_N u_N\right] + \mathbf{b} u_1 = = U_j \sum_{i=1}^{N-2} \mathbf{v}_i + \mathbf{v}_{N-1} u_N \; ; \; u_N \in (0, U_j\rangle \; ; \; j = 1, 2 \tag{35}$$

This is a point of the line segment $X_{N-1}^j X_{N-2}^j$ of the RBC. So, from points of P_b the optimal solution tends in the next sampling period to RBC!

When expressing the distance of an initial point $\mathbf{x}_0 = V_{k,N-k}^{3-j} \in \mathbf{I}_2$ (9) from the RBC by the number k of steps with $u = U_{3-j}$, it is clear from

$$\mathbf{x}_1 = \mathbf{A}\left(U_{3-j} \sum_{i=1}^{k} \mathbf{v}_i + U_j \sum_{i=k+1}^{N} \mathbf{v}_i\right) + \mathbf{b} U_{3-j} = = U_{3-j} \sum_{i=1}^{k-1} \mathbf{v}_i + U_j \sum_{i=k}^{N-1} \mathbf{v}_i \tag{36}$$

that under optimal control with $u_1 = U_{3-j}$ this distance decreases by one sampling period in each control step. So, in general, the aim of the optimal control for $\mathbf{x}_0 \in \mathbf{I}_2$ can be interpreted as: **To reach RBC in the minimum time possible**. This task can be easily solved also without the reachibility sets: RBC is traced out by vertices $X_N{}^j$ computed by N backward steps from the origin under the control $u = U_j$ as

$$X_N^j = \mathbf{b}(-NT) U_j \; ; \; j = 1, 2 \tag{37}$$

The control algorithms can be determined by the requirement to reach RBC in one control step

$$\mathbf{A}\mathbf{x}_0 + \mathbf{b} u_1 = X_{N-1}^j + q\mathbf{v}_N \tag{38}$$

If the computed control signal exceeds the given constraints (2), it has simply to be limited, so that

$$u_1 = sat\{[0 \quad 1][\mathbf{v}_N | \mathbf{b}]^{-1}(X_{N-1}^j - \mathbf{A}\mathbf{x}_0)\} = = sat\{r_0(N) y + r_1(N) \dot{y} + r_c(N)\} \tag{39}$$

9. SPECIAL CASE

In the case of an oscillating undamped system with $a_1 = 0, a_0 = \omega^2$ and the sampling period satisfying

$$K\omega T = \pi \tag{40}$$

(K being positive integer and $\Delta\varphi = \omega T$ the angle traced out by $\mathbf{v}_k$ and $\mathbf{v}_{k+1}$), P_b acquires a regular shape and all its vertices can be expressed in a relatively simple way. As the consequence of (40)

$$\mathbf{v}_{k+iK} = (-1)^i \mathbf{v}_k \tag{41}$$

It is now convenient to introduce strips $P_m{}^j$, $N_m{}^j$ and $Q_M{}^j$ (Fig.3): with m denoting the number of begun half rounds of $\mathbf{v}_N$ around the origin and M denoting the begun full rounds of $\mathbf{v}_N$.

The strips $P_m{}^j$ are defined as a union of all strips $S_N{}^j$ in $\mathbf{I}_n$ $n=m+1$ with Z_N^j constructed using $U_i=U_j$ in (28) (vector $\mathbf{v}_N$ is to the right of p)

$$P_m{}^j = \bigcup_{N=(2m-2)K+1}^{(2m-1)K} S_N^j \tag{42}$$

The strips $N_m{}^j$ are defined as a union of strips $S_N{}^j$ in $\mathbf{I}_n$, $n=m+1$ with Z_N^j constructed using $U_i=U_{3-j}$ in (28) (vector $\mathbf{v}_N$ is to the left of p)

$$N_m{}^j = \bigcup_{N=(m-1)K+1}^{2(m-2)K} S_N^j \tag{43}$$

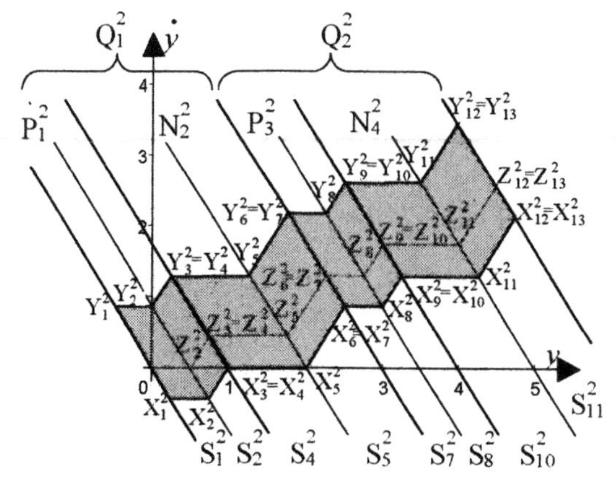

Fig.3. Special case: K=3. P_b corresponding to (58c).

The union of all strips $S_N{}^j$ corresponding to the complete M-th round of $\mathbf{v}_N$ will be denoted as the strip $Q_M{}^j$

$$Q_M{}^j = \bigcup_{N=2(M-1)K+1}^{2MK} S_N^j = P_{2M-1}^j \bigcup N_{2M}^j \tag{44}$$

The strips $P_1{}^j$ and $N_2{}^j$ are separated by a line (33) crossing the point

$$Z_K^j = \mathbf{b}(-KT)U_j - \mathbf{v}_1 U_j = \mathbf{z}_K U_j \tag{45}$$

The lines p_N^j (33) crossing X_N^j in $\mathbf{I}_2$ are described by

$$p = y + \frac{\tan(\omega T/2)}{\omega}\dot{y}; \quad c_N^j = \frac{\cos(\omega T(N-1/2)) - \cos(\omega T/2)}{\omega^2 \cos(\omega T/2)} Uj \tag{46}$$

From (41) and from (28-29) it follows that

$$Z_{2K}^j = Z_K^j - \mathbf{z}_K U_{3-j} \tag{47}$$

and the constant c_{2K}^j of the line crossing Z_{2K}^j and separating $N_2{}^j$ from $P_3{}^j$ and, simultaneously, $Q_1{}^j$ from $Q_2{}^j$ is

$$c_{2K}^j = \frac{\cos\left(\frac{\omega T(2K-1)}{2}\right) - \cos\left(\frac{\omega T}{2}\right)}{\omega^2 \cos\left(\frac{\omega T}{2}\right)}\left(U_j - U_{3-j}\right) \tag{48}$$

In general, the strips $Q_M{}^j$ and $Q_{M+1}{}^j$ are separated by the lines (46) with

$$c_{2MK}^j = Mc_{2K}^j \tag{49}$$

Putting such a line through a point of a strip Q_M^j

$$p(\mathbf{x}_0) + c_{2MK}^j = p(\mathbf{x}_0) + Mc_{2K}^j = 0 \tag{50}$$

and solving it in terms of M results in

$$M = -p(\mathbf{x}_0)/c_{2K}^j \tag{51}$$

which takes in $Q_M{}^j$ values $M^* \in (M, M+1)$. Hence, the index M can be determined as

$$M = \mathrm{int}(M^*) \tag{52}$$

The strips $Q_M{}^j$ are partitioned into the strips $P_m{}^j$ and $N_m{}^j$ by the lines (46) with constants

$$c_{K+2K(M-1)} = c_K^j + (M-1)c_{2K}^j \tag{53}$$

By evaluating the sign of expression (33) with constant (53) one can determine, in which of the two strips the initial point is. Then the computation of the parameter N of the actual strip can be finished iteratively. But, it is also possible to derive an explicit solution based on the solution in P_1^j. Here, writing equation of line (46) crossing a point $\mathbf{x}_0 \in S_N{}^j$ and solving it in terms of N gives

$$N = \mathrm{int}\left\{\frac{1}{2} + \frac{1}{\omega T}\arccos\left[\cos\left(\frac{\omega T}{2}\right)\left(1 + \frac{\omega^2 p(\mathbf{x}_0)}{U_j}\right)\right]\right\} \tag{54}$$

The control signal can be computed by means of (54) and (32) directly, without iterations. The same can be generalized for N_2^j with $N = \overline{N} + K$ and by determining $\overline{N}$ according to (54) with $p(\mathbf{x}_0)/U_j$ being replaced by $p(\mathbf{x}_0 - Z_K^j)/\left[|U_{3-j}|sign(U_j)\right]$.

10. NONUNIQUENESS OF CANONICAL REPRESENTATION

Choice of the sampling period (40) also varies the shape of sets $\mathbf{R}_N$. In $\mathbf{I}_3$ the edges of $\mathbf{R}_N$ parallel to $\mathbf{v}_1$ can be expressed using (12) with m=1 as

$$B_N^j = \mathbf{v}_1 u_1 + U_{3-j}\sum_{i=2}^{K}\mathbf{v}_i + \mathbf{v}_{K+1}U_j + U_j\sum_{i=K+2}^{N}\mathbf{v}_i \tag{55a}$$

$$u_1 \in \langle U_1, U_2 \rangle$$

and the edges parallel to $\mathbf{v}_{K+1}$ with $m=K+1$ as

$$B_N^j = \mathbf{v}_1 U_j + U_{3-j}\sum_{i=2}^{l+1}\mathbf{v}_i + \mathbf{v}_{K+1}u_{K+1} + U_j\sum_{i=K+2}^{N}\mathbf{v}_i \tag{55b}$$

$$u_{K+1} \in \langle U_1, U_2 \rangle$$

Both these edges have one common vertex

$$V_{1,l,N-l-1}^j = \mathbf{v}_1 U_j + U_{3-j}\sum_{i=2}^{K}\mathbf{v}_i + U_j\sum_{i=K+1}^{N}\mathbf{v}_i \tag{56}$$

Changing u_1 from U_j to U_{3-j} the point (55a) moves from the vertex (56) and traces out line segment $(U_{3-j}-U_j)\mathbf{v}_1$. In accordance with (41), a motion of the point (55b) with the length of $(U_j-U_{3-j})\mathbf{v}_1$ corresponds to equivalent change of u_{K+1}. Both edges form together a line segment with the length $2(U_{3-j}-U_j)\mathbf{v}_1$. The point (56) becomes to be an internal point of this line segment. From the relation

$$\mathbf{v}_1 U_j + \mathbf{v}_{K+1}U_j = \mathbf{v}_1(U_j - U_j) = \mathbf{v}_1(u_1 + u_{K+1}) \tag{57}$$

it is obvious that the point (56) can be expressed by any couple u_1, u_{K+1}. Thus, driving this system from the initial point (56) the optimal control is irrelevant to the value $u_1 \in \langle U_1, U_2 \rangle$!

The canonical representation of interior states of edges (55) is also nonunique, but increasing the distance from the point (56) the range of possible combinations (u_1, u_{K+1}) becomes narrower and, finally, the points corresponding to $u_1=u_{K+1}=U_{3-j}$ have a unique canonical representation.

The prolongation of edges parallel to $\mathbf{v}_1$ (Fig.4) brings some degree of freedom into the construction of the proportional zone. The rule (29) for constructing the new vertices of polygonal curves (28) can be modified by any of the possibilities:

a) putting $U_i=U_j$;
b) putting $U_i=U_{3-j}$;
c) putting $U_i=0$; (58)
d) determining U_i by means of $\mathbf{v}_{N-1}$;
e) determining U_i by means of $\mathbf{v}_{N+1}$.

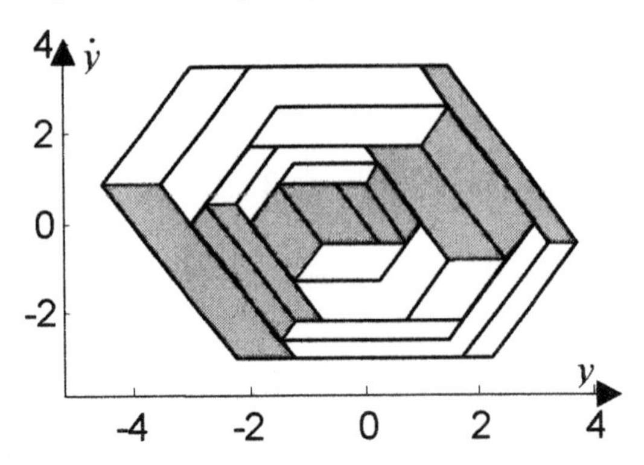

Fig.4. Prolongation of edges of $\mathbf{R}_N$ parallel to $\mathbf{v}_1$.

without changing the shape of the strips S, P, N and Q, since the questionable vectors are parallel to $\mathbf{v}_1$. In Fig.3, P_b is shown for the case (58c) guaranteeing continuity of the control over the phase plane.

CONCLUSIONS

It was shown that choosing sampling period (40) the control algorithm can be expressed explicitly for arbitrary initial point, reasonably simplified and speeded up. Of course, one could ask, if it has a high practical importance, since, in fact, all used functions (like *sin*, etc.) are internally calculated iteratively. So, the only difference between an iterative and the explicit controller is that in the 2nd case the designer is using already available procedures. Once such controller is implemented, the differences are no more important.

Two approaches to the minimum time controller design have been shown: The 1st one based on the reachability sets is generally valid; the 2nd one is restricted just to the initial states $\mathbf{x}_0 \in \mathbf{I}_2$ consisting of two control intervals. At $\mathbf{x}_0 \in \mathbf{I}_n, n>2$, for the sampling period (40) a nonuniqueness of the synthesis occurs. This can be considered as a consequence of the fact that in such points the trajectories corresponding to various control inputs differ only slightly. Such a property holds also in the case of a general sampling period, but it is not so easy to demonstrate. Analogically as in the continuous-data systems (Athans-Falb, 1966), this can be used for the controller simplification based on the 2nd approach, even here the convenience of such solution is yet more stressed by the effect of the time quantization.

The problem of linear PD controller with constrained output and a striplike proportional band P_b is that it cannot sufficiently approximate the optimal P_b traced out by the polygonal curves. If it is optimally tuned for the relatively small initial conditions, overshoot, or even instability occurs for higher initial disturbances. If these are suppressed by modified controller setting, sluggish transient occurs in the vicinity of demanded state.

It is also important to note that the production costs of the minimum time (or of the minimum time pole assignment) controllers may be practically the same as those of the conventional linear PD-controllers. The same holds also for their dynamical properties: Implementing such controllers by standard PCs it is possible to control processes with sampling period in the range of fractions of ms. Windupless I-action and several methods for tuning such controllers have been proposed. So, the only real barrier in using controllers considering the control signal constraints seems to be in the system of the control education and in the strong conservativism of the industrial sector.

ACKNOWLEDGEMENTS

This work has been partially supported by the Slovak Scientific Grant Agency, Grant VG-0145-Hub-Sk1.

REFERENCES

Aström, K. J. , T.Hägglund (1995). *PID controllers : Theory, design, and tuning* - 2. ed., ISA, Research Triangle Park, NC

Athans, M. and P.L.Falb (1966). *Optimal control.* McGraw-Hill, N.York, pp.569-589.

Desoer, C.A. and J.Wing (1961). The minimal time regulator problem for linear sampled-data systems. *J. Franklin Inst.*, **272**, 208-228.

Kuo, B.C. (1970). Discrete-data control systems. *Prentice-Hall*, Englewood Clifs, N.Jersey.

Huba, M. et al. (1987). Digital time-optimal control of nonlinear second-order system. In: *Preprints 10th IFAC World Congress,* Munich, **8**, 29-34.

Huba,M. (1992). Control Algorithms for 2nd Order Minimum Time Systems. *Electrical Engineering Journal* **43**, 233-240. (http://www.kar.elf.stuba.sk/~huba/EE92.pdf)

Huba, M., et al. (1999). Invariant Sets Based Concept of the Pole Assignment Control. In: *ECC'99*, Düsseldorf: VDI/VDE

Huba, M., Bisták, P. (1999). Dynamic Classes in the PID Control. In: *Proc. 1999 American Control Conference*. San Diego: AACC.

Rönnbäck,S. (1996). Nonlinear Dynamic Windup Detection in Anti-Windup Compensators. Preprints CESA '96, Lille, 1014-101

www.elsevier.com/locate/ifac

DESIGN OF (P) I (D) CONTROLLER OPTIMAL PARAMETERS

P. Hudzovič, I. Oravec, D. Sovišová

*Faculty of Electrical Engineering and Information Technology,
Slovak University of Technology, Ilkovičova 3, 81219 Bratislava, Slovakia
e-mail: hudzovic@kasr.elf.stuba.sk*

Abstract: The paper deals with a PID-controller optimal parameters design method. Its focus is on the system of nonlinear equations, which solution enables elimination of all variables but one and elude an iterative approach to solution. *Copyright © 2003 IFAC*

Keywords: PID controller, optimal parameters, solution of nonlinear equations system.

1 INTRODUCTION

Although the history of continuous controller synthesis traces back more than half a century, compensators of this type are still being operated in industrial plants. There is a broad range of methods of their adjusting, varying in quality of the resulting product, calculation requirements or underlying principles. In (Hudzovič, 1999), a controller synthesis method was described, which provides outstanding closed loop parameters. However, to obtain the optimal parameters requires solution to the system of non-linear equations including laborious calculations, which are briefly discussed in the sequel.

2 PRINCIPLE OF THE METHOD

Let the transfer function of the controlled static linear plant be

$$\bar{S}(s) = K\frac{\prod_{i=1}^{m}(1+\bar{t}_i s)}{\prod_{i=1}^{n}(1+\bar{T}_i s)}e^{-\bar{D}s} = K\frac{\sum_{i=0}^{m}\bar{B}_i s^i}{\sum_{i=0}^{n}\bar{A}_i s^i}e^{-\bar{D}s} \quad \begin{matrix}\bar{B}_0 = 1 \\ \bar{A}_0 = 1\end{matrix} \quad m < n \tag{1}$$

Without loss of generality, the transfer function can be normed through the gain K and the time constants $\bar{T}_i$, $\bar{t}_i$ **a** $\bar{D}$ by constants:

$$\tau = \bar{D} + \sum_{i=1}^{n}\bar{T}_i - \sum_{i=1}^{m}\bar{t}_i = \bar{D} + \bar{A}_1 - \bar{B}_1 > 0 \tag{2}$$

which gives the normed expressions:

$$S(s) = \bar{S}(s)/K; \quad T_i = \bar{T}_i/\tau; \quad t_i = \bar{t}_i/\tau \quad D = \bar{D}/\tau \tag{3}$$

and a normed transfer function

$$S(s) = K\frac{\prod_{i=1}^{m}(1+t_i s)}{\prod_{i=1}^{n}(1+T_i s)}e^{-Ds} \tag{4}$$

The task is to design an open-loop with the transfer function $F_0(s)$ of type 1 comprising the normed transfer function (4) and a controller with integral action, which normed transfer function is:

$$R(s) = \frac{\prod_{i=1}^{q}(1+\vartheta_i s)}{Cs} = \frac{\sum_{i=0}^{q} b_i s^i}{Cs} \quad \begin{matrix} b_0 = 1 \\ q \le n-m \end{matrix} \tag{5}$$

where for $q = 2$ the controller is of PID - type , $q = 1$ implies $b_2 = 0$ and the controller is of PI - type and for $q = 0$, $b_2 = b_1 = 0$ of I - type. The actual real numerical values can be obtained using the substitution $\bar{\vartheta} = \tau\vartheta$, or $\bar{C} = KC\tau$.

The algorithm for computing optimal parameters begins by determination of auxiliary coefficients, from the normed transfer function (4):

$$h_k = \delta_{1k}D + \sum_{i=1}^{n}T_i^k - \sum_{i=1}^{m}t_i^k \qquad k = 1,2,3,... \tag{6}$$

where δ_{1k} is the Kronecker symbol, defined as follows: $\delta_{1k} = 1$ if $k = 1$, in other cases $\delta_{1k} = 0$. Using normed parameters implies $h_1 = 1$.

Using the Bôcher formulas, see (Åström and Hägglund, 2000) or (O'Dwyer, 2000), the coefficients h_k allow to compute the Markov's parameters:

$$l_k = \frac{1}{k}\sum_{i=1}^{k}h_i l_{k-i} \qquad l_0 = 1, \qquad k = 1,2,3,... \tag{7}$$

Expressing explicitly the first four of the above parameters yields:

$$\left.\begin{aligned} l_1 &= h_1 = 1 \\ l_2 &= (1+h_2)/2 \\ l_3 &= (1+3h_2+2h_3)/6 \\ l_4 &= (1+6h_2+3h_2^2+8h_3+6h_4)/24 \end{aligned}\right\} \tag{8}$$

By means of the Markov's parameters l_k it is possible to expand the transfer function (4) into a power series:

$$S(s) = \sum_{i=0}^{\infty}(-1)^i l_i s^i \qquad l_0 = 1 \tag{9}$$

Joining (5) and (9), the the open-loop transfer function is:

$$F_0(s) = \frac{1}{Cs}\left[\sum_{i=0}^{q} b_i s^i\right]\left[\sum_{i=0}^{\infty}(-1)^i l_i s^i\right] \tag{10}$$

The underlying idea of the parameter-optimised controller design is to achieve the closest possible approximation of the ideal response (Fig. 1) by the closed-loop step response. The ideal step response can be analytically expressed as

$$r(t) = \begin{cases} 0 & 0 \le t \le fd \\ (t-fd)/(d-fd) & fd \le t \le d \\ 1 & d \le t \le \infty \end{cases} \tag{11}$$

The advantage of this approach consists in that the dynamics of the control circuit is not violated, but adapted to the dynamics of the controlled system.

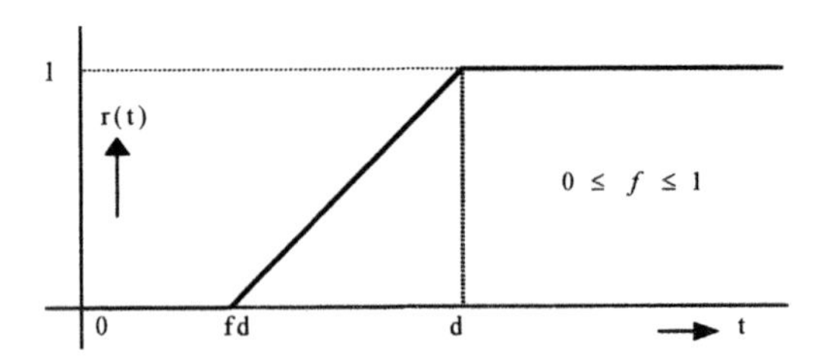

Fig.1. Ideal closed-loop step response

Undoubtedly, the ideal step response has to begin with a zero value, have time delay fd, rise time (1-fd). The values of the parameters d and f are not prescribed, but follow from the actual synthesis.

The closed-loop control corresponds to the step response

$$F_c(s) = \frac{e^{-fds} - e^{-ds}}{(1-f)ds} = 1 - \frac{1}{1+F_0(s)} \tag{12}$$

from which we can easily determine the Laplace transform of the control error:

$$E(s) = 1 - F_c(s) = \frac{(1-f)ds - e^{-fds} + e^{-ds}}{(1-f)ds} = \frac{1}{1+F_0(s)} \tag{13}$$

Expanding this function into an infinite power series and applying a simple manipulation we obtain:

$$\frac{2E(s)}{(1+f)ds} = 1 - \frac{1-f^3}{3(1-f^2)}ds + \frac{1-f^4}{12(1-f^2)}d^2s^2 + \ldots, \tag{14}$$

Combining (10) and (13) yields:

$$\frac{E(s)}{Cs} = \left[1 - (1-b_1-C)s + (l_2-b_1+b_2)s^2 - \ldots\right]^{-1} \tag{15}$$

Comparing the last two expressions result in

$$C = 0.5d(1+f) \tag{16}$$

which substitution into (14) gives:

$$\frac{E(s)}{Cs} = 1 - \frac{2}{3}\frac{1-f^3}{(1-f^2)(1+f)}Cs + \frac{1-f^4}{3(1-f^2)(1+f)^2}C^2s^2 + \ldots \tag{17}$$

or in the preferable form:

$$\frac{E(s)}{Cs} = \begin{bmatrix} 1 + \frac{2}{3}\frac{1+f+f^2}{1+2f+f^2}Cs + \frac{1}{9}\frac{f^4+f^3+4f^2+f+1}{f^4+2f^3+4f^2+2f+1}C^2s^2 - \\ -\frac{2}{135}\frac{f^6+6f^5+12f^4+10f^3+12f^2+6f+1}{f^6+12f^5+15f^4+32f^3+15f^2 12f+1}C^3s^3 + \ldots \end{bmatrix}^{-1} \tag{18}$$

If using the following notation

$$m = 1 - b_1 \tag{19}$$

and

$$a = 3\frac{1+2f+f^2}{1+4f+f^2} \qquad 0 \le f \le 1 \qquad 2 \le a \le 3 \tag{20}$$

and equating coefficients at equal powers of the variable s in square brackets in relations (15) and (18), we obtain the following system of nonlinear equations:

$$am = C \tag{21}$$

$$(3 - 3a + a^2)m^2 - 3m - 3b_2 - 3(l_2 - 1) = 0 \tag{22}$$

$$\begin{aligned} &(2-a)(15-12a+a^2)m^3 - 30l_2m - \\ &- 30b_2 - 30(l_3 - l_2) = 0 \end{aligned} \tag{23}$$

$$\begin{aligned} &(90 - 144a + 81a^2 - 18a^3 + a^4)m^4 - \\ &- 90l_3m - 90l_2b_2 - 90(l_4 - l_3) = 0 \end{aligned} \tag{24}$$

3 COMPUTATION OF OPTIMAL CONTROLLER PARAMETERS

For simplicity we begin with the I-controller. In this case $b_2 = b_1 = 0$ and to specify the controller, it is necessary to know only the coefficient $C = am = a(1-b_1) = a$, which follows from (19) and (21). Thus using relation (22) we have

$$C^2 - 3C + 3(1 - l_2) = 0 \tag{25}$$

from which follows the solution:

$$C^* = 1.5\left[1 + \sqrt{(1+2l_2)/3}\right] \tag{26}$$

Considering PI-controller, the constant $b_2 = 0$. For computing the couple of constants C and $b_1 = 1 - m$, we need to know the value of the parameters a and m, according to the relation (21). We use the equations (22) and (23) to compute them.

This system of two equations with variables a and m can be solved using one of the direct methods. One possibility is to create the objective function - the sum of the squares of the left hand sides of equations. The minimum of this function can be found using some of the gradient methods (Newton-Raphson, conjugate gradient, quasi Newton algorithm), or comparative methods (Nelder and Mead).

It is preferable to eliminate one parameter from one equation, and vary the other until a correct solution is achieved. Most comfortable would be to eliminate one variable and thus to obtain such a condition for the other, which can be solved analytically.

In (Hudzovič, 1999) a procedure is described, in which the variable m is eliminated and for unknown variable a the algebraic equation of the 6th order is obtained. The searched solution is the real root satisfying the condition: $2 \le a \le 3$.

In this paper we propose an equivalent, but more advantageous procedure in which after elimination the parameter a remains one algebraic equation of the 6th order for the variable m and the searched solution $m^* = 1 - b_1^*$ is the real root which satisfies the condition: $0 \le m \le 1$.

The mentioned algebraic equation of 6th order is created as quadratic form:

$$\underline{v}^T (l_2, l_3) \underline{L} \underline{u}(m) = 0 \qquad (27)$$

where the vector of the powers of the parameter m is

$$\underline{u}(m) = \begin{bmatrix} 1 & m & m^2 & m^3 & m^4 & m^5 & m^6 \end{bmatrix}^T, \qquad (28)$$

the vector of the input data

$$\underline{v}(l_2, l_3) = \begin{bmatrix} 1 & l_2 & l_3 & l_2^2 & l_2 l_3 & l_3^2 & l_2^3 \end{bmatrix}^T \qquad (29)$$

and finally it will be the constant matrix of size 7x7:

$$\underline{L} = \begin{bmatrix} 3 & -9 & 94 & -173 & 134 & -49 & 7 \\ -9 & -172 & 201 & -70 & 1 & 0 & 0 \\ 0 & 190 & -190 & 50 & 0 & 0 & 0 \\ 109 & -19 & -5 & 0 & 0 & 0 & 0 \\ -200 & 10 & 0 & 0 & 0 & 0 & 0 \\ 100 & 0 & 0 & 0 & 0 & 0 & 0 \\ -3 & 0 & 0 & 0 & 0 & 0 & 0 \end{bmatrix} \qquad (30)$$

This procedure is more advantageous than the one described in the mentioned paper, because the matrix $\underline{L}$ has many zero elements, which facilitates computation.

After computing the real root of the equation (27), which has to satisfy $0 < m < 1$, the other variable a can be computed using

$$a = \frac{10(l_3 - l_2) + (11 - l_2)m - 11m^2 + m^3}{m(1 - l_2 - m - m^2)} \qquad (31)$$

Considering a PID controller, the nonlinear system of equations (22), (23) and (24) can be again solved using similar procedure as in the previous case: generating the objective function - the sum of squares from left hand sides of equations and finding its minimum using some of the gradient, or comparative methods. It is possible to use the procedure of elimination of variables.

In (Hudzovič, 1999) variables $b2$ and $m = 1 - b_1$ are stepwise eliminated and a quadratic form type (27) is obtained which is the algebraic equation of 38th order. The size of the vector $\underline{v}$ was 43. The size of the matrix $\underline{L}$ was 43x39. Its description would take the whole area assigned for a common article.

Application of MAPLE V4 has proved, that the elimination of b_2 and m is more preferable as it considerably reduces the size of the problem.

The input data vector, which size is 34, will be split into three parts:

$$\underline{v}_1 = \begin{bmatrix} 1 & l_2 & l_3 & l_4 & l_2^2 & l_2 l_3 & l_2 l_4 & l_2^3 & l_3^2 & l_2^2 l_3 & l_3 l_4 \end{bmatrix}^T \qquad (31)$$

$$\underline{v}_2 = \begin{bmatrix} l_2^4 & l_2^2 l_4 & l_2 l_3^2 & l_4^2 & l_2 l_3 l_4 \\ l_3^3 & l_2^5 & l_2^2 l_3^2 & l_2 l_4^2 & l_3^2 l_4 \end{bmatrix}^T \qquad (33)$$

$$\underline{v}_3 = \begin{bmatrix} l_2^3 l_4 & l_2^4 l_3 & l_2^2 l_3 l_4 & l_2 l_3^3 & l_3 l_4^2 & l_2^6 \\ l_2^4 l_4 & l_2^3 l_3^2 & l_2^2 l_4^2 & l_2 l_3^2 l_4 & l_3^4 & l_4^3 \end{bmatrix}^T \qquad (34)$$

whereas the vector of the powers of the variable m will be:

$$\underline{u}(m) = \begin{bmatrix} 1 & m & m^2 & m^3 & m^4 & m^5 & m^6 & m^7 & m^8 \\ m^9 & m^{10} & m^{11} & m^{12} \end{bmatrix}^T = \underline{u}(0:12) \qquad (35)$$

After that the algebraic equation of the 12th order is obtained as quadratic form:

$$\begin{bmatrix} \underline{v}_1^T & \underline{v}_2^T & \underline{v}_3^T \end{bmatrix} \begin{bmatrix} 25\underline{L}_1 & 5\underline{L}_2 & \underline{L}_3 \\ 25\underline{L}_4 & 5\underline{L}_5 & \underline{0} \\ 25\underline{L}_6 & \underline{0} & \underline{0} \end{bmatrix} \underline{u}(m) = 0 \qquad (36)$$

Solving this equation one real root is obtained, which has to satisfy $0 < m < 1$.

Using this root we generate the couple of vectors and its powers:

$$\underline{u}(0:6) = \begin{bmatrix} 1 & m & m^2 & m^3 & m^4 & m^5 & m^6 \end{bmatrix}^T \qquad (37)$$

$$\underline{u}(0:7) = \begin{bmatrix} \underline{u}(0:6) & m^7 \end{bmatrix}^T \qquad (38)$$

in order to be able to compute the other parameter

$$a^* = \frac{5\underline{v}_1^T \underline{N}\underline{u}(0:7)}{m^* \underline{v}_1^T \underline{M}\underline{u}(0:6)} \qquad (39)$$

and finally the last unknown variable is computed using

$$b_2 = (1 - a^* + a^{*2}/3)m^{*2} - m^* + (1 - l_2)/3 \qquad (40)$$

$$\underline{N} = \begin{bmatrix} 50 & -80 & -490 & 820 & -285 & -85 & 40 & 7 \\ -215 & 800 & 220 & -932 & 227 & 73 & 0 & 0 \\ 615 & -1490 & 675 & 77 & -12 & 0 & 0 & 0 \\ -515 & 915 & -300 & -77 & 0 & 0 & 0 & 0 \\ -255 & -45 & -30 & 147 & 0 & 0 & 0 & 0 \\ 105 & 45 & -75 & 0 & 0 & 0 & 0 & 0 \\ 180 & -180 & 0 & 0 & 0 & 0 & 0 & 0 \\ 120 & 30 & 0 & 0 & 0 & 0 & 0 & 0 \\ 55 & 5 & 0 & 0 & 0 & 0 & 0 & 0 \\ -135 & 0 & 0 & 0 & 0 & 0 & 0 & 0 \\ -15 & 0 & 0 & 0 & 0 & 0 & 0 & 0 \end{bmatrix} \qquad (41)$$

The matrices $\underline{N}$ in (41) and $\underline{M}$ in (42) belong of course to the equation (39), whereas the hex of matrixes $\underline{L}_i$ from quadratic form (36) will be in (43) – (47).

$$\underline{M}=\begin{bmatrix} 25 & -1050 & 1675 & -510 & -265 & 105 & 21 \\ 150 & 2025 & -2760 & 580 & 220 & 0 & 0 \\ -200 & -625 & 510 & 105 & 0 & 0 & 0 \\ 250 & -175 & -35 & 0 & 0 & 0 & 0 \\ -225 & -450 & 435 & 0 & 0 & 0 & 0 \\ -25 & 275 & 0 & 0 & 0 & 0 & 0 \\ -75 & 0 & 0 & 0 & 0 & 0 & 0 \\ 75 & 0 & 0 & 0 & 0 & 0 & 0 \\ 25 & 0 & 0 & 0 & 0 & 0 & 0 \\ 0 & 0 & 0 & 0 & 0 & 0 & 0 \\ 0 & 0 & 0 & 0 & 0 & 0 & 0 \end{bmatrix} \qquad (42)$$

$$\underline{L}_1=\begin{bmatrix} 10 & -40 & 460 & -1172 \\ -80 & 230 & -2627 & 5163 \\ 240 & -500 & 477 & 893 \\ -200 & 180 & 1023 & -2106 \\ 70 & -136 & 6733 & -11992 \\ -990 & 3752 & -10998 & 10552 \\ 950 & -2532 & 2412 & -638 \\ 441 & -2769 & -2112 & 6549 \\ 1580 & -5446 & 8602 & -5740 \\ -1593 & 5547 & 1827 & -5259 \\ -2640 & 7652 & -8278 & 3502 \end{bmatrix} \qquad (43)$$

$$\underline{L}_2=\begin{bmatrix} 6330 & -2900 & 79 & 262 \\ -17780 & 595 & 4574 & -1409 \\ -11015 & 8395 & -2267 & -167 \\ 7045 & -1570 & -105 & 44 \\ 34897 & -2036 & -4222 & 640 \\ -13646 & -4398 & 1604 & 210 \\ -174 & 54 & -86 & 0 \\ -16401 & 456 & 906 & 0 \\ 6603 & 714 & -238 & 0 \\ 5286 & 1584 & 0 & 0 \\ -1096 & -894 & 0 & 0 \end{bmatrix} \qquad (44)$$

$$\underline{L}_3=\begin{bmatrix} 355 & -532 & 7 & 18 & 1 \\ -1214 & 473 & 59 & 0 & 0 \\ 649 & 39 & 0 & 0 & 0 \\ -19 & 0 & 0 & 0 & 0 \\ 859 & 0 & 0 & 0 & 0 \\ 0 & 0 & 0 & 0 & 0 \\ 0 & 0 & 0 & 0 & 0 \\ 0 & 0 & 0 & 0 & 0 \\ 0 & 0 & 0 & 0 & 0 \\ 0 & 0 & 0 & 0 & 0 \\ 0 & 0 & 0 & 0 & 0 \end{bmatrix} \qquad (45)$$

$$\underline{L}_4=\begin{bmatrix} 267 & 906 & -153 & -804 \\ 813 & -2400 & 1137 & 162 \\ 733 & -2735 & -615 & 1252 \\ 1120 & -2726 & 2016 & -278 \\ 6 & -966 & -240 & 594 \\ -6 & 1116 & 390 & -714 \\ 369 & -293 & 356 & 36 \\ -207 & -171 & 99 & 0 \\ -543 & 354 & -54 & 0 \\ -567 & 729 & -261 & 0 \\ -467 & 86 & -56 & 0 \end{bmatrix} \qquad (46)$$

$$\underline{L}_5=\begin{bmatrix} 1533 & 0 & 0 & 0 \\ -456 & 0 & 0 & 0 \\ -550 & 0 & 0 & 0 \\ -557 & 0 & 0 & 0 \\ 0 & 0 & 0 & 0 \\ 0 & 0 & 0 & 0 \\ 0 & 0 & 0 & 0 \\ 0 & 0 & 0 & 0 \\ 0 & 0 & 0 & 0 \\ 0 & 0 & 0 & 0 \\ 0 & 0 & 0 & 0 \end{bmatrix} \quad \underline{L}_6=\begin{bmatrix} -426 & 642 & -138 & 0 \\ 387 & 117 & 0 & 0 \\ 666 & -648 & 0 & 0 \\ 210 & 120 & 0 & 0 \\ 147 & -33 & 0 & 0 \\ 9 & 0 & 0 & 0 \\ -27 & 0 & 0 & 0 \\ -210 & 0 & 0 & 0 \\ 27 & 0 & 0 & 0 \\ -90 & 0 & 0 & 0 \\ 10 & 0 & 0 & 0 \\ -9 & 0 & 0 & 0 \end{bmatrix} \qquad (47)$$

Achieved results can be summarised as follows
Using (6) we can determine coefficients h_k of the normed transfer function (4). Markov's parameters l_k can be determined using the Bôcher formulas (7). In case of the I - controller, only the formula (26) is used. Considering a PI controller we have to create the data vector (29) and find the root of the equation (27) $m^* = 1 - b_1^*$. After substituting it into (31) we can find a^* and determine the other parameter C^* using formula (21). Finally, considering a PID controller, we first have to fill up the data vectors from the relations (32), (33) and (34). Then, we can find the solution of (36) m^* ($0 < m^* = 1 - b_1^* < 1$). Substituting it into (39) we compute a^*, after substituting in (40) we obtain b_2 and from (21) the last constant C. The values of the constants have to be converted using the constants K and τ, to obtain correct results.

4 ILLUSTRATION EXAMPLE

Properties of the proposed method will be now compared with the ones of the well known synthesis methods for linear continuous control circuits: Sartorius and Oldenburg's optimal module criterion and Kessler's symmetrical optimum method, which can be used only for systems without transport delay with transfer function which numerator is only zero order polynomial.

In order to objectively appreciate the control quality of the new method, we used the following transfer function of the controlled plant:

$$\bar{S}(s)=\frac{1}{1+16s+81s^2+146s^3+80s^4} \tag{48}$$

The values of optimal controller parameters for I, PI and PID controllers are shown in the Table 1.

The closed-loop time responses to the unit-step input $x(t)$, as well as the time response of the control variable $u(t)$ for I, PI and PID controllers respectively, are shown in Fig.2. - 7.

Table 1 Optimal controller parametres

Controller type:	I	PI		PID		
Time constant:	C	C	θ	C	θ_1	θ_2
Sartorius:	32	13.1	9.4	5.1	7.8	5.5
Kessler:	32	16	8	6	8	5
New method:	42.2	17.5	9.1	6.8	7.8	5.4

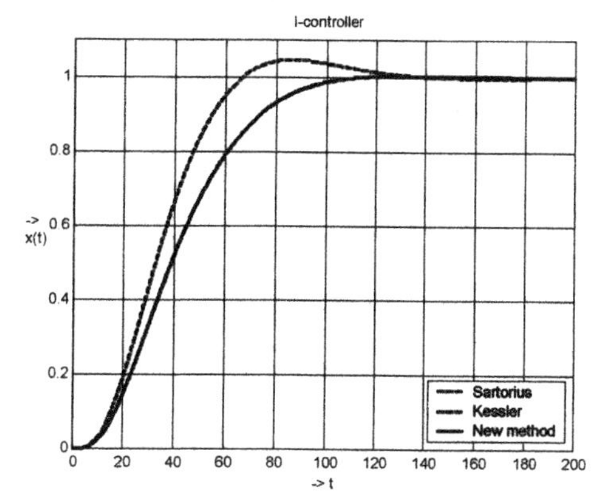

Fig.2. Closed – loop step response (I-controller)

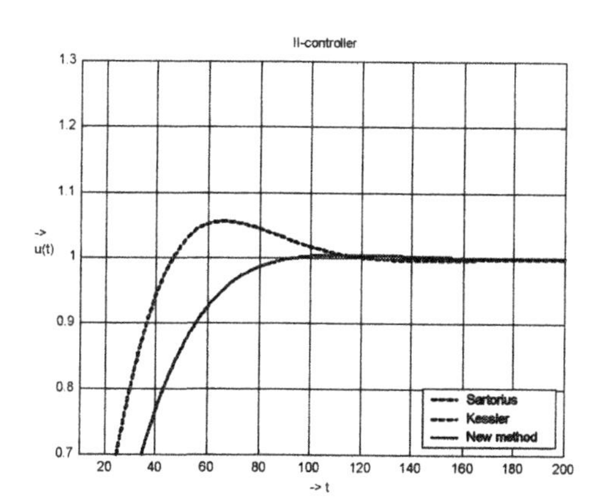

Fig. 3. Time response of the control variable (I- controller)

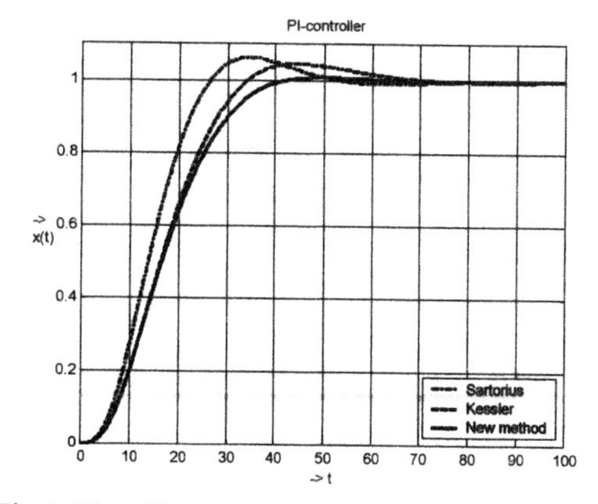

Fig.4. Closed-loop step response (PI-controller)

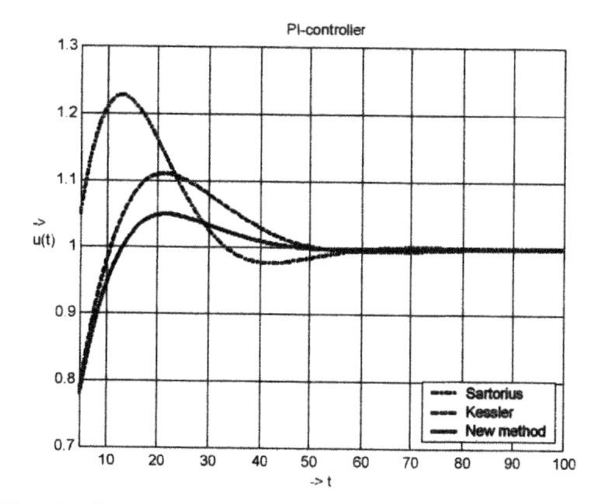

Fig. 5. Time response of the control variable (PI-controller)

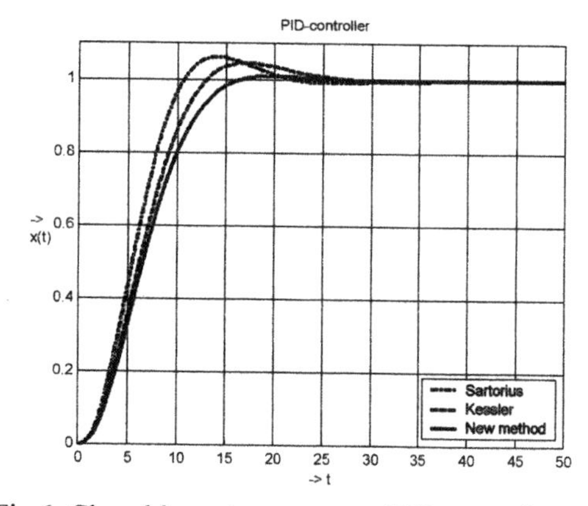

Fig.6. Closed-loop step response (PID-controller)

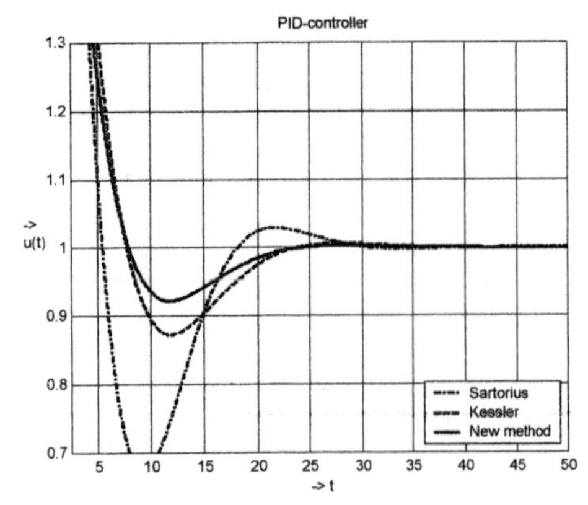

Fig. 7. Time response of the control variable (PID-controller)

By inspection of the figures it is evident, that the parameters of the designed controllers provide responses with a minimum, not overdamped. overshoot.

5 CONCLUSION

The aim of this paper was to present algorithms for solving three nonlinear equations in three unknown parameters. Even though at first sight the algorithm might be found relatively difficult, its programming is relatively simple and needs to be realised just once. Resulting optimal controller parameters are valid for a wide class of controlled plants and give to the control loop outstanding properties.

REFERENCES

Åström, K. J. and T. Hägglund (2000). Benchmark Systems for PID Control, In: *Preprints IFAC Workshop on Digital Control.* Past, Present and Future of PID Control, pp. 181-182

Hudzovič, P. (1999) :Design of Parameter optimized (P)I(D) Controllers, In *Journal of ELECTRICAL ENGINEERING,* VOL. 50, NO. 11-12, 363-375

O'Dwyer, A. (2000). A Summary of PI and PID Controller Tuning Rules for Processes with Time Delay. Part1: PI Controller Tuning rules. In: *Preprints IFAC Workshop on Digital Control.* Past, Present and Future of PID Control, pp. 175-180.

O'Dwyer, A. (2000). A Summary of PI and PID Controller Tuning Rules for Processes with Time Delay. Part1: PID Controller Tuning rules. In: *Preprints IFAC Workshop on Digital Control.* Past, Present and Future of PID Control, pp. 242-247.

Gorez, R., and P. Klán (2000). Nonmodel-based ewplicit Design Relations for PID Controllers. In: *Preprints IFAC Workshop on Digital Control.* Past, Present and Future of PID Control, pp. 141-148.

Klán, P. (2000). Moderní metody nastavení PID regulátorů. Část I: Procesy s přechodovou charakteristikou typu „S". AUTOMA, č. 9.

GENERALIZED ULTIMATE SENSITIVITY CONTROLLER TUNING

Mikuláš Huba and Katarína Žáková

Department of Automation and Control
Faculty of Electrical Engineering and Information Technology
Slovak University of Technology
Ilkovičova 3, 812 19 Bratislava, Slovakia
Email: [huba, katarina.zakova]@elf.stuba.sk

Abstract: The high practical significance of the I_1T_d-plant approximations has been shown by the perhaps most frequently used method for controller tuning by (Ziegler and Nichols, 1942), modified later for sampled data systems by (Takahashi *et al.*, 1971) and by many other authors. This paper extends previous works by developing plant approximations based on the single and double integrator models with dead time. These approximations are then used for tuning of the minimum time pole assignment P- (PD-) controllers designed for single (double) integrator with constrained input.

Keywords: closed-loop identification, delay estimation, P (PD) controllers

1. INTRODUCTION

In the literature it is possible to find various methods for system approximation and controller tuning that are based on measurements at the stability boundary. Perhaps the most widespread one was introduced by (Ziegler and Nichols, 1942) already at the beginning of 40's. They have presented two different methods of the controller setting. While in the first one, the controller parameters are determined on the basis of the transient response measurement, in the "ultimate sensitivity method" it is necessary to bring the system to the stability boundary by raising the gain of a P controller and to measure both the controller critical gain K_u and the period of oscillation P_u. Ziegler and Nichols have designed controllers to guarantee the "quarter amplitude ratio" (quarter amplitude damping). This criterion was also followed by (Parr, 1989) and (Corripio, 1990). Ziegler-Nichols method creates a starting point for numbers of other ultimate cycle tuning rules. Some of them were presented e.g. in (Mantz and Tacconi, 1989), (de Paor, 1993), (Åstrom and Hägglund, 1995). For the sake of simplicity, Ziegler and Nichols have not analyzed the type of time lags involved in the control loop. In this paper, a new method for the plant approximation and the controller tuning of the closed loop is introduced. It is based on the measurement of the period of oscillations and the P- and PD-controller parameters on the stability boundary. For the sake of simplicity, only a dead time delay has been studied.

[1] The work has been partially supported by the Slovak Scientific Grant Agency, Grant No. KG-57-HUBA-Sk1

2. I_1T_D SYSTEM

The single integrator system combined with dead time was analyzed long ago in one of the first textbooks about automatic control (Oldenbourg and Sartorius, 1951).

$$F(s) = \frac{K_S}{s} e^{-T_d s} \tag{1}$$

2.1 *Continuous time case*

Considering a continuous P controller, the closed loop characteristic equation is

$$K_C K_S e^{-T_d s} + s = 0 \qquad (2)$$

The controller design follows a requirement of the fastest monotonic transient response in the presence of the possible maximal controller gain. Such setting can be found by looking for a controller gain extremum that has to satisfy the condition

$$\frac{dK_C}{ds} \stackrel{!}{=} 0 \qquad (3)$$

Solving this equation leads to the optimal controller gain

$$K_{Copt} = \frac{1}{eK_S T_d} \qquad (4)$$

After substituting $s = j\omega$ into (2) one can derive the critical controller gain

$$K_{Ccr} = \frac{\pi}{2K_S T_d} \qquad (5)$$

The corresponding period of oscillation is

$$P_u = \frac{2\pi}{\omega} = 4T_d \qquad (6)$$

For the purpose of identification it is possible to write these relations in the form

$$K_S = \frac{2\pi}{K_{Ccr}\, P_u} \qquad (7)$$

$$T_d = \frac{P_u}{4} \qquad (8)$$

After their substitution into (4) one receives the formula that expresses the relation between the critical and optimal controller gain

$$K_{Copt} = 0.23 K_{Ccr}$$

2.2 *Discrete time case*

Under discrete time P control of the system (1) two principal cases are distinguished. For $t_0 = T_d/T < 1$ (T is a sampling period) one gets the closed loop characteristic equation in the form

$$z\,(z-1) + K_S K_C T\left[(1-t_0)\,z + t_0\right] = 0 \qquad (9)$$

Following the same design criterion as in the continuous case one receives the optimal value of the P controller gain

$$K_{Copt} = \frac{1}{K_S T}\frac{1}{\left(1+\sqrt{t_0}\right)^2}; \quad t_0 = \frac{T_d}{T} < 1 \qquad (10)$$

For $T_d > T$ it is sufficient to consider the transport delay $T_d = kT$ (Žáková and Huba, 2000). Then, the characteristic equation is

$$z^{k+1} - z^k + K_C K_S T = 0 \qquad (11)$$

and the optimal controller gain can be deduced as

$$K_{Copt} = \frac{1}{K_S T}\frac{k^k}{(k+1)^{k+1}}; \quad k = \text{floor}\left(\frac{T_d}{T}\right) \geq 1 \qquad (12)$$

whereby the formula can also be used for non integer ratio T_d/T.

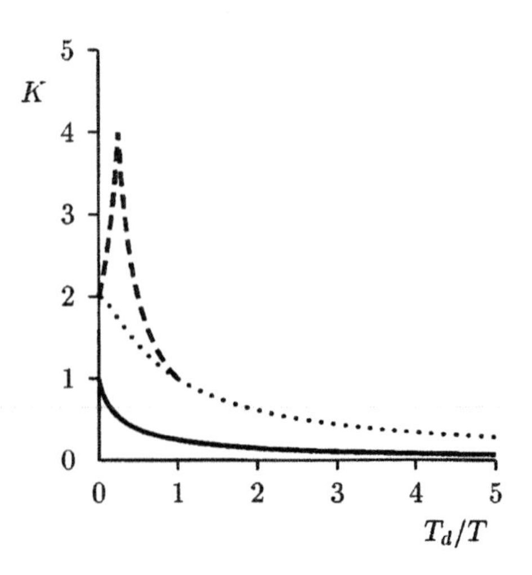

Fig. 1. Optimal and critical controller gain for I_1T_d system. Solid line: K_{opt} for $T_d = kT$; dotted line: K_{cr} for $T_d = kT$; dashed line: K_{cr} for $T_d \leq T/4$ and $T/4 < T_d < T$

The parameters at the stability margin can be derived using the substitution

$$z = e^{j\omega}; \qquad (|z| = 1) \qquad (13)$$

into the equations (9) or (11). The derived formulae can be found in Tab.1.

K_{Ccr}	P_u	
$\frac{2}{K_S T(1-2t_0)}$	$2T$	$t_0 = \frac{T_d}{T} < \frac{1}{4}$
$\frac{1}{K_S T t_0}$	$\frac{2\pi T}{\arccos\frac{2t_0-1}{2t_0}}$	$\frac{1}{4} < t_0 = \frac{T_d}{T} < 1$
$\frac{2}{K_S T}\cos\frac{k\pi}{2k+1}$	$2\,(2k+1)\,T$	$k = \text{floor}(\frac{T_d}{T}) \geq 1$

Tab.1. Critical controller gains

For $t_0 = T_d/T = 1/4$ the critical gain reaches the value $K_{Ccr} = 4/K_S T$ (Fig.1). This leads to the occurrence of the "resonance" effect in the control structure, whereby the amplitude of the output signal excited by a reference step exceeds many times the desired value. Under such conditions it is practically not possible to detect the critical tuning, since either the saturation effect, or the safety requirements become dominant. Similar conclusions have been reached also by (Takahashi *et al.*, 1971) who didn't recommend to use the PID setting for the ratio $t_0 = T_d/T \to 0.25$.

For illustration, the normalized values of all computed critical gains ($K_{cr} = K_{Ccr}K_S T$) are brought in Fig.1.

The formulae for the approximation of system parameters (Tab.2) can be found by substituting (13) into the closed loop characteristic equations (9) and (11). Parameters P_u and K_{Ccr} denote the period of oscillation and the critical controller gain.

K_S	T_d	
$\frac{\sin^2(\frac{\pi}{P_u})}{K_{Ccr}T}$	$\frac{T}{4\sin(\frac{\pi}{P_u})}$	$T_d < T$
$\frac{2\sin(\frac{\pi}{P_u})}{K_{Ccr}T}$	$-\frac{P_u T}{2\pi}\arctan(\varphi)$ $\varphi = -\text{cotan}\left(\frac{\pi}{P_u}\right)$	$T_d > T$

Tab.2. Approximation of I_1T_d system parameters

2.3 *Integral action*

The basic control loop (1st order system + delay + P controller) can be extended by blocks that serve for reconstruction and compensation of an input system disturbance v_i. This creates the integral action of the controller. The new control structure can be seen in Fig.2. The included filter serves for a reduction of possible noise in the structure and to avoid an algebraic loop. In the case when only a linear part of saturation is considered, the whole control structure can be modified to the form that corresponds to PI controller.

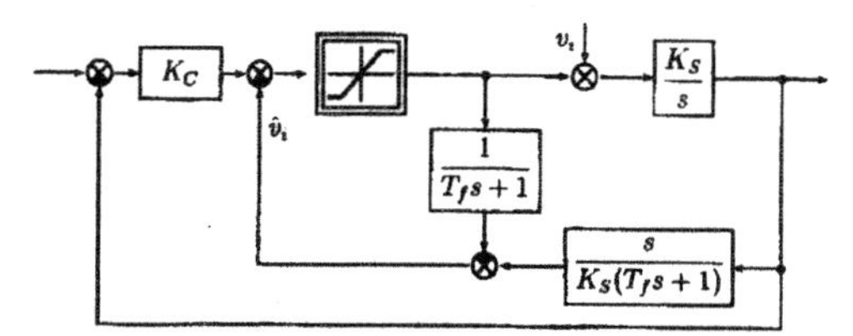

Fig. 2. Reconstruction and compensation of the input system disturbance for I_1T_d system

Again, the optimal loop setting can be looked for from the characteristic equation

$$T_f s^2 + (1 + K_S K_C T_f)\, s e^{-T_d s} + K_S K_C e^{-T_d s} = 0 \quad (14)$$

following the requirement of the dominant triple pole. However, the achieved solutions are complex numbers and the controller and the filter as well are supposed to work with real parameters. Experimentally, it can be shown that the optimal loop dynamics corresponds to parameters, which are derived by approximating the computed complex values by the real part or by the module. The differences between both approximations are negligible and therefore the easier computed real part approximation is taken in the account below. Then, one gets the following parameters for the PI controller setting

$$K_{Copt\ re} = \frac{\sqrt{2}-1}{K_S T_d} e^{-2+\sqrt{2}} \doteq 0.147 K_{Ccr} \quad (15)$$

$$T_{f\ opt\ re} = \frac{2\sqrt{2}+3}{2} T_d \doteq 0.729 P_u \quad (16)$$

However, the question is, if such solution can be still denoted as that one corresponding to the triple real pole.

Discrete equivalent of the controller setting can be derived in a similar way.

3. I_2T_D SYSTEM

The ideas of the previous approximation can be extended for the double integrator with dead time.

3.1 *Continuous time case*

In the case of double integrator system

$$F(s) = \frac{K_S}{s^2} e^{-T_d s} \quad (17)$$

and the PD controller

$$C(s) = K_C (1 + T_D s)$$

the characteristic equation of the closed loop system is

$$s^2 + K_C T_D K_S e^{-T_d s} s + K_C K_S e^{-T_d s} = 0 \quad (18)$$

The optimal setting of PD controller (Huba *et al.*, 1998) follows the same requirements as it was at I_1T_d model - the fastest possible monotonic transient response. Then

$$K_{Copt} = 2\frac{\left(-7+5\sqrt{2}\right)e^{-2+\sqrt{2}}}{K_S T_d^2} \doteq \frac{0.079}{K_S T_d^2}$$

$$T_{Dopt} = \left(3+2\sqrt{2}\right) T_d \doteq 5.828 T_d$$

Similarly as in previous paragraph, the formulae for system approximation can be found. Denoting the controller parameters at the stability margin as K_{Ccr}, T_{Dcr} and the period of oscillation as P_u, the formulae can be expressed using the substitution $s = j\omega$ from the characteristic equation (18). Then

$$K_S = \frac{4\pi^2}{P_u K_{Ccr}\sqrt{4\pi^2 T_{Dcr}^2 + P_u^2}} \quad (19)$$

$$Td = \frac{P_u}{2\pi}\arctan\left(\frac{2\pi}{P_u} T_{Dcr}\right) \quad (20)$$

In Fig.3, the stability boundary of the PD-controller is shown for selected values of T_d.

The main difference in tuning the PD-controller is that one can find infinitely many pairs K_{Ccr}, T_{Dcr} instead of the single critical value of P-controller. The question is, how this can influence the tuning process. On the basis of experimental results it seems that the best choice is to find a critical gain of P controller K_{Ccr}, to set the critical gain of PD controller as $2K_{Ccr}$ and experimentally to find T_{Dcr}.

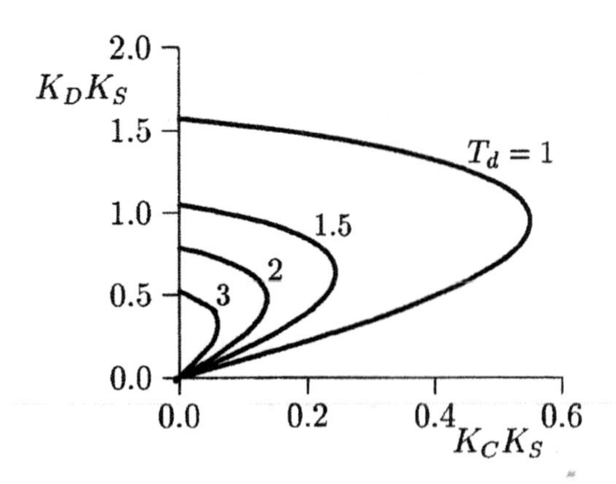

Fig. 3. Critical controller setting for selected values of T_d; $K_D = K_C T_D$

3.2 Discrete time case

The PD controller is considered in the form

$$C(z) = K_C \left(1 + \frac{z-1}{z} T_D\right) \quad (21)$$

For $T_d < T$ ($t_0 = T_d/T < 1$), the discrete counterpart of the system (17) is

$$G(z) = \frac{K_S T^2}{2} \frac{(1-t_0)^2 z^2 + [2t_0(1-t_0)+1]z + t_0^2}{z(z-1)^2} \quad (22)$$

The optimal controller setting can be guaranteed by existence of a triple real pole of the closed loop characteristic equation formed by (21) and (22). Since the solution leads to a 6th order equation, optimal controller parameters can be determined just numerically (see Fig.4). They could be approximated by simple formulae as:

$$K_{Copt} = \frac{2}{K_S T^2} \frac{-1.78t_0^3 + 4.97t_0^2 - 5.75t_0 + 3.5}{100} \quad (23)$$

$$T_{Dopt} = 5.78t_0 + 5.65 \quad (24)$$

whereby $t_0 = T_d/T \in (0,1)$

For $T_d = kT$ ($k = \text{floor}(T_d/T) \geq 1$) the discrete counterpart of the system (17) is as follows

$$G = \frac{K_S T^2}{2} \frac{z+1}{z^k (z-1)^2} \quad (25)$$

The optimal controller setting (Fig.4) can be found from the corresponding close loop characteristic equation

$$z^{k+1}(z-1)^2 + \frac{K_S K_C T^2}{2}\left[(1+T_D) z^2 + z - T_D\right] = 0 \quad (26)$$

as the solution that guarantees a triple real pole. It ensures the requirement of monotonous transient response at the maximal possible velocity

$$K_{Copt} = -\frac{2}{K_S T^2} \frac{h_8 z^{k+1}}{k} \quad (27)$$

$$T_{Dopt} = \frac{k(k+1)(2k - h_2 z) z - k^2 h_1}{2h_8} \quad (28)$$

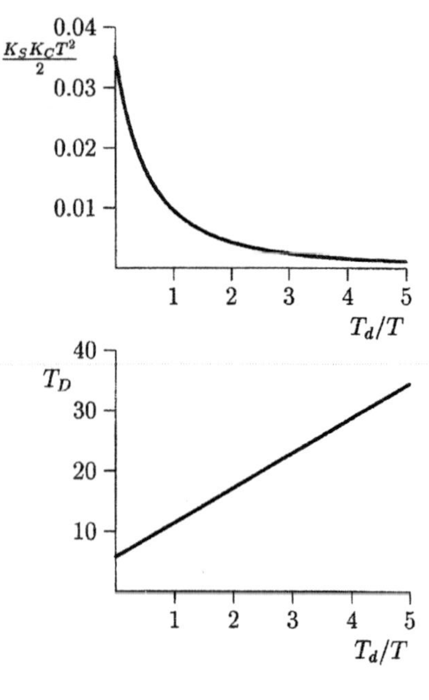

Fig. 4. Optimal controller setting for $I_2 T_d$ model

whereby

$$z = -\frac{1}{4}\frac{6k+6}{k^2+3k+2} + \frac{1}{6}\sqrt{3h_6} + \frac{1}{6}\frac{\sqrt{6h_7}}{\sqrt[4]{(k+1)h_4 h_5}}$$

and

$$h_1 = k - 1; \qquad h_2 = k + 2$$
$$h_3 = 4k^2 h_2^2 (kh_2 + 9) + 135(kh_2 + 1)$$
$$h_4 = \sqrt[3]{h_3 + 3(k+1)\sqrt{3}\sqrt{2h_3 - 27(kh_2+1)}}$$
$$h_5 = 4h_2\sqrt[3]{4}(kh_2+3)^2 + (4kh_2^2 + 15h_2 - 27)h_4 + +2\sqrt[3]{2}h_2 h_4^2$$
$$h_6 = \frac{2h_4\sqrt[3]{2}}{(k+1)h_2} + \frac{4kh_2^2 + 15h_2 - 27}{(k+1)h_2^2} + +\frac{4\sqrt[3]{4}(kh_2+3)^2}{(k+1)h_2 h_4}$$
$$h_7 = \frac{1}{2}h_2\sqrt{h_6}\left[3(4kh_2^2 + 15h_2 - 27)h_4 - h_5\right] + +9\sqrt{3}(2k^2 + 5k + 11)h_4$$
$$h_8 = h_1(k+1)(h_2 z + 6)z^2 - (k+3)(3kh_1 + 2)z + +6 + 2k^2 h_1$$

To derive formulae for the plant approximation the characteristic equation that results from (21) and (22) or (25) respectively has to be rewritten using the substitution (13). The resulting complex equation can be divided into two equations that correspond to its real and imaginary part. Hence the formulae for the system approximation can be found.

For $T_d < T$ one gets

$$K_S = \frac{8\left[1 - T_{Dcr} + 4T_{Dcr}\tau_1\right]\tau_1}{K_{Ccr}T^2(1 - 2t_{0cr})\left[4T_{Dcr}(T_{Dcr}+1)\tau_1 + 1\right]} \quad (29)$$

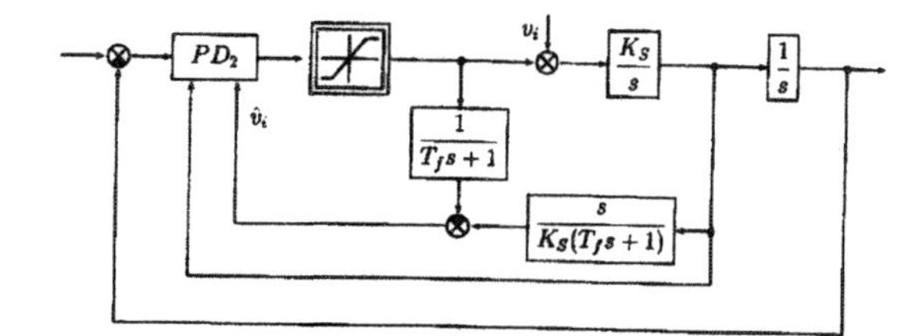

Fig. 5. Reconstruction and compensation of the input system disturbance for I_2T_d system

$$T_d = \frac{T}{4}\frac{4T_{Dcr}\tau_1 + 1 - \sqrt{\tau_2}}{[1 - T_{Dcr} + 4T_{Dcr}\tau_1]\,\tau_1} \qquad (30)$$

$$\begin{aligned}\tau_1 &= \sin^2(\pi T/P_u)\\ \tau_2 &= 64T_{Dcr}^2\tau_1^3 + 32T_{Dcr}(1 - T_{Dcr})\tau_1^2\\ &\quad +4\left(1 - T_{Dcr} + 2T_{Dcr}^2\right)\tau_1 + 1\end{aligned}$$

and for $T_d > T$ it holds

$$K_S = \frac{4\tan\left(\frac{\pi T}{P_u}\right)\sin\frac{\pi T}{P_u}}{K_{Ccr}T^2\sqrt{4T_{Dcr}(1+T_{Dcr})\sin\left(\frac{\pi T}{P_u}\right)+1}} \qquad (31)$$

$$T_d = \frac{[P_u - 4T]\,\pi - P_u\arccos\left(x\sin\left(\frac{\pi T}{P_u}\right)\right)}{2} \qquad (32)$$

$$x = \frac{\left[2(1+T_{Dcr})\cos\left(\frac{2\pi T}{P_u}\right) - 1 - 2T_{Dcr}\right]}{\sqrt{4T_{Dcr}(1+T_{Dcr})\sin\left(\frac{\pi T}{P_u}\right)+1}}$$

3.3 *Integral action*

The integral action of the controller is again based on the reconstruction and compensation of the input system disturbance (Fig.5). In the proportional zone of control, the whole control structure can be modified to the PID controller.

The requirement of the time optimal control algorithm corresponds to the quadruple dominant pole of the characteristic equation. It leads to the following controller setting

$$K_{Copt} = \frac{0.0659}{K_S T_d^2} = 0.0659\frac{\sqrt{39.478\left(\frac{T_{Dcr}}{P_u}\right)^2+1}}{\arctan^2\left(6.283\frac{T_{Dcr}}{P_u}\right)}K_{Ccr} \quad (33)$$

$$T_{f\ opt} = 5.4496T_d = 0.867P_u\arctan\left(6.283\frac{T_{Dcr}}{P_u}\right) \quad (34)$$

$$T_{Dopt} = 4.871T_d = 0.775P_u\arctan\left(6.283\frac{T_{Dcr}}{P_u}\right) \quad (35)$$

In previous, it was derived how to tune the linear PD and PID controller. However, in practice it is recommended to use minimum time pole assignment controllers (MTPA controllers) (Huba and Bisták, 1999) that takes account of the control signal constraints. In the continuous case, the corresponding controller pole (Žáková and Huba, 2000) can be determined as

$$\alpha_e = -\frac{1}{2}K_S K_{Copt} T_{Dopt} \qquad (36)$$

4. EXAMPLE

Let's consider the system described by the following transfer function

$$F(s) = \frac{1}{(s+1)(\alpha s+1)(\alpha^2 s+1)(\alpha^3 s+1)} \qquad (37)$$

where $\alpha = 0.2$. It can be brought to the stability boundary by P controller with the gain $K_{Ccr} = 30.24$. The corresponding period of oscillation is $P_u = 0.562$. Firstly, the system is controlled by PI controller whose output is constrained to $U_{\min} = -2$, $U_{\max} = 2$. At time $t = 8$ a step input disturbance $v_i = 0.5$ occurs.

The controller parameters computed according to (15) and (16) are $K_{Copt} = 4.447$ and $T_{f\ opt} = 0.409$. The simulation results are presented in Fig.6. The quality of process control from the set point step and disturbance rejection can be compared with some other methods. The controller set according to (Parr, 1989) leads to the oscillatory responses. Controller proposed by (Ziegler and Nichols, 1942) has long settling time in comparison with the rest of methods. Quite good results are achieved by (Tyreus and Luyben, 1992) and (Pessen, 1994). The best simulation results are illustrated in Fig.6. The controller based on the derived I_1T_d approximation of the system (15, 16) is able to control the system without overshoot (in step reaction) with the settling time comparable with other methods.

Next, the system was controlled by PID controller that was set according to the approximation by the I_2T_d model derived from measurements at the stability boundary. Following the critical gain of P controller ($K_{Ccr} = 30.24$), the system was brought to the stability margin using the controller setting $K_{Ccr} = 2 * 30.24 = 60.48$ and $T_{Dcr} = 0.021$. The corresponding period of oscillation was $P_u = 0.394$.

The simulation results are presented in Fig.7. They were accomplished using MTPA PID controller (Huba and Bisták, 1999), (Žáková and Huba, 2000) whose pole value is $\alpha = -7.794$. The filter time constant is $T_{f\ opt} = 0.112$. In comparison with PI controller, reasonably faster transient responses are achieved.

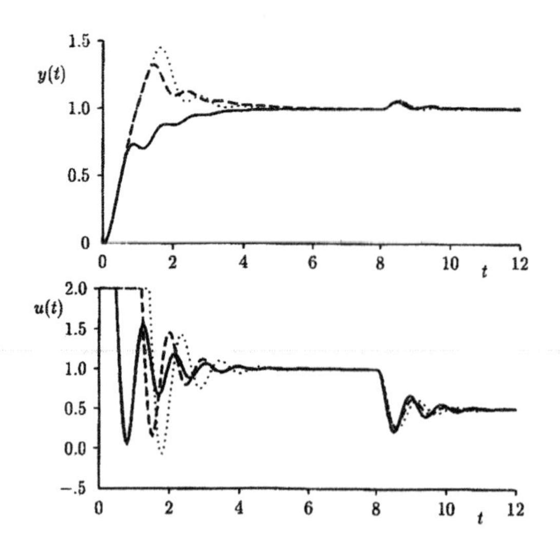

Fig. 6. Transient responses and corresponding control signal for the proposed method - solid line; Tyreus-Luyben method - normal dashed line; Pessen method (dotted line).

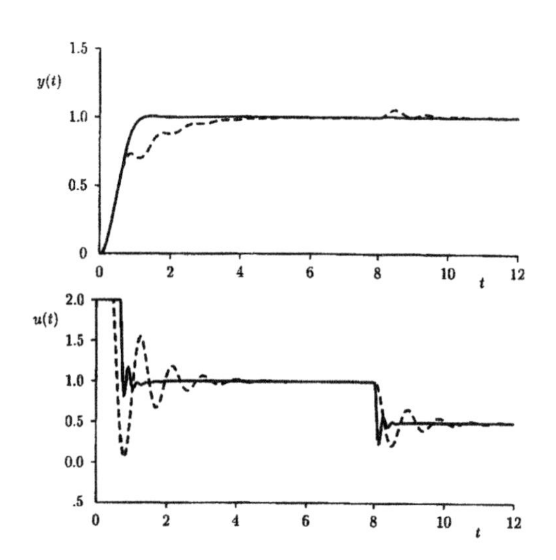

Fig. 7. Output and corresponding control signal (PI controller based on the I_1T_d model - dashed line; PID controller based on the I_2T_d model - solid line)

5. CONCLUSION

A new method for the plant approximation and the controller tuning has been proposed. Although the derived formulae can seem rather complicated in comparison with other methods introduced in the literature, they can be easily implemented by computer. The achieved results confirm their importance. The method can be considered as generalization of the method by Ziegler and Nichols. It is shown that the basic idea of this method can also be extended to higher order approximations. The analytically derived results were illustrated by an example.

6. REFERENCES

Åstrom, K. J. and T. Hägglund (1995). *PID Controllers: Theory, Design, and Tuning.* 2nd ed.. Instrument Society of America.

Corripio, A. B. (1990). *Tuning of Industrial Control Systems.* Instrument Society of America, Research Triangle Park, North Carolina.

de Paor, A. M. (1993). A fiftieth anniversary celebration of the Ziegler-Nichols PID controller. *International Journal of Elect. Enging. Educ.* **30**, 303–316.

Huba, M. and P. Bisták (1999). Dynamic classes in the PID control. In: *Proceedings of the American Control Conference.*

Huba, M., P. Bisták, Z. Skachová and K. Žáková (1998). P- and PD- controllers for i1 and i2 models with dead time. In: *Proceedings of the 6th IEEE Mediterranean Conference on Theory and Practice of Control and Systems* (A. Tornambe, G. Conte and A. M. Perdon, Eds.). World Scientific Publishing Co.. pp. 514–519.

Mantz, R. J. and E. J. Tacconi (1989). Complementary rules to Ziegler and Nichols' rules for a regulating and tracking controller. *International journal of control* **49**(5), 1465–1471.

Oldenbourg, R.C. and H. Sartorius (1951). *Dynamik Selbsttätiger Regelungen.* 2.auflage ed.. R.Oldenbourg-Verlag. München.

Parr, E. A. (1989). *Industrial Control Handbook.* Vol. 3. BSP Professional Books.

Pessen, D. W. (1994). A new look at PID controller tuning. *Transactions of the ASME - Journal of Dynamic Systems, Measurement and Control* **116**, 553–557.

Takahashi, Y., C.S. Chan and D.M. Auslander (1971). Parmetereinstellun bei linearen DDC-algorithmen. *Regelungstechnik und Prozess-Datenverarbeitung* **19**, 237–244.

Tyreus, B. D. and W. L. Luyben (1992). Tuning PI controllers for Integrator/Dead time processes. *Industrial Engineering Chemistry Research* **31**, 2625–2628.

Žáková, K. and M. Huba (2000). PD- controller tuning based on I2Td and I2T1 models. In: *Proceedings of CASYS'2000, 4th Int. Conference on Computing Anticipatory Systems.* Liege, Belgium.

Ziegler, J.G. and N.B. Nichols (1942). Optimum settings for automatic controllers. *Transaction of the A.S.M.E.* pp. 759–768.

ADAPTIVE SELF-TUNING PIDD² ALGORITHM

Mikuláš Alexík

University of Žilina, Dept. of Technical Cybernetics, Slovak Republic

Abstract: This paper describes the PIDD² control algorithm, which has the adaptive self-tuning performance. Algorithm description is extended on antiwindup and state space version of algorithm, which together with PIDD² synthesis was derived by author. The synthesis can be made in a continuous or discrete domain in control loop with proportional third order behavior of controlled plant. However, the algorithm can be applied for plants with higher order and plants with non-minimal phase or transport delay, which is shown in simulation experiments. The identification is followed by synthesis, which is in discrete domain with compensation of sampling interval effect to control loop response quality. Simulation experiments were performed on analogue model of controlled plants as hybrid simulation in real time and are commented in the paper. *Copyright © 2003 IFAC*

Keywords: PIDD² control, sampling interval, control synthesis, simulation.

1. INTRODUCTION

This paper describes continuous and discrete synthesis of PIDD² control algorithm including Antiwindup and state space version of algorithm, which all was derived by author. This algorithm is "the best" of linear algorithms for proportional third order plant, which works with output of controlled process for the control loop depict at Fig. 1. Not only parameters but also structure of the control algorithm outcome from analytic synthesis for demanded quality of control loops responses. However, the algorithm can be applied for plants with higher order with transport delay or non-minimal phase, as is shown in simulation experiments.

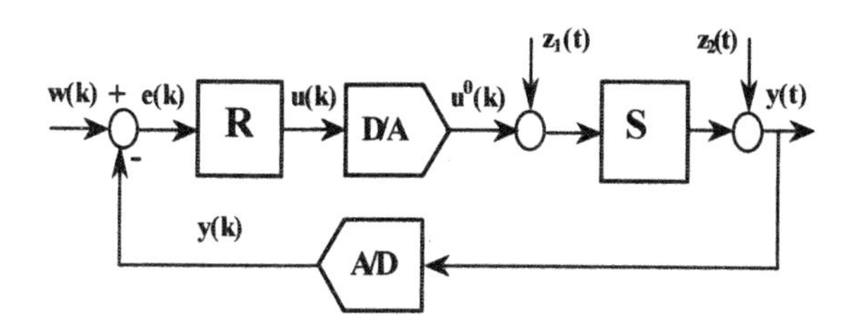

Fig. 1. Block scheme of feedback control loop.

Adaptive self-tuning control loop was executed by continuous identification with controller synthesis following procedure as is shown in Fig. 2. For experimental evaluation and verification of the control algorithm in real time hardware in loop simulation, which replaces real plant with its analogue model (A/D and D/A converters) was used. HIL simulation increases motivation of students and research workers during verification of control algorithms.

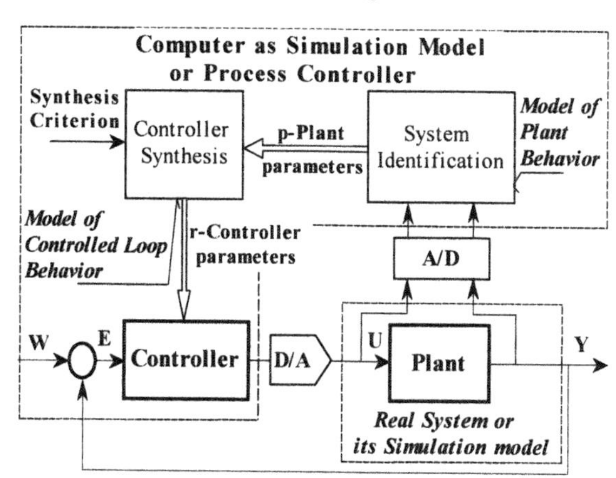

Fig. 2. Hardware in loop simulation of adaptive control loop with on - line identification

Syntheses of adaptive control algorithm from data calculated by continuous identification were described in (Alexík, 1997; Bobál et al, 2000; Kozák, 1994). Synthesis of PID control algorithms was described also in (Vitečková, 1995).

The paper is organised as follows. Section 2 describes demanded quality of control loop responses used by controller synthesis. Section 3 describes continuous and discrete synthesis of PIDD^2 algorithm. Analogue model of the controlled plant used for HIL simulation is described in Section 4. In Section 5 there are described real time simulation experiments according to the block scheme depicted in Fig. 1. The paper ends with conclusion and outlook.

2. DEMANDED QUALITY OF CONTROL LOOP RESPONSE.

The aim of the synthesis of feedback is to propose structure and parameters of the controller. Pole placement control resides in assigning by finding such controller that satisfies the prescribed feedback control dynamics. Consider the best-prescribed dynamics of feedback. The behaviour of the feedback under consideration is described by transient response depicted in Fig. 3. Because all controlled plants are dynamic, it is impossible to consider ideal response (0- in Fig. 5), only responses 1 to 3 for prescribed dynamics of feedback. The best real dynamics of feedback, depicted as response "1" in Fig. 3, is described as the first order transfer function.

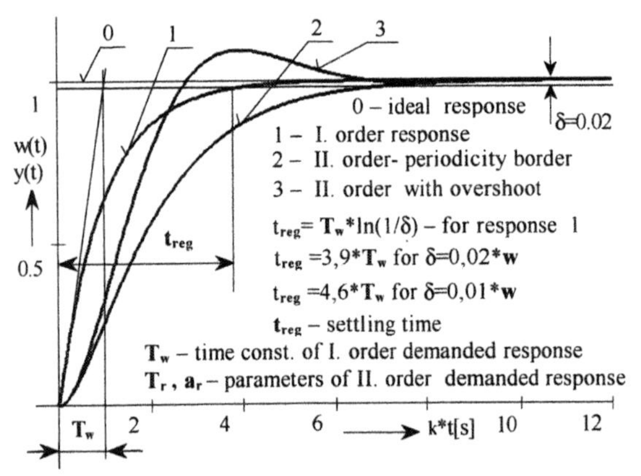

Fig. 3. Transient responses of feedback prescribed dynamics.

3. CONTINUOUS AND DISCRETE PIDD^2 ALGORITHMS.

Consider III. order controlled plant described by equation (1). Three real poles or one real pole and complex poles describe plant dynamics. If we consider prescribed dynamics of feedback as the second order transfer function (case 2 or 3 in Fig. 3), than from control design we derived PID algorithm. If we consider prescribed dynamics of feedback as the first order transfer function (case 1 Fig. 3) than the algorithm PIDD^2 can be carried out according to the equations (2) to (4). The advancement against PID is the possibility to choose the settling time of feedback response by setting time constant "T_w". Synthesis advance for continuous algorithm is clear from equation (2) to (4). Results of this advance are structure and parameters of control algorithm (5).

$$S_3(s)=\frac{K}{(T_3*s+1)*(T^2*s^2+2aT*s+1)} \quad (1)$$

$$Fw(s)=\frac{S_2(s)*R(s)}{1+S_2(s)*R(s)}=\frac{1}{T_w*s+1} \quad (2)$$

$$R(s)=\frac{1}{KT_w\ s}*(T_3*s+1)(T^2*s^2+2aT*s+1) \quad (3)$$

$$R(s)=\frac{2aT+T_3}{K\ T_w}*\left(1+\frac{1}{(T_3+2aT)*s}+\frac{(T^2+T_3 2aT)}{(T_3+2aT)}*s+\frac{T_3T^2}{(T_3+2aT)}*s^2\right) \quad (4)$$

$$R(s)=r_p*\left(1+\frac{1}{T_i*s}+T_d*s+T_{d2}*s^2\right) \quad (5)$$

In this continuous case the influence of time constant T_w to the settling time is described in Figure 3 (if manipulated variable is without constraint) In the case of constraint of manipulated variable Anti-WindUp algorithm will be described for discrete domain. For derivation of discrete PIDD^2 algorithm, let us assume third order plant with real poles. Results will be valid also for plant with one real and complex pole. For conversion of third order transfer function (6) to discrete domain, relations' (7) and (8) are valid.

$$S(s)=\frac{K}{(T_1*s+1)(T_2*s+1)(T_3*s+1)}=\frac{b_1z^2+b_2z+b_3}{z^3+a_1z^2+a_2z+a_3} \quad (6)$$

Calculation from continuous to discrete domain:

$$a_1=-(D_1+D_2+D_3);\quad a_3=-(D_1D_2D_3) \quad (7)$$
$$a_2=(D_1D_2+D_1D_3+D_2D_1);\quad D_i=\exp(T_0/T_i)$$

$$b_1=K\left[1-\frac{T_1^2D_1}{(T_1-T_2)(T_1-T_3)}+\frac{T_2^2D_2}{(T_1-T_2)(T_2-T_3)}-\frac{T_3^2D_3}{(T_1-T_3)(T_2-T_3)}\right] \quad (8)$$

$$b_2=K\left[a_2+\frac{T_2^2(D_1D_3+D_1+D_3)}{(T_1-T_2)(T_2-T_3)}-\frac{T_1^2(D_2D_3+D_2+D_3)}{(T_1-T_2)(T_1-T_3)}-\frac{T_3^2(D_1D_2+D_1+D_2)}{(T_1-T_3)(T_2-T_3)}\right]$$

$$b_3=K\left[a_3+\frac{T_1^2D_2D_3}{(T_1-T_2)(T_1-T_3)}-\frac{T_2^2D_1D_3}{(T_1-T_2)(T_2-T_3)}+\frac{T_3^2D_1D_2}{(T_2-T_3)(T_1-T_3)}\right]$$

The first order prescribed dynamics of feedback in discrete domain can be described as (9). From derived equation (10) of controller Z transfer function of controller can be formally written in form (11). If we compare (11) with Z transfer function of discrete PIDD^2 algorithm (12) then obtained algorithm parameters are in form of (13), which are parameters of controlled plants (6) in discrete domain.

$$F_w(z)=\frac{S_3(z)R(z)}{1+S_3(z)R(z)}=\frac{(1-D_w)}{z-D_w},\ D_w=e^{-(T_0/T_w)} \quad (9)$$

$$R(z)=\frac{Fw(z)}{[1-Fw(z)]*S_3(z)}=$$
$$\frac{(1-D_w)[z^3+a_1z^2+a_2z+a_3]}{(b_1z^2+b_2z+b_3)(z-1)} \quad (10)$$

Equation (8) can be expanded into form :

$$R(z)=\frac{(1-D_w)}{B(z)(z-1)}\left[\begin{matrix}\lim(b_i)z^2(z+1)+S(a_i)z^2(z-1)\\+T_{dA}z(z-1)^2+T_{d2A}(z-1)^3\end{matrix}\right]=$$

$$=\frac{(1-D_w)*S(a_i)z^2}{B(z)}\left[\begin{matrix}1+\frac{0.5\lim(b_i)}{S(a_i)}\frac{z+1}{z-1}+\\+\frac{T_{dA}}{S(a_i)}\frac{z-1}{z}+\\+\frac{T_{d2A}}{S(a_i)}\left(\frac{z-1}{z}\right)^2\end{matrix}\right] \quad (11)$$

$$R(z)=r_p\left[\begin{matrix}1+\frac{T_0}{2T_i}\left(\frac{z+1}{z-1}\right)+\frac{T_d}{T_0}\left(\frac{z-1}{z}\right)+\\+\frac{T_{d2}}{T_0^2}\left(\frac{z-1}{z}\right)^2\end{matrix}\right] \quad (12)$$

Where the expanding constants are : $T_{dA}=(a_2+3a_3)$

$$\lim(b_i)=(1+a_1+a_2+a_3);\quad T_{d2A}=-a_3$$
$$S(a_i)=(0.5-0.5a_1-1.5a_2-2.5a_3) \quad (13)$$

The expressions for proportional gain of controller in equation (11) have to be aligned. If we consider the steady state of feedback then proportional gain of controller can be described by equation (15). Times constants of algorithm by consider (13) are obtained in form (16) and depend on plant parameters and sampling interval only. Proportional gain of control algorithm depends on plant parameters and time constant "T_w" of feedback demanded settling time. Time constant T_w setting necessity constitutes the main weakness of this design.

$$\lim_{z\to1}B(z)=\lim_{z\to1}(b_1z^2+b_2z+b_3)=$$
$$=K(1+a_1+a_2+a_3)=K*\lim(b_i)=$$
$$=K\left[\begin{matrix}1-(D_1+D_2+D_3)+\\+(D_1D_2+D_1D_3+D_2D_3)-D_1D_2D_3\end{matrix}\right] \quad (14)$$

$$r_p=\lim_{z\to1}\left[\frac{(1-D)T_{iB}z^2}{B(z)}\right]=\left[\frac{(1-D_w)}{K}*\frac{S(a_i)}{\lim(b_i)}\right]=$$
$$=r_{pp}*\frac{S(a_i)}{\lim(b_i)} \quad (15)$$

$$T_i=\frac{(0.5-0.5a_1-1.5a_2-2.5a_3)}{(1+a_1+a_2+a_3)}*T_0=\frac{S(a_i)}{\lim(b_i)}*T_0$$
$$T_d=\frac{(a_2+3a_3)*T_0}{(0.5-0.5a_1-1.5a_2-2.5a_3)}=\frac{(a_2+3a_3)}{S(a_i)}*T_0 \quad (16)$$
$$T_{d2}=\frac{-a_3*T_0^2}{(0.5-0.5a_1-1.5a_2-2.5a_3)}=\frac{-a_3}{S(a_i)}*T_0^2$$

For computing between continuous and discrete version of PIDD^2 control algorithm equation (17) to (19) are valid. Especially equations (19) are simple and are as part of nominator of DB(n) algorithm.

$$u(k)=u(k-1)+q_0e(k)+q_1e(k-1)+q_2e(k-2)+$$
$$+q_3e(k-3) \quad (17)$$

$$I=r_i*T_0;\ D=(r_d/T_0);\ D_2=(r_{d2}/T_0^2) \quad (18)$$

$$q_0=(r_p+0.5*I+D+D_2)=(r_{pp}/\lim(b_i);$$
$$q_1=(0.5*I-r_p-2D-3D_2)=q_0*a_1;$$
$$q_2=(D+3D_2)=q_0*a_2;\ q_3=-D_2=q_0*a_3 \quad (19)$$

If $u_d^0(k-1)=[u(k-1)-u^0(k-1)]$, then

$$u(k)=u(k-1)+q_0e(k)+q_1e(k-1)+q_2e(k-2)+$$
$$+q_3e(k-3)-\left(\frac{1}{q_0}\right)\left\langle\begin{matrix}(q_0+q_1)u_d^0(k-1)+\\q_2u_d^0(k-2)+q_3u_d^0(k-3)\end{matrix}\right\rangle \quad (20)$$

State version of PIDD^2 algorithm with AntiWindUp (21) (valid also for PID) derived by author, is better for application as classical AntiWindUp algorithm (20), because there are less arithmetic operation..

$$u(k)=q_0x_3(k)+q_1x_3(k-1)+q_2x_2(k-1)+$$
$$+q_3x_1(k-1) \quad (21)$$
$$x_3(k)=x_3(k-1)+e(k)-(1/q_0)[u(k)-u^0(k)]$$

$u^0(k)$ - Output from D/A converter.

The algorithm in the discrete domain was derived for transfer function (6), which in continuous domain is adequate for more structure of III order transfer function not only for (1) (see Fig. 12 and Fig. 13).

The examples for transfer function with real pole and complex poles are documented at Fig. 4 and Fig.5.

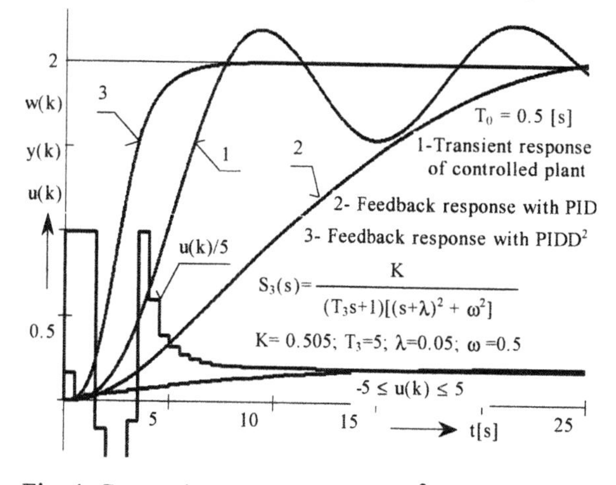

Fig. 4. Comparison of PID and PIDD^2 algorithm.

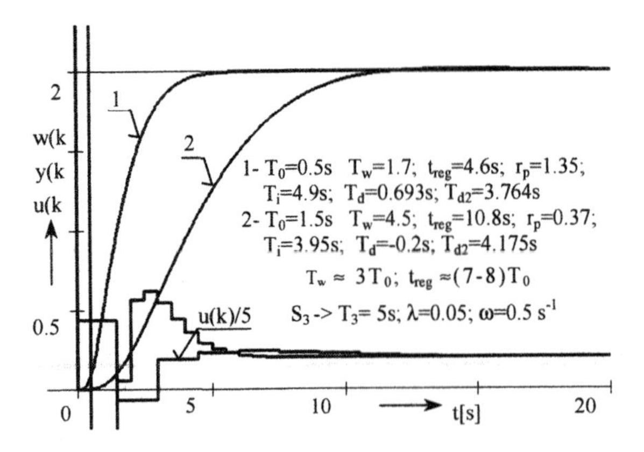

Fig. 5. Transient responses with PIDD^2 algorithm.

4. ANALOGUE MODEL OF THE SYSTEM.

For evaluation of control algorithms, the hybrid simulation in comparison with the digital one is more suitable. In our case, it is the continuous process modelled with the continuous model working in real time. The technical design of the continuous model enables us to model the proportional linear system of the first to the sixth order and to change the parameters and the structure of the controlled system, which is suitable for the evaluation of robustness of the control algorithm. Fig. 6 shows and describes the model. During the transient state, it is possible to change the relative damping for the second order system (S_1) or time constant of the second or first order and the total gain or order of the system from the first to sixth orders.

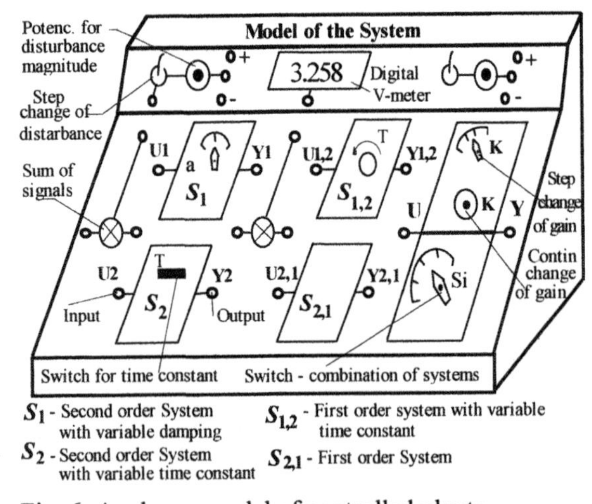

Fig. 6. Analogue model of controlled plants.

It is also possible to connect model as the MIMO system with 2 inputs/outputs. The individual models can be connected to the input of the gain block, marked, as "U" in Fig. 6 and then change the system gain during the transient state in continuous or step manner. The switch *Si* enables to combine individual transfer functions in such a way that the "Y" output has dynamic characteristics of the proportional system until the sixth order. This kind of model of the controlled plant enables us to change individual time constants, gains and structure of the controlled system in a continuous or step-by-step manner

Transient responses of second and fifth order system measured from previously described analogue model of the controlled plant are shown in Fig. 7. Real time simulation experiments described in next Section were mainly realised with fifth order model, which response is depicted as "3" in Fig. 7. All experiments were performed in real time with the application of A/D and D/A converters and the simulations can be repeated when it is demanded.

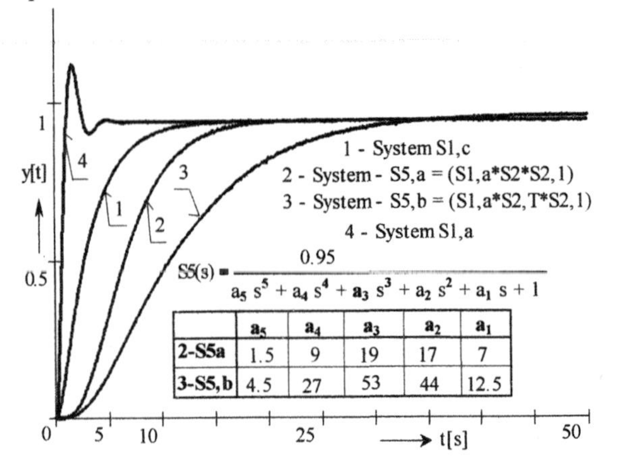

Fig. 7. Transient responses of the analogue model.

5. SIMULATION EXPERIMENTS RESULTS.

Simulation experiments with control algorithms, described in this section, are adaptive control experiments with continuous identification. In principle, the on-line mathematical model of controlled plant and on-line synthesis of control algorithm for demanded quality of control process were suggested. This advance is called Self-Tuning Control (STC) control because the parameters of the control systems are adjusted after the calculation of plant parameters. Identification algorithm REDIC (Kulhavý and Kárny, 1984) was used in all experiments. Comparison of computer simulation and hybrid simulation of adaptive PIDD^2 control, including identified parameter tuning by hybrid simulation, is shown in Fig. 8.

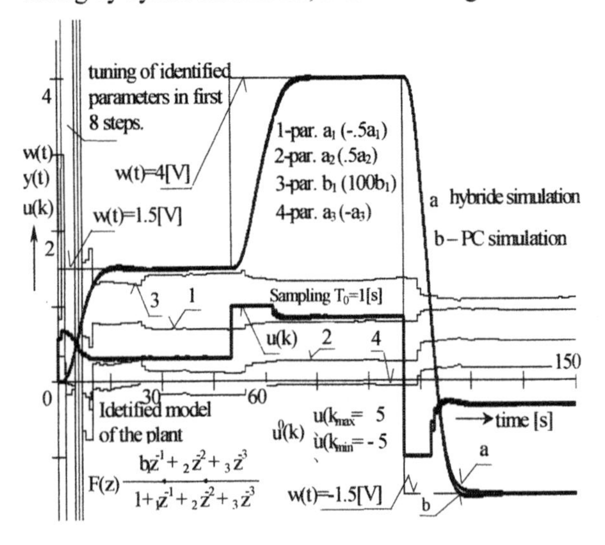

Fig. 8. Comparison of digital and hybrid simulation of adaptive PIDD^2 control

Controlled plant is S5b from Figure 7 and controller synthesis was realised for third order-identified model by algorithm (20).

Also next two Figures show simulation experiments with described algorithm. In the Fig. 9 hybrid simulation with fifth order plant for three different sampling intervals is shown. From Fig. 8 as well as from Fig. 9 is evident, that control algorithm described by equation (17) to (20) enables good quality of controlled loop behavior for several sampling intervals and higher orders of controlled plant as were assumed for controller synthesis. Controller synthesis is independent from sampling interval, as is evident form (17) to (20). However, two problems must be solved completely: continuous identification algorithm, applicable for wide scale of sampling interval and for identification of variable transport delay and convenient continuous calculation of time constant "Tw" which define demanded quality of control loop response.

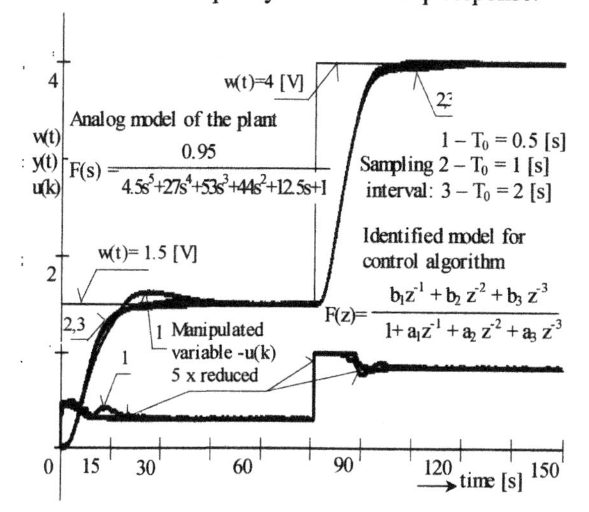

Fig. 9. On –line adaptive control of the fifth order - analogue model +PC +A/D&D/A.

The next simulation experiment shown in Fig. 10 and 11 is also comparison of digital with hybrid simulation of adaptive PIDD^2 control of second and fifth order plant.

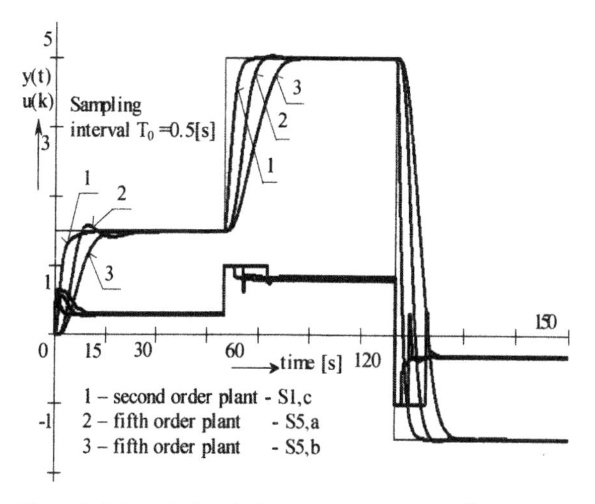

Fig. 10. Digital simulation –adaptive PIDD^2 control of second and fifth order plant

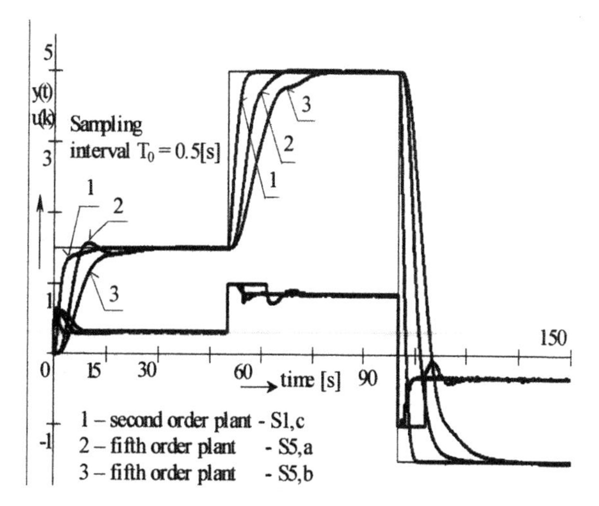

Fig. 11. Hybrid simulation –adaptive PIDD^2 control of second and fifth order plant

As can be seen from both figure algorithms it can be applied also for second order system and transient response quality in the both simulations experiments (digital and hybrid) are very similar. However there is the problem with continuous calculation of time constant "Tw", which is not resulted in a satisfying way and will have to be explained in another subscription.

Specific problems are also with the start of adaptation. In the first seven to several tenth steps (it depends from error of identification, which is computed), PD algorithm with varying gains was applied. In this case, there is a problem with adaptive computing of gains and identification error depending on sampling interval and identified parameter of plant. This problem is not resolved at this time well and coefficient of varying gains and an error of identification when adaptation process is started have to be taken ahead of start of simulation experiment. Simulation experiment with constant gains and constant identification error for start adaptation in all the response are depicted in Fig. 12, where there are hybrid simulation experiments with the same analogue model of plants as in Fig. 11 but with different sampling period.

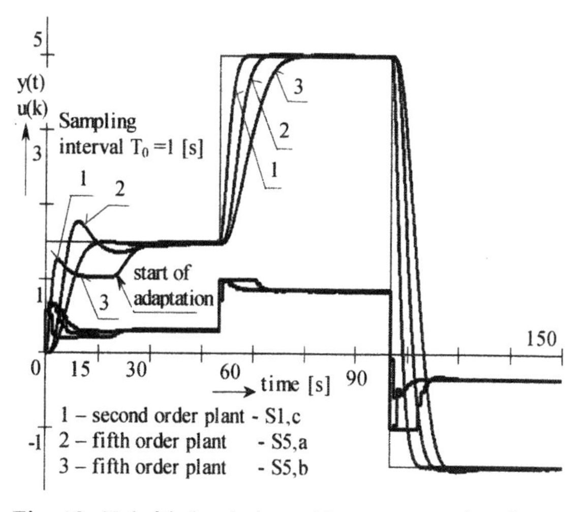

Fig. 12. Hybrid simulation with constant gain of start.

The next simulation experiment shown in Fig. 13 is digital (not hybrid) simulation of adaptive control of plant with non-minimal phase for four different sampling intervals. To control such a type of plant is very difficult, but from Fig. 11 it is evident, that described PIDD^2 adaptive algorithm provides very good quality (close to time-optimal) of controlled loop response. The start of adaptation process (time interval 0-20 [s]) is the weak point of this algorithm.

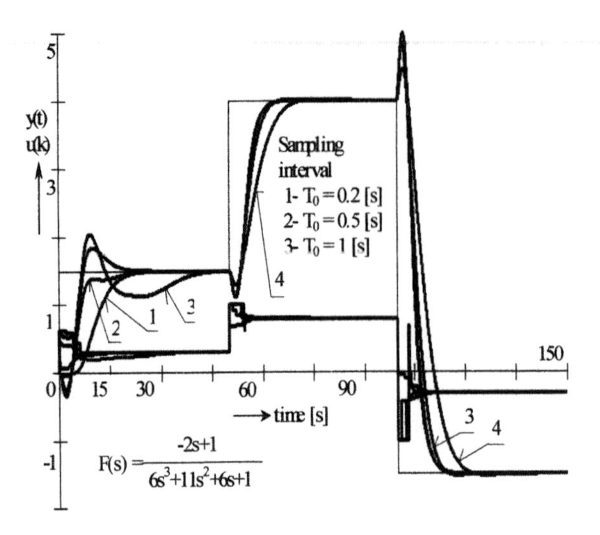

Fig. 13. Digital simulation –adaptive control of plant with non-minimal phase

In the last picture, there is shown discrete simulation of plant with transport delay. There is not adaptive control but in advance synthesis of PIDD^2 algorithm, for adaptive control it is necessary to expand continuous identification by immediate estimating of transport delay. Fig. 14 documented only application possibility of described algorithm also for plant with transport delay.

Simulation experiments clearly emphasize the advantage of hybrid simulation that enables the process of identification of the plant parameters and the adaptation of the controller parameters in a more sophisticated way compared with only a discrete model of the plant and controller.

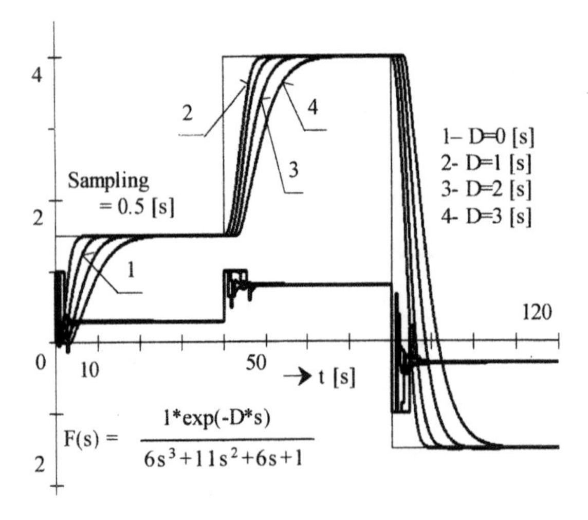

Fig. 14. Control of plant with transport delay.

CONCLUSIONS AND OUTLOOK.

Described analysis and simulation experiments results assign that application of PIDD^2 adaptive algorithm assures better quality of control loop response that classical PID algorithm. Synthesis of algorithm is simple and quality of control loop response is responsible to demanded quality. Described synthesis including continuous identification makes possible adaptive control for wide scale of controlled processes. Simulation experiments with digital models of controlled systems and real time simulation (hybrid simulation) were realised in program environment of ADAPTLAB, developed and realised by author. It is suitable for developing and verification of classical as well as adaptive control algorithms for SISO and MIMO control loops. The program can be used in two basic modes: Simulation and Measurement. The simulation mode works with continuous transfer function set by operator. In the measurement mode the output (input) from model of the plant described in Section 5 is measured with A/D (D/A) converter with sampling interval controlled by real time clock or interrupt from A/D converter. A part of described environment is successfully used in laboratory practice in subjects "Digital Control", "Computer Control" and "Simulation of the Systems" at the University of Žilina.. In future, the author will focus on research of plant with variable transport delay identification, continuous calculation of time constant Tw and start of adaptation problems.

REFERENCES

Alexík M. (1997): Adaptive Self-Tuning PID Algorithm Based on Continuous Synthesis. In: *A Proceeding from the 2nd IFAC Workshop, New Trends in Design of Control Systems.* Smolenice, Slovak Republic, 7-10 September 1997. pp.481-486. .Published for IFAC by Pergamon an Imprit of Elsevier Science, 1998.

Bobál V. et al. (2000). Self -Tuning PID Controllers Based on Dynamics Inversion Method. *In: A Proceeding from IFAC Workshop, Digital Control-Past, Present and Future of PID Control.* University of Catalanya, Terassa, Spain, 2000, pp. 183-188.

Kozák, Š. (1994): Non-linear Self tuning Controller Based upon Laguerre Series Representation. In: *A Proceeding from the 2nd IFAC Workshop, New Trends in Design of Control Systems,* Smolenice,. Slovak Republic September 1994, pp. 432-438.

Kulhavý, R. and Kárny, M. (1984). Tracking of slowly varying parameters by directional forgetting. Paper 14.4/E-4, IFAC World Congress, 1984.

Vitečková M. (1995).: New Approach to Controller Tuning. In: *A Proceeding from the International Conference "micro CAD'95"* Miskolc, Feb. 23, 1995, pp. 18 - 21.

ASPECT – Advanced Control Algorithms on Industrial Controllers

G. Dolanc[1], S. Gerkšič[1], J. Kocijan[1, 4], S. Strmčnik[1], D. Vrančić[1], M. Božiček[2], Z. Marinšek[2], I. Škrjanc[3], S. Blažič[3], A. Stathaki[5], R. King[5], M. Hadjiski[6], K. Boshnakov[6]

[1] Jožef Stefan Institute, Department of Systems and Control, Jamova 39, 1000 Ljubljana, Slovenia
[2] Inea d.o.o., Stegne 11, 1000 Ljubljana, Slovenia
[3] Faculty of Electrical Engineering, Tržaška 25, 1000 Ljubljana, Slovenia
[4] Polytechnic Nova Gorica, Vipavska 13, 5000 Nova gorica, Slovenia
[5] Computer Technology Institute, 11 Akteou and Poulopoulou St., Athens, Greece
[6] University of Chemical Techology and Metallurgy, Kliment Ohridski Blvd. 8, Sofia, Bulgaria
E-mail: gregor.dolanc@ijs.si

Abstract: The advanced control system for control of nonlinear, slowly time variant and delayed processes is described in the paper. The proposed control system performs automatic tuning of fuzzy gain-scheduling controller parameters. The parameters are tuned according to a non-linear process model, which is identified by performing simple experiments on the actual process. The proposed control system (ASPECT) is implemented as a software product and currently runs on several PLC platforms. *Copyright © 2003 IFAC*

Keywords: PID control, Magnitude Optimum, Controller Tuning, Disturbance Rejection, Velocity linearisation

1. INTRODUCTION

There is a large number of SISO processes in process and chemical industries, which are either non-linear, slowly time-variant and/or highly-delayed. It is well known that such processes cannot be efficiently controlled by classical PID controllers, at least not over the entire operating region.

Many research results and theoretical concepts for control of such processes are available. However, relatively large number of scientific publications in this field is not always followed by the corresponding transfer to practice. The reasons are mainly due to a relatively high level of knowledge necessary for implementation, tuning and use of advanced control systems and due to difficult implementation into commercially available hardware equipment and software tools.

The paper presents the so-called ASPECT control system, which has been developed for the mentioned class of processes[1]. This system is implemented within commercially available programmable logic controllers (PLCs).

2. CONCEPT OF THE ASPECT CONTROL SYSTEM

ASPECT control system is implanted as a set of software modules. It consists of three main modules:

---

[1] The ASPECT project was financially supported by EC under contract IST-1999-56407 and co-sponsored by the participating companies INEA, d.o.o., Start Engineering JSCo, and INDELEC Europe S.A.

- "Run-Time" Module (RTM),
- Human Machine Interface module (HMI), and
- Configuration Tool module (CT).

The RTM executes main functions of the control system (control algorithms, tuning of the controller parameters, on-line learning of the process model, estimation of the control quality, coordination of tasks, etc.), and is executed within the programmable logic controller (PLC).

The HMI is implemented within commercially available operator panels or alternatively within personal computer (PC). It is meant for main process signals display and supervision of RTM module status, as well as for the input of operator's commands.

The CT is a programming tool, which runs within a PC. It is used for configuration and adjustment of the RTM module. During the configuration stage, a communication is established between the CT and the RTM.

In the following sections the RTM module structure and its operation is given. The RTM is based on three advanced control algorithms:

1. fuzzy gain-scheduling PID controller,
2. predictive controller with dead time compensation, and
3. neuro-PID controller.

In this paper the attention will be focused on the first algorithm (fuzzy gain-scheduling) only.

Parameters of all three algorithms are tuned automatically according to the non-linear process model. The process model is described by a set of linear local models, which are blended according to the value of scheduling variable. In this way it is possible to describe the operation of non-linear process over the wide range of operation. Non-linear model of the process is obtained experimentally either during regular process operation or with identification. The RTM module is organised as a group of interconnected functional units, called "agents", which are briefly described as follows.

2.1 Control Algorithm agent (CAA)

Main function of the Control Algorithm Agent is the closed-loop control. The main control algorithm is the PID algorithm, upgraded by the fuzzy logic gain scheduling controller (FGSC) (Kocijan et al., 2002) and realised in the form based on velocity-based linearisation approach. FGSC contains the tuning of controller parameters based on non-linear process model. It operates in three different modes:

- AUTO – utilises all functions of the non-linear control algorithm,
- SAFE –PID controller with fixed parameters, and
- MAN – manual controller output operation.

Control algorithm agent runs in three layers, as shown in Figure 1:

<u>Control layer</u> contains the mentioned control algorithm, equipped with additional functions, such as controller output constraint, controller output rate constraint, anti-windup protection, bumpless and conditioned transfer between manual and automatic modes of operation, etc.

<u>Gain-scheduling layer</u> blends between local controller parameters, which correspond to particular local model. Blending is implemented with triangular membership functions and fuzzy logic, according to the scheduling variable (SV), which depends on process input and output and the reference (Kocijan et al., 2002).

<u>Tuning layer</u> tunes controller parameters that belong to the particular local models. The tuning is triggered when a new local model is added or when parameters of the existing local model are modified. Discrete-time local models are converted in the continuous-time form, which is used for controller parameters calculation using magnitude optimum criterion (Vrančić et al., 2001).

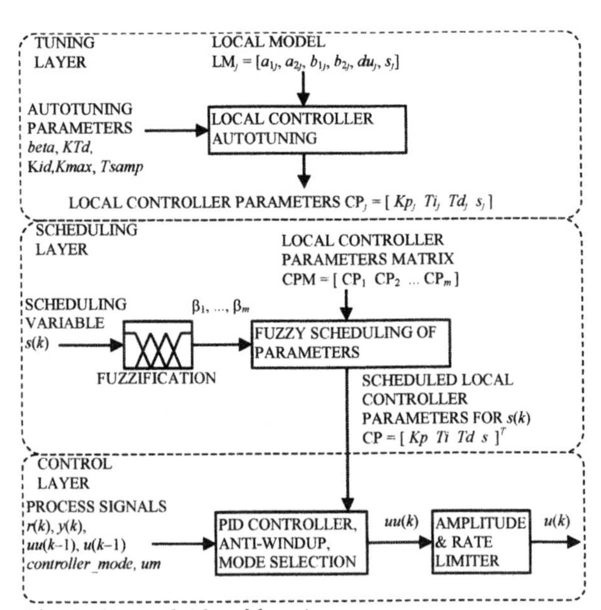

Fig. 1. Control Algorithm Agent

2.2 On-line Learning Agent (OLA)

The task of the On-line Learning Agent is acquiring new local models. Learning is conducted periodically or on demand (by the Operation Supervisor). Since On-line learning is computationally highly demanding, it is pursued as a low priority task in the multitasking operating system. On-line learning performs the following steps (see Figure 2):

Excitation check determines whether the process excitation signal is sufficient in order to conduct the identification.

Identification of local models is performed by using fuzzy instrumental variable identification method (FIV). Result of identification are two sets of local model parameters: the first and the second order linear models with time delay. Fuzzy instrumental variables identification method is initialised by using standard least squares identification method (LS). For more details see Kocijan et al. (2002) and Patent (2002).

Verification/validation of local models is performed by outputs comparison between the corresponding local model and the actual process output in the range of the local model validity. For any local model, verification/validation is performed over three sets of model parameters: initial (existing) model, new model obtained by FIV method, and the new model obtained by the LS method. The model with the best fit according to sum of square errors is chosen as a new model.

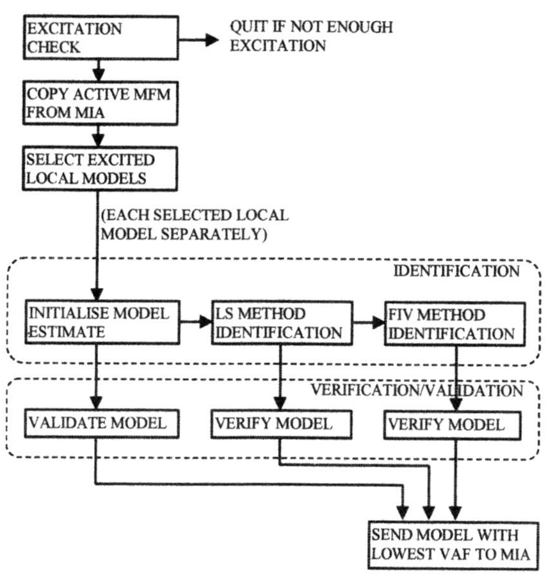

Fig. 2. On-line Learning Agent

2.3 *Model Information Agent (MIA)*

Model Information Agent accepts new local models from OLA and maintains a valid non-linear model, which is a local network model consisting of local linear models. The model equation for the second order local model is

$$\begin{aligned} y(k) &= -a_{1,j}y(k-1) - a_{2,j}y(k-2) + \\ &+ b_{1,j}u(k-1-du_j) + b_{2,j}u(k-2-du_j) + \\ &+ c_{1,j}v(k-1-dv_j) + c_{2,j}v(k-2-dv_j) + r_j \end{aligned},$$

and for the first order model simplifies to

$$\begin{aligned} y(k) &= -a_{1,j}y(k-1) + b_{1,j}u(k-1-du_j) + \\ &+ c_{1,j}v(k-1-dv_j) + r_j \end{aligned},$$

where k is the discrete time index, j is the number of the local model, $y(k)$ is the process output signal (controlled variable, CV), $u(k)$ is the process input signal (manipulated variable, MV), $v(k)$ is the (optional) measured disturbance signal (MD), du is a delay in the MV-CV path, dv is a delay in the MD-CV path, and a_i, b_i, c_i and r_j are parameters of local models.

MIA also keeps status information. The status of each local model is represented by two flags. The first flag denotes if particular local model has been successfully identified at least once after the very first start-up of the system. The second flag signalises if controller parameters have already been tuned according to the corresponding local model. This flag is set by the CAA after successful tuning of the controller parameters to prevent multiple tuning according to unchanged local models. When a new model enters into MIA from OLA, the flag is reset, which means that tuning of the controller parameters is enabled again. MIA supports also an interpolation function for estimation of new local model parameters. These are obtained on the basis of identified local models on adjacent positions.

2.4 *Control Performance Monitor (CPM)*

Control Performance Monitor analyses the real-time signals and estimates a quality of the closed-loop control (control performance) (see Figure 3). If the control performance is poor according to predefined criterion, the Operation Supervisor (OS) can undertake several kind of actions: local model identification, controller parameter tuning or switchover from AUTO to the SAFE mode of operation.

2.5 *Operation Supervisor (OS)*

Operation Supervisor coordinates the operation and interaction of the mentioned functional units and processes interactions of the user/operator. Due to the lack of space, only the main functions are listed:

- coordination and supervision of all units,
- coordination of user interactions,
- processing of data to/from HMI,
- support of several operating regimes,
- processing of configuration data,
- coordination with external signals, and

- manipulation with hardware resources (input and output signals).

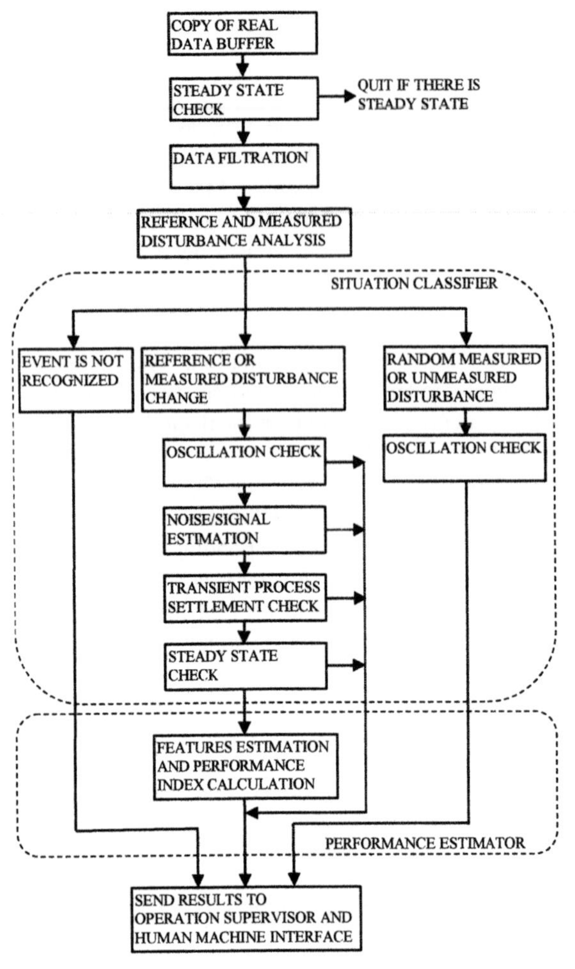

Fig. 3. Estimation of the control performance

3. IMPLEMENTATION

ASPECT control system is implemented as a software, running on several PLC platforms. In this paper an implementation of the run-time module (RTM), within coprocessor module SPAC20 (Mitsubishi floating-point co-processor with 2MB RAM, 40MHz TI TMS 320C32 DSP processor), is presented. HMI is implemented on the operator panel Beijer E700.

RTM is realised as a software unit included in the programming tool "INEA IDR BLOK", which is used for programming and solving special control tasks (DDC) using Mitsubishi PLCs.

4. FIELD TEST

The operation of ASPECT control system has been tested on a hydraulic valve testing line, located in a hydraulic equipment production plant. Simplified scheme of the testing line is shown in Figure 4. The most important control task is to control the pressure on the tested valve by adjusting the speed of pumps P1, P2 and P3. The testing line is non-linear and time variable due to the following reasons:

- steady-state characteristic between the pressure on the valve and the flow rate through the valve (speed of rotation of pumps) is quadratic, i.e. non-linear,
- the opening of the valve can be changed during the test, and
- several different combination of active pumps can be used.

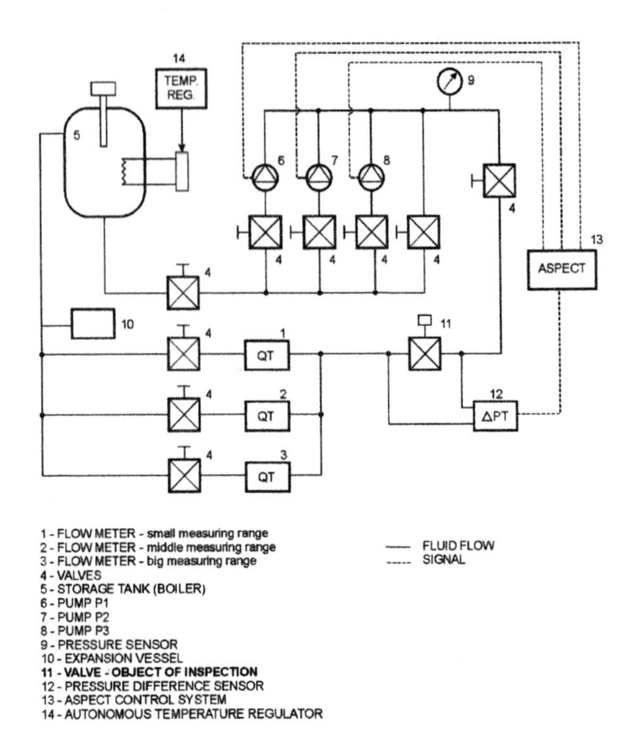

Fig. 4. Scheme of the testing line

Therefore the process steady state gain, i.e. the ratio between the pressure and the speed of the pump is varying with the operating point. Since the process gain and dynamics are crucial for the closed-loop stability, the scheduling variable is selected to be proportional to the pump speed. Process dynamics consists of two main parts: inertia of pump mechanical parts and inertia of the liquid in the pipeline. Both dynamics can be approximated by the first order filters, which sums into the second order process dynamics.

The process was controlled in two different ways: first by using fixed PID controller parameters, and second by using ASPECT control system in AUTO mode, where all functions were activated. Parameters of the fixed PID controller were determined to be optimal at lower pressures on the valve. Controller parameters for AUTO mode were derived automatically using the OLA, by performing experiments, conducted by the OS. Figure 5 shows a closed-loop response of PID control with fixed parameters. Upper graph shows the pressure difference (delta_P) and the set-point (SP). The response of the pressure on the valve was close to optimum but only at lower pressure values. As the pressure increases, the response becomes more oscillatory due to quadratic relation between the pressure and the speed of pump.

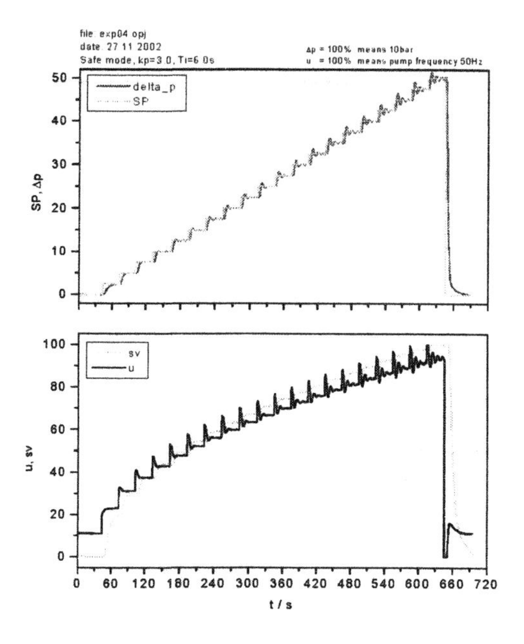

Fig. 5. Control of the pressure when using fixed PID controller parameters

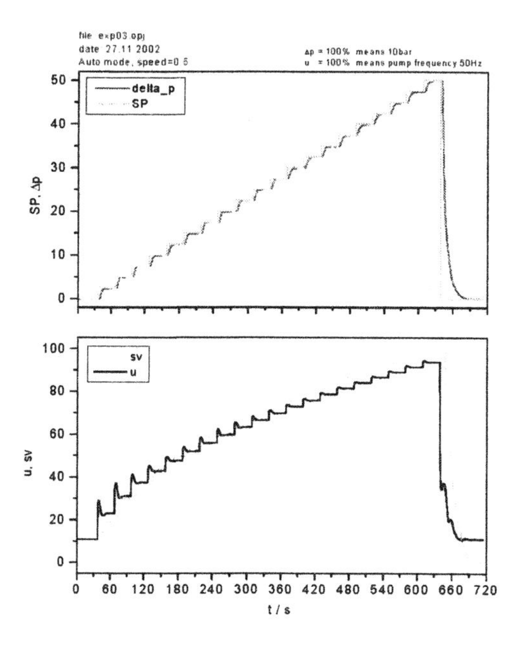

Fig. 6. Control of the pressure when using ASPECT control system

Figure 6 shows the response when ASPECT control system in AUTO mode is used. In comparison to PID controller with fixed parameters, the response is close to optimal in entire region. The PID FGSC gain-scheduling feature of the ASPECT control system adjusts controller parameters in such a manner that the closed-loop transfer function remains stable within entire operating range.

5. CONCLUSIONS

The ASPECT control system has been designed in order to improve the closed-loop performance of the nonlinear, slowly time variant and delayed processes.

Development, design and implementation stages were finished successfully. The operation of the system has been tested by simulations and by experiments on industrial process. Results of experiments have shown advantages over the conventional solutions (e.g. PID controller with fixed parameters).

6. ACKNOWLEDGEMENT

The authors would like to acknowledge the contributions of all other project team members.

ASPECT ©2002 software is property of INEA, d.o.o., Indelec Europe S.A., and Start Engineering JSCo. Patent pending PCT/SI02/00029.

7. REFERENCES

PATENT (2002). Self-tuning controller for non-linear processes described by set of local linear models. PCT/SI02/00029, Slovenia, December 2002.

Kocijan, J., G. Žunič, S. Strmčnik, D. Vrančić, (2002). Fuzzy gain-scheduling control of a gas-liquid separation plant implemented on a PLC. *Int. J. Control*, **75** (14), pp. 1082-1091.

Vrančić, D., S. Strmčnik, and Đ. Juričić, (2001). A magnitude optimum multiple integration tuning method for filtered PID controller. *Automatica*, **37**, pp. 1473-1479.

www.elsevier.com/locate/ifac

POINTS OF VIEW IN CONTROLLER DESIGN BY MEANS OF EXTENDED SYMMETRICAL OPTIMUM METHOD

Stefan Preitl and Radu-Emil Precup

"Politehnica" University of Timisoara, Department of Automation and Ind. Inf.
Bd. V. Parvan 2, RO-300223 Timisoara, Romania
Phone: +40-256-40-3224, -3229, -3230, -3226, Fax: +40-256-403214
E-mail: spreitl@aut.utt.ro, rprecup@aut.utt.ro

Abstract: The paper deals with theoretical and applicative aspects concerning the Symmetrical Optimum method for controller tuning. There are highlighted two approaches regarding the method and a generalization of one of them. Connections with the optimisation methods are given. Due to the applicability of the method in the field of rapid plants, the presented aspects are of permanent actuality.

Keywords: Extension of Symmetrical Optimum method, PI and PID controllers, benchmarks, quadratic performance indices, optimisation problems.

1. INTRODUCTION

In 1958, Kessler has elaborated the basic version of the Symmetrical Optimum (SO) method, and it is characterised by the fact that in the open-loop transfer function (abbreviated t.f.), $H_0(s)$:

$$H_0(s) = H_C(s) H_P(s) , \qquad (1)$$

with $H_C(s)$ – the controller t.f. and $H_P(s)$ – the controlled plant t.f., there is obtained a double pole in the origin. This can correspond to various situations:
- the plant is characterised by an integral component and the controller brings the second one in order to ensure the condition of zero steady-state control error and zero static coefficient,
- the plant does not contain any integral component, but the requirement for zero control error with respect to the ramp modification of the reference input leads to the necessity that the controller should bring the two integral components, and others.

The paper is structured as follows on four parts and presents:
- a short overview on Kessler's version of the SO method (both in the time domain and in the operational domain);
- an application of the SO method to some low order benchmarks given by Åström and Hägglund (1995);
- an extension of the SO method together with connections with integral quadratic performance indices;
- some final conclusions.

It is worthwhile to highlight that the efficiency of controller tuning in terms of the SO method can be proved both in the time domain and in the frequency domain as well (see, for example, (Föllinger, 1985)).

2. A SHORT OVERVIEW ON THE SYMMETRICAL OPTIMUM METHOD

The basic version of the SO method (Kessler, 1958) accepts that the controlled plant has its t.f. expressed as:

$$H_P(s)=\frac{k_P}{(1+sT_\Sigma)\prod_{\nu=1}^{n}(1+sT_\nu)}, \quad (2)$$

or in the case of an included integral component:

$$H_P(s)=\frac{k_P}{s(1+sT_\Sigma)\prod_{\nu=1}^{n}(1+sT_\nu)}, \quad (3)$$

where T_Σ includes the effects of the small time constants of controlled plant, and T_ν represent the large time constants of the plant ($\nu = 1 \ldots n$).

It is well accepted that the quality of the system transients is determined by the shape of the open-loop frequency plots, $H_0(j\omega)$ in the zone of the cut-off frequency ω_c. For ensuring proper control system performance (σ_1 – overshoot, t_s – settling time, t_1 – first settling time, etc.) it is required that the value of ω_c should be as large as possible representing an acceptable value for the phase margin (reserve), φ_r.

For the large time constants of controlled plant T_ν, the t.f. (2) – transferred into the frequency domain – for large values of the frequency ω the following approximation will be accepted:

$$1+j\omega T_\nu \approx j\omega T_\nu ; \quad (4)$$

so, (2) will be approximated by (5):

$$H_P(s)\approx\frac{k_P}{(1+sT_\Sigma)\prod_{\nu=1}^{n}(sT_\nu)}. \quad (5)$$

By using a typical controller (PI, PID, etc.) with the theoretical t.f.:

$$H_C(s)=\frac{\prod_{\mu=1}^{m}(1+s\tau_\mu)}{\tau s}, \quad (6)$$

the open-loop t.f. (1) obtains the form (7):

$$H_0(s)\approx\frac{k_P\prod_{\mu=1}^{m}(1+s\tau_\mu)}{\tau s\,(1+sT_\Sigma)\prod_{\nu=1}^{n}(sT_\nu)}. \quad (7)$$

By doing some computations regarding the equation (7), assuming the condition m = n, it can be expressed in the equivalent form (8):

$$H_0(s)\approx\frac{\prod_{\nu=1}^{n}(1+s\tau_\nu)}{\tau_e\, s\,(1+sT_\Sigma)\prod_{\mu=1}^{n}(s\tau_\mu)}, \quad (8)$$

where:

$$\tau_e=\frac{\tau}{k_P}\cdot\frac{\prod_{\nu=1}^{n}T_\nu}{\prod_{\mu=1}^{n}\tau_\mu}. \quad (9)$$

In Kessler's version the proper control system behaviour is accomplished if there are fulfilled the following three requirements (Kessler, 1958):

- the congruence of the time constants brought by the controller:

$$\tau_1=\tau_2=\ldots=\tau_m=\tau_c ; \quad (10)$$

- a first optimum condition:

$$\tau_c=4\,n\,T_\Sigma ; \quad (11)$$

- a second optimum condition:

$$\tau_e=2\,T_\Sigma . \quad (12)$$

A combined restricted condition (13) will result from the equations (11) and (12):

$$\tau_c=2\,n\,\tau_e . \quad (13)$$

Replacing the conditions (10) in (8) leads to:

$$H_0(s)\approx\frac{(1+\tau_c s)^n}{(1+sT_\Sigma)(\tau_c s)^n\tau_e s}=\frac{[1+1/(s\tau_c)]^n}{\tau_e\, s(1+sT_\Sigma)}. \quad (14)$$

If in the expression of the nominator there are used only the first two terms of the binomial formula in terms of (11) and (12), it will result that the open-loop t.f. (14) will obtain a practical useful form:

$$H_0(s)\approx\frac{1+4T_\Sigma s}{8T_\Sigma^2 s^2(1+T_\Sigma s)} \quad (15)$$

(see (Åström and Hägglund, 1995)).

Accordingly, the closed-loop t.f. with respect to the reference input w can be expressed as:

$$H_w(s)=\frac{b_0+b_1 s}{a_0+a_1 s+a_2 s^2+a_3 s^3}, \quad (16)$$

with $a_0=b_0=1$, $a_1=b_1=4T_\Sigma$, $a_2=8T_\Sigma^2$, $a_3=8T_\Sigma^3$, or, in a factorised version:

$$H_w(s)=\frac{1+4T_\Sigma s}{(1+2T_\Sigma s)(1+2T_\Sigma s+4T_\Sigma^2 s^2)}. \quad (17)$$

However, the obtained dynamic control system performance indices $\sigma_1 \approx 43\%$, $t_s \approx 16.5T_\Sigma$, $t_1 \approx 3.1T_\Sigma$, are seldom acceptable and require the improvement. Referring the t.f. (17), this is usually done in the input-output framework by pre-filtering the reference input with filters that suppress the effect of either only the zero ($-1/4T_\Sigma$) or the pair of complex conjugated poles (pole-zero cancellation, control based on an inverse model of the system).

The main drawback of the SO method is in the relatively small phase margin of the control loop, $\varphi_r \approx 36^\circ$, that is reflected in the increased sensitivity with respect to parametric modifications.

It has to be highlighted that in some papers (see, for example, (Lehoczky, *et al.*, 1977)) the method is considered to belong to both Kessler and Naslin.

3. A PRACTICAL VERSION OF THE SYMMETRICAL OPTIMUM METHOD GIVEN IN (ÅSTRÖM AND HÄGGLUND, 1995)

By starting with the form (16) of the closed-loop t.f. $H_w(s)$, in (Åström and Hägglund, 1995) it is stated that the fulfilment of the conditions (18):

$$2a_0a_2 = a_1^2,\ 2a_1a_3 = a_2^2, \qquad (18)$$

ensures in the t.f. (16) optimal performance guaranteed by the SO method and offer, by design, a compact manner for computing the controller tuning parameters.

Some situations concerning remarkable applications are related to the field of electrical drives, characterized by t.f.s expressed as benchmark type plants:

$$H_P(s) = \frac{k_P}{s(1+sT_\Sigma)}, \qquad (19)$$

or:

$$H_P(s) = \frac{k_P}{s(1+sT_\Sigma)(1+sT_1)}, \qquad (20)$$

where $T_1 >> T_\Sigma$ is the large time constant of the plant.

The use of a PI or of a PID controller having the t.f. (21) and (22), respectively:

$$H_C(s) = \frac{k_c}{s}(1+sT_c), \qquad (21)$$

$$H_C(s) = \frac{k_c}{s}(1+sT_c)(1+sT_c'), \qquad (22)$$

with $T_c' = T_1$ (pole-zero cancellation), simplifies enough the design steps.

By applying the optimisation relations (18), the transfer functions $H_0(s)$ and $H_w(s)$ result in the form of (15) and (17), respectively.

It is to note that the Bode plots of the optimal transfer function $H_0(s)$ (presented in the equation (15)) is symmetrical around the cut-off frequency $\omega_c = 1/(2T_\Sigma)$ (this is the main motivation for the name Symmetrical Optimum in terms of Åström and Hägglund (1995)).

The parameters of the PI controllers and of the PID controllers are computed by means of the compact relations (23):

$$k_c = \frac{1}{8k_P T_\Sigma^2},\ T_c = 4T_\Sigma\ (T_c' = T_1). \qquad (23)$$

It is also worthwhile to remark that the justification pointed out in (Åström and Hägglund, 1995) makes the presentation of the SO method close to the Modulus Optimum method.

Requirements close to those given by the SO method have been formulated also by many other papers including, for example, (Loron, 1997; Lutz and Wendt, 1998; Preitl, 2001), each of the contributors pointing out interesting particular aspects.

4. AN EXTENSION OF THE SYMMETRICAL OPTIMUM METHOD

The authors have presented in (Preitl and Precup, 1999) a generalized form of the equations (18):

$$\beta^{1/2}a_0a_2 = a_1^2,\ \beta^{1/2}a_1a_3 = a_2^2, \qquad (24)$$

where β is a design parameter with the value to be chosen by the developer.

The conditions (24) are applied to the typical situations discussed by Åström and Hägglund (1995), and lead to significantly improved final situations characterized by:
- maintaining the "symmetry" in the sense given in (Åström and Hägglund, 1995), Fig. 1, but with an increased possibility of performance modification; Fig. 2 presents the values of the main control system performance indices as function of the design parameter β;
- compact design relations for the controller tuning parameters, which generalize the relations (23) to the form (25):

$$k_c = \frac{1}{k_P\beta^{3/2}T_\Sigma^2},\ T_c = \beta T_\Sigma\ (T_c' = T_1), \qquad (25)$$

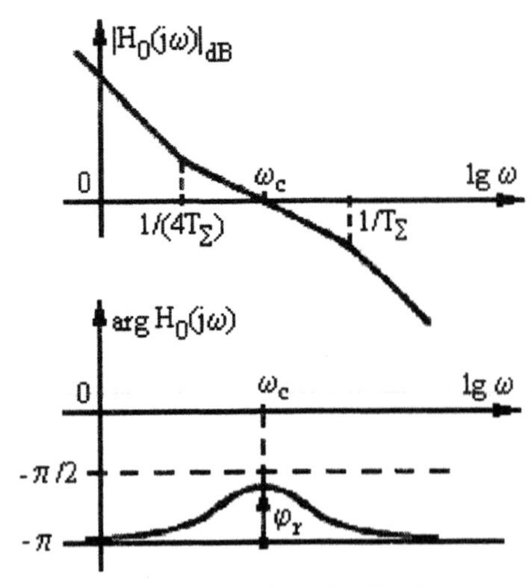

Fig. 1. Optimal open-loop Bode plots.

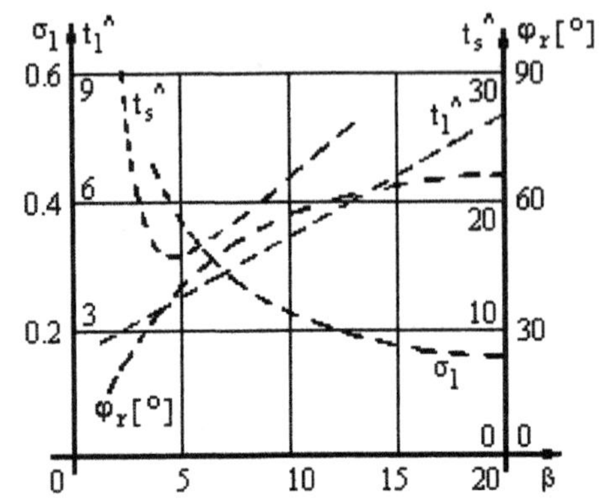

Fig. 2. Performance indices σ_1, $\hat{t}_s$ $(= t_s T_\Sigma)$, $\hat{t}_1$ $(= t_1 T_\Sigma)$ and φ_r [°] versus β.

that can be correlated directly to the desired control system performance indices, Fig. 2;

- the values of β are usually chosen to belong to the recommended interval [4, 20); a decrease of the value of β under the "optimal value" $\beta = 4$ given in (18) leads to the rapid decrease of the phase margin; an increase of the value of β over 20 forces the derivative effect introduced by the controller;
- the improvement of the phase margin φ_r for increasing values of β is of exquisite significance especially in the situations when the parameter k_P (the controlled plant gain) is time-variant.

The authors have presented in (Preitl and Precup, 1999) a method that computes the value of β which ensures – for the variation of k_P in the domain $[k_{Pmin}, k_{Pmax}]$ – a minimum guaranteed phase margin. This is, for example, the situation of variable inertia electrical drives, for which the proposed method offers a good support for controller design.

As the effect of the "optimisation" in terms of the equations (24) the open-loop t.f. $H_0(s)$ and the closed-loop t.f. $H_w(s)$ will obtain the forms (26) and (27), respectively:

$$H_0(s)=\frac{1+\beta T_\Sigma s}{\beta^{3/2} T_\Sigma^2 s^2 (1+T_\Sigma s)}, \tag{26}$$

$$H_w(s)=\frac{1+\beta T_\Sigma s}{\beta^{3/2} T_\Sigma^3 s^3 + \beta^{3/2} T_\Sigma^2 s^2 + \beta T_\Sigma s + 1}. \tag{27}$$

Concerning the design method presented synthetic in this Section there is of interest to point out the connection of the results with those obtained by minimizing the integral quadratic performance indices.

For a large class of control applications there are widely accepted and recommended for use (for example, by Graham and Lathrop (1953), Anderson and Moore (1989)) the following performance indices as cost functions:

- the Integral of Squared Error (ISE) index, I_{2e}:

$$I_{2e} = \int_0^\infty e^2(t)dt, \tag{28}$$

- the generalized ISE index, I_{2g}:

$$I_{2g} = \int_0^\infty [e^2(t) + \tau^2 \dot{e}^2(t)]dt, \tag{29}$$

- the quadratic cross-optimisation index, I_{2c}:

$$I_{2c} = \int_0^\infty [e^2(t) + \rho^2 u^2(t)]dt, \tag{30}$$

where e(t) stands for the control error, u(t) represents the control signal, τ and ρ are weighting parameters available to user's choice.

It can be proved (Preitl and Precup, 2000) that each of these indices can be correlated with the only one value of the design parameter β. So, by using the Parseval relationships the computation of the expressions I_{2e}, I_{2g} and I_{2c} for the dynamic regime characterized by the unit step modification of the reference input w yields:

$$I_{2e} = \frac{\beta T_\Sigma}{2(\beta^{1/2}-1)}, \tag{31}$$

$$I_{2g} = \frac{\tau^2 + \beta T_\Sigma^2}{2T_\Sigma(\beta^{1/2}-1)}, \tag{32}$$

$$I_{2c} = \frac{\beta^{5/2} T_\Sigma^2 + r^2(\beta^{3/2} + \beta - \beta^{1/2} + 1)}{2T_\Sigma \beta^{3/2}(\beta^{1/2}-1)}, \tag{33}$$

with $r = \rho/k_P$.

It can be observed that for a chosen value of τ or ρ the expressions of the cost functions (31) ... (33), depend on the single variable β. Consequently, the three dynamic optimisation problems are transformed into stationary optimisation ones. Therefore, the minimization with respect to β of the expressions (31) ... (33) leads to the minimum values of the quality indices as follows:

$$I_{2e\min} = 2T_\Sigma, \text{ for } \beta = 4, \tag{34}$$

with the graphics $I_{2e} = f(\beta)$ according to Fig. 3;

$$I_{2g\min} = T_\Sigma[1+[1+(\tau/T_\Sigma)^2]^{1/2}], \tag{35}$$

for

$$\beta = 1+[1+(\tau/T_\Sigma)^2]^{1/2}, \tag{36}$$

with the graphics $I_{2g} = f(\beta, \tau)$, detailed for the values (for example) $\tau \in \{0, 0.25, 0.5, 1\}$ (continuous line for $\tau = 0$, dotted line for $\tau = 0.25$, dashed line for $\tau = 0.5$ and dash doted line for $\tau = 1$), according to Fig. 4;

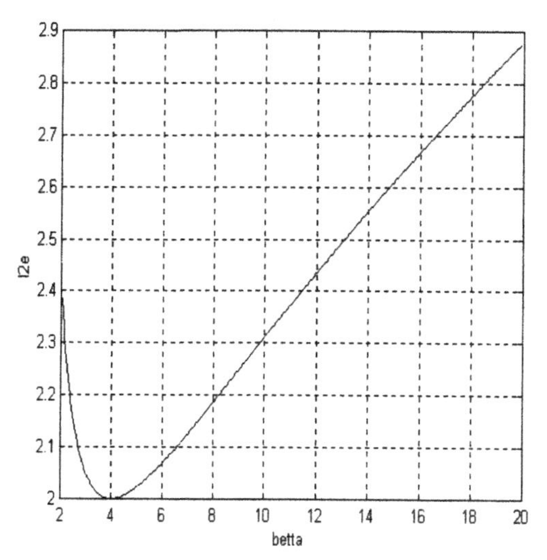

Fig. 3. I_{2e} versus β.

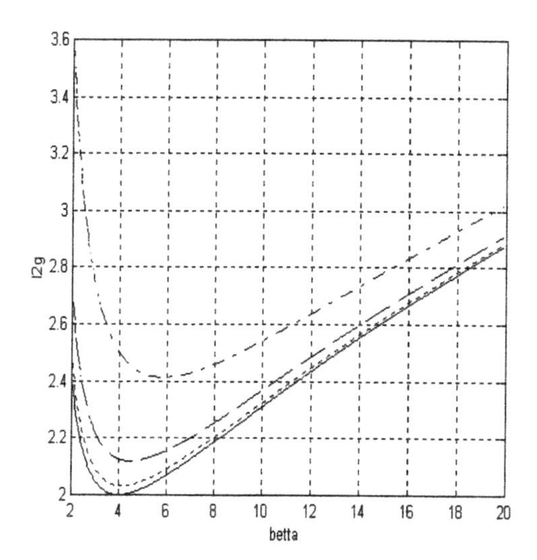

Fig. 4. I_{2g} versus β for $\tau \in \{0, 0.25, 0.5, 1\}$.

$$I_{2c\min} = \frac{T_\Sigma^2 x^5 + r^2(x^3 + x^2 - x + 1)}{2T_\Sigma x^3(x-1)}, \tag{37}$$

with $\beta = x^2$ and x represents the single real solution larger than 4 of the following sixth-order equation:

$$T_\Sigma^2(x^6 - x^5) - r^2(x^4 + 2x^3 - 4x^2 + 6x - 3) = 0, \tag{38}$$

and the graphics $I_{2c} = f(\beta, r)$, detailed for the values (for example) $r \in \{0, 0.25, 0.5, 1\}$ (continuous line for $r = 0$, dotted line for $r = 0.25$, dashed line for $r = 0.5$ and dash doted line for $r = 1$), according to Fig. 5.

These results point out the fact in this case the optimisation procedures based on the mentioned integral quadratic performance indices can be reduced to the proper choice of the value of the parameter β. By the value of β it is reached the characterisation of the quality of the control system by means of the empirical control system performance indices $\{\sigma_1, t_s, t_1, \ldots\}$ which are easily understandable. This represents another exquisite advantage of the method.

It must be outlined that the favourable effects of using the indices I_{2g} and I_{2c}, obtained by increasing the weighting coefficients τ and ρ, respectively, lead to:

- the increase of the value of the indices with respect to the value $I_{2e\min}$ (this is obvious);
- the increase of the value of the parameter β, accompanied by control system performance enhancement, defined by means of empirical control system performance indices σ_1, t_s, t_1 and φ_r.

From the user's point of view can be useful to know the connections between the values of β and those values of τ and ρ, for which the indices I_{2g} and I_{2c}, respectively, obtain their minimum values.

These connections are presented in Fig. 6 for the domain $\beta \in [4, 16]$, in the conditions of a simplified controlled plant having the t.f. (20) with $k_P = 1$ and $T_\Sigma = 1$.

5. CONCLUSIONS

The paper presents points of view concerning two ways for introducing the Symmetrical Optimum method and practical interpretations related to the parameter tuning for conventional PI and PID controllers.

In its second version, presented in (Åström and Hägglund, 1995), there is the possibility for a generalization of the optimisation conditions that enlarges significantly the areas of application and usefulness of the SO method.

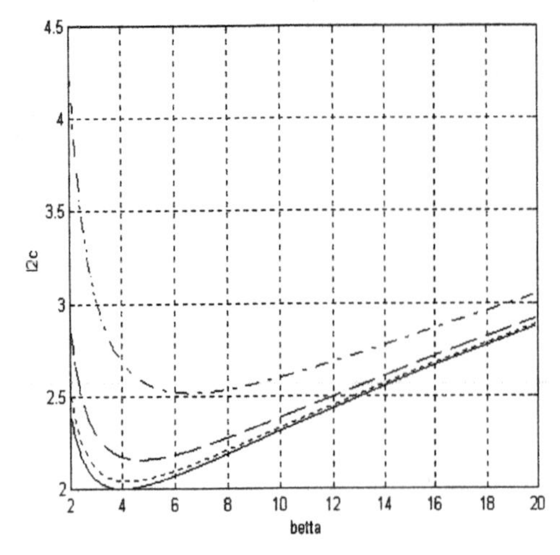

Fig. 5. I_{2c} versus β for $r \in \{0, 0.25, 0.5, 1\}$.

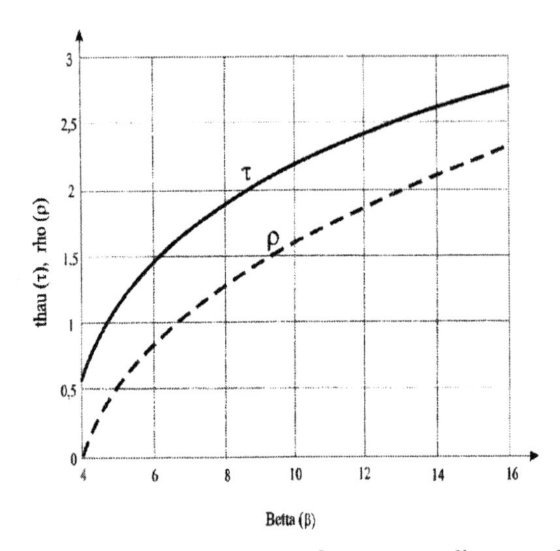

Fig. 6. τ and ρ versus β corresponding to the minimum values of I_{2g} and I_{2c}.

The method introduced and called Extended Symmetrical Optimum (ESO) method has the exquisite advantage in having a single parameter β. The choice of β stays in the hands of the designer. This advantage acts including in the relation with the use of some integral quadratic performance indices (see (31) ... (36)).

The main drawback of the method is in the fact that the designer must have the performance diagrams (Fig. 2) or he must built them. The explicit expressing under analytical forms for all performance diagrams is possible, but difficult.

Although the control solutions are discussed in continuous time, they can be relatively easy implemented in quasi-continuous digital version by using the approaches presented in (Åström and Wittenmark, 1990; Kucera, 1991).

Other aspects derived from the application of the SO method can be of both theoretical and applicative interest.

REFERENCES

Anderson, B.D.O. and J.B. Moore (1989). *Optimal Control. Linear Quadratic Methods.* Prentice-Hall, Englewood Cliffs, NJ.

Åström, K.J. and T. Hägglund (1995). *PID Controllers Theory: Design and Tuning.* Instrument Society of America, Research Triangle Park.

Åström, K.J. and B. Wittenmark (1990). *Computer-Controlled Systems. Theory and Design.* Prentice Hall, Englewood Cliffs, NJ.

Föllinger, O. (1985). *Regelungstechnik,* Elitera Verlag, Berlin.

Graham, D. and R.C. Lathrop (1953). The Synthesis of "Optimum" Transient Response: Criteria and Standard Forms. *Transactions of AIEE. Part II: Applications and Industry,* **72**, 273-288.

Kessler, C. (1958). Das Symetrische Optimum. *Regelungstechnik,* **6**, 395-400, 432-436.

Kucera, V. (1991). *Discrete Linear Control Systems.* Prentice Hall, London.

Lehoczky, J., M. Márkus, and and S. Mucsi (1977). *Szervorendszerek követö szabályozások,* MK, Budapest.

Loron, L. (1997). Tuning of PID Controllers by the Non-symmetrical Optimum Method. *Automatica,* **33**, 103-107.

Lutz, H. and W. Wendt (1998). *Taschenbuch der Regelungstechnik,* Harri Deutsch Verlag, Frankfurt.

Preitl, S. and R.-E. Precup (1999). An Extension of Tuning Relations after Symmetrical Optimum Method for PI and PID Controllers. *Automatica,* **35**, 1731-1736.

Preitl, S. and R.-E. Precup (2000). Cross Optimization Aspects Concerning the Extended Symmetrical Optimum Method. *Preprints of PID'00 IFAC Workshop,* Terrassa, 254-259.

Preitl, Z. (2001). PI and PID Controller Tuning Method for a Class of Systems. *Proceedings of 7th Symposium on Automatic Control and Computer Science,* Iasi, 4 pp.

PID CONTROL OF DISTRIBUTED PARAMETER SYSTEMS

G. Hulkó, C. Belavý

Department of Automation and Measurement, Faculty of Mechanical Engineering, Slovak University of Technology, Nám. Slobody 17, 812 31 Bratislava, Slovak Republic. E-mail:hulko@kam1.vm.stuba.sk

Abstract: In the paper first some new results of distributed parameter systems theory will be indicated. Concept of lumped-input/distributed-output systems will be introduced and distributed parameters feedback control loops will be designed. Further possibilities for PID control of distributed parameter systems, which are considered as lumped-input/distributed-output systems will be demonstrated. *Copyright © 2003 IFAC*

Keywords: distributed parameter systems, lumped-input/distributed-output systems, transfer functions, finite element method, control system synthesis, PID controller.

1. INTRODUCTION

In the engineering practice is paid considerable attention to the motion, whose state variables are fields of variables, or spatial distributed variables. In these cases very frequently is considered the motion of continuum mechanics systems - Distributed Parameter Systems (DPS).

Lately it was found, that DPS is useful to interpret as Lumped-input/Distributed-output Systems (LDS), (Hulkó, 1987, 1990, 1995, 1998). Dynamics of such systems is decomposed to space and time components. LDS enable to design feedback loops for solution of DPS control tasks.

In this paper first some new results of the DPS theory will be indicated, further possibilities of their utilizing for control of continuum mechanics systems will be shown.

Dynamics of continuum mechanics systems as LDS system will be modeled using both lumped and distributed transfer functions and Finite Elements Method (FEM). Control synthesis in time domain will be realized by PID controllers.

2. DISTRIBUTED PARAMETER SYSTEMS

In general DPS are systems whose state or output variables are distributed variables or fields of variables.

In control theory these systems are frequently considered as systems whose dynamics is described by Partial Differential Equations (PDE), (Lions, 1971). In the input-output relation PDE define Distributed-input/Distributed-output Systems (DDS) between distributed input U(x,t) and distributed output variables Y(x,t) at initial and boundary conditions given, see Fig. 1.

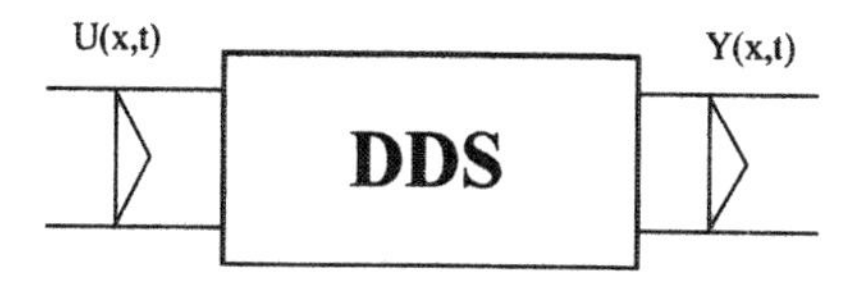

Fig. 1. Distributed-input/distributed-output system

Distributed systems very frequently are found in the engineering practice as LDS, see Fig. 2, having the structure according Fig. 3. When the blocks of this structure are appropriately chosen, DDS block is obtained as a special case of LDS one.

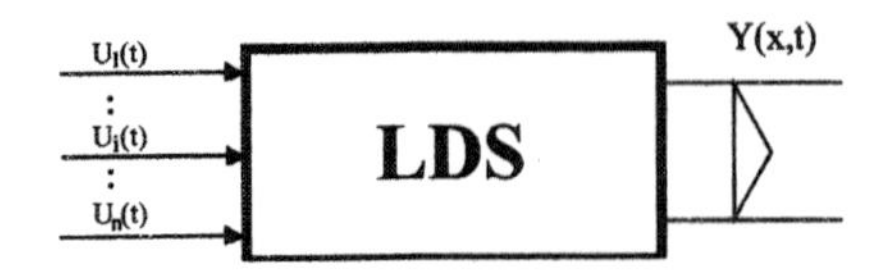

Fig. 2. Lumped-input/distributed-output system: $\{U_i(t)\}_{i=1,n}$ - lumped input variables, Y(x,t) - distributed output variable

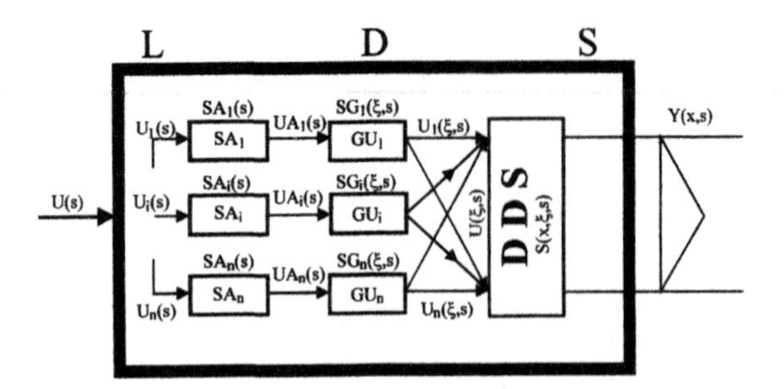

Fig. 3. Lumped-input/distributed-output system structure: $\{SA_i(s)\}_{i=1,n}$ - transfer functions of actuating gears of lumped variables $\{SA_i\}_{i=1,n}$, $\{SG_i(\xi,s)\}_{i=1,n}$ - transfer functions of generators of distributed input variables $\{GU_i\}_{i=1,n}$, $S(x,\xi,s)$- transfer function of distributed-input/ distributed-output system

LDS in this structure offers large possibilities for solving tasks of modelling, control and design of distributed parameter systems in the engineering practice.

3. BASIC CHARACTERISTICS OF LDS DYNAMICS

Let us consider LDS distributed on the interval <0,L>, as shown in Fig. 4. Lumped discrete input variable $U_i(k)$, at unit sampling period, enters through the zero-order hold H_i as $U_i(t)$ into the LDS block. The output of the system will be in the form of distributed variable $Y_i(x,t)$ or $Y_i(x,k)$ respectively. At a point x_i it will be $Y_i(x_i,t)$ or $Y_i(x_i,k)$ respectively. Discrete transfer function between $U_i(k)$ and $Y_i(x_i,k)$ is denoted as $SH_i(x_i,z)$.

When the unit-input variable is used, in the output of the system distributed transient response $HH_i(x,k)$ is got. Its reduced form in the steady-state is expressed as:

$$HHR_i(x,\infty) = \frac{HH_i(x,\infty)}{HH_i(x_i,\infty)} \tag{1}$$

This procedure enables to introduce characteristics for all lumped input variables, i = 1, n of the studied system:

$$\{SH_i(x_i,z)\}_{i=1,n} \tag{2}$$

$$\{HHR_i(x,\infty)\}_{i=1,n} \tag{3}$$

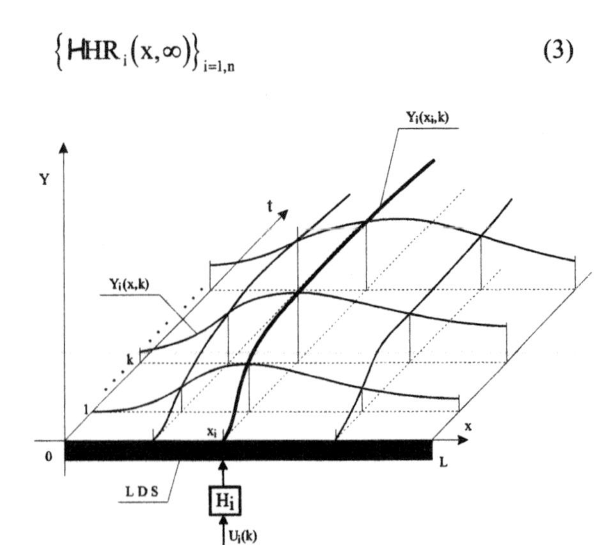

Fig. 4. Lumped-input/distributed-output system with zero-order hold H_i on the input

4. DISTRIBUTED PARAMETER CONTROL LOOPS

For LDS control synthesis, let us introduce distributed parameter feedback control loop, Fig. 5. Let the goal of control is to ensure the steady-state control error to be minimal, i.e. :

$$\min\|E(x,\infty)\| = \min\|W(x,\infty) - Y(x,\infty)\| = \tag{4}$$

$$= \quad \|W(x,\infty) - \breve{Y}(x,\infty)\| = \|\breve{E}(x,\infty)\| \tag{5}$$

where $\|\,.\,\|$ is a norm appropriately chosen.

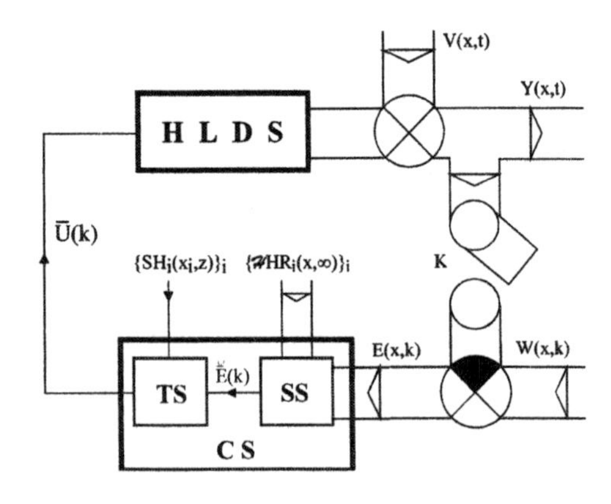

Fig. 5. Distributed parameter feedback control loop: HLDS- LDS system with zero-order holds $\{H_i\}_i$ on the input, CS- control synthesis, TS- control synthesis in time domain, SS- control synthesis in space domain, K- sampling, Y(x,t)- distributed controlled variable, W(x,k)- control variable, V(x,t)- disturbance variable, E(x,k)- control error.

First, an approximation problem (6) in the block of the Space Synthesis (SS) is solved on the set of reduced steady-state distributed transient responses $\left\{HHR_i(x,\infty)\right\}_i$, i=1, n, see Fig. 6.

$$\min_{E_i}\left\|E(x,k)-\sum_{i=1}^{n}E_i(k)HHR_i(x,\infty)\right\| = \\ = \left\|E(x,k)-\sum_{i=1}^{n}\breve{E}_i(k)HHR_i(x,\infty)\right\| \quad (6)$$

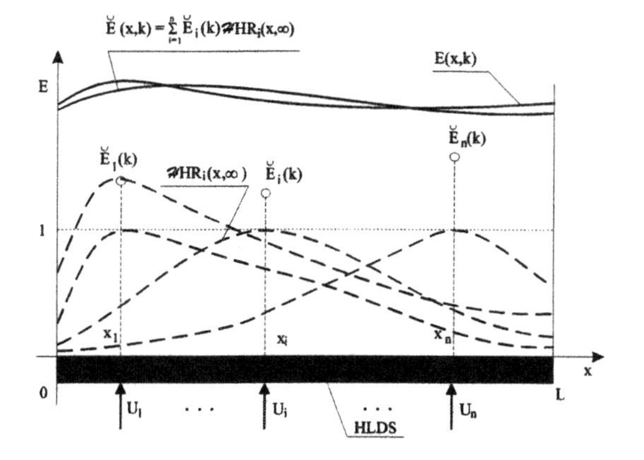

Fig. 6. Approximation problem solution

Further the control errors vector $\left\{\breve{E}_i(k)\right\}_i$ enters into the Time Synthesis block (TS), see Fig. 7.

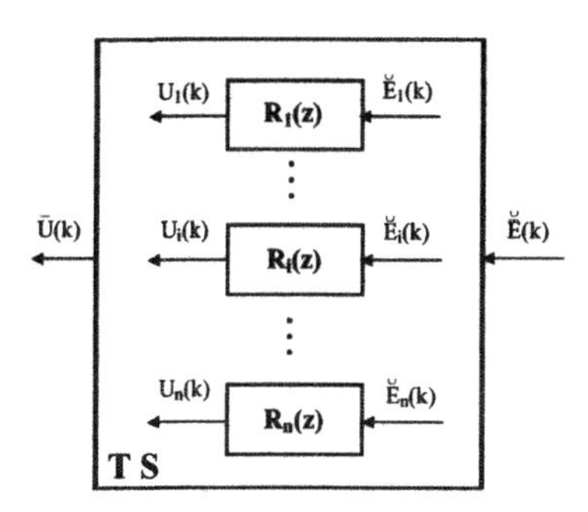

Fig. 7. Time synthesis block

Here vector components $\left\{\breve{E}_i(k)\right\}_i$ are fed through the input of controllers $\{R_i(z)\}_i$ and the sequence of control variables $\overline{U}(k)$ is generated. Tuning of controllers is done for one-parameter control loops $\{SH_i(x_i,z), R_i(z)\}_i$, see Fig. 8.

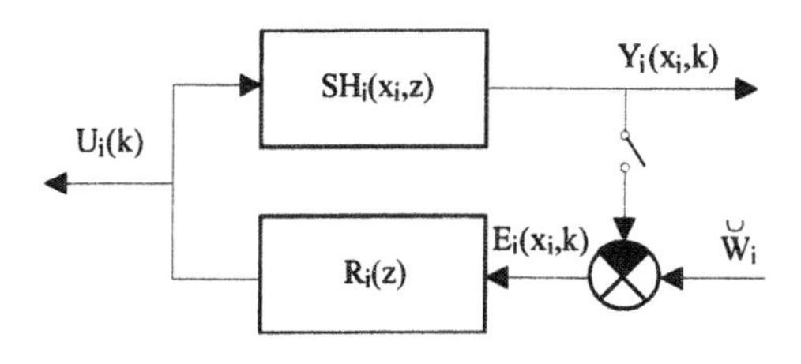

Fig. 8. i-th one-parameter control loop

During the control process for $k\rightarrow\infty$ we get the relation

$$\min_{E_i}\left\|E(x,\infty)-\sum_{i=1}^{n}E_i(\infty)HHR_i(x,\infty)\right\| = \left\|\breve{E}(x,\infty)\right\| \quad (7)$$

thus the control task (4), (5) is accomplished.

Control synthesis technique shows, that for the attainment of required control quality in space domain it is possible to use results of approximation theory.

To design of one-parameter control loops for controllers, $\left\{R_i(z)\right\}_i$ tuning there are results of the control theory of lumped systems at disposal. Thus, for achievement control quality requested in time domain it is necessary to select appropriate control synthesis methods.

Presented control synthesis procedure enables to design also another types of distributed parameter control systems with controllers $\left\{R_i(z)\right\}_i$ in time synthesis block, e.g. see Fig. 9.

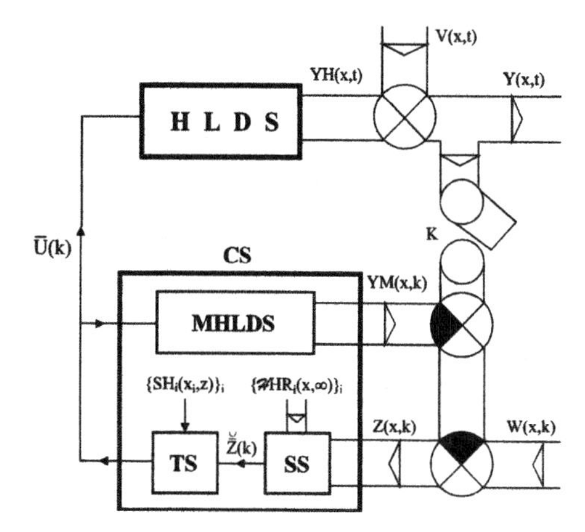

Fig. 9. Distributed parameter internal model control system

Most processes are controlled by proportional-integral-derivative (PID), or *three-term* controllers. The popularity of PID controllers is given mainly both their robust performance in a wide range of operating conditions and functional simplicity.

In one-parameter control loops according Fig. 8, discrete-time versions of PID controllers are used. First, for PID controller with transfer function in s-domain

$$R(s)=K_P+\frac{K_I}{s}+K_D s=K_P\left(1+\frac{1}{T_I s}+T_D s\right) \quad (8)$$

adjustable parameters K_P, K_I, K_D, or T_I, T_D, respectively are determined with usual tuning rules such as Ziegler-Nichols (ZN), integral of time-weight absolute value of the error (ITAE), or root locus method. There are transfer functions $\{S_i(x_i,s)\}_{i=1,n}$, or their discrete-time versions $\{SH_i(x_i,z)\}_{i=1,n}$ are used.

Further, for discrete-time incremental versions of PID controller in the form

$$R(z)=\frac{r_0+r_1z^{-1}+r_2z^{-2}}{1-z^{-1}} \tag{9}$$

parameters are determined for sampling period T as follows:

$$\begin{aligned} r_0 &= K_P\left(1+\frac{T}{2T_I}+\frac{T_D}{T}\right)\\ r_1 &= -K_P\left(1-\frac{T}{2T_I}+\frac{2T_D}{T}\right)\\ r_2 &= K_P\frac{T_D}{T}\end{aligned} \tag{10}$$

5. DISTRIBUTED PARAMETER SYSTEMS CONTROL PROCESS

In previous part we have analyzed the basic functions of distributed parameter control system. Further control processes for various continuum mechanics systems will be presented in MATLAB, with using of software package DPS Toolbox (Hulko et al., 1999).

5.1 Control of smart material structure string deviation

The string with smart material structure is considered as real LDS, see Fig. 10.

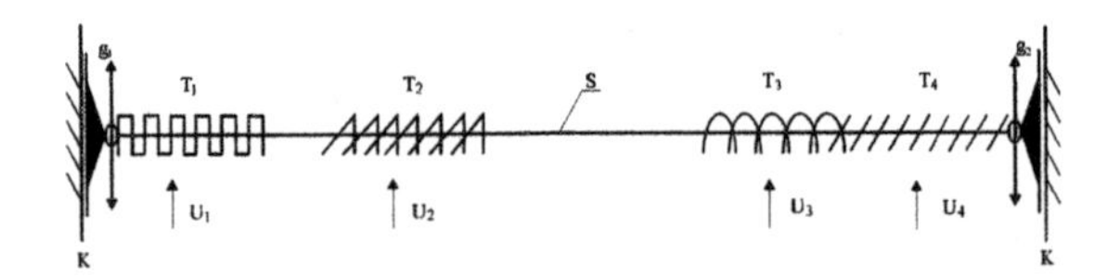

Fig. 10. Oscillating string with smart structure: S - oscillating string, K- frictionless joints, $\{T_i\}_{i=1,4}$- exciting smart structures, $\{U_i\}_{i=1,4}$ - exciting input voltages

Exciting subsystem T_i includes both blocks SA_i and GU_i, see Fig. 3. Dynamics of these blocks represent transfer functions $\{SA_i(s)\}_i$, $\{SG_i(s)\}_i$ and shaping units of distributed input variables $\{T_i(x)\}_i$. Thus, the dynamics T_i is expressed as transfer function $SA_i(s)\ SG_i(s)\ T_i(x)$. The forms of shaping units of distributed input variables in space domain are shown in Fig. 11.

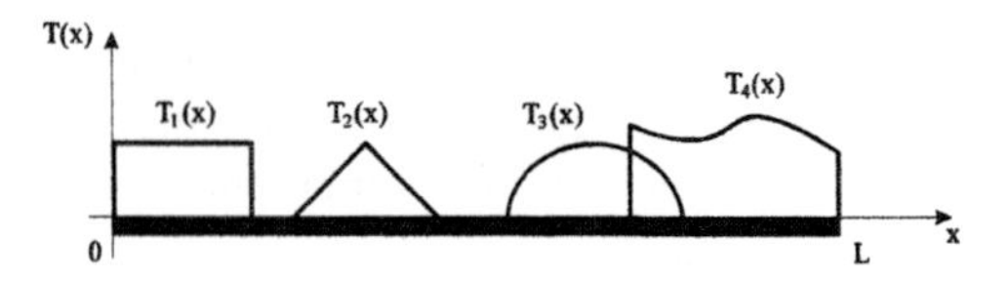

Fig. 11. Layouts and forms of shaping units

The string dynamics is described by distributed transfer function (Butkovskij, 1982):

$$S(x,\xi,s)=\frac{2}{L}\sum_{n=1}^{\infty}\frac{\sin\frac{n\pi x}{L}\sin\frac{n\pi\xi}{L}}{s^2+2\alpha s+\frac{n^2\pi^2a^2}{L^2}} \tag{11}$$

Distributed transfer function $S(x,\xi,s)$ due to a boundary value problem:

$$\frac{\partial^2Y(x,t)}{\partial t^2}+2\alpha\frac{\partial Y(x,t)}{\partial t}-a^2\frac{\partial Y(x,t)}{\partial x^2}=U(x,t) \tag{12}$$

$$Y(x,0)=Y_0(x),\quad \frac{\partial Y(x,0)}{\partial t}=Y_1(x) \tag{13}$$

$$\begin{aligned}&Y(0,t)=g_1(t),\quad Y(L,t)=g_2(t)\\ &0\le x\le L,\quad t\ge 0,\quad a\ne 0\end{aligned} \tag{14}$$

Typical distributed transient responses to particular lumped input variables U_1 - U_4 for given technical parameters of LDS are presented in Fig. 14, where unit variables are: x [m], Amplitude [dm], Time [s].

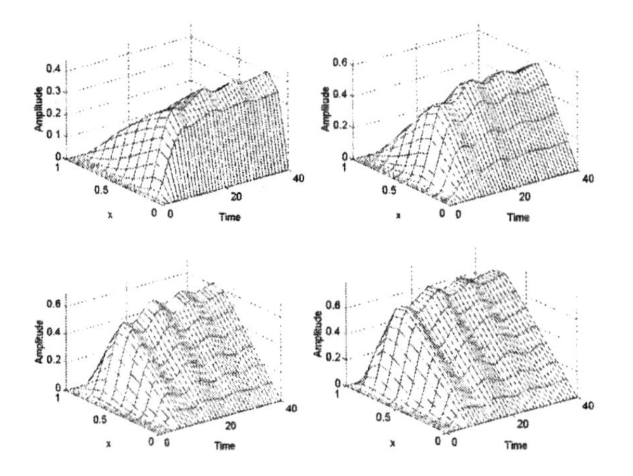

Fig. 12. Distributed transient responses

The control process of string motion in control system with internal model according to the scheme in Fig. 9 with PID controllers in block TS is shown in Fig. 13.

Tuning of controllers was made according ZN rules, where gain margin and associated frequency were determined by MATLAB function *margin*.

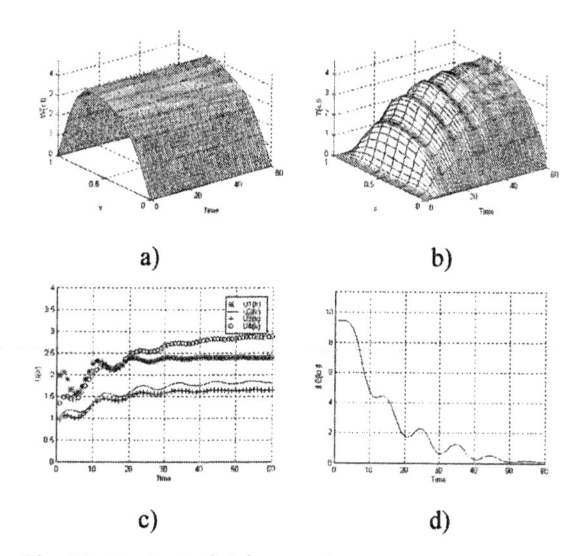

Fig. 13. Control of string motion

a) - distributed reference variable W(x,t)
b) - distributed controlled variable Y(x,t)
c) - control variables $U_i(k)$
d) - quadratic norm of distributed control error

5.2 Control of fixed smart structure membrane motion

Fixed membrane having smart material structure is real LDS, see Fig. 14. The left side the membrane is fixed. Dynamics of excitation subsystems including distributed input variables shaping units in specified subareas $\{\Omega_i\}_{i=1,3}$ are represented by transfer functions: $\{SA_i(s)SG_i(s)\,T_i(x,y)\}_{i=1,3}$. $\{T_i(x,y)\}_{i=1,3}$ have constant forms.

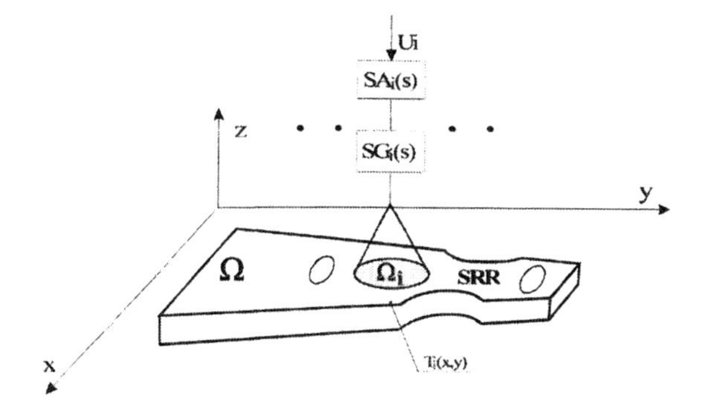

Fig. 14 Fixed membrane with smart material structure

Fixed membrane dynamics is given as a boundary value problem due to hyperbolic PDE in the area Ω.

$$d\frac{\partial^2 Y}{\partial t^2} - \nabla(c\nabla Y) + aY = f \qquad (15)$$

with parameters d, c, a. In fixation points of the membrane Dirichlet boundary conditions, otherwise Neumann ones are to be considered.

Typical results of the control processes of fixed smart structure membrane motion in distributed parameter control systems according Fig.5 with PID controllers in block TS are presented in following figures. Tuning of controllers is based on minimization of the ITAE performance index.

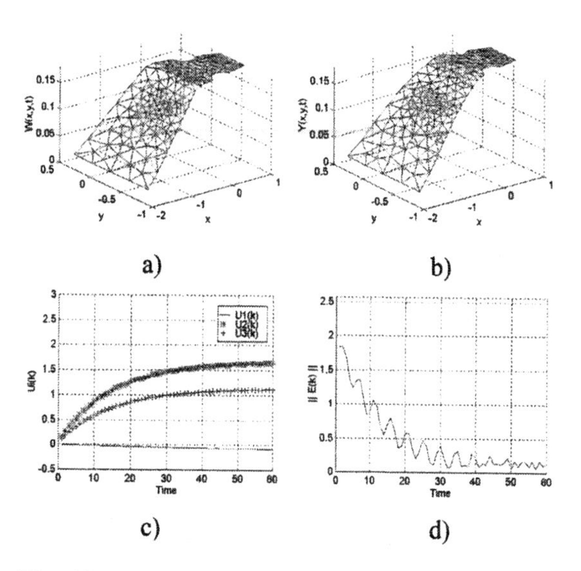

Fig. 15 PID control of fixed membrane motion

a) - distributed reference variable W(x,y,t), $t \to \infty$
b) - distributed controlled variable Y(x,y,t), t=60 s
c) - control variables $U_i(k)$
d) - quadratic norm of distributed control error

5.3 Control of metal plate temperature field

The metal plate with heating units is real LDS, Fig. 16. Dynamics of heating units is given by transfer functions $\{SA_i(s)\ SG_i(s)\ T_i(x,y)\}_i$, where shaping units of distributed input quantities have following forms: $T_1(x,y)$ - constant form, $T_2(x,y)$ - conic form, $T_3(x,y)$ - sinusoidal form, $T_4(x,y)$ - constant form, $T_5(x,y)$ - conic form in specified subareas $\{\Omega_i\}_i$.

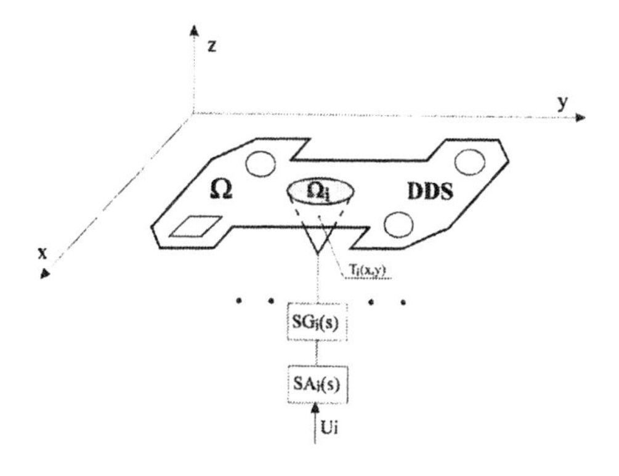

Fig. 16. Metal plate with heating units

Metal plate dynamics as DDS is given in the form of boundary value problem due to parabolic type PDE in an area $\Omega \in E2$:

$$d\frac{\partial Y}{\partial t} - \nabla(c\nabla Y) + aY = f \quad (16)$$

with constants d, c, a and Neumann type boundary conditions

$$\vec{n}(c\nabla Y) + qY = g \quad (17)$$

where $\vec{n}$ is the outward unit normal and q = 0, g = 0.

Mathematical model and dynamic characteristics necessary are determined using the Finite Element Method (FEM). Control synthesis in time domain is based on results of the root locus method, where dominant poles guarantee quality of control. Parameters of controllers were interactively designed using MATLAB GUI function *rltool*. Root locus both for $\{S_i(x_i, s)\}_{i=1}$ and controller (8) is in Fig. 17, where are used constraints: settling time 1000s, peak overshoot 5 %.

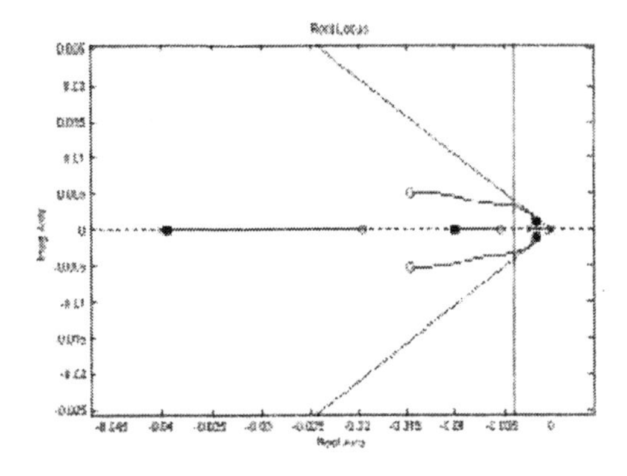

Fig. 17. Root locus for first control loop

Results of the control process are presented in Fig. 18.

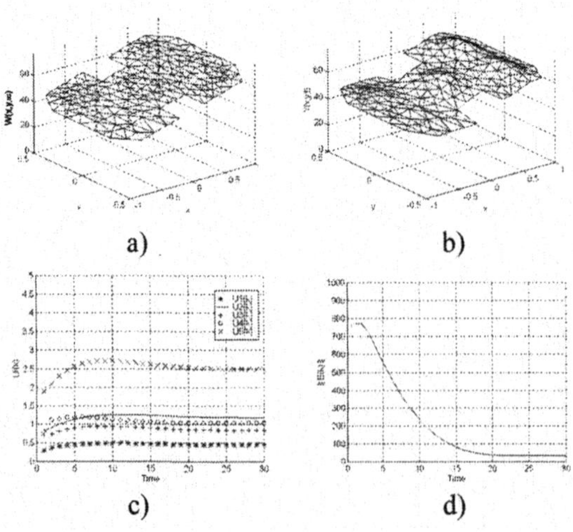

Fig. 18. Control process of metal plate temperature field

a) - distributed reference variable W(x,y,t), $t \to \infty$

b) - distributed controlled variable Y(x,y,t), t=2250 s

c) - control variables $U_i(k)$

d) - quadratic norm of distributed control error

6. CONCLUSION

This paper is dedicated to the original method of modelling and control of distributed parameter systems based on lumped-input/distributed-output systems. Dynamics of such systems is decomposed to space and time components. This representation enables to divide of control task to the time domain and space domain. Control synthesis in space domain is solved as an approximation problem. In time domain discrete-time versions of PID controllers are used. There is possible to use control synthesis methods of lumped systems. Examples of control of DPS are given to illustrate the using approach.

ACKNOWLEDGEMENT

This work has been carried out under the financial support of the VEGA, Slovak Republic project „Control of systems given by numerical structures on complex definition area with demonstrations via Internet" (grant 1/9278/02).

REFERENCES

Butkovskij, A. G. (1982). *Green's Functions and Transfer Functions Handbook.* Ellis Horwood Limited, Publishers Chichester.

Hulkó, G (1987). *Control of Distributed Parameter Systems by means of Multi-Input and Multi-Distributed - Output Systems.* Preprints of 10-th World Congress of IFAC, Munich.

Hulkó, G. *et al.* (1990). *Computer Aided Design of Distributed Parameter Systems of Control.* 11-th World Congress of IFAC, Tallinn.

Hulkó, G. *et al.* (1995). *Identification and Modeling of Lumped-input/distributed-output systems of the Accuracy Requested.* Preprints of European Control Conference ECC 95, Roma.

Hulkó, G. *et al.* (1998). *Modeling, Control and Design of Distributed Parameter Systems with Demonstrations in MATLAB.* Publishing House of STU, Bratislava.

Hulkó, G. *et al.* (1999). *Software package DPSTOOL.* Publishing House of STU, Bratislava.

Lions, J. L. (1971). *Optimal control of systems governed by partial differential equations.* Springer-Verlag.

www.elsevier.com/locate/ifac

A NOVEL APPLICATION OF RELAY FEEDBACK FOR PID AUTO-TUNING

Ioan Nascu[1], **Robin De Keyser**[2]

[1] *Technical University of Cluj Napoca, Department of Automation*
Baritiu 26, Cluj Napoca, 3400, Romania
E-mail: **Ioan.Nascu@aut.utcluj.ro**

[2] *Ghent University, EeSA-department of Electrical Energy, Systems & Automation*
Technologiepark 913, 9052 Gent, Belgium
E-mail: **rdk@autoctrl.rug.ac.be**

Abstract: This paper presents a novel application of the widely used relay-feedback PID-autotuner for stable SISO processes. The proposed method consists of two steps: process identification and controller design. First, a non-iterative procedure is suggested for identification of two points on the process Nyquist curve. A second-order-plus-dead-time model is obtained and then used for PID controller design based on the internal model principle (IMC). For the identification of the two points on the Nyquist curve a pure relay in the feedback loop (as used in standard auto-tuning) and a relay which operates on the integral of the error are used. *Copyright © 2003 IFAC*

Keywords: auto-tuning, PID controller, relay experiment, Nyquist plot .

1. INTRODUCTION

Despite the development of more advanced control strategies, the majority of controllers used in industrial instrumentation are still of the PID type. Their popularity is easy to understand: they have a simple structure, their principle is well understood by instrumentation engineers and their control capabilities have proven to be adequate for most control loops. Moreover, due to process uncertainties, a more sophisticated control scheme is not necessarily more efficient in real-life applications than a well-tuned PID controller. However, it is common that PID controllers are often poorly tuned in practice because the choice of controller parameters requires professional knowledge by the user. To simplify this task and to reduce the time required for it, PID controllers can incorporate *auto-tuning* capabilities, i.e. they are equipped with a mechanism capable of computing the 'correct' parameters automatically when the regulator is connected to the field (Aström, *et al.*, 1984; Aström, *et al.*, 1992; Aström, *et al.*, 1995; Leva, *et al.*, 2002). Auto-tuning is a very desirable feature and almost every industrial PID controller provides it nowadays. These features provide easy-to-use controller tuning and have proven to be well accepted among process engineers.

For the automatic tuning of the PID controllers, several methods have been proposed. Some of these methods are based on identification of one point of the process frequency response, while the others are based on the knowledge of some characteristic parameters of the open-loop process step response. The identification of a point of the process frequency response can be performed either using a proportional regulator, which brings the closed-loop system to the stability boundary, or by a relay forcing the process output to oscillate. Åström and Hägglund (Aström, *et al.*, 1984) report an important and interesting approach. Their method is based on the Ziegler-Nichols frequency domain design formula. A relay connected in a feedback loop with the process is used in order to determine the critical point.

This paper describes the development of an auto-tuning method based on the identification of *two* points on the process Nyquist curve: the intersection with the real negative axis and that with the imaginary negative axis. The use of two relay feedback experiments provides information to determine a four parameters second-order-plus-dead-time (SOPDT) model.

2. THE EXPERIMENTAL SET-UP

A standard PID control system with single input single output, as shown in figure 1, is considered.

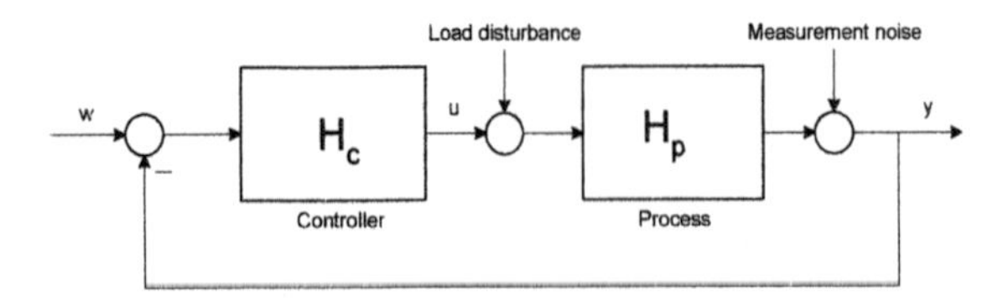

Figure 1. Control system structure

The process H_p is assumed to be linear, stable and proper. The PID controller has a non-interacting structure cascaded with a first order filter:

$$H_c(s) = K_c\left(1 + \frac{1}{sT_i} + sT_d\right)\frac{1}{T_f s + 1} \tag{1}$$

For the purpose of PID-controller design, higher order models are of limited utility even if the process dynamics are theoretically of high order. From the control point of view, complex process models lead to complex controllers. Several papers on PID control are based on the idea of using second order with time delay models. In this paper it is assumed that the process dynamics can be described with reasonable accuracy by a second-order-plus-dead-time (SOPDT) model as:

$$H_p(s) = \frac{k_p e^{-\theta s}}{\tau^2 s^2 + 2\zeta\tau s + 1}, \quad \theta \geq 0 \text{ and } \tau, \zeta > 0 \tag{2}$$

In order to tune the PID controller we will approximate $e^{-\theta s}$ by $(1-\theta s)$ and using the well-known IMC-PID design method we then obtain:

$$\begin{aligned} K_c &= \frac{2\zeta\tau}{k_p(2T_c + \theta)} \\ T_i &= 2\zeta\tau, \quad T_d = \frac{\tau^2}{2\zeta\tau}, \quad T_f = \frac{T_c^2}{2T_c + \theta} \end{aligned} \tag{3}$$

thus resulting in the closed-loop transfer function:

$$H_0(s) = \frac{H_p(s)\cdot H_c(s)}{1 + H_p(s)\cdot H_c(s)} = \frac{e^{-\theta s}}{(T_c s + 1)^2} \tag{4}$$

The parameter T_c is the closed-loop time constant and is a design parameter which can be selected by the user in order to tune the controller 'aggresiveness'.

The parameters of the process transfer function (2) are assumed to be unknown and - in order to identify them - 2 points of the frequency response have to be estimated using two relay feedback experiments: the first experiment uses a relay with hysteresis or an integrating relay (= a relay without hysteresis which has as input the integral of the error) and the second experiment uses a pure relay (without hysteresis). After performing these two experiments the values of the process frequency response at two different frequencies ω_1 and ω_2 are obtained:

$$H_p(j\omega_1) = a_1 + jb_1 \text{ and } H_p(j\omega_2) = a_2 + jb_2 \tag{5}$$

The procedure how to perform these experiments and how to obtain the values for $(a_i, b_i;\ i=1,2)$ is further explained in section 3.

3. THE TUNING METHOD

Note from (2) that $H_p(j\omega)$ can be written as follows:

$$H_p(j\omega) = H'_p(j\omega)H''_p(j\omega), \tag{6}$$

with

$$H'_p(j\omega) = \frac{k_p}{1 - \tau^2\omega^2 + 2\zeta\tau j\omega}, \quad H''_p(j\omega) = (1 - \theta j\omega)$$

If we define - for a certain point on the Nyquist plot - the magnitude, the phase angle and the real and imaginary components of $H_p(j\omega)$, $H'_p(j\omega)$ and $H''_p(j\omega)$ to be M, α, a, b, respectively M', α', a' ,b' and M", α", a", b", then (Ogata, 1995):

$$\begin{aligned} M^2 &= a^2 + b^2 = M'^2 M''^2 = (a'^2 + b'^2)(a''^2 + b''^2) \\ \alpha &= \alpha' + \alpha'' \end{aligned} \tag{7}$$

$$\begin{aligned} tg(\alpha') &= \frac{b'}{a'} = tg(\alpha - \alpha'') = \frac{tg(\alpha) - tg(\alpha'')}{1 + tg(\alpha)tg(\alpha'')} = \\ &= \frac{\frac{b}{a} - \frac{b''}{a''}}{1 + \frac{b}{a}\frac{b''}{a''}} = \frac{a''b - ab''}{aa'' + bb''} \end{aligned} \tag{8}$$

$$b' = a'\frac{a''b - ab''}{aa'' + bb''} \qquad a' = b'\frac{aa'' + bb''}{a''b - ab''} \tag{9}$$

From (9) - and afterwards (7) - it can be written:

$$\begin{aligned} (a'^2 + b'^2) &= a'^2\left[1 + \left(\frac{a''b - ab''}{aa'' + bb''}\right)^2\right] = \\ &= a'^2\frac{(a''^2 + b''^2)(a^2 + b^2)}{(aa'' + bb'')^2} = a'^2\frac{(a''^2 + b''^2)^2(a'^2 + b'^2)}{(aa'' + bb'')^2} \end{aligned}$$

thus leading to:

$$aa'' + bb'' = a'(a''^2 + b''^2) = a'M''^2$$

and using (9) again also to:

$$a''b - ab'' = b'M''^2$$

Thus for the two points obtained from the relay feedback experiments we have:

$$a_1 a_1'' + b_1 b_1'' = a_1' M_1''^2 \quad a_1'' b_1 - a_1 b_1'' = b_1' M_1''^2 \tag{10a}$$

$$a_2 a_2'' + b_2 b_2'' = a_2' M_2''^2 \quad a_2'' b_2 - a_2 b_2'' = b_2' M_2''^2 \tag{10b}$$

The parameters a_i and b_i (i=1, 2) can be obtained using the two relay feedback experiments. In order to identify a point on the process Nyquist curve, a relay connected in a feedback loop with the process is used, forcing the process output to oscillate. Therefore assume that d_i is the relay amplitude and ε_i is the relay hysteresis width. For the given values d_i and ε_i we will obtain oscillation with amplitude h_i and period T_i in the process output.

In the first experiment there are two possibilities: using a relay with hysteresis, a point in the third quadrant on the process Nyquist curve, $P_h(a_1,jb_1)$, is identified (Aström, *et al.*, 1995):

$$\omega_1 = \frac{2\pi}{T_1}, \quad a_1 = \frac{-\pi\sqrt{h_1^2 - \varepsilon_1^2}}{4d_1}, \quad b_1 = -\frac{\pi\varepsilon_1}{4d_1}, \quad M_1^2 = a_1^2 + b_1^2 \tag{11a}$$

or, using an integrating relay ($\varepsilon_1=0$) the point given by the intersection of the process Nyquist curve and the negative imaginary axis, $P_i(a_1,jb_1)$, is identified:

$$\omega_1 = \frac{2\pi}{T_1}, \quad a_1 = 0, \quad b_1 = -\frac{\pi h_1}{4d_1}, \quad M_1^2 = b_1^2 \tag{11b}$$

In the second experiment, using a pure relay ($\varepsilon_2=0$), the point given by the intersection of the process Nyquist curve and the negative real axis, $P_p(a_2,jb_2)$, is identified:

$$\omega_2 = \frac{2\pi}{T_2}, \quad a_2 = \frac{-\pi h_2}{4d_2}, \quad b_2 = 0, \quad M_2^2 = a_2^2 \tag{12}$$

The real and imaginary components of $H_p'(j\omega_i)$ and $H_p''(j\omega_i)$ can be written as:

$$a_i' = \mathrm{Re}\left(H_p'(j\omega_i)\right) = \frac{k_p\left(1-\tau^2\omega_i^2\right)}{\left(1-\tau^2\omega_i^2\right)^2 + \left(2\zeta\tau\omega_i\right)^2} = \frac{\left(1-\tau^2\omega_i^2\right) M_i''^2}{k_p}$$

$$b_i' = \mathrm{Im}\left(H_p'(j\omega_i)\right) = \frac{k_p(-2\zeta\tau\omega_i)}{\left(1-\tau^2\omega_i^2\right)^2 + \left(2\zeta\tau\omega_i\right)^2} = \frac{-2\zeta\tau\omega_i M_i''^2}{k_p}$$

$$a_i'' = \mathrm{Re}\left(H_p''(j\omega_i)\right) = 1 \qquad b_i'' = \mathrm{Im}\left(H_p''(j\omega_i)\right) = -\theta\omega_i \tag{13}$$

Inserting in (10) the relations (13) for a_i', b_i', a_i'', b_i''; $i=1,2$ gives 4 equations with 4 unknown parameters: k_p, θ, τ^2 and $2\zeta\tau$:

$$a_1 - b_1\theta\omega_1 = \frac{\left(1-\tau^2\omega_1^2\right)M_1^2}{k_p} \tag{14.a}$$

$$b_1 + a_1\theta\omega_1 = \frac{-2\zeta\tau\omega_1 M_1^2}{k_p} \tag{14.b}$$

$$a_2 - b_2\theta\omega_2 = \frac{\left(1-\tau^2\omega_2^2\right)M_2^2}{k_p} \tag{14.c}$$

$$b_2 + a_2\theta\omega_2 = \frac{-2\zeta\tau\omega_2 M_2^2}{k_p} \tag{14.d}$$

Using in the first experiment a relay with hysteresis and in the second experiment a pure relay (the 'HR' case) then the first identified point on the process Nyquist curve is in the third quadrant and the second one is the intersection with the negative real axis (thus $b_2=0$). From (14.b) and (14.d) results:

$$\theta = \frac{a_2 b_1}{\omega_1(M_1^2 - a_2 a_1)} \tag{15.a}$$

From (14.c) and (14.a) results:

$$k_p = \frac{M_1^2 M_2^2\left(\omega_2^2 - \omega_1^2\right)}{\omega_2^2 M_2^2\left(a_1 - b_1\theta\omega_1\right) - a_2\omega_1^2 M_1^2} \tag{15.b}$$

Now, (14.a) and (14.b) give:

$$\tau^2 = \frac{1}{\omega_1^2} + \frac{k_p(b_1\theta\omega_1 - a_1)}{\omega_1^2 M_1^2} \tag{15.c}$$

$$2\zeta\tau = -\frac{k_p(b_1 + a_1\theta\omega_1)}{\omega_1 M_1^2} \tag{15.d}$$

These values can then be used in (3) to calculate the PID-parameters.

Using in the first experiment an integrating relay and in the second experiment a pure relay (the 'IR' case) then the first identified point on the process Nyquist curve is the intersection with the negative imaginary axis (thus $a_1=0$) and the second one is the intersection with the negative real axis (thus $b_2=0$). In this case we obtain the following parameters which can then be used in (3) to tune the PID:

$$\theta = \frac{a_2}{\omega_1 b_1}$$

$$k_p = \frac{a_2 b_1^2 \left(\omega_1^2 - \omega_2^2\right)}{b_1\omega_1(a_2\theta\omega_2^2 + b_1\omega_1)}$$

$$\tau^2 = \frac{1}{\omega_1^2}\left(1 + \frac{k_p a_2}{b_1^2}\right) \tag{16}$$

$$2\zeta\tau = -\frac{k_p}{\omega_1 b_1}$$

On the other hand, to perform relay feedback experiments, the process is first brought to steady-state conditions in manual control or with any preliminary tuned PI controller. Measuring steady-state values u_0 and y_0 of the controller and process output, giving then a small perturbation to the control variable and measuring its effect on the process output, the process gain k_p can be easily determined instead of using (15.b) or (16.b). For processes with small nonlinearities better results are then obtained. For processes of first or second order without time delay, the value of θ in the model (2) will be zero. Strictly applying theory, these processes cannot be forced to oscillate by a pure relay. A relay without hysteresis can be used only if the process Nyquist curve crosses the negative real axis. In fact, considering that in any digital controller implementation the sampling process itself introduces a phase lag and that in real situations the process output is filtered, it can be assumed that all the processes in practical cases will oscillate when a relay controller is connected. Unfortunately, in absence of hysteresis, the relay experiment gives small values for the amplitude of oscillation, difficult to measure in the presence of measurement noise. Thus, in case of small amplitude oscillations, in the pure relay experiment a_2 and θ are fixed to zero, k_p is computed from steady-state conditions and τ^2 and $2\zeta\tau$ are computed using measurements only from the first experiment (a_1, b_1, ω_1).

If there are no others restrictions, the amplitude of the relay will be set to $0.1u_0$ and the hysteresis to $0.01y_0$. Better results can be obtained using IR method.

4. EXAMPLES

First the proposed identification method has been tested. An example of the identification with both methods (HR and IR) is shown in Figure 2. The real process transfer function is given by:

$$H_p(s) = \frac{2e^{-50s}}{(80s+1)^3} \tag{17}$$

We identified a second-order-plus-dead-time (SOPDT) model (equation 2). The estimated parameter values are presented in table 1. In this example better results are obtained using an integrating relay (IR case) instead of a relay with hysteresis (HR case). Also, the computations are simpler in the IR case. In both cases better results are obtained determining k_p from steady-state conditions instead of using (15.b) or (16.b). In the IR case (Fig. 2.a) the real process Nyquist plot (solid line) is not too separate from the Nyquist plot of the identified model using steady-state conditions for k_p (dashdot line)

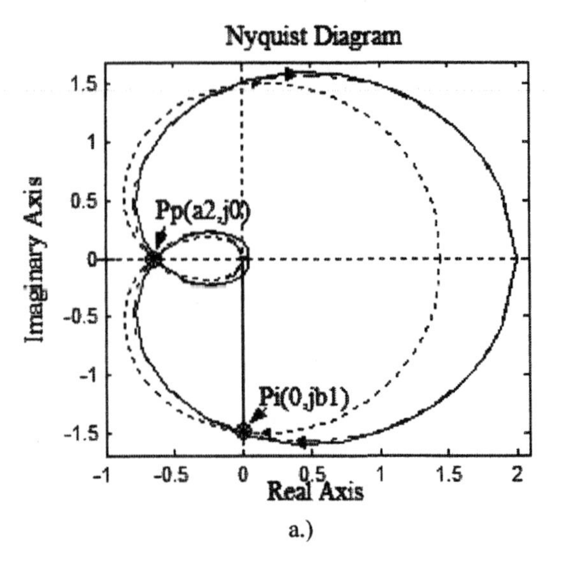

a.)

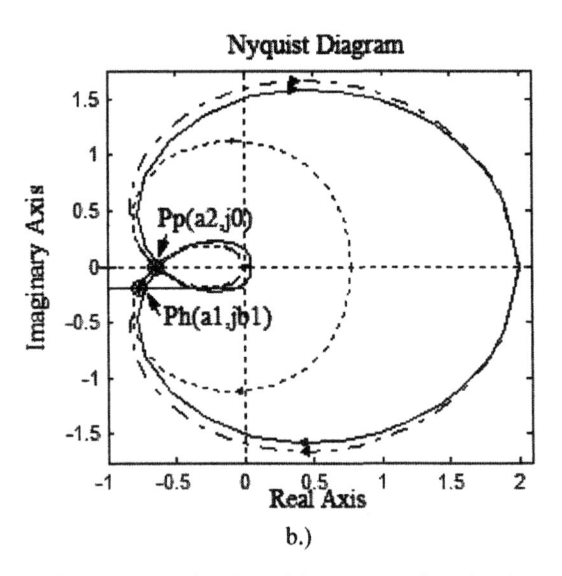

b.)

Figure 2. Nyquist plot of the process given by the transfer function (17) (solid line) and of the process model (2) identified using the two relay experiments IR/HR, with k_p computed from steady-state conditions (dashdot line) respectively from relay experiment (dotted line); a) IR case , b) HR case

Table 1

k_p determination	Method	k_p	θ	$2\zeta\tau$	τ^2
steady-state conditions	HR	2	84	257	20225
	IR	2	78	239	12878
relay-experiment measurements	HR	0.77	84	99	18012
	IR	1.45	78	173	12217

The proposed auto-tuning algorithms have been tested by simulated examples. Different aspects, such as setpoint changes and load disturbances, have been analyzed. The amplitude of the setpoint and the load disturbance is 1. The closed loop step responses using IR method are presented in the left plots and the load disturbance responses in the right plots of Figs. 3/4. The design parameter T_c can be specified by the user or can be chosen automatically as a percent of $2\zeta\tau$ or $2\zeta\tau+\theta$. Here T_c is fixed at $(2\zeta\tau+\theta)/2$ for the upper plots and respectively $(2\zeta\tau+\theta)/10$ for lower plots in Figs. 3/4.

In Fig. 3 the results are presented for processes of order 1 to 4 with no delay:

$$H_p(s) = \frac{2}{\left(\frac{240}{i}s+1\right)^i}, \quad i = 1,\ldots 4. \tag{18}$$

In Fig. 4 the results are presented for the same processes with time delay 50s.

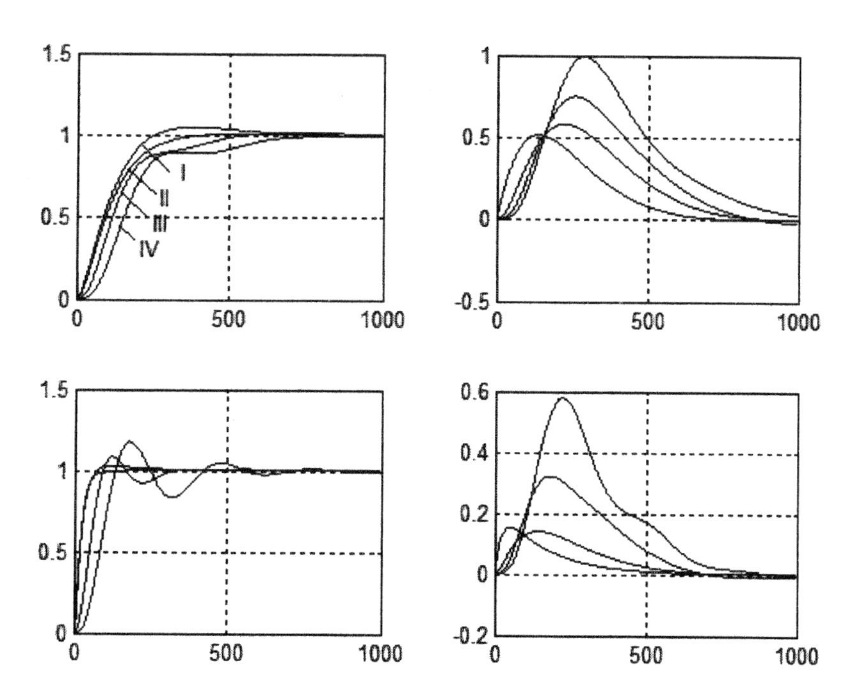

Figure 3. Responses for processes without delay

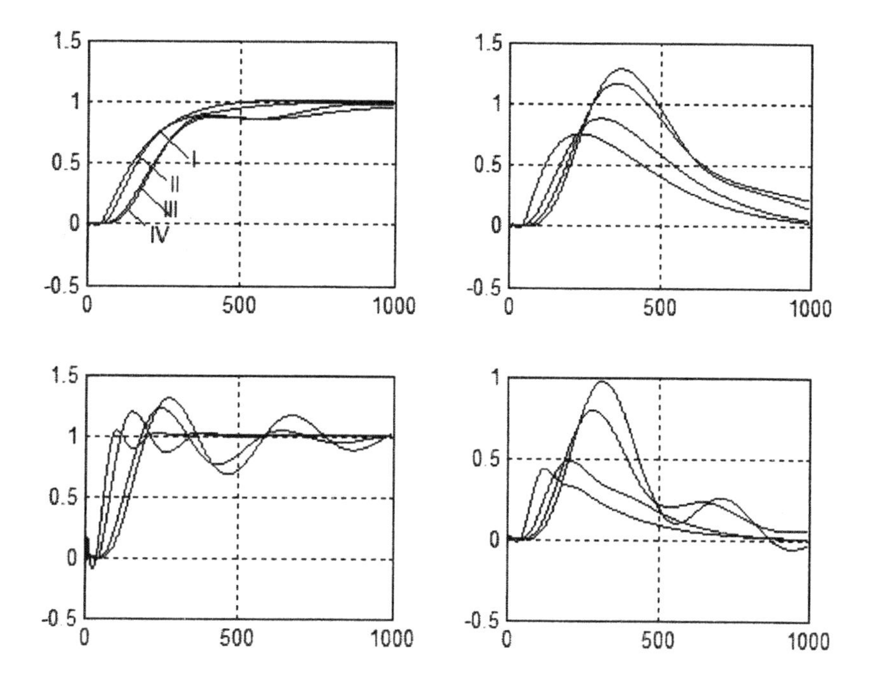

Figure 4. Responses for processes with delay

In figure 5 and 6 the performances of the IR method are compared with two other methods: Ziegler-Nichols method (ZN) and internal model control (IMC). Notice that here IMC can be considered as a kind of reference but this is not an auto-tuning method and it requires the full process model. In figure 5 a second order process given by (18) is considered. The same process with time delay 50 s is considered in figure 6. T_c is fixed at $(2\zeta\tau+\theta)/10$. The closed loop step responses are presented in the upper plots and the load disturbance responses in the lower plots.

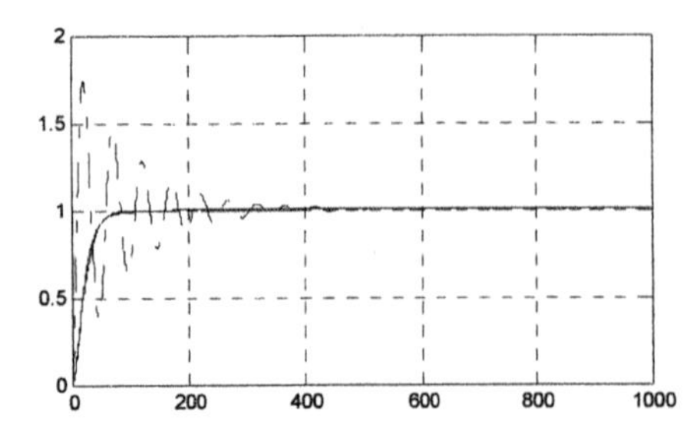

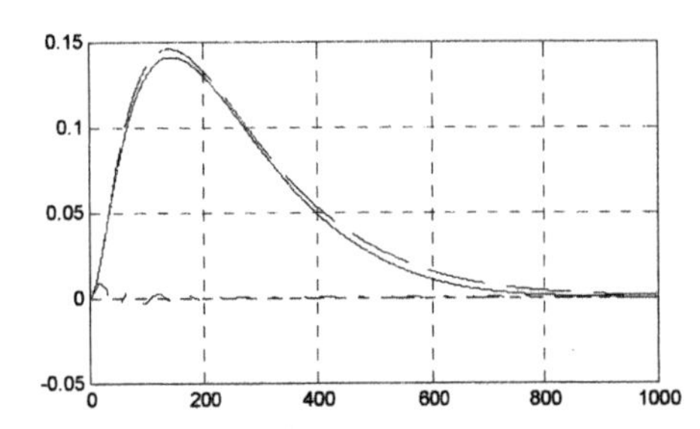

Figure 5. Comparison of the IR(solid line), IMC(dashed) and ZN (dashdot) methods, step response and load disturbance response. Second order process without delay.

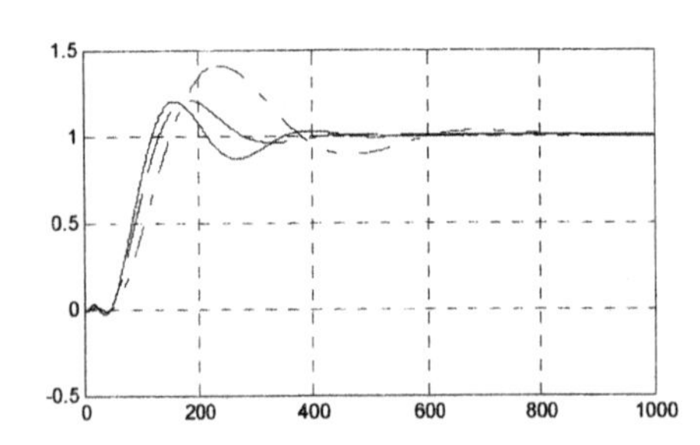

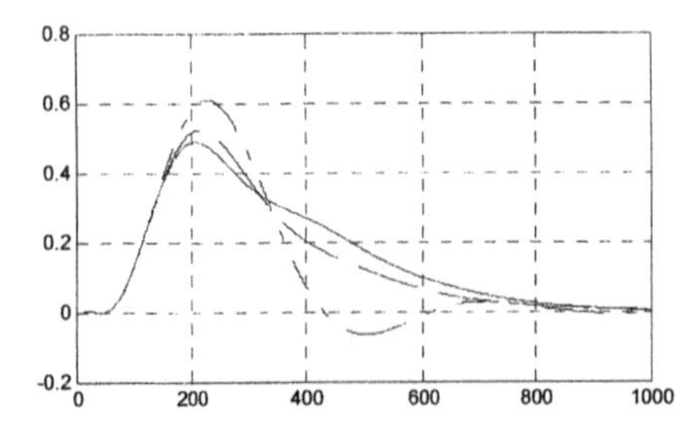

Figure 6. Comparison of the IR(solid line), IMC(dashed) and ZN (dashdot) methods, step response and load disturbance response. Second order process with delay.

5. CONCLUSION

Some relay-based algorithms for auto-tuning of PID controllers have been presented, assuming a simple process model structure and achieving the regulator tuning by identifying two points on the process frequency response and using the IMC-PID design method. This auto-tuning method yields good PID parameters for a specific class of processes (stable processes with self-regulation); it is not a general methodology for arbitrary process models.

Concerning the numerical complexity, the proposed method involves more complicated calculations than Ziegler-Nichols/Åström-Hägglund; however the required experiments are non-iterative and easy to be performed. Experiments and simulation studies have indicated that the presented auto-tuner performs well and can be easily used even by people who are not specialist in automatic control.

REFERENCES

Aström, K.J. and T. Hägglund (1984). Automatic tuning of simple regulators with specifications on phase and amplitude margins, *Automatica*, **Vol.20**, pp.645.

Aström, K.J., C.C. Hang, P. Persson and W.K. Ho (1992). Towards intelligent PID control. *Automatica*, **Vol.28**, pp.1,.

Aström, K.J. and T. Hägglund (1995). *PID Controllers: Theory, Design and Tuning.* Instrument Society of America, Research Triangle Park, NC, USA, 1995.

Leva, A., C. Cox and A. Ruano (2002). Hands-on Pid autotuning: a guide to better utilisation. *IFAC Professional Brief.*

Ogata,K. (1990). *Modern Control Engineering.* Prentice-Hall.

Ho,W.K., C.C. Hang and L.S. Cao (1995). Tuning of PID controllers based on gain and phase margin specifications, *Automatica*, **Vol.31**, No.3, pp.497.

Poulin, E., A. Pomerleau, A. Desbiens and D. Hodouin (1996). Development and evaluation of an auto-tuning and adaptive PID controller, *Automatica*, **Vol.32**, pp.71.

Schei,T.S. (1994). Automatic tuning of PID controllers based on transfer function estimation. *Automatica*, **Vol.30**, pp.1983

A DECOUPLED FUZZY SLIDING MODE APPROACH TO SWING-UP AND STABILIZE AN INVERTED PENDULUM

Mariagrazia Dotoli, Paolo Lino, Biagio Turchiano

Dipartimento di Elettrotecnica ed Elettronica
Politecnico di Bari
Via Re David 200
70125 Bari, Italy

Abstract: In this paper we present a novel decoupled fuzzy sliding mode (DFSM) strategy for swinging-up an inverted pendulum. In the proposed technique the control objective is decomposed into two sub-tasks, i.e., swing-up and stabilization. Accordingly, first a DFSM controller stabilizing the pole is synthesized and optimised via genetic algorithms. Next, a DFSM controller with a piecewise linear sliding manifold is synthesized and optimised, dealing with the task of entering the stabilization zone. Numerical simulations show the effectiveness of the proposed controller for a model encompassing friction as well as a limited control action and a restricted cart travel. *Copyright © 2003 IFAC*

Keywords: nonlinear systems, fuzzy control, sliding mode control, decoupled subsystems, genetic algorithms.

1. INTRODUCTION

Swinging up an inverted pendulum is a well-known benchmark for the investigation of automatic control methodologies. A pole, hinged to a cart moving on a finite track, is swung-up from its pendant equilibrium and stabilized via a motor motioning the cart. Simultaneously, the cart is placed in an objective position on the track: hence, the system is under-actuated. In addition, the control action is limited, since the motor presents a saturation effect.

Numerous studies have been developed on the subject in the scientific literature. Wei, *et al.* (1995) presented a nonlinear technique based on a decomposition of the control law into a sequence of steps, taking into account only the pendulum motion. Wang, *et al.* (1996) applied to the pendulum subsystem the well known parallel distributed decomposition technique, based on a Takagi-Sugeno fuzzy model of the system. Åström and Furuta (2000) as well as Yoshida (1999) proposed energy-based techniques, that work when the track length is unlimited. In such a spirit, recently Chatterjee, *et al.* (2002) proposed a sliding mode control strategy using generalized energy control methods in order to limit the cart position for swinging up a pendulum.

In this paper we introduce a novel Fuzzy Sliding Mode (FSM) technique for swinging-up an inverted pendulum and controlling the connected cart, while minimizing the cart travel. FSM control techniques are hybrid methodologies combining the effectiveness of sliding mode (SM) approaches in controlling nonlinear systems with the immediacy of fuzzy control algorithms. The SM control theory has been successfully established and employed for over a decade for nonlinear systems, see Slotine and Li (1991) for further reading. FSM controllers (Palm, 1994) approximate the nonlinear input/output map of a SM controller by means of a fuzzy inference mechanism (Jantzen, 1999) applied to a linguistic rule base. This results in a smooth control action, reducing the typical SM chattering, and in a restriction of the fuzzy rule table dimension. However, most of the FSM control methodologies presented in the related literature deal with second order nonlinear systems. Recently Lo and Kuo (1998) proposed a method called Decoupled Fuzzy Sliding Mode (DFSM) that deals with fourth order systems and is based on the definition of a control surface encompassing two decoupled controllers designed via two corresponding sliding manifolds. Accordingly, in DFSM control two FSM control objectives are simultaneously taken into account. Moreover, Lo and Kuo (1998) apply the proposed method to the stabilization of an inverted pendulum.

This paper presents a modified DFSM control strategy for swinging-up an inverted pendulum with restricted cart travel. We decompose the control objective into two sub-tasks, namely, swing-up and stabilization: hence, two target zones are defined in the phase plane. First, a DFSM controller stabilizing the pole is designed; next, a similar DFSM controller is synthesized, dealing with the task of entering the stabilization zone. Moreover, a gain scheduler is introduced in order to select the appropriate controller during operation. In particular, following the approach proposed by some of the authors in (Dotoli, *et al.*, 2001), in this paper the swing-up DFSM regulator is defined using a piecewise linear sliding manifold that is bent towards the far off zones of the pole phase plane, in order to let the pole enter the stabilization zone. In addition, we suggest to optimise the controller via Genetic Algorithms (GAs): the interested reader can refer to Dotoli, *et al.* (2002) for a general procedure applying GAs to the optimization of FSM controllers. Several numerical simulations show the effectiveness of the proposed method for an inverted pendulum on a cart including friction and a limited control action, while taking into account restrictions on the cart travel.

The paper is organized as follows. In section 2 the SM and FSM control techniques are briefly reviewed. In section 3 we outline a fourth-order nonlinear model for the inverted pendulum and summarize the control task. Moreover, in section 4 a concise report on the DFSM control technique is given. In section 5 a DFSM controller for the pendulum stabilization is designed first, then a modified DFSM controller is synthesized in order to swing-up the pendulum. Finally, some conclusions are outlined.

2. FUZZY SLIDING MODE CONTROL

In this section we recall the general concepts of SM and FSM control for a second order dynamical system. Consider the following nonlinear second order dynamical system in canonical form:

$$\begin{aligned} \dot{x}_1 &= x_2 \\ \dot{x}_2 &= f(\mathbf{x}) + b(\mathbf{x})u \end{aligned} \tag{1}$$

where $\mathbf{x}=[x_1\ x_2]^T$ is the observable state vector, u is the control input and the system output, $f(\mathbf{x})$ and $b(\mathbf{x})$ are nonlinear functions with $b(\mathbf{x})$ of known bound. In addition, consider a given desired trajectory $x_{1d}(t)$ and the tracking error $e=x_1-x_{1d}$. The basic idea of the SM theory is to force the system, after a finite time reaching phase, to a sliding surface (called sliding line for a second-order system) containing the operating point and defined as follows for (1):

$$s(\mathbf{x}) = \dot{e} + \lambda e = x_2 - x_{2d} + \lambda(x_1 - x_{1d}) = 0 \tag{2}$$

where $x_{2d} = \dot{x}_{1d}$ (representing the desired trajectory for the second state variable) and the sliding constant λ is a strictly positive design parameter. It can be shown that at steady state system (1) follows the desired trajectory once $s(\mathbf{x})=0$, i.e., when the trajectory has reached the sliding line. Hence, in SM control the objective is to design a control law forcing the system onto the sliding surface. Slotine and Li (1991) showed that by selecting the Lyapunov function

$$V = \frac{1}{2}s^2(\mathbf{x}) \tag{3}$$

the sliding line is attractive if the control action is the following:

$$u = \hat{u} - K\mathrm{sign}(s(\mathbf{x})b(\mathbf{x})), \quad K > 0 \tag{4}$$

where 'sign' represents the sign function, and

$$\hat{u} = -b^{-1}(\mathbf{x})(f(\mathbf{x}) - \ddot{x}_{1d} + \lambda\dot{e}). \tag{5}$$

Due to the sign function in (4), the SM controller exhibits chattering, i.e., high frequency switching phenomena. These may be avoided introducing a boundary layer of width Φ, i.e., replacing the sign with a saturation function (Slotine and Li, 1991):

$$u = \hat{u} - K\mathrm{sat}(\Phi^{-1}s(\mathbf{x})b(\mathbf{x})), \quad K, \Phi > 0. \tag{6}$$

However, using a boundary layer can compromise the tracking accuracy, since the tracking error magnitude directly depends on the boundary layer width. A straightforward method reducing the chattering effect is to fuzzify the sliding surface, i.e., combine fuzzy logic algorithms (Jantzen, 1999) with the SM control methodology by replacing K in (5) or (6) with a fuzzy variable K_{fuzz}. The distinctive marks of the resulting FSM technique (Palm, 1994) are the small magnitude of chattering and a low computational effort. It can be shown that FSM is an extension of SM control (Palm, 1994).

The FSM controller rule table is determined on the basis of heuristic conditions in the phase plane, so that the overall control surface yields a SM control law with boundary layer (Slotine and Li, 1991). The procedure does not require a complete identification of the plant and keeps the computational effort relatively low. In the following, the rule table of the FSM regulator is as follows (Lo and Kuo, 1998):

R^1: If s is Negative Big, then u is Positive Big.
R^2: If s is Negative Small, then u is Positive Small.
R^3: If s is Zero, then u is Zero.
R^4: If s is Positive Small, then u is Negative Small.
R^5: If s is Positive Big, then u is Negative Big.

The corresponding membership functions for s and u (Lo and Kuo, 1998) are triangular with completeness equal to 1 (Jantzen, 1999). If the sup-min compositional rule of inference and the center of area defuzzification are adopted (Jantzen, 1999), the FSM emulates a SM with boundary layer (see Lo and Kuo, 1998 and Slotine and Li, 1991).

3. INVERTED PENDULUM ON A CART

In this section we recall the model of an inverted pendulum (see figure 1). A pole, hinged to a cart moving on a track, is balanced upwards by a horizontal force applied to the cart via a DC motor. The cart is simultaneously motioned to an objective position on the track, which is finite. The system observable state vector is $\mathbf{x}=[x_1\ x_2\ x_3\ x_4]^T$, including respectively the cart horizontal distance from the track center, the pole angular distance from the upwards equilibrium point and their derivatives. The force motioning the cart may be expressed as $F=\alpha u$, where u is the input, i.e., the limited motor supply voltage. The system model is (Dotoli, *et al.*, 2001):

$$\begin{aligned} \dot{x}_1 &= x_3 \\ \dot{x}_2 &= x_4 \\ \dot{x}_3 &= f_2(\mathbf{x}) + b_2(\mathbf{x})u \\ \dot{x}_4 &= f_1(\mathbf{x}) + b_1(\mathbf{x})u \end{aligned} \qquad (7)$$

where

$$f_1(\mathbf{x}) = \frac{l\cos x_2(-T_c - \mu x_4^2 \sin x_2) + \mu g \sin x_2 - f_p x_4}{J + \mu l \sin^2 x_2}, \quad (8a)$$

$$b_1(\mathbf{x}) = \frac{l\cos x_2 \alpha}{J + \mu l \sin^2 x_2}, \quad (8b)$$

$$f_2(\mathbf{x}) = \frac{a(-T_c - \mu x_4^2 \sin x_2) + l\cos x_2(\mu g \sin x_2 - f_p x_4)}{J + \mu l \sin^2 x_2}, \quad (8c)$$

$$b_2(\mathbf{x}) = \frac{a\alpha}{J + \mu l \sin^2 x_2}, \quad (8d)$$

with

$$l = \frac{Lm_p}{2(m_c + m_p)}, \quad a = l^2 + \frac{J}{m_c + m_p}, \quad \mu=(m_c+m_p)l.$$

The cart and pole masses are respectively m_c and m_p, g represents the gravity acceleration, L is half the pole length, J is the cart and pole overall moment of inertia with respect to the system center of mass, f_p is the pole rotational friction coefficient and T_c is the horizontal friction acting on the cart, which is a nonlinear function of the cart speed x_3. Note that in (7) $b_1(\mathbf{x})$ and $b_2(\mathbf{x})$ are bounded.

Here, the control task is to swing the pole up from its pendant equilibrium and subsequently stabilize it upwards, while the cart is simultaneously controlled to a position on the track. Hence, the system desired trajectory is $x_{1d}=x_{1d}(t)$, $x_{2d}=0$, $x_{3d}=\dot{x}_{1d}(t)$ and $x_{4d}=0$. There is one control input u, which is bounded: the DC motor saturates.

Remark 1. The inverted pendulum (7) is an under-actuated fourth order system which is not in canonical form. Rather, it includes two second order subsystems, namely, the cart subsystem and the pendulum subsystem, with state vectors respectively $[x_1\ x_3]^T$ and $[x_2\ x_4]^T$, coupled by way of the control input u. For each subsystem the state variables of the other one may be regarded as fictitious uncontrollable inputs.

Remark 2. The pole dynamics, i.e., the second and fourth equations in (7), are affected by the cart only by way of friction T_c. If we neglect this contribution, the pole subsystem is in the form (1). In other words, the main nonlinearities affect the cart subsystem. Hence, to control both the cart and pole with one input, we can balance the rod first, and then the cart.

4. DECOUPLED FUZZY SLIDING MODE

The FSM control technique depicted in Section 2 cannot be applied to a system of the form (7), which is not in canonical form and comprises two coupled subsystems, namely, in the inverted pendulum case, the cart and the pendulum. For such a system Lo and Kuo (1998) proposed the DFSM approach, based on a control surface resulting from two decoupled controllers designed via two corresponding sliding surfaces. In other words, in DFSM control two FSM regulation tasks, addressing the two sub-systems in (7), are taken into account. In the following we briefly review the DFSM technique.

Consider a nonlinear fourth order system expressed in the form (7) with $b_{1,2}(\mathbf{x})$ of known bounds. This system includes two second order subsystems in canonical form with states $[x_1\ x_3]^T$ and $[x_2\ x_4]^T$, respectively. The basic idea of DFSM control is to design a control law such that the single input u simultaneously controls the two subsystems to accomplish the desired performance. To achieve such a goal, the following sliding surfaces are defined (Lo and Kuo, 1998):

$$s_1(\mathbf{x}) = \lambda_1 \cdot (x_2 - x_{2d} - z) + x_4 - x_{4d} = 0 \qquad (9a)$$

$$s_2(\mathbf{x}) = \lambda_2 \cdot (x_1 - x_{1d}) + x_3 - x_{3d} = 0 \qquad (9b)$$

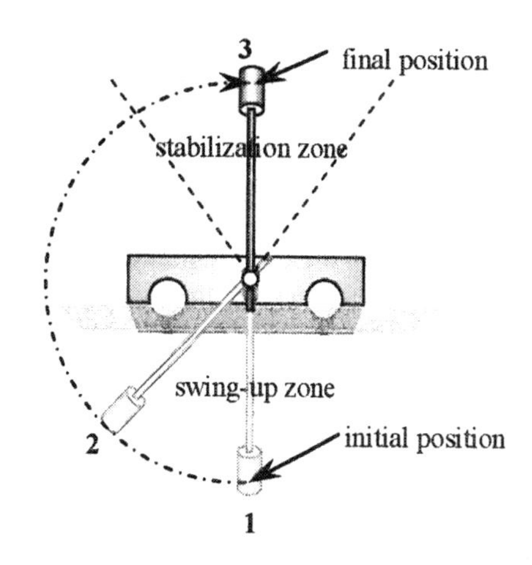

Fig. 1. Pendulum swing-up and stabilization zone.

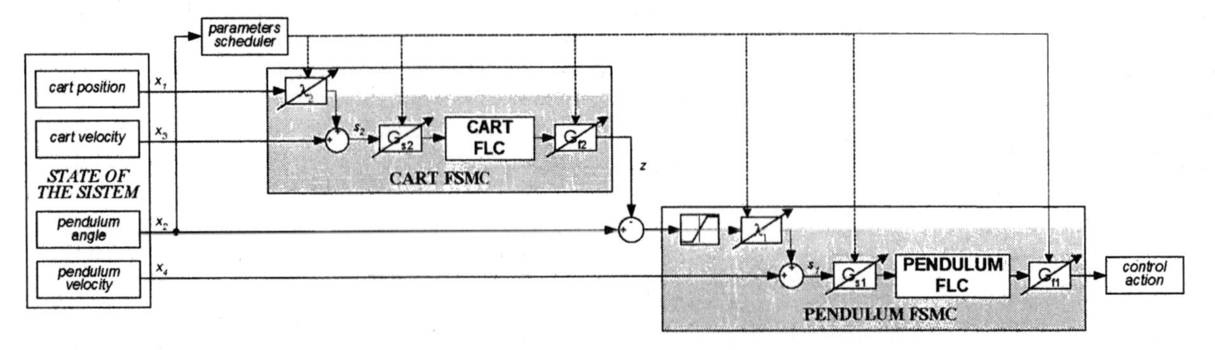

Fig. 2. DFSM controller for stabilization and swing-up of an inverted pendulum with restricted cart travel.

where z is a value proportional to s_2 with the range proper to x_2. A comparison of (9a) with (2) shows the meaning of (9a): the control objective in the first subsystem of (7) changes from $x_2=x_{2d}$ and $x_4=x_{4d}$ to $x_2=x_{2d}+z$ and $x_4=x_{4d}$. On the other hand, (9b) has the same meaning of (2) and its control objectives are $x_1=x_{1d}$ and $x_3=x_{3d}$. Now, let the control law for (9a) be a FSM emulating a SM with boundary layer (6):

$$u_1 = \hat{u}_1 - G_{f_1}\text{sat}(s_1(\mathbf{x})b_1(\mathbf{x})G_{s_1}),\ G_{f1},G_{s1}>0 \quad (10)$$

with

$$\hat{u}_1 = -b_1^{-1}(\mathbf{x})(f_1(\mathbf{x}) - \ddot{x}_{2d} + \lambda_1 x_4 - \lambda_1 \dot{x}_{2d}) \quad (11)$$

and obvious meaning of the other parameters in (10). So, let the control law for (9b) be a FSM mimicking a SM with boundary layer (Lo and Kuo, 1998):

$$z = \text{sat}(s_2 \cdot G_{s_2}) \cdot G_{f_2},\quad 0 < G_{f_2} < 1 \quad (12)$$

where G_{s2} represents the inverse of the width of the boundary layer to s_2, G_{f2} transfers s_2 to the proper range of x_2 and 'sat' indicates the saturation function. Notice that in (12) z is a decaying oscillation signal, since $G_{f2}<1$. Moreover, in (9a) if $s_1=0$ then $x_2=x_{2d}+z$ and $x_4=x_{4d}$.

Now, the control sequence is as follows: when $s_2\neq0$, then $z\neq0$ in (9a) causes (10) to generate a control action that reduces s_2; as s_2 decreases, z decreases too. Hence, at the limit $s_2\rightarrow0$ with $x_1\rightarrow x_{1d}$, then $z\rightarrow0$ with $x_2\rightarrow x_{2d}$, so $s_1\rightarrow0$ and the control objective is achieved (Lo and Kuo, 1998).

Remark 3. Let n be the number of fuzzy sets for each s(**x**) variable attached to the two subsystems. Accordingly, the overall controller comprises n rules of the type described in section 2 for both FSM sub-controllers. Hence, the rule table dimension of the DFSM controller equals 2n. Such a value is minimal when compared with the corresponding typical fuzzy controller rule base dimension, that equals n^2 when n fuzzy labels describe the error of each output in (7).

5. DFSM CONTROL OF THE INVERTED PENDULUM: STABILIZATION AND SWING-UP

The DFSM technique may be implemented for stabilizing the inverted pendulum on a cart (7) with a minimal dimension of the corresponding fuzzy rule table. However, six parameters have to be tuned in the DFSM regulator presented in (Lo and Kuo, 1998): the sliding parameters λ_i (i=1,2) and the input/output normalization gains G_{si} and G_{fi} (i=1,2).

In the following we modify the DFSM technique in order to swing-up the inverted pendulum on a cart. Moreover, we propose the use of GAs for automatic optimisation of the resulting control strategy. In addition, we take into account restrictions such as a finite track and a limited available control action, which are typical for a real system.

5.1 The DFSM controller structure.

It can be shown that a control law implementing equations (9)-(12) is not sufficient to swing the pendulum up and simultaneously control the cart position with the above physical restrictions. Therefore, we decompose the control objective into two sub-tasks, namely, swing-up and stabilization. Hence, two target zones are defined in the phase plane (see figure 1). Accordingly, the controller is defined as follows. First, a DFSM controller stabilizing the pole is designed according to equations (9) to (12); next, a similar DFSM controller is synthesized, dealing with the task of entering the stabilization zone. Finally, a gain scheduler is introduced in order to select the appropriate controller during operation.

The resulting controller structure is depicted in figure 2. The parameters scheduler in figure 2 checks whether the pole is in the stabilization or the swing-up zone and applies the corresponding controller accordingly. In this paper we employ a stabilization zone with range $[-\pi/3,\pi/3]$.

5.2 Genetic optimisation of the DFSM controller.

In the following we propose to automatically tune the parameters of the DFSM controller (see figure 2). Optimisation of the regulator is performed via GAs (Michalewicz, 1996): in other words, we define a genetic search problem whose generic solution is a DFSM controller, implementing in turn stabilization or swing-up. Such a fuzzy system is defined by the input and output membership functions, the fuzzy rules and the mathematical operators performing fuzzification, inference and defuzzification. In principle, it is possible to develop a GA both discovering and optimizing the structure (the rules and operators) as well as the parameters associated to the fuzzy system. However, in order to contain the GA computational complexity and convergence time, this paper concerns the optimization of the parameters of a DFMS controller with respect to a given rule base, as well as to a set of predetermined parameters defining position and shape of the membership functions (see section 2). Hence, the GA optimizes the DFSM controller input/output gains G_{si} (i=1,2) and G_{fi} (i=1,2) as well as the sliding parameters λ_i (i=1,2). Here, all the controller parameters are real values spanning in bounded intervals. This suggested the use of real valued chromosomes as opposed to the traditional binary encoding schema (Michalewicz, 1996). The interested reader can refer to (Dotoli, *et al.*, 2002) for a thorough discussion of automatic optimisation of FSM controllers via GAs.

We propose as GA fitness function (i.e., the design objective to achieve) a two-criteria index, combining with different weights the integral time absolute error of the cart position ($ITAE_{pos}$) with respect to the set point, i.e., $x_{1d}=x_{1d}(t)$ and the ITAE of the pole angle ($ITAE_{angle}$) with respect to the set point, i.e., $x_{2d}=0$. Hence, the fitness measures the performance of each trial DFSM controller generated by the GA and encompassing a limited output. In other words, the GA searches the solution space for DFSM controllers guaranteeing both good set point tracking and low control actions.

5.3 DFSM control for pendulum stabilization and swing up.

The first design step in the proposed procedure is to find a DFSM controller for stabilization according to equations (9) to (12) and optimise it via GAs with the following fitness:

$$Fitness_1=0.5\cdot ITAE_{pos}+0.5\cdot ITAE_{angle}. \quad (13)$$

Then, following the approach proposed by Dotoli, *et al.* (2001), we define the swing-up DFSM regulator using a piecewise linear sliding manifold that is bent in the external regions of the pole error phase plane, in order to let the pole enter the stabilization zone with the physical restrictions previously discussed.

Due to the finite track length and the limited control action, the above DFSM controller cannot swing-up the inverted pendulum from its downwards position, i.e., when the initial error on the angle equals π. In order to swing-up the pole and simultaneously control the cart, we modify the previous DFSM regulator as follows. Since the pole angle value is periodical, the space $(x_2, x_4)\equiv(e_2, e_4)$ is a cylindrical surface of width 2π. Cutting the cylinder along its generatrix passing by the origin determines a discontinuous sliding surface in $(\pi,0)$, see (Dotoli, *et al.*, 2001). Clearly, the proposed control action increases correspondingly to the error value (see section 2); since the actual control action is limited by the DC drive maximum torque, in case of consistent errors on the pole angle the actuator cannot provide the energy required to stabilize the system. In fact, a FSM control law is designed to make the sliding manifold attractive for the trajectory: however, if the initial error on the angle is too large, the trajectory deviates from the switching manifold and spirally converges to a limit cycle around the downwards equilibrium position. In other words, due to the actuator saturation, the pendulum oscillates around its pendant equilibrium.

Now, we modify the sliding surface in order to achieve swing-up with the available control action. The ultimate aim of DFSM control is reaching the desired trajectory through a sliding mode. Hence, in order to ease the reaching phase we suggest to decrease $s_1(\mathbf{x})$, i.e., the vertical distance of the pole trajectory from the sliding surface, when the error exceeds a threshold. The sliding variable is left unaltered for small error values, in order to keep the convergence speed and stability to the upwards equilibrium. As a result, the swing-up initial position $(\pi,0)$ is closer to the modified sliding surface than to the linear switching manifold, and the novel sliding variable is considerably smaller in magnitude than the original one (see figures 3a and 3b). Now, since the control action changes according to the magnitude of the sliding variable (see the FSMC rule table in section 2), the DFSM controller output is correspondingly reduced. Hence, the proposed DFSM with piecewise linear sliding manifold represents an effective alternative to DFSM with linear switching surface in systems with saturated control input. In particular, in this paper we bend the sliding surface for pole errors $e_2\geq 1.86$ [rad].

The above device makes it possible to swing-up the pendulum. Moreover, just like for the DFSM controller operating in the stabilization zone, in order to enhance the controller's performance during swing-up, we optimize it via GAs with the fitness:

$$Fitness_2=0.8\cdot ITAE_{pos}+0.2\cdot ITAE_{angle}. \quad (14)$$

We remark that now the main objective is swinging up the pole, hence in (14) the $ITAE_{angle}$ contribution is multiplied by a higher weight than in (13).

Summing up, we define and optimise two different sets of controller parameters, corresponding to the two operating zones (see figures 1 and 2):

1. zone 1 for $|x_2|<\pi/3$ (stabilization zone) with $\lambda_i=\lambda_{is}$, $G_{fi}=G_{fis}$ and $G_{fi}=G_{fis}$ (i=1,2);
2. zone 2 for $\pi/3<|x_2|<\pi$ (swing-up zone) with $\lambda_i=\lambda_{isu}$, $G_{fi}=G_{fisu}$ and $G_{fi}=G_{fisu}$ (i=1,2).

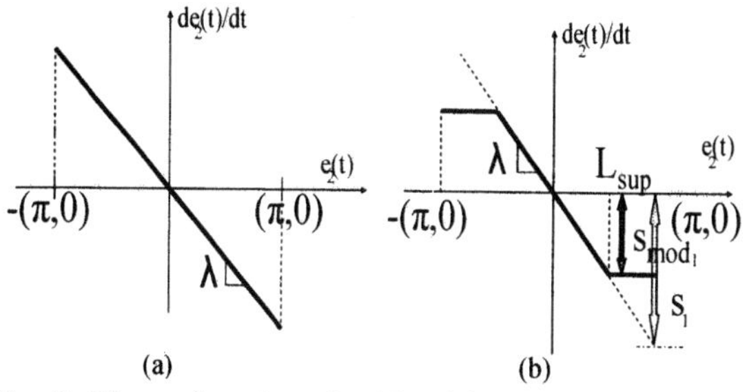

Fig. 3. Phase plane (e_2, e_4) with original (a) and modified (b) sliding surface for the pole subsystem (z=0).

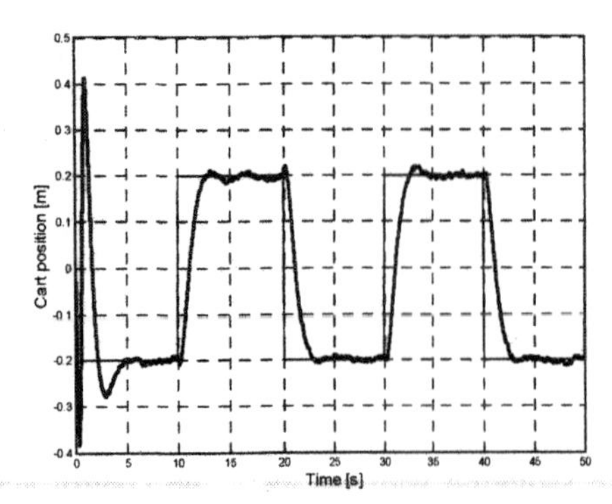

Fig.4. Simulation results for cart position.

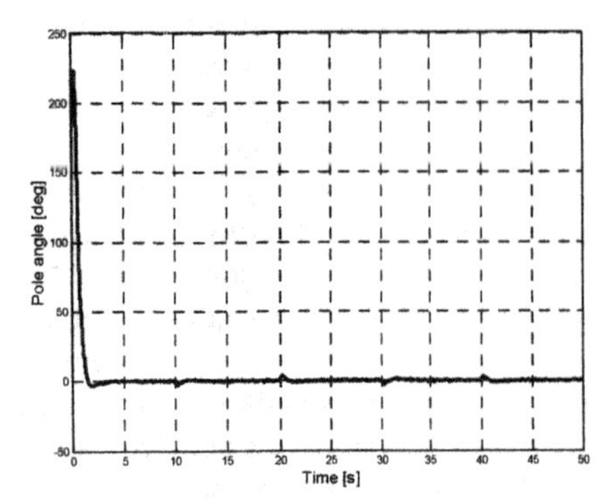

Fig. 5. Simulation results for pole angle.

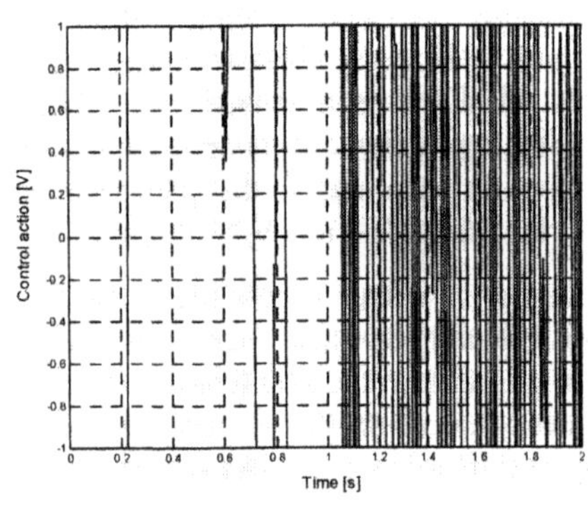

Fig. 6. Simulation results for control action (swing-up).

5.4 Simulation results.

The performance of the proposed DFSM controller is shown in this section. We report some simulation tests carried out on the inverted pendulum in the MATLAB environment, with the objective of swinging-up the pole from its downwards stable equilibrium and simultaneously control the cart position. Figures 4 and 5 display the results for the proposed DFSM control strategy, with a 0.2 m amplitude and 30 s period square wave reference for the cart position. While the cart follows the step reference, the pole is initially swung-up and stabilized (first wave period) and subsequently kept in the upwards position (successive wave periods). The control action during swing-up (figure 6) is kept within the range allowed by the actuator.

CONCLUSIONS

In this paper we propose a novel Decoupled Fuzzy Sliding Mode (DFSM) technique for swinging-up an inverted pendulum and controlling the connected cart, while minimizing the cart travel. The technique is based on a decomposition of the control objective into two sub-tasks, i.e., swing-up and stabilization. Accordingly, on the basis of the pole angular position, we define a gain scheduler selecting the DFSM to control the system. While the stabilization controller is a classical DFSM regulator, we design a swing-up controller based on a piecewise linear sliding manifold that is bent at the extremes of the pole phase plane. The advantage of the approach is in the reduced magnitude of the control action, making it possible to cope with saturation effects. In addition, we suggest the use of genetic optimisation for automatic tuning of the controller parameters. Simulations prove the effectiveness of the proposed technique both for stabilization and swing-up: the upwards pendulum equilibrium is globally stable.

REFERENCES

Åström K.J. and K. Furuta (2000). Swinging up a Pendulum by Energy Control, *Automatica*, **36**, 287-295.

Chatterjee D., A. Patra and H. K. Joglekar (2002). Swing-up and Stabilization of a Cart-Pendulum System under Restricted Cart Length, *Systems & Control Letters*, **47**, 355-364.

Dotoli M., B. Maione, D. Naso and B. Turchiano (2001). Fuzzy Sliding Mode Control for Inverted Pendulum Swing-up with Restricted Travel. In: *Proceedings of the 10th IEEE Conference on Fuzzy Systems*, Melbourne, Australia, 753-756.

Dotoli M., B. Maione and D. Naso (2002). Fuzzy Sliding Mode Controllers Synthesis Through Genetic Optimization. In: *Advances in Computational Intelligence and Learning, Methods and Applications*, Zimmermann H-J., G. Tselentis, M. van Someren and G. Dounias (Eds.), Kluwer Academic, Boston, 331-341.

Jantzen J. (1999). Linear Design Approach to a Fuzzy Controller. In: *Fuzzy logic Control: Advances in Applications*, Verbruggen H. B. and R. Babuska. (Eds.), World Scientific, Singapore, 313-329.

Lo J.-C. and Y.-H. Kuo (1998). Decoupled Fuzzy Sliding-Mode Control, *IEEE Transactions on Fuzzy Systems*, **6**, 426-435.

Michalewicz Z., (1996). *Genetic Algorithms + Data Structures = Evolution Programs*, Springer-Verlag, Berlin.

Palm R., (1994). Robust Control by Fuzzy Sliding Mode, *Automatica*, **30**, 1429-1437.

Slotine J.-J. and W. Li (1991). *Applied Non Linear Control*, Prentice Hall, Englewood Cliffs, NJ.

Yoshida K. (1999). Swing-up Control of an inverted pendulum by energy-based methods. In: *Proceedings of the American Control Conference*, San Diego, California, 4045-4047.

Wang H. O., K. Tanaka and M. F. Griffin (1996). An Approach to Fuzzy Control of Nonlinear Systems: Stability and Design Issues, *IEEE Transactions on Fuzzy Systems*, **4**, 14-23.

Wei Q., W.P. Dayawansa and W.S. Levine (1995). Nonlinear Controller for an Inverted Pendulum Having Restricted Travel, *Automatica*, **31**, 841-850.

OPTIMIZATION OF EXTENDED KALMAN FILTER FOR IMPROVED THRESHOLDING PERFORMANCE

Knut Rapp * **Per-Ole Nyman** *

* *Narvik University College, P.O. Box 385, 8515 Narvik, NORWAY*

Abstract: In this paper it is shown how a simple genetic algorithm can be utilized for tuning the parameters of an extended Kalman filter. Results from applying this genetic algorithm on a specific problem related to signal tracking is shown. Further, the performance index and how to obtain a simple objective function for the genetic algorithm are discussed. Finally it is shown how the genetic algorithm can be modified in order to optimize the filters thresholding performance. *Copyright © 2003 IFAC*

Keywords: Extended Kalman filters, Genetic algorithms, filter design, optimal filtering, search methods

1. INTRODUCTION

Since its introduction in 1960, the Kalman filter technique has grown to be perhaps one of the most often used in the area of process control. The results achieved from the last forty years of research has contributed to a continued expansion of this technique.

In spite of this, the problems of parameter tuning are still not very well investigated. A very limited amount of paper and results seems to be published, but some methods for tuning the filter, including the extended Kalman filter (EFK), are reported see (Powell, 2002), (Oshman and Shaviv, 2000).

Normally it can be assumed that the process measurement noise covariance matrix is known, either from measured values, or from statistics applied on the real process data. However, if the noise covariance is not constant, the filter is to be designed to handle the variations which may occur.

In this paper the problems of tuning and optimizing an EKF used for tracking a periodic signal from a noisy measurement are considered. The signal to be tracked is assumed to be nearly sinusoidal, with varying frequency and amplitude. Further the problem with time varying measurement noise are treated. A simple design method is applied, and the results are discussed.

This filter is applied to the classical problem of controlling the balling drums used in the iron ore industry. For this reason it is assumed that the signal to be tracked has quite low frequency. Normally it will be in the interval (0.015, 0.15)$[\frac{rad}{s}]$ For detail see e.g (Rapp and Nyman, 2003a).

The outline of the paper is as follows: In Section 2, the signal model is described and the EKF equations for this specific model are stated. In section 3, a very brief description of the up today reported results on automatic filter tuning is given. Some results regarding tuning parameters and performance indices are mentioned. It is shown how different choices of the weighting matrix will give filters with different properties. This is illustrated by three different cases. In Section 4, the problem of increasing the filters thresholding performance is discussed, and it is demonstrated how the genetic algorithm may be utilized to achieve this. In Section 5 concluding remarks are stated.

2. SIGNAL MODEL AND EKF EQUATIONS

2.1 *Signal model*

The signal model used by the EKF is a third order state space model, as follows:

$$x_{k+1}^{(1)} = x_k^{(1)} + v_k^{(1)} \tag{1}$$

$$x_{k+1}^{(2)} = x_k^{(1)} + x_k^{(2)} \tag{2}$$

$$x_{k+1}^{(3)} = x_k^3 + v_k^{(2)} \tag{3}$$

$$y_k = x_k^3 \sin x_k^2 + z_k \tag{4}$$

where $x^{(1)}$ is the phase increment (or frequency), $x^{(2)}$ is the phase and $x^{(3)}$ is the amplitude of the oscillation. $\mathbf{v} = [v^{(1)} \quad v^{(2)}]^T$ and z are white, zero mean processes with covariance matrices $\mathbf{Q} = \text{diag}(q_1, q_2)$ and $\mathbf{R_s}$. Locally the state is uniquely determined by the output y, but owing to the factor $\sin(x^{(2)})$ in output equation, this does not hold globally. In fact, a simultaneous change of sign in $x^{(1)}$ and $x^{(2)}$, or a shift of $x^{(2)}$ by any number of periods, does not change the output. However, with a reasonable initialization of the EKF this mild nonuniqueness does in general not cause problems. The choice of the matrix $\mathbf{Q}$ is a compromise between accuracy in steady state and capability to track a changing amplitude or frequency. $\mathbf{R}$ is set equal to the covariance of assumed measurement noise of the true, measured output.

2.2 *EKF equations*

Written in a more compact form, equation (1)-(4) are given by

$$\mathbf{x}_{k+1} = \mathbf{A}\mathbf{x}_k + \mathbf{B}\mathbf{v}_k \tag{5}$$

$$\mathbf{y}_k = g_k(\mathbf{x}_k) + z_k \tag{6}$$

The EKF equations derived from equation (1)-(4) are given by (see e.g. (Chui and Chen, 1999)):

$$\widehat{\mathbf{x}}_{k,k-1} = \mathbf{A}\widehat{\mathbf{x}}_{k-1} \tag{7}$$

$$\widehat{\mathbf{P}}_{k,k-1} = \mathbf{A} \cdot [\mathbf{P}_{k-1,k-1}] \cdot \mathbf{A}^T + \mathbf{B} \cdot \mathbf{Q} \cdot \mathbf{B}^T \tag{8}$$

$$\widehat{\mathbf{x}}_{k,k} = \widehat{\mathbf{x}}_{k,k-1} + \mathbf{K}_k(y_k - g_k(\widehat{\mathbf{x}}_{k,k-1})) \tag{9}$$

$$\widehat{\mathbf{P}}_{k,k} = \widehat{\mathbf{P}}_{k,k-1} - \mathbf{K}_k \left[\frac{\partial g_k}{\partial \mathbf{x}_k}(\widehat{\mathbf{x}}_{k,k-1})\right] \widehat{\mathbf{P}}_{k,k-1} \tag{10}$$

where $\mathbf{A} = \begin{bmatrix} 1 & 0 & 0 \\ 1 & 1 & 0 \\ 0 & 0 & 1 \end{bmatrix}$, $\quad \mathbf{B} = \begin{bmatrix} 1 & 0 \\ 0 & 0 \\ 0 & 1 \end{bmatrix}$

and

$$\mathbf{G}_k = \left[\frac{\partial g_k}{\partial \mathbf{x}_k}\right]_{\mathbf{x}=\mathbf{x}_{k,k-1}} = \left[0 \;\; x_{k,k-1}^{(3)} \cos(x_{k,k-1}^{(2)}) \;\; \sin(x_{k,k-1}^{(2)})\right] \tag{11}$$

The filter gain matrix is given by

$$\mathbf{K}_k = \mathbf{P}_{k,k-1}\left[\mathbf{G}_k(\widehat{\mathbf{x}}_{k,k-1})\right]^T \cdot \left[\left[\mathbf{G}_k(\widehat{\mathbf{x}}_{k,k-1})\right]\mathbf{P}_{k,k-1}\left[\mathbf{G}_k(\widehat{\mathbf{x}}_{k,k-1}) + \mathbf{R}\right]^T\right]^{-1} \tag{12}$$

3. AUTOMATIC FILTER TUNING

3.1 *Automatic tuning methods*

As mentioned in section 1, only a few methods for automatic parameter tuning are reported. In (Powell, 2002) some numerical methods are listed. In these methods the tuning problem is converted to a numerical optimization problem. In particular, Powell discusses and gives examples of how a simplex downhill algorithm can be used to solve the tuning problem when it is posed as a numerical optimization problem. In (Oshman and Shaviv, 2000), the use of genetic algoritms are discussed. In this paper, an genetic algorithm is applied for the numerical optimization.

3.2 *Filter specifications*

This work focuses on the filters ability to track the signal frequency and amplitude. As mentioned in section 1, the signal is assumed to have a quite low frequency.

Three properties are considered to be important:

a) The filter should be able to track the signal even if it is very noisy
b) The variance in the filters estimate is to be as low as possible
c) The filters transient response is to be sufficiently fast

As in almost all set of specifications, this set also include some contradictory wishes. It may be difficult, or even impossible, to obtain very quick transient response and very low variance in the estimate at the same time. An optimally tuned filter will therefore be a trade off between the specifications listed above.

3.3 *Filter performance index*

In this paper $\mathbf{R}$ is assumed known so q_1 and q_2 are the available filter tuning parameters (see (Maybeck, 1979)). In order to apply an numerical optimization algorithm, a suitable performance index is to be available, see (Rapp and

Nyman, 2003b) or (Powell, 2002). A reasonable filter performance index is the (weighted) RMS of the estimation error. As $\mathbf{R}$ is constant the performance index will be a function of the elements in $\mathbf{Q}$ only. The performance index is given by:

$$J(q_1, q_2) = \left[\frac{1}{T+1} \sum_{t=0}^{T} (\hat{\mathbf{e}}_t^T \mathbf{W} \hat{\mathbf{e}}_t) \right] \quad (13)$$

where $\hat{\mathbf{e}} = [(\widehat{x}_1 - x_1), (\widehat{x}_2 - x_2), (\widehat{x}_3 - x_3)]$ is the estimation error vector, T is the final time, which equals the number of samples and $\mathbf{W} = \text{diag}(w_1\ w_2\ w_3)$ is the 3x3 weighting matrix. In all simulations $T = 2000$.

3.4 *The evaluation function*

A basic element in an genetic algorithm is the evaluation function. The evaluation function gives the "fitness value" for each filter candidate to be considered. It is specific for each application, and depends also of the genetic algorithm to be used. The GA used in this work is the free GAOT toolbox, which is design to maximize the evaluation function rather than minimizing it. See (Houck *et al.*, 1996) for a complete description of this algorithm and how it is implemented. A suitable evaluation function candidate is given by

$$E(\mathbf{Q}) = N \left[\sum_{1}^{N} \left(\frac{1}{T+1} \sum_{t=0}^{T} (\hat{\mathbf{e}}_t^T \mathbf{W} \hat{\mathbf{e}}_t) \right) \right]^{-1} \quad (14)$$

where N is the number of Monte Carlo runs used for evaluating each filter candidate. See (Rapp and Nyman, 2003b) for further details.

3.5 *Results from simulations*

The filter is tuned with a signal having amplitude 1 and frequency $6 \cdot 10^{-2}\ \frac{rad}{s}$. The signal noise variances are $\mathbf{R_s} = 1$ and $\mathbf{Q_s} = 10^{-8}$.

Three cases are considered.

1. Equal weight on the frequency- and amplitude error.
2. High weight on the frequency error
3. High weight on the amplitude error

In all three cases the initial population is set to 200 individuals, and it evolves over 100 generations.

In Figure 1, the amplitude and frequency diagram from this filter in case 1 is shown. The dashed line is the true amplitude and the solid-drawn line is the estimate.

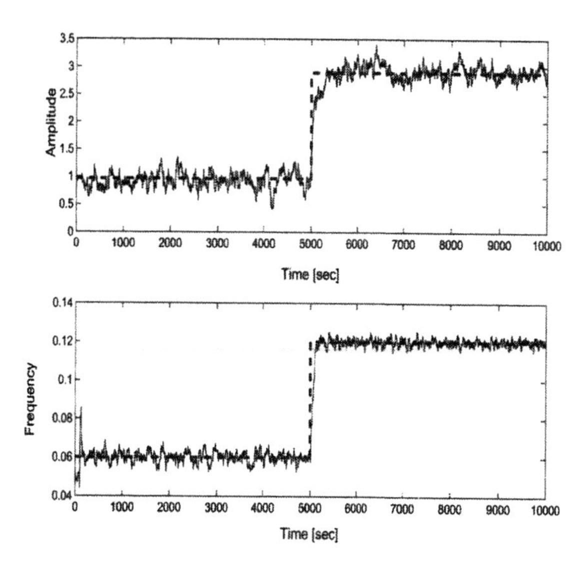

Fig. 1. Amplitude- and frequency diagram 1

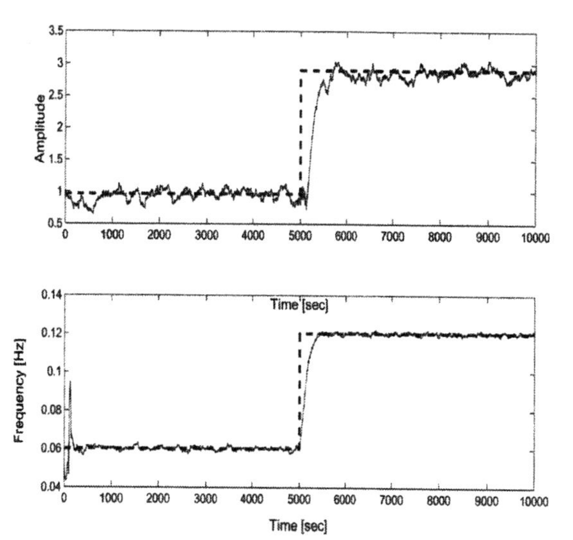

Fig. 2. Amplitude- and frequency diagram 2

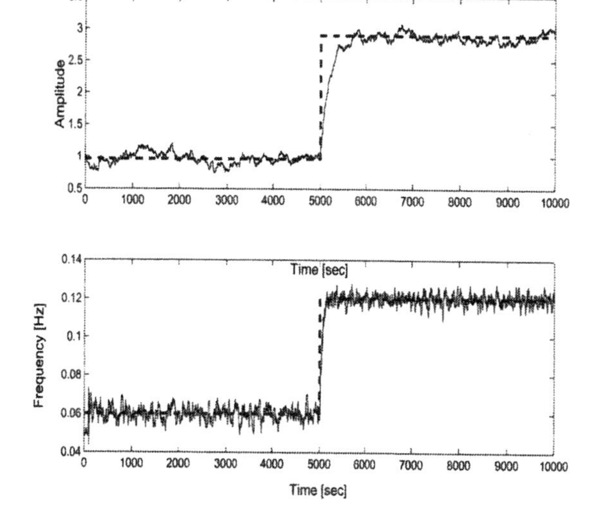

Fig. 3. Amplitude- and frequency diagram 3

In Figures 2 and 3 the amplitude- and frequency diagram for the filters in case 2 and 3 are shown.

As pointed out in section 3.2, the cost of a quick transient response will be a more noisy estimate. This is illustrated in Figure 1 and Figure 2.

In case 1 the filter responds quickly to the changes in both amplitude and frequency. However, the noise in both the amplitude and frequency estimate is considerable.

In case 2 the estimates, and in particular the frequency estimate, have lower variance. This is expected as the estimation error in the frequency is weighted quite heavily. The step response time is, however, increased.

In case 3 the variance in the amplitude estimate is decreased, but as in case two, the step response time is increased.

These cases demonstrates that the genetic algorithm tool is well suited for filter tuning. If some of the properties are more important than others, e.g. low variance in one of the estimated states, then it is very easy to reflect this in the tuning process.

4. OPTIMIZING THE FILTERS THRESHOLDING PERFOMANCE

4.1 *The thresholding phenomenon*

If the filter parameters (q_1 and q_2) are tuned for one value of the signals measurement noise $\mathbf{R_s}$ only, then it is impossible to guarantee the performance if it changes. If $\mathbf{R_s}$ decreases it may be expected that the performance is satisfactory. However, to obtain the optimal EKF for the given model, the filter is to be re-tuned. If $\mathbf{R_s}$ increases, the estimation error may become unacceptable large. In Figure 4 the relative frequency error versus signal-to-noise ratio (SNR) is shown.

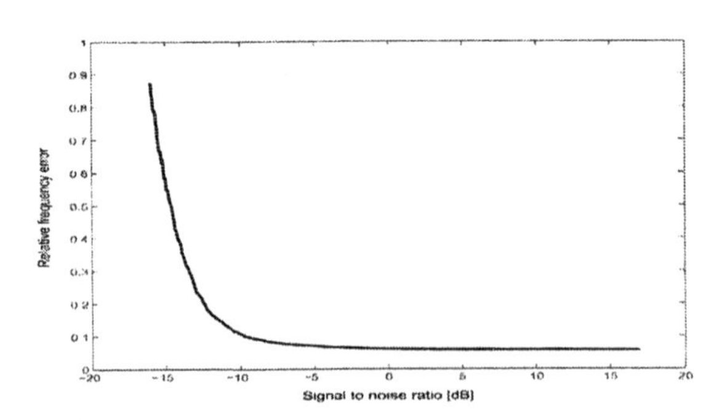

Fig. 4. Frequency error vs. SNR

The signal-to-noise ratio is given by

$$SNR = 10 \log \frac{A^2}{2\text{Var}(v)} \tag{15}$$

where A is the signals amplitude and $\text{Var}(v)$ its variance. As clearly illustrated in Figure 4, the error in the filters frequency estimate starts to increase dramatically for a signal to noise ratio below approximately -12 dB. This phenomenon is refereed to as *the thresholding phenomenon*, see e.g. (Kootsookos and Spanjaard, 1997).

In this paper the term *thresholding performance* refers to the filters ability to estimate a state in a high noise environment, and *thresholding point* refers to the point where the estimation error starts to increase dramatically. Increasing a filters thresholding performance means to move its thresholding point to a lower SNR value.

4.2 *Optimizing criteria*

A optimizing criteria when using the GA should include the following two basic requirements

1) The thresholding point is to be placed at the lowest possible SNR level
2) The estimation error in the low noise area is to be sufficiently low

Even if it is an optimization of the thresholding performance, item number two in the above listing must be included to assure that the resulting filter is useful. Highest possible thresholding performance has no value if the performance in low noise is not satisfactory.

These requirements may be expressed as follows:

1)

$$\min_{\mathbf{Q} \in [\mathbf{Q_0}, \mathbf{Q_1}]} \left(\max_{\mathbf{R_s} \in [\mathbf{R_0}, \mathbf{R_1}]} \| \mathbf{x} - \hat{\mathbf{x}} \|_2 \right) \tag{16}$$

where $[\mathbf{R_0}, \mathbf{R_1}]$ is the interval in which the optimization is carried out and $[\mathbf{Q_0}, \mathbf{Q_1}]$ is the interval for the filter tuning parameter.

2)

$$\max_{\mathbf{R_s} \in [\mathbf{R_0}, \mathbf{R_1}]} \| \mathbf{x} - \hat{\mathbf{x}} \|_2 \leq M \tag{17}$$

where M is the highest allowed value for the 2-norm of the estimation error.

Under the assumption that the estimation error is a monotonous increasing function of the signal measurement noise variance $\mathbf{R_s}$ the evaluation function will be exactly the same as shown in (14). If this assumption is not valid, the evaluation function must be modified. In this paper the following function is applied:

$$E(\mathbf{Q}) = N \left[\sum_1^N \sum_{i=1}^K \left(H_i \frac{1}{T+1} \sum_{t=0}^T E_{t,i} \right) \right]^{-1} \tag{18}$$

where K is the number of points to be considered, H_i is the weight for SNR point number i, N is the number of Monte Carlo runs used for evaluating each filter candidate and $E_{t,i}$ equals $\hat{\mathbf{e}}_t^T \mathbf{W} \hat{\mathbf{e}}_t$ at the i'th SNR point.

4.3 *Results from simulations*

In this section three different cases are considered.

1) the filter is tuned as in the previous section, i.e. with an signal-to-noise ratio SNR equal to -3 dB.
2) the filter is tuned for one single SNR value as in case 1, but now for SNR=-10 dB
3) the filter is tuned for three SNR values; SNR=-10 dB, SNR=-3dB and SNR=0 dB

Case 1

In Figures 5 and 6 the relative amplitude- and frequency error vs. signal-to-noise ratio curves are shown. The solid-drawn line is for an amplitude $A = 1$ and the dashed line for $A = 2$.

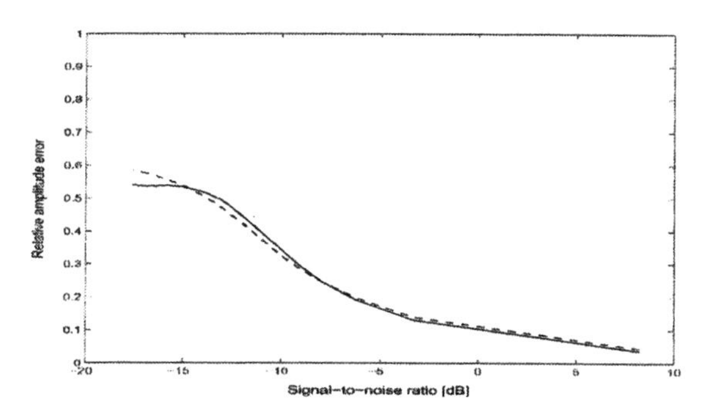

Fig. 5. Amplitude error vs. SNR

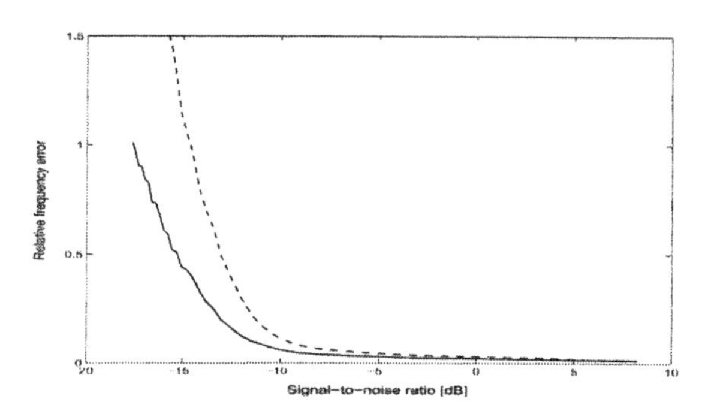

Fig. 6. Frequency error vs. SNR

It should be noted that the relative error may very well exceed 1 (100 %) if e.g. the filter try to track a frequency which is more than twice the signals frequency. This is not very likely to happen in low noise, however, when the SNR is very low this is a common situation.

When comparing Figure 5 and Figure 6, it is observed that thresholding occure in the amplitude estimate on a SNR level where the frequency estimate is still very satisfactory. This suggests that it is the amplitude estimation which is the limiting factor for this filters thresholding performance.

Another interesting observation is that the SNR level where thresholding occure seems to depend on the amplitude.

Case 2

In Figures 7 and 8 the amplitude- and frequency error vs. signal-to-noise ratio curves are shown.

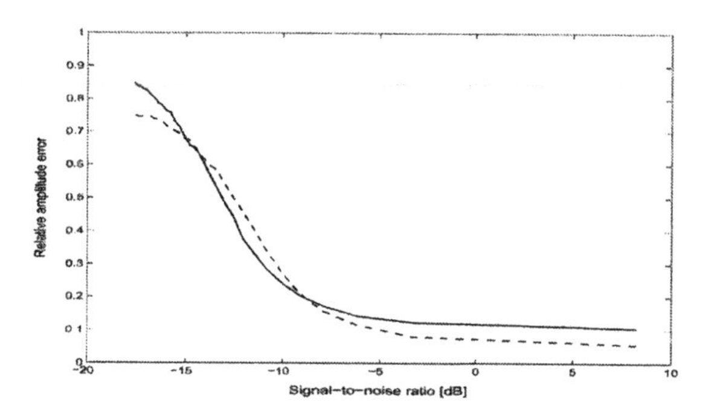

Fig. 7. Amplitude error vs. SNR

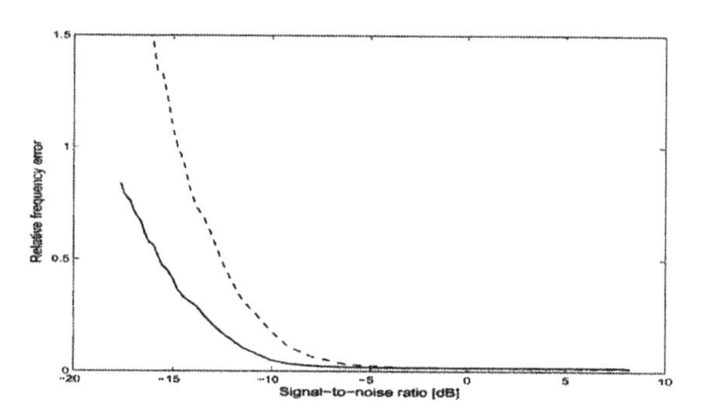

Fig. 8. Frequency error vs. SNR

Compared with the previous case, it is clearly seen that for the amplitude estimation, the thresholding performance has increased considerably. However, the error in low noise area has also increased, in particular when the amplitude is low.

For the frequency estimation the thresholding occure in fact earlier than compared with case 1. This difference is, however, not big.

Case 3

In Figures 9 and 10 the amplitude- and frequency error vs. signal-to-noise ratio curves are shown.

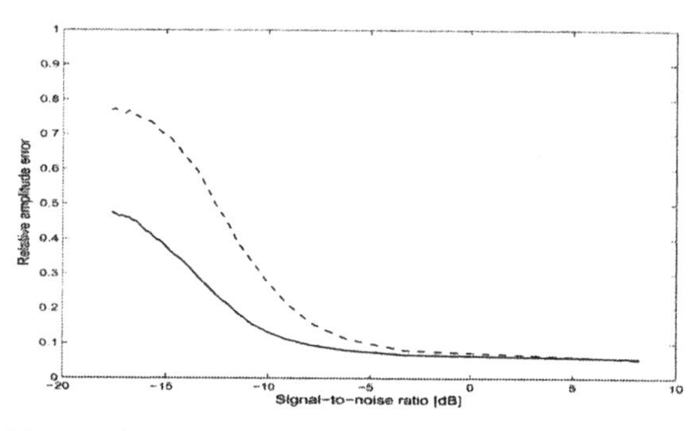

Fig. 9. Amplitude error vs. SNR

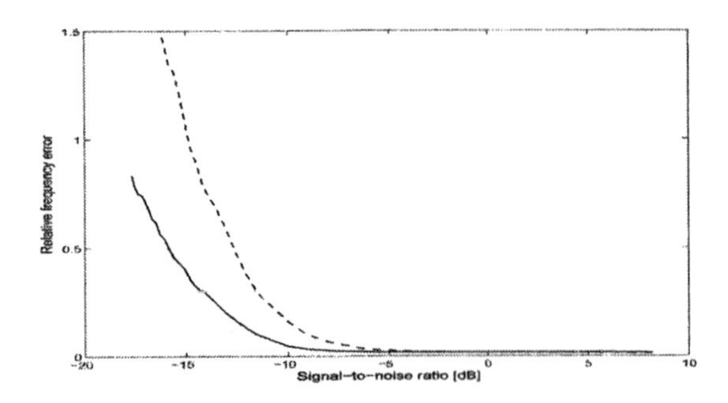

Fig. 10. Frequency error vs. SNR

In this case the amplitude error in low noise area is reduced compared to the previous case. The cost for this is a slightly decreased thresholding performance. For the frequency estimation it is on the same level as before.

4.4 *Discussion*

As demonstrated by the previous cases, the GA may easily be modified to tune a filter with high thresholding performance.

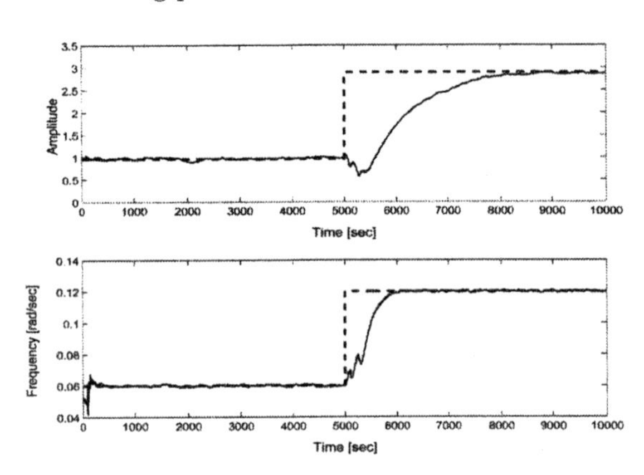

Fig. 11. Amplitude- and frequency diagram

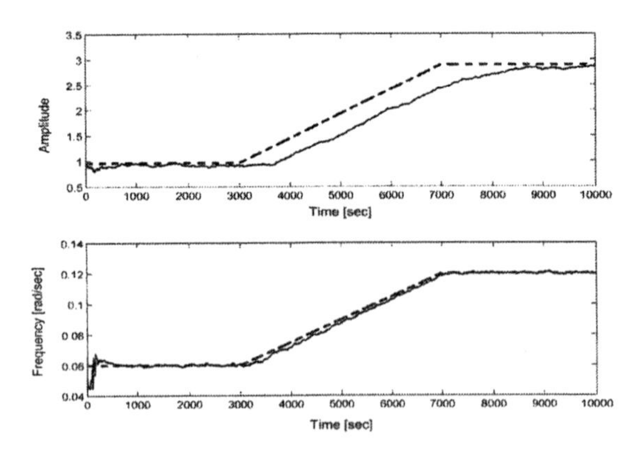

Fig. 12. Amplitude- and frequency diagram

Due to its decreased noise sensitivity, it should be expected that the resulting filter will react slowly on a quick change in frequency or amplitude. For slowly varying amplitude and frequency the performance should be expected to be far better than for rapid changes. This is clearly illustrated in Figure 11 and Figure 12. The filter parameters are equal to those in 4.3 case 3, and the noise variances are as given in section 3.5.

5. CONCLUDING REMARKS

In this paper some aspects regarding optimal tuning of an Extended Kalman filter by use of genetic algorithms has been discussed.

It is demonstrated that genetic algorithms (GA) is a tool well suited for filter tuning.

Further is is demonstrated how the GA can be modified in order to achieve better thresholding performance. This optimization is a trade off between low variance in the estimated states and a quick response. Filters optimized for maximum thresholding performance is only useful if the amplitude and frequency are almost constant, or changes very slowly.

ACKNOWLEDGEMENT

The authors are grateful to LKAB, Sweden, for support both financially and practical during this research.

6. REFERENCES

Chui, C. K. and G. Chen (1999). *Kalman Filtering with Real-Time Applications.* Springer. Berlin.

Houck, C. R., J. A. Joines and M. G. Kay (1996). The genetic algorithm optimization toolbox (gaot) for matlab 5. *http://www.ie.ncsu.edu/mirage/GAToolBox /gaot/.*

Kootsookos, Peter J. and Joanna M. Spanjaard (1997). An extended kalman filter for demodulation of polynomial phase signals. *IEEE Signal Processing Letters.*

Maybeck, Peter S. (1979). *Stochastic Models, Estimation and Control.* Academic Press. New York.

Oshman, Y. and Ilan G. Shaviv (2000). Optimal tuning of a kalman filter using genetic algorithms. *AIAA Paper 2000-4558.*

Powell, T. D. (2002). Automated tuning of an extended kalman filter using the downhill simplex algorithm. *Journal of guidance, control and dynamics* **25**, 901–908.

Rapp, K. and P-. O. Nyman (2003a). Control of the amplitude in a surging balling drum circuit, a new approach to an old problem. *Presented at ECC 2003, Cambridge UK.*

Rapp, K. and P-. O. Nyman (2003b). Genetic algorithm based tuning of an extended kalman filter. *Presented at MMAR 2003, Miedzyzdroje, Poland.*

GENETIC ALGORITHM BASED CONTROLLER DESIGN

Ivan Sekaj

Department of Automatic Control Systems
Faculty of Electrical Engineering and Information Technology
Slovak University of Technology, Ilkovičova 3, 812 19 Bratislava, Slovak Republic
Tel: ++421 2 602 91 585, E-mail: sekaj@kasr.elf.stuba.sk

Abstract: Controller design approach is described, which is based on genetic algorithms. The approach uses an optimisation procedure, where the cost function to be minimised consists of the closed-loop system simulation and a performance index evaluation. The designed system may be complex and practically without any limitations on its type, structure and size. The proposed approach is demonstrated on a robust controller design example. *Copyright © 2003 IFAC*

Keywords: controller design, genetic algorithms, dynamic system simulation, performance index, robust control.

1. INTRODUCTION

In the area of process control we have to design controllers for different types of dynamic systems with specific and sometimes very complex dynamics. For that reason it is necessary to use various controller types and different control loop structures. Modern control theories are able to solve complex tasks, but sometimes the applied approaches are solving only particular problems. Consider a system to be controlled, which has:

- nontrivial structure
- complex static or dynamic behaviour
- nonlinearities
- noise or disturbances
- several inputs / outputs (MIMO system)
- all the above together

In the proposed approach the controller design procedure is based on extensive computer simulations of the entire control loop - with all its specific properties, in combination with a suitable optimisation method. The optimisation method performs a direct search/optimisation in the controller parameter space (Sekaj 1999). The simulation is an essential part of the minimised objective function. It will be shown that due to this approach, the task of the control system parameter design will be transformed into a conventional n-dimensional optimisation problem. The dimension of the search space grows with the number of the designed controller parameters, which can be relatively high. Therefore also the computation effort using "conventional" optimisation methods is growing exponentially. From this reason the powerful genetic algorithm optimisation approach has been used.

2. CONTROLLER DESIGN

2.1 The design principle

The aim of the control system design is to provide required static and dynamic behaviour of the controlled process. Usually, this behaviour is represented in terms of the well-known concepts referred to in the literature: maximum overshoot, settling time, decay rate, steady state error or using different integral performance indexes (Dorf 1990, Kuo 1991, etc.). Without loss of generality let us consider a simple feedback control loop (closed-loop) (Fig. 1) where y is the controlled value, u is the control value, w is the reference signal and e is the control error ($e=w-y$).

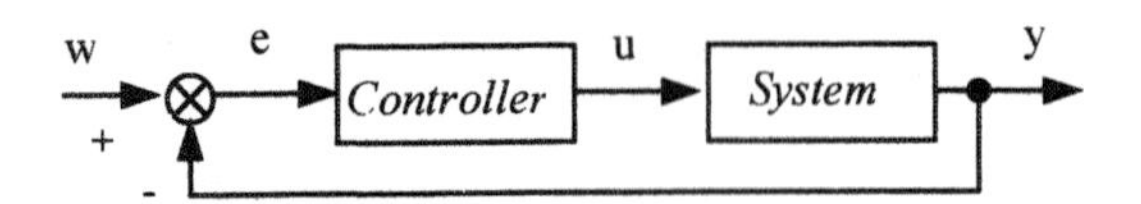

Fig. 1 Simple feedback closed-loop

Consider an appropriate simulation model of this closed-loop to be available. Let us analyse the closed-loop behaviour using the simple integral performance index "Integral of Absolute Control Error", which is defined as

$$I_{AE} = \int_0^T |e(t)| dt \qquad (1)$$

where T is the simulation time. The discrete form of this performance index is

$$I_{AE} = \sum_{k=1}^{N} T_s |e_k| \qquad (2)$$

where T_s is the simulation step size and N is the number of simulation steps.

The controller design principle is actually an optimisation task - search for such controller parameters from the defined parameter space, which minimise the performance index (1) or (2). The cost function (fitness) is a mapping $R^n \rightarrow R$, where n is the number of designed controller parameters. The evaluation of the cost function has two steps. The first step is the computer simulation of the closed-loop response and the second one is the performance index evaluation. In case of designing complex multi-input and multi-output (MIMO) control structures or some other controller types (fuzzy controllers, neuro-controllers, etc.) the dimension n of the search space can be high (more than tens or even hundreds), therefore the cost function may be complex and multi-modal.

A graphical representation of a simple cost function is depicted in Fig.2. It corresponds to a PI controller design example for a feedback loop according to Fig.1. Each point of the surface $G_{IAE}=F(P,I)$ is result of a simulation and performance index (2) evaluation. The optimal PI controller parameters, which are to be find, are the co-ordinates of the global minimum of this surface. This simple 2-D search problem can be solved also via conventional optimisation techniques like bisection, etc. However, in case of more complex control structures with many parameters (practically more than 3 or 4) the search space dimension is growing and the use of conventional optimisation methods, is no more feasible due to high computational requirements. Here the genetic algorithms can be used. A block scheme of a GA-based design is in Fig.3

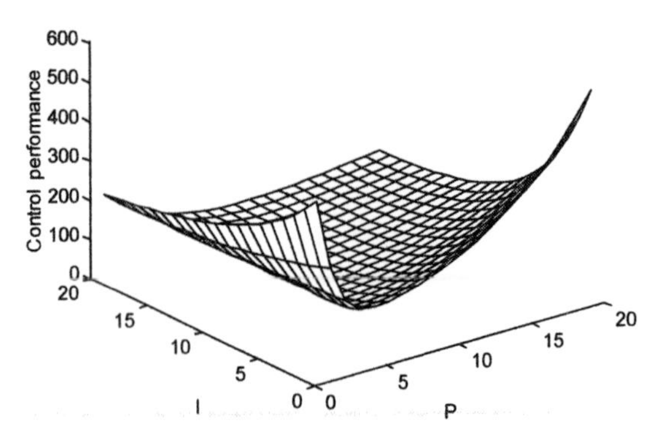

Fig.2 Graphical representation of a cost function using the I_{AE} performance index for a closed loop under PI controller

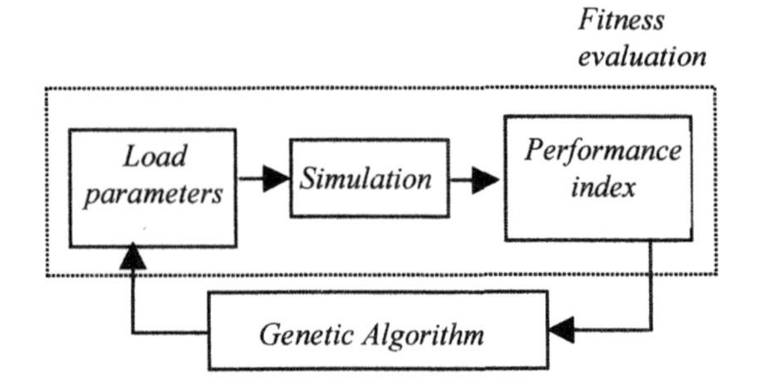

Fig.3 Block scheme of the GA-based controller design

2.2 Genetic algorithms

Genetic algorithms are sufficiently described in literature e.g. (Goldberg 1989, Michalewicz 1996 and others). The GA structure applied in our cases is as follows

1. Initialisation of a population of potential solutions - chromosomes (random, size of the population is between 20 and 50 chromosomes)
2. Fitness evaluation (Fig.3) and testing of the terminating conditions
3. Crossover (random couples, multi-point crossover, 50-75 % of all chromosomes of the population)
4. Mutation (probability 0.5 - 2 % of all genes in population)
5. New population completion (1-3 best chromosomes from previous population + new chromosomes after genetic operations + unchanged chromosomes), jump to the step 2

Note, that also different GA structures and genetic operations can be applied.

The chromosomes in the population are linear strings, which items (genes) are the designed controller parameters. Because the controller parameters are real-number variables and in case of complex problems the number of the searched parameters can be high, in our experiments GA's with real-coded chromosomes, instead of binary-coded ones, have been preferred. For illustration the chromosome of a simple PID controller, which is described in the time domain by the equation

$$u(t) = Pe(t) + I\int e(t)dt + D\frac{de(t)}{dt} \quad (3)$$

where $P \in R$, $I \in R$, $D \in R$ are the proportional, integral and derivative gain respectively, can be in form $ch=\{P,\ I,\ D\}$. Before each simulation, the corresponding chromosome is decoded into the controller parameters of the simulation model and after the simulation the performance index is evaluated.

The definition of the GA terminating conditions is not a simple problem. Based on our experience a favourable terminating method is the definition of the GA generations number. If the solution is still not acceptable, the number of generations is to be increased. Note that in case of complex system designs the controller design based on dynamic process simulations can be a time-consuming and a multi-modal problem and often also a sub-optimal solution can be sufficient.

2.3 The choice of the performance index

Consider that the GA finds the optimal (sub-optimal) solution in the user defined search space of controller parameters. The choice of the performance index has a fundamental effect on the closed-loop (dynamic system) behaviour. Using (1) or (2) normally brings about fast control responses with small overshoots between 2-5%. If it is necessary to damp the overshoot or damp oscillations, it is recommended to insert additional under-integral terms that include absolute values of the first or also the second order control error derivatives

$$J = \int_0^T \alpha|e(t)| + \beta|e'(t)| + \gamma|e''(t)|dt \quad (4)$$

(or output derivatives $|y'(t)|$, $|y''(t)|$) and to increase β and γ with respect to α, where α, β, γ are weight coefficients. In the discrete case the integral is replaced by the sum and the derivative is replaced by the difference. Good results can be obtained also using the performance index

$$J = \alpha\eta + (1-\alpha)t_r \quad (5)$$

where η is the overshoot, t_r is the settling time and $0 < \alpha < 1$ is the weight coefficient. The reference response y_r tracking can be achieved via minimisation of

$$J = \int (y_r(t) - y(t))^2 dt$$

Input energy optimisation can be carried out using the performance index

$$J = \int (\alpha e^2(t) + (1-\alpha)u^2(t))dt$$

where u is the control value. In Fig.4 there are closed-loop responses under a PID, for which the following criterions have been used: a) IAE (2); b) criterion (4) with $\alpha = 1.5, \beta = 1, \gamma = 0$; c) criterion (5) with; $\alpha = 0.25$; d) criterion (5) with $\alpha = 0.75$.

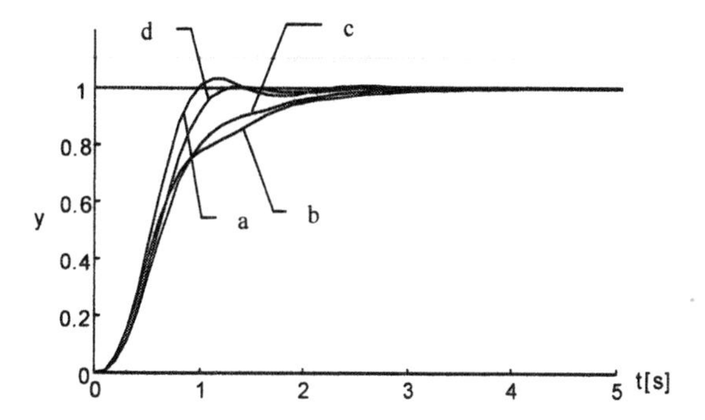

Fig.4 Closed-loop step responses using different performance indexes

Here let us make a short remark about the closed-loop stability. Due to the applied way of minimising the performance index, the closed-loop stability is an implicit attribute of each solution. During the evolution the unstable chromosomes are eliminated because of their high value of performance index. However if necessary, it is possible to include a stability test into each fitness evaluation.

3. ROBUST CONTROLLER DESIGN

An important property of control algorithms is their robustness. One possibility of increasing the controller robustness is via minimising the gain/phase margin in combination with some of the above mentioned integral performance index. Another way, how to increase the controller robustness is to consider separate models for n different operating conditions of the controlled process using the same (robust) controller. The cost function is obtained after evaluation of the closed-loop simulations under all n operating conditions using the equation

$$J = \sum_{i=1}^{n} \int_0^T |e_i(t)|dt \quad (6)$$

It is also recommended to include the measured noise from the real system or other expected disturbances into the simulation model.

Another very powerful robust controller design method is as follows. Consider $c=\{c_1,\ c_2,\ \dots,\ c_q\}$ to be the set of the designed controller parameters and $s=\{s_1,\ s_2,\ \dots,\ s_r\}$ the set of parameters of the controlled object. During the operation, each of the parameters s_i can move within the uncertainty space, which is defined by the matrix with two rows

$$S = \begin{bmatrix} s_{1,\min}, s_{2,\min}, \ldots, s_{r,\min} \\ s_{1,\max}, s_{2,\max}, \ldots, s_{r,\max} \end{bmatrix}$$

where $s_{i,min}$ and $s_{i,max}$ are the minimum and maximum limits of the i-th system parameter. The genetic algorithm mentioned in the part 2.2 uses the following fitness evaluation. In each iteration cycle (generation) n random operating conditions are generated (say $n=10$ etc.). That means n vectors of the system parameters s become random values from the space S. The fitness function is calculated using the performance index (6). In the next generation other n random parameter vectors are generated.

Let us demonstrate the robust design method on an example. Let the nonlinear controlled system be described by the differential equation

$$y'' + a_2 y' + a_1 y + a_0 y^3 = b_0 u$$

The coefficients b_0, a_0, a_1, a_2 are within the intervals $0.5<b_0<5$, $0.5<a_0<5$, $0.5<a_1<5$, $0.5<a_2<5$. Additionally, the simulation model is disturbed at the system output by a white noise. The controller is in form (3) where the parameter limits are $\langle 0;100 \rangle$ for each controller parameter. The used performance index is

$$J = \int_0^T |e(t)| + 0.5|y'(t)|dt$$

In Fig.5 step responses for ten random generated (perturbed) systems from the defined space S with the designed PID controller are depicted.

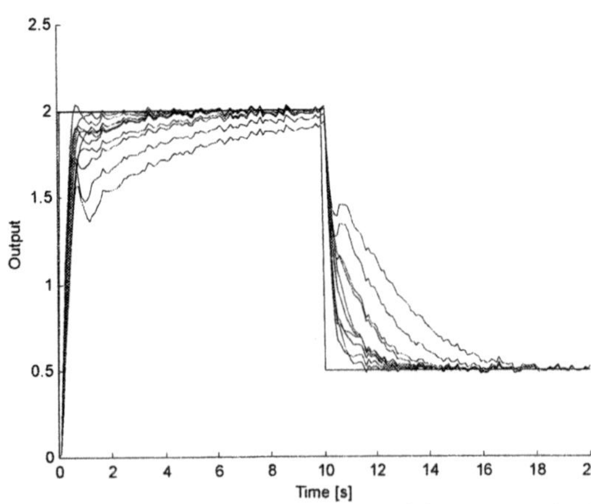

Fig.5 Closed-loop step responses with 10 perturbed systems and with measurement noise

4. CONCLUSION

A genetic algorithm-based optimisation approach for the design of various controller types has been presented. The GA-optimisation procedure is minimising a cost function, which combines system simulation and performance index evaluation. In such a way the dynamic system design with N parameters is transformed into a search problem in the N-dimensional parameter space. The subjects of design/optimisation may be complex control structures, which include systems practically without any limitations on their type and number of their inputs and outputs. The only limitation is the computation time, which is higher in comparison to conventional approaches. The necessary condition for the application of this approach is existence of an appropriate computer model of the controlled system and of the controller.

The presented approach has been successfully applied for controller design of linear, non-linear, stable, unstable, non-minimum-phase systems, systems with saturations, SISO and MIMO systems, for system identification as well as for self-tuning GA-based controllers (Sekaj 2002a). Simulation results have been verified in real-time experiments. Beside controller design tasks the GA-based optimisation approach has been applied also to other process control tasks like static optimisation of processes (Sekaj 2002b).

REFERENCES

Dorf, R.C. (1990): Modern Control Systems, 5th edition, Addison-Wesley publishing Company

Goldberg, D.E. (1989): Genetic Algorithms in Search, Optimisation and Machine Learning, Addisson-Wesley

Kuo B.C. (1991): Automatic Control Systems, Prentice-Hall International Editions

The Math Works Inc.(1999): Matlab User Guide

Mitsukura, Y., Yamamoto, T., Kaneda, M., Fujii, K.(1999): Evolutionary Computation in Designing a PID Control System, Proceedings on the IFAC World Congress 1999, Beijing, 5th-9th july, P.R.China, pp. 497-502

Michalewicz, Z. (1996): Genetic Algorithms + Data Structures = Evolutionary Programs, Springer

Sekaj, I. (1999): Genetic Algorithm-based Control System Design and System Identification, Proceedings on the Int. Conference Mendel '99, June 9-12, Brno, Czech Republic, pp.139-144

Sekaj, I., Foltin, M., Gonos, M. (2002a): Genetic Algorithm Based Adaptive Control of an Electromechanical MIMO System, Proceedings of the GECCO 2002 Conference, July 9-13 , New York, pp. 696

Sekaj, I. (2002b): Genetic Algorithms with Changing Criterion Functions, Intelligent Technologies – Theory and Applications, P.Synčák et al. (Eds), IOS Press, pp.183-188

www.elsevier.com/locate/ifac

NONLINEAR PROCESS IDENTIFICATION AND PREDICTIVE CONTROL BY THE WEIGHTED SUM OF MULTI-MODEL OUTPUTS

U. Schmitz*, R. Haber*, R. Bars**

**Department of Plant and Process Engineering, Laboratory of Process Control, University of Applied Science Cologne, D-50679 Köln, Betzdorfer Str. 2, Germany*
fax: +49-221-8275-2836 and e-mail: ulrich.schmitz@ fh-koln.de, robert.haber@fh-koeln.de
***Department of Automation and Applied Informatics, Budapest University of Technology and Economics, H-1111, Budapest, Goldmann Gy. tér 3., Hungary; fax : +36-1-463-2871.and e-mail : bars@aut.bme.hu*

Abstract: Most industrial processes are nonlinear. In such a case only a nonlinear model valid for the whole working area can ensure a good controller design. The nonlinear process is approximated by a multi-model consisting of the intelligent combination of some linear sub-models. As a very practical way the following identification strategy was used: independent model parameter estimation in the different working points and the calculation of the global valid model output as the weighted sum of the sub-models. As a weighting function the Gaussian function is used. The parameters of the Gaussian function were chosen either without or with optimization of the identification cost function. The global valid nonlinear model was used for model based predictive control. A heat exchanger example illustrates the method. *Copyright © 2003 IFAC*

Keywords: process identification, nonlinear models, predictive control, multi-model, heat exchanger

1. INTRODUCTION

Predictive control is increasingly used in industry because of the good and robust control which can be achieved with such a controller. Most of the predictive control algorithms used in the industry are linear controllers, however most of the processes are nonlinear.

Controlling of a nonlinear process with a linear controller leads often to a small operating area for the controller or to a very slow controller setting in order to achieve a robust control in a large working area. Therefore, the need for nonlinear model based controllers is apparent. In contrary to the linear case, when the predicted controlled variable can be calculated by the sum of a free and a forced response and an analytical minimization of the cost function is possible, with nonlinear process models an iterative optimization of the control cost function is necessary.

In the following one possible solution for a nonlinear model structure, the multi-model approach is presented. This method combines the advantage of the low computational effort with the linear sub-model identification and the complexity of a global valid nonlinear process model which can be used for the prediction. Such a model is then used for a nonlinear predictive controller with iterative optimization.

There are several approaches for identification of a multi-model:

- independent model parameter estimation in the different working points and hard switching between the models (e.g., Pottmann, et.al., 1993),
- independent model parameter estimation in the different working points and linear approximation of the model outputs between the working points (e.g., for control signal blending: Aouf, et.al.2002),
- independent model parameter estimation in the different working points and calculating the global valid model output as the weighted sum of the sub-models (e.g., Johansen and Foss, 1993).

The last approach is applied in this paper. As a weighting function the Gaussian function is used.

The Gaussian function has an unknown parameter (the standard deviation), which has to be optimized. There are two ways to do it: (Nelles; 2001):

- simultaneously estimation of the location of the working points, of the parameters of the linear sub-models and the standard deviation of the Gaussian distribution.
- first selection of the working points, after that estimation of the parameters of the linear sub-models and finally optimization of the standard deviation of the Gaussian distribution.

The second method requires not only less computational efforts, but is most often used in the practice, as prior identified linearized models in different working points show the nonlinear structure of the process. Therefore, this method was used for the modeling, nonlinear identification and predictive control of an electrically heated heat exchanger model.

2. THE NONLINEAR PLANT (HEAT EXCHANGER)

The performance of an electrically heated heat exchanger (Fig. 1) was measured.

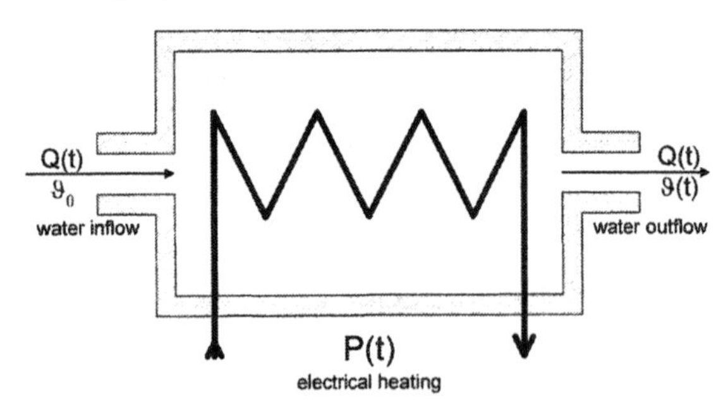

Fig. 1 Electrically heated heat exchanger

The temperature of the water outflow depends on the water inflow and on the electrical heating power. Fig. 2 shows some steps in the heating power with different water flow on this plant.

From the step responses an experimental model has been built, resulting in a nonlinear first-order model between the heating power [P] to the outflow temperature [ϑ], where the time constant and the static gain depends on the inflow [Q] (Haber and Keviczky, 1999). The model was developed by:

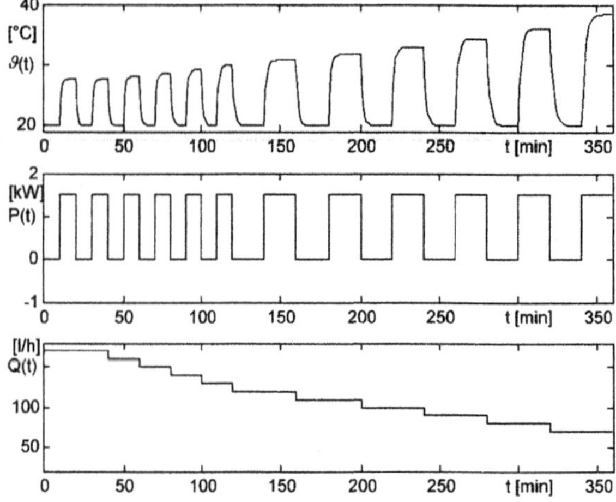

Fig. 2 Temperature step responses of the heat exchanger at different water flows

$$Q(t)c_p\vartheta_o + P(t) = Q(t)c_p\vartheta(t) + \alpha\rho V c_p \frac{d\vartheta(t)}{dt} \quad (1)$$

which leads to the first-order model

$$\frac{\Delta\vartheta(s)}{P(s)} = \frac{K}{1+sT} \quad (2)$$

with

$$K = \frac{1}{c_p Q}$$

and

$$T = \frac{\alpha\rho V}{Q}$$

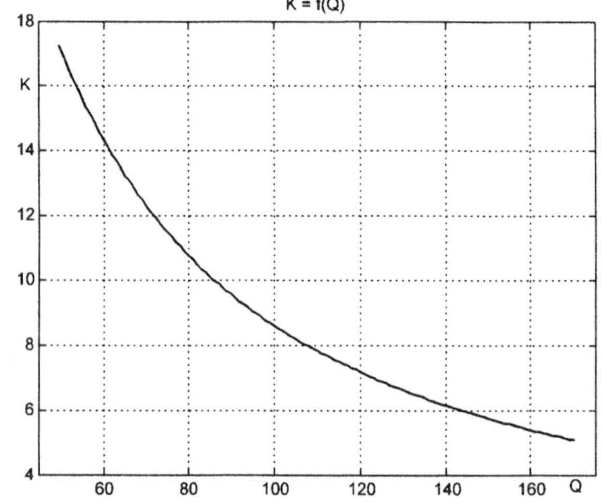

Fig. 3 Dependence of the static gain of the flow

(c_p is the specific heat of the water, ρ is the density of the water and α is an uncertainty factor.)

Figs. 3 and 4 demonstrate how the static gain and the time constant depend on the water flow.

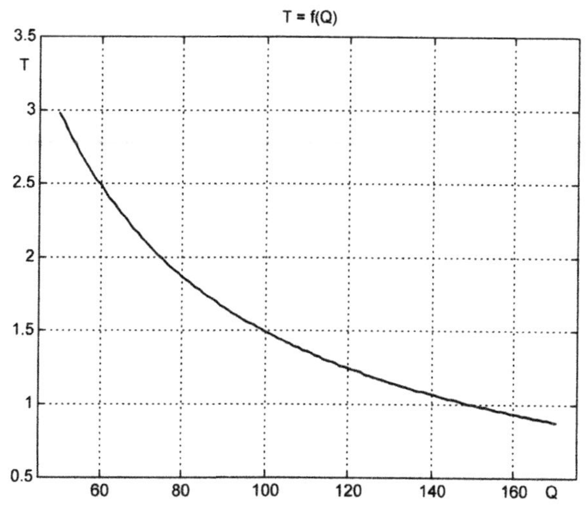

Fig. 4 Dependence of the time constant of the flow

Figures 5 and 6 show the control of this plant for a set point step and a disturbance step in two working points with a linear predictive controller designed for the flow Q=150 m³/h

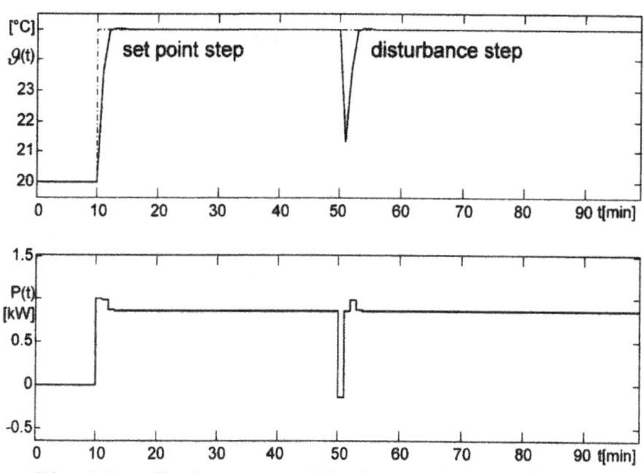

Fig. 5 Predictive control in the working point the controller was designed for (Q=150m³/h)

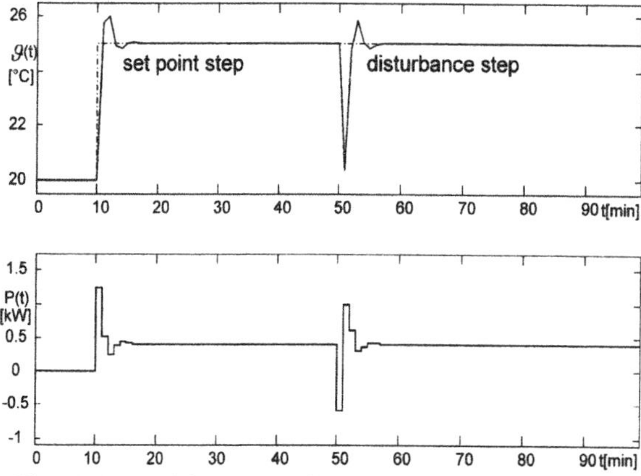

Fig. 6 Control in the working point Q=70 m³/h with the controller designed for the working point Q=150 m³/h

- around the working point the controller was designed for (Fig. 5) and
- in a different working point of Q=70m³/h (Fig. 6).

As it is seen, the control behavior is bad in those working points where the controller was not designed, that means the controller produces a big overshoot and needs a longer time to achieve a steady-state behavior.

3. MULTI-MODEL MODELING APPROACH

The basic idea of the multi-model approach with soft switching between the working points is to combine the output of many linear models, which each are valid for different working points. Fig. 7 demonstrates this idea.

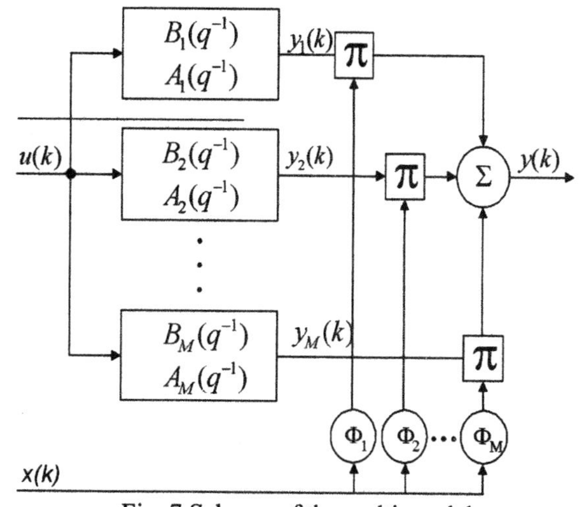

Fig. 7 Scheme of the multi-model

The output of the global model is the weighted sum of the local valid linear models y_j where the weighting factor Φ_i depends on the actual working point Q. Thus the output of the global model is described by:

$$\hat{y}(t) = \sum_{j=1}^{M} \left[y_j(t) \Phi_j(t) \right] \tag{3}$$

The weighting factor Φ_i, for example, can be built by a normalized Gaussian function:

$$\Phi_i(u) = \frac{e^{\left(-\frac{1}{2}\frac{(Q-Q_i)^2}{\sigma_i^2}\right)}}{\sum_{i=1}^{M} e^{\left(-\frac{1}{2}\frac{(Q-Q_i)^2}{\sigma_i^2}\right)}} \tag{4}$$

where Φ_i, is the output of the validity function, Q is the actual working point, Q_i is the working point the local valid linear model is valid for, σ_i is the standard deviation and M is the number of working points considered.

Above model is a special class of the LOLIMOT (LOcal LInear MOdel Tree) (Nelles, 2001), because here the parameters of the linearized models in some working points are estimated independently form each other. Similar idea was presented by Foss and Johansen (1993). This approach was chosen, because with nonlinear plants often linearized models can be estimated in different working points.

Fig. 8 shows the outputs of the normalized Gaussian functions for the heat exchanger model with three local valid linear models in three working points. As it is seen, the sum of all validity functions is 1 for all working points.

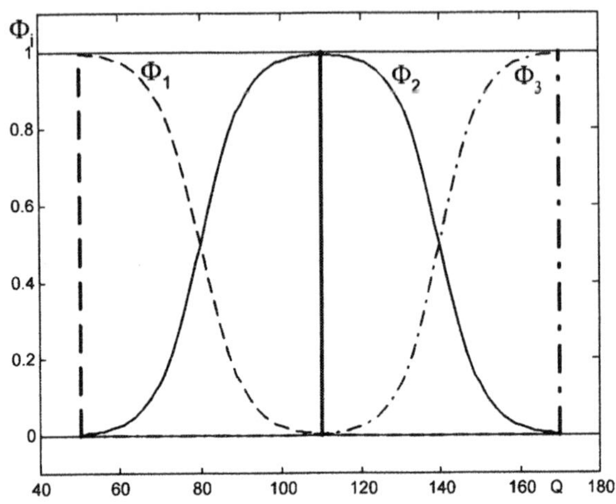

Fig. 8 Outputs of the normalized Gaussian functions with three working points and a standard deviation of σ_i=18.

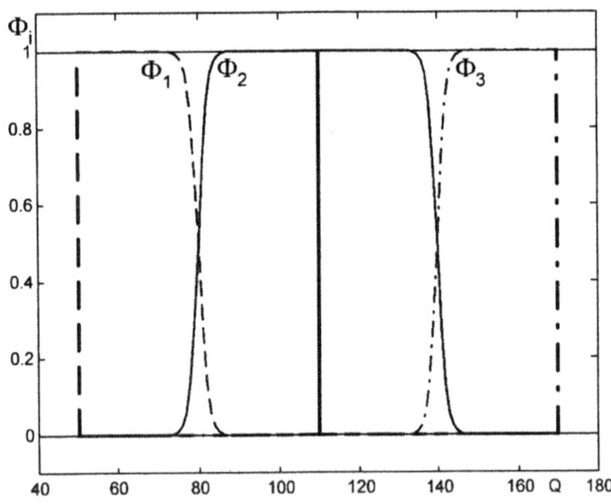

Fig. 9 Outputs of the normalized Gaussian functions with three working points and a standard deviation of σ_i =8.

In the case of normalized Gaussian functions the standard deviation σ_i is a tuning parameter which gives the validity of the local valid linear models around the exact working point. A higher standard deviation makes the validity area of the local valid linear model bigger and the changeover from one local valid linear model to an other is more soft, while a smaller standard deviation makes the area smaller and the changeover from one local valid linear model to an other more hard. Fig. 9 shows the output of the normalized Gaussian functions with a standard deviation of σ_i=8 and Fig. 10 with a standard deviation of σ_i=30.

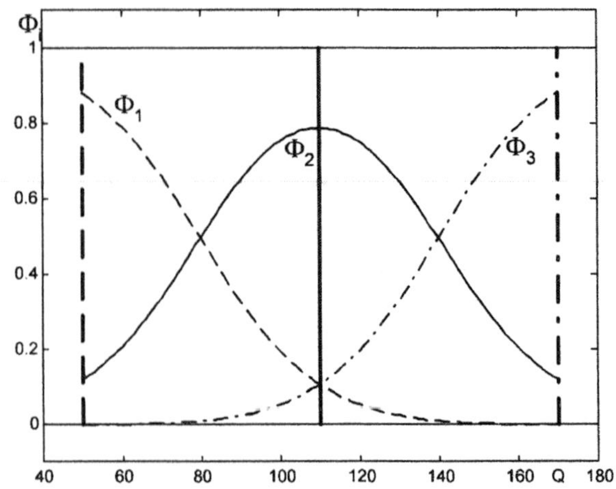

Fig. 10 Outputs of the normalized Gaussian functions with three working points and a standard deviation of σ_i =30.

As described above, the output of the global nonlinear model is the sum of the local valid linear models, multiplied with the local validity function value for the actual working point. For each local working point the nonlinear first-order model can be calculated as a linearized model for the corresponding working point by keeping the flow Q as a constant value which leads to a linear first-order model, valid for the chosen working point (described by the flow Q).

The standard deviation for the normalized Gaussian functions has to be chosen by the user or this parameter can be computed by defining an identification cost function which describes the cost of the predicted model output in proportion to the measured process output with the form:

$$J = \sum_{j=1}^{n} \left[y_{measured} - y_{pred}(P, Q, \Phi) \right] \qquad (5)$$

If this function is written in such a way that the cost depends on the standard deviation then an optimizer can be used to search the best value for the standard deviation.

The signal to be predicted should cover all working points which the global nonlinear model should be valid for. In case of the heat exchanger the flow Q was stepwise changed from Q=170 m^3/h to 50 m^3/h while the heating power was generated with a PRTS (Pseudo Random Ternary Signal). Fig. 11 shows both input signals and the measured temperature of the process.

For an easier optimization the standard deviations were chosen to be equal to each other for the different working points. The optimal standard deviation found by the optimizer was σ_i=24. Fig. 12 shows the output of the normalized Gaussian functions with this standard deviation.

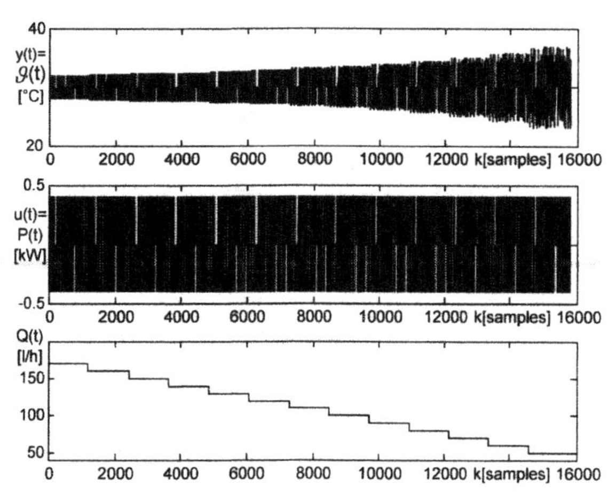

Fig. 11 Test signal for the computation of the optimal standard deviation

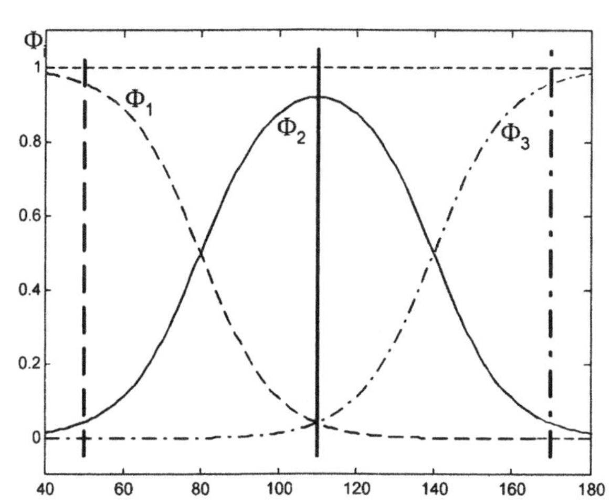

Fig. 12 Outputs of the normalized Gaussian functions with the standard deviation of σ_i =24, found by the optimization routine

Fig. 13 shows the static gains of the linear models in all working points as the output of the nonlinear multi-model with the standard deviation σ_i=24 found by the optimizer.

The nonlinearity of the static gain is still well observable, but also the waveform generated by the multi-model approach with the normalized Gaussian functions is very distinctive at the higher values for the electrical heating power P.

4. PREDICTIVE CONTROL WITH MULTI-MODEL

Using the above multi-model with the analytically derived local valid linear models and the standard deviation σ_i=24 found by the optimization process, a nonlinear predictive controller was designed. The iterative optimization is based in each step on the nonlinear model prediction.

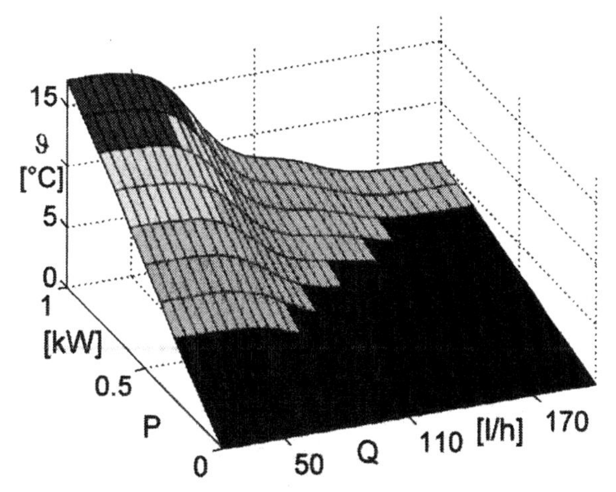

Fig. 13: Surface of the static gains in all working points by the combined output of the global valid nonlinear model.

In order to prevent static errors CARIMA model (Clarke, 1987) was used:

$$y(k) = \frac{B(q^{-1})}{A_\Delta(q^{-1})}\Delta u(k)$$

with

$$A_\Delta(z^{-1}) = \Delta(z^{-1})A(z^{-1}),$$

where

$$\Delta(z^{-1}) = 1 - z^{-1}.$$

Because of the fact that the local valid models are derived independent from the standard deviation and the signal, describing the nonlinear behavior in this method it is possible to use the CARIMA model which works with manipulated signal increments instead of absolute values. The usage of manipulated signal increments is a requirement for predictive control to prevent static errors in the control variable. This is an important and practical difference and extension to the program package LOLIMOT (Nelles, 2001).

The control signal was computed using a non-constrained optimizer and the control was performed in 13 different working points, that means

- for working points where a local valid linear model was fully valid and
- also for some working points between these points, that means where no estimated linearized model was valid.

Fig. 14 shows the result of the control of a set point change with the temperature (upper plot) and the heating power (lower plot) in the working point Q=110 m³/h in which one of the three local valid linear models was fully valid.

Figs 15 and 16 show the control of the heat exchanger in the working points Q=160 m³/h and Q=60 m³/h, respectively. These working points lie between the working points used for the linear identification.

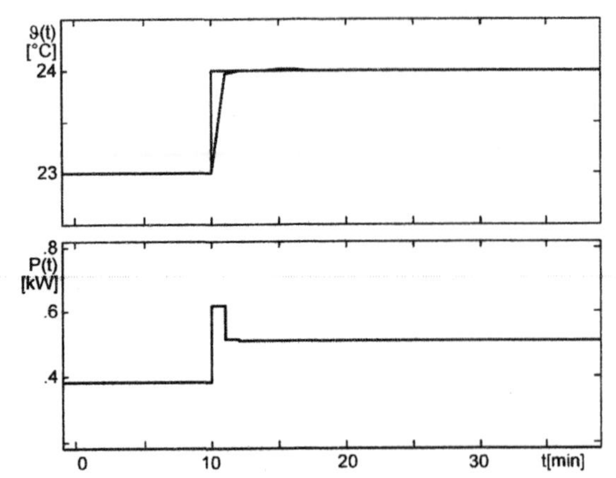

Fig. 14 Control of the heat exchanger in the working point Q=110 m³/h

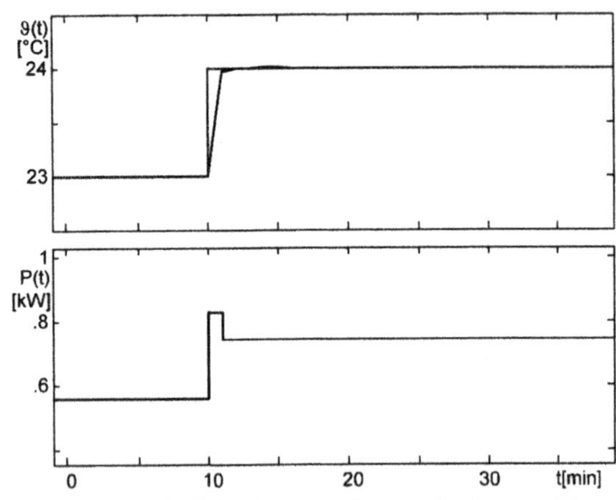

Fig. 15 Control of the heat exchanger in the working point Q=160 m³/h

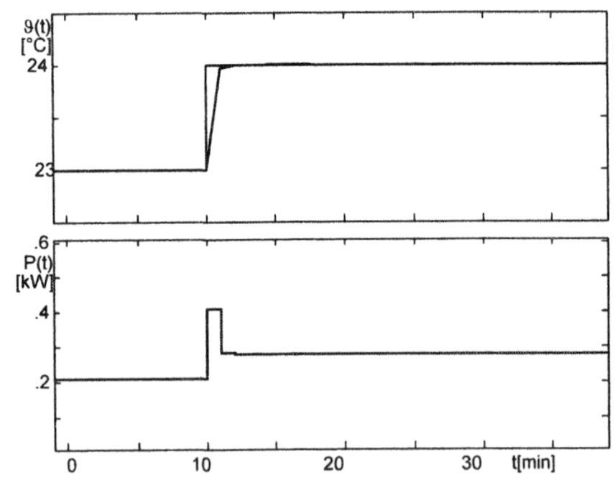

Fig. 16 Control of the heat exchanger in the working point Q=60 m³/h

It can be seen that in all three working points the controlled signal (the temperature) is very similar while the manipulated signal (the heating power) shows differences, e.g. in the initial overshoot and in the dynamic behavior.

5. CONCLUSIONS

A special nonlinear modeling and identification procedure was presented. The process was described by a multi-model with soft switching between the working points. This was realized by combining the outputs of the sub-models as a weighted sum of the sub-model outputs. The weighting factor is a Gaussian function with unknown parameter, which was optimized. The sub-models were estimated by linear dynamical models in three working points.

The global valid nonlinear model was used for the model based predictive control.

A heat exchanger example illustrated the method.

6. ACKNOWLEDGEMENT

The work has been supported earlier by the Ministry of Science and Research of NRW (FRG) in the program „Support of the European Contacts of the Universities / Förderung der Europafähigkeit der Hochschulen" and now is supported by the University of Applied Science Cologne in the program "Advanced Process Identification for Predictive Control" and by the program of EU-Socrates. The third author's (R.B.) work was also supported by the fund of the Hungarian Academy of Sciences for control research and partly by the OTKA fund TO29815. All supports are kindly acknowledged.

7. REFERENCES

Aouf, N., D. G. Bates, I. Postlethwaite, B. Boulet (2002), Scheduling schemes for an integrated flight and propulsion control system, Control Engineering Practice, Vol. 10, p685-696.

Clarke, D.W., C. Mohtadi, P.S. Tuffs (1987)., Generalized predictive control – Part I. The basic algorithm. Automatica, 24, 2, 137-148.

Foss, B. A., Johansen, T. A. (1993), Constructing NARMAX models using ARMAX models, Int. J. Control, Vol. 58, No. 5, p1125-1153

Haber, R., L. Keviczky (1999). Nonlinear System Identification. Input/Output Modelling Approach. Part 1: Parameter Estimation, Part 2: Structure Identification. Kluwer Academic Publisher, Dorchrecht, Netherlands

Johansen, T. A., B. A. Foss (1993), Constructing NARMAX models using ARMAX models, Int. J. Control, Vol. 58, No. 5, p1125-1153

Nelles, O. (2001), Nonlinear system identification, from classical approaches to neural networks and fuzzy models, Springer, Berlin, Germany.

Pottmann, M., H. Unbehauen, D. E. Seborg (1993), Application of a general multi-model approach for identification of highly nonlinear process – a case study, Int. J. Control, Vol. 57, No. 1, p97-120.

www.elsevier.com/locate/ifac

NEURAL NETWORKS FOR REAL TIME CONTROL SYSTEMS.

Čapkovič Miroslav * **Kozák Štefan** **

* *Department of Automatic Control Systems*
Faculty of Electrical Engineering
Slovak University Of Technology
Slovakia
miro@kasr.elf.stuba.sk
** *Department of Automatic Control Systems*
Faculty of Electrical Engineering
Slovak University Of Technology
Slovakia
kozak@kasr.elf.stuba.sk

Abstract: Due to the complexity of neural networks the problem of large computation requirements on computational devices where the network is trained or evaluated and slow network evaluation/training response can appear. The problem is more evident when networks are compared to the conventional algorithms that run with low response time. In general this problem is being solved using by more powerful computational device or parallel systems. This paper deals with a neural network optimized for run on the real time control system to obtain real time responses comparable to the conventional algorithms. Special effort is made for solving of the problem of optimal network structure with reflection to preserve or aproove original network qualities(approximation and generalization capability).

Keywords: Neural networks, real time, function approximation, nonlinear system, network structure optimization.

1. INTRODUCTION

In the last recent years the phenomenon of neural network comes into different scopes of life. Even though the neural network are firstly mention in the seventies, the real applications appeared only after introducing the back-propagation algorithm. At presents, neural networks are used for different computational tasks including relatively simple identification methods to complex task of artificial intelligence. Of course, more network structures and learning algorithm were explored after introducing back-propagation algorithm, but it is still applicable and sufficient to solve the majority of problems introduced in control theory due to their good approximation precision and capability to solve the problems described by (input-output) data relations without exact knowledge of mathematical model. The problem solution by neural network is based on the optimization techniques that optimize the network parameters to reflect best network responses with relation to the solved problem.

1.1 Problems.

As it is well-known the functionality of neural networks is based on the large count of cell structures interconnected into network structure. Such structures brings the power of massive parallelism into computation and thus network can be represented by concurrent working computational units, but other side of this network feature is large computational effort required for network computation algorithms (learning, evaluating). This effort produce long computation time if single computational unit (end-device) is used for performing network algorithms and therefore the usage of the neural networks in real time control systems is not possible. This problem also occurs when the network algorithm is performed on the large data set.

1.2 Problems identification and description.

The problem solved by this paper is optimization of network(structure, learning and evaluating algorithms) to obtain shortest response time **without** affecting the network functionality, which covers particular approximation and generalization abilities of networks optimized by this manner. Because at present the algorithms are *narrowly coupled with computers* or some computation-able devices when they are performed, the optimization of algorithm executable and its transformation for these devices should be taken into the consideration (This coupling will have greater sense when the algorithm is in interaction with the real time system.). These optimizations should covers the optimization on the highest level (the level of algorithm, its task, steps performed by algorithm, ...), middle level (algorithm's representation in the machine-human interface, which generally is a higher programming language) and lowest level (optimal translation of such optimized code into end device executable). Itis identified more major problems of this transformation with relation to the neural networks, which are:

- **Network structure.** In general case the problem is involved by redundant or too complex network structure, which brings multiply computational overhead.
- **Algorithm efficiency.** The problem covers the non efficient logical constructions in the algorithms and also non efficient higher language constructions.
- **Network activation function complexity.** The high computational complexity is product of high time computational complexity of elementary functions used as activation functions.
- **Compiler and device overhead.** The problematic are overheads produced by the higher structural programming language and dynamic architecture of network representation, especially:
  - Overheads produced by parameter copying and stack operations (if any) which are:
    - (1) Program function parameter passing.
    - (2) Program function calling overhead.
  - Overhead produced by accessing internal variables and structured data (represents network parameters) representing network parameters.

The term of *data access overhead* s defined as overhead produced by programming techniques during accessing data structures. Often it is produced by the structural and higher languages used for network representations (N steps needed for accessing the value of variable nested in the structures and arrays instead one - the problem pointer of arithmetics). Term *function and code call overhead* is associated to overhead that is produced by calling pre-compiled libraries or own defined functions. The overhead code is produced by parameters copying and also it can be product of function local variables and jump code.

  - Overheads produced by accessing variables - representing solved problem.

- **High data complexity and redundancy.** The problem is related to the data redundancy in training set and its complexity - the neural network must reflect the complexity of data.

2. PROBLEM SOLUTIONS

2.1 Optimization levels for solving specified problems.

As was mentioned above, major task of this paper is to prepare the network for usage in real-time control systems. For solution of this problem the paper [1] defines performance steps that are introduced here and explicitly explained in the [1]. This article is dedicated to the problem of *network structure optimization.* Each presented optimization technique can be used without dependency with another, but the sequence of optimization levels should be preserved. Five optimization levels for optimizing the networks computation in the computational devices are defined:

2.2 **Optimization level 0**

Pre-process the data by technique of normalization and transformation (linear, nonlinear). It also should contain algorithm for redundant data set elimination. This level can help to optimize the final network structure.

2.3 Optimization level 1

is decomposition of solved problem into the local problems and construction of **local observers.** Each local observer solves the problem on the local area instead of the whole state space. This leads to the network system with more observers and generally simplest network structure. This optimization level covers the design of *local networks* and *switching/gating network*, generally represented by *decision tree.* Major effect of this technique is shortest local networks learning and evaluating of each local observer. Disadvantage is additional processing of switching/gating network - but in the summary the shortest response time can be obtained. The input of this algorithm step is solved problem, the output is divided state space, eventually designed local observers.

2.4 Optimization level 2

solves the problem of **network optimal structure** by algorithm of *structural optimization.* This algorithm covers the suggestion of network optimal structure in relation of network qualitative indicators that are computed from input-output data set. The algorithm cooperates with expert and reasoning system, where the main knowledges of network structure design are stored. The input of this optimization level is initial network structure and solved problem, output is structurally optimized network with respect to the qualitative indicators.

2.5 Optimization level 3

is **elementary math function reduction.** In this level the original node elementary activation function is replaced by less computational expensive function including approximated function, spline approximation or another computational less expensive function. The input of this optimization level is the structure of neural network, solved problem, output is network structure with reduced activation functions.

2.6 Optimization level 4

covers the reduction of **device and algorithm** associated problems. The problem of data access overhead is reduced by the technique of N-nary step data access replacement by 1-nary for variables and 0-nary for constants. The problem of code access overhead is reduced by technique similar to the function in-lining technique. The output of this optimization level is linear code containing only useful code (generally only math instructions.).

2.7 Optimization level 5

is applied on the generated linear code if still some overheads stays here. The algorithm of **linear code optimization and data dependency reduction** is applied on the generated linear code which reduce the multiply occurrence of same chains of instructions operating on the same data. The overhead free linear code can be translated to the device dependent code using specific device natural language.

3. NEURAL NETWORK STRUCTURE OPTIMIZATION

In this paper the algorithm of network structure optimization is introduced (as was presented above in the optimization level 2). The algorithm has role to **search** best structure of network with reflection to the fitness function and network performance indicators. Algorithm starts to search from initial network structure which is passed from previous chain step(s) and generates network structure with better (or best) features as the initial structure. The idea of algorithm originates from genetic algorithms (GA) and evolutionary artificial neural networks (EANN) but as extension it implements fitness function which controls the network structure optimization. Unlike GA, the fitness function $\gamma(\Delta)$ has active role in process of optimization and also in the process of network evaluation.

$$\gamma(\Delta) \tag{1}$$

The fitness function decides from one or more network modification operations applied on the network structure: Especially it choose *where* in network structure (which edge, node) is applied *which* operation. The main idea of appropriate modification selection is evaluation of each network node by some manner and search the worst one (searches for weakest node in the network). In presented implementation this selection is performed by value of "node contribution". This value indicates how large error is produced on the node's output (it is computed by manner of back-propagation algorithm). The algorithm defines basic operations on network structure encapsulated into cloning function δ, including the modification of network structure, learning algorithm and node's activation function. After the application of cloning function, child network is created. This can be repeated more times while the searched structure is found. Because the nature of algorithm, for one network more children can be generated using different operation. The algorithm can generate during runtime large amount of networks - this is a reason why they are placed in the searching tree **T**. The relations between subsequent leafs are relation of "parent-child": Children are created from their parent by applying the structure modification algorithm - cloning function introduced above (fig. 1)

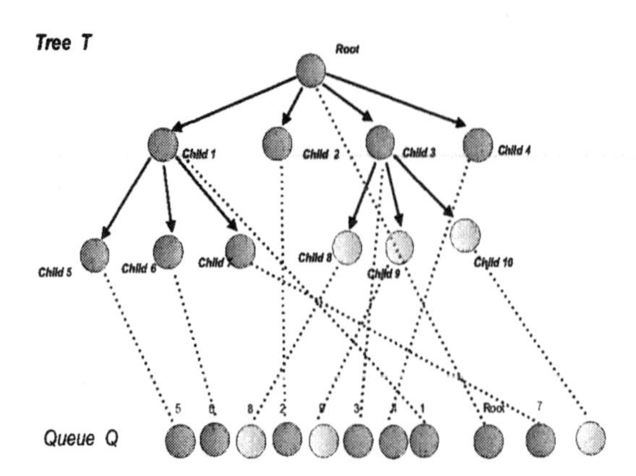

Fig. 1. Relation between searching tree **T** and ordered queue **Q**.

Mathematically the aim of structure optimization is to obtain optimal network structure symbolic representation $\hat{\theta}$, which has optimal (or better) qualitative indicators Δ as the original network θ. The principle of optimal structure computation is shown in the follow steps:

(1) **Modification of the network structure θ by cloning function δ with respect to objective function $\gamma(\Delta)$ The modified structure $\tilde{\theta}$ is obtained:**

$$\tilde{\theta}= \delta(\theta, \gamma, ...) \quad (2)$$

(2) **Linear overhead-free code ξ_L generation for structure $\tilde{\theta}$ (step originating from optimization level 4)**
(3) **Execution of the process of parametric optimization (Network learning) on the modified structure.**

Steps (1) to (3) are repeated while optimal structure $\hat{\theta}$ is obtained. It can be seen that whole optimization process can be broken into two main subprocesses:

- Parametric optimization which correspond to standard network learning algorithm.
- Structural optimization which covers mentioned conventional technique and alternates with it while the optimal structure is obtained.

Before algorithm starts, the start conditions must be met and some data preprocessing should be done (level 0 of optimization chain). First condition of algorithm start is initial network structure θ_{root} design. We can choose from these possibilities:

- The structure is generated from solved problem Π using function composition/decomposition algorithm (see below).
- The structure is obtained from optimizer evaluation system **E** as a most similar from the previous passes.
- The structure is obtained manually.
- The structure is obtained randomly.

When initial network is obtained, the algorithm starts (optimizer elements are explained in subsequent sections):

(1) **Insert the root θ_{root} into searching tree T and ordered queue Q.**
(2) **Using cloning function δ clone the best N_1 (free) nodes of searching tree T and generate new N_2 children.**
It covers the sub-steps:
    (a) For each selected *free* (node was not passed through optimization process) node generate N children [1].
    (b) Insert each generated child into searching tree **T** and ordered queue **Q**.
    (c) Reevaluate each new inserted node in the ordered queue and sort the queue.
(3) **On each new and the best *free* N_2 nodes run parametric optimization and change their state from free to busy.**
(4) **Reevaluate each node which had finished the parametric optimization, mark it as free and reorder the queue.**
(5) **Compare qualitative indicators and if found appropriate (stopping condition can be chosen as small values of time response and large values of approximation and generalization capability), stop the algorithm.**

Algorithm loops between steps 2 and 5 while step 5 satisfies. This algorithm has the following features which are used in the algorithm implementation in distributed environment [1]:

- Grand granurality parallelism (One task equivalent to the network learning).
- Asynchronous event-driven optimizations.

Optimization of neural network is performed by structures of the autonomous suggestion system Ω.

$$\Omega = \Omega(T, Q, E, \delta, \gamma) \quad (3)$$

It keeps the track the optimization of neural network for specific solved problem Π to the final optimal neural network representation. It consists of more elements:

[1] It is 1:N relation.

- Searching tree **T** - symbolic structure that holds the results of particular optimization steps and tracks them (fig. 1). Searching tree is the variable leaf count (because cloning operation can choose to modify the original network by more ways) n-nary optimization tree that has assigned the follows data structures to each leaf:
  - Network structure θ.
  - Network qualitative indicators Δ_θ.
- Ordered queue **Q**, that maps all networks from tree **T** into queue **Q** in sorted order from the network with the best performance indicator to the weakest. Ordering queue is the list data structure with links to the searching tree leafs into specified order. Leafs are ordered from leaf with best value of objective function $\gamma(\Delta)$ to the leaf with the worst value of objective function. The relation between queue and tree can be see on the fig. (1).
- Evaluating system **E** is an extension of expert system that has the role of observer between analyzed problem Π and optimal structure $\hat{\theta}$. It stores information about previous optimization processes that can be later reused in process of the initial network design.

3.1 *Process of cloning*

The operation is initially applied on the root of tree, later on the generated and evaluated leafs and so on. The cloning operation can modify as the network structure as the nature of network learning and evaluation. The definition follows:

1 Definition **(Cloning function.)** *Let's have base of autonomy suggestion system Ω. The cloning function generates the new network $\tilde{\theta}$ from its original copy with respect to the objective function $\Omega.\gamma$ as follows:*

$$\tilde{\theta} = \delta(\Omega, \theta) \tag{4}$$

The general network structure modification operations are:

- Node insertion δ_{ins}.
- Node deletion δ_{del} .
- Node re-wiring δ_{rew} .
- Activation function change δ_{fchg}.

and learning modification:

- Learning function change δ_{learn} .
- Parameter perturbation δ_{pert} .

The operation are explained explicitly below:

2 Definition **(Node insertion δ_{ins}.)** *Let's have disjoint sets of nodes $\mathbf{N_I}$ and $\mathbf{N_O}$ as a subset of network θ set of nodes, $\mathbf{N_I} \subset \theta$ and $\mathbf{N_O} \subset \theta$, $\mathbf{N_I} \not\bigcup \mathbf{N_O}$. Let's have the new node $\mathbf{n_x}$; $\mathbf{n_x} \not\subset \mathbf{N_I}$; $\mathbf{n_x} \not\subset \mathbf{N_O}$ with assigned activation function $\phi_\mathbf{x}$ and η_i inputs, where η_i is count of nodes in N_I.Then the node insertion consists of:*

(1) For each $n_i \subset N_I$ create an edge $e_{i,x}$ from n_i to node n_x.
(2) For each $n_i \subset N_O$ expands its inputs count and create an edge $e_{x,i}$ from n_x to node n_i.

It is preferred to add the node n_x with assigned activation function ϕ_x by algorithm that the network function $\theta(\Pi)$ is not affected (or affected minimally) on this problem Π by this insertion:

$$\theta(\Pi) = \tilde{\theta}(\Pi) \tag{5}$$

After insertion of the node, also the new node n_x takes participation on the network learning.

3 Definition **(Node deletion δ_{del}.)** *Let's have node $n_x \subset \theta$. Let's have the input dependency set N_I of node n_x and output dependency set N_O of node n_x. Node deletion consist of those steps:*

(1) Removal of all edges $e_{i,x}$ where $n_i \subset N_I$.
(2) Removal of all edges $e_{x,i}$ where $n_i \subset N_O$.
(3) Performing nodes input count reduction on each node belong to output dependency set N_O.

It is expected that network preserves its consistency after applying the network modification function. The node deletion and insertion modifies the nodes count in network. The algorithm can also operate on network structure without changing nodes count when it is only adding or deleting existing edges. The following operations do not change node count:

4 Definition **(Node re-wiring δ_{rew}.)** *Let's have node $n_x \subset \theta$. Let's have the input dependency set N_I of node n_x. The node re-wiring refers to two operations:*

- *Input edge insertion.*
- *Input edge deletion.*

In both of operation the input count expansion and reduction must be performed to stay network be consistent and rules mentioned in (def. 2, p.5) and (def. 3, p.5) should be met.

5 Definition **δ_{lchg} (Learning algorithm change.)** *The idea is based on solution of problems with learning convergence and stability. While in standard learning techniques it is used only one learning algorithm in this case more than one learning algorithms are used and algorithm can switch be-*

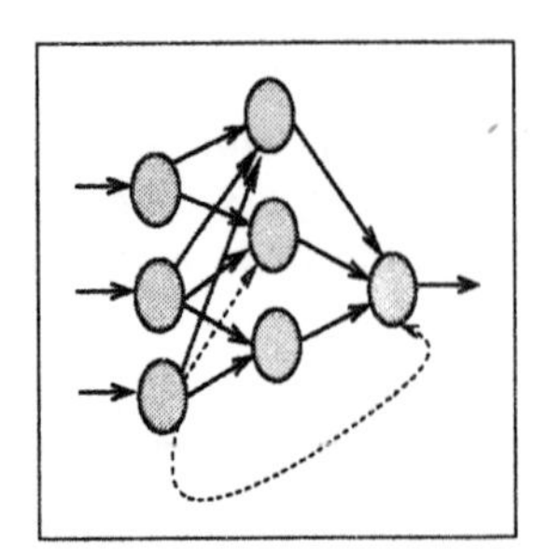

Fig. 2. Example of node re-wiring.

tween them during learning process. The cloning function choose which one is active at the actual moment. The main choosing criteria should be algorithm skills to jump over the local minima, and/or the faster convergence, etc...

The nature of randomness is built into algorithm, specially the random suggestion of network structure and random addition of small values in conventional learning algorithm when it is stranded in local minima.

6 Definition **(Structure and parameter perturbation δ_{pert}.)** *The idea is based on random small perturbation in the original structure of the network using previously defined operations and also random modification of network parameters by small value.*

The cloning function can bring the sense of randomness to the generation of new network structure, whenever the accent of structure generating is placed on the data analysis(initial network suggestion). At this section the criteria how the cloning function can select the operation and node/edge victim is discussed. The overall node contribution σ of node plays a main role. Additionally all operations that modify network structure are requested to leave network consistent.

3.2 Network consistency

To keep network consistent during network modification operation, the following sub-operations must be performed on nodes:

- Node input count expansion.
- Node input count reduction.

The task of node input count expansion or reduction is produced by node insertion and/or deletion. The network consistency is defined as mood of nodes interconnection when all interfaces (inputs of node, outputs of node, parameters of node) are connected into graph (representing network structure) and no subgraph is isolated. The second criteria is to have all interfaces of the node to be connected after such operation. In the case that symmetrical (or symmetrized) functions [1] are used, the function exchange is simple and requires only to expand or to shrink the function symmetrical part. Thus the node function exchange is required only for nodes with non-symmetrical function.

3.3 Neural network

The work [1] introduces a different view on the networks - as to an approximation mathematical graph structure. In comparison to the conventional feed-forward networks they have added extensions which includes:

- Recurrent links in network structure.
- Memory capabilities.
- Parameters assigned to node instead of edges.
- Non-symmetric functions.

Two basic algorithms are defined to operate on the network structure: evaluation algorithm λ_E that evaluates the network output with current network input and its configuration and learning algorithm λ_L that is equivalent of standard learning techniques.

7 Definition **(Network learning algorithm λ_L.)** *Network learning algorithm is an algorithm which optimize the network parameters (It can modify parameters of network and network structure) with respect to the minimal error between training set and network response. Generally, the learning algorithm λ_L can be expressed as follows:*

$$\lambda_L : (\theta, \Delta, \Pi) \rightarrow (\widetilde{\theta}, \widetilde{\Delta}) \tag{6}$$

where $\widetilde{\theta}$ is modified network structure retrieved by learning process and $\widetilde{\Delta}$ are qualitative indicators of modified structure. It can referred to the two basic types of learning algorithms:

- *Gradient-based.*
- *Gradient-free.*

The operation of putting input pattern on the network input and its evaluating is the process of network evaluation. This process can be represented by the follow evaluation algorithm:

8 Definition **(Network evaluation algorithm λ_E.)** *Network evaluation algorithm is an function which maps input of network into scalar output value:*

$$\lambda_E : \lambda_E(\theta, \vec{X_i}) \mapsto Y \tag{7}$$

and for nodes:

$$\forall n_i \in \mathbf{N} \ ; \ \phi_i \in n_i \ ; \ y = \phi_i(\vec{x}, \vec{\alpha}, t); \tag{8}$$

Both algorithms are implemented as symbolic language operate on the network structure.

3.4 Network performance indicators

For network evaluation the work [1] defines terms that differs from standard terminology and adds extra feature used in the proposed algorithms. The basic definition of neural network as a graph structure is presented, and also the basic structures and algorithms used in the process of structural optimization and its implementation. The following attributes evaluate network parameters qualities [1]:

- Network approximation capability μ_A.
- Network generalization capability μ_G.
- Network evaluate time response μ_{T_E}.
- Network learning time response μ_{T_L}

Generally [1] defines the quality indicator as an measurement μ which return value within $< 0,1 >$. The value "1" is assigned to the measured value when the appropriate quality is the best, and "0" when it is poor(est). For better orientation and later possible algorithm modification all qualitative indicators are encapsulated into one qualitative vector that is evaluated into one scalar value. For solved problem Π the qualitative indicator vector is defined as follows:

$$\Gamma(\Pi) = (\mu_m(\vec{X}), \mu_m(\vec{Y}), \mu_c(\vec{X}, \vec{Y}), \quad (9)$$

$$\mu_{D_0}(\vec{\Pi}), \mu_{D_1}(\vec{\Pi}), \mu_{D_2}(\vec{\Pi}), \mu_{\theta^1 X}, \mu_{\theta^1 t}, ...) \quad (10)$$

More can be found in the work [1]. The selection of victim node and edge is based on the node contribution value defined below:

9 Definition **(Overall contribution of node σ.)** *Let's have node n_i in network θ and set of input and output time series $\vec{X}$, $\vec{Y}$. Let's have activation function ϕ_i of node n_i, $n_i \in \mathbf{N}$. We say that the overall contribution of the node n_i (or function ϕ_i) is value*[2]

$$\sigma_i = \sum_{k=0,\dots,K} \partial^k \tilde{\sigma_i} \quad (11)$$

where

$$\partial^k \tilde{\sigma_i} = \sum_{l=1,..,N} \left(\frac{\partial^k \phi(\vec{x_l})}{\partial^k x_l} - \frac{\partial^k y_l}{\partial^k x_l}\right)^2; \ \vec{x_l} \subset \mathbf{X}; \ y_l \subset \mathbf{Y} \quad (12)$$

where K is order of derivates to take into consideration.

[2] We restrict the overall contribution to the second derivate however higher derivates can be used.

3.5 Initial network suggestion

The rough algorithm for initial network structure design is based on algorithm (def.10):

10 Definition **(Function composition/decomposition.)** *It is assumed that for the solved problem Π set of available functions (that can model the problem) Ψ. The algorithm consist of follow steps:*

(1) For solved problem $\Pi_0 = \Pi = (X_0, Y_0)$ find (most) suitable function ϕ_0. Evaluate error ε_0 obtained as follows:

$$\varepsilon_0 = \sum_{\Pi} (Y_i - \phi_0)^2 \quad (13)$$

(2) Generate new problem from previous solved problem as follows:

$$\Pi_1^{parallel} = (X_0, \varepsilon_0) \quad (14)$$

or

$$\Pi_1^{serial} = (\varepsilon_0, Y_0) \quad (15)$$

(3) By these steps we obtain the model:

$$\Pi \sim \phi_0(X_0) + \varepsilon_0 \quad (16)$$

The appropriate composition of function for the actual problem Π_0 is processed by two possible ways as follows:

$$Pi \sim \Phi_1(\Phi_0(X_0)) + \varepsilon_1 \quad (17)$$

by serialization or by the parallelization:

$$\Pi \sim (\phi_0(X_0) + \phi_1(x_0)) + \varepsilon_1 \quad (18)$$

The technique can obtain the appropriate network initial structure assigned to the symbolic representation of this problem (not precise, but good for initial design). This technique can be view as composition constructive method.

3.6 Choosing operation in process of structure modification

The algorithm choose the operation for network modification using the following criteria:

- **Node deletion** when its functionality is identity or near to identity. Other technique is to delete node when its contribution is larger as some threshold

$$\sigma_i > \kappa \quad (19)$$

  Node is deleted so the network function was not affected:

$$\phi_{prev}(\vec{X}) = \phi_{after}(\vec{X}) \quad (20)$$

- **Activation function change** of node i when its overall contribution with replacement function is smaller as of the original contribution.

- **Node insertion.** Node insertion can be performed when it can be found sets of nodes $N_i \subset N, N_o \subset N$ that have overall contribution larger than chosen threshold value κ. The subsets N_i and N_o are choosed accomplished relation that N_o is in output dependency of nodes in N_i. The new node is subsequently inserted with inputs connected to each node in subset N_i and output connected to subset N_o.
- **Node activation function change.** The operation is also chosen when the node contribution exceed some value. The new node activation function is chosen with respect to definition of function composition algorithm - the original node function can be replaced when on the solved problem some other function from available function set has better results as the original function.

4. CONCLUSION

Introduced algorithm is nondeterministic searching algorithm controlled by objective function originating from EANN and GA algorithms. It has two run levels - first is the structure optimization algorithm which run on the top, the second run level is level of conventional used parametric optimization process. It has feature of asynchronous simultaneous event-driven algorithm. The primary result obtained by algorithm is design of optimal network structure, second is building of knowledge base with stored knowledge as the relations between analyzed problems and suggested network structure. These data can later be reused as base of analysis and used for searching of exact mathematical relations describing the solved problem.

ACKNOWLEDGEMENTS

This paper was partially supported by the Slovak Scientific Grant Agency, Grant no. 1/7630/20.

REFERENCES

[1] Čapkovič Miroslav "Neural Networks For Real Control Time Systems, Large And Large Scaled Problems", *Thesis*, Faculty Of Electrotechnics And Informatics, Slovak University Of Technology,2002

MODIFICATION OF THE SMITH PREDICTOR FOR RANDOM DELAYS TREATMENT IN THE NETWORK CONTROL SYSTEMS

[1]Ligušová Jana – [1]Liguš Ján – [2]Barger Pavol

[1]Technical University, Košice, Letná 9/A, 040 01 – Košice – Slovakia
Tel.:++421 95 602 2508, Jana.Ligusova@tuke.sk, Jan.Ligus@tuke.sk
[2]Centre de Recherche en Automatique de Nancy ESSTIN2, rue Jean Lamour54 519
Vandoeuvre – France. Tel: + 33 (0) 3 83 68 51 31, barger@esstin.uhp-nancy.fr

Abstract: This paper discusses analysis and design of control algorithms in Network Control Systems (NCS). The problem analysed is the treatment of systems with random time delays caused by a communication network. The main idea of this paper is to consider network induced delay as a dynamically changeable deadtime. In regard to this assumption, methods and algorithms known from area of control systems with deadtime can be used to compensate influence of network delays in NCS. In our case the method of the Smith predictor was modelled in actuator node. The modelling of NCS is realized by Coloured Petri nets (CPN) and as simulation tool is used Design/CPN software.

Keywords: Network, Control, Petri-nets, Prediction

1. INTRODUCTION

The modern approach in the control systems area is that the connection of sensors, regulators and actuators are not realized in an analogue way, but they are connected into digital communication bus. Mentioned control systems are called network control systems (NCS). NCS are affected by exponential development of communication technologies (protocols and media), where each element of control system represents an independent communication unit directly connected to the communication network, as to the next element of control loop. When considering NCS a major problem consists in the integration of the network-induced delays into the model. These delays caused by the sharing of communication medium can be constant, bounded, or even random.

In general, there are two main approaches in the design of NCS. The first one is the "control approach" in the design of NCS that means to focus on the quality of control algorithm and to design such a control that takes these network delays into consideration. Known approaches and methods sufficiently solve just delays smaller then sampling period (Nilsson, *et al.*, 1998). The second one is the "network approach" in the design of NCS that means to focus on the quality of the communication protocol to prevent the NCS from random network delays. The goal of the paper is to analyse time delays caused by the communication medium and to find appropriate control algorithms which compensate network delays smaller and bigger then sampling period. The paper discusses control approach where the Smith predictor is designed to control a tank system. Modified scheme of the Smith predictor and its implementation in actuator node is presented. Next, in the paper a tank system is described and some simulation results are presented as well. The paper presents the using of Petri nets for dynamic simulation of NCS and as simulation tool Design/CPN is used.

2. DELAYS IN NETWORK CONTROL SYSTEMS

New requirements for control systems which include modularity, control decentralization, integrated diagnostics, fast and easy operation and easy maintenance limit the use of traditional analogue way of interconnection in industrial control. Implementation of a network into the control loop has several advantages as lower cabling price in comparison with the analogue connection, easier installation and maintenance, easy diagnostics of system, increasing of control architecture flexibility, increasing of system reconfiguration etc. But this network interconnection has also some disadvantages as communication constraints, dependability of control from network faults, asynchronous elements of control, unpredictable network faults etc. All of mentioned disadvantages lead to network induced delays. Network induced delays and consecutively their affect to the stability of systems are in general the most discussed problems of NCS and impact the implementation of network into industrial control. Those delays, constant or even random, can not only degrade performance of control system designed without considering of delays but even destabilize the system (Walsh, Bushnel, 1999).

The character of network induced delays depends on chosen network protocol. There are a lot of currently available control networks, but in regard to medium access control mechanism (MAC) control networks can be divided into following three groups:

1. Control networks based on a collision detection mechanism CSMA/CD: Ethernet, Modbus/TCP, Ethernet/IP, EIB, LonWorks, etc.
2. Token passing control networks: Profibus, ControlNet, BACnet, MAP, P-Net, WordFIP etc.
3. Control networks based on a mechanism of arbitration of message priority CSMA/AMP: CAN, DeviceNet, SDS, etc.

Real implementation of NCS into real environment brings other practical reasons for a NCS analysis. The question is how to provide the same quality of control in NCS in case of:

- short-term power failures in communication systems and control network elements,
- disturbances of radio and other types of data transfers in networks,
- changes of program scan durations in regulator, when sampling period is close to scan time,
- treatment of error intervals in communication,
- separated master/slave regulations,
- physical network interruptions,
- communication driver faults and consequent and re-initialisation process,
- routing in WAN networks and control through the Internet, etc.

In general, it is possible to avoid all of these problems by some special types of prevention, for example electric protection, redundant networks etc. This is the reality in plants with a hard real-time or dangerous conditions. Delays with random occurrence can lead to control with non-equidistance steps and continuously to a system nonlinearity. Our goal is not to solve delays in NCS with improvements of networks features, but to solve delays in NCS from the system and control point of view.

The NCS are composed of intelligent sensors (SMART sensors) and actuators (with computation capabilities, time-stamping, communication interfaces etc.), controller units and a communication network as shown on Fig. 1 (Wittenmark, *et al.*, 1995). Delay τ_{sc} represents the network induced delay in direction from the sensor to the controller and the delay τ_{ca} represents the network induced delay in the direction from the controller to the actuator. Delays τ_{sc} and τ_{ca} have different character from the control point of view. In case of delay τ_{sc} the controller can include delay in computation of control value in step k, because together with the measured value the controller will receive timestamp when the value was measured. But the delay τ_{ca} can be included in computation of the control value just in the step $(k+1)$. The control action delay (duration of control value realization) is in general sum of these two delays. Both network induced delays can degrade the performance of control system designed without considering the delay.

The next problem is that in NCS several digital equipments take part in a control loop. The asynchronism of these different clocks in NCS causes a nonstrict periodical sample for measurement and for control application. The network induced delays and multiplicity of clocks make difficult its mathematical formalization.

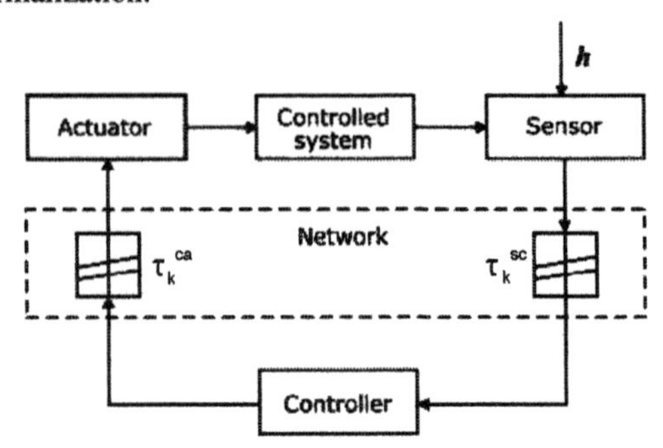

Fig. 1: Scheme of NCS with network induced delays τ_{sc} and τ_{ca}

As mentioned above there are following sources of delays in NCS (Ligusova, 2002):

1. delays caused by activity of individual NCS elements (time necessary for computation, communication synchronization of individual NCS elements, etc.),
2. delays caused by asynchronism of clocks in NCS elements (time necessary for clock synchronization of NCS elements),
3. network induced delays caused by MAC mechanism of communication media - τ_{sc} in direction from sensor to controller and τ_{ca} in direction from controller to actuator (time sharing of communication media, time necessary for signal coding, communication processing etc.),
4. delays caused by unpredictable network faults (time delays those are not caused by the network protocol, but caused by implementation of NCS into a real environment).

The delays can be very variable. Their character can be constant, bounded or random and can go from some milliseconds to several minutes. It is evident that a real communication network shows greater percentage of small delays than big delays. In general it can be said that the frequency of appearance of delays are inversely proportional to their value.

Time delays caused by the network can be, in regard to their duration, classified into following three groups:

- delays smaller than the sampling period of the controlled system $\tau < T$
- delays comparable to the sampling period $\tau = T$
- delays greater than the sampling period $\tau > T$.

The bigger the delay, the more important are potentially its consequences and the system safety can be changed. From the safety point of view the smallest delays could be neglected, but some sequences of small delays are likely to cause bigger delays (Barger *et al.*, 2002).

3. THE IMPLEMENTATION OF THE SMITH PREDICTOR INTO A NETWORK CONTROL SYSTEM

3.1 The Smith predictor and its modification for NCS

When considering the delay of control action in NCS as dynamically changeable deadtime, the Smith predictor can be used for network delays compensation. Scheme of the Smith predictor is shown in Fig. 2. It consists of an ordinary feedback loop (represented by white blocks) plus an inner loop (represented by grey blocks) that introduces two extra blocks in the feedback path. The first term is an estimate of what the process variable would look like in the absence of any disturbances.

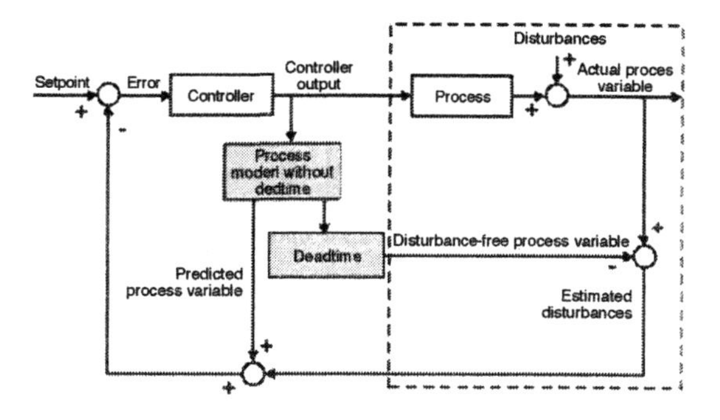

Fig. 2: Modified scheme of the Smith predictor

The mathematical process model used to generate the disturbance-free process variable has two elements connected in series. The first represents all of the process behaviour not attributable to deadtime. The second represents nothing but the deadtime. The deadtime-free element is generally implemented as an ordinary differential or difference equation that includes estimates of all the process gains and time constants. The second element is simply a time delay. The signal that goes into it comes out delayed, but otherwise unchanged. The second term that the Smith strategy introduces into the feedback path is an estimate of what the process variable would look like in the absence of both disturbances and deadtime. Subtracting the disturbance-free process variable from the actual process variable yields to an estimate of the disturbances. By adding this difference to the predicted process variable, Smith created a feedback variable that includes the disturbances, but not the deadtime. In case of NCS just modified scheme of Smith predictor can be used, that represent scheme shown on Fig. 2 without block bordered with dashed line. The reason of this modification is that in case of network delay occurrence we cannot reach the value of actual process variable.

3.2 Implementation of the Smith predictor into NCS

The first goal of our work was to analyse a NCS and to find by analysis appropriate behaviour of NCS elements to ensure stability of the system in case of network faults. In case of a sensor node, there are no algorithms that can improve the quality of control. From theoretical point of view arithmetic average of samples, or geometrical average of samples cannot improve regulation process results because the closed loop with non-equidistance steps is nonlinear system. So, we came to the conclusion that the most significant control alternative in case of network faults is on the focus to actuator and to use some kinds of control variable prediction.

The proposed implementation of the Smith predictor into NCS is shown on Fig. 3. This scheme is realization of Smith predictor for controlled systems without deadtime. We consider just delay of network in NCS as dynamically changeable deadtime. Overall control scheme shown on Fig. 3 uses two control modes. The first mode is used in case of control without network induced delays (lower loop) when controller connected to the network is used and the second mode is used when delays occur in NCS (Smith predictor loop in actuator). It is evident that Smith predictor is realized in actuator node. The arbitration unit between two mentioned modes is a block with the label τ, that represents detection of network faults. In regard to occurrence of network induced delay, an arbitration unit switches between two control modes. Situation on Fig. 3 presents first control mode (control without network induced delays). As shown on Fig. 3 it is normal network control loop where sensor, controller and actuator are connected to control network. But in case of a large network delay occurrence this alternative is not sufficient to ensure stability of NCS, because sensor and controller (Controller 1) nodes will be disconnected from control loop and just control data from actuator have influence to control quality. So in case of network delay occurrence the control will be switched into second control mode that physically represents connection of Smith predictor loop into control. As is shown on Fig. 3 in case of the second mode output of the model y_p is used to calculate control action that is realized on real controlled system. This model output is not sensed but predicted in regard to model behaviour so we need just this software sensor that makes prediction of system output.

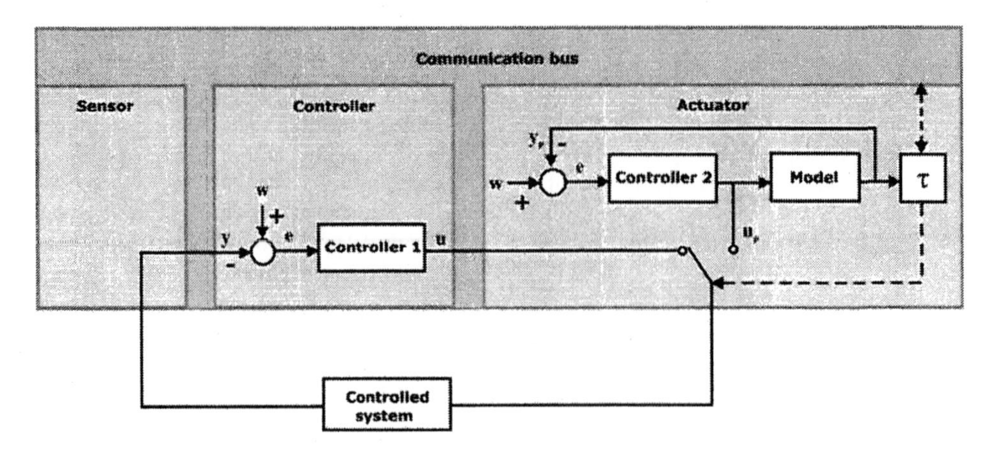

Fig. 3: Implementation of the Smith predictor into Network Control System

From this point thc most important block of the Smith predictor scheme is model, because just a slightest difference between controlled system and model can cause generation of output that can degrade quality of control.

3.3 Block algebra of both control modes – normal control mode and Smith predictor mode

As was mentioned above decision control is used to compensate network delays in NCS, where decision is realized by a delay detection block implemented in actuator node. Decision criteria is the value of the delay on the network that measures the delay detection block. Time, when detection block switches first to second mode is set according to the way of NCS element initialisations. As shown on Fig. 4(a), in case of occurrence of this acceptable delay τ_a normal network control loop is realized without implementation of the Smith predictor. In case of occurrence of bigger delays (Fig. 4(b)) the Smith predictor gets control in NCS. Transfer function of the network closed loop without implementation of the Smith predictor shown on Fig. 4 (a) is as follows:

$$\begin{aligned} e &= w - y \\ u &= eF_{c1} \end{aligned} \quad \rightarrow \quad F = \frac{y}{w} = \frac{F_s F_{c1} F_{\tau ca}}{1 + F_s F_{c1} F_{\tau ca} F_{\tau sc}}$$

where $\tau_{sc} \in (0, \frac{T}{2})$ and $\tau_{ca} \in (0, \frac{T}{2})$. Symbols F_s, F_{c1}, $F_{\tau ca}$ and $F_{\tau sc}$ represent transfer functions of individual blocks on Fig. 4, where F_s is transfer function of controlled system, where F_{c1} is transfer function of controller 1, $F_{\tau ca}$ represents transfer function of network delay block in direction from controller to actuator and $F_{\tau sc}$ represents transfer function of network delay block in direction from sensor to controller.

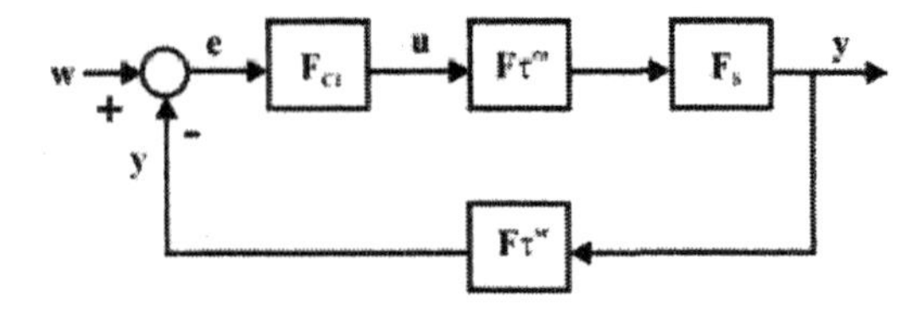

(a)
first control mode (network control mode without implementation of the Smith predictor)

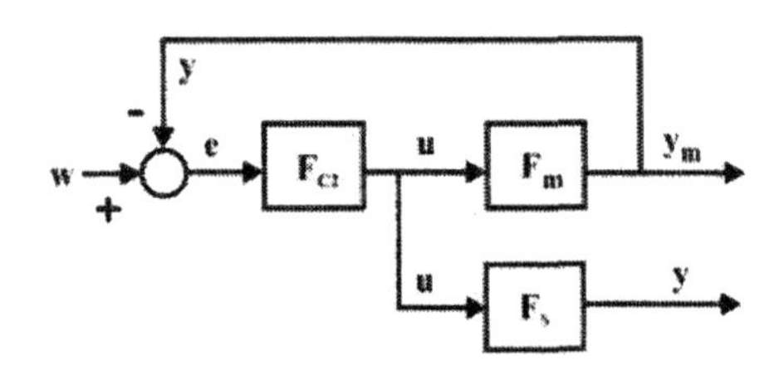

(b)
second control mode (the Smith predictor control mode)

Fig. 4: Block diagrams of first and second control modes

The transfer function of the Smith control loop shown on Fig. 4 (b) is derived as follows:

$$\begin{aligned} e &= w - y_m \\ u &= eF_{c2} \\ y &= uF_s \\ y_m &= uF_m \end{aligned} \quad \rightarrow \quad F = \frac{y}{w} = 1 - \frac{F_s F_m F_{c2}^{\ 2}}{1 + F_{c2} F_m} ,$$

where F_s is transfer function of controlled system, where F_{c2} is transfer function of controller 2 F_m represents the transfer function of the system model used for the Smith prediction.

4. NCS SIMULATION

To test our proposition a model on Fig. 3 was developed. The model includes a sensor, controller, controlled system, actuator and a communication network. Modelling of such a system is not really easy. The most common tools are not adapted for this kind of modelling. From our previous experiences the choice was to develop a model using the Coloured Petri Nets.

The communication network used in this example is a random access type. Any component willing to transmit can do so, if the network is ready. If this is not true then the component can retry the emission later. A message put to transmission is transported through the bus to its destination. This transmission can take either constant or variable time. A number of messages are transported through the network from sensor to the controller and from the controller to the network. Supplementary traffic consists of 3 sources emitting at he same period as the sampling period of the discrete system.

The controller: There are two controllers in our control scheme, but assuming that we have the same model as the controlled system, both controllers have the same regulation constants. The backup controller (controller 2 on Fig. 3) in actuator node is represented as a system of recurrent equations representing a smith predictor composed of a proportional-integration regulator connected with a model of the control system. The delay of two sampling steps is chosen in the Smith predictor. Both controllers are designed as a discrete time system with a regular sampling period fixed at 10 time units (TU). Every 10 TU a new control value is established and prepared for transmission independently on whether a new measure has arrived or not.

A sensor is connected to the controlled process model. Just like the controller this one produces for transmission a new measure value every 10 TU. In order to make the modelled system deterministic, the sensor sampling points are delayed to that of the controller by 1 TU. To simplify the example the sensor disposes always of exact measure without any perturbation. The modeled system is also expressed as a system of recurrent equations. Its sampling period is fixed at 10 TU but it is of little importance in this example. No perturbation on controlled system is considered.

The action element applies instantly and without any degradation the command value on the controlled system. Its study is not a part of the present communication.

The backup control algorithm is used to determine an approximate control value in the case when the communication network did not provide the actuator with the command necessary. This algorithm is a model of a closed loop containing the controller identical to the one in the main closed loop of the system (the Smith predictor), a model of the controlled system and 2 delay blocks of 1 TU each representing the network communication delay. This algorithm is constantly running. No adaptation is considered at this stage of our work. We suppose the actuator knows the desired value to reach.

The study of the network arbitration unit is the main theme of this study. If control message is delayed beyond a limit value τ it switches to the reserve control algorithm running in the actuator. The question is how the value of τ influences the quality of the control. For this reason a series of τ values are tested through a Monte-Carlo simulation.

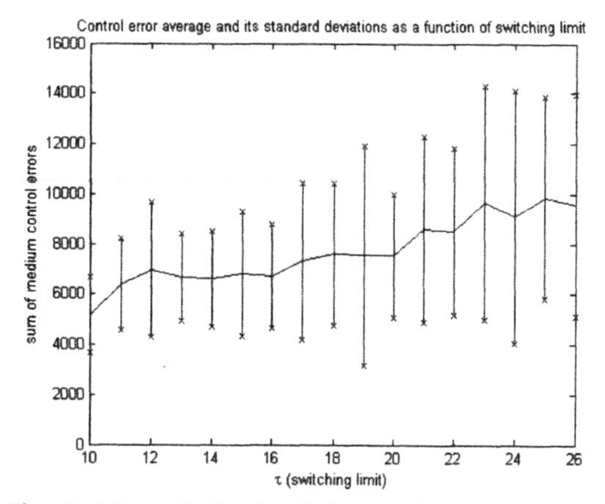

Fig. 5: Monte-Carlo simulation results: control error as function of τ switching limit in arbitration unit

Results in Fig. 5 and Fig. 6 are done in regard to control scheme shown on Fig. 3 under the following conditions: (1) time constant of the controlled system is T=20 TU, (2) sampling period is Ts=10 TU, (3) the range of random network delays is from 1 TU to 17 TU, (4) setting of τ switching limit is in range 10 TU $<= \tau <=$ 26 TU (x axis of Fig.5) (5) for every individual τ value 100 simulations were realized and an average sum of control errors was calculated and displayed (y axis of Fig.5). From the given graph it can be concluded, that increasing the τ value decreases the quality of the control and worsens its estimation. It seems that the dependence is linear. On the following graph (Fig. 6) the relationship between the τ value and the number of the interventions of the auxiliary Smith predictor implemented in the actuator is shown. If the τ value is small, often than The smaller the τ value, the more often the auxiliary Smith predictor is used. The graph shows also that the estimation precision is not dependent on the τ value itself but is always constant.

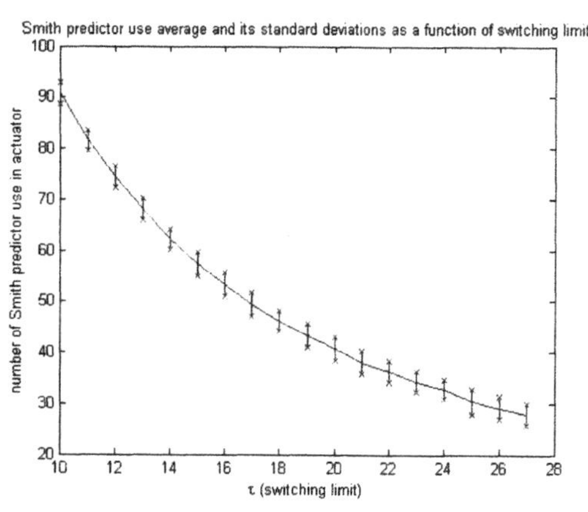

Fig. 6: Monte-Carlo simulation results: The Smith predictor use as function of τ switching limit in arbitration unit

5. CONCLUSION

Considering network delay in described approach improves safety of whole NCS system. There are many important questions yet to be answered for NCS design researchers, and also for designers of NCS elements in industry, too. This paper discussed using of the Smith predictor method for deadtime treatment to compensate network delays in NCS not only smaller then sampling period but bigger then sampling period as well. In regard to assumption, that network induced delay is dynamically changeable deadtime, modification of Smith prediction was implemented into NCS to compensate these delays. The control algorithm is based on decision control, where decision is realized by delay detection block implemented in actuator node. So, in case of NCS with negligible delays normal network control loop is realized shown on Fig. 1 and in case of significant delay occurrence, the Smith predictor realized in actuator node is connected to control loop. It is evident that in case of NCS distribution of intelligence is necessary to ensure stability in case of random network faults. The most important node in NCS design is an actuator node, that has three main functions: (1) standard actuator functions, (2) backup controller and (3) network observer. We recommend to producers of actuators in industry to design hardware and software parts of actuators by way that it will be possible to implement into actuators control shown on Fig. 3. It means to implement into actuator adaptive state-space model of the system, implementation of detection arbitration and switching units into actuator and implementation hot backup of controller parameters. By using of control method described above is possible very effectively compensate network induced delays and network faults, as it is shown in simulation results as well. Similarly, this approach solve transition from non-equidistance control to control with equidistance steps. Of course there are some problems to be solved as to mentioned control method, for example setting of switching limit in detection arbitration unit in the regard to system sampling period as well as sampling period of the model and another important parameters. Petri nets was, because of their capability to model a network (and network delay) used for modelling of discrete systems with non-equidistance steps, that NCS are.

REFERENCES

Barger P., Thiriet J. M., Robert M. Performance and dependability evaluation of distributed dynamical systems, European Conference on System Dependability and Safety (ESRA 2002/lambda-Mu13), pp. 16-22, Lyon (France), 19.-21. March 2002.

Ligušová, J.: Modelling of Network Control Systems Based on Coloured Petri Nets, Intenal scientific conference of TU Košice, Košice, Slovakia, May, 2002

Nilsson J., Bernhardsson B., Wittenmark B. Stochastic analysis and control of real-time systems with random time delays, Automatica 34:1, pp. 57-64, 1998

Walsh G. C., YE H., Bushnel L. Stability analysis of networked control systems, Proceedings American Control Conference, pp. 2867-2880, San Diego, CA, Jun 1999

Wittenmark B., Nilsson J., Torngen M. Timing problems in real-time control systems, Preprints American Control Conference, pp. 2000-2004, Seattle, WA, Jun 1995

SELFTUNING PID CONTROL BASED ON THE CLOSED-LOOP RESPONSE RECOGNITION

Martin Foltin, Ivan Sekaj

Department of Automatic Control Systems, Faculty of Electrical Engineering and Information Technology, Slovak University of Technology, Ilkovičova 3, 812 19 Bratislava, Slovak Republic
Tel : ++421 2 602 91 506 E-mail : foltin@kasr.elf.stuba.sk , sekaj@kasr.elf.stuba.sk,

Abstract: A new type of knowledge adaptive control based on the closed-loop response recognition is proposed. This approach tries to mimic the behaviour of an expert, who has a lot of experience with the particular controlled system. The adaptive system updates the controller parameters according to the changes in system behaviour. The algorithm uses a rule-based system, which evaluates the closed-loop response shape after the decay of the transient mode, and computes the corrections of controller parameters. Because the main problem of such a control structure consists in designing of the rule-base, we focused our interest on the design of an automatic procedure for the rule-base generating. *Copyright © 2003 IFAC*

Keywords: Self tuning, PID controllers, MIMO, Step response

1. INTRODUCTION

To control dynamic processes, which behaviour changes in time or which have nonlinear behaviour, adaptive control systems use to be used. In such cases for control algorithms design commonly exact mathematical models and stability and performance conditions are employed. The results obtained by such an approach are often very good, however sometimes problems may emerge e.g. when the complexity or nonlinearity degree of the model increases or if we have problems with inexact measurements or insufficient a priori information about the process (uncertainty). Thus, sometimes our "exact" mathematical approach and the resulting algorithms cannot be used to control the given process or some phases of it. In that moment it is the task for the human-operator to carry on and to control the process to the required state. The human works with his own model, based on his experience and estimations. He is able to deal with inexact, uncertain, nonnumeric and complex information (he exploits his „natural“ intelligence), which in our case, he is able to apply for the process control. Using the operators „manual adaptation“ of a complex or non-exactly described processes the results are sometimes better than using an „automatic adaptation“. The goal of the following proposed approach is to mimic the human behaviour and to update controller parameters with respect to the required closed-loop response. The underlying idea consists in that after the disturbance occurs we let the transient process decay and then evaluate the close-loop response (Fig. 1.).

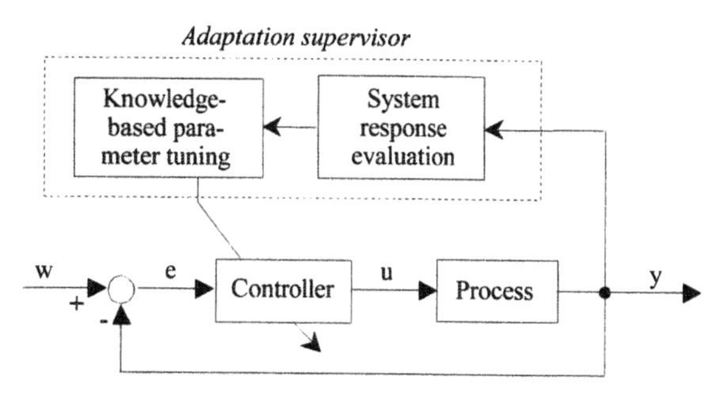

Fig. 1. Block scheme of the rule-based adaptation system of a PID controller.

Using a rule-based mechanism, corrections of actual controller parameters are computed. The aim is to design a tuning algorithm, which will provide a good controller adaptation without any exact a-priori information about the process. Similar methods can be found in (Pfeiffer and Isermann, 1993; DeSilva

1991; Oliveira, *et al.*, 1991). This method is based on a time-response shape recognition using the least-square-error method.

2. THE ADAPTATION MECHANISM

The recognition of the transient response shape is based on the comparison of the time-responses with a set of representative shapes using the least-square-error method. A brief description of this method is as follows. Consider a „nominal system", which is located (when possible) in the centre of the area A in which the system parameters can move during the normal operation

$$A = I_P \times I_I \times I_D,\ A \subset R^3 \quad (1)$$

where I_P, I_I, I_D are the considered intervals for the PID parameters, or it is a system, which can be considered as a typical representative of a class of systems (from point of view system response shape). For such a system we design optimal PID controller parameters P^*, I^*, D^* with respect to selected requirements (performance index etc.) (Sekaj, 1999). Then we start to perform a cycle of systematic changes in the PID controller parameters in a defined universe with defined steps s_P, s_I and s_D. For each set of actual parameters $\{P_d, I_d, D_d\}$ it is realized a simulation of the closed-loop system with the selected disturbance (set point step etc.). The obtained time-response is written into the database together with the corresponding controller parameters set. The number of such records in the database is

$$n = n_P . n_I . n_D \quad (2)$$

where

$$n_P = \frac{I_P}{s_P}, n_I = \frac{I_I}{s_I}, n_D = \frac{I_D}{s_D} \quad (3)$$

For a PI controller the database represents a 2-D table, where each item of the table contains a record of a closed-loop time-response and the set of the corresponding parameters P and I. An example for a particular closed-loop is in the Fig.2, where the marked response represents the desired closed-loop response shape. For the PID case the table is by analogy 3-dimensional.

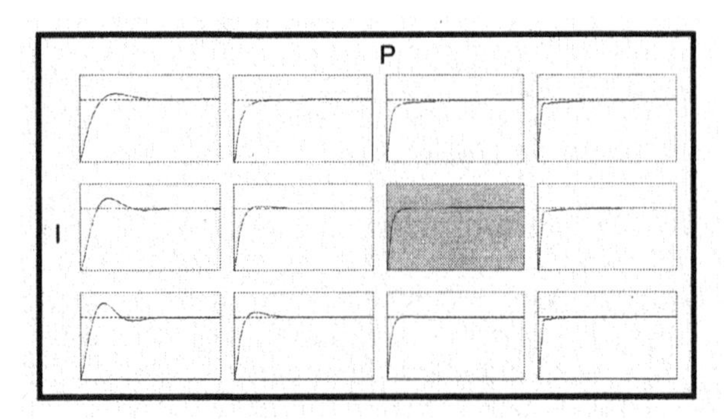

Fig. 2. Database of closed-loop time-responses

After a single-shot generating of such a database all data are ready for the application in the on-line adaptive system. The Adaptation mechanism is as follows. After the detection of some disturbance in the closed-loop we let decay the transient period. Next the acquired time response is normed and then there is searched for the „most similar" time-response from all records in the database, i.e. which minimize the criterion

$$J = \sum_{k=1}^{n} (y_{d,k} - y_k)^2 \rightarrow \min \quad (4)$$

where $y_{d,k}$ is the closed-loop response from the database under the controller parameters P_d, I_d, D_d, d is the ordinal number of the database record with the „most similar" time-response (Fig. 3.) and k is the simulation step.

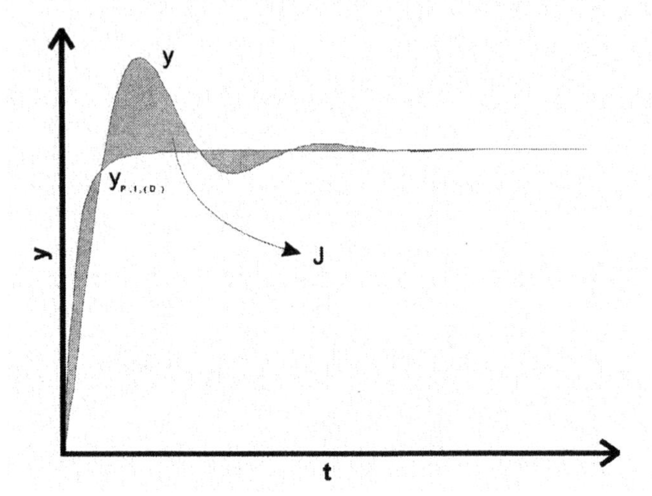

Fig. 3. Evaluation of the "most similar" response

After retrieving of the selected time-response there is loaded the appropriate set $\{P_d, I_d, D_d\}$ from the database or it is interpolated between neighbour items in database.

The last step is the calculation of controller parameter corrections

$$\Delta P = P^* - P_d \quad \Delta I = I^* - I_d \quad \Delta D = D^* - D_d \quad (5)$$

These corrections are used for the actual PID controller updating

$$\begin{aligned} P_{new} &= P_{actual} + \Delta P \\ I_{new} &= I_{actual} + \Delta I \\ D_{new} &= D_{actual} + \Delta D \end{aligned} \quad (6)$$

and the algorithm is waiting for new disturbances and then it is repeating. The main steps of the algorithm are as follows:

1. Detection of a disturbance and waiting for the closed-loop response decay
2. Evaluation of the most similar time response from the database using (4)

3. Interpolation between neighbour PID parameter sets in the database
4. Correction of the actual PID parameter set using (5) and (6)

Despite the fact that this algorithm is not trivial for computation, the response evaluation and parameter correction is calculated only at the moment after decay of the transient period. Therefore there are no problems with the implementation in real-time control.

3. CASE STUDY

An example of the adaptation process of a DC-drive speed is depicted in Fig. 4.

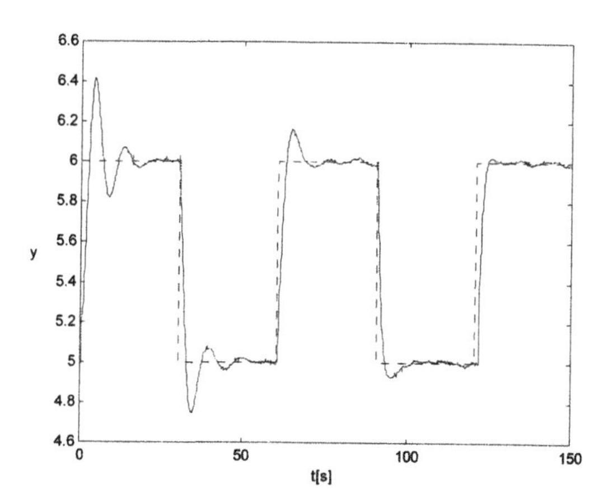

Fig. 4. Adaptation process of the PID controller of a DC-drive speed

Another example in case of a more complex controlled object is as follows. The controlled system consists from two servomotors (two input and two output system –TITO). Between the two motors there are strong interactions. The goal is the independent speed control of both motors. The block scheme of the object with the considered control structure is in Fig.5.

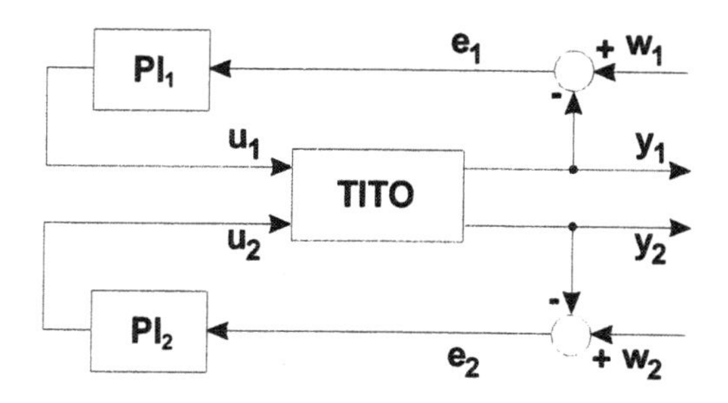

Fig. 5. Block scheme of the closed-loop with 2 servo systems

For the speed control (y_1 and y_2) two independent PI controllers are used PI_1 and PI_2, each with two parameters (P and I). An example of the adaptation process is depicted in Fig. 6.

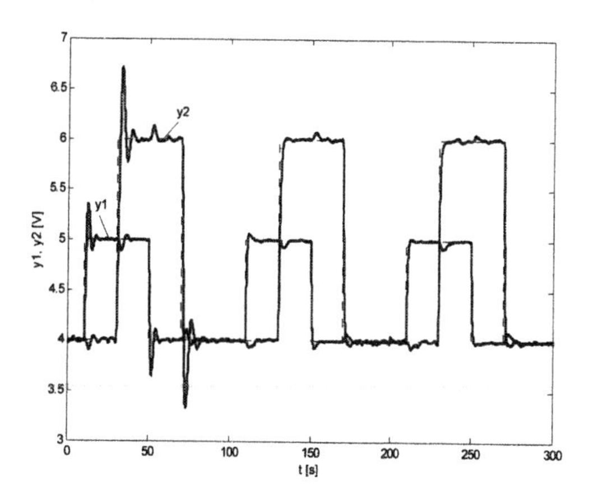

Fig. 6. Adaptation process of the TITO system control

4. CONCLUSION

The described knowledge-based adaptation of a PI / PID controller mimic the tuning of a controller by an experienced human-operator. The advantage of this approach consists in, that it does not require a mathematical model of the controlled process. It is able to adapt the PI / PID controller parameters only after the closed-loop behaviour recognition, after the decay of the transient mode. Another advantage is the possibility to use designed rule-base for a class of similar systems, no matter their time constants are. Experiments in simulation and real-time has shown, that the mentioned adaptation approach is surprising robust and it is working reliable.

Acknowledgement. This work has been supported by the grant No. 1/7630/20 „Intelligent methods of modeling and control" of the Slovak Grant Agency.

REFERENCES

DeSilva, C.W.: An Analytical Framework for Knowledge-Based Tuning of Servo Controllers, Eng.Applic.Artif.Intellig., Vol.4, No.3, 177-189, 1991

Foltin, M.: Design of Knowledge-Based Adaptive Control, thesis, FEI STU Bratislava, 2000, (in slovak)

Oliveira, P., Lima, P, Sentineiro, J.: Fuzzy Supervision on Intelligent Control Systems, ECC 91, Grenoble, France, 1991

Pfeiffer, B.M., Isermann, R.,: Selftuning of classical controllers with fuzzy-logic, Proc. of IMACS Symp."Mathematical and Intelligent models in system Simulation", Brussels 1993

Sekaj, I., Foltin, M.: Adaptive Control Based on the Closed-Loop Response Recognition. VDI Berichte, 2000, Nr. 1761-1805, 82-88.

Sekaj, I., Foltin, M.: Adaptive Control Based on the Closed-Loop Resposne Recognition. Process Control 2001, High Tatras, 188.

Sekaj, I.: Genetic Algorithm-Based Control System Design and System Identification, Conferrence Mendel'99, Brno, Czech Republic, june 9th –12th 1999, pp. 139-144

www.elsevier.com/locate/ifac

INDIRECT ADAPTIVE CONTROL OF NONLINEAR SYSTEMS USING TAKAGI- SUGENO TYPE FUZZY LOGIC

Martin Kratmüller, Ján Murgaš

Department of Automatic Control Systems, Faculty of Electrical Engineering and Information Technology, Slovak University of Technology,
Ilkovičova 3, 812 19 Bratislava, Slovak Republic
Tel : ++421 2 602 91 781
E-mail : mailto:murgas@kasr.elf.stuba.sk,
mailto:kratmuller@kasr.elf.stuba.sk

Abstract: An adaptive fuzzy controller is designed using a collection of fuzzy IF-THEN rules. Parameters of membership functions characterizing the linguistic terms in the fuzzy IF-THEN rules are changing according to some adaptive laws, to control a plant to track a reference trajectory. In this paper, an indirect adaptive fuzzy control design method is developed for general higher order nonlinear continuous systems. The Takagi-Sugeno type fuzzy logic system has been used to approximate the controller. It has been proved that the closed-loop system under this adaptive fuzzy controller is globally stable in the sense that all signals involved are bounded. Finally, the indirect adaptive fuzzy controll method has been applied to control an unstable system. *Copyright © 2003 IFAC*

Keywords: A adaptive fuzzy control, fuzzy control, linear state feedback control, nonlinear systems.

1. INTRODUCTION

Fuzzy logic controllers are generally considered to be applicable to plants that are mathematically poorly understood and where experienced human operators are available.

An adaptive fuzzy system is a fuzzy logic system equipped with an adaptation law. The major advantage of adaptive fuzzy controllers over conventional adaptive controllers is that the adaptive fuzzy controller is capable of incorporating linguistic fuzzy information from human operators. In the indirect adaptive fuzzy control, fuzzy logic systems are used to model the plant. Then the controller is constructed under the assumption that the fuzzy logic system approximates represent the true plant (Fig. 1).

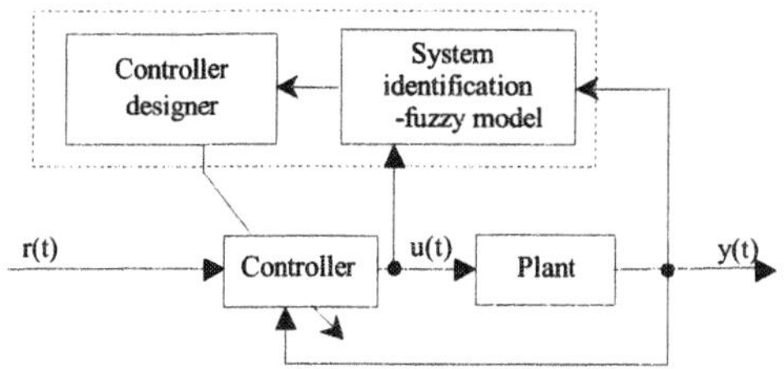

Fig. 1. Indirect adaptive control

In this paper, an alternative indirect adaptive controller is developed. We have use fuzzy logic systems that are different from used in (Chen, *et al.*, 1996; Essounbouli, *et al.*, 2002; Wang, 1993). This form of the fuzzy logic system is called the Takagi-Sugeno type. Those consequent part of the fuzzy rules is based on the locally linear feedback control theory. In Section 2, the considered problem formulation theory is shown. In Section 3, the used fuzzy logic system is described. The main result is presented in Section 4. In Section 5, the proposed indirect adaptive fuzzy controller design method is used to control an unstable system. Some conclusions of this paper are given in Section 6.

2. PROBLEM FORMULATION

Consider the nth-order nonlinear systems of the form

$$x^{(n)} = f\left(x,\dot{x},\ldots,x^{(n-1)}\right) + g\left(x,\dot{x},\ldots,x^{(n-1)}\right)u$$
$$y = x \qquad (1)$$

where f and g is unknown (uncertain) but bounded continous functions and $u \in R$ and $y \in R$ are the input and the output of the system, respectively. Let $\underline{x} = (x, \dot{x}, \ldots, x^{(n-1)})^T \in R^n$ be the state vector of the system which is assumed to be available.

The control objective is to force y to follow a given bounded reference signal y_m under the contraints that all signals involved must be bounded. Hence, a feedback control u based on fuzzy logic systems and an adaptive law for adjusting the paramaters of the fuzzy logic systems are both determined to satisfy the following conditions:

1) The closed-loop system must be globally stable in the sense that all variables, $\underline{x}(t)$, $\underline{\theta}(t)$ and $u(\underline{x} \mid \underline{\theta})$ must be uniformly bounded; that is, $|\underline{x}(t)| \le M_x < \infty$, $|\underline{\theta}(t)| \le M_\theta < \infty$ and $|u(\underline{x} \mid \underline{\theta})| \le M_u < \infty$ for all $t \ge 0$, where M_x, M_θ and M_u are design parameters specified by the designer.
2) The tracking error $e \equiv y - y_m$ should be as small as possible under the constraints in 1).

Then our design objective is to impose an adaptive fuzzy control algorithm so that the following asymptotically stable tracking is achieved

$$e^{(n)} + k_1 e^{(n-1)} + \ldots + k_n e = 0. \qquad (2)$$

The roots of the polynomial $h(s) = s^n + k_1 s^{n-1} + \ldots + k_n$ in the characteristic equation of (2) lie all in the open left-half plane via an adequate choice of coefficients $k_1, k_2, \ldots, k_n$.

3. DESCRIPTION OF THE FUZZY LOGIC SYSTEM

The fuzzy logic system performs a mapping from $U \subset R^n$ to R. We assume $U = U_1 \times \cdots U_n$, where $U_i \subset R$, $i = 1,2,\cdots,n$.

The fuzzy rule base contains a collection of fuzzy IF-THEN rules

$$R^{(l)}: \quad \text{IF } x_1 \text{ is } F_1^l \text{ and ... and } x_n \text{ is } F_n^l$$

$$\text{THEN } \quad y = K_1^l x_1 + K_2^l x_2 + \cdots + K_n^l x_n \qquad (3)$$

where $\underline{x} = (x_1, x_2, \ldots, x_n)^T \in R^n$ and $y \in R$ are the input and the output of the fuzzy system, respectively. F_i^l denotes the fuzzy set in U_i, for l=1,2,...,M. $K_1^l, K_2^l, \cdots, K_n^l$ are constant coefficients of the consequent part of the fuzzy rule. This fuzzy logic system is called the Takagi-Sugeno type. Each fuzzy rule of (3) defines a fuzzy implication $F_1^l \times \cdots \times F_n^l \to y(\underline{x}) = K_1^l x_1 + K_2^l x_2 + \cdots + K_n^l x_n$.

In this paper, we employ the product operation for the fuzzy implication and the *t* norm. The definition of the product operation is the same as the one in (Spooner, and Passino, 1996); besides, the singleton fuzzifier is used. Consequently, the final output value is

$$y(\underline{x}) = \frac{\sum_{l=1}^{M} \left(\prod_{i=1}^{n} \mu_{F_i^l}(x_i) \right) \left(K_1^l x_1 + \cdots + K_n^l x_n \right)}{\sum_{l=1}^{M} \left(\prod_{i=1}^{n} \mu_{F_i^l}(x_i) \right)} \qquad (4)$$

where $\mu_{F_i^l}(x_i)$ are the membership functions for *i*=1,2,...,*n* and *l*=1,2,...,*M*.

If we fix the $\mu_{F_i^l}(x_i)$'s and view the K_i^l's as adjustable parameters, then (4) can be rewritten as

$$y(\underline{x}) = \theta^T \xi(\underline{x}) \qquad (5)$$

where $\theta = (K_1^1, \cdots, K_n^1, K_1^2, \cdots, K_n^2, \cdots, K_n^n)^T$ is a parameter vector and $\xi(\underline{x}) = (\xi_1^1(\underline{x}), \cdots, \xi_n^1(\underline{x}), \xi_1^2(\underline{x}), \cdots, \xi_n^2(\underline{x}), \cdots, \xi_n^n(\underline{x}))^T$ is a regressive vector with the regressor $\xi_j^l(\underline{x})$ defined as (Kratmüller, 2002)

$$\xi_j^l(\underline{x}) = \frac{\left(\prod_{i=1}^{n} \mu_{F_i^l}(x_i) \right) x_j}{\sum_{l=1}^{M} \left(\prod_{i=1}^{n} \mu_{F_i^l}(x_i) \right)}. \qquad (6)$$

Hence, (6) will represent mathematical models of the fuzzy logic system used in the following sections.

4. THE MAIN RESULT

The basic ideas of how to construct an indirect adaptive fuzzy controller to achieve the control objectives defined in Section 2 will be shows. First, let $\underline{e} = (e, \dot{e}, \ldots, e^{(n-1)})^T$ and $\underline{k} = (k_n, \ldots k_1)^T \in R^n$ be such that all roots of the polynomial $h(s) = s^n + k_1 s^{n-1} + \ldots + k_n$ will be shown that are in the open left-hand plane. If the functions f and g are known, then the control law

$$u^* = \frac{1}{g(\underline{x})} \left[-f(\underline{x}) + y_m^{(n)} + \underline{k}^T \underline{e} \right] \qquad (7)$$

applied to (1) results in

$$e^{(n)} + k_1 e^{(n-1)} + \ldots + k_n e = 0 \qquad (8)$$

which implies that $\lim_{t \to \infty} e(t) = 0$ (the main control objective). Since f and g are unknown, the optimal control u^* cannot be implemented. Hence, the fuzzy logic controller will be designed to approximate this optimal control.

We replace f and g in (7) by fuzzy systems $\hat{f}(\underline{x} \mid \underline{\theta}_f)$ and $\hat{g}(\underline{x} \mid \underline{\theta}_g)$, respectively, which are in the form (5). The resulting control law

$$u = \frac{1}{\hat{g}(\underline{x} | \underline{\theta}_g)} \left[-\hat{f}(\underline{x} | \underline{\theta}_f) + y_m^{(n)} + \underline{k}^T \underline{e} \right] \qquad (9)$$

is the so-called certainty equivalent controller (Sastry, and Bodson, 1989; Murgaš, 2000). Applying (9) to (1) and staightforward manipulation, we obtain the error equation

$$e^{(n)} = -\underline{k}^T \underline{e} + \left[\hat{f}(\underline{x} | \underline{\theta}_f) - f(\underline{x}) \right] + \left[\hat{g}(\underline{x} | \underline{\theta}_g) - g(\underline{x}) \right] u \quad (10)$$

or equivalently

$$\dot{\underline{e}} = \Lambda_c \underline{e} + \underline{b}_c\left[\left(\hat{f}(\underline{x}|\underline{\theta}_f) - f(\underline{x})\right) + \left(\hat{g}(\underline{x}|\underline{\theta}_g) - g(\underline{x})\right)u\right] \quad (11)$$

where

$$\Lambda_c = \begin{bmatrix} 0 & 1 & 0 & \cdots & 0 & 0 \\ 0 & 0 & 1 & \cdots & 0 & 0 \\ \vdots & \vdots & \vdots & & \vdots & \vdots \\ 0 & 0 & 0 & \cdots & 0 & 1 \\ -k_n & -k_{n-1} & & \cdots & & -k_1 \end{bmatrix}, \underline{b}_c = \begin{bmatrix} 0 \\ \vdots \\ \vdots \\ 0 \\ 1 \end{bmatrix} \quad (12)$$

Since Λ_c is a stable matrix ($|sI - \Lambda_c| = s^n + k_1 s^{n-1} + \ldots + k_n$ which is stable), we know that there exists a unique positive definite symetric $n \times n$ matrix P which satifies the Lyapunov equation

$$\Lambda_c^T P + P\Lambda_c = -Q \quad (13)$$

where Q is an arbitrary $n \times n$ positive definite matrix. If $V_e = \frac{1}{2}\underline{e}^T P \underline{e}$ then using (11) and (13) we obtain

$$\dot{V}_e = \frac{1}{2}\dot{\underline{e}}^T P \underline{e} + \frac{1}{2}\underline{e}^T P \dot{\underline{e}}$$

$$= -\frac{1}{2}\underline{e}^T Q \underline{e} + \underline{e}^T P \underline{b}_c\left[\left(\hat{f}(\underline{x}|\underline{\theta}_f) - f(\underline{x})\right) + \left(\hat{g}(\underline{x}|\underline{\theta}_g) - g(\underline{x})\right)u\right] \quad (14)$$

The control u is supposed to consist of a fuzzy control u_c and a supervisory control u_s, i.e.,

$$u = u_c + u_s \quad (15)$$

The purpose of this supervisory control u_s is to force $\dot{V}_e \le 0$ when $V_e > \overline{V}$ (which is a constant specified by the designer). Substituing (15) in (1) and using the same manipulation as for obtaining (11), we have the new error equation:

$$\dot{\underline{e}} = \Lambda_c \underline{e} + \underline{b}_c\left[\left(\hat{f}(\underline{x}|\underline{\theta}_f) - f(\underline{x})\right) + \left(\hat{g}(\underline{x}|\underline{\theta}_g) - g(\underline{x})\right)u_c - g(\underline{x})u_s\right] \quad (16)$$

Using (16) and (13) we have

$$\dot{V}_e = -\frac{1}{2}\underline{e}^T Q \underline{e} + \underline{e}^T P \underline{b}_c\left[\left(\hat{f}(\underline{x}|\underline{\theta}_f) - f(\underline{x})\right) + \left(\hat{f}(\underline{x}|\underline{\theta}_f) - f(\underline{x})\right)u_c - g(\underline{x})u_s\right]$$

$$\le -\frac{1}{2}\underline{e}^T Q \underline{e} + \left|\underline{e}^T P \underline{b}_c\right|\left[\left|\hat{f}(\underline{x}|\underline{\theta}_f)\right| + \left|f(\underline{x})\right| + \left|\hat{g}(\underline{x}|\underline{\theta}_g)u_c\right| + \left|g(\underline{x})u_c\right|\right] - \underline{e}^T P \underline{b}_c g(\underline{x}) u_s \quad (17)$$

In order to design u_s such that the right-hand side of (17) is nonpositive we need to know the bounds for f and g, i.e., we have to make the following assumption.

Assumption 1: We can determine functions $f^U(\underline{x})$, $g^U(\underline{x})$ and $g_L(\underline{x})$ such that $|f(\underline{x})| \le f^U(\underline{x})$ and $g_L(\underline{x}) \le g(\underline{x}) \le g^U(\underline{x})$ for $\underline{x} \in U_c$ where $f^U(\underline{x}) < \infty$, $g^U(\underline{x}) < \infty$, and $g_L(\underline{x}) > 0$ $\underline{x} \in U_c$.

Based on f^U, g^U, and g_L and by observing (17), we choose the supervisory control u_s as

$$u_s = I^* \operatorname{sgn}\left(\underline{e}^T P \underline{b}_c\right)\frac{1}{g_L(\underline{x})}\left[\left|\hat{f}(\underline{x}|\underline{\theta}_f)\right| + f^U(\underline{x}) + \left|\hat{g}(\underline{x}|\underline{\theta}_g)u_c\right| + \left|g^U(\underline{x})u_c\right|\right] \quad (18)$$

where

$I^* = 1$ if $V_e > \overline{V}$

$I^* = 0$ if $V_e \le \overline{V}$

Substituting (18) to (17) and considerig the case $V_e > \overline{V}$ yields

$$\dot{V}_e \le -\frac{1}{2}\underline{e}^T Q \underline{e} + \left|\underline{e}^T P \underline{b}_c\right|\left[\left|\hat{f}\right| + |f| + \left|\hat{g}u_c\right| + \left|gu_c\right| - \frac{g}{g_L}\left(\left|\hat{f}\right| + f^U + \left|\hat{g}u_c\right| + \left|g^U u_c\right|\right)\right]$$

$$\le -\frac{1}{2}\underline{e}^T Q \underline{e} \le 0 \quad (19)$$

Next, we will develop an adaptation formula to adjust the parameter vectors θ_f and θ_g. Define the optimal parameter vectors

$$\underline{\theta}_f^* = \arg\min_{\underline{\theta}_f \in \Omega_f}\left[\sup_{\underline{x} \in U_c}\left|\hat{f}(\underline{x}|\underline{\theta}_f) - f(\underline{x})\right|\right]$$

$$\underline{\theta}_g^* = \arg\min_{\underline{\theta}_g \in \Omega_g}\left[\sup_{\underline{x} \in U_c}\left|\hat{g}(\underline{x}|\underline{\theta}_g) - g(\underline{x})\right|\right] \quad (20)$$

where Ω_f and Ω_g are constraint sets for $\underline{\theta}_f$ and $\underline{\theta}_g$, respectively, specified by the designer.

The "minimum approximation error" is defined as follows

$$w = \left(\hat{f}(\underline{x}|\underline{\theta}_f^*) - f(\underline{x})\right) + \left(\hat{g}(\underline{x}|\underline{\theta}_g^*) - g(\underline{x})\right)u_c \quad (21)$$

Then the error equation (16) can be rewritten as

$$\dot{\underline{e}} = \Lambda_c \underline{e} - \underline{b}_c g(\underline{x}) u_s + \underline{b}_c\left[\left(\hat{f}(\underline{x}|\underline{\theta}_f) - \hat{f}(\underline{x}|\underline{\theta}_f^*)\right) + \left(\hat{g}(\underline{x}|\underline{\theta}_g) - \hat{g}(\underline{x}|\underline{\theta}_g^*)\right)u_c + w\right] \quad (22)$$

If we choose $\hat{f}$ and $\hat{g}$ to be fuzzy logic systems in the form (5), then (22) can be rewritten as

$$\dot{\underline{e}} = \Lambda_c \underline{e} - \underline{b}_c g(\underline{x}) u_s + \underline{b}_c w + \underline{b}_c\left[\underline{\phi}_f^T \underline{\xi}(\underline{x}) + \underline{\phi}_g^T \underline{\xi}(\underline{x}) u_c\right] \quad (23)$$

where $\underline{\phi}_f = \underline{\theta}_f - \underline{\theta}_f^*$, $\underline{\phi}_g = \underline{\theta}_g - \underline{\theta}_g^*$. Now consider the candidate Lyapunov function

$$V = \frac{1}{2}\underline{e}^T P \underline{e} + \frac{1}{2\gamma_1}\underline{\phi}_f^T \underline{\phi}_f + \frac{1}{2\gamma_2}\underline{\phi}_g^T \underline{\phi}_g \quad (24)$$

where γ_1 a γ_2 are positive constants. The time derivative of V along the trajectory of (23) is

$$\dot{V} = -\frac{1}{2}\underline{e}^T Q \underline{e} - g(\underline{x})\underline{e}^T P \underline{b}_c u_s + \underline{e}^T P \underline{b}_c w + \frac{1}{\gamma_1}\underline{\phi}_f^T\left[\dot{\underline{\theta}}_f + \gamma_1 \underline{e}^T P \underline{b}_c \underline{\xi}(\underline{x})\right] + \frac{1}{\gamma_2}\underline{\phi}_g^T\left[\dot{\underline{\theta}}_g + \gamma_2 \underline{e}^T P \underline{b}_c \underline{\xi}(\underline{x}) u_c\right] \quad (25)$$

Using (13) and $\dot{\underline{\phi}}_f = \dot{\underline{\theta}}_f$, $\dot{\underline{\phi}}_g = \dot{\underline{\theta}}_g$. From (18) and for $g(\underline{x}) > 0$ we obtain $g(\underline{x})\underline{e}^T P \underline{b}_c u_s \ge 0$. If choosing the adaptive law

$$\dot{\underline{\theta}}_f = -\gamma_1 \underline{e}^T P \underline{b}_c \underline{\xi}(\underline{x}) \quad (26)$$

$$\dot{\underline{\theta}}_g = -\gamma_2 \underline{e}^T P \underline{b}_c \underline{\xi}(\underline{x}) \quad (27)$$

then from (25) results

$$\dot{V} \le -\frac{1}{2}\underline{e}^T Q \underline{e} + \underline{e}^T P \underline{b}_c w \quad (28)$$

$\underline{\theta}_f$ and $\underline{\theta}_g$ are calculated, using the projection algorithm (Goodwin, *et al.*, 2001). The basic idea of the projection algorithm is as follows: if the parameter vector is inside the constraint set or on the boundary of the constraint set but moving toward the inside of the constraint set, then use the simple adaptive law based on the Lyapunov synthesis approach; if the parameter vector is on the boundary of the constraint set but moving toward the outside of the constraint set, then project the gradient vector $\dot{\underline{\theta}}_f$ or $\dot{\underline{\theta}}_g$ onto the supporting hyperplane at $\underline{\theta}_f$ or $\underline{\theta}_g$ to the convex set Ω_f or Ω_g.

Properties of this proposed indirect adaptive fuzzy controller are summarized in the following theorem.

Theorem 1: Consider the plant (1) with the control (15), where u_c is given by (9), u_s by (18), and $\hat{f}$ and $\hat{g}$ by (26) and (27), respectively. Let the parameter vectors $\underline{\theta}_f$ and $\underline{\theta}_g$ be adjusted by the projection algorithm and let Assumption 1 be true. Then, the overall control scheme guarantees the following properties:

1) $|\underline{\theta}_f(t)| \le M_f$, $|\underline{\theta}_g(t)| \le M_g$, all elements of $\underline{\theta}_g$ satisfies $\underline{\theta}_g \ge \varepsilon$ (29)

2) $$|\underline{x}(t)| \le |\underline{y}_m| + \left(\frac{2\bar{V}}{\lambda_{min}}\right)^{\frac{1}{2}} \quad (30)$$

3) $$|u(t)| \le 2M_\theta + \frac{1}{b_L}\left[f^U + |y_m^{(n)}| + |\underline{k}|\left(\frac{2\bar{V}}{\lambda_{min}}\right)^{1/2}\right] \quad (31)$$

for all $t \ge 0$, where λ_{min} is minimum eigenvalue of P, and $\underline{y}_m = (y_m, \dot{y}_m, \ldots, y_m^{(n-1)})^T$.

The proof of Theorem 1 is in (Kratmüller, 2002).

Remark 1: For practical control problems, the state $\underline{x}$ and control u are required to be constrained within certain regions. For given constraints, we can specify the design parameters $\underline{k}, M_f, M_g, \varepsilon$ and $\bar{V}$, based on (30) and (31), such that the state $\underline{x}$ and control u are within the constraint sets. To be able to do this, we need to know some fixed bounds of $|\underline{y}_m|, |y_m^{(n)}|, |f^U(\underline{x})|, g^U(\underline{x})$, and $g_L(\underline{x})$. Since these functions are known to the designer, it should not be difficult to determine these bounds. After these bounds have been determined, we can specify the values on the right-hand sides of (30) and (31) by properly choosing the design parameters. Note that due to (13) and (12) λ_{min} is determined by $\underline{k}$. Therefore, we can specify $\underline{k}$ to achieve a required λ_{min}.

5. SIMULATION

In this section, we apply the indirect adaptive fuzzy controller to a system (Han, *et al.*, 2001)

$$\dot{x}(t) = \frac{1 - e^{-x(t)}}{1 + e^{-x(t)}} + u(t) \quad (32)$$

We want to control the plant to the origin, i.e. to achieve $y_m = 0$. The plant (32) is unstable when uncontrolled because if $u(t) = 0$, then $\dot{x} > 0$ for $x > 0$ and $\dot{x} < 0$ for $x < 0$. We have choosen $\gamma = 1$, $M_x = 3$, $b_L = 0.5 < 1 = b$ and $f^L = 1 \ge (1 + e^{-x(t)})/(1 + e^{-x(t)})$ and have defined six fuzzy sets over interval <-3, 3> labeled N3, N2, N1, P1, P2, P3. The membership functions are

$$\mu_{N3}(x) = \frac{1}{1 + e^{5(x+2)}} \qquad \mu_{N2}(x) = \frac{1}{e^{(x+1.5)^2}}$$

$$\mu_{N1}(x) = \frac{1}{e^{(x+0.5)^2}} \qquad \mu_{P1}(x) = \frac{1}{e^{(x-0.5)^2}}$$

$$\mu_{P2}(x) = \frac{1}{e^{(x-1.5)^2}} \qquad \mu_{P3}(x) = \frac{1}{1 + e^{-5(x-2)}}$$

Choosing $h(s) = s$ and Q=12. We have obtained A=-3, B=1 a P=2. Substituting these values to the algorithm (18), we have got the supervisory control $u_s = I\,\mathrm{sgn}(2e)[|u_c| + 2 + |6e|]$. Finally, we have used SIMULINK and MATLAB TOOLBOX to simulate the closed-loop systém for the initial state $x(0) = 1$. Fig. 2 shows the state $x(t)$ which does not hit the boundary $|x| = 3$. Therefore, the supervisory control u_s never fires. Then we have considered the second case of $M_x = 1$ and kept all other parameters as in simulation in Fig. 2. Simulation result for this case are shown in Fig. 3. We can see that the supervisory control forces the state to remain inside the contraint set $|x| \le 1$.

From (18) we can see that u_s is proportional to the upper bound f^U which is usually very large. Large control actions are undesirable because the implementation cost may increase. Fig. 4 shows the variation of the control inputs for the two cases. We can find out that the variation of u(t) for $M_x = 1$ is stronger. Therefore we have choosen the u_s to operate in the supervisory fashion.

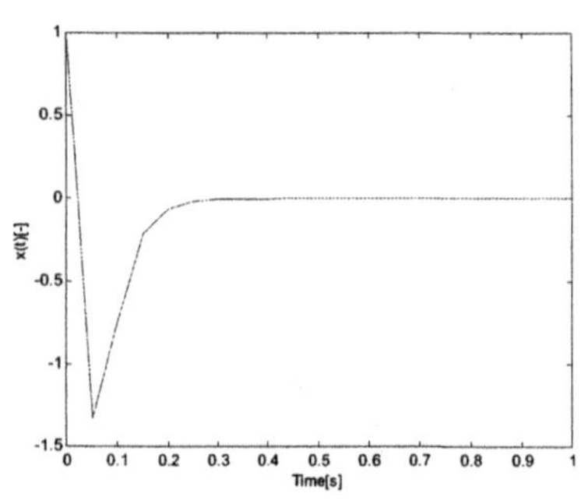

Fig. 2. The system state $x(t)$ using the indirect adaptive fuzzy control with

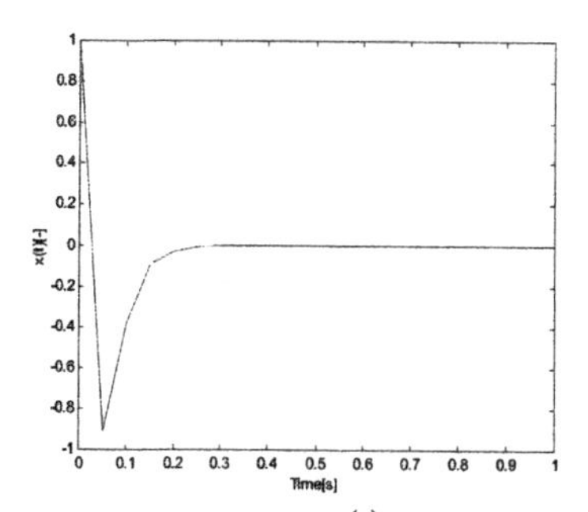

Fig. 3. The system state $x(t)$ using the indirect adaptive fuzzy control with $M_x = 1$

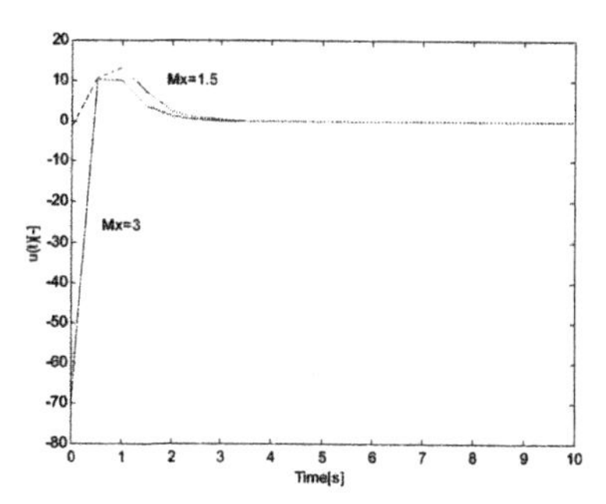

Fig. 4. The control inputs $u = u_c + u_s$ for two cases $M_x = 3$ and $M_x = 1$

6. CONCLUSION

In this paper, the Takagi-Sugeno type fuzzy logic system has been used used in the indirect adaptive fuzzy controller design. The major advantage consist in that the accurate mathematical model of the system is not required to be known. The proposed method can guarantee global stability of the resulting closed-loop system in the sense that all signals involved are uniformly bounded. Besides, the specific formula for the bounds is given too. Finally, the indirect adaptive controller has been used to control an unstable system to the origin. It has also been showed explicitly how the supervisory control forced the state to be within the constraint set.

REFERENCES

Chen, B.S., C. H. Lee, and Y. C. Chang (1996). H^{∞} tracking design of uncertain nonlinear SISO systems: Adaptive fuzzy approach. *IEEE Trans. Fuzzy Syst.*, Vol. 4, pp. 32-43.

Essounbouli, N., A. Hamzaoui and J. Zaytoon (2002). A supervisory robust adaptive fuzzy controller. In IFAC 15th Triennial World Congress, Barcelona, Spain.

Wang, L.X (1993). Stable adaptive fuzzy control of nonlinear systems. *IEEE Trans. Fuzzy Syst.*, Vol. 1, pp. 146-155.

Spooner, J.T. and K. M. Passino (1996). Stable Adaptive Control Using Fuzzy Systems and Neural Networks. *IEEE Trans. Fuzzy Syst*, Vol. 4, pp. 339-359.

Kratmüller, M. (2002). Adaptívne regulatory s fuzzy alebo heuristickým zákonom adaptácie. Diploma, Bratislava (in Slovak).

Sastry, S. and M. Bodson (1989). *Adaptive Control: Stability, Convergence, and Robustnes*. Englewood Cliffs, NJ: Prentice-Hall.

Goodwin, G.C. and D. Q. Mayne (1987). A parameter estimation perspective of continuous time reference adaptive control. *Automatica*, Vol. 23, pp. 57-70.

Han, H., Chun-Yi Su and Y. Stepanenko (2001). Adaptive Control of a class of Nonlinear Systems with Nonlinearly Parametrized Fuzzy Approximators. *IEEE Trans. Fuzzy Syst.*, Vol. 9, pp 315-323.

Murgaš J. (2000). Adaptive control for a class of nonlinear systems. *Journal of EE*, Vol. 51, No. 3-4, pp. 89-93.

INTELLIGENT CONTROL OF CONTINUOUS PROCESS USING NEURO-FUZZY SYSTEM

A. Kachaňák, M. Holiš, M. Holišová, J. Belanský

Slovak University of Technology, Faculty of Mechanical Engineering,
Department of Automation and Measurement,
Nám. Slobody 17, 812 31 Bratislava, Slovakia
Fax : (+421) 524931 and e-mail: Kachanak@kam.vm.stuba.sk

Abstract: The main goal of the paper presented is to outline the possibility of hybrid neuro-fuzzy control systems application for continuous process intelligent control focusing on heating technology. The principle used pointed out the importance of a structured form of the operator's knowledge; further the realization of fuzzy decisions from the supervisory control level as well as the tuning of modified neuro-fuzzy system NEFCON with parametrical learning. The practical use of such approach in the domain of building complexes heating system control is described.

Software realization of the neuro-fuzzy control algorithms for the process optimization on heuristical base used as well as the communication between supervisory SCADA system and the process operator represent an open applied software package making possible its integration to complex monitoring and control system of so called intelligent building. Results attained show possibilities to gain energy savings, thermal comfort improvement and pollution reduction as well.

Keywords: continuous process, intelligent control, qualitative models, neuro-fuzzy system, heating process.

1 INTRODUCTION

At the present time the continuous process control and optimization represents a significant mean to obtain energy savings. As the possibilities of quantitative modeling of complex uncertain processes are limited, their optimization using deterministic or stochastic process models is very difficult and very often impossible. Therefore new methods are to be used – such as the artificial intelligence and qualitative modeling (Kachaňák 2003), further knowledge based or experimental approaches to the process optimization.

The qualitative modeling uses an apriori information concerning the process to be controlled in the form of experience or expert knowledge - e.g. linguistic or fuzzy models, neuro-fuzzy ones. Neuron nets (ANN), evolution algorithms etc. represent another category of qualitative models - as a special class of artificial intelligence systems with unstructured form of experience. Intensive research of qualitative models was strongly supported by the improvement of actual information technologies (machine intelligence). The control of complex continuous systems often realized in the form of distributed multilevel control structures needs new modeling methods. The combination of knowledge based approaches and learning methods using ANN to various forms of neuro-fuzzy systems seems to be of advantage. In ninety's several neuro-fuzzy systems have been presented, such as ANFIS (MANFIS, CANFIS), NEFCOM, FALCON, RFALCON. The practical use of these systems is determined mainly by the characteristics of systems to be optimized and by the ability of the neuro-fuzzy system used to effective learning. In the case, when so called controlled learning (with teacher) cannot be used, it is necessary to use the principle of reinforcement learning (Barto 1998). Such approaches became useful by the realization of hierarchical control levels, characterized by higher complexity or uncertainty of the process being controlled. In this case the realization of a higher control level can bring notable energy savings especially for processes consuming large amount of raw materials and energy.

2 HYBRID NEURO-FUZZY SYSTEM NEFCON

NEFCON (Kruse1993, 1997) is a neuro-fuzzy system based on the architecture named fuzzy perceptron.
This is a special three-layer neuron net. Synaptic connection among individual layers represents here the membership function of individual language values.

Two inputs and one output are assumed, each having three language terms. The fuzzy controller includes five rules in the form:

If x *is* A^1 *and* y *is* A^2 *then* u *is* B,

where x, y are inlets to fuzzy controller representing language variables and u is its outlet.

A^1, A^2, B are particular values of language variables represented by fuzzy sets with membership functions $\mu_1^1, \mu_2^1, \mu_3^1$ for the first inlet, $\mu_1^2, \mu_2^2, \mu_3^2$ for the second inlet and ν_1, ν_2, ν_3 for the outlet.

The perceptron's learning algorithm is based on error back propagation and has two basic parts:

- parametrical learning – changes of form and position of fuzzy sets as well.

- structural learning – rules generation and modification. The main goal of the learning is to make changes of neuro-fuzzy system's parameters improving the control process. Resulting optimal state of the system being controlled can be described by the vector

$$x_{opt} = \left(x_1^{(opt)}, \dots, x_n^{(opt)}\right) \tag{1}$$

where $x_1^{(opt)}, \dots, x_n^{(opt)}$ are state variables representing characteristics of the system.

The layout of fuzzy perceptron in the form of neuron net is shown in Fig. 1.

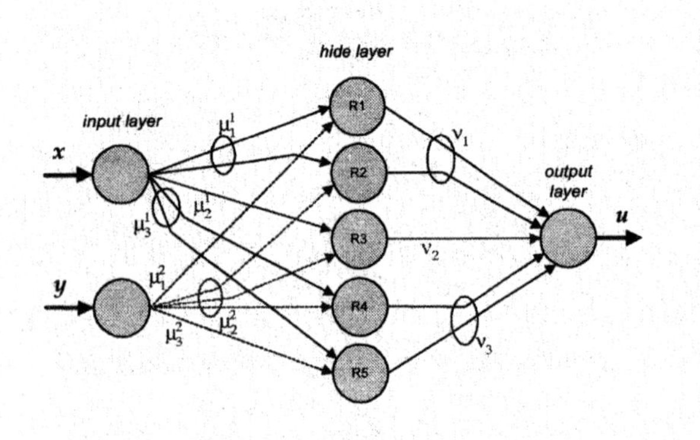

Fig. 1. NEFCON structure

The system controlled is in its optimal state when all its state variables achieve values defined by this vector. Nevertheless the system state can be taken as "optimal" although these values are achieved only approximately - here the possibility of a vague description of system's optimal state using fuzzy sets appears.

Besides the optimal state of the system so called compensation state can be defined, where state variables are not optimal, but are modified in such a way, that the system approaches its optimal state. Similarly, the compensation state can be described by fuzzy sets.

This idea is used for the reinforcement learning for NEFCON, where the reinforcement signal is taken as a fuzzy error describing the state of the system being controlled. Such learning form is then called fuzzy error back propagation.

The learning algorithm of fuzzy sets can be defined as follows:

Assume **S** to be a process with „n" state variables $\xi_i \in X_i$, $i=1,\dots,n$, and one control variable $\eta \in Y$. Each input variable ξ_i is described by triangle fuzzy sets $\mu_j^{(i)}$, $j=1,\dots,p_I$ (with the membership function in the form of a triangle) and the output variable η again by q triangle sets.

For all fuzzy sets there exist limitations $\Psi(\mu_{j_r}^{(i)})$, $\Psi(\upsilon_{j_r})$ constraining the modification of the fuzzy set given.

The algorithm of fuzzy error back propagation for the adaptation of membership functions of the NEFCON system with k rules R_r, $r=1,\dots,k$ repeats following steps:

- designation of NEFCON output value o_η and its application to process **S** to get its new state,

- designation of fuzzy error E following the new process state,

- for each rule R_r the calculation of fuzzy error E_{Rr}

$$E_{Rr} = o_{Rr}.E.\mathrm{sgn}(t_{Rr}) \tag{2}$$

where o_{Rr} is the activated rule expressed as T- norm of rules suppositions ,

- designation of fuzzy sets consequents υ_{jr}:

$$\Delta b_{jr} = \sigma.(c_{jr} - a_{jr}).E_{Rr} \tag{3}$$

$$\Delta a_{jr} = \Delta b_{jr} \tag{4}$$

$$\Delta c_{jr} = \Delta b_{jr} \tag{5}$$

where $\sigma \in \langle 0,1)$ is the learning step. The modification υ_{jr} continues until stopped by the limitation $\Psi(\upsilon_{j_r})$

- designation of fuzzy sets causalities $\mu_{jr}^{(i)}$:

$$\Delta b_{jr}^{(i)} = \sigma.E_{Rr}.(\xi_i - b_{jr}^{(i)}).(c_{jr}^{(i)} - a_{jr}^{(i)}) \tag{6}$$

$$\Delta a_{jr}^{(i)} = -\sigma.E_{Rr}.(b_{jr}^{(i)} - a_{jr}^{(i)}) + \Delta b_{jr}^{(i)} \tag{7}$$

$$\Delta c_{jr}^{(i)} = \sigma.E_{Rr}.(c_{jr}^{(i)} - b_{jr}^{(i)}) + \Delta b_{jr}^{(i)} \tag{8}$$

The modification $\mu_{jr}^{(i)}$ continues until stopped by the limitation $\Psi(\mu_{j_r}^{(i)})$.

This algorithm uses parameters notation of the membership function in the form shown in Fig. 2.

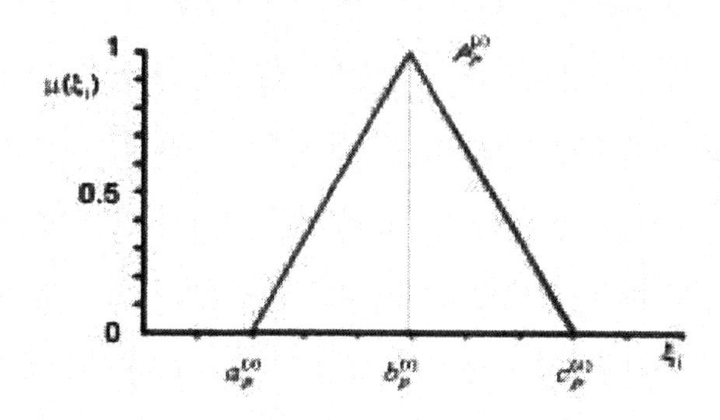

Fig. 2. Parameters of fuzzy sets with triangle membership function

The principle of NEFCON operation is shown in Fig.3.

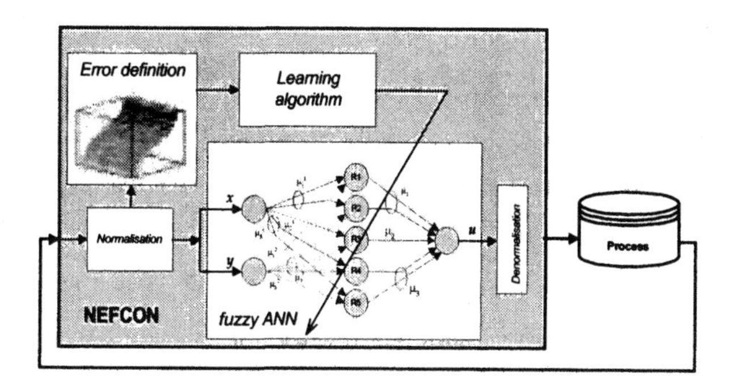

Fig. 3. NEFCON operation

Following the parametrical learning definition the existence of rules base as well as the definition of the number and basic form of fuzzy sets of individual inputs and outputs is necessary.

In the following parametrical learning will be assumed where fuzzy rules by the formulation of the fuzzy system are formed on the base of an apriori information from the expert knowledge base. Theoretical possibility of neuro-fuzzy systems learning algorithm enlargement to fuzzy RULES structures learning exists, but practical experience shows that generally this leads to system complexity enlargement as well as to the system's reliability reduction (Kachaňák, Lehotský 2002).

3 HEATING TECHNOLOGY WITH ZONE EQUITHERMAL CONTROL

Nowadays building heating control systems can be usually found in the form of two level hierarchical control structures. On the first control level standard equithermal zone control is used (Kachaňák, Belanský 1998), as a rule for separate building zones.

This represents in principle dependent control following external temperature and changes of external conditions; undesirable influence of thermal system dead time is eliminated and process control with time-programmable settings of daily or weekly thermal attenuation in separate zones can be done by the operator. Such optimization is based on heuristical principles and it is very subjective depending strongly upon the experience of the operator. The operator mainly tunes the equithermal curve that represents heating characteristics of the building. In principle the optimization of such equithermal control system using neuro-fuzzy system in principle compensates this human factor as an uncertain control factor by a special type of real time expert system with learning ability.

Such system also allows replacing the activity of an inexperienced operator. At the same time it is needful to have the possibility of tuning the fuzzy algorithm depending upon variable operating conditions. By the process optimization designed in such a way it is of advantage to realize the neuro-fuzzy system as a part of adequate SCADA system on a supervisory control level. The necessary condition represents here the existence of such supervisory control level allowing the realization of the process data base with the aim to be used not only for process monitoring but also for its control and optimization. In the case of modern technologies this condition is fulfilled very often. The advantage of keeping the equithermal control on the process control level results in the reliability of control system given and its functionality improvement in suboptimal conditions as well.

In the case of an increasing number of process measuring points (e.g. more than 100) it can be of advantage to use supervisory (monitoring) control level for operator's comfort improvement. Operator's activities will be then limited to the acknowledgment and the elimination of failures and to the storage of synchronous and asynchronous protocols (diaries of events), further parameters settings and settings of daily or weekly attenuations.

The quality of heating process control from the supervisory control level depends directly upon the level of both the experience and knowledge of the operator. Monitored process data (process data base) as well as the knowledge of the operator can be advantageously used for the optimization of heating process following fuzzy principles (Kachaňák, Holiš, Belanský 2000).

4 APPLICATION OF NEURO-FUZZY SYSTEM NEFCON TO THE OPTIMIZATION OF EQUITHERMAL CONTROL

It has been very often confirmed in the practice that quantitative and qualitative knowledge of the process being controlled is more important as technical features of control systems. When the supervisory control level of heat control system is used to process monitoring and to operator's comfort improvement only, then economical gains from such investment will be very problematic. Although for systems having "slow" dynamics (as building heating processes) it is possible to use measured or computed process data as well as heuristic knowledge (experience) of the operator for the off-line supervisory control. Such approach has qualitative character only and it is very subjective. In this context a possibility appears to use for heating processes control at supervisory control level systems with fuzzy logic or real-time expert systems. The process data base besides computing statistics and data integration allows also to compute actual and global trends, to accept measured or introduced by the operator data concerning external weather conditions (wind intensity, sun radiation etc.), as well as internal requirements of heat consumption and to use all information to optimize equithermal control on the process control level. In the following application of fuzzy approach to building heating control optimization will be presented. Good function of equithermal control at the process control level will be assumed as well as the presence of no technical problems relating to the reliability of digital information, control and communication systems. Neuro-fuzzy system was modified for two input and two output variables.

Basic configuration of fuzzy control system is shown in Fig. 4.

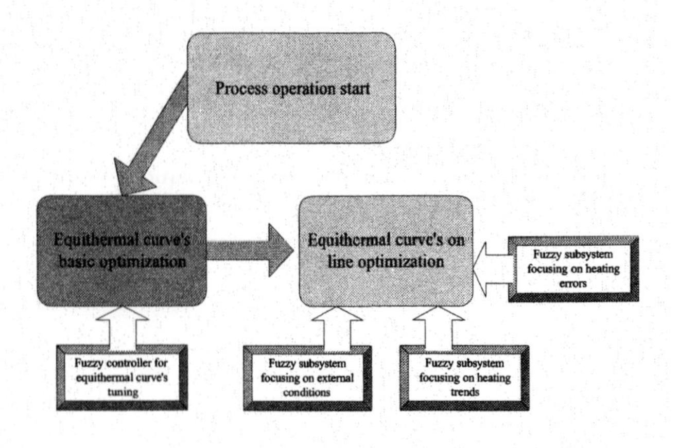

Fig. 4. Control modules layout

By such concept of the system's optimization the use of the fuzzy control implemented to SCADA control system on the supervisory control level seems to be of advantage

The optimization of equithermal control using fuzzy decisions in principle involves the replacement of a human factor as an uncertain control element by a special type of real time expert system. This allows to eliminate control uncertainty and to automate activities of the process operator.

5 RESULTS ATTAINED

The design of heating equithermal control optimization has been verified first by the simulation using applied software package MATLAB SIMULINK and a simplified heating process dynamic model (Kachaňák, Holiš 1999).

Simulations took the equithermal curve approximately (slope and shift). Temperature in the room by the control using this curve is shown in Fig. 5.

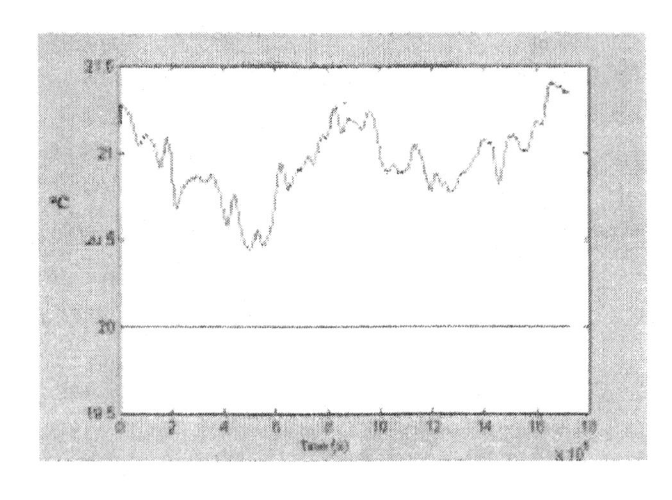

Fig. 5. Room temperature without equithermal curve optimization

Evidently this room is overheated by any external temperature (improper shift), further the room temperature fluctuates following external temperature (improper slope). Room set temperature was 20°C. The course of external temperature is shown in Fig. 6.

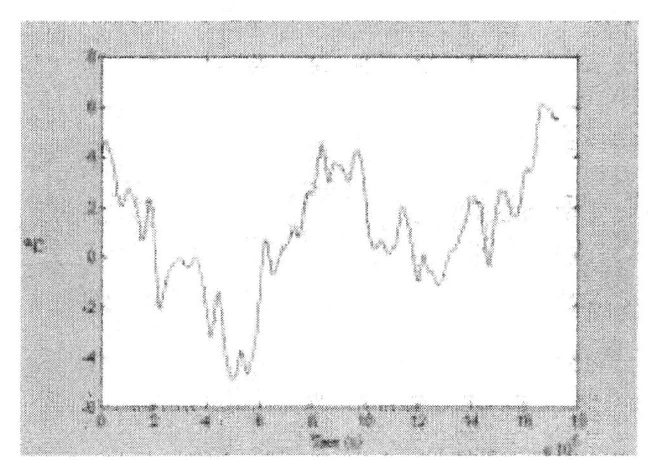

Fig. 6. External temperature

Fuzzy controller proposed is Mamdani type having two inputs:

T internal - thermal comfort

T external – external temperature

and two outputs:

Shift – equithermal curve shift change

Slope - equithermal curve slope change.

The optimization proposed involved a control feedback in the form of internal temperature measurement. SCADA system used (D 2000) allows to get required dada from the process data base with the period of approx. 1hour depending on the dynamics of the process.

Simulation was realized by the same conditions (external temperature, weather conditions) as described above.

The room temperature by basic optimization of equithermal curve using NEFCON is shown in Fig. 7.

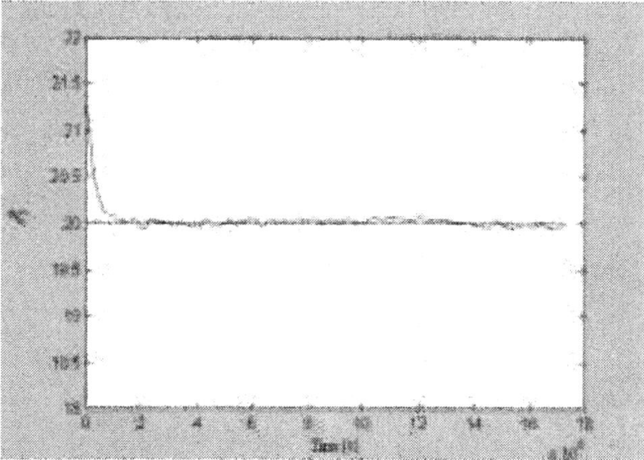

Fig. 7. Room temperature by basic optimization

Heating bodies' temperature is shown in Fig .8.

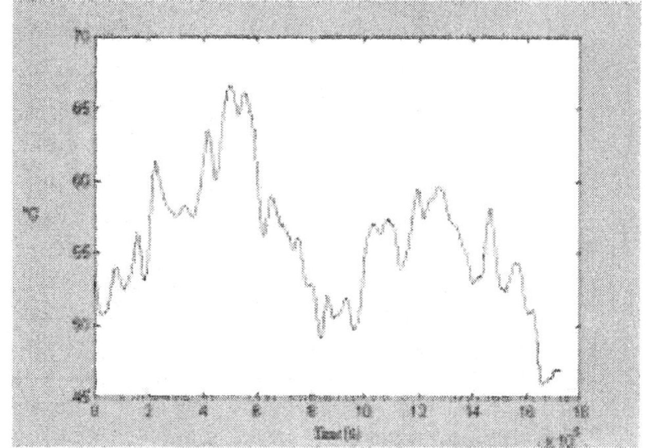

Fig. 8. Heating bodies temperature by basic optimization

Required changes of the equithermal curve's slope and shift are shown in Fig. 9 and Fig. 10 respectively.

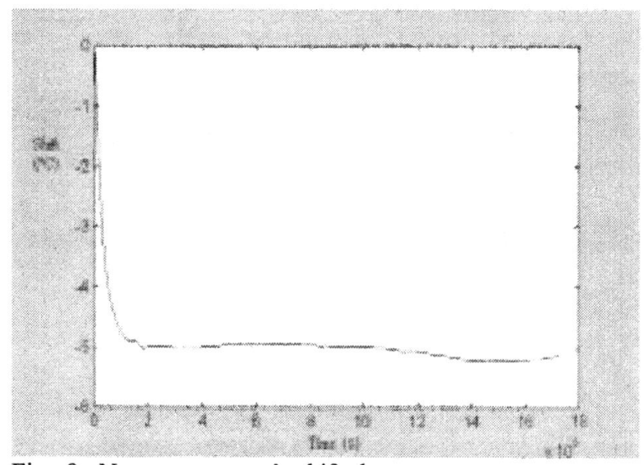

Fig. 9. Necessary curve's shift changes

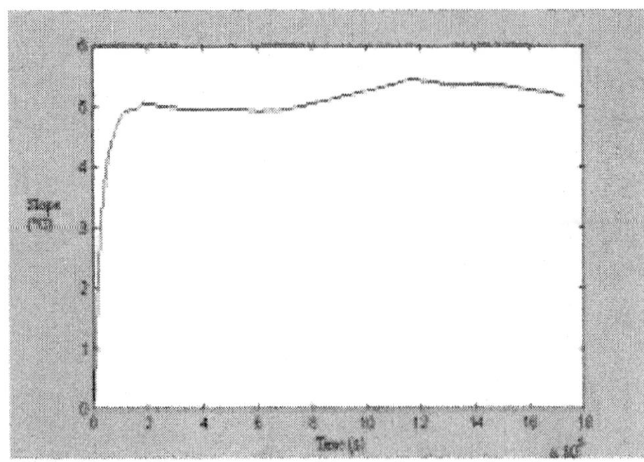

Fig. 10. Necessary equithermal curve's slope changes

This proves the positive influence of the neuro-fuzzy system on the control quality leading to energy savings and pollution reduction as well.

6 CONCLUSION

In the paper brief description of the design of STU buildings heating system's fuzzy control from supervisory control level is presented. Method presented uses historical data from process data base, expert knowledge of process operator and principles of Mamdani's type fuzzy decision as well. Results attained show significant possibilities of heat consumption reduction in building heating systems. The implementation of fuzzy control system at supervisory control level need no changes in equithermal control at process control level.

Fuzzy control algorithm introduced here can be implemented into arbitrary heating control system fulfilling real-time activities requirements.
This control system realization except heat consumption reduction by automatic thermal attenuation also disables any unprofessional intervention of operator in heating system's operation.

ACKONWLEDGMENTS

Results have been attained in the frame of research project VEGA1/9278/02 – Control of Complex Systems at the Department of Automation and Measurement, Faculty of Mechanical Engineering STU at Bratislava Slovakia.

7 REFERENCES

Kruse, Nauck (1993): A Fuzzy Neural Network Learning Fuzzy Control Rules and Membership Functions by Fuzzy Error Back Propagation. Proc. IEEE Int. Conf. On NeuraL Network, San Francisco, 1022-1027

Kruse, Nauck Klawonn (1997): Foundations of Neuro-fuzzy Systems. J. Willey

Sutton, Barto (1998): Reinforcement Learning. A. Bransford book, MIT Press

Kachaňák, Belanský (1998): Fuzzy Control of Building Heating Systems. In Sel. Top. Modeling and Control Vol.1, STU Press Bratislava, 113-118.

Kachaňák, Holiš (1999): Contribution to Fuzzy Control Design of Building Heating Process. Proc. 12th Conf. Process Control 99, Tatranské Matliare, Vol. 2, 152-155

Kachaňák, Holiš, Belanský (2000): Control System Design for Building Heating Process Using Neuro-fuzzy Approach. Proc. IFAC Symp. CSD 2000, Bratislava 560-565

Kachaňák, Lehotský (2002): Analysis of Neuro-fuzzy Systems with Structural and Parametric Learning Characteristic, Proc. Process Control Conf. Kouty nad Desnou, CD ROM

Kachaňák (2003): Quantitative versus Qualitative Modeling. AT&P Journ. No 3, 83-85

Kachaňák, Holiš (2003): Using Neuro-fuzzy System to Heating Process Control AT&P Journ. No 3, 36-38

Holiš (2002): Using Neuro-fuzzy System to Continuous Process Control in Uncertainty Conditions. PhD Thesis, SjF STU Bratislava

www.elsevier.com/locate/ifac

PERFORMANCE IMPROVEMENT PROPERITIES OF NEURAL NETWORKS USING ORTHOGONAL AND WAVELET FUNCTIONS

Ladislav Körösi, Štefan Kozák

Department of Automatic Control Systems,
Faculty of Electrical Engineering and Information Technology,
Slovak University of Technology, Ilkovicova 3, 812 19 Bratislava
e-mail:korosi@kasr.elf.stuba.sk, kozak@kasr.elf.stuba.sk

Abstract: This paper dealt with modelling and control of non-linear dynamic processes by artifiacal neural network with orthogonal and wavelet activation functions. Wavelet decomposition is employed to explore the characteristics of demand time series to extract more patterns from the series, and therefore enhance the forecast capability of neural network based models. Several types of orthogonal functions (Fourier, Legendre, Laguerre and Chebychev) for non-linear process modelling and control are analysed and compared. Simulation results are provided to illustrate high performance of artificial neural networks with the orthogonal activation functions in practical industrial problems. The obtained results are compared and potential shortcomings of the proposed methods for real time industry control are discussed and analysed. *Copyright © 2003 IFAC*

Keywords: neural network, orthogonal functions, self-tuning control, non-linear process modelling, wavelet, forecast

I. INTRODUCTION

In recent years a lot of research work has been carried out on applications of artificial neural networks in non-linear system modelling, identification and control. Artificial neural network model has played a more and more important role in modelling and control of complex industrial processes. As it is well known, the existing control theory and the related control engineering techniques are mainly based on the mathematical models and formulations, which describe the process behaviour and control law. However, even when the adaptive, robust and self-tuning control theory and algorithms are well developed, the complexity of the system inaccuracies due to output measurement, disturbances and uncertainties about the system dynamics make the control problems extremely hard to be solved by conventional control techniques. Thus it is necessary to further explore and investigate some novel control strategies for modelling and control of complex systems.

During the last five years orthonormal functions have been successfully applied in the design of linear and non-linear self-tuning controllers. The major advantage of this approach consists in that the plant can be modelled without any knowledge about its structure, i.e. without any assumptions about the true plant order and time delay. Compared with the ARMA type models, usage of orthonormal series representation based models has several advantages: good approximation of delayed systems, better tolerance to un-modelled dynamics, convenient filter network realization, etc. Orthonormal functions are applicable for continuous-time as well as for discrete-time systems. Consequently, a vast body of work addresses the problem of real system modelling and non-linear control design based on orthogonal, (harmonic, Legendre, Chebychev, Laguerre, wavelet) functions. Thus, the neural network based process control using orthonormal functions is interesting from the practical point of view and has led to many new studies dealing with control problems, e.g. robust stability, performance, controllability, observability, model matching, etc The behavior of a neural network is determined by three basic elements: neuron activation functions, a connecting pattern of neurons and a learning rule. Such learning properties and characteristics permit neural network to identify an unknown process with little a priori knowledge about the dynamics and parameters of the process. The basic principle for the activation function selection consists in achieving a specified target behavior of the dynamic process. The neural network has to mathematically map a wide range of linear and non-linear input-output relations. The commonly used activation functions for neural network design are of various types: sigmoidal functions, radial basis function, Boolean functions, linear functions and combination of the mentioned types.

The proposed paper focuses on modelling and control of continuous-time processes using artificial neural networks with orthonormal activation functions. Based on simulations carried out on many practical models, the implementation possibilities of this approach are outlined. The paper is organized as follows. Section 2 introduces the architecture of neural networks with orthonormal activation functions, section 3 deals with a method of modelling and control of non-linear system using OAFN, Section 4 presents an example, simulation, verification and comparison of orthonormal and conventional neural network for modelling and self-tuning control

II. NEURAL NETWORK STRUCTURES USING WITH ORTHONORMAL ACTIVATION FUNCTIONS

There are many results reported on properties of universal functional approximation for certain classes of feed-forward neural networks. The rigorous proof of universal approximation capabilities of a multi-layer feed-forward neural network with one hidden layer has received much interest during the last decade. All the obtained results assume that the hidden layer activation functions are sigmoidal functions which are either squashing or some non-constant functions, bounded and monotone increasing continuous functions, and the target function belongs to the R^n which is the set of all real-valued continuous functions defined on the topological space R^n.. Each piecewise continuous function satisfying the Dirichlet´s conditions can be successfully modelled by orthonormal functions.. A typical neural structure for approximation of a non-linear function is the three layer structure in Fig.1. The input layer has m nodes with inputs u=[u(1) u(2),...u(m)]. The hidden layer consists of neurons with orthogonal (orthonormal) activation functions. It is assumed that activation functions for these neurons belong to the same class of orthogonal functions and no two neurons have the same order of activation function. The input and output layers consist of linear neurons. The weights between the input and the hidden layers are fixed and depend on the type of orthogonal activation function. The weights for harmonic orthogonal functions are equal to the frequency and for other types of orthonormal functions are unity. The output of the network with orthonormal activation functions is given by the linear combination of activation functions

$$y(u,w)=\sum_{n_1=0}^{N_1-1}\cdots\sum_{n_m=0}^{N_m-1} w_{n_1}..{}_{n_m}\Phi_{n_1}..{}_{n_m}(u)=\Phi^T(u)W \qquad (1)$$

where $u=[u_1\ u_2...u_m]^T$ is an m-dimensional input vector, N_i is the number of neurons associated with the i-th input and W is the vector of weights between hidden and output layers.

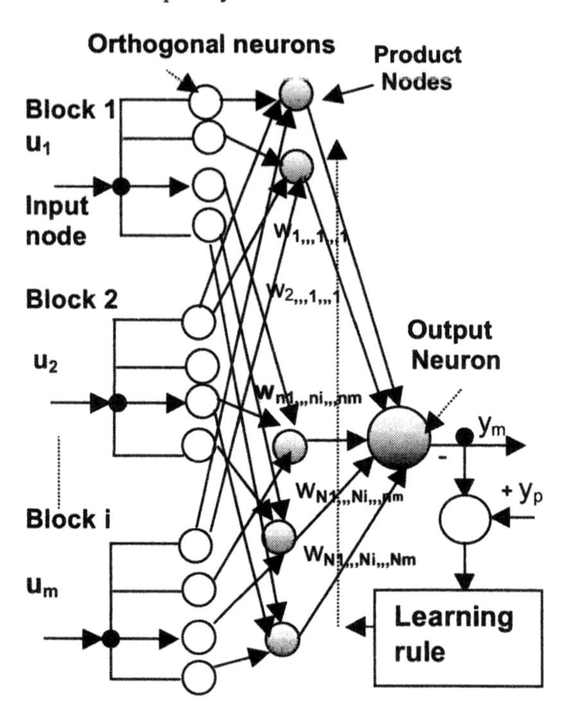

Fig.1 Block scheme of a three-layer neural network with orthogonal activation functions

The orthogonal functions $\phi_{n_1}\cdots_{n_m}(u)$ can be expressed as follows

$$\phi_{n_1}\cdots_{n_m}(u)=\prod_{i=1}^{m}\phi_{n_i}(u_i) \qquad (2)$$

where ϕ_{n_i} are one-dimensional orthogonal functions implemented in each hidden layer, Φ and W are the transformed input vector and the weight vector, respectively, in the n - dimensional network weight space, where n is given as

$$n=\prod_{i=1}^{m}N_i \qquad (3)$$

In the modelling and control system design the following types of orthogonal functions can be used: *Laning – Batin, Hermit, Laguerre, Fourier series, Legendre.*

III. NEURAL NETWORK STRUCTURE USING WAVELET ACTIVATION FUNCTIONS

In this paper the wavelet network will be used as a nonparametric regression estimator. In nonparametric estimation there is a concern for curse of dimensionality. This means that the complexity of the estimator grows quickly with input dimension. The notion of sparse data results from the fact that the functional mapping between the input and output is localized to a small portion of the input space. Exploiting this fact makes is possible to address problems of high input dimension. The wavelet network provides an adaptive discretization of the wavelet transform by choosing influential wavelets based on a given data set. Identifying the important wavelets can also be viewed as eliminating the ones that have no influence on function synthesis. To make this procedure generalizable to large dimensional non-linear systems, a wavelet frame will be used since it is impractical to implement wavelet bases of large dimension (greater than 3).

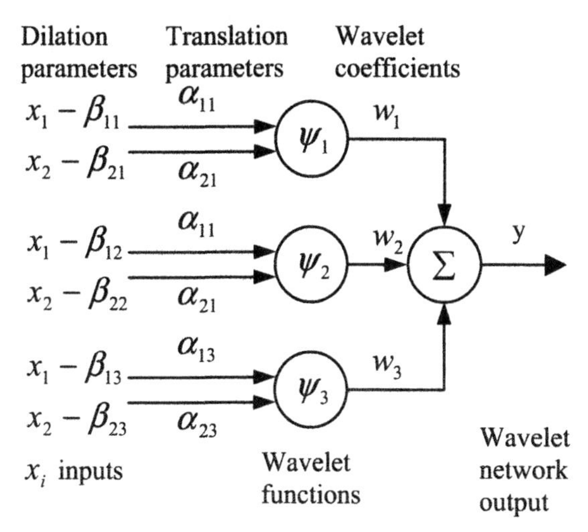

Fig. 2 Two-Input Single-Output Wavelet Network Consisting of Three Wavelets

The similarity between the inverse wavelet transform and neural network motivates the use of a single layer feed-forward network (Fig.2). The network structure is used to accommodate a large number of inputs.

Define input vector

$\Phi = [x_1 \quad x_2]$

From the above model, the wavelet network has the following form:

$$\hat{f}(\Phi) = \sum_{i=1}^{N} w_i \psi(\alpha_i * (\Phi - \beta_i)) \qquad (4)$$

The parameters $w_i \in R$, $\alpha_i \in R^d$, and $\beta_i \in R^d$, are the wavelet coefficient, dilation parameter and translation parameter respectively with d denoting input dimension. The parameter N is the number of wavelets, and '*' denotes component-wise multiplication of vectors. The identification process will be applied to a class of environments that typically exhibit linear regions along with the nonlinearities. To accurately capture this behavior the final model takes the following form:

$$\hat{f}(\Phi) = \sum_{i=1}^{N} w_i \psi(\alpha_i * (\Phi - \beta_i)) + c^T \Phi + b \qquad (5)$$

This thesis will use the Mexican hat wavelet. Consider the one-dimensional radial Mexican hat wavelet:

$$(2.8)\ \psi(x) = (1 - x^2) e^{\frac{-x^2}{2}} \qquad (6)$$

Based on the development in [Kugarajah and Zhang 1995] the *d*-dimensional (*d* being input dimension) wavelet frame becomes:

$$(2.9)\ \psi(x) = (1 - \|x\|^2) e^{\frac{-\|x\|^2}{2}} \qquad (7)$$

where $\|.\|$ denote Euclidean norm.

IV. ALGORITHM FOR MODELLING AND CONTROL OF NON-LINEAR PROCESSES USING ANN WITH OAFN

Today, the control of non-linear processes using the artificial neural networks is extraordinarily interesting from a practical point of view, leading to many new studies and results on control problems such as robust stability and performance.

As neural network modelling is adaptive to parametric and structural variations of the system, artificial neural network with orthonormal activation functions provides an efficient means for robustness maximization when modelling non-linear systems.

In neural networks learning is realized by adapting the network weights in such a manner that the expected value of the mean squared error between the network output an the training output is minimized. The four gradient-type algorithms (the steepest descent , the conjugate , the variable gradient methods and the momentum method) and the RLS (recursive least-square methods) are the most popular training methods. The performance index is

$$J=E\left[(y_p-y_m)^2\right]=E\left[(y_p-\Phi^T w)^2\right]=E(y_p^2)+w^T Rw-2P^T w$$

where y_p is the training process output, w is the vector of estimated weights, $R=E(\Phi\Phi^T)$ is the auto-correlation matrix of orthonormal activation functions and $P=E(y\Phi)$ is the cross -correlation vector. Neural model output y_m is determined as

$$y_m(u,w)=\sum_{n_1=0}^{N-1}\cdots\sum_{n_m=0}^{N_m-1} w_{n_1\cdots n_m}\phi_{n_1\cdots n_m}(u)=\Phi^T(u)W\ \mathrm{T}$$

he auto-correlation matrix R is a diagonal positive definite n-th order matrix with equal entries on the diagonal and with equal eigenvalues. These special properties contribute to a fast parameter convergence during the training mode. The on-line approach is based on updating the weights at each training step toward a negative gradient (updating of the weight vector w and bias vector). A general gradient rule (steepest descent method) for updating the network weights in each discrete step is given by

$$\Delta W(k)=-\alpha\frac{\partial J}{\partial W} \qquad (8)$$

where α is the learning rate and the objective function can be expressed as

$$J(W,k)\equiv\frac{1}{2}e^2=\frac{1}{2}(y_p-y_m)^2$$

and y_p is a process output and y_m is the neural model output. Stability and convergence conditions for orthonormal activation functions can be derived from the inequality $0<\alpha<\dfrac{2}{\phi^T\phi}$ (9)

The on-line algorithm for updating the weights is given by the relation

$$W(k)=W(k-1)+\alpha\left[y_p(k)-y_m(k)\phi(k)\right] \qquad (10)$$

and the fast convergence of the Levenberg-Marquard algorithm is

$$\Delta W(k)=\left[\dot{J}(W)\dot{J}^T(W)+\beta I\right]^{-1}\dot{J}^T(W)e(W) \qquad (11)$$

where $e(.)=(y_p - y_m)$ is the error function, $\dot{J}(W)$ is a Jacobi matrix and β is the learning coefficient.

Recursive least square methods are most popular in identification of parameters of dynamical systems. These methods can be used also for updating the weights of a neural network model. A recursive form of the least square method for updating the weights can be determined on the base of the criterion

$$J(W)=\frac{1}{2}\sum_{i=1}^{k}\varphi^{k-1}\varepsilon^2(i,W) \qquad (12)$$

where φ is the forgetting coefficient $0<\varphi\le 1$, ε is the model error and ϕ is a regression vector composed from orthonormal functions in each node. The recursive form for updating the weights is given

$$W(k)=W(k-1)+V(k)\varepsilon(k)$$

where

$$\varepsilon(k)=y_p(k)-\phi^T(k)W(k-1) \qquad (13)$$

$$V(k)=\frac{P(k-1)\phi(k)}{\varphi+\phi^T(k)P(k-1)\phi(k)}$$

$$P(k)=(P(k-1)-V(k)\phi^T(k)P(k-1))/\phi \qquad (14)$$

The neural adaptive feedback control law is defined $u(k)=g(e(k),\ldots,e(k-1))$ where the control error is $e(k)=r(k)-y_m(k)$, $r(k)$ is the reference signal and y_m is the neural network output. A three-layer computation architecture of the neural controller with an orthonormal activation function can be described by following steps

- The input layer : $z(k)=e(k-i)$, i=0, 1,...,m
- The hidden layer : $s_i(k)=\phi_i(z_i)$, i=0 ,1,...,h

where $\quad \phi(z)=\prod_{i=1}^{m}\phi_i(z) \qquad (15)$

- The output layer

$$u(e,w)=\Phi^T(e)W=$$
$$=\sum_{n_1=0}^{N_1-1}\cdots\sum_{n_m=0}^{N_m-1} w_{n_1}\ldots w_{n_m}s_{n_1}\ldots s_{n_m}(e) \qquad (16)$$

Based on the neural model output y_m the control law determination can be obtained as a result of a direct optimization of the following objective function

$$J(w,e)=\frac{1}{2}\left\{[r(k+1)-y_m(k+1)]^2+\gamma[\Delta u(k)]^2\right\}$$

where $\gamma>0$ is the control effort weighting factor and $\Delta u(k)=u(k)-u(k-1)$ is the control difference and $r(k+1)$ is the reference variable in the $(k+1)$-th step.

$$\frac{\partial J}{\partial u(k)}=\gamma\Delta u(k)\,|+[r(k+1)-y_m(k+1)]\cdot\left[-\frac{\partial y_m(k+1)}{\partial u(k)}\right]$$

Using the steepest descent gradient algorithm, the control variable can be determined as

$$u(k)=u(k-1)-\alpha\frac{\partial J(k)}{\partial u(k)} \tag{17}$$

and $y_M(z,w)=\sum_{n_1=0}^{N_1-1}\cdots\sum_{n_m=0}^{N_m-1} w_{n_1\ldots n_m}\phi_{n_r,\ldots n_m}(z)=\Phi^T(z)W$

where z_I is an m-dimensional vector of model inputs and $\phi_{n_1\ldots n_m}(z)$ are orthonormal functions

$$\phi_{n_1\ldots n_m}(z)=\prod_{i-1}^{m}\phi_{n_i}(z_i)$$

and $z=\begin{bmatrix} y(k-n_y+1),y(k-n_y-2),\ldots,y(k); \\ u(k-n_u+1),u(k-n_u+2,\ldots,u(k) \end{bmatrix}$

Finally,

$$\frac{\partial y_m(k+1|}{\partial u(k)}=\Phi^T(z)W=\sum_{n_1=0}^{N-1}\ldots\sum_{n_m=0}^{N_m-1} w_{n_1}..n_m s_{n_1}..n_m(z)$$

$$s_{n_1\ldots n_m}(z)=\prod_{i-1}^{m}s_{n_i}(z_i)$$

$$s_{n_i}(z_i)=\langle \begin{matrix} \phi_{n_i}(z_i), & ak \quad i<m \\ \phi'_{n_i}(z_i), & ak \quad i=m \end{matrix} \tag{18}$$

where $\phi'_{n_i}(z)$ is a derivative of $\phi_{n_i}(z)$.

Applying the Quasi-Newton method, the control variable can be computed

$$u(k)=u(k-1)-\left(\frac{\partial^2 J}{\partial u^2(k-1}\right)^{-1}\cdot\frac{\partial J}{\partial u(k-1)} \tag{19}$$

The robust version of the proposed control algorithm can be expressed in the form

$$u(k)=u(k-1)-\left(max\left(abs\left(\frac{\partial^2 J}{\partial u^2(k-1)}\right),k\right)\right)^{-1}\frac{\partial J}{\partial u(k-1)}$$

where

$$\frac{\partial^2 J}{\partial u^2(k)}=\gamma+[r(k+1)-y_M(k+1)]\left[-\frac{\partial^2 y_M(k+1)}{\partial u^2(k)}\right]+\left[\frac{\partial y_M(k+1)}{\partial u(k)}\right]^2$$

$$\frac{\partial^2 y_M(k+1)}{\partial u^2(k)}=\sum_{n_1=0}^{(N_1-1)}\cdots\sum_{n_m=0}^{(N_m-1)} w_{n_1\ldots n_m}\delta_{n_r,\ldots n_m}(z)=\Delta^T(z)W$$

$$\delta_{n_1\ldots n_m}(z)=\prod_{i-1}^{m}\delta_{n_i}(z_i)$$

and

$$\delta_{n_i}(z_i)=\langle \begin{matrix} \phi_{n_i}(z_i), & for \quad i<m \\ \phi''_{n_i}(z_i), & for \quad i=m \end{matrix} \tag{20}$$

where $\phi''_{n_i}(z_i)$ is the second derivative of $\phi_{n_i}(z_i)$.

V. CASE STUDY

To evaluate the performance of the orthonormal activation functions a simulation study has been carried out by applying modelling and control to several non-linear systems taken from hydraulic and chemical process. In this paper we demonstrate the proposed approach for the hydraulic system.

Final equations are

$$\dot{S}_s=-\frac{D}{I}S_s+\frac{C_r\eta_m}{I}P_m-\frac{C_r}{I\eta_m}P_p$$

$$\dot{P}_m=-\frac{2BC_r}{V_t}S_s-\frac{2BK_t}{V_t}P_m+\frac{2K_sK_\theta B}{V_t}u(P_s-P_m)^{1/2}$$

After parameter substitution

$$\dot{S}_s=-8{,}2S_S+780P_m--1077{,}6P_P$$

$$\dot{P}_m=-0.6S_s-8.5P_m+64X_s\sqrt{(P_s-P_m)}$$

The time responses of the measured output data (S_S) and the corresponding responses neural models with Chebychev orthonormal activation functions (third order) are depicted in Fig.3 and Fig.4. Time response of the NN control with OAFN is depicted in Fig.5. For simulation six inputs have been used each of them having three orthogonal neurons.

The proposed approach requires a smaller number of iterations giving a faster and a more accurate modelling and control. The Chebychev OAFN has converged to a stabilized error (SSE=43450 for training data and SSE=6.4167 for testing data).

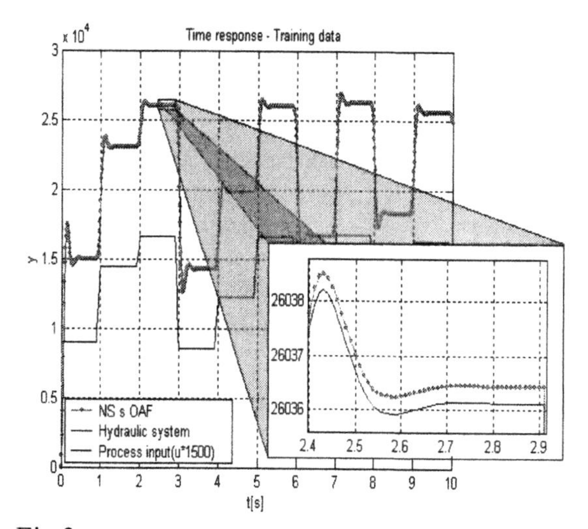

Fig.3 Time responses of the process output S_S (Hydraulic system) and output from the neural model with Chebychev OAFN (NS s OAF) – training data set

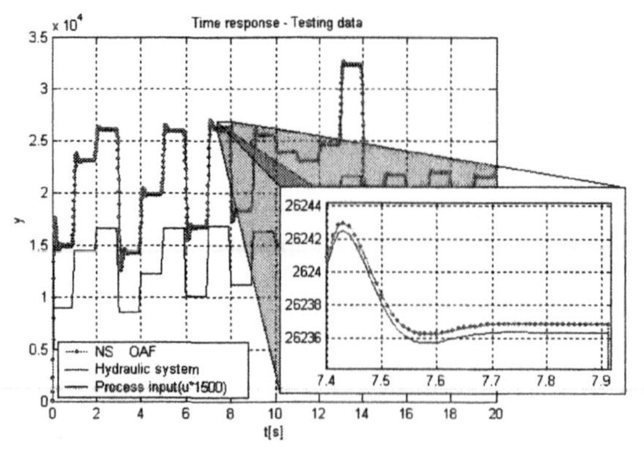

Fig.4 Time responses of the process output S_S (Hydraulic system) and output from the neural model with Chebychev OAFN (NS s OAF) – testing data set

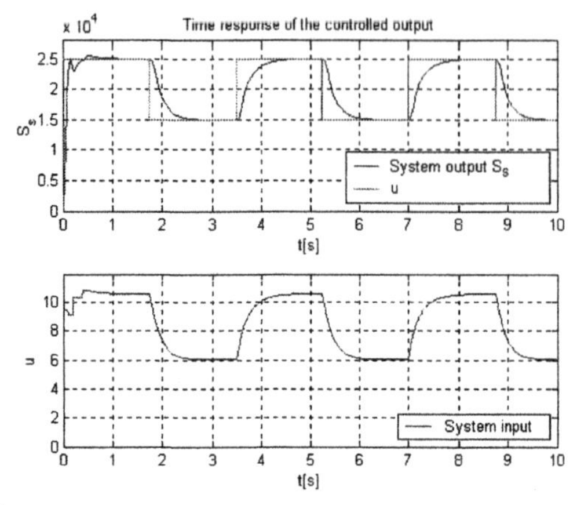

Fig.5 Time responses of the controlled output (S_S) and reference variable under self-tuning controller with Chebychev OAFN

VI. CONCLUSION

This paper addresses the design of a class of artificial neural networks which properties are especially attractive for the self-tuning control. These networks employ the chosen type of orthogonal/ orthonormal functions as activation functions. The proposed approach has several advantages, e.g. a faster and more accurate convergence, smaller number of iterations, etc. The high speed of orthonormal activation function application is highly appropriate for practical real-time usage in self-tuning control structures. The experiments and simulation results prove good properties of OAFN for modelling and self-tuning control in practical problems from industry (chemical industry, gas industry, power plant, etc.).

VII. ACKNOWLEDGMENT

This paper was partially supported by the Slovak Scientific Grant Agency VEGA, Grant No. 7630/20

VIII. REFERENCES

Dumont, G.A., A. Elnaggar and A. Elshafei (1993). Adaptive Predictive Control of Systems with Time-Varying Time Delay. *Int. Journal of Adaptive Control and SignalProcessing*, Vol. 7, No 7, pp. 91-101.

Kozák, Š. (1994) Nonlinear selftuning controller based upon Laguerre series representation. *Proceedings of 1st IFAC Workshop on New Trends in Design of Control Systems,* Smolenice, Slovak Republic, pp. 432-438.

Kozák, Š. (1999). Self-tuning Neural Controller, *Journal of Electrical Engineering,* Vol. 50, No. 11-12, pp. 346-355

LEONDES T. C.: Optimization Techniques – Neural Network Systems Techniques and Applications, ACADEMIC PRESS, 1998, San Diego, California

WAHLBERG B.: Orthonormal Basic Functions Models: A Transformation Analysis, IFAC 1999, 14th Triennial World Congress, Beijing, P. R. China, pp. 355-360

I. Daubechies: "The Wavelet Transform, Time-Frequency Localization and Signal Analysis," *IEEE Transactions on Information Theory*, vol. 36-5, pp. 961-1005, 1990.

C. Chui, *Wavelets: a tutorial in theory and applicaitons*. New York: Academic, 1992.

NONLINEAR MODEL-BASED PREDICTIVE CONTROL

Jana Paulusová, Štefan Kozák, Jakub Grošek

Department of Automatic Control Systems,
Faculty of Electrical Engineering and Information Technology,
Slovak University of Technology,
Ilkovičova 3, 812 19 Bratislava, Slovak Republic
e-mail: paulusj@kasr.elf.stuba.sk, kozak@kasr.elf.stuba.sk

Abstract: In this paper a predictive control designed for the non-linear plant is discussed. The fuzzy-neuro internal model control (FNIMC) and predictive control based on the dynamic matrix control (DMC) algorithm are compared and designed. The proposed DMC and FNIMC used to the control the non-linear process. *Copyright © 2003 IFAC*

Keywords: predictive control, fuzzy-neuro internal model control (FNIMC), nonlinear system, identification

1. INTRODUCTION

The predictive control has become a very important area of research in recent years. The predictive fuzzy-neuro technique is a highly efficient approach in different control applications where conventional strategies call for long lookup tables or complex equation solving.

Originally, model based predictive control (MBPC) algorithms were developed for linear processes, but the basic idea can be transferred to nonlinear systems. The model predictive control (MPC) shows improved performance because the process model allows current computations to consider future dynamic events.

The main idea of MPC is the prediction of the output signal at each sampling instant. The prediction is implicit or explicit depending on the model of the process to be controlled. In the next step a control is selected to bring the predicted process output signal back to the reference signal in the way to minimize the area between the reference and the output signal.

Typical MPC capabilities

- is particularly easy to use for multivariable systems and easily handles process interactions.
- handles time delays, inverse response, as well as other difficult process dynamics.
- utilizes a process model, but no rigid model form is required.
- compensates for the effect of measurable and un-measurable disturbances.
- is posed as optimization problem and is therefore capable of meeting the control objectives by optimizing control effort, and at the same time is capable of handling constraints of all kinds.

Within basic methods, which are essentially referred as predictive control, Dynamic matrix control (DMC) and linear IMC with the added capability of handling non-linear systems are included.

The paper is organized as follows. First, MBPC formulation is briefly introduced in Section 2. Then, the case study is discussed in Section 3. The reliability and effectiveness of the presented methods is shown by application to control the non-linear process in Section 4. A summary and conclusions are given in Section 5.

2. MBPC FORMULATION

MBPC is a general methodology for solving control problems in the time domain. It is based on three main concepts:

1. Explicit use of a model to predict the process output.
2. Computation of a sequence of future control actions by minimizing a given objective function.
3. The use of the receding horizon strategy: only the first control action in the sequence is applied, and the horizons are moved one sample period towards the future, optimisation is repeated.

2.1 IMC Design

The internal model control is very often used control strategy for linear systems and can be used for non-linear systems as well.

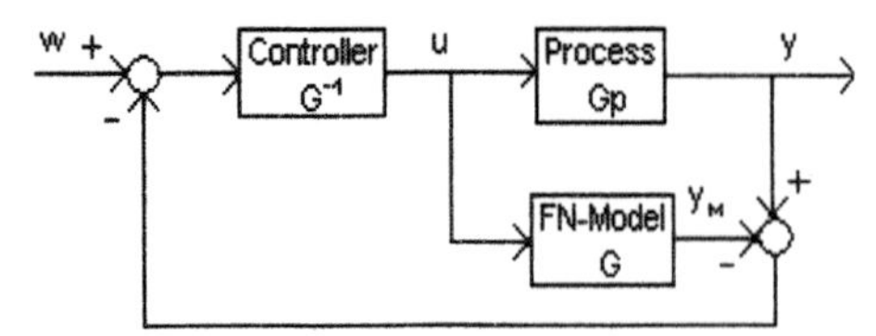

Fig. 1. Internal model control (IMC) scheme

In process control, internal model control (IMC) has gained high popularity due to the good disturbance rejection capabilities and robustness properties of the IMC structure.

The internal model control is very often used control strategy for linear systems and can be used for non-linear systems as well.

One of its conventional forms is shown in Fig.1.

The IMC and the classical feedback control can generate the same loop characteristics. The process model plays an explicit role in the control structure compared to the standard control loop. The IMC structure has some advantages over conventional feedback control loops. If the process is stable, which is true for most industrial processes, the closed loop will be stable for any stable controller. The controller can simply be designed as a feed-forward controller in the IMC scheme.

Since the controller can be designed as an open-loop controller, the ideal choice for the controller is the inverse of the process model.

The IMC design procedure is very simple and reliable.

2.1.1 Fuzzy-neuro model

The model is Takagi-Sugeno fuzzy model (Takagi and Sugeno, 1985), the membership functions are built on the Gaussian distribution curve.

The design of fuzzy model has been proposed from input-output measurement data. Still, compared to other nonlinear techniques, fuzzy models provide a more transparent representation of the identified model.

The parameters of the fuzzy model are adapted by neural network. The fuzzy-neuro model, which is obtained, has very high accuracy.

2.1.2 Predictive inverse model-based fuzzy-neuro control

Main goal of the proposed approach is to implement the predictive model-based control theory, advanced of fuzzy-neuro modeling technique to obtain a model with high accuracy and apply the possibilities of the inverse model-based fuzzy control.

First process outputs are changed with inputs and the same fuzzy-neuro algorithm is used for the obtaining of the inverse model. In the internal model-based scheme the quality of the designed fuzzy-neuro logic controller (FNLC) depends on the accuracy of the inverted fuzzy-neuro model presented by checking error.

2.2 DMC Design

One of the first proposed MBPC methods, and still commercially the most successful one, is DMC. This method was introduced by Cutler (1980).

Model predictive control (MPC) is an optimisation based control methodology that explicitly utilises a dynamic mathematical model of a process to obtain a control signal by minimising an objective function. The model must describe the system well. The future process outputs $y(k+i)$ for $i=1,...,p,$ are predicted over the prediction horizon (H_P) using a model of the process. These values depend on the current process state, and on the future control signals $u(k+i)$ for $i=0,...,m-1,$ over the control horizon (H_C), where $H_C \leq H_P$. The control variable is manipulated only within the control horizon and remains constant afterwards, $u(k+i)=u(k+m-1)$ for $i=H_C,...,H_P-1$.

In Fig. 3 is the basic principle of MBPC.

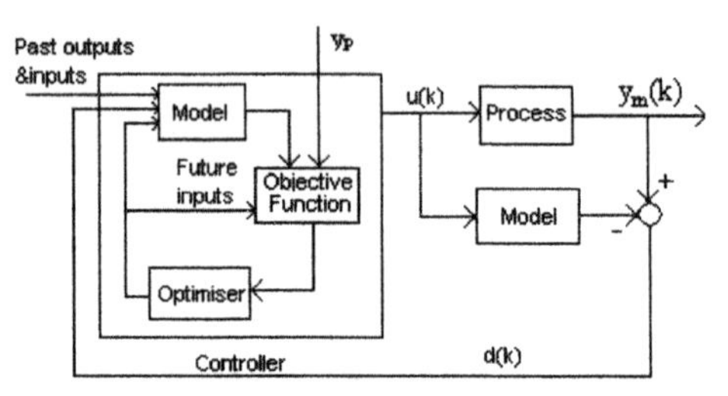

Fig. 2. Block scheme of model predictive control (MPC)

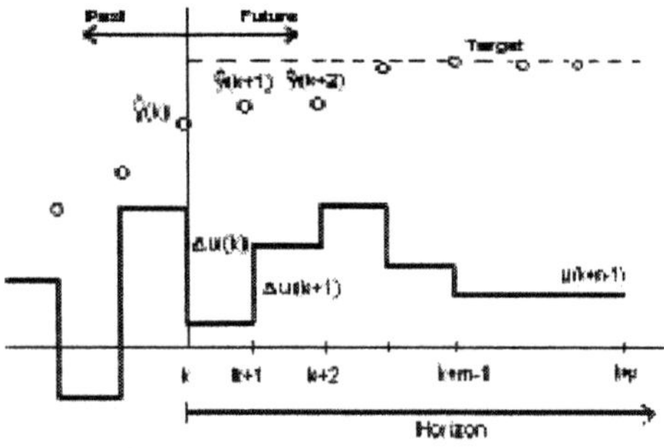

Fig. 3. The principle of DMC

Process interactions and dead-times can be intrinsically handled with model predictive control schemes such as dynamic matrix control (DMC).

The block-scheme of MPC is shown in Fig. 2.

The sequence of future control signals is computed by optimizing a given (cost) function. Often, the system needs to follow a certain reference trajectory defined through set points. In most cases, the difference between system outputs and reference trajectory is used by combination with a cost function on the control effort. A general objective function is the following quadratic form

$$J = \sum_{i=1}^{p} [y_p(k+1 \mid k) - \hat{y}(k+i \mid k)]^2 \gamma_y + \sum_{i=1}^{m} \Delta u(k+i-1 \mid k)^2 \gamma_u \quad (1)$$

where

- y_P is desired set point,
- γ_u and γ_y are weight parameters,
- $\Delta u(k-i)=u(k-i)-u(k-i-1)$ is the change in manipulation variable,
- p is the length of the prediction horizon, m is the length of the control horizon,
- $\hat{y}(k)$ is the process output, at sample instant t is given as

$$\hat{y}(t) = \sum_{i=1}^{\infty} g_i \Delta u(t-i) \quad (2)$$

where g_i are the step response coefficients.
The model employed is a step response of the plant.
The model predictions along the prediction horizon H_P are

$$\hat{y}(t+k \mid t) = \sum_{i=1}^{\infty} g_i \Delta u(t+k-i) + d(t+k \mid t) \quad (3)$$

Disturbances are considered to be constant between sample instants

$$d(t+k \mid t) = y_m(t \mid t) - \sum_{i=1}^{\infty} g_i \Delta u(t+k-i) \quad (4)$$

where

- $y_m(t|t)$ represents the measured value of the process output at time t.

So

$$\hat{y}(t+k \mid t) = \sum_{i=1}^{k} g_i \Delta u(t+k-i) + f(t+k \mid t) \quad (5)$$

where

$$f(t+k \mid t) = y_m(t \mid t) + \sum_{i=1}^{N} (g_{k+1} - g_i) \Delta u(t-i) \quad (6)$$

The prediction of the process output along the length of the prediction horizon, can be written compactly using matrix notation

$$y_P(t) = G \Delta u(t) + f(t) \quad (7)$$

where

$$G = \begin{bmatrix} g_1 & 0 & \cdots & 0 \\ g_2 & g_1 & \cdots & 0 \\ \vdots & \vdots & \ddots & \vdots \\ g_{Hc} & g_{Hc-1} & \cdots & g_1 \\ \vdots & \vdots & \ddots & \vdots \\ g_{Hp} & g_{Hp-1} & \cdots & g_{Hp-Hc+1} \end{bmatrix} \quad (8)$$

Matrix G is called the system's dynamic matrix.
By minimizing objective function the optimal solution is then given in matrix form

$$\Delta u(t) = (G^T \Gamma G + \Lambda)^{-1} G^T \Gamma (y_m(t) - f(t)) \quad (9)$$

where Γ and Λ are matrix of the weight parameters.

2.2.1 Linearization of nonlinear process

The mathematical model of our nonlinear plant is based on observed input-output data. We have done the linearization of nonlinear process with ARX model.
The parameters of the ARX model structure are estimated using the least-squares method. The most used model structure is the simple linear difference equation

$$y(t) + a_1 y(t-1) + \ldots + a_{nA} y(t-n_A) = b_1 u(t-n_K) + \ldots + b_{nB} u(t-n_K-n_B-1) \quad (10)$$

which relates the current output $y(t)$ to a finite number of past outputs $y(t-k)$ and inputs $u(t-k)$.
The structure is thus entirely defined by the three integers n_A, n_B, and n_K. n_A is equal to the number of poles and n_B-1 is the number of zeros, while n_K is the pure time-delay (the dead-time) in the system. For a system under sampled-data control, typically n_K is equal to one if there is no dead-time.
The discrete transfer function model parameters are determined from experimental data via process identification. The discrete function is transformed to continuous transfer function. This average process model is then used directly in the DMC algorithm.

3. CASE STUDY

The mathematical model of nonlinear process is described by nonlinear differential equation

$$\ddot{y}(t) + 0.05\dot{y}(t)[y^2(t) + 1] + 0.12y(t) - 0.2u(t) = 0 \quad (11)$$

The simulation scheme of this process is depicted in Fig. 4.

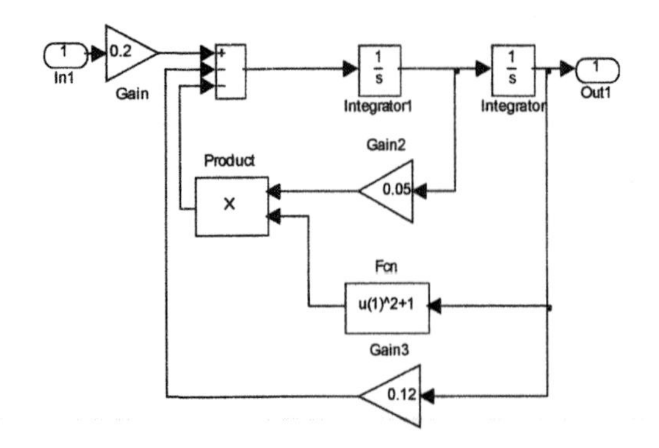

Fig. 4. Simulation scheme of the nonlinear process

4. EXPERIMENTAL RESULTS

For the nonlinear process were proposed several types of controllers, such as FNIMC and DMC.

4.1 IMC Design

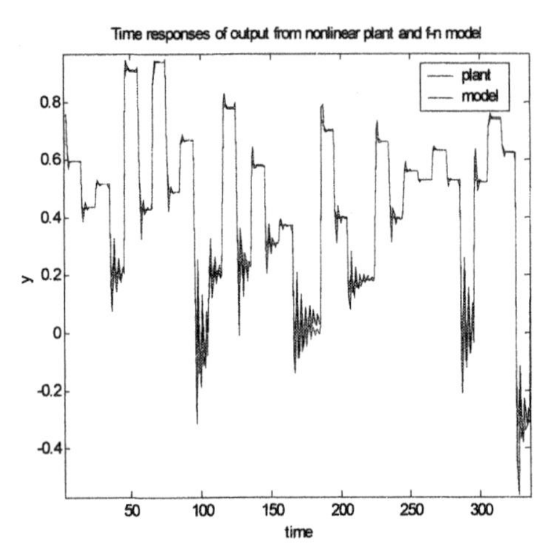

Fig. 5. Time responses of output from the nonlinear plant and the fuzzy-neuro model

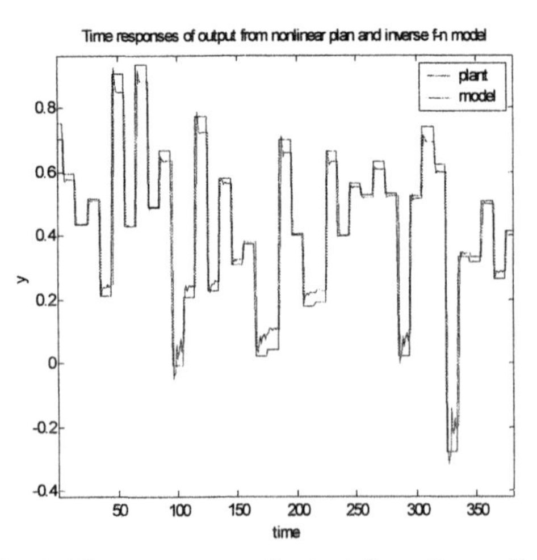

Fig. 6. Time responses of output from the nonlinear model and inverted fuzzy-neuro model

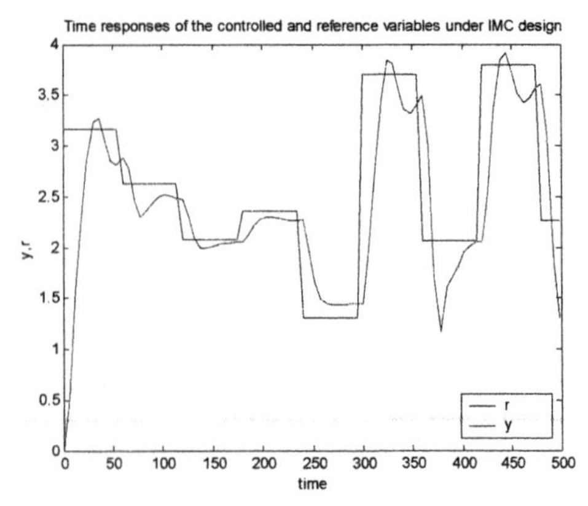

Fig. 7. Time responses of the controlled and reference variables under IMC design

In Fig. 5 and Fig. 6 the fuzzy-neuro model and inverted fuzzy-neuro model accuracy is shown.

Fig. 7 shows the results of IMC applied to the above described process.

4.2 DMC Design

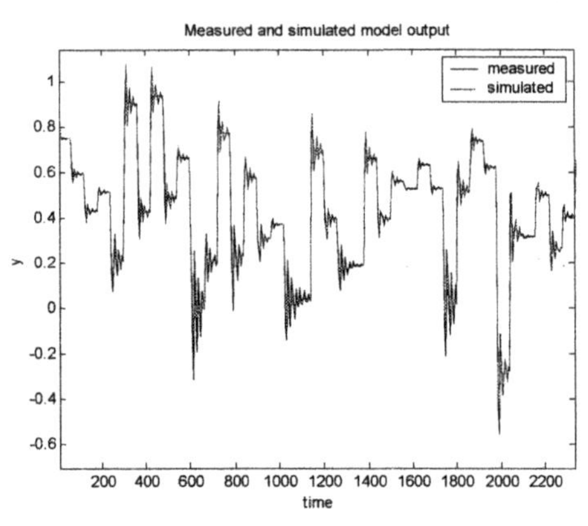

Fig. 8. Time responses of output from the measured and simulated model (ARX model)

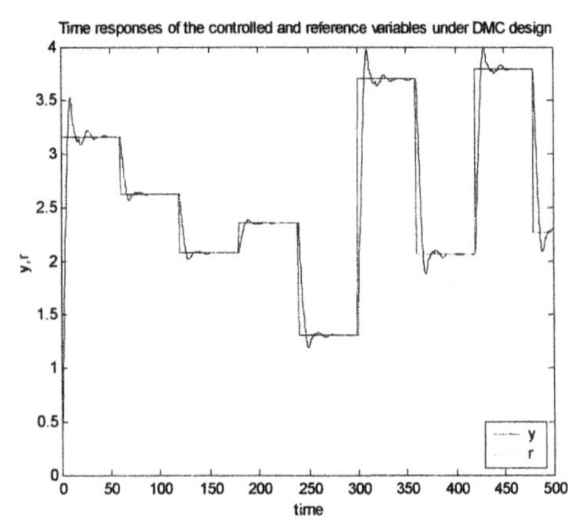

Fig. 9. Time responses of the controlled and reference variables under DMC design (H_p=4, H_C=2)

The linearization of nonlinear process has been done using ARX model.
The parameters of the ARX model structure in polynomial form are

$$A(z) = 1 - 0.3451\, z^{-1} + 0.3785\, z^{-2} - - 0.1637\, z^{-3} + 0.02986\, z^{-4}$$

$$B(z) = 1.756 z^{-1} + 0.9291 z^{-2} + + 0.1266 z^{-3} - 0.1616 z^{-4} \quad (12)$$

The parameters of the discrete function are transformed to continuous transfer function. This average process model is then used directly in the DMC algorithm.
The result of predictive control that is using DMC algorithm (prediction horizon H_P is 4 and control horizon H_C is 2) is shown in Fig. 8.

5. CONCLUSION

In this paper, several types of the predictive controllers were designed and verified for highly nonlinear model of the process. It has been shown that IMC is very effective method, which is based on fuzzy-neuro model and the existing inverse fuzzy-neuro model.
DMC has done with linearization of non-linear process with ARX model.

ACKNOWLEDGEMENT

This paper has been supported by the Slovak Scientific Grant Agency, Grant No.1/0155/03.

6. REFERENCES

Clarke, D. W., Mohdani, C.-Tuffs, P.S. (1989): Generalized predictive control. Part I: The Basic algorithm and part II: Extensions and interpretations. *Automatica, vol.23, No2, 137-160*

Gormandy, B. A., Posttlethwaite, B. E. (2002): Incorporating the Crisp Consequent FRM into Dynamic Matrix Control. *Proceedings of the 10th Mediterranean Conference on Control and Automation-MED 2002, Lisbon, Portugal, July 9-12, 2002*

Fink, A., Nelles, O., Isermann, R. (2002): Nonlinear Internal Model Control for MISO systems based on local linear Neuro-Fuzzy Models. *IFAC, 15th Triennial World Congress, Barcelona, Spain*

Paulusová, J., Kozák, Š. (1999): The comparison of the conventional controllers with fuzzy controllers. *Tatranské Matliare, Slovac Republic, May 31-June 3, 168-171*

Roubos, J. A., Mollov, S., Babuška, R., Verbruggen, H., B. (1999): Fuzzy model-based predictive control using Takagi-Sugeno models. *International Journal of Approximate Reasoning 22, 3-30*

MODELING OF SYSTEMS WITH DELAYS USING HYBRID PETRI NETS

Martina Svádová, Zdeněk Hanzálek

Center for Applied Cybernetics, Czech Technical University,
Karlovo nam. 13, 121 35, Prague 2
fax : +420 2 24357610, e-mail : xsvadova@lab.felk.cvut.cz

Abstract: In practice the systems are not only pure discrete event or pure continuous, but they are combined from discrete and continuous parts together. These systems are called hybrid systems. Hybrid Petri nets (HPNs) allow modeling and analysis of hybrid systems. Since classical HPNs cannot properly model the systems with the delays on continuous product flow, extended Hybrid Petri nets and generalized batches Petri nets (GBPNs) were defined. This paper deals with extended HPNs and GBPNs, which are used in practical examples elaborated in this paper. *Copyright © 2003 IFAC*

Key words: Delay, Hybrid Petri-nets, Extended Hybrid Petri-nets, Batches Petri-nets, Modelling

1. INTRODUCTION

Petri nets (PNs) (Murata, 1989) are used to model and analyze discrete event dynamic systems like manufacturing systems and communication protocols. In continuous Petri nets (CPNs), the markings of places are real numbers and the firing of transitions is a continuous process (David and Alla, 1993). Therefore CPNs allow modeling and analysis of continuous systems. Moreover CPNs can be used to approximate behavior of discrete systems represented by discrete PN with large number of tokens. In such case, analysis of CPNs does not require exhaustive enumeration of the discrete state space.

In practice the systems are not only pure discrete event or pure continuous, but they are combined from discrete and continuous parts together (discrete state variables and continuous state variables). Hybrid Petri nets (HPNs) are one of the tools, which allow modeling of such systems (David and Alla, 1992; David and Alla, 2001). A timed model of HPNs may be used for performance evaluation of the systems. Timed HPNs have inherited various modeling abilities from the previous models related to time. In the classical model adapted by David and Alla (1998), the discrete part of HPN enables to model delay with threshold and the continuous part of HPN models the continuous flows without threshold and without delay.

Beside of that there exist the systems with delays on the continuous product flow (e.g. the product delay on the conveyor, or delay of fluid in pipe). Since there is no threshold of the product volume, the delay on continuous product flow cannot be properly modeled by HPNs. Several authors have proposed extensions of the original HPNs model in order to represent delays on the continuous flows. David and Caramihai (2000) have shown use of extended HPNs, Demongodin (2001) defined generalized batches Petri nets and Brinkman and Blaauboer (2000) assigned delays to continuous places. Extended HPNs and generalized batches Petri nets are used in practical examples elaborated in this text.

This paper is organized as follows: Section 2 briefly presents HPNs, followed by introduction to extended HPNs and generalized batches Petri nets (GBPNs). Section 3 shows various possibilities of modeling systems with the delay on continuous flow. Section 4 presents example with delay on continuous flow: delay in physical layer of communication protocol.

2. DESCRIPTION OF HYBRID PETRI NETS

In this section hybrid Petri nets are defined first. Next basic concepts of extended hybrid Petri nets and generalized batches Petri nets are presented. The instantaneous firing speeds in the continuous part of HPNs are supposed to be constrained by constant values. This model corresponds to classical constant speed continuous Petri Net (CCPN).

A marked timed HPN net is defined by the seven-tuple:

$$<\mathbf{P}, \mathbf{T}, Pre, Post, M_0, V, d> \quad (1)$$

$\mathbf{P} = P \cup CP$	finite and non-empty set of places (P and CP are disjoint)
$\mathbf{T} = T \cup CT$	finite and non-empty set of transitions (T and CT are disjoint)
Pre: $\mathbf{PxT} \rightarrow R^+$ or N	input incidence matrix
Post: $\mathbf{PxT} \rightarrow R^+$ or N	output incidence matrix
M_0: $\mathbf{P} \rightarrow R^+$ or N	vector of initial marking
$V=\{V_1,..,V_m\}$	vector of maximal speeds
$d=\{d_1,..,d_n\}$	vector of timings

In a definition of Pre, Post and M_0, R^+ corresponds to case, when $\mathbf{P} \in CP$ and N represents case, when $\mathbf{P} \in P$.

HPN contains the set of discrete places and transitions (P_i, T_j - where P_i is the discrete place and T_j is the discrete transition) and the set of continuous places and transitions (CP_i, CT_j - where CP_i is continuous place and CT_j is continuous transition). The discrete part and the continuous part influence each other depending on firing rules of discrete and continuous transitions (David and Alla, 1998). The continuous place CP_i can be input or output place of a discrete transition T_j without restriction. T_j is not enabled if the marking of its input place CP_i is lower than $Pre(CP_i, T_j)$. This value represents the threshold of continuous marking. Timed HPN has the delay d_j associated to discrete transitions T_j and the maximal firing speed V_j associated to continuous transition CT_j (calculation of instantaneous firing speed is presented in (David and Alla, 1993). The delay d_j expresses how long the needed quantity (in the case of continuous input place) or the number of tokens (in the case of discrete input places) is reserved. Being enabled during d_j time units the discrete transition T_j is fired.

HPNs can be studied using standard PN analysis techniques such as invariants. Incidence matrix W of HPN is combined from discrete part W^D, continuous part W^C and the part representing influence of continuous places on discrete transitions W^{CD}. The incidence matrix W does not reflect influence of any discrete place P_i on any continuous transition CT_j since $Pre(P_i, CT_j) = Post(P_i, CT_j)$.

$$W = Post - Pre = \begin{pmatrix} W^D & 0 \\ W^{CD} & W^C \end{pmatrix} \quad (2)$$

In HPNs, characteristic vector s of a sequence is a vector for which each element is either non-negative real number corresponding to the firing quantity of a continuous transition CT_j or a positive integer number corresponding to the number of firings of a discrete transition T_j. A marking M can be calculated from initial marking M_0 and its appropriate characteristic vector s using the fundamental equation (3):

$$M = M_0 + W \cdot s \quad (3)$$

2.1 Evolution graph of Hybrid Petri nets

An evolution graph of HPN is an oriented graph consisting of nodes corresponding to IB-states (invariant behavior states) and transitions between these nodes. The IB-state characterizes the state of HPN during certain time. The marking of discrete places and the instantaneous firing speed of continuous transitions remain constant as long as the system is in the same IB-state. The transition of the evolution graph corresponds to event provoking passage from one IB-state to another IB-state. It is labeled by the name of the event, the time elapsed in previous IB-state and the absolute time.

An IB-state consists of two parts in HPNs:

- the discrete part on the left side of the IB-state is given by M_D (the marking of discrete places).
- the continuous part on the right side of the IB-state is given by v_C, the instantaneous speed of continuous transitions and by M_C, the marking of continuous places at the beginning of the IB-state.

Timed HPNs have inherited various modeling abilities from discrete PNs and CPNs, which are related to time. Therefore the discrete part of HPN enables to model delay with threshold and the continuous part of HPN models the continuous flows without threshold and without delay. Beside of that there exist the systems with delays on the continuous product flow (e.g. the product delay on the conveyor, or delay of fluid in pipe). Since there is no threshold of the product volume, the continuous product flow delay cannot be properly modeled by HPNs.

2.2 Extended hybrid Petri nets

The extended HPNs (David and Caramihai, 2000) have the following features in addition to ordinary HPNs:

- an inhibitor arc is allowed
- 0^+ weight of an arc joining a continuous place and discrete transition is allowed
- 0^+ marking of continuous place is allowed

Symbol 0^+ represents infinitely small positive real number. Please notice that due to the second feature we are able to model delays on continuous flows by extended HPNs.

2.3 Generalized Batches Petri nets

Generalized batches Petri nets (GBPNs) (Demongodin, 2001) are another extension of HPNs.

New elements are batch places and batch transitions. Batch place allows to model delays on continuous flows. It is described by the batch function, which is given by speed of transfer, maximal density and length of place $\{V_i, d_{max}, s_i\}$. For each batch place, the set of input transitions consists of only one batch transition and the set of output transitions consists of only one batch transition as well. Therefore there is no structural conflict related to any batch place. The batch transition behaves like the continuous transition.

Important notion in GBPN is internal coherent batch ($ICBp_i$) assigned to the beginning and to the end of the IB-state. The marking of batch place BP_i is given by an ordered sequence of $ICBp_i$. Index *p* of $ICBp_i$ expresses its placing in the ordered sequence of the batch place BP_i. At fixed time t, ICBp is characterized by three continuous variables: length $l_p(t)$, density $d_p(t)$ and position $x_p(t)$ in batch place. If position of the batch is equal to the characterized length of batch place s_i, the batch becomes the output internal coherent batch (OICBp).

3. EXAMPLES

Let us now present a simple example modeled in three distinct ways: by discrete PN, by extended HPN and by GBPN. Fig. 1 represents conveyor with an input buffer B_1 and an output buffer B_2. In initial state, the input buffer B_1 contains 10 kilograms of a product and the output buffer B_2 is empty. The product is transferred via conveyor from B1 to B_2. The conveyor length L is 9 meters and its speed V is 0.5 m/sec. Maximal density D_{max} of the product on the conveyor is 1 kg/m.

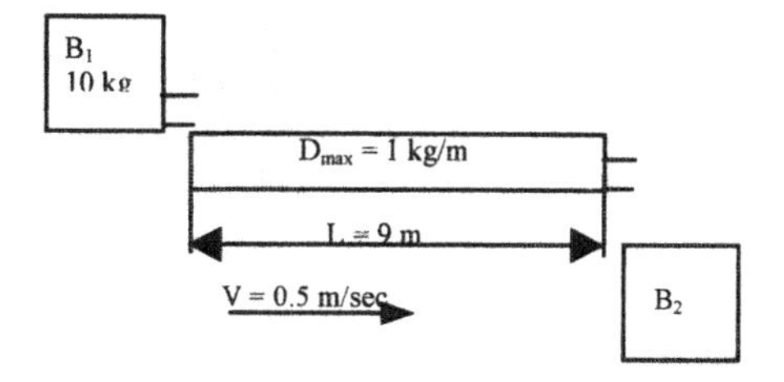

Fig. 1. – Model of conveyor

3.1 Discrete PN model

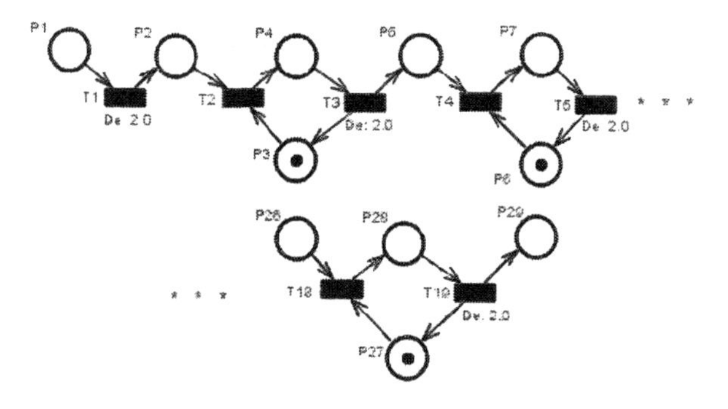

Fig. 2. – Conveyor model by discrete PN

Conveyor model by the discrete PN is shown in Fig. 2. The mode is based on the single server semantics. Input buffer is represented by place P1. Each kilogram of the product is supposed to be a discrete quantity therefore P1 contains 10 tokens. Timing of transition T1, having the threshold Pre(P1,T1)=1, represents a delay needed to put 1 kilogram of the product on the conveyor. T1 fires for the first time when the first kilogram leaves the input buffer B1. Timed transitions T3, T5, ... T19 correspond to the delay on each meter of the conveyor. Places P3, P6, P27 represent that the given part (1 meter) of the conveyor is empty. Places P4, P7,...P28 represent that the given part of the conveyor is full. Between each couple of timed transitions there is one immediate transition (one of T2,T4,... T18), otherwise the timed behavior of the model does not correspond to reality, since the "empty part" of the conveyor (places P3, P6,...P27) is logical and not timed constraint when moving the product from one part to the subsequent one.

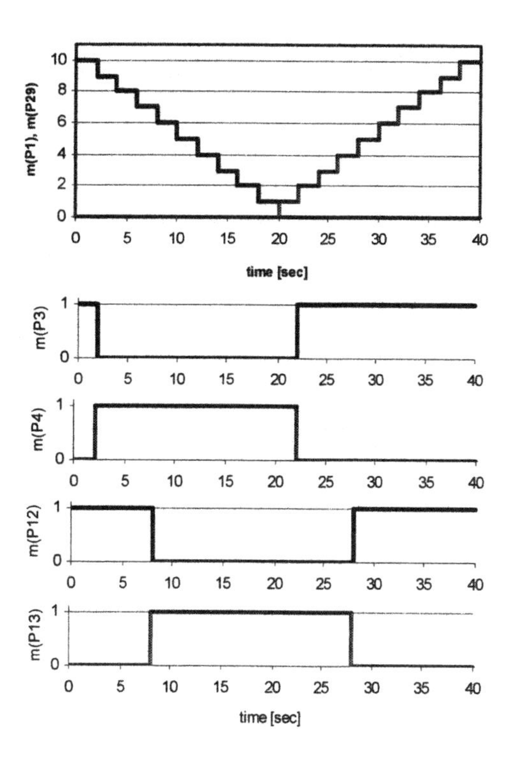

Fig. 3. – The marking evolution of model in Fig. 2

Since the state space of the modeled system is too large, the marking evolution of several places is illustrated in Fig. 3. The first figure represents discharging of the input buffer B_1 (place P1) and filling of the output buffer B_2 (place P29). The input buffer contains 10 parts at time t=0 and one part from input buffer leaves each 2 seconds. The input buffer is empty at time t = 20 sec. At this time, the first part is added to output buffer and each two seconds the next part is added to the output buffer until it contains all 10 parts. Next figures illustrate behavior of the first part (or fourth part respectively) of the conveyor. Marking of P3 (P12) signifies that the first part (or fourth part respectively) is occupied. Since P3 and P4 form P-invariant, the marking of P3 is complement to the marking of P4 - if place P3 contains token, place P4 is empty. The first element becomes occupied from time t = 2 sec to t =22 sec, while the forth element becomes occupied from t = 8 sec to t = 28 sec.

In this example, we have shown also internal behavior of the conveyor, since it distinguishes two delays:

- the discretization of the continuous product quantity by the threshold of T1
- the transport delay on the conveyor (timing of T1, T3,..., T19).

This model is just an approximation of the continuous flow due to the discretization. The model is precise in discrete values of time ($0^+, 2^+, 4^+, \ldots 38^+$).

3.2 *Extended HPN model*

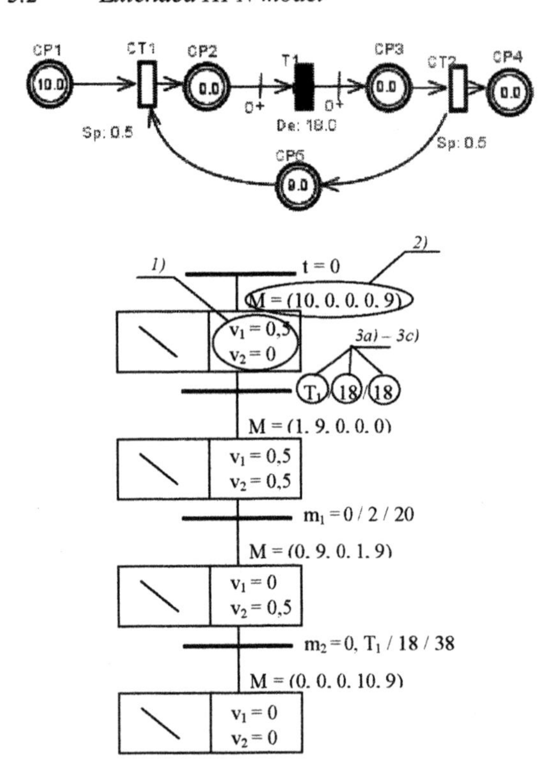

Fig. 4. - Extended HPN model of conveyor and its evolution graph

1) Instantaneous firing speed of T_j^c
2) Marking of P_i^c at the beginning of IB-state
3a) event, which caused passage from one IB-state to the other one
3b) relative time – the time elapsed in previous IB-state
3c) absolute time - the time elapsed from t=0

A conveyor model by extended HPNs is given in Fig. 4. Markings of continuous places CP1 and CP4 correspond to the number of parts in the input buffer B_1 and the output buffer B_2. Speeds of continuous transitions CT1 and CT2 correspond to the speed of the conveyor (the input speed is equal to the output speed). The maximal speeds associated to these transitions correspond to:

$$V1 = V2 = V * D\max = 0.5 * 1 = 0.5 \text{ kg/sec} \quad (4)$$

Place CP5 models the conveyor capacity. The delay associated to the transition T1 corresponds to time, which an infinitely small amount of product spends on the conveyor.

$$d1 = \frac{L}{V} = \frac{9}{0.5} = 18 \text{ sec} \quad (5)$$

The weight 0^+ represents infinitely small threshold of the transition T1. It means, that as soon as the infinitely small quantity of the product is put on the conveyor, the transition T1 becomes enabled and this quantity will reach the end of the conveyor, when time d_1 has elapsed.

3.3 *GBPN model*

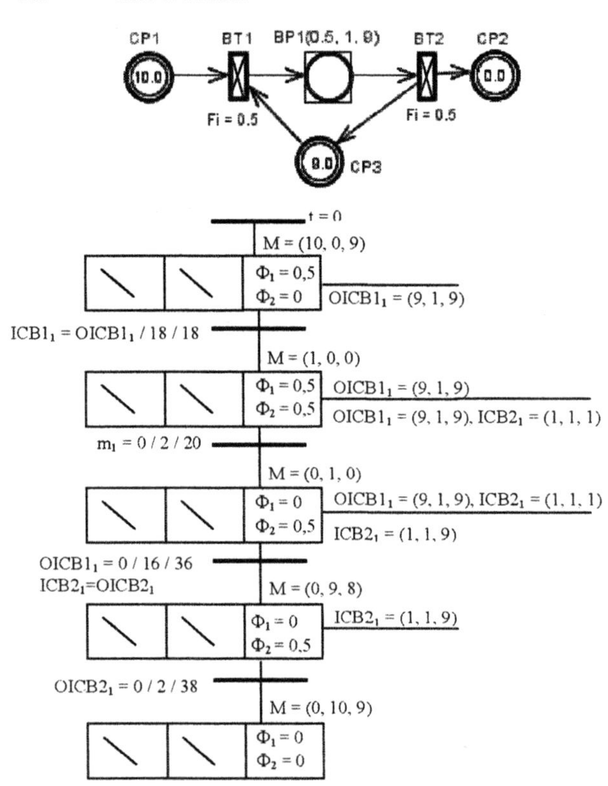

Fig. 5. – GBPN model of conveyor and its evolution graph

A conveyor model by GBPNs is shown in Fig. 5. Continuous places CP1 and CP2 correspond to input and output buffer. Batch place BP1 models the existence of the product on the conveyor. Batch function of BP1, which gives the speed of transfer, maximal density and the length of place, is {0.5, 1, 9}. The maximal speeds Φ_1 and Φ_2 associated to batch transitions BT1 and BT2 are given by equation (4). The behavior is described by the evolution graph in Fig. 5. Each IB-state of GBPN consists of three

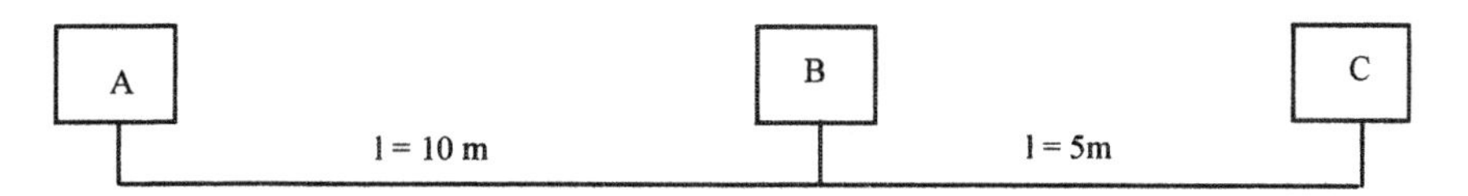

Fig. 6. - Three nodes A, B and C accessing a common bus

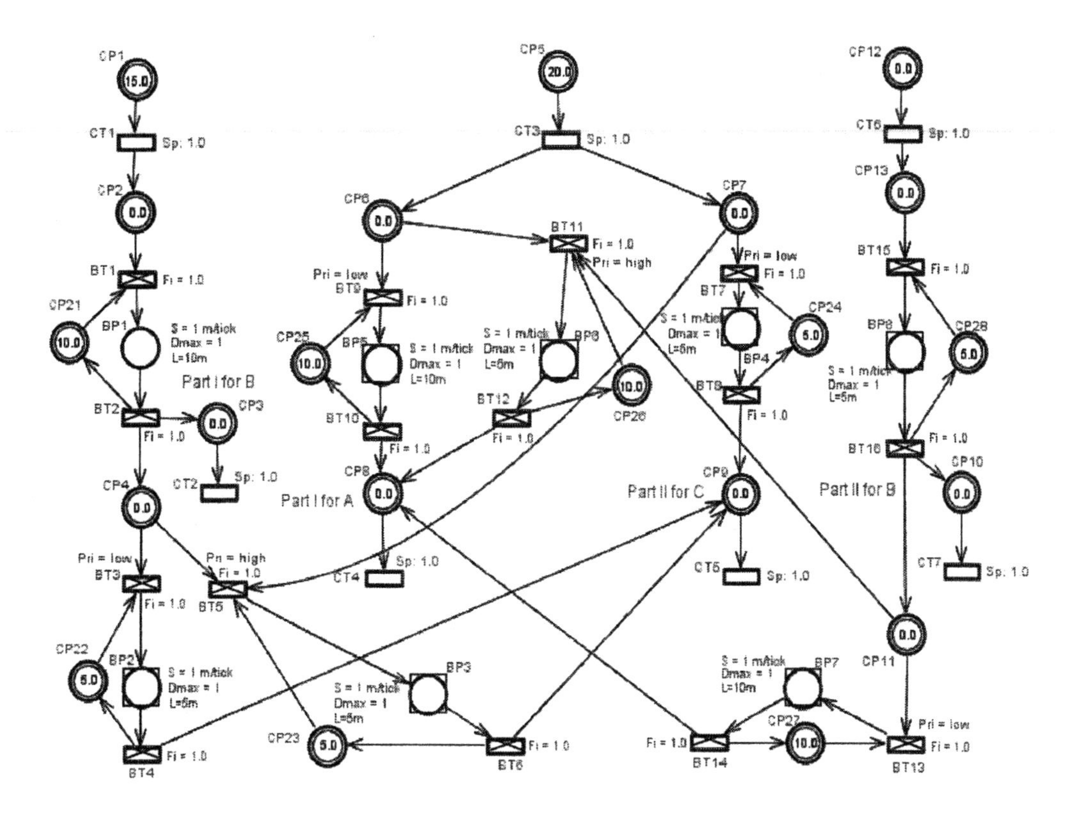

Fig. 7. – Physical layer model by GBPN

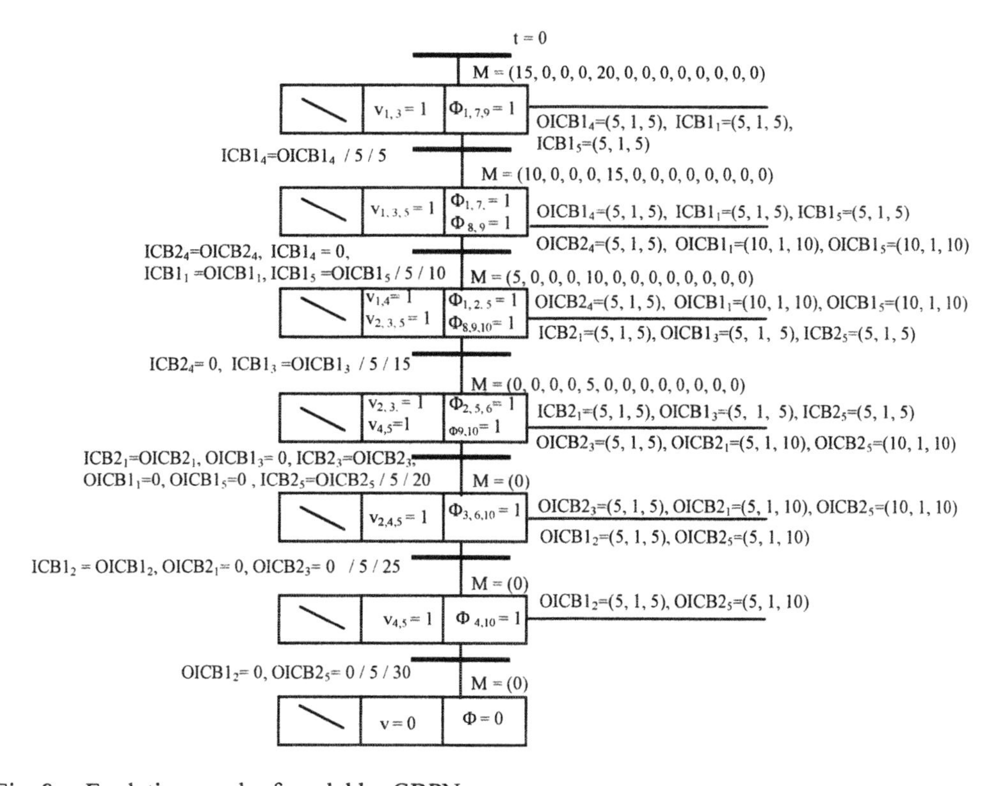

Fig. 8. – Evolution graph of model by GBPN

parts: discrete, continuous and batch part. The batch part contains constant speed of batch transitions during IB-state.

4. INDUSTRIAL EXAMPLE

4.1 Physical layer

Fig. 6 shows three nodes A, B and C accessing a common bus. The bus is one segment consisting of two parts – part I of length 10 meters between stations A and B, part II of length 5 meters between stations B and C. The aim is to model behavior of the physical media and to indicate the traffic on the bus to the Media Access Control sub-layer. In this example the nodes can transmit whenever they want at the speed 1 bit/tick. The signal propagates at the speed 1 meter per tick.

The physical media model by means of the batches Petri Nets is presented in Fig. 7. Continuous transition CT1 represents consecutive transmission of bit sequence from the node A. Batch transitions BT1, BT2, batch place BP1 and continuous place CP21 represent transmission delay on the part I. Traffic on the part I from the left side is indicated to the node B by CP3. Consecutively the bit sequence propagates on the part II with transmission delay 5 (BT3, BT4, BP2 and CP22). Continuous transition CT3 represents transmission of bit sequence from the node B to both directions, to the right and to the left. The transmission delay on part I is 10 ticks (BT9, BT10, BP5 and CP25) and 5 ticks on part II (BT7, BT8, BP4 and CP24). Similarly CT6 represents transmission from the node C with delay 5 ticks on the part II (BT15, BT16, BP8 and CP28) and 10 ticks on the part I (BT13, BT14, BP7 and CP27). When both the bit sequences from the node A and the bit sequences from the node B are propagated on the part II, then they collide and they form one bit sequence. The collided bit sequence is also transmitted with the delay 5 ticks (BT5 BT6, BP3 and CP23). In the same way the collided sequence from nodes B and C on the part I is transmitted with delay 10 (BT11, BT12, BP6 and CP26).
When the system is extended by a new node, the model grows only linearly - each part is represented by one transmission to the right side and one to the left side and one collided transmission to the right side and one collided transmission to the left side (with exception of the end parts).

Fig. 8 shows the evolution graph of GBPN in Fig. 7. Node A starts to transmit 15 bits and node B starts to transmit 20 bits at time 0. The node A detects the bus activity from the right side (corresponding to v_4) during $t \in <10,30)$. The node B detects the bus activity from the left side (corresponding to v_2) during $t \in <10,25)$ and from the right side (corresponding to v_7) there is no bus activity. The node C detects the bus activity from the left side (corresponding to v_5) during $t \in <5,30)$.

5. CONCLUSION

This article shows how the systems with the continuous flows delays can be modeled by extended hybrid Petri Nets. Two extensions have been used in this text – extended HPN by David and Caramihai (2000) and generalized Batches Petri Nets by Demongodin (2001). As illustrated on basic example, both extensions have similar expressive power with respect to modeling of continuous flows.

The physical layer example illustrates how this approach can be used to model the transmission delay on the communication media with bus topology. Indication of the activity on the bus is used by MAC sub-layer in order to decide when the transmission is allowed and when it is not allowed. Since the bit sequences are rather long represented by big amount of tokens, it is attractive to model the physical layer as continuous part of HPN and other layers as discrete part of HPN. The physical layer example in this article illustrates reasonable size of the HPN evolution graph.

This work was supported by the Ministry of Education of the Czech Republic under Project LN00B096.

6. REFERENCES

Brinkman P.L., W.A. Blaauboer (2000). Timed Continuous Petri nets: A Tool for Analysis and Simulation of Discrete Event Systems. In: *European Simulation Symposium*, Ghent.

David, R., H. Alla (1992): In: *Petri Nets and Grafcet.* Prentise Hall, London.

David,R., H. Alla (1993). Autonomous and timed continuous petri nets. In: *Advances in Petri nets*, Springer-Verlag.

David,R., H. Alla (1998). Continuous and Hybrid Nets. In: *Journal of Circuits, Systems and Computers*, Volume **8**, No.1, p. 159-188.

David,R., S.I. Caramihai (2000). Modeling of Delays on Continuous flows Thanks to Extended Hybrid Petri nets. In: *The 4th International conference on Automation of Mixed Processes: Hybrid Dynamic systems*, p.343-350.

David,R., H. Alla (2001). On Hybrid Petri Nets. In: *Discrete event dynamic systems: Theory and applications*, Volume **11**,p. 9-41.

Demongodin, I. (2001): Generalized batches PNs, In: *Discrete event dynamic systems: Theory and applications*, Volume **11**.

Murata T. (1989). Petri Nets: Properties, Analysis and Application. In: *Proceeding of The IEEE*, Volume **77**.

REDUCING COMPLEXITY OF SUPERVISOR SYNTHESIS

Roberto Ziller * Klaus Schneider **

* *University of Karlsruhe, Germany*
** *University of Kaiserslautern, Germany*

Abstract The objective of the supervisory control problem proposed by Ramadge and Wonham is to find a supervisor that constrains a system's behaviour according to a given specification. All its known solutions are of quadratic time complexity wrt. the number of states of the product of the system to be controlled and the specification. We present a new solution that can solve a large class of practical problems in linear time. Experimental results show that the improvement is not only a theoretical issue, but enables us to handle industrial-size problems efficiently. *Copyright © 2003 IFAC*

Keywords: Supervisor synthesis, Ramadge-Wonham, μ-calculus.

1. INTRODUCTION

Many embedded systems used in safety-critical applications consist of reactive real-time controllers, whose design requires automatic tools to improve efficiency and avoid errors made by humans. Modern verification methods (Clarke *et al.*, 1999) allow designers to check a given specification for a controller, but do not support the actual specification process except for providing a simulation trace when an error has been found. Ideally, the specification itself should be generated by a tool that takes the informal requirements of the designer and either outputs a correct specification or rejects them if they cannot be implemented. In general, the result of such a process is a representation of various possible solutions. The present work relates this representation to the notion of *supervisor* and the process of generating it to existing *supervisor synthesis* methods that use finite automata.

A supervisor is an entity that observes the sequence of events in a system and affects the next possible events in order to ensure a desired behaviour. Supervisor synthesis methods and their application vary according to the requirements they can handle. One aspect is whether a method distinguishes events that can be prevented from happening, like turning on an actuator, from those that can not, like a stimulus coming from a sensor. This can be modelled by dividing the event set of an automaton into *controllable* and *uncontrollable* events. A second aspect is whether the requirements may include assertions about deadlocks and livelocks, collectively referred to as *blocking* situations. These can be modelled through an automaton's final states. The latter define a language, and blocking freedom means that a system never gets stuck at a proper prefix of a string in that language. If a model is not concerned with blocking, then any generated prefix will be acceptable and the associated language will be prefix-closed.

A solution that can handle both controllability and blocking is offered by the Ramadge-Wonham (RW) framework (Ramadge and Wonham, 1987; Wonham and Ramadge, 1987; Wonham, 2002; Cassandras and Lafortune, 1999). It includes a method to synthesise a supervisor from two automata, one representing the physically possible behaviour of the system to be controlled, and the other a specification for its desired behaviour. The latter is not yet the product of a formal procedure, but rather a translation of the designer's requirements into a finite automaton. Therefore, it may express the wish for a behaviour of the system that is impossible to implement (for example, preventing a part to be loaded onto a manufacturing machine in a state where this happens automatically). It may also contain blocking situations not realised by the designer. A formal *synthesis procedure* then takes both automata above and generates a third automaton from them. The latter represents the largest implementable and non-blocking subset of the specification. It can therefore replace the original specification, as long as the resulting behaviour is still acceptable, in which case it acts as a supervisor for the system. The strength of the method lies in this formal synthesis capability.

Other methods, some of which were inspired in RW, were developed in parallel. For example, (Aziz *et al.*, 1995) considers the problem of finding a finite state machine (FSM) that, composed with a given FSM, satisfies a given specification. The work is restricted to prefix-closed languages and applies to problems where blocking considerations are not necessary. In (Yevtushenko *et al.*, 2001), a similar problem is cast in terms of solving equations over languages that are again prefix-closed. Failure semantics is used in (Overkamp, 1997) to introduce non-determinism. The method uses prefix-closed languages and cannot handle livelocks. It can handle deadlocks, but there is no distinction between the case where a system stops generating events because it has accomplished some task from that where it just gets stuck in the middle of a task. In (Kumar *et al.*, 1997), RW-synthesis is applied to protocol conversion and extended by a requirement that ensures progress, which also avoids deadlocks. The procedure uses prefix-closed languages and is not concerned with livelocks.

This paper is devoted to the *supervisory control problem* (Ramadge and Wonham, 1987), which considers controllability of events as well as both types of blocking. Given a system with state set $Q_{\mathcal{P}}$ and a specification with state set $Q_{\mathcal{E}}$ that communicate over an event set Σ, the known synthesis algorithms (Wonham and Ramadge, 1987; Kumar and Garg, 1995) have time complexity $O((|Q_{\mathcal{P}}|\,|Q_{\mathcal{E}}|)^2\,|\Sigma|)$ (Ramadge and Wonham, 1989; Cassandras and Lafortune, 1999). An attempt to solve the problem in linear time by analysing both controllability and blocking in the same pass of an algorithm was first made in (Ziller and Cury, 1994), but the solution obtained there was not guaranteed to be livelock-free. Up to now, only the special case of prefix-closed languages (i.e., without considering any type of blocking) could be solved in linear time (Kumar and Garg, 1995).

In this paper, we identify a class of problems that strictly contains the prefix-closed case and show how to solve the problems therein in time $O(|Q_{\mathcal{P}}|\,|Q_{\mathcal{E}}|\,|\Sigma|)$. *This means that many problems concerned with blocking, for which the existing algorithms require quadratic runtime, can now be solved in linear time.* We also argue that the new class contains many problems of practical importance. For the problems outside of it, the new solution automatically falls back into quadratic runtime, but will iterate fewer times than the existing algorithms. Therefore, the result improves all known solutions to the supervisory control problem.

Algorithms depend on some programming or informal language and are thus not well-suited for mathematical manipulation. In this work, we choose a formulation based on μ-calculus, which enables us to first present a new formulation for the algorithm in (Kumar and Garg, 1995) and then modify it to derive our main result. This approach is independent of any such languages and allows us to use well-known results about model checking (Cleaveland *et al.*, 1992) to assess the complexity of the new solution. Further, the equations can be understood by tools originally intended for verification, which are hereby extended to also handle controller synthesis.

The paper is organized as follows: Section 2 presents the Ramadge-Wonham model and states the problem being solved. Section 3 introduces the theoretical tools needed and describes the existing algorithm by means of μ-calculus expressions. Section 4 builds on these expressions to derive the new solution, and Section 5 discusses implementation and shows experimental results.

2. THE RAMADGE-WONHAM FRAMEWORK

The framework parallels continuous systems control theory, in which a system and its controller form a closed loop. There, the feedback signal from the controller enforces a given specification that would not be met by the open-loop behaviour. This foundation on control theory explains the use of the terms *discrete event system* (to designate an event-driven, discrete-space system, in opposition to time-driven, continuous systems) and *plant* (to designate the system to be controlled). It also leads to the assumptions that the latter encompasses the whole possible behaviour of the system (including unwanted situations, like the crash of two robot arms in a manufacturing cell), and that a specification describes a subset of this behaviour that corresponds to the actions wanted to remain executable under control.

The plant is viewed as a system that generates events. It is assumed to have a control input, through which some of the events that could happen in each state can be prevented from occurring. The *supervisor* is an external agent that has the ability to observe the events generated by the plant and to influence its behaviour through the control input, as shown in Figure 1.

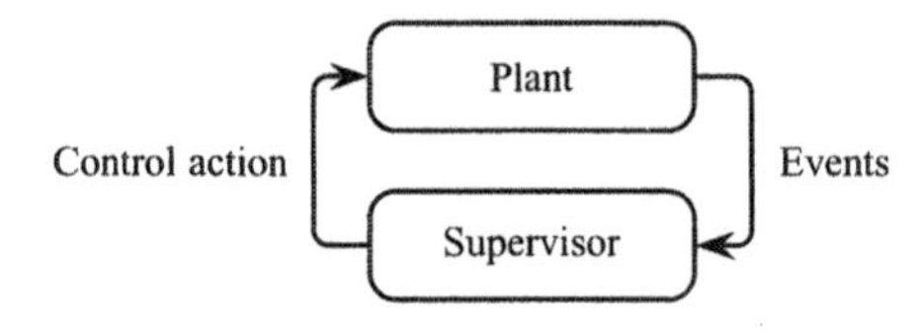

Figure 1. The basic RW-model

Control problems are formulated using language theory and finite automata. A finite automaton is a 5-tuple $\mathcal{A} = \langle \Sigma, Q, \delta, q^0, M \rangle$, where Σ is a set of event labels, Q is a set of states, $\delta \subseteq Q \times \Sigma \times Q$ is a transition relation, and $q^0 \in Q$ is the initial state. The states in the set $M \subseteq Q$ are chosen to mark the completion of tasks by the system and are therefore called *marker states*. Those readers familiar with the RW-literature will recall that δ is traditionally defined as a partial deterministic *function*. We use a *relation* instead because this simplifies notation later on. We write $\delta(q, \sigma, q')$ to signify that $(q, \sigma, q') \in \delta$. In order to ensure the same functionality, we require the relation to be deterministic, that is, $\delta(q, \sigma, q') \wedge \delta(q, \sigma, q'') \Rightarrow q' = q''$.

In the following, it is convenient to define the set of *active events* $\mathsf{act}_{\mathcal{A}}(q)$ as the subset of events for which there is a transition leaving state q:

Definition 1. (Active Events). Given an automaton $\mathcal{A} = \langle \Sigma, Q, \delta, q^0, M \rangle$ and a particular state $q \in Q$, the set of active events of q is:

$$\mathsf{act}_{\mathcal{A}}(q) := \{\sigma \in \Sigma \mid \exists q' \in Q.\ \delta(q, \sigma, q')\}.$$

When a plant and its supervisor are represented by finite automata $\mathcal{A_P}$ and $\mathcal{A_S}$, respectively, the control action of the latter amounts to running these automata in parallel, according to the following definition:

Definition 2. (Automata product). Given two automata $\mathcal{A_P} = \langle \Sigma, Q_\mathcal{P}, \delta_\mathcal{P}, q^0_\mathcal{P}, M_\mathcal{P} \rangle$ and $\mathcal{A_S} = \langle \Sigma, Q_\mathcal{S}, \delta_\mathcal{S}, q^0_\mathcal{S}, M_\mathcal{S} \rangle$, the product $\mathcal{A_P} \times \mathcal{A_S}$ is the automaton $\langle \Sigma, Q_\mathcal{P} \times Q_\mathcal{S}, \delta_{\mathcal{P}\times\mathcal{S}}, (q^0_\mathcal{P}, q^0_\mathcal{S}), M_\mathcal{P} \times M_\mathcal{S} \rangle$, where

$$\delta_{\mathcal{P}\times\mathcal{S}}((p,q), \sigma, (p', q')) \Leftrightarrow \delta_\mathcal{P}(p, \sigma, p') \wedge \delta_\mathcal{S}(q, \sigma, q').$$

Note that if a given transition is present in only one of the states p or q, it will *not* be present in state (p, q), i.e., $\mathsf{act}_{\mathcal{A_P}\times\mathcal{A_S}}((p,q)) = \mathsf{act}_{\mathcal{A_P}}(p) \cap \mathsf{act}_{\mathcal{A_S}}(q)$. The control action of the supervisor enables only the events in $\mathsf{act}_{\mathcal{A_P}\times\mathcal{A_S}}$. Hence, in order to forbid the occurrence of event σ when $\mathcal{A_P} \times \mathcal{A_S}$ is in state (p, q), it suffices to omit σ in state q of $\mathcal{A_S}$.

However, reactive systems may contain events that can not be prevented from occurring. The event set Σ is therefore partitioned into the sets of *controllable events* Σ_c (which the supervisor can disable) and *uncontrollable events* Σ_u (whose occurrence cannot be avoided). This places a condition on the existence supervisors: A specification given by an automaton $\mathcal{A_E}$ can be implemented by a supervisor only if, for every state (p, q) of $\mathcal{A_P} \times \mathcal{A_E}$, every event in $\mathsf{act}_{\mathcal{A_P}}(p) \setminus \mathsf{act}_{\mathcal{A_P}\times\mathcal{A_E}}((p,q))$ is controllable.

Specifications that do not fulfil this requirement are termed *uncontrollable*, because they allow the plant to reach a state in which uncontrollable events can occur and, at the same time, try to forbid the occurrence of one or more of these events in that state. Formally, this means that the product $\mathcal{A_P} \times \mathcal{A_E}$ has one or more *bad states*, which are states (p, q) that fail to satisfy the following condition:

$$\mathsf{act}_{\mathcal{A_P}\times\mathcal{A_E}}((p,q)) \supseteq \mathsf{act}_{\mathcal{A_P}}(p) \cap \Sigma_u. \qquad (1)$$

Analysing the controllability of a specification further requires some language theory: Every automaton $\mathcal{A}$ has an associated *marked language*, denoted $\mathsf{L_m}(\mathcal{A})$, which consists of all event sequences that end up in a marker state, hence representing the tasks the system is able to complete. When δ is extended in the usual way to process strings from Σ^*,

$$\mathsf{L_m}(\mathcal{A}) = \left\{ s \in \Sigma^* : \delta\left(q^0, s, q\right) \wedge q \in M \right\}.$$

Given a specification automaton $\mathcal{A_E}$, the language $K = \mathsf{L_m}(\mathcal{A_E})$ is *controllable* if and only if $\mathcal{A_E}$ has no bad states.

The marked language of plant $\mathcal{A_P}$ under control of supervisor $\mathcal{A_S}$ is equal to $\mathsf{L_m}(\mathcal{A_P}) \cap \mathsf{L_m}(\mathcal{A_S})$, and denoted $\mathsf{L_m}(\mathcal{A_S}/\mathcal{A_P})$. Ramadge and Wonham have shown that, for any plant $\mathcal{A_P}$ and any specification language $K \subseteq \mathsf{L_m}(\mathcal{A_P})$, there exists the *supremal controllable sublanguage* of K, denoted $\mathsf{supC}(K)$. This result is of practical interest: Given that the specification language K is uncontrollable, it is possible to compute $\mathsf{supC}(K)$ and to construct a supervisor $\mathcal{A_S}$ such that $\mathsf{L_m}(\mathcal{A_S}/\mathcal{A_P}) = \mathsf{supC}(K)$. This language can replace the original specification, as long as the resulting behaviour under control is still acceptable.

Another aspect to consider is whether the supervisor always allows the system to make progress towards the completion of some task. This is not the case when the system can (1) reach a state in which no task is finished and no more events can occur (deadlock) or (2) be caught forever within a subset of states, none of which corresponds to a finished task (livelock). A supervisor that avoids these situations is said to be *non-blocking*. A non-blocking automaton is coaccessible, which means that there is at least one path leading from every state to a marker state. Controllability and absence of blocking come together in the problem that is central to the present work:

Definition 3. (Supervisory Control Problem (SCP)). Given a plant $\mathcal{A_P}$, a specification language $K \subseteq \mathsf{L_m}(\mathcal{A_P})$ representing the desired behaviour of $\mathcal{A_P}$ under supervision, and a minimally acceptable behaviour $A_{min} \subseteq K$, find a non-blocking supervisor $\mathcal{A_S}$ such that $A_{min} \subseteq \mathsf{L_m}(\mathcal{A_S}/\mathcal{A_P}) \subseteq K$.

SCP is solvable if and only if $\mathsf{supC}(K) \supseteq A_{min}$, and $\mathsf{supC}(K)$ is its least restrictive solution (Ramadge and Wonham, 1987). A coaccessible automaton $\mathcal{A_S}$ whose marked language is equal to $\mathsf{supC}(K)$ can be computed from the automata $\mathcal{A_P}$ and $\mathcal{A_E}$, with $\mathcal{A_E}$ constructed so that $\mathsf{L_m}(\mathcal{A_E}) = K$. Because the resulting automaton is a supervisor, this computation is often referred to as *supervisor synthesis*.

The reader is referred to (Wonham, 2002; Cassandras and Lafortune, 1999) for further details about the RW-framework.

3. CLASSICAL SUPERVISOR SYNTHESIS

In this section we associate Kripke structures with the automata used in the RW-framework and define a μ-calculus over them. These structures are used in Subsection 3.3 to describe the classical solution in the sense of (Wonham and Ramadge, 1987; Kumar and Garg, 1995). Because the μ-calculus is not usual in this context, we start with a brief review of basic concepts.

3.1 Fixpoint Calculus

Notations for extremal fixpoints of monotone operators have been introduced by different authors (Lassez *et al.*, 1982). In particular, Tarski's work (Tarski, 1955) has been frequently used in verification and synthesis literature (Emerson and Lei, 1986; Thistle and Wonham, 1994). The following is an adaptation of the results found in these sources to suit our needs.

An operator $f : 2^X \to 2^X$ on the powerset 2^X is said to be *monotone* if, for any subsets $x_i, x_j \subseteq X$,

$$x_i \subseteq x_j \Rightarrow f(x_i) \subseteq f(x_j). \tag{2}$$

Such an operator has least and greatest fixpoints, which are the solutions of $x \overset{\mu}{=} f(x)$ and $x \overset{\nu}{=} f(x)$, where the symbols $\overset{\mu}{=}$ and $\overset{\nu}{=}$ indicate the seek for the least and greatest values of x that satisfy these equations. The solutions are denoted $\mu x.f(x)$ and $\nu x.f(x)$, and known to satisfy:

$$\mu x.f(x) = \cap\{x \subseteq X : x = f(x)\} \quad \text{and}$$
$$\nu x.f(x) = \cup\{x \subseteq X : x = f(x)\}.$$

Given that X is finite and f is monotone, the least fixpoint can be found by an iteration starting with $x_0 = \emptyset$ and computing $x_{i+1} = f(x_i)$ until, for some j, $x_j = x_{j-1}$ holds. The greatest fixpoint can be obtained by the same iteration starting with $x_0 = X$. In this paper, the fixpoint operators will be applied to the state set of a Kripke structure.

We shall use a version of the modal μ-calculus, defined on equation systems (Cleaveland *et al.*, 1992; Schneider, 2003) of the form:

$$\begin{cases} x_1 \overset{\sigma_1}{=} \varphi_1 \\ \vdots \\ x_n \overset{\sigma_n}{=} \varphi_n \end{cases}$$

where $\sigma_i \in \{\mu, \nu\}$ for $i \in \{1, \ldots, n\}$. Such a system can be translated into a single fixpoint expression that uses the operators μ and ν, and vice versa. The formulas φ_i may consist of the propositional operators $\neg$, $\wedge$, and $\vee$, and the modal operators EX, AX, EY, and AY. Here, E and A are path quantifiers that specify that the property that follows them in the expression must hold on at least one path (E) or on all paths (A) from some state in a Kripke structure. X and Y are temporal operators that limit the length of the path to the immediate successor or the immediate ancestor states, respectively (see (Clarke *et al.*, 1999; Schneider, 2003) for background on modal operators and temporal logics). As usual, we require that the occurrences of all x_j in every φ_i must occur under an even number of negation symbols.

3.2 Automata and Kripke Structures

A Kripke structure is a finite transition system. As opposed to automata, the transitions of a Kripke structure are not labelled. The transition relation only indicates which states can be reached from a given state, but not why or how. There is also a set of Boolean variables, say, $\mathcal{V}$, and a labelling function that maps each state of the Kripke structure onto a subset of $\mathcal{V}$. The effect of this mapping is to associate with each state a number of Boolean variables which are considered to be true in that state. This is useful to write Boolean expressions representing sets of states of the Kripke structure, e.g. $x_1 \wedge x_2$, which denotes all states in which both variables x_1 and $x_2 \in \mathcal{V}$ are true.

While the definition of a Kripke structure as described above is independent of any automaton, we can also define a special type of structure to suit our specific needs:

Definition 4. (Kripke Structure of an Automaton). Given an automaton $\mathcal{A} = \langle \Sigma, Q, \delta, q^0, M \rangle$ representing the product of a plant and a specification, we define its associated Kripke structure $\mathcal{K}_\mathcal{A} = \langle \mathcal{S}, \mathcal{I}, \mathcal{R}, \mathcal{L} \rangle$ over the Boolean variables $\mathcal{V}_\mathcal{A} := \{x_q \mid q \in Q\} \cup \{x_b, x_m, x_u\}$ as follows:

- $\mathcal{S} := Q \times \{0, 1\}$
- $\mathcal{I} := \{(q^0, 0), (q^0, 1)\}$
- $\mathcal{R}((q, 0), (q', 0)) :\Leftrightarrow \exists \sigma \in \Sigma_u.\delta(q, \sigma, q')$
- $\mathcal{R}((q, 1), (q', 1)) :\Leftrightarrow \exists \sigma \in \Sigma.\delta(q, \sigma, q')$
- $\mathcal{L}((q, 0)) := \{x_q, x_u\} \cup \begin{cases} \{x_b\} & \text{if } q \text{ is bad} \\ \{\} & \text{otherwise} \end{cases}$
- $\mathcal{L}((q, 1)) := \{x_q\} \cup \begin{cases} \{x_m\} & \text{if } q \in M \\ \{\} & \text{if } q \notin M. \end{cases}$

Here, $\Sigma = \Sigma_c \dot\cup \Sigma_u$ (see Section 2), $\mathcal{S}$ is a set of states, $\mathcal{I}$ is the set of initial states, and $\mathcal{R} \subseteq \mathcal{S} \times \mathcal{S}$ relates states $(q, 0)$ and $(q', 0)$ exactly when an uncontrollable event leads from q to q' and states $(q, 1)$ and $(q', 1)$ exactly when there is an event (controllable or not) leading from q to q' in $\mathcal{A}$. This creates a structure with two disconnected substructures, each of which has a copy of the original states in $\mathcal{A}$. Finally, $\mathcal{L}$ assigns a set of labels from $\mathcal{V}_\mathcal{A}$ to each state, thereby enabling us to address sets of states through Boolean expressions. Note that the Kripke structure can be constructed from the automaton in time $O(|Q|\,|\Sigma|)$.

As an example, suppose the automaton in Figure 2 represents the product of some plant and specification. The composite numbers of the states have been replaced by singletons for simplicity. States 3 and 5 are assumed to be bad, and the event set is partitioned into $\Sigma_c = \{\alpha, \lambda\}$ and $\Sigma_u = \{\beta, \gamma\}$. The two halves

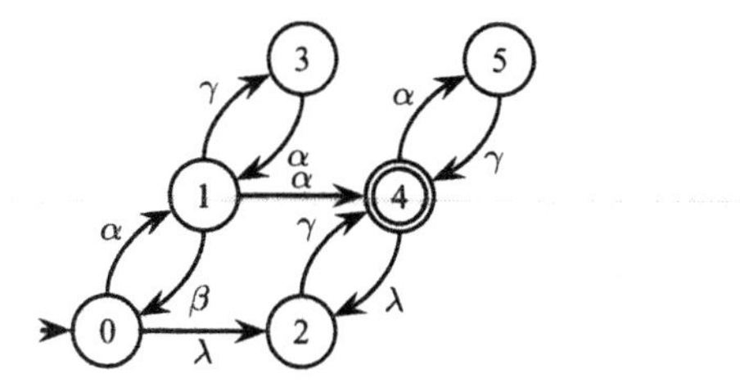

Figure 2. Example automaton

of the associated Kripke structure are shown in Figure 3. Each state (q, i) is labelled with the variable x_q. The left substructure has transitions only where the automaton has uncontrollable transitions, and its states have in common the label x_u. Additionally, the states that correspond to bad states in the automaton have the label x_b. The right substructure reflects all transitions

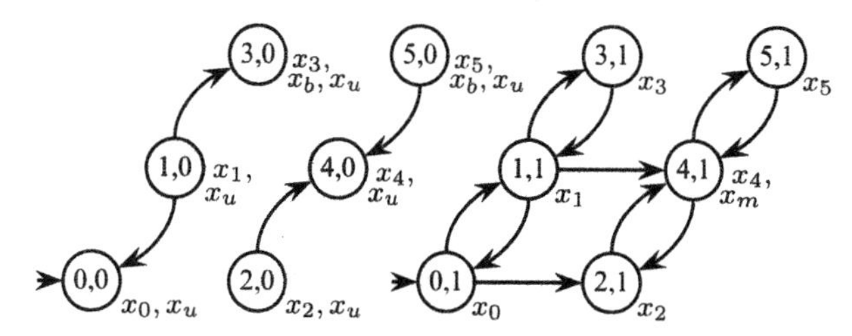

Figure 3. The associated Kripke structure

from the automaton, with the label x_m identifying the state that corresponds to a marker state. Note that x_u distinguishes the states from the two substructures, and that the left side does not know anything about the marker states, while the right side does not know about the bad states.

The following definitions give syntax and semantics of μ-calculus formulas:

Definition 5. (Syntax of μ-Calculus). Given a set of variables $\mathcal{V}$, the set of μ-calculus formulas over $\mathcal{V}$ is defined as the least set $\mathcal{L}_\mu$ that satisfies the following rules:

- $\mathcal{V} \cup \{0, 1\} \subseteq \mathcal{L}_\mu$
- $\neg\varphi, \varphi \vee \psi, \varphi \wedge \psi \in \mathcal{L}_\mu$, provided that $\varphi, \psi \in \mathcal{L}_\mu$
- $\mathsf{EX}\, \varphi, \mathsf{EY}\, \varphi \in \mathcal{L}_\mu$
- $\kappa_\pi(\varphi) \in \mathcal{L}_\mu$, provided that $\varphi \in \mathcal{L}_\mu$
- $\mu x.\varphi \in \mathcal{L}_\mu$, provided that $\varphi \in \mathcal{L}_\mu$.

Definition 5 differs from those usually found in the literature in that it includes the formula $\kappa_\pi(\varphi)$. This allows us to use any monotone state transformer function $\pi : 2^{\mathcal{S}} \to 2^{\mathcal{S}}$ in the computations. In particular, we will define a function π to map states of one of the above substructures to the other.

Definition 6. (Semantics of μ-Calculus). Given a Kripke structure $\mathcal{K} = \langle \mathcal{S}, \mathcal{I}, \mathcal{R}, \mathcal{L} \rangle$ over the variables $\mathcal{V}$, we associate with each formula $\Phi \in \mathcal{L}_\mu$ a set of states $[\![\Phi]\!]_\mathcal{K} \subseteq \mathcal{S}$ by the following rules:

- $[\![0]\!]_\mathcal{K} := \{\}$ and $[\![1]\!]_\mathcal{K} := \mathcal{S}$
- $[\![x]\!]_\mathcal{K} := \{s \in \mathcal{S} \mid x \in \mathcal{L}(s)\}$ for all $x \in \mathcal{V}$
- $[\![\neg\varphi]\!]_\mathcal{K} := \mathcal{S} \setminus [\![\varphi]\!]_\mathcal{K}$
- $[\![\varphi \wedge \psi]\!]_\mathcal{K} := [\![\varphi]\!]_\mathcal{K} \cap [\![\psi]\!]_\mathcal{K}$
- $[\![\varphi \vee \psi]\!]_\mathcal{K} := [\![\varphi]\!]_\mathcal{K} \cup [\![\psi]\!]_\mathcal{K}$
- $[\![\kappa_\pi(\varphi)]\!]_\mathcal{K} := \pi([\![\varphi]\!]_\mathcal{K})$, $\pi : 2^{\mathcal{S}} \to 2^{\mathcal{S}}$ monotone
- $[\![\mathsf{EX}\, \varphi]\!]_\mathcal{K} := \{s \in \mathcal{S} \mid \exists s' \in \mathcal{S}.\mathcal{R}(s, s') \wedge s' \in [\![\varphi]\!]_\mathcal{K}\}$
- $[\![\mathsf{EY}\, \varphi]\!]_\mathcal{K} := \{s' \in \mathcal{S} \mid \exists s \in \mathcal{S}.\mathcal{R}(s, s') \wedge s \in [\![\varphi]\!]_\mathcal{K}\}$
- $[\![\mu x.\varphi]\!]_\mathcal{K} := \bigcap\{\mathcal{Q} \subseteq \mathcal{S} \mid [\![\varphi]\!]_{\mathcal{K}_x^\mathcal{Q}} \subseteq \mathcal{Q}\}$.

The last expression in Definition 6 gives the least set of states $\mathcal{Q} \subseteq \mathcal{S}$ such that $\mathcal{Q} = [\![\varphi]\!]_{\mathcal{K}_x^\mathcal{Q}}$ holds, where $\mathcal{K}_x^\mathcal{Q}$ is the Kripke structure where exactly the states $\mathcal{Q}$ are labelled with the variable x. If φ is a monotonic function of x, then $\mu x.\varphi$ is its least fixpoint (Tarski, 1955). We can also define some further macro operators like $\mathsf{AX}\, \varphi := \neg\mathsf{EX}\, \neg\varphi$, $\mathsf{AY}\, \varphi := \neg\mathsf{EY}\, \neg\varphi$, and $\nu x.\varphi(x) := \neg\mu x.\neg\varphi(\neg x)$. The latter can be shown to be the greatest fixpoint of φ.

Together, Definitions 5 and 6 enable us to write Boolean expressions to denote sets of states, as described informally at the beginning of the current subsection. For example, for the Kripke structure of Figure 3, $[\![x_u \wedge (x_3 \vee x_5)]\!]_{\mathcal{K}_\mathcal{A}} = \{(3,0), (5,0)\}$, and $[\![\mathsf{EX}\, x_m]\!]_{\mathcal{K}_\mathcal{A}} = \{(1,1), (2,1), (5,1)\}$.

In order to apply the above definitions to Kripke structures stemming from automata according to Definition 4, we define $\pi(Q) := \{(q, \neg i) \mid (q, i) \in Q\}$, which is easily verified to be monotone (see condition 2). The function κ_π then toggles the variable x_u that identifies the substructure to which a given state pertains, thereby enabling us to switch from one substructure to the other. For simplicity, we write just κ for κ_π from this point on. Further, we have:

- $[\![x_b]\!]_{\mathcal{K}_\mathcal{A}} := \{(q, 0)\}$ for all q that violate cond. 1
- $[\![x_m]\!]_{\mathcal{K}_\mathcal{A}} := \{(q, 1)\}$ for all $q \in M$
- $[\![x_q]\!]_{\mathcal{K}_\mathcal{A}} := \{q\} \times \{0, 1\}$ for all $q \in Q$
- $[\![x_u]\!]_{\mathcal{K}_\mathcal{A}} := \{(q, 0)\}$ for all $q \in Q$
- $[\![\neg x_u]\!]_{\mathcal{K}_\mathcal{A}} := \{(q, 1)\}$ for all $q \in Q$.

Any formula Φ over the variables $\mathcal{V}_\mathcal{A}$ also describes a subset of the states of $\mathcal{A}$ according to the following projection, which maps states from the Kripke structure back to the originating automaton:

Definition 7. (Kripke Structure State Projection). Given an automaton $\mathcal{A}$, its associated Kripke structure $\mathcal{K}_\mathcal{A}$ and a μ-calculus formula Φ over the variables $\mathcal{V}_\mathcal{A}$, we define the projection of $[\![\Phi]\!]_{\mathcal{K}_\mathcal{A}}$ onto the state set Q of $\mathcal{A}$ as:

$$[\![\Phi]\!]_\mathcal{A} = \{q \in Q \mid (q, 0) \in [\![\Phi]\!]_{\mathcal{K}_\mathcal{A}} \vee (q, 1) \in [\![\Phi]\!]_{\mathcal{K}_\mathcal{A}}\}. \quad (3)$$

With this projection, we can construct an automaton from Φ by restricting the state set of the original automaton $\mathcal{A}$ to $[\![\Phi]\!]_{\mathcal{A}}$. This completes the toolset to convert an automaton into a Kripke structure, compute a subset of states, and translate the result back. For example, the set of states of $\tau \subseteq \delta$ that are accessible only through states pertaining to τ is given by $[\![x_{Ac}]\!]_{\mathcal{A}}$, with:

$$x_{Ac} \stackrel{\mu}{=} \Phi_\tau \wedge (\mathsf{EY}\, x_{Ac} \vee x_{q^0}), \tag{4}$$

where Φ_τ represents the states of τ. Similarly, the set of states of $\tau \subseteq \delta$ that are coaccessible only through states pertaining to τ is given by $[\![x_{Co}]\!]_{\mathcal{A}}$, with:

$$x_{Co} \stackrel{\mu}{=} \Phi_\tau \wedge (\mathsf{EX}\, x_{Co} \vee x_m). \tag{5}$$

Of special interest is the *alternation depth* of a fixpoint expression (Emerson and Lei, 1986; Niwinski, 1986). Roughly speaking, this is the nesting depth of alternating μ and ν-operators whose computation depends on each other. Expressions with a single operator have alternation depth 1 and are also called *alternation-free*. The alternation depth of an equation system is the largest number of blocks of formulas seeking for a least or greatest fixpoint in the equation system that depend on each other to compute. The following result (Schneider, 2003) will be useful to assess the computational complexity of our solution [1]:

Theorem 1. (Complexity of μ-Calculus Model Checking). For every equation system E of alternation depth l, whose *length* $|E|$ is given by the sum of the lengths of the right sides of the equations in the system, and every Kripke structure $\mathcal{K} = \langle \mathcal{S}, \mathcal{I}, \mathcal{R}, \mathcal{L} \rangle$, there is an algorithm to compute its solution in time

$$O\left(\left(\frac{|E|\,|\mathcal{S}|}{l}\right)^{l-1} |\mathcal{R}|\,|E|\right).$$

Corollary 1. A system of equations of constant length and alternation depth l written for a Kripke structure associated to an automaton can be solved in time

$$O\left(|Q|^l\,|\Sigma|\right).$$

Proof. By construction, the Kripke structure from Definition 4 has $|\mathcal{S}| = 2\,|Q|$ and $|\mathcal{R}| \leq 2\,|Q|\,|\Sigma|$. For constant $|E|$, the result follows immediately.

3.3 Solving SCP

We now apply the tools presented in Subsection 3.2 to the classical solutions of SCP (Wonham and Ramadge, 1987; Kumar and Garg, 1995). Our approach is closer to the one in (Kumar and Garg, 1995), because it collects bad and non-coaccessible states instead of eliminating them at each iteration, and collecting states is easier to represent with our Kripke structure. We use Corollary 1 to confirm the quadratic complexity of the algorithm.

The solution for SCP is computed from the product $\mathcal{A}_\mathcal{P} \times \mathcal{A}_\mathcal{E}$ by starting with the initial bad states (those that violate condition 1), and recursively adding the states that can reach some bad state through an uncontrollable transition. When no more bad states can be found, the remaining states are tested for coaccessibility and those found to be non-coaccessible are added to the set of bad states. Because removing bad states can destroy coaccessibility and removing non-coaccessible states can expose new bad states, the algorithm restarts and alternates between these two computations until a fixpoint is reached.

We form the Kripke structure associated to $\mathcal{A}_\mathcal{P} \times \mathcal{A}_\mathcal{E}$ and collect the bad states into the set [2] x_B. The initial value for this set is x_b. We add a state to x_B when it has an uncontrollable transition leading into a state already found to be bad, or non-coaccessible, or both. This requires collecting also the non-coaccessible states, which is done in the set x_N.

The computations use the two substructures of the Kripke structure. Because we are interested only in predecessors of bad states connected to them through uncontrollable transitions, the substructure identified by x_u is the correct choice to collect bad states. On the other hand, coaccessibility requires considering all transitions, and therefore the substructure identified by $\neg x_u$ must be used to compute x_N. When it comes to using the non-coaccessible states in the computation of x_B, we have to switch from one substructure to the other, using the set $\kappa(x_N)$. The above description translates naturally into expression 9 below.

To derive an expression for x_N, we first set $\Phi_\tau = \neg\kappa(x_B)$ in equation 5 to keep only coaccessible states that are not bad:

$$x_{Co} \stackrel{\mu}{=} \neg\kappa(x_B) \wedge (\mathsf{EX}\, x_{Co} \vee x_m). \tag{6}$$

Next, we complement expression 6 to obtain the non-coaccessible states. Note this makes the expression a *greatest* fixpoint:

$$\neg x_{Co} \stackrel{\nu}{=} \kappa(x_B) \vee \mathsf{AX}\, \neg x_{Co} \wedge \neg x_m. \tag{7}$$

While states in $\kappa(x_B)$ are all on the substructure identified by $\neg x_u$, the term $\neg x_m$ introduces unwanted states from x_u. The expression for x_N is therefore obtained by restricting equation 7 to $\neg x_u$. This completes our equation system:

$$\begin{cases} x_N \stackrel{\nu}{=} \kappa(x_B) \vee \neg x_u \wedge \mathsf{AX}\, x_N \wedge \neg x_m & (8) \\ x_B \stackrel{\mu}{=} \mathsf{EX}\,(x_B \vee \kappa(x_N)) \vee x_b. & (9) \end{cases}$$

[1] This holds when the function κ_π can be computed in time $O(\mathcal{R})$, which we assume.

[2] To improve readability, we will write just 'the set x_B' to signify 'the set $[\![x_B]\!]_{\mathcal{A}}$ represented by x_B', as long as the meaning is clear from the context.

The accessible component of the transition relation that results when the bad states and the non-coaccessible states are eliminated can be found by setting $\Phi_\tau = \neg(\kappa(x_B) \vee x_N)$ in equation 4, again restricting the result to $\neg x_u$:

$$x_{Ac} \stackrel{\mu}{=} \neg(\kappa(x_B) \vee x_N) \wedge (\mathsf{EY}\, x_{Ac} \vee x_{q^0} \wedge \neg x_u). \quad (10)$$

The solution for SCP is then given by restricting the automaton $\mathcal{A}_\mathcal{P} \times \mathcal{A}_\mathcal{E}$ to the states $[\![x_{Ac}]\!]_{\mathcal{A}_\mathcal{P} \times \mathcal{A}_\mathcal{E}}$. The complexity of the overall computation is given by Corollary 1. Since the solution has $l = 2$, we get $O(|Q|^2\,|\Sigma|) = O((|Q_\mathcal{P}|\,|Q_\mathcal{E}|)^2\,|\Sigma|)$, which is the known complexity of the original algorithm (Ramadge and Wonham, 1989; Cassandras and Lafortune, 1999). An example of the computation of the fixpoints is given in (Ziller and Schneider, 2003).

4. REDUCING COMPLEXITY

In this section we identify a special class of supervisory control problems that covers many practical situations. These are characterised by the absence of a particular connected component in the product $\mathcal{A}_\mathcal{P} \times \mathcal{A}_\mathcal{E}$, which we call *hidden livelock component*, or HLC for short. This class contains all problems that use prefix-closed languages (which can already be solved in linear time), but also many problems which the algorithms from (Wonham and Ramadge, 1987; Kumar and Garg, 1995) can solve only in quadratic time. Our main result is a set of equations that solves HLC-free problems in linear time. The equations also apply to all other problems, in which case complexity automatically changes to quadratic. Efficiency is nonetheless improved, because the number of iterations is kept smaller.

In the solution presented in Section 3, equations 8 and 9 form a system of alternation depth 2. According to Corollary 1, complexity decreases with the alternation depth. We therefore suggest a new approach to collect non-coaccessible states that leads to a least fixpoint instead of a greatest. We start from the set of states that are initially not coaccessible in $\mathcal{A}_\mathcal{P} \times \mathcal{A}_\mathcal{E}$, which we denote x_n. To compute it, we set $\Phi_\tau = 1$ in equation 5, complement it, and restrict the result to the substructure identified by $\neg x_u$:

$$x_n \stackrel{\nu}{=} \neg x_u \wedge \mathsf{AX}\, x_n \wedge \neg x_m. \quad (11)$$

To compute non-coaccessible states we initialise x_N with x_n and add a new state to x_N if it is not a marker state and all its outgoing transitions lead to states that were already found to be either bad or not coaccessible. This amounts to exchanging equation 8 by equation 12 in our equation system as follows:

$$\begin{cases} x_N \stackrel{\mu}{=} \mathsf{AX}\,(\kappa(x_B) \vee x_N) \vee x_n & (12) \\ x_B \stackrel{\mu}{=} \mathsf{EX}\,(x_B \vee \kappa(x_N)) \vee x_b. & (13) \end{cases}$$

If equation 12 can capture all non-coaccessible states of $\mathcal{S}$, then the solution can be computed from equation 10 as before. Since the equation system above has alternation depth 1, the problem can be solved in time $O(|Q|\,|\Sigma|)$. However, the expression for x_N will fail to capture some non-coaccessible states from $\mathcal{S}$ if they form a component as described in the following:

Definition 8. (Hidden Livelock Component (HLC)). A hidden livelock component is a subset of states of $\mathcal{A}_\mathcal{P} \times \mathcal{A}_\mathcal{E}$ satisfying the following conditions simultaneously:

- they are initially coaccessible, but become not coaccessible after some bad states collected in x_B are removed;
- none of them becomes a bad state;
- they form a connected component.

As an example, consider a specification like the one in Figure 4 (where event labels are not shown because they are irrelevant for the illustration). Suppose that all transitions are controllable and that state 5 is a bad state. Although states 2, 3, and 4 will become not coaccessible after 5 is removed, they will remain undetected because each of them has a transition leading into a state that can not be gathered while solving equations 12 and 13.

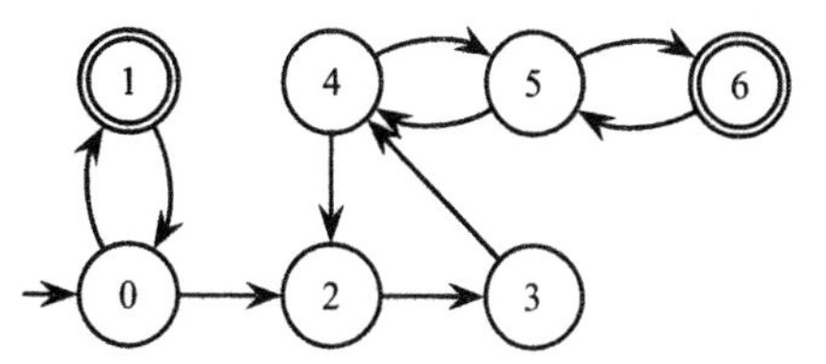

Figure 4. States 2, 3, and 4 form a HLC

Therefore, the states in $(\kappa(x_B) \vee x_N)$ may not be the only states that have to be removed. The only way to test this is to remove them and then check the result for coaccessibility, much like the algorithm from Section 3. By equation 5, the remaining coaccessible states are:

$$x_{Co} \stackrel{\mu}{=} \neg(\kappa(x_B) \vee x_N) \wedge (\mathsf{EX}\, x_{Co} \vee x_m). \quad (14)$$

The complement of this expression, restricted to $\neg x_u$, includes the hidden livelock components exposed by the removal of the bad and the non-coaccessible states found so far. We denote this set by x_H. It is given by equation 15, which completes our equation system for the general solution of SCP. Equation 17 is the same as equation 13, while equation 12 has been augmented by the new set of non-coaccessible states x_H to form equation 16. Note that the first computation of x_H is identical to x_n, so the latter could be dropped. The solution for SCP is again given by restricting $\mathcal{A}_\mathcal{P} \times \mathcal{A}_\mathcal{E}$ to the states given by equation 10.

$$\begin{cases} x_H \stackrel{\nu}{=} (\kappa(x_B) \vee x_N) \vee \neg x_u \wedge \mathsf{AX}\, x_H \wedge \neg x_m & (15) \\ x_N \stackrel{\mu}{=} \mathsf{AX}\,(\kappa(x_B) \vee x_N) \vee x_H & (16) \\ x_B \stackrel{\mu}{=} \mathsf{EX}\,(x_B \vee \kappa(x_N)) \vee x_b & (17) \end{cases}$$

The above system of equations has alternation depth 2. According to Corollary 1, the worst case complexity is thus $O((|Q_\mathcal{P}|\,|Q_\mathcal{E}|)^2\,|\Sigma|)$. To see that this complexity appears only if $\mathcal{A}_\mathcal{P} \times \mathcal{A}_\mathcal{E}$ is not HLC-free, consider that the solution starts by initialising $x_H = 1$ and $x_N = x_B = 0$. x_H is then computed once, which can be done in time $O(|Q_\mathcal{P}|\,|Q_\mathcal{E}|\,|\Sigma|)$. Next, a fixpoint for the subsystem formed by equations 16 and 17 is found. The key point is that x_N and x_B can be computed without reinitialising their values to 0, because they are both least fixpoints. In contrast, equations 8 and 9 required initialising $x_N = 1$ and $x_B = 0$ at each iteration. When x_H is evaluated a second time and there are no HLC's, the iteration stops, including all non-coaccessible states. In the presence of HLC's, the iteration continues automatically, only then yielding a quadratic complexity. This leads to the conclusion that, even in the presence of HLC's, the new solution will be more efficient than the existing ones, because only the non-coaccessible states that form HLC's will be collected in quadratic time. The special case of prefix-closed languages is also done in linear time, since if all states are marker states, there can be no HLC's. Equivalently, the full marking leads to $x_H = x_N = 0$, so it suffices to evaluate x_B.

5. IMPLEMENTATION

This section presents experimental results that illustrate how our new solution improves supervisor synthesis. Traditionally, an automaton is represented in computer memory by transition tables, and writing such programs for the classical and the new solutions would show the expected reduction from quadratic to linear time in a number of examples. However, even if the computational effort becomes linear in the number of states of $\mathcal{A}_\mathcal{P}\times\,\mathcal{A}_\mathcal{E}$, the transition tables of most practical systems would be far too large to fit into the memory of any computer, especially due to the state explosion when computing the product $\mathcal{A}_\mathcal{P} \times \mathcal{A}_\mathcal{E}$.

Therefore, we went one step further and chose a symbolic representation through binary decision diagrams (BDD's) (Bryant, 1986). This requires a new method to find the initial bad states (Ziller, 2002), and also affects computational complexity, since some operations can no longer be done in linear time. This means that we did not stick to linearity, but rather used the new solution to create a program that can handle large state sets and is faster than a similar program based on the classical solution from Section 3. For comparison, the following example has been solved with BDD-based implementations of both solutions.

5.1 The Nim Game

The example is about the synthesis of a winning strategy for the nim game. Given the matches arranged as in Figure 5, each player removes at least one match from one of the rows in each turn. The goal is to force the opponent to take the last match. Events reflecting the moves of the wanted winner are controllable, while those of the opponent are not. The plant encompasses all possible states of the game, while the specification is derived from it by eliminating the state in which the wanted winner would lose. This results in an uncontrollable specification, whose supremal controllable sublanguage is either empty (when there is no winning strategy for the chosen player) or the wanted winning strategy. There is always a winning strategy for one of the players. The reader is referred to (Ziller, 2001; Ziller, 2002) for more details about modelling and solving the example.

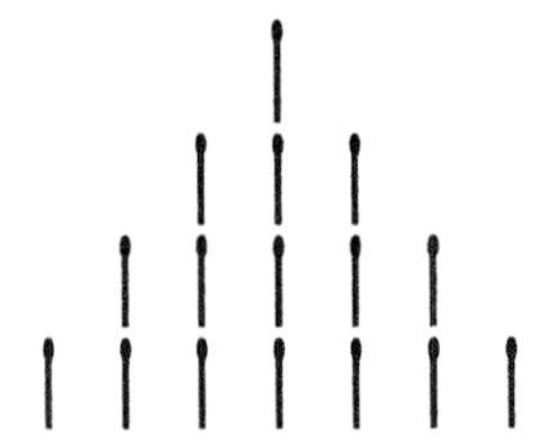

Figure 5. The nim game with 4 rows

Because removed matches do not return to the game, the product of plant and specification is loop-free and hence HLC-free. This means that equations 15 to 17 can be solved in one pass, while the classical solution will require several passes through equations 8 and 9 (see Table 1). Factoring out the effects of the BDD package, this means that this example can be solved in linear time with our new solution, while the existing ones require quadratic time.

We solved the example from 4 to 12 rows. These range from 5920 to 2.82527×10^{14} transitions, thereby reaching the size of real-world problems. The new solution has a clear advantage with respect to time and memory usage; the last case could not even be solved by the classical algorithm within available memory. Time is CPU user time, in seconds, needed to solve equations 8 and 9 (classical) and equations 15 to 17 (new solution). Memory usage is given in the peak number of BDD nodes (1 node = 16 bytes). The programs were written in C++ using CUDD[3] 2.3.1, compiled with GNU g++ 3.2 and run under SUSE Linux 8.1 on a PC with 1 GB RAM and an Athlon XP 2200+ processor.

Table 1. Experimental results

	Classical solution			New solution	
Rows	Time	Node peak	Passes	Time	Node peak
4	0.01	18396	7	0.01	17374
5	0.07	61320	11	0.04	48034
6	0.45	231994	17	0.10	114464
7	3.95	1234576	24	0.40	266742
8	61.8	2498790	31	1.46	591738
9	1618	2587704	39	5.35	1211070
10	53106	6441666	49	49.2	2484482
11	1.86×10^6	32898180	58	1943	2474262
12				180262	3594374

[3] A BDD package by Fabio Somenzi, University of Colorado at Boulder

6. CONCLUSION

The paper presents a new approach to the RW synthesis problem using μ-calculus expressions and refers to results from formal verification to assess its computational complexity. Before this work, only the special class of problems where all states are marked could be solved in linear time. These problems are seen to be strictly contained in the class of HLC-free problems. Our new approach can solve all HLC-free problems in linear time, thereby including many practical examples that the existing algorithms can solve only with quadratic effort. Moreover, an implementation of the new solution based on binary decision diagrams can handle industrial-size problems faster and with far less memory than the corresponding implementation of existing algorithms. Finally, the representation through μ-calculus formulas allows the problem to be solved by verification tools not necessarily conceived with supervisor synthesis in mind.

REFERENCES

Aziz, A., F. Balarin, R.K. Brayton, M.D. Dibenedetto, A. Saldanha and A.L. Sangiovanni-Vincentelli (1995). Supervisory control of finite state machines. In: *Conference on Computer Aided Verification (CAV)* (P. Wolper, Ed.). Vol. 939 of *LNCS*. Springer Verlag. Liege, Belgium. pp. 279–292.

Bryant, R.E. (1986). Graph-based algorithms for boolean function manipulation. *IEEE Trans. on Computers* **C-35**(8), 677–691.

Cassandras, C. G. and S. Lafortune (1999). *Introduction to Discrete Event Systems*. Kluwer Academic Publishers. Boston, U.S.A. ISBN 0-7923-8609-4.

Clarke, Jr, E. M., O. Grumberg and D. A. Peled (1999). *Model Checking*. The MIT Press. London, U.K. ISBN 0-262-03270-8.

Cleaveland, R., M. Klein and B. Steffen (1992). Faster Model Checking for the Modal μ-Calculus. In: *Computer Aided Verification (CAV'92)* (G.v. Bochmann and D.K. Probst, Eds.). Vol. 663 of *LNCS*. Springer-Verlag. Heidelberg, Germany. pp. 410–422.

Emerson, E.A. and C.-L. Lei (1986). Efficient model checking in fragments of the propositional mu-calculus. In: *IEEE Symposium on Logic in Computer Science (LICS)*. IEEE Computer Society Press. Washington, D.C.. pp. 267–278.

Kumar, R. and V. Garg (1995). *Modeling and Control of Logical Discrete Event Systems*. Kluwer Academic Publishers. ISBN 0-7923-9538-7.

Kumar, R., S. Nelvagal and S. I. Marcus (1997). A discrete event systems approach for protocol conversion. *Discrete Event Dynamical Systems* **7**(3), 295–315.

Lassez, J.-L., V.L. Nguyen and E.A. Sonenberg (1982). Fixed point theorems and semantics: a folk tale. *Information Processing Letters* **14**(3), 112–116.

Niwinski, D. (1986). On fixed point clones. In: *International Colloquium on Automata, Languages and Programming (ICALP)*. L. Kott, Ed., vol 226 of *LNCS*, Springer-Verlag. pp. 464–473.

Overkamp, A. (1997). Supervisory control using failure semantics and partial specifications. *IEEE Transactions on Automatic Control* **42**(4), 498–510.

Ramadge, P. J. and W. M. Wonham (1987). Supervisory control of a class of discrete event processes. *SIAM J. of Control and Optimization* **25**(1), 206–230.

Ramadge, P. J. and W. M. Wonham (1989). The control of discrete event systems. *Proceedings of the IEEE* **77**(1), 81–98.

Schneider, K. (2003). *Verification of Reactive Systems – Algorithms and Formal Methods*. EATCS Texts. Springer.

Tarski, A. (1955). A lattice-theoretical fixpoint theorem and its applications. *Pacific J. Math.* **5**(2), 285–309.

Thistle, J. G. and W. M. Wonham (1994). Control of infinite behavior of finite automata. *SIAM J. of Control and Optimization* **32**(4), 1075–1097.

Wonham, W. M. (2002). Notes on control of discrete-event systems. Technical report. Dept. of Electrical and Computer Engineering, University of Toronto. http://www.control.utoronto.ca/DES.

Wonham, W. M. and P. Ramadge (1987). On the supremal controllable sulanguange of a given language. *SIAM J. Control and Optimization* **25**(3), 637–659.

Yevtushenko, N., T. Villa, R. K. Brayton and A. Petrenkoand A. L. Sangiovanni-Vincentelli (2001). Solution of parallel language equations for logic synthesis. In: *Proceedings of the International Conference on Computer-Aided Design*. pp. 103–110.

Ziller, R. and K. Schneider (2003). A μ-Calculus Approach to Supervisor Synthesis. In: *GI/ITG/GMM–Workshop Methoden und Beschreibungssprachen zur Modellierung und Verifikation von Schaltungen und Systemen*. pp. 132–143.

Ziller, R. M. (2001). System Modelling Using Marker States in the RW–Framework. In: *GI/ITG/GMM–Workshop Methoden und Beschreibungssprachen zur Modellierung und Verifikation von Schaltungen und Systemen*. MoPress. pp. 121–130.

Ziller, R. M. (2002). Finding Bad States during Symbolic Supervisor Synthesis. In: *GI/ITG/GMM–Workshop Methoden und Beschreibungssprachen zur Modellierung und Verifikation von Schaltungen und Systemen*. pp. 209–218.

Ziller, R. M. and José E. R. Cury (1994). On the Supremal L-controllable Sublanguage of a non Prefix-Closed Language. *Anais do 10. Congresso Brasileiro de Automática e 6. Congresso Latino-Americano de Controle Automático* **2**, 260–265.

www.elsevier.com/locate/ifac

COALGEBRA AND COINDUCTION IN DECENTRALIZED SUPERVISORY CONTROL

Jan Komenda [1]

*CWI, P.O.Box 94079, 1090 GB Amsterdam,
The Netherlands and
Mathematical Institute, AS CR, Zizkova 22, CZ - 616 62
Brno, Czech Republic*

E-mail: Jan.Komenda@cwi.nl

Abstract: Coalgebraic methods provide new results and insights for the supervisory control of discrete-event systems (DES). In this paper a coalgebraic framework for the decentralized control of DES is proposed. Coobservability, decomposability, and strong decomposability are described by corresponding relations and compared to each other. *Copyright © 2003 IFAC*

Keywords: Decentralized supervisory control, discrete-event systems, coobservability, coalgebra.

1. INTRODUCTION

Discrete-event (dynamical) systems (DES) can be studied using coalgebraic techniques. DES are often represented by automata viewed as a particular algebraic structure. However, they may be also viewed as partial automata, which are coalgebras of a simple functor of the category of sets. Coalgebras are categorial duals of algebras (the corresponding functor operates from a given set rather than to a given set).

This paper presents a formulation of the decentralized control of DES in terms of coalgebra. The basic formalism is the one that has been developed in Rutten (1999) and by the first author in Komenda (2002), i.e. partial automata as models for DES and partial automaton of (partial) languages as the final coalgebra. The main advantage of the use of coalgebra is the naturally algorithmic character of the results, there is a canonical way how to check the properties like decomposability or coobservability by constructing corresponding relations. As an application, the exact relationship between C&P coobservability and decomposability is derived.

The paper is organized as follows. Section 2 recalls the partial automata from Rutten (1999) as the coalgebraic framework for DES represented by automata. In Section 3 after introducing local weak transition structures on partial automata we present an introduction to the decentralized supervisory control. We give an alternative definition of coobservability, which is equivalent to C&P coobservability of and coincides with the original definition of coobservability in the case of two local supervisors. In section 4 relational characterizations of coobservability are given. Section 5 is devoted to the coalgebraic study of decomposability and strong decomposability, that are described by the corresponding relations and compared to coobservability. We show in particular the exact relationship between C&P coobservability and decomposability.

[1] Partially supported by the Grant 201/03/D77 of GACR

2. PARTIAL AUTOMATA

In this section we recall partial automata as coalgebras with a special emphasis on the final coalgebra of partial automata, i.e. partial automaton of partial languages. Let A be an arbitrary set (usually finite and referred to as the set of inputs or events). The empty string will be denoted by ε. Denote by $1 = \{\emptyset\}$ the one element set and by $2 = \{0,1\}$ the set of Booleans. A partial automaton is a pair $S = (S, \langle o,t\rangle)$, where S is a set of states, and a pair of functions $\langle o,t\rangle : S \to 2 \times (1+S)^A$, consists of an output function $o : S \to 2$ and a transition function $S \to (1+S)^A$. The output function o indicates whether a state $s \in S$ is accepting (or terminating) : $o(s) = 1$, denoted also by $s \downarrow$, or not: $o(s) = 0$, denoted by $s \uparrow$. The transition function t associates to each state s in S a function $t(s) : A \to (1+S)$. The set $1+S$ is the disjoint union of S and 1. The meaning of the state transition function is that $t(s)(a) = \emptyset$ iff $t(s)(a)$ is undefined, which means that there is no $a-$transition from the state $s \in S$. $t(s)(a) \in S$ means that $a-$transition from s is possible and we define in this case $t(s)(a) = s_a$, which is denoted mostly by $s \xrightarrow{a} s_a$. This notation can be extended by induction to arbitrary strings in A^*. Assuming that $s \xrightarrow{w} s_w$ has been defined, define $s \xrightarrow{wa}$ iff $t(s_w)(a) \in S$, in which case $s_{wa} = t(s_w)(a)$ and $s \xrightarrow{wa} s_{wa}$.

2.1 *Final automaton of partial languages*

Below we define partial automaton of partial languages over an alphabet (input set) A, denoted by $\mathcal{L} = (\mathcal{L}, \langle o_{\mathcal{L}}, t_{\mathcal{L}}\rangle)$. More formally, $\mathcal{L} = \{\Phi : A^* \to (1+2) \mid dom(\Phi) \neq \emptyset \text{ is prefix-closed}\}$. To each partial language Φ a pair $\langle V, W\rangle$ can be assigned: $W = dom(\Phi)$ and $V = \{w \in A^* \mid \Phi(w) = 1 (\in 2)\}$. Conversely, to a pair $\langle V, W\rangle \in \mathcal{L}$, a function Φ can be assigned : $\Phi(w) = 1$ if $w \in V$, $\Phi(w) = 0$ if $w \in W$ and $w \notin V$ and $\Phi(w)$ is undefined if $w \notin W$. Therefore we can write :

$$\mathcal{L} = \{(V,W) \mid V \subseteq W \subseteq A^*,\ W \neq \emptyset, \text{ and } \bar{W} = W\}.$$

Transition function $t_{\mathcal{L}} : \mathcal{L} \to (1+\mathcal{L})^A$ is defined using the input derivatives. Recall that for any partial language $L = (L^1, L^2) \in \mathcal{L}$, $L_a = (L_a^1, L_a^2)$, where $L_a^i = \{w \in A^* \mid aw \in L^i\}$, $i = 1,2$. If $a \notin L^2$ then L_a is undefined. Given any $L = (L^1, L^2) \in \mathcal{L}$, the partial automaton structure of $\mathcal{L}$ is given by:

$$o_{\mathcal{L}}(L) = \begin{cases} 1 & \text{if } \varepsilon \in L^1 \\ 0 & \text{if } \varepsilon \notin L^1 \end{cases}$$

and

$$t_{\mathcal{L}}(L)(a) = \begin{cases} L_a & \text{if } L_a \text{ is defined} \\ \emptyset & \text{otherwise} \end{cases}.$$

Notice that if L_a is defined, then $L_a^1 \subseteq L_a^2$, $L_a^2 \neq \emptyset$, and L_a^2 is prefix-closed. The following notational conventions will be used: $L \downarrow$ iff $\varepsilon \in L^1$ and $L \xrightarrow{w} L_w$ iff L_w is defined iff $w \in L^2$.

Recall from that $\mathcal{L} = (\mathcal{L}, \langle o_{\mathcal{L}}, t_{\mathcal{L}}\rangle)$ is final among all partial automata: for any partial automaton $S = (S, \langle o,t\rangle)$ there exists a unique homomorphism $l : S \to \mathcal{L}$. Recall that the unique homomorphism l given by finality of $\mathcal{L}$ maps a state $s \in S$ to the partial language $l(s) = (L_s^1, L_s^2) = (\{w \in A^* \mid s \xrightarrow{w} \text{ and } s_w \downarrow\}, \{w \in A^* \mid s \xrightarrow{w}\})$.

Denote the minimal representation of a partial language L by $\langle L\rangle$, i.e. $\langle L\rangle = (DL, \langle o_{\langle L\rangle}, t_{\langle L\rangle}\rangle)$ is a subautomaton of $\mathcal{L}$ generated by L. This means that $o_{\langle L\rangle}$ and $t_{\langle L\rangle}$ are uniquely determined by the corresponding structure of $\mathcal{L}$. The carrier set of this minimal representation of L is denoted by DL, where $DL = \{L_u \mid u \in L^2\}$. Let us call this set the set of derivatives of L. Inclusion of partial languages that corresponds to a simulation relation is always meant componentwise.

3. INTRODUCTION TO DECENTRALIZED SUPERVISORY CONTROL

Decentralized supervisory control is mostly applied because the system has two or more local controllers each receiving different partial observations of the system. Since communication of the local observations is either not possible or possible but costly, the partial observations of the local controllers differ. Decentralized supervisory control consists in considering local controllers $S_1, \ldots, S_n$ and breaking the set of controllable and observable events into locally controllable and locally observable events, denoted by $A_{c,i}$, and $A_{o,i}, i = 1, \ldots, n$ respectively. The natural projections to locally observable events are denoted by $P_i : A^* \to A_{o,i}^*$. The action of P_i is simply to delete events that are not observable by S_i.

The following notation will be used: $\{1, \ldots, n\} = Z_n$, for any $a \in A$: $Z_c^a = \{i \in Z_n : a \in A_{c,i}\}$ and similarly $Z_o^a = \{i \in Z_n : a \in A_{o,i}\}$. Furthermore, we denote $A_c = \cup_{i \in Z_n} A_{c,i}$, $A_o = \cup_{i \in Z_n} A_{o,i}$, $A_{uc} = \cap_{i \in Z_n}(A \setminus A_{c,i})$, and finally $A_{uo} = \cap_{i \in Z_n}(A \setminus A_{o,i})$.

In the following definition we introduce the notion of of weak derivative (transition). Roughly speaking it disregards locally unobservable events.

Definition 3.1. For an event $a \in A$ define $L \overset{P_i(a)}{\Rightarrow}$ if $\exists s \in A^* : P_i(s) = P_i(a)$ and $L \xrightarrow{s} L_s$. Denote in this case $L \overset{P_i(a)}{\Rightarrow} L_s$.

Remark 1. Let us introduce the notation for locally unobservable events $L \stackrel{\varepsilon}{\Rightarrow}_i$ as an abbreviation for $\exists \tau \in A^*$ such that $P_i(\tau) = \varepsilon$ and $L \xrightarrow{\tau}$. We admit $\tau = \varepsilon$, hence $L \stackrel{\varepsilon}{\Rightarrow}_i$ is always true. For $a \in A_{o,i}$ our notation means that there exist $\tau, \tau' \in (A \setminus A_{o,i})^*$ such that $L \xrightarrow{\tau a \tau'} L_{\tau a \tau'}$. This definition can be extended to strings (words in A^*) in the following way:
$L \stackrel{P_i(s)}{\Rightarrow}$ iff $\exists t \in A^* \;:\; P_i(s) = P_i(t)$ and $L \xrightarrow{t}$.
Denote in this case $L \stackrel{P_i(s)}{\Rightarrow} L_t$.

3.1 Coobservability

There are two control architectures for the decentralized supervisory control . The original control architecture is called C&P (conjuctive and permissive.) The local supervisor S_i is then represented as a mapping $\gamma_{C\&P}(S_i, .) : P_i(L(G)) \to \Gamma_i$, where $\Gamma_i = \{C \subseteq A : C \supseteq (A \setminus A_{c,i})\}$ is the set of local control patterns and $\gamma_{C\&P}(S_i, s)$ represents the set of locally enabled events after S_i has observed string $s \in A^*_{o,i}$. The associated control law of the local supervisor S_i is

$$\gamma_{C\&P}(S_i, s) = (A \setminus A_{c,i}) \cup$$
$$\{a \in A_{c,i} : \exists s' \in K^2 \cap P_i^{-1} P_i(s) \text{ and } s'a \in K^2 \}.$$

The control law of the conjunction of local supervisors S_i, $i = 1, \ldots, n$ is given by :

$$(\bigwedge_i \gamma_{C\&P} S_i)(w) = \cap_{i=1}^n \gamma_{C\&P}(S_i, P_i(w)),\; w \in A^*.$$

The necessary and sufficient conditions for a given language to be achieved by a joint action of local supervisors are controllability, $L_m(G)$−closedness, and coobservability. The definition of coobservability from Rudie and Willems (1992) can be extended to n supervisors.

Definition 3.2. (Coobservability.) $K \subseteq L$ is called co-observable with respect to L and $A_{o,i}, i = 1, \ldots, n$ if $(\forall s \in K^2)$, $(\forall a \in A_c$: $sa \in L^2)$ $(\exists i \in Z_c^a)$ such that the following implication holds true:

$$(s' \in K^2 \text{ and } P_i(s) = P_i(s') \text{ and } s'a \in K^2) \Rightarrow$$
$$sa \in K^2.$$

Note that the definition of coobservability has been originally formulated for two local supervisors in Rudie and Wonham (1992). This notion of coobservability is needed for the existence of local supervisors that jointly achieve a given language.

The control policy of local supervisors associated to C&P architecture is called permissive, since the default action is to enable an event whenever a local supervisor has an ambiguity what to do with this event. It should be clear that with the permissive local policy we always achieve all strings in the specification language K, i.e. K is always contained in the language of the closed-loop system. The only concern is the safety, which is expressed by the following definition of C&P coobservability, which states that there always exists a local supervisor that is sure to disable an event resulting in an illegal string. Thus, the following definition of C&P coobservability - Yoo and Lafortune (2002) is much more intuitive then the original definition of coobservability Rudie and Wonham (1992).

Definition 3.3. (*C&P* coobservability.) $K \subseteq L$ is said to be *C&P* co-observable with respect to L and $A_{o,i}, i = 1, \ldots, n$ if for all $s \in K^2$, $a \in A_c$ such that $sa \in L^2 \setminus K^2$

$$\exists i \in Z_c^a \text{ with } (P_i^{-1}(P_i(s))a \cap K^2 = \emptyset.$$

It has been shown in Barrett and Lafortune (2000) that C&P-coobservability coincides for two supervisors with the 'classical' definition of coobservability introduced in Rudie and Wonham (1992). We will show that the definition of coobservability above (definition 3.2) is equivalent to C&P-coobservability and can thus be considered as an extension of the definition given in Rudie and Wonham (1992) to an arbitrary number of local supervisors.

Lemma 2. Coobservability is equivalent to *C&P* coobservability.

4. DECENTRALIZED SUPERVISORY CONTROL AND COALGEBRA

We have presented in Komenda (2002) a coalgebraic approach to the supervisory control of discrete-event systems with partial observations. It is possible to formulate the basic concepts of decentralized supervisory control using coalgebra. First observe that the concept of observational indistinguishability relation can be easily extended to the family of observation indistinguishability relations associated to local observers. For partial automaton S with initial state s_o we define:

Definition 4.1. (Observational indistinguishability relation on S.) A binary relation $Aux_i(S)$, $i \in Z_n$ on S, called an *observational indistinguishability relation*, is the smallest relation such that:

(i) $\langle s_0, s_0 \rangle \in Aux_i(S)$
(ii) If $\langle s, t \rangle \in Aux_i(S)$ then : ($s \stackrel{\varepsilon}{\Rightarrow}_i s'$ for some $s' \in S$ and $t \stackrel{\varepsilon}{\Rightarrow}_i t'$ for some $t' \in S$) $\Rightarrow$ $\langle s', t' \rangle \in Aux_i(S)$

(iii) If $\langle s,t\rangle \in Aux(S)$ then $\forall a \in A_{o,i}$: $(s \xrightarrow{a} s_a$ and $t \xrightarrow{a} t_a) \Rightarrow \langle s_a, t_a\rangle\rangle \in Aux_i(S)$.

Since we need to work with the final automaton of partial languages, and for $K \subseteq L$, it is not in general true then $\langle K\rangle$ is a subautomaton of $\langle L\rangle$, we need also the following concept.

Definition 4.2. A binary relation $Aux_i(K,L) \subseteq (DK \times DL)^2$, $i \in Z_n$, called an *observational indistinguishability relation*, is the smallest relation such that:

(i) $\langle (K,L),(K,L)\rangle \in Aux_i(K,L)$

(ii) If $\langle (M,N),(Q,R)\rangle \in Aux_i(K,L)$ then : $((M,N) \xrightarrow{\varepsilon}_i (M',N')$ for some $(M',N') \in DK \times DL$ and $(Q,R) \xrightarrow{\varepsilon}_i (Q',R')$ for some $(Q',R') \in DK \times DL) \Rightarrow \langle (M',N'),(Q',R')\rangle \in Aux_i(K,L)$

(iii) If $\langle (M,N),(Q,R)\rangle \in Aux_i(K,L)$ then $\forall a \in A_{o,i}$: $((M,N) \xrightarrow{a} (M_a,N_a)$ and $(Q,R) \xrightarrow{a} (Q_a,R_a)) \Rightarrow \langle (M_a,N_a),(Q_a,R_a)\rangle \in Aux_i(K,L)$.

For $\langle (M,N),(Q,R)\rangle \in DK \times DL$ we write $(M,N) \approx^{K,L}_{Aux(i)} (Q,R)$ whenever $\langle (M,N),(Q,R)\rangle \in Aux_i(K,L)$. Similarly as for the centralized $Aux(K)$ we have:

Lemma 3. For given partial languages K,L: $\langle (M,N),(Q,R)\rangle \in Aux_i(K,L)$ iff there exist two strings $s,s' \in K^2$ such that $P_i(s) = P_i(s')$ and $M = K_s$, $N = L_s$, $Q = K_{s'}$, and $R = L_{s'}$.

Now the C&P coobservability can be formulated within the coalgebraic framework of partial automata.

Definition 4.3. (*C&P* Coobservability relation.) Given two (partial) languages K and L, a binary relation $CO(K,L) \subseteq DK \times DL$ is called a *C&P coobservability relation* if for any $\langle M,N\rangle \in CO(K,L)$ the following items hold:

(i) $\forall a \in A$: $M \xrightarrow{a} \Rightarrow \quad N \xrightarrow{a}$ and $\langle M_a,N_a\rangle \in CO(K,L)$

(ii) $\forall a \in A_c$: $N \xrightarrow{a} \Rightarrow \big[(\exists i \in Z^a_c)$ such that $(M' \in DK, N' \in DL$: $(M',N') \approx^{K,L}_{Aux(i)} (M,N)$ and $M' \xrightarrow{a}) \Rightarrow M \xrightarrow{a}\big]$.

For $M \in DK$ and $N \in DL$ we write $M \approx_{CO(K,L)} N$ whenever there exists a C&P coobservability relation $CO(K,L)$ on $DK \times DL$ such that $\langle M,N\rangle \in CO(K,L)$. In order to check whether for a given pair of (partial) languages (K and L), K is C&P coobservable with respect to L and $A_{o,i}, i = 1,\dots,n$, it is sufficient to establish a C&P coobservability relation $O(K,L)$ on $DK \times DL$ such that $\langle K,L\rangle \in O(K,L)$. Indeed, we have:

Theorem 4. A (partial) language K is *C&P* coobservable with respect to L (where $K \subseteq L$) and $A_{o,i}, i = 1,\dots,n$ iff $K \approx_{CO(K,L)} L$.

PROOF. ($\Rightarrow$) Let K be C&P coobservable with respect to L. Denote

$$CO(K,L) = \{\langle K_u, L_u\rangle \in DK \times DL \mid u \in K^2\ \}.$$

Let us show that $CO(K,L)$ is a C&P coobservability relation.
Let $\langle M,N\rangle \in CO(K,L)$. We can assume that $M = K_s$ and $N = L_s$ for $s \in K^2$. We must show that conditions (i) and (ii) are safisfied.
(i) Let $M \xrightarrow{a}$ for $a \in A$. Notice that $K \subseteq L$ implies that for any $u \in K^2$, $K_u \subseteq L_u$. In particular $N \xrightarrow{a}$, because $M = K_s \subseteq L_s = N$ and it follows from the definition of $CO(K,L)$ that $\langle M_a,N_a\rangle \in CO(K,L)$.
(ii) Let $N \xrightarrow{a}$ for $a \in A_c$. Then we have $sa \in L^2$ and recall that $s \in K^2$. Then by C&P coobservability of K with respect to L there exists $i \in Z^a_c$ such that whenever there is a string $s'a \in K^2$ with $P_i(s') = P_i(s)$, then also $sa \in K^2$. Using Lemma 3 this means that there exists $i \in Z^a_c$ such that whenever $(M',N') \approx^{K,L}_{Aux(i)} (M,N)$: $M' \xrightarrow{a}$, then $M \xrightarrow{a}$. Indeed this means that there exist $s',s'' \in A^*$: $M' = K_{s'}$, $N' = L_{s'}$, $M = K_{s''} = K_s$, $N = L_{s''} = L_s$, and $P_i(s'') = P_i(s')$. Note that it can be that $s = s''$. We have $s'' \in K^2$ and $s''a \in L^2$. By applying the coobservability of K, where s'' plays the role of s, it follows that $s''a \in K$, i.e. $K_{s''} = M \xrightarrow{a}$. Hence $CO(K,L)$ is a C&P coobservability relation.

($\Leftarrow$) Let $K \approx_{CO(K,L)} L$. Let us show that K is C&P coobservable with respect to L. For this purpose, let $s \in K^2$ and $a \in A_c$ such that $sa \in L^2$. Then $s \in K^2 \cap L^2$, i.e. $L \xrightarrow{s}$ and $K \xrightarrow{s}$, whence from (i) of definition 4.3 inductively applied $K_s \approx_{CO(K,L)} L_s$. Now, $sa \in L^2$ means that $L_s \xrightarrow{a}$, hence by definition of 4.3 there exists $i \in Z^a_c$ such that whenever $(M',N') \approx^{K,L}_{Aux(i)} (K_s,L_s)$: $\quad M' \xrightarrow{a}$, then $K_s \xrightarrow{a}$. According to Lemma 3 we have for $P_i(s') = P_i(s)$ that $(K_{s'},L_{s'}) \approx^{K,L}_{Aux(i)} (K_s,L_s)$. Also notice that $s'a \in K^2$ is equivalent to $K_{s'} \xrightarrow{a}$. But this means that there exists $i \in Z^a_c$ such that whenever there is a string $s'a \in K^2$ with $P_i(s') = P_i(s)$, then also $sa \in K^2$, i.e. K is C&P coobservable with respect to L and $A_{o,i}, i = 1,\dots,n$. $\square$

5. DECOMPOSABILITY AND STRONG DECOMPOSABILITY

It is known that coobservability is not preserved under unions. Therefore the existence of supremal coobservable sublanguages cannot be guaranted. For this reason, stronger notions of decomposable and strongly decomposable languages have been studied.

Definition 5.1. (Decomposability.) Let $K \subseteq L$ be given partial languages. K is said to be decomposable with respect to L and $A_{o,i}, i = 1, \ldots, n$ if $K^2 = L^2 \cap \bigcap_{i=1}^{n} P_i^{-1}(P_i(K^2))$.

Let us introduce now a binary relation that corresponds to the decomposability.

Definition 5.2. (Decomposability relation.) Given two languages K, L with $K \subseteq L$, a binary relation $D(K, L)$ on $DK \times DL$ is said to be a *decomposability relation* if for any $\langle M, N \rangle \in D(K, L)$ the following items hold:

(i) $\forall a \in A : M \xrightarrow{a} M_a \Rightarrow N \xrightarrow{a} N_a$ and $\langle M_a, N_a \rangle \in D(K, L)$

(ii) $\forall u \in A_{uo} : N \xrightarrow{u} \ \Rightarrow \ M \xrightarrow{u}$

(iii) $\forall a \in A_o : N \xrightarrow{a}$ and $[\forall i \in Z_n : \exists M^{(i)} \in DK$ and $N^{(i)} \in DL$: $(M^{(i)}, N^{(i)}) \approx_{Aux(i)}^{K,L} (M, N)$ and$M^{(i)} \xrightarrow{a})] \Rightarrow M \xrightarrow{a}$.

Using strong and weak transitions, we obtain:

Lemma 5. K is decomposable with respect to L and $A_{o,i}, i = 1, \ldots, n$ iff $\forall w \in A^*$:
$\{ L \xrightarrow{w}$ and $(\forall i = 1, \ldots, n : K \overset{P_i(w)}{\Rightarrow})\} \ \Rightarrow \ K \xrightarrow{w}$.

Theorem 6. A (partial) language K is decomposable with respect to L and $A_{o,i}$, $i = 1, \ldots, n$ (with $K \subseteq L$) iff there exists a decomposability relation $D(K, L) \subseteq DK \times DL$ such that $\langle K, L \rangle \in D(K, L)$.

PROOF. ($\Rightarrow$) Let K be decomposable with respect to L and $A_{o,i}$, $i = 1, \ldots, n$. Denote

$$R = \{\langle K_u, L_u \rangle \mid u \in K^2 \} \subseteq DK \times DL.$$

Let us show that R is indeed a decomposability relation. Assume that $\langle M, N \rangle \in R$. We can assume that $M = K_s$ and $N = L_s$ for $s \in K^2$. We must show that conditions (i)-(iii) of decomposability relations are satisfied.
(i) If for $a \in A$ we have $M \xrightarrow{a}$ (i.e. $a \in M^2$) then $a \in N^2$, i.e. $N \xrightarrow{a}$, because $K \subseteq L$ implies that $M = K_s \subseteq L_s = N$. Moreover, it follows from the definition of R that $\langle M_a, N_a \rangle \in R$.
(ii) Let $u \in A_{uo}$ such that $N \xrightarrow{u}$. Then $L \xrightarrow{su}$, and since $K \xrightarrow{s}$ and $\forall i = 1, \ldots, n :\ \ P_i(su) = P_i(s)$, trivially $\forall i = 1, \ldots, n$ we have $K \overset{P_i(su)}{\Rightarrow}$. Therefore by Lemma 5 $K \xrightarrow{su}$, i.e. $M \xrightarrow{u}$.
(iii) If for $a \in A_o$: $N \xrightarrow{a}$ and $[\forall i \in \{1, \ldots, n\}$: $\exists M^{(i)} \in DK$ and $N^{(i)} \in DL : (M^{(i)}, N^{(i)}) \approx_{Aux(i)}^{K,L} (M, N)$ and $M^{(i)} \xrightarrow{a})]$ then $L \xrightarrow{sa}$ and $\forall i = 1, \ldots, n$ $\exists s_i, s_i' \in K^2$ such that $K_s = K_{s_i}$, $L_s = L_{s_i}$, $M^i = K_{s_i'}$, $N^i = L_{s_i'}$, where $P_i(s_i) = P_i(s_i')$ and $K_{s_i'} \xrightarrow{a}$, hence $K \overset{P_i(s_i a)}{\Rightarrow}$. Recall that $L_{s_i} = L_s \xrightarrow{sa}$. By Lemma 5 we obtain $K \xrightarrow{s_i a}$, i.e. $M = K_{s_i} \xrightarrow{a}$.

($\Leftarrow$) Let there exists a decomposability relation $D(K, L)$ such that $\langle K, L \rangle \in D(K, L)$. We show by induction on the structure of the string $w \in A^*$ that the implication of Lemma 5 holds true. For $w = \varepsilon$ it is trivially true, because K^2 is closed (i.e. $K \xrightarrow{\varepsilon}$). Suppose now that for $w \in A^*$: $L \xrightarrow{w}$ and $(K \overset{P_i(w)}{\Rightarrow} \ \forall i = 1, \ldots, n)$ imply that $K \xrightarrow{w}$.

Let $L \xrightarrow{wa}$ and $K \overset{P_i(wa)}{\Rightarrow}$ $\forall i = 1, \ldots, n$. This implies in particular that $L \xrightarrow{w}$ and $K \overset{P_i(w)}{\Rightarrow}$ $\forall i = 1, \ldots, n$, hence by the induction hypothesis $K \xrightarrow{w}$. Since $\langle K, L \rangle \in D(K, L)$, by inductive application of (i) of the definition of decomposability relation we obtain that $\langle K_w, L_w \rangle \in D(K, L)$. Now suppose first $a \in A_{uo}$ and $L_w \xrightarrow{a}$, then we obtain by (ii) of the definition of decomposability relation $K_w \xrightarrow{a}$, i.e. $K \xrightarrow{wa}$. If $a \in A_o$, then $\forall i$: $K \overset{P_i(wa)}{\Rightarrow}$ means that for all $i = 1, \ldots, n$ there exist $s_i \in A^*$: $(K_{s_i}, L_{s_i}) \approx_{Aux(i)}^{K,L} (K_w, L_w)$ and $K_{s_i} \xrightarrow{a}$. By application of (iii) of decomposability relation we have $K_w \xrightarrow{a}$, i.e. $K \xrightarrow{wa}$. □

Decomposability is related to $C\&P-$coobservability. The following theorem holds.

Theorem 7. Let $A_c \subseteq A_o$, for $i \in Z_n$: $A_{o,i} \cap A_c \subseteq A_{c,i}$, and $K \subseteq L$ be decomposable wrt L and $A_{o,i}$, $i = 1, \ldots, n$. Then K is C&P-coobservable wrt L and $A_{o,i}$, $i \in Z_n$.

PROOF. Let K be decomposable wrt L and $A_{o,i}$, $i = 1, \ldots, n$. We show that if

$$R = \{\langle K_u, L_u \rangle \mid u \in K^2\}$$

is a decomposability relation, it is also a C&P-coobservability relation. Take $M = K_s$ and $N = L_s$ for a $s \in K^2$. Then (i) of coobservability relations trivially holds. Assume by contradiction that (ii) does not hold, i.e. there exists $a \in A_c$ such that $N \xrightarrow{a}$ and $(\forall i \in Z_n \ : \ \ a \in A_{c,i})$ $\exists (M^{(i)}, N^{(i)}) \approx_{Aux(i)}^{K,L} (M, N)$ such that $M^{(i)} \xrightarrow{a}$, while $M \not\xrightarrow{a}$. Recall that $K \xrightarrow{s} M$. The last condition means that $(\forall i \in Z_n \ : \ \ a \in A_{c,i})$: $K \overset{P_i(sa)}{\Rightarrow}$, because by Lemma 3 there exist $s_i, s_i' \in K^2$ not all necessarily different such that $M = K_{s_i}$, $N = L_{s_i}$, $M^i = K_{s_i'}$, $N^i = L_{s_i'}$, with

$P_i(s^i) = P_i(s)$ and $K_{s^i} \xrightarrow{a}$. Now, if $a \notin A_{c,i}$, then $a \notin A_c \cap A_{o,i}$. Since $a \in A_c$, there must be $a \notin A_{o,i}$. This means that $K \xrightarrow{s_i} M$ and $K \overset{P_i(s_i a)}{\Rightarrow}$ (because $M^{(i)} \xrightarrow{a}$). Furthermore, $P_i(sa) = P_i(a)$ for $i \in Z_n$ $a \notin A_{o,i}$. We conclude that for $a \in A_c \subseteq A_o$ we have $L \xrightarrow{s_i a}$ and $\forall i \in Z_n$: $K \overset{P_i(s_i a)}{\Rightarrow}$, i.e. from decomposability of K we have $M = K_{s_i} \xrightarrow{a}$. This is a contradiction. Therefore, we conclude that K is $C\&P-$coobservable. □

Theorem 8. Let $A_c \subseteq A_o$, $K \subseteq L$ is controllable, and $C\&P-$coobservable wrt L and $A_{o,i}$, $i \in Z_n$. Then K is decomposable wrt L and $A_{o,i}$, $i \in Z_n$.

PROOF. Let K be C&P-coobservable wrt L and $A_{o,i}$, $i \in Z_n$. We show that

$$R = \{\langle K_u, L_u\rangle \mid u \in K^2\}$$

is a decomposability relation. Take $M = K_s$ and $N = L_s$ for $s \in K^2$. First of all, (i) of decomposability relations is trivial. Also (ii) easily follows from controllability of K with respect to L and A_{uc}. Indeed, it is sufficient to notice that $A_{uo} = A \setminus A_o \subseteq A \setminus A_c = A_{uc}$. Then (ii) follows from controllability of K. Assume by contradiction that (iii) does not hold, i.e. there exists $a \in A_o$ such that $N \xrightarrow{a}$ and $\forall i \in Z_n$ $\exists (M^{(i)}, N^{(i)}) \approx^{K,L}_{Aux(i)} (M, N)$ such that $M^{(i)} \xrightarrow{a}$, while $M \not\xrightarrow{a}$. We have also $a \in A_c$, because otherwise by controllability we would have $M \xrightarrow{a}$. This means there exists at least one $j \in Z_n$ such that $a \in A_{c,j}$. Since the above condition holds for all $i \in Z_n$, in particular it holds for all $i \in Z_n$ such that $a \in A_{c,i}$. But this is a contradiction with $C\&P-$coobservability. Therefore, we conclude that K is decomposable. □

The set of decomposable sublanguages is not closed under unions either. There is yet a stronger condition, called strong decomposability, which is preserved by arbitrary unions.

Definition 5.3. (Strong Decomposability.) Let $K \subseteq L$ be given partial languages. K is said to be strongly decomposable with respect to L and $A_{o,i}, i = 1, \ldots, n$ if $K^2 = L^2 \cap \cup_{i=1}^{n} P_i^{-1}(P_i(K^2))$.

Now we introduce a binary relation that corresponds to strong decomposability.

Definition 5.4. (Strong decomposability relation.) Given two languages K, L with $K \subseteq L$, a binary relation $SD(K, L)$ on $DK \times DL$ is said to be a *strong decomposability relation* if for any $\langle M, N\rangle \in SD(K, L)$ the following items hold:

(i) $\forall a \in A : M \xrightarrow{a} M_a \Rightarrow N \xrightarrow{a} N_a$ and $\langle M_a, N_a\rangle \in D(K, L)$

(ii) $\forall u \in \cup_{i=1}^{n}(A \setminus A_{o,i}) : N \xrightarrow{u} \Rightarrow M \xrightarrow{u}$

(iii) $\forall a \in \cap_{i=1}^{n} A_{o,i}$: $N \xrightarrow{a}$ and $[\exists i \in Z_n$ and $(\exists M' \in DK, N' \in DL :$ $(M', N') \approx^{K,L}_{Aux(i)} (M, N)$ and $M' \xrightarrow{a})\ ] \Rightarrow M \xrightarrow{a}$.

We have the following theorem:

Theorem 9. A (partial) language K is strongly decomposable with respect to L and $A_{o,i}, i = 1, \ldots, n$ (with $K \subseteq L$) iff there exists a strong decomposability relation $SD(K, L) \subseteq DK \times DL$ such that $\langle K, L\rangle \in SD(K, L)$.

6. CONCLUSION

Decentralized supervisory control of DES has been treated by coalgebraic techniques. Coobservability, decomposability, and strong decomposability have been characterized by appropriate relations in this framework. Exact relationships between properties like C&P coobservability and decomposability have been derived using these relational characterizations.

REFERENCES

G.Barrett and S. Lafortune. Decentralized supervisory control with communicating controllers. *IEEE Trans. on Automatic Control*, 45, p. 1620-1638, 2000.

S.G. Cassandras and S. Lafortune. *Introduction to Discrete Event Systems.* Kluwer Academic Publishers, Dordrecht 1999.

J. Komenda. Coalgebra and coinduction in discrete-event control. Submitted to *Siam Journal on Control and Optimization*, October 2002.

S. Lafortune and E. Chen. The infimal closed and controllable superlanguage and its applications in supervisory control. *IEEE Trans. on Automatic Control*, Vol. 35,N 4, p. 398-405, 1990.

A. Overkamp and J.H. van Schuppen. Maximal solutions in decentralized supervisory control, *SIAM Journal on Control and Optimization*, Vol. 39, No.2, pp. 492-511, 2000.

J.J.M.M. Rutten. Coalgebra, Concurrency, and Control. *Research Report CWI*, SEN3, Amsterdam, 1999.

K. Rudie and W.M. Wonham. Think Globally, Act Locally: Decentralized Supervisory Control. *IEEE Trans. on Automatic Control*, Vol. 37,N 11, p. 1692-1708, 1992.

T.S. Yoo and S. Lafortune. General Architecture for Decentralized Supervisory Contol of Discrete-Event Systems. *Discrete Event Dynamic Systems: Theory and Applications*, 12, 335-377, 2002.

CASE-BASED REASONING APPLIED FOR CAS-DECISION SYSTEM

T.-Tung Dang, B. Frankovič, I. Budinská

Institute of Informatics, Slovak Academy of Sciences
Dúbravská 9, 84507, Bratislava
Slovak Republic
Email: {utrrtung, utrrfran, utrrbudi}@savba.sk

Abstract: This paper presents the CAS-Decision System, whose purpose is to assist the customers in choosing a method for modelling and designing the control system and simulation. Case-based reasoning is used to meet this purpose. We present two methods for searching and extracting the needed information. *Copyright © 2003 IFAC*

Keywords: MAS, Case-based reasoning (CBR), Fuzzy, Decision making.

1. INTRODUCTION

Design of control system for an arbitrary system usually requires a lot of important knowledge or information characterizing the behavior of the system. For example, the user, who is responsible for designing the control system, needs to know whether the given system is continuous or discrete, centralized or decentralized, SISO or MIMO, etc. On the basis of that information, the user can choose an appropriate method for designing the control system for the given system. For each class of systems, the corresponding class of control methods exists, e.g., there are a number of control methods applicable to continuous systems, a number of control or modelling methods applicable to discrete systems, etc. In other words, many control methods are available for different classes of systems. On the other hand, the user might not know everything about each class system and control method; therefore, in order to recognize which method is the most appropriate for the given system, he/she has to contact the experts who are able to resolve this difficulty. Our goal is to build such a system, which replaces the experts and is able to help the user to select a method for modelling and designing the control system and simulation for the given system. The motivation leading to this idea is that the experts are usually unavailable for the user when he/she needs them. In addition, each expert is good at some domains, so each user needs to recognize all the experts when asking for advices. Therefore, creating a common database, which involves all experience or knowledge of the experts, with the purpose to assist the user when he/she meets a new situation, would be a really practical idea.

Due to many advantages of the multi-agent system (MAS) and with respect to the complicated characteristics of the goal given above, the MAS approach seems to be the most feasible manner. There are some significant reasons that motivate us to choose the MAS approach:

Modularity: Each agent is an autonomous module and can work without interventions of the external world. Each agent can have different capabilities or functionalities and through cooperation the agents are able to achieve a variety of goals. From the practical point of view, producing a number of agents (e.g.

software agents - programs) with different capabilities is more effective than creating one agent (e.g. a program), which is able to do everything. In addition, the MAS approach allows separating the original problem solving to a number of sub-problems of a manageable size. For example, many different agents could be used to assist a number of users at one time - there is a significant difference from the traditional centralized approach.

Parallelism: The MAS approach supports parallel processing. A complicated problem could be solved in an acceptable time by using a number of agents, e.g., gathering information from various resources allocated in different places.

Flexibility: The MAS approach is able to react flexibly to each change occurred in the environment. Through cooperation the agents can assist each other to compensate the lack of capability or knowledge. They can share information or own capacity to resolve a newly appeared situation, if one agent is not able to resolve. Beside that, each intelligent agent can do reasoning about with whom and when to cooperate, in order to achieve effective performance.

Of course, there are also several difficult questions associated with the MAS approach, e.g., which types of agents are needed for the use, how many agents are optimal, what is a functionality of each agent, cooperation between agents, etc. These questions will successively be dealt with during the development of this system. This paper focuses only on solving the problem: how the agent system can assist the user in designing the control system for an arbitrary system. This paper is organized as follows: Section 2 describes briefly the architecture of the CAS-Decision System. Section 3 discusses case-based reasoning used in the CAS-Decision System. Section 4 describes briefly agent behavior of each agent in this system. Section 5 discusses some related works.

2. THE CAS-DECISION SYSTEM

The CAS-Decision System (The **C**omputer-**A**id **S**ystem for Support **Decision** Making) is the common project of four academic institutes. This system consists of a number of agents with specific functions and capabilities. The CAS-Decision System is used to assist the user in designing the control system for a set of specific systems. In Figure 1, the architecture of CAS-Decision is shown. A brief description of each agent in this system is presented as follows.

- PA- Personal Agent, that serves as an interface between the user and the CAS-Decision System. The PA's task is to translate the information entered by the user to an appropriate form for agents' execution; and data generated by agents to an understandable form for the user.
- MA - Modelling agent, whose function is to assists the user in choosing the modelling algorithms. On the basis of the information provided by PA, a MA extracts from the database a situation most similar to the current one, loads the modelling algorithm corresponding to it, and then forwards it to the user (through PA).

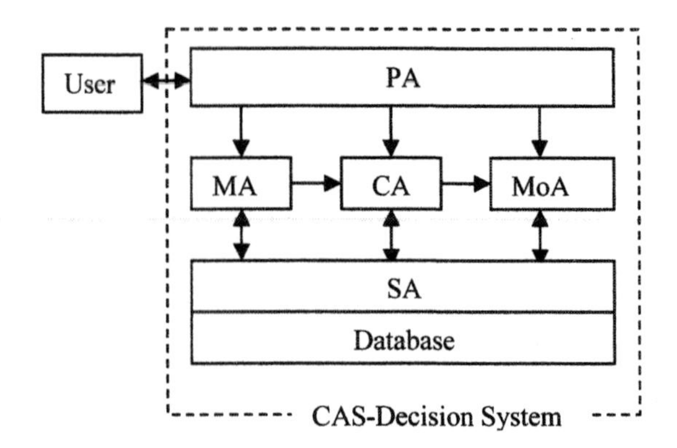

Fig. 1: CAS-Decision architecture

- CA - Controlling agent. Its task is to assist the user in choosing control algorithms, after when the model of the given system is identified (including all its parameters, if the modelling algorithm recommended by MA can do that). Depending on the characteristics of the identified model, CA searches for a similar case in the past and extracts the control method used in this case to recommend the user.
- MoA - Monitoring agent that is used to follow the behaviour of the system, after applying the control algorithm recommended by CA. The results gathered by MoA are then used to evaluate the decisions of MA and CA. The database is updated by the newly achieved results, if the given system appears for the first time or newly achieved results are better than the historical ones.
- SA – Information Search Agent. Its task is to search for specific information from the database and to update or to reorganize the database regularly.
- Database – It involves all historical situations and their corresponding solutions. The database is common and accessible from everywhere. The database structure is also a difficult problem, but to simplify the relational database is considered.

Agents interact as shown in Figure 1. All agents receive information from the user through PA and they can access to the database through SA. Figure 1 also shows the order in which each agent is invoked. MA is activated when the user enters a new situation. CA is activated when important information of the model, which is identified by using the modelling algorithm provided by MA, is known. If the achieved model is not adequate for CA calculation, MA has to repeat the calculation and choose other solution. Each agent is implemented with a capability of doing reasoning. The agents can argue about their choices, e.g. a reason of why the selected algorithm is more appropriate than others, criteria when searching for historical situation similar to the current one, etc. These arguments are useful for the users when they need to make a decision. If the user accepts the control method proposed by CA, MoA tries to simulate the behaviour of the system on the basis of the identified model and the proposed control method.

SA is invoked by MA or CA, when they need to access to some information stored in database. SA uses some search techniques to extract the desired data, e.g. using queries, similarity based search, classification tree, etc. That agent has a task to record and store user's choices and other simulation results achieved.

The next section will describe the agent reasoning in the CAS-Decision System.

3. CASE-BASED REASONING IN THE CAS-DECISION SYSTEM

The basic idea of the CAS-Decision System is to exploit the historical experiences to assist the user to resolve a newly appeared situation. Case-based reasoning (CBR) is suitable for such a purpose, and therefore the agents can apply some knowledge from the CBR domain to fulfil their aims. CBR is built on a common principle: new situations are solved by exploiting and adapting solutions, which were used in similar past situations. CBR usually consists of the following steps:

The most important aspects of historical situations are indexed and stored in a common database. New situations are described by using some basic attributes; afterwards, the similar, existing situations are identified and extracted from the knowledge base. Finally, the previous problem solutions are retrieved and the revised solutions are proposed for the current situation.

The first important requirement for the CAS-Decision System is to build a database for storing information from numerous situations and their solutions. The second requirement is to define a set of rules for classification and extraction of data; and the last one is to design a method for decision making, so that the agents can identify what is best for the user in the current situation. Due to the limited length of this paper, we shall deal with two first problems. The third problem related to decision-making algorithms will be considered in the future.

3.1. Case library for CBR

A case library is a relational database, which involves the important features of each historical situation. Because the CAS-Decision System has to work with different systems and kinds of information, the case library must be generic enough to represent all systems and their associated information. Determining which information is essential for storing is too difficult, since it is impossible to record all information and, on the other hand, because of practical reasons, the database should not be too large.

Other problems related to the database include the format used for data representation, e.g., number, text, graph, etc., and the method used for encoding/decoding these data. In the CAS-Decision System, historical cases are represented by a number of basic attributes, which are indexed in a hierarchical scheme. These attributes are expressed by number or text (the example shown in Figure 2).

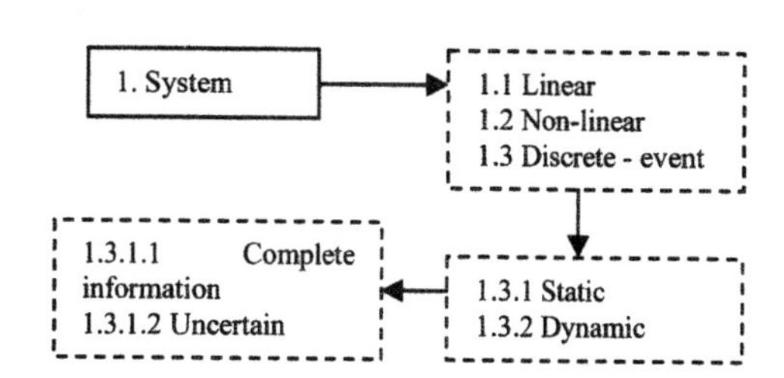

Fig. 2: An example of data representation

The main database consists of two parts. The first part stores a description of all historical situations provided by the experts or added by users during customization of the CAS-Decision System. A description of past situation consists of a number of basic attributes defined in advance. For example, situation ***si*** is a combination of a number of attributes as follows:

Let ***EL*** be a set of all possible attributes.

$$\boldsymbol{si} = \{ el_1 \cup el_2 \cup .... \cup el_n \} \; ; \; \forall i \neq j \in [1,n] \; ; \; el_i \neq el_j \in \boldsymbol{EL}.$$

In this representation, all the attributes are different each from other and they are sorted in a fixed order in order to facilitate search and retrieval.

The second part implies all solutions associated with the cases stored in the first part of the database. Solutions mean which method was used for modelling, designing the control system or simulation.

3.2. Case entry – the user's interface

The manner that the user describes the given situation should be standardized. All the users exploit the same manner for entering the required information. It allows the agents saving a lot of time of classifying and translating input data to an appropriate form. For example, first, the user should enter a type of system like in Figure 2 (*linear, discrete, etc.*), afterwards, the attributes expressing system dynamics, and so on. All these operations are performed by a PA, which provides the user with a feasible interface for this purpose.

3.3. Reasoning algorithms for extracting information

Let us assume that the user sends the CAS-Decision System a description of the current situation (through a PA and in a correct form, which is understandable for agents). The first thing that the agents have to do is to find a similar situation in the library, afterwards, to revise and adapt the solutions applied in the selected case to the current one. Due to the limited scope of this paper, extracting past solutions and their retrieval to adapt to the current case are omitted.

To find a similar case, the agents can use queries to search. Since a relational database is considered to be used, the similar case could be found in a time linear with the number of cases stored in the database. However, the dilemma is how to achieve a case that is identical with or very close to the current one described by the user. Two cases are identical if all their attributes are the same; unfortunately the existence of the identical case is rather rare. In many instances, the agents can find only a case that is "very close" to the target one. Exploiting the similarity degree and an inductive reasoning method can help the agents distinguish situations and choose one of them that is closest to the new instance.

In this paper, two methods proposed to extract information from the database are presented. The first method is based on the similarity degree between two arbitrary situations; the second one is built on the inductive reasoning principle.

The first method works on the following principle: the agents calculate the similarity between two arbitrary situations by comparing all their attributes. Relationships among attributes are evaluated by any number (real or fuzzy) that reflects how much a solution of one situation is useful for the second one, with respect to these attributes. The similarity degree is defined by a combination of those partial relationships.

The second possible method is an inductive reasoning that works as follows: starting with the most important attribute, the agents sort and filter all cases that have the same or *to a certain degree* the same attribute like the target one. This cycle continues with less important attributes, until only one candidate remains. The last case remained is considered as the most similar to the target one.

The important requirement of the both methods is the identification of relationships among attributes. Due to a wide variety of attributes, classifying precisely the dependences among them is impossible. For that reason, the use of a fuzzy classification is preferred, because of its capabilities to express a number of difficult relations among attributes that other classification methods cannot do. For example, the following relation: "an algorithm for modelling system type A *could* be applied to system type B with *the same* or *very good* quality", could be easily expressed by exploiting fuzzy sets.

Let $re(el_1, el_2)$: $\{EL \times EL\} \rightarrow [0,1]$ be a relation expressing the dependence between two attributes $(el_1, el_2) \in \boldsymbol{EL}$. There are some important properties of this relation.

- $re(el_1, el_2) \neq re(el_2, el_1)$, resp. $\neq (1 - re(el_2, el_1))$, i.e. this relation is not symmetric or inverse.
- $re(el_1, el_2) = 1$; when the solution proposed for element el_2 could be applicable to element el_1 without changes. For example, el_1 is a linear system with complete information; el_2 is a linear system with parametrical uncertainty.
- $re(el_1, el_2) = 0$; when both the elements have disjoint domains of effects, i.e. the solution proposed for el_1 is useless for element el_2. For example, el_1 is a discrete event system and el_2 is a linear system.
- $\forall\, el \in \boldsymbol{EL}$; $re(\varnothing, el) = 1$ a $re(el, \varnothing) \geq 0$

The similarity degree between two arbitrary situations $Sim(,)$ is defined as follows:

Let us consider two situations si_1 and si_2 with the same number of attributes.

$si_1=\{el_{1,1}, el_{1,2}, ..., el_{1,n}\}$ and $si_2=\{el_{2,1}, el_{2,2}, ..., el_{2,n}\}$

If the numbers of attributes of each situation are not equal, we can add an empty attribute $\varnothing$ to them. Let us define matrix $W \in [0,1]^{nxn}$ as follows:

$$w_{i,j} = re(el_{1,i}, el_{2,j}) \quad (1)$$

Then,

$$Sim(si_1, si_2) = W^T Q W \quad (2)$$

where $Q=\{q_{i,j}\}^{nxn}$ is a symmetric and positively definite matrix of the same dimension (n x n) as W. Each member $q_{i,j}$ of matrix Q expresses a weight of how much a relation $re(el_{1,i}, el_{2,j})$ can influence the dependence between both situations. Given a target case $si_{current}$ the case that maximizes a function $Sim(si, si_{current})$ is considered as the most similar one, and its solution could be applied or adapted to solve the current situation.

Inductive reasoning method extracts the desired situation by performing a search of a decision tree, which involves all possible historical cases satisfying certain conditions. A decision tree is generated as follows: starting with the most important attribute – propose it is el_1, all cases that satisfy the following condition are added to the decision tree.

$$re(el_1 |_{(si)}, el_1 |_{(si_{current})}) \geq \alpha \quad (3)$$

where coefficient α is a low bound that is used to restrict a set of candidate situations. The classification process continues with less important attributes, until only one candidate solution remains.

After finding a similar situation to the current one, the agents extract the solution that was used in this case (a method for modelling, designing the control system and simulation) and retrieve it appropriately to a solution for the current situation. The user accepts the recommendation and applies it into a practice. The last step that the CAS-Decision System has to do is to observe the real behaviour of the system after the user has applied all the methods proposed by the agents. The results achieved by an observation are intended to evaluate the selected methods and for later use.

4. AGENT BEHAVIOR IN THE CAS-DECISION SYSTEM

This section describes briefly an agent's behaviour by using a CBR algorithm.

An algorithm for assisting the user in modelling, designing the control system and simulation:

Initialization: loading a library of attribute relations, database 1 – descriptions of default cases, database 2 – a set of default solutions.

Input: a description of the current situation and other requirements are entered by the user through PA.

1. MA asks SA to find a similar situation from database 1. Requirements: search and data mining algorithms.
2. After finding solution from database 2 – a method for modelling, MA retrieves the historical solution, adapts it to the current case, then proposes a solution (model with all necessary parameters) to the user, simultaneously sends it to CA.
3. CA receives a (mathematical) model identified by MA and numerical parameters entered by the user. CA works in the same manner like MA, it asks SA to find a similar model stored in database. On the basis of numerical parameters of the model, CA adapts the past solution to the current model and designs the control system.
4. MoA follows the system behavior after applying the designed control system. If all requirements are satisfied, MoA updates the database by newly achieved results → STOP. Otherwise, it requires the MA and CA to repeat their calculation.

One agent can serve many users and can access information in database only through SA. Such approach is used to synchronize an access the database to avoid conflict situations (e.g., one agent performs database updating and at the same time other one extracts information from it). Similarly, all communication between the user and the CAS-Decision System is performed by PA, which provides an appropriate interface to simplify the user work.

5. RELATED WORK

Case-based reasoning has a wide domain of applications. There are some known applications that exploit CBR. Pellucid [Pellucid – IST project, 2002], [Dang *et al.*, 2003] is a similar system to the CAS-Decision System. Its purpose is to exploit historical experience and knowledge of former experts to assist new employees to adapt a new environment. In this system, case retrieval is based on a fuzzy classification as well, but calculation of the similarity degree requires a comparison of each attribute of the first situation with all attributes of the second one. It is different from the method that has been presented here. The next system uses CSB for the same purpose is GMCR (Graph Model for Conflict Resolution methodology) presented by [Ross *et al.*, 2002]. CBR is used to assist in identifying conflict situations and proposing their solutions, but case retrieval is built in a graph model.

Some other works that deal with the same problem as this paper could also be mentioned, e.g., [McSherry, 2002] dealt with the problem of recommendation engineering that requires CBR too. The similarity degree is calculated by a ratio between the numbers of the same attributes and the different ones. [Bhanu and Dong, 2002] presented a fuzzy clustering method, which exploits a K-nearest neighbour search algorithm to extract a similar situation. In this paper the feedback is used to improve extraction accuracy. Further papers which apply fuzzy sets to CBR, should be mentioned, namely, [Dubois *et al.*, 1998], [Rudas, 1999], however, their work focuses more on introducing new operators to express more complicated relationships among situations.

CBR is also used in some other techniques, such as, data mining [Adriaans and Zantinge, 1997], [Giudici *et al.*, 2001], in which CBR is used to find some trends of a large set of data. A similarity is usually evaluated by the distance between two nodes in *n*-dimensional space. A number of CBR systems and their applications can be also found in [Bergmann, 1999].

6. CONCLUSION AND FUTURE WORK

In this paper we have presented a basic architecture and described the behavior of each agent in the CAS-Decision System. CBR is used for recommendation; and two methods – similarity degree and inductive reasoning – have been introduced for searching and extracting the required information from the database. Both methods are easy to implement and they do not require complicated calculation.

The system proposed is currently under development in the environment provided by [Tran *et al.*, 1999, 2000]. A number of modelling, control methods and their required attributes have been collected to create a common database. Each method might require different input data, therefore, the user interface, which serves the user to input data, is designed to be flexible and adaptable to each selected method.

The CAS-Decision System relates to a lot of problems; some of them have already been mentioned in this paper. In the future, the main interest is to focus on cooperation among agents, because there are many different kinds of agents situated in that system, developed by different teams, with different architecture and knowledge. The behavior of each agent also requires detailed investigation, e.g. how agents revise and adapt past solutions to the current situation, how MA can identify a mathematical model on the basis of data entered by the user, which tools are needed for MoA for simulation, etc. This will be the target for future work.

ACKNOWLEDGMENT

This paper is partially supported by APVT and VEGA grant agencies under grants No APVT 51 011602 and VEGA 2/1101/21.

REFERENCES

Adriaans P. and Zantinge D.: Data Mining (1997). *Published by Addison-Westley*, ISBN 0-201-40380-3.

Bhanu B and Dong A. (2002): Concepts learning with fuzzy clustering and relevant feedback. *Engineering Application of AI*, **No. 15**, 123-138.

Bergmann R. (1999): Special issue on case-based reasoning. *Engineering Application of AI*, **No. 12**, 661-759.

Dang T.-Tung, Hluchý L., Budinská I., Nguyen T. G., Laclavík M. Balogh Z. (2003): Knowledge management and data classification in Pellucid. *In Intelligent Information Processing and Web Mining*, Advances in Soft Computing, Springer-Verlag, 563-568.

Dubois D., Esteva F., Garcia P., Godo L., M`antaras R. L., and Prade H. (1998): Fuzzy set modelling in case-based reasoning, *International Journal of Intelligent Systems*, **No. 13**, 345–373.

Giudici P., Heckerman D. and Whittaker J. (2001): Statistical Models for Data Mining. In *Data Mining and Knowledge Discovery*, **Vol. 5 No. 3,** 163-165.

McSherry D. (2002): Recommendation Engineering. *ECAI-02*, 86-90.

Pellucid: http://pellucid.ui.sav.sk/

Ross S., Fang L. and Hipel K. (2002): A case-based reasoning system for conflict resolution: design and implementation. *Engineering Application of AI*, **No. 15**, 369-383.

Rudas J. (1999): Evolutionary operators; new parametric type operator families. *Int. Journal of Fuzzy Systems,* **Vol. 23, No. 2**, 147-166.

Tran D. V., Hluchy L. and Nguyen T. G. (2000): Parallel Program Model for Distributed Systems. In *Proceeding of EURO PVM/MPI – 2000, Lecture Notes in Computer Science,* **Vol. 1908**, 250 – 257.

Tran D. V., Hluchy L. and Nguyen T. G. (1999): Parallel Program Model and Environment. *In Parallel Computing PARCO -99,* 697-704.

www.elsevier.com/locate/ifac

SUPERVISOR EXISTENCE FOR MODULAR DISCRETE-EVENT SYSTEMS

Kurt R. Rohloff and **Stéphane Lafortune** [1]

Department of EECS, University of Michigan
1301 Beal Avenue, Ann Arbor, MI 48109-2122, USA
{krohloff,stephane}@eecs.umich.edu
www.eecs.umich.edu/umdes

Abstract: This paper examines decision problems involving existence conditions for supervisors of modular systems. We investigate the problem of deciding if there exists a supervisor that can augment a modular system to satisfy a given modular global specification. Our system and specification modules are "discrete-event systems" modeled as finite-state automata. The parallel composition operation is used to model the interaction between the various modules. A supervisor for a discrete-event system observes events as they occur in the system and enforces a control action by disabling some controllable events. We find that these types of problems are generally PSPACE-complete for a large class of systems and specifications. *Copyright © 2003 IFAC*

Keywords: Modular and Concurrent Systems, Regular Languages, Specification and Verification, Computational Complexity, Supervision and Control, Discrete Event Systems

1. INTRODUCTION

There has been considerable interest lately in both the discrete-event systems community and the computer science community in concepts related to modular systems. Some systems too complicated to model in a monolithic manner may be easier to model as modular interacting subsystems. Manipulating systems as separate interacting agents has the added advantage of avoiding the "state explosion" problem; when several finite-state systems are combined, the size of the state space of the composed system is potentially exponential in the number of components, so we may wish to keep system models modular whenever possible. Likewise, when supervising or verifying system behavior, there may be several separate specifications and therefore it would be advantageous to keep the specifications modular as well. If we try to combine diverse modular specifications of concurrent systems to form a single monolithic specification, the monolithic specification may be unbearably large due to a similar state explosion problem. These state explosions cause difficulties when attempting to compute solutions to modular supervision problems. Much time is needed to explore the full state space of the composed monolithic system and this hints that these problems are computationally difficult. We investigate the computational complexity of decision problems associated with the supervision of modular systems.

Although there are several ways of specifying modular systems, we make the assumption that the modular systems and specifications discussed in this paper are modeled as deterministic finite-state automata interacting via the parallel compo-

[1] Supported in part by NSF Grant CCR-0082784.

sition operation. This modeling method is generally considered to be the simplest method that is expressive enough to be used for real-world problems, and decision problems related to finite-state deterministic automata are also thought to be relatively easy in general. We show that many questions related to the supervision of these supposedly simple systems are PSPACE-complete, which implies that similar supervision issues for more general models are likewise intractable. PSPACE-complete problems are the most-difficult problems that can be decided using a polynomially bounded amount of memory. The PSPACE problem class includes the NP problem class which is believed to contain problems not solvable in polynomial time.

Although for supervision purposes we find there may not be great savings in time by specifying systems in a modular manner, there is possibly great savings in computation space. This is why researchers began to investigate the use of modules in the first place - to avoid the *state* explosion problem.

The work in this paper was inspired by the automata intersection problem investigated in Kozen (1977) and Rohloff and Lafortune (2002). We reduce automata intersection problems discussed in Rohloff and Lafortune (2002) to supervisor existence problems using polynomial-time many-one reductions. We assume the supervision systems in this paper are "parallel" supervision systems that are realized as deterministic finite-state automata where the supervised system is synthesized by a parallel composition of the supervisor automata with the unsupervised system. See the papers by Cieslak et al. (1988), Lin and Wonham (1988a), Lin and Wonham (1988b), Ramadge and Wonham (1987), Ramadge and Wonham (1989) and Rudie and Wonham (1992) for a sample of major innovative works from the control community on parallel supervision discrete-event systems and the text by Cassandras and Lafortune (1999) for a general introduction to discrete-event system theory. We also investigate decentralized supervision problems where local supervisors make local observations and generate local control actions that are combined globally such as in the papers by Cieslak et al. (1988), Lin and Wonham (1988a), Rudie and Wonham (1992) and Yoo and Lafortune (2002).

There have been several papers from the computer science community that discuss problems similar to ours. The complexity of verification for systems using more complicated models such as temporal logic and alternating tree automata is discussed in Harel et al. (2002), Kupferman et al. (2000) and Vardi and Wolper (1994). The synthesis of distributed systems and supervisors, but under assumptions different from those made in this paper, is discussed in Kupferman and Vardi (2001) and Pnueli and Rosner (1990).

The supervision of modular systems is currently receiving much attention from the control research community. Properties of modular discrete-event systems when the modules have disjoint alphabets are investigated in Queiroz and Cury (2000). Various local specification and concurrent supervision problems, respectively, are investigated in Jiang and Kumar (2000). The supervision of modular systems using specific architectures is discussed in Leduc et al. (2001). With the exception of Gohari and Wonham (2000), which shows NP-hardness results for modular supervision problems, there has been little work investigating the computational complexity of modular supervision. We improve on results in Gohari and Wonham (2000) by showing PSPACE-completeness results. We know of no work besides this paper that places an upper bound on the computational complexity of deciding supervisor existence for modular systems.

In the next section of this paper we present a review of discrete-event system theory. In Section 3 we discuss properties of supervised discrete-event systems. We review results related to the intersection of automata in Section 4. The fifth section demonstrates that several modular supervision problems are PSPACE-complete and Section 6 closes the paper with a discussion of the results.

2. REVIEW OF DISCRETE-EVENT SYSTEMS

Although the notation for automata problems used by researchers in computer science and discrete-event systems is similar, there are subtle differences that need to be distinguished. To aid the reader we briefly review the notation and conventions used in both fields as we reference the literature in both computer science and supervisory control theory. We generally use the notation of computer science theory when we discuss automata intersection problems and we use the notation of discrete-event systems when we discuss issues relating to supervisory control.

For an automaton G, in theoretical computer science, the language *accepted* ($L(G)$) by the automaton G is the set of all strings that lead to a final state starting from the initial state. $L(G)$ is equivalent to the language *marked* ($\mathcal{L}_m(G)$) in discrete-event system theory. The language *generated* in discrete-event system theory ($\mathcal{L}(G)$) is the set of strings whose state transitions are defined by the transition function $\delta^G(\cdot)$ starting from the initial state. Note that we use a script $\mathcal{L}$ for discrete-event systems notation and a regular

L for computer science notation. When $\delta^G(\cdot)$ is a partial function, $\mathcal{L}(G) \subset \Sigma^*$. $\mathcal{L}(G)$ is a prefix-closed language, i.e., it contains all the prefixes of all its strings. $\mathcal{L}_m(G)$ and $L(G)$ are not prefix-closed in general. For a language K, we use the notation $\overline{K}$ to denote the prefix-closure of K which is the set of all the prefixes of all the strings in K. We call an automaton that accepts a prefix-closed language a *prefix-closed automaton.* We also say an automaton is *nonblocking* if the prefix-closure of its marked language is equal to its generated language, i.e., $\overline{\mathcal{L}_m(G)} = \mathcal{L}(G)$.

Given a set of h modules modeled as automata $\{H_1, H_2, ..., H_h\}$, we use the script notation $\mathcal{H}_1^h$ to denote the set $\{H_1, H_2, ..., H_h\}$ and the regular notation H_1^h to denote the parallel composition $H_1 \| H_2 \| ... \| H_h$. H_1^h accepts (generates) a string t if and only if t is accepted (generated) by all automata in $\mathcal{H}_1^h = \{H_1, H_2 ..., H_h\}$. This implies that $\mathcal{L}_m(H_1^h) = \mathcal{L}_m(H_1) \cap ... \cap \mathcal{L}_m(H_h)$.

Similarly, for a set of k languages $\{K_1, K_2, ..., K_k\}$, we use the script notation $\mathcal{K}_1^k$ to denote the set $\{K_1, K_2, ..., K_k\}$ and the regular notation K_1^k to denote the intersection of the languages $K_1 \cap K_2 \cap ... \cap K_k$. We also use the notation $\mathcal{L}(\mathcal{H}_1^h)$ and $\mathcal{L}_m(\mathcal{H}_1^h)$ to denote the sets of languages $\{\mathcal{L}(H_1), \mathcal{L}(H_2), ..., \mathcal{L}(H_h)\}$ and $\{\mathcal{L}_m(H_1), \mathcal{L}_m(H_2), ..., \mathcal{L}_m(H_h)\}$, respectively.

Given a supervisor S and a system G, we denote the composed system of S supervising G as the supervised system S/G. Furthermore, because we assume we are using parallel supervisors realized as finite-state automata, S/G is equivalent to $S \| G$. For the case of multiple supervisors (i.e., decentralized supervision), we assume that an event is disabled if it is disabled by at least one supervisor. For a set of decentralized supervisors $\{S_1, ..., S_s\}$, we adopt a similar notation for $\mathcal{S}_1^s = \{S_1, ..., S_s\}$ and $S_1^s = S_1 \| ... \| S_s$ as seen above for $\mathcal{H}_1^h$ and H_1^h. Hence, a set of supervisors $\mathcal{S}_1^s$ controlling G is equivalent S_1^s/G. A supervisor observes only locally observable events and can disable only locally controllable events, denoted by Σ_{oi} and Σ_{ci}, respectively, for supervisor S_i.

3. PROPERTIES OF SUPERVISED DISCRETE-EVENT SYSTEMS

We now extend the supervisory control theory concepts of controllability, M-closure, and coobservability from Ramadge and Wonham (1987) and Rudie and Wonham (1992) to handle the cases where the systems and specifications are modular.

Let $\mathcal{K}_1^k$ and $\mathcal{M}_1^m$ be sets of languages. Let Σ_{ci} and Σ_{oi} be the locally controllable and observable event sets respectively for $i \in \{1, ..., s\}$. Let $P_i : \Sigma^* \to \Sigma_{oi}^*$ be the natural projection that erases events in $\Sigma \setminus \Sigma_{oi}$. $P_i^{-1} : \Sigma_{oi}^* \to 2^{\Sigma^*}$ is the inverse projection operation which returns all strings in Σ^* that could be projected to a given string in Σ_{oi}^*. Furthermore, let $\Sigma_c = \cup_{i=1}^s \Sigma_{ci}$ and $\Sigma_{uc} = \Sigma \setminus \Sigma_c$

Definition 1. Consider the sets of languages $\mathcal{K}_1^k$ and $\mathcal{M}_1^m$ such that $M_1 = \overline{M_1}, M_2 = \overline{M_2}, ..., M_m = \overline{M_m}$ and the set of uncontrollable events Σ_{uc}. The set of languages $\mathcal{K}_1^k$ is *modular controllable* with respect to $\mathcal{M}_1^m$ and Σ_{uc} if $\overline{K_1^k}\Sigma_{uc} \cap M_1^m \subseteq \overline{K_1^k}$.

Definition 2. Consider the sets of languages $\mathcal{K}_1^k$ and $\mathcal{M}_1^m$ such that $M_1 = \overline{M_1}, M_2 = \overline{M_2}, ..., M_m = \overline{M_m}$. The set of languages $\mathcal{K}_1^k$ is *modular $\mathcal{M}_1^m$-closed* if $K_1^k = \overline{K_1^k} \cap M_1^m$.

Definition 3. Consider the sets of languages $\mathcal{K}_1^k$ and $\mathcal{M}_1^m$ such that $M_1 = \overline{M_1}, M_2 = \overline{M_2}, ..., M_m = \overline{M_m}$ and the sets of locally controllable, Σ_{ci}, and observable, Σ_{oi}, events such that $i \in \{1, ..., s\}$. The set of languages $\mathcal{K}_1^k$ is *modular coobservable* with respect to $\mathcal{M}_1^m$, P_i and Σ_{ci}, $i \in \{1, ..., s\}$ if for all $t \in \overline{K_1^k}$ and for all $\sigma \in \Sigma_c$,

$\left(t\sigma \notin \overline{K_1^k}\right)$ and $(t\sigma \in M_1^m) \Rightarrow$

$\exists i \in \{1, ..., s\}$ such that $P_i^{-1}[P_i(t)]\,\sigma \cap \overline{K_1^k} = \emptyset$ and $\sigma \in \Sigma_{ci}$.

Using these definitions we can demonstrate the following theorem for the existence of supervisors for modular systems.

Theorem 4. For a given set of finite-state automata system modules $\mathcal{G}_1^g$ and a set of finite-state automata specification modules $\mathcal{H}_1^h$ such that H_1^h is nonblocking, there exists a set of partial observation supervisors $\{S_1, S_2, ..., S_s\}$ such that $\mathcal{L}_m(S_1^s/G_1^g) = \mathcal{L}_m(H_1^h)$ and $\mathcal{L}(S_1^s/G_1^g) = \mathcal{L}(H_1^h)$ if and only if the following three conditions hold:

(1) $\mathcal{L}_m(\mathcal{H}_1^h)$ is modular controllable with respect to $\mathcal{L}(\mathcal{G}_1^g)$ and Σ_{uc}.
(2) $\mathcal{L}_m(\mathcal{H}_1^h)$ is modular coobservable with respect to $\mathcal{L}(\mathcal{G}_1^g)$, P_1,...,P_s and Σ_{c1},...,Σ_{cs}.
(3) $\mathcal{L}_m(\mathcal{H}_1^h)$ is modular $\mathcal{L}_m(\mathcal{G}_1^g)$-closed.

The proof of this theorem is a generalization of the proof of the Controllability and Coobservability Theorem discussed in the supervisory control literature (see, e.g., Cassandras and Lafortune (1999)). The proof is constructive and depends on a sample-path argument. This result says that a set of nonblocking supervisors $S_1, S_2, ..., S_s$ exists to achieve a set of modular specifications $\mathcal{H}_1^h$ for a modular system $\mathcal{G}_1^g$ (i.e., $\mathcal{L}_m(S_1^s/G_1^g) = \mathcal{L}_m(H_1^h)$ and $\mathcal{L}(S_1^s/G_1^g) = \mathcal{L}(H_1^h)$) if and only if the system is modular controllable, modular coobservable and modular $\mathcal{L}_m(\mathcal{G}_1^g)$-closed. These properties completely characterize necessary and sufficient conditions for the existence of decentralized supervisors of modular systems. In turn, these properties can play a role in supervisor synthesis when existence conditions are satisfied. Supervisor synthesis for monolithic systems is discussed in Cassandras and Lafortune (1999).

Given $\mathcal{H}_1^h$, deciding if H_1^h is nonblocking is a PSPACE-complete problem. This can be shown using a simple reduction from the automata intersection problem presented in Kozen (1977). However, we may have enough foreknowledge to decide this property holds in a computationally feasible manner and we assume that the modular specifications are given such that H_1^h is nonblocking. If the specification is blocking, no nonblocking supervisors that achieves the specification can exist.

Similarly, the astute reader will note that $\mathcal{L}(H_1^h) \subseteq \mathcal{L}(G_1^g)$ is a necessary condition for $\mathcal{L}_m(S_1^s/G_1^g) = \mathcal{L}_m(H_1^h)$ and for $\mathcal{L}_m(\mathcal{H}_1^h)$ to be modular $\mathcal{L}_m(\mathcal{G}_1^g)$-closed when the system and specification are nonblocking. If $\mathcal{L}_m(H_1^h) \not\subseteq \mathcal{L}_m(G_1^g)$ we can replace the set of specification automata $\mathcal{H}_1^h$ with $\mathcal{H}_1^h \cup \mathcal{G}_1^g$ so that the specification behavior is strictly smaller than the system behavior. $H_1\|...\|H_h\|G_1\|...\|G_g$ would be the monolithic automata marking behavior equivalent to the new specification behavior. This substitution will not alter the computational complexity of the problems we discuss later in this paper.

We can prove a more general theorem concerned only with prefix-closed behavior when we are not concerned with blocking.

Theorem 5. For a given set of prefix-closed finite-state automata system modules $\mathcal{G}_1^g$ and a set of prefix-closed finite-state automata specification modules $\mathcal{H}_1^h$, there exists a set of partial observation supervisors $\{S_1, S_2, ..., S_s\}$ such that $\mathcal{L}(S_1^s/G_1^g) = \mathcal{L}(H_1^h)$ if and only if the following three conditions hold:

(1) $\mathcal{L}(\mathcal{H}_1^h)$ is modular controllable with respect to $\mathcal{L}(\mathcal{G}_1^g)$ and Σ_{uc}.
(2) $\mathcal{L}(\mathcal{H}_1^h)$ is modular coobservable with respect to $\mathcal{L}(\mathcal{G}_1^g)$, P_1,...,P_s and Σ_{c1},...,Σ_{cs}.
(3) $\mathcal{L}(H_1^h) \subseteq \mathcal{L}(G_1^g)$.

As with Theorem 4, if $\mathcal{L}(H_1^h) \not\subseteq \mathcal{L}(G_1^g)$ we can replace $\mathcal{H}_1^h$ with $\mathcal{H}_1^h \cup \mathcal{G}_1^g$ without altering the computational complexity of the problems we discuss later in this paper related to this theorem.

The following proposition will be important when we discuss the computational complexity of supervisor existence.

Proposition 6. The problems of deciding modular controllability, modular coobservability and modular $\mathcal{M}_1^m$-closure for sets of languages specified by finite-state automata are in PSPACE.

PROOF. Proving this proposition relies on a "token" argument similar to that employed by Kozen (1977) for proving that automata intersection emptiness is in PSPACE. Given a set of events Σ_{uc} and two sets of automata $\mathcal{H}_1^h$ and $\mathcal{G}_1^g$, we show that the problem of deciding modular controllability of $\mathcal{L}_m(\mathcal{H}_1^h)$ with respect to $\mathcal{L}(\mathcal{G}_1^g)$ and Σ_{uc} is in PSPACE. Similar proofs exist to show the problems of deciding modular coobservability and modular $\mathcal{M}_1^m$-closure are in PSPACE but are not shown here due to space considerations. Regarding controllability, it is sufficient to show the converse problem of deciding modular non-controllability is in NPSPACE.

A nondeterministic string of events s is generated one event at a time and used to model the state transitions in the finite-state automata in $\mathcal{G}_1^g$ and $\mathcal{H}_1^h$ starting from their respective initial states. The current states of the the automata in $\mathcal{G}_1^g$ and $\mathcal{H}_1^h$ need to be saved and updated as new events are generated. As each new event is generated, we test if $\exists\sigma \in \Sigma_{uc}$ such that

$$[\forall j \in \{1, ..., g\}|(s\sigma \in \mathcal{L}(G_j))]$$
$$\wedge [\forall i \in \{1, ..., h\}|(s \in \mathcal{L}(H_i))]$$
$$\wedge [\exists l \in \{1, ..., h\}|(s\sigma \notin \mathcal{L}_m(H_i))].$$

If this property ever holds then modular controllability does not hold. All of these operations take a polynomial amount of memory with respect to the encodings of $\mathcal{G}_1^g$ and $\mathcal{H}_1^h$. □

Theorem 4 and Proposition 6 result in the following corollary. This result also holds when we do not care about language marking or if the supervisor is nonblocking.

Corollary 7. Given a set of finite-state automata $\mathcal{G}_1^g$, a set of finite-state specification automata $\mathcal{H}_1^h$, sets of observable events $\Sigma_{o1}, ..., \Sigma_{os}$ and sets of controllable events $\Sigma_{c1}, ..., \Sigma_{cs}$, the problem of deciding if there is a set of nonblocking decentralized supervisors S_1^s such that $\mathcal{L}_m(S_1^s/G_1^g) = \mathcal{L}_m(H_1^h)$ and $\mathcal{L}(S_1^s/G_1^g) = \mathcal{L}(H_1^h)$ is in PSPACE.

Now that we have placed an upper bound on the complexity of supervisor existence problems, we show PSPACE-completeness for a large class of problems. First, we need to review that many automata intersection problems are PSPACE-complete.

4. COMPLEXITY OF AUTOMATA INTERSECTION PROBLEMS

Kozen (1977) demonstrates that if given a set $\mathcal{A}_1^a$ of deterministic automata, the problem of deciding if $L(A_1^a) = \emptyset$ is PSPACE-complete. The following theorem is a generalization of Kozen's result to prefix-closed automata that will be used in the next section to show several supervision problems for modular discrete-event systems are PSPACE-complete.

Theorem 8. (Shown by Rohloff and Lafortune (2002).) Given a finite-state automaton B that accepts a prefix-closed language and a set of finite-state automata $\mathcal{A}_1^a$ also accepting prefix-closed languages, the problem of deciding if $L(A_1^a) = L(B)$ is PSPACE-complete.

Due to the nature of the proof in Rohloff and Lafortune (2002), Theorem 8 also holds when we know that $L(B) \neq \emptyset$ and $L(B) \subseteq L(A_1^a)$.

5. EXISTENCE PROBLEMS FOR SUPERVISION OF MODULAR SYSTEMS

In this section we explore the computational complexity of deciding supervisor existence for modular systems under various assumptions. We look at the class of supervisor existence problems where we assume a monolithic prefix-closed specification such that the uncontrolled system allows at least as much behavior as the specification (i.e., for the system automata $\mathcal{G}_1^g$ and the specification automaton H, $\mathcal{L}(H) \subseteq \mathcal{L}(G_1^g)$). We also restrict our attention to architectures with a single full observation supervisor (i.e., $\Sigma = \Sigma_o$). These restrictive assumptions show that we are considering what should be a relatively easy class of problems with one of the simplest possible control structures and one of the simplest possible classes of specifications. This section shows the main result of this paper: even relatively simple classes of supervisor existence problems for modular discrete-event systems are PSPACE-complete.

Theorem 9. The problem of deciding if there is a centralized full-observation supervisor S with controllable event set Σ_c for a set prefix-closed automata $\mathcal{G}_1^g$ and a prefix-closed specification automaton H such that $\mathcal{L}(S/G_1^g) = \mathcal{L}(H)$ is PSPACE-complete.

PROOF. We have already shown in Corollary 7 that this problem is in PSPACE. Suppose $\mathcal{A}_1^a, B$ are prefix-closed automata. We use a many-one polynomial-time mapping to reduce the problem of deciding $L(A_1^a) = L(B)$ to deciding the supervisor existence problem in this theorem. Let $\Sigma_c = \emptyset$. If there exists a centralized full-observation supervisor S such that $\mathcal{L}(S/A_1^a) = \mathcal{L}(B)$, then $L(A_1^a) = L(B)$. If there does not exist a centralized full-observation supervisor such that $\mathcal{L}(S/A_1^a) = \mathcal{L}(B)$, then $L(A_1^a) \neq L(B)$ because no events can be disabled. It should be apparent that this reduction uses a polynomial-time mapping. Since deciding if $L(A_1^a) = L(B)$ is PSPACE-complete from Theorem 8, the supervisor existence problem is also PSPACE-complete. □

This result is particularly disappointing because it shows that a relatively large class of supervisor existence problems involving modular system automata is PSPACE-complete. Due to Theorem 9 it should also be apparent that deciding modular controllability for languages specified by finite-state automata is PSPACE-complete because observability and $\mathcal{L}_m(\mathcal{G})$-closure are implied by full observation and prefix-closure, respectively.

For the case of full control (namely, $\Sigma_c = \Sigma$) and partial observation, we can show using similar proof methods that the supervisor existence problem for a modular system and a monolithic specifications is likewise PSPACE-complete. This implies that deciding both modular observability and modular coobservability for languages specified by deterministic finite-state automata is also PSPACE-complete.

If we do not know that $\mathcal{L}(H_1^h) \neq \emptyset$ or that $\mathcal{L}(H_1^h) \subseteq \mathcal{L}(G_1^g)$, supervisor existence problems remain PSPACE-complete. Likewise, a large class of nonblocking supervisor existence problems for modular systems specified by finite-state automata are also PSPACE-complete due to Theorem 4 because the nonblocking supervisor problems are known to be at least as difficult as prefix-closed specification problems.

6. DISCUSSION

The results in this paper tell us that deciding many supervision problems for modular systems modeled as interacting sets of finite-state automata do not have time-efficient solutions if P $\neq$ PSPACE. There are polynomial time algorithms to decide the monolithic versions of the problems discussed in this paper, but the intuitive generalizations of these algorithms to modular systems take time and space exponential in the number of automata modules. Note that if the number of automata specifying the systems or specifications is bounded, all problems discussed in this paper can be solved in polynomial time.

Despite the negative results regarding the time-complexity of the problems discussed in this paper, we have shown that the problems are in PSPACE. Therefore, there are *always* space-efficient solutions to the problems discussed here for deterministic finite automata modules and other more general systems where we can verify modular controllability, modular coobservability and modular M-closure efficiently in space. These results are in a sense positive in that we can avoid, as far as computation space is concerned, the *state* explosion problem inherent to modular systems. In the worst case the size of the state space of a composed system is exponential in the number of modules, but we only have to store at most a small fraction of those modules in memory to decide supervisor existence.

The results in this paper are also disappointing because (as was mentioned previously), it is generally believed that deterministic finite-state automata problems are fairly simple and the results in this paper together with their proofs indicate that many modular problems using more general system and specification models are also intractable. Many of the PSPACE-completeness results in this paper can be extended to other more complex bounded memory system models such as large classes of bounded Petri nets, temporal logic reactive modules and RAM machines where verification of concurrent behavior can be decided using a polynomial amount of space.

Supervisor synthesis is known to be at least as hard as deciding supervisor existence, so our results show that supervisor synthesis is similarly computationally difficult. We should focus our attention on special cases of interest if we wish to make further progress on developing time-efficient methods for deciding supervisor existence and synthesizing supervisors for modular systems. For instance, it might be helpful to look at specific network architectures or at problems involving systems amenable to a divide-and-conquer approaches.

REFERENCES

C.G. Cassandras and S. Lafortune. *Introduction to Discrete Event Systems*. Kluwer Academic Publishers, Boston, MA, 1999.

R. Cieslak, C. Desclaux, A. Fawaz, and P. Varaiya. Supervisory control of discrete-event processes with partial observations. *IEEE Trans. Auto. Contr.*, 33(3):249–260, March 1988.

P. Gohari and W.M. Wonham. On the complexity of supervisory control design in the RW framework. *IEEE Transactions on Systems, Man and Cybernetics, Part B*, 30(5):643–652, 2000.

D. Harel, O. Kupferman, and M.Y. Vardi. On the complexity of verifying concurrent transition systems. *Information and Computation*, 173: 143–161, 2002.

S. Jiang and R. Kumar. Decentralized control of discrete event systems with specializations to local control and concurrent systems. *IEEE Transactions on Systems, Man and Cybernetics, Part B*, 30(5):653–660, 2000.

D. Kozen. Lower bounds for natural proof systems. In *Proc. 18th Symp. on the Foundations of Computer Science*, pages 254–266, 1977.

O. Kupferman and M. Vardi. Synthesizing distributed systems. In *Proc. 16th IEEE Symp. on Logic in Computer Science*, pages 81–92, 2001.

O. Kupferman, M. Vardi, and P. Wolper. An automata-theoretic approach to branching-time model checking. *Journal of the ACM*, 47(2): 312–360, 2000.

R.J. Leduc, B. Brandin, M. Lawford, and W.M. Wonham. Hierarchical interface-based supervisory control: Serial case. In *Proc. 40th IEEE Conf. on Decision and Control*, pages 4116–4121, 2001.

F. Lin and W. M. Wonham. Decentralized supervisory control of discrete-event systems. *Information Sciences*, 44:199–224, 1988a.

F. Lin and W. M. Wonham. On observability of discrete-event systems. *Information Sciences*, 44:173–198, 1988b.

A. Pnueli and R. Rosner. Distributed reactive systems are hard to synthesize. In *Proc. 31st Symp. on the Foundations of Computer Science*, pages 746–757, 1990.

M. H. Queiroz and J. E. R. Cury. Modular control of composed systems. In *Proc. of 2000 American Control Conference*, 2000.

P.J. Ramadge and W.M. Wonham. Supervisory control of a class of discrete-event processes. *SIAM Journal of Control Optimization*, 25(1): 206–230, 1987.

P.J. Ramadge and W.M. Wonham. The control of discrete-event systems. *Proc. IEEE*, 77(1): 81–98, 1989.

K. Rohloff and S. Lafortune. On the computational complexity of the verification of modular discrete-event systems. In *Proc. 41st IEEE Conf. on Decision and Control*, Las Vegas, Nevada, December 2002.

K. Rudie and W.M. Wonham. Think globally, act locally: Decentralized supervisory control. *IEEE Trans. Auto. Contr.*, 37(11):1692–1708, November 1992.

M. Vardi and P. Wolper. Reasoning about infinite computations. *Information and Computation*, 115:1–37, 1994.

T.-S. Yoo and S. Lafortune. A general architecture for decentralized supervisory control of discrete-event systems. *Journal of Discrete Event Dynamical Systems: Theory and Applications*, 13 (3):335–377, 2002.

AUTOMATED SOLVING THE CONTROL SYNTHESIS PROBLEM FOR THE EXTENDED CLASS OF D.E.D.S.

František Čapkovič *,[1]

* *Institute of Informatics, Slovak Academy of Sciences,*
845 07 Bratislava, Dúbravská cesta 9, Slovak Republic
Phone: ++421-2-5941 1244 , Fax: ++421-2-5477 3271
E-mail: Frantisek.Capkovic@savba.sk
http://www.savba.sk/~utrrcapk/capkhome.htm

Abstract: A generalization of the author's methods published recently is presented in this paper. It extends their validity to the wider class of DEDS (discrete event dynamic systems) to be controlled. Both the previous methods and the innovated one are suitable for the kind of DEDS described by ordinary Petri nets (OPN). However, while the previous methods can be successfully used only in case of DEDS described by the special class of OPN - so called state machines (SM) where each transition has only the single input place and the single output place - the innovated method can be used in case of DEDS described by the wider class of OPN - the bounded OPN (having a limited number of marks in their places) without any restriction on their structure. Thus, the validity of the methods is extended because the class of bounded OPN is undoubtedly wider than the class of OPN represented by SM. Both the original approach and the proposed one can alternatively utilize either ordinary directed graphs (ODG) or bipartite directed graphs (BDG). *Copyright © 2003 IFAC*

Keywords: Discrete-event dynamic systems, directed graphs, bipartite directed graphs, ordinary Petri nets, control synthesis

Petri nets (PN) are often used (Peterson, 1981; Holloway *et al.*, 1997) for DEDS modelling, analysis and control synthesis. Usually, PN places represent DEDS subsystem activities and PN transitions express discrete events occurring in DEDS. Approaches based on OPN allow to use methods of linear algebra for DEDS modelling as well as for the DEDS control synthesis. However, in spite of this fact there exists no general method in analytical terms for the control synthesis of DEDS modelled by OPN. In the recent author publications (Čapkovič, 2000*b*; Čapkovič, 2000*a*; Čapkovič, 2000*c*; Čapkovič, 2002*c*) the methods suitable for SM were presented. SM are the special class of OPN, where each transition has only one input place and only one output place. The control synthesis method utilizing directed graphs (DG) was developed there for such a kind of DEDS. The method (Čapkovič, 2002*a*; Čapkovič, 2002*b*) for SM control synthesis utilizes BDG described e.g. by Diestel (Diestel, 1997). In latter author's works finding the control vectors is simpler and more automated in comparison with the former ones. Because the special class of DEDS described by means of SM is not very wide, this paper is devoted to the idea of extending the validity of

[1] Partially supported by the Slovak Grant Agency for Science (VEGA) under grant # 2/3130/23

the approach to the class of DEDS described by bounded OPN. In comparison with SM it is the multiple wider class of OPN. Before describing the new approach it is necessary to mention the principles of the original approaches that remain the same. We emphasize that in this paper the term vector means by definition the column vector.

1. THE PRINCIPLES OF THE PREVIOUS METHODS

SM can be understood to be the special ODG with the OPN places being the ODG nodes and the OPN transitions being fixed to the ODG edges. In addition to this the nodes are marked in order to express dynamics development. The SM reachability tree can be developed (Čapkovič, 1999) as follows

$$\mathbf{x}(k+1) = \mathbf{\Delta}_k.\mathbf{x}(k) \quad , \quad k = 0, N \qquad (1)$$

where k is the discrete step (the level of the tree); $\mathbf{x}(k) = (\sigma_{p_1}^{(k)}(\gamma), ..., \sigma_{p_n}^{(k)}(\gamma))^T$, $k = 0, N$ is the n-dimensional state vector in the step k; $\sigma_{p_i}^{(k)}(\gamma)$, $i = 1, n$ is the state of the elementary place p_i in the step k. It depends on actual enabling its input transitions. γ symbolizes this dependency. $\mathbf{\Delta}_k = \{\delta_{ij}^{(k)}\}_{n \times n}$, $\delta_{ij}^{(k)} = \gamma_{t_{p_i|p_j}}^{(k)}$, $i = 1, n$, $j = 1, n$. It is the functional matrix where $\gamma_{t_{p_i|p_j}}^{(k)} \in \{0, 1\}$ is the transition function of the PN transition fixed on the edge oriented from p_j to p_i. To avoid problems with computer handling $\mathbf{\Delta}_k$ two simple approaches to control synthesis were proposed. The first of them (Čapkovič, 2000*b*; Čapkovič, 2000*a*; Čapkovič, 2000*c*; Čapkovič, 2002*c*) is based on ODG and the second one (Čapkovič, 2002*a*; Čapkovič, 2002*b*) is based on BDG.

The ODG-based approach operates with the constant matrix $\mathbf{\Delta}$ being the transpose of the adjacency matrix of the ODG representing the SM. The main idea is very simple. To obtain feasible trajectories from a given initial state $\mathbf{x}_0$ to a prescribed terminal state $\mathbf{x}_t$ a special intersection of both the straight-lined reachability tree (developed from $\mathbf{x}_0$ towards $\mathbf{x}_t$) and the backtracking reachability tree (developed from $\mathbf{x}_t$ towards $\mathbf{x}_0$, however, containing the paths oriented towards the terminal state) is performed. The former tree is developed as follows

$$\{\mathbf{x}_1\} = \mathbf{\Delta}.\mathbf{x}_0 \qquad (2)$$

$$\{\mathbf{x}_2\} = \mathbf{\Delta}.\{\mathbf{x}_1\} = \mathbf{\Delta}.(\mathbf{\Delta}.\mathbf{x}_0) = \mathbf{\Delta}^2.\mathbf{x}_0 \qquad (3)$$

$$\dots \quad \dots \quad \dots$$

$$\{\mathbf{x}_N\} = \mathbf{\Delta}.\{\mathbf{x}_{N-1}\} = \mathbf{\Delta}^N.\mathbf{x}_0 \qquad (4)$$

where $\mathbf{x}_N = \mathbf{x}_t$. In general, $\{\mathbf{x}_j\}$ is an aggregate all of the states that are reachable from the previous states. According to graph theory $N \leq (n-1)$. The latter tree is the following

$$\{\mathbf{x}_{N-1}\} = \mathbf{\Delta}^T.\mathbf{x}_N \qquad (5)$$

$$\{\mathbf{x}_{N-2}\} = \mathbf{\Delta}^T.\{\mathbf{x}_{N-1}\} = (\mathbf{\Delta}^T)^2.\mathbf{x}_N \qquad (6)$$

$$\dots \quad \dots \quad \dots$$

$$\{\mathbf{x}_0\} = \mathbf{\Delta}^T.\{\mathbf{x}_1\} = (\mathbf{\Delta}^T)^N.\mathbf{x}_N \qquad (7)$$

Here, $\{\mathbf{x}_j\}$ is an aggregate all of the states from which the next states are reachable. It is clear that $\mathbf{x}_0 \neq \{\mathbf{x}_0\}$ and $\mathbf{x}_N \neq \{\mathbf{x}_N\}$. It is the consequence of the fact that in general, $\mathbf{\Delta}.\mathbf{\Delta}^T \neq \mathbf{I_n}$ as well as $\mathbf{\Delta}^T.\mathbf{\Delta} \neq \mathbf{I}_n$. $\mathbf{I}_n$ is $(n \times n)$ identity matrix. The intersection of the trees is performed as follows

$$\mathbf{M}_1 = (\mathbf{x}_0, {}^1\{\mathbf{x}_1\}, \dots, {}^1\{\mathbf{x}_{N-1}\}, {}^1\{\mathbf{x}_N\}) \qquad (8)$$

$$\mathbf{M}_2 = ({}^2\{\mathbf{x}_0\}, {}^2\{\mathbf{x}_1\}, \dots, {}^2\{\mathbf{x}_{N-1}\}, \mathbf{x}_N) \qquad (9)$$

$$\mathbf{M} = \mathbf{M}_1 \cap \mathbf{M}_2 \qquad (10)$$

$$\mathbf{M} = (\mathbf{x}_0, \{\mathbf{x}_1\}, \dots, \{\mathbf{x}_{N-1}\}, \mathbf{x}_N) \qquad (11)$$

where the matrices $\mathbf{M}_1$, $\mathbf{M}_2$ represent, respectively, the straight-lined tree and the backtracking one. The special intersection both of the trees is performed by means of the column-to-column intersection both of the matrices. Thus, $\{\mathbf{x}_i\} = \min({}^1\{\mathbf{x}_i\}, {}^2\{\mathbf{x}_i\})$, $i = 0, N$ with ${}^1\{\mathbf{x}_0\} = \mathbf{x}_0$, ${}^2\{\mathbf{x}_N\} = \mathbf{x}_N$. In $\mathbf{M}$ the trajectories from the initial state $\mathbf{x}_0$ to the terminal one $\mathbf{x}_N$ are comprehended. $\mathbf{\Delta}_k$ contains relations between the trajectories and the control variables. Thus, the control problem is resolved.

However, OPN in general (SM as well) can be understood to be BDG. Let $S = \{P, T\}$ is the set of the BDG nodes where P is the set of the PN places and T is the set of the PN transitions. Let Δ is the set $S \times S$ of BDG edges. Their occurrence can be expressed by the $((n+m) \times (n+m))$ matrix

$$\mathbf{\Delta} = \begin{pmatrix} \mathbf{\emptyset}_{n \times n} & \mathbf{G}^T \\ \mathbf{F}^T & \mathbf{\emptyset}_{m \times m} \end{pmatrix} \qquad (12)$$

where $\mathbf{\emptyset}_{i \times j}$ in general is the $(i \times j)$ zero matrix; $\mathbf{G}$ is the $(m \times n)$ incidence matrix expressing $T \times P$; $\mathbf{F}$ is the $(n \times m)$ incidence matrix representing $P \times T$. Using the extended state vector $\mathbf{s}_k = (\mathbf{x}_k^T, \mathbf{u}_k^T)^T$ instead of the original one we can use the same algorithm like that used in the previous ODG-based method. Eliminating the zeros blocks, the approach can be decomposed into two algorithms that alternate step-by-step. In such a way we have (in analogy with (8)–(11)) both the system state trajectories and corresponding control strategies

$$\mathbf{X} = {}^1\mathbf{X} \cap {}^2\mathbf{X} \; ; \; \mathbf{U} = {}^1\mathbf{U} \cap {}^2\mathbf{U} \qquad (13)$$

$$\mathbf{X} = (\mathbf{x}_0, \{\mathbf{x}_1\}, \dots, \{\mathbf{x}_{N-1}\}, \mathbf{x}_N) \qquad (14)$$

$$\mathbf{U} = (\{\mathbf{u}_0\}, \{\mathbf{u}_1\}, \dots, \{\mathbf{u}_{N-1}\}) \qquad (15)$$

In some pathological cases (e.g. when a place has no output transition) the approach fails and should be generalized in order to have a possibility to work explicitly in the actual state space.

2. THE PROPOSED APPROACH

The main aim of this paper is to extend the validity of the above methods (based on ODG and BDG) to the bounded OPN with general structure as well as to solve the pathological cases mentioned above in the actual state space. The idea of the proposed approach consists in the fact that the reachability tree (RT) and the corresponding reachability graph (RG) of OPN can be undestood to be the SM. The explicit handling the state space is possible in such a case. Thus, the procedure extending the applicability of the approach consists in the following steps

1. setting a given initial state $\mathbf{x}_0$ of the OPN places (initial marking) to be the root node
2. developing the RT from the root node as well as determining the space of reachable states
3. creating the reachability graph RG by means of mutual connecting the leaves (terminal nodes) occurring repeatedly (i.e. having the same name) with the nonterminal node of the tree with the same name
4. understanding the RG to be the special ODG mentioned above. The nodes represent the entire state vectors during the system dynamics development (in the method above they were only the states of elementary places)
5. understanding the RG to be the SM
6. utilizing one of the methods mentioned above

To quantify the approach and to deal with the problem automatically the following items are necessary: the mathematical model of DEDS in the form of OPN, and the algorithm for enumerating the RT and RG.

2.1 *The mathematical model of OPN*

The simple mathematical model is the following

$$\mathbf{x}_{k+1} = \mathbf{x}_k + \mathbf{B}.\mathbf{u}_k \quad , \quad k = 0, N \tag{16}$$

$$\mathbf{B} = \mathbf{G}^T - \mathbf{F} \tag{17}$$

$$\mathbf{F}.\mathbf{u}_k \leq \mathbf{x}_k \tag{18}$$

where

k is the discrete step of the DEDS dynamics development.

$\mathbf{x}_k = (\sigma_{p_1}^k, ..., \sigma_{p_n}^k)^T$ is the n-dimensional state vector of the system in the step k. Its components $\sigma_{p_i}^k \in \{0, c_{p_i}\}$, $i = 1, n$ express the states of the DEDS elementary subprocesses or operations - 0 (passivity) or $0 < \sigma_{p_i} \leq c_{p_i}$ (activity); c_{p_i} is the capacity of the DEDS subprocess p_i as to its activities.

$\mathbf{u}_k = (\gamma_{t_1}^k, ..., \gamma_{t_m}^k)^T$ is the m-dimensional control vector of the system in the step k. Its components $\gamma_{t_j}^k \in \{0, 1\}$, $j = 1, m$ represent occurring of the DEDS elementary discrete events (e.g. starting or ending the elementary subprocesses or their activities, failures, etc.) - 1 (presence) or 0 (absence) of the corresponding discrete event.

$\mathbf{B}$, $\mathbf{F}$, $\mathbf{G}$ are, respectively, $(n \times m)$, $(n \times m)$ and $(m \times n)$- dimensional structural matrices of constant elements. The matrix $\mathbf{F} = \{f_{ij}\}$; $i = 1, n$, $j = 1, m$; $f_{ij} \in \{0, M_{f_{ij}}\}$ expresses the causal relations among the states of the DEDS and the discrete events occuring during the DEDS operation, where the states are the causes and the events are the consequences - 0 (nonexistence), $M_{f_{ij}} > 0$ (existence and multiplicity) of the corresponding causal relations. The matrix $\mathbf{G} = \{g_{ij}\}$; $i = 1, m$, $j = 1, n$; $g_{ij} \in \{0, M_{g_{ij}}\}$ expresses very analogically the causal relation among the discrete events (the causes) and the DEDS states (the consequences). Both of these matrices are the arcs incidence matrices. The matrix $\mathbf{B}$ is given by means of them according to (17).

$(.)^T$ symbolizes the matrix or vector transpose.

2.2 *The algorithm for finding the RT and RG*

There exists the relatively simple algorithm for generating the RT of OPN with general structure (Peterson, 1981). It was formally described by many other authors too. A simple trial for the computer realization of the algorithm was made e.g. in the Czech research group (CzechResGr, 2001) where all of the reachable states were simultaneously generated too. In the research report (Iordache and Antsaklis, 2002) another approach to finding the RG is described. Utilizing the Czech algorithm (because of its simplicity) the adjacency matrix of the RT (RG as well) and the list of corresponding reachable states can be generated. Thus, any new algorithm needs not be constructed in this paper. The inputs of the algorithm are the PN incidence matrices $\mathbf{G}^T$, $\mathbf{F}$ as well as the PN initial marking (i.e. the initial state vector) $\mathbf{x}_0$. The output of the algorithm is the square matrix $\mathbf{A}_{RT}$ representing the adjacency matrix of the RT in a *quasi-functional* form. Its dimensionality is $(n_{RT} \times n_{RT})$ where the integer n_{RT} denotes the number of the reachable states. The elements of this matrix $a_{i,j}^{RT}$, $i = 1, n_{RT}$; $j = 1, n_{RT}$, are equal either to 0 (when there exists no oriented arc connecting the nodes i, j) or to the positive integer (the ordinal number of the transition through which the oriented arc passes from the node i to node j). It is necessary to say that the *quasi-functional* form of the RG adjacency matrix $\mathbf{A}_{RG} = \mathbf{A}_{RT}$. Another output of the algorithm represents the matrix $\mathbf{X}_{reach}$ which is simultaneously enumerated. Its dimensionality is $(n \times n_{RT})$ and its columns represent the above mentioned space of the OPN reachable states. To utilize the matrix $\mathbf{A}_{RG}$ in the control synthesis algorithm to be proposed its elements must be modified - positive integers must be replaced by the integer 1. Thus, we obtain the (0, 1)-matrix expressing the classical numerical adjacency matrix $\mathbf{A}$. The matrix $\mathbf{\Delta} = \mathbf{A}^T$. Hence, the ODG-based approach can be utilized without any problem. However, any state vector (with dimensionality n_{RT}) represents one of the reachable states. It contains only one nonzero element equal to 1 on the position j corresponding to the j-th column of $\mathbf{X}_{reach}$ in which the actual j-th OPN reachable state vector (with dimensionality n) is placed. Because RT starts from the initial state it needs not be enumerated more. It is sufficient to enumerate only backtracking RT for any given terminal state.

2.3 The modification of the BDG-based approach

Before utilizing the BDG-based control synthesis of the DEDS described by means of the boundary OPN we have to overcome a small problem. Namely, the BDG-based approach handles matrices $\mathbf{F}$, $\mathbf{G}$ as well as the OPN transitions. However, the new approach do not know the matrices. Thus, after enumerating the matrix $\mathbf{A}_{RG}$ we have to disassamble this matrix into the matrices $\mathbf{F}_{RG}$, and $\mathbf{G}_{RG}$. In addition to this the original OPN transitions may offer among the elements of the *quasi-functional* matrix more then once. Consequently, some confusions can occur. To avoid these it is necessary to rename the original transitions in order to obtain the transitions that occure only once. The number of them is T_r (the global number of the elements of the *quasi-functional* matrix). The renaming is performed raw-by-raw in the matrix $\mathbf{A}$ where the nonzero elements are replaced by ordinal integers starting from 1 and finishing at T_r. In such a way the matrix $\mathbf{A}_{T_r}$ is obtained. The dissassambling of the matrix $\mathbf{A}_{T_r}$ into the incidence matrices $\mathbf{F}_{RG}$ and $\mathbf{G}_{RG}$ is given as follows. For $i = 1, n_{RT}$; $j = 1, n_{RT}$ if $A_{RT}(i,j) \neq 0$ and $A_{T_r}(i,j) \neq 0$ we set $T_{rTt}(A_{RT}(i,j), A_{T_r}(i,j)) = 1$, $F_{RG}(i, A_{T_r}(i,j)) = 1$, $G_{RG}(A_{T_r}(i,j), j) = 1$ else we set the elements to be equal to 0. Here, $\mathbf{T}_{rTt}$ is the transformation matrix between the original set of transitions and the fictive one. Hence, $\mathbf{U} = \mathbf{T}_{rT_t}.\mathbf{U}^*$ where the matrix $\mathbf{U}^*$ yields the control strategies (15) computed by means of the set of the fictive transitions.

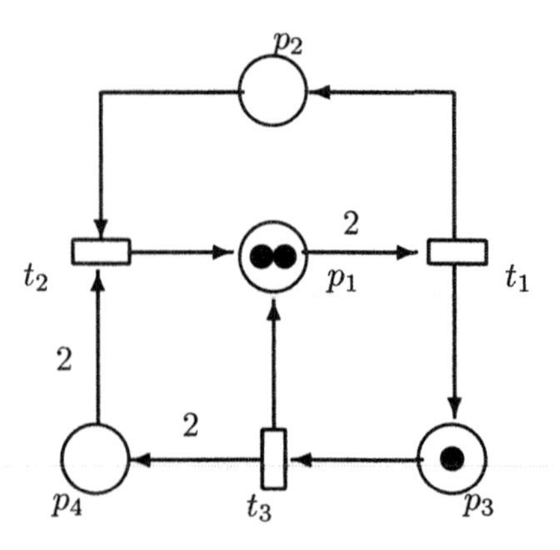

Fig. 1. The OPN-based model of the DEDS

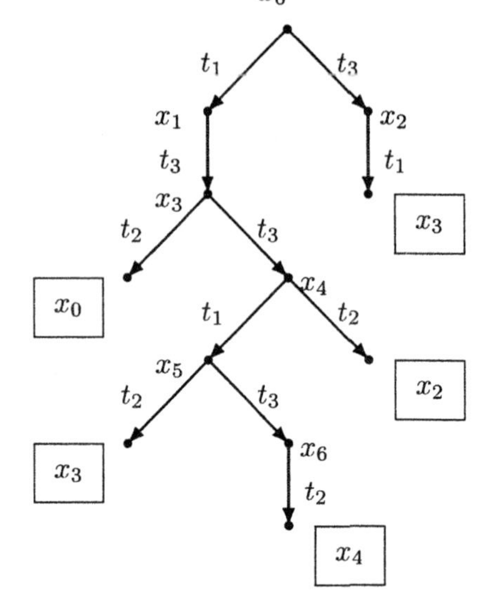

Fig. 2. The corresponding reachability tree

3. THE ILLUSTRATIVE EXAMPLE

Consider the simple case of DEDS leading to the OPN-based model (Pastor *et al.*, 1999) with general structure, even with multiplicity of the oriented arcs. The system can be modelled by means of the OPN presented on the left in Fig. 1. The corresponding RT starting from the initial state $\mathbf{x}_0 = (2, 0, 1, 0)^T$ is given on the right in the Fig. 2 and the RG in the Fig. 3. We can see that the OPN has $n = 4$ places and $m = 3$ transitions. The structural matrices are the following

$$\mathbf{F} = \begin{pmatrix} 2 & 0 & 0 \\ 0 & 1 & 0 \\ 0 & 0 & 1 \\ 0 & 2 & 0 \end{pmatrix} \mathbf{G} = \begin{pmatrix} 0 & 1 & 1 & 0 \\ 1 & 0 & 0 & 0 \\ 1 & 0 & 0 & 2 \end{pmatrix}$$

Connecting mutually the enframed leaves of the same name together and also with the nonterminal node of the same name we obtain the corresponding RG. However, the adjacency matrix $\mathbf{A}_{RT}$ of the RT is the same like the adjacency matrix $\mathbf{A}_{RG}$ of the RG. Its dimensionality is equal to the number of reachable states. In this case

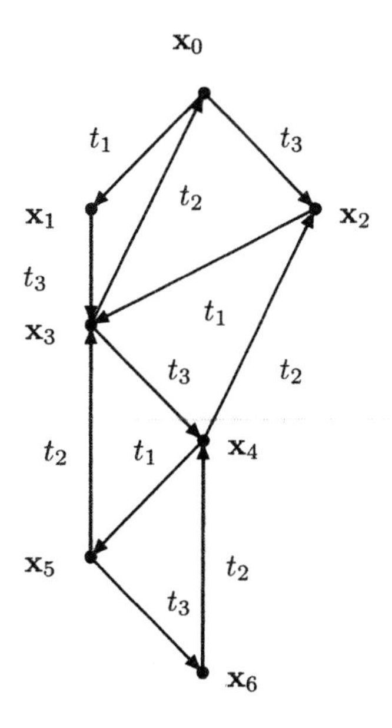

Fig. 3. The reachability graph

$$\mathbf{A}_{RT} = \begin{pmatrix} 0 & 1 & 3 & 0 & 0 & 0 & 0 \\ 0 & 0 & 0 & 3 & 0 & 0 & 0 \\ 0 & 0 & 0 & 1 & 0 & 0 & 0 \\ 2 & 0 & 0 & 0 & 3 & 0 & 0 \\ 0 & 0 & 2 & 0 & 0 & 1 & 0 \\ 0 & 0 & 0 & 2 & 0 & 0 & 3 \\ 0 & 0 & 0 & 0 & 2 & 0 & 0 \end{pmatrix}$$

$$\mathbf{A} = \begin{pmatrix} 0 & 1 & 1 & 0 & 0 & 0 & 0 \\ 0 & 0 & 0 & 1 & 0 & 0 & 0 \\ 0 & 0 & 0 & 1 & 0 & 0 & 0 \\ 1 & 0 & 0 & 0 & 1 & 0 & 0 \\ 0 & 0 & 1 & 0 & 0 & 1 & 0 \\ 0 & 0 & 0 & 1 & 0 & 0 & 1 \\ 0 & 0 & 0 & 0 & 1 & 0 & 0 \end{pmatrix} \quad \mathbf{\Delta} = \mathbf{A}^T \qquad (19)$$

The set of the reachable states can be expressed by means of the following matrix representing (by means of its columns) the vectors in question.

$$\mathbf{X}_{reach} = \begin{pmatrix} 2 & 0 & 3 & 1 & 2 & 0 & 1 \\ 0 & 1 & 0 & 1 & 1 & 2 & 2 \\ 1 & 2 & 0 & 1 & 0 & 1 & 0 \\ 0 & 0 & 2 & 2 & 4 & 4 & 6 \end{pmatrix} \qquad (20)$$

Thus, in the columns $j = 1, 2, ..., n_{RT}$ lie the vectors $\mathbf{x}_0, \mathbf{x}_1, ..., \mathbf{x}_{n_{RT-1}}$. Utilizing the ODG-based method of the control synthesis we have the following solution of the DEDS control synthesis in case when terminal state $\mathbf{x}_t = \mathbf{x}_7 = (1, 2, 0, 6)^T$. It is graphically illustrated in Fig. 4.

$$\mathbf{M}_1 = \begin{pmatrix} 1 & 0 & 0 & 2 & 0 & 0 \\ 0 & 1 & 0 & 0 & 2 & 0 \\ 0 & 1 & 0 & 0 & 4 & 0 \\ 0 & 0 & 2 & 0 & 0 & 8 \\ 0 & 0 & 0 & 2 & 0 & 0 \\ 0 & 0 & 0 & 0 & 2 & 0 \\ 0 & 0 & 0 & 0 & 0 & 2 \end{pmatrix} \quad \mathbf{M}_2 = \begin{pmatrix} 2 & 0 & 0 & 0 & 0 & 0 \\ 0 & 1 & 0 & 0 & 0 & 0 \\ 0 & 1 & 0 & 0 & 0 & 0 \\ 0 & 0 & 1 & 0 & 0 & 0 \\ 3 & 0 & 0 & 1 & 0 & 0 \\ 0 & 2 & 0 & 0 & 1 & 0 \\ 0 & 0 & 1 & 0 & 0 & 1 \end{pmatrix}$$

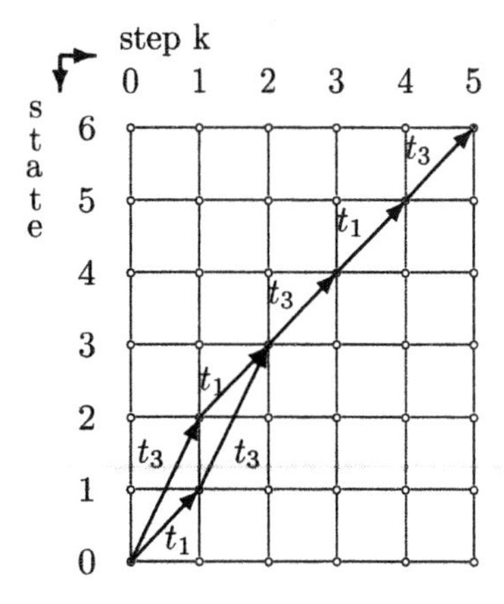

Fig. 4. The solution of the state trajectories

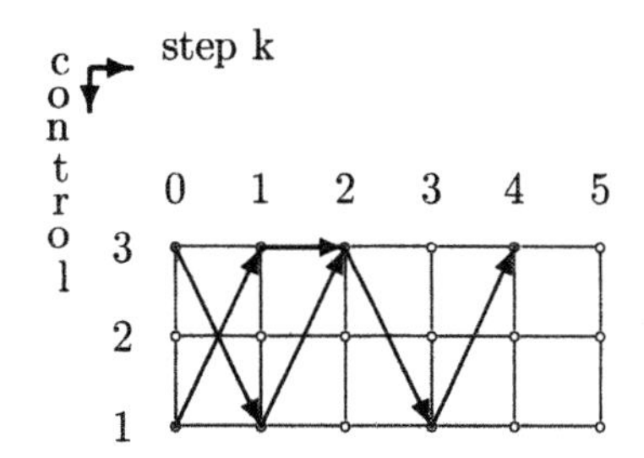

Fig. 5. The solution of the control trajectories

$$\mathbf{M} = \begin{pmatrix} 1 & 0 & 0 & 0 & 0 & 0 \\ 0 & 1 & 0 & 0 & 0 & 0 \\ 0 & 1 & 0 & 0 & 0 & 0 \\ 0 & 0 & 1 & 0 & 0 & 0 \\ 0 & 0 & 0 & 1 & 0 & 0 \\ 0 & 0 & 0 & 0 & 1 & 0 \\ 0 & 0 & 0 & 0 & 0 & 1 \end{pmatrix} \quad \mathbf{A}_{T_r} = \begin{pmatrix} 0 & 1 & 2 & 0 & 0 & 0 & 0 \\ 0 & 0 & 0 & 3 & 0 & 0 & 0 \\ 0 & 0 & 0 & 4 & 0 & 0 & 0 \\ 5 & 0 & 0 & 0 & 6 & 0 & 0 \\ 0 & 0 & 7 & 0 & 0 & 8 & 0 \\ 0 & 0 & 0 & 9 & 0 & 0 & 10 \\ 0 & 0 & 0 & 0 & 11 & 0 & 0 \end{pmatrix}$$

Before the application of the BDG-based approach the renamed set of the fictive transitions has to be defined (their number is 11 in this case - see the above introduced matrix $\mathbf{A}_{T_r}$) and the transformation matrix $\mathbf{T}_{rTt}$ has to be enumerated. Simultaneously, the matrices $\mathbf{F}_{RG}$, $\mathbf{G}_{RG}$ has to be computed.

$$\mathbf{T}_{rTt} = \begin{pmatrix} 1 & 0 & 0 & 1 & 0 & 0 & 0 & 1 & 0 & 0 & 0 \\ 0 & 0 & 0 & 0 & 1 & 0 & 1 & 0 & 1 & 0 & 0 \\ 0 & 1 & 1 & 0 & 0 & 1 & 0 & 0 & 0 & 1 & 0 \end{pmatrix}$$

$$\mathbf{F}_{RG} = \begin{pmatrix} 1 & 1 & 0 & 0 & 0 & 0 & 0 & 0 & 0 & 0 & 0 \\ 0 & 0 & 1 & 0 & 0 & 0 & 0 & 0 & 0 & 0 & 0 \\ 0 & 0 & 0 & 1 & 0 & 0 & 0 & 0 & 0 & 0 & 0 \\ 0 & 0 & 0 & 0 & 1 & 1 & 0 & 0 & 0 & 0 & 0 \\ 0 & 0 & 0 & 0 & 0 & 0 & 1 & 1 & 0 & 0 & 0 \\ 0 & 0 & 0 & 0 & 0 & 0 & 0 & 0 & 1 & 1 & 0 \\ 0 & 0 & 0 & 0 & 0 & 0 & 0 & 0 & 0 & 0 & 1 \end{pmatrix}$$

$$\mathbf{G}^T_{RG} = \begin{pmatrix} 0 & 0 & 0 & 0 & 1 & 0 & 0 & 0 & 0 & 0 & 0 \\ 1 & 0 & 0 & 0 & 0 & 0 & 0 & 0 & 0 & 0 & 0 \\ 0 & 1 & 0 & 0 & 0 & 0 & 1 & 0 & 0 & 0 & 0 \\ 0 & 0 & 1 & 1 & 0 & 0 & 0 & 0 & 1 & 0 & 0 \\ 0 & 0 & 0 & 0 & 0 & 1 & 0 & 0 & 0 & 1 & 0 \\ 0 & 0 & 0 & 0 & 0 & 0 & 0 & 1 & 0 & 0 & 0 \\ 0 & 0 & 0 & 0 & 0 & 0 & 0 & 0 & 0 & 0 & 1 \end{pmatrix}$$

The solution of the state trajectories is the same like that obtained by the ODG-based approach, i.e. ${}^1\mathbf{X} = \mathbf{M}_1$, ${}^2\mathbf{X} = \mathbf{M}_2$, $\mathbf{X} = \mathbf{M}$. The solution of control trajectories (in the space of the fictive transitions) is the following.

$$
{}^1\mathbf{U}^* = \begin{pmatrix} 1&0&0&2&0 \\ 1&0&0&2&0 \\ 0&1&0&0&2 \\ 0&1&0&0&4 \\ 0&0&2&0&0 \\ 0&0&2&0&0 \\ 0&0&0&2&0 \\ 0&0&0&2&0 \\ 0&0&0&0&2 \\ 0&0&0&0&2 \\ 0&0&0&0&0 \end{pmatrix} \quad {}^2\mathbf{U}^* = \begin{pmatrix} 1&0&0&0&0 \\ 1&0&0&0&0 \\ 0&1&0&0&0 \\ 0&1&0&0&0 \\ 0&0&0&0&0 \\ 0&0&1&0&0 \\ 1&0&0&0&0 \\ 2&0&0&1&0 \\ 0&1&0&0&0 \\ 0&1&0&0&1 \\ 0&0&1&0&0 \end{pmatrix}
$$

$$
\mathbf{U}^{*^T} = ({}^1\mathbf{U}^* \cap {}^2\mathbf{U}^*)^T = \begin{pmatrix} 1&1&0&0&0&0&0&0&0&0&0 \\ 0&0&1&1&0&0&0&0&0&0&0 \\ 0&0&0&0&0&1&0&0&0&0&0 \\ 0&0&0&0&0&0&0&1&0&0&0 \\ 0&0&0&0&0&0&0&0&0&1&0 \end{pmatrix}
$$

After the transformation we obtain the solution of the control trajectories in the space of the real transitions. It is given in the Fig. 5.

$$
\mathbf{U} = \mathbf{T}_{rTt}.\mathbf{U}^* = \begin{pmatrix} 1&1&0&1&0 \\ 0&0&0&0&0 \\ 1&1&1&0&1 \end{pmatrix}
$$

4. CONCLUSIONS

The proposed approach presented above seems to be powerful and very useful for the class of DEDS described by the wide class of OPN - the bounded OPN. It yields the procedure of automated solving the control synthesis problem of the class of DEDS. In the first stage the known algorithm is utilized for generation of both the adjacency matrix of the RT starting from a given initial state $\mathbf{x}_0$ and the corresponding space of reachable states $\mathbf{X}_{reach}$. In the second stage the state and/or control trajectories are found for any desired reachable terminal state $\mathbf{x}_t$ by the intersection of both the straight-lined RT and the backtracking RT.

Another problem is how to choose the most suitable trajectory when control task specifications (like criteria, constraints, etc.) are prescribed. Because the specifications for DEDS are usually given in nonanalytical terms (e.g. verbally) a suitable representation of knowledge about them is necessary. Solving such a problem is out of the frame of this paper. In spite of this, it should be very useful to apply the knowledge-based (KB) approach at the choice of the most suitable control. Imbedding the KB approach is a very interesting challenge for the future research.

REFERENCES

Čapkovič, F. (1999). Automated solving of DEDS control problems. In: *Multiple Approaches to Intelligent Systems. LNAI, Vol. 1611 of LNCS* (A. El-Dessouki I. Imam, Y. Kodratoff and M. Ali, Eds.). pp. 735–746. Springer. Berlin-Heidelberg-New York-...-Tokyo.

Čapkovič, F. (2000*a*). Intelligent control of discrete event dynamic systems. In: *Proc. 2000 IEEE Int. Symp. on Intelligent Control* (N.T. Koussoulas P.P. Groumpos and M. Polycarpou, Eds.). pp. 109–114. IEEE Press. Patras, Greece.

Čapkovič, F. (2000*b*). *Modelling and Control of Discrete Event Dynamic Systems. BRICS Report Series, RS-00-26.* University of Aarhus, Denmark. Aarhus, Denmark.

Čapkovič, F. (2000*c*). A solution of DEDS control synthesis problems. In: *Control System Design. Proc. IFAC Conf., Bratislava, Slovak Republic, 18-20 June 2000* (Š. Kozák and M. Huba, Eds.). pp. 343–348. Pergamon, Elsevier Science. Oxford, UK.

Čapkovič, F. (2002*a*). An approach to the control synthesis of state machines. In: *Cybernetics and Systems. Proc. 16th European Meeting on Cybernetics and Systems Research, 2-5 April 2002* (R. Trappl, Ed.). Vol. 1. pp. 81–86. Austrian Society for Cybernetics Studies. Vienna, Austria.

Čapkovič, F. (2002*b*). Control synthesis of a class of DEDS. *KYBERNETES* **31, No. 9/10**, 1274–1281.

Čapkovič, F. (2002*c*). A solution of DEDS control synthesis problems. *Systems Analysis Modelling Simulation* **42, No. 3**, 405–414.

CzechResGr. (2001). http://dce.felk.cvut.cz/cak /Research/index_Research.htm

Diestel, R. (1997). *Graph Theory.* Springer Verlag. New York.

Holloway, L., B. Krogh and A. Giua (1997). A survey of Petri net methods for controlled discrete event systems. *Discrete Event Dynamic Systems: Theory and Applications* **7**, 151–180.

Iordache, M.V. and P.J. Antsaklis (2002). *Software Tools for the Supervisory Control of Petri Nets Based on Place Invariants.* Technical Report ISIS-2002-003. 26 p. University of Notre Dame, USA.

Pastor, E., J. Cortadella and M.A. Pena (1999). Structural methods to improve the symbolic analysis of Petri nets. In: *Application and Theory of Petri Nets 1999. LNCS, Vol. 1639* (S. Donatelli and J. Kleijn, Eds.). pp. 26–45. Springer. Berlin-Heidelberg-New York-...-Tokyo.

Peterson, J.L. (1981). *Petri Net Theory and Modeling the Systems.* Prentice Hall. New York.

TIME EXTENSION OF HIERARCHICAL STATE MACHINES

Jaroslav Fogel *

* *Institute of Informatics Slovak Academy of Sciences, Bratislava, Slovakia*
e-mail : Jaroslav.Fogel@savba.sk

Abstract: We present the time extension of the hierarchical state machines with real-time constructs as clocks, timed guards, and invariants. For timed hierarchical state machines we propose the compositional semantics of traces ensuring that the semantics of the model can be determined from the semantics of its components. Such semantics is proposed with the goal of its successive utilization in the design and verification of the model. At the end of the paper, we discussed the verification problem, and its decomposition to sub-problems with lower complexity as the consequence of the compositional semantics and component refinement using the rules of assume-guarantee reasoning. *Copyright © 2003 IFAC*

Keywords: Hierarchical state machine, timed models, compositional semantics, verification, component refinement.

1. INTRODUCTION

Large industrial processes require models, which inherently express their hierarchical structure, concurrency and dynamics. Mainly, dynamics requires introducing the timing aspects into the model that is important step in the construction of real-time control systems for such processes. As a whole, the model of the control system (which can be done automatically) and the user- supplied model of the controlled plant can be used for the verification of the controller properties concerning mainly the real-time requirements. The specification problem for these real-time applications is more complex since the absolute timing behavior and not only the functional behavior of a system is important. The paper describes a specification language based on the hierarchical state machines (HSM) extended with timing aspects called timed HSM (THSM). We give the compositional semantics of THSM ensuring that the semantics of THSM can be determined from the semantics of its components. Compositionality is also useful when formally analyzing the verification problem. There are several works dealing with HSM as e.g., (Alur *et al.* 1999, Brave and Heymann 1993) where authors describe finite state machines extended with both the hierarchy and concurrency but without timing aspects. Many works devoted to the theory of timed automata (Alur and Dill 1994, Alur 1999), from which we borrow the basic syntax concerning the clocks (time variables). Similar work is done in (David *et al.* 2002), where the authors present real-time extension of UML statecharts, but their semantics is substantially different than ours. From the statecharts syntax we borrow only graphical representation of the hierarchical states (sequential and parallel) and terminology (OR-state, AND-state). The ideas of this work were inspired by work (Alur and Grosu 2000) where authors present the language of hierarchic reactive modules with an observational trace semantics that provides the basis for mode refinement.

The paper is organized as follows. In section 2, we

review THSM and their syntax. In section 3, we give the operational semantics of THSM and in section 4, the compositional semantics of traces of THSM. Finally in section 5 we discuss the verification problem for hierarchical time models, mainly the techniques of compositional proof rules.

2. FORMAL STRUCTURE OF TIMED HSM

Definition 1. A Timed HSM is a structure $THSM = (X, Q, Q_0, E, \delta, \rho, inv, T)$ where:
X is a finite set of timed variables with values from R^+, also called clocks,
Q is a finite set of discrete states,
$Q_0 \subset Q$ is a finite set of the initial states,
E is a finite set of event symbols (input alphabet),
δ is a default entrance function,
ρ is the hierarchy relation on Q,
inv is a function associating with each discrete state $q \in Q$ a convex X-polyhedron[1] called invariant of q,
$T \subseteq Q \times L \times Q$ is a finite set of transitions, where L is a set of labels $l = (\zeta, e, X_0)$, where ζ is a conjunction of atomic constraints on X defining a convex X- polyhedron, called the guard of transition, $e \in E$ is an event symbol, $X_0 \subset X$ is a set of clocks to be reset by taking the transition.
A. States
Q is a finite set of discrete states consisting of the subset Q^+ of sequential or OR- states, the subset $Q^\times$ of parallel or AND- states, and the subset Q^{basic} of basic states. The hierarchical structure of THSM states is represented by the binary relation δ on Q and satisfies the following conditions:

- There exists a unique state r, called the root state of THSM, such that for no state $q \in Q, (q\rho r)$.
- For every state $q \in Q, q \neq r$, there exists a unique state $p \in Q$ such that $(p\rho q)$. The state p is called an immediate super-state of q, whereas q is an immediate sub-state of p.
- A state $q \in Q$ has no immediate sub-states if and only if q is basic $q \in Q^{basic}$.
- If $(p\rho q)$ then either $p \in Q^+ \wedge (q \in Q^+ \vee q \in Q^\times \vee q \in Q^{basic})$ or $p \in Q^\times \wedge (q \in Q^+ \vee q \in Q^{basic})$.

The default entrance function δ is mapping δ: $Q^+ \to Q_0$ where for every $p \in Q^+$ and $q^I \in Q_0$ such that $(p\rho q^I)$, q^I is the default initial sub-state of the super-state p.
A state of THSM is a pair (q, v), where $q \in Q$ is a discrete state and $v \in inv(q)$ is a X-valuation satisfying the invariant of q. X-valuation is a function v: $X \to R^+$ assigning to each clock $x \in X(q)$ a non-negative real value $v(x)$, where $X(q)$ denotes the clocks in the scope of the state q. If discrete state q has not assigned an invariant then the time can progress in q without bounds.
If q is OR- state with immediate sub-states $q_1, ..., q_k$, then

$$inv(q) \supseteq \bigcup_{i=1}^{n} inv'(q_i) \tag{1}$$

where n is a number of such sub-states of q, for which there exists an input transition with label l, in which the set $X_0 \neq 0$ (at least one clock is reset), and
$inv'(q_i) = \bigcup_{j=i}^{n_i} inv(q_j)$
where n_i is a number of sub-states between q_i and q_{i+1} for which $X_0 = 0$.
If there is a finite loop in the OR-state, the invariant of the states from the loop is multiplied by the number of cycles in the loop.
For example, if the constraints of clocks of each sub-state q_i, $i = 1, ..., k$ are X-hyperplanes[2] $x_i < c_i, i = 1, ..., k$ then the invariant of state q will be $inv(q) \equiv x < c$, where $c \geq \sum_{i=1}^{n} c'_i$, where n is as in (1) and $c'_i = max_{i \leq j \leq n_i} c_j$. Figure 1 shows the invariant of OR-state q.

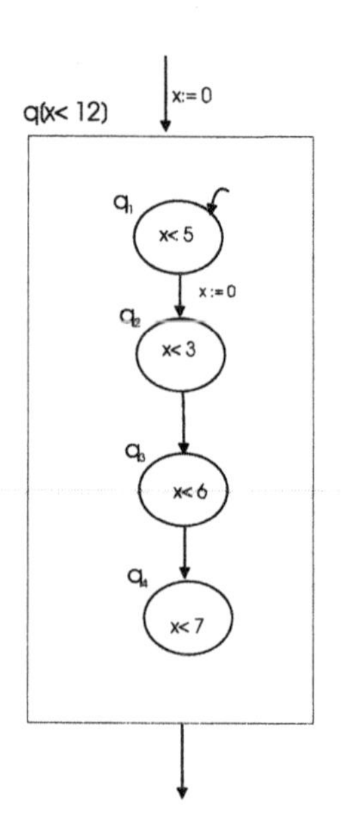

Fig. 1. Invariant of hierarchical OR-state.

If q is AND- state with immediate sub-states $q_1, ..., q_k$, then

$$inv(q) \supseteq \bigcap_{i=1}^{k} inv(q_i). \tag{2}$$

Similarly, if the constraints of clocks are as in the previous case, then for AND-state the invariant of state q can be $inv(q) \equiv x < c$, where $c \geq \min_i(c_i)$.

[1] X- polyhedron is the intersection of an atomic constraints on X, which are in the form $x \sim c$ or $x - y \sim c$, where $x, y \in X, \sim \in (<, \leq, \geq, >)$ and c is a positive integer constant.

[2] X-hyperplane is a set of valuations satisfying an atomic clock constraint.

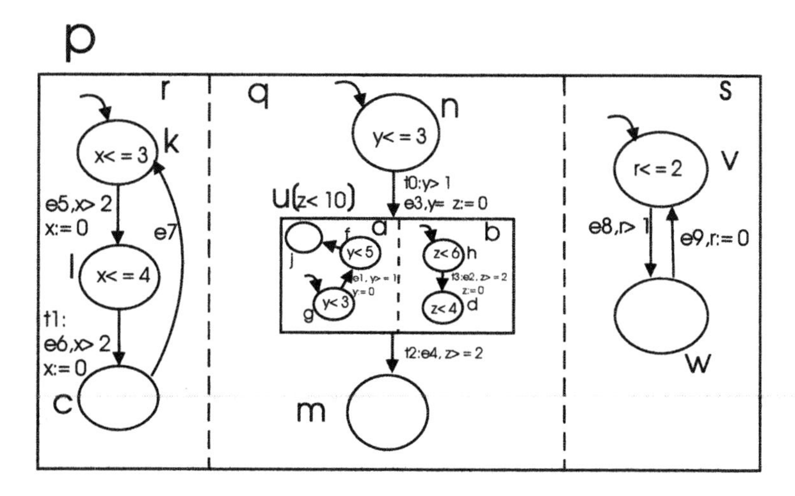

Fig. 2. Timed HSM.

B. Transitions

Consider a state (q, v). Given a transition $t = (q, guard(t), e, X_0, q')$ such that $v \in guard(t)$ and $v' = v[x := 0, \forall x \in X_0] \in inv(q')$, $(q, v) \xrightarrow{e} (q', v')$ is a discrete transition of THSM also called the e-successor of (q, v). The condition $v \in guard(t)$ is also called an enabling condition of transition t. We assume that:

(1) for each transition t with the source state q $guard(t) \cap inv(q) \neq 0$, and
(2) for some outgoing transition t of state q if X-valuation $v \notin inv(q) \Rightarrow v \in guard(t)$.

Condition 1 is necessary for executability of the transition. Condition 2 means that when the invariant of a state is violated some outgoing transition must be enabled.

If q or q' are the hierarchical states the transition t is also called a top-level transition. It connects the final sub-state of q with the initial sub-state of q' given by a default entry function $\delta(q')$. The final sub-state of an OR-state is a state without output transition or if there is a loop within OR-state then a final sub-state is the final state of the loop (it has outgoing transition, which connects it with the initial state of the loop, e.g., states c, m, w from Figure 2). In the case, if q is AND-state with the immediate sub-states, which can be OR-states or basic states, the final states are stated similarly, and taking the transition means a joint exit from each of its orthogonal components. This top level transition is synchronization transition since all orthogonal sub-states of an AND-state must be synchronized on their final states. This requirement is in contradiction with (2) because independent transitions of components may interleave, and interleaving implicitly allows indefinite waiting of a component before achieving synchronization. In spite of that, a synchronized exit from AND- state is the important assumption to introduce the compositional semantics of THSM as, will be shown in the next sections. The above contradiction with (2) can be turned by introducing waiting state as a final state to all "faster" orthogonal components (see sub-state j of a from Figure 2).

If q' is AND- state, the initial states for each sub-state q_i', $i = 1, 2, .., k$ of q' are given by default entry function $\delta(q_i)$, and taking the transition means a fork entrance to each of its orthogonal components. If q and q' are both AND-states, taking the transition means a joint exit from each orthogonal component of q, and fork entrance to each orthogonal component of q'. The transitions between top-level states have a lower priority compared to the inner-level state transitions, if they are enabled simultaneously. For example, the transition $t2$ between states u and m from Figure 2 has lower priority compared to any transition of state u. $t2$ is not taken when event $e4$ occurs even if the enabling condition $z \geq 2$ is fulfilled, since the transition $t3$ is also enabled and has a higher priority compared to $t2$. This assumption gives the necessary condition for the synchronized exit from all orthogonal components of AND-state. The state consistency, which means that the control is always passed back to super-states is ensured by no infinite loops within a sub-state. In contrast with statecharts we do not allow inter-level transitions, i.e., the transitions crossing the borderline of states.

A time transition from (q, v) has the form $(q, v) \xrightarrow{\tau} (q, v + \tau)$ where $\tau \in R^+$, and $v + \tau \in inv(q)$, and it means that system is being in the state q while time elapses. If q is a hierarchical state the time τ is bounded according to the relations (1) or (2) in accordance with its type.

Example.

As an illustration, see Figure 2 showing a partial THSM. The root AND-state p is product of three OR-states r, q, and s. The state q is mapped to lower level sub-states n, u, and m. For instance, sub-state u is AND- state too. The hierarchical structure of THSM can be represented by directed acyclic graph (DAG)(Brave and Heymann 1993). The DAG of THSM from Figure 2 is in Figure

3. An invariant of state is given either in the position of the state or beyond the name of state in the square brackets. The timed transition must satisfy the constraint given by invariant, e.g., the model can be maximally 4 time units in l, so $(l,0) \xrightarrow{4} (l,4)$ is legal transition but $(l,0) \xrightarrow{5} (l,5)$ is not. The discrete transition between states is given by an oriented curve, and can be labeled with the event, the guard condition, and by the reset function. Default initial states of the hierarchical states are marked by small arrow.

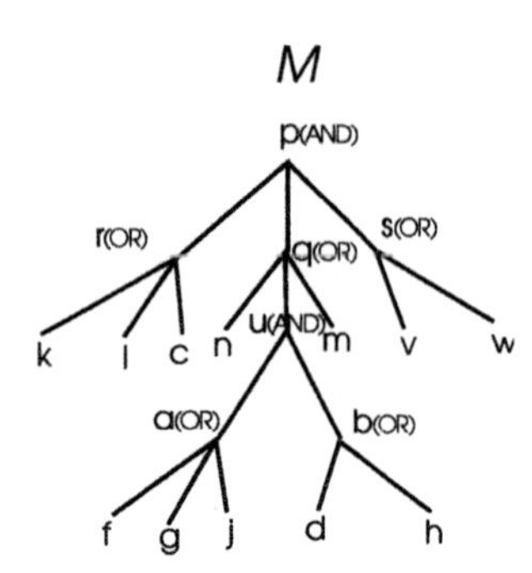

Fig. 3. DAG of THSM M from Figure 2.

3. OPERATIONAL SEMANTICS OF TIMED HSM

Definition 2. Let $s = \langle s_1, s_2, ..., s_n \rangle$ be a tuple of basic states $s_i = (q_i, v)$ where q_i is a basic discrete state and v is a X- valuation satisfying the invariant of q_i, $i = 1, ..., n$ then s will be called a configuration if every pair of discrete states q_i, q_j in s is orthogonal, i.e., their lowest common ancestor[3] is an AND-state.

In Figure 2, the tuple $\langle k, n, v \rangle$ is configuration but $\langle k, f, m, v \rangle$ is not, since the pair of basic states f, m is not orthogonal.

The configuration of state q is full if it is maximal, i.e., it contains maximal number of basic states that the system can be in simultaneously.

The set $\Re_s$ of full configurations for $s = (q, v)$, $q \in Q$ can be computed inductively as follows:

- if q is basic state then $\Re_s = \{\langle s \rangle\}$
- if q with sub-states $q_1, ..., q_k$ is OR- state then $\Re_s = \{\langle \Re_{s_i} \rangle\}$
- if q with sub-states $q_1, ..., q_k$ is AND- state then $\Re_s = \{\langle \Re_{s_1} \times ... \times \Re_{s_k} \rangle\}$.

A. Synchronization

Let $s = \langle s_1, ..., s_i, s_j, ..., s_n \rangle$ and $s' = \langle s_1 ..., s'_i, s'_j, ..., s_n \rangle$ be two configurations, and event $e_i = e_j = e \in l_i \wedge l_j$, where l_i, l_j are labels of transitions t_i, t_j, then the synchronization of discrete transitions t_i, t_j with scopes $\chi(t_i) \neg \rho \chi(t_j)$[4], where $t_i = (s_i, guard(t_i), e, X_i, s'_i)$, $t_j = (s_j, guard(t_j), e, X_j, s'_j)$ yields the transition $t_i \| t_j = (s, guard(t_i) \cap guard(t_j), e, X_i \cup X_j, s')$, where operator $\|$ means simultaneous execution of both transitions. Also event e is called a synchronization event.

B. Interleaving

Let s and s' be two configurations as formerly. Let $e_i \in l_i \neq e_j \in l_j$, i.e., $t_i = (s_i, guard(t_i), e_i, X_i, s'_i)$, $t_j = (s_j, guard(t_j), e_j, X_j, s'_j)$, then the interleaving of transitions t_i, t_j yields the transitions $t_i \| \bot = (s, guard(t_i), e_i, X_i, s')$ with $s_j = s'_j$ for $j \neq i$, and $\bot \| t_j = (s, guard(t_j), e_j, X_j, s')$ with $s_i = s'_i$ for $i \neq j$, where $\bot$ is an empty symbol.

C. Timed transition

Let $s = \langle s_1, ..., s_i, s_j, ..., s_n \rangle$ be a configuration, then time transition $s_i \xrightarrow{\tau} s_i + \tau$, for some $i \in \{1, ..., n\}$ has the result that an amount of time τ passes in the configuration s only if all components can delay τ time units, i.e., $s + \tau = \langle s_1 + \tau, ..., s_i + \tau ..., s_n + \tau \rangle$, and we simply write $s \xrightarrow{\tau} s + \tau$.

D. Traces

A trace of HTA starting from configuration s_0 is a finite (or infinite) sequence $\sigma = s_0 \xrightarrow{\tau_0} s_0 + \tau_0 \xrightarrow{e_0} s_1 \xrightarrow{\tau_1} s_1 + \tau_1 \xrightarrow{e_1} s_2 \xrightarrow{\tau_2} ... \xrightarrow{e_{n-1}} s_n ...$, such that for all $i = 0, 1, 2, ..., s_i + \tau_i$ is the τ_i- successor of s_i, and s_{i+1} is e_i- successor of $s_i + \tau_i$. The set of all traces of HTA will be denoted by Σ.

4. COMPOSITIONAL SEMANTICS OF TRACES

The goal is to define such compositional semantics of traces, which ensures that the semantics of the whole system can be determined from the semantics of its components. In order to show that our semantics is compositional, we need to be able to define the semantics of a hierarchical state only in terms of trace semantics of its sub-states.

In the following lemma we show how traces can be defined with further hierarchical extension of state q.

Lemma 1 (Trace construction). Let $\sigma \in \Sigma_q$ be a trace of state q. Let p be OR-state extension of q $(p \rho q)$ then $\sigma' \in \Sigma_p$ looks like $\sigma' = \sigma_1 \rightarrow \sigma_2 \rightarrow ... \rightarrow \sigma_i \rightarrow \sigma_{i+1} \rightarrow ... \rightarrow \sigma_k$, where σ_j, $j = 1, 2, ..., k$, $j \neq i$ are traces of new components p_j of state p, and $\sigma_i = \sigma$ is an original trace of state q.

[3] The lowest common ancestor (lca) of states q_1 and q_2 is a state q such that q_1, q_2 are sub-states of q, and for every sub-state p of q either p is super-state of q_1, and not super-state of q_2 or opposite.

[4] The scope of transition is denoted as $\chi(t)$, and represents serial lca of the source and target discrete states of transition, and $\chi(t_i) \neg \rho \chi(t_j)$ means that the serial lowest common ancestors of both transitions are not in the hierarchical relation ρ.

The lemma asserts that σ' arises from σ by the concatenation of traces of the individual sub-states of OR- state p. We marked by $\rightarrow$ top-level transitions between sub-states, for which the entrance configuration of trace σ_{i+1} is e_i successor of the exit configuration of trace σ_i.
Consider the case if p is AND-state extension of q arising by adding k orthogonal components p_i, $i = 1, ..., k$ to state q then $\sigma' = s'_0 \stackrel{\tau_0}{\rightarrow} s'_0 + \tau_0 \stackrel{e_0}{\rightarrow} s'_1 \stackrel{\tau_1}{\rightarrow} s'_1 + \tau_1 \stackrel{e_1}{\rightarrow} s'_2 \rightarrow ....$, where s'_i, $i = 1, ..., k$ are the new configurations of state p. Now, new trace σ' cannot be expressed directly as the concatenation of the original trace of q with traces of its super- state extension. The reason is that first, the full configurations of states p and q are different, and second, some new transitions arising with the state extension can be in the synchronization with transitions of state q.
For the special case of THSM called asynchronous THSM, we will show the compositional semantics of traces also for AND- state extension of q. Asynchronous THSM are characterized by sparse interaction between parallel components. That is, there is no synchronization of orthogonal components by means of shared events; all interactions are assumed to be modeled either by high-level transitions or by the transition constraints. We will use the following lemmas.

Lemma 2. With AND- state extension of state q its configuration is extended with a tuple of basic states of the new orthogonal components.
Proof.
It is evident from the construction of full configuration.

Lemma 3. Let p be AND-state extension of state q $(p\rho q)$. If $E_{p-q} \cap E_q = 0$[5] (the set of common events of added orthogonal state components and state q is empty) then σ' is the concatenation of σ with the trace formed by interleaving of transitions t_i, t_j with scopes $\chi(t_i) = p - q$ and $\chi(t_j) = q$ with the difference that now the traces σ' and σ are constructed from the extended state configurations according to Lemma 2.
Proof.
We mark the configuration of p as
$s_p = \langle\langle s_1\rangle, \langle s_2\rangle, ..., \langle s_q\rangle\rangle$, where $\langle s_1\rangle, \langle s_2\rangle, ...$ are configurations of new orthogonal components of state p, and $\langle s_q\rangle$ is the configuration of state q.
We want to prove that the trace $\sigma' = s_{p_0} \stackrel{\tau_{p_0}}{\rightarrow} s_{p_0} + \tau_{p_0} \stackrel{e_{p_0}}{\rightarrow} s_{p_1} \stackrel{\tau_{p_1}}{\rightarrow} s_{p_1} + \tau_{p_1} \stackrel{e_{p_1}}{\rightarrow} s_{p_2} \stackrel{\tau_{p_2}}{\rightarrow} ... \stackrel{e_{p_{m-1}}}{\rightarrow} s_{p_m}$ of state p contains the original trace of state q, $\sigma = s_{q_0} \stackrel{\tau_{q_0}}{\rightarrow} s_{q_0} + \tau_{q_0} \stackrel{e_{q_0}}{\rightarrow} s_{q_1} \stackrel{\tau_{q_1}}{\rightarrow} s_{q_1} + \tau_{q_1} \stackrel{e_{q_1}}{\rightarrow} s_{q_2} \stackrel{\tau_{q_2}}{\rightarrow} ... \stackrel{e_{q_{n-1}}}{\rightarrow} s_{q_n}$, where $m \geq n$.

- The time transition of σ $s_{q_i} \stackrel{\tau_{q_i}}{\rightarrow} s_{q_i} + \tau_{q_i}$ is contained in $s_{p_j} \stackrel{e_{p_j}}{\rightarrow} s_{p_{j+1}} \stackrel{\tau_{p_{j+1}}}{\rightarrow} s_{p_{j+1}} + \tau_{p_{j+1}}$ because $s_{q_i} \subset s_{p_j} \wedge s_{q_i} \subset s_{p_{j+1}}$ and $s_{q_i} + \tau_{q_i} \subset s_{p_{j+1}} + \tau_{p_{j+1}}$, where event $e_{p_j} \notin E_p$.
- The discrete transition of σ $s_{q_{i+1}} \stackrel{e_{q_i}}{\rightarrow} s_{q_{i+2}}$ is contained in $s_{p_{j+1}} \stackrel{\tau_{p_{j+1}}}{\rightarrow} s_{p_{j+1}} + \tau_{p_{j+1}} \stackrel{e_{p_{j+1}}}{\rightarrow} s_{p_{j+2}}$ because $s_{q_{i+1}} \subset s_{p_{j+1}} \wedge s_{q_{i+1}} \subset s_{p_{j+1}} + \tau_{p_{j+1}}$ and $s_{q_{i+2}} \subset s_{p_{j+2}}$, where $e_{p_{j+1}} = e_{q_i}$.

According to the assumption $E_{p-q} \cap E_q = 0$ all discrete state transitions of state p are interleaving, it means that all events $e_{p_i} \in E_{p-q}$ causing the discrete transitions in the trace σ', do not influence the components $\langle s_q\rangle$ of configurations from σ', which are changing in accordance with the configurations from σ. In the case when $E_{p-q} \cap E_q \neq 0$ (the synchronization of orthogonal components by means of shared events), the enabling conditions of transitions from the original trace σ will be different (conjunction of guards of transitions with shared event, see the synchronization case above) than in the case without state extension. It means that in this case the trace σ' will not contain the original trace σ.
The lemmas are used to prove the following theorem:

Theorem 1. The set of traces Σ of THSM can be computed from the set of traces of its sub-states and its discrete and timed transitions.
Proof.
It follows immediately from the preceding lemmas.

4.1 Refinement

The trace semantics allows us to define refinement between THSM. The refinement relation between models captures the notion that two THSMs describe the same system at different levels of detail.

Definition 3. Let
$M1 = (X, Q_1, Q_0^1, E, \delta, \rho, inv_1, T_1)$ and $M2 = (X, Q_2, Q_0^2, E, \delta, \rho, inv_2, T_2)$ be two Timed HSMs such that $Q_1 \subseteq Q_2$, $Q_0^1 \subseteq Q_0^2$, $T_1 \subseteq T_2$, and for all $q \in Q_1 \cap Q_2$, $inv_1(q) = inv_2(q)$ then $M2$ refines $M1$, denoted $M2 \preceq M1$, if $\Sigma_1 \subseteq \Sigma_2$.

Similarly, it can be defined submachine
$M' = (X, Q', Q'_0, E, \delta, \rho, inv', T')$ of THSM
$M = (X, Q, Q_0, E, \delta, \rho, inv, T)$ by the following conditions: $Q' \subseteq Q$, $Q'_0 \subseteq Q_0$ and $T' \subseteq T$ and for all $q \in Q' \cap Q$, $inv'(q) = inv(q)$. It will be marked as $M[M']$.
We show that refinement operator is compositional with respect to the hierarchy relation.

[5] We shall write $z = p - q$ iff z consists of all elements of p that do not appear in q.

Theorem 2. Given THSM M with submachine $M1$, such that
(1) $M1 \preceq N1$, if
THSM N has submachine $N1$, and
(2) the extensions of $M1$ to M and $N1$ to N are equal then $M \preceq N$.
Proof.
We have to prove $M[M1] \preceq N[N1]$. In accordance with the theorem assumption (1) it results that $M[M1] \preceq M[N1]$, and from the assumption (2) results $M[N1] \preceq N[N1]$.
Example.
As an illustration of the theorem we introduce an example of DAG from Figure 3. DAG of $N1$, with state u that is not refined, is given in Figure 4(a). DAG of $M1$, now with state u refined to its sub-states is given in Figure 4(b). N is given in Figure 4(c). You can see that M with DAG given in Figure 3 refines N that is the consequence of the refinement of $N1$ with $M1$.

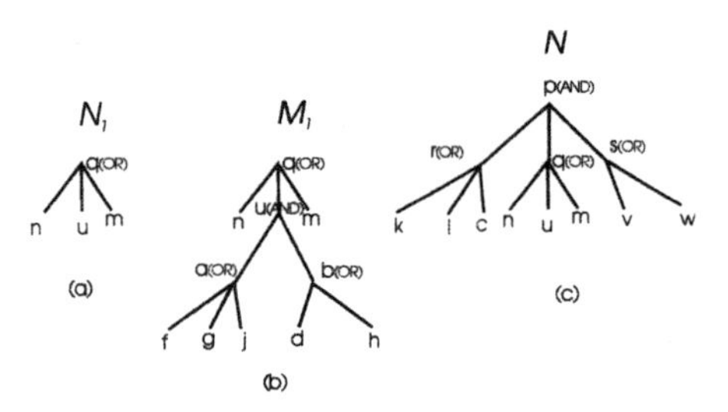

Fig. 4. DAGs of B_1, A_1, B showing the refinment of the hierarchical states.

5. VERIFICATION

The problem of the model verification means to show that the temporal formula given in some specification language, e.g., CTL, LTL is satisfied in the model trajectories or states. For complex models this problem leads to the so called "state explosion problem". In the formal methods there is several techniques solving this problem.
Authors of (Alur and Grosu 2000) proposed automatic verification system based on the principle of assume-guarantee reasoning for the hierarchical hybrid systems. The goal is to decompose complex verification task into subtasks of manageable complexity. The authors demonstrate how the model supports hierarchical, component-based design and analysis. The refinement-checking problem is to prove that the complex component $P_1 \| P_2$ refines simpler component Q. By assume-guarantee reasoning this can be concluded from the two proof obligations $P_1 \| A_2 \preceq A_1 \| Q$ and $A_1 \| P_2 \preceq Q \| A_2$, where abstraction components A_1, A_2 are suitable constraining environment A_2 for P_1, and similarly A_1 for P_2.
The semantics of THSM defined in the paper allows to decompose the proof rule for complex component into proof rules of individual components without abstraction components.

Theorem 3. Let P_1, P_2, Q be THSMs, if $P_1 \preceq Q$ and $P_2 \preceq Q$ are simultaneously replaced in THSM M than $M[P_1 \| P_2] \preceq M[Q]$.
Proof.
In accordance with Theorem 2, it results that $M[P_1] \preceq M[Q]$ and $M[P_2] \preceq M[Q]$ and due to the compositionality of the parallel composition (Theorem 1), it results that
$M[P_1] \| M[P_2] \preceq M[Q] \| M[P_2] \preceq M[Q] \| M[Q] = M[Q]$.
Since it is valid $M[P_1 \| P_2] = M[P_1] \| M[P_2]$ the statement of the theorem is proved.

6. CONCLUSION

We defined the time extension of HSM, and proposed the compositional semantics of THSM. We have shown how the compositional semantics can be exploited in the verification process. In the future, we plan to concentrate our work on the wide class of verification tasks concerning mainly the verification of the temporal formulas logic TCTL expressing the real-time model behaviors.

Acknowledgments
The author is grateful to the SGA for Science VEGA (grant No, 2/1101/22) and APVT agency (grant No, 51 011602) for partial supporting of this work.

REFERENCES

Alur, R. (1999). Timed automata. In: *11th Conference on Computer- Aided Verification.* LNCS 1633, Springer-Verlag. pp. 8–22.
Alur, R. and D. Dill (1994). A theory of timed automata. *Theoretical Computer Science* (126), 183–235.
Alur, R. and R. Grosu (2000). Modular refinement of hierarchic reactive machines. In: *Proceedings of the 27th ACM Symposium on Principles of Programming Languages.* pp. 390–402.
Alur, R., S. Kannan and M. Yannakis (1999). Communicating hierarchical state machines. In: *Proceedings of the 26th International Colloquium on Automata, Languages, and Programming.* LNCS 1644, Springer- Verlag. pp. 169–178.
Brave, Y. and M. Heymann (1993). Control of discrete event systems modeled as hierarchical state machines. *IEEE Transaction on Automatic Control* **38**, 1803–1819.
David, A., M. O. Moller and Wang Yi (2002). Formal Verification of UML Statecharts with Real-Time Extensions. In: *Proceedings of FASE 2002.* LNCS 2306, Springer- Verlag. pp. 218–232.

SOLUTION OF THE MANUFACTURING TRANSPORTATION CONTROL USING PETRI NETS

Branislav Hrúz, Leo Mrafko

Slovak University of Technology, Faculty of Electrical Engineering and Information Technology
Department of Automatic Control Systems
Ilkovičova 3, 812 19 Bratislava, Slovak Republic
Phone: +421 2 60291698, e-mail: hruz@kasr.elf.stuba.sk

Abstract. An analysis of the transportation systems used in the flexible manufacturing systems is introduced in the paper. Based on the analysis a class of the transportation systems is delimited, whose modeling and control is the subject of the paper. Important property of those systems is unidirectional zone movements of the automatic guided vehicles and transport requirements occurring spontaneously during the work of the system. A special group of the colored Petri nets interpreted for control is used for the control specification. The Petri net helps to analyze the function of the transportation system and to write the control program. *Copyright © 2003 IFAC*

Keywords: transportation systems in manufacturing, discrete event dynamic systems, Petri nets, colored Petri nets.

1. INTRODUCTION

A growing attention is paid to transportation problems in the manufacturing systems. As usual in the system engineering there are two problem aspects to be solved, namely how to model transportation subsystems and following-up how to automatically control them.

From the general view the transportation systems can be described as discrete event dynamic systems (DEDS). States are given by positions of the used transportation means and state transitions by passes from some positions to other ones. Consequently, the tools used for modeling and control of DEDS such as Petri nets (Zhou, 1995; Murata, 1989; Abel, 1990; Hrúz, 1994; Čapkovič, 1993; Hrúz, 2003), Grafcet (David and Alla, 1992), reactive flow diagrams (Hrúz, 2003) state-charts (Fogel, 1997) etc. are generally applicable for the transportation systems, as well. As mentioned, such tools serve first for the behavior and function specification of the whole system including its control and second for the specification of the control subsystem in a feedback structure "controlled system ⇔ control system". The models are adapted in a form of so-called synchronized models or models interpreted for control. In such a form they enable to generate easier and more correct the reactive programs performing the automatic control (Hrúz, Niemi, and Virtanen, 1996). This is a way we use in this paper to solve modeling and control of a class of the manufacturing transportation systems.

2. ANALYSIS AND CLASSIFICATION OF TRANSPORTATION CONTROL PROBLEMS

Automatic product transportation systems are usually parts of flexible manufacturing systems (FMS). Vehicles or carts provide the transport of palettes, work-pieces, fixtures, etc. between machining cells, processing centers, servers, robots, testing and measuring devices, buffers, storages, etc. The transport system performance combined with the job scheduling determines the overall efficiency of the manufacturing system (Li, Takamori, and Tadokoro, 1995).

Modeling and control of a transportation system depends to a large extent on the technologic layout and functional requirements. In what follows we consider automatic guided vehicle (AGV) track system. It can be realized as a rail system, inductive wire guidance system or similar (Koff, 1987). As such the tracks are supposed to be fixed unlike as systems with free movements in the respective transportation space. Moreover, in our considerations we assume the tracks are divided into the zones or sections. Such an arrangement enables to apply a thumb rule of the safety, namely that only one vehicle can be in one zone.

The function of the transportation is, of course, subordinated to the work-pieces routings according to the required job scheduling in the particular FMS. The following functions can be distinguished:

a) A job schedule and resulting routes are à priori given and they are repeated cyclically in a manufacturing process. A transportation task within the manufacturing jobs is specified by the transfers from a section to section. The assignments of vehicles to the transfers are fixed.
b) Transportation requests are not given in advance. They occur spontaneously at the beginning and during the manufacturing execution. A vehicle is assigned just for one transfer and after its execution the vehicle is free.

A routing optimisation may be desirable in both cases. In the first case more or less it can be calculated in advance off-line. In the second one it has to be re-calculated during the production run. Transportation durations and/or processing operation durations influence the optimisation in both cases. Most used criteria are the shortest path or minimum transfer time. Dynamic re-planning of the routes or transfers can be relevant mainly in the case b).

As mentioned earlier, a long-time practice is to ensure the transportation safety and the non-collision operation by the track system division into zones or sections allowing one vehicle to occupy a section. Sensors detect the vehicle presence in a section. There can be one sensor in one section or sensors on both ends of a section. We assume a normal stop in the switch or crossing area is not allowed. AGV motions can be controlled via power on/off switching in the rail tracks or by a communication of control signals to the vehicles through rails or by some wireless technique.

An important feature of the manufacturing system transportation is direction of the vehicle movements. There are two possibilities:

1) in all zones or in part of them the bi-directional movements of the vehicles are allowed,
2) in all zones only unidirectional movements are allowed.

Movement freedom has a decisive influence on behavior of a system and on control solution. Usually the first case is intended to obtain a transportation system with many alternative routes and detours while the second one has rather specific connections between respective zones with few alternative connections. The described diverse transportation arrangements require also different and rather distinctive modeling and algorithmic approach in a control solution.

Case a)-1) is solved e.g. in Hrúz, Mrafko, and Bielko, 2003, by means of Petri nets interpreted for control. Case b)-1) is solved in Kim and Tanchoco, 1991, Maza and Castagna, 2002, using so-called reserved time windows and in Hrúz, Mrafko, and Bielko,2002, as a supervisory control problem.

The next sections of the paper deal with the case b) for the transportation requests and with the movement freedom case 2), both in the zoned track transportation systems. A transport requirement contains a section of departure and a goal section. The requirements are transformed by a vehicle routing planner into a sequence of movements through the zones.

3. DESCRIPTION OF THE TECHNOLOGICAL LAYOUT CLASS

A class of the production fixed track transportation systems with strictly unidirectional movements in the zones as indicated above are assumed. During production operations the transportation jobs occur spontaneously depending on the intermediate and partial results of the production. Responsible operators put in required transport jobs during the production process.

Fig. 1 shows a typical arrangement of the transportation system within the delimited class. The system presented in the figure consists of a group of the product checking stations (T0 ÷ T4) and processing centers (Pij) where the defects or malfunctions of the products are removed. Operators in the checking stations decide the next operation and the corresponding next zone where the respective product should be transported. The transportation zones are denoted as Z1, Z2,...,Z41 and the movement switching areas as C1, C2, ..., C13 in Fig. 1. Tab. 1

and Tab. 2 illustrate transport requirements and vehicle assignments for transport jobs. Notation FREE → L1 means a call for a free vehicle. The products are fixed on the palettes. A palette with the product is loaded onto a free vehicle in loading station denoted as L1. This station is combined with checking station TO. Operator calls a free vehicle and then he sends the vehicle with the palette according to the checking result in one of the processing stations P41 through P45 or to some station of P31, P32 or to the testing station T21, etc. U1 and U2 are devices,

Tab. 1. List of transfer requirements.

FIFO stack of transfer requirements
FREE → L1
L1 → T2
FREE → P32
P32 → T1
T1 → U3

which unload palettes from vehicles or load the palettes back into the system. Tab. 3 describes possible transfers between the system stations. The groups of the stations performing identical operations

Tab. 2. Transfers in execution.

Vehicle No	Vehicle status	List of executed transfers given by sequences of zones
5	Busy	Z19 Z15 Z5 Z38 Z34
1	Busy	Z37 Z38 Z34
6	Free	Z27 Z11 Z12 Z13 Z14

Tab. 3. Possible transfer requirements.

Possible transfer requirements
T0 →GP3 or GP4 or T1 or T2 or GU
GP4 → GP3 or T1 or T2 or GU
GP3 → GP4 or T1 or T2 or GU
T2 → T3
T3 → T4
T4 → GP1 or GP4 or GU
GP1 → GP3 or GP4 or T1 or GU
T1 → GP3 or GP4 or T2 or GU or U3
U1 → GP3 or GP4 or T1 or T2
U2 → GP3 or GP4 or T1 or T2
T1 → GP2 or T2
GP2 → GP3 or GP4 or T1

are the following GP1={P11, P12}, GP2={P21, P22}, GP3={P31, P32}, GP4={P41, P42, P43, P44, P45, P46}, GU={U1, U2}.

Special routes are specified for the manipulation with the free vehicles. A free vehicle can be called from any checking or processing station. If a vehicle becomes free it is directed to zone Z1 having capacity of 8 vehicles, or to zone Z30 having capacity of 2 vehicles. The third possibility is to direct the free vehicle on the parking cycle track consisting of the zones Z5 - Z6 - ...- Z15. A command for that is to go to zone Z10. If no job is assigned meanwhile to the

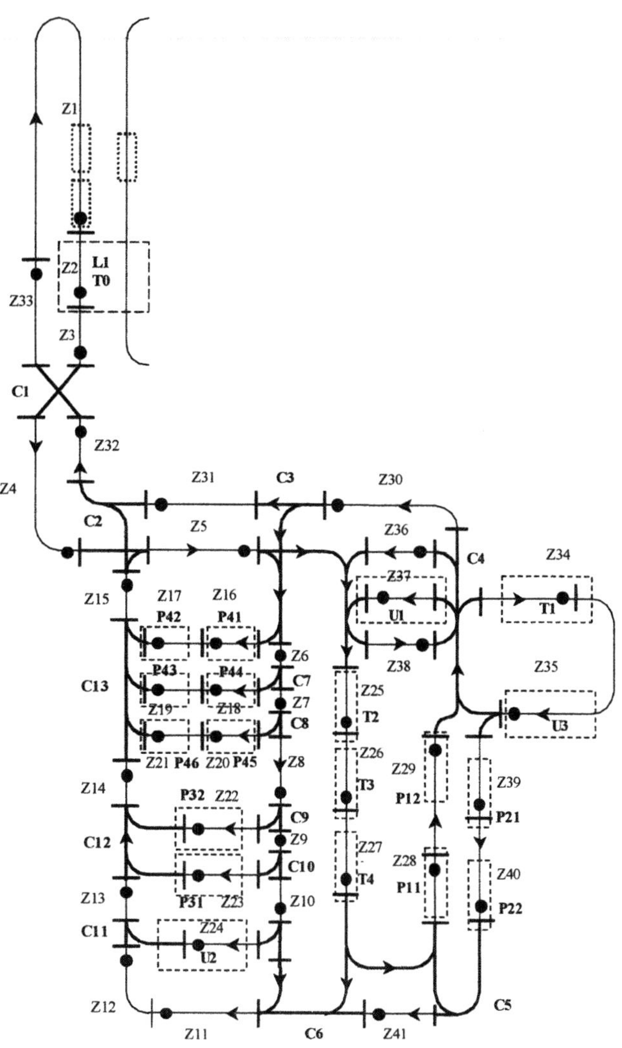

Fig. 1. Technological layout of a transportation system.

vehicle after reaching the goal zone it is sent to zone Z14 etc. The number of the vehicles is determined in the system under study to be 18. The performance analysis of the system with respect to the operation and transportation times and statistics of the jobs is to be analyzed and optimal number of the vehicles calculated. It is not the subject of this paper.

A choice of a route according to an operator's requirement is elaborated by a route planning program. The route planner works according the following rules:

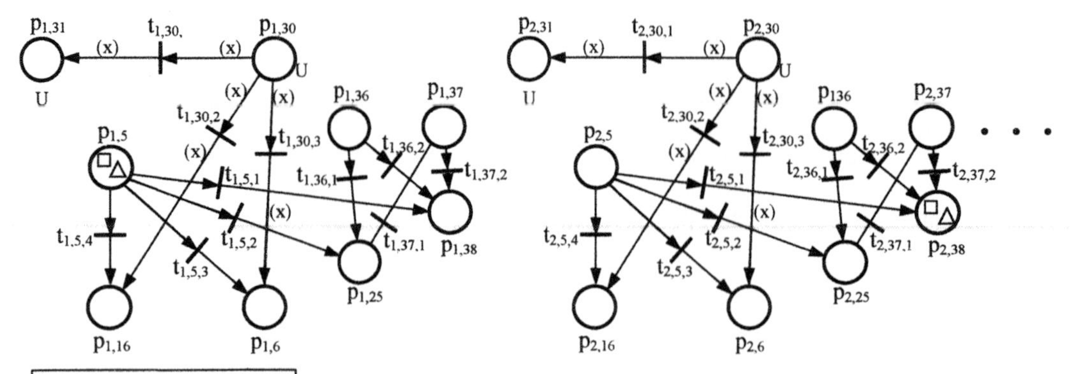

Fig. 2. Colored Petri net for the transportation system consisting of the subnets.

1) A free vehicle is called from the nearest accommodation site of the free vehicles, which has a free vehicle (i.e. from Z1, Z2, and the parking cycle Z5 ÷ Z15).
2) The vehicle is sent to the goal station via the shortest route. If the number of the vehicles on that route is higher than a prescribed value, the detour is chosen if any is specified as reasonable and if it is comparatively free.
3) An unloaded vehicle is sent to a zone for the free vehicles with the following priorities, a) to zone Z1 if there are less than 4 vehicles, b) to zone Z30 if there are less than 2 vehicles, c) to the parking cycle Z5 ÷ Z15 if there are less than 6 vehicles, d) if the preceding numbers are fulfilled the numbers of the vehicles are completed to the maximum values.

4. PETRI NET MODEL OF THE SYSTEM

Color Petri nets interpreted for control are chosen as a suitable tool for the modeling and control design of the transportation system class specified in Section 3. In what follows a construction of the Petri net will be presented for the specified class.

The Petri net consists of the subnets that are not connected. One subnet is constructed for each vehicle. The places of the Petri net correspond to the transportation system zones with a differentiation as for the vehicles. There are different places for different vehicles in the same zone. Determination of the places depends on the location of the sensors detecting the vehicle presence. The vehicle position is unknown if the vehicle is outside of the sensor detecting range. Some of the zones are less than the length of the vehicles so that the vehicle blocks the preceding zone or zones. It should be respected in the control policy.

The switching areas are determined by the situation given by the surrounding zones, the dimensions with respect to the dimension of the vehicles and locations of the vehicle presence detecting sensors. Fig. 2 shows how the Petri nets are created. For the brevity there is depicted in Fig. 2 only a part of the Petri net specifying the control of the vehicle movement through the switching area C3.

Notation of the Petri net places is $p_{i,j}$ where index i is related to vehicle V_i supposing there are vehicles $V_1, V_2, \ldots, V_N$ in the AGV system. Index i distinguishes the i-th Petri net subnet as it is clear from Fig. 2. Index j corresponds to the zone number. E. g. place $p_{1,31}$ belongs to the first subnet and it corresponds to zone 31. Similarly the Petri net transitions are indexed $t_{i,j,k}$ where i, j has the same meaning as for the places and index k is order index of the post-transitions of the place $p_{i,j}$. The used indexing is based on the fact that the subnets in the used Petri net model are the class of the finite state-machine Petri nets.

Two colors are defined in the Petri net, namely □ and Δ. The idea of the colors is the following. The used colored Petri net is interpreted for control in the sense that presence of vehicle V_i in a zone is given by presence of token   in the corresponding place and the respective transitions are equipped with conditions while places with commands both related to the token with color □. If the respective conditions allow it, the token transfers to the following place and a command for the vehicle movement is released. The token with color Δ is transferred after token with color □ if the presence signal of the vehicle is released.

The start signal for the movement of vehicle V_i is denoted by STARTi and for its delayed stop as DLSTOPi.

As explained before each transition in the Petri net is equipped with logic expression binding the transition firing. Expression $e_{i,j,k}$ is assigned to the transition $t_{i,j,k}$. The following logic variable appear in the expressions:

$\alpha_{i,j} = 1$ if $p_{i,j}$ contains token with color □ , otherwise $\alpha_{i,j} = 0$,

$\beta_{i,j} = 1$ if $p_{i,j}$ contains token with color △ , otherwise $\beta_{i,j} = 0$,

$r_{i,j,k}$ is set to value logic 1 for the realized actual route of the i-th vehicle,

$s_{i,j} = 1$ if the vehicle presence sensor is activated, otherwise $s_{i,j} = 0$,

$b_{i,j,k} = 0$ if a vehicle blocks some zones because of the vehicle and zone dimensions. The logic condition system for the switching area C3 and vehicle V_1 is as follows

$$e_{1,5,1} = \left(\alpha_{1,5} \wedge \overline{\alpha_{2,38}} \wedge \ldots \wedge \overline{\alpha_{N,38}} \wedge \beta_{1,5} \wedge r_{1,5,1}\right) \vee \left(\beta_{1,5} \wedge \alpha_{1,38} \wedge s_{1,38} \wedge r_{1,5,1}\right)$$

$$e_{1,5,2} = \left(\alpha_{1,5} \wedge \overline{\alpha_{2,25}} \wedge \ldots \wedge \overline{\alpha_{N,25}} \wedge \beta_{1,5} \wedge r_{1,5,2}\right) \vee \left(\beta_{1,5} \wedge \alpha_{1,25} \wedge s_{1,25} \wedge r_{1,5,2}\right)$$

$$e_{1,5,3} = \left(\alpha_{1,5} \wedge \overline{\alpha_{2,6}} \wedge \ldots \wedge \overline{\alpha_{N,68}} \wedge \beta_{1,5} \wedge r_{1,5,3} \wedge b_{1,5,3}\right) \vee \left(\beta_{1,5} \wedge \alpha_{1,6} \wedge s_{1,6} \wedge r_{1,5,3}\right)$$

$$e_{1,5,4} = \left(\alpha_{1,5} \wedge \overline{\alpha_{2,16}} \wedge \ldots \wedge \overline{\alpha_{N,16}} \wedge \beta_{1,5} \wedge r_{1,5,4}\right) \vee \left(\beta_{1,5} \wedge \alpha_{1,16} \wedge s_{1,16} \wedge r_{1,5,4}\right)$$

$$e_{1,30,1} = \left(\alpha_{1,30} \wedge \overline{\alpha_{2,31}} \wedge \ldots \wedge \overline{\alpha_{N,31}} \wedge \beta_{1,30} \wedge r_{1,30,1}\right) \vee \left(\beta_{1,30} \wedge \alpha_{1,31} \wedge s_{1,31} \wedge r_{1,30,1}\right)$$

$$e_{1,30,2} = \left(\alpha_{1,30} \wedge \overline{\alpha_{2,16}} \wedge \ldots \wedge \overline{\alpha_{N,16}} \wedge \beta_{1,30} \wedge r_{1,30,2}\right) \vee \left(\beta_{1,30} \wedge \alpha_{1,16} \wedge s_{1,16} \wedge r_{1,30,2}\right)$$

$$e_{1,30,3} = \left(\alpha_{1,30} \wedge \overline{\alpha_{2,6}} \wedge \ldots \wedge \overline{\alpha_{N,6}} \wedge \beta_{1,30} \wedge r_{1,30,3} \wedge b_{1,30,3}\right) \vee \left(\beta_{1,30} \wedge \alpha_{1,6} \wedge s_{1,6} \wedge r_{1,30,3}\right)$$

$$e_{1,36,1} = \left(\alpha_{1,36} \wedge \overline{\alpha_{2,25}} \wedge \ldots \wedge \overline{\alpha_{N,25}} \wedge \beta_{1,36} \wedge r_{1,36,1}\right) \vee \left(\beta_{1,36} \wedge \alpha_{1,25} \wedge s_{1,25} \wedge r_{1,36,1}\right)$$

$$e_{1,36,2} = \left(\alpha_{1,36} \wedge \overline{\alpha_{2,38}} \wedge \ldots \wedge \overline{\alpha_{N,38}} \wedge \beta_{1,36} \wedge r_{1,36,2}\right) \vee \left(\beta_{1,36} \wedge \alpha_{1,38} \wedge s_{1,38} \wedge r_{1,36,2}\right)$$

$$e_{1,37,1} = \left(\alpha_{1,36} \wedge \overline{\alpha_{2,25}} \wedge \ldots \wedge \overline{\alpha_{N,25}} \wedge \beta_{1,37} \wedge r_{1,37,1}\right) \vee \left(\beta_{1,37} \wedge \alpha_{1,25} \wedge s_{1,25} \wedge r_{1,37,1}\right)$$

$$e_{1,37,2} = \left(\alpha_{1,37} \wedge \overline{\alpha_{2,38}} \wedge \ldots \wedge \overline{\alpha_{N,38}} \wedge \beta_{1,37} \wedge r_{1,37,2}\right) \vee \left(\beta_{1,37} \wedge \alpha_{1,38} \wedge s_{1,38} \wedge r_{1,37,2}\right)$$

STARTi and DLSTOPi commands are set when token with color □ arrives in any place. When □ arrives in place $p_{1,6}$ the variables $b_{1,5,3}$ and $b_{1,30,3}$ are set to 0. By the token departure they are set to 1.

Expression analog to those given above can be written for each subnet. The commands associated with the places are given in the analog way, as well. A general procedure for the generation of the condition system can be created. In that way a complete system of the equation for the token flow can be obtained, which together with the colored Petri net incorporated in the control program brings about the control of the transportation system.

Petri net subnets concerning individual vehicles can be combined into one colored Petri net, modeling all vehicles in one Petri net structure. The structure is the same, as depicted in Fig. 2, except another token color definitions. A part of the aggregated Petri net is shown in Fig. 3.

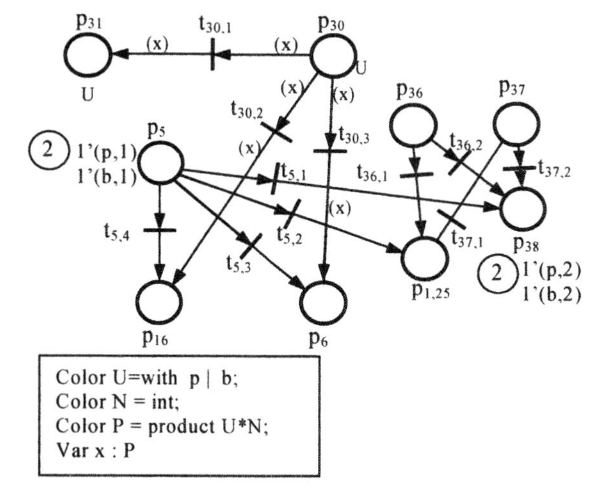

Fig. 3. Colored Petri net for the transportation system.

Two colors are used: U, which correspond to marking used in Fig. 2 (value p corresponds to symbol □ meaning presence of vehicle, value b corresponds to symbol △ meaning blocking of a section by a vehicle. It is also necessary to adjust conditions connected to transitions in order to reflect the new colouring convention. Some of the conditions are simply explainable – e.g. two tokens of different value of the color N cannot meet in one place, i.e. a vehicle cannot enter a section already blocked or occupied by another vehicle.

5. CONCLUSION

Process control of a specified class of a transportation system was solved in the paper using colored Petri nets interpreted for control. A problem of the dead-locks in the system can be solved by means of the reachability graph analysis. Another problem is to analyze the system performance. It depends on the stochastic properties of events and operations in the system. It is to be solved by means of some suitable methods e.g. using stochastic or fuzzy Petri nets. Those problems are a subject of a further research and are beyond the scope of this paper.

6. REFERENCES

Abel, D. (1990). *Petri-Netze für Ingenieure.* Springer-Verlag, Berlin.

Čapkovič, F. (1993). Modelling and justifying discrete production processes by Petri nets. *Computer Integrated Manufacturing Systems*, **Vol. 6**, No 1, pp.27-35.

David, R., and H. Alla (1992). *Petri nets and Grafcet.* Prentice Hall, New York.

Fogel, J. (1997). A statecharts approach to the modeling of discrete manufacturing systems. *Proc. of the 7th Symposium on Computer Aided Control Systems Design*, Gent, April 28-30, 1997.

Hrúz, B. (1994). Discrete event systems modelling and real-time control. *Journal of Electrical Engineering,* **45**,363-370.

Hrúz, B., A. Niemi, and T. Virtanen (1996). Composition of conflict-free Petri net models for control of flexible manufacturing systems. *Proc. of the 13th IFAC World Congress,San Francisco*, **Vol. B**, 37-42.

Hrúz, B., L. Mrafko, and V. Bielko (2002). A comparison of the AGV control solution approaches using Petri nets. *Proc. of the IEEE Conf. on System, Man and Cybernetics*, Hammamet, October 5-9, 2002.

Hrúz, B. (2003). *Modeling and Control of Discrete Event Dynamic Systems with Petri Nets and other Tools.* Publishing House of Slovak University of Technology, Bratislava, 2003.

Kim, C. W. and J. M. A. Tanchoco (1991). Conflict-free shortest-time bidirectional AGV routing. *Int. J. Prod. Res.*, **Vol. 29**, No 12, pp.2377-2391.

Koff, G. A. (1987). Automatic guided vehicle systems: applications, controls and planning. *Material Flow*, **Vol. 4**, 1987, pp. 3-16.

Li, S., T. Takamori, and S. Tadokoro (1995). Scheduling and re-scheduling of AGVs for flexible and agile manufacturing. In Zhou, M.C. (Edit.) *Petri Nets in Flexible and Agile Automation*, Kliwer Academic Pulishers, Boston, pp.189-205.

Maza, S. and P. Castagna (2002). Robust conflict-free routing of bi-directional automated guided vehicles. *Proc. of the IEEE Conf. on System, Man and Cybernetics*, Hammamet, October 5-9, 2002.

Murata, T. (1989). Petri nets: properties, analysis and applications. *Proc. IEEE*, **77**, 541-580.

Zhou, M. C. (1995). *Petri Nets in Flexible and Agile Automation.* Kluwer Academic Publishers, Boston.

ALGORITHMS FOR MODULAR STATE EVOLUTION AND STATE-SPACE ANALYSIS OF PETRI NETS

Jana Flochová, René Boel, George Jiroveanu

Department of Automatic Control Systems, Slovak University of Technology
Ilkovičova 3, 812 19 Bratislava, Slovak republic; SYSTeMS, Universiteit
Gent, Technologiepaark Zwijnaarde 9, 9052 Zwijnaarde Belgium
flochova@kasr.elf.stuba.sk, rene.boel@rug.ac.be, george.jiroveanu@rug.ac.be

Abstract: Computational complexity issues in the field of discrete event system (DES) theory are due to the state space explosion problem. This paper describes algorithms based on the T-time Petri net formalism to provide an abstract representation of the plant that allows computationally tractable, modular analysis and design algorithms for supervisory controllers, avoiding this state space explosion problem. The contribution of this paper consists in the introduction of some decomposition techniques and programs implementing this modular approach to the construction of the reachable state set. These tools are useful for control synthesis and for failure diagnosis of real plants, represented by timed discrete event systems. *Copyright ©2003 IFAC*

Keywords: Discrete-event systems, Petri-nets, Modeling and Simulation, Reachability

1. INTRODUCTION

Discrete event system (DES) models were introduced in the eighties for representing and analyzing automated manufacturing systems, robotics, communication networks, fault detection and recovery mechanisms, traffic control, etc. The activity in these systems is governed by operational rules designed by humans; their dynamics are often characterized by asynchronous occurrences of discrete events (Ho and Cassandras 1983; Cassandras and Lafortune 1999). Today's DES plants have become even more complicated, and intensively interconnected; they are subjected to many constraints concerning safety, reliability, energy consumption, and environment protection. Severe complexity issues in the field of discrete event system theory motivated various modular approaches to the DES model design and analysis, and to model based supervisory control of these plants.

This paper addresses some of these issues. We discuss techniques for representing the state space of a timed DES plant – modeled as a T-time Petri net - having a finite set of state classes. We combine this representation with a modular decomposition of the Petri net, into state machine components interacting via guards and via token synchronized passing (Boel and Jiroveanu 2003). We describe some of the programming tools for implementing a scalable algorithm constructing the set of reachable states for such a plant. These sets of reachable states are important in the supervisory control (Ramadge and Wonham 1987, 1989) and in the fault diagnosis for timed DES.

The outline of the paper is as follows: in the second section we define *T-time controlled Petri nets*; the third section presents modular reachability and state class tree approaches for such a plant. The forth section describes some of the designed algorithms that have been implemented in C++ modules and the PNCONTROL programming library.

2. T-TIME PETRI NETS

Formal definition of a controlled Petri net (Holloway and Krogh 1990; Boel et al. 1995):

Definition 1. A controlled Petri net *PN* (Holloway and Krogh 1990; Boel et al. 1995) is defined as

$$PN = (P,T,F,W,C,B,m_0) \quad (1)$$

where (P,T,F,W,m_0) represents a Petri net:

1. $P = \{p_1, p_2,..., p_n\}$ is a finite non-empty set of places.
2. $T = \{t_1, t_2,..., t_m\}$ is a finite non-empty set of transitions.
3. $F \subset (P\times T)\cup(T\times P)$ is the set of oriented arcs connecting places to transitions, or transition to places.
4. W is the weight function of the arcs.
5. m_0: $P\rightarrow\{0,1,2,3...\}$ is the initial marking.

and where moreover

6. C is the finite set of control places, at most one per transition, connected to transitions via the set of oriented arcs $B \subset C\times T$.

A control u: $C \rightarrow \{0,1\}$ assigns a binary token count to each control place. A transition $t \in T$ in a controlled Petri net can fire if the marking enables the transition (all input places $p\in {}^\circ t$ contain more than $W(p,t)$ tokens) and if the value of $u(c) = 1$, where c is the control place that is connected to transition t according to the model topology ($B(c,t) = 1$). Transition t is said to be control disabled under a control u if $u(c) = 0$ for $B(c,t) = 1$.

The state evolution of Petri net PN is given by:

$$m_{i+1} = m_i + (\boldsymbol{D}^+\text{-}\boldsymbol{D}^-)n_i = m_0 + (\boldsymbol{D}^+\text{-}\boldsymbol{D}^-)x_i \quad (2)$$

where $\boldsymbol{D}^-$, $\boldsymbol{D}^+$ represent the backward and forward incidence matrices (the rows of $W(p,t)$ and of $W(t,p)$ respectively). The vector n_i has exactly one non-zero entry with value 1 indicating which transition fires, $x_i = \sum_{j=1}^{i+1} n_j$ is the firing count vector.

A marking m_i is said to be reachable from a marking m_0 if there exists a sequence of admissible firings that transforms m_0 to m_i. We denote by $R(m_0)$ the set of states reachable from the initial state m_0. $R(M_n)$ is the set of states reachable from the set M_n of allowed initial states. A *PN* is said to be *k-bounded* if $m_i \leq k$ for all $p_i \in P$ and all $m \in R(M_n)$, $k \in$ N. A Petri net *PN* is said to be safe if it is *1-bounded.*

Time Petri net are obtained from Petri nets by associating a (possibly infinite) interval of time delays with each transition.

Definition 2. The *T-time controlled Petri net* (Boel and Stremersch 1998; Boel and Montoya 2000)

$$TCPN = (P,T,F,W,C, B, L,U,m_0) \quad (3)$$

consists of the following entities:

1. $PN = (P,T,F,W,C,B,m_0)$ is a controlled Petri net.
2. Functions L: $T\rightarrow Q\cup\{\infty\}$, U: $T\rightarrow Q\cup\{\infty\}$, $L(t) \leq U(t)$ associate to each transition a static time interval $I(t) = \langle L(t), U(t)\rangle$ (lower and upper bound of time delay).

The dynamical evolution of a T-time Petri net is described as follows. If transition t becomes simultaneously state and control enabled at time τ, then t can and must fire in the interval $\langle\tau+ L(t),\tau+ U(t)\rangle$ (it is forced to fire at time $\tau+ U(t)$), provided t did not become disabled by that time as a result of the firing of another transition). Using a terminology similar to the terminology of timed automata, we say that a local clock is started as soon as transition t becomes state- and control-enabled. The transition t must be executed some time while the clock is in the interval $\langle L(t),U(t)\rangle$. Note that several clocks may be running simultaneously for transition t if the transition is enabled by several tokens (both state- and control enabled). However in practice we encounter only safe T-time Petri nets, where only one clock at a time is running for one transition. This simplifies the notation a lot.

If no control places are present (meaning that clocks controlling the firing of a transition t start to run as soon as t is state enabled) we call the net a T-time Petri net (*TPN*).

3. MODULAR STATE-SPACE ANALYSIS

In order to enumerate all possible trajectories emanating from a given initial condition one has to remember for each enabled transition what the value is of the local clock. This defines the state:

Definition 3. A state of a TPN is a pair $S = (m, I)$ where:

1. m is a marking.
2. I is a firing interval set which is a vector of possible firing times. The number of entries in this vector is given by the number of the transitions enabled by marking m.

A state is reached from the initial state by a given sequence of firing times corresponding to a firing sequence. Since any reachable marking may be reached from the initial marking by different sequence of firing times corresponding to the same firing sequence, the state-space may be infinite. Just as for the Alur and Dill type timed automata (Alur and Dill 1996) one can show that this set of reachable states can be represented finitely, provided the bounds $L(t)$ and $U(t)$ take rational values only. Therefore it makes sense to introduce the following sets of states:

Definition 4. A state class of a TPN associated with a firing sequence s that has been executed in the past, is a pair $C = (m,D)$ in which:

1. *m* is the marking of the class reached from the initial marking by firing sequence s; all states in the class have the same marking.
2. *D* is the firing domain of the class, which is defined as the union of all firing domains of states reachable from the initial state by firing schedules with firing sequence s.

A state class represents the set of all states reachable from the initial state by firing all feasible firing values *corresponding to the same firing sequence.* The procedure of classes enumeration leads to similar result as the set of regions one obtains when one develops the unfolding of the Petri net into a state automaton, adds timing information in the form of intervals depending on clock valuations, and applying Alur and Dill's method for obtaining a finite set of state regions.

Definition 5. A clock stamps state class (CS class) is a 3-tuple $C = (m, D, ST)$ where:

1. *m* is a marking.
2. *D* is a firing domain; for an enabled transition t_i, $D(t_{ij})$ represents the global firing interval of t_{ij}. "Global" means that the values are counted relative to the beginning of the net's execution from the initial CS class $C_0 = (m_0, D_0, ST_0)$.
3. *ST* is the time stamp of the CS class, which is a (global) time interval.

A control- and state enabled transition t_j is said to be firable at CS class C_k if $L_k(t_j) \leq U_k(t_i)$ for all enabled transitions t_i of C_k. The feasible firing interval of $t_j = \langle L_k(t_j), U_k(t_j) \rangle = \langle L(t_j), \min\{U(t_i), \forall t_i$ firable at $C_k\}$. The set of firable transitions at CS class C_k is divided into inherited firable transitions (firable before CS class C_k was reached) and new firable transitions (their clock starts running at the time when C_k is reached).
Figure 1 shows state class trees of a simple time Petri net.

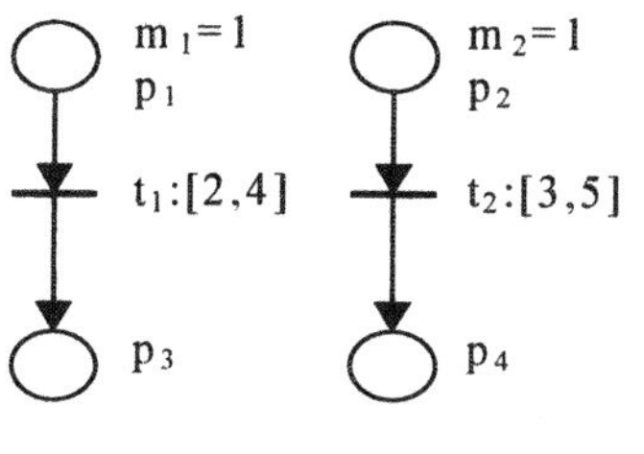

$m_1=1$ p_1 t_1:[2,4] p_3 $m_2=1$ p_2 t_2:[3,5] p_4

Classic state classes

C_0: M_0: (1,1,0,0)
D_0: D_1(t1)=[2,4]
D_0: D_0(t2)=[3,5]
t_1 t_2
C_1: M1: (0,1,1,0) C2: M2: (1,0,0,1)
D_1: $D_1(t_2)$=[0,3] D_2: $D_2(t_1)$=[0,1]
t_2 t_1
C_3: M_3: (0,0,1,1)
D_3: ∅

Clock stamp state classes

C_0: M_0: (1,1,0,0)
ST_0: [0,0]
D_0: $D_1(t_1)$=[2,4]
D_0: $D_0(t_2)$=[3,5]
t1 t2
C_1: M1: (0,1,1,0) C2: M2: (1,0,0,1)
ST_1=[2,4] ST_2=[3,4]
D_1: $D_1(t_2)$=[3,5] D_2: $D_2(t_1)$=[3,4]
t_2 t_1
C_3: M_3: (0,0,1,1) C_3: M_3: (0,0,1,1)
ST_3=[3,5] ST_4=[3,4]
D_3: ∅ D4:∅

Fig. 1 A Simple Time PN and its state class trees

A clock stamps state class can be uniquely mapped into a classic state class. Two state classes C_k and C_l are defined to be equal iff both their markings and their firing domains are equal. The reachability analysis of time Petri nets addresses not only the problem of generating sets of reachable markings, but also the problem of timing property analysis and verification too (Berthomieu and Menasche 1983; Wang et al. 2000).

The computational complexity of generating the set of reachable states, due to the state-space explosion in real-life applications motivated various modular approaches and composition techniques for the reachability tree construction. Untimed algorithms (Valmari 1994, Notomi and Murata 1994, Hudák 1994 etc.) can be used for safe acyclic time controlled Petri nets, where the set of reachable markings is a subset of the set of reachable markings of an untimed net. The timing properties add extra constraints to the description of the plant behavior. In a T-time Petri nets one has to split up marking into several state classes because the further evolution may depend on time when the transition takes place within its intervals of allowed execution.

The algorithms described in the appendices of this paper implement state class tree and clock stamp state class tree design, and synchronous composition of plants represented by several interacting T-time Petri net components. A transition that belongs to several components can only be executed when it is firable in each of these components simultaneously. In our programs we implemented this for Petri nets using the concept synchronic decomposition described e.g. in Hudák (1994).

A modular reachabilty analysis (Boel and Jiroveanu 2003) has been proposed for Time Petri net models consisting of a collection of state machine components that interact via synchronizing transitions or guards. Time guards attached to a transition are logical propositions defined over the state of the entire model involving marking of places in other state machine components. A guarded transition becomes enabled when it is both state enabled and the guard is evaluated *true*. The firing of a transition does not require that the guard remains *true* and it does not consume tokens considered in the expression of the guard.

In order to preserve the branching property (if a state S_k of a state class C_k has a successor in a next state class C_{k+1}, then any state of C_k has a successor in C_{k+1}; Berthomieu and Vernadat 2003) the local state classes have to be split up whenever necessary. Figure 2 illustrates the splitting process. Two *PN* interact via the guard *G* attached to the transition t_2. *G* becomes *true* when a token arrives in the place p_6 (global firing interval [30, 45]), and it starts to be *false* when the token leaves the place in [31, 46]. Both state classes C_1 and C_2 have to be split because some of their tokens missed the *true* evaluation of the guard *g*.

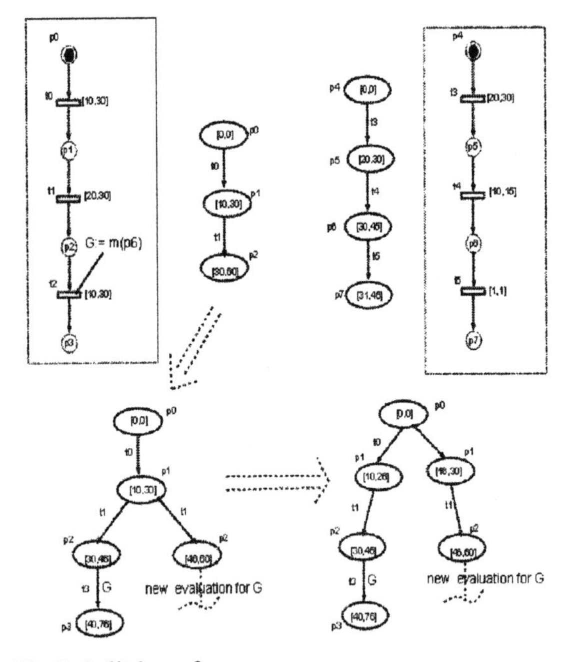

Fig.2 Splitting of processes

The modular construction exploits the concurrency, deriving local reachability trees, while a global trace in the global model is recovered as a feasible combination of local traces of the components.

4. ALGORITHMS

The designed software algorithms were implemented in Matlab and/or written in C++ for OS Linux /Windows.

The Matlab tool PNCONTROL supervisory control tool (Flochová et al. 2001) consists of several modules. The first group of modules is used to edit the Petri nets models, validate these models, to compute the reachability tree and the incidence matrix of the models. The second group of modules implements the methods of finding positive *P* and *T* invariants and the analysis of boundedness, safeness and *L1*-liveness of the process Petri net. The third group of modules is devoted to the supervisory control of DES and to their failure analysis and OFF-line simulation of controlled DES evolution. The basic state class tree and decomposition algorithms have been designed and implemented in C++ for Linux and/or Windows. Some of these new modules have already been integrated in the existing Matlab tool.

The following algorithms have been designed and programmed:

1. Reachability tree (RT) design (Murata 1989; Češka 1994).
2. Modular RT design – synchronic decomposition (T-junction, Hudák 1994).
3. State class trees design (Berthomieu and Menasche 1883; Berthomieu and Diaz 1991)
4. Clock stamp state class tree design for acyclic Petri nets (Wang et al. 2000).
5. Supervisory reachability analysis and backward reachability analysis, supervisor off line design and testing (Yamalidou et al. 1996; Moody and Antsaklis 2000; Flochová and Hrúz 1996).

Algorithms under construction:

6. Modular RT design - *T-junction* for n components.
7. Petri nets with guard interactions and synchronized token passing (Boel and Jiroveanu 2003).
8. Backward firing for T-time PN.
9. Supervisors design (Stremersch and Boel 1998-2001).

5. CONCLUSION

This paper presents some Petri net-based analysis algorithms based on T-time Petri net formalism. The approaches provide particularly simple methods potentially used in supervisory control and model-based failure diagnosis. Forward and backward trees allow the solution of problems of control and observation (including design of supervisory controllers with certain specifications). Moreover they can be used in fault detection algorithms with partial observation. Algorithms that employ the reachability tree and the state class trees analysis are enumerative by nature, but the fact that they can be implemented modularly makes these methods potentially useful for control synthesis of real plants.

Program oriented methods for supervisory control design have been included in the Matlab training tool PNCONTROL and have been used to simulate the real time supervisory control of a discrete event dynamic system, to solve an OFFLINE supervisory control synthesis, and simulate and check the proposed controller of discrete event dynamic systems. The outputs of the programs can be used to solve a real-time supervisory control with help of a quick PLC controller or a quick PN-dedicated controller, which form one of the most commonly used implementation platforms in DES control. The supervisory control theory could be applied for the control synthesis while implementation is performed in the IEC 61131-3 compliant programming environment.

This approaches to the PLC programs design include part time verification techniques, they prevent reaching of forbidden states, and after a systematic design procedure ensure the correctness of the underlying logic in the sense of disabling inputs leading to forbidden possibly dangerous situations, economical damages and dropouts, or avoid some human mistakes. The same algorithms, including the modular reachability analysis, can be used for model-based diagnoser design.

Acknowledgments

This work was supported in part by the Research fellowship for Central and Eastern Europe of the Belgian federal office OSTC, by Slovak grant VG-0155-Koz-Sk1–Intelligent method for modeling and control, and APVT grant 51-011602.

APPENDICES

Brief description of several selected basic algorithms/ program sequences:

Reachability analysis of a safe PN based on clock stamps tree (def.4).

Calculate transitions firable from the initial state class and add initial state class to the database (state classes list), clock stamp $ST_0 = [0,0]$

For all known state classes (while a new state class exists)

Calculate enabled and firable transitions (state, control enabled and time firable).

For all firable transitions:

divide the transitions into inherited and new firable transitions

calculate the feasible firing intervals of the firing transition t_f

calculate the feasible firing intervals of the inherited firable transitions after firing t_f

calculate the feasible firing intervals of the new firable transitions after firing t_f

calculate new marking of the new state class and its clock stamp.

State classes equality checking including state classes with shifted clock stamps and equal marking and relative firing domains.

End for a state class

For a PN with cycles one has to solve the interval intersections of state classes in repeated execution.

For a bounded PN the problem of the multiple enabledness has to be included.

A similar algorithm can be written for the classic state class tree (def. 3) enumeration. The algorithm differs in the computation of transition firing domains according to the distinct definition of state classes. A modification of this classic state class tree using recursive procedures calls for building a new state and its parameters consists of the following steps:

Calculate transitions firable from the initial state class and add initial state class to the database (state classes list).

Starting with initial state class for every new state class calculate enabled and firable transitions, fire each firable transition, compute new marking, transition domains, and recurse down to built the next state class.

T-junction + reachability tree composition/merging (two components, a synchronizing transition) $R(m) \subseteq R(m_1) \times R(m_2))$

Rearrange the incidence matrix (change the place-rows order) according to the components structure

Partitioning of the matrix into components, both components include the synchronizing transition columns

$P_1 \cap P_2 = \varnothing$, $P_1 \cup P_2 = P$,

$T_1 \cap T_2 = \{t'\}$, $T = T_1 \cup T_2$

Calculate the reachability trees of the components

Calculate the initial marking of the composed PN $m_0 = (m_{10}, m_{20})$

Calculate the reachability tree of the composed Petri net:

For both components

For all reachable states

For all enabled transitions

If t is a synchronizing transition

new state = new state of component1, new state of component2

$m_{k+1} = (m_{1k+1}, m_{2k+1})$

If t isn't a synchronizing transition

new state = new state of component1, old state of component2 or old state of component1, new state of component2 – depending on to which component belongs the firing transition

$m_{k+1} = (m_{1k}, m_{2k+1})$ or $= (m_{1k+1}, m_{2k},)$

End transition

End state

End component

The extension for n automata, while any two automata-components N_k, N_l *communicate exactly through one or more synchronizing transitions* $(t'_1, t'_2, \ldots, t'_s)$ *can be described in a similar algorithm.*

The basic supervisory control and backward reachability analysis

Without imposing the return to home states or nodes.
Deadlocks in the state class tree (SCT) are labelled as IA_nodes (inadmissible, forbidden).
Repeat until new forbidden nodes or not allowed edges are found

- *If the edge of the SCT is uncontrollable and its successor is IA_node, the preceding node is labelled as IA_node.*
- *If the edge of the SCT is controllable and its successor is IA_node the edge is not allowed and labelled as NA_edge.*
- *If all the edges going out of a node are not allowed the node is labelled as IA_node*

Imposing the guaranteed return to home states or nodes.
Begin in a home state. Analyse from which admissible nodes it is possible to reach home states using only allowed edges. The admissible nodes from which it is not possible to reach home nodes are converted into forbidden nodes.

REFERENCES

Alur, R. and D.L. Dill (1996). Automata-theoretic verification of real-time systems. *Formal Methods for Real-Time Computing,* Trends in Software Series, John Wiley & Sons Publishers, pp. 55-82. ISBN: 0-471-95835-2.

Berthomieu B. and M. Menasche (1983). An enumerative approach for analyzing time Petri nets. *Information processing 83*. Elsevier Science Publishers, North Holland, pp. 41-46.

Berthomieu B., and M. Diaz (1991). Modeling and verification of Time dependent Systems Using Time Petri nets. *IEEE trans. On software engineering*, **17**, No. 3, pp. 259-273.

Berthomieu B. and François Vernadat (2003) State Class Constructions for Branching Analysis of Time Petri *Nets Lecture Notes in Computer Science,* Springer-Verlag, **2619**, pp. 442-457.

Boel, R.K., Ben-Naoum, L., Van Breusegem, V. (1995). On Forbidden State Problem for a Class of Controlled Petri Nets. In: *IEEE Trans. on Automatic Control*, **40**, No. 10, pp. 1717-1731.

R.K. Boel and G. Stremersch (1998). Forbidden state control synthesis for time Petri net models. In: *Open Problems in Mathematical Systems and Control Theory*. (V.D. Blondel, E.D. Sontag, M. Vidyasagar, J.C. Willems (Ed.)), Springer-Verlag, London, pp. 61-66, ISBN: 1-85233-044-9

Boel, R.K. and F. Montoya (2000). Modular Synthesis of efficient schedules in a timed discrete event plant. In: *Proceedings of the 39th Conference on Decision and Control,* Sydney, Australia, pp. 16-21.

Boel, R.K. and G. Jiroveanu, (2003). Modular Reachability Analysis for Time Petri Nets with Guarded Transitions, *ADHS'03*, St. Malo, France.

Cassandras, C.G. and S. Lafortune (1999): *Introduction to Discrete Event Systems*. Kluwer Academic Publishers, 848 pages, ISBN 0-7923-8609-4.

Češka, M. (1994): *Petriho síte*. Akademické nakladatelství CERM, Brno, Czech republic, 94 pages. ISBN 80-85867-35-4.

Flochová, J. and B. Hrúz (1996). Supervisory control for discrete event dynamic systems based on Petri nets. In: *Proceeding of International conference on Process control*, Horní Bečva, Czech republic, **2**, pp. 80-83.

Flochová, J., Jirsák, R., Boel, R.K. (2001). A Matlab based supervisory controller for discrete event systems. In: *Proceeding of IFAC conference Programmable devices and systems,* Gliwice, Poland, pp. 183-189.

Ho, Y.C. and C. Cassandras (1983). A new Approach to the Analysis of Discrete Event Dynamic Systems. *Automatica*, **19**, pp. 149-167.

Holloway L.E. and B.H. Krogh (1990). Synthesis of Feedback Control Logic for a Class of Controlled Petri Nets. *IEEE Trans. on Automatic Control*, **35**, pp. 514-523.

Hudák, Š. (1994). DE-compositional reachability analysis. *Electrical journal*, **45**, No.11, pp. 424-431.

Moody J.O. and P.J. Antsaklis (2000). Petri net supervisors for DES with uncontrollable and unobservable transitions. *IEEE Trans. on Automatic Control*, **45**, March 2000, pp. 462-476.

Notomi, M. and T. Murata (1994). Hierarchical Reachability Graph of Bounded Petri Nets for Concurrent-Software Analysis. *IEEE Trans. on Software Engineering*, **20**, pp. 325-336.

Ramadge, P. and W.M. Wonham (1997). Supervisory control of a class of discrete event processes. *SIAM J. Control and optimitization*. 25, pp. 206-230.

Ramadge, P.J. and W.M. Wonham (1998). The Control of Discrete Event Systems. In: *Proceeding of the IEEE*, **77**, No.1, pp. 81-98.

Stremersch, G. and R.K. Boel (2001). Decomposition of supervisory control problem for Petri nets. *IEEE Trans. on Automatic Control*, **46**, pp. 1490-1496.

Stremersch, G. (2001). *Supervision of Petri nets*. Kluwer Academic Publishers, 200 pages. ISBN 0-7923-7486-X.

Valmari, A. (1994). Compositional Analysis with Place-Bordered Subnets. *Applications and Theory of Petri nets*, pp. 531-547.

Wang, J., Deng, Y., Xu, G. (2000). Reachability analysis of real time systems using time Petri nets. *IEEE Trans. on Systems,* Man and Cybernetics-Part B, **30**, pp. 725-735.

Yamalidou, K. et al. (1996). Feedback Control of Petri Nets Based on Place Invariants. *Automatica*, **32**, No.1, pp.15-28.

AN EIGENSTRUCTURE BASED METHODOLOGY FOR CONTROLLER REDUCTION

Caroline Chiappa *,**,[1] **Stephanie Chable** **,*
Yann Le Gorrec ***

* *SUPAERO. 10, avenue E.Belin, 31055 Toulouse, France*
** *ONERA, System Control and Flight Dynamics Department. 2, avenue E.Belin, 31055 Toulouse, France*
*** *LAGEP, UCB Lyon 1 - UMR-CNRS-Q-5007 43, Bd du 11 Novembre 1918, 69622 Villeurbanne Cedex France*

Abstract: This paper discusses closed loop controller reduction. The proposed method is based on eigenstructure analysis and equivalent closed loop eigenstructure assignment. In the first step the basic technics used during the design procedure are defined : modal simulation, dominant eigenstructure extraction, eigenstructure assignment using dynamic feedback and controller structuration. From these elementary tools a new design procedure is elaborated, which aims to reduce the order of a given controller (satisfying some closed loop performances). The method is first to determine the significant part of the closed loop eigenstructure of the system controlled by the initial compensator. In the second step a reduced order controller is synthesized using eigenstructure assignment and minimization of the difference (in the form of the Frobenius norm) between both initial and reduced order controllers. The proposed design procedure is applied on an academic example. *Copyright © 2003 IFAC*

Keywords: reduced controller, robust modal control, dominant eigenstructure.

1. INTRODUCTION

Today, the getting of reduced order controller is a coloquid for the automation community. To be modelized, plants need numerous degrees of freedom, which leads to high order state space realization and consequently high order controllers. Indeed, current sophisticated method used for controller synthesis, H_∞, LQG, ..., lead to controller which have the same order, or higher order than the system to be controlled.
However, in relation to implementation problems, low order controller are preferred by industrial designers. Consequently controller order reduction considering closed loop performances is a great challenge.
Two different approaches have been developed. The first approach is based on the controller synthesis from a reduced order model of the system. In the second approach, the controller is synthesized from the initial model of the system. Afterward the synthesized controller can be reduced by various method. These methods are divided into two categories : the *open loop* and the *closed loop* methods.
In the first kind of approach, comparisons between methods of reduction are given in Craig and Su [1990], Madelaine [1998]. The model of the system is approximated by various existing

[1] Corresponding author : Caroline.Chiappa@supaero.fr

methods which are all based on minimization of the modelisation error.

Therefore this kind of reduction needs to be corrected. The use of weighted-frequency improves the reduction by conserving the system transfer with frequency griding. Enns proposed the first step of this method in 1984 Enns [1984a]. Lot of work has followed. For example the use of weighted-frequency has been applied to numerous open loop methods. Enns developed the weighted-balanced reduction Enns [1984a], Enns [1984b], Skelton developed the weighted-cost decomposition reduction Skelton and Hu [1990], Wortelboer introduced the weighted optimal H_2 reduction in general case Wortelboer [1994], and weights are applied to the Hankel optimal approximation reduction Latham and Anderson [1985], Zhou [1993].

Several authors directly apply the model reduction based method to the reduction of the controller. Examples of this open loop approach are given in Prudhomme [1994]. The proposed approaches are based on balanced reduction, optimal reduction, robust parameter reduction, or weighted-frequency reduction.

Another method deal with closed loop consideration during the reduction. Let cite Wang's reduction which is based on the weighted-frequency reduction where perturbation are added on the closed loop transfer function Wang et al. [1998] during the synthesis.

Numerous authors extract the dominant closed loop structure of the nominal controller to reduce it, but Liu and Anderson proposed to reduce the coprime factorization of the nominal controller rather than the nominal controller itself Liu and Anderson [1986]. In Anderson and Liu [1989], they compare their method to the other existing methods. This is the example which will be used at the end of this paper to compare the approach presented here.

Finally, a third approach is to synthesize directly a controller in its reduced form. Lots of studies have been made in this direction, Aouf et al. [2002], Madelaine [1998]. Landau and Karimi directly estimated the reduced order controller parameters with a closed loop identification algorithm Karimi and Landau [2000], Landau et al. [2001]. With this method, the reduced controller keeps the nominal closed loop performances.

This article proposes a method based on the controller reduction (second approach proposed in the introduction). This method is based on the closed loop dominant eigenstructure. The major difference with the previous procedures is the whole dominant eigenstructure of the system closed by the nominal controller is considered instead of the dominant structure of the nominal controller. First, the method used to extract the dominant eigenstructure of the closed loop system is presented. Afterward it is shown how this closed loop eigenstructure can be assigned using reduced order dynamic feedback. Finally, the method is applied on an academic example proposed in Anderson and Liu [1989].

2. THEORY OF ROBUST MODAL CONTROL

The reduction method is based on two main step : dominant eigenstructure extraction using closed loop modal simulation and eigenstructure assignment by dynamic feedback Magni et al. [1998].

2.1 Notations and propositions

Let consider the following linear system, with n states, m inputs and p outputs:

$$\begin{cases} \dot{x} = Ax + Bu \\ y = Cx + Du \end{cases} \tag{1}$$

where x is the state vector, y the vector of measurements and u the vector of inputs, $A \in \mathbb{R}^{n\times n}$, $B \in \mathbb{R}^{n\times m}$, $C \in \mathbb{R}^{p\times n}$ and $D \in \mathbb{R}^{p\times m}$.

where w_i and v_i are the input directions and right eigenvectors associated to the closed loop eigenvalue λ_i

The control feedback is defined by:

$$u(s) = K(s)\, y(s) \tag{2}$$

$K(s)$ is obtained using dynamic eigenstructure assignment defined by Proposition 1.

Proposition 1. The triple $\mathcal{T}_i = (\lambda_i, v_i, w_i)$ satisfying

$$\begin{bmatrix} A - \lambda_i I & B \end{bmatrix} \begin{pmatrix} v_i \\ w_i \end{pmatrix} = 0 \tag{3}$$

is assigned by the dynamic gain $K(s)$ if and only if

$$K(\lambda_i) \begin{bmatrix} C\, v_i + D\, w_i \end{bmatrix} = w_i \tag{4}$$

If the eigenvalue is complex, the equality (4) has to be completed by his conjugated.

The elementary design procedure associate with this proposition is as follows :

- Choose the closed loop eigenvalues λ_i and determine the closed loop admissible eigenvector space using Equation 3. Closed loop eigenvectors are chosen within this subspace using orthogonal projection of open loop eigenvectors. At this step the triple (λ_i, v_i, w_i) is defined.
- Compute $K(s)$ satisfying equation 4. This computation is detailed into section 2.2.

2.2 Resolution technique

The transfer matrix of the gain $K(s)$ is assumed to have the following form each input-output :

$$K_{ij}(s) = \frac{b_{qij}\, s^q + \cdots + b_{1ij}\, s + b_{0ij}}{a_{qij}\, s^q + \cdots + a_{1ij}\, s + a_{0ij}} \qquad (5)$$

Common or different denominators of the matrix $K(s)$ are fixed *a priori.* For choosing the coefficients a_{ijk}, it suffices to choose (see Section 3) the desired roots of each denominator (equal to the roots of an initial controller for example) and to identify the corresponding coefficients. The free parameters for interpolation are the numerator coefficients denoted b_{ijk}.

The structure of the feedback as defined in (5) usually offers a large number of degrees of freedom. For this reason, a quadratic criterion J is considered, for example to keep the controller as close as possible to a initial feedback K_{ref} :

$$J = \sum_{i=1,\ldots,r} \|K_{\text{ref}}(j\omega_i) - K(j\omega_i)\|_F^2 \qquad (6)$$

In order to have the feedback $K(s)$ in (5) written with an explicit dependency on the coefficients b_{ijk}, some notation is introduced. The symbol Ξ_k, denotes unknown coefficients on the kth row of (5)

$$\Xi_k = [b_{0k1} \ldots b_{qk1} \ldots\ldots b_{0kp} \ldots b_{qkp}]$$

With a matrix $X_k(\lambda)$, fixed to determinate $a_{ijk}(\lambda)$, it follows that:

$$K(s) = \sum_{k=1}^{m} e_k \Xi_k X_k(s) \qquad (7)$$

where e_k is the kth vector of the canonical basis of $\mathbb{R}^m$ (*i.e.* a m-dimensional column vector with all entries equal to zero but the kth that is equal to one). (The main result can be found in Magni et al. [1998], Le Gorrec et al. [1998] for derivation, and Magni [2002] for software implementation)

Proposition 2. The problem of computing $K(s)$ satisfying Equation (4) and minimizing Criterion (6) consists of solving for Ξ a LQP problem of the form

$$\Xi A_1 = b_1 \quad ; \quad J = \Xi H \Xi^T + 2\Xi c \qquad (8)$$

where:

$$A_1 = \text{Diag}\{X_1(\lambda_i)(Cv_i{+}Dw_i) \;\; .. X_m(\lambda_i)(Cv_i{+}Dw_i)\}$$

$$H = \text{Diag}\{\sum_{i=1,\ldots,r} X_1(j\omega_i)X_1(j\omega_i)^* \,..\sum_{i=1,\ldots,r} X_m(j\omega_i)X_m(j\omega_i)^*\}$$

$$b_1 = \begin{bmatrix} e_1^T w_1 \\ \vdots \\ e_m^T w_m \end{bmatrix} ; \; c = -\Re \begin{bmatrix} \sum_{i=1,\ldots,r} X_1(j\omega_i)K_{\text{ref}}(j\omega_i)^* e_1 \\ \vdots \\ \sum_{i=1,\ldots,r} X_m(j\omega_i)K_{\text{ref}}(j\omega_i)^* e_m \end{bmatrix}$$

If $r > q$ the matrix H is positive definite. (Dimensions: $A_1 \in \mathbb{R}^{mpq \times m}$, $b_1 \in \mathbb{R}^{m\times 1}$, $H \in \mathbb{R}^{mpq\times mpq}$ and $c \in \mathbb{R}^{mpq\times 1}$.)

3. DESIGN PROCEDURE FOR REDUCED-ORDER CONTROLLERS

In this section we will describe how to combine modal analysis and dynamic eigenstructure assignment to reduce an initial controller order while concurrently satisfying the closed loop performances. The method is based on the fact that the system is entirely defined by its closed loop eigenstructure. First the dominant eigentructure is extracted using modal analysis. In the second step, the dominant eigenstructure of the system will be assigned using reduced order controller which have an a priori fixed structure obtained from the first step of the procedure. Frequency criteria will be minimized to optimize the efficiency of the reduction.

Step 1 : Modal analysis A modal analysis of the closed loop system is performed in order to identify the dominant modes and their associated eigenvectors. Let us recall that the time response of a linear system $\dot{x} = A\,x + B\,u$, $y = C\,x$ is equal to

$$y(t) = \sum_i C\,v_i \int_0^t e^{\lambda_i\,(t-\tau)}\, u_i\, B\, u(\tau)\, d\tau$$

where λ_i and v_i are the already defined eigenvalue and right eigenvectors, u_i are the so-called left eigenvectors respectively. In order to detect the leading modes relative to a given input, it suffices to simulate separately each term

$$C\,v_i \int_0^t e^{\lambda_i\,(t-\tau)}\, u_i\, B\, u(\tau) d\tau \qquad (9)$$

The same analysis is performed by initial controller.

Step 2. Reduction.

- **2.1.** Choice of the controller's poles: this choice is made *a priori* as a subset of the poles of the original controller. The selection is made using the analysis of the dominant eigenvalues obtained with the **Step 1.** At this stage, the coefficients a_{ijk} of equation (5) are all fixed.
- **2.2.** Eigenstructure assignment: the constraints defining the re-assignment of dominant eigenstructure (selected at **Step 1**)

are derived, these constraints correspond to Equations (3) and (4). The constraint's number depends on the desired controller's order. The higher the order, more constraints have to be treated. The equations for eigenstructure assignment, are linear constraints. At this stage, using the notations and the result of Proposition 2, Equation $\Xi A_1 = b_1$ (8) is known.

- **2.3.** Controller structure constraints: Some coefficients of the reduced controller can be fixed. For example, a desired difference of degree between numerator and denominator leads to set the relevant b_{ijk} to zero. So, there is an additional equality constraint and at this stage, the constraints of the form

$$\Xi A'_1 = b'_1$$

are known.

- **2.4.** A quadratic criterion of the form (6) where $K_{\text{ref}}(s)$ is the transfer function of the initial controller is defined. This criterion will fix the degrees of freedom remaining after the treatment of the above constraints. The computed controller will become as close as possible to the initial controller. The choice of the frequencies ω_j depends on the frequency domain features of the initial controller. At this stage, using the result of Proposition 2, the criterion is written in the following form:

$$J = \Xi H \Xi^T + 2\Xi c$$

- **2.5** Solve the LQP problem for Ξ. Then deduce from Ξ the values of the coefficient b_{ijk} to terminate the procedure.

Step 3. Final analysis. Performance is evaluated :

- in the time domain by means of step responses simulations and root locus.
- in the frequency domain by means of Bode plots or singular value analysis.
- in the parametric domain by means of μ analysis.

If some properties are not satisfactory, the procedure of step **2** must be applied again.

remarks : All the needed tools can be found in the Matlab Toolbox "Robust Modal Control *Toolbox*^{TM}" Magni [2000], Magni [2002].

4. APPLICATION

In this section the example proposed by Anderson and Liu [1989] is considered.

4.1 The plant and the initial controller

The plant to be controlled is a four-disk system, represented as a linear, time-invariant, single input and single output, non minimum-phase plant of the eighth order.

The system is defined by his transfer function and the minimal realization of A, B, C is given in Anderson and Liu [1989].

The initial controller is obtained from LQG design. The weighting matrices used for the design are given by $Q = q_1 H'H$, $R = 1$, with $H = [0, 0, 0, 0, 0.55, 11, 1.32, 18]$ and $q_1 = 10^{-6}$. The covariance matrices are $W = q_2 BB'$, $V = 1$ with $q_2 = 2000$.

4.2 Design reduced controller

The design procedure 3 is applied.

Step 1. A modal analysis of the closed loop system is performed in order to identify the dominant modes to replace by eigenstructure assignment and the reduced order controller's poles. The result of this analysis is given in the Table 1.

	poles	residue
controller	$-1.5202 + 0.6509i$	3.2904
	$-0.9410 + 1.6906i$	-3.1861
	$-0.0226 + 0.9993i$	-0.0963
	$-0.3507 + 2.2429i$	0.0072
closed loop	$-0.0477 + 0.0473i$	65.9021
	$-0.0157 + 0.7649i$	-0.3447
	$-0.0278 + 1.4097i$	-0.1274
	$-0.0374 + 1.8496i$	0.0484
	$-0.9902 + 1.6291i$	-0.0019
	$-1.4165 + 0.5720i$	0.0012
	$-0.3567 + 2.2640i$	0.0001
	$-0.0229 + 0.9993i$	-0.0000

Table 1. Residues of the closed loop and controller poles.

Step 2.1 and **2.2** At this stage, the controller's order, the controller's poles and the eigenstructure assignment are chosen. The Table 2 contains these choices.

Note that equivalent **eigenstructure assignment** is directly obtained from the modal analysis done during the **Step 1.** These constraints decrease with the order of the reduction. The rigid modes and the first flexible modes are the most significant modes. When high reduction is necessary, only two (for reduction order equal to 4) or one (for reduction order equal to 3) dominant mode are assigned as the number of degrees of freedom is limited.

Order	Controller's poles	Assignments
7	$-1.5202 + 0.6509i$ $-0.9410 + 1.6906i$ $-0.0226 + 0.9993i$ -0.3	$-0.0477 + 0.0473i$ $-0.0157 + 0.7649i$ $-0.0278 + 1.4097i$
6	$-1.5202 + 0.6509i$ $-0.9410 + 1.6906i$ $-0.0226 + 0.9993i$	$-0.0477 + 0.0473i$ $-0.0157 + 0.7649i$ $-0.0278 + 1.4097i$
5	$-1.5202 + 0.6509i$ $-0.0226 + 0.9993i$ -0.3	$-0.0477 + 0.0473i$ $-0.0157 + 0.7649i$
4	$-1.5202 + 0.6509i$ $-0.0226 + 0.9993i$	$-0.0477 + 0.0473i$ $-0.0157 + 0.7649i$
3	$-1.5202 + 0.6509i$ -0.3	$-0.0477 + 0.0473i$
2	$-0.3 + 0.3i$	$-0.0477 + 0.0473i$
1	-0.3	$-0.0477 + 0.0473i$

Table 2. Reduced order controller and closed loop assignment.

About the **choice of the controller's dynamics**, the same result is obtained. But when uneven order is looked for, the poles of the reduced controller are chosen as a subset of the initial controller poles plus a real pole. The choice of this supplementary pole is based on the bandwidth of the initial controller. The main purpose is to attenuate the high frequencies. Note that in practice the result of the reduction considering the different choices of this real pole is not appreciable.

Step 2.3 Structure constraints: the open loop system has two high frequency flexible modes, so the controller must guarantee roll of properties. This is the reason why the order of the numerator has to be less than the degree of the denominator. The difference between orders is fixed to one for all order reduction excepted for the last, one order reduction ... It is clear that in this case there are not enough degrees of freedom.

Step 2.4 The criterion is defined by (6).

Step 2.5 The reduced order controller is computed.

Step 3. Final analysis Fig. 1 plots all the temporal responses obtained using the 8 different reduced controllers based on modal reduction.

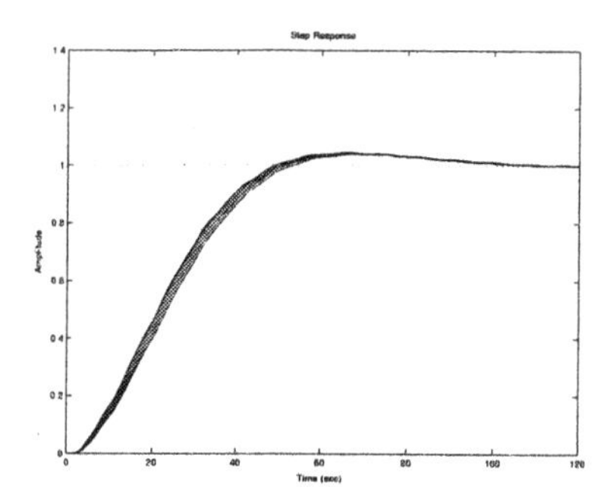

Fig. 1. Time responses with all modal reduced-controller

4.3 Comparison with other methods

The results obtained by this method is compared with the results obtained from the different design procedures. As in Anderson and Liu [1989] the stability of the system closed by the reduced controller is considered.

Order	7	6	5	4	3	2	1
D	S	S	U	U	U	U	-
C	S	U	U	U	U	U	-
RCFU	U	U	S	S	U	S	-
RCFW	S	S	S	S	S	S	-
RMCR	S	S	S	S	S	S	S

Table 3. Stability obtained from the different reduced order controllers. S=Stable, U=Unstable, -=Not available. D : Unweighted Balanced Truncation, C : Unweighted Hankel-norm reduction, RCFU : Unweighted Right Coprime Factorization, RCFW : Weighted Right Coprime Factorization, RMCR : Robut Modal Control Reduction

It appears that the own stability is not sufficient to establish if the reduction is correct. To illustrate this point we can check the results of the controller obtained by *RCFW*, where the reduction order is 3. On Fig. 2 the temporal response of the system closed by this controller is stable but very unsatisfactory. On the contrary the results obtained from the *modal* method are good.

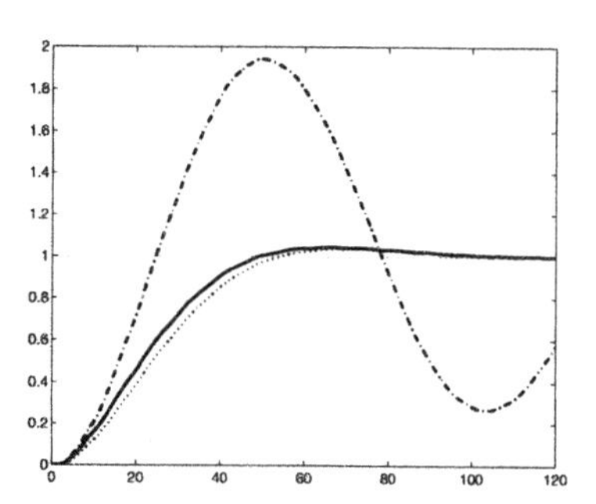

Fig. 2. Temporal responses - initial controller, - modal based controller, - RCFW controller

The proposed method is a closed loop based method which attempts to preserve the temporal and frequency behavior of the initial controller. The *nugap* Zames and EL-Sajjary [1980] obtained between the system closed by our reduced order controller and the initial controller is betwen 0.1 and 0.2 for all order of reduction. This result confirms the last assertion.

5. CONCLUSION

In this paper, we have presented a method of reduction based on dominant eigenstructure assignment. The example of Anderson and Liu [1989] has been chosen as illustrative example. Note this methodology has proved its efficiency for other complex applications with frequency domain and temporal domain schedule of conditions Chable et al. [2002] or in the case of MIMO reductions Chiappa et al. [1998], Chiappa et al. [2001]. The proposed method is very efficient as it respects the intrinsic nature of the considered system. The methodology require, as all other proposed methods, a good knowledge of the system and the initial controller but it is very powerful and simple to apply. Note that computation necessary for the reduction are very simple (LQ formulation of the problem).

REFERENCES

B. D. O. Anderson and Y. Liu. Controller reduction: Concepts and approaches. *IEEE Transactions on Automatic Control*, 34:802–812, 1989.

N. Aouf, B. Boulet, and R.Botez. Model and controller reduction for flexible aircraft preserving robust performance. In *IEEE Transactions on Control Systems Technology*, volume 10, pages 229–237, march 2002.

S. Chable, S. Mahieu, and C. Chiappa. Design and optimization of restricted complexity controller : A modal approach for reduced-order controllers. *European Journal of Control*, 2002.

C. Chiappa, J.F. Magni, C. Döll, and Y. Le Gorrec. Improvement of the robustness of an aircraft autopilot designed by an H_∞ technique. *In Proc. CESA'98 Conference, Nabeul-Hammamet, Tunisia*, 1:1016–1020, April 1998.

C. Chiappa, J.F. Magni, and Y. Le Gorrec. Méthodologie de synthèse de commandes de vol d'un avion souple. partie 2: structuration du correcteur. *APII-JESA, special issue on Robust Control*, 35(1-2):191–208, January 2001.

R.R. Craig and T.J. Su. A review of model reduction methods for structural control design. In J.L. Junkis and C.L. Kirk, editors, *Dynamics of flexible structures in space.* Computational Mechanics Publications / Springer Verlag, 1990.

D. Enns. *Model reduction for control system design.* PhD thesis, Stanford University, 1984a.

D. Enns. Model reduction with balanced realizations: an error bound and frequency weighted generalization. In *Proceedings of the 23rd Conference on Decision and Control*, Las Vegas, USA, December 1984b. IEEE.

A. Karimi and I.D. Landau. Controller order reduction by direct closed loop identification (output matching). In *3rd Rocond IFAC*, 2000.

I.D. Landau, A. Karimi, and A. Constantinescu. Direct controller order reduction by identification in closed loop. *Automatica*, 37:1689–1702, 2001.

G.A. Latham and B.D.O. Anderson. Frequency-weighted optimal hankel norm approximation of stable transfer functions. *Syst. Cont. Lett.*, 5 (4):229–236, 1985.

Y. Le Gorrec, J.F. Magni, C. Döll, and C. Chiappa. A modal multimodel control design approach applied to aircraft autopilot design. *AIAA Journal of Guidance, Control, and Dynamics*, 21(1):77–83, 1998.

Y. Liu and B. D. O. Anderson. Controller reduction via factorization and balancing. 44(2): 507–531, 1986.

B. Madelaine. *Determination d'un modele dynamique pertinent pour la commande: De la reduction la construction.* PhD thesis, ENSAE, 1998.

J.F. Magni. A toolbox for robust multi-model control design RMCT. Anchorage, Alaska, USA, September 2000.

J.F. Magni. *Robust Modal Control with a Toolbox for use with MATLAB.* Kluwer Academic/Plenum Publishers, April 2002.

J.F. Magni, Y. Le Gorrec, and C. Chiappa. A multimodel-based approach to robust and self-scheduled control design. *In Proc. 37th I.E.E.E. Conf. Decision Contr., Tampa, Florida*, pages 3009–3014, 1998.

S. Prudhomme. Reduction et commande: application a des modeles aeroelastiques. Technical report, ONERA, July 1994.

R.E. Skelton and A. Hu. Model reduction with weighted modal cost analysis. In *Proceedings of the Conference of Guidance Navigation and Control.* AIAA, 1990.

G. Wang, V. Sreeram, and W. Q. Liu. Performance preserving controller reduction via additive perturbation of closed loop transfert function. In *AACC*, 1998.

P.M.R. Wortelboer. *Frequency-weighted balanced reduction of closed-loop mechnical servo-systems: Theory and tools.* PhD thesis, Delf University, Netherlands, 1994.

G. Zames and EL-Sajjary. Unstable systems and feedback : The gap metric. *Proceeding of the Allerton Conference*, pages 380–385, Octobre 1980.

K. Zhou. Weighted optimal hankel norm model reduction. In *Proceedings of the 32nd Conference on Decision and Control*, San Antonio, Texas, USA, 1993. IEEE.

PASSIVITY BASED CONTROL OF AUTOMATED DRIVER BASED ON VELOCITY FIELD

A.A. Saifizul* and S. Hosoe**

* *Mechanical Engineering Department, University of Malaya, 50603 Kuala Lumpur, Malaysia*
** *Department of Electronic-Mechanical Engineering, Nagoya University, Furo-cho, Chikusa-ku, Nagoya 464-8501, Japan*

Abstract: The objectives of automated driver are to let vehicles track the desired trajectory in a natural way and to mantain good ride quality. In order to satisfy the above objective, this paper use the control technique based on velocity fields concepts and Passivity Based Control (PBC) method to derive the control law, such that longitudinal, lateral and yaw velocity converge asymtotically to a desired velocity field. Simulation results show the performance of the controller. *Copyright © 2003 IFAC*
Keywords: automated driver, velocity fields, passivity based controller

1. INTRODUCTION

There are two basic issues in vehicle maneuvering: one is the control of the direction of motion of the vehicle; the other is its ability to stabilize its direction of motion against external disturbances. A great number of researches have been reported in the field of automatic steering in highway system (see [1], [2], [3] and the references therein. Almost all the references tended to concentrate on minimizing the lateral deviation from the reference. However minimizing the lateral deviation only is not enough to make vehicle converges to the reference in natural way. Vehicle can converge to the reference within the same period in many ways. Some of them may not be suitable. Thus, we believe that specifying the convergence flow pattern can increase the quality of driving behavior. Very recently [4] represent the desired convergence flow pattern in the form of desired velocity field. In present paper we provide an alternative approach in solving the automatic steering in highway system problem. The new approach is based on velocity field technique and Passivity Based Controller (PBC) method. We believe that this method may cause vehicle tracks the reference in a natural way and maintain good ride quality. The desired velocity field is designed by considering the convergence of vehicle trajectories to the reference (i.e. center of the lane) and also the existence of the controller. Then using the desired velocity field and vehicle dynamics, the velocity error dynamics are generated to obtain passivity conditions for designing the controller gain.

[1] Corresponding author A.A. Saifizul. Tel:(6)03-79674495; e-mail: saifizul@um.edu.my

2. PROBLEM STATEMENT

In this section, we present a description of the problem.

Given a road vehicle equipped with sensor(s) to detect lateral displacement from the reference and with actuators to generate a desired steering action and traction force, design:

- The desired velocity field which may make vehicle converges to the reference when it is completely tracked.
- A controller such that the vehicle longitudinal, lateral and yaw velocity converge and keep tracking with the desired velocity field in the absence of external inputs.

3. VEHICLE MODEL

Consider the dynamics model for vehicle velocity control [5] as

$$\begin{aligned}
\dot{V}_x &= V_y\Omega_z + \frac{2C_{\alpha f}l_1\Omega_z\delta_f}{mV_x} + \frac{2C_{\alpha f}V_y\delta_f}{mV_x} \\
&\quad - \frac{2C_{\alpha f}\delta_f^2}{m} + \frac{F_{xf}}{m} + \frac{R_x}{m} \\
\dot{V}_y &= \frac{2(l_2C_{\alpha r} - l_1C_{\alpha f})\Omega_z}{mV_x} - \frac{2(C_{\alpha r} + C_{\alpha f})V_y}{mV_x} \\
&\quad - V_x\Omega_z + \frac{2C_{\alpha f}\delta_f + F_{xf}\delta_f}{m} + \frac{R_y}{m} \\
\dot{\Omega}_z &= \frac{2l_1C_{\alpha f}\delta_f + l_1F_{xf}\delta_f}{I_z} - \frac{2(l_1^2C_{\alpha f} + l_2^2C_{\alpha r})\Omega_z}{I_zV_x} \\
&\quad - \frac{2(C_{\alpha f}l_1 - C_{\alpha r}l_2)V_y}{I_zV_x}
\end{aligned} \tag{1}$$

where V_x, V_y and Ω_z are the longitudinal velocity, the lateral velocity and the yaw rate, respectively. δ_f is steer angle and F_{xf} is traction force. R_x and R_y are external inputs from environment; other parameters are given in Table 1. In order to make (1) affine in the input (i.e. steer angle δ_f and traction force F_{xf}), let us neglect the second order terms of steer angle δ_f and consider the transformation [6]

$$\begin{aligned}
\delta_T &= 2C_{\alpha f}\frac{V_y + l_1\omega_z}{V_x}\delta_f + F_{xf} \\
F_\delta &= 2C_{\alpha f}\delta_f + F_{xf}\delta_f.
\end{aligned}$$

Then (1) can be rewritten as

$$\begin{aligned}
\dot{V}_x &= V_y\Omega_z + \frac{\delta_T}{m} + \frac{R_x}{m} \\
\dot{V}_y &= \frac{2(l_2C_{\alpha r} - l_1C_{\alpha f})\Omega_z}{mV_x} - \frac{2(C_{\alpha r} + C_{\alpha f})V_y}{mV_x} \\
&\quad - V_x\Omega_z + \frac{F_\delta}{m} + \frac{R_y}{m} \\
\dot{\Omega}_z &= \frac{l_1F_\delta}{I_z} - \frac{2(l_1^2C_{\alpha f} + l_2^2C_{\alpha r})\Omega_z}{I_zV_x} \\
&\quad - \frac{2(C_{\alpha f}l_1 - C_{\alpha r}l_2)V_y}{I_zV_x}.
\end{aligned} \tag{2}$$

Linearizing the system (2) about equilibrium point $V_x = V_{x0}$, $V_y = 0$ and $\Omega_z = 0$, the simplified linear system is expressed as

$$\begin{aligned}
\dot{x} &= Ax + B_1u + B_2d \\
y &= Cx
\end{aligned} \tag{3}$$

where $x = [v_x\ v_y\ \omega]^T$, $u = [\delta_T\ F_\delta]^T$, $d = [R_x\ R_y]^T$ and $y = [y_1\ y_2]^T$;

$$A = \begin{bmatrix} 0 & 0 & 0 \\ 0 & a_{22} & a_{23} \\ 0 & a_{32} & a_{33} \end{bmatrix};\ B_1 = \begin{bmatrix} \frac{1}{m} & 0 \\ 0 & \frac{1}{m} \\ 0 & \frac{l_1}{I_z} \end{bmatrix}$$

$$B_2 = \begin{bmatrix} \frac{1}{m} & 0 \\ 0 & \frac{1}{m} \\ 0 & 0 \end{bmatrix};\ C = \begin{bmatrix} c_{11} & 0 & 0 \\ 0 & c_{22} & c_{23} \end{bmatrix}.$$

Also, a_{ij} in A are defined as follows:

$$\begin{aligned}
a_{22} &= \frac{-2(C_{\alpha r} + C_{\alpha f})}{mV_{x0}} \\
a_{23} &= \frac{2(l_2C_{\alpha r} - l_1C_{\alpha f})}{mV_{x0}} - V_{x0} \\
a_{32} &= \frac{-2(l_1C_{\alpha f} - l_2C_{\alpha r})}{I_zV_{x0}} \\
a_{32} &= \frac{-2(l_1^2C_{\alpha f} + l_2^2C_{\alpha r})}{I_zV_{x0}}.
\end{aligned}$$

Remark: Notice that elements of output matrix C are not specific at this moment. Later they will be determined to satisfy the passivity conditions for designing the controller feedback gain.

4. VELOCITY FIELD

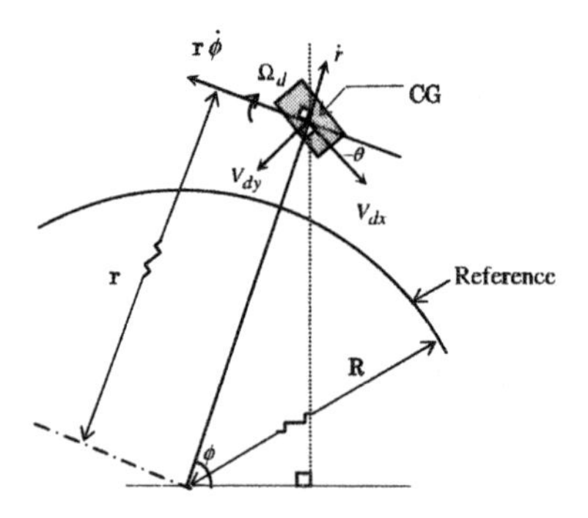

Fig. 1. Schematic diagram of road and vehicle model.

Let V_{dx},V_{dy} and Ω_d represent the local velocity components in the longitudinal, lateral and yaw direction, respectively. y_{cg} is the nearest distance from vehicle center of gravity (CG) to the reference (i.e. $y_{cg} = r - R$) and θ is the angle between longitudinal direction and tangent line to the reference at y_{cg} distance. Other parameters are shown in Figure 1.

We define a desired velocity field by

$$V_d = \begin{bmatrix} V_{dx} \\ V_{dy} \\ \Omega_d \end{bmatrix} = \begin{bmatrix} c_0(1+\frac{y_{cg}}{R}) - \frac{\beta}{Rc_3}\theta \\ c_1 y_{cg} + c_2\theta + \frac{\beta}{Rc_3} \\ c_3\theta + \frac{c_4 c_0}{R} + \frac{c_0}{R} \end{bmatrix}. \quad (4)$$

Parameters c_1, c_2, c_3 and c_4 give some freedom and should be designed based on: convergence of the vehicle to the reference when the desired velocity field V_d in (4) is completely tracked; the existence of feedforward controller which may make vehicle converges to the desired velocity field.

Basically c_0 specify the longitudinal speed and β characterizes the understeer or oversteer behavior. Consider the case where $c_2 > c_0$, vehicle is turning the curve with understeer behavior if $\beta < 0$ and oversteer behavior if $\beta > 0$ and vice versa. Constant terms in second and third row of (4) are introduced to make vehicle travel at constant angle θ along the reference (i.e.$y_{cg} = 0$) and they disappear for straight road case (details explanation for the above statements will be given in Proof in Section 4.1). Intuitively, (4) shows that velocity field are the function of distance y_{cg} and angle θ as shown in Figure 1.

4.1 Convergence Criteria

To guarantee that vehicle can converge to the reference when the desired velocity field is completely tracked, design parameters in (4) need to satisfy some restrictions, which are given by the following Proposition.

Proposition Suppose that

$$c_1 > 0\,, c_3 > 0 \text{ and } c_4 = \frac{-\beta}{c_0(c_0 - c_2)},$$

then vehicle can converge to the reference when the desired velocity field V_d in (4) is completely tracked.

Proof

Referring to Figure 1, by assuming that θ is small, velocity in the normal and tangent direction to the reference and yaw rate can be expressed as

$$\dot{r} = V_x\theta - V_y \quad (5)$$

$$r\dot{\phi} = -V_x - V_y\theta \quad (6)$$

$$\Omega_d = -\dot{\theta} - \dot{\phi}. \quad (7)$$

Consider the case when vehicle completely tracks the desired velocity field, i.e.$V_x = V_{dx}$, $V_y = V_{dy}$ and $\Omega = \Omega_d$. Substituting V_{dx}, V_{dy} in (4) into (5) and (6) as well as neglecting the nonlinear terms, we have

$$\dot{y}_{cg} = -c_1 y_{cg} + (c_0 - c_2)\theta - \frac{\beta}{Rc_3} \quad (8)$$

$$\dot{\phi} = \frac{-c_0}{R.} \quad (9)$$

Then replace (7) by (9) and Ω_d in (4), we obtain

$$\dot{\theta} = -c_3\theta - c_4\frac{c_0}{R.} \quad (10)$$

Let

$$c_4 = \frac{-\beta}{c_0(c_0 - c_2)} \quad (11)$$

as assumed. Then (10) becomes

$$\dot{\theta} = -c_3\theta + \frac{\beta}{(c_0 - c_2)R.} \quad (12)$$

Since $c_1 > 0$ and $c_3 > 0$, from (8) and (12), $[y_{cg}\ \theta]^T$ converge to $[0\ \frac{\beta}{(c_0-c_1)Rc_3}]^T$ as desired.

4.2 Linear case

In the previous section, we have defined V_d in (4) for large range operation (i.e. not only near the equilibrium points). In order to derive the controller based on the simplified linear model (3), it is necessary for us to consider the specific range operation (i.e. near equilibrium points) of velocity field V_d in (4). This can be done by first investigating the equilibrium points of (4).

Consider the case where vehicle moving in straight road and tracks the reference, from (4) and Proof in the Section 4.1, we can obtain equilibrium points as $[c_0\ 0\ 0]$. By subtracting these values from (4), desired velocity field for small range operation near the equilibrium points can be expressed as

$$v_d = \begin{bmatrix} v_{dx} \\ v_{dy} \\ \omega_d \end{bmatrix} = \begin{bmatrix} c_0\frac{y_{cg}}{R} - \frac{\beta}{Rc_3}\theta \\ c_1 y_{cg} + c_2\theta + \frac{\beta}{Rc_3} \\ c_3\theta + \frac{c_4 c_0}{R} + \frac{c_0}{R} \end{bmatrix} \quad (13)$$

5. CONTROLLER SYNTHESIS

In the previous section, we have shown that vehicle can converge to the reference when the desired velocity field V_d in (4), which satisfies some conditions, is completely tracked. The next step is to design a controller so that vehicle longitudinal, lateral and yaw velocity converge and keep tracking with the desired velocity field V_d. For this, we use Passivity Based Control (PBC) method, which breaks up the designing process into 2 stages. This section describes the derivation of velocity error dynamics and the next section concentrates on making velocity error dynamics asymptotically stable in the absence of external inputs.

Let us try to begin this section by representing $\dot{v}_d$ as

$$\dot{v}_d = Av_d + B_1 u_d \tag{14}$$

by some appropriate control input $u_d = [u_{d1}\ u_{d2}]^T$. Differentiating equation (13), we obtain

$$\dot{v}_d = \begin{bmatrix} \frac{c_0}{R}\dot{y}_{cg} - \frac{\beta}{Rc_3}\dot{\theta} \\ c_1\dot{y}_{cg} + c_2\dot{\theta} \\ c_3\dot{\theta} \end{bmatrix}. \tag{15}$$

Thus, equating both sides of the first row of (14) and using (8), (10) and (15), u_{d1} satisfying (14) is obtained as

$$\begin{aligned} u_{d1} &= m\dot{v}_{dx} \\ &= \frac{m}{R}(-c_1 c_0 y_{cg} + \theta(c_0(c_0 - c_2) + \beta)) \\ &\quad -\frac{m}{R}\left(\frac{1 - c_4}{R}\right)\frac{\beta c_0}{c_3}. \end{aligned} \tag{16}$$

On the other hand, to get u_{d2}, parameters c_1, c_2, c_3 and c_4 should satisfy some algebraic conditions. Substituting (8), (10) into (15) and v_{dy} and ω_d into (14), the second and the third row of (14) can be represented as non-homogenous linear system

$$\begin{aligned} &\begin{bmatrix} -c_1^2 - a_{22}c_1 & c_1(c_0 - c_2) - c_3(c_2 + a_{23}) - a_{22}c_2 \\ -a_{32}c_1 & -c_3^2 - a_{32}c_2 - a_{33}c_3 \end{bmatrix} \begin{bmatrix} y_{cg} \\ \theta \end{bmatrix} \\ &+ \frac{1}{R}\begin{bmatrix} -\frac{c_1 - a_{22}}{c_3} & -c_2c_4 - a_{23}(c_4 + 1) \\ -\frac{a_{32}}{c_3} & -c_3c_4 - a_{33}(c_4 + 1) \end{bmatrix} \begin{bmatrix} \beta \\ c_0 \end{bmatrix} \\ &= \begin{bmatrix} \frac{1}{m} \\ \frac{l_1}{I_z} \end{bmatrix} u_{d2}. \end{aligned} \tag{17}$$

Substituting c_4 in (11) into (17) and solving u_{d2} for arbitrarily specified y_{cg}, θ, β and c_0 in (17) by using cross product vector property, we have three determinant equations. From these determinant equations, we derive algebraic conditions for parameters c_1, c_2, and c_3 and they are given by

$$c_1 = \frac{I_z}{l_1 m}a_{32} - a_{22} \tag{18}$$

$$c_2 = \frac{\mu\left(1 - \frac{c_0^2}{\beta}\right) + \frac{I_z}{l_1 m}c_3}{\left(1 - \frac{c_0\mu}{\beta}\right)} \tag{19}$$

where $\mu = \frac{I_z}{l_1 m}a_{33} - a_{23}$.

c_3 is the root of polynomial

$$P_1 {c_3}^2 + P_2 c_3 + c_0 c_1 = 0 \tag{20}$$

where $P_1 = \frac{I_z}{l_1 m}\left(1 - \frac{1}{1 - \frac{c_0\mu}{\beta}}\right)$; $P_2 = \frac{\frac{c_0\mu}{\beta}(c_0 - \mu)}{1 - \frac{c_0\mu}{\beta}}$.

If (18) to (20) and (11) are satisfied, then (17) obviously gives

$$\begin{aligned} u_{d2} &= -m(c_1^2 + a_{22}c_1)y_{cg} \\ &+ m(c_1(c_0 - c_2) - c_2c_3 - a_{22}c_2 - a_{23}c_3)\theta \\ &- m\left(\beta\frac{c_1 + a_{22}}{Rc_3} + c_0\frac{c_2c_4 + a_{23}(c_4 + 1)}{R}\right). \end{aligned} \tag{21}$$

Now, we define the velocity error signals as

$$e = x - v_d. \tag{22}$$

Differentiating (22) and taking into account (3) and (14), the velocity error dynamics can be expressed as

$$\dot{e} = (A + B_1 K)e + B_2 d \tag{23}$$

where $u = u_d + Ke$; $u_d = [u_{d1}\ u_{d2}]^T$ are defined by (16) and (21) and K is feedback controller gain.

6. CONTROL SYSTEM ANALYSIS

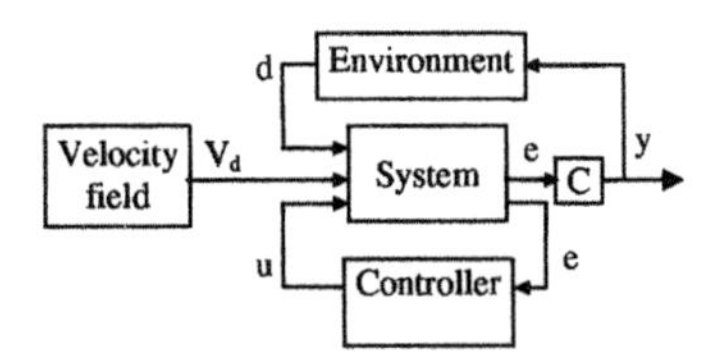

Fig. 2. Block diagram of control system.

Velocity error dynamics which have been derived in the previous chapter should be asymtotically stable in order to make vehicle longitudinal, lateral and yaw velocity converge to the desired velocity field V_d. This can be done by designing the controller gain K using passivity conditions for velocity error dynamics.

From equation (23) and Fig. 2, the velocity error dynamics can be expressed as

$$\begin{aligned} \dot{e} &= (A + B_1 K)e + B_2 d \\ y &= Ce \end{aligned} \tag{24}$$

where

$C = \begin{bmatrix} c_{11} & 0 & 0 \\ 0 & c_{22} & c_{23} \end{bmatrix}$, c_{ij} are design parameters.

Now, we will derive the passivity conditions for (24) and utilize it to obtain gain K. Let us consider the following storage function

$$W(e) = \frac{1}{2}e^T P e \tag{25}$$

$$P > 0 \tag{26}$$

as the kinetic energy of the error dynamics. Differentiating (25) yields

$$\dot{W}(e) = \frac{1}{2}e^T\left\{(A + B_1K)^TP + P(A + B_1K)\right\}e + d^TB_2^TPe. \quad (27)$$

Equation (27) shows that the system (24) is passive w.r.t. the supply rate d^Ty when the following conditions are satisfied:

$$(A + B_1K)^TP + P(A + B_1K) < 0 \quad (28)$$

$$B_2^TP = C \quad (29)$$

Suppose that the external inputs are absent, i.e. $d = 0$. If feedback gain K and matrix P are chosen so that (26), (28) and (29) are satisfied. Then from (27) we have $\dot{W} < 0$, which demonstrates that the velocity error dynamics are asymptotically stable.

However, for error dynamics to be passive with respect to the supply rate d^Ty, equation (29) imposes some restrictions for the selection of elements c_{ij} in C. For this, let us rewrite (26), (28) and (29) as

$$AQ + QA^T + B_1Z + Z^TB_1^T < 0 \quad (30)$$

$$Q > 0 \quad (31)$$

$$B_2^T = CQ \quad (32)$$

where $Q = P^{-1}$, $Z = KQ$ are matrix variables.

Proposition 2 There exist matrix $Q > 0$ only if

$$c_{ij} > 0 \text{ for } i, j = \{1, 2\}$$

where c_{ij} are matrix elements of C in equation (24).

Proof

Observing the structure of matrices B and C, equation (32) can be rewritten as

$$\begin{bmatrix} c_{11} & 0 & 0 \\ 0 & c_{22} & c_{23} \end{bmatrix} \begin{bmatrix} q_{11} & 0 & 0 \\ 0 & q_{22} & q_{23} \\ 0 & q_{23} & q_{33} \end{bmatrix} = \begin{bmatrix} \frac{1}{m} & 0 & 0 \\ 0 & \frac{1}{m} & 0 \end{bmatrix} \cdot (33)$$

Comparing both sides, we obtain

$$c_{11} = \frac{1}{mq_{11}} > 0 \text{ since } q_{11} > 0 \quad (34)$$

$$q_{22} = \frac{1}{mc_{22}} - \frac{c_{23}}{c_{22}}q_{23} \quad (35)$$

$$q_{33} = -\frac{c_{22}}{c_{23}}q_{23} \quad (36)$$

Substituting (35) and (36) into (33) and the fact that the Minor $Q_{11} > 0$, we have

$$\frac{c_{22}}{c_{23}}q_{23}\left(\frac{1}{mc_{22}} - \frac{c_{23}}{c_{22}}q_{23}\right) - q_{23}^2 > 0$$

$$\Rightarrow \frac{q_{23}}{mc_{23}} < 0 \quad (37)$$

Using (35), (36) and (37) we conclude that $c_{22} > 0$ since $q_{22} > 0$ and $q_{33} > 0$.

Notice that, Proof above has shown that only parameter q_{23} in Q is free to be chosen. Thus, we can modify the equality (32) and reduce the passivity conditions to LMI. Substituting (34), (35) and (36) into Q, (31) becomes

$$Q = \begin{bmatrix} \frac{1}{mc_{11}} & 0 & 0 \\ 0 & \frac{1}{mc_{22}} - \frac{c_{23}}{c_{22}}q_{23} & q_{23} \\ 0 & q_{23} & -\frac{c_{22}}{c_{23}}q_{23} \end{bmatrix} \quad (38)$$

We can rewrite (38) as

$$Q_0 - Q_1qQ_1^T > 0 \quad (39)$$

$$\text{where } Q_0 = \begin{bmatrix} \frac{1}{mc_{11}} & 0 & 0 \\ 0 & \frac{1}{mc_{22}} & 0 \\ 0 & 0 & 0 \end{bmatrix};\ q = \frac{c_{23}}{c_{22}}q_{23};$$

$$Q_1 = \begin{bmatrix} 0 & 1 & -\frac{c_{23}}{c_{22}} \end{bmatrix}.$$

Substitute (39) into (30), we have

$$L_0 - L_1qQ_1^T - Q_1qL_1^T + B_1Z + Z^TB_1^T < 0 \quad (40)$$

where $L_1 = AQ$, $L_0 = AQ_0 + Q_0A^T$. Then, (39) and (40) can be expressed as

$$\begin{bmatrix} L_0 - L_1qQ_1^T - Q_1qL_1^T + B_1Z + Z^TB_1^T & 0 \\ 0 & -Q_0 + Q_1qQ_1^T \end{bmatrix} < 0 \quad (41)$$

Now, let us summarize the computation procedures of controller design follow: Step 1, choose c_0 and β by considering the nominal speed and cornering behavior. Step 2, define the desired velocity field v_d by (13). Step 3, compute parameter c_1, c_2, c_3 and c_4 using (11) and (18) to (20). Step 4, determine u_d the feedforward of u by (16) and (21). Step 5, define velocity error dynamics by (24). Step 6, derive the passivity conditions for error dynamics and reduce it to LMI form (41). Step 7, using LMI (41), find q and Z. Then calculate $q23$ using (39) and replace it into Q in (38). Step 8, feedback gain K is obtained as $K = ZQ^{-1}$. Lastly step 9, u is defined as $u = u_d + Ke$.

7. SIMULATION

The new proposed velocity field and controller have been tested in simulation using a nonlinear vehicle model (2). Simulation with initial conditions: $V_x(0) = 15m/s$, $V_y(0) = 0.07m/s$, $\Omega(0) = 0.01rad/s$, $\theta_0 = 0rad$ and radius of curvature $2000m$ are shown in Figure 3 without considering any external disturbances. Given and design parameters for vehicle are given in Table 1. The measurement of y_{cg} and θ are based on equation (5), (6) and (7).

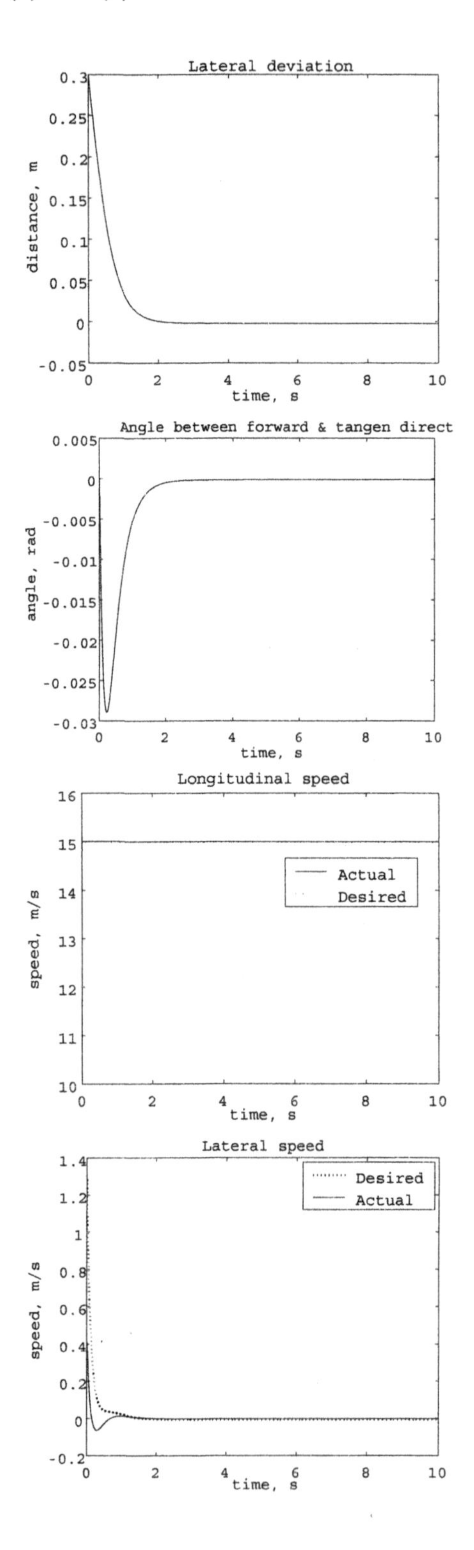

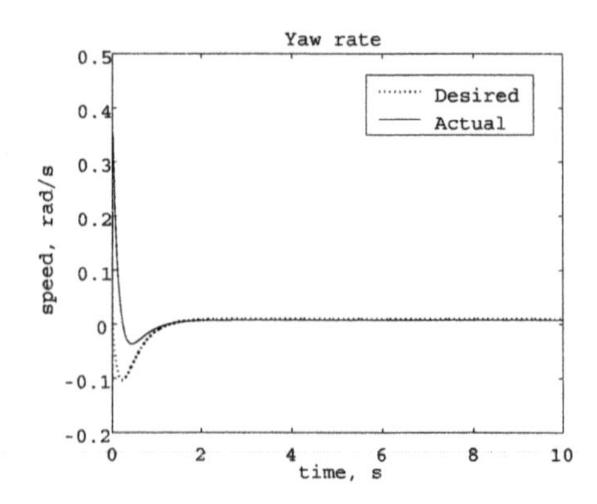

Fig. 3. Simulation results of the controller with $c_0 = 15m/s, \beta = 50m^2/s^2$ and $R = 2000m$

REFERENCES

[1] Shladover, S.E. et al, *Automated vehicle control developments in the PATH program*, IEEE Transactions on Vehicular Technology, Vol. 40, No. 1, February 1991, pp114-130.

[2] H. Peng and M. Tomizuka, *Lateral Control of Front-Wheel-Steering Rubber Tire Vehicles*, PATH reports, Institute of Transportation Studies, University of California at Berkeley, 1990.

[3] Fenton, R.E. and Mayhan, R.J. *Automated highway studies at the Ohio State University-an overview*, IEEE Transactions on Vehicular Technology, Vol. 40, No. 1, February 1991, pp100-113.

[4] T.J. Gordon, M.C. Best and P.J. Dixon, *An automated driver based on convergent vector fields*, Proc IME Part D J. Auto. Eng., Vol. 216, 2002.

[5] J.Y. Wong, *Theory of Ground Vehicles*, John Wiley & Sons, 2nd. Edition.

[6] J.R. Zhang, A. Rachid, and S.J. Xu, *Velocity Controller Design for Automatic Steering of Vehicles*, Proc. ACC, June 25-27, 2001, pp 696-697.

APPENDIX

Table 1. Parameters of vehicle and road model

m	Mass (kg)	2300
I_z	Moment of inertia around z-axis ($kg - m^2$)	3000
l_1	Distance from front tires to CG (m)	1.90
l_2	Distance from rear tires to CG (m)	2.20
$C_{\alpha f,r}$	Cornering stiffness (front, rear tires) (N/rad)	40000, 34000
c_{11}	Elements of output matrix C	0.1
c_{22}		0.03
c_{23}		0.02

www.elsevier.com/locate/ifac

SLIDING MODE CONTROL OF COMPLIANT JOINTS ROBOT

A. B. Sharkawy*, A. Vitko, M. Šavel**, and L. Jurišica*****

**Faculty of Engineering, Assiut University, Assiut 715 16, Egypt*
Fax: +2 088 332 455 and e-mail: asharkawy@hotmail.com
*** Faculty of Mechanical Engineering, Nám. slobody 17, 812 31 Bratislava*
Fax: +421 7 395 315 and e-mail: vitko@kam1.vm.stuba.sk
**** Faculty of Electrical Eng. and IT, Ilkovičova 3, 812 19 Bratislava*
Fax: +421 7 6542 9521 and e-mail: juris@elf.stuba.sk
Slovak University of Technology in Bratislava, SLOVAKIA

Abstract: The lack of knowledge of robot dynamics and joint elasticity for precise tracking control can be overcome by using either robust or adaptive (learning) control. The standard form of adaptive control does not appear to be applicable since the basic assumptions on the system dynamics and non-linear characteristics are rarely satisfied. On the other hand, variable structure control (VSC) is a powerful control method that has its main attractiveness in the invariance to disturbances and parameter variation. The paper presents a robust control for elastic joint robots (EJR) based on variable structure systems (VSS). The controller needs the measurement of position and velocity and the knowledge of the upper bounds of the uncertainties. *Copyright © 2003 IFAC*

Keywords: variable structure systems (VSS), elastic joint robots (EJR), sliding mode.

1. INTRODUCTION

The joint stiffness is mainly due to the use of special low weight gear-boxes, called harmonic drives, which possess very high transmission ratios even they are quite compact and light. A disadvantage of harmonic drives is that compliance or mechanical flexibility is introduced. When elasticities result in eigen frequencies close to the bandwidth of the controlled system, they even cause instabilities. However, including the joint dynamics to the robot model and control rises the order of the dynamics twofold. The complete dynamic model then possesses four states variable for each joint, namely position and speed of the motors and those of the links. From the point of view of control design, the robot manipulator dynamics becomes further complicated if parametric uncertainties and other disturbances are to be taken into account.

Recent works dealing with adaptive control of EJR could be found in (Lozano and Brogliato 1992), (Yuan and Stepanenko 1993), (Brogliato *et al.* 1995). Despite the mathematical achievements of these approaches, the need of the persistency of exiting could not be ignored and even so the convergence of errors may be very slow. For this reason the applicability of the adaptive control doesn't appear at all (Brogliato *et al.* 1998). Variable structure systems (VSS) have been the subject of intensive study for the last few decades. The reader is referred to (Utkin 1977),(DeCarlo *et al.* 1988) for complete details on the theory. The main idea is to force the robot states to slide on predefined hyperplanes called 'sliding surfaces'. This is accomplished by a high frequency switching control among suitable designed alternative control laws. Then the robot dynamics is exclusively determined by the sliding surfaces parameters. Most of the variable structure controllers available in robotic literature are designed for the case of rigid robots. Among those are the studies of (Fu and Liao 1990), (Yao *et al. 1994*), (Kalas and McCorkell 1995), (Myszkorowski 1990). In relation to EJR, implementation of the theory of VSS to the control of EJR could be found in (Ramirez and Spong 1998), (Mrad and Ahmed 1991). The work of (Ramirez and Spong 1998) is based on the idea of feedback linearization within the inner loop. But the controller requires both acceleration and jerk measurements. In (Mrad and Ahmed 1991), a switching control law that guarantees asymptotic stability requires only position and velocity measurements for both motor and link. However the full knowledge of system parameters is needed.

In this work, we introduce an original robust VSC scheme for EJR, which doesn't require neither the measurement of acceleration nor jerk and parameter variation could be easily taken into consideration. Moreover asymptotic stability is ensured regardless of joint elasticity value.

The rest of the paper is organized as follows. Section 2 discusses the dynamic model of EJR and its properties. Section 3 illustrates the relation between the desired motor shaft angle and that of the links, so that joint flexibilties are taken into account. The control scheme is introduced in Section 4. In Section 5 we introduce the boundary layer as a possible technique used for eliminating chattering which is a phenomena associated with VSS. Simulation results are given in Section 6. Section 7 offers our concluding remarks.

2. DYNAMIC MODELLING OF EJR

The dynamic model of an *n*-link manipulator with flexible joints as proposed by M. Spong (Spong 1987):

$$\begin{aligned} &D(q)\ddot{q} + C(q,\dot{q})\dot{q} + G(q) = K(\theta - q) \\ &J\ddot{\theta} + K(\theta - q) = u \end{aligned} \tag{1}$$

in which $D(.)$ is the link inertia matrix, symmetric and positive definite, $C(.)$ is the centrifugal and Coriolis matrix, $G(.)$ is the gravitational vector, K is the drive shaft stiffness matrix, diagonal, J is the motor inertia matrix, diagonal and positive definite, q and θ represent the vectors of link angles, and motor angles respectively. In the next paragraphs we note some of this model's properties and assumptions which will be used in the proposed control scheme.

A useful relationship exists between the time derivative of the inertia matrix $D(q)$ and the Coriolis matrix $C(q,\dot{q})$. For an arbitrary $n \times 1$ vector x we have that

$$x^T[\dot{D}(q) - 2C(q,\dot{q})]\ x = 0 \tag{2}$$

It means that the matrix $[\dot{D}(q) - 2C(q,\dot{q})]$ is skew symmetric, a property which can be derived from the Lagrangian formulation of the manipulator dynamics (Spong 1987).

In most real world problems, one cannot always exactly determine the parameters of a dynamic model. It is however, reasonable to assume that bounds on these parameters do exist. In this work, we assume that only average scalar functions D^o, C^o, G^o, K^o, J^o of $D(q), C(q,\dot{q}), G(q)$, K, J, in (1) are available. The modelling errors resulting from this approximation are bounded by:

$$\begin{aligned} &|\Delta D(q)| \le \delta_D; |\Delta C(q,\dot{q})| \le \delta_C; \\ &|\Delta G(q)| \le \delta_G; |\Delta K| \le \delta_K; |\Delta J| \le \delta_J \end{aligned} \tag{3}$$

where ΔA represents the modelling error of matrix (or vector) A defined by:

$$\Delta A = A - A^o \tag{4}$$

The positive scalar bounds $\delta D, \delta C, \delta G, \delta K, \delta J$ are assumed known functions. It should be noted that this assumption could be easily justified in practice.

3. TRAJECTORY PLANNING

The term 'trajectory planning' is referred here to the definition of a relation between the desired links' trajectories q_d, and those of the joints θ_d. The necessity of this step in the control design is to define the desired joint trajectories θ_d, so that joint flexibility are taken into account. Then, the goal of the control law is to force the actuators position θ to stick to θ_d, and consequently links' motion to q_d.

Let $q_d(t) \in C^4$ denotes a desired link trajectory, which is continuously differentiable up to four times. To define the desired joint angles θ_d, we have adopted an approach similar to that proposed by B. Brogliato *et al* (Brogliato *et al. 1995*). In that approach the right hand side of the first equation (1), which represents dynamics of rigid robots is replaced by the control law of Slotine and Li (Slotine and Li 1988) u_R as follows:

$$\theta_d = K^{o^{-1}} u_R + q_d, \tag{5}$$

where

$u_R = D^o[\ddot{q}_d - 2\lambda\dot{\tilde{q}} - \lambda^2\tilde{q}] + C^o(\dot{q}_d - \lambda\tilde{q}) + G^o$,

λ is positive constant,

$\tilde{q} = q - q_d$ is the vector of tracking errors.

The control signal u_R in (5) is designed for the rigid part of the robot dynamics. To unify this presentation, we use average scalar functions of the matrices D, C, G, instead of the estimated ones in Slotine and Li's algorithm.

4. THE CONTROL DESIGN

In VSS, a high-speed switching control is designed to derive the non-linear plant's trajectory onto a specific and user-chosen surface in the state space, called sliding or switching surface $s(x)$, and to maintain the plant's state trajectory on this surface for all subsequent time. If the control ensures:

$$s_i^T(x)\dot{s}_i(x) \le 0, \qquad i = 1,2,....n, \tag{6}$$

thus a sliding mode exists on the sliding surfaces for all $x \in R^n$ (Brogliato *et al.* 1998), (Utkin 1977). It means that sufficient conditions for rateability / existence of sliding mode are met if for any point in the state space and for all $t \ge t_o$, s and $\dot{s}$ are of opposite sign.

To begin with solving the control problem of EJR, let us choose the following sliding surfaces:

$$s_1 = \dot{q} + \Lambda_1\tilde{q} \text{ and } s_2 = \dot{\theta} + \Lambda_2\tilde{\theta} \tag{7}$$

where matrices Λ_1, Λ_2 have eigenvalues strictly in the right-hand complex plane, $\tilde{q} = q - q_d$ is the vector of linkages tracking errors and $\tilde{\theta} = \theta - \theta_d$ is that of the rotor shafts. For simplicity, we begin with differentiating the sliding surfaces and inserting $D\ddot{q}, J\ddot{q}$ from (1):

$$D\dot{s}_1 = K(\theta - q) + D\Lambda_1\dot{\tilde{q}} - C\dot{q} - G \tag{8}$$

$$J\dot{s}_2 = u + K(q - \theta) + J\Lambda_2\dot{\tilde{\theta}} \tag{9}$$

Now, assume that

$$D\dot{s}_1 = -As_1 = \begin{vmatrix} -a_1\,\text{sgn}(s_{11}) \\ \vdots \\ -a_1\,\text{sgn}(s_{1n}) \end{vmatrix}, \tag{10}$$

and

$$J\dot{s}_2 = -Bs_2 = \begin{vmatrix} -b_1\,\text{sgn}(s_{21}) \\ \vdots \\ -b_n\,\text{sgn}(s_{2n}) \end{vmatrix} \tag{11}$$

where,

$$A = \text{diag}\,[a_i\,\text{sgn}(s_{1i})/s_{1i}],\, a_i(q,\dot{q}) > 0$$
$$B = \text{diag}\,[b_i\,\text{sgn}(s_{2i})/s_{2i}],\, b_i(\theta,\dot{\theta}) > 0$$
$$,i = 1,2\cdots\cdots n,$$

$$\text{sgn}\,(x) = \begin{cases} 1 & x > 0 \\ -1 & x < 0 \\ 0 & x = 0. \end{cases}$$

Consider the candidate Lyapunov function

$$V = \frac{1}{2}s_1^T D\, s_1 + \frac{1}{2}s_2^T J\, s_2 \tag{12}$$

which is a positive semi-definite function, i. e vanishes only when the components of both s_1 and s_2 are equal to zeroes.

Differentiating V with respect to time and using the skew symmetry property, defined in (2), gives

$$\dot{V} = s_1^T D\dot{s}_1 + s_1^T Cs_1 + s_2^T J\dot{s}_2.$$

Substituting from (10), (11)

$$\dot{V} = -s_1^T(A - C)\, s_1 - s_2^T B\, s_2 \tag{13}$$

For $\dot{V}$ to be negative semi-definite the matrices *(A-C)* and B have to be positive definite. For B to be positive definite

$$b_i(\theta,\dot{\theta})\text{sgn}(s_{2i})/s_{2i} > k_{1i} \tag{14}$$

where k_{1i} is the diagonal elements of a positive constant matrix K_1.

The control u that guarantees (14) could be designed as follows. From (9), (11)

$$u + K(q - \theta) + J\Lambda_2\dot{\tilde{\theta}} = -Bs_2 \tag{15}$$

Let the control law be

$$u = T_1 + K^o(\theta - q) - J^o\Lambda_2\dot{\tilde{\theta}} \tag{16}$$

where the control component T_1 is to be determined so that (14) is satisfied. Using parameter bounds given in Section 2, equation (4), and substituting from (16) into (15) yields

$$T_1 + \Delta K(q - \theta) + \Delta J\Lambda_2\dot{\tilde{\theta}} = -Bs_2$$

which in a component wise; see (11):

$$T_{1i} + \sum_{j=1}^{n}[\;\Delta K_{i,j}(q_j - \theta_j) + \Delta J_{i,j}\Lambda_{2i,j}\dot{\tilde{\theta}}_j\;] = -b_i\,\text{sgn}(s_{2i}) \tag{17}$$

From (14) and (17), the control component T_1 can be found as follows:

when $s_{2i} > 0$

$$T_{1i} + \sum_{j=1}^{n}[\;\Delta K_{i,j}(q_j - \theta_j) + \Delta J_{i,j}\Lambda_{2i,j}\dot{\tilde{\theta}}_j\;] = -b_i\,\text{sgn}(s_{2i}) < s_{2i}k_{1i} \tag{18}$$

when $s_{2i} < 0$

$$T_{1i} + \sum_{j=1}^{n}[\;\Delta K_{i,j}(q_j - \theta_j) + \Delta J_{i,j}\Lambda_{2i,j}\dot{\tilde{\theta}}_j\;] = -b_i\,\text{sgn}(s_{2i}) > s_{2i}k_{1i} \tag{19}$$

Combining the above gives

$$\text{sgn}(s_{2i})\left\{\; T_{1i} + \sum_{j=1}^{n}[\;\Delta K_{i,j}(q_j - \theta_j) + \Delta J_{i,j}\Lambda_{2i,j}\dot{\tilde{\theta}}_j\;]\right\} < |s_{2i}|k_{1i} \tag{20}$$

which will be fulfilled if we choose

$$T_{1i} = -\text{sgn}(s_{2i})\left\{ \sum_{j=1}^{n}[\ \delta K_{i,j}(|q_j|+|\theta_j|)+ \delta J_{i,j}\Lambda_{2i,j}\left|\dot{\tilde{\theta}}_j\right| \right\} - s_{2i}k_{1i} \quad (21)$$

where $\delta K, \delta J$ are the bounds of modelling errors as defined in (3). The control component T_1 ensures that the inequality (14), have been satisfied. Consequently, sliding mode will take place on the hyperplane s_{2i} whether such a mode exists on the hyperplane s_{1i} or not. Now, what remains is to design another component T_2 so that the matrix *(A-C)* is positive definite. For *(A-C)* to be positive definite

$$a_i \,\text{sgn}(s_{1i})/s_{1i} > \sum_{j=1}^{n}|C_{i,j}| \quad (22)$$

Following the procedure of steps (15)-(21) to determine the control component T_2, yields

$$T_{2i} = -\text{sgn}(s_{1i})\left\{ \sum_{j=1}^{n}[\ \delta K_{i,j}(|\theta_j|+|q_j|)+ \delta D_{i,j}\Lambda_{1i,j}\left|\dot{\tilde{q}}_j\right| + \delta C_{i,j}|\dot{q}_j|\ ]+\delta G_i \right\} - s_{1i}\sum_{j=1}^{n}|C_{i,j}| \quad (23)$$

The control component T_2 ensures that the inequality (22) has been satisfied, which means the existence of a sliding mode on the hyperplane s_1. Now, the final controller is completed

$$u = T_1 + T_2 + G^o + C^o\dot{q} - J^o\Lambda_2\dot{\tilde{\theta}} - D^o\Lambda_1\dot{\tilde{q}} \quad (24)$$

where T_1 and T_2 are defined in (21) and (23) respectively.

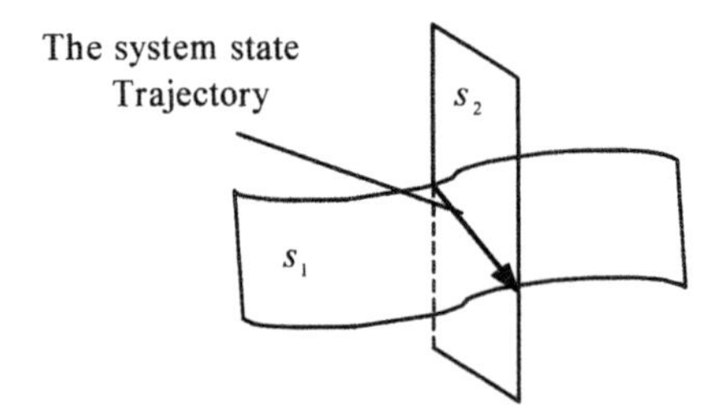

Fig.1. Sliding mode control with multiple sliding surfaces

Consequently, sliding mode will take place at the intersection of the two sliding surfaces (Zhen and Goldenberg 1996). Figure 1 illustrates the control strategy.

Remark 1: Selecting other sliding surfaces could result in other controllers. For example, if we choose sliding surfaces in the form:

$$s_1 = \dot{\tilde{q}} + \Lambda_1\tilde{q} \text{ and } s_2 = \dot{\tilde{\theta}} + \Lambda_2\tilde{\theta} \quad (25)$$

then the control components are

$$T_{1i} = -\text{sgn}(s_{2i})\left\{ \sum_{j=1}^{n}[\ \delta K_{i,j}(|q_j|+|\theta_j|)+ \delta J_{i,j}\Lambda_{2i,j}\left|\dot{\tilde{\theta}}_j\right| + \delta J_{i,j}\left|\ddot{\theta}_{d_j}\right|\ ] \right\} - s_{2i}k_{1i} \quad (26)$$

$$T_{2i} = -\text{sgn}(s_{1i})\left\{ \sum_{j=1}^{n}[\ \delta K_{i,j}(|q_j|+|\theta_j|)+\delta C_{i,j}|\dot{q}_j| + \delta D_{i,j}\Lambda_1\left|\dot{\tilde{q}}\right| + \delta D_{i,j}\left|\ddot{q}_{d_j}\right|\ ]+\delta G \right\} - s_{1i}\sum_{j=1}^{n}|C_{i,j}| \quad (27)$$

and the control law is

$$u = T_1 + T_2 + J^o\ddot{\theta}_d + D^o\ddot{q}_d + G^o + C^o\dot{q} - J^o\Lambda_2\dot{\tilde{\theta}} - D^o\Lambda_1\dot{\tilde{q}} \quad (28)$$

As it is the case in the control law (24), it can be noticed that control components T_1 and T_2 are dependent on the model bounds and the rest of the controller's components in (28) are dependent on average functions for the dynamic model. Each of them is assumed known.

Remark 2: The control laws defined in (24) and (28) are composed of four variables for each joint. They are link position q, velocity $\dot{q}$, actuator position θ and velocity $\dot{\theta}$. This is a common case for most controllers that takes into account joint flexibility of robotic manipulators. Usually, the actuator position measurements are available through encoders fixed to the actuator shaft. Using this signal as the system output, a state observer that delivers estimations for the other states could be found in (Sharkawy 1998).

5. ELIMINATION OF CHATTERING

The control laws (24) and (28) are discontinuous across sliding surfaces, which leads to control chattering. Chattering is undesirable in practice because it involves high control activity and may excite high frequency dynamics neglected in the course of modelling (like unmodelled structural modes, neglected time delays and the like). A possible remedy is to use the boundary layer technique to smoothen out the control discontinuity within a thin boundary layer neighbouring the sliding surface

$$E(t) = \{\ q(t), |s(q,\dot{q})| < \phi\ \}; \qquad \phi > 0$$

where ϕ is the boundary layer thickness. This is achieved by replacing the signum nonlinearity with suitable saturation nonlinearity in the expression of *u*. It means that the term $\text{sgn}(s_i)$ is replaced by s/ϕ inside *E(t),* which leads to tracking within a guaranteed precision (Slotine 1984). Consequently, chattering due to data sampling is eliminated. While such replacement leads to small terminal error, the practical advantages of having smooth control input may be significant.

6. SIMULATION TESTS

In this section, we present numerical tests for the control law (24) with the application of boundary layer. Simulation tests were carried out on a two link planar manipulator. Dynamic equation for such manipulator could be found in (Ramirez and Spong 1998). For such manipulator, the matrices D, C and G in (1) are:

$$D=\begin{pmatrix}\alpha+\beta+2\eta\ \cos\ q_2 & \beta+\eta\ \cos\ q_2\\ \beta+\eta\ \cos\ q_2 & \beta\end{pmatrix},$$

$$C=\begin{pmatrix}-\eta\dot{q}_2\ \sin\ q_2 & -\eta(\dot{q}_1+\dot{q}_2)\sin\ q_2\\ \eta\dot{q}_1\ \sin\ q_2 & 0\end{pmatrix}$$

$$G=\begin{pmatrix}\alpha e\ \cos\ q_1+\eta e\ \cos(q_1+q_2)\\ \eta e\ \cos(q_1+q_2)\end{pmatrix}.$$

in which

$\alpha=r_1^2(m_1+m_2), \beta=m_2r_2^2, \eta=m_2r_1r_2, e=g/r_1$.

The controller matrices D^o, C^o, G^o were chosen as follows:

$$D^o=\begin{pmatrix}\alpha+\beta & \beta\\ \beta & \beta\end{pmatrix},\ G^o=\begin{pmatrix}\eta e\\ \eta e\end{pmatrix},\ C^o=\begin{pmatrix}0 & 0\\ 0 & 0\end{pmatrix},$$

and $\delta D, \delta C, \delta G, \delta J, \delta K$

$$\delta D=\begin{pmatrix}2\eta & \eta\\ \eta & 0\end{pmatrix}, \delta C=\begin{pmatrix}0 & 0\\ 0 & 0\end{pmatrix}, \delta G=\begin{pmatrix}\alpha e\\ 0\end{pmatrix},$$

$$\delta J=0.1J^o, \delta K=0.1K^o.$$

As it may be noticed, they are time independent. With this structure they were implemented in trajectory planning (5). The control law (24) was tested with $K_1=50I$, and $\Lambda_1=\Lambda_2=35I$, where I is the identity matrix. We used the same trajectories and parameter values of Mrad and Ahmed (Mrad and Ahmed 1991). Parameter values were selected as follows:

$K^o=50I\ N.m/rad,\ J^o=0.05I\ kg.m^2,$

$m_1=m_2=5\ kg\ ,\ r_1=r_2=0.025\ m$

The desired links' trajectories are:

$$q_{1d}(t)=-\frac{\pi}{2}+0.3\sin(\pi t),\ q_{2d}(t)=-0.3+0.3\cos(\pi t)$$

Initial conditions were chosen such that $q=\theta=[\pi/2,0], \dot{q}=\dot{\theta}=[0,0]$. A sampling rate of $1000\ Hz$ was incorporated in simulations.

Results of simulations are presented in Figs 2-4. It is shown that the plant trajectories have been converged to the sliding surfaces of links and joints, Fig. 2. Consequently the tracking errors of links have been converged to zero in Fig. 3. The input torques are given in Fig. 4.

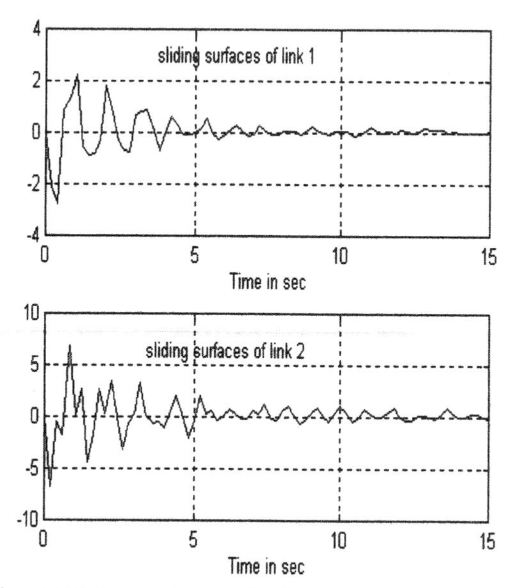

Fig. 2. Sliding surfaces of links

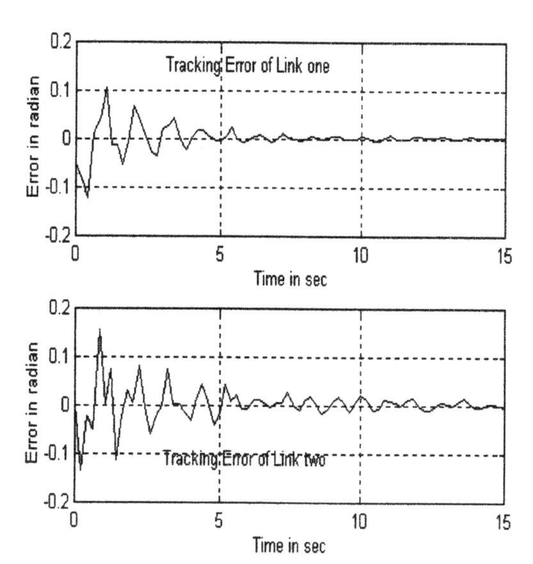

Fig. 3. Tracking error of links

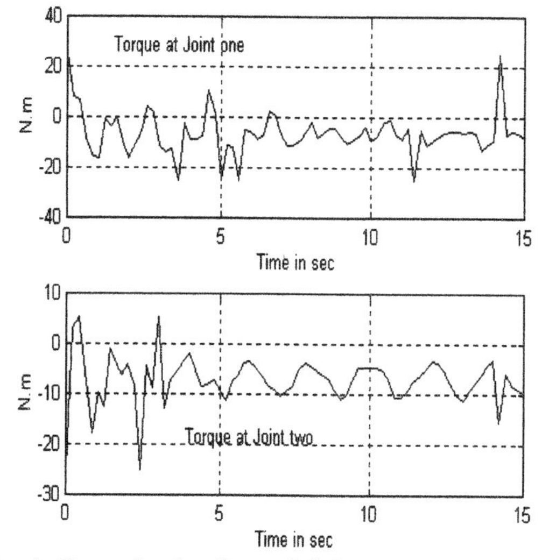

Fig. 4. Control value for each joint

7. CONCLUSIONS

A robust variable structure control scheme for EJR has been presented. The proposed control algorithm has the privilege over the existing VSC algorithms in that exact information about the manipulator parameters is no longer needed. Four feedback signals are needed to synthesise it. They are the position and velocity of the links (after the gearbox) and joints (before the gearbox). A state observer that is able to deliver part of them is discussed in (Sharkawy 1998). The control scheme could be used to derive other control laws by other choices for the sliding hyperplanes. Simulation tests have shown the feasibility of the proposed scheme.

REFERENCES

Brogliato, B., R. Ortega, and R. Lozano (1995). Global Tracking Controllers for Flexible-joint Manipulators: a Comparative Study, *Automatica*, **31**, 941-956.

Brogliato, B., D. Rey, A. Pastore, and J. Barnier (1998). Experimental Comparison of Non-linear controllers for Flexible Joint Manipulators, *Int. J. of Robotics Research*, **17**, 260-281

DeCarlo, A., H. Zak, and P. Matthews. (1988). Variable Structure Control of Non-linear Systems: A Tutorial, *Proceedings of the IEEE*, **76**(3), 212-232.

Fu, L. and T. Liao (1990). Globally Stable Robust Tracking of Non-linear Systems Using Variable Structure Control and with an Application to Robotic Manipulator, *IEEE Trans. on Aut. Conrol.*, **35**, 1345-1350.

Kalas, D. and C. McCorkell (1995). Sliding Mode Robotic Manipulator Trajectory Control in the task Oriented Space", *ISMCR' 95 Proceedings*, pp. 377-382.

Lozano, R. and B. Brogliato (1992). Adaptive Control of Robot Manipulators with Flexible Joints. *IEEE Tran. on Aut. Control*, **37**, 174-181.

Mrad, F. and S. Ahmed (1991). Control of Flexible Joint Robots, *Proceeding of the IEEE Int. Conf. on Robotics and Automat.* Sacramento,pp. 2832-2837.

Myszkorowski, P. (1990). A Class of Robust Controllers For Robot Manipulators, 11*th, IFAC World Congress*, **9**, Tallinn, Etonia, USSR 115 - 120.

Qijie, Z.and S. Chuny (1993). An Adaptive Sliding Mode Scheme for Robot Manipulators, *Int. J. of Contr.*, **57**(2), pp. 261-271.

Ramirez, H. and S., M. Spong (1998). Variable Structural Control of Flexible Joint Manipulators. *Int. J. of Robotics. and Automat.*, **3**, 55-64.

Sharkawy, A. (1998). Control of Flexible Joint Robots, *PhD thesis*, Faculty of Mechanical Engineering, Slovak University of technology in Bratislava.

Slotine, J. (1984). Sliding controller design for non-linear systems, *Int. J. of Control*, **40**(2), 421- 434.

Slotine, J. and W. Li (1988). Adaptive manipulators control: A case study, *IEEE Trans. on Aut. Control*, **33**(11), 995-1003.

Spong, M. (1987). Modelling and Control of Elastic Joint Robots. *Trans. of the ASME*, **109**, 310-319.

Utkin, V. (1977). Variable Structure Systems with Sliding Modes. *IEEE Trans. on Aut. Control*, AC-**22**, 212-222.

Yao, B., S. Chan, and D. Wang (1994). Unified Formulation of Variable Structure Control Scheme for Robot Manipulators, *IEEE Trans. on Aut. Conrol.*, **39**, 372-376.

Yuan, J. and Y. Stepanenko (1993). Composite Adaptive Control of Flexible Joint Robots. *Automatica*, **29**, 677- 888.

Zhen, R. and A. Goldenberg (1996). Variable Structure Hybrid Control of Manipulators in Unconstrained and Constrained Motion, *Trans. of the ASME*, **118**, 327-332.

www.elsevier.com/locate/ifac

NUMERICAL METHOD OF OPTIMAL CONTROL USING COMPACT SUPPORT BASES

Jan Cvejn

Technical University of Liberec, Faculty of Mechatronics
Hálkova 6, 461 17 Liberec, Czech Republic
jan.cvejn@vslib.cz

Abstract: The article describes a numerical method of optimal control computation based on the idea of reduction the original infinite-dimensional problem into finite dimension using systems of base functions. The method can be considered as a generalized discretization. Although many kinds of base systems can be used, a special class of compact support base systems provide a fast way of determining gradient of the criteria functional which can be used with advantage in optimization. *Copyright © 2003 IFAC*

Keywords: optimal control, nonlinear systems, numerical algorithms, gradient methods

1. PROBLEM FORMULATION AND STANDARD APPROACHES

Consider a general optimal control problem:

$$J=\int_0^{t_f} f_0(\boldsymbol{x},\boldsymbol{u},t)dt \rightarrow min \quad (1)$$

$$\frac{d}{dt}\boldsymbol{x}=\boldsymbol{f}(\boldsymbol{x},\boldsymbol{u},t) \quad (2)$$

Vector functions $\boldsymbol{x}$ and $\boldsymbol{u}$ are of order n, m, respectively, $n \geqslant m$. Generalized boundary conditions of the problem are

$$\begin{aligned} &\boldsymbol{x}(0)=\boldsymbol{x}_0 \\ &\psi_j(\boldsymbol{x}(t_f))=0 \, , \, j=1,..r \, , \, r \leqslant n \end{aligned} \quad (3)$$

Functions $f_0, \boldsymbol{f}, \psi_j$ are supposed to be continuously differentiable in arguments $\boldsymbol{x}$ and $\boldsymbol{u}$ and functions $f_0, \boldsymbol{f}, \boldsymbol{u}, \partial f_0/\partial \boldsymbol{u}, \partial f_0/\partial \boldsymbol{x}, \partial \boldsymbol{f}/\partial \boldsymbol{u}, \partial \boldsymbol{f}/\partial \boldsymbol{x}$ integrable in $t \in \langle 0, t_f \rangle$.

The problem is to determine the control process $\boldsymbol{u}(t)$. In literature, usually the following approaches to the problem are described (Polak, Bryson-Ho):

1. Direct numerical minimization of J in functional space of control functions $\boldsymbol{u}(t)$.

2. Discretization of the problem and minimizing J as a real function of finite number of arguments.

3. The third approach consists in transformation of the problem into a set of equations using results of the optimal control theory and solving as a two-point boundary value problem.

In this article, the approaches 1 and 2, based on direct minimization, are discussed.

The discretization approach is especially advantageous if the discrete form of the dynamic system description is available:

$$\begin{aligned} &\boldsymbol{x}_{k+1}-\boldsymbol{x}_k=\boldsymbol{g}_k(\boldsymbol{x}_k,\boldsymbol{u}_k) \\ &k=0\ldots N-1 \end{aligned} \quad (4)$$

In this case, the criteria functional J can be directly minimized as a real function of arguments

$$u_{01}, u_{02}, \ldots u_{N-1,m} \quad (5)$$

The discrete form is natural for technical implementation of control systems that work with constant scanning period. Since the J function depends on finite number of arguments, it can be minimized using many general optimization techniques. There are many theoretical and practical comparisons of various optimization algorithms available (e.g. Fletcher) that enable choosing the most effective method for a given problem. In this case, it is also possible to search for the global extreme, if the used optimization method is designed so.

Unfortunately, for nonlinear systems, explicit discrete form is usually not available. In that case, the discretization is yet possible by numerical computation in each step:

$$x_{k+1} - x_k = \int_{kT}^{(k+1)T} f(x(t), u_k, t)\, dt$$
$$u_k = u(k.T) \tag{6}$$

Although methods of direct minimization in the functional space exist, it is recommended (Polak) to transform the system to the discrete form (4) first, even if it is not explicitly available and has to be evaluated as in (6). In the functional space, the techniques of optimization more often fail to find the actual solution. There are also serious theoretical problems as concerns convergence in this case – for example, some commonly used methods of constrained optimization, including penalty-function methods, need not converge to a solution in an infinite-dimensional space (Polak).

To get a quality discrete control sequence comparable with a continuous control process the scanning period T has to be chosen sufficiently low. Usually, J is then function of hundreds of parameters. Fast optimization of such function J depends on the fact whether its gradient can be determined in an effective manner. Of course, derivations of the criteria function J can be computed approximately as follows:

$$\frac{dJ}{du_{ik}} \approx \frac{J_{+\Delta} - J}{\Delta} \tag{7}$$
$$J_{+\Delta} = J(u_{01}, \ldots u_{ik} + \Delta, \ldots, u_{N-1,m})$$

or

$$\frac{dJ}{du_{ik}} \approx \frac{J_{+\Delta} - J_{-\Delta}}{2\Delta} \tag{8}$$

where Δ is a small positive number. While formula (7) is faster to evaluate, formula (8) is more precise.

Since one computation of the critera value J requires integrating the system differential equations from 0 to t_f, this way of gradient determination is obviously ineffective for higher number of parameters u_{ik}. Fortunately, in the case of optimal control there exist methods how to speed up this computation (Polak). Gradient is used in numerical algorithms of optimization either for direct searching in the direction of the steepest descent, or for successive approximating Hessian matrix of the second derivations. Since direct computation of the Hessian needs $n = N.m$ times more steps than computation of the gradient, it is disadvantageous in many cases.

There exist several ways to deal up with the ending conditions $\psi_j = 0$. Probably the easiest way is to add a special penalty term to the criteria functional:

$$J' = J + K \sum_{j=1}^{r} \psi(t_f)^2 \tag{9}$$

where K is a sufficiently high number. Let $\{K_i\}_1^\infty$ be an increasing sequence of numbers and u_i solution of the optimization problem:

$$J' = \int_0^{t_f} f_0(x, u, t)\, dt + K_i \sum_{j=1}^{r} \psi(t_f)^2 \rightarrow min \tag{10}$$

Then, under certain conditions (Polak), sequence $\{u_i\}_1^\infty$ has limit of u_{opt}, which is the optimal solution of the original problem (1) - (3).

For practical computation, the optimization is carried out in several steps with raising K_i until the ending conditions are fulfilled with sufficient precision. Experiences show that the initial value K_1 should not be set too high – otherwise the computation may take very long.

So far, time t_f was considered as fixed. If the time t_f is not known in forward, it is however possible to transform the problem to a larger problem with a fixed t_f (Polak).

More general idea of reduction the problem to the finite dimension is described in the next paragraph.

2. APPROXIMATING THE CONTROL FUNCTION USING BASE SYSTEMS

Let us choose a value of parameter N and define the set of integrable functions $\{f_i | i=0 \ldots N\}$ such that the optimal control function $u^* \in U$ can be in interval $t \in \langle 0, t_f \rangle$ well approximated by a weighted sum

$$u = \sum_{i=0}^{N} a_i f_i \tag{11}$$

and for each function $u \in U$ that can be expressed in form (11) there exists only one vector of coefficients a_i, i.e. functions f_i are independent.

Set $\{f_i\}$ can be called *base system of order N*.

The system of the base functions $\{f_i\}$ is often chosen so that the functions f_i are orthogonal and the system is *complete* in U for $N \rightarrow \infty$ (e.g. Rektorys), i.e. for any

$u \in U$ there exist coefficients $\{a_j\}_0^\infty$ such that

$$\| u - \sum_{i=0}^{\infty} a_i f_i \|_2 \to 0 \tag{12}$$

Although orthogonality is usually not necessary, only the complete base system can approximate any function $u \in U$ precisely. That is why for numerical computations often the finite base system $\{f_i\}_0^N$ is used such that for $N \to \infty$ holds (12).

Since vector $\boldsymbol{u}$ has m components, the previous expression can be written in the matrix notation

$$\boldsymbol{u}(t) = \boldsymbol{A}.\boldsymbol{b}(t) \tag{13}$$

where $\boldsymbol{A} = (a_{ij})$ is the matrix of weighting coefficients and $\boldsymbol{b}$ is the vector of functions $f_j, j=0\ldots N$.

Substituting (13) into (1) we get J as a function of coefficients a_{ij}:

$$J = \int_0^{t_f} f_0(\boldsymbol{x}, \boldsymbol{A}\,\boldsymbol{b}(t), t)\,dt \tag{14}$$

The previous equation can be written in differential form

$$\frac{dJ}{dt} = f_0(\boldsymbol{x}, \boldsymbol{A}\ \boldsymbol{b}(t), t) \tag{15}$$
$$J(0) = 0$$

This equation can be integrated together with the differential equations of the system and $J(t)$ can be considered as the 0-th component of extended state vector.

$$\frac{d}{dt}\boldsymbol{x}' = \boldsymbol{f}'(\boldsymbol{x}', \boldsymbol{u}, t) \tag{16}$$

where

$$\boldsymbol{f}' = (f_0, .. f_n),\ x_0(0) = 0,\ J = x_0(t_k)\,. \tag{17}$$

Function $J = J(a_{ij})$ can be then numerically minimized to obtain the optimal values of coefficients a_{ij}.

Note that discretization can be considered as a special case of reduction (11). Set $T = t_f / N$ and choose

$$f_i = 1 \text{ for } t \in [i.T, (i+1).T),\ i = 0\ldots N-1$$
$$f_i = 0 \text{ otherwise.} \tag{18}$$

Function $u = \sum_{i=0}^{N-1} a_i f_i$ will be piecewise constant. It is not true discretization however, since $u(t)$ is a function and not a sequence.

The main advantage of the reduction (13) in comparison with the discretization (6) is the fact that in many cases there exist base systems that can better approximate the optimal control function. In consequence, if such systems are used, the number of optimized coefficients can be chosen much lower than in the case of discretization.

Another advantage to the continuous or discrete form is the fact that the function is manipulated in the memory of computer only by a vector of coefficients and does not need to store all the values in time.

Other usable base systems may be for example:

$$f_i = \cos[\frac{\pi}{t_f}.(i+1).t] \tag{19}$$

$$f_i = x^i \tag{20}$$

or $\quad f_i = \Lambda(t - T) \qquad (21)$

where function $\Lambda(t)$ is defined as follows (fig. 1):

$$\Lambda(t) = 0 \quad \text{for} \quad t \in (-\infty, -T] \cup [T, \infty)$$
$$\Lambda(t) = t + T \quad \text{for} \quad t \in (-T, 0)$$
$$\Lambda(t) = T - t \quad \text{for} \quad t \in [0, T) \tag{22}$$

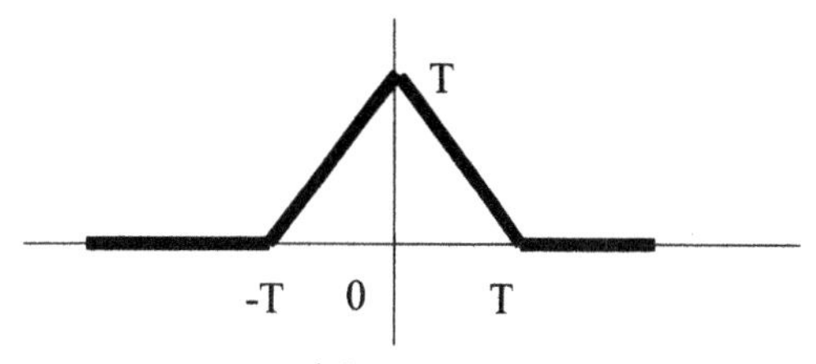

Fig. 1: Function $\Lambda(t)$

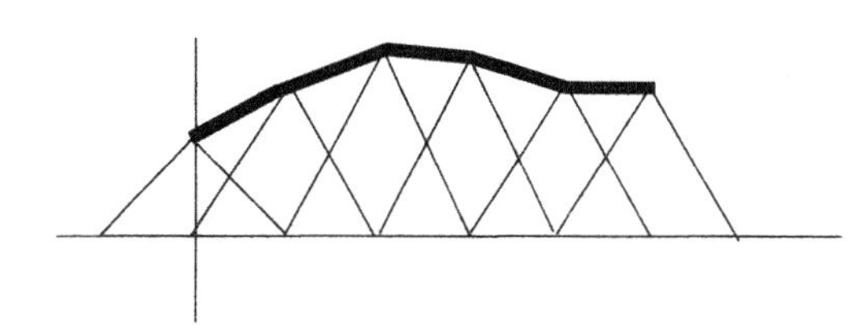

Fig. 2: Approximating a function using $\Lambda(t)$

Let us define a special class of base systems $\{f_i\}$ with the following features:

1. $f_i = f(t - i.T)$, where f is a creating function.
2. $f(t) = 0$ for $t \in R/\Phi$, where $\phi \subset \langle 0, t_f \rangle$ is a closed interval.

Such base systems can be called *compact support bases* since ϕ is a compact subset of interval $\langle 0, t_f \rangle$. The system $\{\Lambda(t - iT)\}$ is a typical example of the compact support base. Fig. 2 shows how a continuous function can be approximated using this base system. These base systems can be also used for numerical solving the differential equations with boundary conditions (Rektorys).

The advantage of the compact support base systems for nonlinear optimal control consists in accelerated computation of the gradient of the functional J as described in the next paragraph.

3. DETERMINING THE GRADIENT OF J

Consider the optimization problem extended by the terminal penalty part:

$$J=\varphi(\boldsymbol{x}(t_f))+\int_0^{t_f} f_0(\boldsymbol{x},\boldsymbol{u},t).dt\rightarrow min \tag{23}$$

Let us define Hamiltonian H as follows:

$$H=f_0+\boldsymbol{p}^T\boldsymbol{f} \tag{24}$$

It can be proven (Polak), that

$$\delta J=\int_0^{t_f}\frac{\partial H}{\partial \boldsymbol{u}}\delta\boldsymbol{u}\,dt \tag{25}$$

while

$$\frac{d}{dt}\boldsymbol{p}^T=-\frac{\partial H}{\partial \boldsymbol{x}} \tag{26}$$

$$\boldsymbol{p}^T(t_f)=\frac{\partial\varphi}{\partial \boldsymbol{x}}(t_f) \tag{27}$$

From (13) follows that

$$\delta\boldsymbol{u}=d\boldsymbol{A}.\boldsymbol{b} \tag{28}$$

Substituting into (25) we get

$$\delta J=\int_0^{t_f}\frac{\partial H}{\partial \boldsymbol{u}}d\boldsymbol{A}\,\boldsymbol{b}(t)\,dt=$$

$$=\sum_{i=1}^{m}\int_0^{t_f}\frac{\partial H}{\partial u_i}d\boldsymbol{A}_i\boldsymbol{b}(t)\,dt \tag{29}$$

Finally,

$$\delta J=\sum_{i=1}^{m}\sum_{j=0}^{N}\int_0^{t_f}\frac{\partial H}{\partial u_i}b_j(t)\,dt\,da_{ij} \tag{30}$$

This means,

$$\frac{dJ}{da_{ij}}=\int_0^{t_f}\frac{\partial H}{\partial u_i}b_j(t)\,dt \tag{31}$$

At first, the system equations are integrated for the current control $\boldsymbol{u}(t)=\boldsymbol{A}.\boldsymbol{b}(t)$ to get the values of $\boldsymbol{x}(t)$. Then the equation (26) can be integrated backwards with terminal condition (27) to obtain values of $\boldsymbol{p}(t)$ and $\partial H/\partial u_i$. This phase of the computation does not depend on the order N. Then the integral (31) is $m(N+1)$ times integrated forwards to compute the partial derivations.

For the compact support base systems, the integral (31) needs to be computed only in interval $\Phi+jT$ since outside it the integrand equals to zero. This feature of the compact support bases is important especially for higher values of N. Moreover, in the case of many compact support base systems, for $N'=2N$ the width of the support Φ' is approximately half width of the original support Φ. In that case, overall computational severity of $dJ/d\boldsymbol{A}$ obviously does not depend on N at all.

For the other base systems, $m(N+1)$ integrals along the whole interval $\langle 0,t_f\rangle$ has to be computed, which is the same work as direct numerical approximation of $dJ/d\boldsymbol{A}$. However, even for that case this method may be more suitable since it is more precise.

Note that the base system $\{\Lambda(t-iT)\}$ has some other useful features. It allows fast computation of the base functions values and enables easy transformation of the weighting coefficients for $N'=2N$ so that the control function keeps the same. In this case,

$$a'_{2i}=2a_i$$
$$a'_{2i+1}=a_i+a_{i+1} \tag{32}$$

Note that values of the coefficients a_i are multiplied by 2 since $T'=t_f/N'=T/2$. The transformation $N\rightarrow N'$ can be used to enhance the solution obtained for the base of order N.

A practical disadvantage of this construction is the need of explicit knowledge of partial derivations of $\boldsymbol{f}$ and f_0. Their numerical determination is ineffective and due to cumulation of numerical errors in integral (31) it can be practically unusable.

4. ALGORITHM IMPLEMENTATION

At this point a complete algorithm optimizing functional (9) based on the conjugate-gradients principle (variant known as Polak-Ribière) can be described, using the transformation of the problem as discussed above.

Suppose that components of the matrix of weighting coefficients $\boldsymbol{A}$ are stored in a vector of parameters $\boldsymbol{X}$ of dimension $n=m(N+1)$.

0. Generate a starting vector of coefficients $\boldsymbol{X}_0$, set $k=0$, $\epsilon=\epsilon_0$, $K=K_0$, $N=N_0$, $\alpha>1$, $\beta>1$.

1. Compute $g_k^T = -[\frac{dJ}{dX}]_{X=X_k}$. If $\|g_k\| < \epsilon$, go to step 4.
2. If $k \bmod n = 0$, let $s_k = g_k$. Otherwise,

$$s_k = g_k + \frac{(g_k^T - g_{k-1}^T) \cdot g_k}{g_{k-1}^T \cdot g_{k-1}^T} \cdot s_{k-1}$$

3. Search for the minimum along the line starting at point X_k in direction s_k with precision proportional to ϵ. Denote this point X_{k+1}. Set $k \leftarrow k+1$ and go to step 1.
4. If $\epsilon \leqslant \epsilon_{min}$, $N \geqslant N_{max}$ and $K \geqslant K_{max}$, end.
5. If $\epsilon > \epsilon_{min}$, set $\epsilon \leftarrow \epsilon / \alpha$.
6. If $K < K_{max}$, set $K \leftarrow \beta . K$.
7. If $N < N_{max}$, set $N \leftarrow 2.N$ and transform the coefficients X_k so that the control function u is the same as for the original N.
8. Set $k = 0$. Go to step 1.

The used conjugate-gradients optimization method is able to find the minimum of a quadratic function in n exact line searches. Its efficiency in nonquadratic problems can be explained by the fact that any continuously differentiable function can be in a neghbourhood of a local extreme approximated by a quadratic function. The method can be easily implemented since it needs only to store the direction vector and the gradient in the last point. That is why it can be used also for the optimization problems with a large number of parameters (e.g. >100), which can be also the case of the optimal control.

For less number of parameters, it is currently recommended to use the quasi-Newton methods (also known as variable-metric methods) that use line searches to approximate the Hessian matrix of the second derivations or directly its inversion (Fletcher).

5. SOLVED EXAMPLE

Finally, the proposed method is demonstrated on a practical example. Consider a mechanical system described by the set of nonlinear equations:

$$m_1 \frac{d^2 x}{dt^2} = k_1 u_1 + m_2 [g.tan(\varphi) + \sin(\varphi)(\frac{d\varphi}{dt})^2]$$
$$L \frac{d^2 \varphi}{dt^2} = -3sin(\varphi) + k_2 u_2 \tag{33}$$

The system moves from the zero starting state at t=0 to the terminal point:

$$x(2) = 1$$
$$\varphi(2) = \frac{dx}{dt}(2) = \frac{d\varphi}{dt}(2) = 0 \tag{34}$$

The criteria of optimality was chosen as quadratic:

$$J = \int_0^2 (u_1^2 + u_2^2)\, dt \tag{35}$$

Base system (21) was used for approximating the control functions.

Fig. 3 shows the optimized control functions of individual elements for N=8, while Fig. 4 shows solution for N=32.

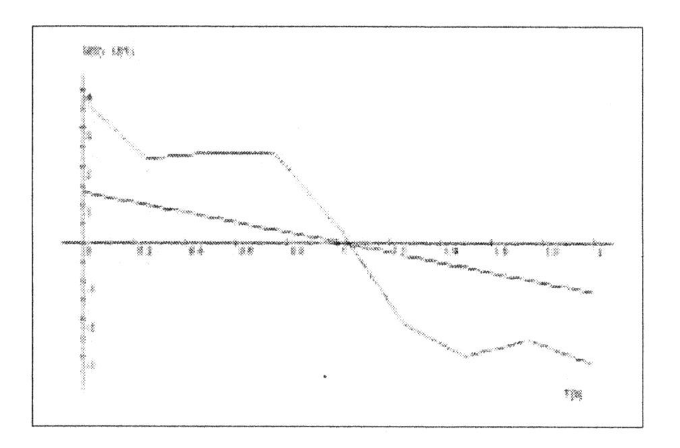

Fig. 3: Results for N=8

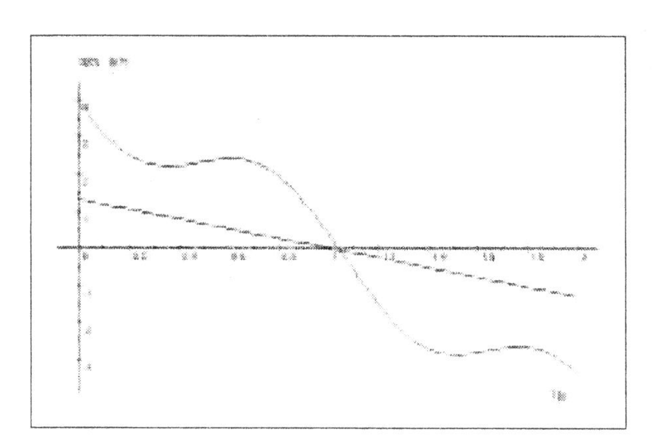

Fig. 4: Results for N=32

REFERENCES

Bryson, A. E., Ho.,Y.C. (1975). *Applied Optimal Control.* Hemisphere corp.

Polak, E. (1971). *Computational Methods In Optimization.* Academic Press, New York.

Fletcher, R. (1987): *Practical Methods of Optimization.* John Wiley & Sons Ltd. 2nd edition.

Rektorys, K. (1980): *Variational Methods in Mathematics, Science and Engineering.* Reidel, Dordrecht. 2nd edition.

THE EXACT VELOCITY LINEARIZATION METHOD

Miroslav Halás, Mikuláš Huba, Katarína Žáková

Department of Automation and Control
Faculty of Electrical Engineering and Information Technology
Slovak University of Technology
Ilkovičova 3, 812 19 Bratislava, Slovakia
E-mail: [halas, huba, katarina.zakova]@elf.stuba.sk

Abstract: The aim of this paper is to present a new method for the nonlinear controller design. The underlying theory evolves from the introduced transfer function of nonlinear systems, that can be used in similar way as transfer functions of classical linear systems. The presented algebra of transfer functions is the same as the linear one. Therefore, the controller design can follow the linear system synthesis. The paper shows, how to design a nonlinear PID controller using the direct method synthesis and also shows that the nonlinear state controller following the pole-placement method is equivalent to the exact linearization method. Since the presented method leads to results identical with the exact linearization method, it has been denoted as the exact velocity linearization method. *Copyright © 2003 IFAC*

Keywords: nonlinear system, velocity linearization, transfer function, PID controller, state controller.

1. INTRODUCTION

This paper tries to formalize some previous results (Leith and Leithead, 1998a,b, 1999) into a transfer function based approach having features similar to the classical linear control design. The proposed method can be classified as the gain scheduling method, or more exactly as the velocity linearization method. The gain scheduling is perhaps one of the most popular nonlinear control design method. It enables linear design methods to be used for nonlinear dynamic systems (Engell, 1995; Khalil, 1996; Lawrence and Rugh, 1995; Leith and Leithead, 2000). The classical gain scheduling allows to describe and modify the nonlinear plant behaviour around an equilibrium operating point. In contrast to this, the velocity linearization allows to do so also for arbitrary operating points and to design the plant behaviour also during the transient responses (Leith and Leithead, 1998a,b).

The paper is organized as follows. In §2, the basic framework for the velocity linearization of SISO nonlinear plant is established. §3 deals with properties of the algebra of nonlinear plant transfer functions. Application in the PID control design is depicted in §4. In §5, an application of the pole placement controller design is discussed. §6 summarizes results of previous chapters and establishes connections to the exact linearization method. Finally, conclusions of the paper are summarized in §7.

2. PRINCIPLE OF THE METHOD

In this section, the exact velocity linearization method, providing the transfer function based approach for nonlinear systems, is introduced.

* The work has been partially supported by the Slovak Scientific Grant Agency, Grant No. KG-57-HUBA-Sk1.

Consider a nonlinear system with dynamics

$$\dot{\mathbf{x}} = \mathbf{F}(\mathbf{x}, u) \quad (1)$$
$$y = G(\mathbf{x}, u)$$

The nonlinear system can be linearized using a series expansion linearization in a given equilibrium operating point $(\mathbf{x}_0, u_0)$

$$\Delta\dot{\mathbf{x}} = \mathbf{A}\Delta\mathbf{x} + \mathbf{b}\Delta u \quad (2)$$
$$\Delta y = \mathbf{c}^T\Delta\mathbf{x} + d\Delta u$$

whereby

$$\mathbf{A} = \nabla_x \mathbf{F}(\mathbf{x}, u); \mathbf{b} = \nabla_u \mathbf{F}(\mathbf{x}, u) \quad (3)$$
$$\mathbf{c}^T = \nabla_x G(\mathbf{x}, u); d = \nabla_u G(\mathbf{x}, u)$$
$$\Delta\mathbf{x} = \mathbf{x} - \mathbf{x}_0; \Delta u = u - u_0$$

In the close vicinity of any equilibrium operating point, the linear theory can be used to design a linear controller. However, the controller transfer function varies according to the chosen equilibrium operating point. The idea of the presented approach is to change controller parameters continuously, so that they permanently maintain values corresponding to the actual operating point, not just the equilibrium operating point. Since the continuous change of parameters is required, differences have to be substituted by derivatives

$$\ddot{\mathbf{x}} = \mathbf{A}\dot{\mathbf{x}} + \mathbf{b}\dot{u} \quad (4)$$
$$\dot{y} = \mathbf{c}^T\dot{\mathbf{x}} + d\dot{u}$$

State space equations (4) represent the velocity form of the nonlinear system (1) and can be also obtained by additional time differentiating of the direct form (1). This velocity form provides an alternative representation of the nonlinear system (Leith and Leithead, 1999) and is valid in the vicinity of any operating point, not just at equilibrium operating points. If the state and input variables are posed as parameters, the velocity form (4) is linear according to the input and output time derivatives. We can, therefore, use standard tools of the linear theory and introduce the transfer function of the nonlinear system (1) as

$$F(s) = \frac{\dot{Y}(s)}{\dot{U}(s)} = \mathbf{c}^T(s\mathbf{I} - \mathbf{A})^{-1}\mathbf{b} + d \quad (5)$$

where $\dot{Y}(s)$ and $\dot{U}(s)$ denote the Laplace transforms of $\dot{y}(t)$ and $\dot{u}(t)$. In general, the transfer function coefficients are not only constants, they can be functions of inputs and states of the nonlinear system. In contrast to the transfer function of a linear system, (5) works with the derivatives of variables and not with the variables themselves. It has to be taken into account in any case. The introduced transfer function (5) is considered as a transfer function of the nonlinear system.

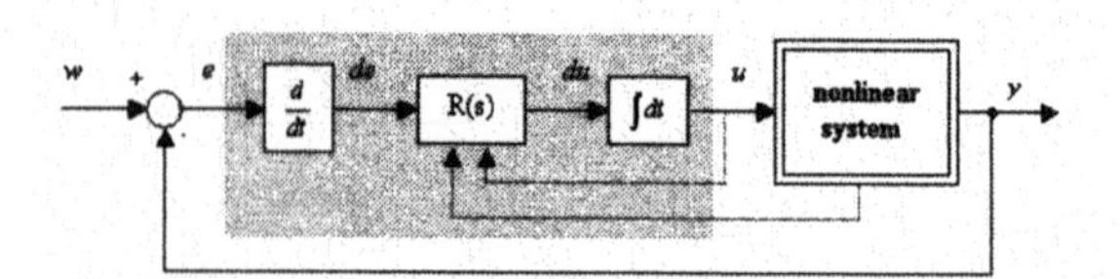

Fig. 1. The controller structure

The described approach uses the entire nonlinear model in its velocity form. It does not deal with the local models and the velocity-based linearization families, which are used by Leith and Leithead (1998a,b, 1999), at all. Therefore, after considering initial conditions, the transfer function (5) characterizes a nonlinear system uniquely.

2.1 *Example*

Consider the nonlinear plant with dynamics

$$\dot{x} = -x^2 + u^2 \quad (6)$$
$$y = x$$

for which the velocity form (4) is

$$\mathbf{A} = (-2x); \mathbf{b} = (2u); \mathbf{c}^T = (1); d = 0 \quad (7)$$

The transfer function can be computed as

$$F(s) = \mathbf{c}^T(s\mathbf{I} - \mathbf{A})^{-1}\mathbf{b} + d = \frac{2u}{s + 2x} \quad (8)$$

Evidently, the transfer function (8) is parametrized by state and input variables of the system (6). The pole is a function of the state x and the gain is a function of the input u. Both, x and u, represent the transfer function parameters. This approach enables linear methods, including a transfer function algebra, to be applied to nonlinear control problems.

The designed controller, respecting the introduced transfer function (5), works with the derivatives of variables and not directly with the variables themselves. Following this, it has to be organized according to Fig. 1. It can be designed by using any linear method. In practice, use of additional derivatives is not very convenient. However, it is often possible to modify the designed controller and to eliminate the input derivative and the output integral from the control structure by an analytic integration.

3. ALGEBRA OF NONLINEAR PLANT TRANSFER FUNCTIONS

Each complex system structure can be divided into three basic connections: series, parallel and feedback connection (Fig. 2). The algebra of nonlinear plant transfer functions is the same as the linear one.

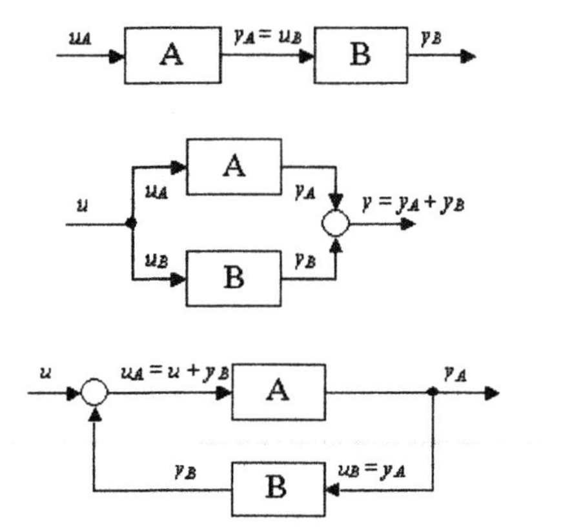

Fig. 2. Basic connections

For series, parallel and feedback connection one obtains:

$$
\begin{aligned}
F(s) &= F_A(s)\,F_B(s) \\
F(s) &= F_A(s) + F_B(s) \\
F(s) &= \frac{F_A(s)}{1 - F_A(s)\,F_B(s)}
\end{aligned} \tag{9}
$$

The validity of (9) can be verified by computing the transfer function from the velocity form of the particular connection. For example, the series connection gives

$$
\begin{aligned}
\dot{\mathbf{x}}_A &= \mathbf{F}_A(\mathbf{x}_A, u_A) \\
\dot{\mathbf{x}}_B &= \mathbf{F}_B(\mathbf{x}_B, u_B) = \mathbf{F}_B(\mathbf{x}_B, y_A) \\
y_B &= G_B(\mathbf{x}_B, u_B) = G_B(\mathbf{x}_B, y_A)
\end{aligned} \tag{10}
$$

where

$$
y_A = G_A(\mathbf{x}_A, u_A) \tag{11}
$$

The velocity form is

$$
\begin{aligned}
\ddot{\mathbf{x}}_A &= \mathbf{A}_A \dot{\mathbf{x}}_\mathbf{A} + \mathbf{b}_A \dot{u}_A \\
\ddot{\mathbf{x}}_B &= \mathbf{A}_B \dot{\mathbf{x}}_\mathbf{B} + \mathbf{b}_B \mathbf{c}_A^T \dot{\mathbf{x}}_\mathbf{A} + \mathbf{b}_B d_A \dot{u}_A \\
\dot{y}_B &= \mathbf{c}_B^T \dot{\mathbf{x}}_\mathbf{B} + d_B \mathbf{c}_A^T \dot{\mathbf{x}}_\mathbf{A} + d_B d_A \dot{u}_A
\end{aligned} \tag{12}
$$

whereby

$$
\begin{aligned}
\mathbf{A}_k &= \nabla_{\mathbf{x}_k} \mathbf{F}_k(\mathbf{x}_k, u_k)\,;\ \mathbf{b}_k = \nabla_{u_k} \mathbf{F}_k(\mathbf{x}_k, u_k) \\
\mathbf{c}_k^T &= \nabla_{\mathbf{x}_k} G_k(\mathbf{x}_k, u_k)\,;\ d_k = \nabla_{u_k} G_k(\mathbf{x}_k, u_k) \\
k &= A \text{ or } B
\end{aligned} \tag{13}
$$

After Laplace transformation of (12) one receives

$$
\begin{aligned}
s\dot{\mathbf{X}}_A(s) &= \mathbf{A}_A \dot{\mathbf{X}}_A(s) + \mathbf{b}_A \dot{U}_A(s) \\
s\dot{\mathbf{X}}_B(s) &= \mathbf{A}_B \dot{\mathbf{X}}_B(s) + \mathbf{b}_B \mathbf{c}_A^T \dot{\mathbf{X}}_A(s) + \mathbf{b}_B d_A \dot{U}_A(s) \\
\dot{Y}_B(s) &= \mathbf{c}_B^T \dot{\mathbf{X}}_B(s) + d_B \mathbf{c}_A^T \dot{\mathbf{X}}_A(s) + d_B d_A \dot{U}_A(s)
\end{aligned} \tag{14}
$$

Finally, the series connection transfer function is

$$
\begin{aligned}
\frac{\dot{Y}_B(s)}{\dot{U}_A(s)} &= \left[\mathbf{c}_B^T (s\mathbf{I} - \mathbf{A}_B)^{-1} \mathbf{b}_B + d_B\right] \cdot \\
&\cdot \left[\mathbf{c}_A^T (s\mathbf{I} - \mathbf{A}_A)^{-1} \mathbf{b}_A + d_A\right] = F_B(s)\,F_A(s)
\end{aligned} \tag{15}
$$

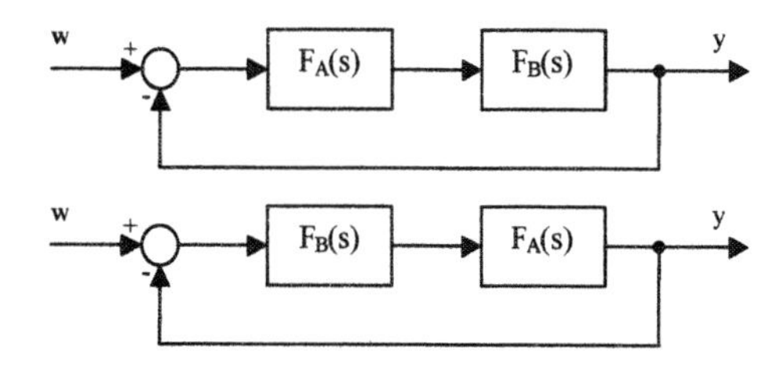

Fig. 3. Feedback with replaced systems

In general, for nonlinear systems the associativity is not valid. However, the introduced transfer function algebra is correct and is the same like the linear one. In each resulting transfer function it is appropriate to replace formally all state and input variables of systems by the state and input variables of the particular systems connection. These facts are depicted in next example.

3.1 *Example*

Consider two nonlinear systems

$$
\begin{aligned}
\dot{x}_A &= -x_A^2 + u_A \\
y_A &= x_A
\end{aligned} \tag{16}
$$

and

$$
\begin{aligned}
\dot{x}_B &= u_B^2 \\
y_B &= x_B + u_B
\end{aligned} \tag{17}
$$

with the corresponding transfer functions

$$
F_A(s) = \frac{1}{s + 2x_A} \tag{18}
$$

$$
F_B(s) = \frac{s + 2u_B}{s} \tag{19}
$$

Both systems are connected in the feedback: firstly, in the order $F_A \rightarrow F_B$ and secondly, in the order $F_B \rightarrow F_A$ (Fig. 3).The final transfer functions of both feedback connections are

$$
F(s) = \frac{F_A(s)\,F_B(s)}{1 + F_A(s)\,F_B(s)} \tag{20}
$$

$$
F(s) = \frac{s + 2u_B}{s(s + 2x_A) + s + 2u_B} \tag{21}
$$

However, it does not mean that systems in Fig. 3 are equivalent. It only means that the final transfer functions of the both connections are computed by the same expression. It is advisable to reflect that parameter u_B has different meaning. Hence, it is appropriate to replace formally single systems state and input variables by state and input variables of the overall feedback structure. In the case of the systems order $F_A \rightarrow F_B$, it obtains $u_B = y_A = x_A$ and the final transfer function can be simplified for $x_A \geq 0$ to the linear transfer function

$$
F_{AB}(s) = \frac{s + 2x_A}{s(s + 2x_A) + s + 2x_A} = \frac{1}{s + 1} \tag{22}
$$

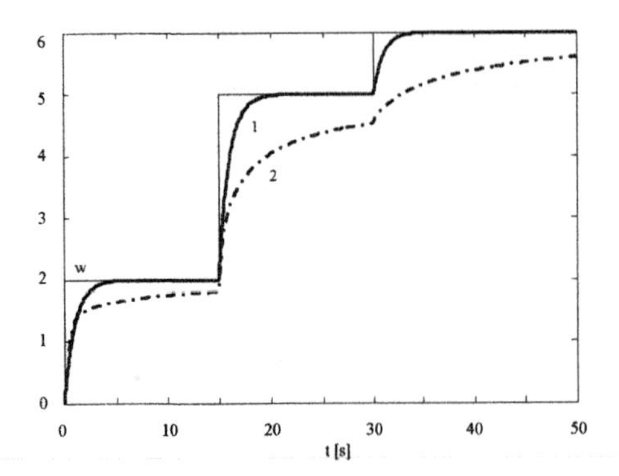

Fig. 4. The feedback step responses

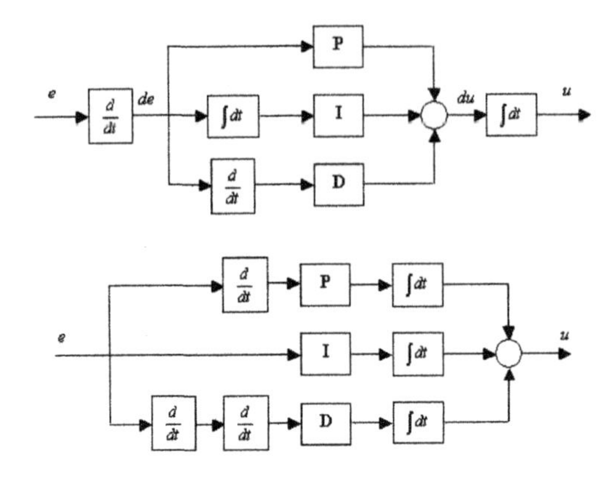

Fig. 5. The PID controller structure

In the case of the order $F_B \rightarrow F_A$, when $u_B = w - y_A = w - x_A$, it cannot be simplified and

$$F_{BA}(s) = \frac{s + 2(w - x_A)}{s(s + 2x_A) + s + 2(w - x_A)} \tag{23}$$

The final step responses (Fig. 4) of systems depicted in (Fig. 3) correspond to their transfer functions (22) and (23). They are linear in the first case and nonlinear in the second one.

4. NONLINEAR PID CONTROLLER

In §2, there has been demonstrated that the designed controller works on the derivatives of variables. Reflecting to the structure (Fig. 1), the PID controller has to be considered according to Fig. 5. Evidently, the I action of the controller can be simplified by cancelling the subsequent derivator and integrator. The undesirable input differentiation of the P and D action can be eliminated by the analytical integration of the controller equation. For example, if the P gain of the designed controller is constant, it leads to

$$\int P\frac{de}{dt}dt = \int Pde = P\int de = Pe \tag{24}$$

Even if the P gain is not constant, it is often possible to solve the integral in an analytic way and to use the simple PID structure depicted in Fig. 6.

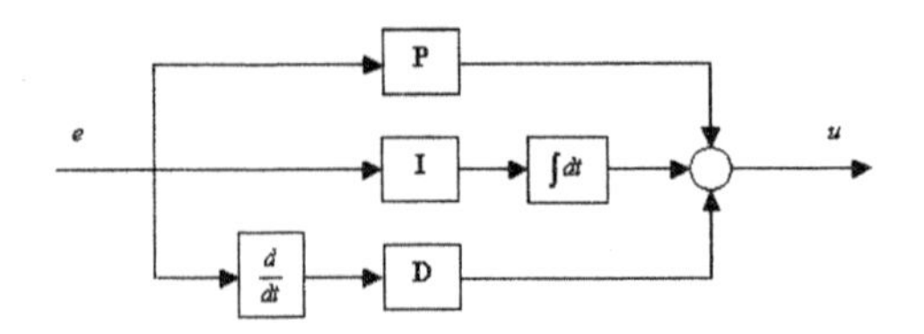

Fig. 6. The simple PID structure

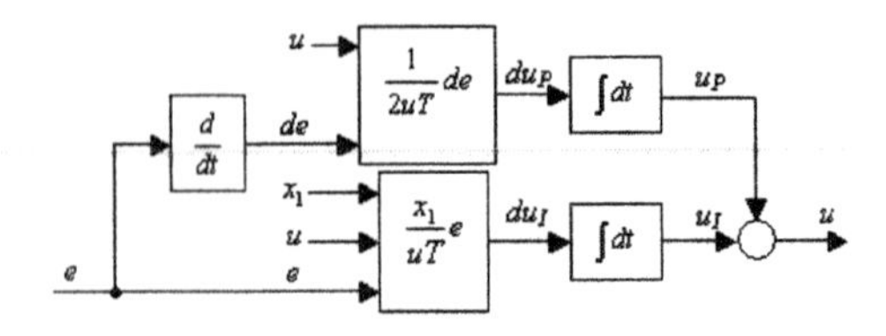

Fig. 7. The nonlinear PI controller

4.1 *Example of PI controller design*

Consider the nonlinear plant (6) with the transfer function (8) and the classical feedback control structure (Fig. 1) which gives

$$G(s) = \frac{R(s)F(s)}{1 + R(s)F(s)} \tag{25}$$

The solution to this equation solved for $R(s)$ is

$$R(s) = \frac{1}{F(s)}\left[\frac{G(s)}{1 - G(s)}\right] \tag{26}$$

If the feedback transfer function is chosen to be $G(s) = \frac{1}{Ts+1}$, the solution represents the PI controller.

$$R(s) = \frac{1}{F(s)}\left[\frac{1}{Ts}\right] = \frac{s + 2x}{2uTs} \tag{27}$$

$$R(s) = \frac{1}{2uT} + \frac{1}{s}\frac{x}{uT} = \frac{\dot{U}(s)}{\dot{E}(s)} \tag{28}$$

The primary controller structure, in its velocity form, is depicted in Fig. 7. Another alternative is to solve the integrals in the nonlinear differential equation

$$\dot{U}(s) = \frac{1}{2uT}\dot{E}(s) + \frac{1}{s}\frac{x}{uT}\dot{E}(s) \tag{29}$$

$$\dot{u} = \frac{1}{2uT}\dot{e} + \int\frac{x}{uT}\dot{e}dt \tag{30}$$

which leaves from the controller transfer function (28) and which represents the velocity form of the designed controller. One follows

$$2u\dot{u} = \frac{1}{T}\dot{e} + \frac{2x}{T}e \tag{31}$$

$$\int 2udu = \int\frac{1}{T}\dot{e}dt + \int\frac{2x}{T}edt \tag{32}$$

$$u^2 = \frac{1}{T}e + \frac{1}{T}2\int xedt \tag{33}$$

$$u = \sqrt{\frac{1}{T}\left(e + 2\int xedt\right)} \tag{34}$$

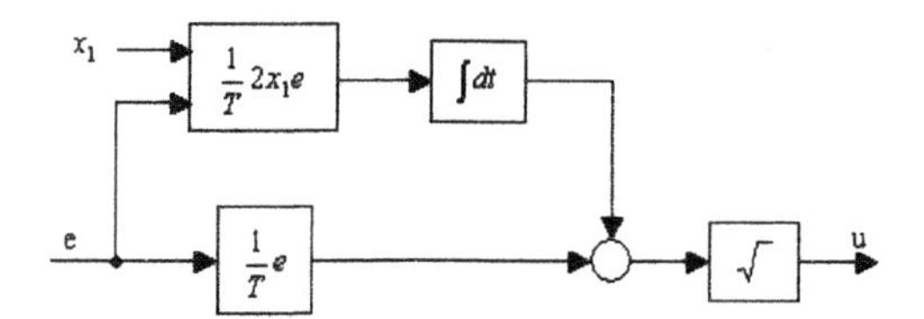

Fig. 8. The simple nonlinear PI controller

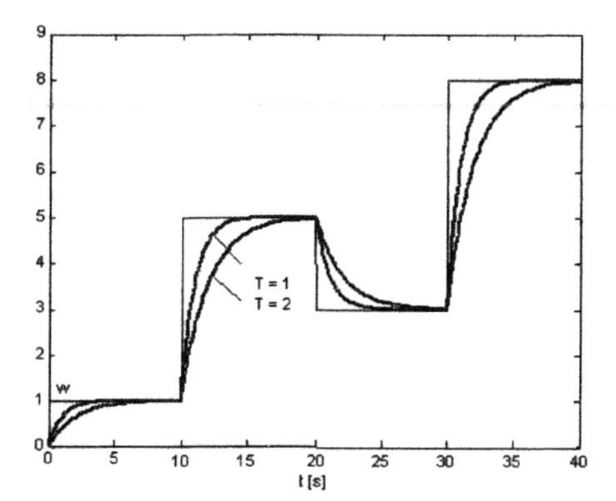

Fig. 9. The output responses

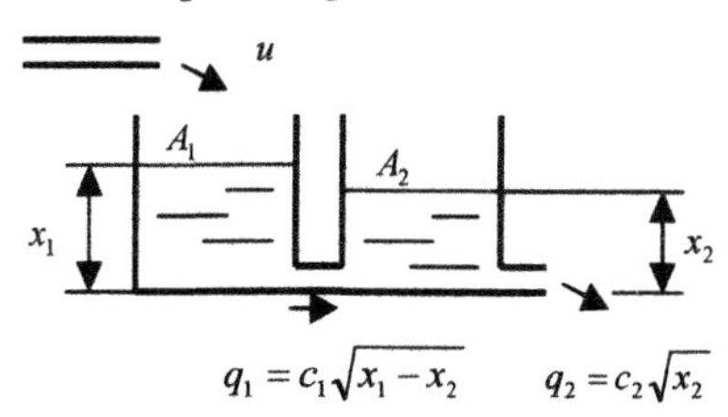

Fig. 10. The two tank system

The final expression represents a nonlinear PI controller (Fig. 8). Both designed structures (Fig. 7 and Fig. 8) are equivalent. It means they have the same transfer function and the same behaviour. Step responses for chosen time constant of feedback structure are shown in Fig.9.

4.2 *PID controller design for a two tank system*

Consider a two tank system depicted in Fig. 10 with dynamics

$$\begin{aligned}\dot{x}_1 &= \frac{1}{A_1}u - \frac{c_1}{A_1}\sqrt{x_1 - x_2}\\ \dot{x}_2 &= \frac{c_1}{A_2}\sqrt{x_1 - x_2} - \frac{c_1}{A_1}\sqrt{x_2}\\ y &= x_2\end{aligned} \tag{35}$$

The transfer function can be computed as

$$F(s) = \frac{\frac{2\sqrt{x_2}}{c_2}}{s^2 a_2 + s a_1 + 1} \tag{36}$$

whereby

$$\begin{aligned}a_2 &= \frac{4A_1A_2}{c_1c_2}\sqrt{x_1 - x_2}\sqrt{x_2}\\ a_1 &= 2\left(\frac{A_1}{c_2} + \frac{A_2}{c_2}\right)\sqrt{x_2} + \frac{2A_1}{c_1}\sqrt{x_1 - x_2}\end{aligned} \tag{37}$$

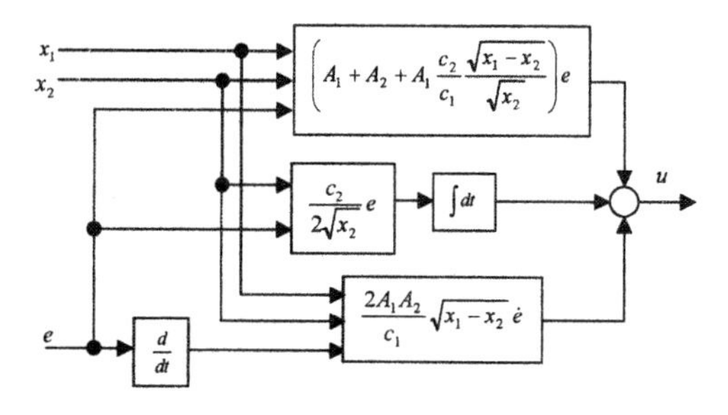

Fig. 11. The nonlinear PID controller

If the transfer function of the feedback is chosen as $G(s) = \frac{1}{Ts+1}$, the solution (26) for $T = 1$ leads to the PID controller with parameters

$$\begin{aligned}P &= A_1 + A_2 + A_1\frac{c_2}{c_1}\frac{\sqrt{x_1 - x_2}}{\sqrt{x_2}}\\ I &= \frac{c_2}{2\sqrt{x_2}}\\ D &= \frac{2A_1A_2}{c_1}\sqrt{x_1 - x_2}\end{aligned} \tag{38}$$

It is always supposed that the full state vector is accessible. The controller structure after the analytical integration is shown in Fig. 11. For $x_1 \geq x_2 > 0$ the time behaviour of the closed loop is equivalent to the linear system $G(s) = \frac{1}{s+1}$. Step responses are the same like in the previous example (Fig. 9).

5. POLE - PLACEMENT METHOD

Consider the nonlinear system

$$\begin{aligned}\dot{x}_1 &= -2x_1 + x_2\\ \dot{x}_2 &= 1 - e^{-x_2} + u\\ y &= x_1\end{aligned} \tag{39}$$

which is taken from Engell (1995), where the gain scheduling around equilibrium operating points was used. Now let us use the introduced approach combined with the pole placement method.

The linearization at $(x_1, x_2)^T$ gives

$$\mathbf{A} = \begin{bmatrix} -2 & 1 \\ 0 & e^{-x_2} \end{bmatrix}; \mathbf{B} = \begin{bmatrix} 0 \\ 1 \end{bmatrix}; \mathbf{C} = \begin{bmatrix} 1 & 0 \end{bmatrix} \tag{40}$$

and the nonlinear plant transfer function is

$$F(s) = \frac{\dot{Y}(s)}{\dot{U}(s)} = \frac{1}{(s+2)(s - e^{-x_2})} \tag{41}$$

If the control law is required in form

$$\dot{u} = r_w\dot{w} - r_{x_1}\dot{x}_1 - r_{x_2}\dot{x}_2 \tag{42}$$

the closed loop matrix will be

$$\begin{aligned}\mathbf{A}_C &= \mathbf{A} - \mathbf{B}\begin{pmatrix} r_{x_1} & r_{x_2} \end{pmatrix}\\ \mathbf{A}_C &= \begin{bmatrix} -2 & 1 \\ -r_{x_1} & e^{-x_2} - r_{x_2} \end{bmatrix}\end{aligned} \tag{43}$$

with the corresponding characteristic polynomial

$$\det(s\mathbf{I} - \mathbf{A}_C) = s^2 + \left(2 - e^{-x_2} + r_{x_2}\right)s - \\ -2e^{-x_2} + 2r_{x_2} + r_{x_1} \quad (44)$$

A comparison of (44) with the characteristic polynomial $s^2 + 8s + 16$ introduced in Engell (1995) gives proportional gains

$$\begin{pmatrix} r_{x_1} \\ r_{x_2} \end{pmatrix} = \begin{pmatrix} 4 \\ 6 + e^{-x_2} \end{pmatrix} \quad (45)$$

$$r_w = -\frac{1}{\mathbf{CA}_C^{-1}\mathbf{B}} = 16 \quad (46)$$

which after substitution to (42) leads to

$$\dot{u} = 16\dot{w} - 4\dot{x}_1 - \left(6 + e^{-x_2}\right)\dot{x}_2 \quad (47)$$

The next step is to integrate this equation according to time, when

$$u = 16w - 4x_1 - 6x_2 + e^{-x_2} - 1 \quad (48)$$

The designed control law is equivalent to the exact linearization of the nonlinear plant (39).

6. EXACT VELOCITY LINEARIZATION METHOD

The control design can be summarized into the following steps:

1. Compute the nonlinear plant transfer function.

2. Use the transfer function algebra to design the controller transfer function, so that the final transfer function of closed loop is linear.

3. Implement the nonlinear controller according to its transfer function.

It is possible to implement the proposed nonlinear controller in its velocity form, or preferably (by solving the integrals) in its analytic form. Both ways are equivalent and yield the same transfer function. Thus, the control design results in a nonlinear closed loop with the prescribed linear transfer function. For example:

$$G(s) = \frac{R(s)F(s)}{1 + R(s)F(s)} = \frac{\dot{Y}(s)}{\dot{W}(s)} \quad (49)$$

It can be shown (Halás, 2002), that if $G(s)$ is a linear transfer function then

$$\frac{\dot{Y}(s)}{\dot{W}(s)} = G(s) \Longleftrightarrow \frac{Y(s)}{W(s)} = G(s) \quad (50)$$

It means the final nonlinear closed loop structure has a time behaviour equivalent to the linear one described by the transfer function $G(s)$. Since under equivalent design requirements, the proposed method gives results identical with the exact linearization method, it can be denoted as the exact velocity linearization method.

7. CONCLUSIONS

In this paper, a new method for the synthesis of nonlinear systems has been presented. Firstly, it introduces the transfer function of the given nonlinear system that can be used in similar way as transfer functions of classical linear systems. Then, based on these transfer functions, a nonlinear controller can be proposed (nonlinear PID controller, state controllers) using methods very well known from the theory of linear systems. In designing a nonlinear PID controller, it is often possible to use classical PID controller structures with nonlinear gain components.

An interesting potential of the method is given by the introduced transfer functions of nonlinear systems and by the fact that it can yield results identical to the exact linearization method. Furthermore, it can be used also in cases, when the exact linearization is not applicable (unstable zero dynamics). However, it is well known that in the velocity form, numerical problems in controllers without I action occur. Therefore, it will be interesting to investigate, under which conditions is the algorithm integrable to a direct form. Or, in a particular case, under which conditions is the velocity form equivalent to the exact linearization.

Moreover, it is always supposed that the full state vector is accessible in the controller design.

8. REFERENCES

Engell, S. (1995). *Entwurf nichtlinearer Regelungen.* R. Oldenbourg Verlag, München; Wien.

Halás, M. (2002). *A Nonlinear Control Algorithm Design.* Diploma Thesis, Department of Automation and Control, Slovak University of Technology, Slovakia (in Slovak).

Khalil, H.K. (1996). *Nonlinear Systems.* Second Edition. Prentice Hall.

Lawrence, D.A. and W.J. Rugh (1995). Gain Scheduling Dynamic Controllers for a Nonlinear Plant. *Automatica,* **31**, 381-390.

Leith, D.J. and W.E. Leithead (1998a). Gain-Scheduled & Nonlinear Systems: Dynamic Analysis by Velocity-Based Linearization Families. *Int. J. Control,* **70**, 289-317.

Leith, D.J. and W.E. Leithead (1998b). Gain-Scheduled Controller Design: An Analytic Framework Directly Incorporating Non-Equilibrium Plant Dynamics. *Int. J. Control,* **70**, 249-269.

Leith, D.J. and W.E. Leithead (1999). Input-output linearization by velocity-based gain-scheduling. *Int. J. Control,* **72**, 229-246.

Leith, D.J. and W.E. Leithead (2000). Survey of Gain-Scheduling Analysis & Design. *Int. J. Control,* **73**, 1001 - 1025.

ALGORITHMS FOR GEOMETRIC CONTROL OF SYSTEMS OVER RINGS

Anna Maria Perdon * **Giovanna Guidone Peroli** *
Massimo Caboara **

* *Dipartimento di Matematica, Università di Ancona via Brecce Bianche, 60131 Ancona, Italy*
** *Università di Genova via Dodecaneso 35, 16146 Genova, Italy*

Abstract: In this paper we describe some algorithms for implementing geometric procedures related to analysis and synthesis of control systems over rings. Software performing symbolic computations allows practical use of the algorithms in particular dealing with delay differential systems. *Copyright © 2003 IFAC*

Keywords: Systems over rings, Geometric approach, Algorithms

1. INTRODUCTION

In recent time, many efforts have been made to extend the so called geometric approach to the class of systems with coefficients in a ring or, simply, systems over rings. The study of this kind of systems is receiving an increasing attention, as they prove to be an efficient tool for investigating and solving design problems concerning continuous-time, delay-differential systems (see (Conte and Perdon, 1997) and the references therein).
The relevance of the geometric approach as design methodology in controlling systems with coefficients in a field relies on the existence of numeric algorithmic procedures to check set theoretic relations and to construct feedbacks. It is more difficult to provide similar results for systems over rings. We have developed algorithms relying on Gröbner basis computation; for other approaches see (Bareiss, 1968), (Malashonok, 1991).

These new procedures, usually acting on vectors and matrices with elements in a ring of polynomials in one ore several indeterminates, cannot be implemented efficiently using numeric algorithms, but they require the use of symbolic computations tools, such as MapleV, Mathematica or CoCoA, see (Capani A., 1995).

In this paper we first present a complete account of the results which, within the geometric approach, allow the solution of many design problems such as that of the Disturbance Decoupling. Then we propose new algorithms to compute feedback solutions to such problems.

In this draft, due to the page limit, only some of the algorithms are explicitly described.

2. PRELIMINARIES

Assume that a delay-differential system Σ_d is described by the equations

$$\mathbf{\Sigma_d} = \begin{cases} \dot{x}(t) = A_0 x(t) + \sum_{i=1}^{a} A_i \delta^i x(t) + \\ \qquad + B_0 u(t) + \sum_{i=1}^{b} + B_i \delta^i u(t) \\ y(t) = C_0 x(t) + \sum_{i=1}^{c} C_i \delta^i x(t) \end{cases} \quad (1)$$

where, denoting by $\mathbf{R}$ the field of real numbers, x belongs to the vector space $\mathbf{R}^n$, u belongs to the input space $\mathbf{R}^m$, y belongs to the output space $\mathbf{R}^p$, A_i, $i = 0, 1, ..., a$, B_i, $i = 0, 1, ..., b$, C_i, $i =$

$0, 1, ..., c$ are matrices of suitable dimensions with entries in $\mathbf{R}$ and δ is the delay operator defined, for any time function $f(t)$, by $\delta f(t) = f(t-h)$, for a given $h \in \mathbf{R}^+$. Replacing formally the delay operator δ by the algebraic indeterminate Δ, it is possible to associate to Σ_d the system Σ over the ring $R = \mathbf{R}[\Delta]$ of real polynomials in one indeterminate defined by the equations

$$\begin{cases} x(t+1) = Ax(t) + Bu(t) \\ y(t) = Cx(t) \end{cases} \quad (2)$$

where x belongs to the free state module $\mathcal{X} = R^n$, u belongs to the free input module $\mathcal{U} = R^m$, y belongs to the free output module $\mathcal{Y} = R^p$ and A, B, C are matrices with entries in $R = \mathbf{R}[\Delta]$ given by $A = \sum_{i=0}^{a} A_i \Delta^i$, $B = \sum_{i=0}^{b} B_i \Delta^i$ and $C = \sum_{i=0}^{c} C_i \Delta^i$. Note that, although they have different meaning, we used the same letters to represent corresponding variables in equations (1) and (2). Actually the systems Σ_d and Σ are quite different objects from a dynamical point of view. The key point is, however, that Σ_d and Σ have the same signal flow graph, so that control problems concerning the input/output behaviour of Σ_d can naturally be formulated in terms of the input/output behaviour of Σ and can possibly be solved in the framework of systems over rings. In turn, solutions found in that framework can be brought back to the original delay-differential framework.
For delay-differential systems with non commensurable delays, the above procedure gives rise to systems over the ring $R = \mathbf{R}[\Delta_1, \Delta_2, ..., \Delta_k]$ of polynomials in several indeterminates with real coefficients. Therefore, in the systems which we are mainly interested in, the ring R can be a Principal Ideal Domain, as in the case of $\mathbf{R}[\Delta]$, or more generally a Noetherian ring, as in the case of $R = \mathbf{R}[\Delta_1, \Delta_2, ..., \Delta_k]$.

3. BASIC GEOMETRIC CONCEPTS

The geometric approach to linear dynamical systems relies upon the notion of controlled invariance and its equivalence to feedback invariance, which is crucial in the applications. For systems over a field, (Basile and Marro, 1992) and (Wonham, 1985), controlled invariance is equivalent to invariance with respect to a static state feedback, while for systems over a ring R the situation is more complex.

Definition 1. Given the system (2) over a ring R, a submodule $\mathcal{V} \subset \mathcal{X}$ is called

i) (A,B)-invariant or controlled invariant if $A\mathcal{V} \subset \mathcal{V} + ImB$;

ii) feedback invariant or $(A+BF)$-invariant if there exists a static R-linear state feedback $F: X \longrightarrow U$ such that $(A+BF)\mathcal{V} \subset \mathcal{V}$.

A feedback law satisfying *ii)* is called a *friend* of $\mathcal{V}$.

If R is a field, i) and ii) are equivalent, but over a ring controlled invariance does not necessarily imply feedback invariance. The crucial property which allows to pass from the geometric property i) to the existence of a feedback is related to the possibility of extending a map defined on $\mathcal{V}$ to a map defined on the whole state module. In particular, the following propositions hold.

Proposition 2. (see (Conte and Perdon, 1995*b*)) Assume that the (A, B)-invariant submodule $\mathcal{V}$ is a direct summand of the state module $\mathcal{X}$ (i.e. there exists a submodule $\mathcal{W}$ such that $\mathcal{X} = \mathcal{W} \oplus \mathcal{V}$). Then, $\mathcal{V}$ is of feedback type.

The above result motivates our interest in the following property.

Proposition 3. (see (Conte and Perdon, 1995*b*)) Over a PID ring R a submodule $\mathcal{V} \subseteq \mathcal{X}$ is a direct summand if and only if the Smith form of any matrix whose columns generate $\mathcal{V}$ is the identity.

In general, if R is not a PID and Proposition 2 does not apply , it is difficult to prove that a submodule is a direct summand but the following result helps us in checking the feedback invariance property. See also (Perdon *et al.*, 2003) for another approach.

Proposition 4. (see (Assan *et al.*, 2002)) Let Σ be a system defined by (2) over a commutative ring R, let $\mathcal{V}$ be a submodule of R^n and let v be an element of $\mathcal{V}$. Denote by $\mathcal{F}_V(v)$ the set of all friends of v, i.e. all $F \in R^{m \times n}$ such that $(A+BF)v \in \mathcal{V}$ and let $\{v_i, i \in \mathcal{I}\}$ be a set of generators for a submodule $\mathcal{V}$ of R^n.
Then $\mathcal{V}$ is feedback invariant if and only if

$$\bigcap_{i \in \mathcal{I}} \mathcal{F}_V(v_i) \neq \emptyset. \quad (3)$$

The above condition is practically computable in many cases which are important for applications as, for instance, when R is the rings of polynomials in one or several indeterminates over the reals. If a controlled invariant submodule is not feedback invariant, a more general property of the same kind is still valid (see (Conte and Perdon, 1995*b*)). In order to describe the situation, assume that $\mathcal{V} \subset \mathcal{X} \subset R^n$ is an (A, B)-invariant submodule generated by the q columns of the matrix V, i.e.

there exist two matrices $L \in R^{q\times q}$ and $M \in R^{m\times q}$ such that

$$AV = VL + BM,$$

and consider the extension Σ_e of the system (2) given by the equations

$$\Sigma_e = \begin{cases} x_e(t+1) = A_e x_e(t) + B_e u_e(t) \\ y(t) = C_e x_e(t) \end{cases}, \quad (4)$$

where

$$x_e = \begin{bmatrix} x(t) \\ x_a(t) \end{bmatrix}, \; u_e = \begin{bmatrix} u(t) \\ u_a(t) \end{bmatrix},$$

$$A_e = \begin{bmatrix} A & 0 \\ 0 & 0_{q\times q} \end{bmatrix}, \; B_e = \begin{bmatrix} B & 0 \\ 0 & I_{q\times q} \end{bmatrix}$$

$$\text{and } C_e = [C \; 0_{q\times q}].$$

The module $\mathcal{V}_e$ generated in $\mathcal{X}_e = R^{n+q}$ by the columns of $V_e = \begin{bmatrix} V \\ I_{q\times q} \end{bmatrix}$ is a direct summand of $\mathcal{X}_e = R^n \bigoplus R^q$ and a controlled invariant submodule for Σ_e, since

$$A_e V_e = \begin{bmatrix} AV \\ 0_{k\times k} \end{bmatrix} = V_e L + B_e \begin{bmatrix} M \\ -L \end{bmatrix}. \quad (5)$$

Then, by Proposition 1, $\mathcal{V}_e$ is feedback invariant and

$$\begin{bmatrix} u(t) \\ u_a(t) \end{bmatrix} = F_e \begin{bmatrix} x(t) \\ x_a(t) \end{bmatrix} + G_e v(t), \quad (6)$$

where

$$F_e = \begin{bmatrix} -FV - M & F \\ -L & 0 \end{bmatrix} \text{ and } G_e = \begin{bmatrix} G \\ G_a \end{bmatrix} \quad (7)$$

is a friend of $\mathcal{V}_e$ for any F.

Remark that the action of F_e can be seen as the action on the original system Σ of a dynamic feedback of the form

$$\begin{cases} x_a(t+1) = A_1 x(t) + A_2 x_a(t) + G_1 v(t) \\ u(t) = Fx(t) + Hx_a(t) + G_2 v(t) \end{cases} \quad (8)$$

where $x_a \in \mathcal{X}_a = R^q$, $v \in R^m$, A_1, A_2, F, H, G_1 and G_2 are matrices of suitable dimensions with entries in the ring R.

Another geometric notion, crucial in dealing with non interacting control problems, is the notion of controllability submodule.

Definition 5. A submodule $\mathcal{R} \subseteq \mathcal{X}$ is said to be a controllability submodule if and only if there exist R-linear maps $F : \mathcal{X} \to \mathcal{U}$, $G : \mathcal{U} \to \mathcal{U}$ such that

$$\mathcal{R} = < A + BF | Im(BG) > .$$

A controllability submodule is (controlled invariant and) feedback invariant. In particular, for every friend F of $\mathcal{R}$ one has

$$\mathcal{R} = < A + BF | ImB \cap \mathcal{R} > .$$

The fact that the one given above is not a purely geometric characterization, but one which relies on the existence of a feedback, makes its use difficult for system with coefficients in a ring. In order to avoid this, a weaker notion was introduced.

Definition 6. (see (Conte and Perdon, 1995*a*)) A submodule $\mathcal{R} \subseteq \mathcal{X}$ is said to be a pre-controllability submodule if the following holds:
i) $\mathcal{R}$ is (A, B)-invariant;
ii) $\mathcal{R}$ coincides with the minimum element $\mathcal{S}_*$ of the family $\mathcal{S}_R$, defined by

$$\mathcal{S}_{\mathcal{R}} = \{\mathcal{S} \subseteq \mathcal{X} \; s.\,t. \; \mathcal{S} = \mathcal{R} \cap (A\mathcal{S} + ImB)\}.$$

The relationship between this notion and that of controllability submodule, is described in (Conte and Perdon, 1995*a*). Here, it is sufficient to recall that a submodule $\mathcal{R} \subseteq \mathcal{X}$ is a controllability submodule if and only if it is a pre-controllability submodule and feedback invariant.

4. THE DISTURBANCE DECOUPLING PROBLEM

Conditions for the solvability of many control problems are formulated in terms of set theoretic relations involving controlled invariant submodules or pre-controllability submodules. For example, let us recall that, for a system Σ described over a Noetherian ring R by equations of the form

$$\begin{cases} x(t+1) = Ax(t) + Bu(t) + Dq(t) \\ y(t) = Cx(t) \end{cases} \quad (9)$$

where $q \in R^k$ denotes a disturbance, the Disturbance Decoupling Problem for Σ consists in finding a feedback law of the form (8) such that, in the compensated system, the disturbance does not appear on the output. Denoting by $\mathcal{V}^*$ the maximum (A, B)-invariant submodule contained in $Ker\ C$, we have

Proposition 7. (see (Conte and Perdon, 1995*b*)) Let Σ be the system defined by equations (9)

over a Noetherian ring R. Then, the Disturbance Decoupling Problem for Σ is solvable if and only if

$$Im\ D \subseteq \mathcal{V}^* + Im\ B, \tag{10}$$

where $\mathcal{V}^*$ is the maximum (A, B)-invariant submodule contained in $Ker\ C$.

Computation of $\mathcal{V}^*$ is necessary in order to check condition (10), but the I.S.A. algorithm for computing $\mathcal{V}^*$, see (Wonham, 1985), does not converge in a finite number of steps in the case of systems over rings. Although a new, different algorithm has been found for systems over a PID in (Assan *et al.*, 1999*b*), the computation of $\mathcal{V}^*$ over more general rings is still an open problem.

However, the solution of the DDP problem can be studied by means of the following alternative result, which makes use of $\mathcal{R}^*_{(BD)}$, the maximum pre-controllability submodule in $Ker\ C$ with respect to the input matrix $(B\ \ D)$.

Proposition 8. (see (Assan *et al.*, 1999*a*)) Given a system Σ, defined over the Noetherian ring R by equations of the form (9), denote by $\mathcal{R}^*_{(BD)}$ the maximum pre-controllability submodule in $Ker\ C$ with respect to the input matrix $(B\ \ D)$. Then, the DDP for Σ is solvable if and only if

$$Im\ D \subseteq \mathcal{R}^*_{(BD)}. \tag{11}$$

Efficient algorithms based on Gröbner basis theory for checking if a given submodule is contained in another one are available over rings of polynomials in several indeterminate over a field. They will be discussed in the following section, where we explicitly present an algorithm to compute the maximum pre-controllability submodule. Other algorithms are available and, due to the pages limit, will be presented in the full paper.
A novel algorithm to compute $\mathcal{R}^*$ by a converging sequence of submodules is proposed in (Assan *et al.*, 1999*a*). In (Assan and Perdon, 1999) we also find a first attempt of implementing such algorithm, relying on the computation of counterimages and intersections of modules over polynomial rings. Since such computations can be very expensive, it is crucial to optimize them. We here propose new and optimized algorithms which compute the substeps of such implementation.

5. ALGORITHMS

In the following we only consider the general case of a noetherian ring: the PID case is much simpler and all the procedures proposed below work without modifications also in that case. Given a system described by equations (2) over a ring R and a submodule $\mathcal{V} \subseteq \mathcal{X}$, spanned by the columns of a matrix V, the controlled invariance property check consists practically in finding matrices L and M satisfying

$$AV = VL + BM, \tag{12}$$

i. e. in solving with respect to the matrices L ed M, the linear system of equations over R

$$[V \mid B] \left[\frac{L}{M}\right] = AV \tag{13}$$

When the coefficients of the system belong, for instance, to a ring of polynomials in one or several indeterminates over the rationals, this can be done by syzygy computations (see any Computer Algebra text, e.g. (Kreuzer and Robbiano, 2000)) or by another technique called *module components elimination*, described in (Caboara M., 2003) and applied to the geometric approach in (Perdon *et al.*, 2003). Both techniques are supported by modern Computer Algebra Systems. See also (Malashonok, 1991) for a survey of other techniques.
In the following, we use the sub procedure `ModImage` and the check of inclusion (11) which are trivial, and the sub procedures `Ker` and `ModCImage`, which can be realized by syzygy computations or by module components elimination. The procedure `SpecialInverse_One` returns, if it exists, one inverse of the argument solving the associate linear system.

```
<ALGORITHM 1:>

Define Is_V_AB_Invariant(A,B,V)
  L:=ColumnVectors(BlockMatrix([[-A*V,V,B]]));
  S:=Gens(Syz(L));
  X0:=Transposed(
       Mat([First(Vi,NumCol(B))|Vi In S]));
  X1:=Transposed(
       Mat([GetFromIToJ(Vi,NumCol(B)+1,
            NumCol(B)+NumCol(V))|Vi In S]));
  X2:=Transposed(
        Mat([Last(Vi,NumCol(V))|Vi In S]));
  TT:=SpecialInverse_One(X0);
  If Not TT[1] Then Return [False]; End;
  K0:=TT[2][1];//K0 is the inverse of X0
  OK:=A*V=V*X1*K0+B*X2*K0;
  Return [True,X1*K0,X2*K0];
End;
```

Assume now that $\mathcal{V}$ is (A, B)-invariant. The existence of the feedback can then be checked by Proposition 3.

If the controlled invariant submodule is not feedback invariant we can compute the extended system and the dynamic feedback (7), as it has been described in Section 3. We refer to (Perdon *et al.*, 2003) for the details and the implementation.

The condition in Proposition 5 needs, to be verified, the computation of $\mathcal{R}^*$. We here propose the following algorithm in CoCoA:

```
                <ALGORITHM 2>

Define SStar(A,B,C)
  KerC:=Ker(C);
  NextS:=B;
  Repeat
    LastS:=NextS;
    NextS:=
      B+ModImage(A,Intersection(LastS,KerC));
  Until NextS=LastS;
  Return LastS;
End;


Define RStar(A,B,C)
  SBKerC:=
  Intersection(SStar(A,B,C),Ker(C));
  NextR:=
  Intersection(SBKerC,ModCImage(A,B));
  Repeat
    LastR:=NextR;
    NextR:=
    Intersection(SBKerC,
                 ModCImage(A,LastR+B));
  Until LastR=NextR;
  Return LastR;
End;
```

Remark 9. Syzygy computations over a computable polynomial ring rely on Gröbner basis computations, hence, they present a quite high worst-case complexity. Practical complexity for given examples is quite difficult to be determined *a priori*. Moreover, Gröbner basis computations are notoriously not stable over the reals, requiring use of unlimited precision rational arithmetic. Nevertheless, modern systems, like CoCoA, see (Kreuzer and Robbiano, 2000), or Singular, see (Greuel *et al.*, 2001), offer quite advanced capabilities in Gröbner basis computations and many cases of practical interest can in fact be worked out. See (Cox *et al.*, 1993) for an easy introduction or (Kreuzer and Robbiano, 2000) for more mathematical detail on Gröbner basis theory.

In (Perdon *et al.*, 2003) we will present several examples where all the computations have been performed with CoCoA 4.1 on a AMD 7 1GHz processor with 750MB RAM running Linux in less than one second CPU time. Here we only report a small example, for space reasons.

Example 10. Let us consider the delay-differential system Σ_d described by the equations

$$\begin{cases} \dot{x}_1(t) = x_1(t) + x_2(t-h_1-h_2)+ \\ \quad +u_1(t-h_1) \\ \dot{x}_2(t) = x_1(t) + x_2(t-h_1-h_2)+ \\ \dot{x}_3(t) = x_4(t) + x_2(t-h_1-h_2)+ \\ \quad +q(t-h_1-h_2) \\ \dot{x}_4(t) = x_2(t-h_2) + x_4(t)+ \\ \quad +u_2(t) + q(t) \\ \dot{x}_5(t) = x_4(t-h_2) + x_2(t-h_1-h_2)+ \\ \quad +u_3(t-h_2) \\ y_1(t) = x_1(t-h_1) - x_2(t) \\ y_2(t) = x_4(t) \end{cases} \tag{14}$$

We can associate to Σ_d the system described over the Noetherian ring $R[\Delta_1, \Delta_2]$ by equations of the form (9), where

$$A = \begin{bmatrix} 1 & \Delta_1\Delta_2 & 0 & 0 & 0 \\ 1 & \Delta_1\Delta_2 & 0 & 0 & 0 \\ 0 & 0 & 0 & \Delta_1 & 0 \\ 0 & \Delta_2 & 0 & 1 & 0 \\ 0 & 0 & 0 & \Delta_2 & 0 \end{bmatrix},$$

$$B = \begin{bmatrix} \Delta_1 & 0 & 0 \\ 0 & 0 & 0 \\ 0 & 0 & 0 \\ 0 & 1 & 0 \\ 0 & 0 & \Delta_2 \end{bmatrix},$$

$$C = \begin{bmatrix} \Delta_1 & -1 & 0 & 0 & 0 \\ 0 & 0 & 0 & 0 & 1 \end{bmatrix}$$

and

$$D = \begin{bmatrix} 0 \\ 0 \\ \Delta_1\Delta_2 \\ 1 \\ 0 \end{bmatrix}.$$

Applying Algorithm 3 we have that that the columns of the matrix

$$V := \begin{bmatrix} 0 & 0 \\ 0 & 0 \\ \Delta_1 & 0 \\ 0 & 1 \\ 0 & 0 \end{bmatrix}$$

generate $\mathcal{R}^*_{(BD)}$, the maximum pre-controllability submodule in $Ker\ C$ with respect to the input matrix $(B\ D)$ and condition (11) is satisfied. Although $\mathcal{R}^*_{(BD)}$ is not a direct summand, condition (3) is satisfied (the computations lasted Cpu times less than 0.5 seconds), therefore $\mathcal{R}^*_{(BD)}$ is feedback invariant. By using Algorithm 1, we obtain

$$L := \begin{bmatrix} 0 & 1 \\ -1 & 0 \end{bmatrix} \text{ and } M := \begin{bmatrix} 0 & 0 \\ 1 & 1 \\ 0 & 1 \end{bmatrix}.$$

Then, a possible feedback is given by (7). Going back to the original delay differential sysytem we have the following state feedback

$$F = \begin{cases} u_1(t) = 0 \\ u_2(t) = -x_3(t) - x_4(t) \\ u_3(t) = -x_4(t) \end{cases}$$

It is easy to verify that the disturbance is decoupled from the output of the compensated system

$$\begin{cases} \dot{x}_1(t) = x_1(t) + x_2(t - h_1 - h_2) \\ \dot{x}_2(t) = x_1(t) + x_2(t - h_1 - h_2) \\ \dot{x}_3(t) = x_4(t - h_1) + q(t - h_1 - h_2) \\ \dot{x}_4(t) = x_2(t - h_2) + x_4(t) \\ \dot{x}_5(t) = 0 \\ y_1(t) = x_1(t - h_1) - x_2(t) \\ y_2(t) = x_4(t) \end{cases} \quad (15)$$

6. CONCLUSIONS

Symbolic computation tools which perform efficiently provide the possibility of implementing basic algorithms for analyzing geometric properties of dynamical systems with coefficients in a ring and for constructing feedback solutions to control problems. Their use allows to develop practical design methodologies for delay differential systems based on the geometric approach.

As a matter of fact, though the geometric approach extended to the rings has been allowing for a couple of decades a deep insight in the properties of several kind of systems, the complete lack of computational techniques prevented such results to be of immediate use: throughout the bibliography concerning this subject it is not easy to find practical examples which are worked out by an application of the theoretical results coming from the geometrical approach. And this happens since, up to now, no computational technique has been developed within the ring context, as it requires a complex knowledge in computational commutative algebra. The results which we here propose represent a first step towards the filling of such a gap: we can now compute the solutions of several design problems, such as DDP, based on the effective computations of fundamental objects of the geometrical approach, and we could consider issues coming from practical design problems, such as robustness of solutions.

7. REFERENCES

Assan, J. and A. M. Perdon (1999). An efficient computation of the solution of the block decoupling problem with coefficient assignment over a ring. *Kybernetika* **35**(6), 765–776.

Assan, J., J. F. Lafay and A. M. Perdon (1999*a*). Computation of maximal pre-controllability submodules over a noetherian ring. *System and Control Letters* **37**(3), 153–161.

Assan, J., J-F Lafay and A. M. Perdon (2002). Feedback invariance and injection invariance for systems over rings. *Proc. 15th IFAC Wordl Congress.*

Assan, J., J.F. Lafay, A.M. Perdon and J.J. Loiseau (1999*b*). Effective computation of maximal controlled invariant submodules over a principal ring. *Proc. 38th IEEE Conference on Decision and Control.*

Bareiss, E.H. (1968). Sylvester identity and multistep integer-preserving gaussian elimination. *Math. Comp.* **22**(103), 565–578.

Basile, G. and G. Marro (1992). *Controlled and Conditioned Invariants in Linear System Theory.* Prentice Hall. Englewood Cliffs, New York.

Caboara M., C. Traverso (2003). Efficient algorithms for basic module operations. Submitted.

Capani A., G. Niesi, L. Robbiano (1995). CoCoA, a system for doing computations in commutative algebra. Available via anonymous ftp from `cocoa.dima.unige.it`.

Conte, G. and A. M. Perdon (1995*a*). The decoupling problem for systems over a ring. *Proc. 34th IEEE Conference on Decision and Control.*

Conte, G. and A. M. Perdon (1995*b*). The disturbance decoupling problem for systems over a ring. *SIAM Journal on Control and Optimization.*

Conte, G. and A. M. Perdon (1997). Noninteracting control problems for delay-differential systems via systems over a ring. *European Journal of Automation* **31**(6), 1059–1076.

Cox, D., J. Little and D. O'Shea (1993). *Ideal, Varieties and Algorithms.* Springer Verlag. Berlin, New York.

Greuel, G. M., G. Pfister and H. Schönemann (2001). SINGULAR 2.0. A Computer Algebra System for Polynomial Computations. Centre for Computer Algebra. University of Kaiserslautern. `http://www.singular.uni-kl.de`.

Kreuzer, M. and L. Robbiano (2000). *Computational Commutative Algebra 1.* Springer Verlag. Berlin, New York.

Malashonok, G.I. (1991). Algorithms for the solution of systems of linear equations in commutative rings. In: *Effective Methods in Algebraic Geometry* (T Mora and C. Traverso, Eds.). pp. 289–298. Birkhauser.

Perdon, A.M., M. Caboara and G. Guidone Peroli (2003). Symbolic computations in the geometric approach to systems over rings. Submitted.

Wonham, M. (1985). *Linear Multivariable Control: a Geometric Approach.* 3rd Ed., Springer Verlag. Englewood Cliffs, New York.

A New Loop-Shaping Procedure for Tuning LQ-PID Regulator

B. S. Suh, S. O. Yun and J. H. Yang

Division of Electrical and Computer Engineering, Hanyang University,
17 Haengdang-Dong, Seongdong-Gu, Seoul, 133-791, Korea
Tel: +82-2-2290-0364, bssuh@hanyang.ac.kr

Abstract: This paper presents a new limiting loop shaping technique for tuning the robust optimal LQ-PID regulators by the suitable weighting factors Q and R in order to satisfy the performance requirements. The technique is developed by pushing all two zeros formed by PID controller closely to a larger pole of the second order plant. As a result, an improved loop shaping is achieved in the high frequencies region on the Bode plot.

Keywords: Loop-Shaping, LQR, Optimal control, PID, Robustness, Tuning.

1. INTRODUCTION

One of the major concerns for the control engineers is to maintain robust stability and good performance to meet the design specifications. For the robustness issue, some robust controller design methods as *LQG-LTR* (Sein, 1979) and H_∞ (Doyle, 1989, Zames, 1981) have been developed.

Although these methods are based on the elaborate mathematical optimization theories, *PID* controllers are still widely used in industry since they are simple and robust, more over familiar to the field engineers. Thus, there is one approach to apply H_∞ control theories to *PID*. Grimble(1990) tried to identify a certain class that H_∞ controllers have a *PID* structure. Maffezzoni and Rocco (1997) derived a model error bound in order to formulate simple rules for a robust tuning of *PID* controller. Panagopoulos (1999) showed that the specifications in terms of maximum sensitivity and maximum complementary sensitivity are related to the weighted H_∞ norm. And Grassi et al.(1996,2001) presented a PID tuning method based on frequency loop-shaping using convex optimization in the sense of the H_∞ norm.

On the while, there is a new *LQ*-approach to utilize the robustness properties of *LQR*. Suh (Suh and Yun,2000, Suh,2000) interpreted Athans' *LQ*-Servo structure(Athans, 1989) as a special type of *PI* controllers with a partial state feedback and solved a loop shaping problem of an output-loop transfer function. He et al. (1998) also suggested an optimal *PI/PID* controller tuning algorithms for the first order plus a time delay process via *LQR* approach to satisfy the design specifications in the domain design.

In this paper, a new LQ-approach based on Shih and Chen's work (Shin and Chen, 1974) is exploited to apply to the conventional *PID* controller. They were able to transform the *PID* tuning into *LQR* design problem by relating the optimal state feedback control law for the second-order system to the *PID* control. Thus the tuning parameters of *PID* controller are determined by the weighting factors Q and R of the cost function.

In this paper, a new limiting loop shaping technique is presented to select the suitable weighting factors Q and R by pushing all two zeros formed by *PID* controller closely to a larger pole of the second order plant. It is able to achieve the improved loop shaping in the high frequencies region on the Bode plot than the previous method (Suh, 2001) for good noise reduction and modelling error. Although this approach is limited to the second order plants, they arise in wide variety applications such as in vibration and structural analysis.

2. PID TUNING FORMULATION BY LQR

Shih and Chen(1974) was able to relate the optimal control law for the second order system to the conventional *PID* control by taking an integral of the output variable as a new state variable in the following way:

Consider the following second order model

$$\frac{d^2y(t)}{dt^2}+2\zeta\,\omega_n\frac{dy(t)}{dt}+\omega_n^2\,y(t)=\omega_n^2\,u(t) \qquad (1)$$

where $y(t)$ is an output variable, $u(t)$ is a control variable, ω_n is a natural frequency, and ζ is a damping ratio. The initial conditions $y(0)$ and $dy(0)/dt$ are specified.

The aim of the control is to drive the output variable to the final steady state values, which is assumed to be $y(\infty)=0$.

Let the state variables $x_p(t)$ be defined as

$$x_p(t)=\begin{bmatrix} y(t) \\ \frac{dy(t)}{dt} \end{bmatrix}=\begin{bmatrix} x_1(t) \\ x_2(t) \end{bmatrix}. \quad (2)$$

Then (1) becomes

$$\frac{dx_p(t)}{dt}=A_p x_p(t)+B_p u(t) \quad (3)$$

$$y(t)=C_p x_p(t) \quad (4)$$

where

$$A_p=\begin{bmatrix} 0 & 1 \\ -\omega_n^2 & -2\zeta\omega_2 \end{bmatrix}, \quad B_p=\begin{bmatrix} 0 \\ \omega_n^2 \end{bmatrix},$$

and

$$C_p=[1 \quad 0] \quad (5)$$

with $x_p(t)$ is specified.

Also, an integrator state vector $x_0(t)$ which is defined as $\frac{dx_0(t)}{dt}=y(t)$ is augmented to the feed forward loop in order to make *PID* structure.

Then the augmented state-space descriptions are

$$x(t)=[x_0(t) \ \vdots \ x_p(t)]^T=\left[\int_0^t y(\tau)d\tau \ \vdots \ y(t) \quad \frac{dy(t)}{dt}\right]^T \quad (6)$$

$$\frac{dx(t)}{dt}=A\,x(t)+B\,u(t) \quad (7)$$

$$y(t)=C\,x(t) \quad (8)$$

where

$$A=\begin{bmatrix} 0 & C_p \\ 0 & A_p \end{bmatrix}=\begin{bmatrix} 0 & 1 & 0 \\ 0 & 0 & 1 \\ 0 & -\omega_n^2 & -2\zeta\omega_n \end{bmatrix}, B=\begin{bmatrix} 0 \\ B_p \end{bmatrix}=\begin{bmatrix} 0 \\ 0 \\ \omega_n^2 \end{bmatrix}$$

and

$$C=[0 \quad C_p]=[0 \quad 1 \quad 0]. \quad (9)$$

For transforming to LQR, consider a cost function is

$$J=\frac{1}{2}\int_0^\infty \left(x(t)^T Q x(t)+u(t)^T R u(t) \right) dt \quad (10)$$

with the assumption that the weighting matrix Q is symmetric and positive semi-definite, and the weighting matrix R is symmetric and positive definite, then the following optimal control law is

$$u(t)=-G\,x(t) \quad (11)$$

where $G=-R^{-1}B^T K$. And $K=K^T$ is a solution matrix of the algebraic Riccati's equation:

$$KA+A^T K+Q-KBR^{-1}B^T K=0 \quad (12)$$

where the design parameters $Q=N^T N$. And $R=\rho I$ is considered.

Let G be decomposed into

$$G=[g_0 \quad g_1 \quad g_2], \quad (13)$$

then the optimal control law of (11) becomes the following *PID* control form:

$$u(t)=-\left(g_0\int_0^t y(\tau)d\tau+g_1 y(t)+g_2\frac{dy(t)}{dt} \right). \quad (14)$$

Consider the conventional *PID* control form:

$$u(t)=-K_p\left(y(t)+\frac{1}{T_i}\int_0^t y(\tau)d\tau+T_d\frac{dy(t)}{dt} \right). \quad (15)$$

Relating the gain components g_i to the proportional gain K_p, integral time T_i, and derivative time T_d gives

$$K_p=g_1, \quad T_i=\frac{g_1}{g_0}, \quad T_d=\frac{g_2}{g_1}. \quad (16)$$

Since the optimal control $u(t)$ of (11) can be considered as the *PID* control form, the *PID* structure shown in Figure 1 and *LQR* structure shown in Figure 2 can be equivalent foe the augmented second order system. The parameters of a *PID* controller K_p, T_i and T_d can be obtained by solving *LQR* design problem.

It is noted that *LQR* is able to guarantee the stability robustness, but it is hard to deal with the performance issues because it does not consist of an output feedback. For this, in this section, a special type of *LQR* structure with an output feedback in the second order system is considered with the augmented state variable description

The appropriate design parameters Q and R of *LQR* are selected to improve the performance issues by a limiting loop shaping technique proposed in the next section.

3. A SELECTION OF WEIGHTING FACTORS Q AND R

In this section, a selection procedure of the design parameters Q and R in the performance index is developed on the basis of the previous singular value matching technique(Suh and Yun,2000) of loop transfer function obtained by braking at the plant output for the improvement of performance robustness. Originally, this technique is an extended version of Athans' matching technique (Athans, 1986) of the loop transfer function broken at the plant input.

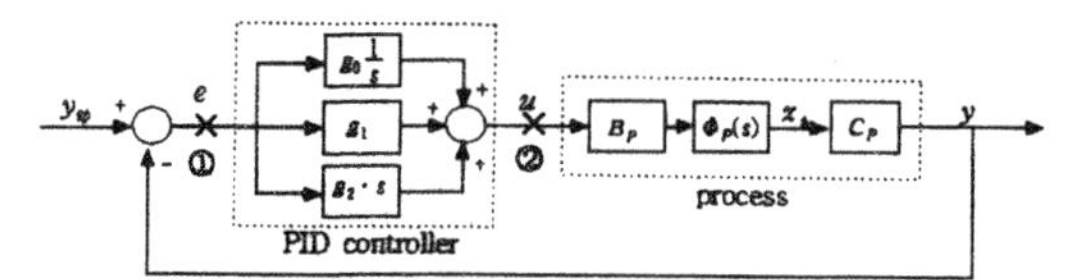

Fig. 1. Block diagram of a PID structure.

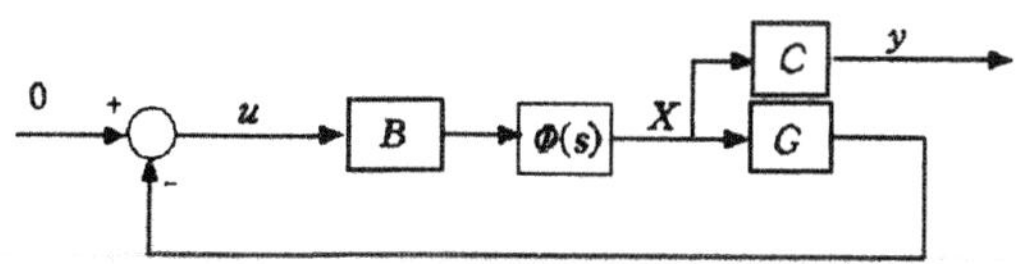
Fig. 2. Block diagram of a LQR.

To deal with the performance issue by the loop shaping method, a loop transfer function $h(s)$ obtained by braking at the plant output point ① in Figure 1 other than plant input point ② in Figure 1 must be considered. A loop transfer function $h(s)$ obtained by braking at the plant output in Figure 1 and $g_L(s)$ obtained by braking at the plant input point in Figure 1 are as follows:

$$h(s) = C_p\,(sI - A_p)^{-1}\,B_p\,(\,g_2 s + g_1 + \frac{g_0}{s})$$
$$= \frac{g_2 s^2 + g_1 s + g_0}{s(s^2 + 2\zeta\,\omega_n s + \omega_n^2)} \qquad (17)$$

$$g_L(s) = (\,g_2 s + g_1 + \frac{g_0}{s})\,C_p\,(sI - A_p)^{-1}\,B_p$$
$$= \frac{g_2 s^2 + g_1 s + g_0}{s(s^2 + 2\zeta\,\omega_n s + \omega_n^2)}. \qquad (18)$$

In SISO systems, the loop transfer function $h(s)$ as the same as $g_L(s)$. Also $g_L(s)$ is the same as the loop transfer function of *LQR*, since $g_{LQ}(s)$ is described as

$$g_{LQ}(s) = G(sI - A)^{-1}B = G\left\{\begin{bmatrix} sI & 0 \\ 0 & sI \end{bmatrix} - \begin{bmatrix} 0 & C_p \\ 0 & A_p \end{bmatrix}\right\}^{-1} B$$
$$= \begin{bmatrix} g_0 & g_1 & g_2 \end{bmatrix} \begin{bmatrix} \frac{1}{s} & \frac{1}{s}C_p(sI - A_p)^{-1} \\ 0 & (sI - A_p)^{-1} \end{bmatrix} \begin{bmatrix} 0 \\ B_p \end{bmatrix}$$
$$= \frac{1}{s} g_0 C_p (sI - A_p)^{-1} B_p + [g_1 \;\; g_2](sI - A_p)^{-1} B_p$$
$$= \frac{g_2 s^2 + g_1 s + g_0}{s(s^2 + 2\zeta\,\omega_n s + \omega_n^2)}. \qquad (19)$$

Let us determine the design parameters Q and R by using the following frequency domain equality:

$$[\,I + g_{LQ}(-j\omega)\,]^T R\,[\,I + g_{LQ}(j\omega)\,]$$
$$= R + g_{OL}(-j\omega)\,g_{OL}(j\omega) \qquad (20)$$

where $g_{OL}(j\omega)$ is an open loop transfer function and $h(s)$ is the same as $g_{LQ}(s)$ in SISO. The (20) can be described

$$[\,I + h(-j\omega)\,]^T R\,[\,I + h(j\omega)\,]$$
$$= R + g_{OL}(-j\omega)\,g_{OL}(j\omega). \qquad (21)$$

The frequency domain characteristics of *LQR* are derived from (20) as

$$\left|\, I + h(j\omega) \,\right| = \sqrt{1 + \frac{1}{\rho}\left|\, N(j\omega I - A)^{-1} B \,\right|^2} \qquad (22)$$

where design parameters $N = [\, n_0 \;\; n_1 \;\; n_2 \,]$.

Far from the cross over frequency of $|\, h(j\omega) \,|$, that is, at low frequencies and at high frequencies on the Bode plot, the (22) can be approximated as

$$\left|\, h(j\omega) \,\right| \approx \frac{1}{\sqrt{\rho}}\left|\, N(j\omega I - A)^{-1}\, B \,\right|. \qquad (23)$$

For the second order system, (23) can be developed into

$$\left|\, h(j\omega) \,\right| = \frac{1}{\sqrt{\rho}}\left|\, \frac{n_2\,\omega_n^2\, s^2 + n_1\,\omega_n^2\, s + n_0\,\omega_n^2}{s(s^2 + 2\zeta\,\omega_n s + \omega_n^2)} \,\right|. \qquad (24)$$

The limiting behaviours of $h(j\omega)$ at low frequencies and high frequencies are approximated as

$$\lim_{\omega \to 0} \left|\, h(j\omega) \,\right| \approx \frac{|\, n_0 \,|}{\omega\sqrt{\rho}} \qquad (25)$$

and

$$\lim_{\omega \to \infty} \left|\, h(j\omega) \,\right| \approx \frac{|\, n_2\,\omega_n^2 \,|}{\omega\sqrt{\rho}}. \qquad (26)$$

Let each the asymptotic lines of $|\, h(j\omega) \,|$ at low frequencies and high frequencies be $l_L(\omega) = \frac{|\, n_0 \,|}{\omega\sqrt{\rho}}$ and $l_H(\omega) = \frac{|\, n_2\,\omega_n^2 \,|}{\omega\sqrt{\rho}}$.

It is well known that the shape of $|\, h(j\omega) \,|$ should not be invade both the command following barrier (or disturbance rejection barrier) "$\alpha(\omega)$" and the sensor noise barrier (or high order modelling error barrier) "$\beta(\omega)$" in Figure 3.

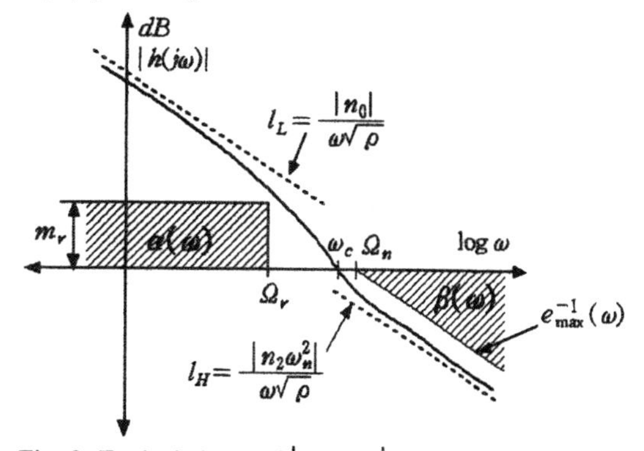

Fig. 3. Typical shape of $|\, h(j\omega) \,|$ with two barriers of the performance requirements

The shape of $|\, h(j\omega) \,|$ at the low frequencies should be existed on the upper than the barrier $\alpha(\omega)$. And the shape of $|\, h(j\omega) \,|$ at the high frequencies should be existed on the lower than the barrier $\beta(\omega)$. That is, the asymptotic lines of $|h(j\omega)|$, $l_L(\omega)$ and $l_H(\omega)$ should be existed on the upper side and the lower side of the barriers $\alpha(\omega)$ and $\beta(\omega)$ each. For this shaping, we can develop a limiting loop shaping technique that the design parameter N plays a role in the shaping of $|h(j\omega)|$ not to be invaded into those two barriers.

Let Ω_r be a boundary frequency of command following barrier $\alpha(\omega)$ and m_r be a height of $\alpha(\omega)$ shown in Figure 3. And let the barrier $\alpha(\omega)$ as

$$\alpha(\omega)=\begin{cases} m_r & , \ \omega\le\Omega_r \\ 0 & , \ \omega>\Omega_r \end{cases}. \tag{27}$$

Since $l_L(\omega)$ should be existed on the upper side of the barrier $\alpha(\omega)$, the asymptotic line of $|h(j\omega)|$ at low frequencies must be satisfied with

$$l_L(\omega)\ :\ \frac{|n_0|}{\omega\sqrt{\rho}} > \alpha(\omega)\ ,\ \omega\le\Omega_r. \tag{28}$$

Solving (28) for the range of the parameter n_0 gives

$$|n_0| > \Omega_r\sqrt{\rho}\cdot m_r. \tag{29}$$

And also, let Ω_n be a boundary frequency of sensor noise barrier $\beta(\omega)$ and $e_{\max}(\omega)$ is the maximum value of modelling error in Figure 3.
And let the barrier $\beta(\omega)$ be

$$\beta(\omega)=\begin{cases} 0 & , \ \omega<\Omega_n \\ e_{\max}^{-1}(\omega) & , \ \omega\ge\Omega_n \end{cases}. \tag{30}$$

Since $l_H(\omega)$ should be existed on the lower side of the barrier $\beta(\omega)$ the asymptotic line of $|h(j\omega)|$ at high frequencies must be satisfied with

$$l_H(\omega)\ :\ \frac{|n_2\omega_n^2|}{\omega\sqrt{\rho}} > \beta(\omega)\ ,\ \omega\ge\Omega_n. \tag{31}$$

Solving (31) for the range of the parameters n_2

$$|n_2| < \frac{\Omega_n\sqrt{\rho}\cdot e_{\max}^{-1}(\omega)}{\omega_n^2}. \tag{32}$$

The ranges of the design parameters n_0 and n_2 which make a satisfactory shape of $|h(j\omega)|$ could be solved with (29) and (32). However, these ranges are too broad to choose values of n_0 and n_2. Not only the constraint between n_0 and n_2, but the range of design parameter n_1 are not considered.

For those problems, the limiting technique made by pushing all two zeros formed by *PID*•controller closely to a larger pole of the second order plant is introduced in the following ways:
Denote poles and zeros of $|h(j\omega)|$ by p_1, p_2, z_1 and z_2. Let p_2 be larger than p_1 as shown in Figure 4.

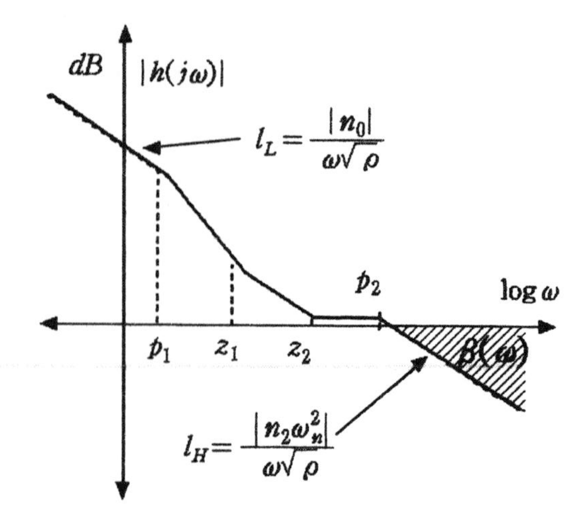

Fig. 4. Shape of $|h(j\omega)|$ with poles and zeros.

Then (24) can be expressed as

$$|h(j\omega)|=\frac{1}{\sqrt{\rho}}\left|\frac{n_2\omega_n^2(s+z_1)(s+z_2)}{s(s+p_1)(s+p_2)}\right|. \tag{33}$$

Since poles of $|h(j\omega)|$, p_1 and p_2 are given by the plant model, the positions of poles on the Bode plot are fixed for a given plant. But two zeros of $|h(j\omega)|$ could be placed (or moved) to any position by selecting the design parameters n_0, n_1 and n_2. Also the suitable shape of $|h(j\omega)|$ could be obtained by controlling the two zeros of $|h(j\omega)|$. Suppose that z_1 and z_2 are moved closely to the larger pole p_2, $l_H(\omega)$ would be farther separated from the sensor noise barrier $\beta(\omega)$, as shown in Figure 5.

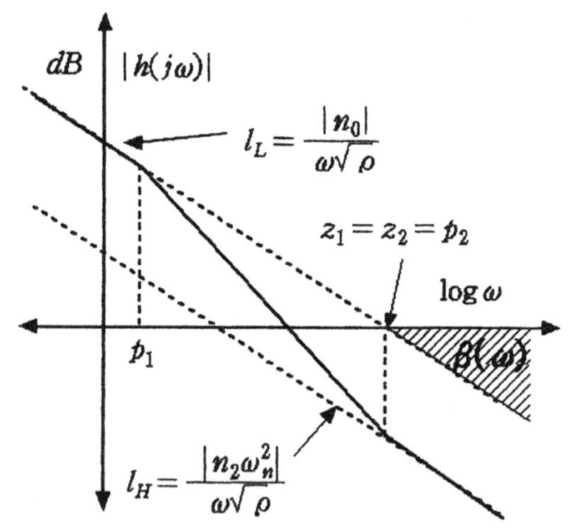

Fig. 5. Shape of $|h(j\omega)|$ taken a limit as $\lim_{z_1,z_2\to p_2}|h(j\omega)|$.

Such the separation can be easily figured out when we consider the general Bode plotting techniques of zeros and poles of transfer function.
Then (33) can be taken a limit in the following way:

$$\lim_{z_1,z_2\to p_2}|h(j\omega)|\approx\frac{1}{\sqrt{\rho}}\left|\frac{n_2\omega_n^2(s^2+2p_2+p_2^2)}{s(s+p_1)(s+p_2)}\right|. \tag{34}$$

Comparing the coefficients of (24) with those of (34), the decomposed parameters are related as

$$N=\left[\begin{array}{lll} n_0 & n_1 & n_2 \end{array}\right]=\left[\begin{array}{lll} n_2 p_2^2 & 2n_2 p_2 & n_2 \end{array}\right]. \quad (35)$$

This implies that the parameter N can be determined by the only one partial parameter n_2. Using (28), (30) and (33), the range of n_2 is finally obtained as

$$\frac{\Omega_r \sqrt{\rho} \cdot m_r}{p_n^2} < |n_2| < \frac{\Omega_n \sqrt{\rho} \cdot e_{\max}^{-1}(\omega)}{\omega_n^2} \quad (36)$$

or

$$\frac{\Omega_r \cdot m_r}{p_n^2} < \frac{|n_2|}{\sqrt{\rho}} < \frac{\Omega_n \cdot e_{\max}^{-1}(\omega)}{\omega_n^2}. \quad (37)$$

Since Ω_r, Ω_n, m_r, p_2, $e_{\max}(\omega)$ and ω_n in (37) are determined from a given plant and its design specifications, n_2 can be selected to satisfy (37) with the appropriate choice of ρ.

Remark 1:
It is remarked that ρ can play a role in the whole shaping of $|h(j\omega)|$ up and down. So, it is effective in adjusting the shaping of $|h(j\omega)|$ not to be invaded into both $\alpha(\omega)$ and $\beta(\omega)$ barriers. Its role is well explained in the Athans' lecture note (Athans, 1986), especially in the cases of the limiting behaviours of $|h(j\omega)|$ at low frequencies and at high frequencies that are represented as (25) and (26).

Remark 2:
It is again remarked that it does not lead to a higher sensitivity when the zeros of loop (33) are moving close to large pole. The sensitivity of loop (33) with respect to zeros z_1 and z_2 is given by

$$\begin{aligned}
S_{h:z_1,z_2} &= \lim_{\Delta z_1 \to 0} \lim_{\Delta z_2 \to 0} \left(\frac{\Delta h / h}{\Delta z_1 / z_1} \cdot \frac{\Delta h / h}{\Delta z_2 / z_2} \right) \\
&= \lim_{\Delta z_1 \to 0} \left(\frac{z_1}{h} \cdot \frac{\Delta h}{\Delta z_1} \right) \lim_{\Delta z_2 \to 0} \left(\frac{z_2}{h} \cdot \frac{\Delta h}{\Delta z_2} \right) \\
&= \left(\frac{z_1}{h} \cdot \frac{\delta h}{\delta z_1} \right) \cdot \left(\frac{z_2}{h} \cdot \frac{\delta h}{\delta z_2} \right) \\
&= \frac{z_1}{s+z_1} \cdot \frac{z_2}{s+z_2} = \frac{z_1 z_2}{s^2+(z_1+z_2)s+z_1 z_2}
\end{aligned}$$

which is a function of the value of s. The sensitivity to changes in zeros z_1 and z_2 is less than 1 for any arbitrary value of s.

Summary of Design Procedure

1. Select n_2 to satisfy (37) for the given design specifications.
2. Calculate N from (35) for the selected n_2.
3. Get the optimal control gain G by solving Riccati's equation parameters $Q = N^T N$ and $R = \rho I$.
4. Solve the parameters of a PID controller K_p, T_i and T_d from (16).

4. SIMULATION RESULTS

Example 1: Let the second order plant $g_p(s)$ and its design specifications be

$$g_p(s) = \frac{1}{s^2+10s+1},$$

$$\alpha(\omega) = \begin{cases} 30\,dB, & \omega \le 1 \\ 0, & \omega > 1 \end{cases}, \quad \beta(\omega) = \begin{cases} 0, & \omega \le 10 \\ 10/\omega, & \omega > 10 \end{cases}.$$

According to the design procedures, the design parameter N and the gain matrix G are obtained as

$$N = [\,342.9643 \quad 69.2929 \quad 3.5000\,],$$
$$G = [\,342.9643 \quad 123.0409 \quad 8.9297\,].$$

For convenience, $\rho = 1$ is taken. Finally, the parameters of *PID* controller become $K_p = 123.0409$, $T_i = 0.3588$ and $T_d = 0.0726$. The simulation results show in Fig. 6 that the curve of closed loop transfer function $C(j\omega)$ formed by the proposed method does not intrude $e_{\max}^{-1}(\omega)$, but the curve of $C(j\omega)$ by Ziegler-Nichols' method intrude $e_{\max}^{-1}(\omega)$. And the comparison of step response of the proposed method and Ziegler-Nichols' one is shown in Figure 7. Such the results imply that the proposed technique is superior to the Ziegler-Nichols' one, with respect to robustness and good command following.

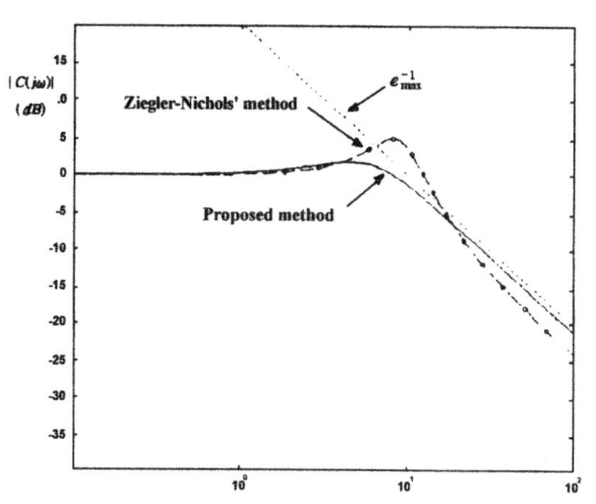

Fig. 6. Shapes of closed loop transfer function for example 1. (The proposed method: solid line, Ziegler-Nichols' method: marked line with o)

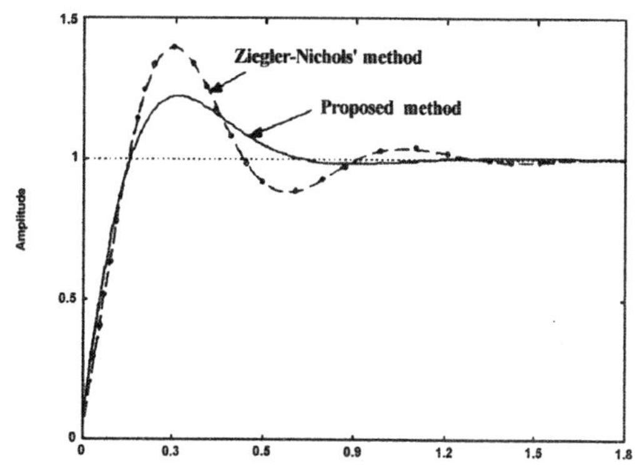

Fig. 7. Closed loop step response of PID controller for example 1. (The proposed method: solid line, Ziegler-Nichols' method: marked line with o)

Example 2: Let the second order unstable system $g_p(s)$ and its specifications be

$$g_p(s) = \frac{10}{s^2 - 9s - 10},$$

$$\alpha(\omega) = \begin{cases} 50\,dB & ,\omega \le 1 \\ 0 & ,\omega > 1 \end{cases}, \beta(\omega) = \begin{cases} 0 & ,\omega \le 10^3 \\ 10^3/\omega & ,\omega > 10^3 \end{cases}.$$

N and G are obtained as

$$N = [\,5630.0 \quad 1126.0 \quad 56.3\,],$$

$$G = [\,5630.0 \quad 1136.9 \quad 59.2\,] \text{ for } \rho = 1.$$

The parameters of a *PID* controller are $K_p = 1136.9$, $T_i = 0.2019$ and $T_d = 0.0521$. And the comparisons of closed loop transfer function and step response of the proposed method and Grassi's method (1996, 2001) are shown in Fig. 8 and Fig. 9. The proposed method is appeared to be better step response than Grassi's one, although it has the more tightened modelling error bound.

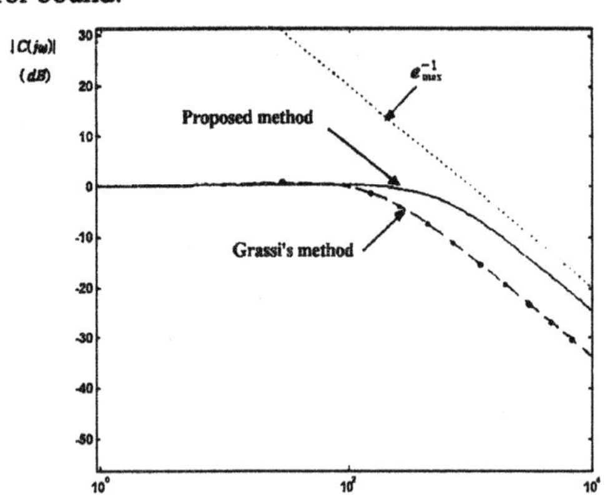

Fig. 8. Shapes of closed loop transfer function for the example 2. (The proposed method: solid line, Grassi' method: marked line with o)

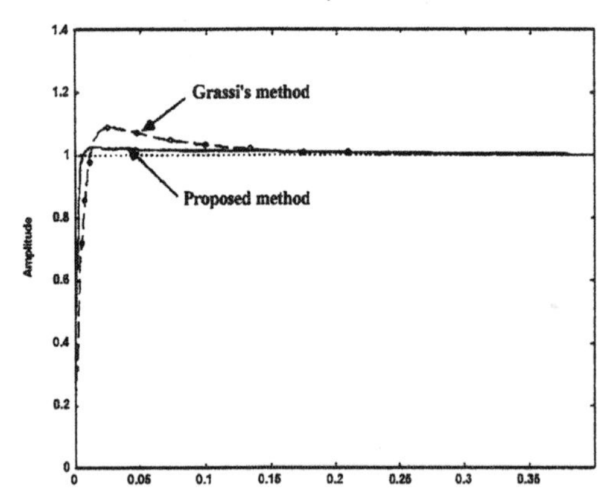

Fig. 9. Closed step response of PID controller for the example 2. (The proposed method: solid line, Grassi' method: marked line with o)

5. CONCLUSION

For the robust optimal tuning of *PID* controller for the second order plant, a new loop shaping technique is developed via *LQR* approach in order to deal with performance issues as well as robustness. The technique is based on pushing all the zeros formed by *PID* controller closely to a larger pole of the plant and is appeared to be effective in handling a loop shaping to satisfy the given design specifications.

REFERENCES

Athans. M. (1986), *Lecture Note on Multivariable Control System*, M.I.T. Ref. No.860224/6234.

Doyle, J., K. Glover, P. P. Khargonekar, and, B. A. Francis (1989), state space solutions to standard H2 and H control problems, *IEEE Trans. Automat. Contr.*, **34**, pp. 831-847.

Grassi, E. and K. S. Tsakalis (1996), PID Controller Tuning by Frequency Loop-Shaping, *Proceedings of the 35th IEEE Conference on Decision & Control*, pp 4776-4781.

Grassi, E., K. S. Tsakalis, S. Dash, S. V. Gaikwad, W. MacArthur, and G. Stein (2001), Integrated System Identification and PID Controller Tuning by Frequency Loop-Shaping, *IEEE Trans. Contr. Syst. Technol.*, **9**, pp.285-294.

Grimble, M. (1990), H^∞ controllers with a PID structure, *J. of Dynamic Syst. Meas. and contr.*, **112**, pp. 325-336.

He, J. B., Q. G. Wang and T. H. Lee (1998), PI/PID Controller Tuning Via LQR Approach, *Proceedings of the 37th IEEE Conference on Decision & Control*, pp 1177-1182.

Mattezzoni, C. and P. Rocco (1997), Robust tuning of PID Regulators Based on Step-Response Identification, *European J. of Contr.*, **3**, pp. 125-136.

Panagopoulos, H. and K. J. Åstrom (1999), PID Control Design and H^∞ Loop Shaping, *Proceedings of the 1999 IEEE International Conference an Control Applications*, pp.103-108.

Sein, G. and J. Doyle (1979), Robustness with Observers, *IEEE Trans. Automat. Contr.*, **24**, pp. 607-611.

Shih, Y. and , C. Chen (1974), On the weighting factors of the quadratic criterion in optimal control, *Int. J. Control*, **19**, pp. 947-955.

Suh, B. S. and S. O. Yun (2000), LQ-servo Design, *Proceedings of The 4th Asia-Pacific Conference on Control & Measurement*, pp. 97-100.

Suh, B. S. (2000), Design of LQ-servo PI Controller Considering Weight, *J. of Korea Institute of Communication Science*, **25**, pp. 570-576.

Suh, B. S. (2001), Robust Optimal tuning of PID regulators for a Second Order System, *XIV International Conference on System Science*, pp.313-319, Wroclaw Poland, September.

Zames. G. (1981), Feedback and Optimal Sensitivity, *IEEE Trans. Automat. Contr.*, **26**, pp. 301-320.

MINIMAL STRUCTURE CONTROL: A PROPOSITION

Amin Nobakhti [1] **, Neil Munro**

Control Systems Centre, UMIST
Manchester M60 1QD, UK
multivariable@IEEE.org

Abstract: This, the first of a two part paper, aims to propose the new concept of Minimal Structure control. The paper outlines and explains why there is a need for this additional class of multivariable controllers and why it is imperative that they be given a separate identify. In addition, two basic tools are presented to aid preliminary analysis and design of Minimal Structure controllers. In the second paper, these tools will be used to design Minimal Structure controllers for two real-life systems to show the powerful nature of this approach to controller design.

Keywords: Multivariable Control, Simply Structured, Minimum Structure

1. INTRODUCTION

Multivariable feedback controller design has undoubtedly been one of the most exciting areas of feedback control in the past three decades. It has continually presented engineers and academics alike, with some of the most challenging control issues and an indication of this can be the sheer speed with which theoretical advancements have been made in this field. This speed, whilst has provided academics with much thrill and stimulation, has been a major downside as far as the industry is concerned. They have simply not been able to keep up to speed with implementation of these controllers to the extend that worryingly large gaps are appearing between the theory and the practice of control engineering. Fortunately in recent years, the academic community has come to acknowledge this and now a major shift of emphasis means that issues with understanding implementation of controllers from an industrial view point have gained a high priority in many lists.

One issue which the industry is particularly concerned with, is the controller structure. On a broad note it is defined as the manner with which the inputs of a system are manipulated to give the outputs. In a single input single output (SISO) system, structure can be directly linked to the system dynamics. In a multivariable system whilst the dynamics of individual channels are still important, an additional consideration from a structural point of view is the cross-coupling in the controller, or the number of non-zero off-diagonal terms. Therefore from here on forth, when the structure of a multivariable controller is referred to, not only dynamical order of each element is in mind, but also the number of non-zero elements. The advantage of defining structure in this way, is that under this definition a 'well structured controller' contains two of the critical factors the industry favors; small number of cross-couplings, and low dynamics for individual elements.

The issue with control structure came in the spot light as a consequence of recent explosion in synthesis based techniques for design of multivariable controllers. What the majority of these techniques have in common is in their tendency to make

[1] The author wishes to thank EPCRS, UMIST and the IEE for supporting this research.

unnecessary use of control dynamics. This often happens when a mode that is dominant in one channel, also appears in another channel without being of much practicality. The ultimate manifestation of this can be seen in standard H_∞ based controllers whose every I/O channel contains the equivalent number of states of the entire system which is being controller. This problem however must not be viewed as disease of modern control. Some classical multivariable techniques [2] may also suffer in the same way . For example in designing dominance based multivariable controllers, what determines the lowest dynamical order of each I_i/O channel [3] , is the order of the highest element. This is to say if, say, element (1,1) is of nth order, by default all other elements in the same column of the controller, i.e. (2,1),...,(n,1) will have n^{th} order transfer functions, whether its needed or in fact its not. This point was illustrated by (Maciejowski, 1989) in his book on multivariable feedback design.

In this light the motives behind this work are self-illuminating. There is a strong need to develop controllers which are not anymore complex than they ought to be, from a practical point of view. As for the nature of this work, the reader should be aware that far from a technical or theoretical paper, this is a paper which aims to promote a new kind of thinking towards design of multivariable controllers and to lay the basis for further theoretical work. What little theory is thus presented here, is merely to show how powerful and intuitive this approach to multivariable controller design can indeed be.

2. CONTROL STRUCTURES

In the previous section a definition of structure was given. This definition is generally accepted within the control community with slight variations. Controller classification based on structure is not a recent trend, however has only been applied on a limited scale. Namely, classification of multivariable controllers into decentralized and centralized controllers. A centralized controller is one where full cross-coupling occurs. The structure of a decentralized controller is however constrained, namely it is either diagonal, or block diagonal.

$$C_d(s) \in \mathbf{C}^{n \times n} = diag \left\{ \begin{array}{c} C_1(s) \in \mathbf{C}^{m_1 \times m_1} \\ \vdots \\ C_k(s) \in \mathbf{C}^{m_k \times m_k} \end{array} \right\} \quad (1)$$

$$where \quad \sum_{i=1}^{k} m_i = n$$

Note that in a decentralized controller there is no restriction on the dynamical order of the non-zero elements. Indeed it has been shown that the optimal (decentralized) controller is in general of infinite order and may be non-unique. A major disadvantage of decentralized control is that in general its performance is dictated by the plant rather than the controller it-self. A decentralized controller will often 'only' yield satisfactory results if the plant is 'only' loosely interaction and poses a diagonal dominant structure to start with. Most multivariable plant cannot be considered as such and this is why the use of decentralized control is limited.

Assuming a decentralized controller is unable to meet the response demands, the only way to meet them, is to use a centralized controller. A centralized controller will almost certainly meet the reference responses however, it leaves a basic question unanswered. What if, in between having to go from a decentralized controller to a centralized one, there is another controller structure which will satisfy the response criteria? Worded differently, one could ask, given an $n \times n$ decentralized controller has n elements and an $n \times n$ centralized controller has n^2 elements, how many of the additional $n(n-1)$ elements are in fact needed to meet the given response criteria? Further, of those needed, how complex do they have to be? Surprisingly its often the case that the majority of performance gained from moving from a decentralized controller, is in fact due to a minority of the additional $n(n-1)$ elements. This leads rather naturally to Minimal Structure control [4] .

A Minimal Structure controller can be defined as such; It is a controller that can deliver the desired responses *closely* with the minimum number of cross-coupling *and* the minimum dynamic on individual elements if the controller. Note the use of the term closely. It is to highlight the fact that often engineering is a process of measuring compromise against gain, and in many places insisting that the controller produce responses exactly equal with the references can be unrealistic and unnecessary specially considering that even of the controller does produce such responses they

[2] Three major Classical multivariable techniques are Diagonal Dominance, Root Locus method and sequential loop closing.

[3] By I_i/O channel, it is meant outputs effected by input i.

[4] The simplicity with which the problem can be outlined should not be deceiving as to how complex the problem actually is. What Minimal Structure ultimately aims to do is to walk a fine line between design and synthesis, which to say the least, is incredibly difficult to formulate theoretically.

will almost certainly change when it is implemented for real. Also note that a controller in this sense can be minimal towards the specifications. It would also be possible to quantify this compromise. For example repose the question this way: "What is the controller with minimum structure that can deliver system responses within 10% of the reference ones. In this case, the 10% represents the maximum deterioration one is prepared to tolerate

A Minimal Structure controller may or may not be a decentralized controller [5], however all Minimal Structure controllers are a subset of centralized controllers. This points towards a possible method of finding a Minimal Structure controller. Namely in a process similar to model order reduction, where-by a complex model is replaced by its equivalent order reduced model, a centralized controller can be reduced to a Minimal Structure controller by identifying elements which can be set to zero, or those that contain unnecessary dynamics, subject to the constraint that the performance is not comprised more than one is prepared to tolerate . Naturally, this process will be called *structure reduction.* Setting out the problem this way has one clear advantage; it allows use of already established design techniques for the initial design part, and new tools will only have to be developed for the second part.

Before proceeding to the next section, an example is given here to stress the critical difference between Minimal Structure control and simply structured control. Consider for example the PI/D controller which is by far the most widely used controller in the industry, and the reason for this favoring has been largely its simple structure. In most cases the PI/D can indeed be considered as a Minimal Structure controller since removal of each term will result in marked performance deterioration [6]. However consider the case where the plant contains a double integrator in the forward path. Then it is no longer necessary to have an I term for zero steady state error to step inputs, and therefore the controller will loose its minimality attribute. Therefore whereas all other controller classifications are solely determined by the controller itself, in the context of Minimal Structure, this is influenced equally by the plant.

[5] A decentralized controller would only be Minimal Structure if it can be shown that not only it is capable of delivering the response demands, but also any further reduction in the dynamical order of each diagonal element would severely deteriorate performance.

[6] Assuming that it, the full controller meets the response demands

3. MINIMAL STRUCTURE CONTROL

In the previous section a two stage structure reduction process was hypothesized for obtaining a Minimal Structure controller. In the first stage a centralized controller would be designed and the second stage would be the structure reduction stage. Two issues are raised in stage one, which can greatly affect the difficulty or the ease of stage two. Namely the technique used to design the centralized controller and the issue of input-output pairing. Both these issues will be expanded upon in the next section.

3.1 Stage 1-Centralized Design

It is obvious how design of the centralized controller, could affect stage two. The more unneeded complexity is introduced to the controller initially, the more they have to be identified and removed in the second stage. It is therefore imperative that the initial design technique produces 'simple' controller, which of course as a criteria would have to meet the response demands. In this work, focus is exclusively made on one multivariable control design technique, namely diagonal dominance, as this is one of the only design techniques that allows the designer explicit control over elements of the multivariable controller. This makes diagonal dominance inherently an easy to use multivariable design technique when there are structural constraints present. Further, being a design technique as opposed to a synthesis method, means it brings with itself all the well rehearsed advantages.

Diagonal dominance (Rosenbrock, 1974), is a design technique that converts a multivariable design problem into several SISO design problems which can be solved using a vast amount of SISO design techniques available. For a plant $G(s) \in \mathbf{C}^{n\times n}$, this involves finding a pre-compensator $K(s) \in \mathbf{C}^{n\times n}$, such that the resulting open loop system $Qd(s) = G(s)K(s)$ satisfies,

$$| q_{ii}(s) | > \sum_{\substack{j=1 \\ j\neq i}}^{n} | q_{ji}(s) | \tag{2}$$

If such $K(s) \in \mathbf{C}^{n\times n}$ is found, for the second stage of the design process, $Qd(s)$ may be replaced by $\acute{Q}_d(s)$ where $\acute{Q}_d(s) = diag\{Qd(s)\}$. Next a diagonal controller $Kc(s)$ is found such that $kc_{ii}(s)\acute{qd}_{ii}(s) \sim ref_i(s)/(1 - ref_i(s))$ for $i = 1, \ldots, n$ where $ref_i(s)$ denotes the reference function for loop i. Note since $\acute{qd}_{ij}(s)kc_{ij}(s) = 0$, $i \neq j$ design of $Kc(s)$ can be broken into n SISO design problems. The final return ratio is thus $Q(s) = G(s)K(s)Kc(s)$ and the multivariable controller is $C(s) = Kc(s)K(s)$. Also note that if $\exists K(s)$ such that 2 is satisfied, then 2 is also satisfied for $Kc(s)Q_d(s)$ since $Kc(s)$ multiplies a column of $Q_d(s)$ by the same gain.

The designer can be given explicit control over the structure of each element of $K(s)$ and $Kc(s)$.[7] This means not only structural constraints are easily seen to (for example limiting the dynamics of each entry to be of no more than second order), it also means one can guarantee that $K(s)$ will not contain RHP zeros or poles, unstable pole-zero cancellation will not happen between $G(s)$ and $K(s)$ and non-proper transfer functions do not appear in $K(s)$ or $Kc(s)$. However, it is easy to see how redundant dynamics could be introduced in even a dominance based controller. The criteria for design of $Kc(s)$ is solely that $kc_{ii}(s)\acute{q}d_{ii}(s) \sim ref_i(s)/(1 - ref_i(s))$ for $i = 1, \ldots, n$. However, when the final $Kc(s)$ transfer function matrix is multiplied by $K(s)$ a whole column of $K(s)$ will be multiplied by the corresponding transfer function in $Kc(s)$.

Input output paring was also mentioned as a factor which could effect the final outcome of this process. It has been suggested that if one is not designing a decentralized controller, the issue of input output paring is not as critical. Strictly speaking this is not a wise engineering approach. Consider for example the RGA (Bristol, 1966) of a decoupling controller. For $G(s)$ the plant and $K(s)$ the pre-compensator, define the column dominance ratio as,

$$CDR = \frac{\sum_{i \neq j} \mid qd_{ij}(jw) \mid}{\mid qd_{jj}(jw) \mid} \tag{3}$$

where $Q_d(s) = G(s)K(s)$. For $CDM \rightarrow 0$, $K(s) \rightarrow D(s)\hat{G(s)}$ where $D_{ij}(s) = 0$ for $i \neq j$. $\Gamma_1(s) = G(s) \otimes \hat{G}^T(s)$ and $\Gamma_2(s) = K(s) \otimes \hat{K}^T(s) = D(s)\hat{G(s)} \otimes inv(D(s)\hat{G(s)})^T(s)$. Since the RGA is not effected by diagonal input or output scaling it follows $\Gamma_2(s) = \hat{G(s)} \otimes inv(\hat{G(s)})^T(s) = G(s)^T \otimes \hat{G(s)} = (G(s) \otimes \hat{G}^T(s))^T = \Gamma_1(s)^T$.[8]

This suggests for a decoupling controller the RGA of the controller is largely dictated by the RGA of the plant and although in general $CDR \not\rightarrow 0$, non-the-less before the controller is actually synthesized a great deal can be inferred from its structure by just looking at the plants RGA and this correlation can be expected to be very high in the lower frequency regions where it is typically easier to achieve very high amounts of dominance. This means if the plants inputs and outputs are ordered such that it will correspond to an RGA with minimized interactions, then the controller will also have minimum amount of interactions. This certainly indicates that I/O pairing is essential in a Minimal Structure control approach. Other reasons for I/O pairing are better over all integrity of the system and issues concerning fault, loop failures etc... The reader can refer to (Maciejowski, 1989) and (Skogestad and Postlethwaite, 1996) for more details.

[7] However, there is no explicit control over elements of C(s).

[8] Where $\otimes$ denotes the element by element shur product.

3.2 Stage 2-Structure Reduction

Using the diagonal dominance approach (or other similar type of design techniques), one is in possession of a centralized controller which can already be clarified as simply structured, however not minimal by our definition. Here, simply structured refers to the fact that the elements of this controller are typically of not higher than a second order structure, as opposed to containing full dynamics of the plant. In order for the minimal controller to be obtained two questions should be asked,

- Are there any cross-coupling channels that can be removed without without deteriorating performance more than the tolerance level?
- Are there any dynamic cross-coupling channels which when replaced by a constant gain will not deteriorate performance more than the tolerance level?[9]

In what follows two basic tools will be proposed to be able to answer these questions which must be viewed as merely a first attempt at this problem.

3.2.1. The Nominal Interactions Numbers Array (NINA) The NINA plot is proposed here as one method of identifying redundant cross-coupling in a controller. Consider for example the effect of removing cross-coupling path $k_{\alpha\beta}$. Denote the diagonal dominant transfer function as $Qd(s)$ before and $\tilde{Q}d(s)$ after setting $k_{\alpha\beta} = 0$. Initially it is assumed that dominance will *not* be effected as a result of setting $k_{\alpha\beta} = 0$.[10] Essentially this means assuming $qd_{i\beta}(s) \stackrel{\circ}{=} \tilde{qd}_{i\beta}(s)$ for $i \neq \beta$. Therefore for $qd_{\beta\beta}(s) \stackrel{\circ}{=} \tilde{qd}_{\beta\beta}(s)$, it is required,

$$\mid k_{\alpha\beta} g_{\beta\alpha} \mid \ll \sum_{\substack{i \neq \alpha \\ i=1}}^{n} \mid k_{i\beta} g_{\beta i} \mid \tag{4}$$

[9] Strictly speaking this second question has been simplified here. What really should be asked is "From those channels which cannot be set to zero, how much dynamics can one remove before the tolerance criteria is violated". Posing the question this way however, would introduce far too much complexity at this early stage.

[10] This assumption will be removed later

This is however assuming that dominance is not effected. Next one needs to establish how much $k_{\alpha\beta}$ contributes to overall achieving of diagonal dominance. Contribution of $k_{\alpha\beta}$ to all other outputs excluding the first is,

$$\sum_{\substack{i \neq \alpha \\ i=1}}^{n} | k_{\alpha\beta} g_{i\alpha} | \quad (5)$$

and therefore provided,

$$| k_{\alpha\beta} | \sum_{\substack{i \neq \beta \\ i=1}}^{n} | g_{i\alpha} | \ll \sum_{\substack{j \neq \alpha \\ j=1}}^{n} | k_{j\beta} | \sum_{\substack{i \neq \beta \\ i=1}}^{n} | g_{ij} | \quad (6)$$

removing $k_{\alpha\beta}$ will not effect diagonal dominance of column 1. (Note changing of $k_{\alpha\beta}$ can only effect column 1).

Inequalities (4) and (6) can be combined in a single graphical criteria, the NINA plot, for determining of potential cross-coupling which can be eliminated. NINA is a frequency dependant array whose elements are defined as follows,

$$NINA(jw) = | K(jw) | \otimes H(jw) \otimes \Omega \quad (7)$$

where,

$$h(jw)_{jk} = \sum_{i=1}^{n} | g_{ij}(jw) | \quad k = 1, \ldots, n \quad (8)$$

$$\hat{\Omega}_{ii} = \sup_{w} \| \, | k_{ij}(jw) \, | h_{ij}(jw) \, \|_{\infty} \quad (9)$$

Elements in NINA which are notably smaller that other elements over the bandwidth , correspond to elements of the pre-compensator which when removed do not considerably effect either the amount of diagonal dominance, nor the response of its corresponding output. Here the bandwidth is denoted by ω_b and is defined by $\omega_b = \{w \mid \bar{\sigma}(G(jw)) = 0.7\bar{\sigma}(G(0))\}$.

We also propose the cumulative NINA (cNINA), which is a slight modification that turns NINA into a numerical, rather than graphical tool. Elements of cNINA are initially defined as follows,

$$cNINA_{ij} = \int_{0}^{\omega_b} NINA_{ij}(jw)dw \quad (10)$$

and using the basic approximation for numerical integration,

$$cNINA_{ij} = \sum_{k=2}^{f} nina_{ij}(jw_k)(w_k - w_{k-1}) \quad (11)$$

where $w_f = \omega_b$. This discretization is also handy since computation is also a discrete process. Note also that although the error of the approximation is large, systematic errors in relative analysis have little significance.

Elements of cNINA indicate the total contribution a given cross-coupling channel makes to the total system response over the control bandwidth. It is arguably more accurate that the basic NINA since it 'averages' out short lived phenomena on the cross-coupling channels. Fuhrer unlike NINA which has to be plotted over w, cNINA returns just a single matrix making analysis easier.

Calculation of cNINA can be more simplified by taking advantage of properties of inequalities. Namely consider dropping the $_{ij}$ subscript and rewriting cNINA as,

$$cNINA = D \otimes U \quad (12)$$

where $u_{ij} = V$, $V = \{w_i \mid i = 1, \ldots, f\}$ and $d_{ij} = \{nina_{ij}(jw_k) \mid k = 1, \ldots, f\}$. Since given $ac < bc \Rightarrow a < b$,

$$sd_{ij} = \sum_{k=1}^{f} d_{ij}(w_k) \quad (13)$$

And deduce that an element which is smallest in SD is also smallest in cNINA, therefore cNINA can be redefined as follows,

$$cNINA_{ij} = \sum_{k=1}^{f} nina_{ij}(jw_k) \quad (14)$$

Which is computationally easier to calculate. Incidentally before this section is ended another use of cNINA is briefly mentioned. If the diagonal elements of cNINA are considerably smaller than the off diagonal terms, it suggest that the plant is not correctly I/O paired. this is so because if an input of a plant has been paired to an output which it does not interact with greatly, then significant cross-coupling activity would be present in the controller and this is exactly what the cNINA has been formulated to show. This difference being with the RGA that, the RGA is incapable of saying anything useful about the controller and the plant simultaneously. Namely for $CDR \to 0$ then $RGA\{Q(s)\} \to I$. Which is logical because if the controller minimized the interactions considerably then one can write $Q(s) = G(s)C(s) \doteq R(s)I$ where $R(s)$ is a diagonal matrix. Since the RGA is independent of scaling, $RGA\{Q(s)\} \doteq RGA\{R(s)I\} = RGA\{I\} = I$. Indeed the most effective way the RGA can be used to identify redundant cross-couplings is applying it to $K(s) \otimes G(s)^T$. Smallest elements of this are the smallest interacting channels on the responses of y_i/e_j, however for only $i = j$. In other words it assumes removing element $k_{\alpha\beta}$ only effects y_β. This is clearly not a trivial assumption.

3.2.2. The Deviation Matrix The standard deviation (std) is a measure of how much a variable varies from its nominal value. A variable which remains constant has a standard deviation of zero and an element which varies rapidly (and has large deviations) has a large std. Thus it is obvious why it is chosen to use the std at this point. The basic std is defined as follows,

$$std = \left(\frac{1}{n} \sum_{i=1}^{n} (x_i - \bar{x})^2 \right)^{\frac{1}{2}} \tag{15}$$

where,

$$\bar{x} = \frac{1}{n} \sum_{i=1}^{n} x_i$$

Note that similar to the variance, the std is not a normalized variable. for example just knowing the std of a variable is 1, is meaningless to asses deviation, unless its mean is also known. There are different way the std can be normalized to overcome this. One method is mean normalization which is defined as follows,

$$std_n = \left(\frac{1}{n} \sum_{i=1}^{n} \left(\frac{x_i - \bar{x}}{\bar{x}} \right)^2 \right)^{\frac{1}{2}} \tag{16}$$

Although this means the variable is now normalized with respect to its mean, however it is not useful for our purposes since it does not take into account the fact that variables are related. In other words this type of normalization is useful for comparing *totally unrelated* variables. In this case, a more practical approach is adopted which takes into account the multivariable nature of the problem and normalized such that the biggest std on each channel of the plant input is 1. Therefore elements of the deviation matrix are defined as follows,

$$dm_{ij} = 1 - sgn \left\{ \frac{v_{ij}}{\Xi_i} \right\} \tag{17}$$

where,

$$v_{ij} = \hat{f} \sum_{k=1}^{f} \mid \hat{g}_{ij}(jw_k) \mid - \overline{\mid \hat{g}_{ij}(jw_k) \mid} v \tag{18}$$

$$\Xi_i = \parallel \xi_i \parallel_\infty \tag{19}$$

$$\xi_i = [v_{i1}, v_{i2} \ldots v_{in}] \tag{20}$$

$$sgn\{x\} = \begin{cases} 1, & x \geq 0.5 \\ 0, & x < 0.5 \end{cases} \tag{21}$$

Interpretation of DM can be as follows. a zero entry means which elements cannot be set to constant and a one entry corresponds to elements which may be a constant. In other word an element of zero corresponds to elements which *cannot* be constant, but a one entry corresponds to elements which *may* be constant.

4. CONCLUDING REMARKS

The main aim and objective of this paper was to point attention towards a gap in the study of multivariable controller design and to raise awareness of possible ways to address this issue. Is it hoped that this work will be a first stepping stone for a ladder which will be extended by future works on this subject. Whether this paper together with its sister paper, do justice to this end, is up to the reader to decide, however the authors feels that the concept, that is of Minimal Structure control is simply to powerful to ignore and not to address.

It is also imperative that this not be viewed as a dominance based subject. Diagonal dominance was chosen here simply to demonstrate the versatility of how one can approach the problem, however, the primary question of Minimal Structure control is technique independent and in fact already works are under way in the Control Systems Centre to develop firmer theoretical basis for this, and also, perhaps more interestingly use modern optimisation tools such as the Genetic Algorithms to solve the problem of Minimal Structure control. That is to say to develop not a two stage design process as outlined here, but to use the GAs to formulate a one-step process. This latter part, namely use of GAs for this effect has shown very promising results which will no doubt be reflected in future publications.

In the second paper (Nobakhti and Munro, 2003), the tools presented in this work will be used to develop Minimal Structure controllers for a four-input four-output reheat furnace and also a two by two automotive gas-turbine engine.

REFERENCES

Bristol, E.H. (1966). On a new measure of interactipon for multivariable process control. *IEEE Transactions on Automatic Control* pp. 43–44.

Maciejowski, J.M. (1989). *Multivariable Feedback Design.* Addison-Wesley Publishing Company.

Nobakhti, A. and N. Munro (2003). Minimal strcuture control: Some examples). *IFAC Confrence on Control Systems Design (CSD'03)-Slovak Republic.*

Rosenbrock, H.H. (1974). *Computer Aided Control System Design.* Academic Press.

Skogestad, S. and Postlethwaite (1996). *Multivariable Feedback Control: Analysis and Design.* John Wilery and Sons.

www.elsevier.com/locate/ifac

MINIMAL STRUCTURE CONTROL: SOME EXAMPLES

Amin Nobakhti [1] **, Neil Munro**

Control Systems Centre, UMIST
Manchester M60 1QD, UK
multivariable@IEEE.org

Abstract: This is the second part of a two part paper on Minimal Structure control. In the first part, the concept of Minimal Structure control was outlined and explained. Further a possible approach was proposed to develop Minimal Structure controllers. This approach included use of two tools also proposed in the first paper. In this paper, this approach will be applied to two real-life examples and will be shown, even given the basic nature of this proposed initial approach, powerful results can be obtained. This approach will be used to to develop Minimal Structure controllers for a four-input four-output reheat furnace and also a two by two automotive gas-turbine engine. *Copyright © 2003 IFAC*

Keywords: Multivariable Control, Simply Structured, Minimum Structure

1. INTRODUCTION

In (Nobakhti and Munro, 2003) a technique was outlined to approach the problem of Minimal Structure control. It is important to emphasize that the proposal merely constituted a basic approach to the problem. It is thus natural to expect that it may be unable to fully unlock the power behind the idea of Minimal Structure control. Non-the-less even with these basic tools, it is shown in this paper how real-life Minimal Structure controllers can be developed.

Two system have been chosen in this paper as real-life test beds. Both are multivariable systems and both have notable interactions. The first system is a 4×4 model of a reheat furnace in a chemical reactor. This model has been the subject of numerous design studies including (Rosenbrock, 1974) and (Patel and Munro, 1982). These design studies ranged from a classical dominance based study to a decentralized controller design. Whilst in both the cases, the plant could be made to behave satisfactorily, the overall performance, specially that of the decentralized controller left much to be desired. A model for the furnace can also be found in (Patel and Munro, 1982).

The second system considered in this paper is a 2×2 model of an Automotive gas-turbines. Automotive gas-turbines were subject of increased research during the past decades, as it was thought they could offer a cleaner and better solution than the internal combustion engine which has effectively remained unchanged for a century. Although the Automotive gas-turbine is only 2×2, it is considerably more interacting than the reheat furnace. It is not possible to design a decentralized controller for this plant which would give an acceptable response by any standards, and even centralized basic dominance-based studies on this system have been faced with some difficulty.

[1] The author wishes to thank EPCRS, UMIST and the IEE for supporting this research.

In what follows two high performance Minimal Structure controllers will developed for these systems.

2. EXAMPLES

Please note the following notation which will be used for the reminder of this paper.

Expression	Definition
$G(s)$	Plant's transfer function
$K(s)$	Pre-compensator (dynamic)
$Km(s)$	minimal pre-compensator
$Kc(s)$	Diagonal controller
$C(s)$	$Kc(s)K(s)$
$Cm(s)$	$Kc(s)Km(s)$
$Q(s)$	$G(s)C(s)$
$Qc(s)$	$Q(s)(I+Q(s))^{-1}$
$Qd(s)$	$G(s)K(s)$
$Qdm(s)$	$G(s)Km(s)$
$Qm(s)$	$G(s)Cm(s)(I+G(s)Cm(s))^{-1}$
DNA	Direct Nyquist Array

Table 1. List of adopted notations

2.1 Gas-Fired Reheat Furnace

A brief introduction was given to this system in the first section. Figure 1 shows the schematics of this system, including all the inputs and outputs to be controlled.

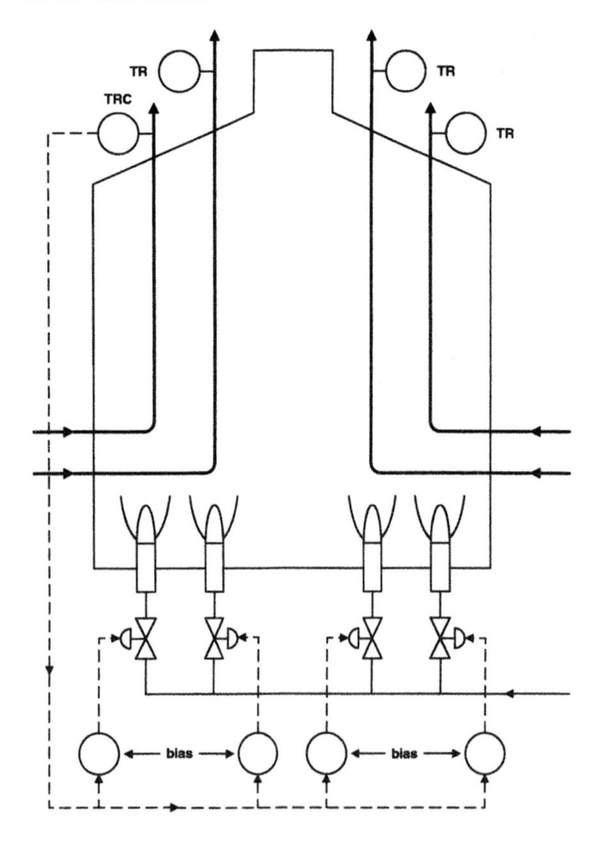

Fig. 1. Schematics of the reheat Furnace

The direct Nyquist Array of $G(s)$ and its step response are shown in Figures 2 and 3 respectively.

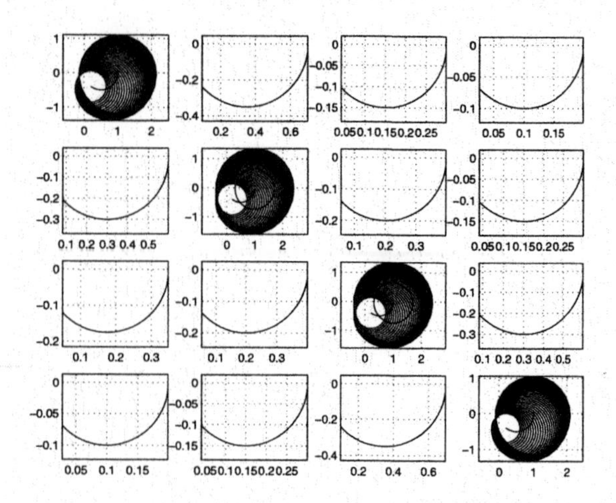

Fig. 2. DNA of $G(s)$ - Reheat Furnace

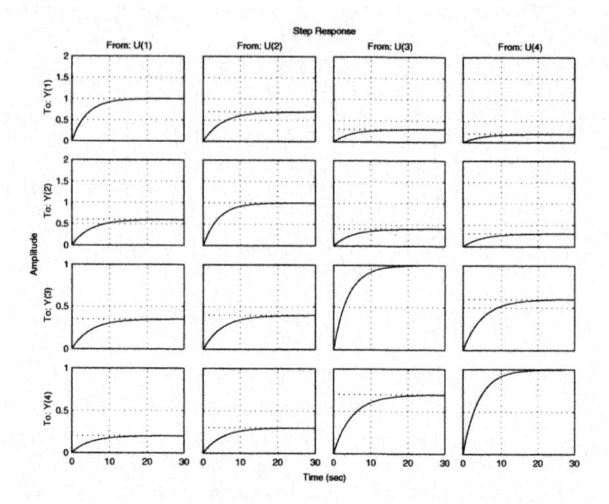

Fig. 3. Step response of $G(s)$ - Reheat Furnace

Using a technique outlined in (Nobakhti, 2002) a dynamic pre-compensator is found for the system to be as follows,

$$K(s) = \begin{pmatrix} \frac{1.6138(s+0.26)}{(s+0.19)} & \frac{-0.8832(s+0.26)}{(s+0.15)} & \frac{-0.13009(s+0.08)}{(s+0.095)} & \frac{0.0081577}{(s+0.1)} \\ \frac{-0.71064(s+0.29)}{(s+0.17)} & \frac{1.7212(s+0.22)}{(s+0.16)} & \frac{-0.28567(s+0.23)}{(s+0.15)} & \frac{-0.15796(s+0.27)}{(s+0.21)} \\ \frac{-0.25505(s+0.225)}{(s+0.14)} & \frac{-0.23234(s+0.3)}{(s+0.24)} & \frac{1.7169(s+0.24)}{(s+0.17)} & \frac{-0.72426(s+0.3)}{(s+0.17)} \\ \frac{0.053284(s+0.4)}{(s+0.1)} & \frac{-0.13565(s+0.25)}{(s+0.17)} & \frac{-0.87672(s+0.3)}{(s+0.172)} & \frac{1.6325(s+0.23)}{(s+0.17)} \end{pmatrix}. \quad (1)$$

The DNA of $Qd(s)$ and its step response are shown in Figures 4 and 5 respectively. The interactions are clearly suppressed by $K(s)$ and the system is almost decoupled. Next a suitable $Kc(s)$ is designed to eliminate the steady state errors and to tune the responses. A good choice was found to be,

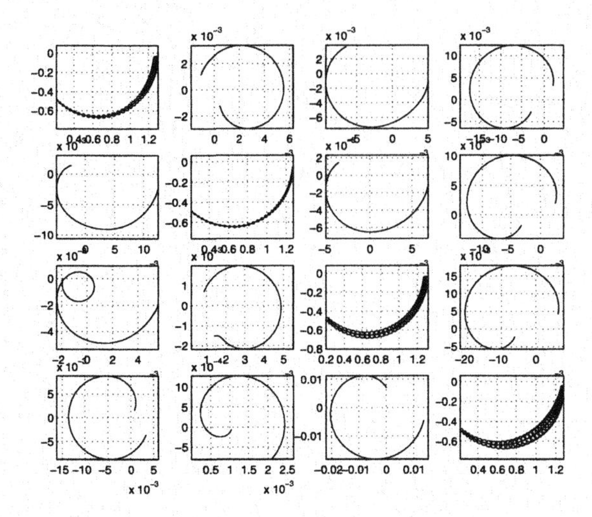

Fig. 4. DNA of $Qd(s)$ - Reheat Furnace

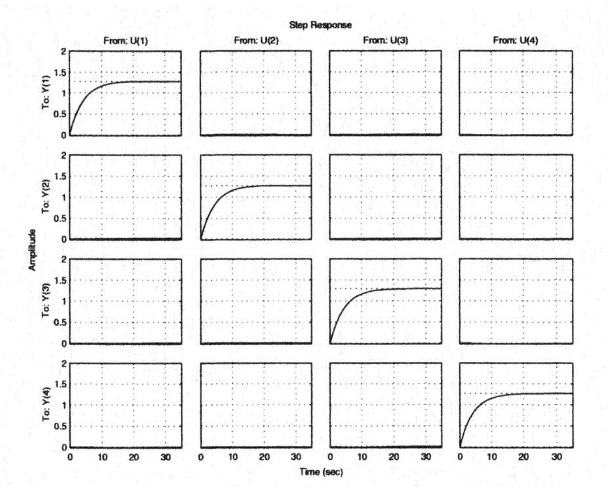

Fig. 5. Step response of $Qd(s)$ - Reheat Furnace

$$kc_{ii}(s) = \frac{3.7s + 0.86}{s} \quad \text{for } i = 1, \ldots, n. \qquad (2)$$

The DNA of $Qc(s)$ and its step response are shown in Figures 6 and 7 respectively. It is seen that the responses of the system are extremely good. The next stage is to find $Km(s)$ via a process of structure reduction as outlined previously. Consider the NINA plot of the system shown in Figure 8 and the cNINA matrix shown in Equation (3).

$$cNINA = \begin{pmatrix} 40.6194 & 28.0308 & 2.4252 & 1.2161 \\ 19.2234 & 40.1804 & 7.4363 & 3.3791 \\ 6.1527 & 4.8661 & 40.1506 & 19.2968 \\ 3.2703 & 3.7701 & 28.9440 & 40.4459 \end{pmatrix} \quad (3)$$

Both of these suggest that at least five elements of $K(s)$ may be set to zero completely, namely; $\{k_{13}, k_{14}, k_{24}, k_{41}, k_{42}\}$. Also note that the diagonal of the cNINA matrix contains the highest entries . This as mentioned in the first paper indicates correct I/O pairing.

Next consider the DM matrix shown in (4). Note its highly interesting structure. It suggests that (with one exception) the only elements which need to be dynamics are the diagonal elements. This matches very well with experimental results from before, which found that closing the loops with just PIs would achieve diagonal dominance. However as mentioned here the emphasis is not on just achieving dominance.

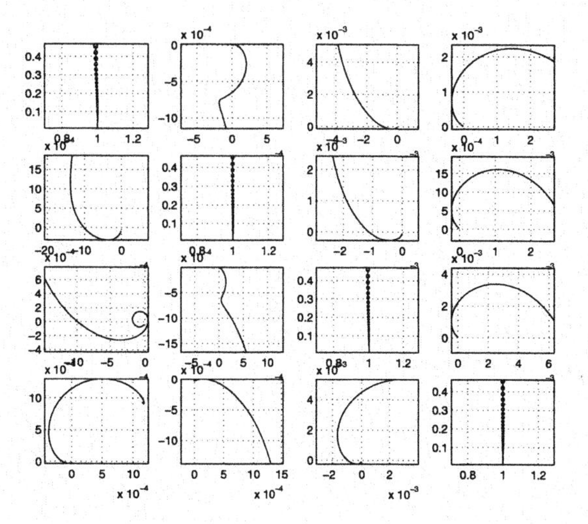

Fig. 6. DNA of $Qc(s)$ - Reheat Furnace

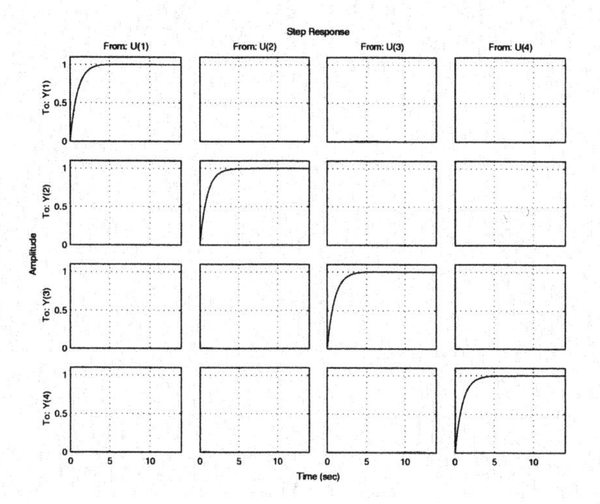

Fig. 7. Step response of $Qc(s)$ - Reheat Furnace

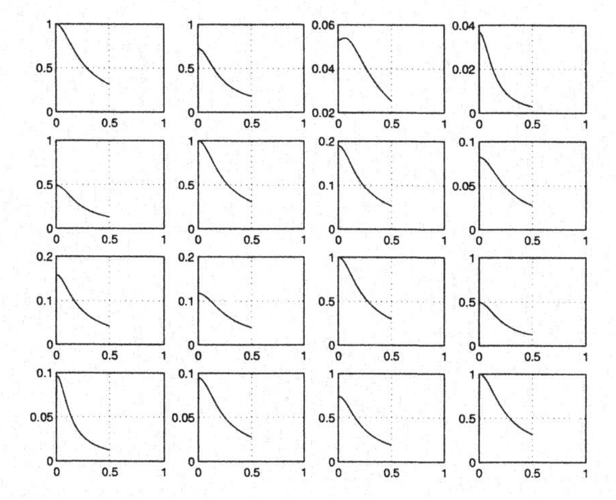

Fig. 8. NINA plot - Reheat Furnace

$$DM = \begin{pmatrix} 0\,1\,1\,1 \\ 1\,0\,1\,1 \\ 1\,1\,0\,0 \\ 1\,1\,1\,0 \end{pmatrix} \tag{4}$$

Consider now subdividing the DM matrix into four 2×2 subsystems as shown in (5). Notice how only the off-diagonal subsystems have 1 entries.

$$DM = \left(\begin{array}{cc|cc} 0 & 1 & 1 & 1 \\ 1 & 0 & 1 & 1 \\ \hline 1 & 1 & 0 & 0 \\ 1 & 1 & 1 & 0 \end{array}\right) \tag{5}$$

This indicates that in fact the elements of these two subsystems can all be set to constant. Since elements $\{k_{13}, k_{14}, k_{24}, k_{41}, k_{42}\}$ are already known to be zero form the NINA analysis, this means there are only three elements to be modified. Namely $\{k_{31}, k_{31}, k_{23}\}$ which were set to $\{-0.249, -0.229, -0.249\}$ respectively. The values were found using the Pseudo-diagonalization algorithm. The resulting, structure reduces $Km(s)$ is therefore,

$$Km(s) = \left(\begin{array}{c|c} Km_{11}(s) & Km_{12} \\ \hline Km_{21} & Km_{22}(s) \end{array}\right) \tag{6}$$

where,

$$Km_{11}(s) = \begin{pmatrix} \dfrac{1.6138(s+0.26)}{(s+0.19)} & \dfrac{-0.8832(s+0.26)}{(s+0.15)} \\ \dfrac{-0.71064(s+0.29)}{(s+0.17)} & \dfrac{1.7212(s+0.22)}{(s+0.16)} \end{pmatrix} \tag{7}$$

$$Km_{22}(s) = \begin{pmatrix} \dfrac{1.7169(s+0.24)}{(s+0.17)} & \dfrac{-0.72426(s+0.3)}{(s+0.17)} \\ \dfrac{-0.87672(s+0.3)}{(s+0.172)} & \dfrac{1.6325(s+0.23)}{(s+0.17)} \end{pmatrix} \tag{8}$$

$$Km_{21}(s) = \begin{pmatrix} -0.229 & -0.249 \\ 0 & 0 \end{pmatrix} \tag{9}$$

$$Km_{12}(s) = \begin{pmatrix} 0 & 0 \\ -0.249 & 0 \end{pmatrix} \tag{10}$$

Figure 9 shows the response of $Qm(s)$ where it can be confirmed that the interactions have increased only slightly (3%), well worth the extra structure that was removed from $K(s)$.

2.2 *Automotive Gas turbine (AGT)*

A brief introduction was given to this system in the first section. Figure 10 shows the schematics of

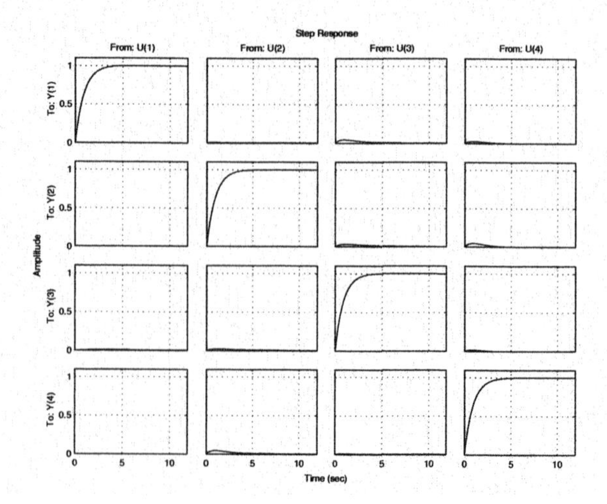

Fig. 9. Step response of $Qm(s)$ - Reheat Furnace

this system, including all the inputs and outputs to be controlled. The Automotive gas-turbine is

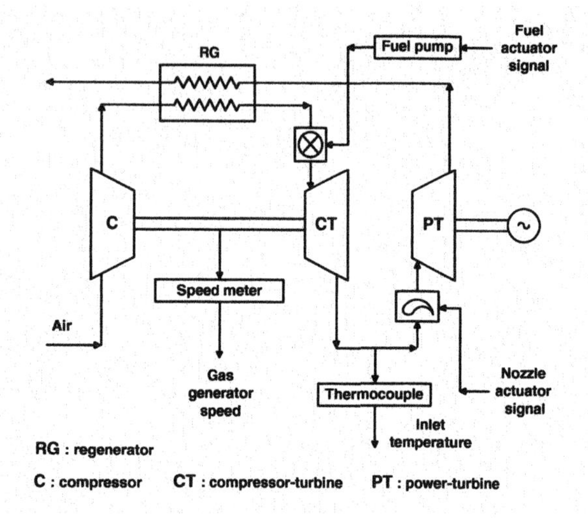

Fig. 10. Schematics of the reheat Furnace

a 2×2 system whose inputs are: fuel pump excitation and nozzle actuator and its outputs are gas generator speed and the inter-turbine temperature. Although it is only of 12 states, it is a particularly difficult system. The DNA of $G(s)$ and its step response are shown in Figures 11 and 12 respectively. The model of the system is as follows,

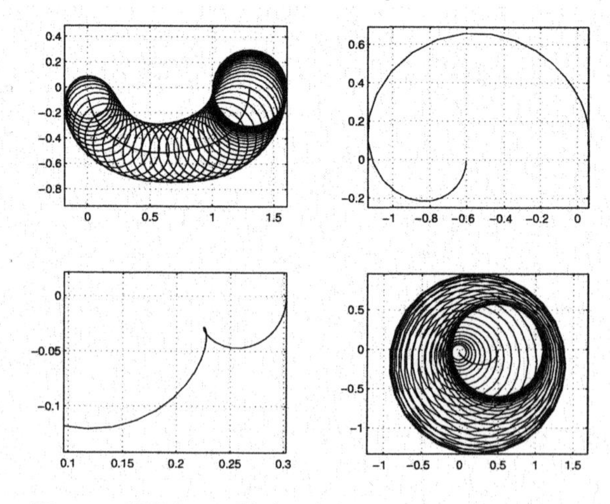

Fig. 11. DNA of $G(s)$ - AGT

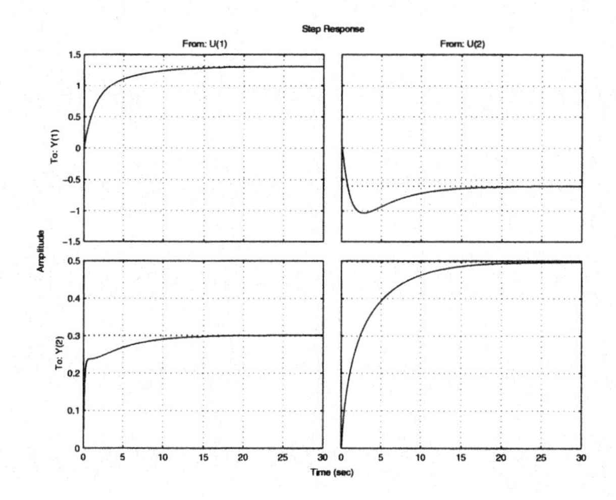

Fig. 12. Step response of $G(s)$ - AGT

$$G(s) = \begin{pmatrix} \frac{0.806s+0.264}{s^2+1.15s+0.202} & \frac{-(15s+1.42)}{s^3+12.8s^2+13.6s+2.36} \\ \frac{1.95s^2+2.12s+0.49}{s^3+9.15s^2+9.4s+1.6} & \frac{7.14s^2+25.8s+9.4}{s^4+20.8s^3+116s^2+111.6s+18.8} \end{pmatrix} \quad (11)$$

Using a technique outlined in (Nobakhti, 2002) a dynamic pre-compensator is found for the system which achieved very high amount of diagonal dominance. The pre-compensator is as follows,

$$K(s) = \begin{pmatrix} \frac{0.049815(s+12)}{(s+34)} & \frac{0.86302(s+0.1)}{(s+0.2)} \\ \frac{-0.40395(s+0.9)}{(s+7)} & \frac{1.5545(s+0.4)(s+10)}{(s+0.23)(s+30)} \end{pmatrix}. \quad (12)$$

The DNA of $Qd(s)$ and its step response are shown in Figures 13 and 14 respectively. The interactions are clearly suppressed by $K(s)$. Note at this point one is not concerned with the shape of the response of the diagonal elements. In design of $K(s)$ the sole factor which is of importance is reduction of interactions to preferably as high as possible, or to at least sufficient levels to satisfy the diagonal dominance criteria. What is *not* important and is the subject of the next stage (loop shaping via design of $Kc(s)$) is the shape of the responses themselves. To find $Km(s)$, $K(s)$ will undergo a structure reduction process as outlined in the text previously. Consider first the NINA plot shown in Figure 15 and the cNINA matrix shown in (13), both of which suggest that no cross-coupling channel can be removed.

$$cNINA = \begin{pmatrix} 38.5 & 13.8 \\ 35.1 & 30 \end{pmatrix} \quad (13)$$

This makes sense since for a 2×2 system if any of the cross-coupling is removed it would result in a triangular structure, and it is easy to prove that a triangular controller can only decouple a triangular system. Now consider the DM matrix shown in (14) which suggest that the whole of the first column of $K(s)$ may be reduced to be just constants.

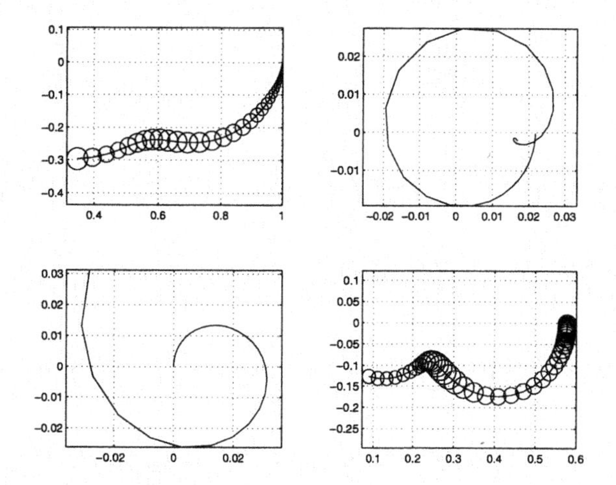

Fig. 13. DNA of $Qd(s)$ - AGT

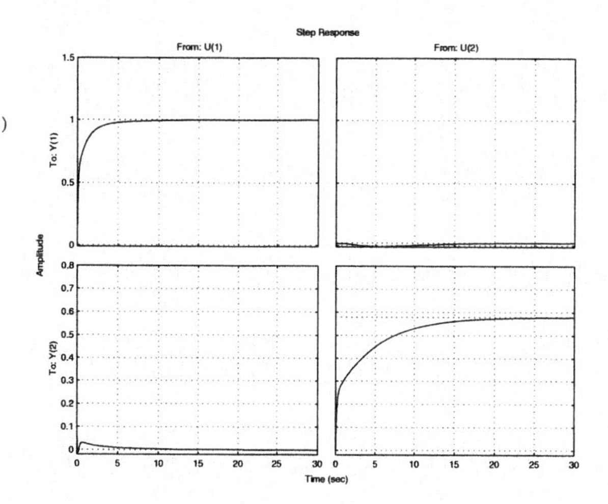

Fig. 14. Step response of $Qd(s)$ - AGT

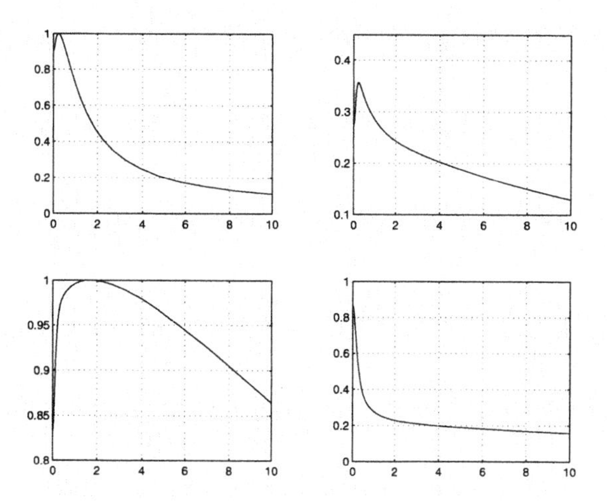

Fig. 15. NINA plot - AGT

$$DM = \begin{pmatrix} 1 & 0 \\ 1 & 0 \end{pmatrix} \quad (14)$$

The first column of $K(s)$ is thus, replaced by $[0.598, \ -0.364]^T$ and hence $Km(s)$ is as follows,

$$Km(s) = \begin{pmatrix} 0.598 & \dfrac{0.86302(s+0.1)}{(s+0.2)} \\ -0.364 & \dfrac{1.5545(s+0.4)(s+10)}{(s+0.23)(s+30)} \end{pmatrix}. \tag{15}$$

Figure 16 shows the response of $Qdm(s)$ where it can be confirmed that indeed very little change is present in the systems responses. Next one might wish to design a $Kc(s)$ to shape the response of each loop into the desired form. For example closing the loops with the $Kc(s)$ shown in Equation (16) will result in the system responses shown in Figure 17.

$$K(s) = \begin{pmatrix} \dfrac{s+1.5}{s} & \\ & \dfrac{12s^2+5.4s+0.6}{s^3+0.66s^2+0.112s} \end{pmatrix} \tag{16}$$

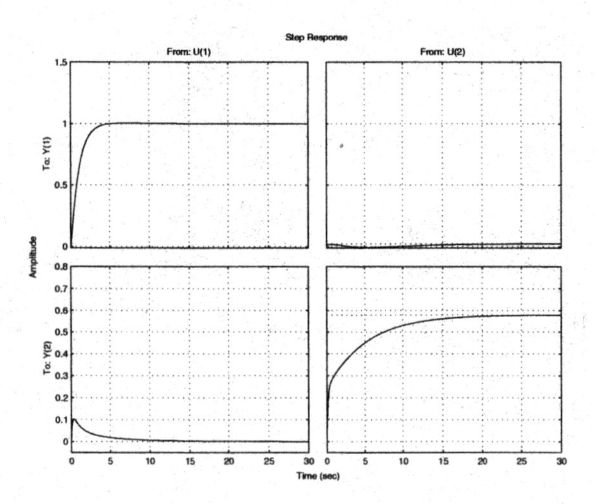

Fig. 16. Step response of $Qdm(s)$ - AGT

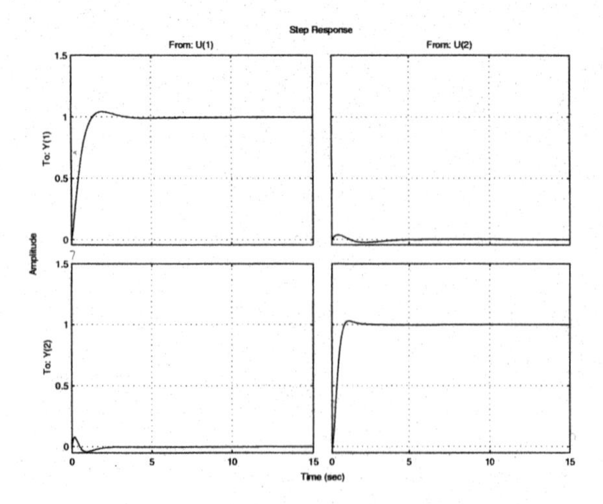

Fig. 17. Step response of $Qm(s)$ - AGT

3. FINAL REMARKS

In this paper, the techniques outlined in the first paper (Nobakhti and Munro, 2003) were successfully applied to two real-lifer systems. Given the elementary nature of the tools used, still the results obtained were interesting and certainly highlighted how effective this approach to design of multivariable controller could be. It is worth reiterating that the purpose of this work, outlined in two papers, is more to lay the foundations for a new kind of thinking towards deign of multivariable controllers, than to present any solid theory as such. Although some theory was indeed presented, the emphasis is more on developing a more industry conscious approach to design of multivariable controllers. The majority of the next generation of control technicians entering service are indeed familiar with most modern approaches to multivariable control and so there is no doubt that these techniques will become widely used and accepted, however during this incubation period, there are real and existing multivariable problems that need to be addressed and solved in a manner which does not deviate much from what the practicing engineer is used to deal with and which also has in mind the issues which are of primary concern to the end users of these systems.

As for the concept of Minimal Structure control, indeed there are many more questions to be asked and many more theories to be developed. As mentioned in the first paper works are under way in the Control Systems Centre to develop firmer theoretical basis for this, and also, perhaps more interestingly use modern optimisation tools such as the Genetic Algorithms to solve the problem of Minimal Structure control. That is to say to develop not a two stage design process, but to use the GAs to formulate a one-step process. Result found in this light will be reflected in due course is future publications.

REFERENCES

Nobakhti, A. (2002). Procedure for systematic design of dynamic pre-compensators. *Proceedings of the 2002 UKACC Conference Control-Sheffield.*

Nobakhti, A. and N. Munro (2003). Minimal strcuture control: A proposition. *IFAC Confrence on Control Systems Design (CSD'03)-Slovak Republic.*

Patel, R.V. and N. Munro (1982). *Multivariable System Theory and Design.* Pergamon Press. Oxford.

Rosenbrock, H.H. (1974). *Computer Aided Control System Design.* Academic Press.

ON ACHIEVABLE BEHAVIORS AND CONTROLLABILITY OF DISCRETE EVENT SYSTEMS IN A BEHAVIORAL FRAMEWORK

Takashi Misaki * **Osamu Kaneko** * **Takao Fujii** *

* *Graduate school of Engineering Science, Osaka University, Machikaneyama, Toyonaka, Osaka, 560-8531, Japan*

Abstract: The purpose of this paper is to investigate the controllability and the achievability of discrete event systems within a behavioral framework. Based on the notion of Willems' behavioral controllability (cf. (Willems, 1989), (Willems, 1991)), we introduce the new concept related to the controllability of discrete event systems. By using the controllability proposed here and the notion related to achievable behaviors (cf. (van der Schaft and Julius, 2002)), we show that the behavioral controllability for a given specification with respect to *language* is equivalent to the existence of a controller so that an interconnected system satisfies the specification *exactly* . A proposed controller here is represented by the intersection of the behavior of a given plant and that of a given (controllable) specification. In order to show the validity of our results, we also clarify that our controllers for a given specification fits the properties of well-known supervisory controllers proposed and developed by Ramadge and Wonham (cf. (Ramadge and Wonham, 1987) *Copyright © 2003 IFAC*

Keywords: Behavioral approach, Discrete event systems, Achievable behaviors, Controllability, Supervisory control

1. INTRODUCTION

In this study, we consider one of the important properties in the discrete event systems theory within a behavioral framework. Particulary, this paper addresses the controllability of discrete event systems within a behavioral framework.

Recently, Willems has proposed *behavioral approach*, which provides a novel viewpoint for dynamical system theory (cf. (Willems, 1989), (Willems, 1991),(Willems, 1997)). In the behavioral approach, a dynamical system is regarded as a set of trajectories that are compatible with the physical laws determining the dynamics of a system. In this approach, the behavior of a system represents intrinsic properties of a dynamical system, while differencial and difference equations play roles as the restrictions to describe the behavior. Thus, the behavioral concept enables us to synthesize and analyze dynamical systems whose behavior can not be represented as solutions of certain equations. On the other hand, in discrete event systems, the dynamics of a system is evolved by the occurence of events (cf. (Cassandras and Lafortune, 1999)). Since such systems take discrete value, asynchronous and nondeterministic, it is difficult to describe their dynamics by using certain equations. Alternatively, it is natural to focus on the set of the trajectories of a system, i.e., the behavior, directly. Thus, it is important and meaningful to formalize and investigate the dynamics of discrete event systems within the behavioral framework. However, little is studied about discrete event systems within a behavioral framework except the works by Raisch et al. (e.g. (Raisch and O'Young, 1998)) or Moor et al. (e.g. (T.Moor and Raisch, 1999)).

By the way, the notion of the controllability is one of the essential properties in dynamical systems. As for the standard system theory, e.g. in the case of system that is finite dimentional, linear and time invariant, the controllability is defined as one of the properties of state variables. However, the role of the state is to store the information of past external trajectories, so the intrinsic properties depending on the state of a system should be clarified with respect to the external behavior directly. As stated above, in discrete event systems there are many cases in which the dynamics are not described by differential or difference equations but by only the sequence of events. From the theoretical point of view, it is interesting to clarify the controllability with respect to the external behavior.

From these reasons, the purpose of this paper is to investigate the controllability and the achievability of discrete event systems within a behavioral framework. Based on the notion of Willems' behavioral controllability (cf. (Willems, 1989), (Willems, 1991)), we introduce the new concept related to the controllability of discrete event systems. By using the controllability proposed here and the notion related to achievable behaviors (cf. (van der Schaft and Julius, 2002)), we show that the behavioral controllability for a given specification with respect to *language* is equivalent to the existence of a controller so that an interconnected system satisfies the specification *exactly* . A proposed controller here is represented by the intersection of the behavior of a given plant and that of a given (controllable) specification. In order to show the validity of our results, we also clarify that our controllers for a given specification fits the properties of well-known supervisory controllers proposed and developed by Ramadge and Wonham (cf. (Ramadge and Wonham, 1987)).

2. BEHAVIORS AND DISCRETE EVENT SYSTEMS

In this section, we introduce some notations, definitions and concepts required throughout this paper. Particulary, we define discrete event systems that are studied in this paper by using the behavioral system theory proposed by Willems (cf. (Willems, 1989), (Willems, 1991)).

Let T denote an additive semi-group. For any abstract set, say $\mathbb{W}$, let $\mathbb{W}^T$ denote the set of all maps from T into $\mathbb{W}$. For $w \in \mathbb{W}^T$ and $t_1, t_2 \in T (t_1 \leq t_2)$, let $w|_{[t_1,t_2]}$ denote the sequence $[w(t_1), w(t_1+1), \cdots, w(t_2)]$. For $w_1, w_2 \in \mathbb{W}^T$ and $\tau \in T$, let $w_1 \wedge_\tau w_2$ denote the concatenation at time τ, i.e. $(w_1 \wedge_\tau w_2)(t) = w_1(t)$ for $t < \tau$ and $(w_1 \wedge_\tau w_2)(t) = w_2(t)$ for $t \geq \tau$.

Before going to defining discrete event systems in the behavioral setting, we introduce a dynamical systems within a behavioral framework as follows.

Definition 2.1. A dynamical system Σ is defined as a triple $\Sigma = (\mathbb{T}, \mathbb{W}, \mathfrak{B})$, where $\mathbb{T}$ is the time axis, $\mathbb{W}$ is the signal space and $\mathfrak{B} \subseteq \mathbb{W}^T$ is the manifest behavior.

In the standard linear differential (or difference) systems, the time axis T is the set of real numbers (integers, respectively). Since we treat discrete event systems in this paper, we assume that T is the set of natural numbers, denoted $\mathbb{N}_0$, throughout this paper. In discrete event systems, $\mathbb{N}_0$ is regarded as the logical time at which an event occurs. Consider a non-empty finite set $\mathbb{A}$, called *the alphabet* or *the event set.* Let $\Box$ denote the blank or the empty string. Here we assume that the set of events include the empty string. Thus, the signal space is $\mathbb{W} := \mathbb{A} \cup \Box$. Moreover, the set of events strings can be described by $\mathbb{W}^{N_0} := (\mathbb{A} \cup \Box)^{N_0}$. Let $\mathfrak{L}$ denote the set of all feasible finite sequences of events, that is to say, $\mathfrak{L} \subseteq W_{[1,\tau]}$ for finite $\tau \in \mathbb{N}_0$. In this paper, our purpose is to construct more generalized theory including formal languages and so on, so we regard that it is no matter whether the behavior is finite or not. Thus, discrete event systems treated in this paper can be formalized as follows:

Definition 2.2. A discrete event system Σ is defined as a triple $\Sigma = (\mathbb{N}_0, \mathbb{W}, \mathfrak{B})$, where $\mathbb{N}_0$ is the logical time axis, $W = \mathbb{A} \cup \{\Box\}$ is the signal space and $\mathfrak{B} \subseteq \mathbb{W}^{N_0}$ is the manifest behavior described by

$$\mathfrak{B} := \{w \in W^{N_0} | \exists \tau \in \mathbb{N}_0 \; s.t. \; w_{[1,\tau]} \in \mathfrak{L} \; and \; w(t) = \Box \; for \; t \in [\tau+1, +\infty]\}, \quad (1)$$

where $\mathbb{A}$ is the event set of Σ and $\mathfrak{L}$ is the set of all feasible finite sequences of events of Σ.

Roughly speaking, the above definition implies that there exists $\tau \in \mathbb{N}_0$ such that $\mathfrak{B}|_{[1,\tau]} = \mathfrak{L}$. Moreover, setting $\tau = \infty$ enables us to consider general discrete event systems, while it should be finite in the usual formal language theory. Let σ^t denote the backward t-shift, i.e. $(\sigma^t f)(\tau) := f(t + \tau)$ for all $\tau \in \mathbb{N}_0$.

Nextly, we introduce the notion of the *latent variables.* In addition to manifest variables, there are many cases in which some auxiliary variables are required to describe a dynamics. Such variables are said to be latent variables. A system with latent variables is defined analogously to Definition 2.2.

Definition 2.3. A dynamical systems with latent variables Σ_a is defined as a quadaple $\Sigma_a = (\mathbb{N}_0, \mathbb{W}, \mathbb{L}, \mathfrak{B}_a)$, where $\mathbb{L}$ is the signal space of latent variables and $\mathfrak{B}_a \subseteq (\mathbb{W} \times \mathbb{L})^{N_0}$ is the full behavior described by

$$\mathfrak{B}_a := \{(w,l) \in (\mathbb{W} \times \mathbb{L})^{N_0} | \exists \tau \in \mathbb{N}_0 \ s.t. \ (w,l)|_{[1,\tau]} \in \mathfrak{L}_a \ and \ (w(t),l(t)) = (\square,\square) \ for \ t \in [\tau+1,+\infty]\}, \quad (2)$$

where $\mathfrak{L}_a$ denotes the set of all feasible finite sequences of events with latent variables, i.e., $\mathfrak{L}_a \subseteq (\mathbb{W} \times \mathbb{L})|_{[1,\tau]}$ for a finite $\tau \in \mathbb{N}_0$.

Next, we introduce the notion of the Willems's behavioral controllability. In the behavioral system theory, the controllability is defined as the connectability of trajectories.

Definition 2.4. A dynamical system $\Sigma = (\mathbb{T}, \mathbb{W}, \mathfrak{B})$ is said to be controllable if for any $w_1, w_2 \in \mathfrak{B}$ there exist $\tau_1, \tau_2 \in \mathbb{T}(\tau_1 \leq \tau_2)$ and $w \in \mathfrak{B}$ such that $w(t) = w_1(t)$ for $t \leq \tau_1$ and $w(t) = w_2(t)$ for $t \geq \tau_2$.

The above definition of controllability implies the existence of free variables yielding the desired trajectory.

Nextly, we explain the notion of control in a behavioral framework (cf. (Willems, 1997)). In this framework, a control is regarded as an *interconnection* of the behaviors of each systems, one is a plant, the other is a controller. In (Kuijper, 1995), the detailed properties of the interconnections were studied for the case of linear and time invariant systems. In this paper, we treat discrete event systems, so we need to introduce the notion of the interconnection from the view of more abstract level.

Definition 2.5. Let $\Sigma_G = (\mathbb{N}_0, \mathbb{W} \times \mathbb{C}, \mathfrak{B}_f)$ denote a given plant with the full behavior. And let $\Sigma_C = (\mathbb{N}_0, \mathbb{C}, \mathfrak{C})$ denote a system with the behavior whose signal space is $\mathbb{C}$. Then, the interconnected system of these two discrete event systems $\Sigma := (\mathbb{N}_0, W, \mathfrak{B}_{cl})$ is said to be the closed loop system, where $\mathfrak{B}_{cl}$ is defined by

$$\mathfrak{B}_{cl} := \{w \in \mathfrak{B} | \exists c \in \mathfrak{C} s.t. (w,c) \in \mathfrak{B}_f\}. \quad (3)$$

Note that the interconnected variable $c \in \mathfrak{C}$ plays a role as a latent variable of the given plant or the closed loop systems.

3. BEHAVIOURAL CONTROLLABILITY OF DISCRETE EVENT SYSTEMS

In this section, along the above preliminaries, we introduce the notion of controllability for discrete event systems within a behavioral framework. And then we provide an equivalent condition for a given specification to be achievable, as our main result of this paper. In the following, we assume that discrete event systems are backward shift invariant, i.e. for the behavior of a system $\Sigma = (\mathbb{N}_0, W, \mathfrak{B})$, $\sigma^\tau \mathfrak{B} \subseteq \mathfrak{B}$ holds for any $\tau \in \mathbb{N}_0$.

Firstly, we introduce our new concept of the controllability of discrete event systems for a given specification. In Definition 2.4, the behavioral controllability can be characterized by the concatenability of all feasible trajectories. In discrete event systems, as stated in (Ramadge and Wonham, 1987), the controllability of the desired languages (i.e. specifications) is equivalent to the existence of a supervisory controller that prohibits undesirable events from occurring. In other words, one of the desirable trajectories can be also prefix of an uncontrollable event. This means that any feasible uncontrollable event is concatenable with some desirable trajectories. By using above notions of concatenability, we define our new concept related to controllability for a given specifications as follows.

Definition 3.1. Let $\Sigma = (\mathbb{N}_0, \mathbb{W}, \mathbb{C}, \mathfrak{B}_f)$ denote a discrete event system with the full behavior. And let $\mathfrak{B}_{spec,f}$ denote a given specification with the full behavior. Then, $\mathfrak{B}_{spec} = \Pi_W(\mathfrak{B}_f)$ is said to be controllable if

$$\{^\forall(w_1,c_1) \in \mathfrak{B}_{spec,f}, ^\forall (w_2,c_2) \in \mathfrak{B}_f \ \text{s.t.} \ \{\exists M,N \in \mathbb{N}_0 \ \text{s.t.} \ \{c_1(M) = c_2(N)\}\}\} \Rightarrow w_1 \wedge_M \sigma^{(M-N)} w_2 \in \mathfrak{B}_{spec} \quad (4)$$

where Π_W is defined as the projection map from $(W \times C)^{N_0}$ onto $\mathbb{W}^{N_0}$.

As stated above, note that Definition 3.2 is concerned with a specification. This standpoint is similar to the work by Ramadge and Wonham in (Ramadge and Wonham, 1987).

Nextly, we apply the notion of a *canonical controller* proposed by van der Schaft in (van der Schaft and Julius, 2002) to discrete event systems. In Definition 2.5, we have introduced a general concept of interconnections in a behavioral framework ((Willems, 1997)). Concerned with behavioral interconnections, the concept of canonical controllers is introduced to achieve a given specification exactly.

Definition 3.2. Let $\Sigma = (\mathbb{N}_0, \mathbb{W}, \mathbb{C}, \mathfrak{B}_f)$ denote a discrete event system with the full behavior. Define $\mathfrak{B} := \Pi_W(\mathfrak{B}_f)$. And let $\mathfrak{B}_{spec} \subseteq \mathfrak{B}$ denote a given specification with respect to the manifest behavior. Define a behavior described by

$$\mathfrak{C} := \{c \in \mathbb{C}^{\mathbb{N}_0} | \exists w \in \mathfrak{B}_{spec} s.t. (w, c) \in \mathfrak{B}_f\} \quad (5)$$

on the signal space $\mathbb{C}$. Then, $\Sigma_c = (\mathbb{N}_0, C, \mathfrak{C})$ is said to be the canonical controller for $\mathfrak{B}_{spec}$.

A canonical controller is the interconnected system of the behavior of a plant and that of a specification through the manifest variable $w \in W^{\mathbb{N}_0}$ as shown in Fig.1 (cf. (van der Schaft and Julius, 2002)). Before going to our main result, we provide the following property, which is important in the sense of not only useful tools for our main results but also self-standing intrinsic result.

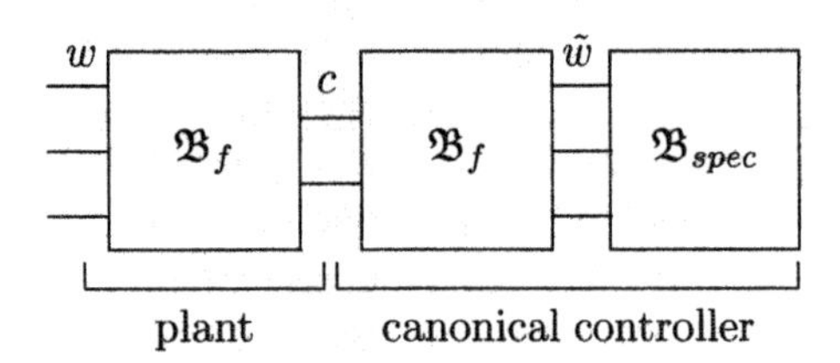

Fig.1 An interconnected system to a canonical controller

Lemma 3.1. Let $\Sigma = (\mathbb{N}_0, \mathbb{W} \times \mathbb{C}, \mathfrak{B}_f)$ denote a discrete event system with the full behavior. And let $\mathfrak{B}_{spec} \subseteq \mathfrak{B}$ denote a given specification with respect to the manifest behavior. Then, there exists a controller $\Sigma_c = (\mathbb{N}_0, \mathbb{C}, \mathfrak{C})$ such that $\mathfrak{B}_{cl} \subseteq \mathfrak{B}_{spec}$.

Proof: Let $\Sigma_c = (\mathbb{N}_0, \mathbb{C}, \mathfrak{C})$ denote a controller for the system Σ, where

$$\mathfrak{C} := \{c \in \mathbb{C}^{\mathbb{N}_0} | \exists w \in \Pi_W(\mathfrak{B}_f) s.t. (w, c) \in \mathfrak{B}_f\} \quad (6)$$

On the other hand, in general, the closed loop behavior $\mathfrak{B}_{cl}$ that consists of the interconnection of $\mathfrak{C}$ and $\mathfrak{B}_f$:

$$\mathfrak{B}_{cl} = \{w \in \Pi_W(\mathfrak{B}_f) | \exists c \in \mathfrak{C} s.t. (w, c) \in \mathfrak{B}_f\}. \quad (7)$$

Now, we introduce a canonical controller $\Sigma_c' := (\mathbb{N}_0, \mathbb{C}, \mathfrak{C}')$ with

$$\mathfrak{C}' := \{c \in \mathbb{C}^{\mathbb{N}_0} | \exists w \in \mathfrak{B}_{spec} s.t. (w, c) \in \mathfrak{B}_f\}. \quad (8)$$

Then, the closed loop behavior $\mathfrak{B}_{cl}'$ that is the interconnection of the canonical controller and a discrete event system is

$$\mathfrak{B}_{cl}' = \{w \in \Pi_W(\mathfrak{B}_f) | \exists c \in \mathfrak{C}' s.t. \\ (w, c) \in \mathfrak{B}_f, w \in \mathfrak{B}_{spec}\} \quad (9)$$

This means that for any manifest variable $w \in \mathfrak{B}_{cl}'$, $w \in \mathfrak{B}_{spec}$ holds. Hence, $\mathfrak{B}_{cl}' \subseteq \mathfrak{B}_{spec}$. (Q.E.D.)

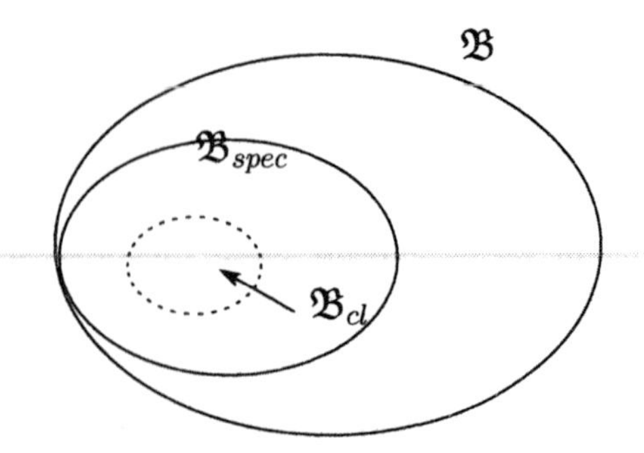

Fig.2 $\mathfrak{B}_{cl} \subseteq \mathfrak{B}_{spec}$

In the above lemma, we can not always guarantee that there exists a controller that satisfies $\mathfrak{B}_{cl} = \mathfrak{B}_{spec}$. In order to achieve this requirement, our controllability stated in Definition 2.4 plays a crucial role as follows.

Lemma 3.2. Let $\Sigma = (\mathbb{N}_0, \mathbb{W} \times \mathbb{C}, \mathfrak{B}_f)$ denote a discrete event system with the full behavior. And let $\mathfrak{B}_{spec} \subseteq \mathfrak{B}$ denote a given specification with respect to the manifest behavior. Then, if $\mathfrak{B}_{spec}$ is controllable, there exists a controller $\Sigma_c = (\mathbb{N}_0, \mathbb{C}, \mathfrak{C})$ such that$\mathfrak{B}_{spec} \subseteq \mathfrak{B}_{cl}$ holds.

Proof: Let $\mathfrak{B}_{spec}$ be controllable, i.e. for any $(w_1, c_1) \in \mathfrak{B}_{spec,f}$ and for any $(w_2, c_2) \in \mathfrak{B}_f$, there exist $M, N \in \mathbb{N}_0$, such that $c_1(M) = c_2(N)$, thus $w_1 \wedge_M \sigma^{-N} w_2 \in \mathfrak{B}_{spec}$ holds. Taking this property into consideration, for a subclass of the behavior of the system $\mathfrak{B}_f' \subseteq \mathfrak{B}_f$, for any $(\tilde{w}_1, \tilde{c}_1) \in \mathfrak{B}_{spec}$ and for any feasible $(\tilde{w}_2, \tilde{c}_2) \in \mathfrak{B}_f'$, there exist $M, N \in \mathbb{N}_0$ such that $\tilde{c}_1(M) = \tilde{c}_2(N)$, $\bar{w} := \tilde{w}_1 \wedge_M \sigma^{-N} \tilde{w}_2 \in \mathfrak{B}_{spec}$ holds. By the assumption, for the behavior $\mathfrak{B}_{spec}$, the property of the backward shift invariant holds, i.e. for any $\tau \in \mathbb{N}_0$, $\sigma^\tau \mathfrak{B}_{spec} \subseteq \mathfrak{B}_{spec}$. Therefore,

$$\bar{w} := \tilde{w}_1 \wedge_M \sigma^{-N} \tilde{w}_2 \in \sigma^{-M} \mathfrak{B}_{spec} \quad (10)$$

that is to say, as $\sigma^M \bar{w} \in \mathfrak{B}_{spec}$, thus we can find $\sigma^{-N} \tilde{w}_2 \in \mathfrak{B}_{spec}$.

From the above discussions, the controllability of the behviour $\mathfrak{B}_{spec}$ is equivalent to saying that any feasible trajectory $w \in \Pi_W(\mathfrak{B}_f)$ for the behavior $\mathfrak{B}_{spec}$ is also $w \in \mathfrak{B}_{spec}$, i.e. the controllability guarantees that $\mathfrak{B}_{spec}$ is closed with regard to the concatenation of two trajectries. Consequently, in the case of that $\mathfrak{B}_{spec}$ is controllable, we can only select a subsystem$\mathfrak{B}_f'$ which cover over the behavior of the specification $\mathfrak{B}_{spec}$. Then, any $w \in \mathfrak{B}_{spec}$ is also $w \in \Pi_W(\mathfrak{B}_f')$, and for this w, there exists $c \in \mathbb{C}^{N_0}$ such that $(w, c) \in \mathfrak{B}_f'$. For such a $c \in \mathbb{C}^{N_0}$, we define the following behavior

$$\mathfrak{C} := \{c \in \Pi_C(\mathfrak{B}'_f) | \exists w \in W^{\mathbf{N}_0} s.t. \\ (w, c) \in \mathfrak{B}'_f\} \quad (11)$$

as a controller for the system $\mathfrak{B}_f$. By using this $\mathfrak{C}$ the closed loop system is as follows

$$\mathfrak{B}_{cl} = \{w \in \Pi_W(\mathfrak{B}'_f) | \exists c \in \mathfrak{C} s.t. \\ (w, c) \in \mathfrak{B}'_f (\subseteq \mathfrak{B}_f)\}. \quad (12)$$

Therefore, for any $w \in \mathfrak{B}_{spec}$, $w \in \mathfrak{B}_{cl}$ holds, i.e. there exists a controller $\mathfrak{B}_c$ such that $\mathfrak{B}_{spec} \subseteq \mathfrak{B}_{cl}$ holds. (Q.E.D.)

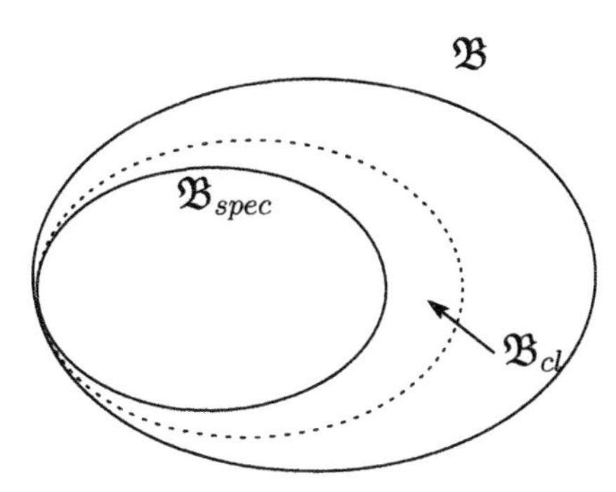

Fig.3 $\mathfrak{B}_{spec} \subseteq \mathfrak{B}_{cl}$

From the above two lemmas, we provide the following theorem .

Theorem 3.1. Let $\Sigma = (\mathbb{N}_0, \mathbb{W} \times \mathbb{C}\mathfrak{B}_f)$ denote a discrete event system with the full behavior. And let $\mathfrak{B}_{spec} \subseteq \mathfrak{B}$ denote a given specification with respect to the manifest behavior. Then, there exists a controller $\Sigma_c = (\mathbb{N}_0, C, \mathfrak{C})$ such that $\mathfrak{B}_{cl} = \mathfrak{B}_{spec}$ if and only if $\mathfrak{B}_{spec}$ is controllable.

Proof: ('If' part) From Lemma 3.1, for a given specification, there exists a canonical controller such that

$$\mathfrak{B}_{cl} \subseteq \mathfrak{B}_{spec} \quad (13)$$

holds. Nextly, for $\mathfrak{B}_f$, it follows from the lemma 3.2 that the controllability of $\mathfrak{B}_{spec}$ enables us to select a subsystem $\mathfrak{B}'_f$ such that $\mathfrak{B}_{spec} \subseteq \Pi_W(\mathfrak{B}'_f)$. By regading this new subsysytem $\mathfrak{B}_f$' as a controller, we have

$$\mathfrak{B}_{spec} \subseteq \mathfrak{B}_{cl} \quad (14)$$

Hence, if $\mathfrak{B}_{spec}$ is controllable, there exists a controller such that $\mathfrak{B}_{cl} = \mathfrak{B}_{cl}$.

('Only if' part) We assume that $\mathfrak{B}_{spec}$ is not controllable, i.e. there exist $(w_1, c_1) \in \mathfrak{B}_{spec,f}$, $(w_2, c_2) \in \mathfrak{B}_f$, and there exist $M, N \in \mathbb{N}_0$ such that $c_1(M) = c_2(N)$ and $w_1 \wedge_M \sigma^{-N} w_2 \notin \mathfrak{B}_{spec}$. On the other hand, since a specification is a subset of the behavior of the system $\mathfrak{B}_f$, $w_1 \wedge_M \sigma^{-N} w_2 \in \Pi_W(\mathfrak{B}_f)$ holds. Now, there exists a controller $\mathfrak{C}$ for the system, and this achieves $\mathfrak{B}_{cl} = \mathfrak{B}_{spec}$, where

$$\mathfrak{B}_{cl} = \{w \in \mathfrak{B}_{spec} | \exists c \in \mathfrak{C} s.t. (w, c) \in \mathfrak{B}_f\} \quad (15)$$

Thus, for any $w_1 \in \mathfrak{B}_{spec}$, $w_1 \in \mathfrak{B}_{cl}$ and $w_1 \in \Pi_W(\mathfrak{B}_{c,f})$ hold. From this, w_2 that is concatenable to$w_1 \in \mathfrak{B}_{spec}$ is that its pair is only the element of the behavior $\mathfrak{B}_c$, so $w_1 \wedge_M \sigma^{-N} w_2 \in \mathfrak{B}_{cl}$ holds. This is in contradiction to the above assumption. (Q.E.D.)

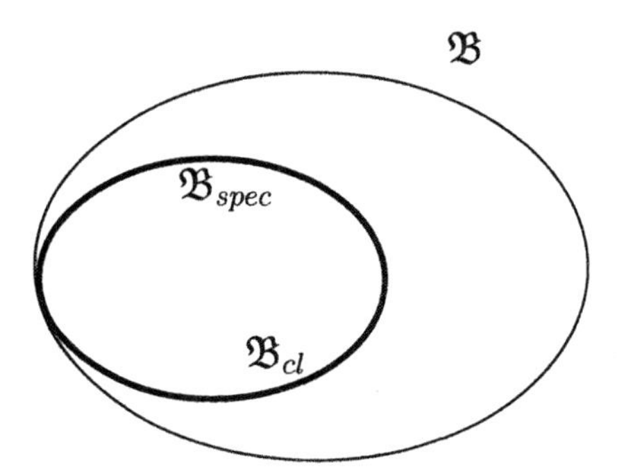

Fig.4 $\mathfrak{B}_{cl} = \mathfrak{B}_{spec}$

As stated in the above section, supervisory controllers for discrete event systems proposed by Ramadge and Wonham restrict undesirable events from occuring for given specifications. Moreover, the supervisor can not restrict uncontrollable events. As for this problem, if specifications guarantee the controllability, there exists a supervisory controller for controllable specifications. On the other hand, controllers treated in this paper (that is to say, canonical controller) can extract only the specification from all behavior of the system, then $\mathfrak{B}_{cl} \subseteq \mathfrak{B}_{spec}$ is achievable automatically. If the specification is controllable, then there exists a controller such that $\mathfrak{B}_{spec} \subseteq \mathfrak{B}_{cl}$ holds. Thus, we can synthesize the controller satisfying $\mathfrak{B}_{cl} = \mathfrak{B}_{spec}$. Such controllers make the closed loop system whose behavior is exactly equivalent to the behavior of specifications. Therefore, we can find the controller treated here is satisfies the property of supervisory controller by Ramadge and Wonham.

4. CONCLUSION

In this manuscript, we have investigated the controllability and the achievability of discrete event systems within a behavioral framework. Based on the notion of Willems' behavioral controllability (cf. (Willems, 1989), (Willems, 1991)), we have introduced the new concept related to the controllability of discrete event systems. By using the controllability proposed here and the notion related to achievable behaviors (cf. (van der Schaft and Julius, 2002) (van der Schaft, 2003)), we have shown that the behavioral controllability for a given specification with respect to *language* is equivalent to the existence of a controller so that an interconnected system satisfies the specification *exactly* . A proposed controller here is represented by the intersection of the behavior of a given plant and that of a given (controllable) specification. In order to show the validity of our results, we have also clarified that our controllers for a given specification fits the properties of well-known supervisory controllers proposed and developed by Ramadge and Wonham (cf. (Ramadge and Wonham, 1987)).

We will show more detailed discussions about this paper in our manuscript ((Kaneko *et al.*, 2003)).

5. REFERENCES

Cassandras, C.G. and S. Lafortune (1999). *Introduction to Discrete Event Systems.* Kluwer Academic Pub. Boston.

Kaneko, O., T. Misaki and T. Fujii (2003). On controllability and interconnections of discrete event systems in a behavioral framework. *Systems and control letters* - **submitted**, –.

Kuijper, M. (1995). Why do stabilizing controllers stabilize. *Automatica* **31**, 621–625.

Raisch, J. and S.D. O'Young (1998). Discrete approximation and supervisory control of continuous systems. *IEEE Trans. on Automat. Contr.* **43**, 569–573.

Ramadge, P.J. and W.M. Wonham (1987). Supervisory control of a class of discrete event processes. *SIAM Journal on Contr. and Optim.* **25**, 206–230.

T.Moor and J. Raisch (1999). Supervisory control of hybrid systems within a behavioral framework. *Systems and control letters* **38**, 157–166.

van der Schaft, A.J. (2003). Achivable behaviors for general systems. *Syst. and Contr. Lett.* **49**, 141–149.

van der Schaft, A.J. and A.A. Julius (2002). Achivable behaviors by composition. *Proc. of 41st Conf. on Dec. and Contr.* **CD-ROM**, 7–12.

Willems, J.C. (1989). Models for dynamics. *Dynamics reported* **2**, 171–269.

Willems, J.C. (1991). Paradigms and puzzles in the theory of dynamical systems. *IEEE. Trans. Automat. Cont.* **36**, 259–294.

Willems, J.C. (1997). On interconnections, control, and feedback. *IEEE. Trans. Automat. Cont.* **42**, 326–339.

PERFORMANCE OF AN ADAPTIVE VOLTAGE REGULATOR ADOPTING DIFFERENT IDENTIFICATION ALGORITHMS

Giuseppe Fusco * **Mario Russo** *

* *Università degli Studi di Cassino*
via G. Di Biasio 43, 03043 Cassino (FR), Italy
e-mail:(fusco,russo)@unicas.it

Abstract: The performance of a previously-proposed self-tuning control scheme for the regulation of nodal voltage in power systems is analyzed. To estimate the parameter variations of the equivalent model representing the power system, three different parameter tracking techniques are adopted and compared. A description of the estimation algorithms and a complete discussion of the implementation issues are reported. Particular attention is paid to assure a good parameter tracking performance as well as to avoid the estimator wind-up. Numerical applications referring to a simulated case study are presented to compare the different performance achievable by the self-tuning regulator in response to both a step reference variation and sudden power system parameter variations. *Copyright © 2003 IFAC*

Keywords: Adaptive Control; Power system control; Self-tuning regulators; Voltage regulation.

1. INTRODUCTION

In electrical power systems, electronic devices are widely used as actuators in nodal voltage control schemes so as to directly and locally regulate the voltage waveform of the busbar at which they are connected. Without loss of generality, this paper refers to a Static Var System with Fixed Capacitor-Thyristor Controlled Reactor, in the remainder denoted with the acronym SVS. Its steady-state operating characteristic is shown in Figure 1. Between points A and B the reactors are switched on partially and behave as a variable reactance. The slope of the characteristic determines the SVS current variation in response to the bus voltage variation.

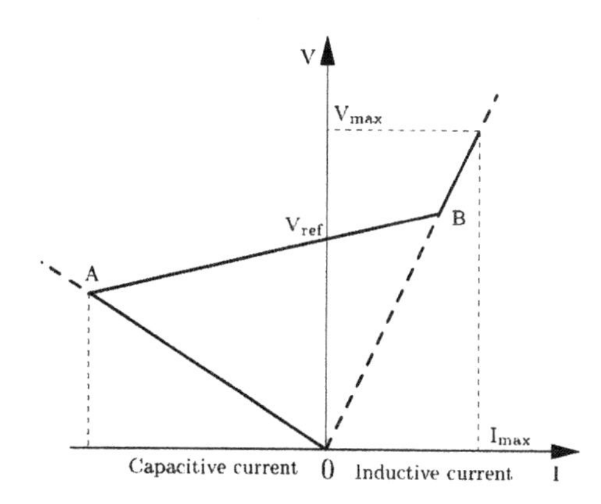

Fig. 1. SVS voltage-current operating characteristic.

Since power system operating conditions continuously and unpredictably change, robust or adaptive approaches are experienced to be attractive in voltage regulation tasks. To this aim, an equivalent model of the power system is necessary. A suitable model adopted to represent an electrical power system is represented by the Thevenin equivalent circuit, both at fundamental and harmonic frequencies (Kundur, 1994). In literature, many methods have been presented to identify the network equivalent impedance using field measurements, see for example (Girgis and McManis, 1989; Girgis *et al.*, 1993; Oliveira *et al.*, 1991; Fusco *et al.*, 2000). They basically refer to two approaches. The first approach uses the existing load current variations and the existing harmonic sources to identify the Thevenin equivalent impedance at each frequency. The second approach uses the switching transients caused by network equipment, e.g. capacitor banks, to identify the network equivalent transfer function by adopting time domain and frequency domain techniques. The design of an adaptive control scheme for the nodal voltage regulation adopting the SVS apparatus has been proposed in (Fusco *et al.*, 2001), while a generalized adaptive approach is presented in (Fusco and Russo, 2003). Figure 2 shows the adopted self-tuning regulation control scheme. Starting from the sampled measurements of the voltage waveform $v(t)$ and of the current $i(t)$ injected by the SVS, two Kalman Filters (KF) estimate the values of the voltage and current phasors at the fundamental frequency. Such values are then used in the identification procedure (CRLS) which estimates the Thevenin equivalent model parameters. The identification algorithm adopts a classical well-known Recursive Least Squares based technique (see for example (Franklin *et al.*, 1998; Wellstead and Zarrop, 1991)), which is extended to take into account physical constraints (Fusco *et al.*, 2000). Finally the estimated parameter values are used in the design of the adaptive voltage regulator (AVR), based on the poles assignment technique.

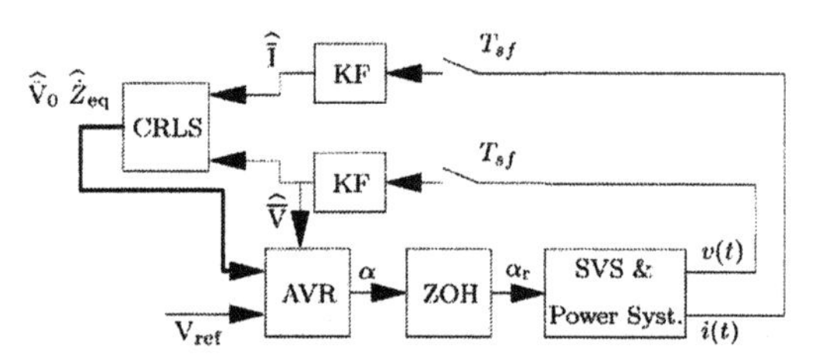

Fig. 2. SVS adaptive voltage regulation control scheme.

This paper analyzes the performance of the described control scheme under three different estimation algorithms. Basically, an adaptive mechanism must be included into the identification algorithm to track the parameter changes of the Thevenin circuit. The first two mechanisms employ a variable forgetting factor technique while the third one adopts a random walk technique. The three different adaptive mechanisms are applied to the constrained recursive least squares technique (block CRLS in Figure 2) to compare the performance of the designed adaptive voltage control in response to parameter variations of the system. Numerical applications referring to a simulated case study are finally presented.

2. SYSTEM MODELING AND IDENTIFICATION PROCEDURE

Let us refer to a power system bus at which the SVS is connected, see Figure 3. The power system

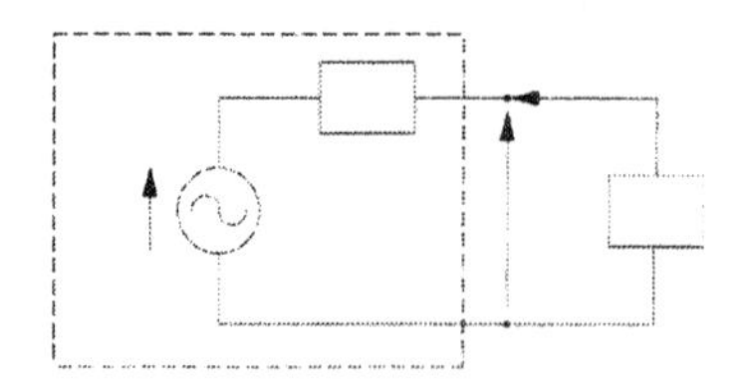

Fig. 3. Power system equivalent electrical circuit and SVS.

is modeled by the Thevenin equivalent circuit at fundamental frequency, as seen from the bus at which the SVS is connected. The Thevenin circuit parameters are the network equivalent impedance $\dot{Z}_{eq} = R + jX$ and the no-load equivalent voltage phasor $\bar{V}_0 = V_{0,r} + jV_{0,i}$. The SVS nodal voltage is represented by the phasor $\bar{V}$. To estimate $\dot{Z}_{eq}$ and $\bar{V}_0$ let's consider a first-order model which includes dynamics only on current phasors (Fusco *et al.*, 2000); it can be described as:

$$\mathbf{y}(k) = \boldsymbol{\phi}(k)\boldsymbol{\Theta} \tag{1}$$

in which $\mathbf{y}(k) = \left(\widehat{V}_r(k) \ \widehat{V}_i(k) \right)$ and

$$\boldsymbol{\phi}(k) = \left(\widehat{I}_r(k) \ \widehat{I}_i(k) \ c \ \widehat{I}_r(k-1) \ \widehat{I}_i(k-1) \right) \tag{2}$$

where $\boldsymbol{\Theta} \in \mathbb{R}^{5\times 2}$ is an unknown matrix whose $i-j$ element is θ_{ij}, and $\widehat{V}_r, \widehat{V}_i, \widehat{I}_r, \widehat{I}_i$ are the real and imaginary components of the SVS phasor voltage and current estimated by the Kalman Filters, see Figure 2. In (2) the quantity c is a constant input; it can be viewed as a scaling factor. Since in steady-state it is

$$I_r(k) = I_r(k-1) = I_r^{\infty}$$
$$I_i(k) = I_i(k-1) = I_i^{\infty}$$

the steady-state value $\mathbf{y}(k) = \mathbf{y}(k-1) = \mathbf{y}^{\infty}$ of the output model (1) is given by:

$$\mathbf{y}^{\infty} = (I_r^{\infty} \ I_i^{\infty} \ c) \begin{pmatrix} \theta_{11}+\theta_{41} & \theta_{12}+\theta_{42} \\ \theta_{21}+\theta_{51} & \theta_{22}+\theta_{52} \\ \theta_{31} & \theta_{32} \end{pmatrix}. \tag{3}$$

At this point, from the circuit shown in Figure 3, in view of equation (3), it is possible to derive the following conditions:

$$\begin{pmatrix} \theta_{11}+\theta_{41} & \theta_{12}+\theta_{42} \\ \theta_{21}+\theta_{51} & \theta_{22}+\theta_{52} \\ \theta_{31} & \theta_{32} \end{pmatrix} = \begin{pmatrix} R & X \\ -X & R \\ \dfrac{V_{0,r}}{c} & \dfrac{V_{0,i}}{c} \end{pmatrix}. \tag{4}$$

Once that N couples of voltage and current estimated phasors are available so as to form the matrices,

$$\mathbf{Y} = \left(\mathbf{y}^{T}(1) \ .... \ \mathbf{y}^{T}(N) \right)^{T} = (\mathbf{Y}_1 \ \mathbf{Y}_2),$$
$$\boldsymbol{\Phi} = \left(\boldsymbol{\phi}^{T}(1) \ .... \ \boldsymbol{\phi}^{T}(N) \right)^{T},$$

the estimated value $\widehat{\boldsymbol{\Theta}}$ of the unknown parameter matrix $\boldsymbol{\Theta}$ can be obtained by solving the following constrained least squares problem:

$$\min_{\widehat{\boldsymbol{\Theta}}} J(\widehat{\boldsymbol{\Theta}}) \tag{5}$$

$$\text{subject to}: \begin{cases} \mathbf{W}_1^{T}\widehat{\boldsymbol{\Theta}}_1 - \mathbf{W}_2^{T}\widehat{\boldsymbol{\Theta}}_2 = 0, \\ \mathbf{W}_2^{T}\widehat{\boldsymbol{\Theta}}_1 - \mathbf{W}_1^{T}\widehat{\boldsymbol{\Theta}}_2 = 0, \end{cases} \tag{6}$$

where

$$J(\widehat{\Theta}) = \epsilon_1^T \epsilon_1 + \epsilon_2^T \epsilon_2, \quad \widehat{\Theta} = \begin{pmatrix} \widehat{\Theta}_1 & \widehat{\Theta}_2 \end{pmatrix},$$

$$\mathbf{W}_1 = (1 \quad 0 \quad 0 \quad 1 \quad 0)^T,$$

$$\mathbf{W}_2 = (0 \quad 1 \quad 0 \quad 0 \quad 1)^T,$$

with

$$(\,\epsilon_1 \; \epsilon_2\,) = \mathbf{Y} - \mathbf{\Phi}\,\widehat{\Theta} = \begin{pmatrix} \mathbf{Y}_1 - \mathbf{\Phi}\,\widehat{\Theta}_1 & \mathbf{Y}_2 - \mathbf{\Phi}\,\widehat{\Theta}_2 \end{pmatrix}.$$

Constraints (6) are forced due to the particular form of (4): they assure the elements of $\widehat{\Theta}_h$ are *physically meaningful*. If enough input vectors $\phi(k)$ are available so that the matrix $\mathbf{\Phi}^T \mathbf{\Phi}$ is nonsingular, then the problem (5) subject to (6) is a convex quadratic programming problem which has a global minimum that can be found by solving the well known first-order conditions on the Lagragian function. According to this, the solution to (5)–(6) is:

$$\begin{aligned} \widehat{\Theta}_1 &= \widehat{\Theta}_1^* - \frac{\lambda_1}{2}\mathbf{P}\mathbf{W}_1 + \frac{\lambda_2}{2}\mathbf{P}\mathbf{W}_2, \\ \widehat{\Theta}_2 &= \widehat{\Theta}_2^* + \frac{\lambda_1}{2}\mathbf{P}\mathbf{W}_2 - \frac{\lambda_2}{2}\mathbf{P}\mathbf{W}_1, \end{aligned} \tag{7}$$

with

$$\widehat{\Theta}_1^* = \mathbf{P}\mathbf{\Phi}^T\mathbf{Y}_1, \quad \widehat{\Theta}_2^* = \mathbf{P}\mathbf{\Phi}^T\mathbf{Y}_2, \tag{8}$$

being $\mathbf{P} = \left(\mathbf{\Phi}^T \mathbf{\Phi}\right)^{-1}$ the covariance matrix and $\lambda_{1,2}$ the Lagrangian multipliers, given by.

$$\begin{aligned} \lambda_1 &= 2\frac{\mathbf{W}_1^T\widehat{\Theta}_1^* - \mathbf{W}_2^T\widehat{\Theta}_2^*}{\mathbf{W}_1^T\mathbf{P}\mathbf{W}_1 + \mathbf{W}_2^T\mathbf{P}\mathbf{W}_2}, \\ \lambda_2 &= 2\frac{\mathbf{W}_2^T\widehat{\Theta}_1^* + \mathbf{W}_1^T\widehat{\Theta}_2^*}{\mathbf{W}_1^T\mathbf{P}\mathbf{W}_1 + \mathbf{W}_2^T\mathbf{P}\mathbf{W}_2}. \end{aligned} \tag{9}$$

In practice, for the identification of the Thevenin circuit parameters, a classical unconstrained recursive least squares can be adopted to estimate $\widehat{\Theta}_1^*$ and $\widehat{\Theta}_2^*$, (see equation 8), which are substituted into (7) to give $\widehat{\Theta}_1$ and $\widehat{\Theta}_2$ through (9). Finally from (4) it is obtained:

$$\widehat{R} = \mathbf{W}_1^T\widehat{\Theta}_1, \qquad \widehat{X} = \mathbf{W}_1^T\widehat{\Theta}_2,$$

$$\widehat{V}_{0,r} = c\,\hat{\theta}_{31}, \qquad \widehat{V}_{0,i} = c\,\hat{\theta}_{32}.$$

3. PARAMETER TRACKING

In actual applications, the constrained RLS algorithm described in the previous Section is required to track the changes of the equivalent circuit parameter values, which usually vary unexpectedly and unpredictably way, due to changes in the supplying network configuration as well as to modified operating conditions. To make the on-line procedure able to track the changes in the equivalent circuit parameters, some adaptive mechanisms must be included into the estimator. Generally, these mechanisms are based on the control of the trace of the covariance matrix $\mathbf{P}(k)$ in the classical RLS algorithm, see Section 2. In this context two main approaches are considered (Astrom and Wittenmark, 1989; Wellstead and Zarrop, 1991).

The first approach arises from the consideration of how the data are weighted in the cost function minimized by RLS algorithms. It is based on forgetting factors which are used to progressively reduce the importance given to past information. Small values of the forgetting factors improve the responsiveness to parameter changes but increase the steady-state variance of the estimator, whereas forgetting factors with values very close to unity reduce both responsiveness and steady-state variance.

The second approach is based on the random walk technique which modulates the matrix $\mathbf{P}(k)$ and prevents it from becoming too small by adding to it a symmetric positive definite matrix $\mathbf{R}(k)$, named Random Walk Matrix. There is, however, considerable arbitrariness in selecting the size of the elements of $\mathbf{R}(k)$. The use of a big-trace random walk matrix causes large excursion in the estimates, whereas smaller values of the elements of the matrix $\mathbf{R}(k)$ allow a smooth transition to the new parameter values but low responsiveness of the algorithm to parameter changes. The drawback of the random walk procedure stands in the detection of the time instant at which the parameter changes occur. To overcome this difficulty the *a posteriori* error, which gives information at each step about the state of the estimator, can be used.

The use of either variable forgetting factors or random walk in RLS algorithms can create some problems in the construction of the covariance matrix. Specifically, if the data vector brings no new information over a long period, the event of covariance matrix blow-up can arise. In order to prevent the estimator wind-up, various solutions have been suggested in literature [7], such as fixing an upper bound on the diagonal elements of the covariance matrix or on the value of its trace, or setting a dead-zone which prevents estimation when new available information is poor. Consequently, some specific mechanisms for preventing estimator wind-up have to be introduced in the recursive estimator procedure.

In the following, three different algorithms are considered in more details. The first two algorithms employ a variable forgetting factor technique: the forgetting factor decreases when sudden changes to the process occur, while it approaches unity in steady-state conditions. The third algorithm adopts a random walk technique.

The first algorithm is derived from (Fortescue *et al.*, 1981): the forgetting factor is chosen at each step so that the estimation is always based on the same overall amount of information: roughly speaking, the "amount of forgetting" is forced to correspond to the amount of new information in the latest measurements. It is shown in (Cordero and Mayne, 1981; Fortescue *et al.*, 1981) that this strategy can be attained by forcing, at each step, the weighted sum of the squares of the *a posteriori* errors to be equal to a constant value Σ_0. In the following the first algorithm is shortly referred to as Variable Forgetting Factor (VFF) algorithm.

The second algorithm is derived from (Wellstead and Sanoff, 1981): the variable forgetting factor is obtained combining a start-up forgetting factor (which, starting from an initial value γ, exponentially approaches unity) with an adaptive forgetting factor (which varies on the basis of a weighted sum of the prediction error). In the following the second algorithm is shortly referred to as Adaptive Forgetting Factor (AFF) algorithm. The similarities and differences in principle between VFF and AFF algorithms have been described in (Wellstead and Sanoff, 1983). In addition, both algorithms update the estimation only when the prediction error is large enough; on the contrary, if it is smaller than a fixed bound, the estimator is switched off; i.e. a dead zone is introduced to prevent estimator wind-up. To switch the estimator on, thus exiting from the dead-zone, the same indicator, the prediction error, is used.

The third algorithm is based on a classical RLS algorithm to update the covariance matrix and adopts a random walk technique to track parameter changes and avoid estimator wind-up. When a parameter change is detected on the basis of the latest values of the estimation errors, matrix $\mathbf{R}(k)$ is chosen so as to assure a big enough trace of the covariance matrix to track the parameter changes. Since the third input c in $\phi(k)$ is a constant bias, see (2), the third diagonal element in $\mathbf{R}(k)$ should be chosen to be much smaller than the other diagonal elements. On the other hand, when the estimation error is small and new available information is poor, to avoid the estimator wind-up, matrix $\mathbf{R}(k)$ is chosen so as to force a constant trace of the covariance matrix (Wellstead and Zarrop, 1991). In the following the third algorithm is shortly referred to as Random Walk (RW) algorithm.

4. SIMULATION CASES STUDY

The adaptive voltage regulation scheme using the described estimation procedures has been tested by time-domain simulations of the power system shown in Figure 4. The simulation is performed in Matlab/Simulink environment. The three-phase 132 kV - 50 Hz system is assumed to be balanced in all its components. In the following, reference is made to the phase voltage peak values. The load L_3 and L_4 are equal to 100 MW and 67 MW, respectively, both with a lagging power factor equal to 0.9; loads are represented by means of shunt resistors and reactances. A 100 MVAR SVS is connected to the bus b_4. An harmonic current generator injecting into bus b_3 a distorted current I_{d3} composed of constant phasors at 5-th and 7-th order harmonic frequencies, is also considered in the power system. Concerning the desired requirements imposed on the adaptive voltage regulation scheme, a desired value of the settling time is chosen equal to 0.140 s with a damping factor equal to 0.9; in addition the desired value of the voltage slope is set equal to 0.005. The three algorithms presented in Section 3 have been adopted in the identification procedure illustrated in Section 2 to identify the equivalent circuit parameters as seen from bus b_4 at fundamental frequency. The simulation studies aim to evidence the performance of the adaptive control scheme in terms of reference step response (regulation task) as well as in terms of (identification task):

- the characteristics of the tracking error (boundness and convergence towards zero) in response to a single sudden system parameter change;
- the responsiveness of the control scheme to adapt to parameter changes;
- the ability to prevent the covariance matrix from blowing-up.

Table I. Steady-state estimations and errors: first simulation case study

	$\lvert\dot{V}_0\rvert$		$\lvert\dot{Z}_{eq}\rvert$		$\angle\dot{Z}_{eq}$	
	[kV]	error [%]	[ohm]	error [%]	[deg]	error [deg]
AFF	112.8	0.4	73.2	0.8	57.3	0.1
VFF	112.7	0.4	74.3	-0.7	57.8	-0.4
RW	112.5	0.6	74.3	-0.7	59.8	-2.4

The first simulation case study shows the adaptive regulator response to a step variation of the voltage reference occurred at time $t = 0.5$ s, see Figure 5. In details, three simulations have been run. In each simulation, the same operating conditions of the power system have been considered while a different estimation algorithm has been adopted. It must be pointed out that in these cases all the three identification algorithms remain inactive during the step response of the adaptive regulator, since no parameter variation in the power system occurs. For this reason only the time evolution of the SVS controlled voltage has been shown. Each of the adopted identification algorithms gives a satisfactory estimate and, consequently, the reported controlled voltage time evolutions are very similar. For the sake of comparison, the steady-state values of the estimated parameters $|\dot{V}_0|$ and $\dot{Z}_{eq}$, together with their corresponding errors are reported in Table I. Finally, it should be noticed from the analysis of Figure 5 that the SVS configuration introduces a small time delay in the response to the reference step (Fusco *et al.*, 2001).

In the second simulation case study it has been investigated the performance of the adaptive reg-

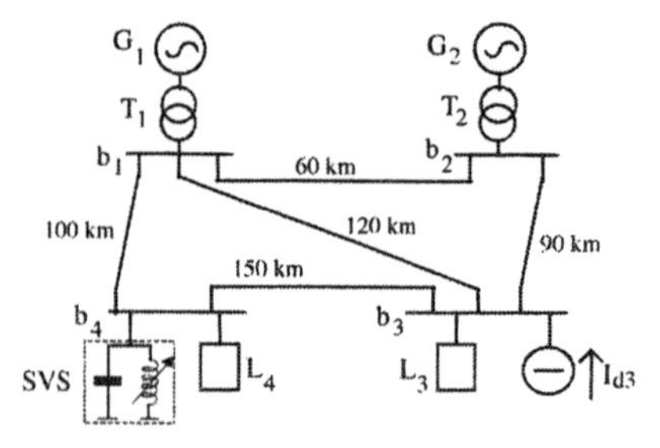

Fig. 4. Simulated power system.

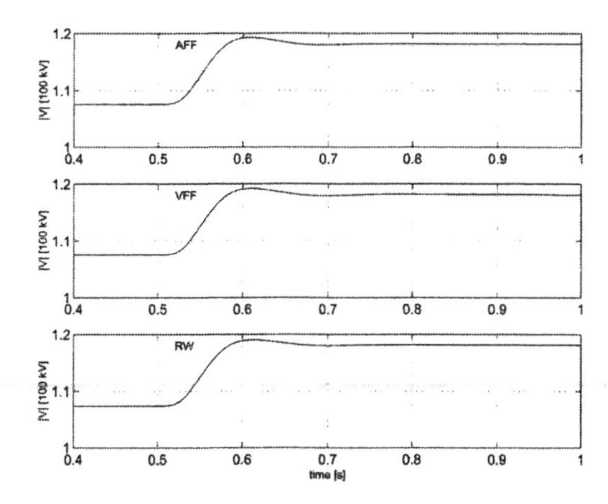

Fig. 5. Time evolution of the controlled SVS peak phase voltage $|\bar{V}|$ for each of the adopted estimation algorithms: step change of the voltage reference.

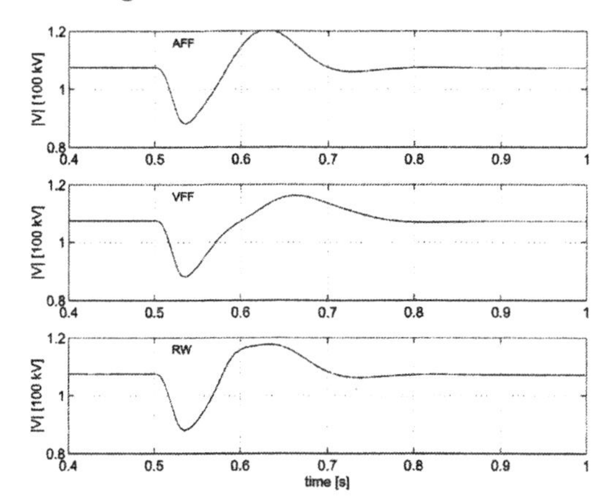

Fig. 6. Time evolution of the controlled SVS peak phase voltage $|\bar{V}|$ for each of the adopted estimation algorithms: step load change.

ulator in response to a sudden system parameter variation, while the voltage reference value is unchanged. In particular, at time $t = 0.5$ s a 50% step increase of the load L_3 at bus b_3 is introduced and, consequently, the SVS voltage suddenly decreases. The time response of the controlled SVS voltage is reported in Figure 6 for each of the three identification algorithms. Each of the depicted time response exhibits a smooth shaping during the transient while ensuring a steady state value very close to the reference one, in spite of the imposed large parameter variation. Concerning the identification task, Figures 7, 8, 9 depict the time evolutions of the estimated parameters of Thevenin equivalent circuit and the covariance matrix trace, respectively, for each of the identification algorithms. The tracking capability of the three algorithms is evidenced by the sudden increase of the trace of $\mathbf{P}(k)$ after the parameter step variation at 0.5 s. At the same time, all three algorithms avoid the estimator wind-up keeping a constant trace when steady-state estimation is reached and new available information is poor. The transient time evolution of the estimated parameters is different for each of the adopted algorithm and strictly dependent on the tuning of the adaptive mechanism. For all the three algorithms

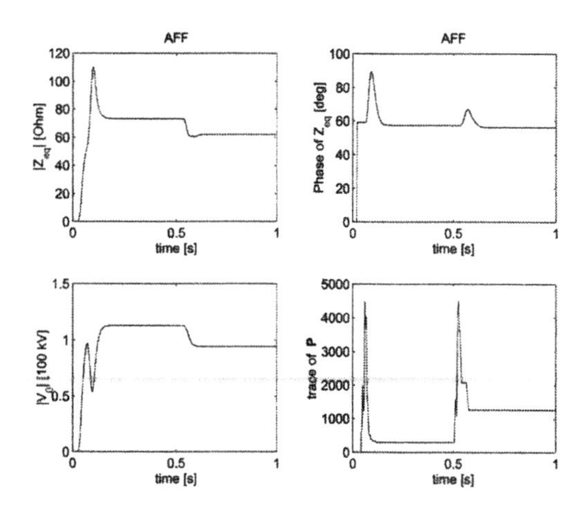

Fig. 7. Time evolution of the estimated Thevenin circuit parameters and of the covariance matrix trace: step load change and AFF algorithm.

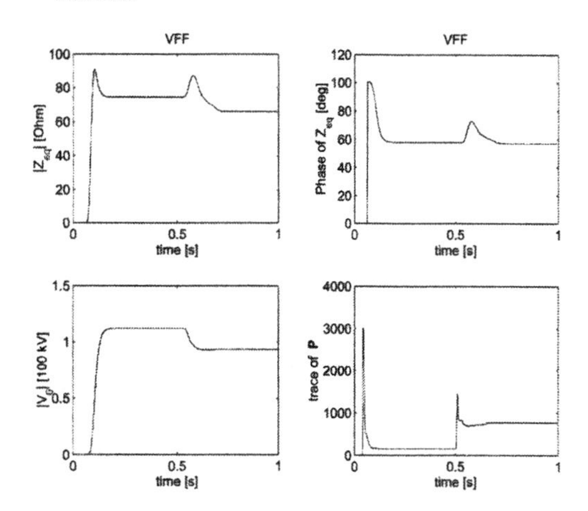

Fig. 8. Time evolution of the estimated Thevenin circuit parameters and of the covariance matrix trace: step load change and VFF algorithm.

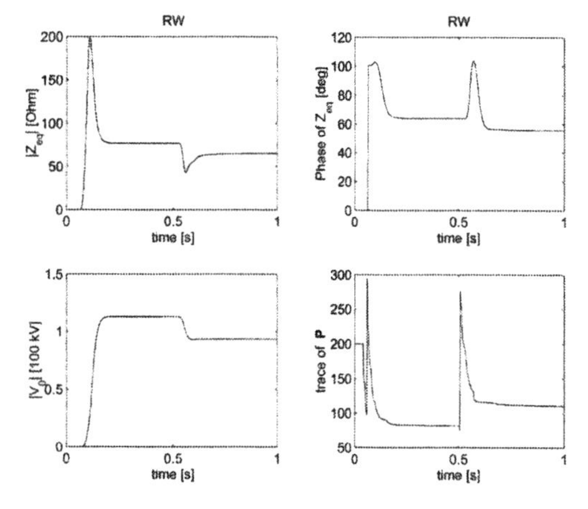

Fig. 9. Time evolution of the estimated Thevenin circuit parameters and of the covariance matrix trace: step load change and RW algorithm.

the achieved tuning guarantees a satisfactory performance. The steady state estimation values and errors are reported in Table II; all the algorithms show very low steady-state variance.

Table II. Steady-state estimations and errors: second simulation case study

	$\|\bar{V}_0\|$		$\|\dot{Z}_{eq}\|$		$\angle \dot{Z}_{eq}$	
		error		error		error
	[kV]	[%]	[ohm]	[%]	[deg]	[deg]
AFF	94.1	-0.3	62.0	2.4	56.2	0.5
VFF	93.7	0.1	65.9	-3.8	57.1	-0.4
RW	93.7	0.1	65.0	-2.4	57.8	-1.1

5. CONCLUSIONS

Three different estimation algorithms have been applied to the peculiar constrained least squares model used in the identification procedure of a SVS adaptive regulation control scheme. The obtained performance of the adaptive regulator under the different estimation algorithms have been compared in some simulation case studies. All the tested algorithms have shown satisfactory behavior; considerations about their applicability and tuning for the identification task have been presented.

6. REFERENCES

Astrom, K. and B. Wittenmark (1989). *Adaptive Control.* Addison-Wesley Publishing Company. New York, USA.

Cordero, A.Osorio and D.Q. Mayne (1981). Deterministic convergence of a self-tuning regulator with variable forgetting factor. *IEE Proc. part.D* **128**(1), 19–23.

Fortescue, T.R., L.S.Kershenbaum and B.E. Ydstie (1981). Implementation of self tuning regulators with variable forgetting factors. *Automatica* **17**(6), 831–835.

Franklin, G.F., J.D. Powell and M.L. Workman (1998). *Digital Control of Dynamic Systems Third Edition.* Addison-Wesley Publishing Company. New York, USA.

Fusco, G., A. Losi and M. Russo (2000). Constrained least squares methods for parameter tracking of power system steady-state equivalent circuits. *IEEE Transactions on Power Delivery* **15**(3), 1073–1080.

Fusco, G., A. Losi and M. Russo (2001). Adaptive voltage regulator design for static var systems. *Control Engineering Practice* **9**(7), 759–767.

Fusco, G. and M. Russo (2003). Self-tuning regulator design for nodal voltage waveform control in electrical power systems. *IEEE Transactions on Control Systems Technology* **11**(2), 258–266.

Girgis, A.A. and R.B. McManis (1989). Frequency domain techniques for modeling distribution or transmission networks using capacitor switching induced transients. *IEEE Transactions on Power Delivery* **4**(3), 1882–1890.

Girgis, A.A., W.H. Quaintance III, J. Qiu and E.B. Makram (1993). A time-domain three-phase power system impedance modeling approach for harmonic filter analysis. *IEEE Transactions on Power Delivery* **8**(2), 504–510.

Kundur, P. (1994). *Power system stability and control.* McGraw-Hill Inc.. New York, USA.

Oliveira, J.C. De, J.W. Resende and M.S. Miskulin (1991). Practical approaches for ac system harmonic impedance measurements. *IEEE Transactions on Power Delivery* **6**(4), 1721–1726.

Wellstead, P.E. and M.B. Zarrop (1991). *Self-tuning systems control and signal processing.* John Wiley & Sons. New York, USA.

Wellstead, P.E. and S.P. Sanoff (1981). Extended self-tuning algorithm. *Int. J. Control* **34**(3), 433–455.

Wellstead, P.E. and S.P. Sanoff (1983). Comments on: Implementation of self-tuning regulators with variable forgetting factor. *Automatica* **19**(3), 345–346.

DESIGN AND SIMULATION TESTING OF THERAPIES FOR A VIRAL DISEASE

Alberto Lima*, **Laura Menini** *,[1]

* *Dip. Informatica, Sistemi e Produzione*
Università di Roma Tor Vergata
Via del Politecnico, 1
00133, Roma, Italy
Email: menini@disp.uniroma2.it

Abstract: The purpose of this paper is to design therapies for a viral disease, and to test their efficacy in simulation. The control input is assumed to be the plasma cell enhancer, which in the related literature is pointed out as the less efficient one. The design is carried out by suitably choosing a nominal trajectory, corresponding to significant nominal initial conditions and to a nominal control input, and by applying an additional closed-loop control, designed by assuming continuous state measurements. Finally, the control law is discretized in order to cope with sampled state measurements, and its application is by means of a suitable generalized holder. *Copyright © 2003 IFAC*

Keywords: Viral disease, LQ control, biological systems.

1. INTRODUCTION

The interest for developing dynamic models of the human immune system has increased in recent years, as witnessed, e.g., by the papers (Asachenkov *et al.*, 1993), (Rundell *et al.*, 1995) and (Mielke and Pandey, 1998). Although realistic models of the immune response are rather complicated, and the identification of their parameters is a very complex task (see, e.g., (Rundell *et al.*, 1998)), there are already some studies on the use of control techniques in order to develop suitable therapies: see (Rundell *et al.*, 1997), (de Souza *et al.*, 2000) and (Stengel *et al.*, 2001). It is stressed that a good understanding on how immune systems work, and to what extent they can be controlled, is of obvious interest for increasing the efficacy of therapies, but can also be of interest in order to develop technological systems that mimic the behaviour of the immune system, as, e.g., it is the idea of (Taheri and Calva, 2001).

In this paper, we follow the study in (Stengel *et al.*, 2001), in which optimal control policies (which do not seem easily computable in real time) were proposed for a simplified model of the human immune system. Here, in order to obtain simpler contro laws, we renounce to true optimality, and we apply a strategy which is based on the linearization of the nonlinear system about a nominal trajectory.

The organization of the paper is as follows: in Section 2, the dynamic model considered is briefly described, in Section 3, after describing in detail the goals of the paper, the design of the proposed control law is dealt with and, finally, some of the simulations showing the performance of the proposed control law are reported in Section 4.

[1] Supported by ASI (grant I/R/277/02) and MIUR (grant COFIN: Matrics)

2. A DYNAMIC MODEL OF THE RESPONSE TO A VIRAL ATTACK

A simple model of the response of an organism subject to a viral disease is given by the following system of differential equations (see (Asachenkov *et al.*, 1993) and (Stengel *et al.*, 2001)):

$$\dot{x}_1(t) = (a_{11} - a_{12}\, x_3(t))\, x_1(t), \tag{1a}$$
$$\dot{x}_2(t) = a_{21}(x_4(t))\, a_{22}\, x_1(t-\tau)\, x_3(t-\tau) - a_{23}\,(x_2(t) - x_2^*) + u(t), \tag{1b}$$
$$\dot{x}_3(t) = a_{31}\, x_2(t) - (a_{32} + a_{33}\, x_1(t))\, x_3(t), \tag{1c}$$
$$\dot{x}_4(t) = a_{41}\, x_1(t) - a_{42}\, x_4(t), \tag{1d}$$

where $t \geq 0$ is time (measured in time units, which can be thought of as days), $x_1(t)$ is the concentration of viruses, $x_2(t)$ is the concentration of plasma cells (carriers and producers of antibodies), $x_3(t)$ is the concentration of antibodies, $x_4(t)$ is the relative characteristic taking into account the health of a damaged organ, $a_{21}(\cdot)$ is a nonlinear function, $u(t)$ is the control input (plasma cell enhancer) and the parameters a_{11}, a_{12}, a_{22}, a_{23}, a_{31}, a_{32}, a_{33}, a_{41}, a_{42}, and x_2^* are physical parameters of the system. The state variable x_4 belongs to the interval $[0,\ 1]$, as a matter of fact, $x_4 = 0$ means that the organ is in perfect health, whereas $x_4 = 1$ means that the organ is completely damaged. The function $a_{21}(\cdot)$ is:

$$a_{21}(x_4) = \alpha^{x_4^\gamma},$$

where $\alpha \approx 0.629 \cdot 10^{-10}$ and $\gamma = 3$. It is a positive decreasing function in the interval $[0,\ 1]$, whose value becomes very small when, for $x_4 \approx 0.55$, the organ is so damaged that it is practically no more able to sustain the growth of plasma cells. Such a function $a_{21}(\cdot)$ is a close approximation of the one considered in (Stengel *et al.*, 2001), but has the advantage of being a differentiable one, thus simplifying the computations in Section 3.1.

The nominal values of the parameters are taken from (Stengel *et al.*, 2001): $a_{11} = 1$, $a_{12} = 1$, $a_{22} = 3$, $a_{23} = 1$, $a_{31} = 1$, $a_{32} = 1.5$, $a_{33} = 0.5$, $a_{41} = 0.5$, $a_{42} = 1$ and $x_2^* = 2$. Notice that all the state variables $x_i(t)$ need to be non-negative, as well as the control input $u(t)$, in order to have a physical meaning. In this respect, the model (1) is well posed, since, as it is easy to verify, if the initial conditions $x_1(\sigma) : [-\tau,\ 0] \to \mathbb{R}$, $x_2(0)$, $x_3(\sigma) : [-\tau,\ 0] \to \mathbb{R}$ and $x_4(0)$ are positive, and $u(t)$ is non-negative, it is guaranteed that the state variables will remain positive for all $t \geq 0$. On the other hand, the constraint $x_4 \leq 1$ is not automatically guaranteed by the simple model (1), so that in this paper the simulations in which such a value will be reached will be stopped at the first time $\bar{t}$ in which $x_4(\bar{t}) = 1$.

The following properties of the open-loop system have been tested for small values of τ, which will be considered null for the control law design (if $\tau < 1/6$ time units, the behaviour of the open-loop system does not change significantly). It can be easily verified that the open-loop system has two equilibria: $x_{e,1} = [0,\ 2,\ 4/3,\ 0]^T$ and $x_{e,2} \approx [1.11,\ 2.05,\ 1,\ 0.55]^T$. The first equilibrium, which corresponds to the health situation, is asymptotically stable, i.e., the organism is able to recover autonomously if the viral attack is small enough. The second equilibrium is unstable, representing a form of cronic disease that cannot be maintained for long, since arbitrarily small perturbations will cause the organism either to recover its health or to attain the limit condition $x_4 = 1$. The value of $x_4 = 1/2$, is taken in this paper as the condition which causes the patient to become aware of his/her illness. Assuming that the organism is in perfect health before the viral attack, hence $x(\sigma) = x_{e,1}$, for all $\sigma \in [-\tau,\ 0)$, the following situations can occur depending on the value of the viral attack $x_1(0^+)$:

- subclinical case: $x_1(0) < x_{sc,\tau} \Rightarrow x_4(t) < 1/2, \forall t \geq 0$ (the patient doesn't even realize he/she is sick);
- clinical case: $x_{sc,\tau} \leq x_1(0) < x_{cr,\tau} \Rightarrow x_4(t) < 1, \forall t \geq 0$ and $\lim_{t\to+\infty} x(t) = x_{e,1}$ (the patient realizes he/she is sick, but the organism is able to recover its health without reaching the limit value $x_4 = 1$);
- cronic case: $x_1(0) = x_{cr,\tau} \Rightarrow x_4(t) < 1, \forall t \geq 0$ and $\lim_{t\to+\infty} x(t) = x_{e,2}$ (a single limit case between clinical and lethal case);
- lethal case: $x_1(0) > x_{cr,\tau} \Rightarrow \exists \bar{t} : x_4(\bar{t}) = 1$ (the organism is not able to recover autonomously, since the damaged organ attains the limit condition $x_4 = 1$);

where the special values of viral attack $x_{sc,\tau}$ and $x_{cr,\tau}$ have the following values when the time-delay is null: $x_{sc,0} \approx 1.85$ and $x_{cr,0} \approx 2.57$.

3. CONTROL DESIGN

The first, obvious, goal of any therapy is to enlarge the domain of attraction of the "healthy" equilibrium $x_{e,1}$, so that to render the organism in closed-loop with the therapy able to recover from viral attacks which would be lethal without therapy. Then, it is also desirable that in the clinical case the convergence to the "healthy" equilibrium $x_{e,1}$ is faster with therapy than without. Moreover, it is desirable that the applied control, i.e., the quantity of drug delivered to the patient, is not unnecessarily high, since any treatment always has related disadvantages. Finally, the proposed control law should be computable in real-time, i.e., based on present or past measurements of the state of the system. All these considerations lead us to the choice of the following rationale to be used for control design: first, in Subsection 3.1, a nominal trajectory $x_n(t)$ is chosen, corresponding to an initial condition $x_{0,n}$ which would be lethal in absence on control, but which is safe enough if the control action can be used. The control action $u_n(t)$, which ensures that such a nominal trajectory is actually followed by the system from the initial condition $x_{0,n}$, will be called open-loop control or nominal control. Then, in Subsection 3.2, the case when the initial condition is different from $x_{0,n}$ will be considered. Letting $\delta x(t) := x(t) - x_n(t)$, a closed-loop control law $\delta u(t)$ will be designed, which ensures that, if the initial condition is "worst" than the nominal one, additional drug is delivered in order to help the organism to reach the nominal (and safe) trajectory, whereas, if the initial condition is "better" than the nominal one, no serious damage is caused to the organism, but, on the contrary, a lighter therapy is delivered, which still helps its recovery (avoiding excess of drug). In Subsection 3.3, the case of sampled measurements is taken into account, and a procedure for time discretization of the proposed control is described.

3.1 *Choice of the nominal trajectory*

In this subsection, as well as in the subsequent one, it is assumed that $\tau = 0$. Then, in order to choose a significant initial condition $x_{0,n}$, the following reasoning is made. We consider as a typical lethal attack the one corresponding to an organism in perfect health (i.e., $x(0^-) = x_{e,1}$) attacked by $x_{n,1}(0^+) = 3$, which is the value chosen as particularly significant by (Stengel *et al.*, 2001), and we assume that no therapy is applied until the time $t_{0,n}$ in which the corresponding motion of the uncontrolled system attains for the first time the condition $x_4(t_{0,n}) = 1/2$. By simulation, it results that, in this conditions, $t_{0,n} \approx 0.5$, and $x(t_{0,n}) = x_{0,n} \approx [2.6,\ 3.9,\ 1.3,\ 0.5]^T$. Then, we consider as nominal initial time $t_{0,n}$ (the time at which the therapy starts in the nominal case) and as nominal initial condition $x(t_{0,n}) = x_{0,n}$.

In order to compute the nominal trajectory, by using, respectively, equations (1a), (1c), (1d) (this last one can be simply integrated) and (1b), the following relationships have been obtained:

$$x_{n,3}(t) = \frac{a_{11}\,x_{n,1}(t) - \dot{x}_{n,1}(t)}{a_{12}\,x_{n,1}(t)}, \tag{2a}$$

$$x_{n,2}(t) = \frac{\dot{x}_{n,3}(t) + (a_{32} + a_{33}\,x_{n,1}(t))\,x_{n,3}(t)}{a_{31}}, \tag{2b}$$

$$x_{n,4}(t) = \mathrm{e}^{a_{42}\,(t-t_{0,n})}x_{0,n,4} + \int_{t_{0,n}}^{t} \mathrm{e}^{a_{42}\,(t-\tau)}a_{41}\,x_{n,1}(\tau)\,\mathrm{d}\tau, \tag{2c}$$

$$u_n(t) = \dot{x}_{n,2}(t) - a_{21}(x_{n,4}(t))\,a_{22}\,x_{n,3}(t)x_{n,3}(t) + a_{23}\,(x_{n,2}(t) - x_2^*), \tag{2d}$$

from which it is clear that, if a suitable $x_{n,1}(t)$ is chosen, all the other entries of $x_n(t) = [x_{n,1}(t),\ x_{n,2}(t),\ x_{n,3}(t),\ x_{n,4}(t)]^T$, and the nominal control input $u_n(t)$ can be computed.

To be admissible, $x_{n,1}(t)$ must satisfy:

$$x_{n,1}(t_{0,n}) = x_{0,n,1}, \tag{3a}$$

$$\dot{x}_{n,1}(t_{0,n}) = (a_{11} - a_{12}\,x_{0,n,3})\,x_{0,n,1}, \tag{3b}$$

$$\ddot{x}_{n,1}(t_{0,n}) = \dot{x}_{n,1}(t_{0,n}) - a_{12}x_{0,n,1}\dot{x}_{n,3}(t_{0,n}), \tag{3c}$$

$$\dot{x}_{n,3}(t_{0,n}) = a_{31}\,x_{0,n,2} - \tag{3d}$$

$$(a_{32} + a_{33}\,x_{0,n,1})\,x_{0,n,3}. \tag{3e}$$

Moreover, it is useful to have some free parameters in the function $x_{n,1}(t)$ in order to ensure the following two desirable properties:
a) $\lim_{t\to+\infty} x_{n,3}(t) = 4/3$ (although the nominal control is thought of as a function defined only in a finite interval $[t_{0,n},\ t_{f,n}]$),
b) $u_n(t) \le 0$, for all $t > t_{0,n}$.

The function $x_{n,1}(t)$ chosen here is the following:

$$x_{n,1}(t) = \frac{A\mathrm{e}^{B\bar{t}}}{1 + C\bar{t} + D\bar{t}^2 + E\bar{t}^3},$$

with $\bar{t} := t - t_{0,n}$, $A = x_{0,n,1}$, $C = -(AB - \dot{x}_1(t_{0,n}))/A$, $D = (\dot{x}_3(t_{0,n}) + C^2)/2$, where $\dot{x}_1(t_{0,n})$ and $\dot{x}_3(t_{0,n})$ are computed by the equations (1) with $\tau = 0$ and the initial condition $x(t_{0,n})$. As for B and E, such parameters have been chosen in order to ensure a suitably fast convergence to the healthy equilibrium of all the state variables, together with properties a) and b) above, and have the following values: $B = 1/3$ and $E = 0.1$. By looking at the plots of the time behaviour of the nominal state variables and input function corresponding to such a choice of $x_{n,1}(t)$, reported in Figure 1, it appears that at time $t_{f,n} = t_{0,n} + 3.2$, the state variables are in a safe condition, from which the organism is able to recover autonomously, hence this is the choice made for the time interval $[t_{0,n},\ t_{f,n}]$ in which to effectively apply the nominal control input.

3.2 *Closed-loop control*

If the viral attack $x_1(0^+)$ is different from $x_{n,1}(0^+)$, or if the initial condition of the organism does not correspond to perfect health, the time t_0 at which $x_4(t_0) = 1/2$ (i.e., the time at which the patient will realize he/she is sick, and at which the therapy will start) is different from $t_{0,n}$. As expected, for small differences in the magnitude of the viral attack from the nominal value and for small differences of the initial condition from the

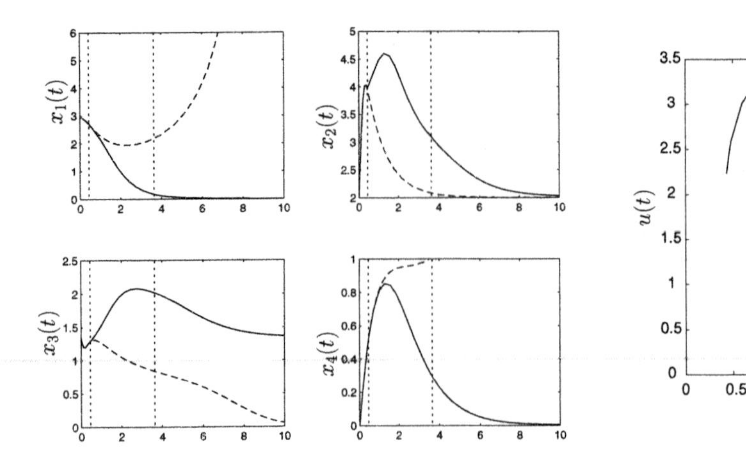

Fig. 1. In the four plots in the left hand side, the time behaviour of the nominal state variables is reported with a continuous line, and the time behaviour of the state variables of the uncontrolled system is reported with a dashed line. Vertical lines are used to emphasize the time interval $[t_{0,n},\ t_{f,n}]$ during which the nominal control (reported in the right hand side plot) is applied.

healthy equilibrium, the nominal control is still a good therapy, i.e., if $t_f = t_0 + 3.2$, the control

$$u(t) = u_n(t - t_0 + t_{0,n}), \quad \forall t \in [t_0,\ t_f],$$

still brings the organism to a state $x(t_f)$ from which it is able to recover autonomously. But, if the magnitude of the viral attack is much bigger than the nominal one, this is not guaranteed, whereas, if the magnitude is significantly smaller (but still in the clinical case, otherwise the therapy does not even start) the applied control input could be too large, i.e., it could lead to an unreasonable increment of the plasma cell values, which is not needed for the reaction to a small viral attack. For such reasons, in this section a closed-loop control action $\delta_u(t)$ is designed, to be added to the nominal control action, in order to enlarge the region of attraction of the healthy equilibrium (i.e., to be able to cure patients attacked by $x_1(0^+) > 3$) and to reduce the amount of drug delivered in case of lighter viral attacks. In particular, the system (1) has been linearized about the nominal trajectory, obtaining:

$$\Delta\dot{x}(t) = A(t)\Delta x(t) + B(t)\Delta u(t), \quad t \in [t_{0,n},\ t_{f,n}],$$

with $\Delta x(t) \approx \delta x(t - t_{0,n} + t_0)$ and $\Delta u(t) \approx \delta u(t - t_{0,n} + t_0)$, for all $t \in [t_{0,n},\ t_{f,n}]$, whereas $\delta x(t) := x(t) - x_n(t - t_0 + t_{0,n})$ and $\delta u(t) := u(t) - u_n(t - t_0 + t_{0,n})$, for all $t \in [t_0,\ t_f]$, with $t_f := t_0 + t_{f,n} - t_{0,n}$. Then, the time varying gain-matrix $K(t)$ which minimizes the quadratic performance index

$$J = \int_{t_{0,n}}^{t_{f,n}} \Delta x^T(t) Q \Delta x^T(t) + R\,\Delta u^2(t) dt + \Delta x^T(t_{f,n}) P_f \Delta x^T(t_{f,n}),$$

with $Q,\ P_f \in \mathbb{R}^{4\times 4}$, $Q\ P_f \geq 0$, $R \in \mathbb{R}$, $R > 0$, under the control law $\Delta u(t) = -K(t)\Delta x(t)$, for the linearized system, has been computed, according to the following well-known formula (Dorato *et al.*, 1995):

$$K(t) = \frac{1}{R} B^T(t) P(t),$$

where $P(t) \in \mathbb{R}^{4\times 4}$ is the solution from $P(t_{f,n}) = P_f$ of the Riccati differential equation:

$$-\dot{P}(t) = -\frac{1}{R} P(t) B(t) B^T(t) P(t) + A^T(t) P(t) + P(t) A(t) + Q.$$

Finally, after several simulations performed in order to choose the following matrices:

$$Q = \text{diag}(1,\ 0,\ 0,\ 250),\ R = 1/50,\ P_f = I_4,$$

the following control law has been applied to the original nonlinear system (1)

$$u(t) = \begin{cases} \Pi\Big(u_n(\eta(t)) - K(\eta(t))\delta x(t)\Big), \ \forall t \in (t_0,\ t_f], \\ 0,\ \forall t \in (0,\ t_0] \text{ and } \forall t > t_f, \end{cases}$$

with $\eta(t) := t - t_0 + t_{0,n}$, and

$$\Pi(z) := \begin{cases} z \text{ if } z \geq 0, \\ 0 \text{ if } z < 0, \end{cases}$$

obtaining satisfactory results in simulation, as described in Subsection 4.1.

3.3 *Time discretization*

In order to adapt the proposed control law to sampled measurements, it has been assumed that the state of the system is measured once per time unit, i.e., at times t_0, $t_1 = t_0 + 1$, $t_2 = t_0 + 2$, $t_3 = t_0 + 3$, and that, at each measurement time, on the basis of the available measurements, 6 doses of drug have to be computed to be delivered at equally spaced times up to next measurement. The proposed algorithm is to compute $u(t_i)$ according to the described continuous-time control law, and to deliver a total amount of drug equal

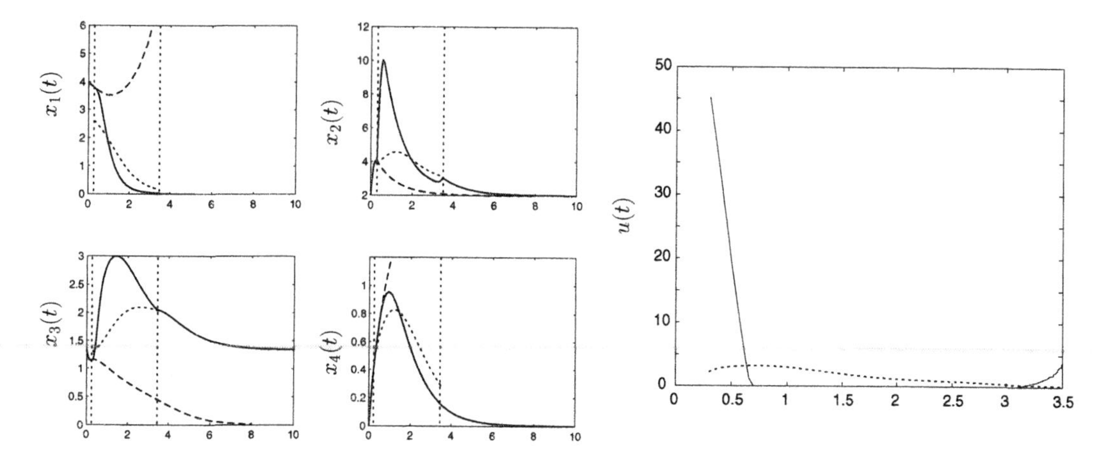

Fig. 2. The controlled system under the viral attack $x_1(0^+) = 4$. In the four plots in the left hand side, the time behaviour of the system controlled with the proposed control law is reported with continuous lines, and, for comparison, the nominal state variables (translated) are reported with dotted lines and the state variables of the uncontrolled system are reported with dashed lines. Vertical lines are used to emphasize the time interval $[t_n, t_f]$. In the right hand side plot, the applied control is reported with a continuous line and the (translated) nominal control with a dotted line.

to 1/4 of the amount that would be delivered if a zero-order holder was used up to time t_{i+1}, divided into 6 "impulsive" doses having linearly decreasing behaviour (see Fig. 4).

4. SIMULATIONS

In this section, simulation results for the case of $\tau = 0$ and with the nominal values for the physical parameters of the system are reported. The simulations performed for small non-null values of the time delay τ, and for perturbed values of the physical parameters, are not reported for lack of space, but they also showed satisfactory results.

4.1 Continuous-time control

In Fig. 2, the case of a very serious viral attack, corresponding to $x_1(0^+) = 4$ is considered. It can be seen that, despite the large differences with the nominal trajectory (translated in time, in order to emphasize the role of the feedback control law), the proposed control is able to bring the organism to a safe condition, from which it can recover quickly. Notice the cusp in x_2 at time t_f (due to the non null value of the input right before t_f), that can be easily reduced with a different P_f.

In Fig. 3, the case of a clinical, but light, viral attack, corresponding to $x_1(0^+) = 2.5$ is considered. It can be seem from the reported plots that, right after the time t_0, no control is needed, but, after about one time unit, some drug is delivered, which helps the organism.

4.2 Discrete-time control

In the most serious viral attacks, the discrete-time control performs even better than the corresponding continuous-time one, as it is natural since, in choosing the multiplying factor of 1/4, we were conservative. On the other hand, the performance loss in the clinical less serious cases is acceptable, as it is shown in Fig. 4.

5. CONCLUSIONS

With respects to other control laws presented in the literature, the following steps towards the applicability of the proposed control law have been performed: (1) the proposed control law can be computed in real time, based on the measurements of the state, (2) the possibility that such measurements are available only at the sampling times has been considered, (3) the fact that the therapy could be activated only after the patient realizes that he/she is sick has been taken into account. The main step that remains to be done is the design of output feedback control laws.

The method presented here can be used both in the case of a different available control input, e.g., a direct virus killer or an antibody enhancer (with a minor modification in the procedure reported in Subsection 3.1, which becomes partially numeric), and in the case when two or more scalar inputs are available. The choice made in this paper, that of having a single control input, namely the plasma cell enhancer, ensures the simplicity of the choice of the nominal trajectory (having two or more scalar inputs would give more degrees of freedom, but would have rendered the process in Section 3.1 more complex), and takes into account that (see (Stengel *et al.*, 2001)) such an input is the most

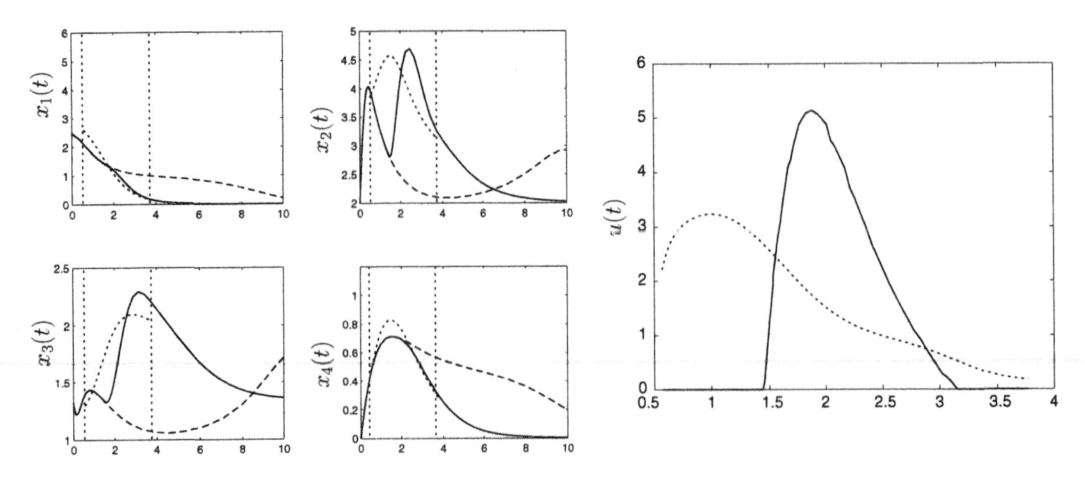

Fig. 3. The controlled system under the viral attack $x_1(0^+) = 2.5$. The same notations of Fig.2 are used.

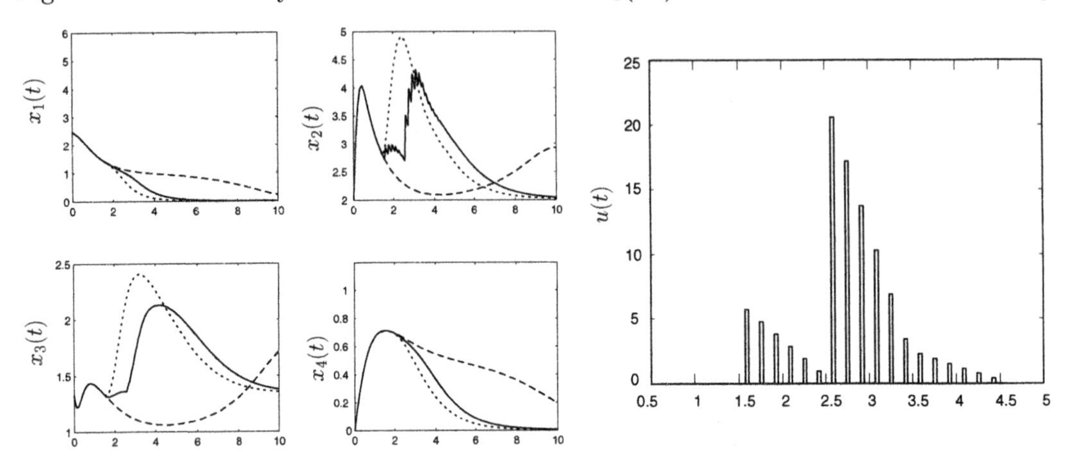

Fig. 4. The system controlled with discrete-time control under $x_1(0^+) = 2.5$. In the four plots in the left hand side, the time behaviour of the system controlled with discrete-time control is reported with continuous lines, and, for comparison, the state variables of the system controlled with continuous-time control are reported with dotted lines, and, the state variables of the uncontrolled system are reported with dashed lines. Vertical lines are used to emphasize the time interval $[t_n,\ t_f]$ during which the control is applied. In the right hand side plot, the applied discrete-time control is reported.

difficult to use, having the least direct action on the organ's health.

REFERENCES

Asachenkov, A., G. Marchuk, R. Mohler and S. Zuev (1993). *Disease Dynamics*. Birkhäuser.

de Souza, J.A.M.F., M.A. Leonel Caetano and T. Yoneyama (2000). Optimal control theory applied to the anti-viral treatment of aids. In: *Proc. of the Conference on Decision and Control*. Sidney, Australia. pp. 4839–4844.

Dorato, P., C. Abdallah and V. Cerone (1995). *Linear Quadratic Control: An Introduction*. Prentice–Hall. Englewood Cliffs, NJ.

Mielke, A. and R.B. Pandey (1998). A computer simulation study of cell population in a fuzzy interaction model for mutating hiv. *Physica A* **251**, 430–438.

Rundell, A., H. HogenEsch and R. DeCarlo (1995). Enhanced modeling pf the immune system to incorporate natural killer cells and memory. In: *Proc. of the American Control Conference*. Seattle, WA. pp. 255–259.

Rundell, A., R. DeCarlo, P. Doerschuk and H. HogenEsch (1998). Parameter identification for an autonomous 11th order nonlinear model of a physiological process. In: *Proc. of the American Control Conference*. Philadelphia, PA. pp. 3585–3589.

Rundell, A., R. DeCarlo, V. Balakrishnan and H. HogenEsch (1997). Investigations into treatment strategies aiding the immune response to haemophilus influenzae. In: *Proc. of the American Control Conference*. Albuquerque, NM. pp. 208–212.

Stengel, R. F., R. Ghigliazza, N. Kulkharni and O. Laplace (2001). Optimal Control of a Viral Disease. In: *Proc. of the American Control Conference*. Arlington, VA.

Taheri, S.A. and G. Calva (2001). Imitating the human immune system capabilities for multi-agent federation formation. In: *Proc. of the Int. Symposium on Intelligent Control*. Mexico City, Mexico. pp. 25–30.

POLYNOMIAL PARAMETRIZATION OF MULTIRATE DEAD-BEAT RIPPLE-FREE CONTROLLERS [1]

Jiří Mošna * Jiří Melichar ** Pavel Pešek ***

University of West Bohemia, Univerzitní 8, 306 14 Plzeň, Czech Republic. E-mail: * mosna@kky.zcu.cz
** melichar@kky.zcu.cz
*** pesek@kky.zcu.cz

Abstract: In this paper, a solution is given to the polynomial parametrization of the set of admissible causal multirate dead-beat ripple-free controllers, if simultaneously, the desired region of closed-loop poles is chosen. The admissible set of all causal DBRF controllers is obtained through the solution of multirate Diophantine equation, which is polynomially transformable to the standard Diophantine equation. The parametrization of the set of DBRF controllers is carried out in a finite dimensional space of polynomials that transforms the minimum-step control sequence into the control sequence corresponding to the controller with a given degree of freedom. *Copyright © 2003 IFAC*

Keywords: multirate sampled-data systems, dead-beat control, Diophantine equation

1. INTRODUCTION

The dead-beat (DB) control of single-rate sampled-data systems with zero steady-state error required only at sampling time instants after a finite settling time (Ackermann, 1985; Zafiriou and Morari, 1985) proved to be of disputable practical applicability due to often appearing intersample ripples, analysed e.g. in (Urikura and Nagata, 1987). Necessary and sufficient conditions for dead-beat ripple-free (DBRF) control are derived for error feedback controllers e.g. in (Sirisena, 1985; Jetto, 1994) and for state feedback controllers in (Urikura and Nagata, 1987). The increased number of steps has introduced the additional degree of freedom into dead-beat control problems and enabled formulation and solution to various optimization problems (Isermann, 1981; Barbargires and Karybakas, 1994; Fikar and Kučera, 2000; Rao and Rawlings, 2000; Mošna *et al.*, 2001). Often, in practice, it is desired to sample certain variables at a faster rate to achieve an improved performance and obtain the extra degrees of freedom for manipulating control variable. More recently, the relevance of systems with multirate sampling has been recognised and a large number of papers has appeared (Grasselli *et al.*, 1995). In this paper, a solution is given to the polynomial parametrization of DBRF controllers for multirate sampled-data systems. We restrict ourselves to linear servosystems, where the sampling period is an integer multiple of the control period and only a non-pathological sampling condition is required to guarrantee controllability and observability (Chen, 1995). In contrast to literature, it is adopted a polynomial approach to determine multirate discretised models. The admissible set of all causal DBRF controllers is obtained through the solution of multirate Diophantine equation (MDE), which is polynomially transformable to the standard Diophantine equation (DE). The optimization is conditioned by a new way of parametrization of a competitive

[1] This work was supported by the Ministry of Education of Czech Republic - Project No: MSM 235200004 and Grant Agency of the Czech Republic - Project No: 102/01/1347

set of controllers in a finite dimensional space of polynomials that transforms the minimum-step control sequence into the control sequence corresponding to the controller with a given degree of freedom. This approach also reflects stabilization properties, assured through the specification of the region Ω^+ containing admissible roots of the closed-loop polynomial. Mutual relations between the minimum number of control steps, sampling steps and the degree of freedom of admissible set of DBRF controllers with regard to the chosen region Ω^+ are analysed.

2. PLANT DISCRETIZATION

Consider the position of a linear position servo system described by its Laplace transform

$$\varphi(s) = \frac{b(s)}{sa(s)}u(s) + \frac{c(s)}{sa(s)} \tag{1}$$

where u is input and $a(s)$,$b(s)$ are coprime polynomials with $b(0) \neq 0$ and $\delta b \leq \delta a$ ($\delta = degree$). The polynomial $c(s)$, $\delta c \leq \delta a$, reflects a general initial condition. Nominally, it is supposed that the servo system is in a steady-state starting position $\varphi = w_{bgn}$ characterised by the relation $c(s) = a(s)w_{bgn}$. Suppose that the linear servo system in cascade with the zero-order hold is controlled with a control period h at time instants τ_k, $k = 0, 1, \ldots$. Further we suppose that the sampling of the plant output is carried out repeatedly after r control steps with the sampling period $H = rh$ in time instants t_k, synchronized with control time instants τ_k so that it holds $t_k = \tau_{rk}$. Our basic aim is to transfer the system in a finite number N of sampling steps from a nominal steady-state starting position $\varphi(t_o) = w_{bgn}$ to a given steady-state final position $\varphi(t) = w_{end}$, for $t \geq t_N$. Supposing that the sampling frequency is not in resonance with any of damped natural frequencies of the plant, the aim can be attained if and only if $\varphi(t_k) = w_{end}$ and $u_k = 0$ for $k \geq Nr$, where u_k is the value of input variable at time $t \in (\tau_k, \tau_{k+1})$.

The discrete-time plant discretised by the multirate sampling control with the period h is obtained in the coprime form

$$\varphi(z) = \frac{b_h(z)}{(z-1)a_h(z)}u(z) + \frac{zc_h(z)}{(z-1)a_h(z)} \tag{2}$$

where $a_h(0) \neq 0$ and the initial condition $c_h(z)$, $\delta c_h \leq \delta a_h$, is nominally given by $c_h(z) = a_h(z)w_{bgn}$. Further we denote the sequence of position values in sampling instants $t_k = \tau_{rk}$ as φ_H^o and as φ_H^i the sequence of position values in time instants $\tau_{rk+i} = t_k + i * h$, where $k = 0, 1, \cdots$, i.e. this sequence is shifted against φ_H^o by $i * h$. Its z-transform will be denoted $\varphi_H^i(q) = \mathcal{Z}\{\varphi^i\}_{z=q}$, whereas the symbol q is used for the z-transform variable to stress that the transformation is applied to the selected sequences with the sampling period H. Analogical procedure can be applied to selected sequences for input vlaues u_H^i. Then it holds

$$\varphi(z) = \sum_{i=0}^{r-1} z^{-i}\varphi_H^i(z^r), \quad u(z) = \sum_{i=0}^{r-1} z^{-i}u_H^i(z^r) \tag{3}$$

From the discrete-time plant (2), the discrete-time plant describing the sequence of position values φ_H^o, timed by the sampling period $H = rh$, can easily be obtained through the monic transformation polynomial $t(z)$, defined as a polynomial with the least possible degree such that $t(z)a_h(z) = a_H(z^r)$. After the substitution $q = z^r$ we obtain the polynomial $a_H(q)$. The set of polynomials $t(z)$ is always non-empty and properties of $t(z)$ are preserved even if $t(z)$ is multiplied by any non-zero number. Notice that the transformation of polynomials' product can be obtained as the product of their transformation polynomials. From the transformation $a_H(z^r) = t(z)a_h(z)$ it is obvious that the degree of $t(z)$ is an invariant given by $\delta t = (r-1)\delta a_h$. The polynomial $t(z)$ is unique and can be found through the solution of the set of $\delta t(z)$ linear algebraic equations via equating the corresponding coefficients. If, however, $r = 2^k$ the polynomial $t(z)$ can be determined more precisely and by a numerically far less demanding procedure. The transformation of the polynomial $(z-1)$ is achieved through its multiplication by the polynomial $t^{\odot}(z) = \sum_{i=0}^{r-1} z^i$ and thus it holds $t^{\odot}(z)(z-1) = (z^r - 1)$.

Expanding numerators in the plant model (2) by the polynomial $t^{\odot}(z)t(z)$ we get

$$\begin{aligned} t^{\odot}(z)t(z)b_h(z) &= \sum_{i=0}^{r-1} z^i b_H^i(z^r); \\ zt^{\odot}(z)t(z)c_h(z) &= \sum_{i=0}^{r-1} z^i \bar{c}_H^i(z^r) \end{aligned} \tag{4}$$

The expanded denominator has the form

$$t^{\odot}(z)(z-1)t(z)a_h(z) = (z^r - 1)a_H(z^r) \tag{5}$$

Substituting (3)–(5) into the plant (2) we obtain

$$\begin{aligned} \sum_{i=0}^{r-1} z^{-i}\varphi_H^i(z^r) &= \sum_{i=0}^{r-1} z^i \frac{\bar{c}_H^i(z^r)}{(z^r-1)a_H(z^r)} + \\ &+ \sum_{i=0}^{r-1}\sum_{j=0}^{r-1} z^{i-j} \frac{b_H^i(z^r)}{(z^r-1)a_H(z^r)} u_H^j(z^r) \end{aligned} \tag{6}$$

Comparing the left-hand and right-hand side of the identity (6) we obtain for $\varphi_H^o(z^r)$

$$\begin{aligned} \varphi_H^o(z^r) = &\sum_{i=0}^{r-1} \frac{b_H^i(z^r)}{(z^r-1)a_H(z^r)} u_H^i(z^r) + \\ &+ \frac{\bar{c}_H^0(z^r)}{(z^r-1)a_H(z^r)} \end{aligned} \tag{7}$$

As it always holds $\bar{c}_H^o(0) = 0$ we can generally write $\bar{c}_H^o(z^r) = z^r c_H^o(z^r)$ and after the substitution of $q = z^r$, the sequence of position values φ_H^o has the form

$$\varphi_H^o(q) = \frac{\boldsymbol{b}_H(q)}{(q-1)a_H(q)}\boldsymbol{u}_H(q)' + \frac{qc_H^o(q)}{(q-1)a_H(q)} \quad (8)$$

where $\boldsymbol{b}_H(q)$ resp. $\boldsymbol{u}_H(q)$ is r-dimensional row vector built up from the elements $b_H^i(q)$ resp. $u_H^i(q)$, for $i = 0, s, r-1$. It can be easily proved that under accepted suppositions it holds $\delta a_H(q) = \delta a_h(z)$, $\delta b_H^i(q) \leq \delta a_H(q)$, $\delta c_H^o(q) \leq \delta a_H(q)$, and, nominally, $c_H^o(q) = a_H(q)$.

In order to realize the desired DBRF transition, we aim at the design of an error causal feedback multirate digital controller

$$\begin{aligned} \boldsymbol{u}_H(q) &= \frac{\boldsymbol{b}_D(q)}{a_D(q)}e_H(q); \\ \boldsymbol{b}_D(q) &= [\ b_D^o(q), s, b_D^{r-1}(q)\ ] \\ \delta b_D^i(q) &\leq \delta a_D(q), \quad i = 0, s, r-1 \end{aligned} \quad (9)$$

where $e_H(q)$ is z-transform of the sequence of control error values $e(t) = w(t) - \varphi(t)$ at sampling time instants t_k. Polynomials in the row vector $\boldsymbol{b}_D(q)$ and the polynomial $a_D(q)$, cancelled by their greatest common divisor $gcd(a_D, \boldsymbol{b}_D)$, are denoted $\boldsymbol{b}_{Dc}(q)$ and $a_{Dc}(q)$ and define a coprime transfer function of the controller in (9). Naturally, our final aim will be the determination of the set of all DBRF controllers realizing the desired transition and simultaneously respecting a chosen measure of closed-loop stability. The linear parametrization of such set may then constitute a basic prerequisite for optimization.

3. MULTIRATE DIOPHANTINE EQUATION

Considering nominal initial conditions, the nominal transition can be realized through the control sequence generated by the open-loop DBRF controller.The feedback asserts its influence for non-nominal initial conditions and the controller can provide a desired measure of closed-loop stability, however, at the expense of bulkiness of obtainable nominal DBRF transitions. To make clear this relationships, we turn our attention to the specification of the set of all the control sequences that realize the transition from a general initial condition to the zero steady-state $w_{end} = 0$. The transition can always be carried out in $\delta a_h + 1$ control steps, such sequence is unique and thus there always exists at least one solution for the sequence with the length $\eta \geq \delta a_h + 1$.

Denote the z-transform of the position $\varphi(\tau_k)$ and the η-step control sequence u_k as

$$\varphi(z) = \frac{x(z)}{z^{\eta-1}} \qquad u(z) = \frac{y(z)}{z^{\eta-1}} \qquad (10)$$

where $x(z)$ and $y(z)$ are polynomials with the degree at most $\eta - 1$. Thus for $\eta \geq \delta a_h + 1$, there can be found to each $c_h(z)$, $\delta c_h \leq \delta a_h$ such polynomials $x(z)$ and $y(z)$ that

$$\frac{x(z)}{z^{\eta-1}} = \frac{b_h(z)}{(z-1)a_h(z)}\frac{y(z)}{z^{\eta-1}} + \frac{zc_h(z)}{(z-1)a_h(z)} \quad (11)$$

The identity (11) will come true if and only if

$$(z-1)a_h(z)x(z) + b_h(z)y(z) = z^{\eta}c_h(z) \quad (12)$$

Solving DE (12) with respect to the unknown $x(z)$ and $y(z)$, it is possible to find for the given initial condition $c_h(z)$ and for $\delta y < \eta$ just all the η-step control sequences assuring $\varphi(\tau_k) = 0$ for $k \geq \eta$. Under the condition $\delta y < \eta$, the solution of (12) always exists for $\eta \geq \delta a_h + 1$ and has $\eta_f = \eta - \delta a_h - 1$ degrees of freedom.

Analogically we may proceed for the discrete-time plant (8) describing the sequence of position values φ_H^o. The condition $\varphi(t_k) = 0$ for $k \geq N_c$ is equivalent to the zero steady-state condition $\varphi(t) = 0$ for $t \geq t_{N_c}$. Necessary and sufficient condition for the existence of control sequences realizing the transition from a general initial condition to the zero steady-state is given by MDE

$$(q-1)a_H(q)x^o(q) + \boldsymbol{b}_H(q)\boldsymbol{y}(q)' = q^{N_c}c^o(q) \quad (13)$$

where the polynomial $x^o(q)$ defines $\varphi_H^o(q) = \frac{x^o(q)}{q^{N_c-1}}$, row vector $\boldsymbol{y}(q)$ of polynomials $y^i(q)$ defines $u_H^i(q) = \frac{y^i(q)}{q^{N_c-1}}$ for $i = 0, 1, s, r-1$ and $c^o(q) = c_H^o(q)$ (see (7)). If $\delta \boldsymbol{y} < N$, the solutions of (13) specify the set of all rN-step DBRF control sequences, which is equivalent to the set of DBRF control sequences specified by DE (12) for $\eta = rN_c$. Thus every solution of the MDE (13) can be obtained also from DE (12) for $\delta y(z) \leq \eta$. For equivalent solutions it obviously holds

$$x(z) = \sum_{i=0}^{r-1} z^i x^i(z^r); \quad y(z) = \sum_{i=0}^{r-1} z^i y^i(z^r) \quad (14)$$

For an arbitrary $\delta c^o \leq \delta a_H$ and under the condition $N_c r \geq \delta a_H + 1$, MDE has always a solution for which $\delta \boldsymbol{y}(q) < N_c$ and such solution has $\eta_f = rN_c - \delta a_h - 1$ degrees of freedom.

If a DBRF control sequence is to be generated by a causal multirate error feedback controller (9), such controller cannot anticipate non-zero control error and thus the nominal output of the controller is constrained by the condition $\delta \boldsymbol{y}(q) \leq \delta x^o(q)$. Such solutions can be obtained from MDE (13) for $c^o(q) = q^{\delta a_H - \delta c_H^o} c_H^o(q)$ in the form

$$\begin{aligned} \varphi(q) &= \frac{1}{q^{\delta a_H - \delta c_H^o}} \frac{x^o(q)}{q^{N_c-1}} \\ \boldsymbol{u}(q) &= \frac{1}{q^{\delta a_H - \delta c_H^o}} \frac{\boldsymbol{y}(q)}{z^{N_c-1}} \end{aligned} \quad (15)$$

and for all searched solutions with $\delta \boldsymbol{y}(q) < N_c$ it holds $\delta x^o(q) = N_c - 1$.

Each solution of MDE (13) can be obtained through the solution of DE (12), if only the polynomial of an initial condition $c^o(q)$ is transformed to $c_h(z)$ by means of the transformation implicitly given by (4). Since the polynomials $c^o(q)$, $\delta c^o(q) \leq \delta a_H(q)$ and $c_h(z)$, $\delta c_h(z) \leq \delta a_h(z)$ can be viewed as only formally different expressions for the same initial state, the linear transformation between spaces of polynomials is always regular.

4. MULTIRATE DBRF STABILIZATION

Consider the model (8) and the multirate controller (9). Changing an initial position w_{bgn} by a constant Δw, the error response $e(q)$ and the control response $\boldsymbol{u}(q)$ are described by

$$e(q) = \frac{q a_H(q)}{a_C(q)} a_{Dc}(q) \Delta w \tag{16}$$

$$\boldsymbol{u}(q) = \frac{q a_H(q)}{a_C(q)} \boldsymbol{b}_{Dc}(q) \Delta w \tag{17}$$

$$a_C(q) = (q-1) a_H(q) a_{Dc}(q) + \boldsymbol{b}_H(q) \boldsymbol{b}'_{Dc}(q) \tag{18}$$

where $a_C(q)$ is the closed-loop pole polynomial.

The control system realizes DBRF position change in the finite number of steps if and only if the denominators of fractions (16), (17) after the reduction are equal to q^{N_κ}. Since it always holds $gcd(a_H, q) = 1$ and $gcd(\ \boldsymbol{b}_{Dc},\ a_{Dc}) = 1$, this can be attained only if the closed-loop polynomial has the form $a_C(z) = q^{N_\kappa} a_H^\times(q)$, where $gcd(a_H^\times, a_H) = a_H^\times(q)$ or equivalently if

$$(q-1) \cdot a_H \cdot a_{Dc} + \boldsymbol{b}_H \cdot \boldsymbol{b}'_{Dc} = q^{N_\kappa} \cdot a_H^\times(q) \tag{19}$$

Provided that $gcd(a_H, \boldsymbol{b}_H) = 1$, the solution of MDE (19) for unknown polynomials a_{Dc} and $\boldsymbol{b}_{Dc}$ exists for sufficiently large N_κ.

The roots of the factor $a_H^\times$ of the closed-loop polynomial a_C determine the measure of the control system stability, however, the condition of stability may restrict the competitive set of controllers in the optimization of the desired DBRF position change response. The controller is said to be generally admissible, provided it produces a polynomial a_C, the roots of which are laid out in a region Ω_H^+ of the complex plane. The region Ω_H^+ is supposed to be a transform of the region Ω^+ conveyed by the function $\exp(\omega H)$.

As an admissible controller we denote the causal controller (9), described by the polynomials b_{Dc} and a_{Dc} satisfying the identity (19), where

$$N_\kappa \leq N \quad \wedge \quad (a_H^\times, a_H^+) = a_H^\times(q) \tag{20}$$

and $a_H^+(q) = gcd(a_H(q))$ has all roots in Ω_H^+. Its complement $a_H^-(q)$ is the polynomial for which it holds $a_H(q) = a_H^+(q) a_H^-(q)$. Hereafter, we aim at finding the set of all admissible controllers $\mathcal{D}_H(\Omega^+, N)$. Complying with (20), it is obvious that for $\Omega_1^+ \subseteq \Omega_2^+$ and $N_1 \leq N_2$ the relation $\mathcal{D}_H(\Omega_1^+, N_1) \subseteq \mathcal{D}_H(\Omega_2^+, N_2)$ can be deduced. Remark that by tightening up the requirements on stability conditions and the settling time, we constrain the set of admissible controllers.

5. ADMISSIBLE SET OF CONTROLLERS

In order to parametrize the set of admissible controllers $\mathcal{D}_H(\Omega^+, N)$, it is advisable to admit the divisibility of the controller polynomials $a_d(q)$ and $\mathbf{b}_d(q)$ by the polynomial

$$gcd(a_D, \boldsymbol{b}_D) = q^{N - N_\kappa} \frac{a_H^+(q)}{a_H^\times(q)} \tag{21}$$

Multiplying the identity (19) by the factor (21) and using of the validity of $a_D = gcd(a_D, \boldsymbol{b}_D) a_{Dc}$ and $\boldsymbol{b}_d = gcd(a_D, \boldsymbol{b}_D) \boldsymbol{b}_{Dc}$, we get

$$\begin{aligned}(q-1) a_H(q) a_D(q) + \boldsymbol{b}_H(q) \boldsymbol{b}'_D(q) = \\ = q^{N - \delta a_H^-} q^{\delta a_H^-} a_H^+(q)\end{aligned} \tag{22}$$

If the controller $D(q) = \boldsymbol{b}_D(q)/a_D(q)$ is admissible, its coprime form after the extension by the factor (21) satisfies the identity (22) and vice versa. Namely, if it holds (22) then it also holds $a_{cl}(q) gcd(a_D, \boldsymbol{b}_D) = q^N a_H^+(q)$. The set of admissible controllers $\mathcal{D}_H(\Omega^+, N)$ is generated by all causal solutions of MDE (22) for the unknown polynomials $a_D(q)$ and $\boldsymbol{b}_D(q)$. All causal solutions for which $\delta \boldsymbol{b}_D(q) \leq \delta a_D(q)$ can be obtained from MDE (13) after the substitution $c^o(q) = q^{\delta a_H^-} a_H^+(q)$ and $N_c = N - \delta a_H^-$. Then, it follows that the equation (22) has solution if $r(N - \delta a_H^-) \geq \delta a_H + 1$. Thus we obtain

$$rN \geq \eta_o = \delta a_H^- + r \delta a_H^- + 1 \tag{23}$$

$$N \geq N_o = \delta a_H^- + \text{ceil}\left\{\frac{\delta a_H + 1}{r}\right\} \tag{24}$$

where η_o denotes the minimum number of control steps, N_o denotes the minimum number of sampling steps and the function ceil$\{\bullet\}$ returns the smallest integer greater than or equal to its numeric argument. Then the solution has $\eta_f = r(N - \delta a_H^-) - \delta a_H - 1$ degrees of freedom and for all those solutions it holds $\delta a_D(q) = N - \delta a_H^- - 1$.

In order to find the set of all causal controllers satisfying MDE (22) we convert the solution of MDE (13) to the solution of DE (12). It is sufficient to use (4) and find to the polynomial $c^o(q) = q^{\delta a_H^-} a_H^+(q)$ the corresponding polynomial $c_h(z)$. Introducing polynomials $t^\oplus(z)$ and $t^\ominus(z)$, for which $t^\oplus a_h^+(z) = a_H^+(z^r)$ and $t^\ominus a_h^-(z) = a_H^-(z^r)$, the polynomial $t(z)$ for $a_h(z)$ will be in the form $t(z) = t^{\oplus}(z)t^{\ominus}(z)$. From (4) it follows that $c_h(z) = c^{\ominus}(z)a_h^+(z)$. After the substitution into (4) we get the condition for $c^{\ominus}(z)$

$$zt^{\odot}(z)t^{\ominus}(z)c^{\ominus}(z) = z^{r(\delta a_H^- + 1)} + \sum_{i=1}^{r-1} z^i c_i^{\ominus}(z^r) \quad (25)$$

where $c_i^{\ominus}(z^r)$ are arbitrary polynomials. The searched polynomial $c^{\ominus}(z)$ always exists and its degree is δa_H^-. The polynomial is uniquely given by the solution of $\delta a_h^- + 1$ equations for $\delta a_h^- + 1$ coefficients. Using the relation $rN_c = r(N - \delta a_h^-) = \eta_f + \delta a_h + 1$, the solution of MDE (22) can be converted to the solution of DE

$$(z-1)\cdot a_h \cdot a_d + b_h \cdot b_d = z^{\eta_f} \cdot z^{\delta a_h + 1} \cdot c^{\ominus} \cdot a_h^+ \quad (26)$$

where the searched solutions are given as

$$\begin{aligned} b_d(z) &= \sum_{i=0}^{r-1} z^i b_D^i(z^r) \\ a_d(z) &= a_D(z^r) + \sum_{i=1}^{r-1} z^i a_D^i(z^r) \end{aligned} \quad (27)$$

The causality condition of the multirate controller takes on the form $\delta b_d(z) \leq \delta a_d(z)$. The equation (26) has the unique causal solution $a_{do}(z)$, $b_{do}(z)$ if and only if $\eta_f = 0$. The set of all causal solutions can be parametrized in a linear η_f-dimensional space of polynomials $\beta(z)$, $\delta\beta < \eta_f = r(N - \delta a_h^-) - \delta a_h^+ - 1$ as

$$\begin{aligned} b_d(z) &= z^{\eta_f} b_{do} + (z-1)a_h(z)\beta(z) \\ a_d(z) &= z^{\eta_f} a_{do} - b_h(z)\beta(z) \end{aligned} \quad (28)$$

In each solution of (26) the polynomial $a_h^+(z)$ must be a factor of the polynomial $b_d(z)$ and therefore its causal solution can be expressed as

$$b_d(z) = a_h^+(z)\alpha(z) \quad (29)$$

$$\alpha(z) = z^{\eta_f}\alpha_o(z) + (z-1)a_h^-(z)\beta(z) \quad (30)$$

$$a_d(z) = z^{\eta_f} a_{do} - b_h(z)\beta(z) \quad (31)$$

where $\alpha_o(z)$ and $a_{do}(z)$ are solutions of DE

$$(z-1)\cdot a_h^- \cdot a_{do} + b_h \cdot \alpha_o = z^{\delta a_h + 1} \cdot c^{\ominus} \quad (32)$$

with $\delta a_{do} = \delta a_h$. Under accepted assumptions, such solution always exists and is unique. Then it holds $\delta\alpha_o \leq \delta a_h^-$. The relations (29) and (31) together with (27) parametrize the set of admissible controllers $\mathcal{D}_H(\Omega^+, N)$ for $N \geq N_o$ in η_f-dimensional space of polynomials β, having the degree at most $\eta_f - 1$.

Multiplying (22) by the polynomial $a_H^-(q)\Delta w$, we obtain DE (13) for open-loop DBRF transitions from the nominal initial condition $c^o(q) = a_H(q)\Delta w$. By comparison it follows that

$$u_H(q) = \frac{a_H^-(q)\boldsymbol{b}(q)}{q^{N-1}}\Delta w \quad (33)$$

According to (3) and with the usage of $a_H^-(z^r) = t^{\ominus}(z)a_h^-(z)$, we obtain after the substitution (30) the parametrized DBRF control sequence

$$\begin{aligned} u(z) = & \frac{a_h(z)t^{\ominus}(z)\alpha_o(z)}{z^{\eta_o - 1}}\Delta w + \\ & + \frac{a_h(z)(z-1)t^{\ominus}(z)a_h^-(z)}{z^{\eta_o}}\frac{\beta(z)}{z^{\eta_f - 1}}\Delta w \end{aligned} \quad (34)$$

For $\beta(z) = 0$ we obtain the minimum-step control sequence with the length η_0. For $a_h^+ = a_h$ the minimum-step controller produces the minimum-step open-loop control sequence independently on the sampling period. For $r > \delta a_H + 1$ the minimum-step controller produces in each sampling step the non-zero control segment with the length utmost $\delta a_H + 1$ whilst the remaining control have zero value. Remark that there can exist more controllers realizing the transition at most in N_o sampling steps and they build up a space having $r\text{ceiling}\{(\delta a_h + 1)/r\} - \delta a_h - 1$ degrees of freedom. Turning our attention back to DE (12) which specifies all DBRF control sequences, we find out that for the nominal initial condition $c_h(z) = a_h(z)\Delta w$ the polynomial $y(z)$ of the nominal control sequence always contains the polynomial a_h as its factor. In comparison with (34) it is obvious that the DBRF control sequence, nominally generated by the controller, always contains in the polynomial $y(z)$ an additional polynomial $t^{\ominus}(z)$ as its factor. This condition of feedback stabilization cuts down the number of admissible nominal DBRF transitions compared to the open-loop control. The only exception arises if $a_h^+ = a_h$, i.e. in case that there is no need to compensate for the dynamics of the servo system.

6. EXAMPLE

Consider the servo system described by

$$G = \frac{b(s)}{sa(s)} = \frac{1}{0.03s^4 + 0.07s^3 + 0.6s^2 + s} \quad (35)$$

Choosing the control period h and discretising the plant with the zero-order hold, we get the discrete-time plant discretized by the multirate sampling control described by the discrete transfer function

$$G_h = \frac{b_h(z)}{(z-1)a_h(z)} \quad (36)$$

with stable polynomial a_h. Taking into account the admissible measure of the closed-loop system stability defined by the set Ω, only two border choices of the factor a_h^+ will be considered: $a_h^+ = a_h$ and $a_h^+ = 1$. In the first case, all the stable poles are acceptable for the closed-loop polynomial, in the second case no open-loop poles are acceptable for the closed-loop polynomial. In all experiments the sampling period is $H = 0.4sec$ and the desired transition is $N = 7$. Control frequence is considered as r-multiple of the sampling frequence in succesive steps $r = 1, 2, 4, 8$. The discrete-time plant (2) is determined with the usage of polynomials $t^{\oplus}(z)$, $t^{\ominus}(z)$. The set of admissible DBRF controllers $\mathcal{D}_H(\Omega^+, N)$ in the form (28) is parametrized in the space of polynomials $\beta(z)$ having the degree $\delta\beta < \eta_f$, where $\eta_f = rN - \eta_0$ is the degree of freedom in the space of admissible controllers. For this purpose, it is necessary to find via the solution of DE (28) the admissible minimum-step controller realizing the admissible DBRF transition in η_0 control steps (24) and thus in N_o sampling steps (25). The minimal possible time period of the transition in the set of admissible controllers is equal to $T_{min} = h\eta_o$. Prior to the determination of the solution of DE it is necessary to determine for the respective r and $t^{\ominus}(z)$ the polynomial $c^{\ominus}(z)$ from (25). For the monitored cases with $a_h^+ = a_h$ resp. $a_h^+ = 1$ it holds $t^{\oplus}(z) = t(z)$ and $t^{\ominus}(z) = 1$ resp. $t^{\oplus}(z) = 1$ and $t^{\ominus}(z) = t(z)$, where $t(z)$ transforms the polynomial $a_h(z)$. The values characterizing parametric spaces of controllers are in Table 1.

		$a_h^+ = a_h$				$a_h^+ = 1$			
r	h	N_o	η_o	T_{min}	η_f	N_o	η_o	T_{min}	η_f
1	0.4	4	4	1.6	3	7	7	2.8	0
2	0.2	2	4	0.8	10	5	10	2	4
4	0.1	1	4	0.4	24	4	16	1.6	12
8	0.05	1	4	0.2	52	4	28	1.4	28

Table 1. Values characterizing parametric spaces of controllers

The step responses of the minimum-step linear servo system for $r = 8$ and $a^+ = a$ resp. $a^+ = 1$ are illustrated in Fig.1.

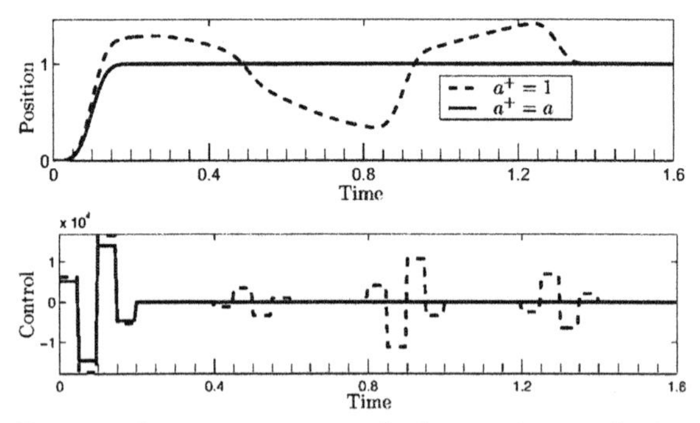

Fig. 1. Step responses of the position of the minimum-step linear servo system

7. CONCLUSIONS

The problem of multirate DBRF position control of a linear servosystem with the chosen stability region Ω^+ has been formulated. The method for the parametrization of the space of all admissible multirate controllers realizing the desired transition with the given number of sampling steps N in the η_f-dimensional space of polynomials was developed. The degree of freedom η_f is given by the difference between the number of dispensable control steps rN and the minimum number of control steps η_o, which is strongly influenced by the requirements on closed-loop stability. Each pole of the servosystem outside the region Ω^+ extends the minimum transition time by one sampling step, i.e. by r control steps. The increase of the number of control steps r within the sampling period H, under the fixed Ω^+ and N, leads to the increase of the dimension of the parametric space. Projecting the parametric space into control sequences (34) enables straightforward multirate optimization in the sense of minimum effort criterion, minimum fuel consumption and the like (Fikar and Kučera, 2000; Mošna *et al.*, 2001).

REFERENCES

Ackermann, J. (1985). *Sampled-Data Control Systems.* Springer-Verlag. Berlin.

Barbargires, C. A. and C. A. Karybakas (1994). Ripple-free dead-beat control of DC servo motors. In: *Proc. 2nd IEEE Mediterranean Symposium on New Directions in Control and Automation.* Chania, Crete. pp. 469–476.

Chen, T. (1995). *Optimal sampled-data control systems.* Springer. London.

Fikar, M. and V. Kučera (2000). On minimum finite lenght control problem. *Int. J. Control* **73**(2), 152–158.

Grasselli, G. M., L. Jetto and S. Longhi (1995). Ripple-free dead-beat tracking for multirate sampled-data systems. *Intrenational Journal of Control* **61**(6), 1437–1455.

Isermann, R. (1981). *Digital Control Systems.* Springer–Verlag.

Jetto, L. (1994). Deadbeat ripple-free tracking with disturbance rejection: A polynomial approach. *IEEE Trans. Automat. Control* pp. 1459–1764.

Mošna, J., J. Melichar and P. Pešek (2001). Minimum effort dead–beat control of linear servomechanisms with ripple–free response. *Optimization and Applications* pp. 127–144.

Rao, C. V. and J. B. Rawlings (2000). Linear programming and model predictive control. *Journal of Process Control* **10**, 283–289.

Sirisena, H. R. (1985). Ripple-free deadbeat control of SISO discrete systems. *IEEE Trans. Automat. Control* pp. 168–170.

Urikura, S. and A. Nagata (1987). Ripple-free deadbeat control for sampled-data systems. *IEEE Trans. Automat. Control* pp. 474–482.

Zafiriou, E. and M. Morari (1985). Digital controllers for SISO systems: a review and a new algorithm. *Internat. J. Control* pp. 855–876.

SATURATION AVOIDANCE CONTROL OF A MAGNETIC LEVITATION SYSTEM WITH ENHANCED PERFORMANCES [1]

M. Canale * S. Malan *

* *Dip. di Automatica e Informatica, Politecnico di Torino,
Corso Duca degli Abruzzi 24, 10129 Torino, Italy.
massimo.canale@polito.it, stefano.malan@polito.it*

Abstract: The problem of designing a robust control system for a magnetic ball levitator is considered, avoiding actuator saturation. Two different design procedures are proposed and compared, taking into account both time and frequency domain performances. In particular, it will be shown how the performances obtained by a simple cascade controller can be improved by the joint employment of a predictive control law. *Copyright © 2003 IFAC*

Keywords: Control applications, Constraint control, Robust control, Saturation Avoidance.

1. INTRODUCTION

Control design of linear systems with actuator saturation makes up one of the most important design problems to be faced in practice (see e.g. IJRN95, 1995; Tarbouriech and G. Garcia Eds., 1997; Stoorvogel and Sannuti, 2000). Indeed, in several applications the system to be controlled works in the neighborhood of some operating condition where a linearized model can provide a good approximation, but severe constraints on actuator activity are present. This kind of problems are particularly relevant in magnetic levitation systems, that have been extensively studied by researcher for their simple but unstable and nonlinear behaviour (see, for example, Cho *et al.*, 1993; Trumper *et al.*, 1997; Oliveira *et al.*, 1999; Yang and Tateishi, 2001). In this paper the attention will be focused on a magnetic ball levitator used to perform research and educational experiments (see Carabelli and Malan, 1996).

The aim of this work is to design a control system able to achieve, around a given equilibrium point, robust stability and a certain level of both time and frequency domain nominal performances, avoiding input saturation. In order to describe appropriately the plant behaviour, taking into account parameter uncertainty and/or unmodelled dynamics, a linear model set can be suitably employed. The control requirements can be obtained by a simple linear controller designed using standard H_∞ techniques. The main drawback of this approach relies on the fact that, for satisfying the limitations on the command signal, the frequency domain system performances may degradate. These performances may be recovered by using a slightly more complicated control structure which employs the combined action of the linear compensator and a predictive control law.

The proposed design procedure is tested using model data related to a laboratory levitator at Politecnico di Torino and its effectiveness is analyzed by means of simulated data.

2. LEVITATOR SYSTEM DESCRIPTION

Figure 1 shows (on the left) a sketch of the physical layout of the magnetic ball levitator and (on the right) the corresponding block diagram containing related parts and signals. In particular, u is the command

[1] This research was supported in part by funds of Ministero dell'Istruzione, dell'Università e della Ricerca under the Project "Robustness and optimization techniques for control of uncertain systems".

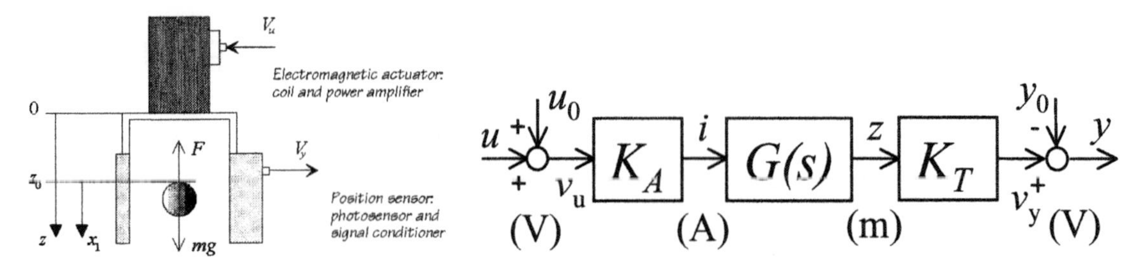

Fig. 1. Schematic and block diagram of magnetic ball levitator.

signal, u_0 the equilibrium command, v_u the actuator input, i the current power signal, z the ball position, v_y the transducer output, y is the output signal, y_0 the equilibrium output.

The actuator is composed by a power transconductance amplifier loaded with electromagnetic coil impedance. The local current feedback assures a high frequency dynamic that can be neglected in the model. Thus the transfer function between the controller output signal v_u and the electromagnetic current i is well approximated by

$$\frac{I(s)}{V_u(s)} = K_A = 0.40 \text{ AV}^{-1} \tag{1}$$

here assumed as a pure constant with an overload protection corresponding to maximum allowed command of 7 V (beyond this value the current is set to zero). Since the magnetic force may be only attractive the lower command is limited to 0 V. The measured points of the actuator static characteristic are depicted in Figure 2, together with its linear approximation given by K_A. These saturation limits ($0 \div 7$ V) are the ones that will be considered for the design procedure described in Section 3.

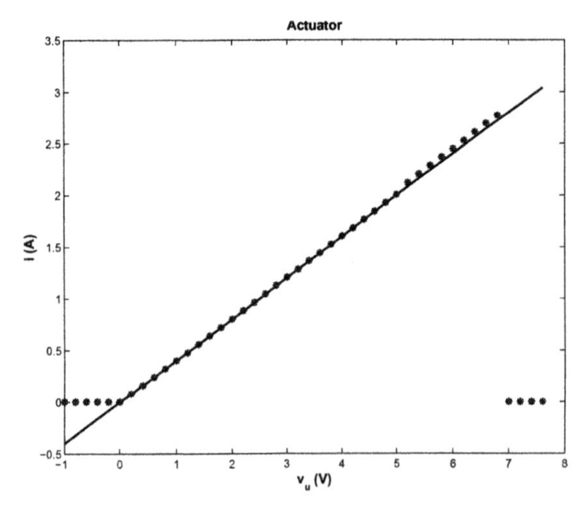

Fig. 2. Actuator static characteristic.

The transducer is a differential optical position sensor that gives out a signal v_y proportional to the vertical displacement z of the levitating ball from the central position fixed at $z_C = 6.27 \cdot 10^{-2}$ m. It is characterized by the nonlinear measured behaviour depicted in Figure 3 and it is well approximated, in the range $[z_C - 0.01,\ z_C + 0.01]$ m, by the linear approximation

$$\frac{V_y(s)}{Z(s)} = K_T = 652 \text{ V m}^{-1} \tag{2}$$

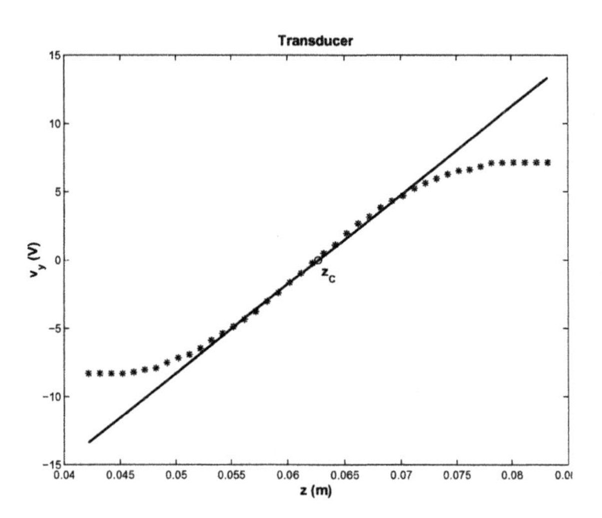

Fig. 3. Transducer static characteristic.

The plant to be controlled is composed by a levitating ball on which the electromagnetic force is exerted. The electromagnetic force F is nonlinear both in the coil current i and in the distance z between coil and ball:

$$F(i, z) = k\frac{i^2}{z^2} \tag{3}$$

with $k = 3.48 \cdot 10^{-4}$ Nm^{2}A^{-2} experimentally obtained around the desired equilibrium position.

The equilibrium of the forces acting on the ball

$$m\ddot{z} = mg - F(i, z) \tag{4}$$

brings to the following state space equations, linearized around the equilibrium position (close to the transducer central position z_C) $z_0 = 6 \cdot 10^{-2}$ m , corresponding to an equilibrium voltage output signal $y_0 = -1.76$ V, for an equilibrium current $i_0 = 0.62$ A corresponding to an equilibrium voltage command $u_0 = 1.54$ V:

$$\begin{cases} \dot{x}_1(t) = x_2(t) \\ \dot{x}_2(t) = \frac{h_x}{m}\, x_1(t) - \frac{h_i}{m}\, K_A\, u(t) \end{cases} \quad (5)$$

$$y(t) = K_T\, x_1(t) \quad (6)$$

where $x_1 = z - z_0$, $x_2 = \dot{z}$. The corresponding transfer function is

$$M(s) = \frac{Y(s)}{U(s)} = K_A \cdot G(s) \cdot K_T = = K_A \cdot \frac{-h_i}{m\, s^2 - h_x} \cdot K_T \quad (7)$$

with ball mass $m = 0.02$ kg, magnetic flux constant $h_i = 0.12$ kg m s^{-2} A^{-1} and negative stiffness $h_x = 6.54$ kg s^{-2}.

The described model does not approximate sufficiently well the plant behaviour. To take into account the difference between model and plant, around the chosen equilibrium position, an additive model set has been considered:

$$P = M + \Delta \text{ with } |\Delta(j\omega)| \leq W_\Delta(j\omega), \omega \in \Re^+ \quad (8)$$

The experimentally measured maximum additive perturbation Δ is such that $W_\Delta(j\omega) = 0.035$, $\forall \omega \in \Re^+$.

3. CONTROL SYSTEM DESIGN

The considered control design requirements are the following:

- robust stability;
- command signal v_u in the linear behaviour range: no saturation;
- optimize sensitivity function frequency response: minimize low frequency output disturbances effect and maximize the bandwidth;
- optimize the time response performance to a step reference signal: zero steady state following error, minimize rise time, overshoot and settling time.

The strong requirement to avoid saturation is due to the desire of having an almost linear behaviour of the controlled levitator. Moreover, note that, beyond the upper saturation limit of 7 V, the plant is switched off for safety reasons (see Figure 2), not allowing the experiment prosecution. Of course, this problem can be avoided by imposing a command maximum value slightly less than 7 V in the control algorithm software, but this is not a way to overcome the saturation problem.

The design of the control system can be performed by means of standard H_∞ techniques using a cascade controller $C(s)$ with negative unitary feedback structure. They allow to consider both robust stability requirements and nominal system performances. In particular, supposing that the plant P to be controlled is described by the additive model set (8), the control requirements may be recast into the following H_∞ optimization problem, which takes into account a certain performance level on the nominal sensitivity function subject to a robust stability condition, with respect to the additive perturbation Δ:

$$C(s) = \arg\min_{C(s)} \left\| \frac{W_S(s)}{1 + M(s)C(s)} \right\|_\infty \qquad s.t. \left\| \frac{W_E(s) \cdot C(s)}{1 + M(s)C(s)} \right\|_\infty < 1 \quad (9)$$

where $W_S(s)$ is a suitable frequency dependent weighting function used to shape the nominal sensitivity, while $W_E(s)$ is a rational low order function overbounding the model uncertainty Δ as tightly as possible. Moreover, the maximum value of the magnitude of the control variable $u(t)$ can be considered in the design by appropriate choices of the same weighting function $W_E(s)$ which penalizes the control sensitivity function $R(s) = \frac{C(s)}{1 + M(s)C(s)}$.

Given this context, two typical situations may occur such that the overall system performances could be limited.

Situation 1.
The weighting function W_E used to overbound the model uncertainty is such that the maximum command amplitude requirements are met. As a consequence, the input variable may present limitations on its full potentiality in terms of maximum allowed effort leading to slower responses of the output variable.

Situation 2.
The weighting function W_E used to overbound the model uncertainty is such that the maximum command amplitude exceeds the prescribed bounds. In this case a more stringent weighting function $W_E^{'}$ has to be employed to meet the command specifications. The immediate consequence is that the sensitivity functions requirements have to be relaxed resulting in frequency domain performance limitation in terms of bandwidth amplitude and output disturbance rejection.

In order to enhance the time response performances and to couple in an effective way both time and frequency domain specifications an improved control structure is proposed which employs a nominal Model Predictive Control (MPC) law together with the linear cascade controller C.

In a time domain design context, the satisfaction of hard input constraint can be achieved, for the nominal model M, by computing the control moves $\bar{u}$, in a receding horizon fashion, by solving (e.g.) the following MPC optimization problem at each time step:

$$\min_{U}\left\{\sum_{i=1}^{h_p}\|y(k+i|k)-r(k)\|^2+\sum_{i=0}^{h_c}\|\bar{u}(k+i|k)\|^2\right\}$$

$$U=[\bar{u}(k|k),\ldots,\bar{u}(k+h_c-1|k)]$$

$$\textbf{s.t. } U\in\mathcal{U}=\{\bar{u}\in U\ :\ 0>\bar{u}_{min}\leq\bar{u}\leq\bar{u}_{max}>0\} \tag{10}$$

where $r(k)$ is a prescribed reference signal, $\|\cdot\|$ is the euclidean norm and h_p and h_c are respectively the prediction and the control horizons satisfying $h_c < h_p \leq \infty$. As a matter of facts, in order to guarantee system stability, a further constraint has to be added to the optimization problem (10) (see for more details the recent survey Mayne *et al.*, 2000).

Now consider the control structure depicted on Figure 4 where the controller C can be designed using the optimization problem (9), the signal $\bar{u}$ is the input sequence computed according to (10) and $\bar{y}$ is the output of model M for the input $\bar{u}$.

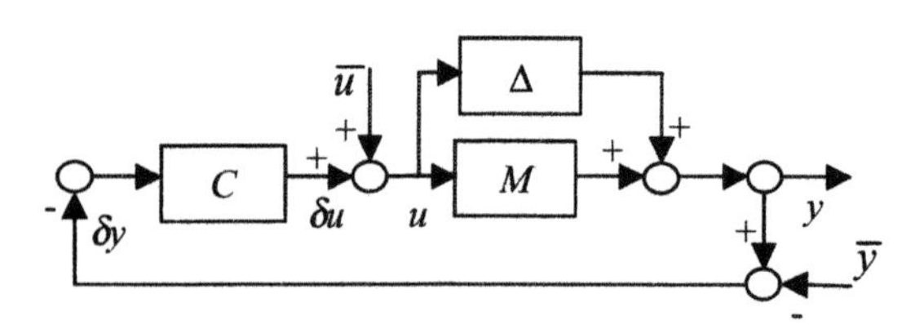

Fig. 4. Enhanced control structure.

This way, the system input u which has to be bounded is the sum of the two components $\bar{u}$ (the MPC one) and δu (the contribution of controller C):

$$u=\bar{u}+\delta u \tag{11}$$

Note that the MPC contribution $\bar{u}$ acts simply as a bounded exogenous signal in the control structure of Figure 4. This way, no feedback loops are closed by the MPC control law. Moreover, as also signal $\bar{y}$ is bounded, the stability of the overall system is guaranteed by controller C.

The enhanced control structure depicted in Figure 4 allows to overcome both the performance limitations pointed out in Situation 1 and 2 when the controller C has been designed according to the H_∞ criteria (9). The improvements that can be achieved by such structure can be summarized as follows:

Situation 1.

The MPC input $\bar{u}$ may be designed so that the amplitude and the effort characteristics of the overall signal u may be improved so to allow the full exploitation of its potentiality.

Situation 2.

The design of the MPC input $\bar{u}$ may be performed so that the required bound on the control variable u is met by preserving as much as possible (i.e. without strengthening the weight W_E) the original H_∞ design.

It has to be noted that both in Situation 1 and 2, the command limitation requirements have to be satisfied. In order to accomplish this aim the two components $\bar{u}$ and δu must fulfill, for all the systems P in the set (8), the following relation

$$\begin{gathered}|u|\leq u_{max}\\ \Updownarrow\\ |\bar{u}+\delta u|\leq u_{max}\\ \Uparrow\\ |\bar{u}|+|\delta u|\leq u_{max}\end{gathered} \tag{12}$$

In order to satisfy (12), the amplitude of the $\bar{u}$ and δu components may be designed such that:

$$\begin{aligned}&|\delta u|\leq\varepsilon\cdot u_{max}\\ &\qquad\qquad\qquad\qquad 0\leq\varepsilon\leq 1\\ &|\bar{u}|\leq(1-\varepsilon)\cdot u_{max}\end{aligned} \tag{13}$$

Of course, if ε tends to zero, the feedback component δu is severely limited, allowing a major contribution from the MPC control. On the contrary, if ε tends to one, the MPC component $\bar{u}$ is severely limited, allowing a major contribution from the feedback control. A suitable choice of the ε value is obtained by trials in order to reach a reasonable tradeoff between time domain and frequency domain performance requirements.

The limitation on $\bar{u}$ may be achieved and "tuned" by the MPC design procedure (10), whereas, for the limitation on δu, the "robust" computation procedure introduced in Reinelt and Canale (2001) may be conveniently adapted.

It has to be remarked that the weighting function $W_E(s)$ which penalizes the control sensitivity function $R(s)$ in the optimization problem (9), the control limitations $\bar{u}_{min}$ and $\bar{u}_{max}$ in the MPC problem (10) and the bound u_{max} in the inequalities (12) and (13) are suitably chosen according to the saturation limits ($0\div 7$ V) of signal v_u, taking also into account the equilibrium voltage command $u_0 = 1.54$ V.

4. SIMULATION RESULTS

The simulation results have been carried out analyzing the system behaviour in face of a step reference of amplitude -1.4 V, corresponding to a ball position change of about $1\cdot 10^{-2}$ m above the equilibrium $z_0 = 6\cdot 10^{-2}$ m.

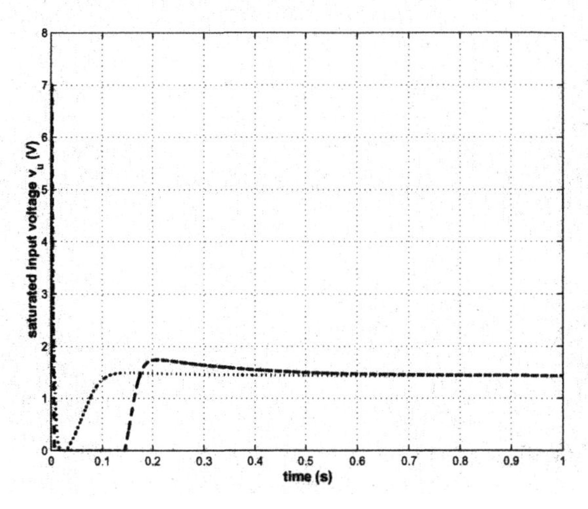

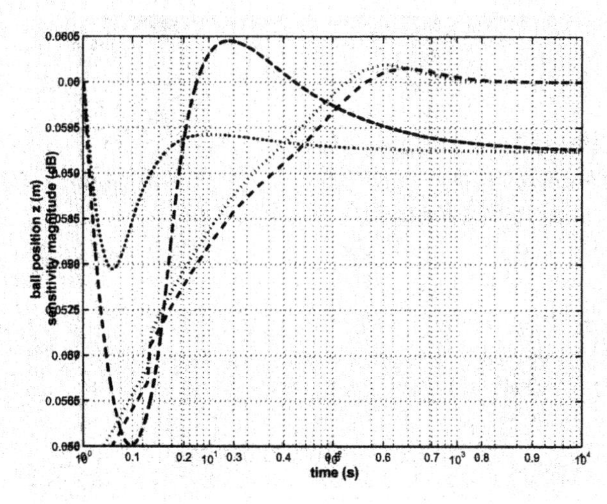

Fig. 5. Left: Saturated input signal using controllers C_1 (dashed), C_2 (dotted). Right: Ball position using controllers C_1 (dashed), C_2 (dotted).

The H_∞ design (9) has been performed by means of the following weighting function for the nominal sensitivity:

$$W_S(s) = 0.83\frac{s+100}{s} \quad (14)$$

It allows to take into account zero steady state following error for a step reference signals, a bandwidth amplitude larger than 100 rad/s, a resonance peak less than 2 dB allowing good stability margin properties, and a disturbance attenuation level of about 40 dB for $\omega \leq 1$ rad/s.

Moreover, as already mentioned in Section 2, it has been computed that the model uncertainty related to the system can be described by a weighting function $W_E(s)$ as follows:

$$W_E(s) = 0.035 \quad (15)$$

Using the weights (14) and (15) an H_∞ controller $C_1(s)$ has been designed and the related simulated behaviours of the input voltage v_u and the ball position z are reported in Figure 5 with dashed lines. Moreover, in Figure 6, the magnitude of the achieved sensitivity function is reported using dashed tract.

The results presented in Figure 5 show that, using controller $C_1(s)$, the control input v_u exceeds both the upper and lower prescribed limitations resulting in a degradate performance of the output response. Note that the output variation is inside the linear behaviour zone $z_C \pm 0.01$ m of the transducer (see Figure 3), so the performance degradation is essentially due to the input saturation. This kind of behaviour can be take back to the Situation 2 described in Section 3. Then, in order to meet the bound specifications for the input signal v_u, a new H_∞ controller $C_2(s)$ has been designed using a more stringent weight $W_E'(s) = 0.075$.

In Figure 5, it is shown (dotted lines) the time domain performance improvement (a better settling time and less undershoot and overshoot) obtained by controller $C_2(s)$ due to its saturation avoidance characteristics. On the other hand, the new design produces, as expected, a degradation on the achieved frequency performances as shown in Figure 6 (dotted line), where the magnitude of the sensitivity function is higher than the one obtained with controller $C_1(s)$. In particular, such performance degradation is reflected on a slight increase of the resonance peak and on a lower disturbance attenuation level, even if the performance specifications are yet met.

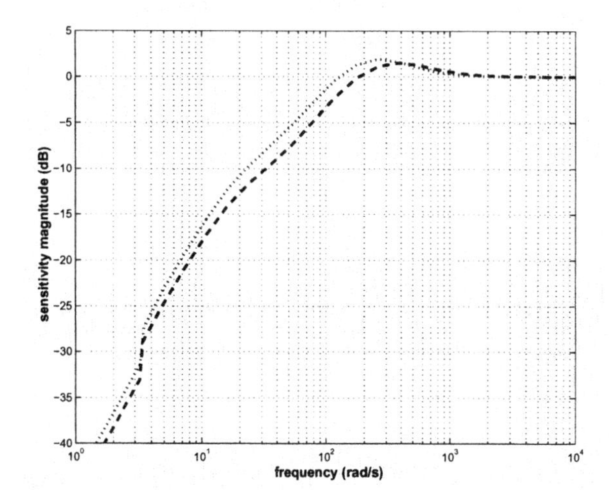

Fig. 6. Nominal sensitivity magnitude for controllers C_1 (dashed) and C_2 (dotted).

In order to enhance the overall performance level, according to the design procedure introduced in Section 3, a predictive nominal control law has been designed

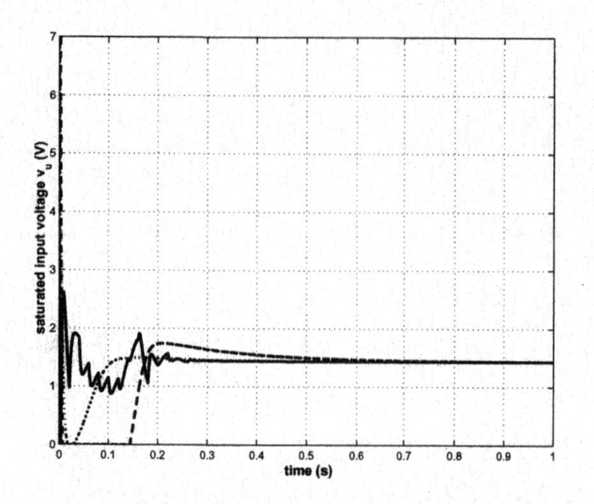

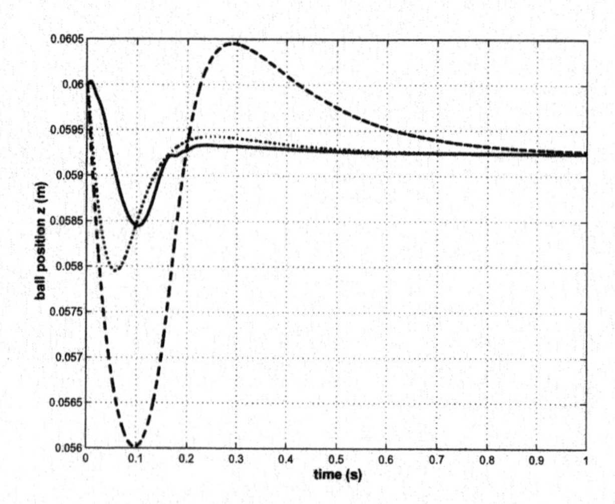

Fig. 7. Left: Saturated input signal using controllers $C_1 + MPC$ (solid), C_1 (dashed) and C_2 (dotted). Right: Ball position using controllers $C_1 + MPC$ (solid), C_1 (dashed) and C_2 (dotted).

using standard techniques for unstable systems. Using the "maximal output computation" results introduced in Reinelt and Canale (2001), such predictive control law can be used jointly with the action of the linear controller $C_1(s)$ without exceeding the limitations on the control variable v_u.

In Figure 7 the results obtained by using the combined action of controllers $C_1(s)$ and MPC are shown in solid lines. As it can be noted, the time domain performances are clearly improved with respect to the ones achieved by using the penalized controller $C_2(s)$ in terms of both peak response and settling time. Moreover, as shown in Figure 6, the usage of controller $C_1(s)$ allows the achievement of a slight improvement of the frequency domain characteristics of the controlled system.

5. CONCLUSION

The problem of designing a robust control system for a magnetic ball levitator has been considered, taking into account the necessity to avoid actuator saturation. Two different standard design procedures have been proposed and compared by simulation, considering both time and frequency domain performances. In particular, it has been shown how the degradate performances obtained by a simple cascade controller, that avoids input saturation, can be improved by the joint employment of a predictive control law.

6. ACKNOWLEDGEMENTS

The authors would like to thank Marco Colangelo, former student at Politecnico di Torino, whose master thesis originated the present work.

7. REFERENCES

Carabelli, S. and S. Malan, (1996). "Low order H_∞ controller design for a magnetic ball levitator", *Proc. World Automation Congress*, pp.101-106, Montpellier (France).

Cho, D., Y. Kato, D. Spilman, (1993). "Sliding mode and classical controllers in magnetic levitation systems", *IEEE Control System Magazine*, vol.13, pp.42-48.

IJRN95 (1995). Special issue on saturating actuators. *International Journal of Robust and Nonlinear Control.*

Mayne, D. Q., J. B. Rawlings, C. V. Rao and P. O. M. Skokaert (2000). Constrained model predictive control: Stability and optimality. *Automatica* **36**, 789–814.

Oliveira, V.A., E.F. Costa, J.B Vargas, (1999). "Digital implementation of a magnetic suspension control system for laboratory experiments", *IEEE Trans. on Education*, vol.42, pp.315-322, 1999.

Reinelt, W. and M. Canale, (2001). "Robust control of SISO systems subject to hard input constraints", *Proc. European Control Conference*, pp.1536-1541, Porto (Portugal).

Stoorvogel, A. A. and P. Sannuti (2000). *Control of Linear Systems with Regulation and Input Constraints*. Springer Verlag. New York.

Tarbouriech, S. and G. Garcia Eds. (1997). *Control of Uncertain Systems with Bounded Inputs*. Springer Verlag. New York.

Trumper, D.L., S.M.Olson, P.K. Subrahmanyan, (1997) "Linearizing control of magnetic suspension systems", *IEEE Trans. on Control Systems Technology*, pp.427-438.

Yang, Z.J. and M. Tateishi, (2001). "Adaptive robust nonlinear control of a magnetic levitation system", *Automatica*, vol.37, pp.1125-1131.

IDENTIFICATION OF WIENER SYSTEMS WITH MULTISEGMENT PIECEWISE-LINEAR NONLINEARITIES

Jozef Vörös

Slovak Technical University
Faculty of Electrical Engineering
Department of Automation and Control, Ilkovicova 3
812 19 Bratislava, Slovakia
e-mail: voros@nov1.kar.elf.stuba.sk

Abstract. The paper deals with the description of block-oriented nonlinear dynamic systems having multisegment piecewise-linear nonlinearities and with their identification using the Wiener model. A new form of multisegment nonlinearity representation and multiple application of a decomposition technique provides a special form of Wiener model with the minimum number of parameters. The model is linear in all the parameters to be estimated and is used in the proposed identification method. An illustrative example is included. *Copyright © 2003 IFAC*

Keywords: Nonlinear system; identification; multisegment piecewise-linear nonlinearities; Wiener model.

1. INTRODUCTION

Many nonlinear dynamic systems may be represented by the Wiener model with a linear dynamic system followed by a nonlinear static block (Haber, Keviczky, 1999). The linear block of Wiener model is typically described by its transfer function or a FIR model, e.g., (Menold et al., 1997). The characteristics of nonlinear blocks are generally approximated by polynomials of proper degree or splines, e.g., (Pajunen, 1992), (Norquay et al., 1998), (Zhu, 1999).

In previous applications dealing with block-oriented nonlinear dynamic models with multisegment piecewise-linear characteristics either a series of piecewise linear models were used to approximate the given nonlinear system over an operating range (Billings, Voon, 1987), or the static nonlinearity of Wiener model was approximated with a piecewise-linear function (Wigren, 1993). In these cases, specific a priori knowledge on the nonlinear characteristics is required to determine proper intervals of the approximation or to choose appropriate grid points.

This paper deals with modeling and identification of nonlinear dynamic systems, which can be represented by the Wiener model with the nonlinear static block containing the multisegment piecewise-linear function shown in Fig. 1. Static piecewise-linear nonlinearities in cascade with linear dynamic systems are often encountered in control systems. They allow to describe processes with different gains in different input intervals.

A decomposition of nonlinear system description based on multiple application of the key term separation principle is presented (Vörös, 1995, 2001). The aim of decomposition is to reduce the appearance of nonlinearly joined parameters in the system description. This provides a special form of Wiener model with the minimum number of parameters. The derived nonlinear model is linear in all the nonlinear and linear block parameters to be estimated. The parameter estimation algorithm is an iterative one, with the internal variables' estimation. To demonstrate the feasibility of the identification method, an illustrative example is included.

2. MULTISEGMENT PIECEWISE-LINEAR NONLINEARITY DESCRIPTION

For the description of the multisegment piecewise-linear characteristic depicted in Fig. 1, the key term separation principle can be applied parallel for more terms to obtain such a description, that simplifies the estimation of parameters involved (Vörös, 2002).

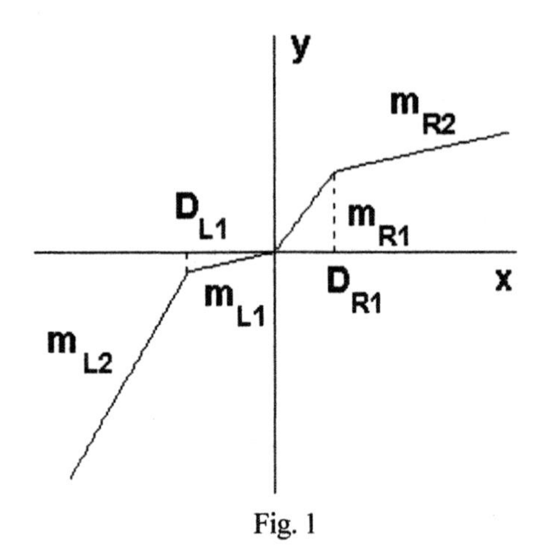

Fig. 1

The output of assumed nonlinearity y(t) according to Fig. 1 depends on the sign and magnitude of input x(t) and can be written as

$$y(t) = \begin{cases} m_{R1}\, x(t), & \text{if } 0 \le x(t) \le D_{R1}, \\ m_{R2}\,[x(t) - D_{R1}] + m_{R1}.D_{R1}, & \text{if } x(t) > D_{R1}, \end{cases} \tag{2.1}$$

$$y(t) = \begin{cases} m_{L1}\, x(t), & \text{if } D_{L1} \le x(t) < 0, \\ m_{L2}\,[x(t) - D_{L1}] + m_{L1}.D_{L1}, & \text{if } x(t) < D_{L1}, \end{cases} \tag{2.2}$$

where $|m_{R1}| < \infty$, $|m_{R2}| < \infty$ are the corresponding segment slopes and $0 \le D_{R1} < \infty$ is the constant for the positive inputs, $|m_{L1}| < \infty$, $|m_{L2}| < \infty$ are the corresponding segment slopes and $-\infty < D_{L1} \le 0$ is the constant for the negative inputs (Tao and Tian, 1998).

The switching function h(t), or more precisely h[x(t)], introduced in (Kung and Womack, 1984) is defined as

$$h(t) = h[x(t)] = \begin{cases} 0, & \text{if } x(t) > 0, \\ 1, & \text{if } x(t) < 0. \end{cases} \tag{2.3}$$

However, this can be extended as follows:

$$h[x(t)-D] = \begin{cases} 0, & \text{if } x(t) - D > 0, \\ 1, & \text{if } x(t) - D < 0, \end{cases} \tag{2.4}$$

to switch between two sets of input values x(t). Now using this form of switching function, the nonlinear characteristic in Fig. 1 can be described in the following input-output form

$$\begin{aligned} y(t) &= m_{R1}\, h[-x(t)]\, x(t) + \\ &+ (m_{R2} - m_{R1})\, h[D_{R1} - x(t)]\, [x(t) - D_{R1}] + \\ &+ m_{L1}\, h[x(t)]\, x(t) + \\ &+ (m_{L2} - m_{L1})\, h[x(t) - D_{L1}]\, [x(t) - D_{L1}]. \end{aligned} \tag{2.5}$$

As the equation (2.5) describing the nonlinearity output y(t) contains parameter combinations, it is not appropriate for the parameter estimation and the above mentioned decomposition technique can be applied to the separation of parameters.

The mapping given by (2.5) can be considered as the composite mapping

$$\begin{aligned} y(t) &= m_{R1}\, h[-x(t)]\, x(t) + f_1(t)\, x(t) - D_{R1}\, f_1(t) + \\ &+ m_{L1}\, h[x(t)]\, x(t) + f_2(t)\, x(t) - D_{L1}\, f_2(t), \end{aligned} \tag{2.6}$$

where the inner mappings are

$$f_1(t) = (m_{R2} - m_{R1})\, h[D_{R1} - x(t)], \tag{2.7}$$

$$f_2(t) = (m_{L2} - m_{L1})\, h[x(t) - D_{L1}]. \tag{2.8}$$

Choosing $f_1(t)$ and $f_2(t)$ as the key terms and after proper half-substituting (2.7) and (2.8) into (2.6), i.e., (2.7) only for $f_1(t)$ in the second term on the right-hand side of (2.6) and (2.8) only for $f_2(t)$ in the fifth term on the right-hand side of (2.6), the following expression is obtained:

$$\begin{aligned} y(t) &= m_{R1}\, h[-x(t)]\, x(t) + \\ &+ (m_{R2} - m_{R1})\, h[D_{R1} - x(t)]\, x(t) - D_{R1}\, f_1(t) + \\ &+ m_{L1}\, h[x(t)]\, x(t) + \\ &+ (m_{L2} - m_{L1})\, h[x(t) - D_{L1}]\, x(t) - D_{L1}\, f_2(t). \end{aligned} \tag{2.9}$$

Now the original mapping (2.5), describing the multisegment piecewise-linear characteristic of Fig. 1, is replaced by the equation (2.9) and those of (2.7) and (2.8) defining the variables $f_1(t)$ and $f_2(t)$, respectively.

The proposed description for the multisegment piecewise-linear characteristic (2.5) can be generalized as follows:

$$\begin{aligned} y(t) &= \sum_{i=1}^{n_R} (m_{R,i} - m_{R,i-1})\, h[D_{R,i-1} - x(t)]\, [x(t) - D_{R,i-1}] + \\ &+ \sum_{j=1}^{n_L} (m_{L,j} - m_{L,j-1})\, h[x(t) - D_{L,j-1}]\, [x(t) - D_{L,j-1}], \end{aligned} \tag{2.10}$$

where $|m_{R,i}| < \infty$, $|m_{L,j}| < \infty$ are the segment slopes, $0 \le D_{R,i} < D_{R,i+1} < \infty$ are the constants for the positive inputs, while $-\infty < D_{L,j+1} < D_{L,j} \leq 0$ are the constants for the negative inputs. Then (2.10) can be considered as a composite mapping and consequently decomposed in the same way as described above for (2.5). To coincide with (2.5), we can put $m_{R,0} = D_{R,0} = m_{L,0} = D_{L,0} = 0$.

3. WIENER MODEL

The Wiener model is given by a linear dynamic system followed by a static nonlinearity block and is shown in Fig. 2. The difference equation model of its linear block can be given as

$$x(t) = A(q^{-1})\, u(t) + [1 - B(q^{-1})]\, x(t), \tag{3.1}$$

where u(t) and x(t) are the inputs and outputs, respectively, $A(q^{-1})$ and $B(q^{-1})$ are scalar polynomials in the unit delay operator q^{-1}

$$A(q^{-1}) = a_0 + a_1 q^{-1} + \dots + a_m q^{-m}, \tag{3.2}$$
$$B(q^{-1}) = 1 + b_1 q^{-1} + \dots + b_n q^{-n}. \tag{3.3}$$

The output of linear block is identical with the input of nonlinear one. However, a direct substitution of x(t) from (3.1) into (2.9) would result in a very complex expression. Therefore, the key term separation principle can be applied again to simplify the model equation.

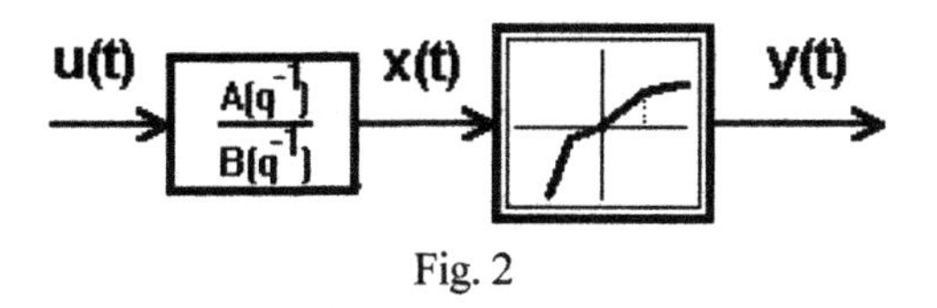

Fig. 2

The equation (2.9) describing the multisegment piecewise-linear nonlinearity can be rewritten as follows:

$$\begin{aligned} y(t) = {} & m_{R1}\, x(t) + (m_{L1} - m_{R1})\, h[x(t)]\, x(t) + \\ & + (m_{R2} - m_{R1})\, h[D_{R1} - x(t)]\, x(t) - D_{R1}\, f_1(t) + \\ & + (m_{L2} - m_{L1})\, h[x(t) - D_{L1}]\, x(t) - D_{L1}\, f_2(t). \end{aligned} \tag{3.4}$$

Assuming that $m_{R1} = 1$ (this is always possible in the given system description), we separate the variable x(t) as the key term of the nonlinear mapping (3.4). Then after the substitution of x(t) from (3.1) into (3.4) for the key term only, the system output is given in the form

$$\begin{aligned} y(t) = {} & A(q^{-1})u(t) + [1 - B(q^{-1})]x(t) + m_1\, h[x(t)]\, x(t) + \\ & + (m_{R2} - m_{R1})\, h[D_{R1} - x(t)]\, x(t) - D_{R1}\, f_1(t) + \\ & + (m_{L2} - m_{L1})\, h[x(t) - D_{L1}]\, x(t) - D_{L1}\, f_2(t), \end{aligned} \tag{3.5}$$

where $m_1 = m_{L1} - 1$. The equation (3.5) and those of (3.1), (2.7) and (2.8) defining the internal variables x(t), $f_1(t)$, and $f_2(t)$ represent a special form of Wiener model with multisegment piecewise-linear nonlinearity as the result of multiple application of the key term separation principle. The model has the minimum number of parameters and all of them enter the expressions linearly, except D_{R1} and D_{L1}, which appear both linearly and nonlinearly.

The Wiener model with multisegment piecewise-linear nonlinearity given by (3.5) can be put into a concise form

$$y(t) = \phi^T(t).\Theta_o, \tag{3.6}$$

where the data vector is defined as

$$\begin{aligned} \phi^T(t) = [& u(t), \dots, u(t-m), -x(t-1), \dots, -x(t-n), \\ & h[x(t)]x(t), h[D_{R1} - x(t)]x(t), -f_1(t), \\ & h[x(t) - D_{L1}]\, x(t), -f_2(t)], \end{aligned} \tag{3.7}$$

and the vector of parameters is

$$\begin{aligned} \Theta_o^T = [& a_0, \dots, a_m, b_1, \dots, b_n, \\ & m_1, m_{R2} - m_{R1}, D_{R1}, m_{L2} - m_{L1}, D_{L1}]. \end{aligned} \tag{3.8}$$

In the same way, the general case of multisegment piecewise linear nonlinearity can be incorporated into the Wiener model. After series of decompositions of (2.10) and the following half-substitution for the separated internal variable x(t), the resulting form of Wiener model will be linear in all the parameters to be estimated.

4. IDENTIFICATION ALGORITHM

An identification algorithm using the derived special form of Wiener model was developed enabling the simultaneous estimation of all the model parameters using input/output records. As the internal variables x(t), $f_1(t)$, and $f_2(t)$ are unmeasurable and therefore must be estimated, we have to consider an iterative estimation process, as in the case of Wiener model with polynomial nonlinearities (Vörös, 1995).

The proposed iterative algorithm is based on the use of the preceding estimates of model parameters for the estimation of unmeasurable internal variables. Assigning as

$${}^s x(t) = {}^s A(q^{-1})\, u(t) + [1 - {}^s B(q^{-1})]\, {}^s x(t), \tag{4.1}$$
$${}^s f_1(t) = ({}^s m_{R2} - {}^s m_{R1})\, h[{}^s D_{R1} - {}^s x(t)], \tag{4.2}$$
$${}^s f_2(t) = ({}^s m_{L2} - {}^s m_{L1})\, h[{}^s x(t) - {}^s D_{L1}], \tag{4.3}$$

the estimates of internal variables in the s-th step, the error to be minimized is gained from (3.6) in the vector form

$$e(t) = y(t) - {}^s\phi^T(t).{}^{s+1}\Theta, \tag{4.4}$$

where ${}^s\phi(t)$ is the data vector with the corresponding estimates of internal variables according to (4.1)-(4.3) and ${}^{s+1}\Theta$ is the (s+1)-th estimate of the parameter vector.

The various steps in the iterative procedure may be now stated as follows:

i) Minimizing a proper criterion based on (4.4) the estimates of both linear and nonlinear block parameters ${}^{s+1}\Theta$ are yielded using the data vector ${}^{s}\phi(t)$ with the s-th estimates of internal variables.
ii) Using (4.1)-(4.3) the estimates of internal variables ${}^{s+1}x(t)$, ${}^{s+1}f_1(t)$, and ${}^{s+1}f_2(t)$ are evaluated by means of the recent estimates of model parameters.
iii) If the estimation criterion is met, the procedure ends; else it continues repeating steps i) and ii).

Note that in the first step nonzero initial values of ${}^{1}D_{R1}$ and ${}^{1}D_{L1}$ have to be used to start up the iterative algorithm.

The identification algorithm based on the derived special form of Wiener model enables the direct estimation of all the model parameters using input/output records. The same steps can be applied to the Wiener model with the general description of multisegment piecewise-linear nonlinearity according to (2.10).

5. EXAMPLE

The proposed method for the identification of nonlinear dynamic systems with multisegment piecewise-linear nonlinearities using Wiener model was implemented and tested by means of MATLAB packages. Several systems were simulated and the estimations of all the model parameters (those of linear and nonlinear blocks) were carried out on the basis of input and output records as well as the estimated internal variables.

To illustrate the feasibility of the identification method using the described technique, the following example shows the results of parameter estimation process for the Wiener system with a multisegment piecewise-linear nonlinearity. The linear dynamic system is given by the equation

$$[1 - 0.2q^{-1} + 0.35q^{-2}]\, x(t) = [q^{-1} + 0.5q^{-2}]\, u(t)\,.$$

and the nonlinear block is given by the characteristic shown in Fig. 3. The 4-segment nonlinearity is characterized by the following parameters:

$m_1 = -0.8$
$m_{R2} - m_{R1} = -0.6$
$m_{L2} - m_{L1} = 0.9$
$D_{R1} = 0.5$
$D_{L1} = -0.4$

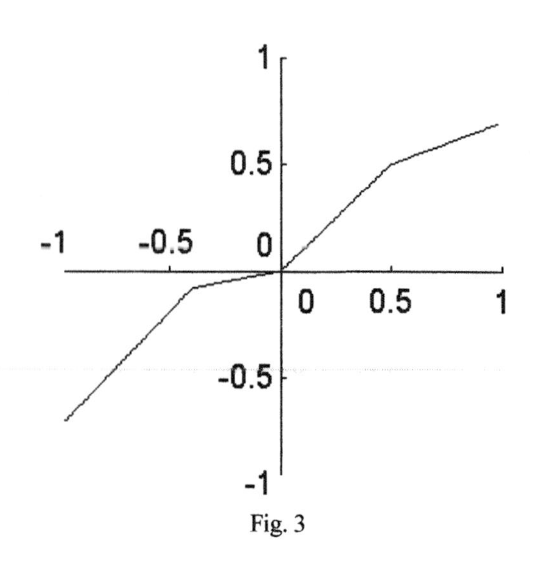

Fig. 3

The least squares method has been employed for the simultaneous estimation of both linear and nonlinear block parameters. The initial values of the parameters were chosen zero, except ${}^{1}D_{R1} = 0.1$, ${}^{1}D_{L1} = -0.1$. The identification was carried out with 2000 samples, using a random input with $|u(t)| \leq 2$. The values of parameter estimates are included in Table 1 and Table 2, where MSEx and MSEy are the mean square errors for the internal variable x(t) and the output y(t), respectively. The real values of the system parameters are written in the bottom row. The parameters' estimates are equal to the real values after 10 iterations.

Note, that the possible high values of $D_{R,i}$ and low values of $D_{L,j}$ in the early steps of the iterative procedure might require input signals with higher/lower amplitude, i.e., higher than the largest $D_{i,j}$, or lower than the smallest $D_{L,j}$, to ensure sufficient excitation for the estimation algorithm. However, proper limits for these parameters during the estimation process can overcome this problem.

6. CONCLUSIONS

A new approach to the parameter identification of nonlinear dynamic systems having multisegment piecewise-linear nonlinearities using the Wiener model has been presented. Multiple application of the decomposition technique based on the key term separation principle has provided a new and universal form of multisegment nonlinearity representation, which has led to a special form of Wiener model with the minimum number of parameters. All the model parameters to be estimated appear explicitly in the model description and they can be estimated simultaneously. A priori knowledge is restricted to proper limits for $D_{R,i}$ and $D_{L,j}$.

The proposed iterative method of model parameters' estimation uses records of measured input/output data and estimated internal variables. The presented example of identification process demonstrates the feasibility and good convergence properties of the proposed technique for Wiener models with multisegment piecewise-linear characteristics. This means that the identified models provide adequate representations of the given type of nonlinear dynamic systems. The proposed form of model suggests that it can also be incorporated into proper recursive schemes and used in adaptive control algorithms.

ACKNOWLEDGMENT

The author gratefully acknowledges financial support from the Slovak Scientific Grant Agency.

REFERENCES

Billings, S. A. and W. S. Voon (1987). Piecewise linear identification of non-linear systems. *Int. J. Control*, Vol. **46**, No. 1, pp. 215-235.

Haber, R. and L. Keviczky (1999). *Nonlinear System Identification - Input-Output Modeling Approach.* Kluwer Academic Publishers, Dorchrecht Boston London.

Kung, M. C. and B. F. Womack (1984). Discrete time adaptive control of linear dynamic systems with a two-segment piecewise-linear asymmetric nonlinearity. *IEEE Trans. Automatic Control*, Vol. AC-29, No. **2**, pp. 1170-172.

Menold, P.H., F. Allgöwer, and R.K. Pearson (1997). Nonlinear structure identification of chemical processes. *Computers & Chemical Engineering*, Vol. **21**, Suppl., pp. S137-S142.

Norquay, S.J., L. Palazoglu and J.A. Romagnoli (1998). Model predictive control based on Wiener models. *Chemical Engineering Science*, Vol. **53**, No. 1, pp. 75-84.

Pajunen, G.A. (1992). Adaptive control of Wiener type nonlinear systems. *Automatica*, Vol. **28**, No. 4, pp. 781-785.

Tao, G. and M. Tian (1998). Discrete-time adaptive control of systems with multisegment piecewise-linear nonlinearities. *IEEE Trans. Automatic Control*, Vol. **43**, No. 5, pp. 719-723.

Vörös, J. (1995). Identification of nonlinear dynamic systems using extended Hammerstein and Wiener models. *Control-Theory and Advanced Technology*, Vol. **10**, No. 4, Part 2, pp. 1203-1212.

Vörös, J. (2001). Parameter identification of Wiener systems with discontinuous nonlinearities. Systems & Control Letters, Vol. **44**, No. 5, pp. 363-372.

Vörös, J. (2002). Modeling and parameter identification of systems with multisegment piecewise-linear characteristics. IEEE Trans. Automatic Control, Vol. **47**, No. 1, pp. 184-188.

Wigren, T. (1993). Recursive prediction error identification using the nonlinear Wiener model. *Automatica*, Vol. **29**, No. 4, pp. 1011-1025.

Zhu, Y. (1999). Parametric Wiener model identification for control. *Proceedings 14th IFAC World Congress*, Beijing, Vol. H, pp. 37-42.

Table 1

It.	a_1	a_2	b_1	b_2	m_1	m_{R2}-m_{R1}	m_{L2}-m_{L1}	D_{R1}	D_{L1}
1	0.7072	0.3776	-0.1729	0.3273	-0.9169	0.0067	0.9409	0.1000	-0.1000
2	0.9951	0.5096	-0.2691	0.4819	-1.1327	-0.6644	1.1202	0.1000	-0.2621
3	1.1885	0.6586	-0.2151	0.4434	-0.9530	-0.6237	0.7611	0.2145	-0.2565
4	1.0124	0.5751	-0.1345	0.2961	-0.6646	-0.4052	0.6483	0.2808	-0.3638
5	1.0065	0.5159	-0.1872	0.3375	-0.8003	-0.5439	0.8936	0.5815	-0.5495
6	0.9870	0.4944	-0.1941	0.3412	-0.6864	-0.5838	0.8016	0.5513	-0.4041
7	0.9930	0.4923	-0.2039	0.3520	-0.8054	-0.6020	0.9161	0.5134	-0.4494
8	0.9989	0.4984	-0.2021	0.3528	-0.7886	-0.6031	0.8902	0.4982	-0.3929
9	0.9994	0.5005	-0.1997	0.3504	-0.7996	-0.5997	0.8995	0.4975	-0.4042
10	0.9998	0.5005	-0.1997	0.3501	-0.7999	-0.6001	0.9000	0.5003	-0.4002
r	1.0000	0.5000	-0.2000	0.3500	-0.8000	-0.6000	0.9000	0.5000	-0.4000

Table 2

It.	MSEx	MSEy
1	0.04068959	0.00993635
2	0.01633690	0.02646342
3	0.04399294	0.01044311
4	0.00117435	0.00285225
5	0.00007029	0.00543578
6	0.00019076	0.00064808
7	0.00002367	0.00065903
8	0.00000383	0.00004513
9	0.00000020	0.00000470
10	0.00000003	0.00000003

www.elsevier.com/locate/ifac

ROBUST MODEL PREDICTIVE CONTROL OF LINEAR TIME INVARIANT SYSTEMS WITH DISTURBANCES

J. Zeman* B. Rohá l'-Ilkiv*

* *Department of Automation and Measurement*
Faculty of Mechanical Engineering, Slovak University of Technology
Nám. Slobody 17, 812 31 Bratislava, Slovakia
fax: +421 2 52495315 and e-mail: zeman@kam.vm.stuba.sk

Abstract: In recent few years authors investigate methods for establishing and solving stability of model predictive control (MPC) with linear time-invariant systems. In this paper a modification of a predictive control scheme to ensure robust feasibility and stability for a case of discrete time linear systems with disturbances is submitted. A new cost function, which penalizes a deviation of state trajectories from a robust invariant set and deviation from some ideal control law is presented. The cost allows formulation of robustly stable MPC that can be solved via min-max optimization using a single linear program. The theory of robust invariant sets is investigated and applied. Further an algorithm for finding robust invariant and feasible sets, that are used in the new linear cost function, is presented. These suggested robust invariant sets have suitable properties and are less computational intensive. A min-max formulation of MPC than involves minimization of the maximum of disturbance influence, which range over a given polytope of possible disturbance realizations, is presented. In general, the solution of this problem is computationally demanding with large number of unknown parameters. The appropriate use of the performance criteria with respect to the receding horizon strategy, low complexity robust invariant sets and feedback policy with consideration of only the extreme realizations of disturbance signals can decrease this number of parameters. An efficient solution of the robust min-max optimization defined for mixed $1/\infty$ norm is given. *Copyright © 2003 IFAC*

Keywords: Stable robust model predictive control, Robust invariant and feasible sets, Linear cost function, Feedback min-max optimization

1. INTRODUCTION

The actual main motivation of researchers in area of control is development of robust control theories that can guarantee closed-loop stability in the presence of disturbances. The solutions of this problem strongly depend on the uncertainty description, which can assists researcher with the controller design procedure. In recent few years there have been established two main streams of the research in this area, H-infinity control and model predictive control. The H-infinity control creates a controller that minimizes the effect of worst possible disturbance through minimization of infinity norm of a sensitive function. The robust MPC theory tries directly incorporate disturbance signals or model uncertainties into MPC design procedures.

Initially the robust stabilizing controllers were designed for linear unconstrained systems. For example, in Konstantopulos and Antsaklis (1996) there was designed linear quadratic controller intended for state feedback policy. The robustness of such controller was accomplished by deriving bounds on parametric uncertainties that could be stabilized by the derived nominal feedback. The predictive controllers that explicitly consider the process and model uncertainties, when determining the optimal control policies, are called robust predictive controllers. The main concept of such controllers is similar to the idea of H-infinity controllers and consists in the minimization of "worst" disturbance effect to the process behavior. Frequently, the optimal control action is achieved from solution of a min-max optimization problem that minimizes the "worst"

performance objective function for given uncertainty description.
For constrained uncertain systems, researchers obviously prove stability by employing the infinity horizon problem formulation and establish the stability arguments analogously to techniques which are used for proving stability of nominal MPC schemes, i.e., based on terminal costs, terminal stability constraints, local optimal control policies etc. Authors (Lee and Cooley, 1997) and (Lee and Yu, 1997) have proposed the infinity-horizon objective function together with a min-max optimization procedure for obtaining optimal control actions. Father, in (Lee and Kouvaritakis, 1990) and (Mayne and Schroeder, 1997) there is designed a min-max control law that steers the current state into an invariant region in which some stabilizing state feedback guarantees convergence to the origin. Wang (2002) has used the terminal cost function as a basic part of stability condition in the robust predictive control design procedure. In this way the optimal control actions are obtained as a min-max minimization task of an augmented performance function.
The main disadvantage of the min-max based control tasks is their huge computational intensity. Many attempts have been made with approximations of solution of min-max tasks to decrease their computational intensity. Bemporad et. all (2001) proposed an explicit (off-line) solution of the min-max problem rather than the on-line computation of a standard linear or quadratic programming.
Due the various assumption and approximation made, it is difficult to directly compare different min-max MPC schemes. The robust MPC schemes can be classified in two main categories: open and closed loop schemes. In the open loop scheme the single control sequence or sequence of perturbations of a stabilizing control law is used to minimize the worst case cost. On the other hand the closed loop min-max control law minimizes the worst case cost over a sequence of feedback control laws.

2. PROBLEM STATEMENT

This paper deals with a numerically tractable procedure for robust model predictive control (RMPC) when constrains on input and states are presented. Consider a regulation problem to the origin of a discrete time linear time-invariant plant with bounded state disturbances. It is further supposed that system states are measurable and

$$x_{k+1} = Ax_k + Bu_k + w_k \qquad (1)$$

where $x_k \in R^n$ is the state vector, $u_k \in R^m$ is the manipulated input and $w_k \in W$ is a persistence disturbance that only takes values in the polytope $W \subset R^n$ and (A, B) is stabilizable. Upper and lower bounds $-x_{\min} \le x \le x_{\max}$, $-u_{\min} \le u \le u_{\max}$ construct in the state space the polytopes X, U, $x_k \in X, u_k \in U$, which are bounded and contains the origin.
Since the disturbance is presented, it is impossible to drive the state to the origin, so the standard MPC schemes must be modified. An obvious possibility is to ignore the disturbance, but the resulting performance of such control can be poor. To incorporate explicitly the disturbance to the solution we may consider a min-max formulation of MPC problem.
In this paper, an MPC modification that is based on min-max optimization of cost function, in which is included the notion of state feedback is considered. As it was mentioned, the origin of the disturbed system is no more $x = 0$, but some neighborhood of this origin, the set T. This target robust invariant set T must be contained in the admissible set X of system states. Now, the goal is to obtain the feedback control law $u_k = u(x_k)$ such that states are robustly steered to the target set T. Simultaneously the given state and input constraints are satisfied, while some worst case cost is minimized.

3. CONTROL LAW

Let as consider a dual mode control law similar to Scokaert and Mayne (1998). For the dual mode control it is necessary to design the *inner* and *outer* controller as follows

$$\begin{aligned} u_{k+i} &= u(x_k), \quad i = 0, \ldots, N-1, \quad x \notin T \\ u_{k+i} &= -Kx_{k+i}, \quad \forall i \ge N, \quad x \in T \end{aligned} \qquad (2)$$

The outer controller operates when the state is outside the invariant set and steers the state of the system to the target robust invariant set T. The inner controller operates when the state is in the robust control invariant set T. As the inner controller, the state feedback control law $u_{k+i} = -Kx_{k+i}$, $\forall i \ge N$ will be further considered. The closed loop properties (eigenvalues, eigenvectors) of the state feedback matrix K are very important for the consistent construction of the invariant target set T. Frequently as the inner controller the linear quadratic gain K is used

$$K = (B^T PB + R)^{-1} B^T PA \qquad (3)$$

As the outer controller following feedback min-max MPC scheme will be used. Let w_k^l, at time k is the possible l-th realization of the disturbance. Let the vector $w^l := \{w_0^l, w_1^l, \ldots, w_{N-1}^l\}$ denotes the disturbance sequences over the finite horizon N, $k = 0,1,\ldots, N-1$ where $l \in L$ indexes these realizations. Further, let $u^l := \{u_0^l, u_1^l, \ldots, u_{N-1}^l\}$ denotes a control sequence associated with the l-th realization of disturbance and $x^l := \{x_0^l, x_1^l, \ldots, x_N^l\}$ represents the sequence of solutions of the following model equation

$$x_{k+1}^l = Ax_k^l + Bu_k^l + w_k^l, \quad l \in L \tag{4}$$

Let as assume $x_0^l = x_0$ the same for all realizations of the disturbance signal and that the state x_0 is measurable at every time k. Then the optimization problem is as follows

$$u(x) = \min_u \max_{l \in L} \left[\sum_{i=1}^N L(x_{k+i}^l, u_{k+i-1}^l) \right] \tag{5}$$

where

$$u(x) = u^1, u^2, \ldots, u^l, \quad l \in L \tag{6}$$

subject to the state, input, stability and "causality" constrains:

$$\begin{gathered} x_{k+i}^l \in X, \quad i = 1, \ldots, N, \quad \forall l \in L \\ u_{k+i-1}^l \in U, \quad i = 1, \ldots, N, \quad \forall l \in L \\ x_{k+N}^l \in T, \quad l \in L \end{gathered} \tag{7}$$

$$x_{k+i}^1 = x_{k+i}^2 \Rightarrow u_{k+i}^1 = u_{k+i}^2, \quad \forall l \in L \tag{8}$$

The notion of feedback is than introduced into the control optimization by allowing a different control sequence for each realization of the disturbance. The causality constraints (8) associate each predicted state at time k with a single control input. This fact reduces the degree of freedom and makes such control actions independent of the control and disturbance sequences taken to reach the current state. As we can see, min-max optimization problem (5) with defined constrains (7,8) is due to the infinite number L of disturbance realization computationally infeasible.

The problem (5) is solvable, when the linear model, convex cost and polytopic bounded disturbances are used.. The solution is to consider only the extreme realizations of the disturbance signal (Mayne and Schroeder,1997), that take values at the vertices of disturbance set W. The resulting finite subset of indexes $L_v \subset L$ enables us design the next finite dimensional min-max optimization problem

$$u(x) = \min_u \max_{l \in L_v} \left[\sum_{i=1}^N L(x_{k+i}^l, u_{k+i-1}^l) \right] \tag{9}$$

where

$$\begin{gathered} u(x) = u^1, u^2, \ldots, u^l, \quad l \in L_v \\ x_{k+i}^l \in X, \quad i = 1, \ldots, N, \quad \forall l \in L_v \\ u_{k+i-1}^l \in U, \quad k = 1, \ldots, N, \quad \forall l \in L_v \\ x_{k+N}^l \in T, \quad l \in L_v \end{gathered} \tag{10}$$

$$x_{k+i}^1 = x_{kk+i}^2 \Rightarrow u_{k+i}^1 = u_{k+i}^2, \quad \forall l \in L_v \tag{11}$$

Now the causality constrains can be explained through the associating the same control input with each vertex of resulting disturbance realization. This issue rapidly decreases the number of unknown parameters. To illustrate this issue, we may consider the nature of very simple Example 2. Without consideration of *causality constrains*, we have to compute v^N unknown input signals, where the variable v denotes the number of vertices of the disturbance polytope W and N is the length of prediction horizon. For given Example 2, $v = 4$ and $N = 3$, so the number of different disturbance realizations and the number of unknown input signals is equal to 4^3. The used observation (11) reduces this number of control inputs to $1 + v^1 + \ldots + v^{N-1}$, so in the Example 2, there was really computed only $1 + 4 + 4^2$ values of control inputs.

4. LINEAR COST FUNCTION

The quadratic performance cost function is mostly used in the classical implementation schemes of MPC usually penalizing the deviation between the predicted states and the origin of the state-space. As we have mentioned in the previous section, the presence of disturbance changes this formulation, so new cost function is needed. Kerrigan and Maciejowski (2002) proposed a new linear cost function, which is less computational intensive and includes the notation of disturbance as

$$\begin{gathered} L(x_{k+i}^l, u_{k+i-1}^l) = \min_{y \in T} \left\| Q(x_{k+i}^l - y_{k+i}^l) \right\|_\infty + \\ + \left\| R(u_{k+i-1}^l - Kx_{k+i-1}^l) \right\|_\infty \end{gathered} \tag{12}$$

The new cost function can be interpreted in the similar fashion as the classical cost used in the standard MPC theory. The first term of the cost penalizes the deviation of the state trajectory from the disturbance invariant set T and the second term implements the idea of penalizing only the deviation from some ideal input signal. It is further assumed that the ideal input signal is generated by a suitable chosen LQ feedback gain. It would be interesting to extend this approach also with the idea of pre-stabilizing predictions introduced in Rossiter et. all (1998).

The solution of the task (12) is based on minimization of mixed $1/\infty$ norm, namely $1-$norm with respect to the time and $\infty-$ norm with respect to the space. The practical aspects of the solution will be mentioned in section 5.

5. CONSTRUCTION OF ROBUST INVARIANT AND FEASIBLE SETS

Disturbance positive invariant and feasible sets of states are such sets that given the state feedback control law $u_k = -Kx_k$, their satisfy the input and state constraints and make feedback trajectory remain inside the set with the presence of the disturbance. Thus, if the initial state x_0 occur in the robust invariant set, the feedback control law $u_k = -Kx_k$ keeps all subsequent states in the set, despite the disturbance. As we have mentioned in the previous section appropriate choice of K has the fundamental effect for the size and shape of such sets. In general, control theory describes two types of invariant sets, in the context of shape, ellipsoidal and polyhedral sets. In this paper we notice only the way of construction polyhedral invariant sets. Let us start with the design of the set T as the minimal disturbance set according to Mayne (1997)

$$T = \sum_{i=0}^{\infty} (A - BK)^i W \tag{13}$$

where the states in the set T must satisfy

$$T \subseteq X, \quad -KT \subseteq U \tag{14}$$

The evaluation of the set T in this form is an infinite dimensional problem, but with the appropriate choice of K it is possible to find an integer $s \geq n$ such that $(A - BK)^s = 0$. The inner controller is used as soon as the state enters T. For all initial states $x_k \in T$, holds

$$\begin{aligned} x_{k+1} &= (A - BK)x_k + w_k \\ x_{k+1} &= \Phi x_k + w_k \end{aligned} \tag{15}$$

where the controller K ensures, that beyond the step $j \geq s$, all $x_j, w_j \in R^n$ satisfy equation $(A - BK)^s x_j = 0$ and also condition $(A - BK)^s w_j = 0$. Than we may evaluate the performance of system (15), as $x_{k+j} = w_{k+j-1} + \Phi w_{k-j-2} + \ldots + \Phi^{s-1} w_{k-j-s}$. Hence, we can write for the set T following equation

$$\begin{aligned} T &= \sum_{i=0}^{s-1} (A - BK)^i W \\ T &\subseteq X, -KT \subseteq U \end{aligned} \tag{16}$$

The shape of such polyhedral set is rather complex, due to many inequality constrains. This condition in a fact largely increase the number of inequalities (10) in the minimization task of the cost function (9), but on the other hand such sets are quite optimal with respect to optimal ellipsoidal invariant ones.

In order to overcome the difficulty with complex shape of the previous designed invariant sets, we can propose a more simple shape polyhedron - low complexity polyhedron - and than try to inscribe the polyhedron within a polyhedron that is defined with the input variable constraints.

First, we need to recall some well-known results concerning convex polyhedral and invariant sets, namely the definition of polyhedral sets, the definition of positive invariant set, the Extended Farkas lemma for inscribing a polyhedron into another polyhedron and a lemma giving the conditions for positive invariance.

A polyhedral set in the state space is an intersection of half-spaces and can be defined as follows

$$\begin{aligned} S(\mathrm{G}, g_1, g_2) &= \{x \in R^n; -g_1 \leq Gx \leq g_2\} \\ G &\in R^{v \times n}, g_1, g_2 \in R^v \end{aligned} \tag{17}$$

According to selected triple G, g_1, g_2 the polyhedral set S can be bounded (polytope), unbounded, including or not including the origin point. In particular, if $g_1, g_2 > 0$ the origin $x = 0$ belongs to the interior of polyhedral set.

Definition 1. The set G is positively invariant set of the system

$$\begin{aligned} x_{k+1} &= (A - BK)x_k \\ x_{k+1} &= \Phi x_k \end{aligned} \tag{18}$$

if and only if for any initial state $x_0 \in G$ complete trajectory of state vector x_k remains inside G.

Lemma 1. (Extended Farkas lemma). Given two polyhedra $S_1(\mathrm{G}_1, \mathrm{g}_1) = \{x \in R^n; G_1 x \leq g_1\}$ and $S_2(\mathrm{G}_2, \mathrm{g}_2) = \{x \in R^n; G_2 x \leq g_2\}$ then $S_1 \subseteq S_2$ if and only if there exist a non-negative matrix H such that $HG_1 = G_2$ and $Hg_1 \leq g_2$.

Proposition 1. (Condition for positive invariance). The polyhedral set $S(\mathrm{G}, g_1, g_2) = \{x \in R^n; Gx \geq g_1, Gx \leq g_2\}$ where $G \in R^{v \times n}$, rank $G = v$ and $g_1, g_2 > 0$ is positive invariant of system (18) if and only if there exist a matrix $H \in R^{v \times v}$ such that

$$G\Phi = HG \tag{19}$$

and

$$\begin{bmatrix} H^+ - I & -H^- \\ -H^- & H^+ - I \end{bmatrix} \begin{bmatrix} g_1 \\ g_2 \end{bmatrix} \leq 0 \tag{20}$$

Remark 1. The Proposition 1 is extension of general definition for positive invariance of convex sets (Definition 1) and Farkas lemma (Lemma 1.). Generally, Proposition 1 says that the set G is positively invariant if every state included in the set G is included in the set $G\Phi$ too. The set $G\Phi$ contains all subsequent states from the set G for system (18).

Bistoris (1988) established the invariant conditions for polyhedral sets and Dam and Nieuwenhuis (1995) proposed a linear algorithm for constructing such positively invariant and feasible sets. Kouvaritakis et. all (2002) further extended this approach. First we will introduce this approach and then we will try to adapt this method for finding the disturbance invariant and feasible set. T

An Invariant polyhedron can be constructed using invariant half-spaces corresponding to real eigenvectors of closed loop system (18). If the closed loop system matrix Φ is diagonalizable, all eigenvalues are real and inside the unit circle, the described invariant set is a polytope

$$P(Z,z_1,z_2)=\{x\in R^n;-z_1\le Zx\le z_2\} \quad (21)$$
$$Z\in R^{nxn}, z_1,z_2\in R^n$$

For the polytope P the definition matrix H, defined by invariance condition (19), always exists as

$$H = Z\Phi Z^{-1} \quad (22)$$

The polytope $P(Z,z_1,z_2)$ is positively invariant if for calculated matrix H the relation (20) holds for all positives z_1,z_2. This is always true under the assumption that all eigenvalues are real and negative. Considering that the definition matrix H is defined in advance, the task is then to determine z_1,z_2 in such manner, that the resulting polytope may obtain desired properties.
The input constrains define in the state space a non-symmetrical polyhedral set of admissible states

$$D(K,u_1,u_2)=\{x\in R^n;-u_1\le Kx\le u_2\} \quad (23)$$
$$K\in R^{nxn}, u_1,u_2\in R^n$$

It is possible to find condition under which the invariant polytope (21) can be inscribed into the unbounded polyhedron (23), induced in the state space with the used constrains, and achieve positive parameters z_1,z_2 maximizing the polytope volume.
Using following definitions:

$$P^*(Z^*,z^*)=\{x\in R^n; Z^*x\le z^*\}, Z^*=\begin{bmatrix} Z \\ -Z\end{bmatrix}, z^*=\begin{bmatrix} z_1 \\ z_2\end{bmatrix} \quad (24)$$

$$D^*(K^*,u^*)=\{x\in R^n; K^*x\le u^*\}, K^*=\begin{bmatrix} K \\ -K\end{bmatrix}, u^*=\begin{bmatrix} u_1 \\ u_2\end{bmatrix} \quad (25)$$

the condition for inscribing the invariant polytope (24) to the input constrains polyhedron (25) provides the following Proposition:

Proposition 2. Given the definition matrix P^* of the polytope $P^*(Z^*,z^*)$ and the polyhedron $D^*(K^*,u^*)$, then $P^*\subseteq D^*$ if and only if there exist a non-negative vector z^* such that

$$\begin{bmatrix} M^+ & -M^- \\ -M^- & M^+\end{bmatrix}\begin{bmatrix} z_1 \\ z_2\end{bmatrix}\le\begin{bmatrix} u_1 \\ u_2\end{bmatrix} \quad (26)$$

where

$$M = KZ^{-1} \quad (27)$$

and matrixes M^+,M^- are defined as $h^+_{i,j}=\max(h_{i,j},0)$ respectively $h^-_{i,j}=\min(h_{i,j},0)$.
Under these assumptions (19-27) we can construct a polytope P satisfying conditions for positive invariance of the Proposition 1 and feasibility conditions of the Proposition 2. The design of such polyhedron can be accomplished using the following algorithm for maximizing polytope volume P from Kouvaritakis et all. (2002):

Algorithm 1:

1. Set x_0,choose ε to be to zero $(\varepsilon\to 0)$, define F,f and x in the following way
$$F=\begin{bmatrix} M^+ & -M^- \\ -M^- & M^+ \\ H^+-I & H^- \\ -H^- & H^+-I\end{bmatrix},\ f=\begin{bmatrix} u_1 \\ u_2 \\ 0 \\ 0\end{bmatrix},\ x=\begin{bmatrix} z_1 \\ z_2\end{bmatrix}$$
2. Minimize
$$t_{\min}=\min t$$
subject to
$$Fx\le f+t.I$$
$$x\ge x_0-tI$$
where x is assumed as a free variable
3. Set $x_0=x$
4. repeat step 2.,3., until $|t_{\min}|<\varepsilon$
5. finally $\hat{x}=x$

Consider now that the invariant and feasible polytope (24) is known for the closed loop system (18), than we can try to achieve a modification of the set P^* when the disturbance in (15) is presented. Using (24) and substituting for x (15) we can write

$$Z^*(\Phi x_k+w_k)\le z^* \quad (28)$$

or also

$$Z^*\Phi x_k\le z^*-Zw_k \quad (29)$$

Equation (29) is the expression of polytope P in the subsequent step. Solving following $2n$ optimization problems:

$$\min[z_i^*-Z_i^*w],\quad i=1,2,\ldots,2n \quad (30)$$

subject to

$$w\in W \quad (31)$$

where the symbol z_i^* denotes the i-th component of the vector z^* and the symbol Z_i^* denotes the i-th row vector of matrix Z^*, we obtain a new right hand site of equation (29). In the case that z_i^* is constant, the minimum of the problem is obtained by solution of:

$$\max_{w\in W}[Z_i^*w],\quad i=1,2,\ldots,2n \quad (32)$$

With the symbol v_i^* we will further denote the maximum of the i-th problem of (32). We can further conclude that the presence of disturbance changes the borders of the polytope P in the following manner

$$Z^{*}\Phi x \le z^{*} - v^{*} \quad (33)$$

$$v^{*} = \left(v_1^{*}, v_2^{*}, \ldots, v_{2n}^{*}\right)^T$$

Now, when we were able to obtain the expression for modified polytope P , in the view of Remark 1 we can try to summarize our conclusions:

Proposition 3. (Conditions for Disturbance positive invariance). The polyhedral set $P(Z, z_1, z_2) = P^{*}(Z^{*}, z^{*}) = \{x \in R^n; Z^{*}x \le z^{*}\}$ where $Z^{*} \in R^{2nxn}$ and $z^{*} > 0$ is disturbance positive invariant for system (15) if and only if there exist a matrix $H \in R^{vxv}$ such that

$$Z\Phi = HZ \quad (34)$$

and

$$\begin{bmatrix} H^{+} - I & -H^{-} \\ -H^{-} & H^{+} - I \end{bmatrix} \begin{bmatrix} z_1 \\ z_2 \end{bmatrix} \le -v^{*} \quad (35)$$

where v^{*} is solution of maximization problem:

$$\max\lfloor Z_i^{*} w \rfloor, \quad i = 1,2,\ldots,2n \quad (36)$$

Under the given Proposition 3 it is possible to achieve a disturbance invariant and feasible polytope P . That means to design the algorithm for maximizing polytope P within the given constraints:

Algorithm 2:

1. Solve the optimization problem of minimizing
$$v^{*} = \max\lfloor Z_i^{*} w \rfloor, \quad i = 1,2,\ldots,2n$$
subject to
$$w \in W$$
2. Set x_0, choose ε to be to zero $(\varepsilon \to 0)$, define F, f and x in the following way
$$F = \begin{bmatrix} M^{+} & -M^{-} \\ -M^{-} & M^{+} \\ H^{+} - I & H^{-} \\ -H^{-} & H^{+} - I \end{bmatrix}, \quad f = \begin{bmatrix} u_1 \\ u_2 \\ -v^{*} \end{bmatrix}, \quad x = \begin{bmatrix} z_1 \\ z_2 \end{bmatrix}$$
3. Minimize
$$t_{\min} = \min t$$
subject to
$$Fx \le f + t.I$$
$$x \ge x_0 - tI$$
where x is assumed as a free variable
4. Set $x_0 = x$
5. repeat step 2.,3., until $|t_{\min}| < \varepsilon$
6. finally $\hat{x} = x$

6. SOLUTON OF MPC WITH MIXED $1/\infty$ NORM

The classical implementation of MPC uses quadratic cost functions, which lead to quadratic programming problem. In (Keriggan and Maciejowski,2002) there was introduced a new linear cost function, which makes the problem of a disturbances presence tractable. In this chapter we will try to describe how this problem can be solved using a linear programming, if the stage cost (3) is used.

Denote with $J(x_0, u^l, w^l)$ the realization of cost (12) on the predicted horizon N, where x_0 is the initial state and u^l is the optimal sequence of input signal associated with a given disturbance realisation w^l:

$$J(x_0, u^l, w^l) = \sum_{i=1}^{N} L(x_{k+i}^l, u_{k+i-1}^l) \quad (37)$$

Then the min-max optimisation through all disturbance sequences $l \in L$ can be written in a following shorter form

$$\max_{u \in C(x)} \min_{l \in L_v} J(x_0, u^l, w^l) \quad (38)$$

Using the standard approach of (Bemporad et. all,2001) we can rewrite (37) as a linear program. The sum of components of any vector $\{\varepsilon_1^x, \ldots, \varepsilon_N^x, \varepsilon_1^u, \ldots, \varepsilon_N^u\}$, which must satisfied conditions

$$\begin{aligned} -1_n \varepsilon_i^x &\le Q(x_{k+i} - y_{k+i}), \quad i = 1,2,\ldots,N \\ -1_n \varepsilon_i^x &\le -Q(x_{k+i} - y_{k+i}), \quad i = 1,2,\ldots,N \\ -1\varepsilon_i^u &\le R(u_{k+i-1} - Kx_{k+i-1}), \quad i = 1,2,\ldots,N \\ -1\varepsilon_i^u &\le -R(u_{k+i-1} - Kx_{k+i-1}), \quad i = 1,2,\ldots,N \end{aligned} \quad (39)$$

represents the upper bound on $J(x_0, u, w)$, where $1_n = [1, \ldots, 1]^T$ is the unit vector with appropriate length. It can be proven, that the vector $z = \{\varepsilon_1^x, \ldots, \varepsilon_N^x, \varepsilon_1^u, \ldots, \varepsilon_N^u, u_k, \ldots, u_{k+N-1}, y_{k+1}, \ldots, y_{k+N}\}$, which includes slack and unknown variables with appropriate length, $s = (n+m)*N + m*N + n*N$, satisfies equation (39) and simultaneously minimizes $J(z) = \varepsilon_1^x + \ldots + \varepsilon_N^l + \varepsilon_1^u + \ldots + \varepsilon_N^u$. After this manner the vector solves the original problem $J(x_0, u, w)$ and the same optimum is achieved. Therefore, the problem (39) can be reformulated as the following LP program:

$$\min_z \{z \in C(x), J(x, u^l, w^l) \le \gamma, \quad \forall l \in L_v\} \quad (40)$$

The polytope $C(x)$ is implicitly defined by subsequent constrains

$$-1_n \varepsilon(l)_i^x \le \pm Q(x_{k+i}^l - y_{k+i}^l), \quad i = 1,2,\ldots,N, \quad \forall l \in L_v$$

$$-1\varepsilon(l)_i^u \le \pm R(u_{k+i-1}^l - Kx_{k+i-1}^l), \quad i = 1,2,\ldots,N,$$

$$
\begin{aligned}
x_{k+i}^{l} \in X, \quad i=1,\dots,N, \quad \forall l \in L_v \\
u_{k+i-1}^{l} \in U, \quad k=1,\dots,N, \quad \forall l \in L_v \\
x_{k+N}^{l} \in T, \quad \forall l \in L_v \\
x_k^{l_1} = x_k^{l_2} \Rightarrow u_k^{l_1} = u_k^{l_2}
\end{aligned} \tag{41}
$$

where the $\pm$ means that the constraint is duplicated for each sign.

7. EXAMPLES

In this section a simulation example with the Algorithm 2 for finding the robust invariant and feasible polytope is provided. The second example will demonstrate the performance of control with the min-max MPC algorithm proposed in the previous section.

Example 1:
Consider following discrete time invariant system with disturbance

$$x_{k+1} = \begin{bmatrix} 0.9350 & 0.5190 \\ 0.3840 & 0.8310 \end{bmatrix} x_k + \begin{bmatrix} -1.4460 \\ -0.7010 \end{bmatrix} u_k + \begin{bmatrix} 1 & 0 \\ 0 & 1 \end{bmatrix} w_k \tag{42}$$

$$X = \{x \in R^2, \|x\| \le 5\}, U = \{u \in R, \|u\| \le 5\} \tag{43}$$

The system (42) is stabilizing with LQ feedback gain $K = [-0.4978 \quad -0.4588]$. The Figures 1 and 2 illustrate the changes of the nominal invariant and feasible set when the disturbance is presented. As we can see the volume of nominal invariant polytope is decreasing while the volume of disturbance polytope is increasing. Figure 1 shows the robust invariant and feasible set when the presented disturbance takes values from the polytope $W = (w \in R^2, \|w\| \le 1.5)$. We can also compare the invariant and feasible set without disturbance. The Algorithm 2 always tries to inscribe the achieved shape within the borders due to feasibility. The Figure 2 shows the same situation for different value of the uncertain polytope $W = (w \in R^2, \|w\| \le 2.5)$

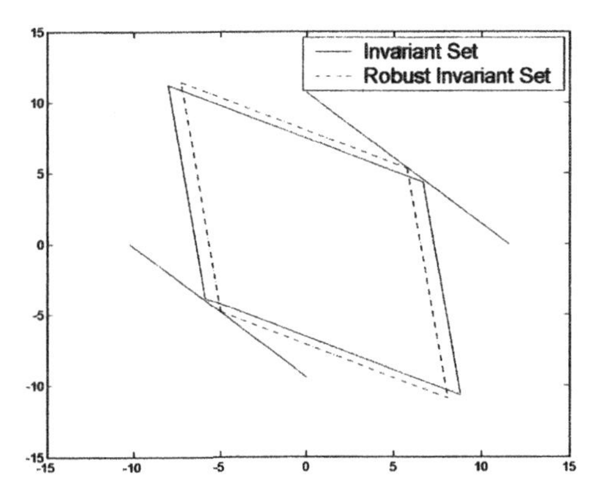

Fig. 1. Robust invariant and feasible set $\|w\| \le 1.5$

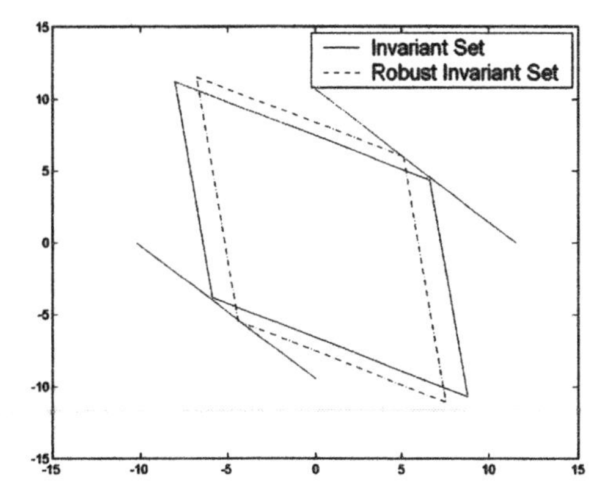

Fig. 1. Robust invariant and feasible set $\|w\| \le 2.5$

Example 2:
Consider the system (42) as in the Example 1 with constrains,

$$X = \{x \in R^2, \|x_i\| \le 5\} U = \{u \in R, \|u\| \le 3\} \tag{44}$$
$$W = (w \in R^2, \|w_i\| \le 0.5)$$

and the same LQ feedback gain K. Applying the Algorithm 2, we find the following target set T

$$T = \left\{ \begin{bmatrix} -3.4114 \\ -4.3503 \end{bmatrix} \le \begin{bmatrix} -0.8858 & -1.1262 \\ -0.4641 & -0.9920 \end{bmatrix} \le \begin{bmatrix} 3.4114 \\ 4.3503 \end{bmatrix} \right\} \tag{45}$$

Consider the stage cost in form as

$$L(x_{k+i}^l, u_{k+i-1}^l) = \min_{y \in T} \| Q(x_{k+i}^l - y_{k+i}^l) \|_\infty + \| R(u_{k+i-1}^l - K x_{k+i-1}^l) \|_\infty \tag{46}$$

, where the $Q = I$ and $R = 1$, set control horizon to $N = 3$ and minimize the cost

$$\max_{u \in C(x),\, l \in L_v} \min J(x_0, u^l, w^l) \tag{47}$$

The results are depicted at Figure 3 and 4 .At Figure 3 the closed loop states are simulated from the initial state $x_0 = [6 \quad 5]$ with $w_k = \pm 0.5/k$. At Figure 4 corresponding input signals are drawn. As it was expected, in spite of the disturbance the control law drives the state to the end-region without the violation of constrains.

ACKNOWLEDGEMENT

The authors wish to thank Slovak grant agency VEGA for financial support – project no. 1/9278/02

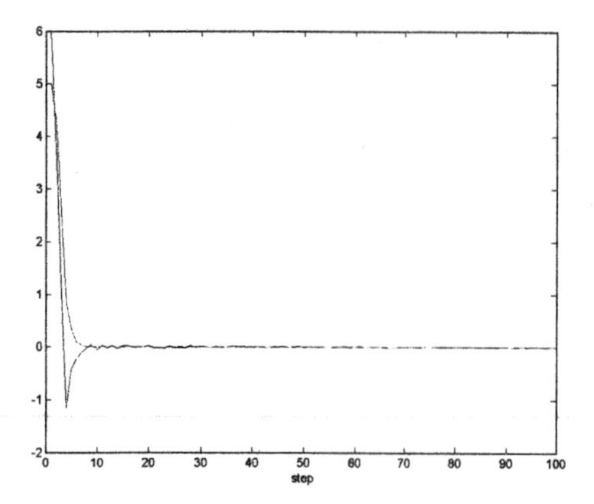

Fig 3. Closed loop response of the second example to a random disturbance signal

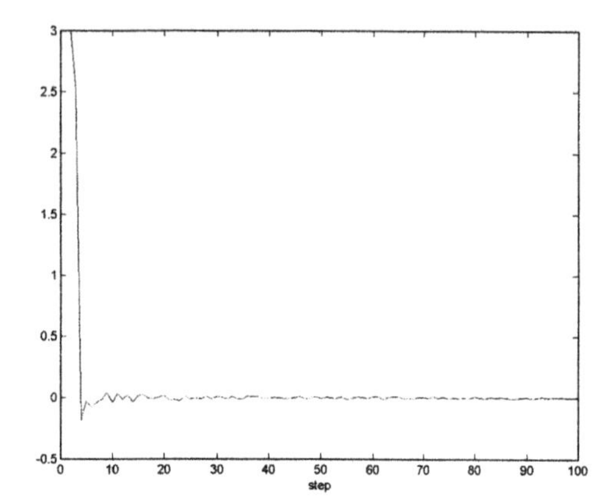

Fig 4. Input signal of the second example

REFERENCES

Bemporad A., M. Morari, V. Dua, E.N Pistokopulos (2001) The explicit linear quadratic regulator for constrained systems, Automatica, 38, pp. 4-19

Bitsoris, G. (1988) On the positive invariance of polyhedral sets for discrete-time systems. *Systems and Control Letters* 11, pp. 243-248.

Dam A.A., J.W. Nieuwenhuis (1995) A linear programming algorithm for invariant polyhedral sets of discrete-time linear systems. *Systems and Control Letters* 25, 337-341.

Farison J., S Kolla, (1990), Stability robustness bounds for discrete time linear regulator., *Int. J System Science*, 22(11), pp. 2285-2298

Kerrigan E. C., J.M Maciejowski (2002), Feedback min-max model predictive control using a single linear program: Robust stability and explicit solution, *Technical report CUED/F-INFENG/TR.440*, Department of Engineering University of Cambridge, United Kingdom

Konstantopulos I., P. Antsaklis (1996), Optimal design of robust controllers for unconstrained discrete time systems, *International Journal of control*, 65(1), pp.71-91

Kouvaritakis, B., M. Cannon, A. Karas, B. Rohal-Ilkiv , C. Belavy (2002) Asymmetric constrains with polyhedral sets in MPC with application to coupled tanks system, *IEEE Conference on Decisions and Control.* Las Vegas, Nevada, USA, 2002

Kwakernaak H., R. Sivan (1972) *Linear Optimal Control Systems*, John Wiley and Sons, New York

Lee J.H., B.L. Cooley (1997) Stable min-max control for state-space systems with bounded input matrix, *In Proceedings of American Control Conference, Albuquerque*, pp. 2945-2949

Lee J.H., Z. Yu (1997) Worst-case formulation of model predictive control with bounded parameters, *Automatica,* 33(5), pp. 763-781

Lee J.H,B. Kouvaritakis (1990) A linear programming approach to constrained predictive control, *IEEE Transaction on Automatic control*, Submitted

Mayne D., Schroeder W. (1997), Robust time-optimal control of constrained linear systems, *Automatica*, 33(12), pp. 814-824

Rossiter, J.A., B Kouvaritakis, , M.J. Rice, (1998) A numerically robust state-space approach to stable-predictive control strategies. Automatica, 34(1), pp. 65-75

Wang Y, (2002) Robust Model Predictive Control, PhD Thesis, *University of Wisconsin-Medison*

ROBUST DESIGN OF PID BASED ACC S&G SYSTEMS

M. Canale * S. Malan *

* *Dip. di Automatica e Informatica, Politecnico di Torino,*
Corso Duca degli Abruzzi 24, 10129 Torino, Italy.
massimo.canale@polito.it, stefano.malan@polito.it

Abstract: In this paper an ACC Stop and Go control design procedure is proposed by taking into account generic performance requirements. It will be described that safeness and comfort features can be considered in the control scheme by suitably exploiting the speed reference generation strategy. It will be shown how the vehicle longitudinal dynamics can be characterized and identified by using standard vehicle documentation and field recorded data by means of interval plant families. Using this plant description, very simple P and PI robust controller can be used to regulate the vehicle dynamics according to a prescribed speed reference profile. *Copyright © 2003 IFAC*

Keywords: PID control, Vehicle dynamic modelling and identification, Control applications, Robust control.

1. INTRODUCTION

In modern vehicle control, the need to assist the driver in different driving tasks originated from safety and comfort reasons and, in particular, the longitudinal control problem has been an important matter of research since the nineties.

The Adaptive Cruise Control (ACC) problem was afforded in many papers by considering either vehicle independence or vehicle cooperation by means of communication systems (see for example Wang and Rajamani, 2002; Lu *et al.*, 2002; Kato *et al.*, 2002, and the reference therein). More recently, some papers started to consider the ACC problem at low speeds, that is to say with Stop and Go features (ACC/S&G) for an urban scenery or, in general, in non urban traffic jam (see for example Marsden *et al.*, 2001).

In this paper, the problem of the longitudinal vehicle control design is considered. In particular, it will described how safe and comfortable acceleration profiles can be suitably computed using the standard vehicle data employed by the control algorithm. The problem on how to take into account the nonlinear longitudinal behaviour in an effective way for the control design will be also discussed. Finally it will be shown how robust low complexity controllers (i.e. P, PI) can be suitably designed considering simple performance requirements.

2. ACC/S&G PROBLEM DESCRIPTION

The controlled vehicle moves in an urban scenery, either strictly following, along the lane, the preceding vehicles at a target relative distance $\overline{d}_R$ or at a target velocity $\overline{v}_F$ in absence of a preceding vehicle. According to the preceding traffic conditions, the controlled vehicle has to accelerate or decelerate and even to stop and restart automatically, in safety conditions. The considered control structure is depicted in Figure 1: the block named "Vehicle" represents the physical system, equipped with obstacle detection radar sensor, vehicle sensors and real time data acquisition system, and is the object of the modelling and identification problem that will be described in Section 4; the other three blocks represent the control architecture. The signals in Figure 1 are:

Plant commands

(1) α: throttle opening angle;
(2) β: brake circuit pressure;

Controlled outputs

(3) v_F: controlled vehicle speed;
(4) d_R: relative distance between the controlled vehicle and the preceding one;

Measured output

(5) v_R: relative speed between the controlled vehicle and the preceding one;

Control signals

(6) a_{ref}, a_c, a_{rq}: respectively reference, compensated and requested accelerations.

The Reference Generator block simulates the human driver actions and its aim is to generate, on the basis of the on line acquired vehicle and environment data,

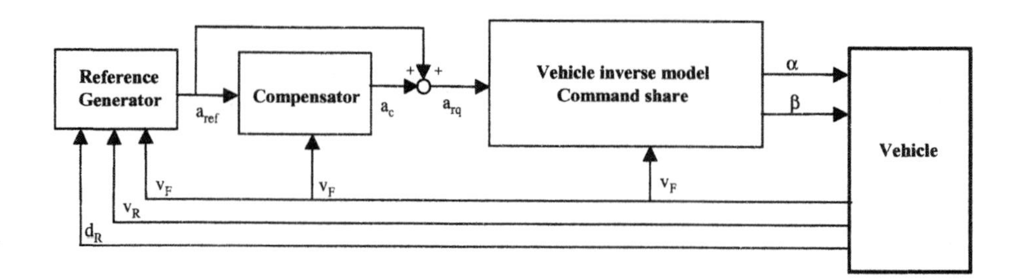

Fig. 1. Control structure.

an acceleration profile a_{ref} in agreement with safety and comfort requirements. As discussed in Canale *et al.* (2002), also user preferences can be taken into account in the Reference Generator design by tuning its parameters according to the driving style. In general a_{ref} is a function $f(\cdot)$ of the relative speed v_R and the relative distance d_R with respect to the preceding vehicle and of the controlled vehicle speed v_F, i.e $a_{ref} = f(v_R, d_R, v_F)$, (see, for example, Burnham *et al.*, 1974; Joannou and Chien, 1993). In Section 3, it will be described in details the structure of this block and shown how the relation $a_{ref} = f(v_R, d_R, v_F)$ can be "defined" for the ACC/S&G task.

The Compensator block is essentially made up by a Proportional plus Integral or by a simple Proportional algorithm used to compensate the error between the reference vehicle speed $v_{ref} = \int a_{ref}$ and the measured vehicle speed v_F and its design will be described in details in Section 5.

The Vehicle Inverse Model / Command Share block generates the throttle angle α or the brake pressure β commands, from the requested acceleration a_{rq}. Note that this block is essentially a variable static gain that takes into account the main static non linearities of the vehicle, but it does not require a design procedure and it will not further considered.

In order to perform the control design procedure in a more effective way, the considered Stop and Go manoeuvre may be divided into three main phases:

GO corresponding to the transient required to approach the subsequent CRUISE / TRACK phase; note that no distance tracking is possible, in this phase, due to its transient characteristics.

CRUISE / TRACK corresponding to the situation in which the vehicle is performing a constant speed following, in absence of a preceding vehicle, or a relative distance following, in presence of the preceding vehicle, respectively.

STOP corresponding to the situation requiring a deceleration leading to stop the vehicle.

The typical behaviours of the signals involved in the control problem recorded during manual driving are reported in Figure 2. They show in a more quantitative way the different phases in which the Stop and Go manoeuvre has been divided.

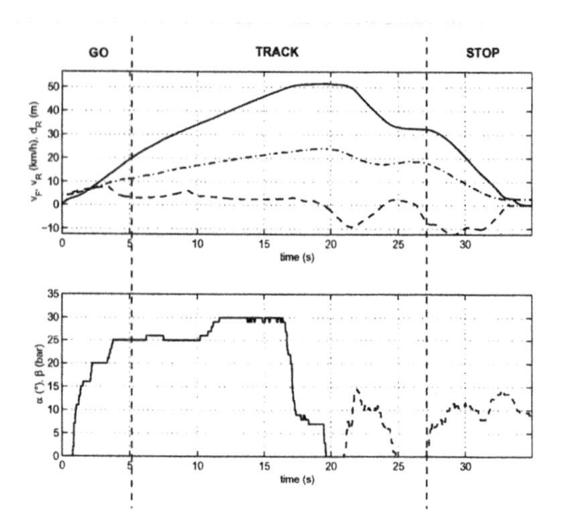

Fig. 2. Acquired signals. Above: v_F (km/h) solid, v_R (km/h) dashed, d_R (m) dash-dotted. Below: α (°) solid, β (bar) dashed.

3. REFERENCE GENERATOR DESIGN

The reference generator design plays a key role for the longitudinal vehicle control in the ACC/S&G driving task. In principle, the acceleration profile must be generated taking into account safety and comfort of the required driving manoeuvres.

It appears reasonable to differentiate the reference generation according to each one of the Stop and Go phases outlined in Section 2. The switching among the three different phases is handled by a supervision logic (here not described) that, on the basis of the acquired signals, enables the correct generation procedure.

GO Considering this phase, the reference generation structure of Figure 3 can be suitably exploited. As it

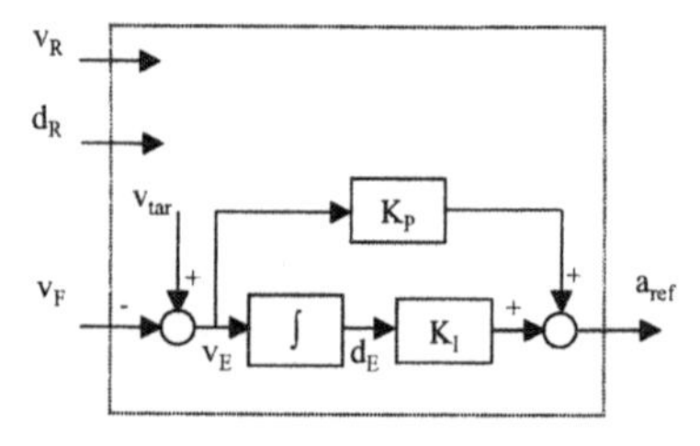

Fig. 3. GO and CRUISE phases reference generator.

can be seen, only the measured signal v_F is used for the reference generation while the other two acquired signals d_R and v_R are only considered for switching logic purposes. The required acceleration profile a_{ref}

can then be obtained by fixing the reference generator "internal" input as $v_{tar} = v_F^{\alpha}$ where v_F^{α} is the vehicle speed obtained by applying to the vehicle a ramp throttle signal α with a certain slope s_{α}. Typical values of s_{α} observed during manual driving lie in the interval $30 \div 70$ $°/s$. Note that signal v_F^{α} can be precomputed and stored. For more details on this procedure see also the description in Canale *et al.* (2002).

This way, the reference generation for the GO phase forces the vehicle longitudinal control to follow the speed reference v_{tar} until a CRUISE or TRACK phase is reached. The parameters K_P and K_I can be tuned taking into account the comfort characteristics of the acceleration profile. To this end, let us consider the scheme depicted in Figure 4. This scheme can be considered as a reasonable linear approximation of the real behaviour of the controlled longitudinal vehicle dynamics. Coefficient $\eta < 1$ represents, as a first approximation, the rate of the speed that has been really rendered to the vehicle, detracted the friction losses and aerodynamic effects. The transfer function

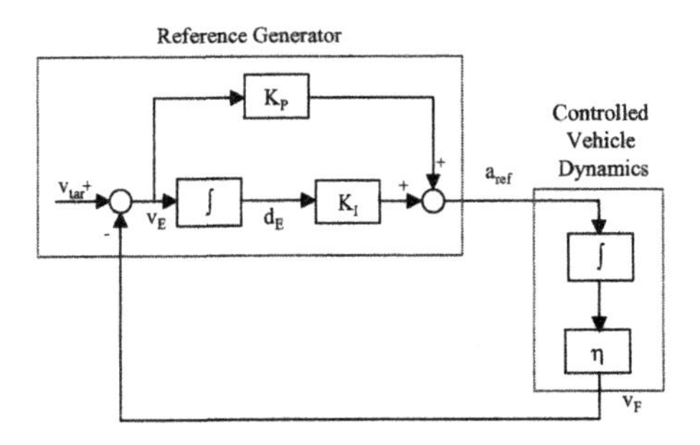

Fig. 4. Approximated controlled longitudinal vehicle dynamics and reference generation scheme.

$G_{ref}(s)$ between the input v_{tar} and the output of the reference generator a_{ref} is given by:

$$G_{ref}(s) = \eta s \frac{K_P s + K_I}{s^2 + K_P s + K_I} \qquad (1)$$

This way, in order to obtain the smoothest and the fastest acceleration profile, the poles of the transfer function (1) have to coincide, leading to the following relation between K_I and K_P parameters $K_I = \frac{K_P^2}{4}$. The value of K_P may be chosen to tune the response speed of a_{ref}. Typical values for K_P range from 0.5 to 0.7 s^{-1}. Consequently K_I values range from 0.0625 to 0.1225 s^{-2}.

CRUISE Regarding this phase all the reasoning made in the previous item are still valid with the only difference that, after the GO phase transient, v_{tar} is set to a constant value selected by the driver according to the urban speed limitation: $v_{tar} \leq 70$ km/h.

TRACK This phase is characterized by the tracking of the target distance d_S:

$$d_S = s_0 + T_H \cdot v_F \qquad (2)$$

where $T_H \triangleq d_R/v_F$ is the headway time, v_F is the vehicle speed and s_0 can be regarded as a sort of a fixed safety distance. In this case the reference generation structure showed in Figure 5 may be adopted.

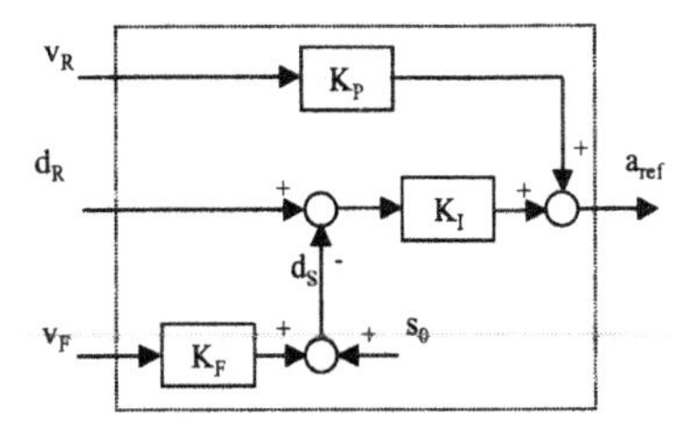

Fig. 5. TRACK phase reference generator.

Now, considering the scheme in Figure 5 and equation (2), the most natural way to tune this feature is to set $K_F = T_H$. In manual driving, typical values of T_H are such that: $1.0 \leq T_H \leq 1.6$ s.

The values of K_P and K_I may be computed in order to ensure comfort characteristics to the generated acceleration profile by following similar reasoning as done in GO and CRUISE phases.

STOP During this phase a deceleration is required whose intensity depends on the way the preceding vehicle is going to be approached or, in other words, to the emergency level induced by the approaching manoeuvre and measured by the collision time $T_C \triangleq d_R/v_R$. Due to the smoothness of the acceleration profiles generated by both the schemes in Figure 3 and 5, in dangerous situations, it may not suffice to avoid collision. This way, a purely kinematics approach of the form:

$$a_{ref} = \gamma \cdot \frac{v_R^2}{2(d_R - d_S)} \qquad (3)$$

may be used, where $\gamma = \gamma(T_C) \geq 1$ is a coefficient measuring the level of emergence required by the braking manoeuvre (Persson *et al.*, 1999).

As already noted in Section 2, the tuning of the reference generator parameters can be also worked out considering driver personalization features. The details on how driver characteristics can be studied and taken into account in the reference generation design are discussed in Canale and Malan (2002) and in Canale *et al.* (2002) respectively.

4. VEHICLE LONGITUDINAL DYNAMICS MODELLING AND IDENTIFICATION

In order to design the longitudinal control strategy a suitable model of the vehicle behaviour has to be identified. According to the different Stop & Go situations described in Section 2 and the control structure described in Figure 1 two different models for the longitudinal dynamics will be taken into account. First, the relation G_{α} between the throttle angle α and the vehicle speed v_F will be considered for describing both the GO and CRUISE/TRACK working situation.

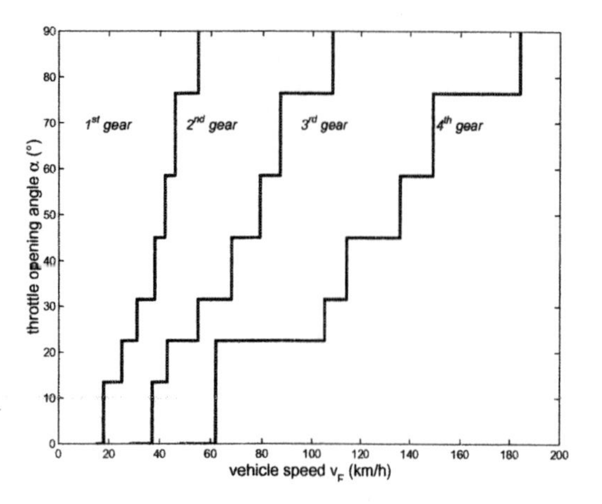

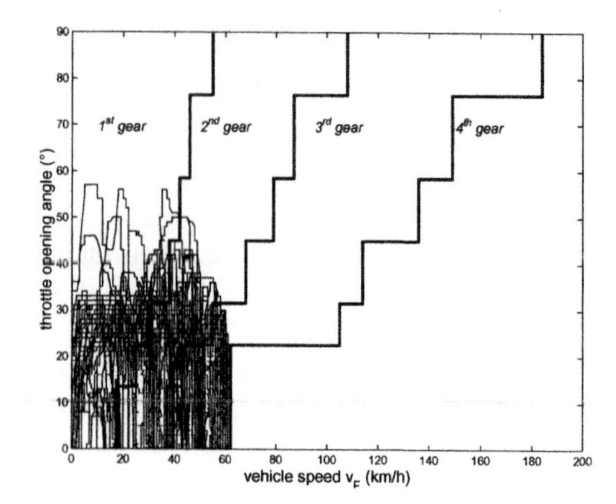

Fig. 6. Gear map (left) and covering of the operating points on it (right).

On the other side, the relation G_β between the brake circuit pressure β and the vehicle speed v_F will be used to describe the vehicle dynamics in the STOP manoeuvres. The relation G_α is, actually, non linear due to the engine torque-acceleration mapping and the effects of the inserted gear on the vehicle speed. This way, given the non linear nature of the problem and the different situations in which the vehicle operates during the GO and CRUISE/TRACK phases, it appears reasonable to consider a set of different linear models for the synthesis of the longitudinal control strategy. In Hunt *et al.* (1996) and in Wang (1992), the modelling of the longitudinal dynamics is performed by considering a different linear model for each of the inserted gear. Following this idea, the key problem is to find such working situations by using real data and the standard vehicle data sheets. To this end, the map describing the influence of the relation between α and v_F on the inserted gear, depicted in Figure 6 on the left, can be suitably employed. Moreover, in order to detect the most significant operating points for the considered Stop and Go scenery, the data collected and employed in the driver behaviour study described in Canale and Malan (2002) have been reported on this "throttle-speed" gear map. The covering of all the acquired data is shown in Figure 6 on the right.

Once the operating points have been selected, the problem of performing suitable experiments in order to identify the vehicle dynamic behaviour in such points has to be considered. In Wang (1992) it is shown how "chirp" signals can be conveniently used to this end. In order to excite the necessary frequency range, the usage of such signals in the identification phase requires very long journeys. Moreover, in order to avoid influences of the road slope on the measured speed, the identification experiments have to be performed on totally "flat" roads. This fact introduces a severe limitation in performing field experiment as test roads, with such required characteristics, were not available. Consequently, to investigate anyhow the vehicle dynamics characteristics, local step responses in the considered operating points have been considered. In Figure 7 it is shown a typical behaviour of such responses.

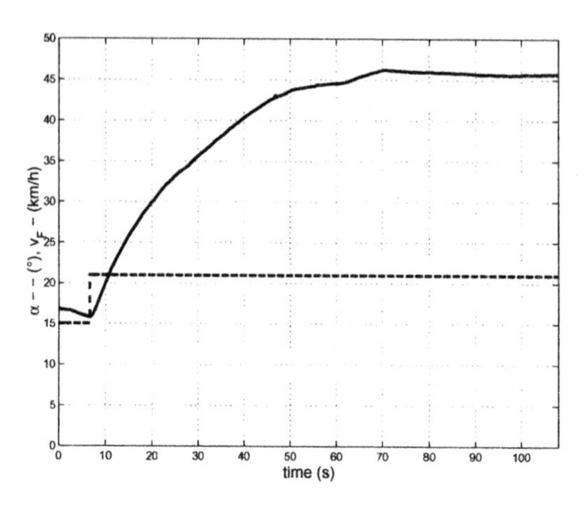

Fig. 7. Local step response for the G_α relation.

It can be easily recognized that the relation between the throttle opening angle and the vehicle speed is essentially a first order dynamic (see also Hunt *et al.*, 1996). This way, the vehicle behaviour when the throttle command is activated can be described by the following first order model:

$$G_\alpha(s) = \frac{K_\alpha}{1 + \tau_\alpha s} \tag{4}$$

Standard identification procedures gave the following results for the values of parameters K_α and τ_α:

$$4 \leq K_\alpha \leq 5 \; km \; h^{-1}/^\circ \;\; 25 \leq \tau_\alpha \leq 35 \; s$$

These obtained intervals cover the behaviour of the vehicle longitudinal dynamic at the different working situations given by the inserted gear.

Regarding the modeling in brake activation circumstances (i.e. STOP manoeuvres) the situation is simpler than in the throttle case as it has been observed (Persson *et al.*, 1999) that the connection between the brake circuit pressure β and the vehicle acceleration $\dot{v}_F$ is linear. This fact has been confirmed by the field recorded data. Thus the relation G_β can be modeled as an integrator

$$G_\beta(s) = \frac{K_\beta}{s} \tag{5}$$

On the basis of the same acquired data the following bounds for parameter K_β have been found:

$$-0.025 \leq K_\beta \leq -0.018 \; km \; h^{-1}/bar$$

As before, this obtained interval covers the behaviour of the vehicle longitudinal dynamic at the different working situations given by different speed.

The proposed models (4) and (5) are linear ones with interval parameters and accounts for the longitudinal vehicle dynamic at different inserted gears. The aim of the described modelling procedure is to obtain models useful to design robust linear compensators, as shown in Section 5, taking into account the main dynamic of the plant. Other negligible dynamics and the non linearities, arising mainly when the vehicle begins to move, have not yet been considered.

5. COMPENSATOR DESIGN

The Compensator block of Figure 1 was designed on the basis of the longitudinal dynamic vehicle models described in Section 4 taking into account few performance requirements.

The generic structure of the Compensator block is depicted in Figure 8 where the reference acceleration a_{ref} is integrated and compared with the vehicle velocity v_F to generate a velocity error v_E. This error is elaborated by the block $C(s)$ to generate the compensated acceleration a_c. The compensator transfer function $C(s)$ was designed independently for the GO and CRUISE/TRACK phases and for the STOP phase due to the different dynamic models, described in Section 4, characterizing these phases. Given the interval form of the uncertainty description of the plant models (4) and (5) and requiring a robust solution of the design problem, the methodologies described in Malan *et al.* (1997) by an algorithmic point of view and applied to a PID compensator design in Canale *et al.* (1999) have been used.

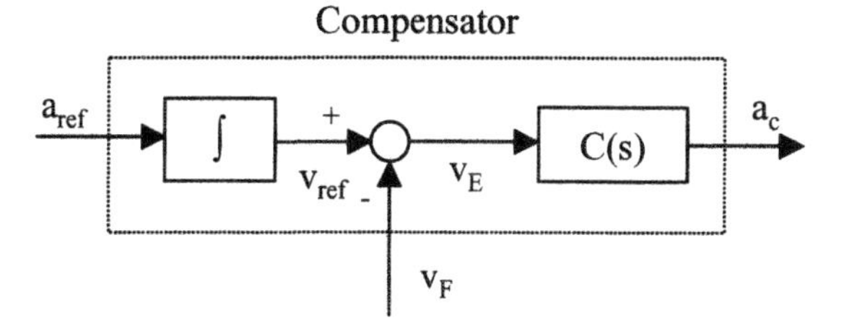

Fig. 8. Compensator structure.

GO and **CRUISE/TRACK** Model (4) was considered together with the following performance specification:

- stability;
- steady state delay to a ramp reference input of about 0.5 s corresponding to a steady state error of 0.5 m/s for a unitary ramp speed reference.
- maximum allowed acceleration of about 3 m/s^2 corresponding to a maximum α control effort of 90 °;

A simple PI controller can be used to satisfy the performance requirements:

$$C(s) = \frac{K_c\,(1 + \tau_c s)}{s} \tag{6}$$

Given the very simple plant model and performance requirements, minimum phase compensators suffice to find a suitable solution to the design problem. For this reason only positive values of parameters K_c and τ_c have been considered. Applying the procedures described in Malan *et al.* (1997), the set of design parameters that robustly meets the design specification has been computed. Such set is shown in Figure 9, where the parameters belonging to the dark gray areas do not satisfy the requirements, while the ones belonging to the complementary areas (white and light gray) do satisfy the specifications.

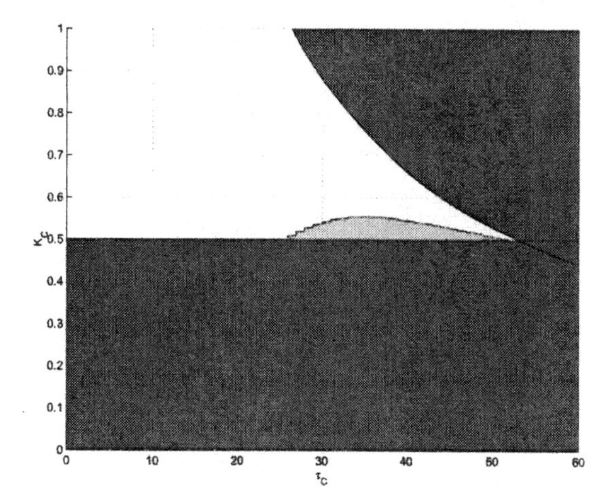

Fig. 9. Design parameter set ensuring robust performances.

In order to confirm the obtained results, the parameters belonging to the light gray area of Figure 9 have been used in simulation, confirming the robust satisfaction of the control requirements. This light gray area is the ones that, in simulation, gives the best transient response, not explicitly taken into account, as specification, in the compensator design procedure.

STOP Model (5) was considered together with the following performance specification:

- stability;
- steady state delay to a ramp reference input of about 0.5 s corresponding to a steady state error of 0.5 m/s for a unitary ramp speed reference.
- maximum allowed deceleration of about 4 m/s^2 corresponding to a maximum β control effort of 60 bar.

A very simple P controller can be used to satisfy the performance requirements:

$$C(s) = K_c \tag{7}$$

Due to the negative value of gain K_β in equation (5) together with the presence of a negative gain in the Vehicle Inverse Model – Command Share block of Figure 1, stability is obtained with a positive gain

K_c. A minimum value for K_c can be easily computed according to the steady state error specification, while the requirement on the maximum control effort limits the maximum value of the same parameter.

The switching among the two different controllers is handled by the same supervision logic (here not described) that handles the Reference Generator algorithm switching and that, on the basis of the requested acceleration a_{rq} and of the acquired signals, enables the correct α or β command generation through the Vehicle Inverse Model / Command Share block.

In order to show the effectiveness of the proposed ACC/S&G system, in Figure 10 it is reported a recorded vehicle speed v_F sequence and it is compared with the one generated by simulating the behaviour of the overall controlled system. Such simulation has been performed by setting the reference generator parameters according to the personalization criteria proposed in Canale *et al.* (2002), while the compensators parameters are chosen inside the light gray area of Figure 9 for the GO and TRACK phases and inside the computed interval of K_c for the STOP phase.

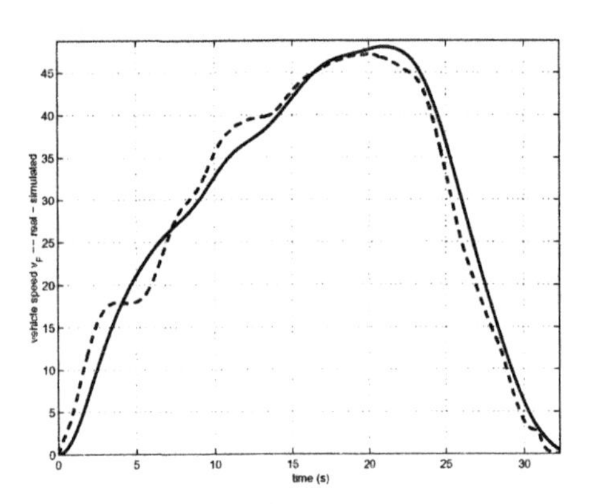

Fig. 10. Real (dashed) and simulated (solid) behaviour of v_F (km/h).

Note that only very slight differences between real and simulated v_F signals can be noticed. These differences can be explained by considering the fact that the automatic driving system takes only into account the first car in front of the controlled vehicle, by means of the radar measurements. On the contrary, the human driver has a more complete information by taking into account the behaviour of a greater number of preceding cars. This fact is a drawback of ACC equipped only with radar sensor and could be overcome by means of different hardware equipments including, for example, a video camera sensor.

6. CONCLUSIONS

The problem of designing an ACC Stop & Go system taking into account the driver behaviour has been afforded. It has been shown how suitable speed reference profiles can be simply generated satisfying safeness and comfort characteristics. Moreover the problem of characterizing the longitudinal vehicle dynamics in an effective way for the control algorithm design has been solved using standard vehicle documentation and real acquired data. Then it has been demonstrated by simulated data that very simple PI and P control strategies can be used to regulate the vehicle dynamics according to the required speed reference profile.

7. REFERENCES

Burnham, G.O., J. Seo and G.A. Bekey (1974). Identification of human drivers models in car following. *IEEE Transactions on Automatic Control* **19**(6), 911–915.

Canale, M. and S. Malan (2002). Analysis and classification of human driving behaviour in an urban environment. *Cognition Technology & Work* **4**(4), 197–206.

Canale, M., G. Fiorio, S. Malan and M. Taragna (1999). Robust tuning of low order controllers via uncertainty model identification. *European Journal of Control* **5**(2), 316–328.

Canale, M., S. Malan and V. Murdocco (2002). Personalization of ACC Stop and Go task based on human driver behaviour analysis. In: *Proc. 15th IFAC World Congress on Automatic Control.* Barcelona.

Hunt, K.J., J. C. Kalkkuhl, H. Fritz and T.A. Johansen (1996). Constructive empirical modelling of longitudinal vehicle dynamics using local model networks. *Control Engineering Practice* **4**(2), 167–178.

Joannou, P.A. and C.C. Chien (1993). Autonomous intelligent cruise control. *IEEE Transactions on Vehicular Technology* **42**(4), 657–672.

Kato, S., S. Tsugawa, K. Tokuda, T. Matsui and H. Fujii (2002). Vehicle control algorithms for cooperative driving with automated vehicles and intervehicle communications. *IEEE Transactions on intelligent transportation systems* **3**(3), 155–161.

Lu, X.Y., J.K. Hedrick and M. Drew (2002). ACC/CACC - control design, stability and robust performance. In: *Proc. American Control Conference*. Anchorage, USA. pp. 4327–4332.

Malan, S., M. Milanese and M. Taragna (1997). Robust analysis and design of control systems using interval arithmetic. *Automatica*.

Marsden, G., M. Brackstone and M. McDonald (2001). Assessment of the stop and go function using real driving behaviour. In: *Proc. ADAS 2001 Conference*. Birmingham, U.K.. pp. 76–80.

Persson, M., F. Botling, E. Hesslow and R. Johansson (1999). Stop and go controllers for adaptive cruise control. In: *Proc. IEEE Conference on Control Applications*. pp. 1692–1697.

Wang, J. and R. Rajamani (2002). Adaptive cruise control system design and its impact on highway traffic flow. In: *Proc. American Control Conference*. Anchorage, USA. pp. 3690–3695.

Wang, W. (1992). Modeling scheme for vehicle longitudinal control. In: *Proc. 31st Conference on Decision and Control*. Tucson, AZ. pp. 549–554.

ROBUST OUTPUT FEEDBACK CONTROL SYNTHESIS: LMI APPROACH

Vojtech Veselý [1]

Dept. of Automatic Control Systems,
Faculty of Electrical Engineering and Information Technology,
Slovak University of Technology, Bratislava, Slovak Republic.
E-mail: vesely@ kasr.elf.stuba.sk

Abstract: The paper addresses the problem of robust output feedback controller design with a guaranteed cost and parameter dependent Lyapunov function quadratic stability for linear continuous time polytopic systems. The proposed design methods lead to a non-iterative LMI based algorithm. Numerical example is given to illustrate the design procedure. *Copyright © 2003 IFAC*

Keywords: Quadratic Stability, Parameter Dependent Quadratic Stability, Robust Controller, Output Feedback

1. INTRODUCTION

Robustness has been recognized as a key issue in the analysis and design of control systems for the last two decades. During the last decades numerous papers dealing with the design of static robust output feedback control schemes to stabilize uncertain systems have been published (Benton and Smith, 1999; El Ghaoui and Balakrishnan, 1994; Geromel, De Souza and Skelton, 1998; Li Yu and Jian Chu, 1999; Mehdi, Al Hamid and Perrin, 1996; Pogyeon, Young Soo Moon and Wook Hyun Kwon, 1999; Tuan, Apkarian, Hosoe and Tuy, 2000; Veselý, 2002; Elaiw and Gyurkovics, 2003). Various approaches have been used to study the two aspects of the robust stabilization problem, namely conditions under which the linear system described in state space can be stabilized via output feedback and the respective procedure to obtain a stabilizing or robustly stabilizing control law.

The necessary and sufficient conditions to stabilize the linear continuous time invariant system via static output feedback can be found in (Kučera, and De Souza, 1995; Veselý, 2001). In the above and other papers, the authors basically conclude that despite the availability of many approaches and numerical algorithms the static output feedback problem is still open.

Recently, it has been shown that an extremely wide array of robust controller design problems can be reduced to the problem of finding a feasible point under a Biaffine Matrix Inequality (BMI) constraint. The BMI has been introduced in (Goh, Safonov and Papavassilopoulos, 1995). In this paper, the BMI problem of robust controller design with output feedback is reduced to a LMI problem (Boyd and et al, 1994).The theory of Linear Matrix Inequalities has been used to design robust output feedback controllers in (Benton and Smith, 1999; El Ghaoui and Balakrishnan, 1994; Li Yu and Jian Chu, 1999; Tuan, Apkarian, Hosoe, and Tuy, 2000; Veselý, 2002). Most of the above works present iterative algorithms in which a set of LMI problems are repeated until certain convergence criteria are met. The V-K iteration algorithm proposed in (El Ghaoui and Balakrishnan, 1994) is based on an alternative solution of two convex LMI optimization problems obtained by fixing the Lyapunov matrix or the gain controller matrix.

[1] The work has been partially supported by Slovak Scientific Grant Agency, Grant N 1/0158/03

This algorithm is guaranteed to converge, but not necessarily, to the global optimum of the problem depending on the starting conditions.
The main criticism formulated by control engineers against modern robust analysis and design methods for linear systems concerns lack of efficient easy-to-use and systematic numerical tools. This is especially true when analyzing robust stability as affected by highly structured uncertainty with BMI, for which no polynomial-time algorithm has been proposed so far (Henrion, Alzelier and Peaucelle, 2002).
This paper is concerned with the class of uncertain linear systems that can be described as.

$$\dot{x}(t) = (A_0 + A_1\theta_1 + \ldots + A_p\theta_k)x(t) \quad (1)$$

where $\theta = [\theta_1 ... \theta_p] \in R^p$ is a vector of uncertain and possibly time varying real parameters.
The system represented by(1) is a polytope of linear affine systems which can be described by a list of its vertices

$$\dot{x}(t) = A_{ci}x(t), \quad i = 1, 2, ..., N \quad (2)$$

where $N = 2^p$.
The system represented by (2) is quadratically stable if and only if there is a common Lyapunov matrix $P > 0$ such that

$$A_{ci}^T P + PA_{ci} < 0, \quad i = 1, 2, ..., N \quad (3)$$

A weakness of quadratic stability is that it guards against arbitrary fast parameter variations. As a result, this test tends to be conservative for constant or slow-varying parameters θ for polytopic systems. To reduce conservatism when (1) is affine in θ and the system parameters are time invariant, in (Gahinet, Apkarian and Chilali, 1996) a parameter-dependent Lyapunov functions $P(\theta)$ has been used in the form

$$P(\theta) = P_0 + \theta_1 P_1 + ... + \theta_p P_p \quad (4)$$

Robust controller design with guaranteed cost and affine quadratic stability has been proposed in (Veselý, 2002). Other types of parameter-dependent Lyapunov functions have been proposed in (De Oliviera, Bernussou, and Geromel, 1999) for stability analysis of linear discrete time systems and for analysis and the design of continuous time systems with affine type uncertainties in (Shaked, 2001; Henrion, Alzelier and Peaucelle, 2002; Takahashi, Ramos, and Peres, 2002). In this paper, we pursue the idea of (Takahashi, Ramos, and Peres, 2002) and introduce a new robust controller LMI design procedure with less conservative results and guaranteed cost. The proposed approach allows to reduce important class of BMI problems to LMI ones. For guaranteed cost and system (2) this leads to a non iterative LMI based algorithm. The proposed design procedure guarantees with sufficient conditions robust parameter dependant Lyapunov function quadratic stability (PDQS) for closed loop systems.
The paper is organized as follows. In Section 2 the problem formulation and some preliminary results are brought. The main results are given in Section 3. In Section 4 the obtained theoretical results are applied. We have used the standard notation. A real symmetric positive (negative) definite matrix is denoted by $P > 0$ $(P < 0)$. Much of the notation and terminology follows the references of (Kučera, and De Souza, 1995; and Gahinet, Apkarian and Chilali, 1996).

2. PRELIMINARIES AND PROBLEM FORMULATION

Consider the following affine linear time invariant continuous time uncertain system

$$\dot{x}(t) = A(\theta)x(t) + B(\theta)u(t) \quad (5)$$
$$y(t) = C(\theta)x(t), \quad x(0) = x_0$$

where $x(t) \in R^n$ is the plant state; $u(t) \in R^m$ is the control input; $y(t) \in R^l$ is the output of system; $A(\theta), B(\theta), C(\theta)$ are matrices of appropriate dimensions and

$$A(\theta) = A_0 + A_1\theta_1 + ... + A_p\theta_p$$
$$B(\theta) = B_0 + A_1\theta_1 + ... + B_p\theta_p$$
$$C(\theta) = C_0 + C_1\theta_1 + ... + C_p\theta_p$$

Note that, in order to keep the polytope affine property, the matrix $B(\theta)$ or $C(\theta)$ must be precisely known. In the following we assume that $C(\theta)$ is known and equal to the matrix C. In general, a polytope description of uncertainties results in a less conservative controller design than other characterizations of uncertainty (Boyd and et al, 1994). However, as the number of uncertain parameters increases, the number of vertices increases exponentially, and the design time increases exponentially too. The system represented by (5) is a polytope of linear systems. The linear matrix inequality approach requires the system (5) to be described by a list of its vertices, i.e., in the form

$$\{(A_{v1}, B_{v1}, C_{v1}), ..., (A_{vN}, B_{vN}, C_{vN})\} \quad (6)$$

Consequently, the system (6) is static output feedback quadratically stabilizable if and only if there is a Lyapunov matrix $P > 0$ and a feedback matrix F such that

$$(A_{vi} + B_{vi}FC_{vi})^T P + \quad (7)$$
$$P(A_{vi} + B_{vi}FC_{vi}) < 0, \quad i = 1, 2, ..., N$$

If (7) holds for $P > 0$ and some F, then the vertices of the polytope (6) are said to be simultaneously quadratically stabilized by F. It is well known (Boyd and et al, 1994) that if P is a common Lyapunov matrix for the vertices of the polytope (6), it serves as a common Lyapunov function for the uncertain system (5) for all admissible uncertainties $\theta_i \in < \underline{\theta_i}, \overline{\theta_i} >, i = 1, 2, ..., p$. In (6), each vertex is computed for a different permutation of the p variables θ_i, alternatively taken at their maximum and minimum values.
Recently, a new type of the parameter dependent Lyapunov function (PDLF) has been introduced (Takahashi, Ramos, and Peres, 2002) in the form

$$P(\alpha) = \sum_{i=1}^{N} P_i\alpha_i \quad \sum_{i=1}^{N} \alpha_i = 1 \quad \alpha_i \geq 0 \quad (8)$$

The robust stability conditions for the closed-loop polytopic system with output feedback algorithm

$$u = FCx \quad (9)$$

$$A_c = \sum_{i=1}^{N} A_{ci}\alpha_i = \sum_{i=1}^{N} (A_{vi} + B_{vi}FC)\alpha_i \quad (10)$$

are given by following lemma (Takahashi, Ramos, and Peres, 2002).
Lemma 1.
Suppose there exist positive definite Lyapunov matrices $P_j, j = 1, 2, ..., N$ such that

$$A_{ci}^T P_{vi} + P_{vi}A_{ci} < -I \quad (11)$$

$$A_{ck}^T P_{vj} + P_{vj}A_{ck} + A_{cj}^T P_{vk} + \quad (12)$$

$$P_{vk}A_{cj} < \frac{2}{N-1}I; i = 1, 2, ..., N$$

$$k = 1, 2, ..., N-1; j = k+1, ..., N$$

then

$$P(\alpha) = \sum_{i=1}^{N} P_{vi}\alpha_i \quad \sum_{i=1}^{N} \alpha_i = 1 \quad \alpha_i \geq 0 \quad (13)$$

is a parameter dependent Lyapunov function for any

$$A_c = \sum_{i=1}^{N} A_{ci}\alpha_i \quad (14)$$

□

The following performance index is associated with the system (5)

$$J = \int_0^\infty (x(t)^T Qx(t) + u(t)^T Ru(t))dt \quad (15)$$

where $Q = Q^T \geq 0, R = R^T > 0$ are matrices of compatible dimensions.

The problem studied in this paper can be formulated as follows: for a continuous time system described by (5) design a static output feedback controller with the gain matrix F and the control algorithm

$$u(t) = Fy(t) = FCx(t) \quad (16)$$

so that the closed loop system

$$\dot{x} = (A(\alpha) + B(\alpha)FC)x(t) = A_c x(t) \quad (17)$$

where

$$A(\alpha) = \sum_{i=1}^{N} A_{vi}\alpha_i \quad ...$$

is PDQS with guaranteed cost.
Definition 2. Consider the system (5). If there exists a control law u^* and a positive scalar J^* such that the closed loop system (17) is stable and the closed loop value cost function (15) satisfies $J \leq J^*$, then J^* is said to be the guaranteed cost and u^* is said to be the guaranteed cost control law for system (5). □

3. MAIN RESULTS

In this paragraph we present a novel procedure to design a static output feedback controller for polytopic continuous time linear systems (5) which ensures the guaranteed cost and PDQS of the closed loop system. One of the main result is summarized in the following theorem.
Theorem 1. Consider the closed loop polytopic system (17). Then the following statements are equivalent.
a. The polytopic system (17) is static output feedback simultaneously PDQS stabilizable with a guaranteed cost

$$\int_0^\infty (x^T Qx + u^T Ru)dt \leq \max_i x_0^T P_{vi}x_0 = J^* (18)$$

when $P_{vi} > 0 \quad i = 1, 2, ..., N$.
b. There exist symmetric matrices $P_{vj} > 0, j = 1, 2, ..., N, R > 0, Q > 0, M > 0$ and a matrix F such that the following inequalities hold

$$(A_{vi} + B_{vi}FC)^T P_{vi} + P_{vi}(A_{vi} + B_{vi}FC) + (19)$$

$$Q + C^T F^T R_1 FC \leq -M$$

for $i = 1, 2, ..., N$.

$$A_{ck}^T P_{vj} + P_{vj}A_{ck} + A_{cj}TP_{vk} \quad (20)$$

$$+P_{vk}A_{cj} + C^T F^T RFC \leq \frac{2}{N-1}M$$

where

$$k = 1, 2, ..., N-1; j = k+1, ..., N$$

$R_1 = \frac{N+1}{2N}R$ and M is positive definite matrix which helps relaxing the conservatism of results.
c. There exist matrices $P_{vj} > 0, R > 0, Q > 0, M > 0$ and a matrix F such that the following inequalities hold

$$(A_{vi} + B_{vi}FC + P_{vi})^T(A_{vi} + B_{vi}FC+ \quad (21)$$
$$P_{vi}) - (A_{vi} + B_{vi}FC)^T(A_{vi} + B_{vi}FC)-$$
$$P_{vi}P_{vi} + Q + C^TF^TR_1FC \leq -M$$
$$i = 1, 2, ..., N$$

and

$$(A_{vk} + B_{vk}FC + P_{vj})^T(A_{vk} + B_{vk}FC+$$
$$P_{vj}) + (A_{vj} + B_{vj}FC + P_{vk})^T(A_{vj}+$$
$$B_{vj}FC + P_{vk}) - [A_{vk}^TA_{vk} + A_{vj}^TA_{vj}+ \quad (22)$$
$$(A_{vk}^TB_{vk} + A_{vj}^TB_{vj})FC + ((A_{vk}^TB_{vk} + A_{vj}^T$$
$$B_{vj})FC)^T] - P_{vk}P_{vk} - P_{vj}P_{vj} + C^T$$
$$F^T(R - B_{vk}^TB_{vk} - B_{vj}^TB_{vj})FC \leq \frac{2}{N-1}M$$

along with the condition

$$R - B_{vk}^TB_{vk} - B_{vj}^TB_{vj} > 0 \quad (23)$$

where

$$k = 1, 2, ..., N-1; j = k+1, ..., N$$

Proof. For a polytopic system (17) and a Lyapunov function $V = x^TP(\alpha)x$ it is well known that there is a guaranteed cost if the following inequality holds

$$A_c^TP(\alpha) + P(\alpha)A_c + Q + C^TF^TRFC \leq 0 (24)$$

Assume, the inequalities (19) and (20) to hold, then using (24) we obtain

$$\sum_{i=1}^{N}(A_{ci}^TP_{vi} + P_{vi}A_{ci} + Q + C^TF^TR_1 \quad (25)$$
$$FC)\alpha_i^2 + \sum_{k=1}^{N-1}\sum_{j=k+1}^{N}(A_{ck}^TP_{vj} + P_{vj}A_{ck}+$$
$$A_{cj}^TP_{vk} + PvkA_{cj} + C^TF^TRFC)\alpha_k\alpha_j$$

Hence,

$$\alpha_i^2 \in < \frac{1}{N}, 1 >, \quad \sum_{k=1}^{N-1}\sum_{j=k+1}^{N}\alpha_k\alpha_j \leq \frac{N-1}{2N}$$

and for the worst case we obtain

$$\sum_{i=1}^{N}(A_{ci}^TP_{vi} + P_{vi}A_{ci})\alpha_i^2+$$
$$\sum_{k=1}^{N-1}\sum_{j=k+1}^{N}(A_{ck}^TP_{vj} + P_{vj}A_{ck}+$$
$$A_{cj}^TP_{vk} + P_{vk}A_{cj})\alpha_k\alpha_j + Q + C^TF^TRFC \leq$$
$$A_c^TP(\alpha) + P(\alpha)A_c + Q + C^TF^TRFC \quad (26)$$

Substituting the right hand side of (21) and (22) to (25) we obtain

$$\sum_{i=1}^{N}(-M)\alpha_i^2 + \sum_{k=1}^{N-1}\sum_{j=k+1}^{N}\frac{2}{N-1}M\alpha_k\alpha_j$$

or

$$-M(\sum_{i=1}^{N}(N-1)\alpha_i^2 + \sum_{k=1}^{N-1}\sum_{j=k+1}^{N}2\alpha_k\alpha_j) =$$
$$-M\sum_{i=1}^{N-1}\sum_{j=i+1}^{N}(\alpha_i - \alpha_j)^2 < 0 \quad (27)$$

Inequality (27) proves stability of closed loop polytopic systems (17) with PDLF and equation (26) proves the equivalence of the first and second statements.The proof of the equivalence of the second and third statement ((19) and (21)) is based on the following equation

$$(A_{vi} + B_{vi}FC + P_{vi})^T(A_{vi} + B_{vi}FC + P_{vi}) =$$
$$(A_{vi} + B_{vi}FC)^T(A_{vi} + B_{vi}FC) + (A_{vi}+ (28)$$
$$B_{vi}FC)^TP_{vi} + P_{vi}(A_{vi} + B_{vi}FC) + P_{vi}P_{vi}$$

Substituting for the first and second terms in (19) from (28) the inequality (21) has been obtained. The equivalence of the second and third statements is then evident. □

For LMI solution of inequalities (21) and (22) with (23), the term " $-PP$" has to be replaced using either of the two proposed possibilities.

(1) $$-P_{vi}P_{vi} \leq -P_{vi}D_{vi} - D_{vi}P_{vi}+ \quad (29)$$
$$D_{vi}D_{vi} = F_{vi}$$

where $D_{vi} = D_{vi}^T > 0$ is the initial value of P_{vi}.

(2) $$-P_{vi}P_{vi} \leq (\varrho^2 - \varrho_1^2)I - Z_i = F_{zi} \quad (30)$$

where

$$\varrho_1 I < P_{vi} < \varrho I \quad i = 1, 2, ..., N$$
$$\begin{bmatrix} Z_i & P \\ P & I \end{bmatrix} > 0$$
$$0 \leq Z_i \leq \varrho^2$$

$Z_i = Z_i^T > 0$ is any positive definite matrix, a new LMI variable that satisfies condition (30).

From (21) and (29) the following LMI condition results

$$\begin{bmatrix} L_i & C^T F^T & (A_{ci} + P_{vi})^T \\ FC & -(R_{1i})^{-1} & 0 \\ A_{ci} + P_{vi} & 0 & -I \end{bmatrix} < 0$$

$$0 < P_{vi} < \varrho I \quad i = 1, 2, ..., N \qquad (31)$$

where

$$R_{1i} = R_1 - B_{vi}^T B_{vi} > 0$$

$$L_i = -[A_{vi}^T A_{vi} + A_{vi}^T B_{vi} FC +$$

$$(A_{vi}^T B_{vi} FC)^T] + Q + M + F_{vi}$$

$i = 1, 2, ..., N$. Equation (22) yields to the following LMI problem

$$\begin{bmatrix} L_{kj} & C^T F^T & S_{kj} & S_{jk} \\ FC & -(R_{kj})^{-1} & 0 & 0 \\ S_{kj}^T & 0 & -I & 0 \\ S_{jk}^T & 0 & 0 & -I \end{bmatrix} < 0$$

$$L_{kj} = -[A_{vk}^T A_{vk} + A_{vj}^T A_{vj} + \qquad (32)$$

$$(A_{vk}^T B_{vk} + A_{vj}^T B_{vj}) FC + ((A_{vk}^T B_{vk} + A_{vj}^T$$

$$B_{vj}) FC)^T] + F_{vk} + F_{vj} - \frac{2}{N-1} M$$

$$S_{kj} = (A_{vk} + B_{vk} FC + P_{vj})^T$$

$$S_{jk} = (A_{vj} + B_{vj} FC + P_{vk})^T$$

with condition

$$R_{kj} = R - B_{vk}^T B_{vk} - B_{vj}^T B_{vj} > 0$$

and

$$k = 1, 2, ..., N-1 \quad j = 1, 2, ..., N$$

If the LMI solutions to (31) and (32) are feasible with respect to matrices $F, P_{vi}, i = 1, 2, ..., N$ and M, then the uncertain system (5) is parameter dependent quadratically stable with a guaranteed cost control law

$$u = Fy$$

whereby

$$J^* = \max_i x_0^T P_{vi} x_0$$

is the guaranteed cost for the uncertain closed loop system.

4. EXAMPLE

This example has been borrowed from (Benton and Smith,1999) to demonstrate the use of the algorithm given by (31) and (32). It is known that the presented system is static output feedback stabilizable. Let (A, B, C) in (1) be defined as

$$A = \begin{bmatrix} -0.036 & 0.0271 & 0.0188 & -0.4555 \\ 0.0482 & -1.010 & 0.0024 & -4.0208 \\ 0.1002 & q_1(t) & -0.707 & q_2(t) \\ 0 & 0 & 1 & 0 \end{bmatrix}$$

$$B = \begin{bmatrix} 0.4422 & 0.1761 \\ q_3(t) & -7.59222 \\ -5.520 & 4.490 \\ 0 & 0 \end{bmatrix} \qquad C = \begin{bmatrix} 0 & 1 & 0 & 0 \end{bmatrix}$$

with parameters bounds $-0.6319 \leq q_1(t) \leq 1.3681, 1.22 \leq q_2(t) \leq 1.420$, and $2.7446 \leq q_3(t) \leq 4.3446$. Find a stabilizing output feedback matrix F. The four vertices are calculated. The nominal model of (A, B) is given by the above matrices by taking the entries $A(3,2) = 0.3681, A(3,4) = 1.32$ and $B(2,1) = 3.5446$. The affine model uncertainty (5) (A_1, A_2, B_1, B_2) are matrices with the following entries $A_1(3,2) = 1, A_2(3,4) = 0.1$ and $B_1(2,1) = 0.8, B_2 = 0$ with $\theta_i \in < -1, 1 >, i = 1, 2$. Other entries of the above uncertain matrices are equal to zero. The nominal model is unstable with eigenvalues:

$$eig\{-2.0516, 0.2529 \pm 0.3247i, -0.2078\}$$

Let the structure of F be defined as

$$F^T = [F(1,1) \quad F(2,1)]$$

- Eqs. (29),(31), (32).
  For $Q = q * I, q = .00001, R = r * I$, r=246.2 $\varrho = 100$ and $\theta_m = 1$ the following results are obtained.

$$F^T = [-1.502 \quad 2.726]$$

$$maxeig(CL) = -.0683; \max_i \lambda_M(P_{vi}) = 98.02$$

- For $\theta_m = 2, r = 264.2$ the results are as follows

$$F^T = [-1.3167 \quad 2.6816]$$

$$maxeig(CL) = -.0676; \lambda_M(P_{vi}) = 98.72$$

- For $\theta_m = 2.5, r = 273.86$the results are as follows

$$F^T = [-1.2879 \quad 3.0857]$$

$$maxeig(CL) = -.0684; \lambda_M(P_{vi}) = 99.83$$

- For $\theta_m = 1, r = 246.2, \varrho = 100$ and varied q the following results are obtained.
  a. $q = .1$

$$F^T = [-1.1071 \quad 2.699]$$

$$maxeig(CL) = -.0719; \lambda_M(P_{vi}) = 98.16$$

Note, that eigenvalues of matrix M are $eig(M) = [.9984; .8902; .6963; .1222]$. Hence, they are different from the identity matrix as supposed in Lemma 1 (Takahashi, Ramos, and Peres, 2002). We can conclude that using the matrix M in the design procedure less conservative results can be obtained. b. $q = 1$

$$F^T = [-1.5033 \quad 2.8373]$$

$$maxeig(CL) = -.0689; \lambda_M(P_{vi}) = 99.34$$

c. $q = 6$

$$F^T = [-2.3635 \quad 2.8861]$$

$$maxeig(CL) = -.0616; \lambda_M(P_{vi}) = 99.92$$

All LMI solutions are feasible. Note that for the above parameters the V-K iterative method does not provide any reasonable solutions. Usually, a repeated procedure generates less conservative results than the first one. The convergence of the above special repeated procedure has not been proven yet, however if the reasoning of (El Ghaoui and Balakrishnan, 1994)that V-K iterative procedure is guaranteed to converge is taken into account we can conclude that the proposed algorithm is guaranteed to converge but not necessarily to the global optimum of the problem, depending on the starting conditions.

5. CONCLUSIONS

In this paper, we have proposed a novel procedure to design robust output feedback controller for linear systems with affine parameter uncertainty. A feasible solution to the proposed output feedback controller design procedure guarantees with sufficient conditions the parameter dependent Lyapunov function quadratic stability and guaranteed cost. The proposed design procedure pursues the idea of (Takahashi, Ramos, and Peres, 2002). Examples show that the proposed approach gives less conservative results than those that could be obtained by a quadratic design procedure.

6. REFERENCES

Benton R. E. JR. and Smith D.: A Non Iterative LMI Based Algorithm For Robust Static Output Feedback Stabilization, (1999) *Int.J.Control*, 72, 14, 1322-1330.

Boyd S., El Ghaoui L.,Feron E., and Balakrishnan V.: Linear Matrix Inequalities in System and Control Theory, (1994), SIAM, 115, Philadelphia.

El Ghaoui L. and Balakrishnan V.: Synthesis of Fixed Structure Controllers Via Numerical Optimization, (1994), In: *Proc. 33rd Conf. on Decision and Control*, Lake Buena Vista, FL 2678-2683.

Elaiw A.M. and Gyurkovics E.: Stabilizing Receding Horizon Control of Sampled-Data Nonlinear Systems via their Aproximate Discrete - Time Models, (2003), In:*IFAC Workshop on Control Applicattions of Optimisation*, Visegrad, 56-61.

Gahinet P., Apkarian P., and Chilali M.: Affine Parameter Dependent Lyapunov Functions and Real Parametric Uncertainty, (1996), *IEEE Trans.on AC*, 41, 3, 436-442.

Geromel J. C., De Souza C. C., and Skelton R. E.: Static Output Feedback Controllers: Stability and Convexity, (1998), *IEEE Trans. on AC*, 43, 1, 120-125.

Goh K. C., Safonov M. G., and Papavassilopoulos G. P.: Global Optimization For the Biaffine Matrix Inequality Problem, (1995), *Journal of Global Optimization*, 7, 365-380.

Henrion D., Alzelier D. and Peaucelle D. : Positive Polynomial Matrices and Improved Robustness Conditions, (2002) In: 15th Triennial World Congres, Barcelona , CD.

Kozáková A.: Robust Decentralized Control of Complex Systems in Frequancy Domain. In: 2nd IFAC Workoshop on NTDCS, Elsevier Kindlington,Uk, 1999

Kučera V. and De Souza C. E.: A necessary and Sufficient Conditions for Output Feedback Stabilizability, (1995), *Automatica* 31, 9, 1357-1359.

Li Yu and Jian Chu: An LMI Approach to Guaranteed Cost of Linear Uncertain Time Delay Systems, (1999), *Automatica*, 35, 1155-1159.

De Oliviera M.C. Bernussou J. and Geromel J.C.: A new discrete-time robust stability condition. *Systems and Control Letters*, 1999, 37, 261-265.

Shaked U.: Improved LMI Representations for the Analysis and the Design of Continuous-Time Systems with Polytopic Type Uncertainty *IEEE Trans on AC* , 2001, Vol46, N4, 652-656.

Takahashi R.H.C. , Ramos D.C.W. and Peres P.L.D. : Robust Control Synthesis via a Genetic Algorithm and LMIS, (2002) In: 15th Triennial World Congress, Barcelona, CD.

Tuan H.D., Apkarian P., Hosoe S., and Tuy H. : D.C. Optimization Approach to Robust Control: Feasibility Problems, (2000), *Int.J. Control*, 73, 2, 89-104.

Veselý V.: Static Output Feedback Controller Design, (2001), *Kybernetika*, 37, 2, 205-221.

Veselý V.: Design of Robust Output Affine Quadratic Controller, (2002), In: 15th Triennial World Congress, Barcelona, CD.

B-BAC: THE ROBUST APPROACH TO THE NONLINEAR ADAPTIVE CONTROL OF INDUSTRIAL PROCESSES

Jacek Czeczot

Institute of Automatic Control, Technical University of Silesia
ul. Akademicka 16, 44-100 Gliwice, Poland, jczeczot@ia.polsl.gliwice.pl

Abstract: In this paper the generalised and robust approach to the nonlinear adaptive control of industrial processes (so called: B-BAC – Balance-Based Adaptive Control) is presented. It is shown how to apply the generalised dynamic equation, describing a controlled variable, for the B-BAController design. The adaptability of the control law results from the fact that all the process nonlinearities (e.g. due to kinetics) are replaced by the only one time varying term, which is then estimated as one parameter. The possibility of the B-BAC application is illustrated on two different industrial examples: the continuous fermentation process and the heat exchanger. The simulation results prove very good control performance and the robustness of the B-BAC in comparison with the conventional PI controller. *Copyright © 2003 IFAC*

Keywords: Robust control, Model-based control, Adaptive control, Least-squares estimation, Biotechnology, Heat exchangers.

1. INTRODUCTION

For the last decades the methodology of the model-based nonlinear control has become the subject of growing interest due to its ability to employ the process nonlinearity directly in the control law (Richalet, 1993; Joshi *et al.*, 1997; Metzger, 2000). This approach usually provides significant improvements over linear control techniques but there is one limitation: we have to know the nonlinear model of the process. This model can be physical or empirical but in both cases when the significant improvement of the control performance is expected, the adaptability should be employed in the control law to ensure its robustness. The adaptability itself usually consists in the estimation of the model parameters, included in the control law, and therefore the complete form of the mathematical model must be given. However, in majority of cases, when physical model is employed, there is a large uncertainty on the nonlinear part of the model. Moreover, the model parameters vary in time and due to this fact usually there is a need to apply some sophisticated multiparameter identification techniques on-line. Consequently, in such cases the very well known difficulties with identifiability and with the estimation convergence arise.

The approach, suggested in this paper and called Balance-Based Adaptive Controller (B-BAC), combines simplicity and generality, that are characteristic of the classical PID controller, with very good control performance and robustness resulting partially from its adaptability and feedforward action and partially from the characteristic properties of the methodology itself. The B-BAC is dedicated to control a wide class of technological processes and its generality follows from the fact that the control law is derived on the basis of the simplified part of a nonlinear physical model of a process, namely on the general balance-based dynamic equation describing a controlled variable. In this equation the nonlinearity, resulting from a number of reactions and/or heat exchange phenomena taking place due to a process, is represented by the only one time varying term. The value of this term is unmeasurable and thus it must be estimated on-line. However, since there is only this one parameter to estimate, it can be easily managed by the recursive least-squares procedure. This approach allows us to avoid common difficulties either with large uncertainty on the reaction and/or heat exchange kinetics and nonlinearity as well as with the multiparameter identification. Moreover, there is no longer a need to know the complete form of the physical model of the process. Only its part, describing the controlled variable, must be given in the simplified form.

This paper is organised as follows. First, the theoretical approach to the B-BAC methodology is presented. Then, the generality of this approach is shown by the B-BAC application to two different examples of the industrial processes: the continuous fermentation process and the heat exchanger. Simulation results prove that in both cases our control methodology ensures much better control performance than the classical PI controller, which nowadays is still in use in the majority of industrial control applications. Concluding remarks complete the paper.

2. THEORETICAL APPROACH TO THE B-BAC METHODOLOGY

As it was said in the previous Section, the B-BAC is dedicated to control a wide range of technological processes for which it is possible to define the control goal in the following way: one of the parameters characterising a process, defined here as Y(t) and called the controlled variable, should be kept equal to its pre-defined set-point Y_{sp}. Y(t) can be chosen as one of state variables (a component concentration or the temperature) or as a combination of two or more state variables. In a process a number of isothermal or nonisothermal biochemical reactions and/or heat exchange phenomena with unknown kinetics can take place. A process itself takes place in a tank with time varying volume V(t) [m^3].

The dynamical behavior of Y(t) can be described by the following well known general ordinary differential equation written on the basis of the mass or of the energy balance considerations:

$$\frac{dY(t)}{dt} = \frac{1}{V(t)} \underline{F}^T(t)\underline{Y}_F(t) - R_Y(t) \quad (1)$$

The vector product $\underline{F}^T(t)\underline{Y}_F(t)$ represents mass or energy fluxes incoming to or outcoming from the reactor tank. The elements of the vector $\underline{F}(t)$ are the combination of the volumetric flow rates and, consequently, the vector $\underline{Y}_F(t)$ is the corresponding vector to $\underline{F}(t)$ and its elements are the combination of the inlet values of Y(t) and of the value of Y(t) itself. $R_Y(t)$ is a positive or negative time varying term with an unknown expression form. It represents "one global reaction" including all reversible and/or irreversible reactions or heat exchange and/or production with unknown and nonlinear kinetics that influence the value of Y(t). Let us note that in the case when Y(t) is a state variable, the equation (1) has a generalized and simplified form of a state equation describing Y(t) and taken directly from a mathematical model of a process. However, if Y(t) is a combination of two or more state variables, a number of state equations from a mathematical model must be combined together and rearranged to the form of the equation (1).

Once the equation (1) has been obtained, it can be a basis for the B-BAC under the following assumptions:

- the control variable must be chosen as one of the elements of the vectors $\underline{F}(t)$ or $\underline{Y}_F(t)$,
- the other elements of the vectors $\underline{F}(t)$ and $\underline{Y}_F(t)$ as well as the value of Y(t) must be measurable on-line at least at discrete moments of time or they should be known by choice of the user.

If the above requirements are met, at this stage we can apply the same methodology as for the 'model reference linearising control' (Isidori, 1989; Bastin and Dochain, 1990). For our control goal let us assume the following stable first-order closed loop dynamics:

$$\frac{dY(t)}{dt} = \lambda\left(Y_{sp} - Y(t)\right) \quad (2)$$

where $\lambda>0$ is the tuning parameter. After combining the equations (1) and (2) we can obtain the following equation:

$$\underline{F}^T(t)\underline{Y}_F(t) = \lambda V(t)\left(Y_{sp} - Y(t)\right) + V(t)R_Y(t) \quad (3)$$

Once the control variable has been chosen, the above equation can be rearranged to obtain the control law describing its value. This control law has the form of the 'model reference linearising controller' and is very well known in the bibliography.

In order to provide the adaptability to the control law resulting from the equation (3) there is a need to estimate the value of the nonmeasurable term $R_Y(t)$. This value can be estimated on-line at discrete moments of time by the recursive least-squares method with the forgetting factor α. The estimation procedure is also based on the discretised form of the simplified model (1) (Czeczot, 1997, 1998). When we replace $R_Y(t)$ by its discrete time estimate $\hat{R}_Y^i$ in the equation (3), it can be rewritten in the following discrete time form with i – discretisation instant:

$$\underline{F}^{T,i}\underline{Y}_F^i = \lambda V^i\left(Y_{sp} - Y^i\right) + V^i\hat{R}_Y^i \tag{4}$$

The equation (4) is a basis for the B-BAC. There is only a need to rearrange it to obtain the control law in the form that allows us to calculate the value of the control variable.

3. CASE STUDIES

In this Section it is shown how to apply the suggested B-BAC methodology for the adaptive control of industrial processes. We consider two test-examples: the continuous fermentation process and the heat exchanger. These examples have been carefully selected to prove the generality of the approach presented in the previous Section. Let us note that although only these two cases are studied in this paper, they are the representative selection of a wide range of industrial processes for which our B-BAC methodology can be applied to ensure robust control and to improve the control performance.

3.1 Example 1: continuous fermentation process

The continuous fermentation process (for instance bacterial production of amino acids, e.g. lysine) is considered as the first real world example. Its mathematical model (Bastin and Dochain, 1990) consists of three nonlinear state equations describing respectively the biomass concentration X [g/L], the substrate concentration S [g/L] and the outcoming product concentration P [g/L] :

$$\frac{dX(t)}{dt} = \mu(t)X(t) - \frac{F(t)}{V}X(t) \tag{5a}$$

$$\frac{dS(t)}{dt} = \frac{F(t)}{V}\left(S_{in}(t) - S(t)\right) - k_1\mu(t)X(t) - k_2\nu(t)X(t) \tag{5b}$$

$$\frac{dP(t)}{dt} = \nu(t)X(t) - \frac{F(t)}{V}P(t), \tag{5c}$$

The Monod-type model (Monod, 1949) is applied for describing the specific growth rate $\mu(t)$ [1/h] and the parabolic expression defines the specific production rate $\nu(t)$ [1/h] :

$$\mu(t) = \mu_{max}\frac{S(t)}{K_M + S(t)}, \tag{5d}$$

$$\nu(t) = \begin{cases} S(t)\left(\nu_0 - \nu_1 S(t)\right) & 0 \le S(t) \le \dfrac{\nu_0}{\nu_1} \\ 0 & S(t) > \dfrac{\nu_0}{\nu_1} \end{cases} \tag{5e}$$

In the model (5) F represents the volumetric flow rate [L/h], V – the volume of the bioreactor tank [L], S_{in} – the inlet substrate concentration [g/L], k_1, k_2 – the yield coefficients [-], μ_{max} – the maximum specific growth rate [1/h], K_M – the saturation constant [g/L] and ν_0, ν_1 –constant parameters.

In practical applications of such a system the yield-productivity conflict occurs and thus there is a need to manage this problem by properly designed control loop. It was shown in (Bastin and Dochain, 1990; Van Impre and Bastin, 1995) that the effective control of the system (5) consists in regulating the outlet substrate concentration S(t) during the bioreactor activity at the properly chosen set-point S_{sp}. Notice that this control goal is the same as the one defined for B-BAC in the Section 1.

Let us note that the state equation (5b) can be directly simplified to the form of the general equation (1). For this system there are two possible control variables: the volumetric flow rate F(t) and the inlet substrate concentration $S_{in}(t)$, and the choice of the control variable determines the structure of the vectors $\underline{F}(t)$ and $\underline{Y}_F(t)$ in the following way:

- for the control variable F(t): $\underline{F}(t) = [F(t)]$, $\underline{Y}_F(t) = [S_{in}(t) - S(t)]$,
- for the control variable $S_{in}(t)$: $\underline{F}^T(t) = [F(t), -F(t)]$, $\underline{Y}_F^T(t) = [S_{in}(t), S(t)]$.

For both cases the controlled variable Y(t) = S(t) and the volume of the reactor tank V(t) = V = const. The nonlinear term of the equation (5b), describing the reaction taking place inside the bioreactor tank, is expressed by the following equation :

$$R_Y(t) = \left(k_1\mu(t) + k_2\nu(t)\right)X(t), \tag{6}$$

but this expression is considered to be unknown for the B-BAC derivation.

In (Bastin and Dochain, 1990; Bastin, 1991; Dochain, 1991, 1992) it is shown how to derive the sophisticated estimation methods to obtain the values of the time varying model parameters. Those methods base on the complete form of the nonlinear model (5) or on the reduced order model of the process. In (Czeczot, 1998, 1999) it was shown how to design two nonlinear model-based adaptive predictive controllers with the application of the substrate consumption rate estimated on-line as one term. That approach is very similar to the B-BAC methodology and it ensures the same good control performance. However, it is less general and therefore it is interesting to show the possibility of the B-BAC application to control the system (5).

It can be seen that the equation (5b), written in the form of the equation (1), then can be easily discretised and rearranged to the form of the equation (4). When we apply the estimation procedure for the on-line estimation of the time varying term $R_Y(t)$ and then replace this term by its estimate $\hat{R}_Y^i$, we can obtain the control law in the form of the equation (4), which is a basis for the B-BAC. The final form of the B-BAC is dependent on the choice of the control variable:

- for the control variable F(t)

$$F^i = \frac{\lambda V\left(S_{sp} - S^i\right) + V\hat{R}_Y^i}{S_{in}^i - S^i} \tag{7}$$

- for the control variable $S_{in}(t)$

$$S_{in}^i = \frac{\lambda V\left(S_{sp} - S^i\right) + V\hat{R}_Y^i + F^i S^i}{F^i} \tag{8}$$

3.2 EXAMPLE 2: heat exchange process

The mathematical model of the heat exchanger, considered in this paper as the second example, consists of two nonlinear state equations:

$$\frac{dT_1(t)}{dt} = \frac{F_1}{V_1}\left(T_{1,in}(t) - T_1(t)\right) - \frac{hA_c\left(T_1(t) - T_2(t)\right)}{\rho_1 cp_1 V_1} \tag{9a}$$

$$\frac{dT_2(t)}{dt} = \frac{F_2}{V_2}\left(T_{2,in}(t) - T_2(t)\right) + \frac{hA_c\left(T_1(t) - T_2(t)\right)}{\rho_2 cp_2 V_2} \tag{9b}$$

where T_1 [K] represents the outlet temperature in the first circuit, T_2 [K] – the outlet temperature in the second circuit, $T_{1,in}$ [K] – the inlet temperature in the first circuit, $T_{2,in}$ [K] – the inlet temperature in the second circuit, F_1 [m^3/h] – the volumetric flow rate in the first circuit, F_2 [m^3/h] – the volumetric flow rate in the second circuit, ρ_1 [kg/m^3] – the density of the medium in the first circuit, ρ_2 [kg/m^3] – the density of the medium in the second circuit, cp_1 [J/kg K] – the specific heat of the medium in the first circuit, cp_2 [J/kg K] – the specific heat of the medium in the second circuit, V_1 [m^3] – the volume of the heat exchanger in the first circuit, V_2 [m^3] – the volume of the heat exchanger in the second circuit, h [W/m^2 K] – the overall heat transfer coefficient and A_c [m^2] – the heat transfer area.

This model was successfully used to describe the most important dynamical properties of the heat exchanger working as a part of the pilot heat distribution plant developed at our laboratory in Gliwice (Laszczyk and Pasek, 1995).

Let us consider the control problem to keep the temperature $T_2(t)$ equal to its set value T_{sp} by manipulating the flow rate $F_2(t)$. Thus our B-BAC can be based directly on the simplified state equation (9b) written in the form of the equation (1). Once the control variable $F_2(t)$ has been chosen, the vectors $\underline{F}(t)$ and $\underline{Y}_F(t)$ have the following form:

- $\underline{F}(t) = \left[F_2(t)\right]$, $\underline{Y}_F(t) = \left[T_{2,in}(t) - T_2(t)\right]$.

The controlled variable $Y(t) = T_2(t)$ and the volume $V(t) = V_2 =$ const. The nonlinear heat exchange term of the equation (9b), describing the heat exchange phenomena and considered to be unknown for our B-BAC methodology, is described by the following expression :

$$R_Y(t) = -\frac{hA_c\left(T_1(t) - T_2(t)\right)}{\rho_2 cp_2 V_2} \tag{10}$$

In this case the equation (9b), written in the form of the equation (1), can also be discretised and easily rearranged to the form of the equation (4). After applying the estimation procedure for the on-line estimation of the time varying term $R_Y(t)$ we can then replace this term by its estimate $\hat{R}_Y^i$ and finally we can obtain the control law in the form of the equation (4), which is a basis for the B-BAC. After some simple rearrangements we can obtain the final B-BAC law:

$$F_2^i = \frac{\lambda V_2\left(T_{sp} - T_2^i\right) + V_2\hat{R}^i}{T_{2,in}^i - T_2^i} \tag{11}$$

4. SIMULATION RESULTS

In this section the very good control performance and the robustness of the suggested B-BAC methodology are proved by the computer simulation. Both examples with the values of the model parameters given below and described by the complete nonlinear models (5) and (9) are considered as the real world systems to be controlled. The controllers (7), (8), and (11), derived in the Section 3 in line with our B-BAC methodology, are used in control loop for successful control of these systems. The selected simulation results of their control performance are presented in the paper and this performance can be compared with the classical PI controller, which is still in use in the vast majority of automatic control loops in the process industries (≈ 90%) (Seborg, 1999).

4.1 Example 1 - continuation

For the continuous fermentation process, described by the nonlinear model (5), the following values of the parameters have been chosen for simulation: $\mu_{max} = 0.35$, $K_M = k_1 = 0.4$, $k_2 = 0.05$, $\nu_0 = 15$, $\nu_1 = 5$, $V = 1$. The steady state of the system is characterized by the following values of the parameters: $S_{in} = 60$, $F = 0.2$, $X(0) = 29.1$, $S(0) = 0.53$, $P(0) = 956{,}6$. For both controllers (7) and (8) the sampling time for the estimation and for the substrate control has been chosen as $T_R = 5$ [min]. The estimate initial value $\hat{R}_Y^0 = 10$, the forgetting factor $\alpha = 0.1$, $P^0 = 1000$ and the tuning parameter $\lambda = 1$.

The control performance of the suggested B-BAControllers (7) and (8) can be respectively investigated in the Fig. 1 and 2. In both cases the controllers are able to keep the process (5) stable, even in the presence of the step changes applied on the set-point S_{sp} as well as on the indicated disturbing parameters. In the comparison with the conventional PI controller both B-BAControllers ensure much shorter regulation time and much smaller overregulations of the controlled variable resulted from the perturbation changes. Moreover, in the Fig. 3 it can be seen that, in opposite to the PI controller application, only the application of the suggested B-BAC (8) allows the regulation error to be reduced to zero in the steady state.

4.2 Example 2- continuation

In the case of the heat exchange process, described by the nonlinear model (9), the simulation runs have been carried out for the following values of the model parameters: $V_1 = V_2 = 0.01$, $hA_c = 100000$, $\rho_1 = \rho_2 = 1000$, $cp_1 = cp_2 = 4200$. In the steady state $T_1(0) = 327.12$, $T_2(0) = 302.88$, $T_{1,in} = 330$, $T_{2,in} = 300$, $F_1 = F_2 = 0.2$. The sampling time both for the controller (11) and for the estimation procedure $T_R = 15$ [min]. The estimate initial value $\hat{R}_Y^0 = 57.7$, the forgetting factor $\alpha = 0.1$, $P^0 = 1000$ and the tuning parameter $\lambda = 5$.

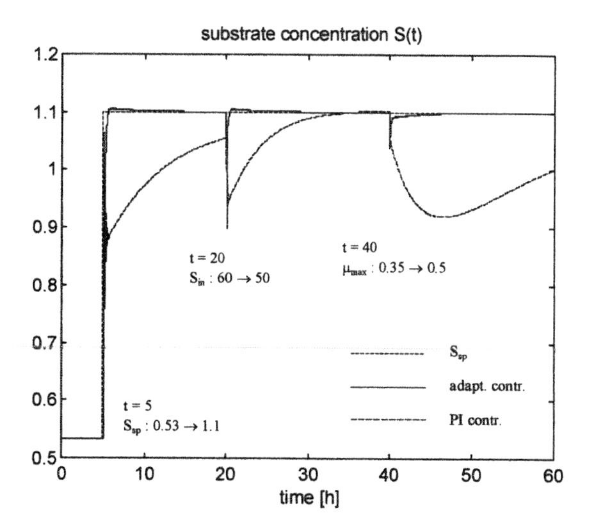

Fig. 1. B-BAControl of the continuous fermentation process with F(t) as the control variable

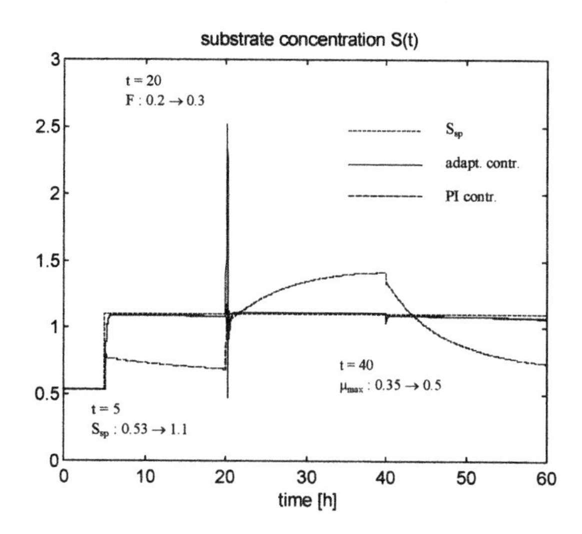

Fig. 2. B-BAControl of the continuous fermentation process with $S_{in}(t)$ as the control variable

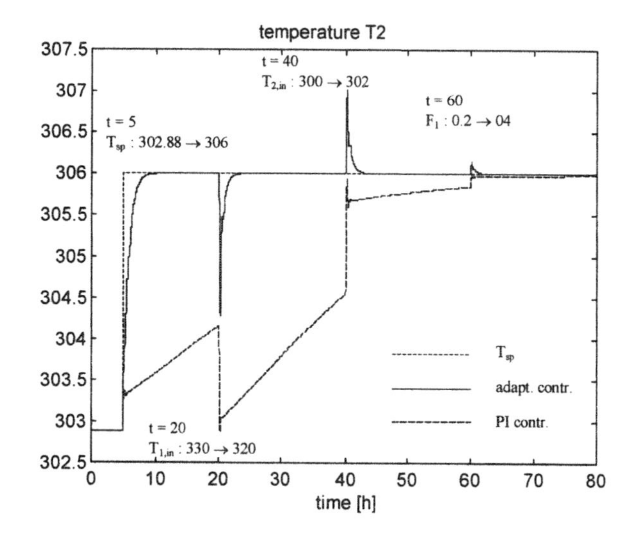

Fig. 3. B-BAControl of the heat exchange process

Fig. 3 shows the control performance of the controller (11) in the presence of the step changes applied on the set-point T_{sp} and on the indicated disturbing parameters. It can be seen that the PI controller provides unacceptable regulation time, even if it ensures the stable behaviour of the control loop. The application of the B-BAC (11) results in much shorter regulation time, which, in consequence, allows the regulation error to be reduced to zero in the steady state.

Let us note that for the process (9) it is assumed that there is no heat loss between the exchanger tank and the surrounding temperature. This assumption is very realistic because such devices are usually very good (almost perfectly) insulated against heat loss. However, the term, describing heat loss, could be easily included in the model (9). Yet, due to the robustness of the B-BAC methodology, this modification would not change the form of the control law (11) because this term would be included as a part of the unknown term $R_Y(t)$ estimated on-line.

5. CONCLUSIONS

In this paper we have presented the generalised approach to the nonlinear adaptive control strategy: Balance-Based Adaptive Controller (B-BAC). The key property of this approach consists in the fact that there is no necessity to have the knowledge of the nonlinear kinetics of a process. Once controlled and control variables have been chosen, the general dynamic equation (1), describing a controlled variable, can be written and then used as a basis for the B-BAC derivation. In this equation all the process kinetics are replaced by one time varying term. This term can be easily estimated on the basis of the same equation (1) by the recursive least-squares method to provide the adaptability of the B-BAC. Let us note that there is no need to assume any form of a nonlinear expression describing this term. There is also no need to know a complete mathematical model of a process. Since there is only this one parameter to estimate, we avoid the common difficulties with the multivariable identification. The influence of the measurement noise can be easily decreased by adjusting the value of the forgetting factor.

The generality of the B-BAC methodology is demonstrated in this paper with the application to the control of two practical examples. The simulation results are very promising because in both cases the control performance of our B-BAC is much better than of the classical PI controller. Thus our methodology can be suggested for the application in the control loops in the process industry due to its generality, robustness and very good control performance.

Finally, last but not the least, let us emphasise on the another important property of the suggested B-BAC methodology in comparison with the others adaptive control approaches – simplicity. In consequence, it allows us to implement this methodology very easily as the PC-based virtual controller or as a stand-alone PLC-based controller.

Acknowledgements. This work was supported by Polish Committee of Scientific Research (KBN) under grant 4T 11A 01924.

REFERENCES

Bastin G. (1991). Nonlinear and adaptive control in biotechnology. A tutorial. European Control Conference ECC'91, Grenoble, France.

Bastin G., Dochain D. (1990). *On-line estimation and adaptive control of bioreactors.* Elsevier Science Publishers B.V.

Czeczot J. (1997). *The application of the substrate consumption rate to the monitoring and control of the biological processes of water purification.* Ph. D. Thesis, Technical University of Silesia, Gliwice, Poland (in polish).

Czeczot J. (1998). Model-based Adaptive Predictive Control of Fed-Batch Fermentation Process with the Substrate Consumption Rate Application, IFAC Workshop on Adaptive Systems in Control and Signal Processing, University of Strathclyde, Glasgow, Scotland, UK, pp. 357 –362.

Czeczot J. (1999). Substrate consumption rate application for adaptive control of continuous bioreactor – noisy case study. *Archive of Control Sciences*, **9**, No. ¾, pp. 33 – 52.

Dochain D. (1991). Design of adaptive controllers for non-linear stirred tank bioreactors: extension to the MIMO situation. *J. Proc. Cont.*, **1**.

Dochain D. (1992). Adaptive control algorithms for non-minimum phase nonlinear bioreactors. *Computers Chem. Engng.*, **16**, No. 5, pp. 449-469.

Isidori A. (1989). *Nonlinear control systems.* New York, Springer-Verlag.

Joshi N.V., Murugan P., Rhinehart R.R. (1997). Experimental comparison of control strategies. *Control Eng. Practice*, **5**, No. 7, pp. 885-896.

Laszczyk P., Pasek K. (1995). Modelling and simulation of the pilot heat distribution plant. System Analysis, Modelling, Simulation. Vol. 18-19, pp. 245-248.

Metzger M. (2000). *Modelling, simulation and control of continuous processes.* Edition of Jacek Skalmierski Computer Studio, Gliwice, Poland.

Monod J. (1949). The growth of bacterial cultures. *Ann. Rev. Microbial*, **3**, pp. 371-394.

Richalet J. (1993). Industrial applications of model based predictive control. *Automatica*, **29**, No. 5, pp. 1251-1274.

Seborg D.E. (1999). A perspective on advanced strategies for process control. *ATP*, **41**, No. 11, pp. 13 – 31.

Van Impre J.F., Bastin G. (1995). Optimal adaptive control of fed-batch fermentation processes. *Control Eng. Practice*, **3**, No. 7, pp. 939-954.

www.elsevier.com/locate/ifac

ROBUST CONTROL OF UNSTABLE NON-MINIMUM PHASE MIMO UNCERTAIN SYSTEM

Ilan Rusnak
RAFAEL, P.O.Box, 2250, Haifa 31021, Israel.

Abstract: The robustness of an algorithm for control of discrete multi-input multi-output linear uncertain systems is presented. This algorithm is based on the state and parameters observability form and a certainty equivalence controller. The algorithm requires an estimate of the order of the dominant-low frequency part of the system. The effect of the high order dynamics is modeled and tested by introduction of unmodelled dynamics. The robustness is checked by simulation on a MIMO unstable nonminimum phase system. The simulations demonstrate that the algorithm preserves stability and performance for large amounts of unmodelled dynamics *Copyright © 2003 IFAC*

Keywords: Robust control, Adaptive control, MIMO systems, Non-minimum phase systems, Uncertain linear systems.

1. Introduction

The control of uncertain systems is covered by the adaptive control theory. The main goal of the "first generation" adaptive controllers has been to maintain stability and performance in terms of the steady state tracking error. These algorithms require the knowledge of the low or high frequency gain, the relative degree, and more. The transient, e.g., the performance on finite time interval, is not covered. Lately, the issue of optimizing the performance has emerged. The problem of adaptive optimal control on finite time interval has been posed and a so called "candidate adaptive controller" is presented (Bitmead, *at al.*, 1990).
A control of continuous deterministic single input single output uncertain system based on the state and parameters canonical form is dealt with in (Rusnak, *et al.*, 1992). This algorithm requires only the knowledge of the order of the system. In (Rusnak, 1996) this algorithm had been applied to the control of unstable nonminimum phase plant. This reference includes a discussion and review of existing adaptive algorithms as well. That algorithm has been generalized for MIMO systems in (Rusnak, 2000a). The algorithm is achieved by the introduction of the State and Parameters Observability Form for MIMO systems. This new representation of MIMO linear time-invariant systems enables application of existing tools for estimation and control of discrete linear time-varying systems. The algorithm requires the knowledge of the order of the system. However, for any real system there exists only an estimate of

system. In other words, there are always unmodeled dynamics. The case when the order of the system is unknown and only an upper bound on the system order is known, has been dealt and a control algorithm is presented in (Rusnak, 2000b).

The objective of this paper is to approach the problem of unknown system order of from different direction. Here we apply the algorithm with an estimate of the system order (which in practice is always smaller than the actual order). Specifically, we demonstrate the robustness of the algorithm (Rusnak, 2000a) for the case when for the algorithm we assume an estimate of the order, say an n-th order system while the actual system is of higher order, that is there are unmodeled dynamics. The robustness is demonstrated on a fourth order system. The algorithm assumes only a 2nd order MIMO system while the actual system is a 4th order. The simulations demonstrate the robustness, in terms of stability and performance with respect to unmodeled dynamics.

2. The Control Algorithm

For the completeness of presentation this section presents the algorithm (Rusnak, 2000a) whose robustness is considered in this paper.

2a. Simultaneous State and Parameters Estimation

For completeness, we present here the derivation of the SPOF and state relevant performance results. We assume that only the order of the system and the number of inputs and outputs are known.
We consider an n-th order discrete multi-input multi-output linear time-invariant system in the observer canonical form

$$x(t+1) = Ax(t) + Bu(t), x(t_0) = x_0, \quad (1)$$
$$y(t) = Cx(t), t \geq t_0,$$

where the input $u(t) \in l_2^m[t_0,t_1]$; $A \in R^{nxn}$;

$C \in R^{pxn}$; $B \in R^{nxm}$; $x(t) \in R^{nx1}$ is the state; and $y(t) \in R^{p}$ is the output; m is the number of inputs; and p is the number of outputs. We assume that the system (1) is observable. The initial state x_0 is unknown. The unknown parameters are the entries of the matrices A, B, and C, expressed in terms of the coefficients of the transfer matrix

$$H(z) = C\,(zI-A)^{-1}\,B = \frac{1}{d(z)}\,N(z)$$

$$d(z) = z^n + a_1 z^{n-1} + \ldots + a_n \tag{2}$$

$$N(z)=N_1 z^{n-1}+N_2 z^{n-2}+ +N_n, N_i \in R^{pxm},\ i=1,2,...n.$$

That is, we have n+nmp unknown parameters.
In order to derive the SPOF we first represent the system (1) in the block-observer form (Kailath, 1980) as detailed next. That is

$$x_0(t+1) = A_0 x_0(t) + B_0 u(t)\ , x_0(t_0) = x_{00}\ , \tag{3}$$
$$y(t) = C_0 x_0(t), t \geq t_0,$$

where $A_0 \in R^{npxnp}$; $C_0 \in R^{pxnp}$; $B_0 \in R^{npxm}$; and $x_0(t) \in R^{npx1}$ is the state of the block observer form. We have

$$H(z) = C_0\,(zI-A_0)^{-1}\,B_0 = \frac{1}{d(z)}\,N(z) \tag{4}$$

$$A_o = \begin{bmatrix} -a_1 I_p & I_p & 0 & \ldots & 0 \\ -a_2 I_p & 0 & I_p & \ldots & 0 \\ \vdots & \vdots & \vdots & \ddots & \vdots \\ \vdots & \vdots & \vdots & \ldots & I_p \\ -a_n I_p & 0 & 0 & \ldots & 0 \end{bmatrix}, B_o = \begin{bmatrix} N_1 \\ N_2 \\ \vdots \\ \vdots \\ N_n \end{bmatrix}, \tag{5}$$

$$C_o = \begin{bmatrix} I_p & 0 & 0 & \cdots & 0 \end{bmatrix}$$

This representation is not minimal, however it is observable (Kailath, 1980). To derive the state and parameters observability form, let us choose the Brunovsk form (A_B, C_0)

$$A_B = \begin{bmatrix} 0 & I_p & 0 & \ldots & 0 \\ 0 & 0 & I_p & \ldots & 0 \\ \vdots & \vdots & \vdots & \ddots & \vdots \\ \vdots & \vdots & \vdots & \ldots & I_p \\ 0 & 0 & 0 & \ldots & 0 \end{bmatrix}, \tag{6}$$

and (A_B, C_0) is observable. Then, there exist $H_0 \in R^{npxp}$ such that $A_0 = A_B + H_0 C_0$ (Kailath, 1980, Chen, 1970), where

$$H_o = \begin{bmatrix} -a_1 I_p \\ -a_2 I_p \\ \vdots \\ \vdots \\ -a_n I_p \end{bmatrix}, H_o = h \otimes I_p, \tag{7}$$

where we denote $h = -[a_1\ a_2\ \ldots\ a_{n-1}\ a_n]^T$ and $\otimes$is the Kronecker product. Further we can write

$$x_o(t+1) = (A_B + \overline{H}_o C_o) x_o(t) + \delta H y(t) + \overline{B}_o u(t) + \delta B u(t) \tag{8}$$

where $\delta H = H_o - \overline{H}_o$, $\delta B = B_o - \overline{B}_o$, and $\overline{H}_o$, and $\overline{B}_o$ are some estimates of H_0 and B_0 (Their nominal value, the average value, the most probable value). We can rewrite

$$\delta H y(t) = \begin{bmatrix} y(t) & 0 & 0 & \cdots & 0 \\ 0 & y(t) & 0 & \cdots & 0 \\ \vdots & \vdots & \ddots & \cdots & \vdots \\ 0 & 0 & \cdots & y(t) & 0 \\ 0 & 0 & \cdots & \cdots & y(t) \end{bmatrix} \delta h$$

$$= \text{block diagonal } (y(t))\delta h = bd(y(t))\ \delta h\ , \tag{9}$$

$$\delta H y(t) = \begin{bmatrix} u^T(t) & 0 & 0 & \cdots & 0 \\ 0 & u^T(t) & 0 & \cdots & 0 \\ \vdots & \vdots & \ddots & \cdots & \vdots \\ 0 & 0 & \cdots & u^T(t) & 0 \\ 0 & 0 & \cdots & \cdots & u^T(t) \end{bmatrix} \delta b$$

$$= \text{block diagonal}(u^T(t))\delta b = bd(u^T(t))\delta b, \tag{10}$$

and where

$$\delta b = \begin{bmatrix} \delta B_{o1}{}^T \\ \delta B_{o2}{}^T \\ \\ \delta B_{on}{}^T \end{bmatrix} \in R^{nmp\times 1}, \tag{11}$$

where r=nm and δB_{oi} , i=1,2,... ,r, are the rows of δB_0. That is, we have

$$x_o(t+1) = (A_B + \overline{H}_o C_o) x_o(t) + bd(y(t))\delta h + bd(u(t))\delta b + \overline{B}_o u(t) \tag{12}$$

where $\delta h = h - h_0$, and $\delta b = b - b_0$.
Furthermore, we have time invariant system, so that we have

$$\delta h(t+1) = \delta h(t), \quad \delta h(t_0) = h - h_o, \tag{13}$$
$$\delta b(t+1) = \delta b(t), \quad \delta b(t_0) = b - b_o.$$

In equation (12) we have exactly n+nmp parameters, in δh and δb. Equations (12,13) are a different representation of equation (1). Now we can write (12,13) in the augmented form, the *state and parameters observability form*, as

$$\begin{bmatrix} x_o(t+1) \\ \delta h(t+1) \\ \delta b(t+1) \end{bmatrix} = \begin{bmatrix} A_B + H_o C_o & bd(y(t)) & bd(u^T(t)) \\ 0 & I & 0 \\ 0 & 0 & I \end{bmatrix} \begin{bmatrix} x_o(t) \\ \delta h(t) \\ \delta b(t) \end{bmatrix} + \begin{bmatrix} \overline{B}_o \\ 0 \\ 0 \end{bmatrix} u(t) \tag{14}$$

$$y(t) = \begin{bmatrix} C_o & 0 & 0 \end{bmatrix} \begin{bmatrix} x_o(t) \\ \delta h(t) \\ \delta b(t) \end{bmatrix}, t \geq t_0,$$

which will be written as a time varying system

$$\mathbf{X}(t+1) = \mathbf{A(t)}\mathbf{X}(t) + \mathbf{B}u(t), X(t_0) = X_0 \tag{15}$$
$$y(t) = \mathbf{C}\mathbf{X}(t)$$

where

$$\mathbf{X}(t) = \begin{bmatrix} x_o(t) \\ \delta h(t) \\ \delta b(t) \end{bmatrix}, \mathbf{A}(t) = \begin{bmatrix} A_B + \overline{H}_o C_o & bd(y(t)) & bd(u^T(t)) \\ 0 & I & 0 \\ 0 & 0 & I \end{bmatrix}, \tag{16}$$

$$\mathbf{B} = \begin{bmatrix} \overline{B}_o \\ 0 \\ 0 \end{bmatrix}, \mathbf{C}^T = \begin{bmatrix} C_o^T \\ 0 \\ 0 \end{bmatrix}, X(t_0) = \begin{bmatrix} x_o(t_0) \\ 0 \\ 0 \end{bmatrix}.$$

The design of an estimator for a time-varying system is not straightforward. Here we use the estimator is

$$\hat{\mathbf{X}}(t+1) = \mathbf{A}(t)\hat{\mathbf{X}}(t) + \mathbf{B}u(t) + \mathbf{K}(t)[y(t+1) - \mathbf{C}\mathbf{A}(t)\hat{\mathbf{X}}(t) - \mathbf{C}\mathbf{B}u(t)], \hat{\mathbf{X}}(t_0) = \hat{\mathbf{X}}_0 \tag{17a}$$

where

$$\mathbf{Q}(t) = \mathbf{A}(t-1)[I - \mathbf{K}(t-1)\,\mathbf{C}]\mathbf{Q}(t-1)\mathbf{A}(t-1)^T + \mathbf{V}_1, \quad \mathbf{Q}(t_0) = \mathbf{Q}_0 \tag{17b}$$

$$\mathbf{K}(t) = \mathbf{Q}(t)\mathbf{C}^T\,[\,\mathbf{C}\,\mathbf{Q}(t)\,\mathbf{C}^T + V_2]^{-1}.$$

where $V_1 \geq 0$, $V_2 > 0$ and $Q_0 \geq 0$ are tuning parameters. These equations are easily solved since up to time t, $\mathbf{A}(.)$, $\mathbf{B}$ and $\mathbf{C}$ are known and the iteration goes forward. Moreover, since $(\mathbf{A}(t),\mathbf{c})$ is uniformly completely observable, by imposing the required input to the plant, and we set during the tuning $(\mathbf{A}(t),V_1^{1/2})$ to be uniformly completely controllable the estimation errors converge exponentially to zero (Kwakernaak and Sivan, 1972). For further details see proposition 1 in the next section and the related proofs in the references.

2b. The Control

In this section we present controller based on the SPOF (15) and the certainty equivalence principle. Then in proposition 1 the conditions for global BIBO stability of the control algorithm and global exponential stability of the state transition matrix are stated. The certainty equivalence control law is given by

$$u(t) = -\hat{K}_c(t)\hat{x}_o(t) + \hat{K}_o(t)v_e(t), \tag{18}$$

where $v_e(t) \in R^{m\times 1}$ is an external input, $\hat{K}_o(t)$ is a gain that sets the gain between the external input and the output, and $\hat{K}_c(t)$ is a certainty equivalence controller gain given by

$$\hat{K}_c(t) = K_c(\hat{A}_o, \hat{B}_o, C_o) = K_c(A_0 + \hat{H}(t)C_o, \hat{B}(t), C_o) \tag{19}$$

where $\hat{x}_o(t)$, $\hat{H}(t) = H_o + \delta\hat{H}$ and $\hat{B}(t) = B_o + \delta\hat{B}$ are from (17a). The certainty equivalence controller gain is such that

$K_c(A,B,C) = K_c$, and $A\text{-}BK_c$ is exponentially stable.(20)

That is, the certainty equivalence controller K_c, for the exact values of the parameters of the system, is a stabilizing controller (exponentially stable).

The following can be stated:

Proposition 1: If:

i) $K_c = K_c(A,B,C)$ is bounded and with finite incremental gain function of the variables A, B, C;

ii) $A\text{-}BK_c$ is exponentially stable;

iii) the external input $v_e(t)$ is bounded and such that $(\mathbf{A}(t),\mathbf{C})$ is uniformly completely observable; and

iv) the tuning matrix V_1 is such that $(\mathbf{A}^T(t),V_1^{1/2})$ is uniformly completely controllable; then:

I) The system (1) with the observer-estimator (17) and the certainty equivalence feedback (18) is BIBO stable;

II) The system transition matrix is

exponentially stable;

III) Almost always exists a bounded input $v_e(t)$ such that $(\mathbf{A}(t), V_2^{1/2})$ is uniformly completely observable.

Proof: See for SISO system (Rusnak, 1999). The proof of MIMO is exactly the same.

Here we used for the controller gain the solution of the Riccati equation.

$$\hat{A}_o^T S \hat{A}_o - S - \hat{A}_o^T S \hat{B}_o (R + \hat{B}_o^T S \hat{B}_o)^{-1} \hat{B}_o^T S \hat{A}_o + Q = 0,$$
$$K_c = (R + \hat{B}_o^T S \hat{B}_o)^{-1} \hat{B}_o^T S \hat{A}_o,$$
$$K_o = \left[C_o (I - \hat{A}_o + \hat{B}_o K_c)^{-1} \hat{B}_o\right]^{-1}. \quad (21)$$

This solution does not satisfy the finite incremental gain requirement when the estimated plant approaches unobservability. This had been dealt with by bounding the gains.

3. Example

The block diagram of the algorithm is presented in figure 1. To examine the robustness of the proposed controller we used a 4th order 2 input and 2 outputs system. Although the order of the plant is n=4, the

is only n=2.

The model of the plant and the unmodelled dynamics is,

$$\begin{bmatrix} Y_1(z) \\ Y_2(z) \end{bmatrix} = \frac{1}{d(z)} \begin{bmatrix} N_{11}(z) & N_{12}(z) \\ N_{21}(z) & N_{22}(z) \end{bmatrix} \begin{bmatrix} \Gamma_1(z) & 0 \\ 0 & \Gamma_2(z) \end{bmatrix} \begin{bmatrix} U_1(z) \\ U_2(z) \end{bmatrix} \quad (22)$$

where the plant is

$$\begin{aligned} d(z) &= z^2 - 1z + 1.1 \\ N_{11}(z) &= 1z + 2 \\ N_{12}(z) &= 1.5z - 2.5 \\ N_{21}(z) &= 3z - 4 \\ N_{22}(z) &= 3.5z + 4.5 \end{aligned} \quad (23)$$

This is the discretized representation of the continuous plant

$$\begin{aligned} d(s) &= s^2 - 0.95s + 105.0 = (s - 0.4750 + 10.2359i)(s - 0.4750 - 10.2359i) \\ N_{11}(s) &= -5.5(s - 57) \\ N_{12}(s) &= 23.2(s - 4.5) \\ N_{21}(s) &= 40.7(s - 2.6) \\ N_{22}(s) &= 5.1(s - 166) \end{aligned} \quad (24)$$

for sampling period of T=0.1sec. Notice the two unstable complex roots and the zeros are in the RHP. The unmodeled dynamics in the s-domain are

$$\Gamma_1(s) = \frac{1}{s\gamma_1 + 1}$$
$$\Gamma_2(s) = \frac{1}{s\gamma_2 + 1} \quad (25).$$

Small values of γ_1 and γ_2 represent small amount of unmodeled dynamics. As these values get larger it represent a higher degree on unmoddeled dynamics. For the following examples the initial conditions of the plant are zero. The estimation algorithm assumes a 50% error in the knowledge of the actual parameters of the low frequency (and no knowledge on the unmodeled dynamics). Further

$$\begin{aligned} V_1 &= 10^{-2} \text{diag}(10,10,10,10,1,1,1,1,1,1,1,1) \\ V_2(s) &= 10^{-1} \text{diag}(1,1) \\ Q_o &= \text{diag}(1,1,1,1,1,1,1,1,1,1,1,1) \\ Q_c &= 10^4 \text{diag}(1,1,1,1) \\ R &= \text{diag}(1,1) \end{aligned} \quad (26)$$

Figure 2 presents the external inputs (desired outputs) of the plant. The first two seconds are forced inputs that impose the requirement that the input to the plant should be persistently exciting (guarantees that the augmented system will be uniformly completely observable, as required by proposition 1). Then the input is zero, until in the first input a step is applied at t=3sec and in the second input at t=4sec. All simulations were preformed for this exact input. The reason for this is to show that the response changes only very slightly.

Figures 3 to 5 present the performance of the controller for low amount of unmodeled dynamics, they are for γ_1 = 1/250 and γ_2 = 1/260 (approximately at 40Hz).

Figure 3 presents the estimated parameters of the plant versus time. One can see that the algorithm converges quickly to the correct values of the parameters (23).

Figure 4 presents the response of the closed loop for the external input of figure 2. The upper plots show the response for the whole period of 5sec, and the lower figures present the last 2.5sec. One can see the cross-coupling between the two channels.

Figure 5 shows the input of the plant. One can see that during the first period of operation the input does not exceed high values.

From these figures one can see that the algorithm performs well if the order of the amount of unmodeled dynamics is small.

Figures 6 and 8 present the performance of the controller for large amount of unmodeled dynamics, they are for γ_1 = 1/25 and γ_2 = 1/26 (approximately at 4Hz). Notice that this is a large amount of uncertainty. The unstable poles of the plant are at 1.6Hz. Such plant (order n=2) does not reflect the system. In spite of it, the simulation demonstrates that the algorithm preserves stability and the reduction in performance acceptable.

Figure 6 presents the estimated parameters of the plant. One can see that the algorithm does not converge to the correct values of the 2nd order plant parameters.

Figure 7 presents the response of the closed loop for the external input of figure 2 in the same format as figure 5.

Figure 8 shows the input of the plant. One can see that during the first period of operation, when the parameters are converging, the input does not exceed high values.

From these figures one can see that the algorithm performs well even in the presence of large amount of unmodeled dynamics.

Especially, comparison of figures 3,4,5 to figures 6,7,8, respectively, reveals that there are only minor changes due to the larger unmodeled dynamics.

4. Conclusions

A robust control algorithm based on the state and parameters observability form has been applied to an uncertain MIMO system. This algorithm needs only the knowledge of the order of the system, thus it is inherently robust with respect to parameters values. Specifically it is robust with respect to the low or high frequency gain and the relative degree. The issue when the order of the system is unknown or too high to be implemented is addressed by introduction of unmodeled dynamics. The simulations demonstrate the robustness in terms of stability and performance of the algorithm with respect to unmodeled dynamics.

5. References

Bitmead, R.R., Gevers, M. and Wertz, V.:" Adaptive Optimal Control, Prentice Hall, Int., 1990.

Chen, C.T.: *Linear System Theory and Design,* Holt, Rinehart and Winston, Inc., New York, 1970.

Kailath, T. (1980). *Linear Systems*, Prentice-Hall, Inc. Englewood Cliffs, N.J..

Kwakernaak, and Sivan, R. (1972). *Linear Optimal Control*, John Wiley & Sons, Inc..

Rusnak, I., Steinberg, M., Guez, A. and BarKana, I.. (1992). "On-Line Identification and Control of Linearized Aircraft Dynamics," IEEE AES Magazine, July 1992, pp. 56-60.

Rusnak, I. (1996)." Control for Unstable Non-Minimum Phase Uncertain Dynamic Vehicle," IEEE Trans. on AES, Vol. 32, AES-32, No. 3, July 1996, pp. 945-951.

Rusnak, I. (1999)." Optimal Adaptive Tracking of Uncertain Stochastic Discrete Systems," The 7th IEEE Mediterranean Conference on Control & Automation, MED 99, 28-30 June 1999, Haifa, Israel.

Rusnak, I. (2000a). "Control of MIMO Uncertain Systems," CSD 2000, IFAC Conference on Control Systems Design, June 18-20, 2000, Bratislava, Slovak Republic.

Rusnak, I. (2000b). "Control of Unknown Order Uncertain Systems, " ROCOND 2000, 3rd IFAC Symposium on Robust Control Design, Prague, Czech Republic, June 21-23, 2000.

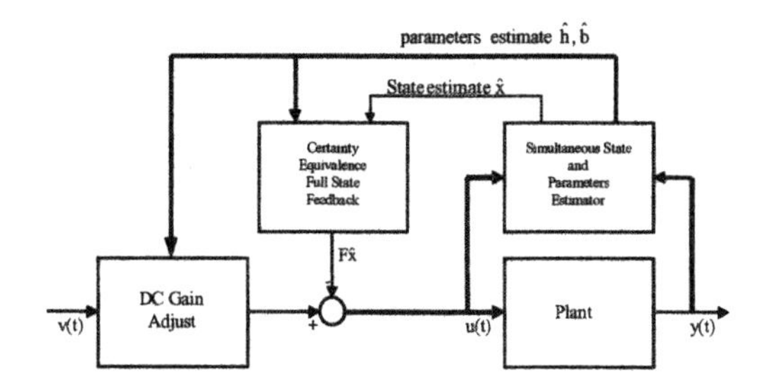

Figure 1: The block diagram of the controller for linear uncertain system.

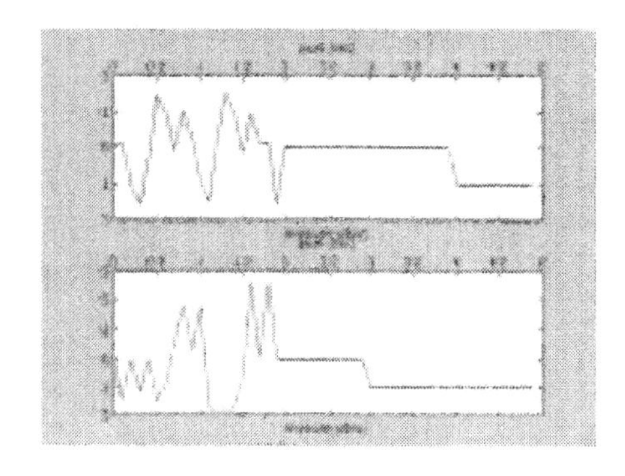

Figure 2: The external input-desired output of the closed loop.

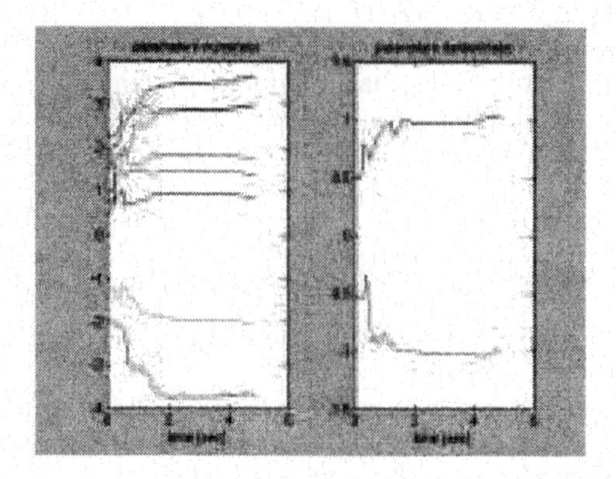

Figure 3: The estimated parameters of the plant versus time controller for low amount of unmodeled dynamics.

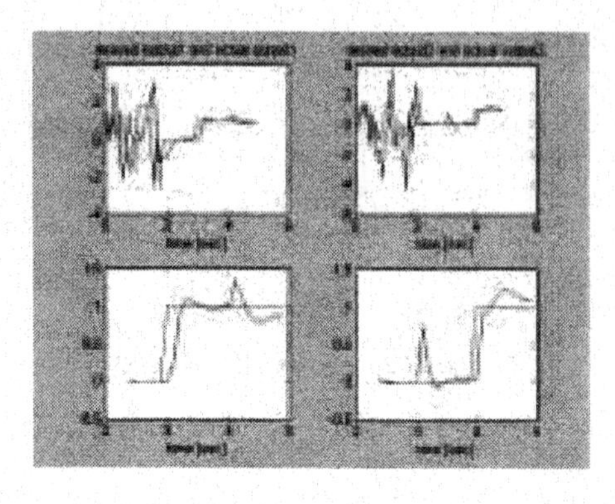

Figure 4:The response of the closed loop controller for low amount of unmodeled dynamics.

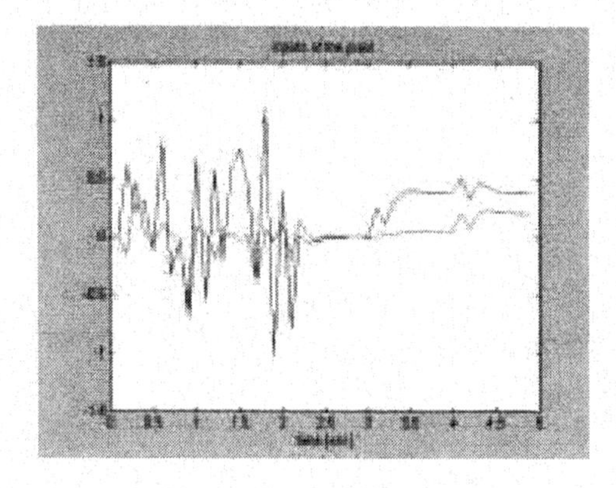

Figure 5: The input of the plant controller for low amount of unmodeled dynamics.

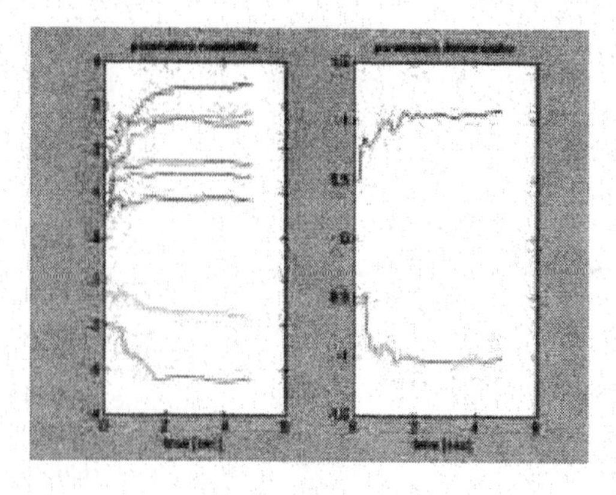

Figure 6: The estimated parameters for large amount of unmodeled dynamics.

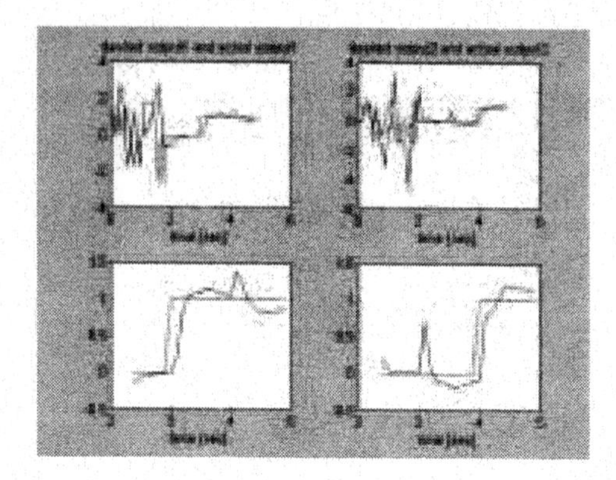

Figure 7: The response of the closed loop controller for large amount of unmodeled dynamics.

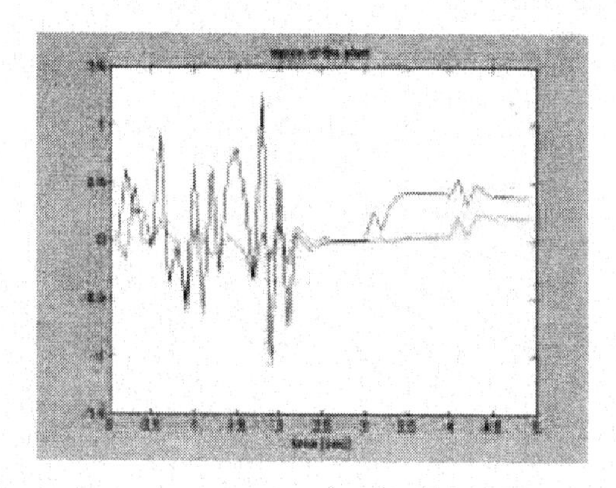

Figure 8: The input of the plant controller for large amount of unmodeled dynamics.

REGULATION OF CANALS IRRIGATION : MULTIVARIABLE BOUNDARY CONTROL APPROACH BY INTERNAL MODEL.

DOS SANTOS Valérie * TOURÉ Youssoufi *

* *Laboratoire de Vision et Robotique, Bourges, France*

Abstract: This paper presents a direct approach control synthesis using the shallow water partial differential equations (pde) description. The purpose is to give an alternative solution for the standard regulation of canals irrigation control problem. This regulation control problem is stated as a boundary control design with a particular form of Internal Model Control (IMC) structure. Moreover, the internal model takes into account the dynamic of the gate of canal, and we consider the inhomogeneous case i.e. all parameters of the equilibrium point are space dependent. Simulation results are given using a nonlinear model of the canal irrigation. *Copyright © 2003 IFAC*

Keywords: Shallow water equation, canals irrigation, semigroup, multivariable boundary control.

1. INTRODUCTION

The regulation of canals irrigation presents an economic and environment interest and many researches are done in this area. These works have been done with different approaches concerning the class of models and the class of control synthesis ([1]).

This paper presents the class of the partial differential equations description of the canal by shallow water equations also called Saint Venant equations. The regulation problem is addressed by the direct methods, which means that the control design is based directly on the infinite dimensional system theory ([3], [6], [8], [9]).
In this work, the internal model control structure is used for the synthesis of a robust control of a monovariable and multivariable boundary control. This internal model boundary control (IMBC), which is introduced in [11] for parabolic system with an exponentially stable semigroup, is studied here for an hyperbolic stable system.

In the next section, the control problem is stated as a boundary control of a linear system. The nonlinear system and the linear boundary control system are given.
In section three, the abstract boundary control system is stated with infinite dimensional systems state space representation. The well-posedness of the open loop system is done with a semigroup approach following Fattorini's abstract boundary control system approach ([4]). Then, we show that the open loop system has a stable semigroup.
The IMBC structure is studied as an extended state space system of a closed loop system with an integral type feedback control. Then, the closed loop can be viewed as a bounded perturbation of open loop system by the control parameters ([6], [9]). Previous stability results ([6], [9], [11]) are used to give sufficient conditions for control synthesis parameters.
In the last section, some simulation results are given on a stable tracking problem of the water-

flow and height, around an equilibrium state according to this multivariable control synthesis.

2. THE CANAL REGULATION PROBLEM : A BOUNDARY CONTROL SYSTEM.

We consider the following class of canal represented in Fig. 1, where $Q(x,t)$ denotes the water-flow, $Z(x,t)$ the water height in the canal, and L the length of the canal part to be controlled between the upstream reservoir ($x = 0$) and the downstream ($x = L$) of the canal (Fig. 1 and 2). $U_0(t)$ and $U_L(t)$ denotes the gate control level at abscisse 0 and L, respectively.

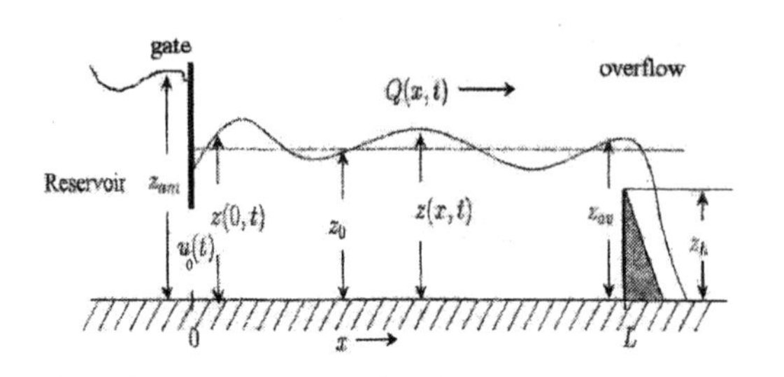

Fig. 1. Canal scheme : one gate

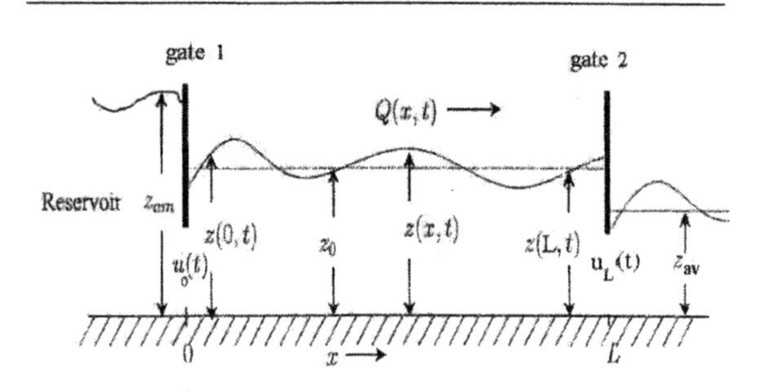

Fig. 2. Canal scheme : two gates

The regulation problem addresses the stabilization of the water-flow and/or the height, around the equilibrium behavior denoted (z_e, q_e). Hence a linearized model can be involved to describe the deviations around the equilibrium behavior.

2.1 Modelling

The canal is supposed to have a sufficient length, L, such that an uniform behaviour can be assumed, in the lateral direction. The shallow water's model for a rectangular canal are then the following nonlinear pde :

$$\partial_t Q = -\partial_x\left(\frac{Q^2}{bz} + \frac{1}{2}gbZ^2\right) + gbZ(I - J), \quad (1)$$

$$\partial_t Z = -\partial_x \frac{Q}{b}, \quad (2)$$

$$Z(x,0) = Z_0(x), \quad Q(x,0) = Q_0(x), \quad (3)$$

Different boundary conditions allow us to consider two cases of the control problem :

Case a *monovariable control* : The equation of the upstream boundary is given by

$$Q(0,t) = U_0(t)\Psi_1(Z(0,t)), \quad (4)$$

The other boundary condition is a downstream overflow (Fig. 1) :

$$Z(L,t) = \Psi_2(Q(L,t)), \quad (5)$$

Note that the output to be controlled is the level at $x_0 = L$.

Case b *multivariable control* : The equation of the upstream boundary remains as described in equation(4). A second control variable appears in the downstream condition with an additional gate (Fig. 2):

$$Q(L,t) = U_L(t)\Psi_3(Z(L,t)), \quad (6)$$

where $\Psi_1(Z) = K_1\sqrt{2g(z_{am} - Z)}$

$$\Psi_2(Q) = \left(\frac{Q^2}{2gK_2^2}\right)^{1/3} + h_s, \quad \Psi_3(Z) = K_2\sqrt{2g(Z - z_{av})}$$

and $U(t)$ is the control, b is the width, I the slope of the canal bottom, and J the slope's rubbing, expressed with the Manning-Strickler expression :

$$J = \frac{n^2Q^2}{(bZ)^2R^{4/3}}, \quad R = \frac{bZ}{b + 2Z}. \quad (7)$$

2.2 A regulation model

The previous system (1)-(6) is linearized around the equilibrium state :

$$\partial_x q_e = 0$$

$$\partial_x z_e = gbz_e \frac{I + J_e + \frac{4}{3}J_e\frac{1}{1+2z_e/b}}{gbz_e - q_e^2/bz_e^2}, \quad (8)$$

and the fluvial case is assumed :

$$z_e > \sqrt[3]{q_e^2/(gb^2)}. \quad (9)$$

Note that, q_e is constant but z_e is space dependent.
The linearized system is given by :

$$\begin{aligned}\partial_t\xi(t) &= (\partial_t z(t) \;\; \partial_t q(t))^t \\ &= A_1(x)\partial_x\xi(x) + A_2(x)\xi(x) \quad (10)\\ \xi(x,0) &= \xi_0(x) \\ q(0,t) &= u_{0,e}\partial_z\Psi_1(z_e(0,t))z(0,t) + u_0(t)\Psi_1(z_e(0,t))\end{aligned}$$

Case a $\quad z(L,t) = \partial_q\Psi_2(q_e)q(L,t)$

Case b

$$q(L,t) = u_{L,e}\partial_z\Psi_3(z_e(L,t))z(L,t)+u_L(t)\Psi_3(z_e(L,t))$$

where $u_{0,e}$ and $u_{L,e}$ are the gate control level corresponding to the equilibrium point and

$$A_1(x) = \begin{pmatrix} 0 & -a_1(x) \\ -a_2(x) & -a_3(x) \end{pmatrix},$$

$$A_2(x) = \begin{pmatrix} 0 & 0 \\ a_4(x) & -a_5(x) \end{pmatrix},$$

with

$$a_1(x) = 1/b, \quad a_2(x) = gbz_e(x) - \frac{q_e^2}{bz_e^2(x)},$$

$$a_4(x) = gb(I + J_e(x) + \frac{4}{3}\frac{1}{1+2z_e(x)/b}),$$

$$a_3(x) = \frac{2q_e}{bz_e(x)} \quad a_5(x) = \frac{2gbJ_e(x)z_e(x)}{q_e}.$$

Now, the control problem is to find the variations of the control $u_0(t)$ at the boundary $x = 0$ (case a) and $u_L(t)$ at the boundary $x = L$ (case b), such that the output variations at the boundary $x = L$ (measured variable), become zero or track a reference no persistent signal $r(t)$, a stable step response of a non oscillatory single system, for example.

3. CONTROL SYNTHESIS : THE IMBC STRUCTURE

The state space representation allows us to use the semi-group approach which is well suited for infinite dimensional systems. The abstract boundary control system is stated first ([4]) and the extended control system is stated in the IMBC structure according to an integral control law.

3.1 The abstract boundary control system

The linearized boundary control model can be formulated as follow :

$$\partial_t\xi(t) = A_d(x)\xi(t), \quad x \in \Omega =]0,L[, \quad t > 0 \quad (11)$$

$$F_b\xi(t) = B_bu(t), \quad on \ \Gamma = \partial\Omega, t > 0$$

$$\xi(x,0) = \xi_0(x) \quad (12)$$

where $A_d(x) = A_1(x)\partial_x + A_2(x)$.

The output is measured at $x_0 = L$:

$$y(t) = C\xi(t), \quad t \geq 0 \quad (13)$$

$$C = \begin{pmatrix} \frac{1}{2\mu}\int_{x_0-\mu}^{x_0+\mu} 1_{x_0\pm\mu} & 0 \end{pmatrix}, \quad \mu > 0$$

The abstract boundary control system follows the change of variables and operators ([1], [3], [4]).
- Consider the operator A defined as :

$$D(A) = \{\varphi \in D(A_d) : F_b\varphi = 0\} = D(A_d) \cap Ker(F_b)$$

and $A\varphi = A_d\varphi \ \forall\varphi \in D(A)$. A is assumed closed and densely defined in $X = L_2(0,L)$.
- Consider the change of variables :

$$\xi(t) = \varphi(t) + Du(t) \quad \forall t \geq 0 \quad (14)$$

where D is a bounded operator from U the control space, to X the state space, such that

$$Du \in D(A_d)$$
$$F_b\ (Du(t)) = B_bu(t) \ \forall u(t) \in U$$

Moreover this operator D can be chosen such that it leaves the operator A_d unchanged (i.e. $Im(D) \subset Ker(A_d)$).
Then the system (11)-(12) is equivalent to :

$$\dot\varphi(t) = A\varphi(t) - D\dot u(t), \quad \varphi(t) \in D(A), \quad t > 0$$
$$\varphi(0) = \xi(0) - Du(0)$$

which has the classical solution :

$$\varphi(t) = T_A(t)\varphi_0 - \int_0^t T_A(t-s)D\dot u(s)ds$$

where $\dot u$ is assumed to be a continuous time function and one suppose that A is an infinitesimal generator of a C_0 semi-group $T_A(t)$ such that the solution $\varphi(t) = T_A(t)\varphi_0$ exists and belongs to $D(A)$.

In the **case a**, we have $U = R$ and :

$$D(A_d) = \{\xi \in X : \xi' \in X \text{ and } z(L) = \partial_q\Psi_2(q_e)q(L)\},$$
$$Ker(F_b) = \{\xi \in X : \xi' \in X \text{ and } q(0) = u_{0,e}\partial_z\Psi_1(z_e(0))z(0)\} \quad (15)$$

In the **case b**, we have $U = R^2$ and:

$$D(A_d) = \{\xi \in X : \xi' \in X\}$$
$$Ker(F_b) = \{\xi \in X : \xi' \in X, \ q(0) = u_{0,e}\partial_z\Psi_1(z_e(0))z(0), \ q(L) = u_{L,e}\partial_z\Psi_3(z_e(L))z(L)\} \quad (16)$$

Proposition 1. The open loop system is well-posed, i.e. it is a generator of a C_0-semi-group.

PROOF. Let $A = A_1\partial_x + A_2$. It can be proved that A is a bounded linear operator. It is clearly linear, and it is bounded by hypothesis (9). Indeed, let

$$\|A\| = \sup_{\|\xi\|=1} \|A\xi\|, \|\xi\| = \|\xi\|_{L^2} + \|\partial_x \xi\|_{L^2}.$$

So,

$$\|A\xi\|^2 = \int_0^L (A\xi)^t A\xi$$

$$= \int_0^L (A_1\partial_x\xi)^t A_1\partial_x\xi + (A_2\xi)^t A_2\xi$$

$$+(A_1\partial_x\xi)^t A_2\xi + (A_2\xi)^t A_1\partial_x\xi$$

$$\|A_1\partial_x\xi\|^2 = \int_0^L a_1^2(\partial_x q)^2 + (a_2\partial_x z + a_3\partial_x q)^2$$

$$\|A_2\xi\|^2 = \int_0^L (a_4 z - a_5 q)^2 \leq \int_0^L (a_4 z)^2 + (a_5 q)^2$$

Using Cauchy-Minkowski inequalities, with the fact that $\|\xi\| = 1$, z_e bounded, Q_e constant, each integral is bounded . □

Proposition 2. The open loop system has a stable semigroup.

PROOF. Recall that the open loop abstract boundary control system is

$$\dot{\varphi}(t) = A\varphi(t) \;\; t > 0$$

$$\varphi(0) = \varphi_0 \;\; in \;\; D(A)$$

and $\varphi(t) = T_A(t)\varphi_0$ (according to proposition 1 where $T_A(t)$ is the C_0 semi-group generated by the operator A defined in this section :

$$A(x) = A_1(x)\partial_x + A_2(x).$$

Let us consider the two parts of this operator

- A_2 is semi-definite negative, following the fact that its spectral set is :

$$\sigma(A_2) = \{0\} \cup \{-a_5(x)/a_5(x) > 0 \;\; \forall \;\; 0 \leq x \leq L\}.$$

- The spectral set of $A_1\partial_x$ is defined as follows ([6]) :

$$\sigma(A_1\partial_x) = \sigma_p(A_1\partial_x)$$

$$= \{\mu_n : \mu_n = \mu + \frac{2i\pi n}{L}, n \in Z\},$$

so the real part of each eigenvalues verifies the two equations :

$$-a_1 Q' = \mu z, \qquad -a_2 z' - a_3 q' = \mu q. \qquad (17)$$

The first one gives :

$$-a_1(q(x) - q(x')) = \int_{x'}^{x} \mu z, \forall x, x' \in (0, L) \quad (18)$$

According to the boundary conditions in $D(A)$, z and q have opposite sign at $x = 0$, and the same sign at $x = L$.

So, there's two possibilities,

(1) z is null for at least one x, and q is null too in this point,
(2) q is null for at least one x, but z is not necessary null in this point).

Let us consider the two possibilities :

(1) let $x' \in (0, L)$, the smaller x, not null, so that $z(x') = 0$ and q has a constant sign in the new interval obtained, then

$$a_1 q(0) = \int_0^{x'} \mu(x) z(x) dx$$

and $z < 0$ over $[0, x')$, so $q > 0$ over $[0, x')$ and

$$\int_0^{x'} \mu(x) z(x) dx > 0 \Rightarrow \mu z \geq 0 \;\; p.p.$$

$$\Rightarrow \mu \leq 0 \;\; p.p.; \;\; \mu \neq 0$$

over $[0, x')$, since μ has constant sign. It would be the same if $z > 0$ and $q < 0$.

(2) In the same way, let $x' \in (0, L)$, the smaller x so that $q(x') = 0$. Since z has a constant sign over this interval, the calculations are the same.

Now, suppose that there is an $x = \alpha \in]0, L[$, such that $\mu(\alpha)$ is null, then it is an extremum of μ and so $\mu'(\alpha) = 0$. Moreover, it implies that if $\mu^{(n)}(\alpha) = 0$ then $(q')^{(n)}(\alpha) = 0$ and $(z')^{(n)}(\alpha) = 0$. Using those relations and (17) in

$$\mu(x) = \frac{-q'(x)}{bz(x)}$$

contradictions appear. Consequently, $\mu < 0, \forall x \in (0, L)$, so

$$\Re e(\sigma(A_1\partial_x)) < 0.$$

Moreover, since $A_1\partial_x$ is bounded ([10]),

$$\Re e(\sigma(A_1\partial_x)) < 0 \Rightarrow \langle A_1\partial_x\varphi, \varphi\rangle \leq 0.$$

So for all $\varphi \in D(A)$,

$$\langle A_2\varphi, \varphi\rangle \leq 0$$

and

$$\langle A_1\partial_x\varphi, \varphi\rangle \leq 0.$$

Now let $V(t)$ be the following Lyapunov-LaSalle function,

$$V(t) = \frac{1}{2}\|\varphi(t)\|^2_{L_2(0,L)} = \frac{1}{2}\|T_A(t)\varphi_0\|^2_{L_2(0,L)},$$

then

$$\dot{V}(t) = \langle A_1\partial_x\varphi, \varphi\rangle + \langle A_2\varphi, \varphi\rangle$$

and $\dot{V}(t) \leq 0$.

According to the Lyapunov approach, the open loop system is stable system. □

The control objective can be now achieved by including a simple control law in the IMBC control structure.

3.2 The IMBC structure

This control structure is a particular case of the classical IMC structure since it contains an internal feedback on the linear system. Moreover the control acts simultaneously on the linear control system and the nonlinear model.

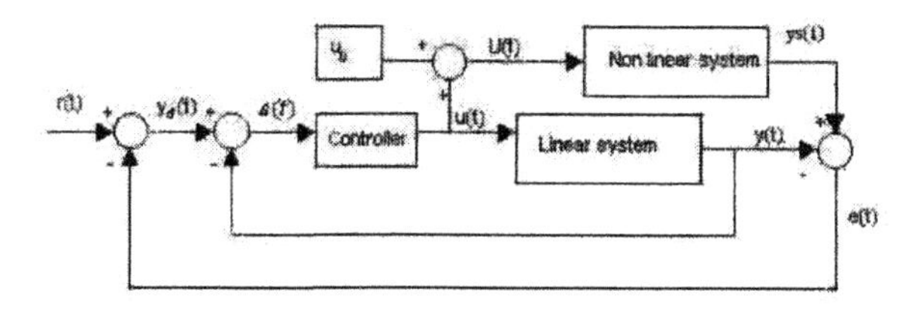

Fig. 3. IMBC structure

The control law is chosen as an integral type feed back control

$$u(t) = \alpha\kappa\xi(t) \tag{19}$$

with $\dot{\xi}(t) = \varepsilon(t)$ and where

$$\varepsilon(t) = r(t) - y(t) - e(t)$$

which can be used with a perfect model (i.e. $e(t) \equiv 0, \forall t$) for control synthesis :
if $e(t) \equiv 0 \quad y(t) = y_s(t), \;\; \varepsilon(t) = r(t) - y_s(t)$

if $e(t) \neq 0 \quad \varepsilon(t) = r(t) - y(t) - e(t) = r(t) - y(t) - (y_s(t) - y(t)) = r(t) - y_s(t)$.

The IMBC state space :
Let

$$\dot{\zeta}(t) = r(t) - C\varphi(t) - CDu(t)$$

then using (14)

$$\dot{\zeta}(t) = r(t) - C\varphi(t) - \alpha\kappa CD\zeta(t) \tag{20}$$

Let $x_a(t) = (\phi(t) \;\; \zeta(t))^t$, the extended IMBC state space system is

$$\left\{ \begin{array}{c} \dot{x_a}(t) = A(\alpha)x_a(t) + B(\alpha)r(t) \\ x_a(0) = x_{a0} \end{array} \right. \tag{21}$$

where

$$A(\alpha) = \begin{pmatrix} A & 0 \\ -G & 0 \end{pmatrix} + \alpha \begin{pmatrix} D\kappa C & 0 \\ 0 & -\kappa CD \end{pmatrix} \tag{22}$$

$$+\alpha^2 \begin{pmatrix} 0 & \kappa^2 DCD \\ 0 & 0 \end{pmatrix},$$

$$B(\alpha) = \begin{pmatrix} -\alpha D\kappa \\ 1 \end{pmatrix}.$$

$A(\alpha)$ can be viewed as a bounded perturbation of A ([6]) :

$$A(\alpha) = A_e + \alpha A_e^{(1)} + \alpha^2 A_e^{(2)},$$

where $A_e^{(1)}$ and $A_e^{(2)}$ are bounded operators.

3.3 Closed loop : stability (and regulation) results

The stability of the closed loop operator $A(\alpha)$ can be achieved using a stable perturbation gain for α and κ.

Proposition 3. A sufficient condition of the closed loop system stability is given by :

$$0 < \alpha < \min_{\lambda \in \Gamma}(a\|R(\lambda, A_e)\| + 1)^{-1}$$

with $a = \max(\|A_e^{(1)}\|, \|A_e^{(2)}\|), \;\; \Gamma \in \rho(A_e)$,

$$\Re(\sigma(-\kappa CD)) < 0.$$

PROOF. see [11] □

4. SIMULATION RESULTS

To take into account a more realistic dynamic of the vanne, a second order system is written :

$$\ddot{u}(t) + 2\aleph\omega_n\dot{u}(t) + \omega_n^2 u(t) = k\omega_n^2 v(t) \tag{23}$$

$$\Leftrightarrow \left\{ \begin{array}{c} \dot{u}_1 = u_2 \\ \dot{u}_2 = -2\aleph\omega_n u_2 - \omega_n^2 u_1 + k\omega_n^2 v(t) \end{array} \right.$$

The reference signal is a step response (range= 0.13) of a first order system. For the shallow water model used, the output variable is Z ([7]). The variation of the height and the control in monovariable and in multivariable are as follows :

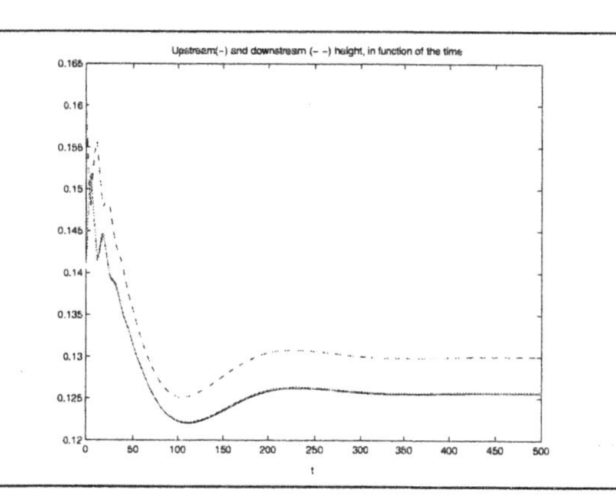

Fig. 4. Height regulation in multivariable

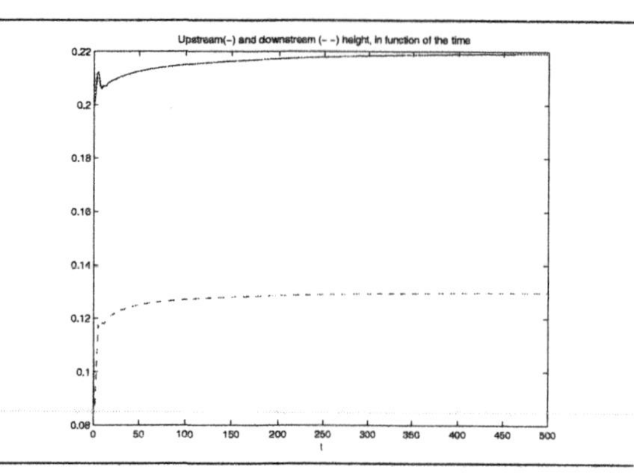

Fig. 5. Height regulation in monovariable

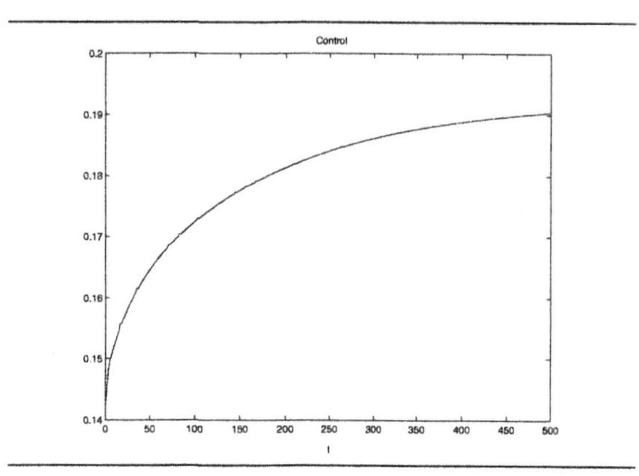

Fig. 6. Control in monovariable

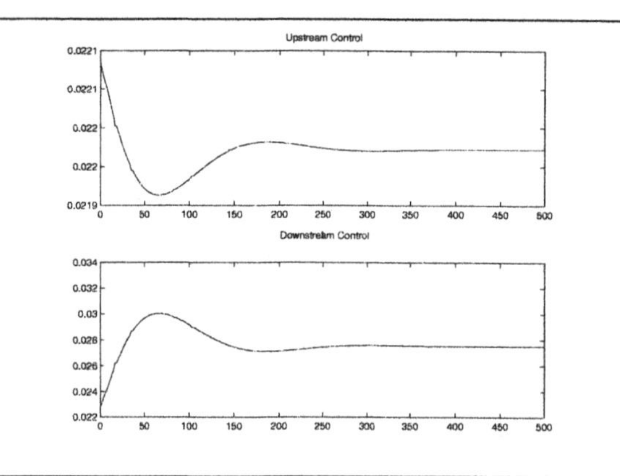

Fig. 7. Control in multivariable

5. CONCLUSION

The above simulation results show the well-suitability of the infinite dimensional system approach for regulation of the canals. The robustness of the IMBC is also an important property, since the control which is synthesis on a linearized system, acts well on the nonlinear.
Moreover, notice that multivariable approach enables, more easily, to take into account the more realistic situation with lateral leak or supply and the connection with more numerous canals with their regulation.

REFERENCES

[1] CHEN M.-L., Commande optimale et robuste des équations aux dérivées partielles, *Thèse de l'Institut National Polytechnique de Grenoble,* Juillet 2001.

[2] CHEN P. & QIN H., Controllability of Linear Systems in Banach Spaces, *Systems & Control Letters*, 45 p. 155-161, 2002.

[3] CURTAIN R.F. & ZWART H., An introduction to Infinite Dimensional Linear Systems.*Springer Verlag,* 1995

[4] FATTORINI H.O., Boundary Control Systems, *SIAM J. Control*, Vol. 6, No. 3, 1968.

[5] JOSSERAND L. & TOURE Y., PI-controller in IMC structure for Distribued Parameter System.*Proc. if the CESA IMACS/IEEE-SMC Multiconference*, Lille, France, Vol. 2, p. 1168-1172, July, 9-12, 2000.

[6] KATO T., Pertubation Theory for Linear Operators.*Springer Verlag,* Berlin, 1966.

[7] KURGANOV A. and LEVY D., Central-Upwind Schemes for the Saint-Venant System.

[8] PAZY A., Semigroups of Linear Operators and Applications to Partial Differential Equations, *Applied Mathematical Sciences, Springer Verlag*, Vol.44., 1983.

[9] POHJOLAINEN S.A., Robust multivariables PI-controller for Infinite Dimensional Systems,*IEEE Trans. Auomat.Contr., AC(27), 17-30,*1982.

[10] TRIGGIANI R., On the stability problem in Banach space, *J. of Math. Anal. and Appl., (52); p. 383-403*, 1975.

[11] TOURE Y. & JOSSERAND L.,An extension of IMC to Boundary Control of Distributed Parameter Systems.*Proc. of the IEEE SMC-CCS, 3, p. 2426-2431, Orlando, FL.*, 1997.

ROBUST OUTPUT FEEDBACK DESIGN OF DISCRETE-TIME SYSTEMS – LINEAR MATRIX INEQUALITY METHODS

Danica Rosinová, Vojtech Veselý

Department of Automatic Control Systems
Slovak University of Technology
Faculty of Electrical Engineering and IT
Bratislava, Slovakia
E-mail: rosinova@kasr.elf.stuba.sk , vesely@kasr.elf.stuba.sk

Abstract
Two novel linear matrix inequality (LMI) - based procedures to receive stabilizing robust output feedback gain are presented, one of them being modification of previous results of (Oliveira et al.,1999). The proposed robust control law stabilizes the respective uncertain discrete-time system described by polytopic model with guaranteed cost. The obtained results are compared with other LMI results from literature and illustrated on the example. *Copyright © 2003 IFAC*

Keywords: discrete-time systems, robust stability, polytopic uncertainties, output feedback, parameter-dependent Lyapunov function, LMI approach, guaranteed cost

1. INTRODUCTION

Robust control of linear systems has attracted considerable interest lately and various approaches for analysis and control design for uncertain linear systems have been adopted (e.g. Oliveira et.al.,1999, Crusius and Trofino,1999, Henrion et al.,2002, Takahashi et al.,2002). This paper deals with robust control design in discrete time-domain via static output feedback using LMI approach.

The studied problem comprises two issues: robust stabilization and static output feedback design. In general, the *static output feedback* design yields a non-convex formulation, therefore requiring non-polynomial (NP) hard algorithms as it is using bilinear matrix inequalities (BMI). To avoid computational burden, other approach resorts to the solutions based on convex optimization. This is done either using iterative procedures where the question of convergence remains open or adding the supplementary conditions to the output feedback problem so that it is restricted to a convex one. In the latter case obviously the results are more or less conservative, however computationally attractive (Sharav-Shapiro et al.,1998, Crusius and Trofino,1999, Vesely,2001, Rosinova, Vesely and Kucera, 2000, Henrion et al.,2002). The present effort focuses on finding the ways to relax the conservatism and to develop simple computationally efficient algorithms based on standard software tools like LMI solvers to obtain the required result – stabilizing output feedback gain matrix.

The frequent approach that is used to study *robust stabilization* of uncertain systems is based on quadratic stability notion. To reduce the conservatism of quadratic stability approach the parameter-dependent Lyapunov function has been introduced and the respective stability condition has been developed in different forms (Oliveira et.al.,1999, Feron at al.,1996, Henrion et al.,2002, Takahashi et al.,2002). The LMI condition and the respective design procedure have been proposed for robust state feedback control, however in the case of output feedback the problems mentioned above still remain.

In this paper several methods of robust stabilizing control design are after some modification compared with the novel robust output feedback control design procedure provided by the authors. The respective methods are briefly characterised and their properties are demonstrated on illustrative examples. The stability margin is considered as well as performance index (in the sense of guaranteed cost).

2. PROBLEM FORMULATION AND PRELIMINARIES

Consider a linear discrete-time uncertain dynamic system

$$\begin{aligned} x(k+1) &= (A+\delta A)x(k) + (B+\delta B)u(k) \\ y(k) &= Cx(k) \end{aligned} \tag{1}$$

where $x(k) \in R^n, u(k) \in R^m, y(k) \in R^l$ are state, control and output vectors respectively; A,B,C are known constant matrices of appropriate dimensions and $\delta A, \delta B$ are matrices of uncertainties of appropriate dimensions. Uncertainties are considered to be of the affine type

$$\delta A = \sum_{j=1}^{p} \varepsilon_j \bar{A}_j,\ \delta B = \sum_{j=1}^{p} \varepsilon_j \bar{B}_j \tag{2}$$

where $\underline{\varepsilon}_j \le \varepsilon_j \le \bar{\varepsilon}_j$ are unknown parameters; $\bar{A}_j, \bar{B}_j, j = 1,2,...,p$ are constant matrices of the corresponding dimensions. The affine parameter dependent model (1), (2) can be readily converted into a polytopic one and described by a list of its vertices

$$\{(A_1, B_1, C), (A_2, B_2, C), ..., (A_N, B_N, C)\}$$

$$N = 2^p. \tag{3}$$

The considered control law is static output feedback

$$u(k) = KCx(k). \tag{4}$$

Then the uncertain closed-loop polytopic system is described by

$$x(k+1) = A_C(\alpha)x(k) \tag{5}$$

where

$$A_C(\alpha) \in \left\{ \sum_{i=1}^{N} \alpha_i (A_i + B_i KC), \sum_{i=1}^{N} \alpha_i = 1, \alpha_i \ge 0 \right\}. \tag{6}$$

Consider a quadratic cost function associated with the uncertain system (1), (2),(4) or, alternatively, (5), (6).

$$J = \sum_{k=0}^{\infty} [x(k)^T Qx(k) + u(k)^T Ru(k)]. \tag{7}$$

where Q, R are symmetric positive definite matrices, $Q \in R^{n \times n}, R \in R^{m \times m}$.

The control aim is to determine conditions and the corresponding controller design procedure for output feedback stabilization of the uncertain system (1), (2), (4) (or alternatively (5), (6)) and to achieve the guaranteed cost (i.e. to guarantee the performance index upper limit).

Firstly, several notions are specified that are used in the following. The *quadratic stability* is equivalent to the existence of one Lyapunov function for the whole set of system models that describes the uncertain system. The polytopic system is quadratically stable if and only if there exists a Lyapunov function for all vertices of the respective polytope describing the uncertain system. The following lemma summarizes the quadratic stability condition for a closed loop uncertain polytopic system with output feedback.

Lemma 1

System (1) with uncertainties (2) and control law (4), or equivalently the polytopic system (5), (6) is quadratically stable if and only if there exists a symmetric positive definite matrix P such that

$$(A_i + B_i KC)^T P(A_i + B_i KC) - P < 0 \tag{8}$$

for i=1,2,...,N.

The following equivalence provides a very useful tool to transform Lyapunov-type matrix inequality into LMI avoiding the products of the respective matrices P and A. This is achieved by the introduction of an additional matrix G without restricting it to any special form.

Lemma 2 (Oliveira et al.,1999)

The following conditions are equivalent:

(i) There exists a symmetric matrix $P>0$ such that

$$A^T PA - P < 0 \tag{9}$$

(ii) There exist a symmetric matrix P and a matrix G such that

$$\begin{bmatrix} -P & A^T G^T \\ GA & -G - G^T + P \end{bmatrix} < 0. \tag{10}$$

The above lemma can be readily used also when parameter-dependent Lyapunov function is applied in order to reduce the conservatism of quadratic stability approach. The parameter-dependent Lyapunov function $P(\alpha)$ and the respective stability condition is considered in compliance with (Oliveira et al.,1999).

Definition 1 (according to Oliveira et al.,1999)

System (5) is robustly stable in the convex uncertainty domain (6) with parameter-dependent Lyapunov function if and only if there exists a matrix $P(\alpha) = P(\alpha)^T > 0$ such that

$$A_C(\alpha)^T P(\alpha) A_C(\alpha) - P(\alpha) < 0 \tag{11}$$

for all α such that $A_C(\alpha)$ is given by (6).

Now let us introduce several notions concerning the concept of *guaranteed cost*, that is considered here in the sense respective to LQ approach.

The notion of guaranteed cost J_0 represents the cost function value for the closed loop system $J \le J_0$ for all admissible uncertainties and considered initial conditions.

The following result provides the basis for further developments in the next section.

Lemma 3

Consider the nominal system (1) without uncertainties, and cost function (7). The following statements are equivalent

(i) System (1) without uncertainties is static output feedback quadratically stabilizable by (4) with guaranteed cost

$$J \le J_0 = x_0^T P x_0 \qquad (12)$$

where P is a real symmetric positive definite matrix, $x_0 = x(0)$ is initial value of the state vector $x(k)$.

(ii) The following inequality holds for some real symmetric positive definite matrix P and a matrix K

$$(A+BKC)^T P(A+BKC) - P + C^T K^T RKC + Q \le 0 \qquad (13)$$

Frequently, the maximal eigenvalue of P, denote it $\lambda_M(P)$, is considered to evaluate the right hand side of (12).

Considering the parameter-dependent Lyapunov function in the form $P(\alpha) = \sum_{i=1}^{N} a_i P_i$ for uncertain system we will use $\max_i\{\lambda_M(P_i)\}$ to evaluate the cost function.

(Obviously $\lambda_M P(\alpha) \le \max_i\{\lambda_M(P_i)\}$).

3. ROBUST OUTPUT FEEDBACK CONTROL DESIGN – MAIN RESULTS

In this section LMI based design procedures are presented and compared with other existing LMI control design methods (Crusius and Trofino,1999, Henrion et al.,2002). Firstly, the original procedure developed by the authors is given in 3.1. Secondly, in 3.2, the iterative procedure is proposed of so-called V-K iteration type (Ghaoui and Balakrishnan, 1994). This V-K iteration procedure is based on the results of (Oliveira et al.,1999), see Lemma 2, and modify them to receive stabilizing *output* feedback. Thirdly, in 3.3, the procedures of (Crusius and Trofino,1999) and (Henrion et al.,2002) are considered for comparison, the latter slightly modified to the iterative one to relax the necessity of appropriate input data choice.

3.1. Output feedback stabilization procedure with guaranteed cost

A novel condition for output feedback stabilization of uncertain polytopic system (5), (6) and the respective LMI – based control design procedure is presented. To obtain LMI formulation the sufficient stability condition is considered and the parameter-dependent Lyapunov function is applied including extra degree of freedom to avoid too conservative results.

We start with several results useful for further developments.

Lemma 4

The following two statements are equivalent for the given system matrices A, B, C of the corresponding dimensions.

(i) There exist a matrix $P = P^T > 0$ and K of appropriate dimension so that

$$(A+BKC)^T P(A+BKC) - P + Q + C^T K^T RKC < 0 \qquad (14)$$

(ii) There exist a matrix $P = P^T > 0$ and K of appropriate dimension so that

$$\begin{bmatrix} -P+Q & (A+BKC)^T & C^T K^T \\ A+BKC & -P^{-1} & 0 \\ KC & 0 & -R^{-1} \end{bmatrix} < 0. \qquad (15)$$

Since (15) is not in the LMI form due to P^{-1} we will use the following inequality to substitute for P^{-1}.

$$-P^{-1} \le -\frac{2}{\rho} D + \frac{1}{\rho^2} DPD \qquad (16)$$

for any matrices $P=P^T>0$, $D=D^T$ and real scalar $\rho > 0$.

Inequality (16) follows from

$$(D-\rho P^{-1})^T \frac{P}{\rho}(D-\rho P^{-1}) \ge 0. \qquad (17)$$

Free scalar parameter ρ is introduced to extend the degree of freedom.

The following theorem (Rosinova and Vesely, 2003) provides a way to LMI based output feedback design.

Theorem 1

Uncertain system (5) is static output feedback robustly stabilizable with guaranteed cost with respect to cost function (7) if for some $D_i = D_i^T$ there exist a feedback gain matrix K, symmetric matrices P_i and a matrix Z satisfying LMI

$$\begin{bmatrix} -P_i+Q & (A_i+B_iKC)^T & 0 & C^TK^T \\ A_i+B_iKC & -\frac{2}{\rho}D_i & \frac{1}{\rho}D_i^TZ & 0 \\ 0 & \frac{1}{\rho}Z^TD_i & -Z-Z^T+P_i & 0 \\ KC & 0 & 0 & -R^{-1} \end{bmatrix} < 0$$

for $i=1,...,N$. (18)

The proof is ommited (it can be found in (Rosinova and Vesely, 2003).

The above Theorem 1 provides sufficient condition for robust stability with guaranteed cost. If (18) gives feasible solution for unknown P_i, Z and K, the resulting output feedback control guarantees the robust stability of uncertain system (1), (2) and value of cost function limited by $\lambda_M P(\alpha)$.

3.2. *V-K iterative procedure for output feedback stabilizing control*

In (Oliveira et al., 1999) the sufficient stability condition (see Lemma 2 above) was developed using parameter dependent Lyapunov function and the respective LMI formulation to find state feedback stabilizing control was provided. In this section we use the former result to build iterative procedure to receive ***output*** feedback control, the stability condition is extended to include performance index (7). The respective sufficient stability condition with guaranteed cost is then in the form

$$\begin{bmatrix} -P_i+Q & (A_i+B_iKC)^TG^T & C^TK^T \\ G(A_i+B_iKC) & -G-G^T+P_i & 0 \\ KC & 0 & -R^{-1} \end{bmatrix} < 0. \tag{19}$$

Since besides P_i both matrices K and G are unknown we propose the iterative procedure to convert the problem into LMI formulation. The V-K iteration approach is used that was declared to have good convergence properties (Ghaoui and Balakrishnan, 1994).

Algorithm V-K

1. Initialization:
   - stability test for vertex system matrices A_i,
   - for unstable A_i choose multiplier a_i so that $A_{pi}=a_iA_i$ is stable, for stable A_i set $A_{pi}=A_i$.
   - set maximal number of iterations *max* and prescribed error ε
2. Set $j \leftarrow 0$. Compute initial value G_0 from the following LMI (with unknown P_i and G_0)

$$\begin{bmatrix} -P_i+Q & A_{pi}^TG_0^T & 0 \\ G_0^TA_{pi} & -G_0-G_0^T+P_i & 0 \\ 0 & 0 & -R^{-1} \end{bmatrix} < 0 \tag{20}$$

3. Set $j \leftarrow j+1$. Compute $K^{(j)}$ and $P_i^{(j)}$ from LMI (21)

$$\begin{bmatrix} -P_i^{(j)}+Q & (A_i+B_iK_{(j)}C)^TG_{(j-1)}^T & C^TK_{(j)}^T \\ G_{(j-1)}(A_i+B_iK_{(j)}C) & -G_{(j-1)}-G_{(j-1)}^T+P_i^{(j)} & 0 \\ K_{(j)}C & 0 & -R^{-1} \end{bmatrix} < 0 \tag{21}$$

   compute $G_{(j)}$ from LMI (22) (with unknown $G_{(j)}$ and $P_i^{(j)}$)

$$\begin{bmatrix} -P_i^{(j)}+Q & (A_i+B_iK_{(j)}C)^TG_{(j)}^T & C^TK_{(j)}^T \\ G_{(j)}(A_i+B_iK_{(j)}C) & -G_{(j)}-G_{(j)}^T+P_i^{(j)} & 0 \\ K_{(j)}C & 0 & -R^{-1} \end{bmatrix} < 0 \tag{22}$$

4. Check terminal condition ($j < \max$) and ($\|G_{(j)}-G_{(j-1)}\|/\|G_{(j)}\| < \varepsilon$)

   if it does not hold, repeat step 3. else end.

Though Algorithm V-K is iterative it provides rather good qualities in practical examples.

3.3. *Existing procedures for output feedback stabilizing gains – (Crusius and Trofino, Henrion et al.)*

To compare our results from sections 3.1. and 3.2. with the existing ones we briefly recall the respective results of (Crusius and Trofino, 1999) and (Henrion et al., 2002). The latter method we turned into iterative form to relax its sensitivity to the initial input data.

Algorithm C-T (Crusius and Trofino, 1999)

- Solve the following LMI for unknown matrices F and W of appropriate dimensions, the W being symmetric (corresponding to P^{-1}).

$$\begin{bmatrix} -W & WA_i^T-C^TF^TB_i^T \\ A_iW-B_iFC & -W \end{bmatrix} < 0,\ i=1,2,...,N$$

$$W > 0 \tag{23}$$

$$MC = CW \tag{24}$$

- Compute the corresponding output feedback gain matrix

$$K = FM^{-1}. \tag{25}$$

The above algorithm can be used under the assumption that the equation (24) can be met which is not always the case and limit the use of it. Algorithm C-T is computationally rather efficient and does not require any iteration or initial data choice. However it is based on quadratic stability and since feedback gain K is computed from (24), (25), there is no obvious way to use parameter-dependent Lyapunov function – the varying W_i would be required for A_i in such case.

The other way to compute stabilizing output feedback gain was developed in (Henrion et al., 2002), based on the sufficient LMI stability condition. In a discrete-time case it is described by

$$\begin{bmatrix} F^T A_c + A_c^T F + P_i & -A_c^T - F^T \\ -A_c - F & 2I - P_i \end{bmatrix} > 0,$$

$$i=1,2,...,N, \quad A_c = A_i + B_i KC \qquad (26)$$

for a given stable matrix F and unknown K and P_i of appropriate dimensions.

Since it is not clear how to choose the stable matrix F, while this choice is important for the result, we propose the following iterative variant of (26).

Algorithm H

1. Initialization – choice of initial stable matrix F_0 :
   - test a stability of mean value of vertex system matrices A_i, $A_0 = \sum_{i=1}^{N} A_i / N$, for unstable A_0 choose multiplier a_0 so that $F_0 = a_0 A_0$ is stable, for stable A_0 set $F_0 = A_0$;
   - determine the required precision ε and maximal number of iterations **iter**; $j \leftarrow 0$

2. Compute P_i and K from (26) having $F = F_j$

   Set $F_{j+1} = \frac{1}{N} \sum_{i=1}^{N} (A_i + B_i KC)$

3. $j \leftarrow j+1$;

   repeat step 2. until $\|F_{j+1} - F_j\| / \|F_j\| < \varepsilon$ or j>**iter**.

Condition (26) can be extended to include cost function matrices Q and R analogously as (10) was extended to (19). However in examples it then provides much more difficult to obtain feasible solution in comparison to algorithms described in sections 3.1. and 3.2. even in the case that the iterative variant has been adopted for the extended condition, analogous to the Algorithm H. Therefore in the illustrative example we tested the mere stabilizing output design method given in Algorithm H.

The algorithms C-T and H provide the stabilizing output feedback gain (if the respective LMI's are feasible), however they do not involve the performance index. To evaluate and compare the quality of obtained results we therefore consider Lemma 3 together with (10) and (11). The guaranteed cost for closed loop system is then evaluated using parameter-dependent Lyapunov function $P(\alpha) = \sum_{i=1}^{N} a_i P_i$, where P_i are solutions of LMI (unknown P_i and G)

$$\begin{bmatrix} -P_i + Q + C^T K^T RKC & (A_i + B_i KC)^T G^T \\ G(A_i + B_i KC) & -G^T - G + P_i \end{bmatrix} < 0,$$

$$i=1,2,...,N. \qquad (27)$$

Then $\max_i \{\lambda_M(P_i)\}$ provides the upper bound on cost function (see last paragraph in Section 2).

4. EXAMPLE

The results developed in previous section are illustrated on the following example.

Example
Consider uncertain system (1), (2) with matrices

$$A = \begin{bmatrix} .7118 & .0736 & .1262 \\ .7200 & .6462 & 2.3432 \\ 0 & 0 & .6388 \end{bmatrix} \quad B = \begin{bmatrix} .0122 & .0412 \\ .3548 & .1230 \\ .2015 & .2301 \end{bmatrix}$$

- nominal model

$$\tilde{A}_1 = \begin{bmatrix} .08 & .006 & .01 \\ .07 & .03 & .1 \\ 0 & 0 & .03 \end{bmatrix} \quad \tilde{B}_1 = \begin{bmatrix} 0 & .001 \\ .012 & 0 \\ .007 & .004 \end{bmatrix}$$

$$\tilde{A}_2 = \begin{bmatrix} 0 & .0004 & 0 \\ .1 & 0 & .12 \\ 0 & 0 & 0 \end{bmatrix} \quad \tilde{B}_2 = \begin{bmatrix} 0 & 0 \\ .02 & 0 \\ .014 & .02 \end{bmatrix}$$

$$C = \begin{bmatrix} 0 & 1 & 0 \\ 0 & 0 & 1 \end{bmatrix} \quad \text{with } -1 \le \varepsilon_i \le 1, i = 1,2.$$

The vertices of the corresponding polytopic model are

$$A_1 = \begin{bmatrix} .7918 & .0792 & .1362 \\ .6900 & .6762 & 2.3232 \\ 0 & 0 & .6688 \end{bmatrix} \quad B_1 = \begin{bmatrix} .0122 & .0422 \\ .3468 & .1230 \\ .1945 & .2141 \end{bmatrix}$$

$$A_2 = \begin{bmatrix} .6318 & .0680 & .1162 \\ .7500 & .6162 & 2.3632 \\ 0 & 0 & .6088 \end{bmatrix} \quad B_2 = \begin{bmatrix} .0122 & .0402 \\ .3628 & .1230 \\ .2085 & .2461 \end{bmatrix}$$

$$A_3 = \begin{bmatrix} .7918 & .0800 & .1362 \\ .8900 & .6762 & 2.5632 \\ 0 & 0 & .6688 \end{bmatrix} \quad B_3 = \begin{bmatrix} .0122 & .0422 \\ .3868 & .1230 \\ .2225 & .2541 \end{bmatrix}$$

$$A_4 = \begin{bmatrix} .6318 & .0672 & .1162 \\ .5500 & .6162 & 2.1232 \\ 0 & 0 & .6088 \end{bmatrix} \quad B_4 = \begin{bmatrix} .0122 & .0402 \\ .3228 & .1230 \\ .1805 & .2061 \end{bmatrix}$$

$$C = \begin{bmatrix} 0 & 1 & 0 \\ 0 & 0 & 1 \end{bmatrix}.$$

The spectral radii for A_1, A_2, A_3, A_4 are respectively $\rho_1 = 0{,}9748; \rho_2 = 0{,}8499; \rho_3 = 1{,}007; \rho_4 = 0{,}8164$; i.e. the vertex (A_3, B_3, C) without control corresponds to unstable system.

The quadratic cost function matrices $Q = 0{,}1 * I_n$ and $R = 0{,}1 * I_m$ are considered.

The results obtained using the stabilizing robust output feedback design methods described in Section 3 are summarized in the following table.

Method	output feedback gain matrix K	upper bound on guarant. cost $\hat{J}_0$
Novel method 3.1.	$\begin{bmatrix} -0{,}9638 & -4{,}1954 \\ 0{,}2867 & -0{,}4189 \end{bmatrix}$	13,251 (**8,04** from (18))
AlgorithmV-K iteration method 3.2.	$\begin{bmatrix} -2{,}6665 & -7{,}4946 \\ 1{,}7799 & 3{,}6662 \end{bmatrix}$	35,745
Algorithm C-T section 3.3.	$\begin{bmatrix} -1{,}5682 & -5{,}777 \\ 0{,}2971 & 0{,}4155 \end{bmatrix}$	19,898
Algorithm H section 3.3. (7 iterations)	$\begin{bmatrix} -2{,}8228 & -7{,}3502 \\ 1{,}3722 & 3{,}9961 \end{bmatrix}$	39,964

Table 1.

The results in Table 1 shows that the method 3.1. "tailor" the output feedback gain to minimize the considered cost function value. Simulation results then shows less oscillations than it is using other method results. The simulation results are omitted due to space reasons. All the tested methods provide in this case stabilizing output feedback gains, the respective LMI's provide feasible solutions.

5. CONCLUSION

The novel robust output feedback design procedures were proposed and studied in comparison with previous results. The proposed control design scheme includes the terms corresponding to performance index, therefore the resulting control law both robustly stabilize the uncertain system and tends to minimize the chosen quadratic cost function. The obtained results indicate the potential qualities of studied methods concerning both robust stability and performance measured by quadratic cost function.

REFERENCES

Boyd,S., ElGhaoui,L., Feron,E. and Balakrishnan V., (1994), Linear Matrix Inequalities in System and Control. *SIAM*, Vol.15, Philadelphia.

Crusius,C.A.R., and Trofino, A., (1999), LMI Conditions for Output Feedback Control Problems. *IEEE Trans. Aut. Control*, 44, 1053-1057.

El Ghaoui,L., and Balakrishnan,V., (1994), Synthesis of Fixed Structure Controllers Via Numerical Optimization. In: *Proc. of the 33rd Conference on Decision and Control*, Lake Buena Vista (FL), 169-175.

Feron,E., Apkarian,P. and Gahinet,P., (1996), Analysis and Synthesis of Robust Control Systems via Parameter-Dependent Lyapunov Functions. *IEEE Trans. Aut. Control*, 41, 1041-1046.

Goh,K.C., Safonov,M.G., Papavassilopoulos, G.P., (1995), Global Optimization for the Biaffine Matrix Inequality Problem. *Journal of Global Optimization*, 7, 365-380.

Henrion,D., Arzelier,D. and Peaucelle,D., (2002), Positive Polynomial Matrices and Improved LMI Robustness Conditions. In: IFAC 15th Triennial World Congress, Barcelona, CD.

Oliveira,M.C., Bernussou,J. and Geromel,J.C., (1999), A New Discrete-Time Robust Stability Condition. *Systems and Control Letters*, 37, 261-265.

Rosinova,D., Vesely,V. and Kucera,V., (2000), A Necessary and sufficient condition for Output Feedback Stabilization of Linear Discrete-Time Systems. In: *Proceedings from the IFAC Conference Control Systems Design*, 18-20 June, Bratislava, Slovakia, 171-174.

Rosinova,D. and Vesely,V. (2003), Robust Output Feedback Design of Linear Discrete-time systems – LMI approach. In: 4th IFAC Symposium on Robust Control Design, June 25-27, Milan, Italy, CD.

Sharav-Shapiro,N.Z., Palmor,Z.J., Steinberg, A.S., (1998), Output stabilizing robust control for discrete uncertain systems, *Automatica*, **34**, 731-739.

Takahashi,R.H.C., Ramos,D.C.W. and Peres, P.L.D., (2002), Robust Control Synthesis via a Genetic Algorithm and LMIS. In: IFAC 15th Triennial World Congress, Barcelona, CD.

Veselý,V, (2001), Static output feedback controller design. *Kybernetika*, **37**, 2001, 205-221.

AN ALTERNATIVE TO CIRCUMVENT HJI COMPUTATION DIFFICULTY IN NONLINEAR ROBUSTNESS ANALYSIS

Assia Henni and Houria Siguerdidjane

Supelec, Service Automatique, Plateau de Moulon, 91192 Gif-sur-Yvette, France
email:
assia.henni@supelec.fr, houria.siguerdidjane@supelec.fr

Abstract: In this paper, it is proposed a possible alternative way to overcome the difficulty encountered by the NLMI characterization of robustness analysis for nonlinear uncertain systems. Indeed, as it has been shown in the literature, the NLMI characterization is not anyhow an easy task, it is as hard as Lyapunov stability analysis. So, it is here used the notion of the dissipation in order to reformulate the stability robustness problem. A judicious way to design a robust nonlinear control to circumvent the NLMI Computation difficulty is proposed. In order to illustrate this approach, an application is given. It deals with the stabilization of a spacecraft during vertically descent on the surface of a planet with bounded uncertainties. *Copyright © 2003 IFAC*

Keywords: Robustness, nonlinear control systems,control system analysis, control system synthesis, uncertain dynamic systems.

1. INTRODUCTION

In this paper, robustness analysis for a class of uncertain nonlinear systems with bounded uncertainties is considered. On the basis of $\mathcal{L}_2$-gain analysis and small-gain theorem for nonlinear systems , the problem of robustness stability may be characterized by a partial differential inequality, by means of a Hamilton-Jacobi inequality (HJI) which is satisfied by a storage function V associated to the closed-loop system. In (Lu and Doyle, 1994; Lu and Doyle, 1997), the HJI conditions are expressed in terms of nonlinear matrix inequalities (NLMI) which are convex. The NLMI conditions involve neither a finite number of unknowns nor finite number of constraints. In (Huang and Lu, 1996; Lu and Doyle, 1997), NLMI computations have been proposed. It is proved that the NLMI characterizations for robustness analysis is *as hard as Lyapunov stability analysis.*

Therefore, we herein propose an easier way to overcome this difficulty by using the notion of dissipation. It is known that a system with positive storage function is dissipative with respect to the so-called supply rate and thus is robustly stable. So then, the problem of the stability robustness can be reformulated. In order to find a robust nonlinear control, a judicious way consists of checking a Lyapunov function and feedback control law that render the closed loop system dissipative. Let us mention that, this idea has not been exploited in the literature, at least to our best knowledge.

Let us consider the following control affine system:

$$\begin{cases} \dot{x} = f(x) + g(x)u \\ y = h(x) \end{cases} \tag{1}$$

where $x \in \mathbf{R^n}$ is the state vector, $u \in \mathbf{R^m}$ is the input vector and $y \in \mathbf{R^p}$ is the output vector.

It is assumed that : $f, g, h \in C^0$ are vectors or matrices valued functions and $f(0) = 0$, $h(0) = 0$. The system evolves on a convex open subset $\mathbf{X} \subset \mathbf{R^n}$ containing the origin. Thus, $0 \in \mathbf{R^n}$ is the equilibrium of the system with $u = 0$.

Definition 1.1
System (1) with the initial condition $x(0) = 0$ is said to have $\mathcal{L}_2$-gain less than or equal to 1 if

$$\int_0^T \|y(t)\|^2\,dt \le \int_0^T \|u(t)\|^2\,dt \qquad (2)$$

for all $T \ge 0$ and $u(t) \in \mathcal{L}_2^e(\mathbf{R}^+)$. Where $\mathcal{L}_2^e(\mathbf{R}^+)$ is the extended space of $\mathcal{L}_2(\mathbf{R}^+)$ which is defined as the set of all vector-valued functions $u(t)$ on $\mathbf{R}^+$ such that $\|u(t)\|_2 := (\int_0^\infty \|u(t)\|^2\,dt)^{1/2} < \infty$.

The following theorem characterizes $\mathcal{L}_2$-gains for a class of nonlinear systems which are asymptotically stable in terms of the Hamilton-Jacobi inequality.

Theorem 1.1
Consider system G given by (1), it is asymptotically stable and has $\mathcal{L}_2$-gain ≤ 1 if there exists a $\mathbf{C}^1$ positive definite function $V : \mathbf{X} \to \mathbf{R}^+$ such that Hamilton-Jacobi inequality (3), holds for all $x \in \mathbf{X}$.

$$\begin{aligned} H(V,x) := {} & \frac{\partial V}{\partial x}(x)f(x) + h^T(x)h(x) \\ & + \frac{1}{2}\frac{\partial V}{\partial x}(x)g(x)g^T(x)\frac{\partial^T V}{\partial x}(x) \le 0 \end{aligned} \qquad (3)$$

The upperscript T denotes the transposition sign.

Note that condition (3) in theorem (1.1) is expressed in terms of the so-called NLMI in (Lu and Doyle, 1994; Lu and Doyle, 1997). This characterization is affine in $V(x)$.

The solutions of this inequality form a convex set but it is not an easy task to employ it in numerical computations. In (Huang and Lu, 1996; Lu and Doyle, 1997), the authors propose computational issues using finite differences and finite elements approximations. However, theses schemes are useful tools in the case of low dimensional problems only. Moreover, the solutions depend on the initial conditions, which nevertheless is not obvious. With the NLMI characterization, it is shown that the NLMI computation for robustness analysis is *as hard as Lyapunov stability analysis* (Lu and Doyle, 1997).

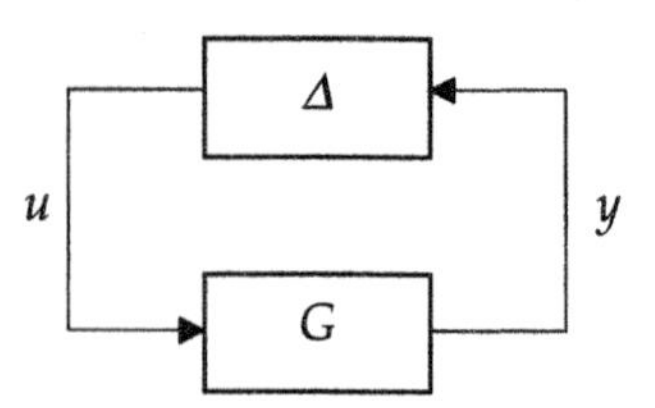

Fig. 1. The uncertain system

2. ROBUSTNESS ANALYSIS AND SYNTHESIS FOR UNCERTAIN NONLINEAR SYSTEMS

Consider the uncertain nonlinear system described in Fig.1 as a feedback system, where G is the nominal system and has realization similar to system (1). Δ is the uncertainty system. It belongs to a bounded-norm set:

$$\mathbf{B\Delta} := \{\Delta \;:\; \Delta \text{ is causal stable and has } \mathcal{L}_2\text{-gain} \le 1\}$$

Definition 2.1
The uncertain nonlinear system is robustly stable if for each $\Delta \in \mathbf{B\Delta}$, the feedback system is well-posed and is asymptotically stable around the origin.

Consider the robustness stability analysis in the case where the uncertainty Δ is unstructured perturbation having the following input affine realization :

$$\Delta : \quad \begin{cases} \dot{\xi} = f_\Delta(\xi) + g_\Delta(\xi)y \\ u = h_\Delta(\xi) \end{cases} \qquad (4)$$

where ξ evolves in $\mathbf{X_0}$, with $f_\Delta(0) = 0$, $h_\Delta(0) = 0$.

The uncertainty Δ has $\mathcal{L}_2$-gain ≤ 1 if and only if there exists a positive definite $\mathbf{C}^1$ storage function $\Theta : \mathbf{X_0} \to \mathbf{R}^+$ for system (4) with respect to the supply rate " $\frac{1}{2}\|y\|^2 - \frac{1}{2}\|u\|^2$ " such that inequality (5) holds, for all $\xi \in \mathbf{X_0}$.

$$\begin{aligned} \Psi(\Theta,\xi) := {} & \frac{\partial \Theta}{\partial \xi}(\xi)f_\Delta(\xi) + h_\Delta^T(\xi)h_\Delta(\xi) \\ & + \frac{1}{2}\frac{\partial \Theta}{\partial \xi}(x)g_\Delta(\xi)g_\Delta^T(\xi)\frac{\partial^T \Theta}{\partial \xi}(\xi) \le 0 \end{aligned} \qquad (5)$$

It is shown that inequality (5) is equivalent to the following dissipation inequality (van der Schaft, 1992).

$$\dot{\Theta}(\xi) \le \frac{1}{2}\|y\|^2 - \frac{1}{2}\|u\|^2 \qquad (6)$$

for all $y \in \mathbf{R^p}$ with $\Theta(0) = 0$.

Theorem 2.1
The uncertain nonlinear system described in Fig.1 is ***robustly stable*** *if there exists a positive definite* $\mathbf{C}^1$ *function* $V : \mathbf{X} \to \mathbf{R}^+$ *satisfying inequality (3).*

Proof 2.1
As for the system uncertainty Δ*, Hamilton-Jacobi inequality (3) can be rewritten as the following dissipation inequality :*

$$\dot{V}(x) \leq \frac{1}{2}\|u\|^2 - \frac{1}{2}\|y\|^2 \qquad (7)$$

for all $u \in \mathbf{R^m}$ *with* $V(0) = 0$ *and* $y = h(x)$.

Let us define a positive definite function W *on* $\mathbf{X} \times \mathbf{X_0}$ *as*

$$W(x, \xi) = V(x) + \Theta(\xi) \qquad (8)$$

So from (6) and (7), it follows that

$$\dot{W}(x, \xi) \leq \dot{V}(x) + \dot{\Theta}(\xi) \leq 0 \qquad (9)$$

Furthemore, if $\dot{W}(x, \xi) = 0$*, then* $x = 0$ *and* $\xi = 0$.

Thus, by LaSalle invariance principle, $W : \mathbf{X} \times \mathbf{X_0} \to \mathbf{R}^+$ *is a Lyapunov function for the nonlinear uncertain system. Therefore, the system is robustly stable.*

So, from the equivalence between Hamilton-Jacobi inequality (3) and dissipation inequality (7), it is shown, in short manner, that the robustness stability analysis is based on the analysis of the dissipation inequality of the nominal system (Henni and Siguerdidjane, 2003).

Thus, it would be more judicious to take into account the dissipation inequality in the synthesis of the feedback control law in order to render the uncertain nonlinear system robustly stable.

In particular, for a Single Input-Single Output systems, inequality (7) becomes :

$$\dot{V} - \frac{1}{2}u^2 + \frac{1}{2}y^2 \leq 0 \qquad (10)$$

Keep in mind that our goal is to find a robust nonlinear control u such that the closed loop system is asymptotically stable at the operating point x = 0.

So, one may check a Lyapunov function and feedback control law that render the closed loop system dissipative.

Thus, the problem of the stability robustness synthesis may be treated as follows :

First of all, one may check a positive definite storage function V. In this case, the storage function will play the role of Lyapunov function for the stability.

In a second step, the desired feedback control law u will be obtained such that differential dissipation inequality (10) holds.

As application, let us consider a design of robust nonlinear control for a spacecraft during vertically descent on the surface of a planet. The system uncertainty is treated as parametric uncertainties on the rate of ejection per unit time and the relative ejection velocity of the gases in the thruster.

3. A SPACECRAFT DURING VERTICALLY DESCENT ON THE SURFACE OF A PLANET

The problem of smooth descent toward the surface of a planet which exhibits noneligible atmospherique resistance has been already treated by methods based on the differential algebraic approach in (Sira-Ramirez, 1992) or by sliding mode control in (Sira-Ramirez, 1990). This last induces an exponential stability.

Its dynamics are :

$$\begin{cases} \dot{x}_1 = x_2 \\ \dot{x}_2 = g - \dfrac{\gamma}{x_3}x_2^2 - \dfrac{\alpha\sigma}{x_3}u \\ \dot{x}_3 = -\alpha u \\ \\ y = x_1 \end{cases} \qquad (11)$$

where x_1 is the position on the vertical axis oriented downwards, x_2 is the downwards velocity and x_3 represents the combined mass of the vehicle and the residual fuel . The constant α represents rate of ejection per unit time, the constant σ is the relative ejection velocity of the gases in the thruster and $\alpha\sigma$ is the maximum thrust of the braking engine while γ is a positive quantity representing atmospheric resistance.

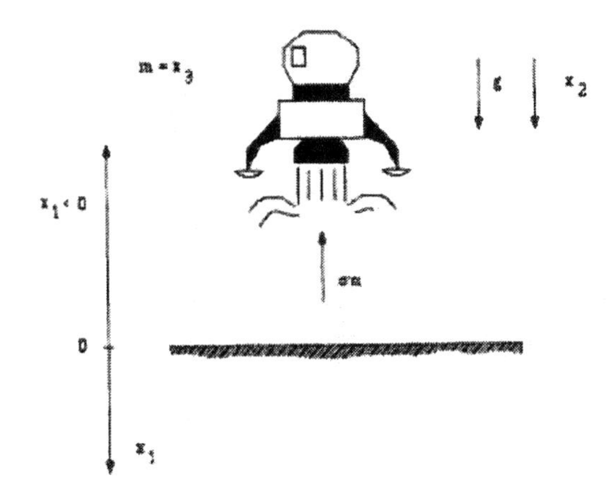

Fig. 2. Vertically controlled descent on the surface of a planet.

3.1 Robust nonlinear control design

Our aim is to stabilize the system around a given position which simulates a descent of the module on a planet.

Let us choose the control Lyapunov function (CLF) as :

$$V = \frac{1}{2}(x_1 + \frac{x_2 x_3}{\sigma} - (1+a)x_3)^2 \qquad (12)$$

We strictly assign the CLF by taking:

$$u = -\frac{1}{\alpha(a - \frac{x_2}{\sigma})}(\frac{x_1}{2} + x_2 + \frac{g}{\sigma}x_3 - \frac{\gamma}{\sigma}x_2^2) \\ -\frac{1}{\alpha(a - \frac{x_2}{\sigma})}(\frac{2x_2 x_3}{\sigma} - (1+a)x_3) \qquad (13)$$

Let us now define H as being the operator of dissipation such that :

$$H = \dot{V} - \frac{1}{2}u^2 + \frac{1}{2}y^2 \qquad (14)$$

According to inequality (10), this operator must be negative definite during the stabilization phase and zero for the next one.

3.2 Simulation results

Numerical simulations have been performed with the following parameters :

$$\sigma = 200\ m/s \qquad \alpha = 50\ kg/s$$

$$g = 3.72\ m/s^2 \qquad \gamma = 1\ kg/m.$$

Figures $3-8$ depict the behavior of the state variables, the dissipation operator, control input and Lyapunov function derivative under the designed controller proposed in Section (3.1) during descent maneuver.

Herein, it has been shown that the convergence of the nonlinear system towards an equilibrium point, for instance, requires less control effort than the control law based on the differential algebraic approach designed in (Sira-Ramirez, 1992).

In order to bring more results and to extend those given in (Sira-Ramirez, 1992), numerical simulations have been also performed for system under parametric uncertainties $\Delta\alpha = 50\%$ and $\Delta\gamma = -50\%$ as that is illustrated in figures $9-10$ and $11-12$.

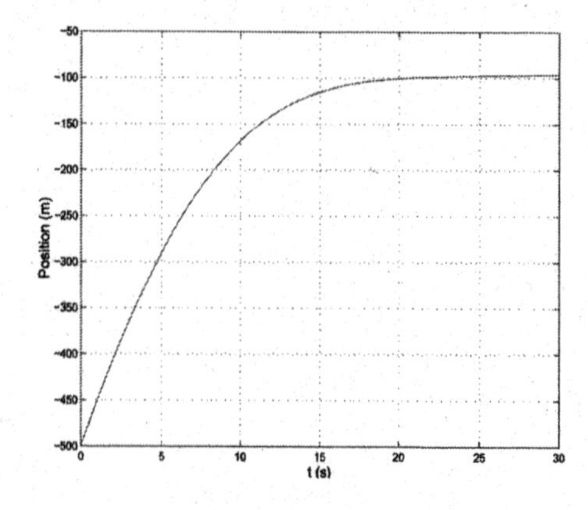

Fig. 3. Vertical position (in m) vs time (in s)

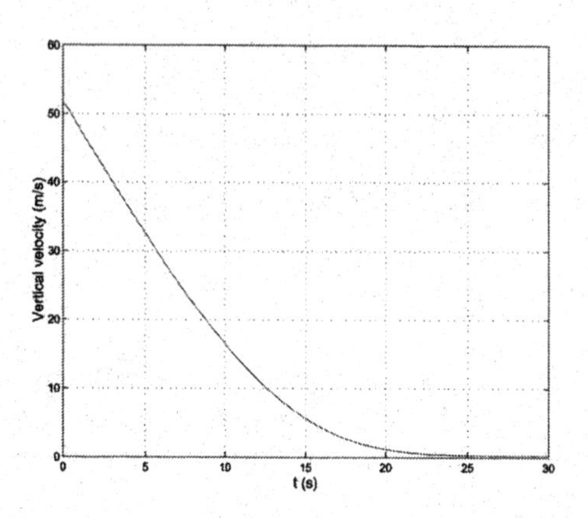

Fig. 4. Vertical velocity (in m/s) vs time (in s)

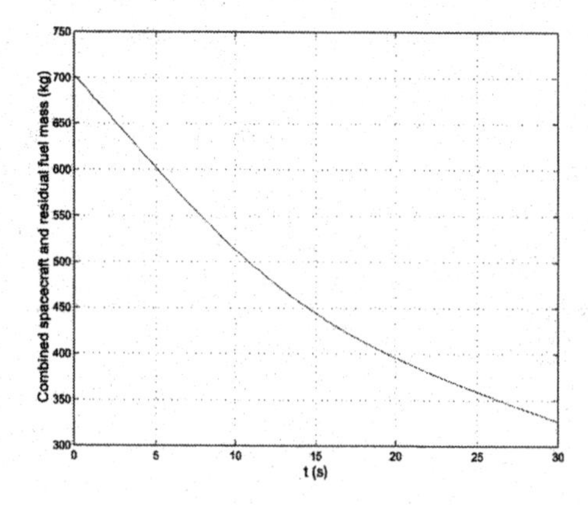

Fig. 5. Combined spacecraft and residual fuel mass (in kg) vs time (in s)

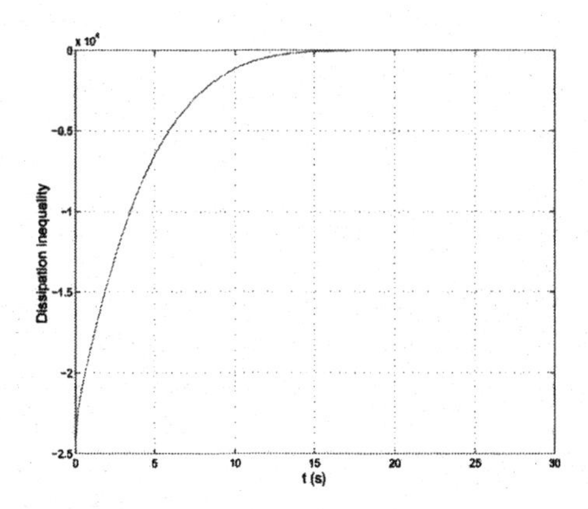

Fig. 6. Dissipation operator versus time (in s)

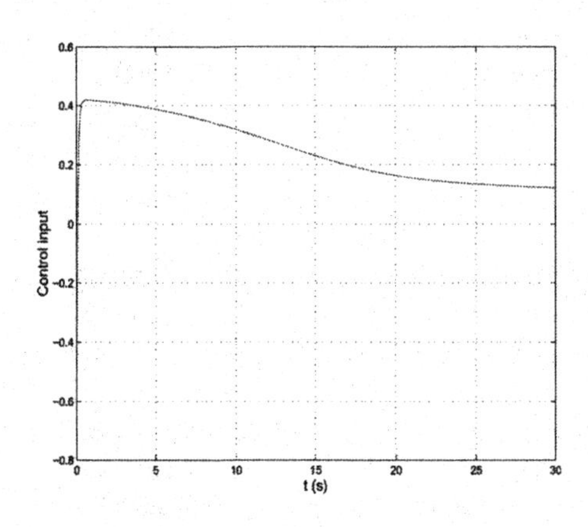

Fig. 7. Control input versus time (in s)

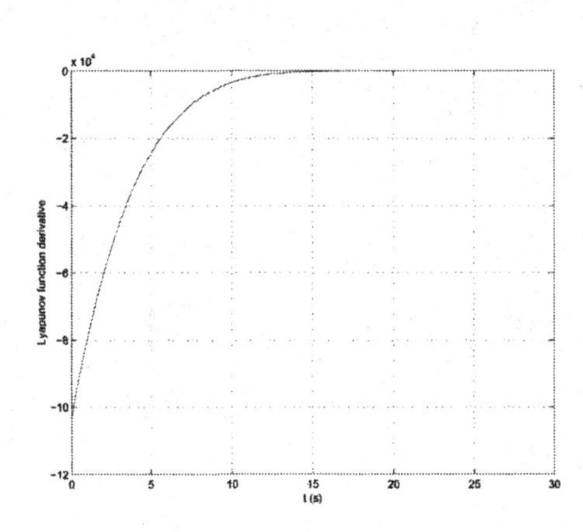

Fig. 8. Lyapunov function derivative versus time (in s)

3.3 System performance with parametric uncertainty $\Delta\alpha = 50\%$

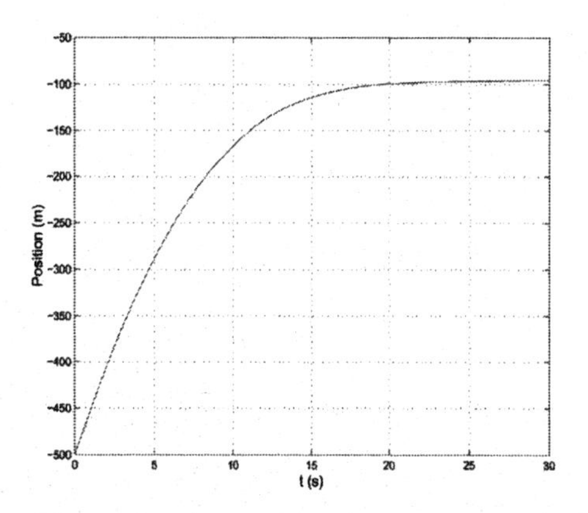

Fig. 9. Vertical position (in m) versus time (in s) with $\Delta\alpha = 50\%$

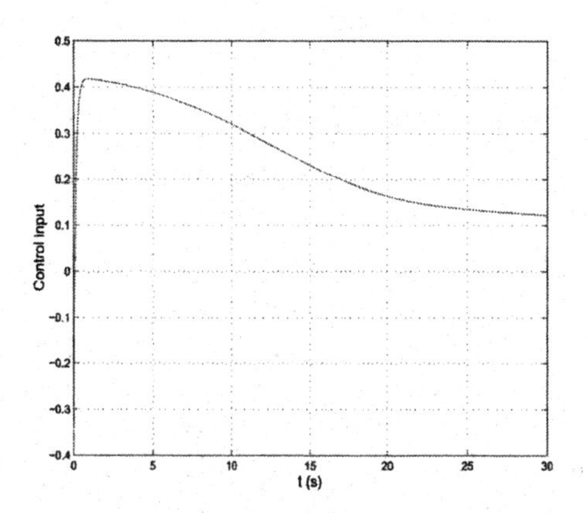

Fig. 10. Control input versus time (in s) with $\Delta\alpha = 50\%$

3.4 System performance with parametric uncertainty $\Delta\gamma = -50\%$

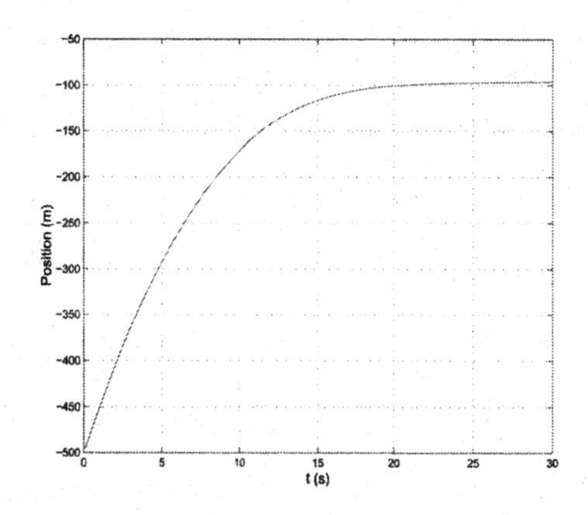

Fig. 11. Vertical position (in m) versus time (in s) with $\Delta\gamma = -50\%$

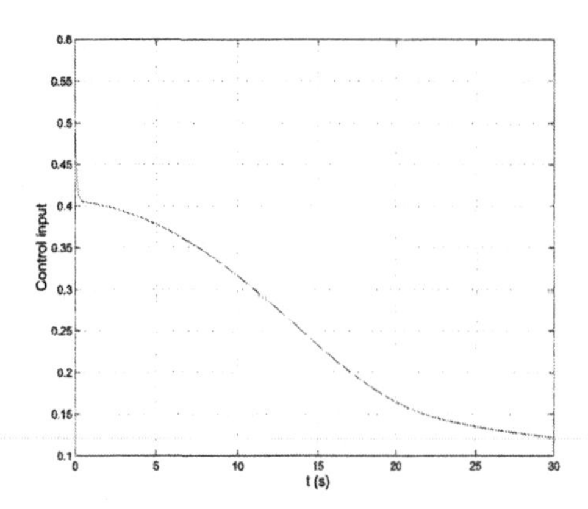

Fig. 12. Control input versus time (in s) with $\Delta\gamma = -50\%$

3.5 Discussion

The required performance are expressed in terms of the overshoot (by means that no overshoot is desirable) and the tracking error accuracy. The numerical simulations have been performed for both nominal system and under uncertainties $\Delta\alpha = 50\%$ and $\Delta\gamma = -50\%$.

It may be observed that concerning the position, the overshoot and the accuracy are within the required performance. Similarly, the control of the vertically descent of the spacecraft designed by our approach requires less control effort than the control law based on the differential algebraic approach designed in (Sira-Ramirez, 1992) in the case of the system under uncertainties.

4. CONCLUSION

In this paper, a possible new idea for robustness analysis and synthesis for uncertain nonlinear systems has been proposed. In order to find a robust nonlinear control, it has been shown that one may check a Lyapunov function and feedback control law that render the closed loop system dissipative.

As application and in order to point out the fact that this approach brings interesting results and more insight to some problems, stabilization of a spacecraft during vertically descent on the surface of a planet has been described.

For nominal system, the obtained simulation results are in concordance with the required performance. It is shown that the convergence of the nonlinear system towards an equilibrium point, requires less control effort than the control law based on the differential algebraic approach.

In summary, this paper emphasize that it is indeed possible to render a class of uncertain nonlinear systems robustly stable without solving any HJI thus to circumvent the computation difficulty of this one. This idea has never been exploited in the literature, to our best knowledge.

REFERENCES

Henni, A. and H. Siguerdidjane (2003). A possible new way for robustness analysis and synthesis of uncertain nonlinear system. *Short paper submitted to the IEEE Conference on Decision and Control.*

Huang, D. J. and W.M. Lu (1996). Nonlinear optimal control: Alternatives to hamilton-jacobi equation. *Proceedings of IEEE Conference on Decision and Control.*

Lu, W.M. and J.C. Doyle (1994). Robustness analysis and synthesis for nonlinear uncertain systems. *Proceedings of the 33rd Conference on Decision and Control* pp. 787–792.

Lu, W.M. and J.C. Doyle (1997). Robustness analysis and synthesis for nonlinear uncertain systems. *IEEE Transactions on Automatic Control* **42**(12), 1654–1662.

Sira-Ramirez, H. (1990). A variable structure control approach to the problem of soft landing on a planet. *Contr. Ttheory Adv. Technol.* **6**(1), 53–73.

Sira-Ramirez, H. (1992). The differential algebraic approach in nonlinear dynamical feedback controlled landing maneuvers. *IEEE Transactions on Automatic Control* **37**(4), 518–524.

van der Schaft, A. J. (1992). L $_2$-gain analysis of nonlinear systems and nonlinear state feedback H_∞ control. *IEEE Transactions on Automatic Control* **37**(6), 770–784.

QUADRATICALLY STABILIZED DISCRETE-TIME ROBUST LQ CONTROL

Dušan Krokavec, Anna Filasová

Technical University of Košice,
Faculty of Electrical Engineering and Informatics
Department of Cybernetics and Artificial Intelligence
Letná 9/B, 042 00 Košice, Slovak Republic
e-mail: krokavec@tuke.sk, filasova@tuke.sk

Abstract: The purpose of the paper is to present a class of algorithms used to solve the optimization problem concerning with discrete-time LQ control for linear systems with norm-bounded parameter uncertainties as well as a method for the problem reducing to a standard formulation but with the necessary modifications of the performance index weighting matrices. Some generalized considerations for the algorithm procedures are given and the problem of equivalent control is outlined. The procedure results the constant feedback for the linear controller defined in terms of recursive solution of a Riccati equation. *Copyright © 2003 IFAC*

Keywords: Discrete-time systems, LQR control methods, robust control, uncertain linear systems.

1. INTRODUCTION

All real systems operate in a stochastic environment where they are subject to unknown disturbances and in addition, the controller has to rely, in practice, on imperfect measurements. One of the principal reasons for introducing feedback into a control system is to obtain relative insensibility to changes in plant parameters and to disturbances. It is well-known fact that linear quadratic (LQ) optimal control yields a stable closed-loop system and the minimal value of the criterion for any initial (non-zero) condition. The disadvantage of applying LQ controllers, however, is that the performance of this controllers, based upon nominal parameter values, is severally degraded in the presence of parameter deviations. In order to overcome this, some information about system parameter uncertainties can be used in properly chosen performance index weighting matrices and the principle can be combined with the optimal stochastic state estimation based on the famous recursive Kalman filters.

In the last years many significant results have spurred interest in problem of determining a linear quadratic optimal control law for a linear system containing time-varying uncertainty; e.g. see (Neto, Dion and Dugard, 1992) and references therein. One approach to the problem of finding the optimal results is technique based on norm bounded uncertainties.

In this note one class of algorithms to solve the optimization problem concerning with discrete-time LQ control for linear systems with norm-bounded parameter uncertainty in the state and input matrices is considered. An interesting point is that presented robust LQ control has the essential structure of a standard LQ control but with the necessary modifications of the performance index weighting matrices to account for the parameter uncertainty. Used principle can be viewed as dual problem to those methods applied for uncertain discrete-time systems in (de Souza, Fu and Xie, 1993), (Xie, Soh and de Souza, 1994).

2. LQ CONTROL PRINCIPLE

The systems under consideration are uncertain discrete-time linear dynamic systems

$$\mathbf{x}(i+1) = \\ = (\mathbf{F}+\Delta\mathbf{F}(i))\mathbf{x}(i) + (\mathbf{G}+\Delta\mathbf{G}(i))\mathbf{u}(i) \quad (1)$$

with system measurement given by

$$\mathbf{y}(i) = \mathbf{C}\mathbf{x}(i) \quad (2)$$

where $\mathbf{x}(i) \in I\!R^n$, $\mathbf{u}(i) \in I\!R^r$, $\mathbf{y}(i) \in I\!R^m$, matrices $\mathbf{F} \in I\!R^{n\times n}$, $\mathbf{G} \in I\!R^{n\times r}$, $\mathbf{C} \in I\!R^{m\times n}$ are finite valued, and $\Delta\mathbf{F}(i)$, $\Delta\mathbf{G}(i)$ are unknown matrices which represent time-varying parametric uncertainties. It is assumed that considered uncertainty matrices to be of the form

$$\left[\,\Delta\mathbf{F}(i)\ \Delta\mathbf{G}(i)\,\right] = \mathbf{M}\mathbf{H}(i)\left[\,\mathbf{N}_1\ \mathbf{N}_2\,\right] \quad (3)$$

$$\mathbf{H}(i)\mathbf{H}^T(i) \le \mathbf{I} \quad (4)$$

$\mathbf{N}_1 \in I\!R^{h\times n}$, $\mathbf{N}_2 \in I\!R^{h\times r}$ and $\mathbf{M} \in I\!R^{n\times h}$ are known constant matrices which specify which elements of the nominal matrices $\mathbf{F}$ and $\mathbf{G}$ are affected by uncertain parameters in unknown matrix $\mathbf{H}(i) \in I\!R^{h\times h}$.

For system (1), (2) the optimal control design task is, in general, to determine the control

$$\mathbf{u}(i) = -\mathbf{K}(i)\mathbf{x}(i) \quad (5)$$

that minimizes the quadratic cost function

$$J_N = \mathbf{x}^T(N)\mathbf{Q}^{\bullet}\mathbf{x}(N) + \sum_{i=0}^{N-1}\mathbf{x}^T(i)\mathbf{J}_J(i)\mathbf{x}(i) \quad (6)$$

$$\mathbf{J}_J(i) = \mathbf{Q}_0 + \mathbf{J}_{JK}(i) \quad (7)$$

$$\mathbf{J}_{JK}(i) = \\ = -\mathbf{S}_0\mathbf{K}(i) - \mathbf{K}^T(i)\mathbf{S}_0^T + \mathbf{K}^T(i)\mathbf{R}_0\mathbf{K}(i) \quad (8)$$

where N is finite, $\mathbf{Q}_0 \in I\!R^{n\times n}$ and $\mathbf{Q}^{\bullet} \in I\!R^{n\times n}$ are symmetric positive semi-definite matrices, $\mathbf{R}_0 \in I\!R^{m\times m}$ is a symmetric positive definite matrix, and $\mathbf{K}(i) \in I\!R^{n\times r}$ is the optimal control gain matrix.

For the best obtainable Lyapunov function

$$V(\mathbf{x}(i)) = \mathbf{x}^T(i)\mathbf{P}(i-1)\mathbf{x}(i) \quad (9)$$

($\mathbf{P}(-1) = \mathbf{P}(0)$ and $\mathbf{P}(i-1) \in I\!R^{n\times n}$ is a symmetric positive definite matrix) the difference $\Delta V(\mathbf{x}(i)) = V(\mathbf{x}(i+1)) - V(\mathbf{x}(i))$ is given by

$$\Delta V(\mathbf{x}(i)) = \mathbf{x}^T(t)(\mathbf{J}_{V0}(i) + \mathbf{J}_{VK}(i))\mathbf{x}(i) \quad (10)$$

$$\mathbf{J}_{V0}(i) = -\mathbf{P}(i-1) + \mathbf{F}^T\mathbf{P}(i)\mathbf{F} + \\ +\Delta\mathbf{F}^T(i)\mathbf{P}(i)\mathbf{F} + \mathbf{F}^T\mathbf{P}(i)\Delta\mathbf{F}(i) + \\ +\Delta\mathbf{F}^T(i)\mathbf{P}(i)\Delta\mathbf{F}(i) \quad (11)$$

$$\mathbf{J}_{VK}(i) = \\ = -(\mathbf{F}+\Delta\mathbf{F}(i))^T\mathbf{P}(i)(\mathbf{G}+\Delta\mathbf{G}(i))\mathbf{K}(i) - \\ -\mathbf{K}^T(i)(\mathbf{G}+\Delta\mathbf{G}(i))^T\mathbf{P}(i)(\mathbf{F}+\Delta\mathbf{F}(i)) + \\ +\mathbf{K}^T(i)(\mathbf{G}+\Delta\mathbf{G}(i))^T\mathbf{P}(i)(\mathbf{G}+\Delta\mathbf{G}(i))\mathbf{K}(i) \quad (12)$$

and the Lyapunov function at the step N is

$$V(\mathbf{x}(N)) = \sum_{i=0}^{N-1}\Delta V(\mathbf{x}(i)) = \\ = \mathbf{x}^T(N)\mathbf{P}(N-1)\mathbf{x}(N) - \mathbf{x}^T(0)\mathbf{P}(0)\mathbf{x}(0) \quad (13)$$

$$V(\mathbf{x}(N)) = \sum_{i=0}^{N-1}\mathbf{x}^T(i)\begin{pmatrix}\mathbf{J}_{V0}(i)+\\ +\mathbf{J}_{VK}(i)\end{pmatrix}\mathbf{x}(i) \quad (14)$$

respectively. Adding (14) and subtracting (13) to (6) the cost function for control law specified in equation (5) can be brought to the form

$$J_N = \mathbf{x}^T(0)\mathbf{P}(0)\mathbf{x}(0) + \sum_{i=0}^{n-1}\mathbf{x}^T(i)\mathbf{J}(i)\mathbf{x}(i) \quad (15)$$

$$\mathbf{P}(N-1) = \mathbf{Q}^{\bullet} \quad (16)$$

$$\mathbf{J}(i) = \mathbf{Q}_0 + \mathbf{J}_{JK}(i) + \mathbf{J}_{V0}(i) + \mathbf{J}_{VK}(i) \quad (17)$$

3. QUADRATIC STABILIZATION

Denote the gain independent elements $\mathbf{J}_0(i) = \mathbf{Q}_0 + \mathbf{J}_{V0}(i)$ of (17) as $\mathbf{J}_0(i) = \mathbf{J}_{01}(i) + \mathbf{J}_{02}(i)$, where

$$\mathbf{J}_{01}(i) = \mathbf{Q}_0 - \mathbf{P}(i-1) + \mathbf{F}^T\mathbf{P}(i)\mathbf{F} + \\ +\Delta\mathbf{F}^T(i)\mathbf{P}(i)\mathbf{F} + \mathbf{F}^T\mathbf{P}(i)\Delta\mathbf{F}(i) \quad (18)$$

$$\mathbf{J}_{02}(i) = \Delta\mathbf{F}^T(i)\mathbf{P}(i)\Delta\mathbf{F}(i) \quad (19)$$

Let at the first approximation step the second order differences are neglected. Since immediately follows from (3)

$$\Delta\mathbf{F}(i) = \mathbf{M}\mathbf{H}(i)\mathbf{N}_1 \quad (20)$$

using condition (4) and the identity

$$\mathbf{A}\mathbf{B}^T + \mathbf{B}\mathbf{A}^T \le \mathbf{A}\mathbf{A}^T + \mathbf{B}\mathbf{B}^T \quad (21)$$

(18) takes the form

$$\mathbf{J}_{01}(i) = \mathbf{Q}_0 - \\ -\mathbf{P}(i-1) + \frac{\sqrt{\varepsilon}}{\sqrt{\varepsilon}}(\mathbf{M}\mathbf{H}(i)\mathbf{N}_1)^T\mathbf{P}(i)\mathbf{F} + \\ +\mathbf{F}^T\mathbf{P}(i)\mathbf{F} + \frac{\sqrt{\varepsilon}}{\sqrt{\varepsilon}}\mathbf{F}^T\mathbf{P}(i)\mathbf{M}\mathbf{H}(i)\mathbf{N}_1 \le \\ \le \mathbf{Q}_0 - \mathbf{P}(i-1) + \frac{1}{\varepsilon}\mathbf{N}_1^T\mathbf{N}_1 + \\ +\mathbf{F}^T(\mathbf{P}(i) + \mathbf{P}(i)\varepsilon\mathbf{M}\mathbf{M}^T\mathbf{P}(i))\mathbf{F} \quad (22)$$

Denoting

$$\mathbf{Q}^* = \frac{1}{\varepsilon}\mathbf{N}_1^T\mathbf{N}_1, \quad \mathbf{M}^* = \varepsilon\mathbf{M}\mathbf{M}^T \\ \mathbf{Q}_1 = \mathbf{Q}_0 + \mathbf{Q}^* \tag{23}$$

since $\mathbf{J}_{01}(i)$ is a positive semi-definite matrix for steady-state solution be

$$\mathbf{P} - \mathbf{Q}_1 \le \mathbf{F}^T(\mathbf{P} + \mathbf{P}\mathbf{M}^*\mathbf{P})\mathbf{F} \tag{24}$$

$$\mathbf{F}(\mathbf{P} - \mathbf{Q}_1)^{-1}\mathbf{F}^T \le (\mathbf{P} + \mathbf{P}\mathbf{M}^*\mathbf{P})^{-1} \tag{25}$$

Using the identity

$$(\mathbf{A} + \mathbf{B}\mathbf{D}\mathbf{B}^T)^{-1} = \mathbf{A}^{-1} - \\ -\mathbf{A}^{-1}\mathbf{B}(\mathbf{D}^{-1} + \mathbf{B}^T\mathbf{A}^{-1}\mathbf{B})^{-1}\mathbf{B}^T\mathbf{A}^{-1} \tag{26}$$

an upper bound of (25) is

$$(\mathbf{P} + \mathbf{P}\mathbf{M}^*\mathbf{P})^{-1} = \\ = \mathbf{P}^{-1} - (\mathbf{M}^{-*} + \mathbf{P})^{-1} < \mathbf{P}^{-1} - \mathbf{M}^* \tag{27}$$

$$\mathbf{F}(\mathbf{P}^{-1} - \mathbf{P}^{-1}(\mathbf{P}^{-1} - \mathbf{Q}_1^{-1})^{-1}\mathbf{P}^{-1})\mathbf{F}^T < \\ < \mathbf{P}^{-1} - \mathbf{M}^* \tag{28}$$

Thus

$$\mathbf{F}\mathbf{P}^{-1}\mathbf{F}^T - \mathbf{P}^{-1} + \mathbf{M}^* + \\ +\mathbf{F}\mathbf{P}^{-1}\mathbf{Q}_1(\mathbf{P} - \mathbf{Q}_1)^{-1}\mathbf{F}^T = \\ = \mathbf{F}\mathbf{P}^{-1}\mathbf{F}^T - \mathbf{P}^{-1} + \mathbf{M}^* + \\ +\mathbf{F}(\mathbf{P} - \mathbf{Q}_1)^{-1}\mathbf{Q}_1\mathbf{P}^{-1}\mathbf{F}^T < 0 \tag{29}$$

and for

$$\Delta\mathbf{F}_e^T = \frac{1}{2}\mathbf{Q}_1(\mathbf{P} - \mathbf{Q}_1)^{-1}\mathbf{F}^T \tag{30}$$

(29) takes the form

$$\mathbf{F}\mathbf{P}^{-1}\mathbf{F}^T - \mathbf{P}^{-1} + \mathbf{M}^* + \\ +\mathbf{F}\mathbf{P}^{-1}\Delta\mathbf{F}_e + \Delta\mathbf{F}_e^T\mathbf{P}^{-1}\mathbf{F}^T < 0 \tag{31}$$

$$(\mathbf{F} + \Delta\mathbf{F}_e)\mathbf{P}^{-1}(\mathbf{F} + \Delta\mathbf{F}_e)^T - \mathbf{P}^{-1} + \\ +\mathbf{M}^* - \Delta\mathbf{F}_e\mathbf{P}^{-1}\Delta\mathbf{F}_e^T < 0 \tag{32}$$

respectively. Since $\mathbf{P}$ is positive definite, it is evident that for $\mathbf{M}^* \le \Delta\mathbf{F}_e\mathbf{P}^{-1}\Delta\mathbf{F}_e^T$ equivalent dual autonomous system with the system matrix $(\mathbf{F} + \Delta\mathbf{F}_e)^T$ is quadratically stabile. Then for $\mathbf{F}_e = \mathbf{F} + \Delta\mathbf{F}_e$ takes $\mathbf{J}_0(i) = \mathbf{Q}_0 + \mathbf{J}_{V0}(i)$ form

$$\mathbf{J}_0(i) = \mathbf{Q}_0 - \mathbf{P}(i-1) + \\ +(\mathbf{F}_e + (\Delta\mathbf{F}(i) - \Delta\mathbf{F}_e))^T\mathbf{P}(i)(\mathbf{F}_e + (\Delta\mathbf{F}(i) - \Delta\mathbf{F}_e)) \tag{33}$$

Since (30) approximates first order differences in (33) the rest task is to obtain second order difference upper-bounds in steady-state. Thus

$$\mathbf{J}_{0a} = (\Delta\mathbf{F}(i) - \Delta\mathbf{F}_e)^T\mathbf{P}(i)(\Delta\mathbf{F}(i) - \Delta\mathbf{F}_e) = \\ = \Delta\mathbf{F}_e^T\mathbf{P}(i)\Delta\mathbf{F}_e - \frac{\sqrt{\varepsilon}}{\sqrt{\varepsilon}}\Delta\mathbf{F}_e^T\mathbf{P}(i)\mathbf{M}\mathbf{H}(i)\mathbf{N}_1 - \\ -\frac{\sqrt{\varepsilon}}{\sqrt{\varepsilon}}(\mathbf{M}\mathbf{H}(i)\mathbf{N}_1)^T\mathbf{P}(i)\Delta\mathbf{F}_e + \Delta\mathbf{F}^T(i)\mathbf{P}(i)\Delta\mathbf{F}(i) \tag{34}$$

and with (24) at steady-state be

$$\mathbf{J}_{0a} \le \\ \le \Delta\mathbf{F}_e^T(\mathbf{P} + \mathbf{P}\varepsilon\mathbf{M}\mathbf{M}^T\mathbf{P})\Delta\mathbf{F}_e + \frac{1}{\varepsilon}\mathbf{N}_1^T\mathbf{N}_1 \le \\ \le \Delta\mathbf{F}_e^T\mathbf{F}^{-T}(\mathbf{P} - \mathbf{Q}_1)\mathbf{F}^{-1}\Delta\mathbf{F}_e + \mathbf{Q}^* = \\ = \mathbf{Q}^* + \frac{1}{4}\mathbf{Q}_1(\mathbf{P} - \mathbf{Q}_1)^{-1}\mathbf{Q}_1 = \mathbf{Q}_e \tag{35}$$

Thus, the upper-bound of (33) may be written as

$$\mathbf{J}_0 = \mathbf{Q}_e - \mathbf{P}(i-1) + \mathbf{F}_e^T\mathbf{P}(i)\mathbf{F}_e \tag{36}$$

4. REST UNCERTAINTY BOUNDS

Separate the gain dependent elements $\mathbf{J}_{KC}(i) = \mathbf{J}_{JK}(i) + \mathbf{J}_{VK}(i)$ of (17) as $\mathbf{J}_{KC}(i) = -\mathbf{J}_K(i) - -\mathbf{J}^K(i) + \mathbf{J}_K^K(i)$, where

$$\mathbf{J}_K(i) = \mathbf{J}^{KT}(i) = \\ = (\mathbf{S}_0 + \mathbf{F} + \Delta\mathbf{F}(i))^T\mathbf{P}(i)(\mathbf{G} + \Delta\mathbf{G}(i))\mathbf{K}(i) \tag{37}$$

$$\mathbf{J}_K^K(i) = \mathbf{K}^T(i)(\mathbf{R}_0 + \mathbf{G} + \Delta\mathbf{G}(i))^T. \\ .\mathbf{P}(i)(\mathbf{G} + \Delta\mathbf{G}(i))\mathbf{K}(i) \tag{38}$$

Starting with (3) one can obtain

$$\Delta\mathbf{G}(i) = \Delta\mathbf{F}(i)\mathbf{N}_1^{-1}\mathbf{N}_2 \tag{39}$$

$$\Delta\mathbf{G}_e = \Delta\mathbf{F}_e\mathbf{N}_1^{-1}\mathbf{N}_2 \tag{40}$$

where $\mathbf{N}_1^{-1}$ is the Penrose inverse of $\mathbf{N}_1$. This gives for (37), using (34), (35) under the same assumptions as above, formulae

$$\mathbf{J}_K(i) \le (\mathbf{S}_0 + \mathbf{F}_e^T\mathbf{P}(i)\mathbf{G}_e)\mathbf{K}(i) + \\ +(\Delta\mathbf{F}(i) - \Delta\mathbf{F}_e)^T\mathbf{P}(i). \\ .(\Delta\mathbf{F}(i) - \Delta\mathbf{F}_e)\mathbf{N}_1^{-1}\mathbf{N}_2\mathbf{K}(i) \le \\ \le (\mathbf{S}_0 + \mathbf{F}_e^T\mathbf{P}(i)\mathbf{G}_e)\mathbf{K}(i) + \\ +(\mathbf{Q}^* + \frac{1}{4}\mathbf{Q}_1(\mathbf{P} - \mathbf{Q}_1)^{-1}\mathbf{Q}_1)\mathbf{N}_1^{-1}\mathbf{N}_2\mathbf{K}(i) \tag{41}$$

which is equivalent to

$$\mathbf{J}_K(i) \le (\mathbf{S}_e + \mathbf{F}_e^T\mathbf{P}(i)\mathbf{G}_e)\mathbf{K}(i) \tag{42}$$

$$\mathbf{S}_e = \mathbf{S}_0 + \\ +(\mathbf{Q}^* + \frac{1}{4}\mathbf{Q}_1(\mathbf{P} - \mathbf{Q}_1)^{-1}\mathbf{Q}_1)\mathbf{N}_1^{-1}\mathbf{N}_2 \tag{43}$$

By the same way (38) may be rewritten as

$$\mathbf{J}_K^K(i) \le \mathbf{K}^T(i)(\mathbf{R}_0 + \mathbf{G}_e^T\mathbf{P}(i)\mathbf{G}_e)\mathbf{K}(i) + \\ +\mathbf{K}^T(i)\mathbf{N}_2^T\mathbf{N}_1^{-T}(\Delta\mathbf{F}(i) - \Delta\mathbf{F}_e)^T\mathbf{P}(i). \\ .(\Delta\mathbf{F}(i) - \Delta\mathbf{F}(i))\mathbf{N}_1^{-1}\mathbf{N}_2\mathbf{K}(i) \tag{44}$$

and with (34), (35) acceptation

$$\mathbf{J}_K^K(i) \le \mathbf{K}^T(i)(\mathbf{R}_e + \mathbf{G}_e^T\mathbf{P}(i)\mathbf{G}_e)\mathbf{K}(i) \tag{45}$$

$$\mathbf{R}_e = \mathbf{R}_0 + \\ +\mathbf{N}_2^T\mathbf{N}_1^{-T}(\mathbf{Q}^* + \frac{1}{4}\mathbf{Q}_1(\mathbf{P} - \mathbf{Q}_1)^{-1}\mathbf{Q}_1)\mathbf{N}_1^{-1}\mathbf{N}_2 \tag{46}$$

5. QUADRATIC OPTIMIZATION

Accepting all obtained approximation the resulting upper-bound of (17) may be written as

$$\begin{aligned} \mathbf{J}(i) \leq \mathbf{Q}_e - \mathbf{P}(i-1) + \mathbf{F}_e^T\mathbf{P}(i)\mathbf{F}_e - \\ -(\mathbf{S}_e + \mathbf{F}_e^T\mathbf{P}(i)\mathbf{G}_e)\mathbf{K}(i) - \\ -\mathbf{K}^T(i)(\mathbf{S}_e + \mathbf{F}_e^T\mathbf{P}(i)\mathbf{G}_e)^T + \\ +\mathbf{K}^T(i)(\mathbf{R}_e + \mathbf{G}_e^T\mathbf{P}(i)\mathbf{G}_e)\mathbf{K}(i) = \mathbf{J}_u(i) \end{aligned} \quad (47)$$

where

$$\mathbf{F}_e = \mathbf{F} + \Delta\mathbf{F}_e,\ \Delta\mathbf{F}_e = \frac{1}{2}\mathbf{F}(\mathbf{P}-\mathbf{Q}_1)^{-1}\mathbf{Q}_1 \quad (48)$$

$$\mathbf{G}_e = \mathbf{G} + \Delta\mathbf{G}_e,\ \Delta\mathbf{G}_e = \Delta\mathbf{F}_e\mathbf{N}_1^{-1}\mathbf{N}_2 \quad (49)$$

$$\mathbf{Q}_1 = \mathbf{Q}_0 + \mathbf{Q}^*,\ \mathbf{Q}_{1e} = \frac{1}{4}\mathbf{Q}_1(\mathbf{P} - \mathbf{Q}_1)^{-1}\mathbf{Q}_1 \quad (50)$$

$$\begin{aligned} \mathbf{Q}^* = \frac{1}{\varepsilon}\mathbf{N}_1^T\mathbf{N}_1 \quad \mathbf{R}^* = \frac{1}{\varepsilon}\mathbf{N}_2^T\mathbf{N}_2 \\ \mathbf{S}^* = \frac{1}{\varepsilon}\mathbf{N}_1^T\mathbf{N}_2 \quad \mathbf{M}^* = \varepsilon\mathbf{M}\mathbf{M}^T \end{aligned} \quad (51)$$

$$\mathbf{Q}_e = \mathbf{Q}_0 + \mathbf{Q}^* + \mathbf{Q}_{1e} \quad (52)$$

$$\mathbf{S}_e = \mathbf{S}_0 + \mathbf{S}^* + \mathbf{Q}_{1e}\mathbf{N}_1^{-1}\mathbf{N}_2 \quad (53)$$

$$\mathbf{R}_e = \mathbf{R}_0 + \mathbf{R}^* + \mathbf{N}_2^T\mathbf{N}_1^{-T}\mathbf{Q}_{1e}\mathbf{N}_1^{-1}\mathbf{N}_2 \quad (54)$$

$\mathbf{J}_u(i)$ is a symmetric upper bound of the weighting matrix of the modified quadratic criterion (15) for LQ control of uncertain systems (1), (2), (3), (4), $\mathbf{N}_1^{-1}$ is the Penrose inverse of $\mathbf{N}_1$, $(\mathbf{Q}_0, \mathbf{R}_0, \mathbf{S}_0)$ are nominal performance index weighting matrices and $\varepsilon \in (0,\ 1 >$ is a design parameter.

The minimal value of (47) is obtained if

$$\begin{aligned} \mathbf{K}^T(i)(\mathbf{S}_e + \mathbf{F}_e^T\mathbf{P}(i)\mathbf{G}_e)^T + \\ +\mathbf{K}^T(i)(\mathbf{R}_e + \mathbf{G}_e^T\mathbf{P}(i)\mathbf{G}_e)\mathbf{K}(i) = \mathbf{0} \end{aligned} \quad (55)$$

i.e. gain matrix $\mathbf{K}(i)$ of the control law

$$\mathbf{u}(i) = -\mathbf{K}(i)\mathbf{x}(i) \quad (56)$$

is given by the term

$$\begin{aligned} \mathbf{K}(i) = \\ = (\mathbf{R}_e + \mathbf{G}_e^T\mathbf{P}(i)\mathbf{G}_e)^{-1}(\mathbf{G}_e^T\mathbf{P}(i)\mathbf{F}_e + \mathbf{S}_e^T) \end{aligned} \quad (57)$$

and $\mathbf{P}(i)$ is a solution of the Riccati equation

$$\begin{aligned} \mathbf{0} = \mathbf{Q}_e - \mathbf{P}(i-1) + \mathbf{F}_e^T\mathbf{P}(i)\mathbf{F}_e - \\ -(\mathbf{S}_e + \mathbf{F}_e^T\mathbf{P}(i)\mathbf{G}_e)\mathbf{K}(i) \end{aligned} \quad (58)$$

The constant gain state feedback controller for infinity control time and control law

$$\mathbf{u}(i) = -\mathbf{K}\mathbf{x}(i) \quad (59)$$

is given by the steady-solution solution of (57), (58), i.e. by the solution of the algebraic Riccati equation.

Presented LQ control is robust for defined structure of uncertainty in system and input matrices.

6. ILLUSTRATIVE EXAMPLE

The dynamic system of a helicopter in a vertical plain (Schmitendorf, 1988), (Krokavec and Filasová, 2000) for defined range of operating condition is described for sampling period $T_s = 0.1s$ by the discrete-time state-space equations (1), (3) and (4) where

$$\mathbf{F} = \begin{bmatrix} 0.9964 & 0.0026 & -0.0004 & -0.0460 \\ 0.0045 & 0.9038 & -0.0188 & -0.3834 \\ 0.0097 & 0.0263 & 0.9379 & 0.1223 \\ 0.0005 & 0.0014 & 0.0968 & 1.0063 \end{bmatrix}$$

$$\Delta\mathbf{F}(i) = \begin{bmatrix} 0 & 0 & 0 & 0 \\ 0 & 0 & 0 & 0 \\ 0 & f_{32} & 0 & f_{34} \\ 0 & 0 & 0 & 0 \end{bmatrix}$$

$$\mathbf{G} = \begin{bmatrix} 0.0444 & 0.0167 \\ 0.2932 & -0.7252 \\ -0.5298 & 0.4726 \\ -0.0268 & 0.0241 \end{bmatrix},\ \Delta\mathbf{G}(i) = \begin{bmatrix} 0 & 0 \\ g_{21} & 0 \\ g_{31} & 0 \\ 0 & 0 \end{bmatrix}$$

and $|f_{32}| \leq 0.0219$, $|f_{34}| \leq 0.1203$, $|g_{21}| \leq 0.1966$, $|g_{31}| < 0.0028$. Autonomous nominal system is unstable with system matrix eigenvalues $\lambda_1 = 0.8203$, $\lambda_2 = 0.9673$, $\lambda_{3,4} = 1.0284 \pm 0.0102i$.

With $\mathbf{Q}_0 = \mathbf{I}$, $\mathbf{R}_0 = \mathbf{I}$, $\mathbf{S}_0 = \mathbf{0}$ the steady-state solution of Riccati equation for nominal system parameters is obtained as

$$\mathbf{P}_N = \begin{bmatrix} 22.7202 & 1.0600 & 0.5927 & -9.2225 \\ 1.0600 & 2.5246 & 1.2060 & -0.3770 \\ 0.5627 & 1.2060 & 3.0532 & 1.6414 \\ -9.2225 & -0.3770 & 1.6414 & 18.9653 \end{bmatrix}$$

the optimal control gain matrix is

$$\mathbf{K}_N = \begin{bmatrix} 0.7111 & -0.1662 & -0.7227 & -0.9722 \\ 0.0902 & -0.5560 & 0.0877 & 0.5419 \end{bmatrix}$$

and the eigenvalues of the nominal closed-loop system are $\lambda_1 = 0.3471$, $\lambda_2 = 0.9294$, $\lambda_{3,4} = 0.8585 \pm 0.0362i$.

One can verify that parametric uncertainties with above given structures significantly change the dominant closed-loop eigenvalue type.

To use presented technique let

$$\begin{bmatrix} \Delta\mathbf{F}(i) & \Delta\mathbf{G}(i) \end{bmatrix} = \mathbf{M}\mathbf{H}(i)\begin{bmatrix} \mathbf{N}_1 & \mathbf{N}_2 \end{bmatrix}$$

where

$$\mathbf{N}_1 = \begin{bmatrix} 0 & 0 & 0 & 0 \\ 0 & 0.0219 & 0 & 0.1203 \end{bmatrix}$$

$$\mathbf{N}_2 = \begin{bmatrix} 0.1966 & 0 \\ 0.0028 & 0 \end{bmatrix},\ \mathbf{M} = \begin{bmatrix} 0 & 0 \\ 1 & 0 \\ 0 & 1 \\ 0 & 0 \end{bmatrix}$$

Using $\varepsilon = 0.1$ the matrices of the uncertainty bounds are

$$\mathbf{Q}^* = \frac{1}{\varepsilon}\mathbf{N}_1^T\mathbf{N}_1 = \begin{bmatrix} 0 & 0 & 0 & 0 \\ 0 & 0.0048 & 0 & 0.0263 \\ 0 & 0 & 0 & 0 \\ 0 & 0.0263 & 0 & 0.1447 \end{bmatrix}$$

$$\mathbf{S}^* = \frac{1}{\varepsilon}\mathbf{N}_1^T\mathbf{N}_2 = \begin{bmatrix} 0 & 0 \\ 0.0006 & 0 \\ 0 & 0 \\ 0.0034 & 0 \end{bmatrix}$$

$$\mathbf{R}^* = \frac{1}{\varepsilon}\mathbf{N}_2^T\mathbf{N}_2 = \begin{bmatrix} 0.3866 & 0 \\ 0 & 0 \end{bmatrix}$$

$$\mathbf{M}^* = \varepsilon\mathbf{M}\mathbf{M}^T = \begin{bmatrix} 0 & 0 & 0 & 0 \\ 0 & 0.1 & 0 & 0 \\ 0 & 0 & 0.1 & 0 \\ 0 & 0 & 0 & 0 \end{bmatrix}$$

Starting with $\mathbf{Q}_0 = \mathbf{I}$, $\mathbf{R}_0 = \mathbf{I}$, $\mathbf{S}_0 = \mathbf{0}$ and using $\mathbf{P}_N$ as a value for matrix $\mathbf{Q}_{1e}$ initialization, the Riccati equation (58) has positive definite solution

$$\mathbf{P} = \begin{bmatrix} 43.6425 & 2.5918 & 1.8503 & -17.4444 \\ 2.5918 & 4.1461 & 2.2133 & -0.4052 \\ 1.8503 & 2.2133 & 5.2467 & 2.7281 \\ -17.4444 & -0.4052 & 2.7281 & 35.7182 \end{bmatrix}$$

where a simple iterative procedure was designed to obtain this using the standard MATLAB function *dare*($\mathbf{F},\mathbf{G},\mathbf{Q},\mathbf{R},\mathbf{S}$).

Thus, the modified optimal criterion parameters are

$$\mathbf{Q}_{1e} = \begin{bmatrix} 0.0080 & -0.0024 & -0.0051 & 0.0050 \\ -0.0024 & 0.1351 & -0.0731 & 0.0109 \\ -0.0051 & -0.0731 & 0.1065 & -0.0155 \\ 0.0050 & 0.0109 & -0.0155 & 0.0142 \end{bmatrix}$$

$$\mathbf{Q}_e = \begin{bmatrix} 1.0080 & -0.0024 & -0.0051 & 0.0050 \\ -0.0024 & 1.1399 & -0.0731 & 0.0373 \\ -0.0051 & -0.0731 & 1.1065 & -0.0155 \\ 0.0050 & 0.0373 & -0.0155 & 1.1589 \end{bmatrix}$$

$$\mathbf{S}_e = \begin{bmatrix} 0.0001 & 0 \\ 0.0014 & 0 \\ -0.0006 & 0 \\ 0.0037 & 0 \end{bmatrix}, \quad \mathbf{R}_e = \begin{bmatrix} 1.3866 & 0 \\ 0 & 1 \end{bmatrix}$$

and the structures of equivalent matrix differences are of the form

$$\Delta\mathbf{F}_e = \begin{bmatrix} 0.0156 & -0.0045 & -0.0095 & 0.0089 \\ -0.0076 & 0.2406 & -0.1259 & 0.0104 \\ -0.0084 & -0.1286 & 0.1929 & -0.0254 \\ 0.0079 & -0.0008 & -0.0035 & 0.0215 \end{bmatrix}$$

$$\mathbf{G}_e = \begin{bmatrix} 0.0002 & 0 \\ 0.0012 & 0 \\ -0.0011 & 0 \\ 0.0005 & 0 \end{bmatrix}$$

Then, from (57), optimal robust control law is $\mathbf{u}(t) = -\mathbf{K}\mathbf{x}(t)$ with

$$\mathbf{K} = \begin{bmatrix} 0.8378 & -0.2291 & -0.9292 & -1.0947 \\ 0.1040 & -0.8873 & 0.1716 & 0.6109 \end{bmatrix}$$

The eigenvalues of the closed-loop system for this robust feedback gain matrix and the nominal system parameters $\mathbf{F}, \mathbf{G}$ are $\lambda_1 = 0.0203$, $\lambda_2 = 0.7359$, $\lambda_{3,4} = 0.9277 \pm 0.0363i$ and shifts of the real part of the dominant eigenvalue for

$$\mathbf{H}(i) = \begin{bmatrix} a & 0 \\ 0 & b \end{bmatrix}, \quad a, b = -1, 0, 1$$

are given in Table 1.

$+0$	-0	$0+$	$0-$
0.0148	-0.0185	-0.0073	0.0031
$++$	$+-$	$-+$	$--$
0.0176	0.0136	-0.0084	-0.0455

Table 1. Dominant eigenvalue shift

One can verify that the matrix $\mathbf{M}^* - \Delta\mathbf{F}_e\mathbf{P}^{-1}\Delta\mathbf{F}_e^T$ is non-positive definite.

7. CONCLUDING REMARKS

The paper presents the basic design principle of a simple robust LQ controller for discrete-time linear multi-variable dynamic systems with norm-bounded uncertainties. The exposed problems was reduced to a standard formulation but with the necessary modifications of the performance index weighting matrices to account for the parameter uncertainty. Resulting control is robust for defined structure of uncertainties in system and input matrices.

Presented applications can be considered as a task concerned the class of quadratic stabilization control problems where the optimal solutions for uncertain discrete-time systems were formulated. The proposed method present some new design features and generalizations. It should be emphasized that the advantage offered by the proposed approach is in its computational simplicity.

An example is presented to demonstrate the role of uncertainty upper-bounds in the design procedure resulting in the performance index weighting matrices modification and, as consequence, in quadratic stabilization of equivalent dual autonomous system.

ACKNOWLEDGEMENTS

The work presented in this paper was supported by Grant Agency of Ministry of Education and Academy of Science of Slovak Republic VEGA under Grant No. 1/9028/02.

8. REFERENCES

Anderson, B.D.O. and J.B. Moore (1990). *Optimal Control. Linear Quadratic Methods.* Prentice Hall, Englewood Cliffs.

Filasová, A. (1997). Robust controller design for large-scale uncertain dynamic systems. In: *Preprints of the 2nd IFAC Workshop on New Trends in Design of Control Systems (Eds. Kozák, Š., Huba, M.).* Smolenice, Slovak Republic, pp. 427-432.

Krokavec, D. (1998). Robust state estimation for structured noise uncertainty. In: *Proceedings of the 12th Conference Process Control '99 (Eds. Mikleš, J., Dvoran, J., Krejči, S., Fikar, M.).* Tatranské Matliare, Slovak Republic, pp. 235-239.

Krokavec, D. and A. Filasová (2000). Unmatched uncertainties in robust LQ control. In: *Proceedings of the IFAC Symposium Robust Control Design 2000 (Eds. Kučera, V., Šebek, M.), Prague, Czech Republic.* Pergamon, Elsevier Science.

Krokavec, D. and A. Filasová (2002). *Optimal Stochastic Systems.* Elfa, Košice (in Slovak).

Neto, A.T., J.M. Dion and L. Dugard (1992). Robustness bounds for LQ regulators. *IEEE Trans. Automat. Contr.*, **37**, pp. 1373-1377.

Schmitendorf, W.E. (1988). Designing stabilizing controllers for uncertain systems using Riccati equation approach. *IEEE Trans. Automat. Contr.*, **33**, pp. 376-379.

de Souza, C.E., M. Fu and L. Xie (1993) H_∞ analysis and synthesis of discrete-time systems with time-varying uncertainty. *IEEE Trans. Automat. Contr.*, **38**, pp. 459-462.

Xie, L., C.E. de Souza and M. Fu (1991) H_∞ estimation for discrete-time linear uncertain systems. *Int. J. Robust and Nonlinear Contr.*, **1**, pp. 111-123.

Xie, L., Y.C. Soh and C.E. de Souza (1994) Robust Kalman filtering for uncertain discrete-time systems. *IEEE Trans. Automat. Contr.*, **39**, pp. 1310-1314.

HYBRID IMC DEAD-BEAT CONTROLLER DESIGN IN DELTA DOMAIN

Zs. Preitl*, R. Bars*, I. Vajk*, R. Haber**

**Department of Automation and Applied Informatics, Budapest University of Technology and Economics, MTA-BME Control Research Group*
H-1111, Budapest, Goldmann Gy. tér 3., Hungary; fax : +36-1-463-2871
e-mail : preitl@aut.bme.hu, bars@aut.bme.hu, vajk@aut.bme.hu, robert.haber@fh-koeln.de
***Department of Plant and Process Engineering, University of Applied Science Cologne, Germany*

Abstract: The paper presents a control system design technique in delta domain for IMC (Internal Model Control) structure. Also a generalised form for delta domain Dead-beat control algorithm is given, a hybrid implementation of the controller (Z-domain combined with delta domain) is presented and its architecture is compared with the pure delta-domain implementation. The effect of placing the limitations in the IMC structure is also studied. The hybrid control structure is compared with a similar Dead-beat controller in Z-domain for second and third order plants (benchmarks). An illustrative sensitivity analysis between the hybrid control system and the Z-domain system has been performed.

Keywords: Controller design, delta domain, generalised hybrid Dead-beat controller, sensitivity analysis

1. INTRODUCTION

Sampling in digital control systems may cause unwanted behaviour in some cases. Two aspects are to be highlighted:

- on one hand using the well-known shift operator (q)

$$qx_k = x_{k+1} \quad \text{or} \quad q^{-1}x_k = x_{k-1} \qquad (1)$$

a model is obtained which is much different from the d/dt time operator model.

- on the other hand the reduction of the stability region to the unit circle centred in the origin imposes a more precise description of the parameters (poles, zeros), often with 4 - 6 digits at least. In this sense the shift operator representation is more sensitive to changes in parameters than the continuous time representation. Also, if the h sampling time is too small, the use of the digital controller can lead to system sensitivity to the parameter changes [Istepanian *et. al.* 2001].

One way to overcome these disadvantages is the use of the *delta transformation* introduced in [Middleton *et. al.* 1990] by means of:

$$\delta = \frac{q-1}{h} \quad \text{or} \quad \gamma = \frac{z-1}{h} \qquad (2)$$

where q^{-1} is the shift operator in time domain, z^{-1} is its associated complex variable and γ is the variable associated to the delta operator.

Two properties of the delta transformation are very useful in controller design:
i) One that reflects the fact that in the case of the delta transformation the stability domain expands as the sampling period gets smaller;
ii) A second one that characterizes the way of pole transposition – particularly of the dominant complex conjugate poles of a second order continuous system - from Z-domain to delta.

For example in case of a second order continuous system with damping coefficient ξ=0.707 when using Z transformation the well-known „heart-shaped" territory limited inside the unit circle is obtained (see Fig.1). For the delta representation for h=1 it is shifted in the circle centered in (-1,0), and for h= 0.1 it becomes 10 times larger. By reducing the sampling period even more the poles of the system in delta domain get closer to the continuous poles, especially the dominant ones. This remark sustaines the idea of controller design techniques in delta domain being close to the design methods from continuous time.

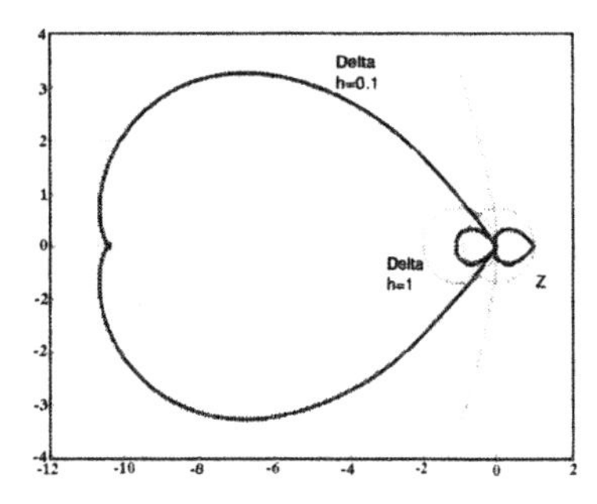

Fig.1 Dominant complex-conjugate poles locus in continuous (ξ=0.707), delta and Z domain

In [Goodwin *et. al.* 2001] there are presented different design techniques for the delta-controllers; they can be grouped into two different categories:

- one where the controller design is in continuous time, according to the continuous model of the plant followed by the replace of the variable s with the corresponding delta-domain variable γ;
- another where the direct controller design is in the delta domain, based on the delta model of the process; one remarkable advantage of this direct design is that delta domain model of the process already contains the zero-order-hold (ZOH). In this context, for a given continuous transfer function (abbreviated t.f.) $H(s)$, the delta t.f. $H(\gamma)$ is obtained using *generalised operational transform* [Middleton *et. al.* 1990] of form:

$$H(\gamma)=\frac{\gamma}{1+h\gamma}T[L^{-1}\{\frac{1}{s}H(s)\}] \tag{3}$$

The paper presents a hybrid design technique based on IMC control structure based on the delta model of the plant extended with the dead time represented directly in discrete domain (Z).

At first dead-beat control using IMC structure and a proposed hybrid delta and Z domain implementation is presented. Then the effect of limitations in the hybrid IMC structure is analysed. Finally in case of a second order plant an illustrative sensitivity analysis points out some useful conclusions, which sustains the viability of the proposed method.

Benchmark type plant model was used for the controller design. Simulation results exemplify the good results of the study.

2. DEAD-BEAT CONTROL USING IMC STRUCTURE AND HYBRID DELTA DOMAIN AND Z-DOMAIN IMPLEMENTATION

Let the t.f. of a continuous plant be of form:

$$H_P(s)=\frac{B_C(s)}{A_C(s)}\cdot e^{-s\cdot T_H} \tag{4}$$

Sampling it with a ZOH using a sampling time h the discrete t.f. is obtained

$$H_P(z)=\frac{B_D(z)}{A_D(z)}\cdot z^{-d} \quad \text{where} \quad d=T_H/h. \tag{5}$$

The corresponding delta t.f., $H_P(\gamma)$, results in form of:

$$H_P(\gamma)=\frac{B(\gamma)}{A(\gamma)}\cdot\frac{1}{(h\gamma+1)^d} \tag{6}$$

The zeros of the plant can be decomposed into cancelling zeros B^+ (zeros inside the "heart-shape") and non-cancelling zeros B^- (zeros outside the "heart-shape", Fig.1, including inverse unstable zeros as well), it results:

$$B(\gamma)=B^+(\gamma)\cdot B^-(\gamma) \tag{7}$$

taking into account the requirement of zero steady-state error, the following condition is imposed:

$$B^-(0)=1 \tag{8}$$

This condition results from the final value theorem expressed in delta domain:

$$\lim_{t\to\infty} f(t)=\lim_{\gamma\to 0}\{\gamma F(\gamma)\} \tag{9}$$

If the design of a Dead-beat controller (abbreviated DB-controller) in delta domain is wanted, then for the closed loop t.f. $H_W(\gamma)$, must be imposed the form (10):

$$H_W(\gamma)=\frac{H_C\cdot H_P}{1+H_C\cdot H_P}=B^-(\gamma)\cdot\frac{1}{(h\gamma+1)^d}\cdot\frac{1}{(h\gamma+1)^n} \tag{10}$$

where n is defined as:

$$n=1+\deg(B^-) \tag{11}$$

Form (10), of the imposed closed loop t.f., results from the considerations that first of all the minimum number of sample times in which DB behaviour can be achieved is n (see (11) for definition of n). Secondly, also the dead time must be taken into account when imposing certain closed loop behaviour; and last but not least the non-cancelling zero has to be excluded from the controller's expression in order to avoid intersampling oscillations and also zero steady-state error.

If a one degree-of-freedom (1-DOF) controller is used, calculating the general expression of the controller from equation (10), it results in form of:

$$H_C(\gamma)=\frac{A(h\gamma+1)^d}{B^+[(h\gamma+1)^{d+n}-B^-]} \tag{12}$$

(Restrictions in the control signal have not been taken into account.) The main disadvantage is that when using small sampling period and the dead time is high, the expression of the controller becomes complicated, the powers of the dead time in the denominator result in very small numbers, thus the use of the delta transformation looses its advantage.

One alternative, which eliminates this disadvantage and combines both delta models and Z-domain representation using an IMC structure, is presented in Fig. 2.

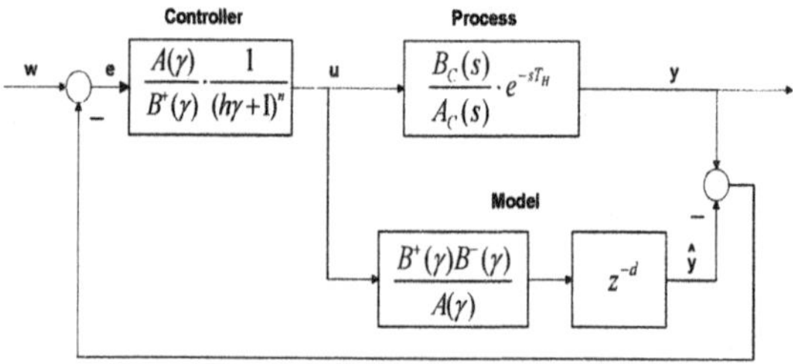

Fig.2 Hybrid IMC structure

As is presented in literature (see [Lunze, 1997], [Datta 1998] *et. al.*), if the plant is open-loop stable, then the Internal Model Control (IMC) structures are very attractive for some control applications. One of the main advantages of the IMC structure, see [Farkas *et.al.* 2002] consists in the fact that if there is no mismatch between the plant and the model (they have the same transfer function) the controller is actually open loop. This makes the design relatively simple. It is also shown that the classical IMC structure is actually equivalent to the Youla-Kucera parametrization of all stabilizing controllers.

In case of delta domain design the use of a hybrid structure, proposed in Fig.2 could be advantageous. Such a structure combines both *Z*-domain representation and delta-domain representation and so it combines the advantages of the delta parametrization and the possibility of dealing easily with the dead time by introducing it in *Z*-domain.

Following this way, the controller can be adapted to different values of the dead time by changing the value of *d* in the model. Thus this controller is very sensitive to mismatch in the dead time (as well as pure *Z* - domain controllers of the kind).

The controller is obtained by imposing to the closed loop transfer function the same condition as in (10). Accordingly, the controller t.f., $H_C(\gamma)$, obtains the expression:

$$H_C(\gamma) = \frac{A}{B^+} \cdot \frac{1}{(h\gamma + 1)^n} \qquad (13)$$

and the internal model of the plant is hybrid, composed of a part in delta model and another, in the *Z*-domain, expression of the dead time is presented in Fig.3.

It has be outlined again, that the plant must be in this case stable.

2.1. DEAD-BEAT CONTROL IN CASE OF A LOW (SECOND) ORDER PLANT

Let the t.f. of the continuous plant be:

$$H_P(s) = \frac{A_C}{(T_{1S}s + 1)(T_{2S}s + 1)} \cdot e^{-sT_H} \qquad (14)$$

Sampling with ZOH (Zero Order Hold) results in a delta t.f. having the expression (15):

$$H_P(\gamma) = A\frac{\tau\gamma + 1}{(T_1\gamma + 1)(T_2\gamma + 1)} \cdot \frac{1}{(h\gamma + 1)^d} \qquad (15)$$

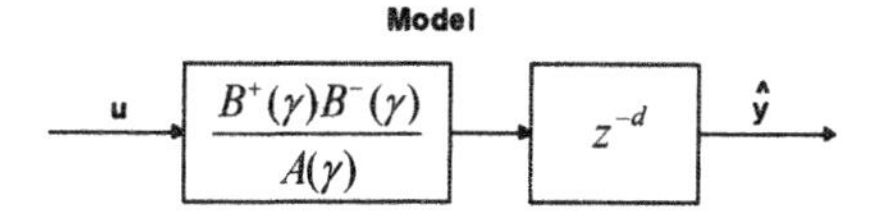

Fig.3. Internal hybrid model of the plant

Proceeding similarly in the described way the expression of the controller results based on relations (11), (13) and Fig.3.

As a numerical example, let the t.f. of the plant with a dead time $T_H = 0.1$ be (16). Having a sampling period of h=0.05 it leads to a discrete dead time d=2.

$$H_P(s) = \frac{1}{(0.67s + 1)(0.33s + 1)} \cdot e^{-0.1S} \qquad (16)$$

Based on rel.(3), the delta t.f. of the continuous plant results in form of:

$$H_P(\gamma) = \frac{0.0259\gamma + 1}{(0.6953\gamma + 1)(0.3556\gamma + 1)} \cdot \frac{1}{(0.05\gamma + 1)^2} \qquad (17)$$

According to (8), the zero can be decomposed into:

$$B^+(\gamma) = 1 \quad \text{and} \quad B^-(\gamma) = 0.0259\gamma + 1 \qquad (18)$$

Implementing the control structure according to the hybrid architecture developed, the controller results in form of:

$$H_C(\gamma) = \frac{(0.6953\gamma + 1)(0.3556\gamma + 1)}{(0.05\gamma + 1)^2} \qquad (19)$$

Finally, the plant model would be added derived from its delta t.f. as were presented in Fig.2.

For a comparison of the solution let be analysed the *Z*-domain DB controller obtained on the same basis. For the considered plant, the Z t.f. result as:

$$H_P(z) = \frac{0.0052z + 0.0049}{z^2 - 1.7875z + 0.7976} \cdot \frac{1}{z^2} \qquad (20)$$

The developed DB controller in *Z*-domain, $H_C(\gamma)$,is:

$$H_C(z) = \frac{z^2 - 1.7875z + 0.7976}{0.0101z^2} \qquad (21)$$

It has to be mentioned that the same IMC structure has been used as in delta-domain, but using a controller only in *Z*-domain. There are no significant differences in the system responses but the control signal is quite high (Fig. 4).

By increasing the sampling period *h* (e.g. h=0.1 or greater) the control signal will decrease.

2.2. MODIFICATION OF DEAD-TIME IN THE HYBRID ARCHITECTURE

A great advantage of the proposed hybrid architecture (Fig.2) is that when modifying the value of the dead time T_H , the controller does not need to be redesigned, only a simple parameter change in the plant model in *Z*-domain solve in a favourable way the problem, the solution is very simple to implement.

For example, let be the same plant with a different dead time (T_H=0.2 and corresponding, d=4):

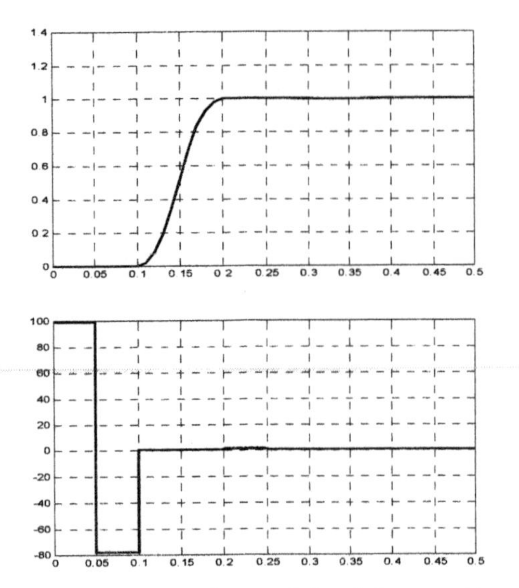

Fig.4. Output and control signals in case of DB-control

$$H_P(s) = \frac{1}{(0.67s+1)(0.33s+1)} \cdot e^{-0.2s} \qquad (22)$$

The IMC delta controller resulting is also (19), and the plant model would change as in Fig.5.

In this case the system's step response obtained through simulation is presented in Fig.6. It is to be remarked that there are no changes is the behavior of the system, only the effect of the increased dead-time appears. No other implementation problem – due for example of a pure delta-domain representation – is to be found.

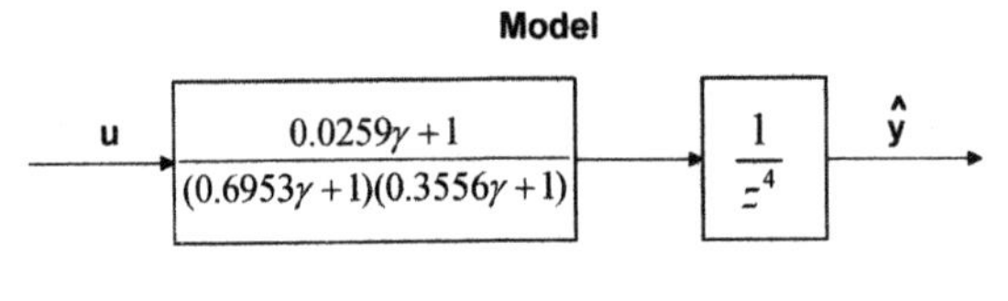

Fig.5 Hybrid plant model for d=4

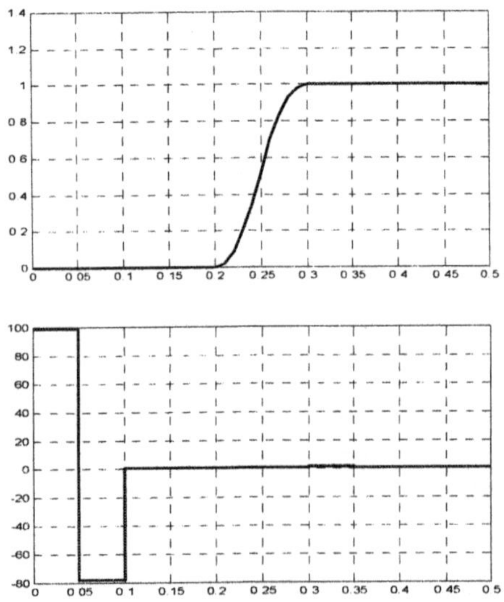

Fig.6 Hybrid process model for d=4

3. EFFECT OF LIMITATIONS IN THE HYBRID IMC STRUCTURE

A significant binding of control with practice is represented by the introduction of limitations (saturation) in the controller's structure. The convenient placement of the limiting element represents a very important practical problem. In this context the effect of the place of the control signal limitation within IMC structure has been also analysed. In this idea let be considered two significant cases:

- one, when the limitation is inside the IMC structure (Fig.7)
- the second one, when the limitation is outside the IMC structure (Fig.8)

Taking for plant (16) and imposing a limitation of +/- 10 units (correlated with a concrete application these values can be changed adequately) at the output of the controller we obtain for the first case a system response without overshoot (σ_1=0), (Fig.9) and in the second case a response with overshoot (Fig.10). In both cases the settling time is the same.

Focused on comparing only the system outputs for both cases (Fig.11) it can be seen that when the limitation is incorporated inside the IMC structure (Fig.7) not only there is no overshoot but the control signal does not oscillate so much between the extreme values (limits). Similar investigations related to the location of the saturation have been considered in [Bars *et.al.* 1995].

It has to be also mentioned that the results are almost the same for hybrid IMC as well as for pure *Z*-domain IMC structure but the design in delta seems to be more user friendly. It can be seen that when the limitation is incorporated inside the IMC structure (Fig.7) not only there is no overshoot but the control signal does not oscillate so much between the extreme values (limits).

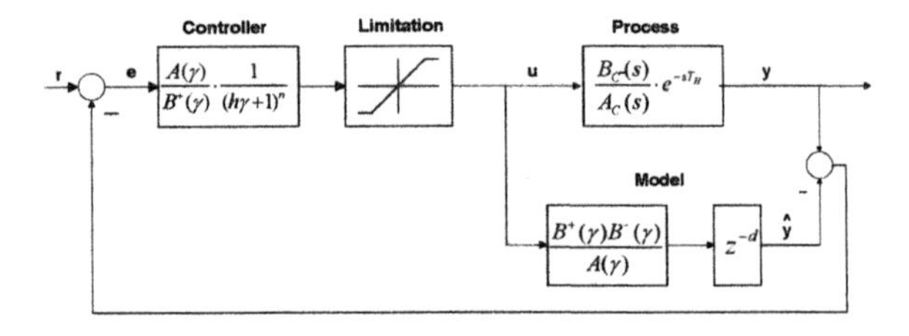

Fig.7 Hybrid IMC structure with limitation inside the model

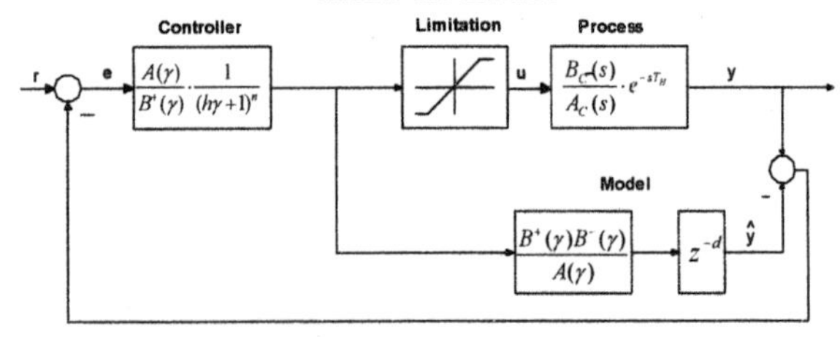

Fig.8 Hybrid IMC structure with limitation outside the model

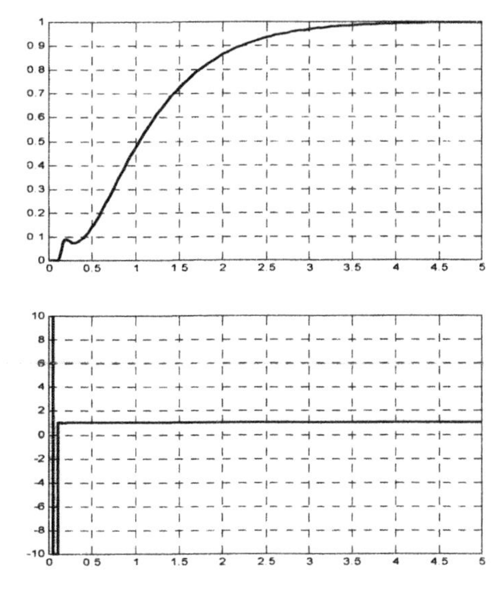

Fig.9 Step response and control signal in case of limitation inside the IMC structure

4. ILLUSTRATIVE SENSITIVITY ANALYSIS IN CASE OF A SECOND ORDER PLANT

In ideal case there is no mismatch between the plant and the identified model. In this case the control is actually open loop for the IMC structure. For example let the original delta transfer function of the plant be as presented in Fig.5, the coefficients given with four digits represent perfect match of model-plant. For comparison suppose there are mismatches between the identified model and the actual plant, for example the model's coefficients are established with only three digits. Also suppose also the *Z*-domain DB controller given by (20). The delta model of the plant is (having three digits coefficients):

$$H_M(\gamma) = \frac{0.025\gamma + 1}{0.247\gamma^2 + 1.050\gamma + 1} \cdot \frac{1}{(0.05\gamma + 1)^2} \quad (23)$$

and *Z*-domain model is (the same three digits coefficients convention has been followed):

$$H_M(z) = \frac{0.005z + 0.004}{z^2 - 1.787z + 0.797} \cdot \frac{1}{z^2} \quad (24)$$

According to the IMC structure the internal model would be different from the plant, and also the actual controller part would be designed accordingly. This way the delta controller results in form of:

$$H_C(\gamma) = \frac{0.247\gamma^2 + 1.05\gamma + 1}{(0.05\gamma + 1)^2} = \frac{(0.694\gamma + 1)(0.355\gamma + 1)}{(0.05\gamma + 1)^2} \quad (25)$$

and in *Z*-domain the controller would be:

$$H_C(z) = \frac{z^2 - 1.787z + 0.797}{0.0101 z^2} \quad (26)$$

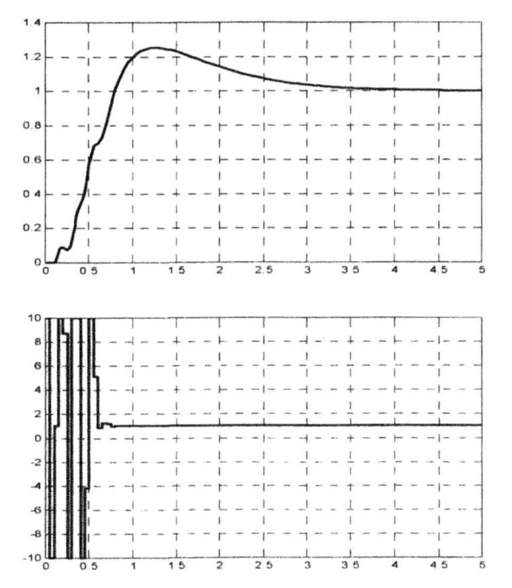

Fig.10 Step response and control signal in case of limitation outside the IMC structure

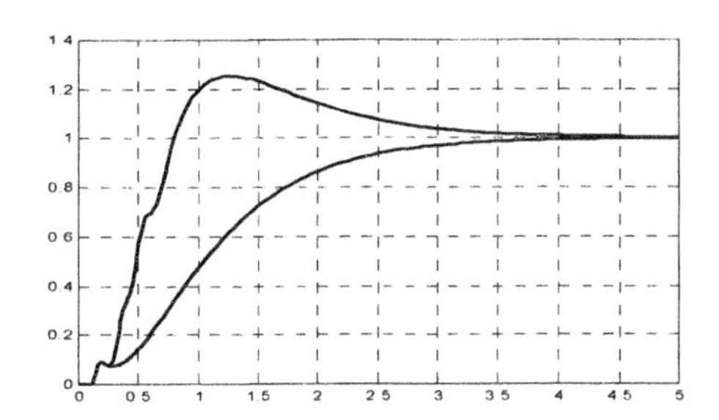

Fig.11 Comparison of step responses

In ideal case, as mentioned, there is no mismatch and there are no disturbances, and so the control is open loop. Suppose the case – interesting from the implementation point of view - when there are no disturbances, but there is the above mentioned mismatch between the plant and model. Taking the open loop structure (Fig.12) let be compared the delta-domain structure with the *Z*-domain structure.

Comparing the outputs and the control signals for delta (line) and *Z*-domain (dot-line) it can be noticed that the *Z*-domain representation is more sensitive to parameter mismatch, reflected in the static error (Fig.13). Operating in open loop actually represents reference signal tracking.

Using the IMC structure this static error disappears since an integrating effect is introduced due to the structure. Comparing the step responses and the control signals of the two systems in closed loop it results that the pure *Z*-domain structure (dot-line) is more sensitive to parameter mismatch than the hybrid structure (line) (Fig.14), but the static error is zero, due to the integrating effect.

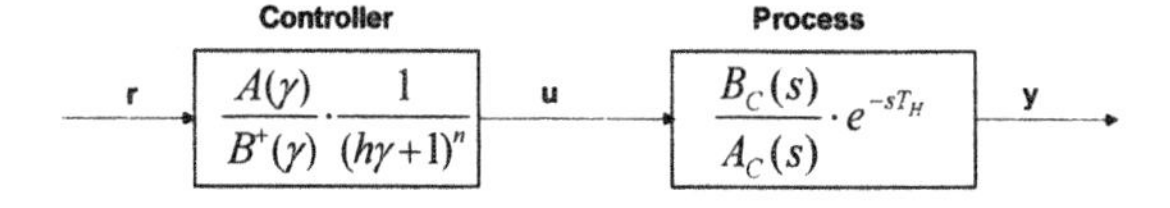

Fig.12 Open loop structure with delta domain controller

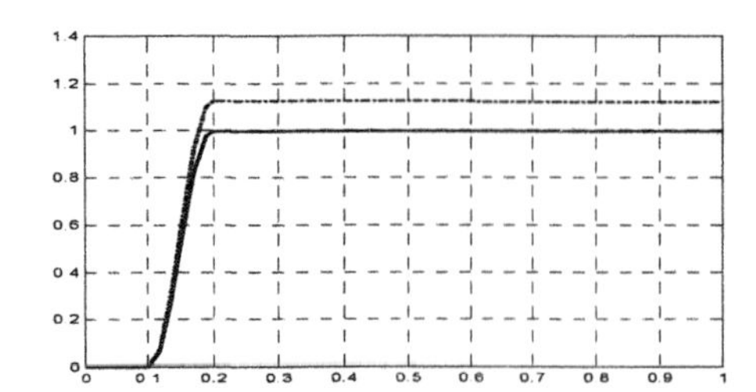

Fig.13 Open loop behaviour with delta and Z-domain controllers

5. CONCLUSIONS

The paper presents a new approach to control system design based on the Internal Model Control (IMC) in delta domain, with a mixed representation of the plant model within the IMC controller. The method is based on the dual representation in delta and Z discrete domain of the plant which has the advantage of the delta parametrization and the possibility of dealing easily with the dead time by introducing it in Z-domain. The dynamics of the system is described in the delta domain.

A great advantage of the proposed hybrid architecture is that when modifying the value of the dead time T_H, the controller does not need to be redesigned, only a simple parameter change in the plant model in Z-domain solve in a favourable way the problem, the solution is very simple to implement.

In order to sustain the close connection of control to the practical cases the effect of placing the limitation element was studied. Two significant cases have been considered:

- one, when the limitation is inside the IMC structure (Fig.7);
- the second one, when the limitation is outside the IMC structure (Fig.8).

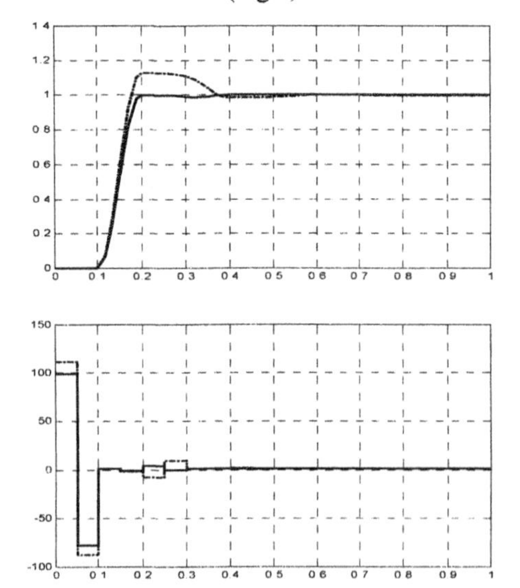

Fig.14 Sensitivity analysis between hybrid IMC structure and pure Z-domain structure (step responses and control signal)

Finally an illustrative analysis – based on simulation – of the sensitivity of the IMC structure to changes in parameter representation has been performed. Such an analysis is useful in case of delta representation of the models versus the discrete Z-domain representation.

The analysis highlights the fact that the use of the hybrid IMC controller – in delta and Z domain – referring to the plant's dead time stays for a viable alternative of controller design based on IMC structure.

The theoretical results have been verified through simulation on low order plants – benchmark types – frequently applied for verification of control solutions. The comparison criteria taken into account in the paper was the difference in the step responses under same conditions obtained through simulation.

ACKNOWLEDGMENT

The authors' work from BUTE was supported by the fund of the Hungarian Academy of Sciences for control research and partly by the OTKA fund T042741. The research was also supported by the bilateral EU-Socrates-Erasmus cooperation between the two universities. All supports are kindly acknowledged.

REFERENCES

Åström, K.J., Hägglund, T., (2000), *Benchmark systems for PID control,* PID'00 IFAC Workshop on Digital Control, Preprints, Terrassa (Spain), pp.181-182.

Bars, R., Habermayer, M., (1995), *Investigation of Saturation Effect in Linear one-step-ahead Predictive Control Algorithms,* IFAC Workshop on Control Applications of Optimization, Haifa, Postprints, pp.41-46.

Datta, A., (1998), *Adaptive Internal Model Control,* Springer Verlag.

Farkas, I., Vajk, I., (2002), *Internal Model-Based Controller for a Solar Plant,* IFAC 15th Triennial World Congress, Barcelona, Spain

Goodwin, G.C., Graebe, S.F., Salgado, M.E., (2001), *Control System Design,* Prentice Hall.

Istepanian, R.H.S., Whidborne, J.F., (2001), *Digital Controller Implementation and Fragility,* Springer Verlag.

Lunze, J., (1997), *Regelungstechnik - 2* Springer Verlag.

Middleton, R.H., Goodwin, G.C., (1990), *Digital Control and Estimation, A Unified Approach,* Prentice Hall.

EXTREMAL TRANSFER FUNCTIONS IN ROBUST CONTROL SYSTEM DESIGN

Ľubomír Grman, Vojtech Veselý

Department of Automatic Control Systems,
Faculty of Electrical Engineering and Information Technology,
Slovak University of Technology
Ilkovičova 3, 812 19 Bratislava, Slovak Republic
e-mail: grman@kasr.elf.stuba.sk, vesely@kasr.elf.stuba.sk

Abstract: The paper addresses the design of a robust output feedback controller for SISO systems using extremal transfer functions and the classical control theory approach. It focuses on robust stabilization of uncertain plants, which belong to linear or multilinear uncertain systems. A survey on extremal transfer functions is given and it is shown that for the proposed robust controller design procedure the classical linear control theory can be applied providing necessary and sufficient or sufficient stability conditions. *Copyright © 2003 IFAC*

Keywords: robust control, extremal transfer function, interval control system and multilinear interval system

1. INTRODUCTION

Robustness has been recognized as a key issue in the analysis and design of control systems for the last two decades.

The main criticism formulated by control engineers against modern robust analysis and design methods for linear systems concerns the lack of efficient, easy-to-use and systematic numerical tools. Indeed, a lot of analysis techniques and most of the design techniques for uncertain systems boil down to non-convex bilinear matrix inequality (BMI) problems, for which no polynomial-time algorithm has been proposed so far (Henrion, *et al.*, 2002). This is especially true when analysing robust stability or designing robust controllers for MIMO systems affected by highly structured uncertainties, or when seeking a low order or a given order robust controller.

In this paper we focus on the problem of robust stabilization of an uncertain single input – single output plant, which belongs to linear or multilinear interval systems. A survey on extremal transfer functions is given and it is shown that for the robust controller design procedure the classical linear control theory can be applied with necessary and sufficient or sufficient stability conditions.

Even though significant progress has been made recently in the field of analysing robust stability for parametric uncertain systems, the robust controller design procedure is still an open problem. Indeed, in (Bhattacharyya, *et al.*, 1995) it is pointed out that a significant deficiency of the control theory is lack of no conservative robust controller design methods. Recent developments in the robust control of systems with parametric uncertainty have been inspired by the Kharitonov Theorem (Kharitonov, 1979). By means of this theorem it is sufficient to determine stability only of four Kharitonov polynomials. Kharitonov's theorem has been generalized for the control problem (Chapellat, Bhattacharyya, 1989). The Generalized Kharitonov Theorem shows that for a compensator to robustly stabilize the system it is sufficient if it stabilizes a prescribed set of line segments in the plant parameter space. Under special conditions on compensator it is sufficient to stabilize the Kharitonov vertices. The next important substantial progress in the robust analysis of parameter stability is the Edge Theorem (Bartlett, *et al.*, 1988). The Edge Theorem allows to constructively determine the root space of a family of linearly parametrized systems. There are situations when several linear interval systems are connected in series. In such a case the global control object is considered to belong to a multilinear systems class (Chapellat *et al.*, 1994). The main tool to approach this problem is the Mapping Theorem (Hollot, Xu, 1989) which shows that the image set of multilinear interval polynomials is contained in the convex hull of the vertices.

2. PROBLEM STATEMENT

Consider the transfer function

$$G(s)=\frac{P_1(s)}{P_2(s)} \tag{1}$$

where $P_1(s)$, $P_2(s)$ are linear or multilinear interval polynomials with respect to parametric uncertainty described in the parameter box uncertainty Q.

The problem studied in this paper can be formulated as follows: For a continuous time system described by the transfer function (1) the robust controller

$$R(s)=\frac{F_1(s)}{F_2(s)} \tag{2}$$

is to be designed with fixed polynomials $F_1(s)$, $F_2(s)$ such that for the closed loop system

$$G_c(s)=\frac{P_1(s)F_1(s)}{1+P_2(s)F_2(s)} \tag{3}$$

robust stability (RS) and a specified robust performance (RP) are guaranteed.

3. EXTREMAL TRANSFER FUNCTIONS

3.1 Linear interval case

We will deal with characteristic polynomials of the form

$$A(s)=F_1(s)P_1(s)+F_2(s)P_2(s) \tag{4}$$

where

$$P_i(s)=p_{0,i}+p_{1,i}s+\cdots+p_{n_i,i}s^{n_i},\ i=1,2 \tag{5}$$

Each $P_i(s)$ is a linear interval polynomial specified by the intervals

$$p_{j,i}\in\left\langle \underline{p}_{j,i},\ \overline{p}_{j,i}\right\rangle,\ i=1,2,\ j=0,1,\ldots,n_i \tag{6}$$

The corresponding parameter box is then

$$Q_i=\left\{p_i:\ \underline{p}_{j,i}<p_{j,i}<\overline{p}_{j,i},\ i=1,2;\ j=0,1,\ldots,n_i\right\} \tag{7}$$

and the global parameter uncertainty box is given

$$Q=Q_1\times Q_2 \tag{8}$$

where $p_i^T=\left\lfloor p_{0,i}\ p_{1,i}\cdots p_{n_i,i}\right\rfloor,\ i=1,2$.

Following assumptions about the linear interval polynomials are considered:

1) Elements of $p_i, i=1,2$ perturb independently of each other. Equivalently, Q is a n_1+n_2 axis parallel rectangular box.
2) All characteristic polynomials (4) are of the same degree.

According to (Bhattacharyya, *et al.*, 1995) the stability problem of (4) can be solved using the Generalized Kharitonov Theorem.

Theorem 1. (Chapellat, Bhattacharyya, 1989)
For a given $F(s)=(F_1(s),F_2(s))$ of real polynomials:

1) $F(s)$ stabilizes the linear interval polynomials $P(s)=(P_1(s),P_2(s))$ for all $p\in Q$ if and only if the controller stabilizes the extremal transfer function

$$G_E(s)=\left\{\frac{K_1(s)}{S_2(s)}\cup\frac{S_1(s)}{K_2(s)}\right\} \tag{9}$$

2) Moreover, if the polynomials of the controller $F_i(s)$, $i=1,2$ are of the form

$$F_i(s)=s^{t_i}(a_is+b_i)U_i(s)Z_i(s) \tag{10}$$

then it is sufficient that the controller $F(s)$ stabilizes the Kharitonov transfer function

$$G_K(s)=\frac{K_1(s)}{K_2(s)} \tag{11}$$

3) Finally, stabilizing (11) is not sufficient to stabilize $P(s)=(P_1(s),P_2(s))$ when the controller polynomials $F_i(s)$, $i=1,2$, do not satisfy the condition (10)

□

where

$$K_i(s)=\left\{K_i^1(s),\ K_i^2(s),\ K_i^3(s),\ K_i^4(s)\right\} \tag{12}$$

stand for Kharitonov polynomials (Kharitonov, 1979) corresponding to each $P_i(s)$

$$S_i(s)=\left\{\left\lfloor K_i^1,\ K_i^2\right\rfloor,\left\lfloor K_i^1,\ K_i^3\right\rfloor,\left\lfloor K_i^2,\ K_i^4\right\rfloor,\left\lfloor K_i^3,\ K_i^4\right\rfloor\right\} \tag{13}$$

are the four Kharitonov segments for corresponding $P_i(s)$; $U_i(s)$ is an anti-Hurwitz polynomial, $Z_i(s)$ are even or odd polynomials, a_i, b_i are real positive numbers and $t_i\geq 0$.

Note: $S_i^1=\lambda K_i^1(s)+(1-\lambda)K_i^2(s)$, $\lambda\in\langle 0,1\rangle$

3.2 Linear affine case

Let the transfer function (1) can be rewritten in the following affine form

$$G(s)=\frac{P_1(s)}{P_2(s)}=\frac{P_{0,1}(s)+\sum_{i=1}^{p}P_{i,1}(s)q_i}{P_{0,2}(s)+\sum_{i=1}^{p}P_{i,2}(s)q_i} \tag{14}$$

where $P_{j,1}(s), P_{j,2}(s)$ for $j = 0, 1, 2, ..., p$ are real polynomials with constant parameters and the uncertain parameter q_i belongs to the interval $q_i \in \langle \underline{q_i}, \overline{q_i} \rangle$, $i = 1, 2, ..., p$.

The system represented by (14) is a polytope of linear systems, which can be described by a list of its vertices

$$G_{Vj}(s) = \frac{P_{V1,j}(s)}{P_{V2,j}(s)}, \; j = 1,2,\ldots,N, \; N = 2^p \qquad (15)$$

computed for different permutations of the p variable q_i, $i = 1, 2, ..., p$ alternatively taken at their maximum $\overline{q_i}$ and minimum $\underline{q_i}$.

The characteristic polynomial of polytopic system with controller (2) is given as follows

$$A_{Vj}(s) = F_1(s)P_{V1,j}(s) + F_2(s)P_{V2,j}(s), \; j = 1,2,\ldots,N \quad (16)$$

A polytopic family of characteristic polynomials can be represented as the convex hull

$$\Delta(s) = \sum_{j=1}^{N} \lambda_j A_{Vj}, \; \lambda_j \geq 0, \; \sum_{j=1}^{N} \lambda_j = 1 \qquad (17)$$

We make the assumption that all polynomials have the same degree.

Theorem 2. Edge theorem (Bartlett, *et al.*, 1988)
Let Q be a p-dimensional polytope that is its vertices and edges describe the convex hull (17). Then the boundary of $R(Q)$ is contained in the root space of the exposed edges of Q.

□

Due to Theorem 2 the characteristic polynomials (16) will be stable if and only if the following set of segments are stable

$$E(s) = \{\lambda A_{Vi}(s) + (1-\lambda)A_{Vj}(s)\}, \; i, j = 1,2\ldots, p2^{p-1} \quad (18)$$

Both i and j has to be taken as the vertices number of corresponding edges.
Substituting (16) to (18) after some manipulation one obtains the following extremal transfer function for affine system (14)

$$G_P(s) = \frac{\lambda P_{V1,i} + (1-\lambda)P_{V1,j}}{\lambda P_{V2,i} + (1-\lambda)P_{V2,j}}, \; \lambda \in \langle 0, 1 \rangle \qquad (19)$$

Lemma 1.
The controller (2) stabilizes the affine system (14) for all $q \in Q$ if and only if the controller stabilizes the extremal transfer function for polytopic system (19) for all $\lambda \in \langle 0, 1 \rangle$.

□

The problem addressed in the Theorem 1 deals with a polytope and therefore using the Edge theorem can solve it. In general, the sets of extremal transfer functions (9) and (19) are quite different. While the number of $G_E(s)$ is equal to 32 ($G_K(s)$ - 16) the number of extremal transfer functions (19) depends exponentially on the number of uncertain parameters q_i, $i = 1, 2, ..., p$ and equal to $p2^{p-1}$.

3.3 Multilinear case

Let the uncertain plant transfer function (1) be

$$G(s) = \frac{P_{11}(s)P_{12}(s)\cdots P_{1n}(s)}{P_{21}(s)P_{22}(s)\cdots P_{2d}(s)} \qquad (20)$$

where $P_{ij}(s) = p_0^{ij} + p_1^{ij}s + \cdots + p_{n_{ij}}^{ij}s^{n_{ij}}$.

Each $P_{ij}(s)$, $i = 1, 2, ..., n(d)$ belongs to a linear interval polynomial specified as

$$p_k^{ij} \in \langle \underline{p}_k^{ij}, \overline{p}_k^{ij} \rangle, \; i = 1,2, \; j = 0,1,\ldots,n(d), \; k = 0,1,\ldots,n_{ij}$$

with independently varying parameters. Let $K_{ij}(s)$ and $S_{ij}(s)$ denote the respective Kharitonov polynomials and Kharitonov segments of corresponding real interval polynomial $P_{ij}(s)$.
An extremal transfer function is given as follows (Chapellat, *et al.*, 1994)

$$M_E(s) = \left\{ \frac{S_{11}(s)S_{12}(s)\cdots S_{1n}(s)}{K_{21}(s)K_{22}(s)\cdots K_{2d}(s)} \cup \frac{K_{11}(s)K_{12}(s)\cdots K_{1n}(s)}{S_{21}(s)S_{22}(s)\cdots S_{2d}(s)} \right\} \qquad (21)$$

Lemma 2.
The controller (2) stabilizes the multilinear system (20) for the whole uncertainty box if and only if the controller stabilizes the extremal transfer function (21).

□

Note that the number of extremal transfer function of (21) is $2.4^d.4^n$.

Consider the transfer function of multilinear interval system to be a ratio of multilinear polynomials with independent parameters. A proper stable system with transfer function of the form (1) will be considered where

$$P_1(s) = \sum_{i=1}^{m} C_i(s) \prod_{j=1}^{r_i} D_{ij}(s) \qquad (22)$$

$$P_2(s) = \sum_{i=1}^{m} H_i(s) \prod_{j=1}^{k_i} L_{ij}(s)$$

with $C_i(s)$ and $H_i(s)$ being fixed polynomials and the $D_{ij}(s)$ and $L_{ij}(s)$ being independent real linear interval polynomials.
Let d and l denote the sets of coefficients of the corresponding interval polynomials. d and l vary in a prescribed uncertainty box $d \in \Lambda$ and $l \in \Pi$.
$P_1(s)$ and $P_2(s)$ are coprime polynomials over the box $Q = \Pi \times \Lambda$ and it is assumed that $P_2(s) \neq 0$ for all $l \in \Pi$ and $s = j\omega$, $\omega > 0$. Introduce the Kharitonov polynomials and segments associated with $L_{ij}(s)$ and $D_{ij}(s)$, respectively.
The extremal transfer function of (22) is given as follows (Bhattacharyya, *et al.*, 1995)

$$M_{SE}(s)=\left\{\frac{\Gamma_K(s)}{\Omega_E(s)}\cup\frac{\Gamma_E(s)}{\Omega_K(s)}\right\} \quad (23)$$

where

$$\Gamma_E^l(s)=\bigcup_{l=1}^{m}\Gamma_E^l(s)$$

$$\Gamma_E^l(s)=\sum_{i=1}^{m}C_i(s)\prod_{j=1}^{r_i}N_{Lij}(s)+C_i(s)\prod_{j=1}^{r_k}T_{Lij}(s) \quad (24)$$

$$\Gamma_K(s)=\sum_{i=1}^{m}C_i(s)\prod_{j=1}^{k_i}N_{Lij}(s)$$

The same equations hold for $\Omega_E(s)$ and $\Omega_K(s)$ with polynomial $P_2(s)$.

Lemma 3.
The controller (2) stabilizes the multilinear system (22) for all (d, l) if and only if the controller stabilizes the extremal transfer function (23).

□

Now, consider a multilinear polytopic system where the entries of the uncertainty vector $q^T=[q_1,\ldots,q_p]$ in the transfer function (1) are in multilinear form.
The closed-loop characteristic polynomial is given as follows (4)

$$A(s)=F_1(s)P_1(s,q)+F_2(s)P_2(s,q) \quad (25)$$

Let the uncertain parameters $q_i\in\langle \underline{q_i},\overline{q_i}\rangle$, $i = 1,2,\ldots,p$ belong to a p-dimensional uncertain parameter box Q with $N=2^p$ vertices and $p2^{p-1}$ edges.
Denote the characteristic polynomials in corresponding vertices of Q as follows

$$A_v(s,q)=\{A(s): q_i=\underline{q}_i \text{ or } q_i=\overline{q}_i, i=1,2,\cdots,p\}= \\ =\{v_1(s),\ldots,v_N(s)\} \quad (26)$$

Let $\Delta(s)$ denote the convex hull of the vertex polynomials $A_v(s,q)$

$$\Delta(s)=\sum_{i=1}^{N}\lambda_i v_i(s),\ \sum_{i=1}^{N}\lambda_i=1 \quad (27)$$

where $\lambda_i\in\langle 0, 1\rangle$.
Under the assumptions that

a) for any $q\in Q$, the polynomials (25) and (26) are of the same degree,
b) for any $s=j\omega$, $\omega>0$, $\Delta(s)\neq 0$ in (27),
c) there exists at least one $q^*\in Q$ such that (25) is stable,

the characteristic polynomial (25) is stable if the convex hull (27) and equivalently the sets of characteristic polynomial edges

$$E(s)=\{\lambda v_i(s)+(1-\lambda)v_j(s): v_i(s),v_j(s)\in A_v(s,q)\} \quad (28)$$

are stable where $\lambda\in\langle 0, 1\rangle$.
With respect to (25), the vertex characteristic polynomials of (26) can be rewritten as follows

$$v_i=F_1(s)v_{P1i}+F_2(s)v_{P2i},\ i=1,2,\ldots,N \quad (29)$$

where $v_{P1i}(s)$, $v_{P2i}(s)$ are the vertex polynomials of $P_1(s,q)$, $P_2(s,q)$, respectively.
A simple manipulation of the entries of $E(s)$ yields

$$E_K(s)=F_1(s)[\lambda v_{P1i}+(1-\lambda)v_{P2j}]+ \\ +F_2(s)[\lambda v_{P1i}+(1-\lambda)v_{P2j}] \quad (30)$$

With respect to (30), the extremal transfer function of the multilinear polytopic system is as follows

$$M_{SP}(s)=\frac{\lambda v_{P1i}(s)+(1-\lambda)v_{P1j}(s)}{\lambda v_{P2i}(s)+(1-\lambda)v_{P2j}(s)} \quad (31)$$

where $i\neq j\ i,j=1,2,\ldots,\frac{2^p!}{2(2^p-2)!}$.

Note that the Mapping Theorem (Hollot, Xu, 1989) shows that the image set of a multilinear interval polynomial (25) is contained in the convex hull of the vertices of Q (26). A sufficient condition for the entire image set to exclude zero (Zero Exclusion Principle) is that the convex hull excludes zero. This suggests that stability of the multilinear set (25) can be guaranteed by solving the stability of the convex hull of the vertex polynomials (27).

Lemma 4.
The controller (2) stabilizes the multilinear polytopic system with characteristic polynomial (25) for all $q\in Q$ if the controller (2) stabilizes the extremal transfer function (31).

□

Note that if Q is not an axis parallel box or the dependency on parameters in the characteristic polynomial (25) is not multilinear, the above lemma does not hold.

4. EXAMPLES

Example 1

As a real example we have considered the problem of robust controller design to control the speed of two serially connected small DC motors.
The controlled process has been identified in three working points using the ARMAX model. The interval transfer function of the process is of the form

$$G(s)=\frac{P_1(s)}{P_2(s)}=\frac{p_{12}s^2+p_{11}s+p_{10}}{p_{22}s^2+p_{21}s+p_{20}} \quad (32)$$

where $p_{22}=1$, $p_{21}\in[1.83\ 2.25]$, $p_{20}\in[0.64\ 0.67]$, $p_{12}\in[0.011\ 0.025]$, $p_{11}\in[-0.58\ -0.32]$, $p_{10}\in[1.54\ 2]$.

First design method

The robust controller design has been carried out using the Nejmark D-curve method (Nejmark, 1978). The method is based on splitting the space of parameters to regions with an equal number of unstable roots of the characteristic equation. In this method the curve for P, I, D gain have been obtained.

Fig. 1 depicts the D-curve for choosing the integration coefficient I of PID controller.

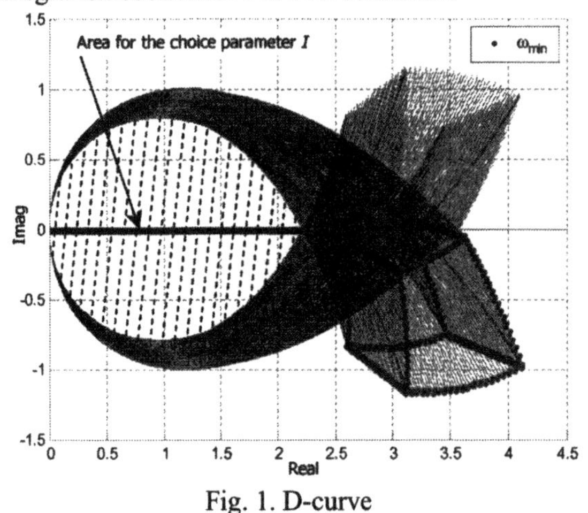

Fig. 1. D-curve

The robust PID controller designed using this method for 32 extremal transfer functions (9) is of the form

$$R(s)=P+\frac{I}{s}+Ds=1.3+\frac{0.6}{s}+0.5s \tag{33}$$

the degree of stability $\alpha = 0.633$ has been achieved.
Note: The degree of stability with a negative sign is equal to the maximum eigenvalue of the stable closed loop system defined by the extremal transfer function and a PID controller.

Second design method

The controller design is carried out by means of the criterion of the minimum of integral of the squared error (I_{SE}) and the criterion of minimum integral of squared error multiplied by time (I_{TSE}). The integral performance criterions provide information about the control process on the basis of integral error for all time values.
The algorithm for calculating I_{SE} designed by Nekolný (Nekolný, 1961) comes from the Parseval's integral in form

$$I_{SE}=\int_0^{\infty}[e(t)]^2\,dt=\frac{1}{2\pi j}\int_{c-j\infty}^{c+j\infty}E(s)E(-s)ds \tag{34}$$

where $E(s)$ is the Laplace transform of the tracking error

$$E(s)=\frac{D(s)}{A(s)}=\frac{d_{n-1}s^{n-1}+\cdots+d_1s+d_0}{a_ns^n+\cdots+a_1s+a_0} \tag{35}$$

The integral of the square error multiplied by time can be written as

$$I_{TSE}=\int_0^{\infty}te^2(t)dt \tag{36}$$

For the integral (36), formulas in the closed form have been derived.

Design of unknown controller parameters have been carried out using *minimax problem* formulation

$$\min_x \max_{\{F_i\}}\{F_i(x)\} \tag{37}$$

which is realized by the *fminimax* function in the Matlab Optimization Toolbox. This function minimizes the worst-case value of a set of multivariable functions.
The robust PI controllers designed by the I_{SE} and I_{TSE} criterions, respectively, using extremal transfer functions $G_P(s)$ (19) are of the form

$$R(s)=P+\frac{I}{s}=0.86+\frac{0.24}{s}\quad (I_{SE}) \tag{38}$$

$$R(s)=P+\frac{I}{s}=1.05+\frac{0.23}{s}\quad (I_{TSE}) \tag{39}$$

We achieved the degree of stability $\alpha = 0.209$ for the I_{SE} criterion and $\alpha = 0.17$ for the I_{TSE} criterion.

Example 2

In this example the interval plant proposed by Hollot and Yang (1990) has been considered

$$G(s)=\frac{P_1(s)}{P_2(s)}=\frac{p_{10}}{(s+0.1)(s+0.2)(s+25)(s+75)} \tag{40}$$

where $p_{10}\in[1;\,5000]$.
As it can be observed, this plant has only one uncertainty parameter.

Third design method

The robust controller design is carried out by Bode diagrams (Kuo, 1991). There are 404 extremal transfer functions derived for linear interval uncertainty. In the frequency domain a robust PI controller for a required phase margin $\Delta\varphi = 50°$ has been designed in the form

$$R(s)=K\left(1+\frac{1}{T_is}\right)=\frac{0.0376s+0.00104}{s} \tag{41}$$

where $K = 0.0376$ and $T_i = 36.2$ [s].

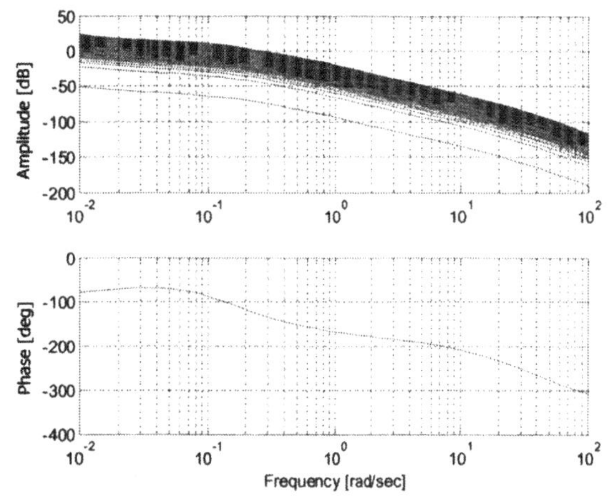

Fig. 2. Bode diagrams of the open loop system: $G_P(s).R(s)$

Fig. 2 depicts Bode diagrams of the open loop system. The achieved gain margin ΔK and phase margin $\Delta\varphi$ are:

$$\begin{aligned} \Delta K &= 34.2 \text{ dB} \\ \Delta\varphi &= 49.5° \end{aligned} \tag{42}$$

From the Bode diagrams it is possible to see that the designed robust controller guarantees stability and performance for linear interval transfer function with uncertainties.

Fourth design method

We will design a robust controller for a linear system (40) described by polynomial matrices (Henrion *et al.*, 2002). The design problem amounts to finding a dynamical output-feedback controller with a transfer function $F_2^{-1}(s)F_1(s)$ such that the closed-loop denominator matrix

$$A(s) = F_1(s)P_1(s) + F_2(s)P_2(s) \tag{43}$$

is robustly stable for all admissible uncertainties.

Let us assume that the static feedback matrix K satisfies structural LMI constraints and the controller polynomial matrices $F_1(s) = F_{10} + F_{11}s + \ldots$ and $F_2(s) = F_{20} + F_{21}s + \ldots$ entering linearly in polynomial matrix $A(s)$ have a prescribed structure, which we denote by the LMI

$$G(A) \geq 0. \tag{44}$$

For a PID controller the coefficients $F_1(s)$ and $F_2(s)$ will be

$$\frac{F_1(s)}{F_2(s)} = K_P + \frac{K_I}{s} + K_D s \tag{45}$$

where $F_{20} = 0$, $F_{21} = 1$, $F_{22} = 0$ and $F_{10} = K_I$, $F_{11} = K_P$, $F_{12} = K_D$.

Under these assumptions the following lemma (Henrion *et al.*, 2002) can be formulated

Lemma 5.

The transfer function in the affine form (14) with uncertainty q_i, $i = 1,\ldots,p$ is robustly stabilizable by a constrained output feedback controller $F_1(s)$, $F_2(s)$ if for a stable nominal characteristic polynomial $D(s)$ of the same degree as polynomial matrices $A_i(s) = P_{1i}(s)F_1(s) + P_{2i}(s)F_2(s)$, there exist some matrices $P_i = P_i^T$ satisfying the LMI

$$D^T A_i + A_i^T D - H(P_i) > 0,\ i = 1,\ldots,m \tag{46}$$

with the additional LMI constraint (44).

□

In our case a nominal controller is $R_n(s) = 0.05 + 0.002/s + 0.001s$ and with help of SeDuMi we have obtained the robustly stabilizing PID controller

$$R(s) = P + \frac{I}{s} + Ds = 0.71 + \frac{0.14}{s} + 2.54s \tag{47}$$

The achieved degree of stability is $\alpha = 0.003935$.

5. CONCLUSION

The main aim of this paper has been to present a survey of extremal transfer functions and robust controller design using classical control theory approach. The proposed robust controller design procedures with extremal transfer functions guarantee specificified performance and stability with necessary and sufficient or sufficient stability conditions.

REFERENCES

Bartlett, A.C., Hollot, C.V. and Lin, H. (1988): Root Location of an Entire Polytope of Polynomials: It Suffices to Check the Edges. *Mathematics of Control, Signals and Systems*, vol. 1.

Bhattacharyya, S.P., Chapellat, H. and Keel, L.H. (1995): Robust Control: The Parametric Approach, *Prentice Hall, Englewood Cliffs, N.J.*

Chapellat, H., Bhattacharyya, S.P. (1989): A Generalization of Kharitonov's Theorem: Robust Stability Interval Plants. *IEEE Transaction on Automatic Control*, AC-34, No. 3, pp. 306-311.

Chapellat, H., Keel, L.H. and Bhattacharyya, S.P. (1994): Extremal Robustness Properties of Multilinear Interval systems. *Automatica*, 30, No. 6, pp. 1037-1042.

Henrion, D., Arzelier, D. and Peaucelle, D. (2002): Positive Polynomial Matrices and Improved LMI Robustness Conditions, In: *15th Triennial World Congress of the International Federation of Automatic Control*, CD.

Henrion, D. and Bachalier, D. (2001): Low-order Robust Controller Design for Interval Plants. *Int. J. Control*, vol 74, No. 1, pp. 1-9.

Hollot, C.V. and Xu, Z.I. (1989): When is the Image of a Multilinear Function a Polytope? A Conjecture. In: *Proceedings of the 28th IEEE Conference on Decision and Control*, Tampa, FL, pp. 1890-1891.

Kharitonov, V. L. (1979): Asymptotic Stability of an Equilibrium Position of a Family of Systems of Linear Differential Equations. *Differential Equations*, vol. 14, pp. 148.

Kuo, B. C. (1991): Automatic Control Systems. Prentice Hall, pp. 672-677.

Nejmark, J. J. (1978): Dynamic systems and control process. *Nauka*, Moskva. (in Russian)

Nekolný, J. (1961): Simultaneous check stability and quality of control. *ČSAV*, Praha, pp. 35-58. (in Czech).

www.elsevier.com/locate/ifac

A NEW METHODOLOGY FOR FREQUENCY DOMAIN DESIGN OF ROBUST DECENTRALIZED CONTROLLERS

Alena Kozáková [*], **Vojtech Veselý** [*]

[*] *Department of Automatic Control Systems,*
Faculty of Electrical Engineering and Information Technology
Slovak University of Technology,
Ilkovičova 3, 812 19 Bratislava, Slovak Republic
kozakova@kasr.elf.stuba.sk

Abstract: The paper proposes an original frequency domain approach to decentralized controller design for performance applied within a robust design strategy. The novelty of the proposed approach consists in assessing the influence of interactions by means of characteristic functions/loci of the plant interaction matrix, and further using them to modify mathematical models of subsystems thus defining the so called „equivalent subsystems". The independent design carried out for equivalent subsystems provides local controllers that guarantee a prespecified performance of the full system. The developed design procedure applied to the nominal model yields local controllers that provide robust properties within a specified range of uncertainty. Theoretical conclusions are supported by an example. *Copyright © 2003 IFAC*

Keywords: decentralized control, Nyquist's criterion, spectral Nyquist plot, characteristic loci, robustness

1 INTRODUCTION

Industrial plants are complex systems typical by multiple inputs and multiple outputs (MIMO systems). Usually they arise as interconnection of a finite number of subsystems. If strong interactions within the plant are to be compensated for then multivariable controllers are used. However, there may be practical reasons that make restrictions on controller structure necessary or reasonable. In an extreme case, the controller is split into several local feedbacks and becomes a decentralized controller. Compared with centralized full-controller systems such controller structure constraints bring about certain performance deterioration. However, this drawback is weighted against important benefits such as hardware simplicity, operation simplicity and reliability improvement as well as design simplicity [2,7]. Due to them, decentralized control (DC) design techniques remain probably the most popular among control engineers, in particular the frequency domain ones which provide insightful solutions and link to the classical control theory.

Major multivariable frequency-response Nyquist-type design techniques were developed in the late 60′ and throughout the 70's; almost simultaneously with the non-interacting Characteristic locus (CL) technique [3,4]. Development of decentralized control (DC) techniques dates back to the 70′; the research, though not so excessive, is still going on.

The DC design has two main steps: first, a suitable control structure (pairing inputs with outputs) has to be selected; then, local controllers are designed for individual subsystems. There are three general approaches to the local controller design: simultaneous, sequential and independent designs [2, 7].

In the independent design, effect of interactions in the full system is assessed, and transformed into bounds for individual subsystems which are to be considered in the local controller design in order to provide stability and a required performance of the full system.

In the sequel the control structure selection is supposed to be already completed. To account for plant interactions, an original approach is proposed based upon representing interactions by means of their characteristic loci (eigenvalue loci) [1,3,4,5] and further using them to modify mathematical models of individual subsystems. For such modified subsystems is applied then. Designed local controllers guarantee fulfilment of performance requirements imposed on the full system; moreover, for a specified range of operating conditions the proposed design procedure applied to the nominal model guarantees closed-loop robustness. Theoretical results are supported by an example.

The paper is organized as follows: theoretical preliminaries of the proposed technique are surveyed in Section 2, problem formulation is in Section 3, main results along with the proposed design procedure and robustness study are presented in Section 4 and verified on an example in Section 5. Conclusions are given at the end of the paper.

2 PRELIMINARIES

Consider a MIMO system *G(s)* and the controller *R(s)* in a standard feedback configuration (Fig. 1)

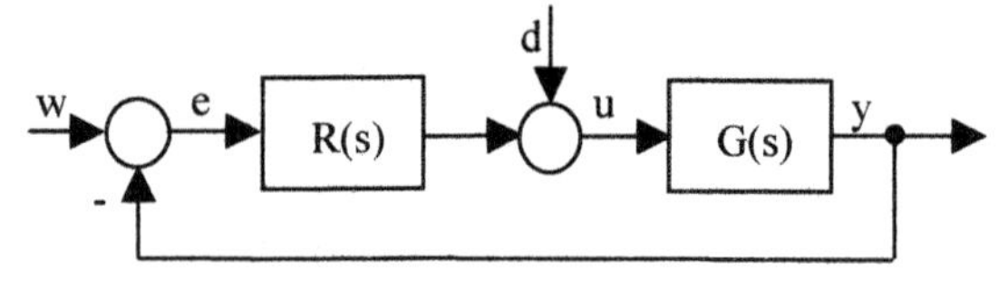

Fig. 1 Standard feedback configuration

where $G(s) \in R^{m \times l}$ and $R(s) \in R^{l \times m}$ are transfer function matrices and w, u, y, e, d are respectively vectors of reference, control, output, control error and disturbance of compatible dimensions. Hereafter, only square matrices will be considered, i.e. $m=l$.

The necessary and sufficient condition for the closed-loop stability provides the generalized Nyquist stability theorem based upon the concept of the system return difference [5, 7] defined as

$$F(s) = [I + Q(s)] \tag{1}$$

where

$F(s) \in R^{m \times m}$ is the system return-difference matrix

$Q(s) \in R^{m \times m}$ is the (open) loop transfer function matrix, for the system in Fig.1 $Q(s) = G(s)R(s)$

$H(s) \in R^{m \times m}$ is the closed-loop transfer function matrix, $H(s) = Q(s)[I + Q(s)]^{-1}$

R denotes the field of rational functions in s.

The *Nyquist D-contour* is a large contour in the complex plane consisting of the imaginary axis $s = j\omega$ and an infinite semi-circle into the right-half plane. It has to avoid locations where *Q(s)* has $j\omega$-axis poles (e.g. if *R(s)* includes integrators) by making small indentations around these points such as to include them to the left-half plane. Thus, unstable poles of *Q(s)* will be considered those in the *open* right-half plane. *Nyquist plot* of a complex function $g(s)$ is the image of the Nyquist under *g(s)*, whereby the *D*-contour is traversed clockwise; $N[k, g(s)]$ denotes number of anticlockwise encirclements of the point *(k, j0)* by the Nyquist plot of *g(s)*.

The characteristic polynomial of the system in Fig.1 is

$$\det F(s) = \det[I + Q(s)] = \det[I + G(s)R(s)] \tag{2}$$

If *Q(s)* has n_q unstable poles, the closed-loop stability can be determined using the generalized Nyquist stability criterion [1, 3, 4, 5, 7].

<u>*Theorem 1*</u> *(Generalized Nyquist Stability Theorem)*

The feedback system in Fig. 1 is stable if and only if

a. $\det F(s) \neq 0$

b. $N[0, \det F(s)] = n_q \quad (3)$

where n_q *is the number of its open-loop unstable poles.* □

As rational functions in s form a field, standard arithmetical manipulations with transfer functions can be carried out. Thus, eigenvalues of a square matrix $Q(s) \in R^{m \times m}$ (whose elements are rational functions in s) are themselves rational functions in s. They are called *characteristic functions* [3, 4, 5]. The m characteristic functions $q_i(s),\ i = 1,\dots,m$ of $Q(s)$ are given by

$$\det[q_i(s)I_m - Q(s)] = 0 \qquad i = 1,\dots,m \tag{4}$$

Then

$$\det F(s) = \det[I + Q(s)] = \prod_{i=1}^{m} [1 + q_i(s)] \tag{5}$$

Characteristic loci (CL) $q_i(j\omega),\ i = 1,2,\dots,m$ are the set of loci in the complex plane traced out by the characteristic functions of *Q(s)* as s traverses the Nyquist *D*-contour. This set is called the *spectral Nyquist plot* [1]. The *degree* of the spectral Nyquist plot is the sum of anticlockwise encirclements with respect to a specified point in the complex plane, contributed by the characteristic loci of $Q(s)$. A stability test analogous to *Theorem 1* has been derived in terms of the CL's [1, 3, 4, 5, 7].

Theorem 2

The closed-loop system with the open-loop transfer function Q(s) is stable if and only if $det[I+Q(s)] \neq 0$ and the degree of the spectral Nyquist plot of Q(s) with respect to (-1 ,0j) fulfils the following condition (with n_q being the number of unstable poles of Q(s))

$$\sum_{i=1}^{m} N[-1,\, q_i(s)] = n_q \tag{6a}$$

or alternatively, if considering the spectral Nyquist plot of [I+Q(s)]

$$\sum_{i=1}^{m} N\{0, [1+\, q_i(s)]\} = n_q \tag{6b}$$

□

Remarks:

1. In the sequel, whenever possible, matrices and their corresponding characteristic transfer functions will be denoted by like capital and small letters, respectively.
2. The CL's are not uniquely defined since whenever two particular loci cross, the identity of the correct continued path is lost. Furthermore, each locus does not in general form a closed contour. Fortunately, as the underlying characterristic functions are piecewise analytic, it is possible to concatenate the *m* CL's of $[I+Q(s)]$ to form one or more close contours [1].
3. Theorems 1 and 2 are equivalent, therefore

$$N\{0, det[I+Q(s)]\} = \sum_{i=1}^{m} N\{0, [1+\, q_i(s)]\} = n_q \tag{7}$$

In general, for points other than (0,0j), degrees of the respective plots are different [1].

3 PROBLEM FORMULATION

Consider a complex system with *m* subsystems *(m=l)*, described by a square transfer matrix $G(s) \in R^{m\times m}$ that can be split into the diagonal and off-diagonal parts. The transfer matrix collecting diagonal entries of *G(s)* is the model of decoupled (isolated) subsystems; interactions between subsystems are represented by the off-diagonal entries, i.e.

$$G(s) = G_d(s) + G_m(s) \tag{8}$$

where $G_d(s) = diag\{G_i(s)\}_{i=1,\dots,m}$

whereby $det\, G_d(s) \neq 0 \quad \forall s$

and $G_m(s) = G(s) - G_d(s)$

For the system (8), a decentralized controller is to be designed

$$R(s) = diag\,\{R_i(s)\}_{i=1,\dots,m}, \quad det\, R(s) \neq 0 \quad \forall s \tag{9}$$

where $R_i(s)$ is transfer function of the *i-th* subsystem local controller.

In compliance with the independent design framework, the point at issue is to appropriately assess the effect of interactions and to translate it into constraints for local controller designs to guarantee stability of the full closed-loop system and its required performance.

4 MAIN RESULTS

4.1 *Theoretical Background*

The proposed decentralized control design technique is based upon factorising the closed-loop characteristic polynomial of the full system (2) under the decentralized controller (9) in the following way

$$\begin{aligned} det\, F(s) &= det\{I + R(s)[(G_d(s) + G_m(s)]\} = \\ &= det\, R(s)\, det[R^{-1}(s) + G_d(s) + G_m(s)] \end{aligned} \tag{10}$$

Existence of $R^{-1}(s)$ is implied by the assumption that $det\, R(s) \neq 0$. Denote

$$F_1(s) = R^{-1}(s) + G_d(s) + G_m(s) \tag{11}$$

Then, with respect to (8)-(10), the following corollary holds.

Corollary 1

A closed-loop system comprising the system (9) and the decentralized controller (10) is stable if and only if

a. $det\, F_1(s) \neq 0$

b. $N[0,\, det\, F_1(s)] + N[0,\, det\, R(s)] = n_q \quad (12)$

If R(s) is stable, $N\{0,\, det[R(s)]\} = 0$ *and the encirclement condition (13b) reduces to*

$$\begin{aligned} &N[0,\, det\, F_1(s)] = \\ &= N\{0,\, det[R^{-1}(s) + G_d(s) + G_m(s)]\} = n_q \end{aligned} \tag{13}$$

□

As the term $[R^{-1}(s) + G_d(s)]$ in (11) is diagonal and includes all necessary information on subsystem dynamics, it is possible to specify a required performance of individual subsystems using an appropriately chosen diagonal matrix *P(s)*:

$$R^{-1}(s) + G_d(s) \overset{!}{=} P(s) \tag{14}$$

where $P(s) = diag\{\, p_i(s)\}_{i=1,\dots,m}$

From (15) results

$$I + R(s)[G_d(s) - P(s)] = 0 \tag{15}$$

or, on the subsystem level

$$1 + R_i(s) G_i^{eq}(s) = 0 \qquad i = 1,2,\dots,m \tag{16}$$

where

$$G_i^{eq}(s) = G_i(s) - p_i(s) \qquad i = 1,2,\dots,m \tag{17}$$

$G_i^{eq}(s)$ denotes the *i-th* subsystem transfer function modified by $p_i(s)$ and will be called the transfer function of the *i-th equivalent subsystem* or simply the *equivalent transfer function*. Similarly, (17) will be denoted the *i-th equivalent characteristic equation.*

Modifying (11) using (14) we obtain

$$det\, F_I(s) = det[P(s) + G_m(s)] \qquad (18)$$

Considering (18) we can formulate the encirclement stability conditions for the closed-loop system under a decentralized controller in terms of the spectral Nyquist plot of $F_I(s)$.

Corollary 2

A closed-loop system comprising the system (8) and a stable decentralized controller (9) is stable if

1. *there exists such a diagonal matrix* $P(s) = diag\{p_i(s)\}_{i=1,...,m}$ *that each equivalent subsystem* $G_i^{eq}(s) = G_i(s) - p_i(s),\ i = 1,...,m$ *can be stabilized by its related local controller* $R_i(s)$*, i.e. each equivalent characteristic polynomial*

   $CLCP_i^{eq} = 1 + R_i(s)G_i^{eq}(s) \qquad i = 1,2,...,m$

   has stable roots;

2. *any of the following conditions is satisfied*

   $$N[0,\ det[P(s) + G_m(s)] = n_m \qquad (19a)$$

   $$\sum_{i=1}^{m} N[0,\ m_i(s)] = n_m \qquad (19b)$$

   where $m_i(s),\ i = 1,...,m$ *are characteristic functions of* $M(s) = P(s) + G_m(s)$*;* n_m *is the number of its unstable poles.* □

Thus far, only stability in the DC design has been considered. Any additional performance requirements need to be included in the design by a suitable choice of $P(s) = diag\{p_i(s)\}_{i=1,...,m}$. Next, the choice of $p_i(s)$ is discussed in detail.

4.2 Development of the Decentralized Controller Design Technique

For the independent design the $p_i(s),\ i = 1,...,m$ actually represent bounds for local controller designs. To guarantee closed-loop stability of the full system they should be chosen such as to appropriately account for the interaction term $G_m(s)$.

According to (4), characteristic functions of $G_m(s)$ are defined

$$det[g_i(s)I - G_m(s)] = 0 \qquad i = 1,...,m \qquad (20)$$

For identical entries in the diagonal of $P(s)$, substituting (14) in (18) and equating to zero actually defines the m characteristic functions of $[-G_m(s)]$

$$det[p_i(s)I + G_m] = 0 \qquad i = 1,...,m \qquad (21)$$

To be able to include the performance issue in the design, the following considerations about $P(s) = p(s)I$ motivated by (20) and (21) have been adopted:

1. If $p(s) = p_\ell(s) = -g_\ell(s)$ for fixed $\ell \in \{1,...,m\}$ then

$$detF_I(s) = \prod_{i=1}^{m}[p(s) + g_i(s)] = \prod_{i=1}^{m}[-g_\ell(s) + g_i(s)] = 0 \qquad (22)$$

and the closed-loop system is not stable (it is on the stability boundary);

2. If $p(s) = p_\ell[s(\alpha)] = -g_\ell[s(\alpha)]$, fixed $\ell \in \{1,..,m\}$ (23)

where we consider the generalized frequency $s = -\alpha + j\omega,\ \alpha \geq 0;\ \omega \in (-\infty, \infty)$ and $s(\alpha)$ indicates the particular case when $\alpha > 0$, then

$$\begin{aligned} detF_I[s(\alpha)] &= \prod_{i=1}^{m}\{p_\ell[s(\alpha)] + g_i[s(\alpha)]\} = \\ &= \prod_{i=1}^{m}\{-g_\ell[s(\alpha)] + g_i[s(\alpha)]\} = 0 \end{aligned} \qquad (24)$$

Thus, as $det\, F_I[s(\alpha)] = 0$, the modified closed-loop system is exactly on the stability boundary "shifted to $(-\alpha)$", hence it is stable with the decay rate α. The equivalent subsystems transfer functions are

$$G_{i\ell}^{eq}[s(\alpha)] = G_i[s(\alpha)] - p_\ell[s(\alpha)] \quad i = 1,2,..., m \qquad (25)$$

It is noteworthy that a closed-loop system stable with a decay rate $\alpha \geq 0$ it is necessarily stable also for $\alpha = 0$.

According to (19b) and considering (22) and (23), the encirclement condition for the closed-loop stability in terms of the spectral Nyquist plot of $F_I(s)$ is then

$$\begin{aligned} &\sum_{i=1}^{m} N\{0,\ [p_\ell[s(\alpha)] + g_i(s)]\} = \\ &= \sum_{i=1}^{m} N\{0,\ m_{i\ell}^{eq}[s(\alpha)]\} = n_m \end{aligned} \qquad (26)$$

where

$$\begin{aligned} m_{i\ell}^{eq}[s(\alpha)] &= \{p_\ell[s(\alpha)] + g_i(s)\} = \\ &= \{-g_\ell[s(\alpha)] + g_i(s)\}, \qquad i = 1,...,m \end{aligned}$$

will be called *equivalent characteristic functions* of $M(s) = [P(s) + G_m(s)]$; n_m is the number of unstable poles of $M(s)$.

If for any $s = -\alpha + j\omega,\ \alpha \geq 0,\ \omega \in (-\infty, \infty)$

$$\begin{aligned} detF_I(s) &= \prod_{i=1}^{m}\{p_\ell[s(\alpha)] + g_i(s)\} = \\ &= \prod_{i=1}^{m}\{-g_\ell[s(\alpha)] + g_i(s)\} = 0 \end{aligned} \qquad (27)$$

i.e. if at any frequency, say ω_I, the particular $p_\ell(-\alpha + j\omega_I)$ and any characteristic locus $g_i(\omega_I),\ i \in \{1,...,m\}$ happen to cross, the closed loop system is not stable. The main theoretical results from the preceding section are summarized next.

Definition 1 (Set of stable characteristic functions)
$p_\ell[s(\alpha)]$ will be called a stable characteristic function if $\forall g_i(s),\ i=1,\dots,m$,
$s=-\alpha+j\omega,\ \alpha\geq 0,\ \omega\in(-\infty,\infty)$

$$1.\ \{p_\ell[s(\alpha)]+g_i(s)\}\neq 0;$$

$$2.\ \sum_{i=1}^{m} N\{0,\ m_{i\ell}^{eq}[s(\alpha)]\}=n_m$$

where $m_{i\ell}^{eq}[s(\alpha)]=\{p_\ell[s(\alpha)]+g_i(s)\}$ and n_m is the number of unstable poles of $M(s)=[P(s)+G_m(s)]$.
The set of all stable characteristic functions for a system will be denoted P_S. □

Lemma 1
The closed-loop system comprising the system (8) and a stable decentralized controller (9) is stable if

1. $p_\ell[s(\alpha)]=-g_\ell[s(\alpha)]$, *fixed* $\ell\in\{1,\dots,m\}$, $s=-\alpha+j\omega,\ \alpha\geq 0,\ \omega\in(-\infty,\infty)$ *belongs* to P_S and
2. the closed-loop characteristic polynomials of all equivalent subsystems (equivalent characteristic polynomials
   $CLCP_i^{eq}=1+R_{ii}(s)G_i^{eq}(s),\ i=1,2,\dots,m$
   are stable. □

Proof of Lemma 1 results from previous considerations.

4.3 Decentralized Controller Design Procedure

1. Partition the controlled system into subsystems (diagonal part) and interactions (off-diagonal part)
   a. $G(s)=G_d(s)+G_m(s)$
   b. specify $\alpha>0$;
2. Find the characteristic functions of $G_m(s)$: $g_i(s),\ i=1,\dots,m$
3. Choose $p_\ell[s(\alpha)]=-g_\ell[s(\alpha)]$; $p_\ell[s(\alpha)]\in P_S$ by examining equivalent characteristic loci $m_{i\ell}^{eq}[s(\alpha)]=\{p_\ell[s(\alpha)]+g_i(s)\},\ i=1,\dots,m$
   If no such $p_\ell[s(\alpha)]\in P_S$ can be found, no stabilizing decentralized controller can be designed using this approach, and the procedure stops; else
4. Design local controllers $R_i(s)$ for all m equivalent subsystems
   $G_i^{eq}[s(\alpha)]=G_i[s(\alpha)]-p_\ell[s(\alpha)]\quad i=1,\dots,m;$
   using any suitable design technique, e.g. the Neymark D-partition method [6].

4.4 Robust Controller Design Procedure

The developed DC design procedure is applicable within the decentralized robust controller design strategy developed in [2]. In the robust approach the DC is to be designed for the uncertain system

$$\tilde{G}(s)=G_D(s)+G_m(s)+\Delta(s) \tag{28}$$

where the diagonal matrix $G_D(s)$ is considered as the nominal model, while interconnections along with the uncertainty are included in the generalized multiplicative perturbation matrix

$$L_R(s)=[\tilde{G}(s)-G_D(s)]G_D^{-1}(s) \tag{29}$$

The global uncertain system $\tilde{G}(s)$ is to be controlled using the controller (9), which has to guarantee closed-loop stability of the nominal (diagonal) system (nominal stability)

$$H_D(s)=G_D(s)R(s)[I+G_D(s)R(s)]^{-1} \tag{30}$$

as well as of the overall uncertain system (robust stability)

$$\tilde{H}(s)=\tilde{G}(s)R(s)[I+\tilde{G}(s)R(s)]^{-1} \tag{31}$$

The robust DC design strategy evolves from the sufficient condition for robust stability of systems under DC (SCRS-DC) [2]

$$\|L_R H_D\|<1 \qquad \forall\omega \tag{32}$$

Basically, the DC design for robust stability consists in breaking down (32) into local bounds for subsystems, which if fulfilled, guarantee closed-loop stability of the overall uncertain system (31). For the standard unity feedback configuration applied in each local loop the subsystem constraints are as follows

$$\left|\frac{G_{ii}(j\omega)R_i(j\omega)}{1+G_{ii}(j\omega)R_i(j\omega)}\right|<M_0(\omega)\quad i=1,\dots,m;\quad\forall\omega \tag{33}$$

$$M_0(\omega)=\frac{1}{\|[\tilde{G}(j\omega)-G_D(j\omega)]G_D^{-1}(j\omega)\|} \tag{34}$$

$M_0(\omega)$ does not depend on the controller, it actually assesses uncertainty and interactions within the system. The expression on the l..h.s. in (33) is the local closed loop magnitude $|H_i(j\omega)|$, hence (33) becomes

$$|H_i(j\omega)|<M_0(\omega)\quad i=1,\dots,m;\quad\forall\omega \tag{35}$$

Put simply, a system under a decentralized controller is robustly stable if there are such local controllers, which provide stability of related isolated subsystems and simultaneously comply with (33). Graphical interpretation of (33) provides useful robust stability criteria, the simplest-to-use being the following.

Magnitude robust stability criterion [2]:
Let $G_D(s)$ and $\tilde{G}(s)$ have the same number of unstable poles. The overall uncertain closed-loop

system (31) is guaranteed to be stable in face of specified uncertainty (both structured and/or unstructured) as long as there are such local controllers $R_i(s),\ i=1,\dots,m$ which stabilize related subsystems and simultaneously ensure that over the given frequency range, the $|H_i(j\omega)|$-versus-ω plot is upper-bounded by the $M_0(\omega)$-versus-ω plot.

This robust stability criterion enables an at-a-glance testing of stability of the global uncertain system as well as of individual subsystems. In addition, information on subsystem performance in terms of bandwidth and resonance peak is provided. Besides testing robust stability, the Magnitude criterion can be used as a tool for heuristic tuning of local controllers, whereby stability of the global system is improved by increasing relative stability measures of subsystems.

The proposed robust controller design procedure is illustrated on an example.

5 EXAMPLE

The proposed design technique has been applied for the design of a robust decentralized controller for a TITO system

$$G(s)=\begin{bmatrix} G_{11}(s) & G_{12}(s) \\ G_{21}(s) & G_{22}(s) \end{bmatrix}$$

The process is given by two transfer function matrices $\tilde{G}_1$ and $\tilde{G}_2$ considered as perturbed models, defining the operating range of the system.

1st operating point - $\tilde{G}_1(s)$:

$$\tilde{G}_{11}(s) = \frac{0.0172s^2-0.1040s+0.4487}{s^3+2.232s^2+2.098s+0.6122}$$

$$\tilde{G}_{12}(s) = \frac{0.012s^2-0.0346s-0.1197}{s^3+2.741s^2+1.881s+.5701}$$

$$\tilde{G}_{21}(s) = \frac{0.0203s^2-0.0535s+1.114}{s^3+4.129s^2+6.228s+3.795}$$

$$\tilde{G}_{22}(s) = \frac{0.0174s^2-0.1067s+0.3996}{s^3+3.257s^2+1.809s+0.3414}$$

2nd operating point - $\tilde{G}_2(s)$:

$$\tilde{G}_{11}(s) = \frac{0.0163s^2-0.0995s+0.4272}{s^3+2.193s^2+2.049s+0.6091}$$

$$\tilde{G}_{12}(s) = \frac{0.0191s^2-0.0404s-0.1014}{s^3+2.367s^2+1.686s+0.5165}$$

$$\tilde{G}_{21}(s) = \frac{0.0062s^2-0.0148s+0.9213}{s^3+3.724s^2+5.402s+3.3}$$

$$\tilde{G}_{22}(s) = \frac{0.041s^2-0.1437s+0.4843}{s^3+3.772s^2+2.21s+0.433}$$

The designed controller has to guarantee a required performance of the nominal model and robust stability of the global system over the operating range. The nominal model *G(s)* has been obtained as a model of mean parameter values [7]. The chosen decay rate is $\alpha = 0.1$

Equivalent characteristic loci for $\ell = 1$, are in Fig. 2. Obviously, $p_l(s,\alpha)\in P_S$ as $n_m = 0$.

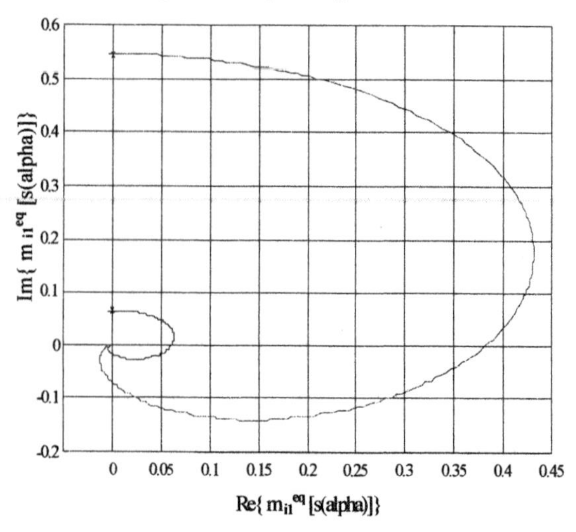

Fig. 2 Equivalent characteristic loci ($\ell = 1$ and $\alpha = 0.1$)

For the related equivalent subsystems, local PI controllers were designed in form

$$R(s) = r_0 + \frac{r_1}{s}$$

applying the Neymark D-partition method to both equivalent characteristic equations (Fig. 3).

$$R_i^{-1}(s)+\{G_i[s(\alpha)]-p_l[s(\alpha)]\}=0 \qquad i=1,2$$

1st equivalent subsystem

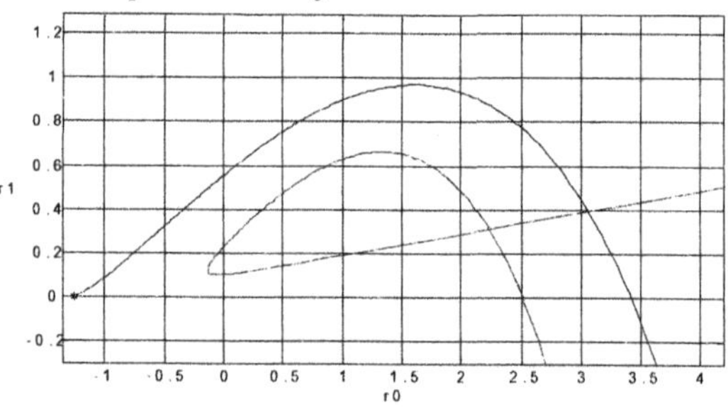

2nd equivalent subsystem

Fig. 3 Neymark D-contours for both equivalent subsystems, $\alpha = \{0, 0.1\}$

For both equivalent subsystems, parameters of local PI controllers have been chosen from the boundary of the closed areas in Fig. 3, which correspond to the closed-loop decay rate of the full system $\alpha = 0.1$. The upper contours correspond to $\alpha = 0$, the inside of the closed areas to $\alpha > 0.1$. Design results are summarized in Tab 1.

Subsystem	Controller	Achieved decay rate α
1	$R_1 = 1.2740 + \frac{0.2149}{s}$	0.1007
2	$R_2 = 0.9222 + \frac{0.1429}{s}$	

Tab. 1 Results of the local controller design

The whole closed-loop eigenvalue set is

$$\Lambda = \{-0.1007 \pm 0.0091j;\ -0.2499 \pm 0.4113j;\ -0.367 \pm 0.7166j; -0.4163 \pm 0.3871j;\ -1.0801 \pm 0.9296j; -1.3172; -1.7043;\ -1.8145; -2.9918\}$$

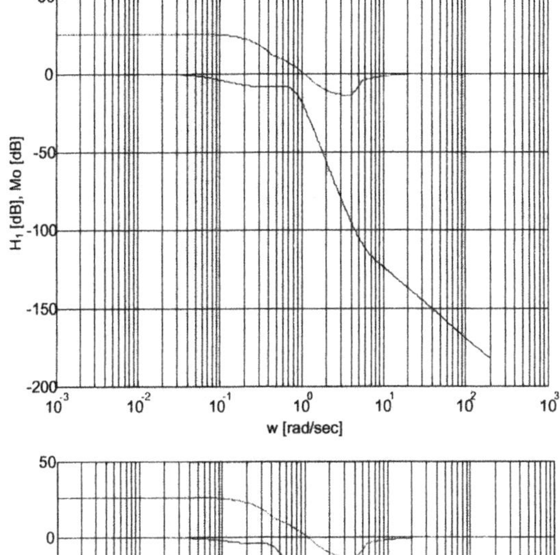

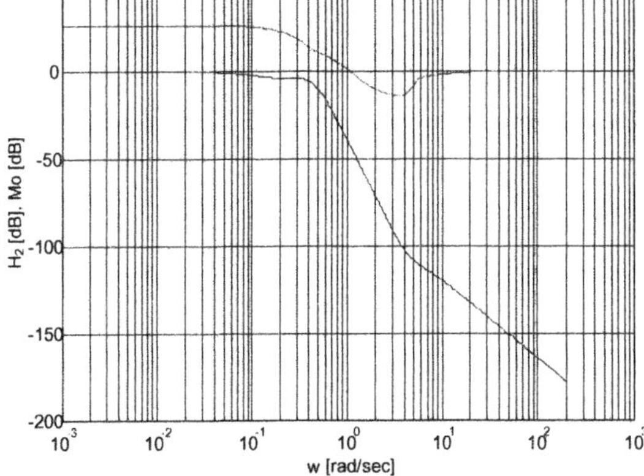

Fig. 4 Magnitude robust stability test for both subsystems

The robust stability test using the Magnitude criterion proves robust stability of the full system over the whole operating range specified by transfer matrices in two operating points.

CONCLUSION

In this paper a novel frequency-domain approach to the decentralized controller design for performance has been proposed. It is based on including plant interactions in the design of local controllers through their characteristic functions, modified so as to achieve a required closed-loop performance in terms of the full system decay rate. The independent design is carried out for the so called „equivalent subsystems" - models of individual decoupled subsystems modified using characteristic functions of the plant interaction matrix. Local controllers designed for equivalent subsystems guarantee fulfilment of performance requirements imposed on the full system. When applied within the robust control strategy [2] the developed design procedure applied to the nominal model yields local controllers that provide robust properties over a specified range of uncertainty. Theoretical results are supported with results obtained by solving several examples one of which is included.

ACKNOWLEDGEMENT

This work has been supported by the Scientific Grant Agency of the Ministry of Education of the Slovak Republic and the Slovak Academy of Sciences under Grant No. 1/0158/03.

REFERENCES

[1] DeCarlo, R.A., Saeks, R. "*Interconnected dynamical systems*", Marcel Dekker, Inc., 1981.

[2] Kozáková, A.: Robust Decentralized Control of Complex Systems in the Frequency Domain. *2nd IFAC Workshop on New Trends in Design of Control Systems, 7-10 Sept. 1997, Smolenice, Slovakia*, Elsevier, Kidlington, UK, 1998.

[3] MacFarlane, A.G.J., Belletrutti, J.J. "The Characteristic Locus Design Method", *Automatica*, **9**, pp. 575-588 (1973).

[4] MacFarlane, .G.J., Kouvaritakis, B.: A design technique for linear multivariable feedback systems. *Int. Journal of Control*, **25**, No.1 (1977), 837-874.

[5] MacFarlane, A.G.J., Postlethwaite, I.: The Generalized Nyquist Stability Criterion and Multivariable Root Loci. *Int. Journal of Control*, **25**, No. 1 (1977), 81-127.

[6] Neymark, J.J. "Dynamical systems and controlled processes", Nauka, Moscow, 1978 (in Russian).

[7] Skogestad, S., Postlethwaite, I. "*Multivariable Fedback Control: Analysis and Design*", John Wiley & Sons, 1996.

whole operating range specified by transfer functions in two operating points.

CONCLUSIONS

... subsystems, parameters of local ... have been chosen from the nominal ... in Fig. ... which correspond to the ...

www.elsevier.com/locate/ifac

AFFINE ROBUST CONTROLLER DESIGN FOR LINEAR PARAMETRIC UNCERTAIN SYSTEMS WITH OUTPUT FEEDBACK

M. Hypiusová, V. Veselý

Department of Automatic Control Systems,
Faculty of Electrical Engineering and Information Technology,
Slovak University of Technology, Ilkovičova 3, 812 19 Bratislava, Slovakia.
E-mail: hypiusov@kasr.elf.stuba.sk, vesely@kasr.elf.stuba.sk

Abstract: The paper deals with a new procedure for robust output feedback controller design providing affine quadratic stability. The developed procedure leads to a non-iterative LMI based algorithm. The proposed affine robust controller applied to a plant consisting of two co-operating DC servomotors was evaluated using affine quadratic and quadratic stability test. *Copyright © 2003 IFAC*

Keywords: Robust control, Linear matrix inequality, Affine quadratic stability, Output feedback.

1. INTRODUCTION

For many real processes a controller design has to cope with the effect of uncertainties, which very often causes poor performance or even instability of the closed-loop systems. Considering these facts, the stability and performance robustness are of great practical importance and have attracted a lot of interest for several decades.

The output feedback problem is one of the most important open questions of control engineering. In a simple way, the problem can be formulated as follows: for a given complex linear system a robust controller with output feedback is to be found, which would provide some desirable characteristics to the closed-loop system, or determine that such a feedback does not exist.

During the last two decades numerous papers dealing with the design of robust output feedback control schemes to stabilize such systems have been published (Benton and Smith, 1999; Goh, *et al.*, 1996; Iwasaki, *et al.*, 1994; Kose and Jabbari, 1999; Li Yu and Jian Chu, 1999; Veselý, 2000). In the above papers, the authors basically conclude that the problem of static output feedback is still open despite the availability of many approaches and numerical algorithms.

The theory of linear matrix inequalities (LMI) (Boyd, *et al.*, 1994) has been used in several design methods (Benton and Smith, 1999; Li Yu and Jian Chu, 1999). Most of above works present iterative algorithms in which a set of equations or a set of LMIs is repeated until certain convergence criteria are met. The authors in (Kose and Jabbari, 1999) have studied conditions under which the designed output feedback controllers can be divided into two stages and the dynamic output feedback can be obtained. Authors in (Benton and Smith, 1999) have proposed an LMI based algorithm, which does not require iteration of LMI problems. The goal is to eliminate the need for iteration by an appropriate choice of initializing state feedback matrix. The proposed algorithm can be used to robustly stabilize a polytopic system via static output feedback.

In this paper one approach to the robust controller design has been developed. The new necessary and sufficient conditions to stabilize continuous time systems via static output feedback are used to design the robust controller. The proposed method is based on the results of (Gahinet, *et al.*, 1996) to design a robust output feedback controller for linear continuous-time systems, which ensures affine quadratic closed loop stability. The proposed method for polytope systems leads to a non-iterative LMI based algorithm. For this method the designer can prescribe the structure of output feedback controller.

The paper is organized as follows. In Section 2 the problem formulation and some preliminary results are presented. The main results are given in Section 3. In Section 4 the obtained theoretical results are applied to an example. We use a standard notation. A real symmetric positive (negative) definite matrix P is denoted by $P > 0\ (P < 0)$. Much of the notation and terminology complies with the references (Benton and Smith, 1999; Boyd, *et al.*, 1994; Kučera and De Souza, 1995).

2. PROBLEM FORMULATION AND PRELIMINARIES

In the context of robustness analysis and synthesis of robust controllers for linear time invariant systems the following uncertain model is commonly used

$$\begin{aligned}\dot{x}(t) &= (A+\delta A)x(t)+(B+\delta B)u(t)\\ y(t) &= Cx(t)\end{aligned} \tag{1}$$

where $x \in R^n$, $u \in R^m$ and $y \in R^l$ are state, control and output vectors, respectively; A, B and C are known matrices of appropriate dimensions; δA, δB are unknown but norm bounded uncertainties. In the next development the matrix affine type uncertain structure will be used

$$\begin{gathered}\delta A = \sum_{i=1}^{p}\varepsilon_i A_i \quad \delta B = \sum_{i=1}^{s}\varepsilon_i B_i\\ \underline{\varepsilon}_i \le \varepsilon_i \le \overline{\varepsilon}_i\end{gathered} \tag{2}$$

where A_i, B_i are known matrices; ε_i are unknown parameters. In general, the polytope characterization of uncertainties results in less conservative controller designs than using other characterizations of uncertainty (Boyd, *et al.*, 1994).

The problem studied in this paper can be formulated as follows. For linear continuous-time invariant system described by (1) a robust static output feedback controller is to be designed with the control algorithm

$$u = FCx \tag{3}$$

such that the closed loop system

$$\dot{x} = (A+BFC)x + (\delta A + \delta BFC)x \tag{4}$$

is stable for all admissible uncertainties described by (2) and simultaneously guaranteeing the suboptimal solution to the performance index

$$J = \int_0^\infty \left(x^T Qx + u^T Ru\right)dt \tag{5}$$

where $Q = Q^T \ge 0$ and $R = R^T \ge 0$ are matrices of compatible dimensions. The nominal model of the system (1) is

$$\begin{aligned}\dot{x}(t) &= Ax(t)+Bu(t)\\ y(t) &= Cx(t)\end{aligned} \tag{6}$$

The following lemma is well known (Lankaster, 1969).

Lemma 1. Suppose $P > 0$ to be the solution of the following Lyapunov matrix equation

$$A^T P + PA + Q = 0 \tag{7}$$

Then A is stable iff $Q > 0$.

If such P exists, we say that the matrix A is quadratically stable. A linear time invariant system is stable if and only if it is quadratically stable. It is possible, however, that e.g. linear polytopic systems can be stable without being quadratically stable (Boyd, *et al.*, 1994).

3. OUTPUT FEEDBACK CONTROLLER DESIGN

In this paragraph we present a new procedure to design a static output feedback controller for continuous-time systems guaranteeing affine quadratic closed loop stability. This section is concerned with a class of uncertain polytope linear systems that can be described by state-space equations of the form

$$\begin{aligned}\dot{x}(t) &= A(\varepsilon)x(t)+B(\varepsilon)u(t) \qquad x(0)=x_0\\ y(t) &= C(\varepsilon)x(t)\end{aligned} \tag{8}$$

where

$$\begin{aligned}A(\varepsilon) &= A_0 + A_1\varepsilon_1 + A_2\varepsilon_2 + \ldots + A_p\varepsilon_p\\ B(\varepsilon) &= B_0 + B_1\varepsilon_1 + B_2\varepsilon_2 + \ldots + B_p\varepsilon_p\\ C(\varepsilon) &= C_0 + C_1\varepsilon_1 + C_2\varepsilon_2 + \ldots + C_p\varepsilon_p\end{aligned}$$

and $\varepsilon = \lfloor \varepsilon_1, \varepsilon_2, \ldots, \varepsilon_p \rfloor$ is a vector of uncertain and possibly time-varying real parameters; $A_0, B_0, C_0, A_1, B_1, C_1, \ldots$ are known fixed matrices. Consider that either the matrix $B(\varepsilon)$ or $C(\varepsilon)$ is varied and the other matrix B or C, respectively, is constant.

We also assume that lower and upper bounds for both the parameter values and rates of variation are available. Specifically,

$$\varepsilon_i \in \langle \underline{\varepsilon}_i, \overline{\varepsilon}_i \rangle \text{ and } \dot{\varepsilon}_i \in \langle \underline{r}_i, \overline{r}_i \rangle,\ i = 1,2,\ldots,p \quad (9)$$

where $\underline{\varepsilon}_i, \overline{\varepsilon}_i, \underline{r}_i, \overline{r}_i$ are known constants. In the sequel

$$\begin{aligned} \Omega &= \{(\omega_1, \omega_2, \ldots, \omega_p) : \omega_i \in \{\underline{\varepsilon}_i, \overline{\varepsilon}_i\}\} \\ \mathrm{T} &= \{(\tau_1, \tau_2, \ldots, \tau_p) : \tau_i \in \{\underline{r}_i, \overline{r}_i\}\} \end{aligned} \quad (10)$$

denote the set of 2^{2p} vertices or corners of this parameter box.

The closed loop system (8) with the control algorithm

$$u = FC(\varepsilon)x \quad (11)$$

is given as follows

$$\dot{x} = (A(\varepsilon) + B(\varepsilon)FC(\varepsilon))x = A_c(\varepsilon)x \quad (12)$$

The following definition has been introduced in (Gahinet, *et al.*, 1996).

Definition 1. The linear system (12) is affine quadratically stable if there exist p+1 symmetric matrices $P_0, P_1, \ldots, P_p$ such that

$$V = x^T P(\varepsilon)x > 0 \quad (13)$$

$$\frac{dV}{dt} = x^T \left(A_c(\varepsilon)^T P(\varepsilon) + P(\varepsilon)A_c(\varepsilon) + \frac{dP(\varepsilon)}{dt} \right) x < 0 \quad (14)$$

hold for all admissible trajectories of the parameter vector ε with conditions (9) where

$$P(\varepsilon) = P_0 + P_1\varepsilon_1 + P_2\varepsilon_2 + \ldots + P_p\varepsilon_p > 0 .$$

Note that the quadratic stability corresponds to the case when $P_1 = P_2 = \ldots = P_p = 0$.

Next, we assume that $A(\varepsilon)$ affinely depends on the vector ε as in (8) and that ε and $\dot{\varepsilon}$ range in the polytope defined by (9). The sufficient stability conditions for affine quadratic stability of the closed loop system (12) are given in the next Theorem (Gahinet, *et al.*, 1996).

Theorem 1. Consider a linear system governed by (12). Let Ω and T denote sets of vertices of the parameter box (10) and let

$$\varepsilon_m = \left[\frac{\underline{\varepsilon}_1 + \overline{\varepsilon}_1}{2}, \ldots, \frac{\underline{\varepsilon}_p + \overline{\varepsilon}_p}{2} \right] \quad (15)$$

denote the average value of the parameter vector.

A system is affine quadratically stable if $A_c(\varepsilon_m)$ is stable and if there exist p+1 symmetric matrices $P_0, \ldots, P_p$ such that $P(\varepsilon) = P_0 + P_1\varepsilon_1 + P_2\varepsilon_2 + \ldots + P_p\varepsilon_p > 0$ satisfies

$$A_c(\varepsilon)^T P(\varepsilon) + P(\varepsilon)A_c(\varepsilon) + \sum_{i=1}^{p} P_i\tau_i + \sum_{i=1}^{p} \varepsilon_i^2 M_i < 0 \quad (16)$$

for all $(\omega, \tau) \in \Omega \times \mathrm{T}$ and

$$A_{c_i}^T P_i + P_i A_{c_i} + M_i \geq 0 \text{ for } i = 1,2,\ldots,p \quad (17)$$

where $A_{c_i} = A_i + B_i FC_i$ and $M_i = M_i^T \geq 0$ are some nonnegative definite matrices.

Using matrices M_i in inequalities (16) and (17) the conservatism of the affine quadratic stability test we can be reduced.

The next theorem is the main result of our paper and gives sufficient conditions for output feedback stabilizability of the closed-loop system (12). We consider $M_1 = \ldots = M_p = 0$ for the controller synthesis and some $M_i = M_i^T \geq 0$ for the closed loop system analysis using affine quadratic stability test.

Theorem 2. The sufficient condition for the output feedback affine quadratic stabilizability of the closed loop system (12) is given as follows.

1. $A_c(\varepsilon_m)$ is stable.
2. There exist $R = R^T > 0$, $Q = Q^T > 0$ such that

$$\begin{aligned} &A(\varepsilon)^T P(\varepsilon) + P(\varepsilon)A(\varepsilon) - P(\varepsilon)B(\varepsilon)R^{-1}B(\varepsilon)^T P(\varepsilon) - \\ &- C(\varepsilon)^T F^T RFC(\varepsilon) + \sum_{k=1}^{p} P_k\tau_k + Q < 0 \\ &\left(B(\varepsilon)^T P(\varepsilon) + RFC(\varepsilon)\right)\Phi_\varepsilon^{-1}\left(B(\varepsilon)^T P(\varepsilon) + RFC(\varepsilon)\right) - \\ &- R < 0 \end{aligned} \quad (18)$$

hold for all $(\omega, \tau) \in \Omega \times \mathrm{T}$ where

$$\Phi_\varepsilon = -\Big(A(\varepsilon)^T P(\varepsilon) + P(\varepsilon)A(\varepsilon) - P(\varepsilon)B(\varepsilon)R^{-1}B(\varepsilon)^T P(\varepsilon) - \\ - C(\varepsilon)^T F^T RFC(\varepsilon) + \sum_{k=1}^{p} P_k\tau_k + Q\Big)$$

Proof of this theorem is omitted.

The system represented by (8) is a polytope of linear systems described by a set of its vertices, i.e. in the form

$$\{(D_1, E_1, C), (D_2, E_2, C), \ldots, (D_N, E_N, C)\} \quad (19)$$

where

$$N = 2^p$$
$$D_1 = A_0 + \underline{\varepsilon}_1 A_1 + \underline{\varepsilon}_2 A_2 + \ldots$$

$$E_1 = B_0 + \underline{\varepsilon}_1 B_1 + \underline{\varepsilon}_2 B_2 + \ldots$$

Equations (16), (17) and (18) have given impulse for the following algorithm for computation of the output feedback gain matrix F.

Algorithm:

1. Compute ARI

$$D_i^T P_0 + P_0 D_i - P_0 B_i R^{-1} B_i^T P_0 + Q \le 0 \qquad i = 1,2,\ldots,N \tag{20}$$

with respect to $P_0 = P_0^T > 0$.

2. Compute the matrix P_1 from the following inequalities

$$D_i^T (P_0 + \varepsilon_1 P_1) + (P_0 + \varepsilon_1 P_1) D_i + \tau_1 P_1 -$$
$$- (P_0 + \varepsilon_1 P_1) E_i R^{-1} E_i^T (P_0 + \varepsilon_1 P_1) + Q \le 0$$
$$i = 1,2,\ldots,N$$
$$P_0 + \varepsilon_1 P_1 \ge 0$$

for all $(\varepsilon_1, \tau_1) \in \Omega \times \mathrm{T}$ and $A_1^T P_1 + P_1 A_1 \ge 0$

3. Compute the matrix P_2 from the following inequalities

$$D_i^T (P_0 + \varepsilon_1 P_1 + \varepsilon_2 P_2) + (P_0 + \varepsilon_1 P_1 + \varepsilon_2 P_2) D_i -$$
$$- (P_0 + \varepsilon_1 P_1 + \varepsilon_2 P_2) E_i R^{-1} E_i^T (P_0 + \varepsilon_1 P_1 + \varepsilon_2 P_2) +$$
$$+ \tau_1 P_1 + \tau_2 P_2 + Q \le 0$$

for all $(\varepsilon_1, \tau_1, \varepsilon_2, \tau_2) \in \Omega \times \mathrm{T}$, $i = 1,2,\ldots,N$ and

$$P_0 + \varepsilon_1 P_1 + \varepsilon_2 P_2 > 0, \quad A_2^T P_2 + P_2 A_2 \ge 0$$
$$\vdots$$

Last Step: Compute the output feedback gain matrix F

$$\left(B(\varepsilon)^T P(\varepsilon) + RFC\right) \Phi_\varepsilon^{-1} \left(B(\varepsilon)^T P(\varepsilon) + RFC\right) - R < 0 \tag{21}$$

where

$$\Phi_\varepsilon = -\Big(A(\varepsilon)^T P(\varepsilon) + P(\varepsilon) A(\varepsilon) -$$
$$- P(\varepsilon) B(\varepsilon) R^{-1} B(\varepsilon)^T P(\varepsilon) + \sum_{k=1}^{p} P_k \tau_k + Q\Big)$$

for all $(\varepsilon, \tau) \in \Omega \times \mathrm{T}$.

Note that if the solutions to (20) and (21) are feasible for $P_1 = P_2 = \ldots = 0$ and the closed loop system (12) is quadratically stable in all N vertices of the polytope then the system is quadratically stable for all uncertainties given by the vector ε.

4. EXAMPLE

Consider the mathematical model of a plant (1) and (2) where $\varepsilon_i \in \langle -1, 1 \rangle$, $i = 1,2$

$$A_0 = \begin{bmatrix} 0 & -0.215 & 0 & 0 & 0 & 0 & 0 & 0 \\ 1 & -1.014 & 0 & 0 & 0 & 0 & 0 & 0 \\ 0 & 0 & 0 & -0.261 & 0 & 0 & 0 & 0 \\ 0 & 0 & 1 & -0.911 & 0 & 0 & 0 & 0 \\ 0 & 0 & 0 & 0 & 0 & -0.164 & 0 & 0 \\ 0 & 0 & 0 & 0 & 1 & -0.814 & 0 & 0 \\ 0 & 0 & 0 & 0 & 0 & 0 & 0 & -0.228 \\ 0 & 0 & 0 & 0 & 0 & 0 & 1 & -0.822 \end{bmatrix}$$

$$A_1 = \begin{bmatrix} 0 & -0.0250 & 0 & 0 & 0 & 0 & 0 & 0 \\ 0 & -0.1395 & 0 & 0 & 0 & 0 & 0 & 0 \\ 0 & 0 & 0 & -0.0938 & 0 & 0 & 0 & 0 \\ 0 & 0 & 0 & -0.2911 & 0 & 0 & 0 & 0 \\ 0 & 0 & 0 & 0 & 0 & 0.0188 & 0 & 0 \\ 0 & 0 & 0 & 0 & 0 & 0.0208 & 0 & 0 \\ 0 & 0 & 0 & 0 & 0 & 0 & 0 & -0.0333 \\ 0 & 0 & 0 & 0 & 0 & 0 & 0 & -0.1173 \end{bmatrix}$$

$$A_2 = \begin{bmatrix} 0 & 0.0125 & 0 & 0 & 0 & 0 & 0 & 0 \\ 0 & 0.0594 & 0 & 0 & 0 & 0 & 0 & 0 \\ 0 & 0 & 0 & 0.0116 & 0 & 0 & 0 & 0 \\ 0 & 0 & 0 & 0.0308 & 0 & 0 & 0 & 0 \\ 0 & 0 & 0 & 0 & 0 & -0.0156 & 0 & 0 \\ 0 & 0 & 0 & 0 & 0 & -0.0565 & 0 & 0 \\ 0 & 0 & 0 & 0 & 0 & 0 & 0 & 0.0434 \\ 0 & 0 & 0 & 0 & 0 & 0 & 0 & 0.1258 \end{bmatrix}$$

$$B_0 = \begin{bmatrix} 0.3148 & 0 \\ 0.0478 & 0 \\ 0 & -0.1028 \\ 0 & -0.0091 \\ -0.0841 & 0 \\ -0.0287 & 0 \\ 0 & 0.3676 \\ 0 & 0.2448 \end{bmatrix} \quad B_1 = \begin{bmatrix} 0.0625 & 0 \\ -0.0798 & 0 \\ 0 & -0.0462 \\ 0 & 0.0449 \\ 0.0016 & 0 \\ 0.0072 & 0 \\ 0 & 0.0770 \\ 0 & -0.0050 \end{bmatrix}$$

$$B_2 = \begin{bmatrix} -0.0094 & 0 \\ 0.0151 & 0 \\ 0 & 0.0019 \\ 0 & -0.0030 \\ -0.0121 & 0 \\ -0.0030 & 0 \\ 0 & -0.0640 \\ 0 & 0.0189 \end{bmatrix} \quad C^T = \begin{bmatrix} 0 & 0 \\ 1 & 0 \\ 0 & 0 \\ 1 & 0 \\ 0 & 0 \\ 0 & 1 \\ 0 & 0 \\ 0 & 1 \end{bmatrix}$$

In case of a PI controller design the extended state space model is to be used. This extension is carried out as follows

$$A_I = \begin{bmatrix} A & 0 \\ C & 0 \end{bmatrix} \quad B_I = \begin{bmatrix} B \\ 0 \end{bmatrix} \quad C_I = \begin{bmatrix} C & 0 \\ 0 & I \end{bmatrix}$$

$$\delta A_I = \begin{bmatrix} \delta A & 0 \\ 0 & 0 \end{bmatrix} \quad \delta B_I = \begin{bmatrix} \delta B \\ 0 \end{bmatrix}$$

In case of decentralized control with 2 inputs and 4 outputs, application of the static output feedback yields the matrix F in the following form

$$F = \begin{bmatrix} F_{11} & 0 & F_{13} & 0 \\ 0 & F_{22} & 0 & F_{24} \end{bmatrix}$$

and the ith obtained transfer function of the PI subcontroller is as follows

$$G_{Rij} = F_{ii} + \frac{F_{ii+2}}{s}$$

Some of the calculated controller parameters for various settings of the Q and R matrices along with the maximum close loop eigenvalues are listed in the Table 1, where $\varepsilon_i = 1$, $Q = qI$ and $R = rI$. As the controllers stabilize the polytope vertices and one point inside the polytope, they ensure the affine quadratic stability for the closed loop system with uncertainties.

Table 2 summarizes robust analysis results of closed loop systems under affine quadratic controllers. Quadratic stability test (QS) and affine quadratic stability test (AQS) were applied. Presented QS and AQS results indicate percentages by which it is possible to extend the maximal value of the uncertainty parameters $\varepsilon_{lm} = \max\|\varepsilon_i\|$ still preserving either closed loop quadratic stability or closed loop affine quadratic stability, respectively.

For the third controller QS is less than 100 %, which means that the closed loop quadratic stability is satisfied only for $\varepsilon_{lm} = 0.8213$.

The results show that the affine controllers give good performance and larger region of affine quadratic stability. Further it is possible to see that the affine quadratic stability test is less conservative than the quadratic stability test.

5. CONCLUSION

The main aim of this paper has been to present a new method for solving the problem of designing robust controllers with output feedback via non-iterative LMI approach guaranteeing the affine quadratic stability of the uncertain closed loop system. Application of this approach to a real laboratory model has resulted in a high robustness measure.

Table 1. Controllers designed for various robustness and performance requirements

q	r	F	*Max. close loop eigenvalues*
100	1	$\begin{bmatrix} 1.410 & 0 & 0.676 & 0 \\ 0 & 4.559 & 0 & 1.572 \end{bmatrix}$	-0.27±0.57i
1	5	$\begin{bmatrix} 0.375 & 0 & 0.402 & 0 \\ 0 & 0.491 & 0 & 0.486 \end{bmatrix}$	-0.13±0.37i
0.5	1	$\begin{bmatrix} 0.762 & 0 & 0.935 & 0 \\ 0 & 0.696 & 0 & 0.777 \end{bmatrix}$	-0.12±0.55i

Table 2. Robust analysis of affine controllers by quadratic stability test and affine quadratic stability test

F	QS [%]	AQS [%]
$\begin{bmatrix} 1.410 & 0 & 0.676 & 0 \\ 0 & 4.559 & 0 & 1.572 \end{bmatrix}$	173.14	246.19
$\begin{bmatrix} 0.375 & 0 & 0.402 & 0 \\ 0 & 0.491 & 0 & 0.486 \end{bmatrix}$	102.15	245.31
$\begin{bmatrix} 0.762 & 0 & 0.935 & 0 \\ 0 & 0.696 & 0 & 0.777 \end{bmatrix}$	82.13	246.09

REFERENCES

Benton, I.E. and D. Smith (1999). A non iterative LMI based algorithm for robust static output feedback stabilization. *IJC*, Vol. 72, pp. 1322-1330.

Boyd, S., L. El Ghaoui, E. Feron and V. Balakrishnam (1994). Linear Matrix Inequalities in System and Control Theory. *SIAM*, 15.

Gahinet, P., P. Apkarian and M. Chilali (1996). Affine parameter-dependent Lyapunov functions and real parametric uncertainty, *IEEE Trans. on AC*, Vol. 42, pp. 436-442.

Goh, K.C., M.G. Safonov and J.H. Ly (1996). Robust synthesis via bilinear matrix inequalities. *Int. Journal of Robust and Nonlinear Control*, Vol. 6, pp. 1079-1095.

Iwasaki, T., R.E. Skelton and J.C Geromel (1994). Linear quadratic suboptimal control with static output feedback. *Systems and Control Letters*, Vol. 23, pp. 421-430.

Kose, I.E. and F. Jabbari (1999). Robust control of linear systems with real parametric uncertainty. *Automatica*, Vol. 35, pp. 679-687.

Kučera, V. and C.E. De Souza (1995). A necessary and sufficient conditions for output feedback stabilizability. *Automatica*, Vol. 31, pp. 1357-1359.

Lankaster, P. (1969). Theory of matrices. *Academic press*, New-York, London.

Li Yu and Jian Chu (1999). An LMI approach to guaranteed cost control of linear uncertain time-delay systems. *Automatica*, Vol. 35, pp. 1155-1159.

Veselý, V. (2000). Output robust controller design for linear parametric uncertain systems, In: *The 3rd IFAC Symposium on Robust Control Design*, Praha, Czech Republic.

SLIDING MODE CONTROLLERS FOR THE REGULATION OF DC/DC POWER CONVERTERS

Rui Esteves Araújo*, Américo Vicente Leite, Diamantino Silva Freitas***

* *Faculty of Engineering of the University of Porto*
Dr. Roberto Frias, 4200-465 Porto , Portugal
** Polytechnic Institute of Bragança
Campus de Sta. Apolónia,
Apartado 134, 5301-587 Bragança, Portugal

Abstract: Sliding mode controllers are derived for the control of the average output voltage in DC/DC power converters. The controller design is carried out on the basis of well-known bilinear models of such circuits. A cascaded control structure is chosen for ease of control realization and to exploit the motion separation property of this power converter. The performance of the proposed sliding mode controllers is tested for the buck and boost converter type. The numerical simulations will demonstrate the efficiency of sliding mode techniques in this field as a powerful alternative to other existing methods. *Copyright © 2003 IFAC*

Keywords: Robust control, Sliding mode control, Nonlinear control, Output feedback stabilization.

1. INTRODUCTION

Switched mode power converters lie at the heart of DC power supplies, bringing the advantages of high efficiency and low mass. They can be modeled as variable structure systems because of the abrupt topological changes suffered by the circuit controlled by a discontinuous control action. They constitute a natural field of application of sliding mode controllers. Sliding mode control is more than a promising technique in the field of automatic control. Moreover, with time it is gaining increasing importance as a universal tool for the robust control of linear and nonlinear systems. It permits the realization of very robust and simple regulators of low cost. Several application of this technique have been developed, mainly in the field of electrical motors control (Utkin, 1993, Araújo, *et al.*, 2000). In fact, since that the power converter is a nonlinear system with variable structure behaviour, the sliding mode control has became the natural solution for direct implementation of its control.

Traditionally, the control problems of DC/DC power converters are solved by using linear controllers that control the average of the output voltage by adjusting the duty cycle of the converter through PWM techniques. Frequently, the state average models of DC/DC power converters are used to derive all the necessary transfer functions to design a controller by using linear control techniques. This is a simple and useful method but it has some very distinct drawbacks like that only the equivalent behavior is represented (in sense that the ripple is excluded), and the control is designed around the steady state solution (equilibrium point), which leaves most of the state-space out of consideration. The regulation is often achieved by a linear controller, showing good performances when disturbances are very small. On the contrary, when disturbances are large, the results of regulation become unsatisfactory. On other hand, since stability cannot be guaranteed more than in a small region around the steady-state operating point, several add-on circuits like: up-ramp,

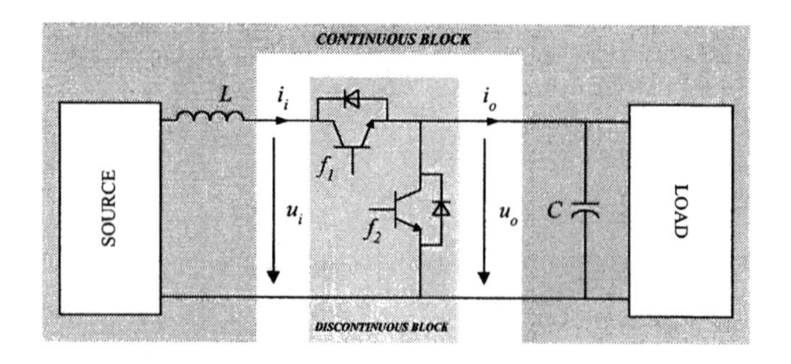

FIG. 1. Functional decomposition of DC/DC converters.

duty cycle limitation, clock, etc., must be added to support the controller.

The sliding mode technique offers an alternative way to implement a control action that exploits the inherent variable structure nature of the power converter. In particular, this technique offers several advantages in DC/DC power converters: - Robustness even for large load and supply variations, - Good dynamic response and - Simple implementation. Although initial ideas of sliding mode control have already been proposal in the 1970 decade, industrial applications are still very rare. Almost all implementations have seen made only for research investigations. Sliding mode control of switched power supplies was first treated by Venkataramanan, (1985) after the survey paper of Utkin (1977), and more recently by Sira-Ramirez and co-authors from a passivity viewpoint (Sira-Ramirez H., *et al.*, 1997). The topic was also treated in the context of the control of the Cúk converter in Oppenheimer M., *et al.*, (1996), where the authors demonstrated the possibility of applying a switching surface that is a linear combination of several state variables. Connections of sliding mode controllers with DC/DC power converters involving the natural time-scale separation properties was first treated by Sira-Ramirez (1988) from a geometric viewpoint.

In this paper, we will propose controllers based on the sliding mode technique, which don't employ PWM. The objective is to present a systematic approach of sliding mode controllers for the control of DC/DC power converter. In addition, we intend to put in evidence the most distinctive feature of a sliding mode controller, that is its robustness to parameter uncertainty and external disturbances. These features make this control technique a valid alternative in industrial applications. Finally, we wish to show how the hardware implementation is much easier for sliding mode control techniques when compared with classical control approaches.

This paper is organized as follows. Section 2 presents the basic concepts of this design methodology for DC/DC converters. The basic idea of sliding mode control is also suitable for other power converters. Section 3 develops the sliding mode controllers and demonstrates, for Buck and Boost converter cases, the simplicity of sliding mode control. The simulations' results are presented in section 4. Section 5 contains the main conclusions.

2. DESIGN CONCEPTS OF DC/DC POWER CONVERTERS

The results of this section, regarding the hybrid nature of DC/DC power converters, extend the work found in Araújo (2001), where only the Buck-type DC/DC power converter is treated. It must be noted that the hybrid nature of the power converter is used to structure the controller. The term "hybrid" has many meanings, one of which is: the dynamic evaluation of the converter depends on coupling between variables that take values in a continuum and variables that take values in a finite or countable set.

2.1 Decomposition in Sub-systems and Control Circuit Design

The controlled system of a power converter is constituted of three subsystems: the DC power supply, the static converter and the load as described in Figure 1. Having in account its hybrid nature and independently of the power converter type (buck, boost, Cúk, etc.), it is possible to make a functional decomposition into two blocks: a discontinuous block and continuous one. The first block comprises the switching elements. Its structure largely depends on the power converter type. On the other hand, the continuous block is independent of the static converter and comprises the power supply, load and reactive components. The reactive components are an inductor and a capacitor that work for intermediate storage of energy. It is clear from Figure1 that the system involves inter-action between discrete and continuous dynamics.

The decomposition of the controlled system allows defining a certain systematic approach about control systems design (see Figure 2). The key idea of the design concept for the power converters presented in this work is the cascaded control structure. According to this idea, any controller should consist of two loops: an inner current loop lies with the discontinuous block and an outer voltage control loop lies with the continuous one. This control structure is chosen for ease of control realization and to exploit the time-scale and functional separation properties of power converters. Such characteristic results from the desirable time-scale separation properties of the input filter and the RC output circuit time constant and the hybrid nature of power converters. In fact, for all switched power supplies used in practice, the motion rate of the current is much faster than the motion rate of the output voltage. In this context, we use the sliding mode approach for the control of the inner current loop and the voltage control loop is realized with standard linear control techniques.

The sliding mode control techniques are well suited for controlling static converters because of their natural switching structure. The design consists of two major phases. The first is the selection of a switching surface, so that the system, when restricted to this surface responds in the desired manner. The second is the design of a control law, which satisfies a set of sufficient conditions for the existence and reachability of a sliding mode. The aim of the next sub-section is to formulate the control problem.

2.2 *Application of sliding mode control to the power electronics*

In general terms the control problem can be formulated as follows:

Given a switching period T>0, determine a feasible control input u^v, so that a controlled variable follows as closely as possible a reference signal, despite influence of disturbances, measurement errors and parameter variations in the system.

The control problem as formulated leads to a number of issues. One is to select a switching surface so that the system produces a desired behaviour when restricted to it. Another is to choose the control law u^v that can induce a sliding motion close to the sliding surface of the system.

Usually the DC/DC power converters can be described with a unified state-space formulation in the form of a hybrid system defined on $\Re^n$.

$$\dot{x} = Ax + Bu^v x \tag{1}$$

where $x \in \Re^n$ is the state vector, $A \in \Re^{n \times n}$ and $B \in \Re^{n \times n}$ are matrices with constant values and u^v is an m-dimensional vector that lives in a finite set

$$u^v \in \mathcal{U} = \{u^1, u^2, \cdots, u^N\} \subset \Re^m, \quad N \geq 2 \tag{2}$$

where each element u^v can be one between the 2^m different vectors, such as:

$$u^v = \left\{ \underset{1}{\begin{bmatrix} 0 \\ \vdots \\ 0 \end{bmatrix}} \underset{2}{\begin{bmatrix} 0 \\ \vdots \\ 1 \end{bmatrix}} \cdots \underset{2^m}{\begin{bmatrix} 1 \\ \vdots \\ 1 \end{bmatrix}} \right\}, \quad i = 1, \cdots, 2^m. \tag{3}$$

The switching instants are determined by appropriate surfaces (that will be determined later), which are chosen to achieve a desired dynamic response. Between two commutations, the control signal u^v is a constant vector.

Note that the differential equation (1) describing the generic converter corresponds to a bilinear system due to the multiplicative character of the control variable u^v with respect to the state vector. Bilinear systems have been found in diverse processes and fields and many control strategies have been developed in the past, such as the quadratic feedback control proposed by Gutman, (1981). Here, we will focus on a class of the controllers designed by sliding mode techniques (Ping-Chen, Y., *et al.*, 2000). It must be noticed that for the DC/DC converters the control signal u^v only takes values from the discrete set $u^v \in \mathcal{U} = \{0, 1\}$.

Now, for the system (1), we advocate a discontinuous control as

$$u^v = \frac{1}{2}(1 - sign\ (s)) \tag{4}$$

where s is a switching function in the sense of sliding mode theory, see Utkin (1992). This control action is a direct large signal control approach and DC/DC converter is self-oscillating. Explicitly, this means that the switching frequency depends on the rate of change of function s and on the amplitude of the hysteresis band. This problem can be unacceptable if the range of variations becomes too high.

3. SLIDING MODE CONTROLLERS FOR DC/DC POWER CONVERTERS

3.1 *The model and controller for the Buck converter*

Consider the buck-type converter circuit shown in Figure 3. The differential equations that describe the circuit are given in (5).

where x_1 and x_2 represent the input inductor current and the output capacitor voltage variables, respectively. The positive quantity V_{cc} represents the constant voltage value of the power supply.

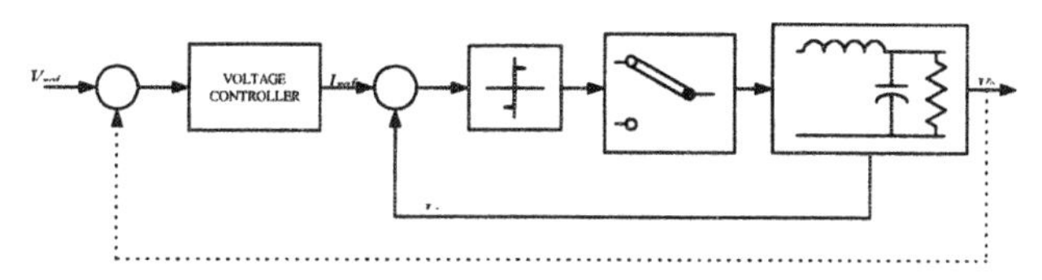

FIG. 2. Cascade control structure of DC/DC converters.

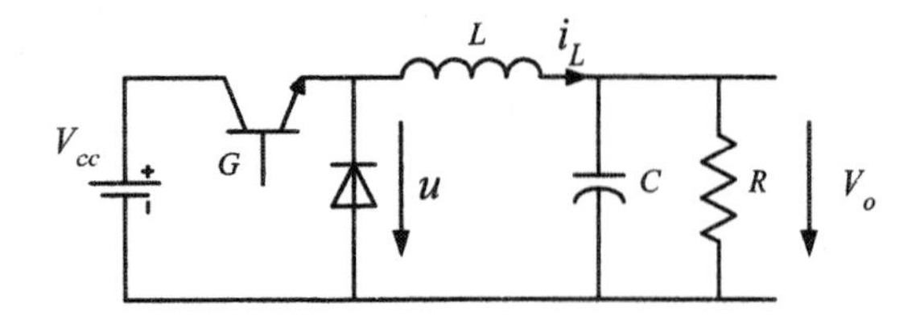

FIG. 3. Buck converter circuit.

$$\begin{aligned}\dot{x}_1 &= -\frac{1}{L}x_2 + \frac{V_{cc}}{L}u^v \\ \dot{x}_2 &= \frac{1}{C}x_1 - \frac{1}{RC}x_2\end{aligned} \quad . \tag{5}$$

The variable u^v denotes the switch state, acting as a control input.

We now consider the problem of constructing an appropriate control law that may ensure an asymptotic voltage tracking. Suppose it is desired to regulate the output capacitor voltage to a constant value. This is expressed mathematically by

$$\lim_{t\to\infty} (V_0 - V_{ref}) = 0, \qquad \dot{V}_0 = 0 . \tag{6}$$

Now, corresponding to this objective for the output voltage, the required input current may be represented by a function $I_{ref}(t)$, to be determined in thefollowing.

This methodology is justified by the inspection of the model of the system in the following way: first, it is verified that the inductor current (state variable x_1) is manipulated by means of the control variable u^v (first equation of 5). Second, taking care of the purpose of control the following can be derived from equation (6):

$$\dot{V}_0 = 0 \Leftrightarrow \dot{x}_2 = 0 \Rightarrow I_{ref} = \frac{V_{ref}}{R} . \tag{7}$$

Since the objective of the controller is to drive V_o to some constant value, it is clear that the inductor current will play a central role. Consequently, the first objective of the controller is to guarantee that the state variable x_1 follows the trajectory established in equation (7). Then, the first loop (of greater dynamics) is controlled through the establishment of a function of commutation defined as:

$$s = I_L - I_{ref} \tag{8}$$

where s is the surface variable and the superscript *ref* indicate the reference or set-point. The set point value is determined by the upper level controller. On s=0, the inductor current will converge to its reference. In order to enforce sliding mode in the surface s in (8), a control action based on (4) is used. The condition for sliding mode to exist is derived from $s\dot{s} < 0$. Thus, we have

$$s\dot{s} < 0 \Leftrightarrow s(\dot{I}_L - \dot{I}_{ref}) < 0 \Leftrightarrow s\left(-\frac{x_2}{L} + \frac{V_{cc}}{L}u^v\right) < 0 \tag{9}$$

taking care of equation (9) that defines the control action, results in two situations:

if $s > 0 \Rightarrow u^v = 0$, as it is necessary to guarantee $\dot{s} < 0$ for the existence condition to be true, it is enough then to guarantee that $x_2 > 0$;

if $s < 0 \Rightarrow u^v = 1$, as it is necessary to guarantee $\dot{s} > 0$, for the existence condition to be true it is enough that $x_2 < V_{cc}$. In conclusion, the domain of attraction of the specified surface is verified if

$$0 < x_2 < V_{cc} . \tag{10}$$

This mathematical result has immediate physical interpretation: the output voltage of this static converter is always lesser than that of the source.

We now follow on to verify if the control objective is fulfilled. The dynamics of the output voltage of the converter in sliding mode is given by:

$$\dot{V}_o = -\frac{1}{RC}V_o + \frac{1}{RC}V_{ref} \tag{11}$$

and the solution of this first order differential equation (11) is:

$$V_o(t) = V_{ref} + (V_o(t_a) - V_{ref})\, e^{-(t-t_a)/RC} . \tag{12}$$

The second term of the solution will decay rapidly to zero. Under this condition $\lim_{t\to\infty} (V_0 - V_{ref}) = 0$ is true, see (6).

3.2 Model and Controller Design for the Boost type converter

Consider now the boost converter circuit shown in Figure 4. The converter model is described by the following set of differential equations, with variables defined as before:

$$\begin{aligned}\dot{x}_1 &= -(1 - u^v)\frac{1}{L}x_2 + \frac{V_{cc}}{L} \\ \dot{x}_2 &= (1 - u^v)\frac{1}{C}x_1 - \frac{1}{RC}x_2\end{aligned} \tag{13}$$

Following the same procedure as in the previous case, a desired current is obtained from the outer voltage loop as

$$I_{ref} = \frac{V_{ref}^2}{RV_{cc}} \tag{14}$$

where V_{ref} is the desired output voltage. The switching function for the inner current control is defined , like in (8):

$$s = I_L - I_{ref} . \tag{15}$$

This choice is, as usual, motivated by the motion

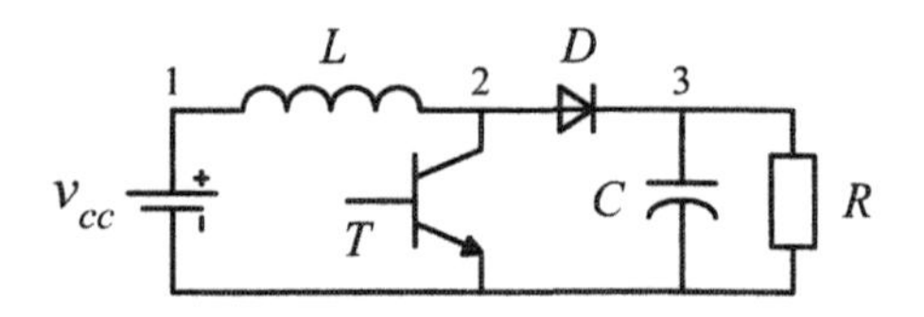

FIG. 4. Boost converter circuit.

separation property of power converters. This means that the dynamics of the inductor current is much faster than the dynamics of the output voltage. In order to enforce the inductor current to track the desired value, the control action can be similar to (4). Now, the condition that guarantee the existence of sliding mode is found by applying the convergence condition $s\,\dot{s} < 0$ to the system (13), given

$$x_2 > V_{cc}\,. \tag{16}$$

Intuitively, (16) implies that the output voltage will always be greater than the input voltage.

In the ideal case (infinite switching frequency) the control action can be replaced by its mean value u_{eq}, the equivalent control, which is a continuous variable, is derived by formally solving $\dot{s} = \dot{x}_1 = 0$. From (13), proceeding like Utkin (1992), we obtain for the control input u^v:

$$u_{eq} = 1 - \frac{V_{cc}}{x_2}\,. \tag{17}$$

The last expression for the equivalent control, determines the behaviour of the nominal system restricted to the switching surface. Now, we will see the equivalent equation of the output voltage in the sliding mode. Introducing (17) into the dynamic model of the boost converter, we obtain the output voltage:

$$\dot{x}_2 = -\frac{1}{RC}\left(x_2 - \frac{V_{ref}^2}{x_2}\right). \tag{18}$$

This equation can be solved explicitly as

$$x_2 = \sqrt{\left(V_{ref}^2 + \left(x_2^2(t_0) - V_{ref}^2\right)\right)e^{-2(t-t_0)/RC}} \tag{19}$$

where t_0 is the reaching instant to the switching surface and $x_2(t_0)$ is the output voltage at time t_0. We can see from (19) that x_2 tends to V_{ref} asymptotically as t goes to infinity. Then, there exists a control action u^v [that satisfying (4)] for the converter model (13) such that

$$\lim_{t\to\infty} (V_0 - V_{ref}) = 0\,. \tag{20}$$

Consequently, there exists a finite time $t > t_0$ such that $V_0(t) = V_{ref}$, $\forall\, t \geq t_0$. Furthermore, regulation can be achieved with a suitable selection of u^v only.

4. SIMULATIONS RESULTS

Simulations were performed for the closed-loop behavior of a buck and boost circuits regulated by means of the sliding mode controllers. The proposed control scheme for the buck converter, illustrated in Figure 5, that also represents the basis of the hardware, was investigated by means of simulations done using the Vector Control Signal Processing Blockset for Simulink environment (Araújo, *et al.*, 1997, Araújo, *et al.*, 1998). The numerical experiment consists of analyzing the performance of the controller in tracking of the trajectory of reference of the voltage intended to the output of the converter. Thus, the initial reference of the voltage is of 10 V and for the time instant t= 0,05 s the reference is subjected to a step variation towards the value of 15 V. For the presented results, we considered the following parameters:

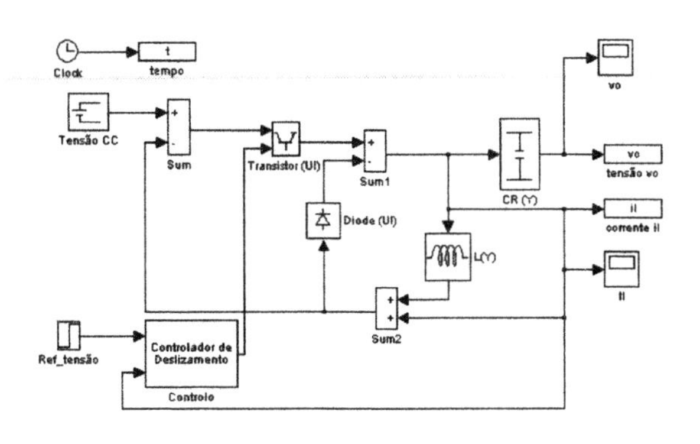

FIG. 5. Simulink model of a sliding mode buck DC/DC converter.

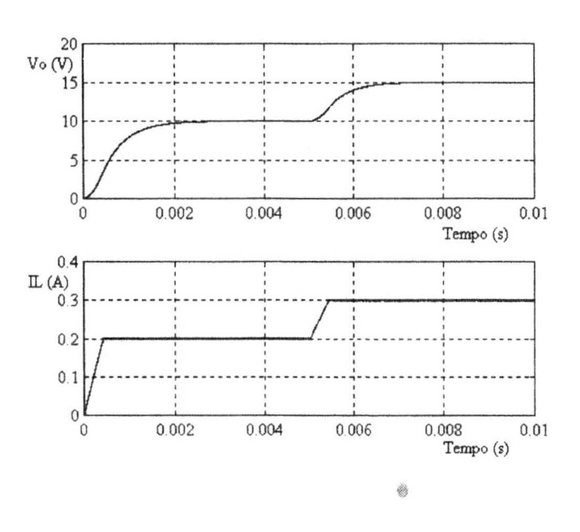

FIG. 6. Output voltage and inductor current waveforms in the buck converter with a sliding control scheme.

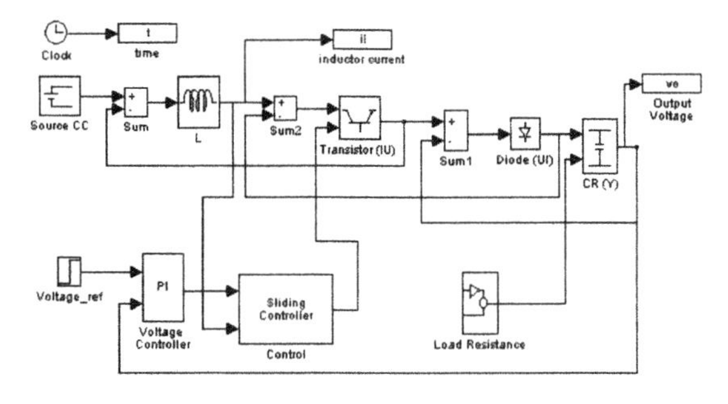

FIG. 7 Simulink model of a sliding mode boost DC/DC converter.

$V_{cc} = 20\ \text{V}$, $R = 50\ \Omega$, $C = 10\ \mu\text{F}$, $L = 40\ \text{mH}$.

Figure 6 shows the simulation results of the proposed control scheme for the buck DC/DC converter. From its analysis a good functioning may be verified. Notice that the inductor current and the output voltage converge rapidly to their references values.

A boost circuit with a feedback control circuit shown in Figure 7 was simulated to verify the main concept of load regulation. The circuit parameter values were taken to be:

$V_{cc} = 10\ \text{V}$, $R = 40\ \Omega$, $C = 10\ \mu\text{F}$, $L = 3.5\ \text{mH}$.

The desired output voltage was set to 20 V. Because unknown load resistance variations are normally considered as the main source that affect the behavior of the closed-loop performance of the controlled converter, the load resistance was stepped in a large range during the experiment. The robustness to load resistance uncertainties is verified, when an unmodeled sudden change in the load resistance was set to 80 per cent of its nominal value. As can be seen from Figure 8 the controller manages rapidly to restore the desired steady-state conditions immediately after the load perturbation disappears. Note that the output voltage converges rapidly to their reference value.

In conclusion, as is proven in the two simulations examples, the proposed methodology to design sliding mode controllers for DC/DC power converters achieves the desired direct regulation of the output voltage. This type of controller also exhibits a good degree of robustness with respect to the load disturbance.

5. CONCLUSIONS

This paper has presented the design of sliding mode controllers for DC/DC power converters. The design methodology provides evidence that the proposed sliding mode controller is capable of regulating the output voltage of the buck and boost converters. The advantages of sliding controllers over a classical control alternative lie in the hardware simplicity and closed loop robustness. The main objectives of this work have therefore been satisfied: - sliding mode control has been applied in the inner current loop that is similar for all types of power converters and the controllers don't introduce any additional complexity. Such possibilities were derived from the desirable time-scale separation properties between the L input filter and the RC output circuit time constant. On the other hand, the simplicity makes this approach attractive for implementation since the switch state is determined only from sign information about an affined functional of the converter state. Simulations results confirmed the effectiveness of the proposed control approaches.

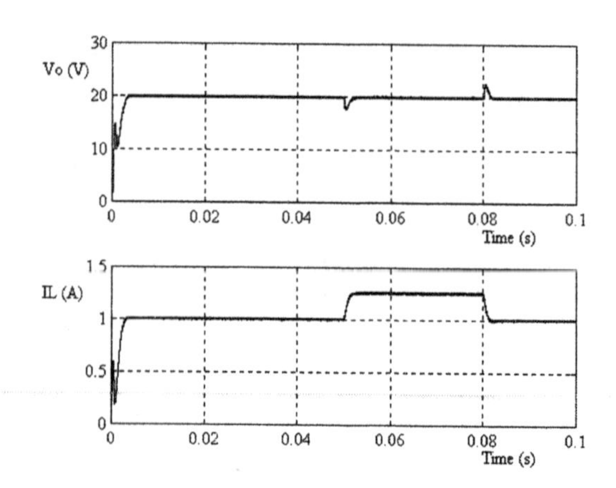

FIG. 8. Robustness test of the sliding mode controller in a boost converter to load disturbance.

REFERENCES

Araújo, R.E., A.V. Leite, and D.S. Freitas, (1997). The Vector Control Signal Processing Blockset for Use with Matlab and Simulink, *Proceedings of the IEEE International Symposium on Industrial Electronics.*

Araújo, R.E., A.V. Leite., and D.S. Freitas, (1998). The development of vector control signal processing blockset for Simulink: philosophy and implementation, in *Proceeding sof the 24th Annual Conf. Of the IEEE Industial Electronics Society, Aachen, Germany.*

Araújo, R.E. and D.S. Freitas, (2000). Non-linear control of an induction motor: sliding mode theory leads to robust and simple solution, *Int. Journal of Adaptive control and Signal Processing*, vol. 14, pp. 331-353.

Araújo, R.E., (2001). *Modelação, identificação e controlo do motor de indução trifásico*, Ph D. Thesis (in Portuguese), FEUP.

Gutman, P.O., (1981). Stabilizing controllers for bilinear systems" *IEEE Trans. Automatic. Control*, vol. 26, 1981, pp. 917-922.

Oppenheimer M., I. Husain, M. Elbuluk, and J.A. Abreu-Garcia. (1996) Sliding Mode Control of the Cúk Converter in *Proceedings of the 1996 Power Electronics Society Conference*, 0-7803-3500-7/96, pp. 1519-1526.

Ping-Chen, Y., J.L. Chang and K. M. Lai, (2000). Stability Analysis and Bang-Bang sliding control of a Class of Single-Input Bilinear Systems," *IEEE Trans. Automatic. Control*, vol. 45, nº 11, pp. 2150-2154.

Sira-Ramirez, H. and M. Ilic-Spong, (1988). A Geometric Approach to the Feedback Control of Switchmode Dc to DC Power Supplies, *IEEE Trans. on Circuits and Systems*, vol. 35, nº 10, pp. 1291-1298

Sira-Ramirez H., R. A. Perez-Moreno, R. Ortega and M. Garcia Esteban, (1997). *Automatica*, vol. 33, nº 4, pp. 499-513.

Utkin V., (1977) Variable Structure Systems with Sliding Modes" *IEEE Trans. Automatic. Control*, vol. 22, pp. 212-222.

Utkin V., (1992)., *Sliding Modes in Control and Optimization*, Springer-Verlag, Berlin,

Utkin, V., (1993), Sliding Mode Control design Principles and Applications to Electric Drives. *IEEE Trans. Industrial Electronics*, vol. 40, nº 1, pp. 23-36

Venkataramanan V., A. Sabanovic and S. Cúk, (1985). Sliding mode control of DC to DC converters, in *Proceedings IECON 85*, pp. 251-258.

www.elsevier.com/locate/ifac

A NAVIGATION ALGORITHM FOR A MOBILE ROBOT CHASING A MOVING LANDMARK IN AN UNKNOWN ENVIRONMENT

Fabrizio Romanelli, Francesco Martinelli, Luca Zaccarian

Dipartimento di Informatica Sistemi e Produzione
Università di Roma Tor Vergata
I-00133 Rome, Italy
e-mail: fab.romanelli@libero.it,
martinelli@disp.uniroma2.it,
zack@disp.uniroma2.it

Abstract: In this paper we describe a control paradigm for the navigation of a mobile robot toward a moving landmark in an unknown indoor environment with obstacles. The overall control paradigm has been successfully tested on a real integrated navigation system which comprises a webcam and a mobile platform equipped with odometric, sonar and bumper sensors. The webcam is used for landmark recognition while sonar, bumper and odometric sensors are used for navigation purposes. In the paper we present a hardware description of the considered experimental system and discuss the proposed control algorithms. Guidelines regarding both hardware and control algorithms are drawn to improve the efficiency of the overall system. *Copyright © 2003 IFAC*

Keywords: Image Processing, Image Recognition, Obstacle Avoidance, Obstacle Detection, Robot Navigation Systems

1. INTRODUCTION

Mobile robot navigation is a well investigated research area, at the borderline between Automatic Control and Artificial Intelligence, greatly appealing both for practical applications and challenging theoretical problems. A vast literature deals with the problem of finding a trajectory in an environment containing obstacles (see Latombe (1991) and references therein). The basic problem may become much more involved if we consider uncertainties regarding the environment or the obstacles, if the position of the obstacles may vary in the environment or if other robots are present in the workspace and could collaborate to solve some assigned tasks (Latombe (1991), see also Gentili and Martinelli (2001) and references therein for motion planning of robot formations). Additional complications arise if one takes into account robot kinematic and/or dynamic constraints. The presence of kinematic constraints, in particular nonholonomic constraints (see, e.g., Latombe (1991) and Murray et al. (1994)), may require to completely re-design the robot trajectory. Dealing with uncertainties on the environment is often a difficult task, which is usually solved through sensor fusion algorithms. The uncertainty on the environment requires smart localization algorithms and dynamic trajectory design and control to avoid unknown obstacles appearing only during

[1] This work has been partially supported by ASI and MIUR (cofin: MATRICS).

the execution of the mission (see Grandoni et al. (2001) for a comprehensive tutorial). For mobile robot navigation in a partially unknown environment, it is frequent to adopt control schemes which attempt to recognize known features in the environment (called *landmarks*). These reference objects are usually in a fixed and known position of the environment and can be used for localization and navigation purposes.

Many applications have been developed to direct mobile robots navigation toward landmarks, using 3D images detecting systems (Se et al. (2001)) or dedicated cameras (Nourbakhsh et al. (1997)) or with landmark extraction learning and detecting systems (Trahanias et al. (1997)). In this paper we address a slightly different problem. In our scenario the landmark is a moving object and the robot attempts to maintain a predefined distance from it. The environment is completely unknown and the robot follows the landmark while employing the on board sensory systems to detect and avoid unknown obstacles along its path.

The choice of acquiring 2D images from a webcam (see the experience in Coleman et al. (2002)) greatly reduces the possibilities to implement accurate localization and mapping algorithms (such as those in Se et al. (2001)). However, the present experience, with the environmental detection of obstacles effected by the robot sonars and bumpers, proves that a webcam is still sufficient to obtain satisfactory results on guided navigation toward landmarks in an environment with unknown obstacles.

Our integrated navigation system is divided into two subsystems: a *visual subsystem* to capture, process and localize the landmark, and an *obstacle avoidance subsystem* to avoid obstacle collisions and to proceed forward along the desired direction by effecting a successful obstacle avoidance strategy. In the visual subsystem, based on the acquisition and processing of the image arising from the webcam, the localization algorithm provides information about the desired forward motion direction. In particular, the robot turret, where the webcam is mounted, is suitably rotated with the goal of centering the landmark in the webcam viewfield. Consequently, the turret orientation provides the required information about the desired forward direction. In the obstacle avoidance subsystem, to avoid collision with the obstacles we have developed an obstacle avoidance algorithm, using the robot proximity sensors (bumpers) and long distance sensors (sonars). Based on the combination of the two subsystems commented above, the robot is able to proceed toward the landmark in a completely autonomous manner.

2. LANDMARK EXTRACTION AND RECOGNITION FOR NAVIGATION

2.1 Overview

The developed experiment works with the parallel execution of two programs: one to control the webcam and the image processing algorithms, and the other one to control the navigation system. The software that manages the *visual system* works as follows: the first step is to acquire the image from the webcam. Subsequently, such an image resides in the PC memory to be suitably processed by digital filters. The filtering step consists in successive transformations, the first one being the conversion into greyscale in order to delete the useless information given by the image colors. The second transformation consists in extracting the contours of the image using the *Sobel* filter (see, e.g., Castleman (2000)), which emphasizes the image discontinuities. To decrease the workload, the filtered image is finally transformed from grayscale into black and white. The following step consists in recognizing the landmark within the filtered image so that in the subsequent one the coordinates of the landmark center are estimated. Finally, based on the landmark position estimate, the robot turret (consequently, the webcam) is rotated to center the landmark on the webcam viewfield.

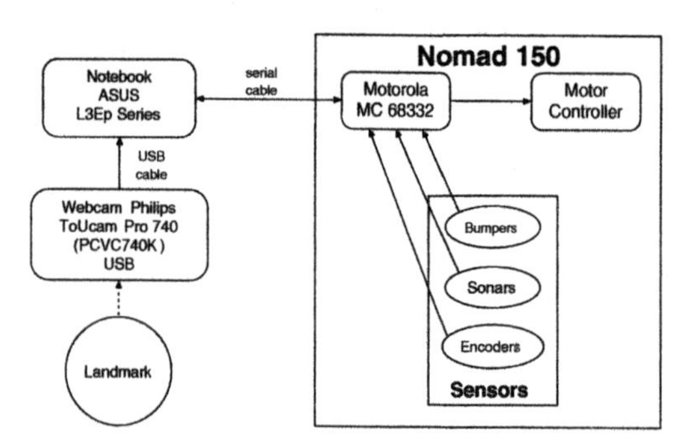

Fig. 1. Block diagram of the experimental device.

The software that controls the robot *navigation system* runs in parallel to the previous one. This software is able to communicate with the robot hardware which monitors the status of all the robot sensors and commands the desired position, speed and acceleration of the robot. During the robot navigation we use the information acquired from the on-board sensors (encoders, sonars and bumpers) and the information acquired from the visual system management software (the turret angle) to determine the forward motion path. In particular, the landmark direction is determined from the turret and the wheel base angle and the possible presence of obstacles is determined from the sonar and bumper data. If an obstacle is located, the robot can maneuver around it by way of the information provided by the obstacle avoidance algorithm. Finally, when the robot reaches the desired distance from the landmark, the program stops.

2.2 Image processing and landmark recognition

In this section we will explain how the acquired image is processed by the visual system management software.

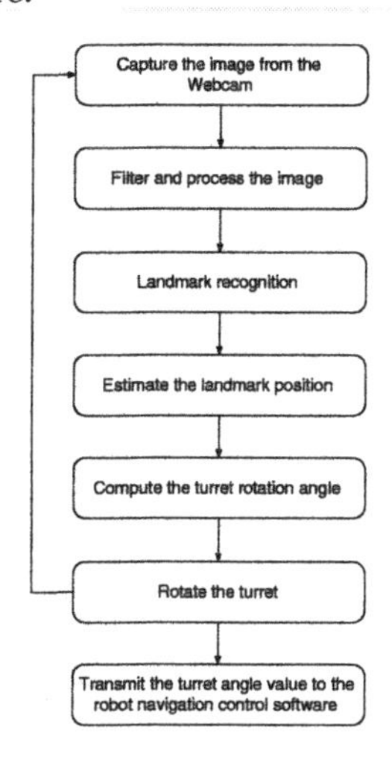

Fig. 2. The visual system management software flowchart.

After the acquisition of the image, the first task is to convert it from the native webcam palette to the greyscale color space. In this step we delete the color information from the source image, put it into memory and convert it from the native color space representation into the 8 bits per pixel greyscale representation. Consequently, the data size of the resulting image is significantly reduced. In Figure 3, we report an example of the acquired image, after the conversion into greyscale.

Fig. 3. An example of the raw greyscale image.

The next step is to extract the contours from the image using the *Sobel* filter (see, e.g., Castleman (2000)). After this filtering action, the contours of the image are highly emphasized, so that the landmark recognition algorithm is facilitated. Figure 4 shows the image resulting from the application of the Sobel filter to the image in Figure 3. Now the landmark detection is an easy task.

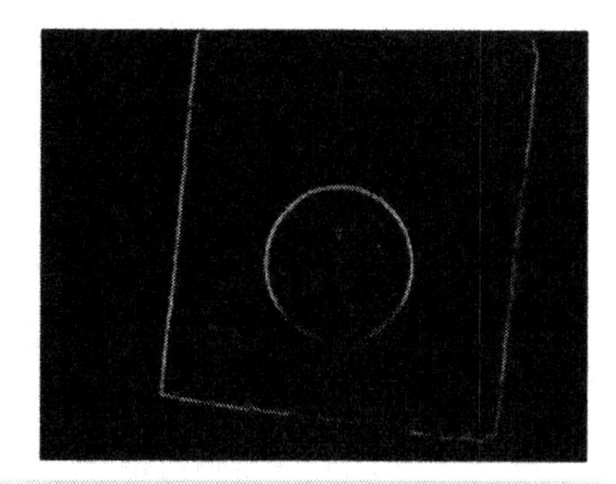

Fig. 4. Edge detected image.

To further emphasize the image contours we apply now a transformation from the greyscale color space to the bidimensional color space (black and white), obtaining the result in figure 5.

Fig. 5. Black and white image.

From the filtered image of Figure 5 we need to estimate the *landmark* position. The landmark recognition algorithm focuses the attention on a subimage consisting of all the pixels whose distance from the image border is greater than a given predefined value. Starting from the pixel on the upper left corner of this subimage, the algorithm selects equally spaced points of the subimage (thereby defining a grid of such a subimage) and analyzes each of these points as a possible candidate for being the center of the landmark. To guarantee as much as possible a uniform execution time, the pixels are processed in pairs, where each pair is composed of one pixel on the left half image and of the symmetric pixel with respect to the image center. For each pixel of coordinates (x_c, y_c), the algorithm sets the circle radius to r_c and checks the following conditions to determine whether or not it corresponds to the landmark center:

$$(X_i, Y_i) \in \text{circle } \forall i = 0, \ldots, 180 \quad (1)$$

$$X_i = x_c + r_c \cos\theta_i$$

$$Y_i = y_c - r_c \sin\theta_i,$$

where $\theta_i = \frac{\pi}{180} i$. This procedure is repeated for r_c ranging from 34 to 150 pixel, to be able to identify circles of different sizes. Circles that correspond to smaller or larger radii won't be recognized by the algorithm. The corresponding maximum and minimum distances between the robot and the landmark can be estimated as 2.10 and 0.5 meters, respectively. Note that in equation (1) only 180 points are checked around the candidate center to guarantee a sufficiently fast processing time on each pixel. Moreover, additional heuristic strategies have been implemented to speed up the search process when certain conditions indicate preliminarily that the candidate pixel cannot be the landmark center. When a circle is detected, the algorithm returns the coordinates of the center to be used in the next step of the program.

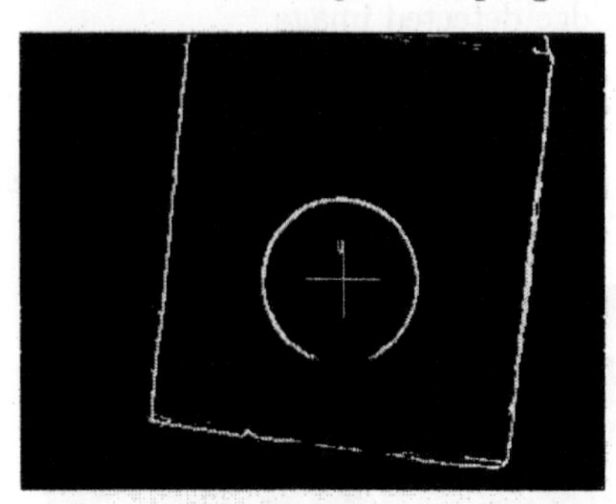

Fig. 6. The *landmark* detected in the image.

Now we have the *landmark* coordinates with respect to the image center (Figure 6 shows the detected *landmark* and its center is marked with a cross). Indeed, the only coordinate that we need to know is the x_c coordinate, because the control goal is to follow the *landmark* horizontal motion. Therefore, we need information about the horizontal distance between the landmark center and the image center. Based on this information, the robot turret can be suitably rotated to align the landmark with the image center. Note that the larger the horizontal distance is, the larger the commanded rotation will be.

2.3 Obstacle avoidance algorithm

The preliminary step of the obstacle avoidance algorithm consists in establishing the connection between the personal computer and the robot hardware, and in setting up the operating speed and acceleration of the robot, both for the forward (translational) motion, for the steering motion of the wheels and for the rotational motion of the turret. Subsequently, the robot is ready to navigate and the control over the turret angle is transferred to the visual control system software, which updates the navigation parameters.

After this preliminary step (the initialization phase), the obstacle avoidance algorithm enters the main loop. Here, based on the turret rotation angle provided by the visual management software and based on the wheel base orientation, a subset of sonars are selected, which cover an angle of view of 90° in the forward direction. Based on this subset of sonars, information is derived about the

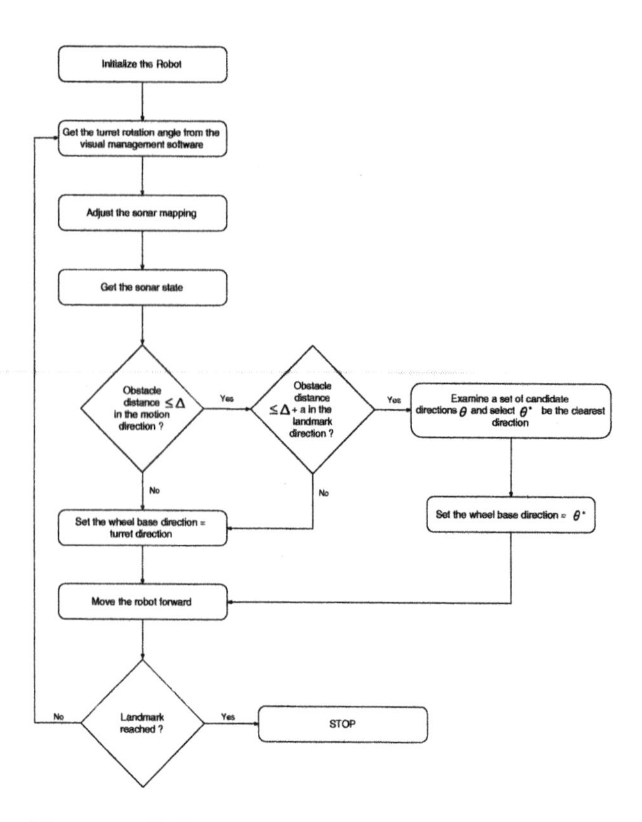

Fig. 7. Robot navigation control software flowchart.

distance between the robot and possible obstacles located along its forward path. If information is retrieved about an obstacle closer than a minimum allowable distance Δ (which is an adjustable threshold of the algorithm), the robot stops its motion and searches for the clearest direction to follow in order to avoid the obstacle. Such a direction is then followed by the robot until the straight path between the robot and the landmark is clear from obstacles. Subsequently, the forward direction is updated again and the algorithm goes back to its original state. During the avoidance phase, the algorithm tolerates a smaller distance from the obstacle (which we denote $\Delta + a$, with $a < 0$) so that the avoidance phase is facilitated in the cases where the obstacle is placed between the robot and the landmark. Indeed, to guarantee a smooth obstacle avoidance maneuver, after detecting the obstacle at a distance Δ it is desirable that the robot slowly deviates from its original direction, thereby inevitably getting closer to the detected obstacle. While iterating the above algorithm, the robot keeps approaching the landmark. When a prescribed distance $d \in (d_m, d_M)$ from the landmark is reached, the program stops.

3. EXPERIMENTAL APPLICATION

3.1 Hardware description

The integrated navigation system consists of the robot *Nomad 150* from Nomadic Technologies, of a USB Webcam Philips *ToUcam Pro 740* and of a laptop computer running Linux (distribution Mandrake 9.0).

The *Nomad 150* is a three degree of freedom cylindrical robot. Its wheel base can translate forward and backward and rotate clockwise and counterclockwise by way of three actuated wheels. The robot turret can rotate independently from the wheel base. The maximum rotational speed of the turret (ω_t) and of the wheel base (ω_w) are 60° per second; the maximum translational speed (v_r) of the wheel base is 61 *cm* per second. The *Nomad 150* on board sensing system consists of 16 sonar sensors located on the upper sensor ring, capable of detecting obstacles within the range [43-648] cm, and of 20 independent pressure sensitive sensors (bumpers) located on two rings on the lower part of the robot, which are able to detect contact with an object. Odometric sensors placed on each of the three wheels allow to permanently monitor the relative position of the robot with respect to the initial configuration (see Borenstein et al. (1997) for odometry and landmark navigation positioning systems).

The webcam *Philips ToUcam Pro 740* consists of a CCD sensor with a maximum resolution of 640x480 pixel (VGA resolution); the integrated lens is 6 mm f 2.0 H33^o (horizontal angle of view), and the maximum framerate is 30 fps. The image captured from the webcam, at a bit depth of 24 bits, is coded in the YUV420P color space: the native webcam palette. The selection of such a webcam has been primarily dictated by the fact that it can be programmed in C++ in linux environment using the *Video 4 Linux* library, like the *Nomad 150*, which can be programmed in C++ using the libraries provided by Nomadic Technologies.

The laptop PC used is an ASUS notebook equipped with a 2 GHz Intel Pentium 4, and 256 MB of RAM. Mandrake Linux 9.0 has been installed on this PC to host all the application software.

Fig. 8. The complete experimental setup.

The *Nomad 150* serves as a chassis for the entire system (see Figure 8): the webcam is aligned with the robot sonar number 0; the notebook is placed on the top of the robot turret. The connections between the webcam and the PC consist of a USB cable; moreover, the serial port of the PC is connected to the *host* port of the robot by a serial cable.

The landmark used in the application consists of a black sphere on a white background support; we selected a sphere for the landmark because its shape is invariant for all possible orientations.

3.2 Experimental results

Experimental tests were conducted in an indoor environment: in the hallways of the Engineering building of the University of Rome, Tor Vergata. Lighting conditions vary throughout the hallway, therefore these tests were challenging: indeed, artificial lighting corresponds to an average lighting level of 190 lux (resulting from minimal values of 135 lux up to maximum values of 250 lux).

The webcam has been initialized at a 320x240 pixel resolution, capturing images at 30 fps. This resolution allows to detect the *landmark* (whose radius is $D = 17{,}5$ cm) between a minimum distance $d_m = 0,50$ m and a maximum distance $d_M = 2,10$ m. The initial distance of the *Landmark* was approximately $1,20$ m and the landmark was moved by an operator parallel to the floor at an approximate speed of 25 *cm* per second, moving away from the robot. The *Landmark* motion was intentionally misaligned with the robot motion, so that a persistent turret rotation was required for the webcam to track the landmark. During the experiment, the landmark motion was adjusted to keep the robot/landmark distance approximately constant. Since the translational speed v_r of the wheel base, the rotational speed ω_w of the wheel base and the rotational speed ω_t of the turret have been fixed as $v_r = 20$ cm/s, $\omega_t = 30$ $degree/s$, $\omega_w = 30$ $degree/s$. The components of the istantaneous speed of the *landmark* can be estimated around 20 *cm* per second in the robot motion direction and around 15 *cm* per second in the orthogonal direction.

Since the step computational time of the visual management software (namely, the average time interval necessary to perform the landmark detection from a grabbed frame) corresponds to approximately 0,5 seconds, the sideways motion of the landmark with respect to the webcam orientation (i.e., the turret orientation) was limited by the limited field of view of the webcam itself (this corresponds to 33 degrees in the horizontal direction - H33^0). Indeed, based on the current robot/landmark distance d, and on the landmark diameter D, such a limitation can be quantified, as represented in Figure 9, through an estimation of the maximum transverse displacement (s) allowable during the maximum visual management program step computational time:

$$s = d \tan(33^0/2) - \frac{D}{2\cos(33^0/2)} \quad (2)$$

Based on condition (2) if the landmark is centered on the webcam field of view at each program step, an upper bound for the transverse motion of the landmark can be estimated by evaluating (2) with $d = 1,2\ m$ (the average landmark distance) and $D = 17,5\ cm$ (the landmark diameter). The resulting upper bound is $s_M \simeq 25\ cm$.

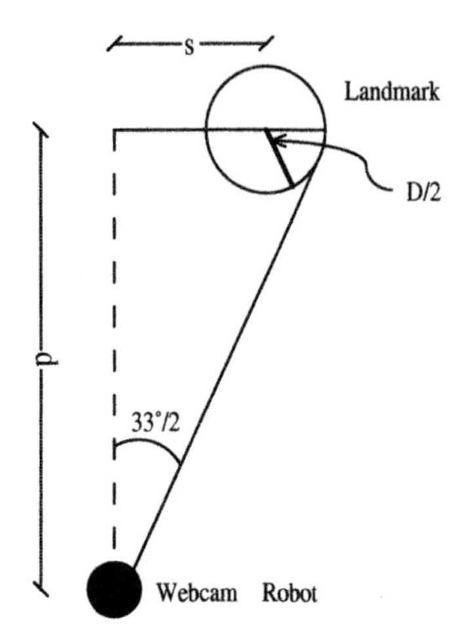

Fig. 9. System scheme.

Consequently, since the step computational time of the visual management software is estimated around 0,5 seconds, the maximum transverse speed of the landmark can be fixed as 50 *cm* per second. In our experiments we used a transverse speed of approximately 15 *cm* per second. During the experiment, an obstacle has been placed between the robot and the landmark to test the obstacle avoidance software. The resulting performance showed successful detection and avoidance by way of the robot which, after detecting the obstacle presence at a (predefined) distance $\Delta = 71\ cm$, has been able to redirect the motion toward a clear lateral path, while keeping the turret oriented towards the landmark. Subsequently, the robot successfully recovered the straight direction towards the landmark, as soon as the corresponding path was clear from obstacles.

Several additional tests were conducted by varying the robot/*landmark* distance and the landmark speed, according to the previously described strategy. In all the experiments, the robot successfully followed the landmark moved by the operator, covering an overall distance of approximately 10 m along a curved hallway direction while avoiding collisions with the walls and with several unknown obstacles.

4. CONCLUSIONS

In this paper we described experimental results on navigation of a mobile robot toward a moving landmark in an unknown indoor environment with obstacles via visual feedback. Experimental results demonstrated the suitability of integrated systems for mobile robots navigation based on the combined action of visual feedback and obstacle avoidance algorithms, even when employing low cost optical devices.

REFERENCES

J. Borenstein, L. Feng, H.R. Everett, and D. Wehe. Mobile robot positioning - sensors and techniques. *Journal of Robotic Systems, Special issue on mobile robots*, 14(4):213–249, 1997.

Kennet R. Castleman. *Digital Image Processing.* Prentice Hall, 2000.

D. Coleman, D. Creel, D. Drinka, N. Holifield, W. Mantzel, A. Saidi, and M. Zuffoletti. Implementation of an autonomous aerial reconnaissance system. Technical report, University of Texas at Austin, IEEE Robot Team Aerial Robotics Project, 2002.

F. Gentili and F. Martinelli. Optimal paths for robot group formations based on dynamic programming. *International Journal of Robotics and Automation*, 16(4), 2001.

F. Grandoni, A. Martinelli, F. Martinelli, S. Nicosia, and P. Valigi. Sensor fusion for robot localization. In B. Siciliano, A. Bicchi, S. Nicosia, and P. Valigi, editors, *Articulated and Mobile Robotics for Services and Technologies (RAMSETE)*, Lecture Notes in Control and Information Sciences, pages 251–273. Springer-Verlag, London, 2001.

Jean-Claude Latombe. *Robot Motion Planning.* Kluwer Academic Publishers, 1991.

R.M. Murray, Z. Li, and S.S. Sastry. *A mathematical introduction to robot manipulation.* CRC Press, 1994.

I.R. Nourbakhsh, D. Andre, C. Tomasi, and M.R. Genesereth. Obstacle avoidance via depth from focus. *Robotics and autonomous systems*, pages 151–158, 1997.

S. Se, D. Lowe, and J. Little. Local and global localization for mobile robots using visual landmarks. *IEEE/RSJ International conference on Intelligent Robots and Systems, pp 414-420*, 2001.

P.E. Trahanias, S. Velissaris, and T. Garavelos. Visual landmark extraction and recognition for autonomous robot navigation. In *IEEE/RSJ International Conference on Intelligent Robots and Systems*, pages 1036–1042, 1997.

CONTROLLING OF PENSION FUND INVESTMENT BY USING BELLMAN'S OPTIMALITY PRINCIPLE

Miroslav Šimandl * **Marek Lešek** **

* *simandl@kky.zcu.cz, University of West Bohemia, Pilsen*
** *marek.lesek@email.cz, University of West Bohemia, Pilsen*

Abstract: This paper analyzes the financial risk in a contribution defined pension fund in the Czech republic. The Bellman's optimality principle is used to derive the best allocation of a pension fund asset in the two-asset world. The principal results concern the suitability of the optimal pension fund strategy and the large variability of the level of achievement in pension fund asset in the case of variable rates of assets return. *Copyright © 2003 IFAC*

Keywords: Pension fund, Mathematical modeling, Bellman's optimal principle

1. INTRODUCTION

The present work analyzes the financial risk in contribution defined pension fund. The financial risks of contribution defined pension fund are analyzed for investing in a two-level system of distributing investment. In this work, the best way of investing between low and high-risk investment is suggested. For computing the optimal investment strategy for multiplicative rates of return dynamical programming is used.

The use of dynamic programming is not new in the pension fund area. (Sung and Haberman, 1994) use Bellman's optimality principle to minimize simultaneously the contribution risk and the solvency risk in a defined benefit pension scheme and derive the optimal contribution rate. (Cairns, 1997) in a continuous time framework and (Owadally, 1998) a discrete time framework apply this principle to a defined benefit pension scheme in order to derive the optimal contribution rate and the optimal allocation decision in a two-asset world.

(Vigna and Haberman, 2001) use Bellman's optimality principle in contribution defined pension fund. The optimal allocation decision in two-asset world was derived. The net replacement ratio was used as level of reference asset. The asset of contribution defined pension fund is valorized by exponential function.

This paper is focused on deriving optimal strategy for allocation decision in a two-asset world for multiplication valorized contribution defined pension fund in Czech republic.

In the work (Šimandl, 1998), the Mathematical Model of Pension Fund in the Czech Republic was firstly established. This basic model was extended and adapted by other author for a better description of reality. We will use the modification which was described in work (Šimandl and Lešek, 2002). The main change, that was done, allows the model to invest assets to two groups of capital - high risk (stocks etc.) and low risk (bonds etc.).

The paper has the following structure. Section 2 is denoted to describe a mathematical model of a contribution defined pension fund which was designed in work (Šimandl and Lešek, 2002). The contents of the section 3 is focused on a problem of dynamic programming. In the first part it is defined problem of dynamic programming for the best allocation of pension fund asset in two-asset world and in the second part the Bellman's optimality principle is used for solving the problem of dynamical programming. In section 4 it is given a illustration example of applying the Bellman's optimal principle to the best allocation of pension fund asset and the brief summary is done in section 5.

2. THE MATHEMATICAL MODEL OF PENSION FUND IN THE CZECH REPUBLIC

(Šimandl and Lešek, 2002) describe the design of pension fund mathematical model in two-asset world in state-space form. The state $\mathbf{x}(t)$ is described by set of stochastic non-linear equations.

$$x_1(t+1) = f_1(x_1(t) + w_1(t)) \tag{1}$$

$$x_2(t+1) = x_2(t) + w_2(t) \tag{2}$$

$$x_3(t+1) = x_3(t) + w_3(t) \tag{3}$$

$$x_4(t+1) = x_1(t)x_4(t) + u_k(t) \tag{4}$$

$$\begin{aligned} x_5(t+1) = & [(1-s(t))\nu(t) + s(t)\lambda(t)]. \\ & .[(x_3(t) - x_2(t))x_4(t) + \\ & + x_5(t) + u_p(t)] \end{aligned} \tag{5}$$

In following part, equations of pension fund mathematical model will be described.

Equation (1) describes an evolution of the portion of clients who stay in the pension fund at time $t+1$ (except new incoming) to number of clients at time t, where $f_1(\ldots)$ represents saturation function that holds value of expression $x_1(t)+w(t)$ in define interval $\langle 0, 1\rangle$ and is valid for all t, that $f_1(x_1(t) + w_1(t)) = 0$, if $x_1(t) + w_1(t) < 0$ and at the same time $f_1(x_1(t) + w_1(t)) = 1$, if $x_1(t) + w_1(t) > 1$ and $w_1(t)$ is a random variable with mean μ_1 and variance σ_1^2.

Equation (2) describes a development of an average value of benefit pay-off per client in the pension fund (this amount already include money that clients carry out to other pension funds if they are transferring), where $w_2(t)$ is a random variable with mean μ_2 and variance σ_2^2.

The next equation, (3), represents a evolution of an average value of contribution per client including a contribution which the State supports the supplementary pension secure, where $w_3(t)$ is a random variable with mean μ_3 and variance σ_3^2.

Equation (4) represents a dynamical development of number of the pension fund clients, where $u_k(t) = u_{k_1} + u_{k_2}$ is number of clients that join the pension fund at time t and u_{k_1} is number of the clients which have not already contributed and u_{k_2} is number of the clients which have contributed to the other pension fund and save some money.

The last equation, (5), describes a dynamical evolution of the size of the pension fund assets, where $u_p(t)$ is a size of the amount which clients bring to the pension fund from other pension fund in which they have saved.

We assume that annual rates of returns $\nu(t)$ (for low-risk investment) and $\lambda(t)$ (for high-risk investment) from asset ($\bar{x}_5(t)$) are normally distributed random variables.

$$\nu(t) \approx N(\overline{\nu}, \omega_1^2) \quad \text{a} \quad \lambda(t) \approx N(\overline{\lambda}, \omega_2^2) \tag{6}$$

where

$$\overline{\nu} \leq \overline{\lambda}, \quad \omega_1^2 \leq \omega_2^2$$

and it is also assumed that the random variables ($\nu(t)$ and $\lambda(t)$) and pension fund asset ($x_5(t)$) are independent

$$\begin{aligned} cov(\nu(t), \lambda(g)) &= 0 \ \forall \ t, g \\ cov(x_5(t), \nu(g)) &= 0 \ \forall \ g \geq t \\ cov(x_5(t), \lambda(g)) &= 0 \ \forall \ g \geq t \end{aligned} \tag{7}$$

The fifth equation of mathematical model contains the investment strategy and the Bellman's optimal principle will be applied to it. The states $x_2(t)$, $x_3(t)$ and $x_4(t)$ are measurable then it is possible to arrange the equation for fifth state ($x_5(t)$) to following form:

$$\begin{aligned} x_5(t+1) = & [x_5(t) + c(t)]. \\ & .[(1-s(t))\nu(t) + s(t)\lambda(t)] \end{aligned} \tag{8}$$

where

$$c(t) = (x_3(t) - x_2(t))x_4(t) + u_p(t)$$

The output equation is designed in virtue of fact, that a part of the states ($x_2(t)$, $x_3(t)$, $x_4(t)$ and $x_5(t)$) is directly measurable. Hence, the output equation is in linear form.

$$\mathbf{y}(t) = \mathbf{C}.\mathbf{x}(t) \tag{9}$$

where C will be in form:

$$\mathbf{C} = \begin{bmatrix} 0 & 1 & 0 & 0 & 0 & 0 \\ 0 & 0 & 1 & 0 & 0 & 0 \\ 0 & 0 & 0 & 1 & 0 & 0 \\ 0 & 0 & 0 & 0 & 1 & 0 \end{bmatrix} \tag{10}$$

and $\mathbf{x}(t)$ is the state vector.

3. PROBLEM OF DYNAMIC PROGRAMMING

We start with the documents (Sung and Haberman, 1994) and (Owadally, 1998), where the "costs" function was defined for benefit defined pension fund, and (Vigna and Haberman, 2001) where is defined "costs" function for contribution defined pension fund in form

$$C(t) = \theta_1 (x_5(t) - \tilde{x}_5(t))^2 \tag{11}$$

where $\theta_1 > 0$ is weight coefficient and $\tilde{x}_5(t)$ is expected asset of a pension fund, which the fund should own at time t. The amount of expected asset will be counted on voted rate of return, which is defined as arithmetical average of chosen rates of return for investment to high-risk $(\overline{\lambda})$ and low-risk $(\overline{\nu})$ asset and it is defined by equation

$$\tilde{x}_5(t) = (\tilde{x}_5(t-1) + c(t-1))\frac{1}{2}(\overline{\nu} + \overline{\lambda}).$$

The set of expected assets $\tilde{x}_5(t)$ is known in advance and it is fixed at each discrete time moment $\{\tilde{x}_5(t)\}_{t=1,2,\ldots,N}$ and $\tilde{x}_5(0) = x_5(0)$

The border condition from "costs" function at time N is

$$C(N) = \theta_0 (x_5(N) - \tilde{x}_5(N))^2 \tag{12}$$

For pension fund managers, it is more important to reach final target than particular targets. This principle is applied in weight coefficients θ_1 and θ_0, where condition for weight coefficients is $\theta_1 \leq \theta_0$. Summary of future cost at time t is defined by equation, which count all future values of the criteria functions multiplied by a discount factor (γ):

$$G(t) = \sum_{b=t}^{N} \gamma^{b-t} C(b) \tag{13}$$

where γ is the inter-temporal discount factor, which can be seen as a "psychological" discount rate or as a risk discount rate (Cairns, 1997).

For solving the problem of optimal investment distribution, it is necessary to define set $I(t)$ as set of all available information at time t:

$$I(t) = \{\mathbf{y}(0), \mathbf{y}(1), \ldots, \mathbf{y}(t), s(0), s(1), \ldots, s(t-1)\}$$

Criteria function at time t can be defined as

$$J(I(t)) = \min_{\pi_t} E[G(t)|I(t)], \tag{14}$$

$$t = 0, 1, \ldots, N$$

where π_t represents the set of future investment strategies

$$\pi_t = \{\{s(t_b)\}_{t_b=t,t+1,\ldots,N-1} : 0 \leq s(t_b) \leq 1\} = \{\{s(t), s(t+1), \ldots, s(t(N-1))\} : 0 \leq s(t_b) \leq 1\}$$

The objective of mathematical programming is to find investment strategy π_t, which would get minimal average value of future cost $G(t)$.

3.1 Bellman's optimal principle

Bellman's optimal principle, which is given in (Bellman and Kabala, 1965), has for above noticed problem the Bellman's equation in form

$$J(I(t)) = \min_{\pi_t} E\left[\sum_{s=t}^{N} \gamma C(s)|I(t)\right] = \min_{s(t)} [C(t) + \gamma E[J(I(t+1)|I(t))]] \tag{15}$$

We notice that $\{\nu(t)\}$ and $\{\lambda(t)\}$ are independent, $x_5(t)$ is Marcovian process and it is valid for conditional density of a probability that

$$p[x_5(t+1)|I(t)] = p[x_5(t+1)|\mathbf{y}(t)]$$

and also

$$p[x_5(t+1), x_5(t+2), \ldots, x_5(N)|I(t)] = p[x_5(t+1), x_5(t+2), \ldots, x_5(N)|\mathbf{y}(t)]$$

It follows from the previous equation that conditional densities of probability are in form

$$p(G(t)|I(t)) = p(G(t)|\mathbf{y}(t)) \tag{16}$$

and

$$J(x_5(t), t) = \min_{s(t)} [C(t) + \gamma E[J(x_5(t+1), t+1)| |\mathbf{y}(t))]], \; t = 0, 1, \ldots, t_{N-1} \tag{17}$$

On the basis of the asset equation (8) and the fact, that $\{\nu(t)\}$, $\{\lambda(t)\}$ and $x_5(t)$ are independent (7), it is possible to derive the conditioned mean $E[x_5(t+1)|\mathbf{y}(t)]$, the conditioned mean of second power $E[x_5^2(t+1)|\mathbf{y}(t)]$ and the conditioned variation $var[x_5(t+1)|\mathbf{y}(t)]$. It easy to verify that

$$E[x_5(t+1)|\mathbf{y}(t)] = (x_5(t) + c(t)) [(1 - s(t))\overline{\nu} + s(t)\overline{\lambda}] \tag{18}$$

$$E[x_5^2(t+1)|\mathbf{y}(t)] = (x_5(t) + c(t))^2 [(1 - s(t))^2(\overline{\nu}^2 + \omega_1^2) + s(t)^2(\overline{\lambda}^2 + \omega_2^2) + 2s(t)(1 - s(t))\overline{\nu}\overline{\lambda}] \tag{19}$$

$$var[x_5(t+1)|\mathbf{y}(t)] = (x_5(t) + c(t))^2 [(1 - s(t))^2\omega_1^2 + s^2(t)\omega_2^2] \tag{20}$$

Without loss of generality it is possible to assume that θ_0 is multiple of θ_1. If it is assumed that $\theta_1 = 1$ then it is possible to define $\theta_0 = \theta$, where

$\theta \geq 1$. The dynamic programming problem is transformed to:

$$J(x_5(t),t) = \min_{s(t)}[(x_5(t) - \tilde{x}_5(t))^2 + \\ + \gamma E[J(x_5(t+1), t+1)|\mathbf{y}(t)]] \quad (21)$$

with boundary condition $t = N$:

$$J(x_5(N), N) = \theta(x_5(N) - \tilde{x}_5(N))^2 \quad (22)$$

where $\tilde{x}_5(N)$ is expected amount of pension fund asset from time $t = 0$ when the cost $c(t)$ and the amount of profit from high-risk investment (with mean $\overline{\lambda}$) and amount of profit from low-risk investment (with mean $\overline{\nu}$) to time $t = N$. We notice, that the realized pension fund target ($\tilde{x}_5(N)$), where half of asset is invested to high-risk capital and second half is invest to low-risk capital, is in form:

$$\tilde{x}_5(N) = (\tilde{x}_5(0) + c(0))\left(\frac{\overline{\nu}+\overline{\lambda}}{2}\right)^N + \\ + \sum_{k=1}^{N-1} c(N-k)\left(\frac{\overline{\nu}+\overline{\lambda}}{2}\right)^k. \quad (23)$$

3.2 Solution of the dynamic programming problem

We assume that it is possible to verify a hypothesis. This hypothesis expects to find and solve the minimal of equation (21) as LQ problem. We expect the solution in form:

$$J(x_5(t),t) = P(t)x_5^2(t) - 2Q(t)x_5(t) + \\ + R(t) \quad (24)$$

For boundary condition (22) at time N the values of coefficient ($P(t)$, $Q(t)$ a $R(t)$) are acceptable for solving expectations (24)

$$P(N) = \theta \quad Q(N) = \theta\tilde{x}_5(N) \quad (25)$$

$$R(N) = \theta\tilde{x}_5^2(N)$$

Assuming that the hypothesis is satisfied for $t+1$. We rewrite the equation (24) to this time moment.

$$J(x_5(t+1), t+1) = P(t+1)x_5^2(t+1) - \\ - 2Q(t+1)x_5(t+1) + \\ + R(t+1) \quad (26)$$

It has to be shown that the hypothesis is satisfied for time t.

Using (19), (20) and (26) it is not difficult to verify that conditioned mean at time $t+1$ is equal to second degree polynomial with coefficients $L(t)$, $M(t)$ and $N(t)$ and unknown parameter is value of optimal distribution of invested asset $s(t)$

$$E[J(x_5(t+1), t+1)|\mathbf{y}(t)] = L(t)s^2(t) + \\ + M(t)s(t) + N(t) = \\ = \psi(s(t)) \quad (27)$$

where

$$L(t) = P(t+1)(x_5(t) + c(t))^2 \\ (\overline{\nu}^2 + \omega_1^2 + \overline{\lambda}^2 + \omega_2^2 - 2\overline{\nu}\overline{\lambda}) \quad (28)$$

$$M(t) = 2P(t+1)(x_5(t) + c(t))^2(\overline{\nu}\overline{\lambda} - \overline{\nu}^2 - \omega_1^2) - \\ - 2Q(t+1)(x_5(t) + c(t))(\overline{\lambda} - \overline{\nu}) \quad (29)$$

$$N(t) = P(t+1)(x_5(t) + c(t))^2(\overline{\nu}^2 + \omega_1^2) - \\ - 2Q(t+1)(x_5(t) + c(t))\overline{\nu} + R(t+1) \quad (30)$$

From equation (27) it is possible to retrieve unique global extreme $\psi^*(t)$ and it is assumed that it is valid $L(t) > 0$ for $\forall t$, then the extreme is unique global minimum

$$\psi(s^*(t)) = Z^*(t)$$

and

$$s^*(t) = -\frac{M(t)}{2L(t)} \quad (31)$$

$$Z^*(t) = N(t) - \frac{M^2(t)}{4L(t)} \quad (32)$$

Substitution (28), (29) and (30) to (32) we get

$$Z^*(t) = P'(t)x_5^2(t) + Q'(t)x_5(t) + R'(t) \quad (33)$$

where

$$P'(t) = HP(t+1) \quad (34)$$

$$Q'(t) = 2Hc(t)P(t+1) + 2KQ(t+1) \quad (35)$$

$$R'(t) = Hc^2(t)P(t+1) + 2Kc(t)Q(t+1) - \\ - \frac{Q(t+1}{DP(t+1)}(\overline{\lambda} - \overline{\nu})^2 \quad (36)$$

where

$$D = \overline{\nu}^2 + \omega_1^2 + \overline{\lambda}^2 + \omega_2^2 - 2\overline{\nu}\overline{\lambda} \quad (37)$$

$$H = \overline{\nu}^2 + \omega_1^2 - \frac{1}{D}(\overline{\nu}\overline{\lambda} - \overline{\nu}^2 - \omega_1^2) \quad (38)$$

$$K = \frac{1}{D}(\overline{\nu}\overline{\lambda} - \overline{\nu}^2 - \omega_1^2)(\overline{\lambda} - \overline{\nu}) - \overline{\nu} \quad (39)$$

Equation (17) becomes now

$$J(x_5(t),t) = \min_{s(t)}[C(t) + \\ + \gamma E[J(x_5(t+1), t+1)|\mathbf{y}(t)]] = \\ = \min_{s(t)}[(x_5(t) - \tilde{x}_5(t))^2 +$$

$$+\gamma E[J(x_5(t+1),t+1)|\mathbf{y}(t)]] =$$
$$= (x_5(t) - \tilde{x}_5(t))^2 + \gamma Z^*(t) =$$
$$= P(t)x_5^2(t) - 2Q(t)x_5(t) + R(t) \quad (40)$$

where

$$P(t) = 1 + \gamma P'(t) \quad (41)$$
$$Q(t) = \tilde{x}_5(t) - 0.5\gamma Q'(t) \quad (42)$$
$$R(t) = \tilde{x}_5^2(t) + \gamma R'(t) \quad (43)$$

where P'(t), Q'(t) and R'(t) are given by (34) and (35) respectively by (36) above. The hypothesis which, finds solution of equation (24) as second degree polynomial, was proved.

3.3 Optimal investment strategy

It is now possible to determine the optimal investment strategy ($s^*(t)$) by substituting (28) and (29) to (31). This leads to

$$s^*(t) = \frac{Q(t+1)V}{P(t+1)(x_5(t)+c)D} - \frac{W}{D} \quad (44)$$

where the sequences $\{P(t)\}_{t=1,\ldots,N}$, $\{Q(t)\}_{t=1,\ldots,N}$ and $\{R(t)\}_{t=1,\ldots,N}$ are given recursively by

$$P(t) = 1 + \gamma HP(t+1) \quad (45)$$
$$Q(t) = \tilde{x}_5(t) - \gamma c(t)HP(t+1) -$$
$$- \gamma KQ(t+1) \quad (46)$$
$$R(t) = \tilde{x}_5^2(t) + \gamma c^2(t)HP(t+1) +$$
$$+ 2\gamma c(t)KQ(t+1) + R(t+1) -$$
$$- \frac{Q(t+1)}{DP(t+1)}(\overline{\lambda} - \overline{\nu})^2 \quad (47)$$

where the boundary conditions $P(N)$, $Q(N)$ and $R(N)$ are given by (25), D, H and K are given by (37), (38) and (39) above and V and W are given by

$$V = \overline{\lambda} - \overline{\nu} \quad (48)$$
$$W = \overline{\nu}\overline{\lambda} - \overline{\nu}^2 - \omega_1^2 \quad (49)$$

4. ILLUSTRATION EXAMPLE

In the illustration example, it is assumed that we have a pension fund with only one client who saves a constant amount during whole period and finishes saving in strictly defined term.

The optimal distribution of asset depends on many parameters. Firstly, it is necessary to decide on length of time invested period. In this illustration, the time period for managing asset was set to 10, 20, 30 and 40 years.

For simplification, we made assumption, that client will not transfer any finances from the other pension funds, so it is assumed that $x_5(0) = 0$ and $u_p(t) = 0$ for $\forall t$. The inter-temporal discount factor was defined as $\gamma = 0.95$.

The amount of contributions (costs=0) is constant and it is equal to c = 420 Kč, where the client contributes 300 Kč and amount 120 Kč is paid as state contribution for supporting supplementary pension insurance.

The rates of return assume the following values. The mean for low-risk investment ($\overline{\nu}$) is 4% and for high-risk ($\overline{\lambda}$) it is 6%. The standard deviation for low-risk investment (ω_1) is 2.5% and for high-risk (ω_2) it is 20%.

In the figure (1), behaviour of optimal investment strategy $s(t)$ is shown. This simulation scenario is very close to the current situation in the Czech republic, where it is possible to invest with a low rate of return in a low-risk asset. On the other hand, if pension fund management wants to get higher rate of returns, they have to invest in a very risky asset.

In this case, the optimal investment strategy has a strictly descending characteristic. The differences between a real asset $x_5(t)$and a target asset $\tilde{x}_5(t)$ depends on time of service. In all cases the real asset is higher then the target asset.

5. CONCLUSION

This paper was focused on utilization of the mathematical model of the contribution defined pension fund in the Czech republic. The structure of mathematical model has been already given in (Šimandl and Lešek, 2002). The method of the best allocation of a pension fund pension fund asset in two-asset world was introduced.

Analysis of behaviour of the pension fund investment for distribution of capital to low-risk investment and high-risk investment was performed. It is possible to expect that pension fund managers know the values for each type of rates of return. And we proved that on basis, we can automatically allocate capital between high-risk and low-risk investment.

In the work (Vigna and Haberman, 2001), it is mentioned that managers in the Great Britain firstly invested in high-risk investments and at the end of service period, they moved capital to low-risk investments. Simulation results show the same strategies in how to best allocate in a two-asset world are similar to the empirical behaviour that pension fund managers in actual pension fund employ.

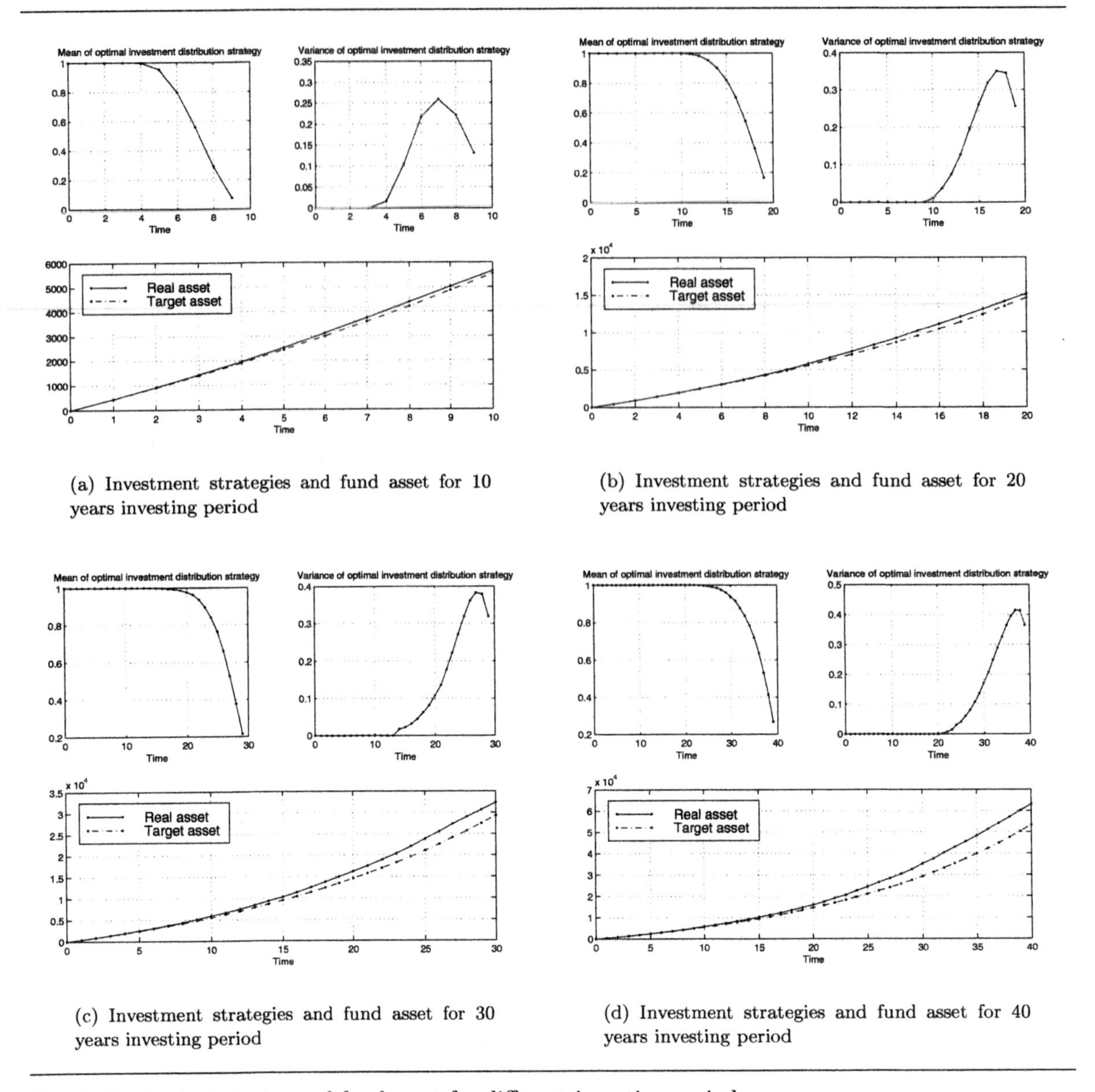

(a) Investment strategies and fund asset for 10 years investing period

(b) Investment strategies and fund asset for 20 years investing period

(c) Investment strategies and fund asset for 30 years investing period

(d) Investment strategies and fund asset for 40 years investing period

Fig. 1. Optimal strategies and fund asset for different investing period

6. ACKNOWLEDGMENT

The work was supported by the Grant Agency of the Czech Republic, project GA ČR 102/01/0021 and by the Ministry of Education, Youth and Sports of Czech Republic MSM 2352 00004.

REFERENCES

Bellman, R. and R. Kabala (1965). *Dynamic programming and modern control theory*. Academic press. New York.

Cairns, A.J. (1997). A comparition of optimal and stochatic control strategies for continuous-time pension fund models. AFIR'97.

Owadally, M.J. (1998). The dynamics and control of pension funding. PhD thesis. The City University, London.

Sung, J.H. and S. Haberman (1994). Dynamic approaches to pension funding. *Insurance: Mathematic and Economics.*

Vigna, E. and S. Haberman (2001). Optimal investment strategy for defined cotribution pension schemes. *Insurance: Mathematic and Economics.*

Šimandl, M. (1998). *Návrh matematických modelů pro návrh některých parametrů penzijního fondu.* Západočeská univerzita. Plzeň. (in Czech).

Šimandl, M. and M. Lešek (2001). Design of structrure and state estimation of pension fund model. In: *Nostradamus, 4th Internacional Conference on Prediction and Nonlinear Dynamics.* Tomas Bata University in Zlín.

Šimandl, M. and M. Lešek (2002). *Rozložení investic penzijního fondu s využitím Bellmanova principu optimality.* Západočeská univerzita. Plzeň. (in Czech).

www.elsevier.com/locate/ifac

A NONLINEAR FLOW CONTROLLER FOR ATM NETWORKS

Andrzej Bartoszewicz

Institute of Automatic Control, Technical University of Łódź,
18/22 Stefanowskiego St., 90-924 Łódź, Poland,
andpbart@ck-sg.p.lodz.pl.

Abstract: In this paper congestion control in Asynchronous Transfer Mode networks is considered. A new nonlinear flow control strategy for a single connection is proposed. The strategy combines the relay type controller with Smith predictor used for the delay compensation. The control scheme proposed in the paper is particularly suitable for loss sensitive applications. It guarantees full link utilisation and no cell loss in the controlled connection. Furthermore, transmission rates generated by the controller are limited according to the maximum available bandwidth of the connection. Finally, the paper shows that a special case of a sliding mode is induced by the proposed strategy in the controlled system. *Copyright © 2003 IFAC*

Keywords: Flow control, ATM networks, time delay systems, Smith principle, sliding mode control.

1. INTRODUCTION

Voice, video and data transmission through the high speed telecommunication networks plays the crucial role for proper management of today's information based society. The asynchronous transfer mode (ATM) technology, well suited for these purposes, is an important standard in the design and implementation of broadband integrated services digital networks (B-ISDN). The ATM networks are connection oriented, i.e. a virtual circuit is established between each source and destination for the connection lifetime. After setting up a virtual circuit, data is sent in relatively short, fixed size packets, usually called cells. Each data cell is 53 bytes long and consists of 48 bytes of transmitted information and a 5 byte long header. The small fixed cell size reduces delay variation which could be particularly harmful for multimedia traffic.

In order to properly serve diverse needs of different users the ATM Forum defines the following five service categories:

1. Constant bit rate (CBR) service category provides the bandwidth which is always available to its user. This category is used by real time service. Typical examples of the service are television, telephone, etc.
2. Variable bit rate (VBR) is designed for both real and non-real time applications. An example of such real time application is video conferencing, and multimedia email is an example of non-real time VBR service.
3. Available bit rate (ABR) is a service category whose rate depends on the available bandwidth. Users should adjust their flow rate according to the feedback information received from the network. This category is particularly important for congestion avoidance and full resource utilisation. Electronic mail is an example of this service category.
4. Unspecified bit rate (UBR) category is used to send data on the first in first out (FIFO) basis, using the capacity not consumed by other services. No initial commitment is made to a UBR source and no feedback concerning congestion is provided. This type of service can be used for background file transfer.
5. Guaranteed frame rate (GFR) is a service intended for non-real time applications with little rate requirements. No feedback control protocol is applied in this service category. An example of this service is frame relay interworking.

As stated above ABR is the only service category using feedback information to control source flow rate. Therefore, ABR plays the crucial role for congestion control and effective resource utilisation. The problem of ABR flow control in ATM networks is considered in this paper.

The difficulty of the ABR flow control is mainly caused by long propagation delays in the network. If congestion occurs at a specific node, information about this condition must be conveyed to the source. Transmitting this information involves feedback propagation delay. After this information has been received by the source, it can be used to adjust the flow rate. However, the adjusted flow rate will start to affect the congested node only after forward propagation delay. Though, both forward and feedback propagation delays depend on the distance between the source and the congested node the sum of these two delays, usually called round trip time (RTT), does not depend on the congested node location in the network.

The ABR flow rate control has recently been studied in several papers. A valuable survey of earlier congestion control mechanisms is given by Jain (1996). Izmailov (1995) considered a single connection controlled by a linear regulator whose output signal is generated according to the several states of the buffer measured at different time instants. Asymptotic stability, nonoscillatory system behaviour and locally optimal rate of convergence have been proved. Chong et. al. (1998) proposed and thoroughly studied the performance of a simple queue length based flow control algorithm with dynamic queue threshold adjustment. Lengliz and Kamoun (2000) introduced a proportional plus derivative controller which is computationally efficient and can be easily implemented in ATM networks. Imer et. al. (2001) gave a brief, excellent tutorial exposition of the ABR control problem and presented new stochastic and deterministic control algorithms. Another interesting approach to the problem of flow rate control in communication networks has been proposed by Quet et al. (2002). The authors considered a single bottleneck multi-source ATM network and applied minimisation of an H-infinity norm to the design of a flow rate controller. The proposed controller guarantees stability robustness to uncertain and time-varying propagation delays in various channels. Adaptive control strategies for ABR flow regulation have been proposed by Laberteaux and Rohrs (2002). Their strategies reduce convergence time and improve queue length management. Also a neural network controller for ABR service in ATM networks has recently been proposed. Jagannathan and Talluri (2002) showed that their neural network controller can guarantee stability of the closed loop system and the desired quality of service (QoS).

Due to the significant propagation delays which are critical for the closed loop performance, several researchers applied Smith principle to control ABR flow in communication networks. Mascolo (1999) considered the single connection congestion control problem in a general packet switching network. He used the deterministic fluid model approximation of packet flow and applied transfer functions to describe the network dynamics. The designed controller was applied to the ABR traffic control in ATM network and compared with ERICA standard. Furthermore, Mascolo showed that Transmission Control Protocol / Internet Protocol (TCP/IP) implements a Smith predictor to control network congestion. In the next paper (Mascolo 2000) the same author applied the Smith principle to the network supporting multiple ABR connections with different propagation delays. The proposed control algorithm guarantees no cell loss, full and fair network utilisation, and ensures exponential convergence of queue levels to stationary values without oscillations or overshoots. Gomez-Stern et. al. (2002) further studied the ABR flow control using Smith principle. They proposed a proportional-integral (PI) controller which helps to reduce the average queue level and its sensitivity to the available bandwidth. Saturation issues in the system were handled using anti-wind up techniques.

In this paper the ABR flow control in ATM networks is considered. Similarly to the approach introduced in (Mascolo 1997; Mascolo 1999; Mascolo 2000; Gomez-Stern et. al. 2002; Grieco and Mascolo 2002) Smith principle is applied, however as opposed to those papers a new nonlinear control strategy is proposed. This strategy combines Smith principle with the conventional on–off controller. The strategy guarantees full link utilisation and no cell loss in the controlled virtual circuit. As a result, the need of cell retransmission is eliminated. Furthermore, transmission rates generated by the algorithm are limited by an arbitrary value greater than the maximum available bandwidth of the controlled connection.

The remainder of this paper is organised as follows. The model of the network considered in the paper is introduced in Section 2. Then the main result, i.e. the new nonlinear control system is presented and discussed in Section 3. Finally, Section 4 comprises conclusions of the paper.

2. NETWORK MODEL

In this paper the ABR flow control in a virtual circuit is considered. The circuit consists of a single source, a number of nodes connected by bi-directional links and a single destination. It is assumed that there is one bottleneck node in the circuit. In other words only one node has more cells to send than the link connecting it to the next node can actually carry. The rate of cell outflow from the bottleneck node depends on the available bandwidth which will be modelled as an a'priori unknown function of time. The source sends data cells (at the rate determined by the controller) and resource management cells. The resource management cells are processed by the nodes on the priority basis, i.e. they are not queued but sent to the next node without delay. These cells carry information about the network condition. After reaching the destination they are immediately sent back to the source, along the same path they arrived. The information carried by the resource management cells is used by the controller to adjust the source rate.

Further in this paper data transfer in modelled as an incompressible fluid flow. The following notation is used: t denotes time and RTT stands for the round trip time which is equal to the sum of forward and backward propagation delays denoted as T_f and T_b respectively.

$$RTT = T_f + T_b \tag{1}$$

Furthermore, $x(t)$ denotes the bottleneck node queue length, x_d the demand value of $x(t)$ and $d(t)$ represents the bandwidth available for the controlled connection at time t. After setting up the connection, the buffer of the bottleneck node is empty, so for any time $t < T_f$ the queue length

$$x(t) = 0 \tag{2}$$

On the other hand for $t \geq T_f$

$$x(t) = \int_{T_f}^{t} \left[a\left(\tau - T_f\right) - h(\tau)\right] d\tau \tag{3}$$

where $a(t)$ is the source rate, and $h(t)$ represents the bandwidth which is actually consumed by the bottleneck node at time t. If the queue length at this node $x(t)$ is greater than zero, then the entire available bandwidth is consumed $h(t) = d(t)$. Otherwise, i.e. when $x(t) = 0$, then $h(t)$ is determined by the rate of data arrival at the node. In this case, the available bandwidth may not be fully utilised. Consequently, for any time t

$$0 \leq h(t) \leq d(t) \leq d_{max} \tag{4}$$

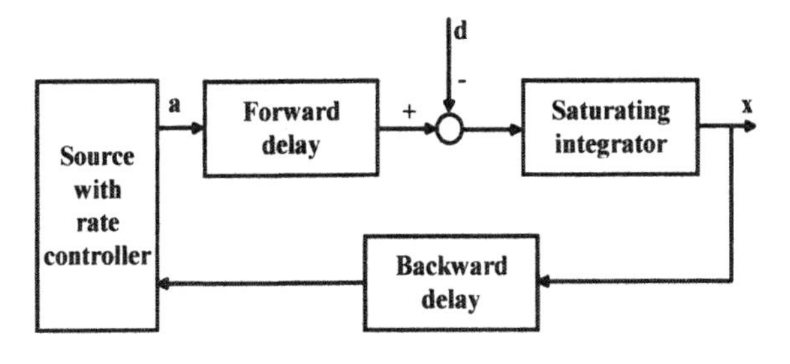

Figure 1 Control system

The block diagram of the flow control system considered in this paper is shown in Figure 1. Since a nonlinear on–off type controller is designed, the source rate $a(t)$ is changed between the two values: a_{max} and zero. It is assumed that the following two conditions are satisfied

$$a_{max} > d_{max} \quad \text{and} \quad a_{max} \geq x_d \cdot RTT^{-1} \tag{5}$$

The source rate controller should assure that the bottleneck buffer overflow is avoided and the node has always enough data to send. The first condition implies that data cells are not lost and there is no need for their retransmission, while the latter one assures full network utilisation which is highly desirable for economic reasons. In the next section it will be shown how to achieve these favourable properties of the considered connection.

3. PROPOSED CONTROLLER

In this section the following control algorithm is proposed:

(i) The source rate during the time interval from zero up to RTT is chosen in such a way that

$$0 \leq a(t) \leq a_{max} \tag{6}$$

$$\int_{t=0}^{RTT} a(\tau) d\tau = x_d \tag{7}$$

(ii) For any time $t \geq RTT$ the source rate is determined as follows

$$a(t) = \frac{1}{2} a_{max} + \\ + \frac{1}{2} a_{max} \cdot sgn\left[x_d - x\left(t - T_b\right) - \int_{t-RTT}^{t} a(\tau) d\tau\right] \tag{8}$$

Notice that the proposed algorithm leaves some freedom when selecting the source rate at the beginning of the control action. It requires only that the total number of cells sent from the initial time $t_0 = 0$ to the time instant RTT equals x_d. Consequently, the source rate during this interval could be chosen as

$$a(t) = \frac{x_d}{RTT} = \text{const.} \tag{9}$$

In the sequel two theorems presenting basic properties of the proposed control strategy are introduced.

Theorem 1

If relations (4) – (8) are satisfied, then the queue length is always upper bounded by its demand value

$x(t) \leq x_d$ (10)

Proof

Let us first notice that for any time $t \leq RTT$ equations (6) and (7) directly imply that relation (10) is satisfied. On the other hand, since in this paper data transfer is modelled as an incompressible fluid flow, from relations (4) and (6) and (8) one concludes that the queue length $x(t)$ is a continuous function of time. Consequently, at time $t \geq RTT$ the length can increase only if

$$a(t - T_f) = a_{max} \tag{11}$$

which implies

$$x_d - x(t - RTT) - \int_{t-RTT-T_f}^{t-T_f} a(\tau)d\tau > 0 \tag{12}$$

and consequently

$$\begin{aligned} x_d - x(t - RTT) - \int_{t-RTT-T_f}^{t-T_f} a(\tau)d\tau &= \\ = x_d - \int_0^{t-RTT-T_f} a(\tau)d\tau + \int_0^{t-RTT} h(\tau)d\tau &+ \\ - \int_{t-RTT-T_f}^{t-T_f} a(\tau)d\tau &= \\ = x_d - \int_0^{t-T_f} a(\tau)d\tau + \int_0^{t} h(\tau)d\tau - \int_{t-RTT}^{t} h(\tau)d\tau &= \\ = x_d - x(t) - \int_{t-RTT}^{t} h(\tau)d\tau &> 0 \end{aligned} \tag{13}$$

It follows from relations (4) and (13) that the queue length can increase at time t, only if

$$x(t) < x_d \tag{14}$$

i.e. only when its current value is smaller than the demand value x_d. As a consequence $x(t)$ will never exceed x_d. This conclusion ends the proof of Theorem 1.

This theorem shows that selecting x_d smaller than or equal to the buffer capacity, one can assure no cell loss in the controlled virtual circuit. This is particularly important for loss sensitive traffic like Web pages or file transfer. Another desired property of a suitably designed flow control system is full link utilisation. The next theorem shows that the proper choice of the buffer capacity actually ensures this property.

Theorem 2

If relations (4), (7) and (8) are satisfied, and the following inequality

$$x_d > RTT \cdot d_{max} + \delta \tag{15}$$

(where δ is a positive constant) holds, then for any time $t \geq RTT + T_f$ the queue length is greater than δ. In particular, if $x_d > RTT \cdot d_{max}$, then the queue length is greater than zero.

Proof

Let us first notice that the queue length at time $t = RTT + T_f$

$$\begin{aligned} x\left(RTT + T_f\right) &= \int_{T_f}^{RTT+T_f} \left[a\left(\tau - T_f\right) - h(\tau)\right]d\tau = \\ &= \int_0^{RTT} a(\tau)d\tau - \int_{T_f}^{RTT+T_f} h(\tau)d\tau = \\ &= x_d - \int_{T_f}^{RTT+T_f} h(\tau)d\tau \geq \\ &\geq x_d - RTT \cdot d_{max} > \delta \end{aligned} \tag{16}$$

is greater than δ. Furthermore, recall that $x(t)$ is a continuous function of time. This function will increase at time $t \geq RTT + T_f$ if the argument of the sign function in equation (8) at time $t - T_f$ is positive, i.e. when

$$\sigma\left(t - T_f\right) = x_d - x(t - RTT) - \int_{t-RTT-T_f}^{t-T_f} a(\tau)d\tau > 0 \tag{17}$$

On the other hand

$$\begin{aligned} \sigma\left(t - T_f\right) &= x_d - x(t - RTT) - \int_{t-RTT-T_f}^{t-T_f} a(\tau)d\tau = \\ &= x_d - x(t) + \int_{t-RTT-T_f}^{t-T_f} a(\tau)d\tau + \\ &- \int_{t-RTT}^{t} h(\tau)d\tau - \int_{t-RTT-T_f}^{t-T_f} a(\tau)d\tau = \\ &= x_d - x(t) - \int_{t-RTT}^{t} h(\tau)d\tau \end{aligned} \tag{18}$$

and the following evaluation can be made

$$\sigma\left(t-T_f\right) \geq x_d - x(t) - RTT \cdot d_{max} > \delta - x(t) \quad (19)$$

It follows from relation (19) that the queue would always grow if its length $x(t)$ was smaller than δ, and consequently that the length never drops below this value. This conclusion ends the proof.

Theorem 2 shows that using the strategy proposed in this paper one can always assure full link utilisation, provided the bottleneck node buffer capacity is greater than the product $RTT \cdot d_{max}$. This shows that our strategy offers essentially better performance than the similar algorithms proposed earlier in the literature (Mascolo 1997; Mascolo 1999).

Theorem 3
If relations (4) – (8) are satisfied and the available bandwidth is not equal to zero, then for any time greater than or equal RTT a stable sliding mode takes place on the surface

$$\sigma(t) = x_d - x(t - T_b) - \int_{t-RTT}^{t} a(\tau)d\tau = 0 \quad (20)$$

Proof
Let us begin the proof calculating

$$\sigma(RTT) = x_d - x(RTT - T_b) - \int_{0}^{RTT} a(\tau)d\tau = 0 \quad (21)$$

Furthermore, for any time $t \geq RTT$

$$\begin{aligned}\sigma(t) &= x_d - x(t - T_b) - \int_{t-RTT}^{t} a(\tau)d\tau = \\ &= x_d - \int_{0}^{t-T_b-T_f} a(\tau)d\tau + \int_{0}^{t-T_b} h(\tau)d\tau + \\ &\quad - \int_{t-RTT}^{t} a(\tau)d\tau = \\ &= x_d - \int_{0}^{t} a(\tau)d\tau + \int_{0}^{t-T_b} h(\tau)d\tau = \\ &= \int_{0}^{t-T_b} h(\tau)d\tau - \int_{RTT}^{t} a(\tau)d\tau\end{aligned} \quad (22)$$

Now in order to prove the theorem the sign of the following product will be checked

$$\begin{aligned}\sigma(t)\dot{\sigma}(t) &= \sigma(t)\left[h(t - T_b) - a(t)\right] = \\ &= \sigma(t)h(t - T_b) - \frac{\sigma(t)a_{max}}{2} + \\ &\quad - \frac{\sigma(t)a_{max}\, sgn[\sigma(t)]}{2}\end{aligned} \quad (23)$$

This product is negative for any $\sigma(t) \neq 0$, which implies that this variable having reached zero at time RTT cannot deviate from this value anymore. This conclusion finally proves existence of the stable sliding mode on the surface $\sigma = 0$.

Now Theorem 3 will be used to find a relation between the bandwidth previously consumed by the controlled connection and the average current flow rate.

First the following definition is introduced.

Definition
The equivalent value of the discontinuous control $a(t)$ is defined as a continuous signal $a_{av}(t)$ whose effect on the system is the same as the effect of the original discontinuous control $a(t)$.

Theorem 4
If the assumptions of Theorem 3 are satisfied, then for any time greater than or equal RTT, the equivalent source rate at time t is equal to the bandwidth actually consumed at time $t - T_b$.

Proof
Theorem 3 implies that for any time $t \geq RTT$, $\sigma(t) = 0$. Consequently, it follows directly from equation (22) that for $t \geq RTT$

$$h(t - T_b) = a_{av}(t) \quad (24)$$

which ends the proof.

Finally let us notice that the steady state value of the queue length, i.e. the queue length when the bandwidth available for the controlled connection $d(t) = d_{ss}$ is constant, can be expressed as follows

$$x_{ss} = x_d - RTT \cdot d_{ss} \quad (25)$$

It can be seen from relation (25) that decreasing the demand value x_d helps to obtain better quality of service, i.e. smaller cell delay time and smaller variation of this time. Therefore, this relation can be used to achieve appropriate trade-off between the required quality of service and the degree of network utilisation.

4. CONCLUSIONS

In this paper a new nonlinear flow control strategy for ATM networks has been proposed. The strategy combines Smith predictor with the conventional relay controller. The strategy is well suited for loss sensitive applications since it guarantees that the bottleneck buffer capacity will never be exceeded in the controlled virtual connection. Furthermore, the strategy helps to achieve full resource utilisation. This favourable performance is achieved without using excessive flow rate. The rate is always limited, and the upper bound on this rate can be chosen at any level greater than the maximum bandwidth available for the controlled connection.

The control algorithm proposed in this paper could also be used for delay sensitive applications. It has been demonstrated that, if necessary, the quality of service can be appropriately improved. This is possible since decreasing the queue length reference value x_d, one can easily lower the total transmission delay. However, this better quality of service is obtained at the cost of smaller resource utilisation in the controlled connection.

Finally, let us point out that detailed study of the proposed control strategy reveals existence of a stable sliding mode in the system. The sliding variable is defined as the difference between the demand value of the queue length and its delayed measured value augmented by a corrective term. This corrective term takes into account the number of cells sent by the source during the last round trip time. The sliding mode control strategy proposed in the paper seems to be a feasible option for a general case of more complex time delay systems.

REFERENCES

Bartoszewicz A., Lipka G. (2002). Zastosowanie predyktora Smith'a z elementem nieliniowym do kontroli przeciążeń w szybkich sieciach transmisji danych ATM. *X Konferencja Sieci i Systemy Informatyczne*, Łódź, pp. 27-35.

Chong S., Nagarajan R., Wang Y. T. (1998). First-order rate-based flow control with dynamic queue threshold for high-speed wide-area ATM networks. *Computer Networks and ISDN Systems*, **29**, pp. 2201-2212.

Gómez-Stern F., Fornés J. M., Rubio F. R. (2002). Dead-time compensation for ABR traffic control over ATM networks. *Control Engineering Practice*, **10**, pp. 481-491.

Grieco L., Mascolo S. (2002). Smith's predictor and feedforward disturbance compensation for ATM congestion control. *Proceedings of the 41st IEEE Conference on Decision and Control*, Las Vegas, pp. 987-992.

Imer O. C., Compans S., Basar T., Srikant R. (2001). Available bit rate congestion control in ATM networks. *IEEE Control Systems Magazine*, pp. 38-56.

Izmailov R. (1995). Adaptive feedback control algorithms for large data transfers in high-speed networks. *IEEE Transactions on Automatic Control*, **40**, pp. 1469-1471.

Jagannathan S., Talluri J. (2002). Predictive congestion control of ATM networks: multiple sources/single buffer scenario. *Automatica* **38**, pp. 815-820.

Jain R. (1996). Congestion control and traffic management in ATM networks: recent advances and a survey. *Computer Networks and ISDN Systems*, **28**, pp. 1723-1738.

Kulkarni L. A., Li S. (1998). Performance analysis of a rate-based feedback control scheme. *IEEE/ACM Transaction on Networking*, **6**, pp. 797-810.

Laberteaux K. P., Rohrs Ch. E., Antsaklis P. J. (2002). A Practical Controller for Explicit rate Congestion Control. *IEEE Transactions on Automatic Control*, **47**, pp. 960-978.

Lengliz I., Kamoun F. (2000). A rate-based flow control method for ABR service in ATM networks. *Computer Networks*, **34**, pp. 129-138.

Mascolo S. (1997). Smith's principle for congestion control in high speed ATM networks. *Proceedings of the 36th IEEE Conference on Decision and Control*, San Diego, pp. 4595-4600.

Mascolo S. (1999). Congestion control in high-speed communication networks using the Smith principle. *Automatica*, **35**, pp. 1921-1935.

Mascolo S. (2000). Smith's principle for congestion control in high-speed data networks. *IEEE Transactions on Automatic Control*, **45**, pp. 358-364.

Niculescu S. I. (2000). On delay robustness analysis of a simple control algorithm in high-speed networks. *Automatica*, **38**, 885-889.

Quet P. F., Ataslar B., Iftar A., Özbay H., Kalyanaraman S., Kang T. (2002). Rate-based flow controllers for communication networks in the presence of uncertain time-varying multiple time-delays. *Automatica*, pp. 917-928.

FPQSPN AND ITS APPLICATION IN ANALYSIS OF ETHERNET

Josef Čapek

Czech Technical University in Prague, FEL, Department of Control Engineering, Karlovo nám. 13, 121 35, Prague 2, Czech Republic

Abstract: FPQSPN is a high level Petri net model allowing simplified analysis of systems loaded by finitely large population. FPQSPN is nearly as graphical and easy to implement as low-level nets models. Here we present the definition of FPQSPN and we present its application to throughput analysis of Ethernet. Complete accurate stochastic Petri net model of Ethernet is shown and its simulation results are presented. *Copyright © 2003 IFAC*

Keywords: Ethernet, Petri-nets, modelling, simulation, communication protocols, Queuing theory

1. INTRODUCTION

Performance analysis of communication protocols is an effort to quantify their properties and represent them using numerical measures. It is also the development and study of mathematical models that predict the performance of networks.

Models are frequent sources of mistakes. Their construction or actually their confidence is getting more and more important. For obtaining of confident results we need transparent and easy to validate modelling and analysis methods.

We introduce a complete stochastic Petri net model of a widely used non-deterministic communication protocol and we explain how we increased its confidence.

2. FPQSPN

Finite Population Queuing Systems Petri Nets (FPQSPN) is a model of *high level Petri nets* suitable for modelling of systems, where several identical functional elements (the population) share some resources. The main reason leading in definition of FPQSPN was to develop a transparent and easy to implement model that enables simple modelling of finitely large population in queuing systems and that is nearly as graphical and transparent as low-level nets models. A FPQSPN diagram typically comprises a sub-diagram of population and a sub-diagram representing shared resources. The shared resources may represent a *server*, for example a call-centre or a communication channel, and rest of the model represents the *customers*. The customers are, for example, nodes accessing the communication channel or simply users requesting the call-centre.

Definition

In FPQSPN, places and transitions are divided into two groups: *shared objects* and *non-shared objects*. In spite of this fact, FPQSPN models are bi-partite graphs. There are rules for placing of arcs between places and transitions that are both shared or non-shared objects or one of them is a shared object and the other one is a non-shared object. Graphically, the shared objects are distinguished by double line. The arcs are drawn by single line regardless of objects they connect. Formally, FPQSPN is a 10-tuple ($\mathbf{P}$, $\mathbf{T}$, I, O, M_0, so, tt, $E(t)$, t_2, k) such that:

P = $\{P_i \mid i \neq j, i \in IN, j \in IN\}$ is an un-ordered finite set of places,

T = $\{T_i \mid i \neq j, i \in IN, j \in IN\}$ is an un-ordered finite set of transitions,

$I \subseteq$ $\mathbf{T} \times \mathbf{P}$ is an input function,

$O \subseteq$ $\mathbf{T} \times \mathbf{P}$ is an output function,

M_0: $\mathbf{P} \rightarrow IN^0$ is an initial marking,

so: $\mathbf{T}$, $\mathbf{P}$ $\rightarrow$ {shared, non-shared} determines whether the object is shared object or not,

This work has been supported by the Ministry of Education of the Czech Republic under Project no. J04/98:212300013 tt: T → {zero, deterministic, *set of probability distribution functions*} determines whether the transition is immediate, deterministic or stochastic ...,

E(t): **T** → *IR* is the mean of transition's time t,

t_2: **T** → *IR* meaning of the parameter depends on *tt* – e.g. zero, variance, minimum ...

k k ∈ IN0 is number of customers (the size of population),

$\mathbf{P} \cup \mathbf{T} \neq \varnothing$ and $\mathbf{P} \cap \mathbf{T} = \varnothing$.

P and **T** are sets where each element is labelled by a unique integer number i which is greater than zero. Generally, the numbers i must not necessarily form any sequence. Note, that **P** and **T** are un-ordered sets and it is *eventually possible* to order them by i.

Unfolding Algorithm

Unfolding algorithm inputs folded FPQSPN model as a set of places, transitions, and arcs and outputs its unfolded version. It is de-facto, a low-level stochastic Petri net model where the population is represented by k replications *the non-shared* sub-diagram.

1. For each object that is a non-shared place P_i or a non-shared transition T_i, create a set of k places $\mathbf{P}_k = \{P_n | n = i + |\mathbf{P}|*j, j = [0, k\text{-}1]\}$ respectively a set of k transitions $\mathbf{T}_k = \{T_n | n = i + |\mathbf{T}|*j, j = [0, k\text{-}1]\}$.
2. For each object that is a shared P_i or T_i, create P_i respectively T_i.
3. For each arc between non-shared P_i and T_i, create a set of arcs and connect their sources and destinations to objects in $\mathbf{P}_k$ and $\mathbf{T}_k$ created in point 1.
4. For each arc between a non-shared object and a shared object, create k arcs having the same index numbers on the shared object side. Non-shared side connect to objects in $\mathbf{P}_k$ or $\mathbf{T}_k$ created in pt. 1.
5. For each arc between shared P_i and T_i, create arc between P_i and T_i.

Time Aspects

FPQSPN are Petri nets with time. Speaking about them, we should briefly recall some elements of Petri nets temporal semantics.

There are *immediate transitions* and *timed transitions* in FPQSPN. While an immediate transition may be either *disabled* or *enabled*, for timed transitions we distinguish even three states: *disabled*, *enabled* or *fireable*. A transition is *enabled* if all its input places are 'marked enough'. Otherwise it is *disabled*. The *fireable* state means that a transition may be fired (an immediate transition is *fireable* if it is *enabled*). The rule for becoming an enabled timed transition *fireable* depends on temporal semantics. In FPQSPN, the delay t of a *timed transition* T means the time between the moment when the transition T becomes *enabled* and the moment when it becomes *fireable*. In other words, T becomes *fireable* after it has been continuously *enabled* for the period t. Enabled *timed transitions* do not reserve tokens in their input places.

"An important issue that arises at every transition firing when timed transitions are used in a model is how to manage the timers of all the transition that do not fire." (Marsan *et al.*, 1995). For convenience, we recall three ways of managing of timers during transition firing: *Resamplig*: at each and every transition firing, the timers of all the timed transitions are discarded. *Enabling memory*: at each transition firing, the timers of all timed transitions that are disabled are discarded, while the timers of transitions that are enabled hold their values. When a disabled timed transition becomes enabled again, value of its timer is discarded. *Age memory*: at each transition firing, the timers of all the timed transition hold their present values. When a disabled transitions becomes enabled again, its timer starts at the value it was holding at the moment, when it was disabled. *Enabling memory* concept is used in FPQSPN.

Enabling degree means, how many times a transition is enabled at given marking M or equivalently how many times the enabling condition is fulfilled. Several possibilities were mentioned in (David, Alla, 1992). FPQSPN use *single server semantics*, which means that a timed transition can be *fireable*, *enabled* or *disabled* at one instant (e.g., *enabled* transition cannot be *fireable* and so on).

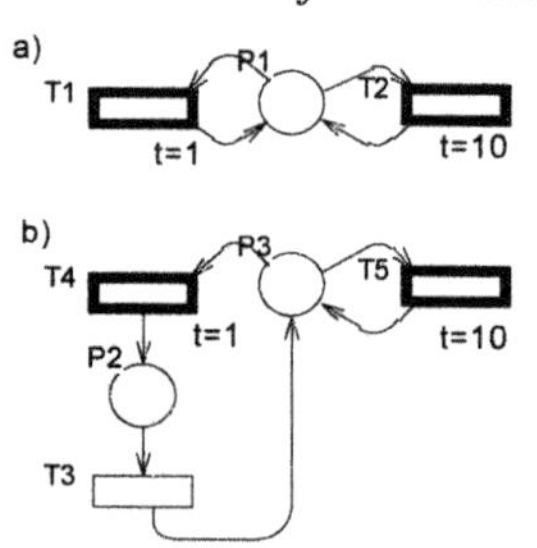

Fig 1 a self loop and a loop

A self-loop transition T_1 (a transition in self-loop – see Fig 1a) doesn't reset timed transitions that are in conflict with T_1 (it does not restart their timers). Two models in Fig 1 behave completely differently: while T_1 in Fig 1a, which has shorter time, is fired approximately ten times more often than T_2, transition T_5 in Fig 1b is never fired. Firing of a transition is an atomic immediate action that represents consuming of tokens in input places and producing ones in output places. It means, that the tokens are produced in output places at exactly the same time as they are consumed in input places so we cannot say whether the tokens are consumed earlier and produced later or whether they are consumed later and produced earlier - they are produced and consumed at the same moment. Hence, firing of T_1 in Fig 1a does not reset timer of T_2 so both of them fire according to their times. Thanks to what was said, it holds that the token count in a directed circuit of a marked graph is invariant in time (model's virtual time) and it is also invariant under any marking: $M(\boldsymbol{C}) = M_0(\boldsymbol{C})$ for each directed circuit $\boldsymbol{C}$. This is in concordance to what is known (e.g. Murata, 1989) about marked graphs.

There are no transitions priorities in FPQSPN. When there are two or more *fireable* transitions at one instant, each of them fires with the same probability. When an immediate transition and a timed (non-fireable) transition are enabled, the *immediate* transition fires first. We would only emphasize that it is not because of priorities but it is because the *immediate* transition becomes *fireable* immediately but the timed transition becomes *fireable* after its *delay*.

In figures in the rest of paper we graphically distinguish transitions probability distribution functions: Outlines of *immediate transitions* are drawn using thin solid line, non-immediate transitions are drawn by thick line. Deterministic transitions (time *t* is a constant) are not filled. Transitions having exponential distribution are filled with forward diagonal lines. Transitions having uniform distribution are filled using vertical lines.

Example

Les us take a look at the model of critical section in Fig 2. We would point out here that places and transitions that are non-shared objects (T_1, T_2 T_3, P_1 and P_2) appear *k*-times after the unfolding (see Fig 3). We excuse for not to explain meaning of all objects.

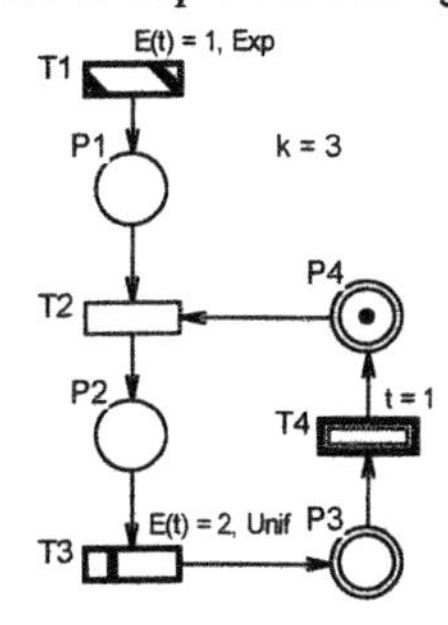

Fig 2 FPQSPN model of critical section

Unfolded models are used during simulation. Thanks to this, it is for example possible to evaluate how many times particular users actually enter the critical section - (firing of T_2, T_6, T_{10}) and how many times the critical section entrance is requested (T_1, T_5, T_9).

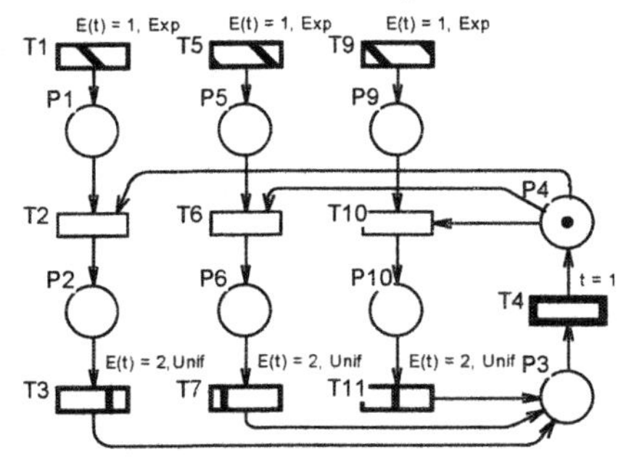

Fig 3 Unfolded version of the model

3. PERFORMANCE MEASURES OF COMMUNICATION PROTOCOLS

From performance analysis point of view, there are at least four main elements considered in the models: *Communication protocol*, *network topology*, *traffic model*, and *performance measures*. In all models, we consider *star* network topology. *Traffic model* is often called the *arrival process*. In some cases, the traffic is assumed to be produced by a very large number of customers – we speak about *infinite population queuing systems* and sometimes it is not. Then we speak about *finite population queuing systems*. Finite population queuing systems are more realistic but their analysis is usually more difficult. Many theoretical works e.g. (Kleinrock, Tobagi, 1975) are based on assumption of infinitely large population. They take advantage of the "*large number law*", which states, that with a very high probability the demand of whole population at any time instant approximately equals to the sum of the average demands. Thanks to this, all packet arrivals can be approximated by a single arrival process (e.g. the Poisson process).

Network or network device is a general open queuing system (see Fig 4). Depending on the specific network model or device operation, it has its *arrival process*, a *departure process*, a *recycle process*, a *queue size* and a *service discipline*. $T(n)$ in Fig 4 is transmission time of nth packet. There are two parameters related to load:

Input load (I) is average number of new jobs arriving to the system from outside of its boundary over a time interval (Kleinrock, Tobagi, 1975).

Offered load (G) is the average number of jobs (or executions of MAC algorithm) offered to the queuing system over T. Note that G, which equals to λ/T (λ is mean arrival rate), represents also the average number of users trying to transmit per T. In retrial models (See Fig 4), G comprises not only I but also users that were previously unsuccessful.

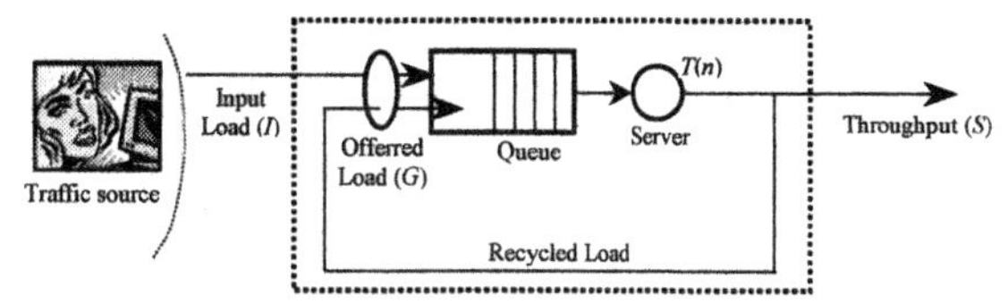

Fig 4 A general queuing model

Throughput (S), is defined as a fraction of the nominal network bandwidth that is used for carrying successfully transmitted data (the *channel utilisation*). It is average number of successfully transmitted packets that leave the queuing system's boundary during T.

Construction of models and parameters evaluation

First, we find out how long the important channel actions last and we assign delays to the *timed* transitions that represent the actions. Firings of particular transitions represent events (e.g. Sinclair, 2002) arising during simulation. Some channel properties are enumerated as ratios between *firing counts* (FC) of particular transitions. Some parameters are enumerated as transition *firing frequencies* (FF). *Firing frequency of i-th transition* (FF_i) equals to FC_i divided by the simulation length.

If channel parameters are normalised to some parameter (as we have already mentioned, it is usually T or $E(T)$), transitions' times are recalculated. Firing *frequencies represent average numbers of occurrences of the monitored events per* T or $E(T)$:

FF_5 corresponds to S and FF_1 represents number of packet arrivals per T. Clearly $FF_5/FF_1 = FC_5/FC_1$ represents the *ratio of successful packets* (the ratio between what is put on channel and what we get).

The result of simulation task is called the *firing count vector* resp. *firing frequency vector* and the result of a series of simulation tasks (simulation sequence) is called the *firing count matrix* resp. the *firing frequency matrix.*

4. ETHERNET

Here we present a confident model of Ethernet. It was developed by a step-by step approach based on modelling techniques known in queuing analysis and Perti net theory. Details can be found in (Josef Čapek 2001).

Objects P_1, P_2, P_3, P_4, P_5, P_6, P_7, P_{11}, T_2, T_3, T_4, T_5, T_6, T_7, T_8, T_9 in Fig 5 represent model of communication channel. There are two meanings of *channel state:* the *actual channel state* given by marking of P_2, P_3 and P_4 and the *channel state as it is visible* by the nodes.

Places P_2, P_3, P_4 represent three actual channel states: P_2 represents channel *idle* state (*free state*), P_3 represents channel *busy* and P_4 represents *collision.* Places P_2, P_3, and P_4 form a conservative component. Stochastic time transition T_1 represents a Poisson packet arrival process. Time of T_5 represents T and time of T_6 represents delay between arrival time of the last packet participating in a collision and the moment, when the collision is resolved. Time of T_6 also equals approximately to *jam time.* After a packet is offered to idle channel (a token appears in P_1), T_2 is fired and produces a token in P_3, which resides there until T_5 is fired or until a new packet arrives, whichever comes first. In latter case, T_3 is fired and the channel changes its state to *collision.* If no other packet arrives during firing time of T_6 the collision disappears and the channel returns to its idle state as well as after firing of T_5. Otherwise, T_4 is fired and T_6 is temporarily disabled. Its time is re-assigned after firing of T_4 (see *enabling memory*).

A token in place P_{10} represents *visibly busy* channel and when there is no token there the channel is *visibly idle.* The channel becomes visible busy a time units after an arbitrary node enters its transmission (firing of T_2) and it becomes visible idle a time units after the currently transmitting node finishes its transmission (firing of T_5). If there is a collision on channel, the channel becomes visibly idle a time units after the collision disappears (firing of T_6). Note, that the propagation delay a expressed in *time units* means the channel propagation delay measured in seconds divided by $E(T)$.

In *Markov Modulated Retrial Poisson Packet Arrivals Process* (P_{12}, P_{13}, P_{14}, P_{16}, P_{17} and T_1, T_{10}, T_{11}, T_{12}, T_{13}), there is an input queue of capacity given by initial marking of P_{13}. Packets arrive only when node's input queue P_{12} is not full (non-zero marking of P_{13}). Un-transmitted packets are either waiting in queue P_{12} or they are waiting in retrial buffer P_{16}. If the channel is visibly active at the moment when the node transmits, it reschedules the packet to a random time given by BEB algorithm (Molle, 1994). When the channel becomes visibly idle the node immediately transmits its frame. T_{14} fires when the channel is visibly idle and the node has a packet in its *transmission queue* P_{14}. Left part of Fig 5 represents the BEB algorithm. Places $\{P_{20} \ldots P_{29}\}$ represent a retrial counter of a currently transmitted packet. At arbitrary moment, only one of them is marked by a token. The stochastic time transitions with uniform distribution $\{T_{20} \ldots T_{29}\}$ increase the retrial counter and the transitions $\{T_{30} \ldots T_{39}\}$ reset the counter to its initial value each time a packet is successfully transmitted (T_{13} is fired) or when a packet is dropped (firing of T_{15}).

$E(t)$ of T_{20} equals to the time slot length r. The time of T_{21} equals to 2r, the time of T_{22} equals to 4r and so on up to T_{29}, which time equals to 512r. When a token resides in P_{20}, a packet is retransmitted during one time slot, when a token resides in P_{22}, a packet is retransmitted during four time slots and so on. For example, if $E(T)$ equals to 10 240 bits (1 280 bytes), $E(t)$ of T_{20} equals approximately to 0.05.

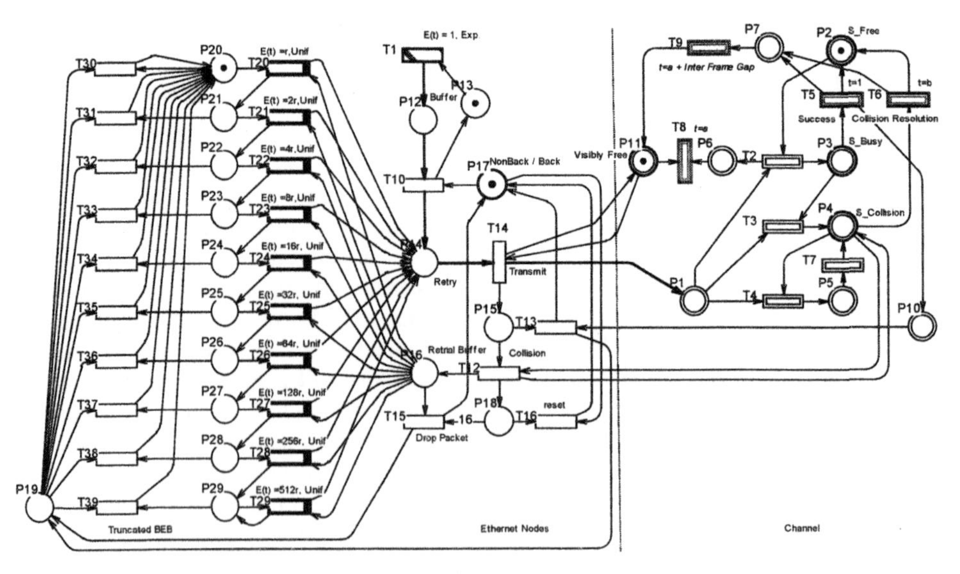

Fig 5 FPQSPN model of Ethernet viewed as retrial M/D/1/∞/k system

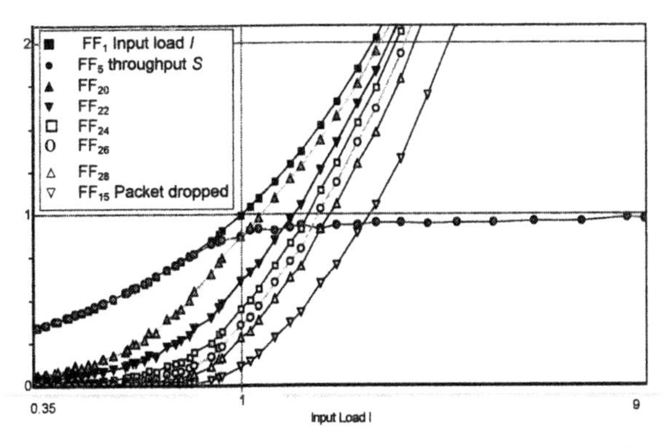

Fig 6 1500 nodes Ethernet network: firing frequencies of selected transitions as functions of *I*

We present results of simulation sequences for the following two network configurations: 1500 and 5000 nodes and variable network load. In each of the 50 simulation tasks of the sequences the simulator changes the E(*t*) of T_1, runs the simulation and evaluates transition firing frequencies.

Selected firing frequencies of the network comprising 1500 nodes are plotted in Fig 6 as functions of *I* (FF_1). In average, FF_{20}, FF_{22}, FF_{24}, FF_{26} and FF_{28} show how often are the packets retransmitted at least ones, at least three times … FF_{28} represents the average number of packets that were retransmitted nine or more than nine times. For small *I* the number of retrials is small. For large values of *I*, FF_{20} approaches to FF_1, which means that almost all packets are retransmitted. $FF_5 + FF_{15} = FF_1$. A packet that arrives to the system is either successfully transmitted or it is dropped (neglecting simulation stop errors). Note, that Knowledge of the average number of packet retransmissions allows us to compute an average time a packet spends in the system.

Firing frequencies of a network comprising 5000 nodes are plotted in Fig 6 as functions of *I* (FF_1).

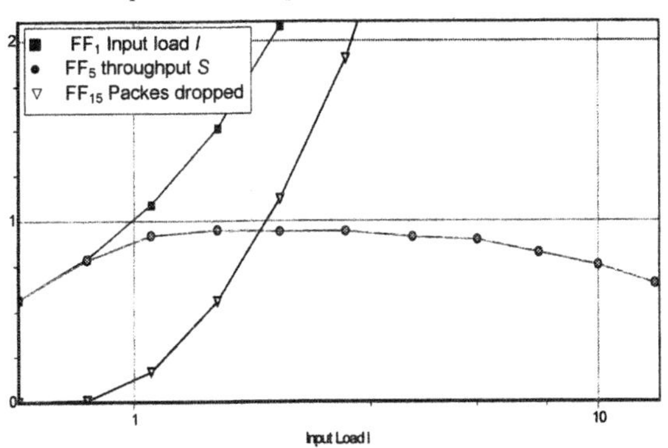

Fig 7 5000 nodes Ethernet network: *I*, *S* and number of dropped packets plotted as functions of *I*

Ethernet is usually used in technologies, where throughput is the most important parameter (*best effort* strategy) regardless of provided quality. From that reason we use the following stability definition: *"the channel is stable if the throughput S is a non-decreasing function of the input load I"*. For 1 500 nodes, Ethernet channel *behaves stable* (*S* is a non-decreasing function of *I*). For 5 000 nodes, the channel is not stable (See Fig 7) because there is a notable throughput degradation for large values of *I*. Fact that Ethernet behaves stable in case of 1500 nodes is surprising because Ethernet is known to be *theoretically unstable* for more than 1024 nodes. Ethernet seems to be stable for 1500 nodes because there is a difference between stability as it is known from theory of systems and 'stable behaviour of channel', which means that *S* does not significantly decrease when *I* is increased. Ethernet is unstable for 1500 nodes, but from practical point of view it behaves stable.

BEB provides CSMA/CD Ethernet networks 'good degree of stability' and the reason why we do not use collision domains comprising about 1000 nodes is usually insufficient channel bandwidth and large packet delays. Although big CSMA/CD networks are stable from the MAC point of view, they are unusable from the users point of view.

5. STPNPLAY

STPNPlay is a multithreaded MDI tool for modelling and analysis of stochastic Petri net models. STPNPlay comprises three functional elements: *Petri Net Editor*, *Petri net Simulator* and *Matrix Module*.

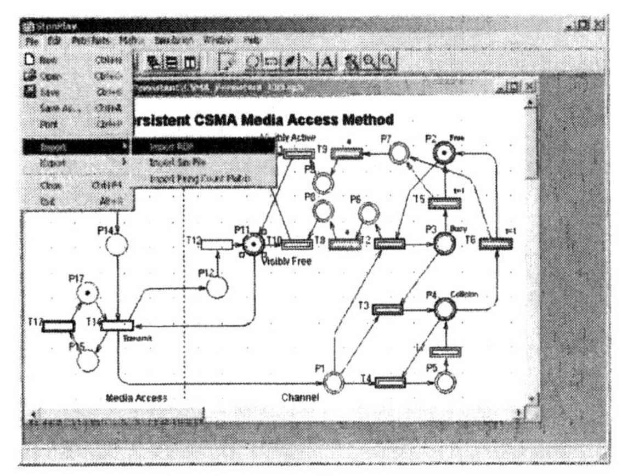

Fig 8 STPN Play: *Petri net editor*

STPNPlay editor allows drawing of places, transitions, arcs and two types of auxiliary objects: lines and texts. Editor provides continuous zoom to fit the Petri net diagram into any area, undelete of deleted objects, export diagrams in various graphical formats like *bmp*, *wmf* or *emf*.

In the *simulator* the user can simulate the model, set-up the simulation parameters and display results in charts. The user can also import and export the unfolded model. Multiple simulations having same or different priorities can be run simultaneously (each simulation process is a Win32 thread).

Several types of simulation output are supported (see Fig 10): *firing sequence*, *marking sequence*, *firing frequency matrix*, *firing count matrix*, *network firing frequency matrix* (average values of results generated by the population of customers) and *network firing count matrix*.

Simulation can be automatically stopped at one of the following conditions: length of firing sequence, maximal simulation time or combination of both. Confidence interval conditions are planned to be implemented in version 1.0.

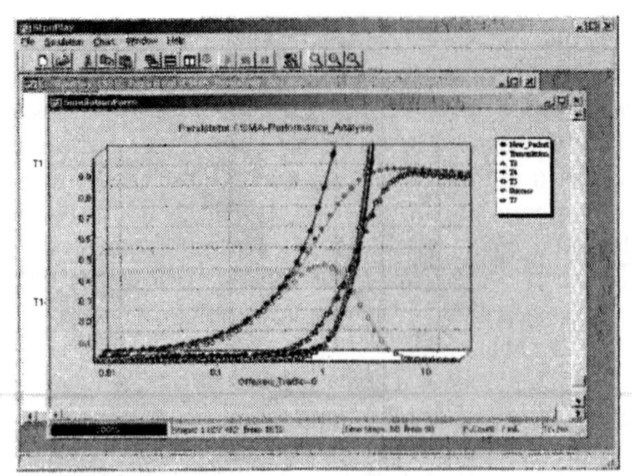

Fig 9 STPNPlay: 3D-Chart in logarithmic coordinates

Simulated result is displayed in chart, which can be interactively zoomed or copied in clip-board in various formats. The user can select up to 8 characteristics. X-axis represents either *simulated time*, *simulation task number* or a FF. Y-axis represents a FF or the *simulated time*. Simulated results as well as Petri net diagrams can be exported in various formats (*csv* or *txt*...).

The simulator allows automated execution of *simulation sequences* (sequences of simulation tasks), which is helpful in construction of various characteristics plotted as functions of some system parameters. In each task of a simulation sequence, the simulator changes a parameter of the model and performs the simulation. Sequence of parameter values forms either an arithmetic or geometric progression so as it is possible to plot the simulation results either in linear or logarithm coordinates.

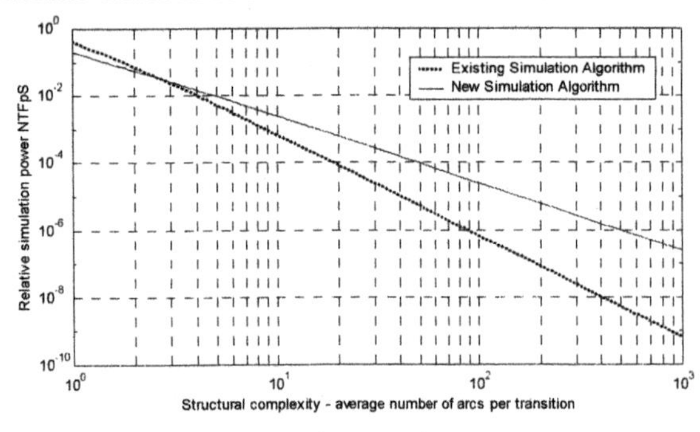

Fig 10 STPNPlay: Simulation speed

STPNPlay uses new, efficient discrete-event token player simulation algorithm. Its speed is independent of the number of places and transitions in the model. Complexity of simulated models is still an issue in simulations - simulation power of STPNPlay's algorithm is less dependent on structural complexity than existing algorithms (Marsan at al. 1995, Chiola 1991). When we define the structural complexity as the average number of arcs per transition (Josef Čapek 2001) and when we define the simulation power as the number of transitions fired per second (NTFpS), computational complexity of StpnPlay is On^2 while computational complexity of the existing algorithms is On^3 (Josef Čapek 2001) (See Fig 10). For 2 GHz PC, the NTFpS of STPNPlay simulator is up to several millions. For example, the model of 5 000 nodes network comprises about 135 000 transitions and 90 000 places and a simulation comprising 50 different simulations (Fig 7) takes only several hours.

6. CONCLUSION

FPQSPN is a high-level Petri net class suitable for analysis of models of systems with finite population. In contrast to other high-level Petri net classes, FPQSPN is easy to implement and it is nearly as transparent as low-level Petri nets. We demonstrate its application to throughput analysis of Ethernet.

Presented confident model of Ethernet (Josef Čapek 2001) was developed by a step-by step approach based on modelling techniques known in queuing analysis and Perti net theory. Its simulation results were compared to results in literature.

Models were drawn and simulations were made in *STPNPlay*. STPNPlay (Čapek, Hanzálek 2001) is a powerful Petri net modelling and simulation tool supporting portability of both models and simulation results to other tools and platforms. It uses very FAST and efficient discrete-event token player. The power of the tool excels mainly for heavy simulations of large models. StpnPlay is available for free on the internet: HTTP://dce.felk.cvut.cz/capekj/StpnPlay.

One of the current works of authors laboratory is the application of FPQSPN in analysis of analysis of wire-less communication standards.

7. REFERENCES

R. David, H. Alla, "Petri Nets and Grafcet" Prentice Hall International (UK) Ltd, 1992, ISBN: 0-13-327537-X

Tadao Murata: "Petri Nets: Properties, Analysis and Applications" Proceedings of the IEEE, vol. 77, No. 4, April 1989.

Molle M. "A New Binary Logarithmic Arbitration Method for Ethernet" Computer Systems Research Institute, University of Toronto, Toronto Canada M5S 1A1. Technical Report CSRI-298, April 1994.

Josef Čapek, "Petri Net Simulation of Non-deterministic MAC layers of Computer Communication Networks Dissertation Thesis" Czech Technical University in Prague, Department of control Engineering, 2001

Sinclair B., 2002, Simulation of Computer Systems and Computer Networks: A Process Oriented Approach Rice University, Houston, Texas, January 8

G. Chiola, "Simulation framework for timed and stochastic Petri nets" International Journal on Computer Simulation 1(2): pp. 153-168, 1991.

Čapek, J. & Hanzálek Z. (2001) StpnPlay a modeling and simulation Petri net tool, Proc. of PNPM2001, Kemper P. (Ed.) univ. of Dortmund, Germany, Sep. 2001.

Kleinrock, L. and Tobagi, F. A., "Packet Switching in radio Channels: Part I—Carrier Sense Multiple Access Modes and Their Throughput Delay Characteristic", IEEE Trans. Comm., Corn-23, Dec. 1975, pp. 1400.

Marsan M. A., Balbo G., Conte G. Donatelli S. and Franceschinis G., 1995, Modelling with Generalised Stochastic Petri Nets, John Wiley & Sons Ltd. ISBN 0 471 93059 8

www.elsevier.com/locate/ifac

ROBUST CONTROL OF pH PROCESS ON THE BASIS OF THE B-BAC MATHODOLOGY

J. Czeczot

Institute of Automatic Control, Technical University of Silesia
ul. Akademicka 16, 44-100 Gliwice, Poland, jczeczot@ia.polsl.gliwice.pl

Abstract: This paper deals with the new B-BAC (Balance-Based Adaptive Control) approach to the adaptive control of the neutralisation process. The control law is based on the simplified and very general experimental dynamic model of the process, which has a form of a classical balance equation with only one time-varying parameter, representing all the nonlinearities of the process. The value of this parameter is estimated on-line by the recursive least-squares procedure. Both the control law and the estimation procedure need only the measurement of pH and of flow rates. Moreover, there is no need to assume any description approximating the nonlinearities of the process. The results of the simulation experiments with application of our controller to the complete nonlinear model prove its very good control performance and robustness. *Copyright © 2003 IFAC*

Keywords: Robust control, pH control, Adaptive control, Model-based control, Least-squares estimation.

1. INTRODUCTION

Control of pH has been a difficult and challenging problem due to time-varying and nonlinear characteristics of pH processes. Since even a well-tuned classical PID controller cannot provide the satisfactory performance of the neutralisation process (Henson and Seborg, 1997), a large number of the sophisticated nonlinear adaptive controllers have been developed and can be find in the literature (see e.g. Buchholt and Kummel, 1979; Piovoso and Wiliams, 1985; Loh *et al.*, 1995; Graebe *et al.*, 1996; Garrido *et al.*, 1997; Henson and Seborg, 1997; Stebel, 2002).

The key problem of the pH control consists in the fact that the deterministic part of the mathematical model of the process describes only its hydrodynamic aspect. All the nonlinearities, and especially the relationship between the acid and base concentrations and the pH, is described by the strongly nonlinear expression with a large number of parameters. Moreover, in the practice there is always a large uncertainty not only on the form of this nonlinear expression but also on the values of these parameters. Thus, these kind of mathematical models are useless for the model-based approach to the controller design.

The main contribution of this paper is to suggest the new B-BAC control strategy for the control of pH. This strategy is based on a very simple, general and rather experimental model of the process. This model has a form of the classical dynamic balance equation, describing the hydrodynamic aspect of the process, with only one time-varying parameter, representing all the nonlinearities of the process. The control law, derived on the basis of this model, has also very simple form and needs only the measurement data of the values of pH and of the flow rates. Its adaptability results from the on-line estimation of the parameter, that represents the nonlinearities of the process, as one time-varying term. Its robust and very good control performance follows from the very general form of the simplified model of the process, which is a basis for the controller design, and from its adaptability and feedforward action.

This paper is organised as follows. First, the complete nonlinear mathematical model of the pH process is given. This model is considered as the real-world example of the pH process to be controlled during the simulation experiments. Then, the control problem is formulated and the B-BAController is derived on the basis of the simplified and general model of the process. The very good control performance and robustness of this controller are proved by the simulation results, which, together with the concluding remarks, complete the paper.

2. DYNAMIC MODEL OF THE PROCESS

As it was said before, the simple neutralisation process, shown in the Fig. 1, is considered in this paper as the real-world example.

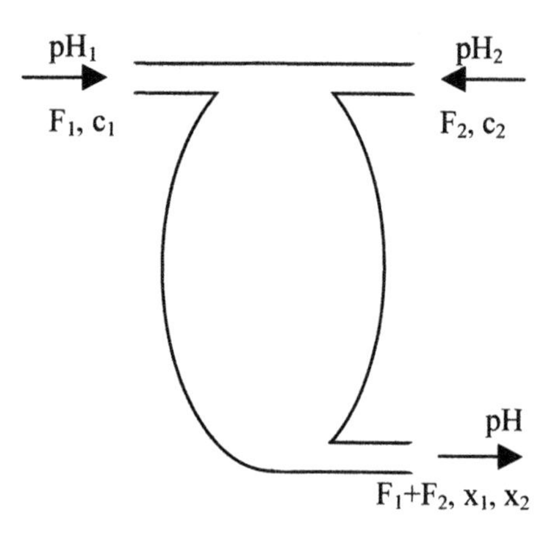

Fig. 1. Simplified diagram of the neutralisation process

There are two inlets to the reactor tank: one is the acid inlet flow with the flow rate F_1 [L/min] and with the inlet acid concentration c_1 [mole/L] and the other is the base inlet flow with the flow rate F_2 [L/min] and with the inlet base concentration c_2 [mole/L]. The values of pH_1 and pH_2 correspond to the value of pH for the acid and base inlet flows, respectively. The neutralisation process takes place in the perfectly mixed reactor tank of the constant volume V [L] and the process results in the outlet acid concentration x_1 [mole/L] and the outlet base concentration x_2 [mole/L] and, in consequence, in the value of pH at the outlet.

This simple neutralisation process can be successfully described by the McAvoy dynamic model (McAvoy, 1972). It consists of two dynamic equations, describing the concentrations x_1 and x_2:

$$\frac{dx_1}{dt} = \frac{F_1}{V}(c_1 - x_1) - \frac{F_2}{V}x_1 \qquad (1a)$$

$$\frac{dx_2}{dt} = \frac{F_2}{V}(c_2 - x_2) - \frac{F_1}{V}x_2, \qquad (1b)$$

and of three algebraic equations, respectively describing the values of pH_1, pH_2 and pH:

$$10^{-3pH_1} + 10^{-2pH_1}K_a + 10^{-pH_1}(-K_a c_1 - K_w) - \\ -K_a K_w = 0 \qquad (1c)$$

$$10^{-3pH_2} + 10^{-2pH_2}(K_a + c_2) + 10^{-pH_2}(K_a c_2 - K_w) - \\ -K_a K_w = 0 \qquad (1d)$$

$$10^{-3pH} + 10^{-2pH}(K_a + x_2) + \\ + 10^{-pH}(K_a(x_2 - x_1) - K_w) - K_a K_w = 0 \qquad (1e)$$

where $K_a = 1.8*10^{-5}$ – acetic acid equilibrium constant and $K_w = 10^{-14}$ – water equilibrium constant.

Let us remind that although the dynamic equations (1a-b) are bilinear, the very strong nonlinearity of the process follows from the form of the equations (1c-e), describing the relationship between the concentrations of the acid and base and the corresponding value of pH, which is the only measurable parameter.

3. B-BAC CONTROL OF THE NEUTRALISATION PROCESS

The B-BAC (Balance-Based Adaptive Control) methodology (Czeczot, 2001, 2002) is dedicated to control a wide range of technological processes due to its generality, simplicity and robustness. In this paper we present the possibility of the B-BAC application to control the neutralisation process described in the previous Section.

3.1 Formulation of the control problem

As the control goal let us consider the problem of controlling the outlet pH value at a prescribed set point value pH_{sp} (e.g. $pH_{sp} = 7.0$) by acting on the base flow rate F_2 as the control variable under the following conditions:

- the state variables x_1, x_2 as well as the inlet concentrations c_1, c_2 are not measurable and unknown,
- only the values of pH_1, pH_2, pH, F_1, F_2 and V are known, either by on-line measurement or by choice of the user,

- the form of the nonlinear equations (1c-e) are unknown or, at least, there is a large uncertainty on it.

As it can be seen, these conditions are very realistic and it is difficult to develop the sophisticated control system for such a process. Thus, these conditions should be considered as significant limitations during the controller design.

3.2 Balance-Based Adaptive Controller

The choice of the B-BAC methodology is not accidental. It is possible to manage all the limitations listed above due to the generality and robustness of the B-BAC approach.

First let us suggest the general dynamic equation, describing the controlled variable pH:

$$\frac{dpH}{dt} = \frac{F_1}{V}(pH_1 - pH) + \frac{F_2}{V}(pH_2 - pH) - R \quad (2)$$

This equation has no specific physical meaning. It is based on the form of classical balance equations with the time-varying term R representing all nonlinearities of the process. Thus, the equation (2) describes the "balance of pH concentration". Let us note that the equation (2) can be always satisfied since it is always possible to find the value of R that satisfies this equation at the particular moment of time. When we assume that the value of R can vary in time, it is obvious that we can assume that the equation (2) can be satisfied at each moment of time by appropriate choice of the current value of R. This proves that the general dynamic equation (2) describes the dynamic of pH for our process with high accuracy. Now we only need to manage the problem how to find the correct value of R and how to use this equation as a basis for our control law. For this purpose let us rewrite the equation (2) in the discrete-time form with i denoting the discretisation instant:

$$\left(\frac{dpH}{dt}\right)_i = \frac{F_1^i}{V}(pH_1^i - pH^i) + \frac{F_2^i}{V}(pH_2^i - pH^i) - R^i \quad (3)$$

Now, at this stage of the controller design we can apply the same methodology as in the case of the "model reference linearising control" (Isidori, 1989; Bastin and Dochain, 1990). For our control goal let us assume the following stable first-order closed loop dynamics at a discrete moment of time:

$$\left(\frac{dpH}{dt}\right)_i = \lambda(pH_{sp} - pH^i) \quad (4)$$

where λ is the tuning parameter. After combining the equations (3) and (4) we obtain the following equation:

$$\lambda V(pH_{sp} - pH^i) = F_1^i(pH_1^i - pH^i) + F_2^i(pH_2^i - pH^i) - VR^i \quad (5)$$

The equation (5) is a basis for the final B-BAControl law and thus, after rearranging, we obtain the final form of the B-BAController in the discrete-time form for the neutralisation process:

$$F_2^i = \frac{\lambda V(pH_{sp} - pH^i) - F_1^i(pH_1^i - pH^i) + V\hat{R}^i}{pH_2^i - pH^i} \quad (6)$$

where $\hat{R}^i$ is the on-line estimate of the unknown parameter R. The estimation procedure is based on the recursive least-squares method, which can be applied due to the fact that the equation (3) is linear with respect to the unknown parameter R. Moreover, there is only one parameter that must be estimated, which allows us to avoid all the difficulties following from the on-line multiparameter identification. Thus, we can suggest the equations describing the estimate $\hat{R}^i$ at the discrete moments of time (Czeczot, 1997, 1998, 2001):

$$\hat{R}_Y^i = \hat{R}_Y^{i-1} - VT_R P^i(y^i + V^i T_R \hat{R}_Y^{i-1}) \quad (7)$$

with

$$P^i = \frac{P^{i-1}}{\alpha}\left(1 - \frac{V^2 T_R^2 P^{i-1}}{\alpha + V^2 T_R^2 P^{i-1}}\right) \quad (7a)$$

$$y^i = V(pH^i - pH^{i-1}) - F_1^i(pH_1^i - pH^i)T_R - F_2^i(pH_2^i - pH^i)T_R \quad (7b)$$

where T_R [min] – sampling time, α - forgetting factor.

Let us note once again that all the parameters needed for calculation of the control law (6) are measurable or known by choice of the user except the value of $\hat{R}^i$, which has to be estimated on-line but on the basis of the same measurable or known parameters. Thus we derived the control law on the basis of the B-BAC methodology despite of all the limitations presented in the previous Subsection.

4. SIMULATION RESULTS

The neutralisation process (1) is considered as the real world system that is to be controlled and it has been simulated under the following steady-state conditions: $c_1 = 0.008$, $c_2 = 0.008$, $F_1 = 0.3$, $F_2 = 0.3$, $V = 3$. The initial values of the state variables $x_1(0) = 0.004$ and $x_2(0) = 0.004$.

A large number of simulation experiments have been carried out to investigate the control performance and the robustness of the B-BAController (6). The most interesting and representative results are presented in this Section in the Fig. 2, 3 and 4. These Figures show the time evolution of the controlled variable pH (Fig. 2), of the control variable F_2 (Fig. 3) and of the estimate for the parameter R (Fig. 4), and they correspond to the results of the same simulation experiment.

The solid line in the Fig. 2, 3, 4 correspond to the case when it is assumed that all the variables, namely pH_1, pH_2, pH and F_1, are measurable on-line, which, of course, is in line with the limitations presented in the Section 3. The experiment is performed as follows. For the first time of 2 [min] the process is operated in open loop and only the estimation procedure (7) with the forgetting factor $\alpha = 0.95$ and with the sampling time $T_R = 0.6$ [sec] is carried out. It is the preliminary open-loop stage of the estimation that is necessary to avoid the transient at this stage of the estimation due to the improper choice of the initial value $\hat{R}^0$. Since the true value of R is unknown (it is impossible to calculate it on the basis of the model of the process (1), even in the case of computer simulation), it is difficult to find the initial value $\hat{R}^0$ that allows us to avoid the significant transient and, in consequence, to ensure good control performance. Thus, it is important to run the estimation procedure in open loop until the estimate converges and after that to close the control loop.

In our case the control loop is closed at t = 2 with the set point $pH_{sp} = 7$ and then, one by one, the indicated step changes of the disturbing parameters are applied to the process. The sampling time for the controller is the same as for the estimation procedure ($T_R = 0.6$ [sec]) and its tuning parameter $\lambda = 0.2$.

Fig. 2 shows very good control performance of the B-BAController (6). Once the control loop is closed, the controlled variable pH is regulated at its set point pH_{sp} in a very short time and without significant oscillations. The controller keeps the process stable, even in the presence of the disturbances step changes, due to its adaptability and feedforward action. The corresponding time evolution of the control variable F_2 is presented in the Fig. 3 (solid line). Let us note that this variable varies smoothly and slowly, which is fully acceptable in the practice. The estimation accuracy for this case can be seen in Fig. 4 (solid line) and, although the true value of the estimated parameter R is unknown, it can be seen that the estimate convergence time is very short, which ensures very good adaptability of the suggested controller (6).

The second simulation experiment has been performed to investigate the robustness of the suggested B-BAController (6). This controller is implemented in the control loop under the same simulation conditions as described above for our first experiment. The only difference follows from the fact that the second experiment has been carried out under more restrictive limitations in comparison to the ones described in the Section 3. Namely, we additionally assumed that there is a large uncertainty on the value of the disturbing variable F_1 because its value is not accessible by the on-line measurement. The current value of this variable is needed both for the calculation of the control law (6) and for the estimation procedure (7). Thus, due to the lack of the measurement data of this parameter, for these calculations we used the value of $F_1 = 0.35$, estimated somehow off-line, and we assumed that this value is constant during the second experiment. Let us note that the restrictive limitation follows not only from the fact that the value of F_1 is assumed to be constant but also in the fact that there is a 12.5% error on its value in the steady state.

The results of this experiment are depicted in the Fig. 2, 3 and 4, drawn with the dotted line. Although the value of F_1 is unknown and assumed to be constant, the control performance of our controller is still very good. Note that there is only a slight difference at the moment when the step change of F_1 is applied to the system, but even in this case the set point pH_{sp} is reached and tracked, yet with more significant oscillations. This is a direct consequence of the robustness of the B-BAControl law (6) introduced both by the form of the controller, based on the very general form of the process model (2), and by its adaptability. Note that this very good control performance and, in consequence, the robustness have been obtained due to a bias on the value of the estimate of the parameter R (Fig. 4).

5. CONCLUDING REMARKS

In this paper the new B-BAC approach to the adaptive control of the neutralisation process is suggested. It is shown how to derive the model-based control law on the basis of the very general, experimental and simplified model of the process, without any knowledge not only about the process nonlinearities but also about the complete form of the mathematical model of the process. This control law has been evaluated by simulation with application to the complete nonlinear mathematical model of the neutralisation process given in Section 2. Simulation results prove very a good control performance of the suggested controller in the presence of the changes of the set point value and of the disturbing parameters.

The adaptability of the suggested controller follows from the on-line estimation of the unknown value of the time-varying parameter, representing all nonlinearities of the process. Since there is only one parameter to estimate, we avoid the common difficulties with the multivariable identification and, in consequence, we ensure very good convergence and properties of the estimation procedure.

In this paper the control performance is tested without any measurement noise, which, of course, always exists in the practical applications. Let us note that it is possible to decrease the influence of the measurement noise by the adjusting of the forgetting factor α and by using the low-pass filters. Since it is very important to investigate the influence of the measurement noise on the control performance, it will be studied in the nearest future.

The robustness of the B-BAController (6) is also tested in this paper and it is proved by simulation because, despite of the error on the value of one of the disturbing parameter, the control performance is still very good.

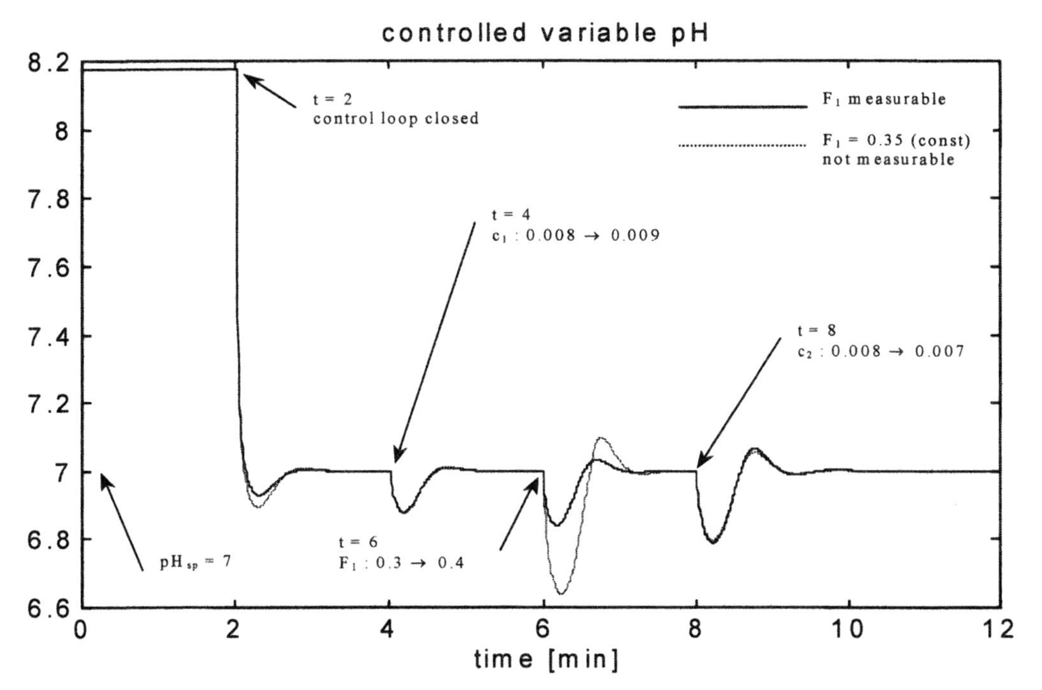

Fig. 2. B-BAC robust control: the controlled variable pH, closed loop experiment

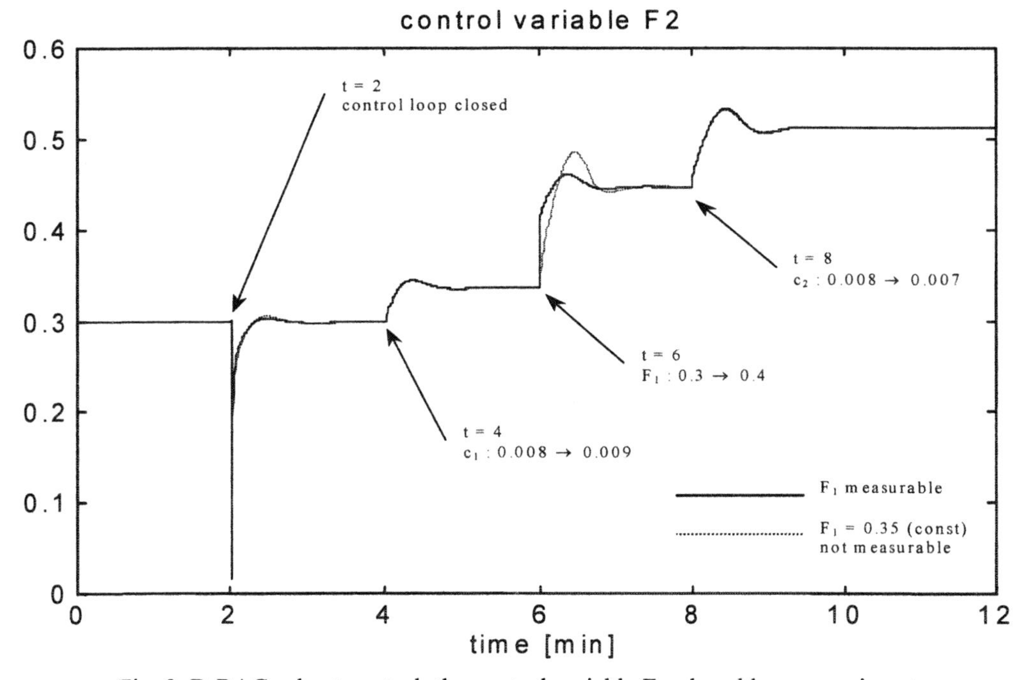

Fig. 3. B-BAC robust control: the control variable F_2, closed loop experiment

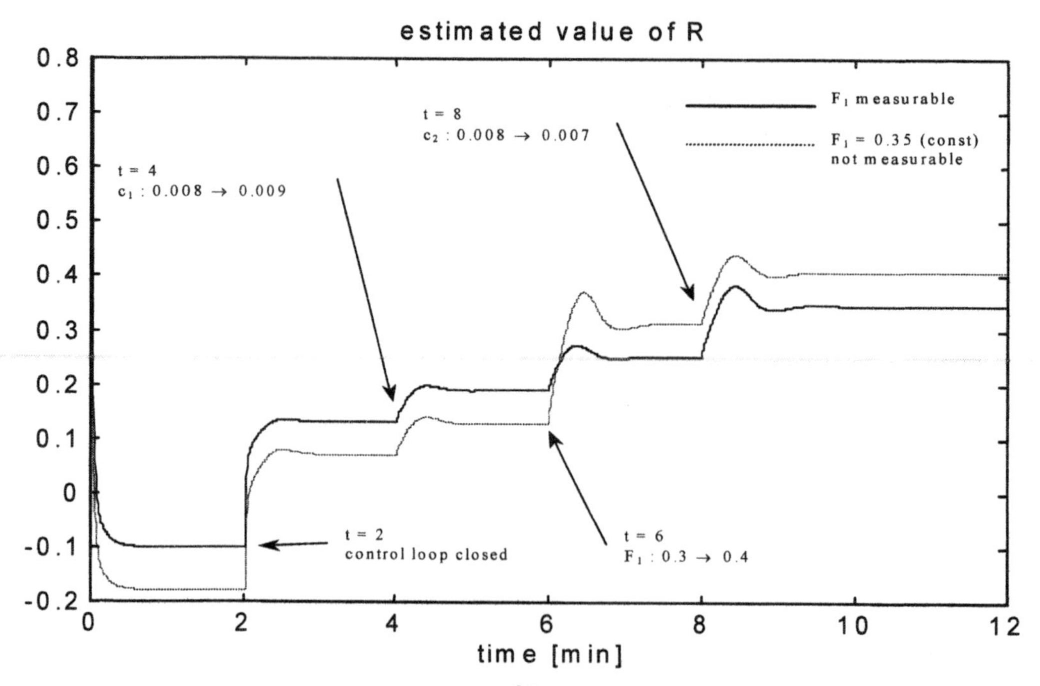

Fig. 4. B-BAC robust control: the estimate $\hat{R}^i$ of the parameter R, closed loop experiment

To summarize, let us state that the suggested B-BAController is a very promising alternative for the successful control of the very nonlinear neutralisation process. The form of the control law is based on the very general and simplified model of the process, which ensures its robustness and generality. Moreover, the simplicity of the control law, in opposite to other more sophisticated model-based approaches, allows us to expect a significant number of practical applications in the future. For our part we are also planning to implement this B-BAController for the control of the neutralisation pilot plant, developed at our laboratory in Gliwice (Metzger and Choinski, 2001).

Acknowledgements. This work was supported by Polish Committee of Scientific Research (KBN) under grant 4T 11A 01924.

REFERENCES

Bastin G., Dochain D. (1990). *On-line estimation and adaptive control of bioreactors.* Elsevier Science Publishers B.V..

Buchholt F., Kummel M. (1979). Self-tuning control of a pH-neutralisation process. *Automatica*, **15**, p. 665.

Czeczot J. (1997). The application of the substrate consumption rate to the monitoring and control of the biological processes of water purification. Ph. D. Thesis, Technical University of Silesia, Gliwice, Poland (in polish).

Czeczot J. (1998). Model-based Adaptive Predictive Control of Fed-Batch Fermentation Process with the Substrate Consumption Rate Application, *IFAC Workshop on Adaptive Systems in Control and Signal Processing*, University of Strathclyde, Glasgow, Scotland, UK, pp. 357 –362.

Czeczot J. (2001). Balance-based adaptive control of the heat exchange process. *7th IEEE International Conference on Methods and Models in Automation and Robotics, MMAR 2001*, Miedzyzdroje, Poland.

Czeczot J. (2002). Robust Balance-Based Adaptive Control of the CSTR. *8th IEEE International Conference on Methods and Models in Automation and Robotics MMAR 2002*, Szczecin, Poland, pp. 327-332.

Garrido R., Adroer M., Poch M. (1997). Wastewater neutralization control based in fuzzylogic: simulation results. *Ind. Eng. Chem. Res.*, **36**, pp. 1665 – 1674.

Graebe S.F., Seron M.M., Goodwin G.C. (1996). Nonlinear tracking and input disturbance rejection with application to pH control. *J. Proc. Cont.*, **6**, No. 2/3, pp. 195-202.

Henson M.A., Seborg D.E. (1997). Adaptive input-output linearization of a pH neutralisation process. *Int. J. of Adaptive Control and Signal Proc.*, **11**, pp 171-200.

Isidori A. (1989). *Non-linear control systems.* Springer, Berlin.

Loh A., Looi K., Fong K. (1995). Neural network modelling and control strategies for a pH process. *J. Proc. Control*, **5**, No. 6, pp. 355 – 362.

McAvoy T.J. (1972). Dynamic models for pH and other fast equilibrium systems. *Int. Eng. Chem. Process Des. Develop.*, **11**, pp. 630 – 631.

Metzger M., Choinski D. (2001). Neutralisation pilot plant. Activity Report, Silesian University of Technology, Institute of Automatic Control, Gliwice, 2001, pp. 67 – 69.

Piovoso M., Williams E.I., duPont de Nemours & Co Inc (1985). Self-tuning pH control: a difficult problem, an effective solution. In Tech.

Stebel K. (2002). Polynomial approximation approach to modelling and control of pH process. 15th IFAC World Congress, Barcelona, Spain, July 21-26.

www.elsevier.com/locate/ifac

MICROCONTROLLER BASED DATALOGGER FOR TRAFFIC INFORMATION COLLECTION IN PARKING AREAS

Luciano Spinello[1], Luca Spaghetti[1], Francesco Martinelli[1], Valentino Di Salvo[2]

[1] *Dipartimento di Informatica Sistemi e Produzione*
Università di Roma Tor Vergata
I-00133 Rome, Italy
e-mail: `lucianospinello@universitor.it`,
`lucaspaghetti@fastwebnet.it`,
`martinelli@disp.uniroma2.it`

[2] *Traffic Engineer*
e-mail: `valentinoambrogio.disalvo@fastwebnet.it`

Abstract: In this paper an automatic device to collect data regarding open parking areas is described. It is based on a loop vehicle detector and a datalogger where collected data are recorded for subsequent processing operations on a standard PC. The device is controlled through a commercial microcontroller unit. The circuit scheme and firmware design are discussed in detail in the paper. Experimental results are provided in the paper together with a discussion of related realizations available in the technical literature. *Copyright © 2003 IFAC*

Keywords: Traffic Data Collection, Microprocessor control, Data logging, Sensors.

1. INTRODUCTION

For statistical purposes, parking companies regularly collect data regarding the occupancy state of public parking areas. Each park slot should be monitored on a basis of $24h$ per day. This kind of study is usually realized by a manual procedure, which is very expensive and subject to many errors.

For this reason, it is of interest to design and realize a device to be placed in each park slot to be monitored, performing the human work in a more precise and efficient way.

However, the requirements specified for this device are very restrictive:

- it should be easy to place it in a standard car parking slot;
- it should have an autonomy of 7 days (night and day) using a common battery;
- it should be sufficiently robust to environment disturbances;
- it should be realized with commercial and low cost components.

At present, such a kind of device is not available on the market. A large part of traffic data collection devices have been realized to perform flow measurements, especially in a situation of heavy traffic conditions. It would be very hard to adapt

[1] This work has been partially supported by ASI and MIUR (cofin: MATRICS).

these devices to our purposes, because the requirements are completely different. If a traffic data collection device accidentally skips the passage of a few vehicles, the relative error committed is usually negligible; in fact this kind of devices is used on highways where the typical traffic flow ranges in the order of 2500 and more vehicles per hour. On the contrary, the error could be dramatic if the requested park slot device skips a vehicle which takes busy the park slot for several days. In addition, a parking measurement device is a finite state machine, unlike a flow traffic counter, since operations performed depend on the park slot state (free or busy).

A problem similar to the one addressed in this paper, even if in a domain completely different, is considered in Luharuka and Krishnamurti (2002). The objective in Luharuka and Krishnamurti (2002) is the realization of a portable datalogger for physiological sensing. The requirements of this device are an autonomy of 12h and an high reliability. The projected device comprises a commercial GSR (galvanic skin response) sensor which is able to provide output data like heart rate, temperature and blood pressure. An onboard datalogger containing a great amount of eeprom guarantees the recording of 7-20 MB data corresponding to 12 hours logging report. Such data can be downloaded on a PC at the end of a recording session.

Many similarities can be observed between the device described in this paper and the one presented in Luharuka and Krishnamurti (2002). They are based on the same control unit family microprocessor (MPU Microchip PIC16XXXX), a very high quality and low cost component; they both use a battery as unique power supply, with voltages conversions where needed.

It would be really appealing (and it is actually considered as a future development of our project) the possibility of realizing power independent measurement devices connected to a central data collection system, like the traffic information collection system realized by Koito industries and described in Matsuda (1999). In the system described in Matsuda (1999), detectors are powered by small solar cells and communicate detected data to a central device through free-space optical transmission links. Vehicle detectors are based on ultrasonic sensors. This system was installed in February 1999 along the national route 246 to collect traffic information regarding vehicles running east and west between Tokyo and Shizuoka. In this area a commercial power supply is not available.

The paper is organized as follows: in Section 2 we discuss the hardware design, in Section 3 we describe the control algorithm and its implementation, in Section 4 we present some experimental results, finally, in Section 5, we sketch some concluding remarks.

2. HARDWARE DESIGN

2.1 A sensor survey.

The first step to realize the requested device is the selection of a sensor or a proper combination of sensors with high reliability, acceptable costs and low energy consumption. The following sensors have been considered: inductive/capacitive proximity switches, photoelectric sensors, microwave sensors, vibration based monitoring sensors, ultrasonic proximity switches, loop vehicle detectors. Videocameras have not been considered mainly for privacy violation reasons.

For different reasons, the loop vehicle detector has appeared as the most promising sensor. It is characterized by medium energy consumption (65mA), an optimal sensing range and an adjustable sensibility. This kind of sensor is commonly used in automatic switched gate parking areas to detect the presence of vehicles which approach to an exit gate. The sensor is small for dimension and weight but the loop requires a surface of about 1 m^2.

2.2 Model concept

A long time has been spent to understand which sensors and which components should have been considered and in which way they should have been disposed into the parking slot. At the end of these investigations, the *Inductive model* has been selected, which is based on a loop vehicle detector sensor, located in a parking slot as illustrated in figure 1.

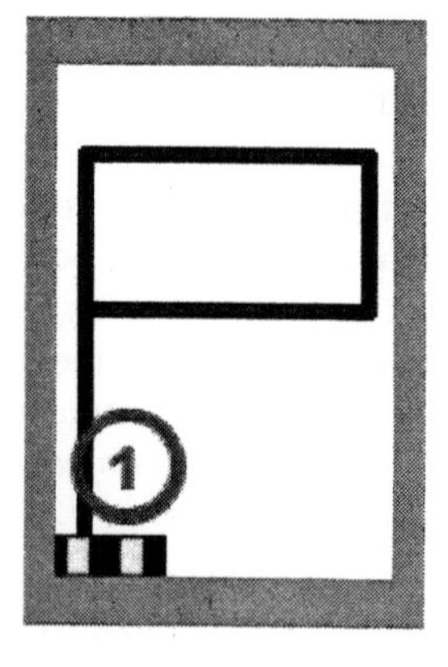

Fig. 1. Sensor placement in a park slot (inductive model)

Data are recorded in a HoboEvent Datalogger produced by the OnSet Computers. The overall device is controlled by a MPU Microchip PIC16F84, which operates as a two state finite machine, one state corresponding to *Empty parking* and the other corresponding to *Busy parking.* Energy is given by a unique commercial 12V battery.

The control device, sensor and datalogger units, are located inside a rigid box placed near the sidewalk, while the loop of the vehicle detector has to be located in the middle of the parking slot under a special PVC covering. The vehicle detector sensor is always on during the overall measurement session. The control algorithm works according to the following logic: when a vehicle enters a park slot, its presence is immediately detected but the entrance event is registered in the datalogger only if the vehicle stays in the park slot for at least 20 s, as explained below.

2.3 The FG1 detector

In this subsection we provide a detailed description of the sensor used in the realized device, the FG1 vehicle loop detector produced by FAAC (see Fig. 2). It is commonly used for gate automation systems and produced for this kind of applications.

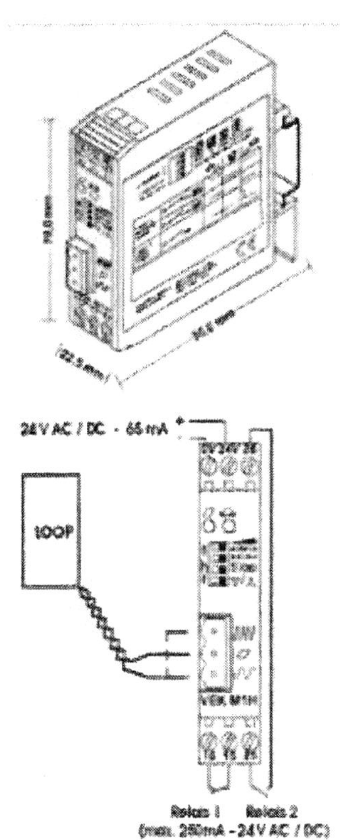

Fig. 2. Detector FG1

The operation of the FG1 Detector is based on a Hartley type HF-circuit, where the oscillator inductor is the loop. When a metallic body (like a car) is over the loop, the inductance and consequently the circuit natural oscillation frequency are changed. A PLL discriminator transforms this frequency variation in a digital signal. By properly tuning some adjustable parameters, it is possible to have precise measurements, ignoring frequency variations caused by environment disturbances. The FG1 Detector would require a 24V power supply but it correctly works using a 12V battery.

Two different outputs, corresponding to two different electro-mechanical relays, give respectively the signal of *busy loop* and of *loop in exit transition.*

Different parameters can be set to obtain the desired behavior: the vehicle loop detector sensitivity, the circuit frequency oscillation and the operation principles of the output relays.

The vehicle loop detector sensitivity is a really important parameter to be selected. The precision of the measurements depends on its good tuning and it has been set only after a large campaign of experiments. The circuit frequency oscillation characterizes the velocity of the sensor of detecting a state transition. For parking purposes a low setting of this parameter is acceptable and allows to save energy. The operation principles of the output relays can be set in two ways. In the first operation way, if the loop is busy, the corresponding relays is closed, while it would remain open in the second operation way. The other relay (the one corresponding to the exit transition output) stays closed for 300 ms after the vehicle has left the park slot.

2.4 Data recording

Data provided by the sensor are stored in a stand alone (i.e. non-integrated) datalogger, which is an autonomous device designed to record a large quantity of data and to transfer them on a standard PC throughout the serial port.

The datalogger used in our device is the *HoboEvent* produced by OnSet Computers. This datalogger is produced for environmental data collection, it has been used for example as a component in a rain gauge. It is not commonly used in automation devices, but it is appealing for its functionality, low cost, and robustness, especially with respect to environment agents. It is able to record up to 8000 events, comprising date and time in the following format: dd/mm/yy, hh/mm/ss. To record an event, it is necessary to shorten a contact between two metal wires coming out the rigid box containing the device. The Hoboevent is a completely autonomous device working with a proper CR 2032-lithium battery, for a time of 1 year of activity. It is possible to set some parameters of the Hoboevent through a special software package provided by the producer. One of the most important parameters is the option *ignore event,* through which it is possible to have the Hoboevent ignore events for a given amount of time after the contact shortening. This parameter has been used in our project, as explained below.

2.5 *Circuit design*

All the described components have been assembled as in the circuit represented in Fig. 3.

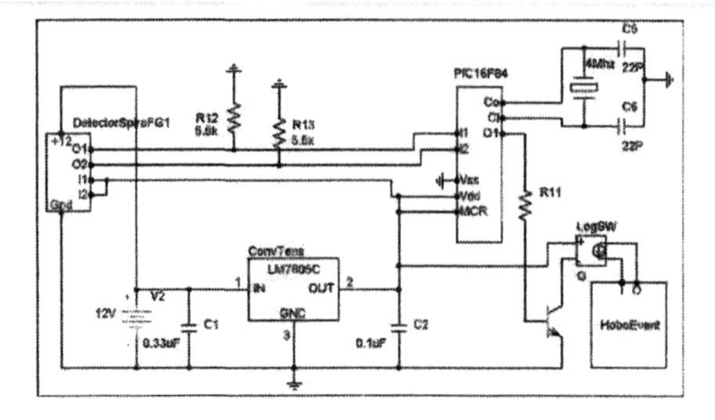

Fig. 3. The circuit scheme

Both the FG1 detector and the micro-controller circuit PIC 16F84 are powered by a 12V battery, through a proper voltage conversion. The voltage conversion is performed through an integrated transformer LM7805, which realizes a linear conversion voltage from a maximum of 30V to 5V stabilized. The two electrolytic condensers connected with ground are suggested by the producer (Texas Instruments) to stabilize the output and to reduce the overshoot phenomenon.

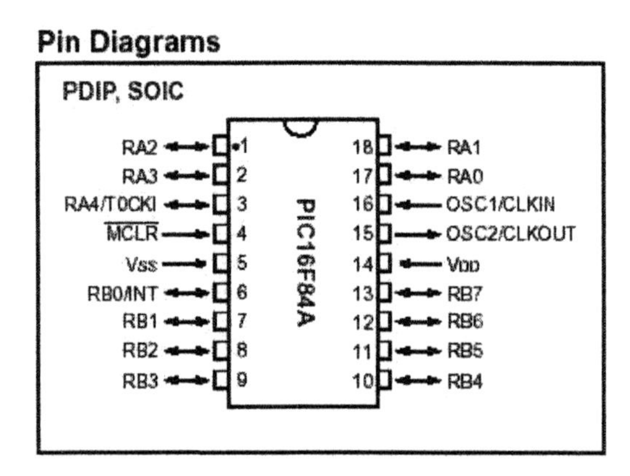

Fig. 4. Pin diagram

The two outputs of the FG1 detector are connected to the inputs of the PIC. The PIC requires a 4MHz quartz oscillator, considering that it is equipped with a prescaler, it actually works at 1MHz. The two condensers connected with mass complete the oscillating circuit. The PIC is obviously connected to the ground and to the 5V power supply. The MCRL pin works as a reset pin: when the voltage is high on this pin, the PIC starts to execute the firmware loaded in its flash memory. One of the output of the PIC should be connected to the relay contact of the Hoboevent. Actually, to prevent an excessive load to the PIC controller, this output pin is connected to the base pin of a BJT NPN which is used as an electronic switch.

Since the selected datalogger uses an electromechanical relay, the option *ignore event* of this unit has been set to 1-2 s to prevent the registration of fictitious events due the bouncing phenomenon existing in this type of relays. In fact, when the contact of the relay is closed, the relay shutter, which shorts the two output wires of the datalogger, could oscillate and generate a sequence of fictitious events. It is particularly important to take care of this undesirable phenomenon as the component gets older and older.

Two leds are placed on the control card. They give a fast representation of the controller state: if the green led is on, the park slot is free while, if the red led is on, the park slot is busy. Finally, if the green led is blinking the device is in the decisional waiting loop.

3. FIRMWARE DESIGN

The operation of the PIC16F84 control device follows the behavior of a two state Mealy automaton (see Fig. 5).

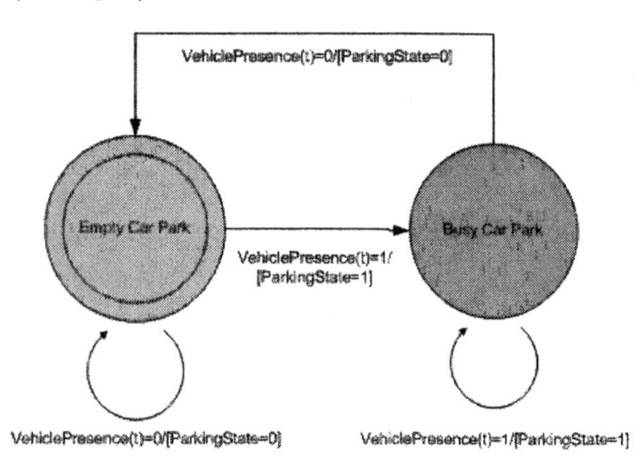

Fig. 5. Mealy automaton

The operation of the automaton represented in Fig. 5 is depicted in Table 1, where $S(t)$ represents the state of the automaton at time t and can assume two values: *Busy Car Park*, shortly *Busy* and *Empty Car Park*, shortly *Empty*. The input of the automaton at time t is denoted as *VehiclePresence(t)*, shortly *VPresence(t)*, while the output of the automaton is the *ParkingState*, shortly *PState*.

S(t)	VPresence(t) = 1	VPresence(t) = 0
Busy	Busy; PState=1	Empty, PState =0
Empty	Busy; PState=1	Empty, PState =0

Table 1.

The most important module of the firmware is the *state selector*, represented in the flow diagram of Fig. 6.

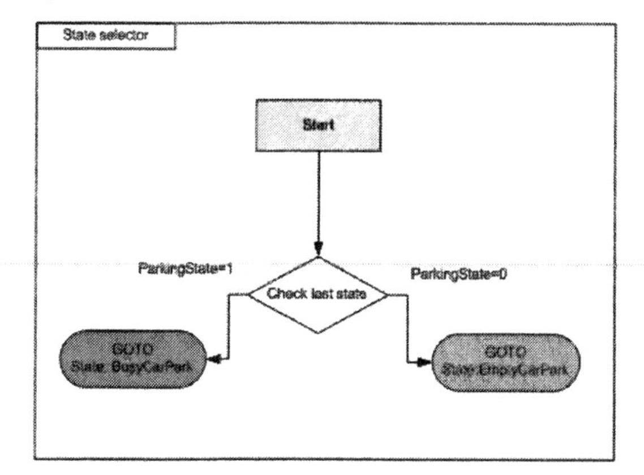

Fig. 6. State selector

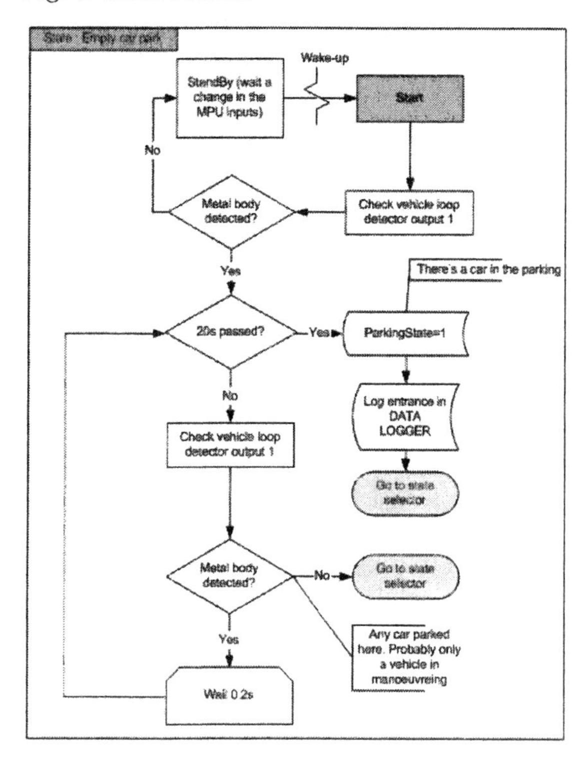

Fig. 7. Flow diagram for the state of empty parking

Such a kind of control paradigm, based on a Mealy automaton, is necessary since the datalogger is not able of detecting the type of the event but only records the closure of its relay.

The control algorithms implemented in the firmware are represented in the flow diagrams of Figures 7 and 8 for the Empty and for the Busy states, respectively. The idea of the procedure is as follows:

Empty State When a vehicle is entering the park slot:

- the Occupancy State is 0 (Empty State);
- the loop detects a metallic mass (i.e. Relays 1 is closed);
- the controller starts a 20s waiting period;

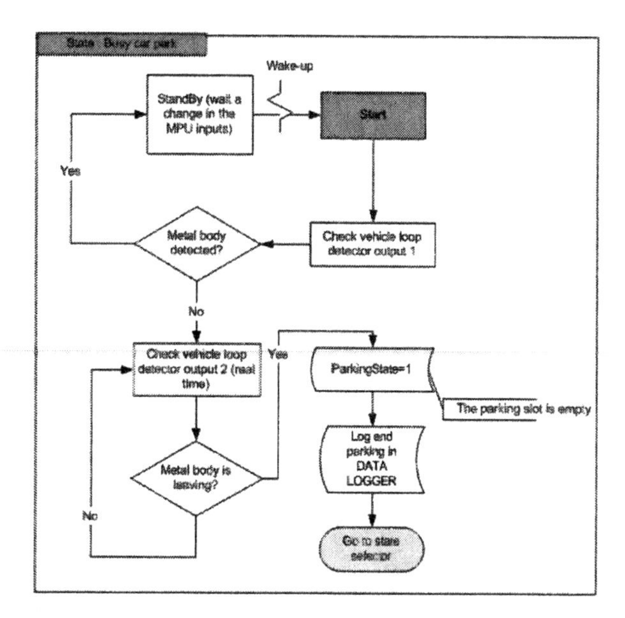

Fig. 8. Flow diagram for the state of busy parking

- if the vehicle remains for at last 20s,
  then an event is registered in the datalogger;
  the Occupancy State is set to 1 (Busy State);
  else no event is registered in the datalogger.

Busy State When the vehicle is leaving the park slot:

- the Occupancy State is 1 (Busy State);
- the loop detects the absence of metallic masses (i.e. Relays 1 is closed);
- the loop detects an outgoing metallic mass (i.e. Relays 2 is closed for 300ms);
- an event is registered in the datalogger;
- the Occupancy State is set to 0 (Empty State).

The part of the algorithm regarding the detection of loop in exit transition (which uses Relays 2 output) has been developed to avoid improper operations of the sensor.

Since the realized device should operate 24h per day for about one week, it is of interest to contain as much as possible power consumption. To do this, the control circuit remains active only for short time periods: 20s when a vehicle is entering the park slot and only for the time the vehicle requires to free the park slot. For the remaining time, the PIC is maintained in standby, in low power consumption mode, where the needed current is 1000 times smaller (from 2mA to 2μA).

3.1 The firmware programming language

The PIC is programmed through a RISC (Reduced Instruction Set Computer) assembler language specific of the Microchip.

It comprises 35 1-byte instructions (called *op, operation*) which are executed by the processor in one or two clock cycles. Since the processor works at 1 MHz, the execution of each instruction requires 1 or 2 μs. This is a negligible speed if compared to the clock frequency of modern processors, however it is completely sufficient to ensure the real time application we are interested in.

The PIC also comprises Ram and eeprom which makes it suitable in many application domains.

All processor family 16FXXX may be programmed by cheap serial port programmers and the assembler may be compiled using software available on the Microchip website. The firmware (compiled assembler) is then uploaded through a free software in the PIC flash memory.

4. RESULTS OF THE PROTOTYPE TESTING

The Sos.T.A. device (represented in Fig. 9) is characterized by a valuable precision in data detection. The large quantity of experiments showed that the device is able to recognize a parking event even if only half of the vehicle covers the parking slot area. If the sensor is located in the middle of the park slot, erroneous or improper vehicle positions can be over-passed. In a real urban scenario, where vehicles are positioned quite randomly with respect to the parking slots, the Sos.T.A. device correctly detected all the parking events, apart from a negligible number of errors mainly due to misplacements of the loop sensor or to battery exhaustion.

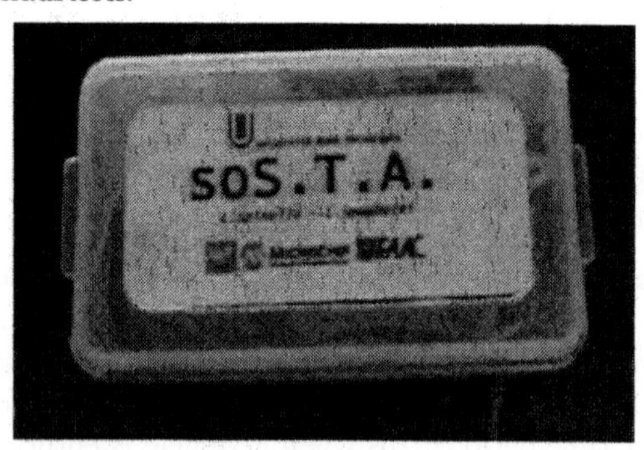

Fig. 9. The realized device

To reduce the detection error as much as possible, it is extremely important to select a proper value for the sensitivity of the loop sensor: a proper setting allows to exactly discriminate between park slot busy and empty and, in addition, to reject disturbances like motor-cycles standing or passing through the parking area.

Data can be analyzed through a processing data software developed in Borland CPP Builder on Win9x/NT/XP platforms.

In figure 10 it is possible to distinguish the interval of times where the parking slot has been busy, which are the black rectangles. At the base of each rectangle it is possible to read the starting and the ending times of each parking event. For convenience, a text box provides a list of all data collected. The realized software can manage more windows at the same time, allowing a fast comparison of data collected, and it is equipped with zoom and pan capabilities.

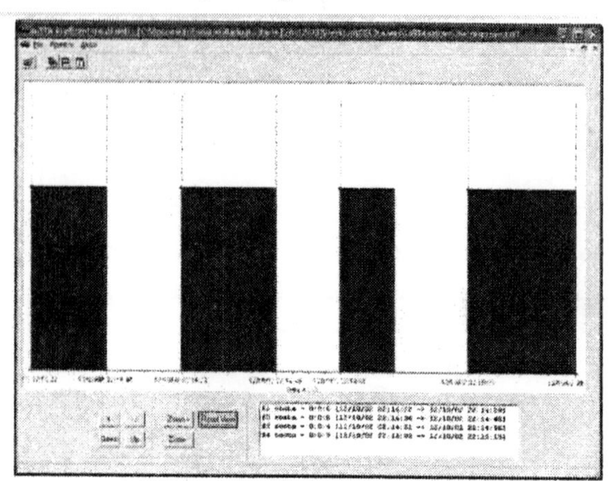

Fig. 10. The visualization window

5. CONCLUSIONS

In this paper a microcontroller based datalogger for parking data collection has been described. A loop vehicle detector has been selected as the most suitable sensor for this application. The described device satisfies all the specified requirements, namely the robustness, the low costs and the installation simplicity. As a future work, it is of interest to integrate the product and to develop a solar cell powered system connected to a data collection network.

REFERENCES

R. Luharuka and S. Krishnamurti. A microcontroller based datalogger for physiological sensing. In *IEEE instrumentation and measurement technology conferece*, Anchorage, Alaska, May 2002.

M. Matsuda. Traffic information collection systems powered by solar cells. In *1999 IEEE/IEEJ/JSAI International Conference on Intelligent Transportation Systems*, pages 870–873, 1999.

STRATEGY AND PHILOSOPHY OF DISTRICT HEATING SYSTEMS CONTROL

Jaroslav BALÁTĚ, Bronislav CHRAMCOV, Michal PRINC

*Institute of Information Technologies,
Faculty of Technology, Tomas Bata University in Zlín,
Mostní 5139, 760 01 Zlín, Czech Republic,
E-mail: balate@ft.utb.cz, chramcov@ft.utb.cz, princ@ft.utb.cz*

Abstract: The District Heating Systems (DHS) are being developed in large cities in accordance with their growth. The DHS are formed by enlarging networks of heat distribution to consumers and at the same time they interconnect the heat sources gradually built. DHS is used in larger cities of some European countries e.g. in Germany, France, Denmark, Finland, Sweden, Netherlands, Czech Republic, Poland and others. Production technology of heat by means of combined production of power and heat (CHP) is an important way to increasing of thermal efficiency of closed thermal loop. The paper shows the system access to the control of extensive DHS - controlled plant. It concerns automatic control of technological string "production, transport + distribution, consumption" of extensive district heating and that is the contribution of this paper. Control by means of advanced control algorithms is a tool (up to now neglect) for decreasing cost of energy and increasing the level of environment protection. *Copyright © 2003 IFAC*

Keywords: District Heating, Control Algorithms, Control System, Hierarchical Control, Combined Heat and Power – CHP, Hydrothermal Power Systems.

1. INTRODUCTION

Cogeneration of Power and Heat is an important way to increasing of thermal efficiency of closed thermal loop. The experiences in design of Control Strategy for Extensive District Heating System in the towns Brno and Prague in Czech Republic are summarised in the paper. It involves the connection of main author's operational experiences gained during many years of his work in the Power and Heating Plant and his further scientific - research activities on the technical university in co-operation with his colleagues, PhD students and with his students. The design of control strategy shows the basic concept of control methods of the district heating system of a specific locations in town Brno and in town Prague. Each district heating system has its specific features and therefore it is necessary to create a philosophy of control for each of them. From the point of view of control, this philosophy consists of both general regularities and special features of a specific locality. The idea of a system approach to design of technological string control "production, transport + distribution, consumption" resulted from the specific solution of the way of control in real time and also from short-time preparation of district heating operation in region of Brno, Czech Republic during the last decade of 20th century. The knowledge of operation and experience motivated the author to creating the methods and conception of control of this technological string.

At present it is known from the literature that the problems of optimum control of combined heat and power production (CHP) in sources and also systems of automatic heat consumption control are solved only separately. Very few attention is paid to analysis of static and dynamic behaviour of heat networks and utilization of these features for operation control of these networks. There are no any works dealing with elimination of transport delay in transport of heat in heat networks. Publications dealing with system approach to control of the technological string as a controlled plant are missing at all. This fact was the motivated cause for solving a new problem.

2. CHARACTERISATION OF CONTROL OF DISTRICT HEATING SYSTEM (DHSC)

The function of DHS is to ensure permanently all economically justified and socially necessary demands for supply of energy to all consumers namely at minimum costs and when respecting further important aspects e. g. of environment. The demands of consumers on the delivered energy can be specified by the following requirements:

- to provide delivery of justified quantity of energy to all consumers according to their needs varying in time,
- to provide stable and uninterrupted delivery of energy,
- to provide permanent delivery of energy within the limits of specified quality indexes (e.g. temperature of hot water in hot-water piping, steam pressure and temperature in steam piping or possibly frequency and voltage of electric energy at combined production of electric energy and heat etc.)

It is necessary to harmonise the demands of consumers on energy with the requirement of maximum operating economy of the whole DHS i.e. both in the production and delivery of heat with emphasis on complying with ecological indexes in the respective locality.

Economical control of the subsystem of heat energy production consists of two tasks i.e. of *economical distribution of load between separate co-operating sources in circle heat network including economical load distribution between separate production units inside the sources and further also of determination of the suitable composition* of co-operating sources including determination of suitable composition of co-operating units inside the source as well as *in determination of suitable economically based starting and switching off the production units or possibly also the whole sources.*

The important fact is the production technology using *combined production of power and heat.*

Economical operation of the heat distribution network subsystem, heat exchanger stations and consumer's equipment require heat output control in three levels of hierarchy:

- control in the source of heat (on the input into the heat network; or possibly the forced limitation of heat supply);
- control in the transfer station;
- control in the system of consumers.

Economical control in the system of consumers is fully in competence of consumers. Control of heat exchanger stations operation is from the point of view of automatic control quite autonomous; however here it is necessary to think over also further possibilities namely at large consumers.

For heat supply control in DHS the heat supply control in the source of heat is decisive i. e. on the input into the heat network.

The system of district heating - for centralised heat supply is by its character large extensive system. There exists a great number of dependent variables (e.g. relative increases of outputs of separate production units of sources); it is necessary to fix their value always for a certain value of one or more (of vector) independent variables (e.g. change in demand for energy supply to consumers, influence of change in combined production of heat and electric energy, influence of weather conditions, changes of parameters of fuel supplied to the sources etc.).

In this paper there is considered the circular steam heat network supplied by several heat sources. In these sources there are heat exchangers steam-hot water for supplying heat to consumers in autonomous (local) radial networks.

3. BRIEF CHARACTERISTIC OF SPECIFIC LOCALITIES OF DHS

3.1 DHS in Brno

According to the analysis the supposed locality has approx. 400 thousand permanently resident people namely in approx. 164 thousand flats. The total heat input has been 3500 MW_t. This total heat consumption is supplied approx. 30% from Heat Supply System. It is a circle steam network with local radial hot-water networks in each cooperating heat source. Delivery of heat by means of hot-water piping from the nuclear power plant situated in a distance of 42 km was considered (Balátě, 1988).

The following terms are used in further text:

- *District Heating System (DHS)*, system with co-operating Power and Heating Plant and Heating Plants - to be understood as present heat sources and heat networks in the supposed locality,
- *Integrated District Heating System (IDHS)* - to be understood as co-operation of DHS with the heat source in nuclear power plant situated in distance about 40 km; according to the design it had to supply heat using the heat feeder to the blocks of flats in housing estates of the town or possibly to the other connected consumers.

Principal scheme of the design to build up an Integrated District Heating System is on Fig.1. The design including the perspective heat network development in supposed locality and by using the heat source in nuclear power plant is shown on Fig. 2.

The used abbreviations have the following meaning: TBŠ - Power and Heating Plant, Š - TBŠ - Peak Boilers in TBŠ, further VSB, VČM, VBS - Heating Plants, ZT KPS, ZT 1BS, ZT Zetor – Power and Heating Plants in industrial factories, TKO - Incinerating Plant of Solid Municipal Waste, JE - Nuclear Power Plant.

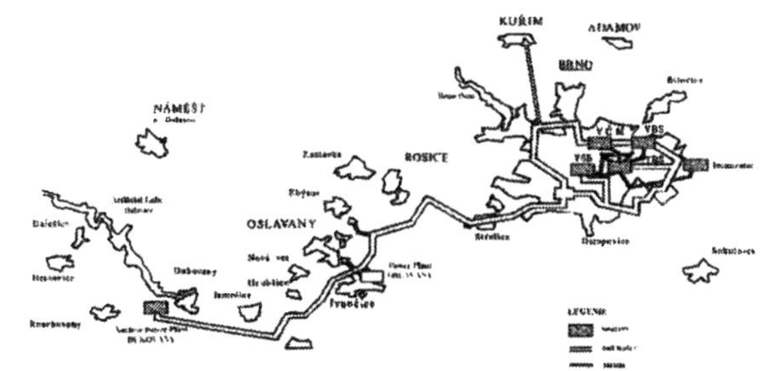

Fig. 1. Integrated District Heating System.

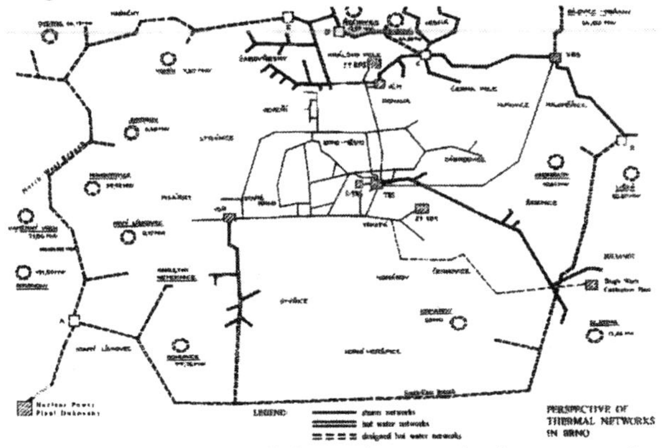

Fig. 2. Perspective of thermal networks in supposed locality.

It is obvious to see from these figures that IDHS forms the original (present) sources which are interconnected by circle steam network and the heat feeder from the nuclear power plant in which had to supply heat in hot water as heat - transfer medium to blocks of flats in new housing estates on the outskirts of town. These housing estates are at present equipped by gas boiler rooms but these should have been reconstructed to exchange stations connected to *north-west and south-east branches of hot water by-pass piping* of the supposed town. The heat sources TBŠ, VSB, VBS and also VČM have their own heat exchangers which deliver heat to their own autonomous hot-water systems. The aim was to interconnect these autonomous hot-water systems with branches of hot-water by-pass piping. Thus the extensive hot-water network had to come into being which would be equipped by sector closures for the possibility to realise changes in network configuration in dependence to the heat consumption in blocks of flats.

3.2 DHS in Prague

It is a hot-water network with installed heat output of 1995 MW_t and electrical output of 138 MW_e. The CPH source cooperating from the distance of 36 km is connected to the original DHS. It is an analogical problem as above-mentioned way of control (Balátě, 1996).

4. PHILOSOPHY OF CONTROL METHOD

Philosophy of control is according to author's opinion a basic view of control (i.e. perhaps also by a manual method on a base of operators's experience) and this method (way) of control has to be made objective by means of utilization of objective methods of control supported by control methods using the method of automatic and automated control or using the method of advanced control algorithms.

From the discussion with the operating organisation in locality of DHS in Brno ensued that the control will be realised as follows:

1. Short-time and long-time preparation in dependence to the development trend of weather, co-operation with the superior energy dispatching and with the heat supply dispatching of operating organisation.
2. Operational control in real time in dependence to immediate heat consumption.

The aim of the control has been fixed:

To utilise the source of heat in nuclear power plant as much as possible; to supply the rest of necessary heat from the existing sources of DHS with the requirement of minimum heat consumption in fuel and fulfilling ecological indexes.

For fulfilling the aims of control the control system should ensure activities illustrated by the algorithms of control in the scheme on the Fig. 3. The separate blocks represent activities which concern both the preparation of heat production i.e. covering the predicted course of daily heat supply diagram and control of technological string in real time.

The Incinerating Plant TKO has a specific standing. The task of the incinerating plant is the liquidation of solid municipal waste and therefore the heat production is a secondary product. It will be a duty of IDHS dispatcher to withdraw the heat produced in the incinerating plant and to include it to the bottom part of HSDD in steam i. e. to the area of basic load. The supplied outputs produced in Power and Heating Plants of Factories will be located in the same way.

Long-time and short-time preparation includes the activities specified in paragraphs No. 1, 2, 3 and 4. Operational control in real time includes the activities according to paragraphs No. 5, 6, 7, 8, 9 and 10. The sequence and reassuming the separate activities is obvious from the Fig. 3.

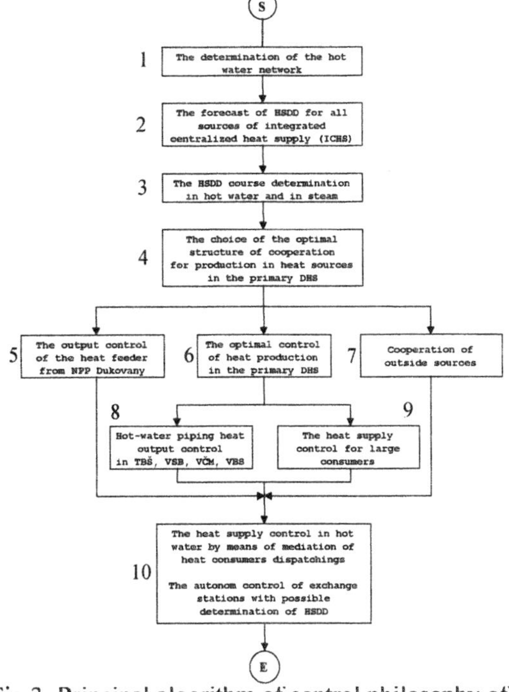

Fig.3. Principal algorithm of control philosophy of district heating of Brno.

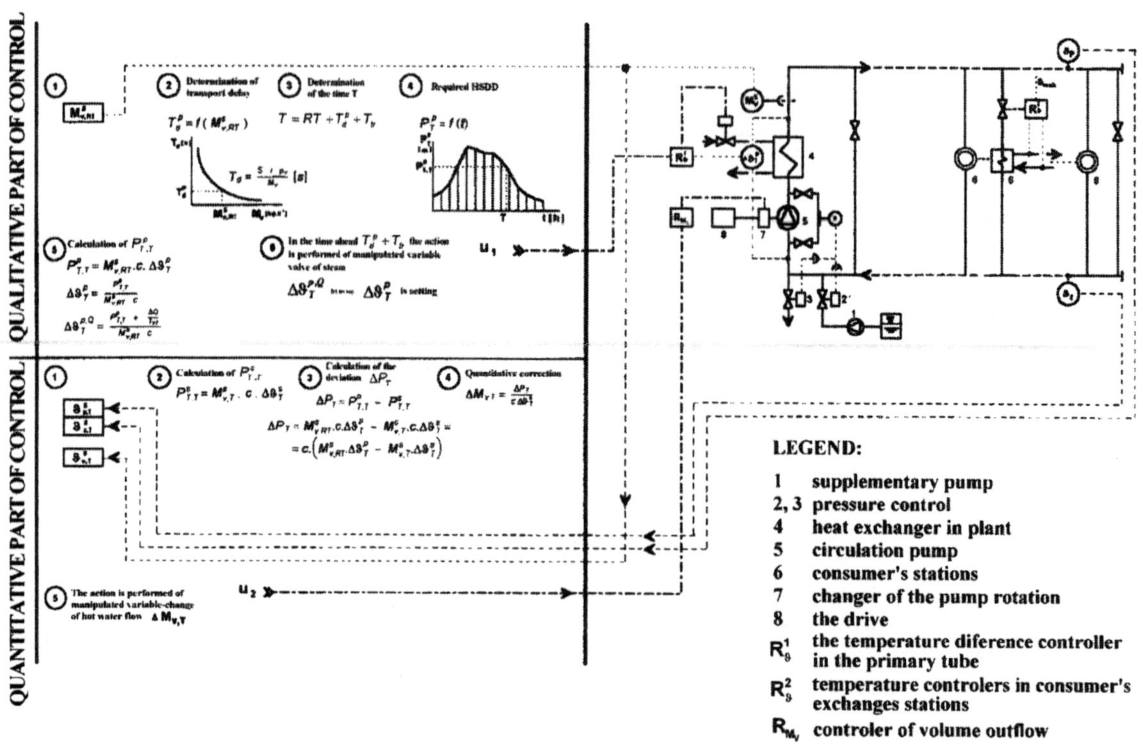

Fig. 4. The qualitative-quantitative method of hot-water piping heat output control (Balátě and Sysala, 1998).

5. SUMMARY OF THE METHODS USED FOR SOLUTION OF AUTOMATED CONTROL OF IDHS

The following methods has been or are solved in our laboratory:

1. For heat production:

1.1. Optimisation of heat production in production plant by the method of linear or non-linear programming.

1.2. Making up criterion of regime optimum for co-operation of power and heating plants and heating plants by heat supplying in district heating system namely from the point of view of costs decreasing of produced heat and power and of increasing the environment protection.

1.3. Making up algorithm for selecting optimum operational composition of sources of DHS including elaborating methods of calculating costs of a change of DHS operational composition or operational variant in separate sources.

1.4. Analysis of static and dynamic features of main technologic equipment of DHS sources.

2. For heat distribution:

2.1. Analysis of static features of heat networks and their utilisation for heat distribution control namely:- steam networks, - hot-water networks.

2.2. Making up algorithm for qualitative-quantitative method of control of heat delivery by hot-water piping using continuously calculated prediction of heat supply daily diagram. The function of the algorithm is obvious from the Fig.4.

2.3. Automation of operation of consumer's heat exchangers including visualisation of technologic process.

3. For control of the whole DHS:

3.1. Co-ordination of sub-system control of the DHS i. e. of production plants and distribution of heat.

3.2. Utilisation of prediction calculation of course of daily diagram of heat supply for control of DHS.

3.3. Design of function of Heat Supply Dispatching including its co-operation with Heat Supply Dispatchings of operation organisations or possibly with direct large consumers.

3.4. Design of instrumentation of Heat Supply Dispatching by discrete distributed control system covering corresponding hierarchy of control.

The following part of this paper brings near the analysis of the static features of heat networks and their use for heat distribution control and of the prediction calculation of the course of heat supply daily diagram.

6. STATIC ANALYSIS OF HEAT NETWORKS OPERATION

As a tool for the analysis of behaviour of the heat networks, a calculation program for simulation of pressure, through-flow and temperature relations in distributing steam and hot water networks was created. For the calculation of the mass flow in the separate branches of a circular piping system a method called *'circle flow'* was used. It is an analogy with the method of loop current, which is very often used in electrical engineering. The mass flow corresponds here to the electric current, the pressure is proportional to the voltage and the hydraulic resistance corresponds to the electric resistance. If we apply the *1st and 2nd Kirchhoff's Law* to the defined network, we obtain a system of non-linear algebraic equations. The procedure is shown on a simple network having four points of junction and six branches (see Fig. 5)

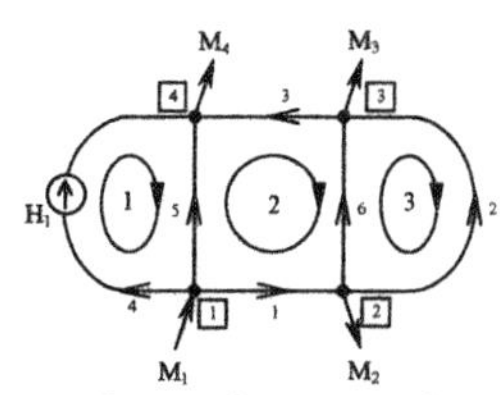

Fig. 5. Scheme of sample network.

We mark the unknown mass flow of the liquid in the i-th branch by symbol x_i, the taken away/delivered quantity in the i-th point of junction by symbol M_i and the active pressure in the i-th branch by H_i. From 1st Kirchhoff's Law it results: The sum of all flow quantities for a determined point of junction equals zero. Therefore, we can write the following equations of balance:

Point of junction 1 $\quad -x_1 - x_4 - x_5 + M_1 = 0,$ (1a)

Point of junction 2 $\quad x_1 - x_2 - x_6 - M_2 = 0,$ (1b)

Point of junction 3 $\quad x_2 - x_3 + x_6 - M_3 = 0.$ (1c)

The pressure losses in branches Δp arising at flowing of the liquid through the circular piping can be calculated according to the following formula

$$\Delta p = \left(\frac{8L}{\pi^2 d^5 \rho \left(2\log\left(3.72\frac{d}{k}\right)\right)^2} + \frac{8\sum\xi}{\pi^2 d^4 \rho} \right) x^2 = r.x^2 \qquad (2)$$

where: L [m] - length of straight sectors of the piping, d [m] - inside diameter of the piping, ρ [kg.m^{-3}]- specific mass of the flowing liquid, k [m] - absolute roughness of the inside surface of the piping, Σξ [-] - sum of the coefficients of the inside resistance of the piping, x [kg.s^{-1}] - mass flow through the piping, r [Pa.s^2.kg^{-2}] - hydraulic resistance of the piping.

For the chosen loops 1, 2 and 3 (see Fig. 5) we can write according to the formula (2) and according to the 2nd Kirchhoff's Law the following equations for the pressure losses in branches:

Loop 1 $\quad r_4|x_4|x_4 - r_5|x_5|x_5 - H_1 = 0,$ (3a)

Loop 2 $\quad -r_1|x_1|x_1 - r_3|x_3|x_3 + r_5|x_5|x_5 - r_6|x_6|x_6 = 0,$ (3b)

Loop 3 $\quad -r_2|x_2|x_2 + r_6|x_6|x_6 = 0.$ (3c)

To solve the system of non-linear equations represented in our case by equation (1) and (3) we shall use the Newton's iterative method.

The calculation runs on the basis of determined network parameters, source delivery and the consumer's demands. The results of calculation can be shown on the screen in the form of tables or simple column graphs, see Fig.7. Presented data are results of the calculation, which was worked out for the heat network described in the Fig.6.

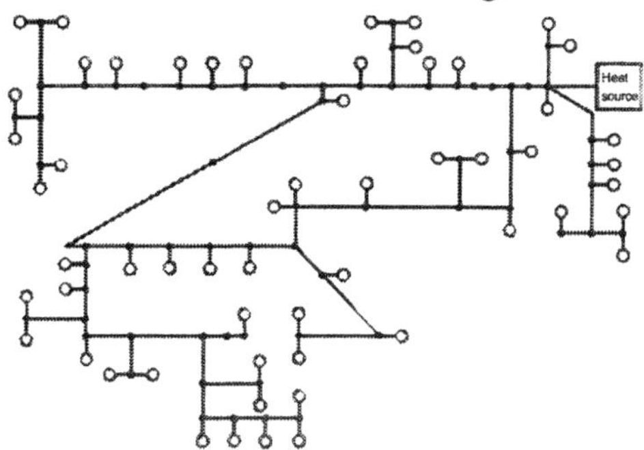

Fig.6. Network topology

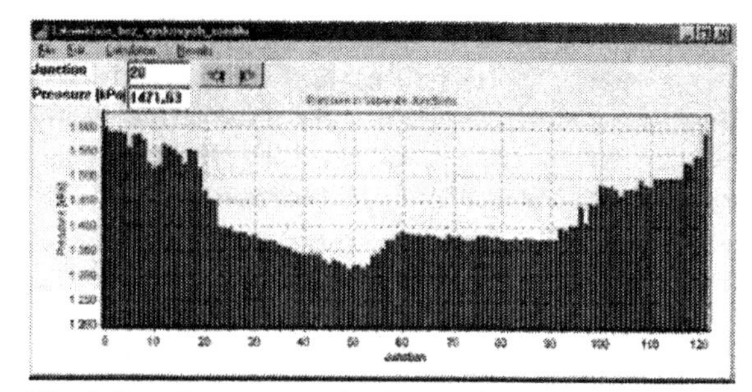

Fig.7. Results presentation in graphs.

The practical importance of the program is not only for design of distributing networks, but above all in operation of district heating systems. The possibility of simulation of relations in the network for various operating conditions, i.e. for instantaneous output of sources or for instantaneous consumptions is very useful for dispatchers in power and heating plants. This can be a contribution not only from technical aspects but also from economic ones.

7. PREDICTION OF COURSE OF HEAT SUPPLY DAILY DIAGRAM

An improvement of technological process control level can be achieved by time series analysis in order to predict their future behaviour. We can find an application of this prediction also by the control in the Centralised Heat Supply System (CHSS), especially for control of hot-water piping heat output. The forecast is determined for two ways of using:

1. The predictions of whole heat supply daily diagram (HSDD). Forecast of HSDD (see Fig.8) of whole CHSS is utilised for the purpose of heat production control and thus for the purpose of the optimal distribution loading between cooperative production sources and production units inside this sources.

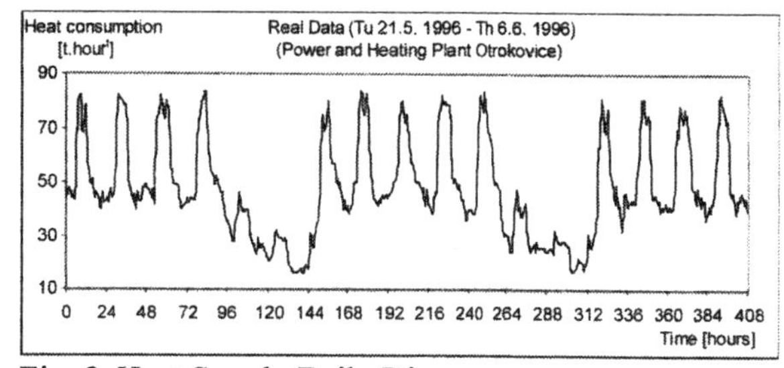

Fig. 8. Heat Supply Daily Diagram.

2. The prediction is determined for continuous acquisition of necessary heat output ahead of the time. It depends on transport delay, namely in the range of 2 up to 16 hours depending on distance of heat sources from consumers, and it is different for each locality.

The methods of forecast. Most of methods which solve the prediction of HSDD are based on mass data processing. But this method has a big disadvantage. It consists of real data out of date. From this point of view it is available to use the forecast methods according to the methodology of Box –Jenkins (1976). This method works with fixed number of values, which are updated for each sampling period.

The Box-Jenkins (BJ) method. This methodology is based on the correlation analysis of time series and it works with stochastic models, which enable to give a true picture of trend component and also of periodic components. The course of time series of HSDD contains two periodic components (daily and weekly period). But general model according to BJ enables to describe only one periodic component. We can propose two possible approaches to calculation of forecast to describe both periodic components.

- **The method that uses the model with double filtration**
- **The method – superposition of models**

First we introduce simplified form (5) of general model according to BJ for the next using when there is used substitution in the form (4). We can find more detailed analysis of general model in work.

$$F=\Phi_P^{-1}(B^s)\cdot\varphi_p^{-1}(B)\cdot\Theta_Q(B^s)\cdot\theta_q(B)\cdot\nabla_s^{-D}\cdot\nabla^{-d} \quad (4)$$

$$z_t = F\cdot a_t \quad (5)$$

The method that uses model with double filtration. We can describe model with double filtration through the substitution (4). The model in the form (6) is the result of it.

$$z_t = F\cdot\nabla_{s^*}^{D^*}\cdot a_t \quad (6)$$

where: D – degree of seasonal difference – daily (in equation 4), D* - degree of seasonal difference – weekly, s – daily period (in equation 4), s* - weekly period

It is important to adhere to the general plan for using the method that uses model with double filtration for calculation of HSDD prediction.

The model in the form (6) enables to describe the HSDD course (i.e. it describes daily periodic component and also weekly one). It can be used for analysis and prediction of following regular influence of calendar (Saturday, Sunday).

The method – superposition of models. We can use second method i.e. superposition of models for elimination of regular influence of calendar. This method is being used on two models in the form (5). These models are discerned by means of symbols * and **. The time series marked by symbol * is series of values of HSDD outputs in every hour (the sampling period is 1 hour). The time series marked by means of symbol ** is series of values of heat consumption per day (the sampling period is 1 day). The plan of calculating prediction by means of the method of superposition of models is shown on the Fig. 9.

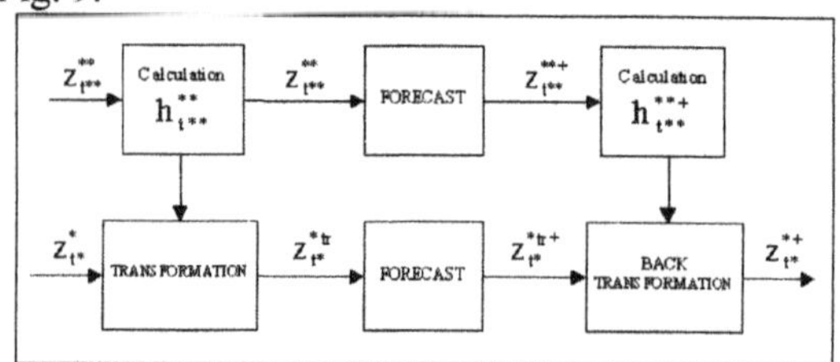

Fig. 9. Superposition of models - plan of calculating prediction

Results of Calculation of HSDD. Pursuant to the mentioned theory and literature a program was created in Matlab. This program enables to choose available mathematical statistical model for calculation of prediction of HSDD course and it is used for forecast of HSDD, which is an essential part of control of DHS by means of qualitative-quantitative method. All testing and calculations are based on lot of real data. These data were obtained in specific locality. Examples of results of the forecast calculations are shown on the next figure.

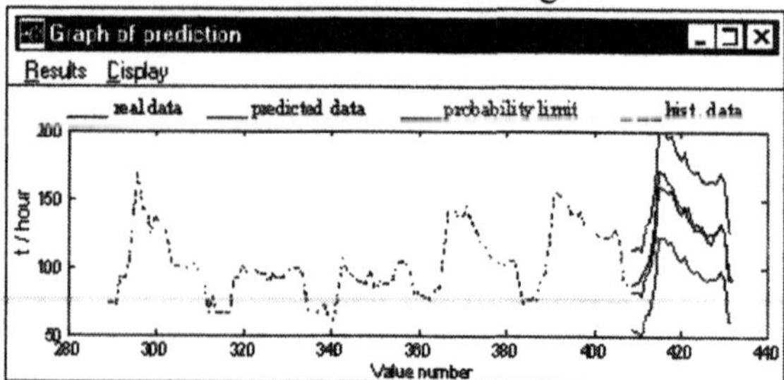

Fig. 10. Results of prediction of HSDD.

8. CONCLUSION

The complicated process of technological string of heat production, transfer and distribution of heat or possibly of heat consumption, as it was mentioned above, requires the system approach when solving the concept of automated control by means of advanced control algorithms. We believe that the intellectual process when determining the philosophy of control can be interesting and useful for the technical public. It is a demonstration of necessity to know technology for designers of control system.

It is necessary to stress that each DHS has its specific features which have to be solved specifically and individually.

We can inform readers that this idea of the presented paper was the base of the proposal in the frame of 5th Framework Programme of EC called: *Cost Effective and Environmental Compatible Operation of District Heating Systems by means of Advanced Control Algorithms.*

REFERENCES

Selected research works in the field of District Heating Control:

Balátě, J. at al. (1988). *The study of district heating control in Brno*, Research work No: VZ-HSZ-070 in Czech, TU Brno, Faculty of Mechanical Engineering, Brno.

Balátě, J. (1996). *District heating in Prague - Design of Philosophy of Control*, Research Work No: KAŘT-PTAS/1996 in Czech, TU Brno, Faculty of Technology in Zlín, Zlín.

Special appreciation with the gold medal of the patent No: 279253 - CZ:

Balátě, J. (1997). *Controle de la production thermique dans circuits de chaleur et méthode utilisée.* 46th World Exhibition of Innovation, Research and New Technology, Brussels EUREKA´97, Brussel, Belgium.

Balátě, J. and T. Sysala (1998). The Way of District Heating Output Control by Means of Hot Water Piping. In: *Preprints: 5th IFAC Workshop on Algorithms and Architectures for Real-Time Control – AARTC ´98Cancun*, Mexico.

Box, G.E.P. and G.M. Jenkins (1976). *Time Series Analysis, Forecasting and Control.* Holden Day, San Francisco.

www.elsevier.com/locate/ifac

CONTROL SYSTEMS DESIGN FOR PRACTICAL TEACHING

Juan J. Gude Prego, Evaristo Kahoraho Bukubiye

Departamento de Automática. Universidad de Deusto
Avda. de las Universidades, 24. 48.007 Bilbao (Spain)
Phone: +34-94 4139000. Fax: +34-94 4139101
jgude@eside.deusto.es

Abstract: In this work, we present our experience in designing laboratory prototypes for practical teaching in control engineering at the Faculty of Engineering (ESIDE) at the University of Deusto. These prototypes have been designed in order to achieve a practical training that is more closely related industrial practice. For this design, we first analyse the technical profile generally demanded by enterprises. The professional qualification of our students in the fields of Industrial Instrumentation, Systems and Automation is analysed. Developed prototypes are described and main differences between using commercial educational equipment and using real industrial equipment are analysed.

Keywords: Control Education, Control Theory, PID Controllers, Control Algorithms, Computer Aided Control Systems, Monitored Control Systems.

1. INTRODUCTION

In this paper we present our experience in the design of control systems for training in Control Engineering.

The four laboratory prototypes that have been designed have an industrial focus and create a bridge between theory and industrial practice. The equipment developed is typical of processes in industry: an ion exchanger, a steam boiler, a heat exchanger and a pH control of an effluent. These prototypes were designed by the Department of Control and were developed jointly with MIESA Engineering (Spain). The prototypes are used in laboratory classes in the fields of Automation and Control in Automation Engineering studies of the University of Deusto.

The prototypes and their use in laboratory classes have been designed as a whole and with a "top-down" strategy. In other words, they are designed from the point of view of both the final qualification required by industry and the knowledge required to complete the formal education.
This design provides an integration of all the technologies acquired by the students in their theoretical-practical education.

2. THE TECHNICAL PROFILE DEMANDED BY INDUSTRY IN AN AUTOMATION ENGINEER

The acquisition and development of equipment for the laboratories of the Department of Control, which includes the prototypes described in this communication, has been conducted with the technical profile demanded in the field of Industrial Automation in the Basque Country (Kahoraho, *et al.*, 1997).

Industry demands that the University trains professionals qualified to improve and/or innovate industrial processes. Consequently, Engineering study programmes must be technologically up-to-date, and special attention must be paid on practical training. Specifically, the Automation Engineering programme should include topics about the new Automation technologies.

An Automation Engineer's knowledge should extend from the conceptualisation and the design of complex automation systems to their maintenance, including integration and implementation of advanced manufacturing systems.

In our opinion, knowledge acquired in ESIDE encompasses the following three profiles and their corresponding study programmes in Control and Industrial Engineering:

- *Process Engineer*. This profile is achieved with a deep theoretical and practical training in Process Control, Electric Engines and Distributed Automation.
- *Manufacturing Engineer*. Most of the knowledge is obtained through subjects such as Automation Design, Industrial Instrumentation, Communications and Industrial Computer Science.
- *Implementation Engineer*. This skill is obtained through subjects such as Industrial Automation.

This is explained in detail in (Kahoraho, *et al.*, 2000).

3. PROFESSIONAL QUALIFICATIONS OF OUR STUDENTS IN THE FIELDS OF INSTRUMENTATION, SYSTEMS AND AUTOMATION

The Department of Control and Electronics provides Practical Education in the following technologies:

Industrial Automation education has two levels. In the first level, students are trained in designing basic automatic devices, assembling pneumatic, hydraulic and electric devices. In the second level, training is focused on PLCs programming and designing of automation systems based on PLCs and the automatic devices previously mentioned.

The abilities achieved in the field of Industrial Instrumentation include the following topics:

- Conceptualisation of instrumentation loops. Familiarisation with different instrumentation loops. Ability to implement instrumentation loops.
- Interpretation and realization of piping and instrumentation diagrams (P&ID's).
- Selection and sizing of instrumentation.
- Industrial instrumentation calibration.

In Industrial Computer Science, the practical training is based on programming data acquisition cards for PCs and applications development on C language for serial communication, the study of SCADA systems and the development of supervision and control applications, and fieldbus communications.
This is completed with lectures on real time systems and industrial communication networks.

The control skills include:

- Control System Analysis.
- Analysis of the influence of the loop parameters on the control loop performance
- Tuning of the conventional control loop structure.

Apart from that, the focus is on design and implementation of DDC control algorithms, as well as on advanced algorithms such as optimal, robust, adaptive and predictive control.

Students also must be able to use Matlab and its toolboxes.

4. DEVELOPED PROTOTYPES

The Department of Control's goal is developing real industrial systems to improve practical education and to reduce the classical dissociation between the University formal education and industrial practice.

4.1 Description of the equipment

The prototypes that have been developed are:

- An ion exchanger,
- A steam boiler,
- A heat exchanger and,
- A pH control of an effluent.

See the heat exchanger prototype in figure 1.

The main characteristics are the real industrial instrumentation and controllers used in their construction.

Fig. 1. Heat exchanger prototype with all its components: industrial instrumentation and three optional control modes.

The equipment includes a PC, Ethernet connection to the Internet, a *Siemens Sipart DR 24* or *DR 19* industrial regulator, a *Siemens S7-300* PLC with a *Profibus* fieldbus communications module and two *Advantech* data acquisition cards to record the values measured by the equipment's instrumentation and to activate pumps and valves.
These prototypes have circuits with pressure security valves and security elements.

In these prototypes, three different technologies for industrial plant control have been implemented:

a) Control by Siemens *Sipart DR 24 or DR19* Industrial Controller.
b) Computer control using *Matlab* to develop and implement the control algorithms. Drives that permit direct access to inputs and outputs of the data acquisition cards from *Matlab* have been developed.
c) Control by the PLC, where the aforementioned control algorithms have been implemented. The PC uses the fieldbus Profibus to communicate to the PLC.

The first control mode has display recorders to visualize the different variables in the process.
The last two control modes have a graphic user interface by the SCADA *Factory Link (Monitor Pro)*. This equipment has a switch to select the use of one or another of these control modes implemented with different technologies.

4.2 Description of the processes

Ion exchanger. The prototype simulates the process of industrial residues filtration or dissolution micro-filtration.

In figure 2 the diagram of the process is shown. The system consists of a water dump D-001 that simulates the liquid to be filtered, an impulsion pump B-001 with its frequency inverter SIC-001, an ion exchanger module simulated by the valve R-001, a control valve PV-001, pressure and flow transmitters PT-001, PT-002 and FT-001, and a pressure and a flow controller.

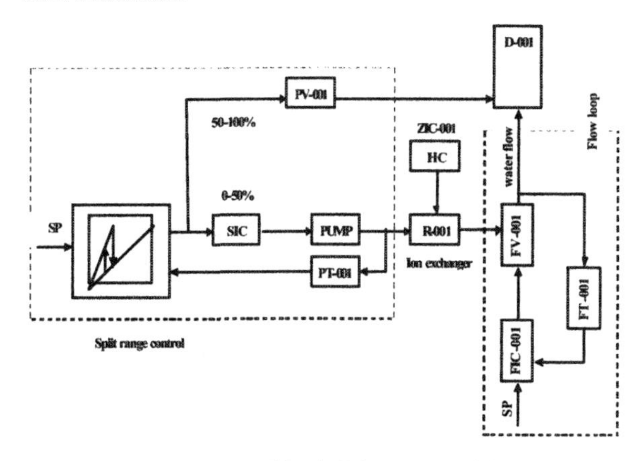

Fig. 2. Ion Exchanger Block Diagram and its control structures.

There are two control loops: a pressure loop in order to simulate the pressure differences downstream and upstream the membrane (valve R-001) and a flow loop to control the flow that returns to the dump. Load perturbations can be generated by opening or closing the valve R-001.

The flow loop is a feedback control and in the pressure loop a split range or duplex control has been implemented. This last kind of control has two possible actions in order to increase or decrease the controlled variable (pressure measured by PT-001): to activate the pump so that the water goes to the main dump through the ion exchanger, and to activate a bypass so that the water goes to the main dump without passing through the module.

Steam boiler. This process simulates a steam boiler level control. For security reasons, we have eliminated the heating process, so that simulation is a simplified process, although the same control structure as the real process has been maintained.

B-001 pump sends water from the main dump D-001 to the BL-001 boiler; because of the pump's nonlinearity, flow control is done by the valve FV-001. The FT-001 flowmeter measures the feed flow, being a part of the cascade loop.

The fictitious extraction of steam is made by the B-002 pump. Its power is controlled by the SIC-002 frequency inverter. The FT-002 Coriolis flowmeter measures the extraction mass flow, being a part of the feedforward loop.

Modifying the setting of the frequency inverter, load perturbations can be generated.

It is a multiloop control of a single variable. A feedback loop, a feedforward control and a cascade control are implemented, as shown in figure 3.

We use a feedforward loop to correct changes in the extraction flow before they affect the boiler level and a cascade loop to correct the effect of the changes in the feed flow.

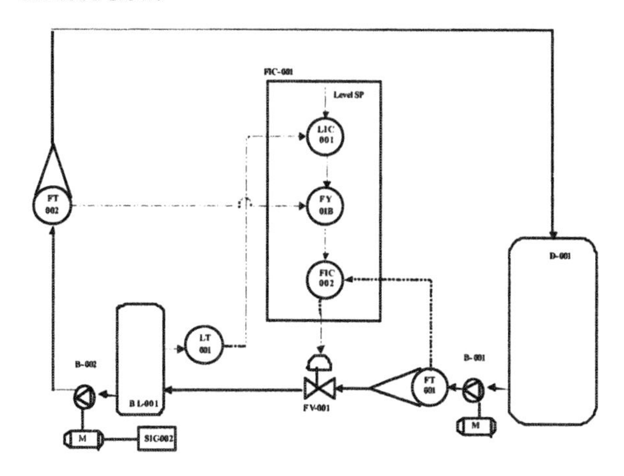

Fig. 3. Steam Boiler Prototype's diagram.

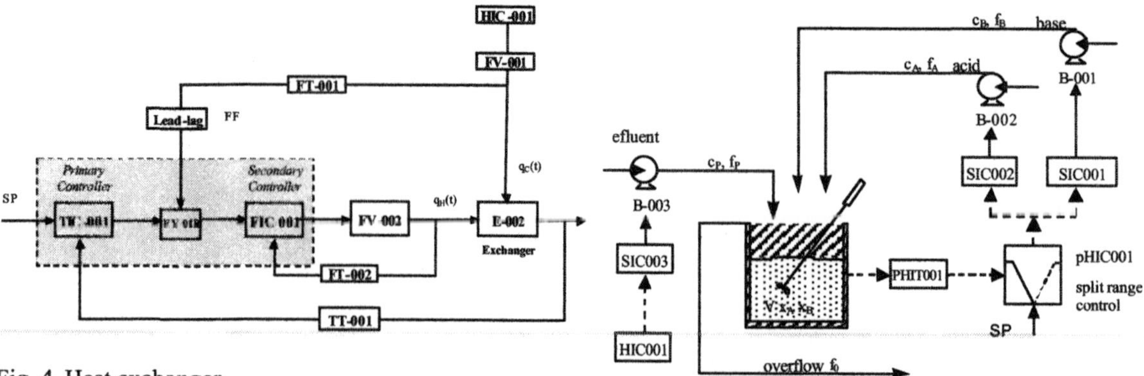

Fig. 4. Heat exchanger.

Fig. 5. pH Control of an effluent.

Heat exchanger. This is a closed circuit system composed of a cold water dump (main dump), a hot water dump with a thermo resistance to maintain the hot water temperature, and a heat exchanger.
The closed circuits that are followed by the cold and the hot water are the following: a pump sends the cold fluid from main dump to the exchanger, where the exchange of energy is produced. Lastly, an air cooler dissipates the heat from this fluid, which is returned to the main dump. Another pump sends the heating fluid to the exchanger. After the loss of heat energy, the fluid returns to the dump.

In the block diagram in figure 4, we can see the structure of the process control system, which includes the following sensors and actuators:

- PT-100 temperature sensors, used to measure the temperature of the hot fluid and the cold fluid inlet and outlet to the exchanger.
- Electric valves that permit the cooling fluid and the heating fluid to pass into the exchanger
- Flowmeters that measure the flow of hot and cold water upon leaving the respective valves

We can produce load changes modifying the cold water flow.

A feedback loop, a cascade control and a feedforward control are implemented. It's a multiloop control of a single variable.
Cascade loop tries to correct the effect of the heating fluid flow changes and feedforward loop tries to correct load changes before they affect the controlled variable.

pH control of an effluent. The process consists of controlling the pH of an effluent with a concentration c_P and a flow f_P that is fed into an agitation tank with a variable volume V.
The tank has an overflow, so that when the maximum volume is exceeded, the volume remains constant and the outlet flow is f_0.
The composition of the effluent is modified using two additives: a strong base (NaOH) with a concentration c_B and a flow f_B and a strong acid (ClH) concentration c_A and a flow f_A.

Three pumps send the efluent, the base and the acid to the agitation tank.

To control this process we use a split range control to modify the setting of frequency inverters that control the speed of the pumps that add the base or the acid. The effluent's flow can also be modified by manual control in order to generate load changes, as indicated in the diagram in figure 5.

4.3 Graphical interface

A graphic user interface has been developed. See the main screen of the Steam Boiler in figure 6. This interface was designed with the SCADA Factory Link.

The interface designed allows one to visualize the process variables and to interact with the process:

It allows one:

- To change SP.
- To use different working modes for the controllers (manual, automatic).
- To change the PID parameters of the controllers.
- To introduce load perturbations.

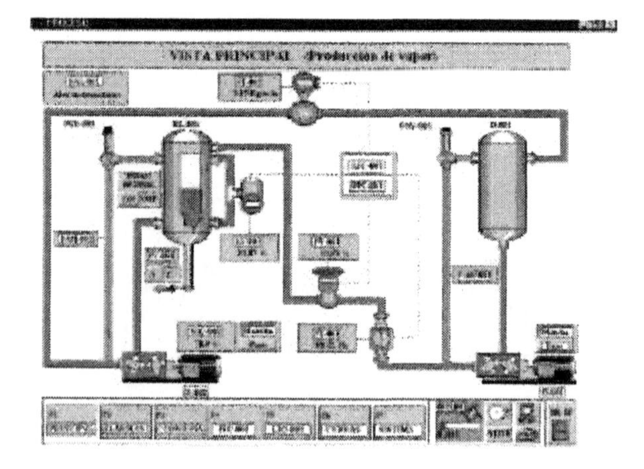

Fig. 6. Graphic user interface of the Steam Boiler.

The application starts with the main screen, where a process diagram with animated graphs appears, as in figure 6. In this screen, you can visualize all the process variables and you are able to start the process.

Additional screens include:

- An alarms screen.
- A curves screen. The user can select the variable she/he wishes to represent in a graph.

Each loop has its own screen, where the user can change PID parameters, SP, %OP,...

The application allows one to tune the controllers by two methods: the open loop method and the relay method. See (Kiong, *et al.*, 1999; Lopez, *et al.*, 1967 and Rovira, *et al.*, 1969)

4.4 Use of industrial communications and the new technologies

An application of remote supervision via Ethernet by the SCADA Factory Link or WinCC has been developed. This is the development of a communication application between PCs based on local area networks such as Ethernet. In this way, anyone can control as well as supervise the process from any location.

This prototype has several technologies for industrial plant control, communications and supervision, as shown in figure 7.

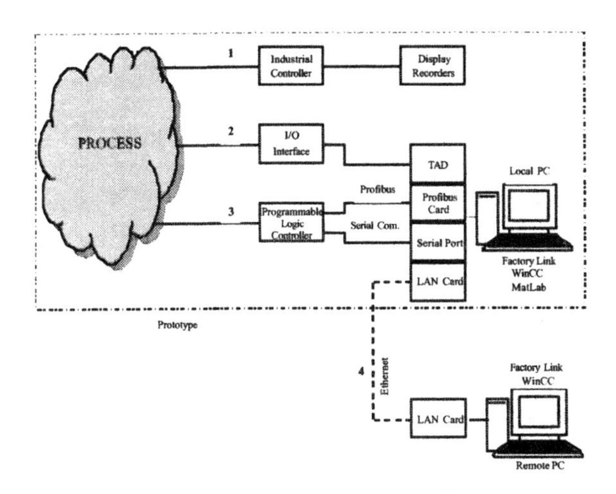

Fig.7. Diagram of the communication, and supervision applications.

4.5 Scope of the prototypes

These prototypes are used in the practical teaching of the following subjects: Industrial Instrumentation, Control Systems, Linear Systems and Industrial Communication Networks, as shown in (Gude and Kahoraho, 2002a; Riera and Kahoraho, 2001; Gude and Riera, 2002b; Revilla, 2001).

The laboratory lessons start from the modelling of the processes and are designed to familiarise the students with industrial control conceptualisation. The students learn to validate the model. More information about the modelling of these processes can be obtained in (Wellstead, 1979; Smith, 1985; Shinskey, 1988; Franks, 1972 and Balchem 1995).

The practical education is aimed at qualifying students for programming, handling and tuning the industrial PID controllers and to become familiar with the conventional control structures. For this theoretical-practical concepts see (Astrom, 1995; Smith, 1985 and Kiong, 1999).

Nearly of the 90% of the industrial applications can be controlled by a PID controller and the problem difficulties consist in tuning correctly its parameters (see Astrom, 1995), so that our effort is centred in achieving that our students learn PID tuning methods.

For the basic education, we consider that it is enough if our students learn to apply the open and closed loop Ziegler-Nichols method and to implement a three element control: feedback, cascade and feedforward control for the Steam Boiler and the Heat Exchanger prototypes.
With the ion exchanger prototype, students learn to apply the open loop Ziegler-Nichols method in order to tune the flow loop and the relay method to tune the pressure loop.
The use of the pH control prototype consists in familiarising the students with the problem of the control of a non-linear process. They apply the relay tuning method in order to obtain PID parameters.
With these two last prototypes they familiarise with the split range control structure.

On the other hand, it is important that they can be able to apply directly tuning formulae and later to make a readjustment of the PID parameters in order to improve the system performance. They make an analysis of the influence of the loop parameters on the control loop response, so that they can check the effect in the system response when they change PID parameters.

With Heat Exchanger prototypes, students can perform one, two or three-element control, i.e. feedback, feedback and cascade, feedback, cascade and feedforward control, and check that the control is improved. The order of this structure changes in the Steam Boiler prototype. Two-element control corresponds to feedback and feedforward control and three-element control corresponds to feedback, feedforward and cascade control.

Students learn to design control algorithms and their computer-based implementation. They use MatLab for this purpose. See (Kahoraho and Pérez, 1998; Astrom and Wittenmark, 1997)

The use of several technologies for control of industrial plants is introduced. Thus, They can learn to develop their own supervision application of the process.

These prototypes permit the introduction of an educational tool such as *Matlab* in an industrial prototype.

A remote supervision of a plant application has been developed as a communication application between PCs based on local area networks. Communication applications between PLCs and between PCs and PLCs based on Profibus fieldbus technology. This field is focused on the development of communication applications between PCs and PLCs based on local area. Students develop their own applications for remote supervision of the process.

The use of real industrial instrumentation in the construction of the prototypes allows us to study its dynamic behaviour and calibration and to work with P&IDs. Students learn to program MatLab applications in order to obtain calibration curves and the study of some transducer's dynamic response.

5. CONCLUSIONS

In this paper, we explained our reflections in the process of designing four laboratory prototypes for teaching Control Engineering.

The main advantage achieved by using this equipment instead of commercial educational equipment is that commercial equipment does not allow enough experience/familiarisation with measures and results obtained in real industrial processes. Thereby, we decided to complete the classical didactic equipment with real industrial processes.

This equipment implies the reduction of the classical dissociation between the University formal education and industry practice.

The same equipment is being developed for advanced control engineering, such as adaptive control, identification and robust control for the future.

REFERENCES

Astrom, K. and B. Wittenmark (1997). Computer Controlled Systems: Theory and Design. Prentice Hall, 3rd Edition.

Astrom, K. and T. Hagglund (1995). PID Controllers: Theory, Design and Tuning. ISA, 2nd Edition.

Balchen, J.G. and K.I. Mumme (1995). Process Control System: Structures and Applications. Blackie Academic & Professional.

Franks, F. (1972). Modelling and simulation in chemical engineering. John Wiley & Sons.

Gude, J.J. and E. Kahoraho (2002a). Prácticas de Sistemas Lineales. University of Deusto, Bilbao (Spain).

Gude, J.J. and I. Riera (2002b). Prácticas de Instrumentación Electrónica. University of Deusto, Bilbao (Spain).

Kahoraho, E., N. Pérez, J. Revilla and J. García (2000). Professional training difficulties in engineering schools – ESIDE experience. *3rd International Conference on Quality, Reliability & Maintenance QRM2000*, pp. 263-268. University of Oxford, UK.

Kahoraho E and N. Pérez (1998). Control Digital. University of Deusto, Bilbao (Spain).

Kahoraho, E., N. Pérez and M.D. Rubio (1997). Memoria del proyecto de adquisición de equipamiento para los laboratorios. Internal documentation. University of Deusto, Bilbao (Spain).

Kiong, T.K., W. Quing-Guo and H.C. Chieh (1999). Advances in PID Control. Advances in Industrial Control. Springer.

López, A.M., J.A. Miller, P.W. Murrill, C.L. Smith. (1967). A comparison of open loop techniques for tuning controllers. Control Engineering, **vol. 14** (12).

Revilla, J. (2001). Prácticas de Redes de Comunicaciones Industriales. University of Deusto, Bilbao (Spain).

Riera, I., E. Kahoraho (2001). Prácticas de Sistemas de Control. University of Deusto, Bilbao

Rovira A.A., P.W. Murrill, C.L. Smith. (1969). Tuning Controllers for setpoint changes. Instruments and Control Systems.

Shinskey, F.G. (1988). Process Control Systems. Mc Graw-Hill, 3rd edition.

Smith, C.A., A.B. Corripio (1985). Principles and Practice of Automatic Process Control. John Wiley & Sons.

Wellstead, P. (1979). Introduction to physical system modelling. Academic Press.

VERIFIABLE-BY-INSPECTION FAULT-TOLERANT CONTROL SOFTWARE FOR A DEPENDABLE CHEMICAL PROCESS USING ANALYTICAL REDUNDANCY

G. Thiele, Th. Laarz, G. Schulz-Ekloff, D. Brunßen, J. Westerkamp
R. Neimeier, L. Renner, E. Wendland

Universität Bremen, FB1
Institut für Automatisierungstechnik
Otto-Hahn-Allee, 28359 Bremen
thiele@iat.uni-bremen.de

Abstract: Verification and certification of real-time software is of growing importance in industry, especially for safety-critical applications in order to guarantee software (SW) correctness. If the real-time tasks are designed in Function Block Diagram (FBD) language using the international standard IEC 61131-3 for programmable logic controllers (PLCs), "verification by inspection" becomes an attractive alternative to formal methods.
As a case-study, the fault-tolerant control software for a dependable chemical process (CSTR) will be chosen considering the dependability functions for sensor faults distributed over a three-level controller hierarchy. Different from earlier papers, this contribution deals with practical implementation problems using commercial PLCs, i.e. the impact of the integration of analytical redundancy as fault-tolerance-oriented components on the computation time, robustification of analytical redundancy and HW/SW-in-the-loop simulation using the conventional terminal-oriented PLC-I/O. Furthermore, a comment is given on the impact of verifiability-by-inspection on standardizability of function-blocks. *Copyright © 2003 IFAC*

Keywords: Fault-tolerant software, function block, fuzzy control, hierarchical control, process control, reliability, graceful degradation, safety, sensor failure, verifiability.

1. INTRODUCTION

Verification-by-inspection of real-time software (Halang, 2000; Thiele et al., 2001) are of growing interest for embedded systems in safety-critical applications (Blanke et al., 2000; Isermann et al., 2000). This is especially true for software including fault-tolerance-oriented components (Thiele et al., 2001; Thiele et al., 2003), e.g. for fault detection and isolation (FDI).
Different from other papers this paper deals with implementation aspects of the problem, i.e. of the design of a three-level controller-hierarchy, the reconfiguration aspect of the fuzzy-controller, overcoming the impacts of the restrictions of conventional PLCs on HW/SW-in-the-loop simulation by definition of I/O-FBs, qualitative assessment of the computation-time-problem, and robustification and supervision of analytic redundancy. Furthermore, a comment on the impact of verifiability-by-inspection on standardizability of FBs are given.
In the paper the design and implementation of a verifiable task for the tracking control of a chemical CSTR-process (Vielhaben et al., 2001) will be presented using a hierarchically structured FBD and suitable dependability functions, e.g. using an "one-

observer approach" (Jakubek et al., 2000) as analytical redundancy.
Especially the design of the fault-identification and inference (FII) component for process dependability (highest level of a three-level controller hierarchy) will be presented.
Results on controller-task schedulability and robustification of the analytical redundancy are given, and practical aspects of the implementation on a commercial μPLC are discussed.

2. VERIFICATION BY INSPECTION

2.1 Verification of FBDs

Assuming the existence of a library of already verified FBs and functions, also the verification of FBDs can be accomplished through verification-by-inspection with respect to the FB-types (FB-classes), the FB-connections, the FB-parameters and the feasibility of the FB-computing-sequence (Thiele et al., 2001).

2.2 Structuring of FBDs

Although complexity is a barrier for the transparency of design of FBDs, as it is for formal methods, the possibility of using "derived FBs" (IEC, 1999) and of hierarchical FBD-structuring proves to be suitable for verifiability-by-inspection for complex problems as well.
As a conceptual architecture for controller tasks a three-level hierarchy (Blanke et al., 2000) as FBD-structure is chosen. This structure is well-suited for the assignment of dependability functions, i.e. analytical redundancy (level 2), supervision of CAN-sensor communication and of observer stability (level 3), inference of the corresponding dependability pattern (level 3), and fault accommodation and reconfiguration (level 2) using the available effectors in the basic control-loop (level 1), due to the actual dependability pattern.

2.3 Derived FBs

A useful instrument keeping verifiability-by-inspection for more complex FBDs, is the use of problem-oriented "derived FBs" (IEC, 1999). A typical example is the fuzzy controller (FUZZY_REG), as a popular version of a qualitative controller, in Figure 2. This strategy of problem-oriented design has to be given precedence over the standardizability of FBs (Münnemann et al., 2002) if verifiability-by-inspection is imperative.

3. DEPENDABLE PROCESS STATES

The intention of integrating fault-tolerance-oriented components into the controller-SW is to maintain the reliability of the process. In order to avoid shut-down as long as possible, the set of reliable states can eventually be enlarged in the sense of graceful degradation (Musa, 1999). With respect to Figure 1,

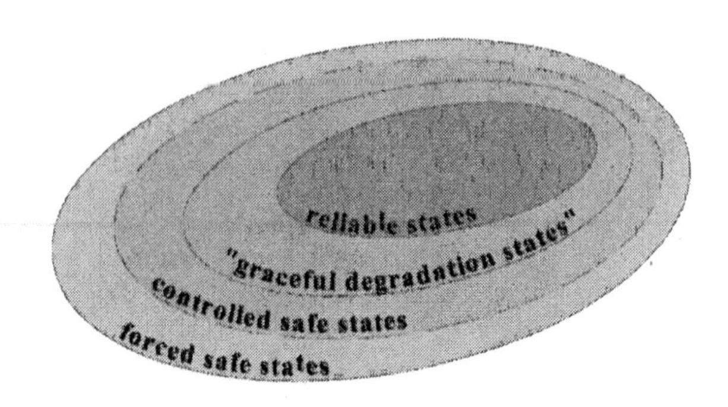

Fig. 1. Dependable process states

there is a "safety-reserve" (Steinhorst, 1999) available, as long as the set of safe states is "larger" than the set of graceful degradation states. The set of "controlled safe states" should be a subset of the set of "forced safe states" serving to maintain safety if the feedback-controller itself is unable to do so. A special safety function or even a special protection system (IEC, 1998) has to guarantee safety in such cases (Thiele et al., 2001).
Note that the states in Figure 1 are of hybrid nature (Blanke et al., 2000), i.e. these states comprise the continuous state components of the technical process as well as discrete components describing, e.g., the operating states of the sensors and of the analytical redundancy.

4. DISTRIBUTION OF DEPENDABILITY FUNCTIONS

Figure 2 illustrates a task-design for the fuzzy-control (Preuß, 1993) of a chemical process.
The distribution of dependability functions over the components of this three-level architecture has to be chosen in accordance to the feasibility of the FB-computing-sequence. A consequence of this is a necessary partitioning of level 2 in the (conceptual) components "fault detection and isolation (FDI)" and "fault accommodation and reconfiguration (FAR)" and the assignment of analytical redundancy as a dependability function to FDI. Although the assignment of analytical redundancy to FAR seems meaningful at first glance, the additional detection of "measurement uncertainties" (Benitez-Perez et al., 1997) by the observers besides output reconstruction makes the assignment to FDI reasonable.
Level 3 (fault identification and inference, FII) assembles all the relevant information of level 2 which can eventually be generated already by detectors of level 1 (e.g. the detection of CAN-communication failures or the receipt of faults from self-validating sensors (Benitez-Perez et al., 2002)

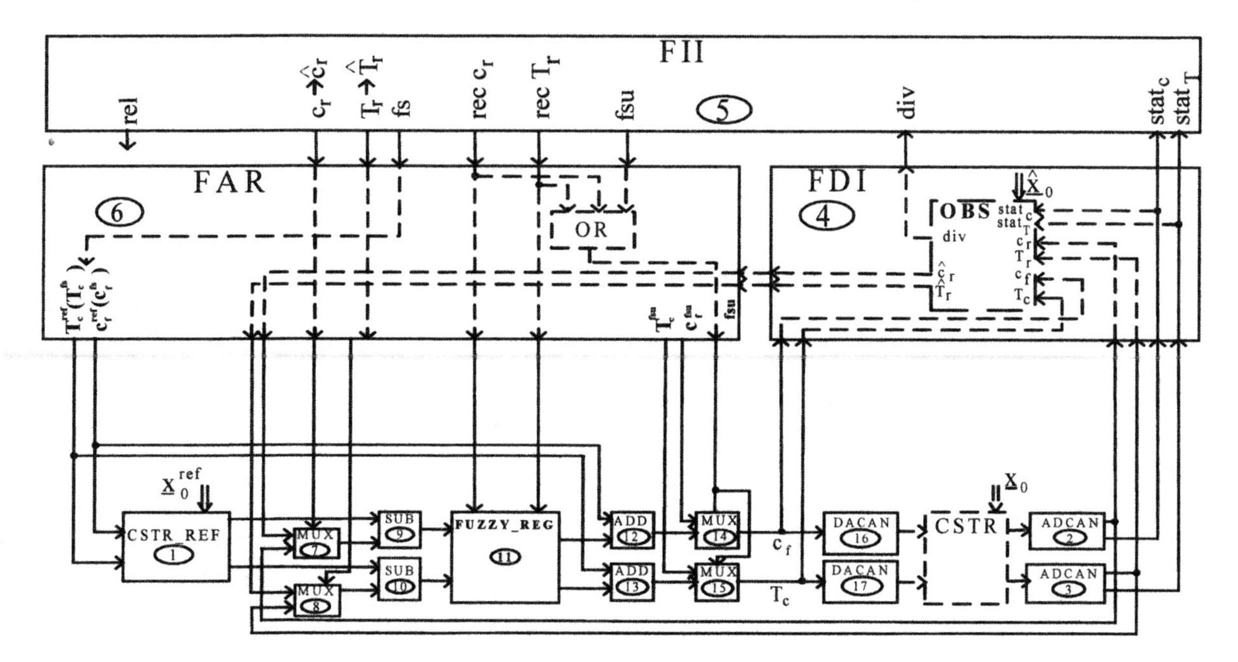

Fig. 2. Conceptual design of a FBD-controller task with fault-tolerance-oriented components and feasible computing-sequence 1 - 17 (*rel* = reliable, *fs* = fail-safe, *fsu* = fail-safe uncontrolled, *div* = divergent, *rec* T_r = reconfiguration for T_r, $stat_T$ = status of temperature measurement)

by FBs of class ADCAN).

A typical example of the "rules" of an FII-component is given in Figure 3 using the process-states presented in Figure 1. The rules have been derived for its application in corresponding exceptional situations from the expert's knowledge, and can be deduced by inspection again from the FBD given in Figure 3, e.g. in the case of "AND-FUNCTION 5.2":

WHEN <T_r-sensor fails> **AND** <c_r-sensor is operational> **AND** <analytic redundancy is non-divergent> **THEN** <substitute T_r for its reconstructed equivalent $\hat{T}_r$>.

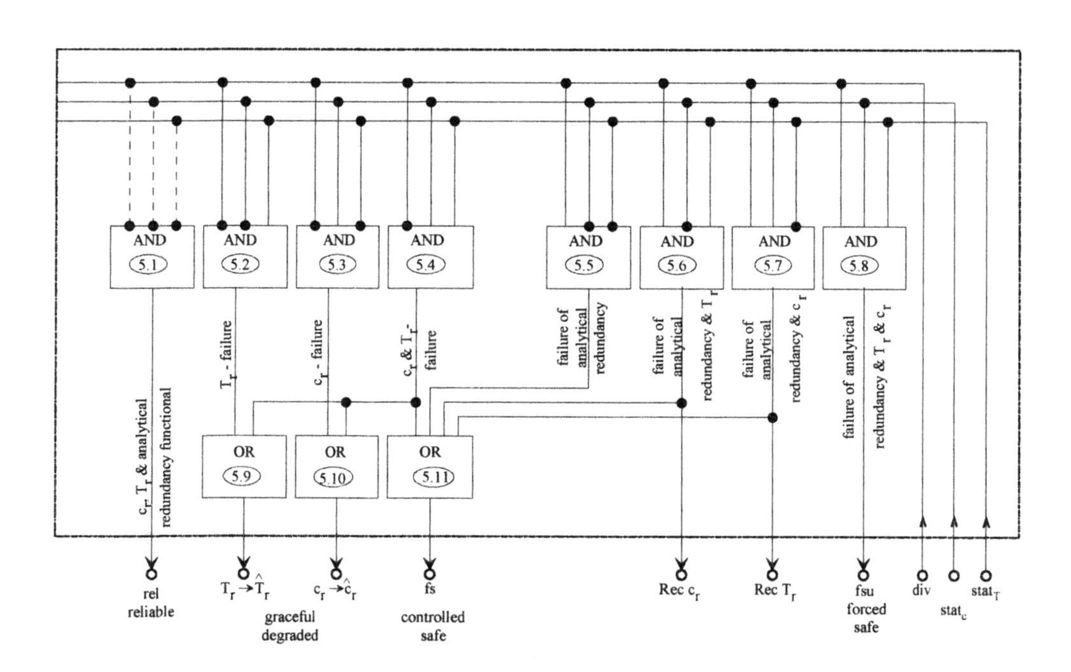

Fig. 3. FBD-design for the FII-component (level 3) of a hierarchical controller

The inference-results of level 3 will then be transferred back to level 2 (FAR) which represents the interface to the effectors (of level 1). These are represented in Figure 2, on the one hand, by the two multiplexers in the feedback path of the controller (i.e. 7 & 8) for the eventual switching to reconstructed measurements and, on the other hand, by the two multiplexers in the control-signal paths (i.e. 14 & 15) realising the forced safety using an inherent safe trajectory of the autonomous process.

In the case of the failure of the analytical redundancy and (only) one sensor, a possible respective reconfiguration of the fuzzy controller is inferred and mapped by level 2 to level 1, if implemented there. If so, a respective one-input fuzzy-controller has to be provided for these cases instead of the nominal 2-input fuzzy-controller, in order to keep a safety-reserve.

5. THE COMPUTING-TIME PROBLEM

In Figure 4 different scenarios ("matrix lines" 1..3) using different real (ß=1) and simulated (ß=2) sampling periods ("matrix columns" 1..3) are analysed with respect to schedulability (an experiment will be referenced by its matrix-position (e.g. 1,1) in the following discussion).
The control-task for the basic control-loop with simulated CSTR is schedulable for Ts=1 sec, ß=2 (on a μPLC based on a M 68331 micro-controller, 25 MHz), i.e. in stretched time (1,3). Therefore it will, in principle, be possible to schedule the task also for real-time, if the communication part "C" is small enough (1,1). From this result, it can be deduced, that the controller in the case of the real CSTR, based on a 1-observer-solution (2,1), using the reconfigurable observer of Figure 5 (Thiele et al., 2003), should be schedulable again if the observer computing-time (which should be dominated by the integration-time of the non-linear observation-model) is nearly the same as the simulation-time of the CSTR-model.
But it is also clear, that a 2-observer-solution, even if the CSTR is not simulated, will not be schedulable in real-time at all (3,1).

Also, using a larger sample-time is not sufficient for schedulability of the 2-observer-solution, because simulation-times of the observation-model will double by doubling of sample-time, e.g., for the non-linear continuous observer (3,2).

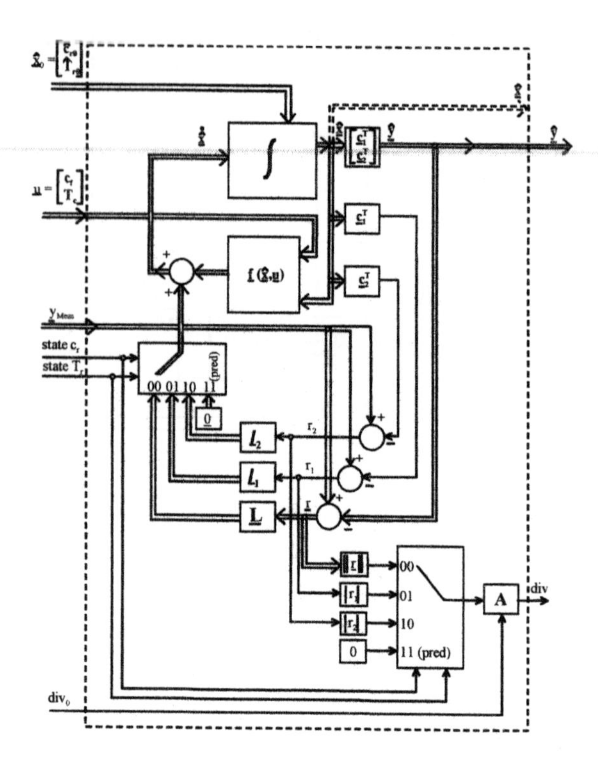

Fig. 5. Re-configurable observer (A = analysis and assessment of the residuals, *pred* = prediction, div_0 = re-initialisation of divergence detection)

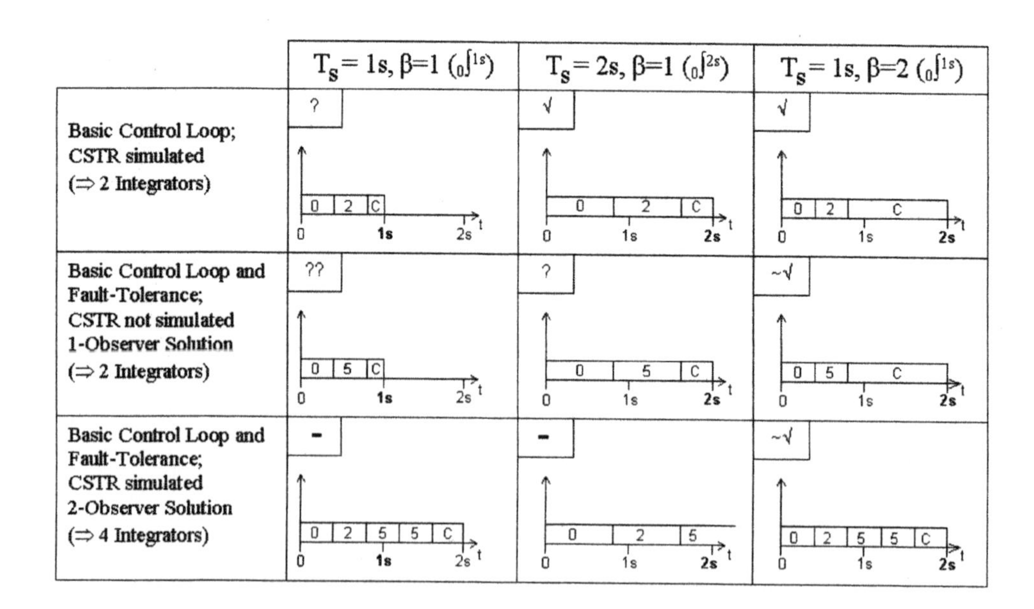

Fig. 4. Qualitative comparison of computing times for different fault-tolerant control-system scenarios ($T_{s,\ sim} = \beta \cdot T_s$ (transformed sample time), *"0"* (Reference trajectory FB), *"2"* (CSTR FB), *"5"* (Observer FB), *"C"* (μPLC ↔ PC communication-time))

6. REDUNDANCY ROBUSTIFICATION

If exception handling (GI-FG 4.4.2, 1998) were available in ST (Structured Text, IEC 61131) as the programming language used for implementing the observer as a basic FB, the "answer" to a possible instability exception of the observer could be its re-initialisation, using the last and best plausible states available, in the sense of graceful degradation. If re-initialisation fails shut-down has to be initiated. Therefore, observer-divergence, e.g. induced by the measurement-approximation over the sampling interval, must be avoided as far as possible.

One way of robustising the analytic redundancy is the application of some of the available measurements to avoid linearisation of the observer, as a usual source of its restriction to local stability. In this case the error-equation of the observer will be of

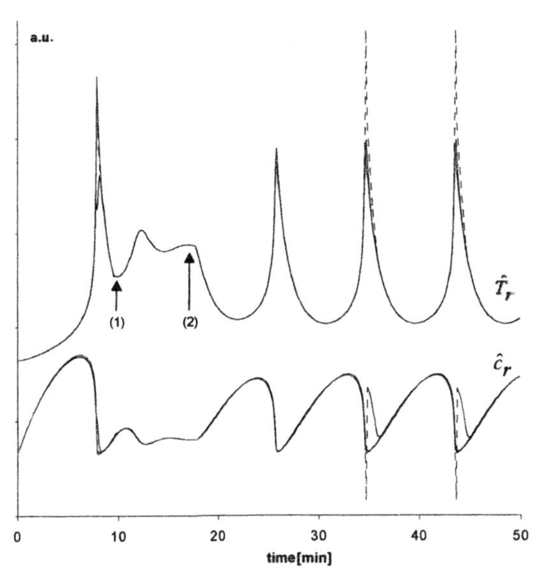

Fig. 6. Linearised observer (---) and non-linearised observer (—) changing to steady state (1) and to limit cycle (2) using cooling temperature as control variable

the type

$$\dot{\mathbf{e}} = \mathbf{F}(\mathbf{y})\mathbf{e}; \mathbf{e} = \mathbf{x} - \hat{\mathbf{x}}\,; \mathbf{y} = \textit{known measurements} \quad (1)$$

with constant eigenvalues of $\mathbf{F}(\mathbf{y})$ in the negative half-plane (Che et al., 1993). Unfortunately, sometimes such an observer-type does not exist for all measurement combinations so that an one-observer approach is not feasible or the achieved increase of performance quality is, again, already exhausted by using prediction-mode and switching to a safe state of the observer for some measurement patterns. In Figure 6, a comparision of a linearised and a non-linearised type of a single measurement observer using concentration c_r is given. The linearised observer shows so called "bursting phenomena", which are well known from recursive parameter estimation (Anderson, 1985), if the operating mode is changed from a safe state back to normal operating mode ((2) in Figure 6).

7. PRACTICAL IMPLEMENTATION PROBLEMS

7.1 HW-/SW-in-the-loop simulation and PLC-"terminals"

In order to test the fault-tolerant PLC-controller, the CSTR-plant (Figure 2) has "only" to be substituted for a (derived) FB, simulating the non-linear process model, in principle. Clearly, for HW-in-the-loop simulation, the FBs of class DACAN and ADCAN have to be "inversely mirrored", by FBs of class ADCAN and DACAN, respectively, in the "simulation-FB". But furthermore, there must not be any restrictions on the definition of the FB-computing-sequence, i.e. the initial values of the process-model have to be put out via DACAN-FBs before reading them by the ADCAN-FBs of the controller. If, instead, the conventional PLC-principle is kept, i.e. to read all input-terminals first before executing the FBD, followed by writing all output-terminals, simulating the process-model on the same PLC is not possible. Special I/O-FBs have to be designed for this purpose, instead, using, e.g., Remote Transmission Request (RTR) of CAN-Bus protocol directly.

7.2 *HW-/SW-in-the-loop simulation and pre-emptive multitasking*

An even more elegant concept of HW-/SW-in-the-loop-simulation, from the point of keeping the controller-task in the true application without any modification, uses a separate task of higher priority for the execution of the derived FB which simulates the plant. Unfortunately, some PLC-platforms featuring IEC 61131 do not implement pre-emptive multitasking yet, and, as a consequence, ignore the PRIORITY-attribute, although this would suffice already in this case of a single controller.

7.3 *Configuration of FBs and verification-by-inspection*

Unfortunately the standard IEC 61131 only allows the information-flow of FBs from left to right. This is not advantageous with respect to verifiability-by-inspection because connection configurations become unnecessarily complicated. Also, the use of an "auto-router" for connection-configuration is not helpful for realizing verifiability-by-inspection because this forces connections which cannot, in principle, be problem-oriented.

8. CONCLUSIONS

The FBD-language of the international standard IEC 61131 turns out to be a promising tool for the design of fault-tolerant controllers due to verifiability-by-inspection which is imperative for systems of high safety integrity level, e.g. SIL 3 or SIL 4 (IEC, 1998).

Concepts as hierarchical controller-design, re-design of the analytic redundancy for robustification of re-configurable observers as analytic redundancy, HW-/SW-in-the-loop test, and of testing the schedulability of the controller-task are easily understood and therefore applicable by control-engineers as well as by process-engineers.

A promising controller-platform is the CoDeSys system of the CoDeSys Automation Alliance (3S, 2001). Unfortunately, not all manufacturers implement all necessary features, e.g. pre-emptive multi-tasking.

Also a minor modifications of the standard IEC 61131-3 would be helpful to ease verifiability-by-inspection for "safety-critical" applications, e.g. permitting the FB-information flow not only from "left to right", but also from "right to left".

REFERENCES

Anderson, B.D.O. (1985). Adaptive Systems, Lack of Persistency of Excitation and Bursting Phenomena. *Automatica*, Vol. 21, No. 3, 247-258.

Benitez-Perez, H., H.A. Thompson and P.J. Fleming (1997). Implementation of a smart sensor using analytical redundancy techniques. In: *Proc. SAFEPROCESS 97* (R. J. Patton, J. Chen, Eds.), pp. 569-574. Pergamon, Elsevier Science Ltd., Oxford.

Benitez-Perez, H., F. Garcia-Nocetti (2002). Reconfigurable distributed control using smart peripheral elements. *Control Engineering Practice,* in press, available online, 19 Nov. 2002, Elsevier Science Ltd., Oxford.

Blanke, M., C.W. Frei, F. Kraus, R.J. Patton, M. Staroswiecki (2000). What is fault-tolerant control? In: *Proc. SAFEPROCESS 2000* (A.M. Edelmayer, C. Banyasz, Eds.), pp. 40-51. Pergamon, Elsevier Science Ltd., Oxford.

Che, G., M. Köhne (1993). Design of non-linear state controllers and observers for the case-study CSTR. *VDI-Bericht*, Vol. 1026, pp. 235-249. VDI-Verlag, Düsseldorf.

Filkorn, T., M. Hölzlein, P. Warkentin, M. Weiss (1999). Formal verification of PLC-programs. In: Proc. *14th IFAC World Congress, Beijing*, pp. 503-508.

Gertler, J.J. (1998). *Fault detection and diagnosis in engineering systems.* Marcel Dekker, New York.

GI-FG 4.4.2 (1998). *PEARL 90 language report.* ftp://ftp.irt.uni-hannover.de/pub/pearl/report.pdf

Halang, W.A. (2000). Programmable control on the safety integrity levels 3 and 4. In: *Proc. 4th International Symposium Programmable Electronic Systems in Safety Related Applications* (TÜV Nord, EWICS TC7, TÜV Rheinland, Eds.).

IEC (1998). *IEC 61508-3. Functional safety of electric / electronic / programmable electronic safety-related systems, Part 3: Software requirements.* International Standard, IEC, Geneva.

IEC (1999). *IEC 61131-3. Programmable controllers, Part 3: Programming languages.* International Standard, 2nd Ed. IEC, Geneva.

Isermann, R., R. Schwarz, S. Stölzl (2000). Fault-tolerant drive-by-wire systems - concepts and realisations -. In: *Proc. SAFEPROCESS 2000* (A.M. Edelmayer, C. Banyasz, Eds.), pp. 1-15. Pergamon, Elsevier Science Ltd., Oxford.

Jakubek, S., H.P. Jörgl (2000). Sensor fault diagnosis in a turbo-charged combustion engine. In: *Proc. SAFEPROCESS 2000* (A.M. Edelmayer, C. Banyasz, Eds.), pp. 560-565. Pergamon, Elsevier Science Ltd., Oxford.

Kaaniche, M., J.-C. Laprie, J.-P. Blanquart (2002). A framework for dependability engineering of critical computing systems. *Safety Science*, Vol. 40, 731-752. Pergamon, Elsevier Science Ltd., Oxford.

Münnemann, A., U. Enste, U. Epple (2002). Hybrid modeling of complex process control function blocks. In: *Modelling, analysis, and design of hybrid systems* (S. Engell, G. Frehse, E. Schnieder, Eds.), LNCIS, Vol. 279, pp. 53-65. Springer, Berlin.

Musa, J. (1999). *Software reliability engineering.* McGraw-Hill, New York.

Preuß, H.P. (1993). Fuzzy control in process automation. *International Journal of System Science*, Vol. 24, 1849-1861.

Van Schrick, D., P.C. Müller (2000). Reliability models for sensor fault detection with state-estimator schemes. In: *Issues of fault diagnosis for dynamic systems* (R.J. Patton, P.M. Frank, R.N. Clark, Eds.), pp. 219-244. Springer, Berlin.

Smart Software Solutions (3S) (2001). *Handbook of PLC-program development using CoDeSys 2.2.* CoDeSys for Automation Alliance.

Steinhorst, W. (1999). Safety-technical systems. Reliability and safety of controlled and uncontrolled systems. Vieweg, Braunschweig.

Thiele, G., R. Neimeier, L. Renner, E. Wendland, G. Schulz-Ekloff (2001). On the development of certifiable real-time software using function-block diagrams in automation. In: *Proc. PEARL 2001, Echtzeit - Kommunikation und Ethernet / Internet* (P. Holleczek, B. Vogel-Heuser, Eds.), pp. 77-86. Springer, Berlin.

Thiele, G., Th. Laarz, R. Neimeier, G. Schulz-Ekloff, L. Renner, E. Wendland (2003). Development of verifiable-by-inspection real-time control software for a dependable chemical process. In: *Proc. SAFEPROCESS 2003* (M. Staroswiecki, N. E. Wu, Eds.), pp. 993-998. IFAC CD-ROM.

Vielhaben, T., G. Schulz-Ekloff, R. Neimeier, G. Thiele (2001). Design of fuzzy control for maintaining a CSTR in the steep section of its non-linear characteristic. *Chemical Engineering and Technology*, Vol. 24, 205-212.

www.elsevier.com/locate/ifac

WIRELESS COMMUNICATION STANDARDS FOR PROCESS AUTOMATION

Zdeněk Bradáč, František Bradáč, Petr Fiedler, Petr Cach, Zdeněk Rejda, Miloš Prokop

Brno University of Technology,
Brno, Czech Republic,
bradac@feec.vutbr.cz,
bradac@fme.vutbr.cz,
fiedler@feec.vutbr.cz,
cach@feec.vutbr.cz,
xrejda00@stud.feec.vutbr.cz,
xproko07@stud.feec.vutbr.cz

Abstract: The paper shows the challenges associated with attempts to utilize Bluetooth communication interface for automation purposes like control and data acquisition. The paper describes major challenges, selected approach, first tests, first experiences with Bluetooth on Linux and first prototypes of embedded devices used for testing. Moreover the paper discusses some specific challenges associated with utilization of wireless communication techniques (with focus to Bluetooth) in automation. Finally the paper presents ideas (formal methods) that could be used to analyze, and improve reliability of such system, and approach (time triggered approach with clock synchronization) that could be used to improve the deterministicity of wireless communication channels.

Keywords: control applications, remote control, communication networks, communication protocols, data acquisition

1. INTRODUCTION

Bluetooth technology is the first widespread PAN (Personal Area Network) technology. The primary focus of this technology is to consumer electronics. However there are possible applications in the field of automation too. Bluetooth itself does not support industrial applications so to gain some experience it was necessary to do some experiments and development. The paper describes the effort to develop Bluetooth platform for a new Bluetooth applications.

To test basic features of Bluetooth communication modules a Linux OS was used. The Linux was the only platform that provides free source codes of Bluetooth protocols, so it was possible to analyze the Bluetooth libraries in the form of source codes and port the code to embedded platforms. The Bluetooth part of Linux is called BlueZ. The BlueZ modules contain Bluetooth stack and support Bluetooth modules from all major manufacturers. The BlueZ enables to immediately start experiments at various protocol levels. As the embedded applications need to support the whole stack, it was necessary to start at the HCI level. The utilities that are part of the BlueZ allow to analyze and to monitor the Bluetooth communication over RS-232 and USB in the form of basic communication packets.

For the initial experiments we have chosen standard Bluetooth modules with USB interface made by MSI. The experiments continued using Comtec Teleca SE modules that are equipped with both USB and RS-232 interfaces. The first modules are based on Cambridge Silicon Radio chipset, while the Comtec modules are based on Ericsson chipset. We found the compatibility among modules from different manufacturers to be very good. The initial testing was done with manufacturer-supplied software for Windows 2000. The Linux workstation equipped with a Bluetooth module was used to send requests to the modules on PCs with Windows 2000. These experiments were performed to gain some experience with HCI interface/protocol. However we found that the latest BlueZ libraries still do not have support for all Bluetooth features and profiles. We have performed our tests using BlueZ version 2.0, which for example does not provide support for LAN profile that would enable wireless connection to standard LAN.

The experience gained with Linux OS and utilization of HCI interface over USB enabled to create sample data acquisition application. In this application there is one master data acquisition node and up-to three slave sensoric nodes. The master station is in a form of PC and the slave stations are based on simple microcontrollers. The sample application thus forms a single Bluetooth piconet.

2. LINUX BASED TESTING OF BLUETOOTH

Implementation of Bluetooth stack based just on the written specification is very difficult so the existence of Linux with free (open source) Bluetooth stack is very helpful. The Bluetooth addendum for Linux called BlueZ is part of the Linux kernel and is part of some recent distributions. We have tested RedHat 7.2 and Suse 8 Linux distributions.

2.1 BlueZ

BlueZ is a software package for Linux, which contains Bluetooth stack. The BlueZ consists of:

- BlueZ Core – core of Bluetooth stack
- HCI UART, USB, PCMCIA and virtual HCI drivers
- L2CAPmodule – implementation of L2CAP Layer of Bluetooth stack
- SCO module – implementation and control of SCO packets
- HCIDUMP – monitoring tool
- Configuration and test tools

The available tools and drivers support Bluetooth modules connected using both standard interfaces (USB and RS-232). The tools allow to set-up and configure the Bluetooth devices, including full set-up of various communication parameters.

The effort to analyze source codes was not very efficient as the source code is not written on a fully platform independent manner and there are many Linux specific parts of code. The major advantage of the Linux based approach was in the HCIDUMP utility that enabled to monitor the flow of commands and data to/from Bluetooth modules. HCIDUMP thus enabled to have hands-on experience with Bluetooth without the need to fully understand to the complex Bluetooth specification.

2.2 HCIDUMP

HCIDUMP application allows to monitor every part of communication, including all levels of protocol stack (e.g. HCI, L2CAP, ...), in various modes. The data can be displayed in raw form, ASCII form, or in hexadecimal form. The application allows to display the data including various labels that help to identify and describe various parts of the protocol. The HCIDUMP is a great tool for everyone who needs to understand the Bluetooth at the protocol level.

3. EXPERIMENTAL APPLICATIONS

The experiments have confirmed that modules from different vendors utilize the same HCI protocol and the same HCI commands. Thus it enables to be manufacturer independent without the need to re-write the software when there is a need to change a module vendor. If the software utilizes only standard commands then the interoperability should be flawless. Some chipset manufacturers extend the standard HCI commands to support non-standard features (e.g. Ericsson Specific HCI-commands), however to bring the modules to operational state these non-standard extensions are not needed.

Based on the first experiment there has been created a model software application, which interconnected two personal computers using Bluetooth. The modules were controlled directly using HCI commands without utilization of third party software. All the newly created software libraries were written in ANSI C to ensure future portability to embedded world. As the embedded application will build on RS-232 interface the experiments were performed using this interface too.

3.1 The experimental applications

The experimental application was targeted to run in the Windows 2000 environment, while utilizing the newly created Bluetooth libraries. The libraries were written in strict ANSI C as the source code was from the beginning intended for porting to MCS51 embedded platform.

The experimental application runs on Windows 2000 and directly controls the Comtec Teleca modules using RS-232. The control is performed strictly only by HCI commands sent over RS-232.

The application itself allows configuring and parameterizing of communication modules. Application was built according our needs and wishes that stem out from our experience with HCIDUMP.

This application is used for:

- Testing of protocols at the HCI level
- Monitoring of data transmissions – REQUEST/ANSWER
- Testing of correctness of implementation of basic HCI commands
- Implementation of basic L2CAP functions for transmission of data packet

Moreover the application supports commands for creation of a piconet from available nodes. There have been selected a subset of necessary commands. These commands are necessary for configuration and setup of modules, inclusion of module into the communication network, setup of communication between modules and commands that enable to transmit useful data. Implemented commands include commands for search and identification of available Bluetooth nodes.

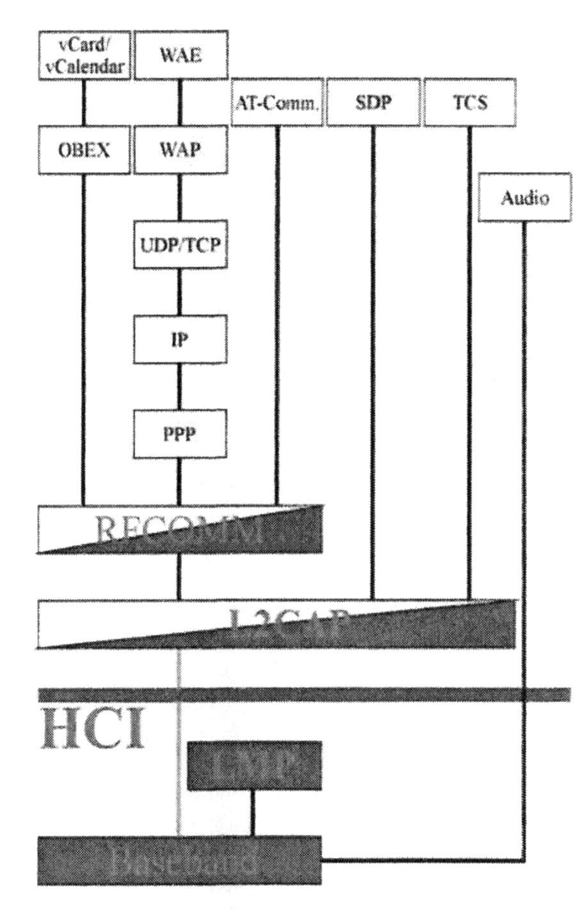

Fig. 1. Implementation of Bluetooth stack

Implemented basic HCI commands are:

- HCI_RESET
- HCI_READ_BD_ADDR
- HCI_CHANGE_NAME
- HCI_WRITE_SCAN_ENABLE
- HCI_INQUIRY
- HCI_REMOTE_NAME_REQUEST1

HCI commands needed for transmission of data packets and commands needed by L2CAP layer:

- HCI_GetDataMessage
- HCI_SendDataMessage
- HCI_SendDataPacket

Each HCI command is defined by Bluetooth specification. All used data types are also defined by the Bluetooth Standard. Fig. 1. shows implemented parts of Bluetooth stack.

3.2 Bluetooth stack terminology

Baseband

Layers „Baseband" and „Link Control Layer" enable physical interconnections between different devices in frame of basic interconnection so called PICONET. Layer is responsible for basic synchronization and communication control based on frequency hopping sequence. Layer also controls two different connection types called SCO „Synchronous Connection Oriented" and ACL „Asynchronous Connectionless". Layer enables simultaneous communication on SCO and ACL over one multiplexed RF link.

LMP – Link Manager Protocol

LMP layer is responsible for session establishment between Bluetooth devices. Layer is also responsible to negotiate length of packets. Sleep modes, power management, link management and generation of cryptographic keys for authentification and cryptography are also controlled by this layer.

HCI – Host Controller Interface

HCI layer provides uniform interface and uniform access method to Bluetooth hardware. Layer includes command interface, link management and HW conditions monitor. Layer also contains control and events registers.

L2CAP – Logical Link Control and Adaptation Protocol

Layer provides services for higher-level super-ordinate layers for establishment of connection-oriented and connectionless-oriented data transfers. Layer performs following services:

- Multiplexing: L2CAP supports multiplexing of different protocols defined by Bluetooth stack mainly SDP, RFCOM and TCS Binary.
- Segmentation and Reassembly: L2CAP is responsible for fragmentation and reassembly of data frames to packets with maximal allowed packet length (Maximum Transmission Unit - MTU). Longer data frames have to be fragmented to packets with allowed length before sending. On the receive side the received packets have to be joined to the original data frames.
- Quality of Services: Task assures parameters like transfer rate and delay.

- Groups: Task implements access rights for Piconet group mapping.

RFCOMM
RFCOMM protocol is an instrument for serial port emulation. Protocol is suitable for applications utilizing classical asynchronous serial interface for data transfers. RFCOMM layer is equipped by RS-232 emulation and includes transfer control signals.

SDP – Service Discovery Protocol
SDP protocol defines methods, which the client station can use to obtain information about Bluetooth server services. Services define mechanisms, how the client station can search server services without apriory knowledge of them. Layer defines services like search of new services in the network and detection of terminated services.

TCS Binary – Telephony Control –Binary
TCS Binary protocol is bit-oriented protocol to define control, establishment and transfer of voice and data packets between Bluetooth stations.

Audio
Bluetooth specification defines methods for audio transfers between Bluetooth devices. Audio data transfers don't utilize services of L2CAP layer. Audio data transfers are directly utilizing Baseband layer.

3.3 Bluetooth stack independent protocols

Telephony Control – AT Commands
Bluetooth standard supports set of modem AT command to assure data transfer over emulated serial port. This layer is utilizing RFCOMM services for serial port emulation.

PPP - Point-to-Point Protocol
PPP protocol is packet oriented protocol and therefore it is necessary to use methods for packet data stream conversion to serial data stream. RFCOMM services are utilized for PPP data connection.

UDP/TCP –IP Protocols
Set of these protocols enables interconnection with other devices connected to Internet. In this case, the functionality of the Bluetooth device can be described as an Internet Bridge. Set of TCP/IP/PPP protocols are utilized for all Internet bridge accesses in framework of Bluetooth 1.0 standard. In the future the OBEX protocol will be utilized.

WAP – Wireless Application Protocol
WAP protocol is wireless protocol for sharing of Internet services over different mobile communication devices. Bluetooth wireless standard can be used as wireless network for transfer of data from WAP master to the WAP client.

OBEX Protocol
OBEX protocol is an optional protocol of application layer to assure infrared communication. Protocol utilizes client-server architecture and it's independent of transfer API. OBEX protocol utilizes RFCOMM layer as a main transportation layer.

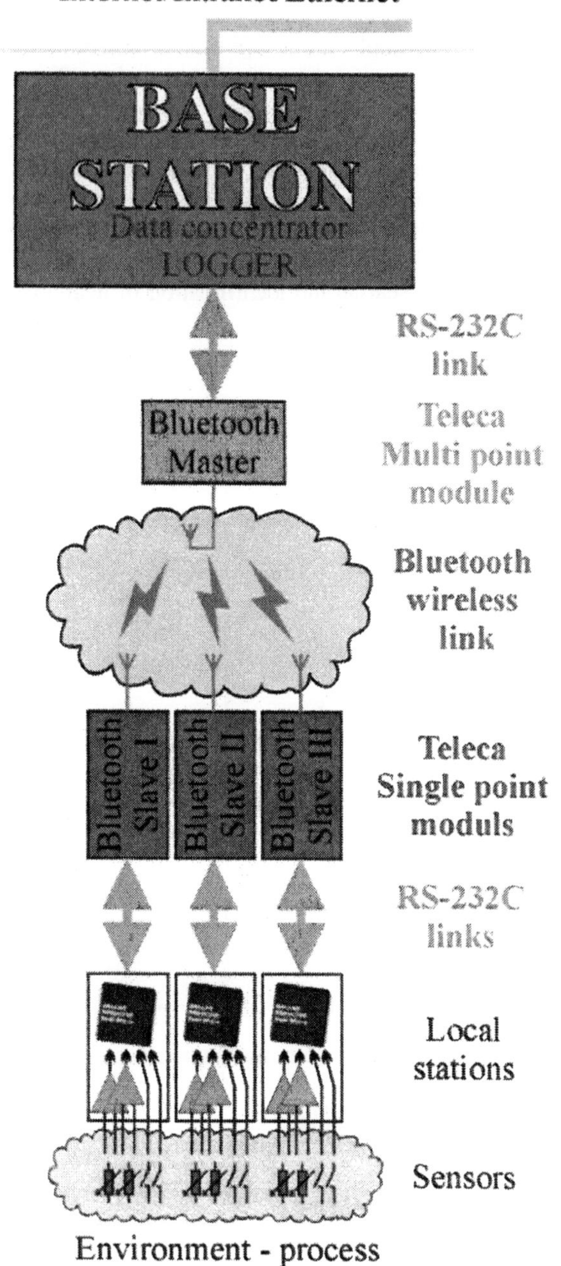

Fig. 2. Experimental Bluetooth application

4. EMBEDDED APPLICATION OF BLUETOOTH

The above-described experiments with wireless communication were foundation for creation of embedded application that would enable to acquire data from wirelessly connected sensors. Mentioned embedded structure is shown on Fig. 2.

A single data-logger gathers information from sensors with wireless interface. The data logger then makes these data available over Internet as it is equipped with Ethernet interface with support of TCP/IP protocols. The master station can be a PC, embedded PC or special data acquisition device (Data concentrator Sensvision™ with Net+Arm CPU). The master station is a piconet Master and it has to be equipped by Bluetooth module with Multipoint capability. The local stations are embedded devices based on MCS51 family microprocessors. The first prototypes are based on Atmel/Temic AT89C51CC01.

4.1 Construction of local sensoric station

Local station is based on microcontroller AT89C51CC01. This microcontroller is equipped with external data memory for communication buffer and data structures needed by Bluetooth communication libraries. The Comtec Teleca Bluetooth module is connected to the embedded system using RS-232 interface, the communication speed over the RS-232 is 57 600 kbps. Each local station is equipped by two analog inputs with 10-bit resolution. Such resolution is suitable enough as this local station serves for testing of the suitability of the selected approach.

5. CHALLENGES ASSOCIATED WITH DISTRIBUTED WIRELESS COMMUNICATION STRUCTURES

The wireless solutions due to their wireless nature bring new challenges, especially in the field of reliability and security. The wireless technologies are prone to jamming, which negatively influences the reliability, predictability and deterministicity. Moreover unauthorized reception/transmission of data is difficult to detect so the wireless technologies have to deal with the security issues much more than it is necessary with classical fieldbuses.

The reliability of wireless systems as for temporary or permanent failures can be analyzed and solved using formal methods, while the deterministicity of the wireless control systems can be assured using time triggered control approach.

5.1 Formal model/specification of hierarchical decentralised control system

Many control systems for industrial controls of manufacturing or production are built as distributed control system. There are used PLCs or other microprocessor based control systems, intelligent sensors and actors connected together by industrial serial fieldbuses or wireless communication buses. These systems are often connected to a high-level network together with workstations, operator stations, database servers etc. In near future a huge amount of new systems will be based on that scheme. These systems are going to be larger and more complex, but more difficult and sensitive to faults. Very important problem is to achieve better safety, functionality and reliability. One of subparts to achieve fault-tolerance of a system is to create a mathematical model of packet's routing a program module mapping in hierarchical decentralized control systems. The mathematical model has to be applicable for solving of optimization problem - routing problem and program module mapping. There are several criterions which are corresponding with a miscellaneous configurations and requests. The formal specification used for mathematical model's creation has to fill requests for easy and very fast solving of routing problem, because the routing problem is only one of many sub-problems like optimal program module mapping, resources administration, in-time tuning of system and etc., which are generating a huge and complex optimization problem. Definition of sufficient formal specification will consequently enable optimal routing and module mapping in the framework of hierarchical decentralized system. It will also enable to solve optimization tasks focused to balanced load of buses, time optimal transfers and other optimization functions based on different requirements and criteria.

5.2. Graphical model/specification of hierarchical decentralized control system

It is possible to accept hypotheses, that whole decentralized control system can be modeled by oriented weighted chart. Formal specification of hierarchical decentralized control system has to be extended to model of different transfer and functional capabilities of serial buses and of resource description of control system elements. This needs to carry new requirements on weighting and functional limitation of arcs of graph presentation. Formal specification of hierarchical decentralized control system allows using infinity value as weighting of arcs. There are problems with discontinuities at enumerating of optimization problems mainly with enumerating of criterial functionals. Based on above-mentioned hypotheses is possible to introduce global model, which consists of parts presenting serial communication nets, distributed peripherals and embedded peripherals.

5.3. Mathematical model/specification of hierarchical decentralised control system

Mathematical model of optimal routing problem as well as optimal module mapping problem are based on formal description of graphic chart presentation of hierarchical decentralized control system by set of linear equations with constrains. Let input to the mathematical model is given by set of two graphs, first graph let describes HW resources of system and second one let describes SW requirement of program modules. The weighted oriented graph modeling a decentralized control system can be transformed into a mathematical representation by LP (linear programming). Some optimization problems of decentralized control system require to use more complex methods like MILP (mixed integer linear programming) methods, etc.

5.4. Bluetooth vs Real-time

The suitability of Bluetooth technology for real-time automation purposes is questionable. At present time the valid Bluetooth specification is version 1.1. The present Bluetooth supports various transmission modes (synchronous or asynchronous; symmetric or asymmetric), however functionality that would provide predictable QoS (Quality of Services) is expected to be added to Bluetooth version 2.0. Also the communication delays with Bluetooth may be varying which makes the utilization of Bluetooth for hard-real time purposes not so straightforward. For applications that require deterministic performance the application has to utilize some advanced control techniques. For timely data acquisition it is wise to utilize clock synchronization methods to achieve unified time base among distributed data acquisition nodes.

5.5. Clock synchronization and time-triggered approach

Without clock synchronization it is not possible to guarantee ability to reconstruct state of the controlled system using data acquired from various distributed sensors using non-deterministic communication channel. With uniform time base it is possible to attach a time stamp to each acquired data. Such time stamp allows to relate data acquired from distributed sensors without a need for deterministic communication channel. Ability to reliably reconstruct state of the system is necessary for some control algorithms and it is vital for successful post-mortem analysis in case of failures. Moreover uniform time base is the first step to time-triggered control, which is very reliable control approach that is able to guarantee deterministic behavior of systems in applications that utilize non-deterministic communication channels.

Bluetooth technology allows to transfer TCP/IP protocol. Over the TCP/IP, there are defined clock synchronization protocols (e.g. NTP, SNTP) that allow to achieve very precise and accurate clock synchronization with a common time base. The time base can be synchronized either locally (limited difference among clock values indicated by distributed nodes at the same time), or the whole system can be synchronized to global time (limited difference between global time and clock value of any distributed node).

With clock synchronization and realistic requirements on the Bluetooth technology, even without QoS it is possible to successfully utilize Bluetooth for automation purposes, including real-time applications.
The QoS is expected to appear in future versions of Bluetooth standards.

6. CONCLUSION

Wireless automation systems utilizing Bluetooth technology can be quite successful in applications where wired fieldbuses do not represent optimal solution. Bluetooth and other PAN wireless technologies can be a good means for short-range wireless data transfer, however lack of automation profiles makes interoperability of Bluetooth automation devices difficult. Moreover reliability and security of each particular Bluetooth application has to be addressed.

7. ACKNOWLEDGEMENTS

The paper presents research and development that is supported by Ministry of Trade and Industry (FD-K/104), Ministry of Education (FRVŠ G1/2224) and Grant agency of the Czech Republic (GAČR 102/03/1097). Without kind support of above-mentioned agencies the research and development would not be possible.

REFERENCES

FIEDLER P., CACH P. Internet – technologies that are missing, 1st Intl Workshop on real-Time LANs in the Internet Age in frame of the 14th Euromicro Intl Conference on Real-Time Systems, Vienna, Austria, June 18, 2002

BRADÁČ, Z., BRADÁČ, F., ZEZULKA, F. Short range radio data transfer for process automation In Contribution of IFAC PDS2003. Programmable Devices and Systems PDS2003. Ostrava: VSB Ostrava, 2003, s. 37 – 40

BRADÁČ, Z., BRADÁČ, F. Packet routing problem – LP problem In International Conference Cybernetics and Informatics. Cybernetics and Informatics. Trebisov: STU Bratislava, 2002, s. 95 - 96, ISBN 80-227-1749-5

BLUETOOTH SIG 2001. Specification of the Bluetooth System, Volume 1 and Volume 2, Core, Version 1.1. Bluetooth SIG, February 22 2001, http://www.bluetooth.com

www.elsevier.com/locate/ifac

SAMPLED TIME FLOW CONTROL FOR WIDE AREA CONNECTION ORIENTED COMMUNICATION NETWORKS

Andrzej Bartoszewicz

Institute of Automatic Control, Technical University of Łódź,
18/22 Stefanowskiego St., 90-924 Łódź, Poland,
andpbart@ck-sg.p.lodz.pl.

Abstract: This paper presents a new, sampled time flow control algorithm for fast connection oriented communication networks. The algorithm employs a simple, proportional controller together with Smith predictor applied for delay compensation. It has been proved that an appropriate selection of the bottleneck buffer capacity assures no cell loss and the maximum possible throughput of the controlled virtual circuit. Moreover, transmission rates generated by the algorithm are always non-negative and bounded by the maximum bandwidth available for the controlled connection. These properties of the proposed algorithm enable its direct implementation in the network environment. The steady state queue length of the bottleneck node buffer is estimated.

Keywords: Sampled time flow control, high-speed communication networks, time delay systems, Smith principle.

1. INTRODUCTION

Flow control in fast communication networks has recently been studied by several researchers (Izmailov 1995; Mascolo 1997; Chong et al. 1998; Kulkarni and Li 1998; Mascolo 1999; Lengliz and Kamoun 2000; Mascolo 2000; Imer et al. 2001; Gómez-Stern et al. 2002; Grieco and Mascolo 2002; Jagannathan and Talluri 2002; Laberteaux et al. 2002; Niculescu 2002; Quet et al. 2002). The main purpose of the proposed control schemes is to assure low packet loss factor, high throughput and the quality of service required by the particular application. The difficulty of the task is mainly caused by the significant propagation delays involved not only in the process of data transmission, but also in the feedback information transfer. If congestion occurs at a specific node, information about this condition must be conveyed to the source. Transmitting this information involves feedback propagation delay. After this information has been received by the source, it can be used to adjust the flow rate. However, the adjusted flow rate will begin to affect the congested node only after forward propagation delay. The sum of the two delays, usually called the round trip time (RTT), does not depend on the congested node location in the connection oriented network and plays an important role in the design of control protocols for the wide area networks.

Early flow control strategies for fast communication networks primarily used simple, relay type controllers. However, direct implementation of an on-off controller in the time delay system such as a communication network may cause significant undesirable oscillations of the congested node buffer occupancy level. Therefore, more sophisticated control schemes were introduced. Izmailov (1995) proposed a linear regulator whose output signal is generated according to the several states of the buffer measured at different time instants. He considered a single connection and proved asymptotic stability, nonoscillatory system behaviour and locally optimal rate of convergence. Chong et. al. (1998) proposed and thoroughly studied the performance of a simple queue length based flow control algorithm with dynamic queue threshold adjustment. Lengliz and Kamoun (2000) introduced a proportional plus derivative controller which is computationally efficient and can be easily implemented in asynchronous transfer mode networks. Imer et. al. (2001) gave a brief, excellent tutorial exposition of the flow control problem and presented new stochastic and deterministic control algorithms. Another interesting approach to the problem of flow rate control in communication networks has been proposed by Quet et al. In the recent paper (Quet et al. 2002) the authors considered a single bottleneck multi-source network and applied minimisation of an H-infinity norm to the design of a flow rate controller. The proposed controller guarantees stability robustness to uncertain and time-varying propagation delays in various channels. Adaptive control strategies for the available bit rate service flow regulation have been proposed by Laberteaux and Rohrs (2002). Their strategies reduce convergence time and improve queue length management. Also a neural network controller for communication networks has recently been proposed. Jagannathan and Talluri (2002) showed that their neural network controller can guarantee stability of the closed loop system and the desired quality of service.

As it has already been mentioned the significant propagation delays are critical for the closed loop performance of the wide area networks. Therefore, several researchers applied the Smith principle to solve the problem of flow control in high speed communication networks (Mascolo 1997; Mascolo 1999; Gómez-Stern et al. 2002; Grieco and Mascolo 2002). In the seminal paper Mascolo (1999) considered the single connection congestion control problem in a general packet switching network. He used the deterministic fluid model approximation of packet flow and applied transfer functions to describe the network dynamics. The designed controller was applied to the available bit rate (ABR) traffic control in asynchronous transfer mode (ATM) networks and compared with ERICA standard. Furthermore, Mascolo showed that Transmission Control Protocol / Internet Protocol (TCP/IP) implements a Smith predictor to control network congestion. In the next paper (Mascolo 2000) the same author applied the Smith principle to the network supporting multiple ABR connections with different propagation delays. The proposed control algorithm guarantees no cell loss, full and fair network utilisation, and ensures exponential convergence of queue levels to stationary values without oscillations or overshoots. Gomez-Stern et. al. (2002) further studied the ABR flow control using Smith principle. They proposed a proportional-integral (PI) controller which helps to reduce the average queue level and its sensitivity to the available bandwidth variations. Saturation issues in the system were handled using anti-wind up techniques. However, in all the papers mentioned above (Mascolo 1997; Mascolo 1999; Mascolo 2000; Gómez-Stern et al. 2002; Grieco and Mascolo 2002), only the continuous time control schemes were proposed. It was assumed that the feedback information about the state of the network was always available, and that the source rate could be modified at any time instant. Unfortunately, this is not the case in a typical communication network, and therefore a more precise analysis and design of the control system can help to obtain an improved performance and better resource utilisation.

In this paper the problem of flow control in wide area communication networks is considered. Similarly to the approach introduced in the papers (Mascolo 1997; Mascolo 1999; Mascolo 2000; Gómez-Stern et al. 2002; Grieco and Mascolo 2002) the Smith principle is applied, however as opposed to those papers here the fact that the source rate can only be changed at discrete time instants is directly taken into account. In other words, the problem of sampling in the considered system is explicitly addressed. In this paper an algorithm which combines Smith principle with the discrete time proportional controller is proposed. It is proved that the algorithm generates always nonnegative and bounded flow rate and ensures no cell loss in the controlled virtual circuit. Furthermore, a necessary and sufficient condition for full resource utilisation is derived.

The remainder of this paper is organised as follows. The model of the network used throughout the paper is introduced in Section 2. Then the main result, i.e. the new sampled time control system is presented and discussed in Section 3. In this section important properties of the system are proved. Finally, Section 4 comprises conclusions of the paper.

2. NETWORK MODEL

In this paper the problem of flow control in a single virtual circuit of a wide area communication network is considered. The circuit consists of a single source, a number of nodes connected by bi-directional links and a single destination. It is assumed that there is a single bottleneck node in the circuit. In other words only one node has more cells to send than the link connecting it to the next node can actually carry. The rate of cell outflow from the bottleneck node depends on the available bandwidth which will be modelled as an a'priori unknown function of time. The source sends resource management cells at regularly spaced time intervals. These cells are processed by the nodes on the priority basis, i.e. they are not queued but sent to the next node without delay. They carry information about the network condition. After reaching the destination they are immediately sent back to the source, along the same path they arrived. The information carried by the resource management cells is used by the controller to adjust the source rate.

Further in this paper data transfer in modelled as an incompressible fluid flow. The following notation is used: t denotes time, T represents the control period, and RTT stands for the round trip time which is equal to the sum of forward and backward propagation delays denoted as T_f and T_b respectively.

$$RTT = T_f + T_b \tag{1}$$

Furthermore, $x(t)$ denotes the bottleneck node queue length, x_d the demand value of $x(t)$ and $d(t)$ represents the bandwidth available for the controlled connection at time t. After setting up the connection, the buffer of the bottleneck node is empty, so for any time $t < T_f$ the queue length

$$x(t) = 0 \tag{2}$$

On the other hand for $t \geq T_f$

$$x(t) = \int_{T_f}^{t} \left[a\left(\tau - T_f\right) - h(\tau)\right] d\tau \tag{3}$$

where $a(t)$ is the source rate, and $h(t)$ represents the bandwidth which is actually consumed by the bottleneck node at time t. If the queue length at this node $x(t)$ is greater than zero, then the entire available bandwidth is consumed $h(t) = d(t)$. Otherwise, i.e. when $x(t) = 0$, then $h(t)$ is determined by the rate of data arrival at the node. In this case, the available bandwidth may not be fully utilised. Consequently, for any time t

$$0 \leq h(t) \leq d(t) \leq d_{max} \tag{4}$$

Since in this paper a sampled time control system is designed, the source rate $a(t)$ can be changed only at regularly spaced time instants $t = n \cdot T$, where n is a nonnegative integer. Consequently,

$$\forall t \in \left(n \cdot T, (n+1)T\right),\ a(t) = a(n \cdot T) = \text{const.} \tag{5}$$

Further, it is assumed that the round trip time is a multiple of the control period, i.e.

$$\exists\, n_R \in \{2, 3, \ldots\}, \text{ such that } RTT = n_R \cdot T \tag{6}$$

This assumption can be easily satisfied, since backward delay T_b can always be appropriately augmented by the source controller with an extra delay. This delay, even in the worst case, is shorter than the control period T.

The source rate controller should assure that the bottleneck buffer overflow will be avoided and the node will always have enough data to send. The first condition implies that data cells are not lost and there is no need for their retransmission, while the latter one assures full network utilisation which is highly desirable for economic reasons. In the next section it will be shown that using sampled time controller these favourable properties of the considered connection can be achieved.

3. CONTROL STRATEGY

In this section the following control algorithm is proposed:

(i) The source rate during the time interval from zero up to RTT is chosen in such a way that $a(t)$ is nonnegative and

$$\int_{t=0}^{RTT} a(\tau) d\tau = x_d \tag{7}$$

(ii) For any time $t \geq RTT$ the source rate is determined as follows

$$a(t) = a(nT) = \\ = K\left[x_d - x\left(nT - T_b\right) - \int_{nT-RTT}^{nT} a(\tau) d\tau\right] \tag{8}$$

where n is the biggest integer such that $nT \leq t$ and K represents the constant gain of the proportional controller.

Notice that the proposed algorithm leaves some degree of freedom when selecting the source rate at the beginning of the control action. It requires only that the total number of cells sent from the initial time $t_0 = 0$ to the time instant RTT equals x_d. Consequently, the source rate during this interval could for example be chosen as

$$a(t) = \frac{x_d}{RTT} = \text{const.} \tag{9}$$

In the sequel two important theorems presenting basic properties of the proposed control strategy are introduced. For that purpose the queue length $x(t)$ is first estimated.

Let δ be a real number such that

$$0 < \delta \leq T \tag{10}$$

Then for any time $t \geq RTT + T_f$ the queue length can be expressed as

$$x(t) = x\left(nT + T_f + \delta\right) = \\ = x\left(nT + T_f\right) + \delta \cdot a(nT) - \int_{nT+T_f}^{nT+T_f+\delta} h(\tau) d\tau =$$

$$= x\left(nT + T_f\right) +$$

$$\delta \cdot K\left[x_d - x\left(nT - T_b\right) - \int_{nT-RTT}^{nT} a(\tau)d\tau\right]$$

$$- \int_{nT+T_f}^{nT+T_f+\delta} h(\tau)d\tau =$$

$$= x\left(nT + T_f\right) + \delta \cdot K\left[x_d - \int_0^{nT-RTT} a(\tau)d\tau +\right.$$

$$\left.+ \int_0^{nT-T_b} h(\tau)d\tau - \int_{nT-RTT}^{nT} a(\tau)d\tau\right] - \int_{nT+T_f}^{nT+T_f+\delta} h(\tau)d\tau =$$

$$= x\left(nT + T_f\right) + \delta \cdot K\left[x_d - \int_0^{nT} a(\tau)d\tau +\right.$$

$$\left.+ \int_0^{nT+T_f} h(\tau)d\tau - \int_{nT-T_b}^{nT+T_f} h(\tau)d\tau\right] - \int_{nT+T_f}^{nT+T_f+\delta} h(\tau)d\tau = \tag{11}$$

$$= x\left(nT + T_f\right) +$$

$$+ \delta \cdot K\left[x_d - x\left(nT + T_f\right) - \int_{nT-T_b}^{nT+T_f} h(\tau)d\tau\right] +$$

$$- \int_{nT+T_f}^{nT+T_f+\delta} h(\tau)d\tau$$

Theorem 1
If relations (4) – (8) hold and the controller gain K satisfies the following inequalities

$$0 < K \le T^{-1} \tag{12}$$

then the queue length is always upper bounded by its demand value

$$x(t) \le x_d \tag{13}$$

Proof
Since for $t \le RTT$, $a(t)$ is a nonnegative function of time, it follows directly from equation (7) that for any time $t \le RTT$ relation (13) is satisfied. On the other hand, the product $\delta \cdot K \le 1$. Consequently, using equation (11), for any time $t \ge RTT$ one gets

$$x(t) = x\left(nT + T_f + \delta\right) =$$

$$x\left(nT + T_f\right) + \delta \cdot K\left[x_d - x\left(nT + T_f\right) - \int_{nT-T_b}^{nT+T_f} h(\tau)d\tau\right] +$$

$$- \int_{nT+T_f}^{nT+T_f+\delta} h(\tau)d\tau \le$$

$$\le x\left(nT + T_f\right) + \delta \cdot K\left[x_d - x\left(nT + T_f\right)\right] \le x_d \tag{14}$$

This conclusion ends the proof.

This theorem shows that selecting x_d smaller than or equal to the buffer capacity, one can assure no cell loss in the controlled virtual circuit. This is particularly important for loss sensitive traffic like Web pages or file transfer. Another desired property of a suitably designed flow control system is full link utilisation. The next theorem shows that the proper choice of the buffer capacity actually ensures this property.

Theorem 2
If relations (4) – (8) are satisfied, and the following inequality

$$x_d > (RTT + 1/K) \cdot d_{max} \tag{15}$$

holds, then for any time $t \ge RTT + T_f$ the queue length is greater than zero.

Proof
Let us first calculate the queue length at time $t = RTT + T_f$

$$x\left(RTT + T_f\right) = \int_{T_f}^{RTT+T_f} \left[a\left(\tau - T_f\right) - h(\tau)\right]d\tau =$$

$$= \int_0^{RTT} a(\tau)d\tau - \int_{T_f}^{RTT+T_f} h(\tau)d\tau =$$

$$= x_d - \int_{T_f}^{RTT+T_f} h(\tau)d\tau \ge \tag{16}$$

$$\ge x_d - RTT \cdot d_{max} > K^{-1} \cdot d_{max} > 0$$

It can be clearly seen from relations (16) that the queue length at time $t = RTT + T_f$, $x(RTT + T_f)$ is greater than zero. On the other hand, at any time $t \ge RTT + T_f$ the queue length satisfies the following relations

$$x(t) = x\left(nT + T_f + \delta\right) \ge$$

$$\ge x\left(nT + T_f\right) +$$

$$+ \delta \cdot K\left[x_d - x\left(nT + T_f\right) - RTT \cdot d_{max}\right] + \tag{17}$$

$$- \delta \cdot d_{max} =$$

$$= x\left(nT + T_f\right) +$$

$$+ \delta \cdot K\left[x_d - x\left(nT + T_f\right) - RTT \cdot d_{max} - \frac{d_{max}}{K}\right] >$$

$$> x\left(nT + T_f\right) \cdot \left(1 - \delta \cdot K\right) \geq 0$$

Relation (17) shows that the queue length is always greater than zero. This conclusion ends the proof.

Theorem 2 shows that using the strategy proposed in this paper one can always assure full link utilisation, provided the bottleneck node buffer capacity is greater than $(RTT + 1/K) \cdot d_{max}$. Taking into account condition (12) it can be easily concluded that the minimum buffer capacity guaranteeing full resource utilisation equals $(RTT + T) \cdot d_{max}$, i.e. it is equal to the product of the controlled connection maximum bandwidth and the sum of the transmission delay and the control period.

Any control strategy can actually be implemented in the network environment, only if it generates the flow rate which is always non-negative and bounded. These properties of the strategy proposed in the paper will now be formulated as two theorems and formally proved.

Theorem 3
If relations (4) – (8) and (12) hold then the flow rate $a(t)$ is always non-negative.

Proof
For any time $t \leq RTT$, the flow rate $a(t)$ is nonnegative by assumption. On the other hand, in order to prove this property for $t > RTT$ it is essential to show that

$$\alpha(nT) = x\left(nT - T_b\right) + \int_{nT-RTT}^{nT} a(\tau)d\tau \tag{18}$$

is always smaller than or equal to the demand value of the queue length x_d. For that purpose one can calculate

$$\alpha(nT) = x\left(nT - T_b\right) + \int_{nT-RTT}^{nT} a(\tau)d\tau =$$

$$= \int_{0}^{nT-RTT} a(\tau)d\tau + \int_{nT-RTT}^{nT} a(\tau)d\tau - \int_{0}^{nT-T_b} h(\tau)d\tau =$$

$$= \int_{0}^{(n-1)T-RTT} a(\tau)d\tau + \int_{(n-1)T-RTT}^{nT-RTT} a(\tau)d\tau + \int_{nT-RTT}^{nT} a(\tau)d\tau +$$

$$- \int_{0}^{(n-1)T-T_b} h(\tau)d\tau - \int_{(n-1)T-T_b}^{nT-T_b} h(\tau)d\tau =$$

$$= x\left[(n-1)T - T_b\right] +$$

$$\int_{(n-1)T-RTT}^{(n-1)T} a(\tau)d\tau + \int_{(n-1)T}^{nT} a(\tau)d\tau - \int_{(n-1)T-T_b}^{nT-T_b} h(\tau)d\tau =$$

$$= x\left[(n-1)T - T_b\right] + \int_{(n-1)T-RTT}^{(n-1)T} a(\tau)d\tau + \tag{19}$$

$$+ KT\left[x_d - x\left[(n-1)T - T_b\right] - \int_{(n-1)T-RTT}^{(n-1)T} a(\tau)d\tau\right] +$$

$$- \int_{(n-1)T-T_b}^{nT-T_b} h(\tau)d\tau$$

and consequently

$$\alpha(nT) \leq x\left[(n-1)T - T_b\right] + \int_{(n-1)T-RTT}^{(n-1)T} a(\tau)d\tau +$$

$$+ KT\left[x_d - x\left[(n-1)T - T_b\right] - \int_{(n-1)T-RTT}^{(n-1)T} a(\tau)d\tau\right] = \tag{20}$$

$$= \alpha\left[(n-1)T\right] + KT\left\{x_d - \alpha\left[(n-1)T\right]\right\}$$

Since it follows from equations (3) and (7) that

$$\alpha\left(n_R \cdot T = RTT\right) =$$

$$= x\left(RTT - T_b\right) + \int_{0}^{RTT} a(\tau)d\tau = x_d \tag{21}$$

taking into account inequality (20), one concludes that for any time nT

$$\alpha(nT) \leq x_d \tag{22}$$

and consequently for any time t the flow rate $a(t)$ is nonnegative. This statement ends the proof of the theorem.

Theorem 3 shows that the strategy proposed in this paper can be directly implemented in the network environment.

In the sequel it will be proved that the flow rate is not only nonnegative but also upper bounded by the maximum available bandwidth d_{max}.

Theorem 4
If relations (4) – (8), (12) and (15) hold then for any time $t > RTT$ the flow rate $a(t)$ is always bounded by the maximum bandwidth available for the controlled connection, i.e.

$$a(t) \le d_{max} \quad (23)$$

Proof
First recall equation (21) which states that $\alpha(n_R \cdot T = RTT) = x_d$. Furthermore, let

$$\gamma = x_d - (RTT + 1/K) \cdot d_{max} \quad (24)$$

Then from equation (19) for any time $t > RTT$

$$\begin{aligned} \alpha(nT) &= \alpha[(n-1)T] - KT\alpha[(n-1)T] + \\ &+ KTx_d - \int_{(n-1)T-T_b}^{nT-T_b} h(\tau)d\tau \ge \\ &\ge \alpha[(n-1)T] - KT\alpha[(n-1)T] + \\ &+ KT\left(RTTd_{max} + \frac{1}{K}d_{max} + \gamma\right) - Td_{max} = \\ &= \alpha[(n-1)T] - KT\{\alpha[(n-1)T] - RTTd_{max} - \gamma\} \end{aligned} \quad (25)$$

This relation together with equation (21) imply that after the time instant $t = RTT$ the variable $\alpha(n \cdot T)$ will always be greater than or equal to $\gamma + RTT \cdot d_{max}$. Consequently, the flow rate

$$\begin{aligned} a(t) &= a(nT) = \\ &= K\left[RTTd_{max} + \frac{1}{K}d_{max} + \gamma - \alpha(nT)\right] \le d_{max} \end{aligned} \quad (26)$$

will never exceed the maximum bandwidth available for the controlled connection. This conclusion ends the proof.

Finally notice that the steady state value of the queue length, when the strategy proposed in this paper is applied, can be expressed as

$$x_{ss} = x_d - (RTT + 1/K) \cdot d_{ss} > 0 \quad (27)$$

It can be seen from relation (27) that the steady state value of the queue length x_{ss} is always greater than or equal to $x_d - (RTT + T) \cdot d_{ss}$.

4. CONCLUSIONS

In this paper a new sampled time flow control strategy for wide area high speed communication networks has been proposed. The strategy combines Smith predictor with the proportional discrete time controller. It has been proved that the proposed solution assures full link utilisation and no cell loss in the controlled connection. This favourable performance is achieved without using excessive flow rate. The flow rate is always nonnegative and bounded by the maximum bandwidth available for the controlled connection.

REFERENCES

Chong S., Nagarajan R., Wang Y. T. (1998). First-order rate-based flow control with dynamic queue threshold for high-speed wide-area ATM networks. *Computer Networks and ISDN Systems*, **29**, pp. 2201-2212.

Gómez-Stern F., Fornés J. M., Rubio F. R. (2002). Dead-time compensation for ABR traffic control over ATM networks. *Control Engineering Practice*, **10**, pp. 481-491.

Grieco L., Mascolo S. (2002). Smith's predictor and feedforward disturbance compensation for ATM congestion control. *Proc. of the 41st IEEE CDC*, Las Vegas, pp. 987-992.

Imer O. C., Compans S., Basar T., Srikant R. (2001). Available bit rate congestion control in ATM networks. *IEEE Control Systems Magazine*, pp. 38-56.

Izmailov R. (1995). Adaptive feedback control algorithms for large data transfers in high-speed networks. *IEEE Transactions on Automatic Control*, **40**, pp. 1469-1471.

Jagannathan S., Talluri J. (2002). Predictive congestion control of ATM networks: multiple sources/single buffer scenario. *Automatica* **38**, pp. 815-820.

Jain R. (1996). Congestion control and traffic management in ATM networks: recent advances and a survey. *Computer Networks and ISDN Systems*, **28**, pp. 1723-1738.

Kulkarni L. A., Li S. (1998). Performance analysis of a rate-based feedback control scheme. *IEEE/ACM Transaction on Networking*, **6**, pp. 797-810.

Laberteaux K. P., Rohrs Ch. E., Antsaklis P. J. (2002). A Practical Controller for Explicit rate Congestion Control. *IEEE Transactions on Automatic Control*, **47**, pp. 960-978.

Lengliz I., Kamoun F. (2000). A rate-based flow control method for ABR service in ATM networks. *Computer Networks*, **34**, pp. 129-138.

Mascolo S. (1997). Smith's principle for congestion control in high speed ATM networks. *Proceedings of the 36th IEEE Conference on Decision and Control*, San Diego, pp. 4595-4600.

Mascolo S. (1999). Congestion control in high-speed communication networks using the Smith principle. *Automatica*, **35**, pp. 1921-1935.

Mascolo S. (2000). Smith's principle for congestion control in high-speed data networks. *IEEE Transactions on Automatic Control*, **45**, pp. 358-364.

Niculescu S. I. (2000). On delay robustness analysis of a simple control algorithm in high-speed networks. *Automatica*, **38**, 885-889.

Quet P. F., Ataslar B., Iftar A., Özbay H., Kalyanaraman S., Kang T. (2002). Rate-based flow controllers for communication networks in the presence of uncertain time-varying multiple time-delays. *Automatica*, pp. 917-928.

www.elsevier.com/locate/ifac

BRINGING ETHERNET TO THE SENSOR LEVEL

Petr Cach, Petr Fiedler, Zdeněk Bradáč, Zdeněk Rejda, Miloš Prokop

Brno University of Technology,
Brno, Czech Republic
cach@feec.vutbr.cz,
fiedler@feec.vutbr.cz,
bradac@feec.vutbr.cz,
xrejda00@stud.feec.vutbr.cz,
xproko07@stud.feec.vutbr.cz

Abstract: This paper deals with the utilisation of Ethernet based system directly on the sensor bus level. It describes several problems encountered when designing a sensor with embedded Ethernet interface. A realisation of such sensor is described. *Copyright © 2003 IFAC*

Keywords: Ethernet, communication networks, communication protocols, data acquisition

1. INTRODUCTION

The original concept of the Ethernet was developed during seventies by consortium of three companies – Xerox, Intel and Digital. Originally the Ethernet was intended to be used as communication technology in local networks, interconnecting personal computers, servers, printers and other devices. The concept of the Ethernet proved to be very successful and after several improvements (such as increasing it's speed, changing physical medium, etc.) it is the most used technology in local networks.

Few years ago the Ethernet started it's move into a new niche. The already wide base of Ethernet systems installed on the office level led naturally to extension into the manufacturing level. Ethernet as the physical and data link layer makes it possible to use a variety of TCP/IP protocols. Many components available on the market, low price and broad experience with LANs made good start for industrial Ethernet. Despite many initial objections (it is non-deterministic, unreliable, etc.) the industrial Ethernet is now generally accepted and used as an alternative of the traditional industrial network systems. The currently available industrial Ethernet systems are generally intended to connect the factory-floor level and enterprise level, or to interconnect intelligent devices (such as PLC, remote I/Os, iPC, ...) on the factory-floor level.

The only area not addressed by the industrial Ethernet systems is the sensor bus level. In this area the traditional fieldbus systems are dominating (e.g. DeviceNet, AS-i). The integration with a higher-level system is matter of gateways.

As far as possible, all devices and subsystems in the plant shall communicate using the same protocol in order to eliminate gateways. Linking from one protocol to another always results in poorer integration than if a single protocol is used because differences in semantics require mapping of data. So a unified communication protocol in the plant is still preferable.

2. ETHERNET IN A SENSOR

To avoid using of such gateways a sensor with embedded Ethernet interface can be used. To build such sensor, three main issues have to be addressed:

- the powering of such sensor
- an Ethernet connector with suitable protection against water and dust
- the size of components

Let's start with the power supply. Unfortunately the Ethernet interface has higher power consumption compared to other fieldbuses. Moreover, it is unfeasible to install a separate power supply with each sensor, not mentioning the need for additional power connector.

Some of the traditional sensor buses offer the possibility to power the connected device directly using the communication cable, using either signal lines (AS-i) or separate wires (DeviceNET). Similar solution is also available for Ethernet. The standard is being developed by IEEE consortium as the 802.3af [1]. This initiative was originally driven by the need of VoIP (Voice over IP), but can be used to power any device with power consumption less than 12.5W.

The power is inserted either in the unused pairs of the CAT5 Ethernet cable (Midspan solution) or directly to the signal lines. Part of the 802.3af standard allows the Power Sourcing Equipment (PSE) to automatically recognize if the connected device is 802.3af compatible. If not, the PSE will not put power on the cable so incompatible devices won't be damaged. The standard 802.3af also provides SNMP management of the PSE (often called an Injector) and can alert a network management system if a device is in trouble.

Because Ethernet spans to areas, which were occupied by traditional fieldbus systems it is necessary to define suitable connection technology satisfying the protection standards (IP67) of these areas. The RJ45 connector is de facto standard in an office environment. It is also suitable for applications in an IP20 environment, which requires mounting and cabling of the equipment in a protected environment. For harsh environments however the higher protection class is required. There are basically two alternative solutions to this problem: a modified RJ45 connector as well as a M12 (micro-style) connector.

There exist at least three RJ45 style IP67 solutions. Each of them is developed and supported by different group or company ([2], [3], [4]). As a result there are three solutions incompatible with each other.

The M12 connector has been used for long time industrial applications. This connector is a standard not only for sensor connections but also for the various fieldbus systems. The essential feature of this connector is the very low failure rate and hence the high reliability at the manufacturing plant. Another very important asset is the smaller footprint of the M12 connector compared to a sealed RJ45 solution. The M12 style connector [5] for industrial Ethernet application is proposed by ODVA (Open Device Vendor Association) for Ethernet/IP protocol and accepted by IAONA (Industrial Automation Open Networking Alliance). This type of connector is more suitable for Ethernet sensor than the RJ45 solutions.

The hardware of the sensor with Ethernet interface has to include three main parts: a processor with sufficient performance to handle high data rates of Ethernet and TCP/IP communication, adequate amount of memory – both RAM for data and ROM for program and an Ethernet Controller. In the worst case this means a system with four chips and complex printed circuit board and of course bigger size of the whole system. In a sensor the available space for the electronic components is very limited, so the size of the system is crucial. To fit the electronics into the limited space it is necessary to use some solution, which integrates all the components on one chip. Such integrated solution means lower price, lower complexity and higher reliability.

3. REALISATION OF A UNIVERSAL MODULE

In order to be able to quickly develop a range of sensors with embedded Ethernet, a universal sensor module has been developed at the authors department. This module was developed in co-operation with BD Sensors company.

The module is based on microprocessor IP2022 [8]. This microprocessor is very fast RISC processor (up to 120 MHz), which has embedded 64kB Flash and 20kB RAM memory. Although it doesn't have integrated hardware Ethernet controller on the chip, it has some special peripherials, which in combination with the high speed of the processor enables to process 10Mbit Ethernet protocol in software. These features make it the best choice for this highly integrated purpose. There are other systems on the market, which integrate some or the all parts in one chip (e.g. Dallas 80C400, IPC@CHIP, NET+ARM), but none of them fulfils all the needs.

Although the IP2002 integrates 10-bit A/D converter on the chip, the sensor module uses external 16/24-bit A/D converter AD7714. The converter features five pseudo-differential (or three fully differential) input channels, programmable gain for each input channel and low pass filter.

To extend the memory space of the module a serial flash memory was added. It has the size of 512kB and can be used to either store measured data, content of a web server or program code for the processor.

The module is 802.3af compatible and is powered over the Ethernet cable.

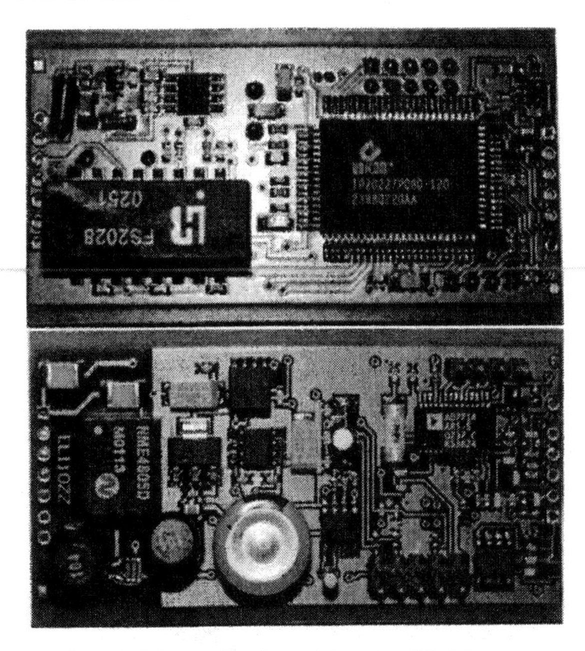

Fig. 1. The realised module, size 67x35mm

4. SOFTWARE OF THE MODULE

One of the benefits of the Ethernet and TCP/IP protocols is the possibility to use several application protocols above the TCP/IP. There is wide range of available solutions. It is possible to use some standard Internet protocol – http, ftp, smtp – or some special industrial application protocol – Modbus/TCP, Ethernet/IP, ProfiNet, OPC-DX, etc. The situation with industrial application protocols is similar to the situation in the area of traditional fieldbuses. There are several of them proposed by different organisations or manufactures. Many of them are proprietary or semi-open, usually derived from traditional fieldbuses. There is not yet one generally accepted standard – war of fieldbuses continues on the new level.

In the module, there are used two application protocols. The first one is the http protocol – the module contains embedded web server. The second protocol is an industrial application protocol Modbus/TCP. The Modbus was used as it is completely free to use without any licence fee and is widely used and supported by OPC and DDE.

Because the module has available flash memory, the software of the module provides option to store measured data in the flash memory. The database system is configurable over the web server. A user can define two databases with different properties – size, sampling time, etc. A so-called spare function is available – the user specifies a condition and when the condition is met the system will store only some of the measured data (it can spare the space in the memory). The web server can be also used to display the stored data, either in the form of a table or graph (uses Java applet). The web server can be also easily used to provide data as a xml file, which can be very easily processed by standard database tools.

5. CONCLUSION

Based on the previously described module a pressure sensor was realised. This sensor can be used either standalone, or integrated in some higher control system.

The disadvantage of the described solution is its limitation to only 10 Mb Ethernet version. The 100 Mb version would require Ethernet controller implemented in hardware, no current microprocessor can handle it in software. However the availability of similar system with embedded 100 Mb Ethernet can be only matter of time.

Currently only http and Modbus/TCP protocols are available for the sensor. However the system offers potentiality to further development. There are two industrial protocols being developed, which can aspire on the general industrial application protocol – a protocol developed by OPC Foundation based on the xml services – OPC-XML and protocol developed by IDA (Interface for Distributed Automation).

6. ACKNOWLEDGEMENTS

The paper presents research and development that is supported by Ministry of Trade and Industry (FD-K/104), Ministry of Education (FRVŠ G1/2224) and Grant agency of the Czech Republic (GAČR 102/03/1097). Without kind support of above-mentioned agencies the research and development would not be possible.

REFERENCES

IEEE P802.3af, DTE Power via MDI Task Force, http://grouper.ieee.org/groups/802/3/af/.

YAMAICHI ELECTRONICS, www.yamaichi.de/Y-con/

SIEMON, www.siemon.com

LUMBERG, www.lumbergusa.com

OPEN DEVICE VENDOR ASSOCIATION, www.odva.com

INDUSTRIAL AUTOMATION OPEN NETWORKING ALLIANCE, www.iaona.com

UBICOM, www.ubicom.com

IDA, www.ida-foundation.com

CACH P., FIEDLER P., ZEZULKA F., Sensor/Actuator web oriented interface, Proceedings of PDS2001 IFAC Workshop, Gliwice, Poland, 2001

FIEDLER P., CACH P. Internet – technologies that are missing, 1st Intl Workshop on real-Time LANs in the Internet Age in frame of the 14th Euromicro Intl Conference on Real-Time Systems, Vienna, Austria, June 18, 2002

www.elsevier.com/locate/ifac

POLICY ANALYSIS IN ECONOMICS, AGENT-BASED TRANSACTIONS

Jaroslav Zajac

Department of Economics and Management, FEI, Slovak University of Technology, Ilkovicova 3, 81209 Bratislava, jzajac@elf.stuba.sk

Abstract: The market and hierarchies framework is built on next categories: opportunism, bounded rationality and asset specificity, and market failures under given conditions, can be reduced by means of hierarchical agent-organizations. The benefits of hierarchy flow from the fact that it attenuates opportunism, attenuates the problem stemming from bounded rationality, and reduces bargaining cost deriving by assets specificity, is to explain why fee agents have chosen to renounce part of their freedom of action.

Keywords: Efficient parameters, Agent relations, Asymmetric information.

1 INTRODUCTION

The institutional comparison can be interpreted as the evolution of agents, if it were then possible to claim that heterogeneous endowments can cause hierarchical relations, there would remain the need agent action. The problem of economic power, such an ability can by influencing the subjective elements of the voluntary choice of agents, the system of objective constraints that defines the agent decision-making sets. The analysis of the mechanism through which social interaction modifies constraints on agents results leads then to discuss economic power. Economy can be constructed that are populated with agents who determine their interactions with other agents and with their environment on the basis of internalized social norms, internal behavioral rules, and data acquired on the basis of experience (Tesfatsion, 2001). Agents continually adapt their behavior in response self-organization. These natural selection pressures induce agents to engage in continual experimentation with new rules of behavior, all subsequent events in these worlds can be initiated and driven by agent-agent and agent-environment interactions without further outside intervention. Each agent's participation decision thus induces widespread externalities on all other agents, and by choosing whether to participate, an agent affect

2 AGENT-BASED ECONOMICS AND CONTROL SYSTEM

Each agent has several possible values for $\alpha \in [0,1]$, α is the profit elasticity of the preference score, $(1-\alpha)$ is the elasticity with respect to trust., this contributes to a change in the scores they assign to different agents. We consider populations of agents indexed on $[0,1]$, continuous distribution functions $F(x)$ and $G(y)$, and their respective densities by $f(x)$ *and* $g(y)$. And allow for asymmetries in the sizes of the populations and let λ^m *and* λ^w denote the measure of two group of agents (Bloch, Ryder, 2000), and the measure of matched agents is given by $M(\lambda^m,\lambda^w)$, where M is increasing in both its arguments and: $M(\lambda^m,\lambda^w) \le \min(\lambda^m,\lambda^w)$. A search equilibrium is a stationary strategy profile (σ_x,σ_y) such that for almost all one group of agents x in $[0,1]$, all strategies σ_x^i:

$$EU_x(\sigma_x,\sigma_y) \ge EU_x(\sigma_x^i,\sigma_y),$$

and for almost other group of agents y in $[0,1]$, all strategies σ_y^i:

$$EU_y(\sigma_y,\sigma_x) \geq EU_y(\sigma_y^i,\sigma_x).$$

Since both players announce a choice, if one of the agents reject the match, the other agents is indifferent between accepting and rejecting. This equilibrium is characterized by the formation of two sequences of subintervals

$I^n = (a^n, a^{n-1})$ *and* $J^n = (b^n, b^{n-1})$ *of* $[0,1]$,

the subintervals are defined:

$$a^n(1-\delta) = \delta \frac{M(\lambda^m,\lambda^w)}{\lambda^w} \int_{a^n}^{a^{n-1}} (x-a^n) f(x) dx,$$

and

$$b^n(1-\delta) = \delta \frac{M(\lambda^m,\lambda^w)}{\lambda^m} \int_{b^n}^{b^{n-1}} (y-b^n) g(y) dy.$$

3 POLICY ANALYSIS

The concept of a causal parameter within a well-posed model is defined in an economic setting that respects the constraints imposed by preferences, endowments, and social interactions through markets. We study a price adjustment process with *m* consumers and *n* commodities. Each agent-consumer takes prices and solves the optimization problem:

$$\max_{x_{i1},\ldots,x_{in}} U_i(x_{i1},\ldots,x_{in}) \quad i=1,\ldots,m,$$

subject to her budget constraint:

$$\sum_{j=1}^{n} p_j x_{ij} \leq \sum_{j=1}^{n} p_j w_{ij}, \quad i=1,\ldots,m$$

where U is a utility function of agent-consumer i, x_{ij} the amount of good j consumed by agent-consumer i, p_j the price of good j and w_{ij} the initial endowment of good j owned by consumer i. Aggregate excess demand for good j is:

$$z(p) = \sum_{i=1}^{m} x_{ij}(p,w_i) - \sum_{i=1}^{m} w_{ij}, \quad j=1,\ldots,n,$$

and stability of the competitive equilibrium is determined by the properties of the aggregate excess demand as well as the specific form of the adjustment rule (Tuinstra, 2000). From the homogeneity we have that if p^* is the equilibrium price, then so is αp^* for any $\alpha > 0$. Assume $g(p) = c$ can be as $p_k = h_k(p_1,\ldots,p_{k-1},p_{k+1},\ldots,p_n)$ for some k, and analyze the global dynamics we have to specify utility functions and endowments:

$$U_i(x_{i1},\ldots,x_{in}) = \left(\sum_{j=1}^{n} \alpha_{ij} x_{ij}^p\right)^{1/p}, \quad p<0, \quad i=1,\ldots,n$$

The corresponding agent demand functions are:

$$x_{ij} = (p_1,\ldots,p_n) = \frac{(\alpha_{ij}/p_j)^\sigma}{\sum_{k=1}^{n} p_k (\alpha_{ik}/p_k)^\sigma} \sum_{j=1}^{n} p_j w_{ij},$$

$$i,j = 1,\ldots,n,$$

with $\sigma = 1/(1-p)$ the elasticity of substitution, with prices p and in preference parameters α_{ik}, and initial endowment w_i. The price adjustment process now becomes:

$$p_{i,t+1} = p_{it}\left(1+\lambda\left[\sum_{i=1}^{n} \frac{p_{it}(\alpha_{ii}/p_{it})^\sigma}{\sum_{j=1}^{n} p_{jt}(\alpha_{ij}/p_{jt})^\sigma} - 1\right]\right)$$

Let the utility maximizing consumption bundle be given by:

$$x = d^U(p,w) = \arg\max_{x\in\Re}\{U(x) | p'x \leq p'w\},$$

and consider the problem of maximizing $U(Mx)$ subject to the budget constraint $p'x \leq p'(M'w)$, where $M^{-1} = M'$, this problem is equivalent with maximizing $U(Mx)$

subject to $(Mp)' Mx \leq (Mp)' w$ with the solution:

$$x = M'd(Mp,w).$$

Let $q = Fp$, the price adjustment process becomes $q_{t+1} = F f F^{-1}(q_t) = h(q_t)$. Differentiating with respect to q gives:

$$M\frac{\partial h}{\partial q'}(q) = \frac{\partial h}{\partial q'}(Mq)M.$$

The agent-planner faces the following Lagrangian:

$$L^W = S(G, x(p,G)) - px(p,G) + (1+\lambda+\tau)[(p-p^0)x(pG) - C(G)]'$$

the exercise tax follows:

$$\frac{p-p^0}{p} = -\frac{\lambda+\tau}{1+\lambda+\tau}\frac{1}{\eta},$$

where η is the price elasticity of demand. Let us assume that the agent-planner considers the following groups, which has a vested interest in provided good and in the taxed private good. The agent-tax players are finance the incomes, and the beneficiaries of the budget surplus which results if the exercise tax, supporters of the agent-spender and the agent-taxer. The agent-planner employs the following support function:

$$\Omega^P = \sigma_{VL}(S(G,x) - px) - \sigma_{GT}(I+J) + \sigma_{BS}(R-C) + \sigma_S U + \sigma_T V + \hbar,$$

where $\hbar$ is a sum of terms which are for the choice of the agent-planner's instruments, σ_{VL} is weights given for vested-interest agents, σ_{GT} for the general agent-tax players, σ_{BS} for the beneficiaries of the budget surplus and σ_S, σ_T for the agent-supporters of the agent-government. This leads to the following Hamiltonian:

$$H^P = \begin{Bmatrix} \sigma_{VL} S(G(\theta)) + \sigma_S U(\theta) \\ -\sigma_{GT}[U(\theta) + \psi(e(\theta)) - wG(\theta)] - \\ \sigma_{BS} C(\theta, e(\theta) G(\theta)) \end{Bmatrix} f(\theta)$$

$$+\tau(\theta)[R - C(\theta, e(\theta), G(\theta))] + \mu(\theta)[h(\theta, e(\theta), G(\theta))] + cons\tan t$$

,

where constant abbreviates the sum of all terms which are constant for the agent-spender's optimization , resulting both from the agent-taxer's and agent-consumers' decisions. The above result depends on the agent-planner's lack of information and on the separation of the agent-taxer and the agent-spender problems, where we have the following Lagrangian function:

$$L^P = \sigma_{VL}[S(G, x(p, G)) - px(p, G)] +$$
$$(\sigma_{BS} + \tau)[(p - p^0)x(p, G) - C(G)] + H^{\cdot}$$

The tax-related marginal condition is:

$$\frac{p - p^0}{p} = -\frac{\sigma_{BS} - \sigma_{VL} + \tau}{\sigma_{BS} + \tau}\frac{1}{\eta},$$

since the agent-spender's income is paid from general taxation, the agent-tax players influence the agent-spender's incentive pay.

4 GROWTH AND WELFARE AND DESIGN

There is a continuum of length of agent-worker/consumers who maximize the utility function, with consumption good C and the amount of they supply L:

$$V = \int_0^\infty e^{-pt}\frac{C^{1-\sigma}}{1-\sigma}h(L)dt \ ,$$

where σ is positive and different from one, p is positive, $h(L)$ is C and the following increasing in consumption and decreasing in labor (Pelloni, Waldmann, 2000):

$$h(L) > 0$$

$(1-\sigma)h'(L) < 0$, with budget constraint:

$$I = r(1-\tau_k - \tau_k^l)K + WL - (1-t_c)C - t_a\overline{K} - t_c\overline{C} + \tau_k^l r\overline{K},$$

where I is investment. Agent-households derive their income be renting the capital stock K and by supplying labor L to agent-firms, taking the interest rate r and the wage W, and capital income is taxed at a rate $\tau_k + \tau_k^l$. Consumption is subsidized at a rate τ_c, and there lump sum taxes proportional to average capital, $t_a\overline{K}$ and to average consumption $t_c\overline{C}$, and there is a lump sum subsidy proportional to average capital income $\tau_k^l r K$. The necessary and sufficient conditions for maximizing utility is respected and:

$$C^{-\sigma}h(L) = (1-t_c)\lambda$$

$$\frac{h'(L)C^{1-\sigma}}{(\sigma-1)} = \lambda W$$

$$\frac{\dot\lambda}{\lambda} = p - (1 - \tau_k^l - \tau_k)r$$

$$\lim_{t\to\infty} K\lambda e^{-pt} = 0.$$

We assume that the production set at the agent-firm level in labor L and capital K, and each agent acts in isolation taking the actions of other agents. Competition among agent-firms are paid their marginal products, that is:

$$Y = rK + WL$$

$$W = Kf'(L)$$

$$r = f(L) - f'(L)L = r(L).$$

We have been that the agent-government consumption G , the flow agent-government budget constraint can be written:

$$G + t_c C + \tau_k^l r\overline{K} = (\tau_k^l + \tau_k)rK + t_a + t_c\overline{C},$$

since average and aggregate variables are the same:

$$G = (\tau_k r + t_a)K.$$

For local indeterminacy it is sufficient that, if the initial consumption to capital ratio is slightly higher than the balanced growth consumption to capital ratio. Then the rate of growth of consumption is lower than the rate of growth of capital, so the consumption to capital ratio return to their steady-state values. The analysis of transitional dynamics is used to show that the signs of the effects of tax policies on the balanced growth equilibrium depend on whether the no-tax balanced growth labor supply $\tilde{L}$. Assuming the economy is always in steady state, the indirect utility function V is:

$$V = \int_0^\infty \frac{C_0^{1-\sigma}h}{1-\sigma}e^{(-p+g(1-\sigma))t}dt$$

$$= \frac{\frac{C_0^{1-\sigma}}{1-\sigma}h}{(p - g(1-\sigma))} = \frac{\frac{h^{2-\sigma}K_0^{1-\sigma}}{(1-\sigma)}\left(\frac{(\sigma-1)}{(1-t_c)h'}f'\right)^{1-\sigma}}{\left((\sigma-1)\frac{h}{(1-t_c)h'}f' - f + r(1-\tau_k^l) + \tau_a\right)}$$

The growth of consumption and capital are:

$$\frac{r(1-\tau_k^l) - p}{\sigma} = f - (\sigma-1)\frac{h}{h'}f',$$

and

$$N(\hat{L}) - \frac{r\tau_k^l}{\sigma} = 0,$$

we find the following effect on labor supply:

$$\frac{d\hat{L}}{d\tau_k^l} = \frac{r(\tilde{L})}{\sigma N'(\tilde{L})}.$$

When $\tau_k = 0,\ \tau_c = 0,\ \tau_a = 0$ we derive:

$$\frac{\partial(\log((1-\sigma)V))}{(1-\sigma)\partial\tau_k^l} = \frac{r}{(1-\sigma)\left((\sigma-1)\frac{h}{h'} - \tilde{L}\right)f'},$$

$$\frac{\partial(\log((1-\sigma)V))}{(1-\sigma)\partial L} > 0.$$

Even if the balanced growth path is locally indeterminate, it is possible to demonstrate an effect of taxes on long run growth. The steady-state effects on growth and welfare of a tax on capital income τ_k, whose proceeds are used to pay for expenditures $G = r\tau_k$, and the equality between the rates of growth of consumption and capital becomes:

$$\frac{(1-\tau_k)r - p}{\sigma} = f - (\sigma-1)\frac{h}{h'}f' - r\tau_k.$$

A small wasted tax on capital increases the rate of growth and welfare in the balanced growth equilibrium, and can be written:

$$N(\hat{L}) - \frac{\tau_k(1-\sigma)r}{\sigma} = 0,$$

around the equilibrium $L = \tilde{L},\ T = 0$ we find:

$$\frac{d\hat{L}}{d\tau_k} = \frac{(1-\sigma)r(\tilde{L})}{\sigma N'(\tilde{L})},$$

at the market equilibrium would be better off if increase the rate of interest and induce faster capital accumulation, for an agent to supply labor. By differentiating around the equilibrium we get:

$$\frac{d\hat{L}}{d\tau_a} = \frac{1}{-N'(\tilde{L})},$$

and the positive effect of a wasted lump sum tax on labour supply and growth can be understood as a combination of a direct wealth effect and an indirect substitution effect. Suppose that the agent-government subsidizes consumption at the rate τ_c paying for the policy with a lump sum tax, the conditions becomes:

$$\frac{r-p}{\sigma} = f + \frac{(1-\sigma)h}{(1-\tau_c)h'}f', \text{ and}$$

$$N(\hat{L}) - \frac{\tau_c(1-\sigma)}{(1-\tau_c)}\frac{h}{h'}f' = 0,$$

and differentiating around:

$$\frac{d\hat{L}}{d\tau_c} = \frac{(\sigma-1)\frac{h(\tilde{L})}{h'(\tilde{L})}f'(\tilde{L})}{-N'(\tilde{L})}.$$

We shown that a small amount of capital taxation, will increase both the rate of balanced growth and balanced growth welfare whenever the balanced growth path is locally indeterminate.

5 CONCLUSION

In real decision-making processes, the agent institutional context poses a number of constraints, every choice is constrained, and free voluntary and coerced relations are compatible. The analysis of the mechanism of control system design through which social interaction modifies constraints on agents and produces economic results leads to competition on both sides of the relation prevents each agent from enjoying an unfair return for functions performed.

REFERENCES

Bloch, F., Ryder, H. 2000. Two-sided search, marriages, and matchmakers. International Economic Review vol. 41, no. 1, 93-115.

Bos, D. 2000. Earmarked taxation: welfare versus political support. Journal of Public Economics 75, 439-462.

Heckman, J., J. 2000. Causal parameters and policy analysis in economics: a twentieth century retrospective. The Quarterly Journal of Economics, Feb., 45-97.

Klos, T. B., Nooteboom, B. Agent-based computational transaction cost economics. Journal of Economic Dynamics & Control 25, 503-526.

Palermo, G. 2000. Economic Power and the Firm in New Institutional Economics: Two Conflicting Problems. Journal of Economic Issues, vol. XXXIV, no. 3, 573-601.

Pelloni, A., Waldmann, R. 2000. Can waste improve welfare? Journal of Public Economics 77, 45-79.

Tesfatsion, L. 2001. Introduction to the special issue on agent-based computational economics. Journal of Economic Dynamics & Control 25, 281-293.

Tuinstra, J. 2000. A discrete and symmetric price adjustment process on the simplex. Journal of Economic Dynamics & Control 24, 881-907.

Tuinstra, J. 2000. The emergence of political business cycles in a two-sector general equilibrium model. European Journal of Political Economy 16, 509-534.

Vicary, S. 2000. Donations to a public good in a large economy. European Economic Review 44, 609-618.

GO-CAD: VIRTUAL REALITY SIMULATION

Tomáš Šebo, Igor Hantuch, Igor Hantuch jr.

Department of Automatic Control Systems,
Faculty of Electrical Engineering and Information Technology,
Slovak University of Technology, Ilkovičova 3, 812 19 Bratislava
e-mail:hantuch@.elf.stuba.sk

Abstract: GO-CAD brings some novel functions and features to the intelligent building design and virtual reality simulation. This paper presents a novel methodology simulation during lighting simulation of a building to acquire data about the mutual interaction between a human – a virtual reality visitor represented by the so-called "avatar", and a 3D building model and distance designing via Internet performed by a distributed multi-professional team. To demonstrate possibilities of the GO-CAD based modelling in the virtual reality, a lighting model has been chosen. *Copyright © 2003 IFAC*

Keywords: virtual reality simulation, avater in 3D interactive inteligent building, multi-professional IB design via Internet, SW-agent programming.

1 WHAT IS GO-CAD ... ?

GO-CAD is a novel original SW tool developed jointly at the Department of Automation and Control of the Faculty of Electrical Engineering and Information Technology and the Department of Building Construction of the Faculty of Civil Engineering, both from the Slovak University of Technology in Bratislava. GO-CAD allows:

- during lighting simulation of a building to acquire data about the mutual interaction between a human – a virtual reality visitor represented by the so-called "avatar", and a 3D building model
- distance designing via Internet performed by a distributed multi-professional team

GO-CAD brings some novel functions and features to the intelligent building design and virtual reality simulation.

Unlike present SW products, Go CAD enables acquisition of data arising in the mutual interaction „Avatar – environment". Besides, it enables to control more effectively collective design processes as well as designing and collective co-operation of various professionals, using Internet.

2 WHAT DOES GO-CAD ENABLE?

- loading dynamic segments into the model
- control of avatar's motion
- acquisition of data from dynamic segments (new)
- monitoring avatar's motion:
  - in drawings
  - inside the object – house - from the avatar's viewpoint
- multi-user virtual reality design in a unbounded 3D space
- access to the project via Internet

The most simple way – it is necessary to have installed a HTML browser with Java support, the situations are displayed in VRML using a plug-in browser module.

3 HOW IS GO-CAD PROGRAMMED ?

GO-CAD has been developed using the following SW environments:

- JAVA (control and monitoring algorithms)
- VRML (3D visualization of the project)
- PHP (dynamic HTML pages generating)
- HTML (final representation)

4 HOW DOES GO-CAD OPERATE... AN EXPERIMENT

GO-CAD deals with simulation of functions of the so-called " intelligent buildings" (IB) in the civil engineering.
First, a building model is to be generated using an arbitrary CAD tool (e.g. ArchiCAD).
GO-CAD allows to build-in the so-called dynamic segments in the building model, such as moving building elements (e.g. doors and windows) responding to a contact with avatar (Fig. 1 opening the window. The major group consists of elements designed within the project part „Electrical equipment and mains". In case of an „intelligent building" even this project layer use to be intricate. Beside motion scanners, which respond to a avatar's movements and/or presence, light switches, and switch-controlled lights used in the experiment, it is possible to use a variety of equipment elements during the design.
As dynamic segments respond in the real time, special requirements use to be imposed on them (e.g. a minimum memory size).
In GO-CAD, human's motion within the building is specified by the avatar's time-trajectory.
In the virtual reality, GO-CAD provides interaction between a human or avatar, and the model by means of dynamic objects responding to avatar's presence (motion scanners, active segments – doors, windows which open on avatar's clicking etc.).

In our case study, the model includes switchers and lighting control lights. Various ways of their wiring and provided benefits are subject matter of the simulation.

Prior to starting an experiment the following items are to be specified:

- avatar's motion
- light switching

The following is being evaluated:

- consumption of electricity for lighting
- lighting elements switching number (service life)

Experiment outcomes

- selection of a more effective design option
- modelling for supporting decision taking during the design

A family house has been chosen used for living and running business. Motion trajectories during the day have been designed for two different cases:

- for a house dweller (leaving in the morning and returning back in the afternoon) and
- for a worker of an company resident in the house (coming in the morning and leaving in the afternoon), considering also the workplace visitors.

Light switching is realized:

- by rooms – the light is switched on only in the room or a part of it, where the avatar has activated motion scanners
- by floors – the light illuminates only the floor where the avatar is present
- central switching – is the avatar in the house, the whole house is being illuminated

5 GO-CAD'S OPERATION MODES

GO-CAD operates in two basic modes:

- designing performed by an individual
- operation in a multi-user environment

The designer can create a model of a building with lighting, whereby the amount of electricity necessary for lighting is being monitored. To create an illusion of a visitor's motion in the house, dynamic segments are entered in the model, which, when activated (e.g. by clicking a door), open and the visitor can pass to a next room. In this way, the visitor can get into the whole house (if he has entrance authorization).

Fig. 1 Dynamic opening of a sliding door

In this building, the light is turned on immediately on visitor's entering the room (Fig. 1). This is enabled due to a novel feature of the virtual space – the possibility to scan avatar's actions (in our case realized by fitting scanners for identification of movements) followed by modelling and simulation by means of a mathematical model, based upon this principle.

Fig. 2 Turning on the light on entering a room

To be able to simply compare the benefits of the presence-based lighting control with the hitherto used light control methods, four types of lighting control have been implemented in the building. The first type is the most energy demanding one; it consisted in lighting up the whole building on crossing the doorway. In this case, all rooms in the building are lighted up. The second type consists in lighting whole individual floors, which is about half less energy demanding compared with the first approach. In this case, only a half of the building is lighted up. The third approach consists in lighting up whole rooms, and finally the fourth one in lighting up just those spaces where persons are present, i.e. it is the less energy demanding method.

All four methods are being evaluated in a control center; for individual control types, the consumed energy is charted in a dynamic chart. The chart shows the amount of energy consumed under individual control methods during e.g. 200s. The consumed power sampling and its depiction in the chart are in progress every 500 ms (Fig. 5). Every room in the building has a different energy intensiveness of illumination and on walking through individual rooms, the light is being turned on and off depending on how a scanner in each particular room evaluates the person's presence and which regulation type it uses. Out of the depicted chart it is possible to simply and quickly select the most appropriate method.

GO-CAD is designed to enable implementation (import) of IB models generated under any CAD environment (e.g. ArchiCAD, AutoCAD, ...). By converting a project e.g. from ArchiCAD into GO-CAD we obtain the basic project of the house, which includes all visible parts. In this stage, it is possible to view the house from any side and to walk through individual rooms.

Avatar's motion can be controlled in two ways:

- manually (like in computer games)
- using a pre-specified trajectory

In the house, we can view every nook and cranny, visually "verify" all its properties based on own requirements, or to rely on the demo created for the purpose of this presentation. In the former case, the visitor himself specifies which door to open, which room to enter and, conversely, which room he does not visit. In Fig.3 on the left there is depicted a situation on visitor's opening the door. This action has been evoked by simply clicking the door. As soon as the visitor performed this action, the door opened smoothly and he could continue exploring the "virtual world".

Fig. 3 The doorway: the door on the left was opened on customer's request, on the right, the customer chose a demo programmed by the presentation authors.

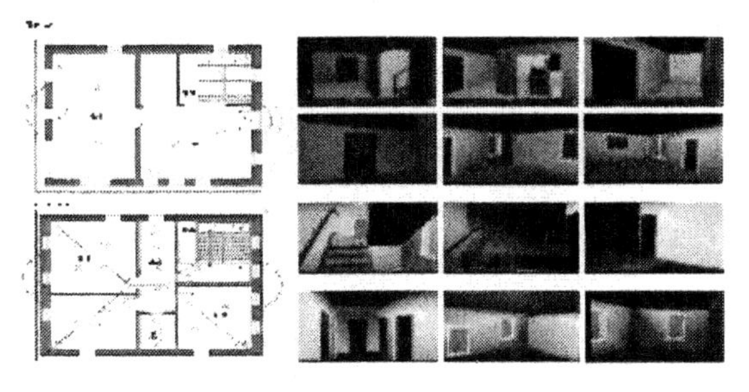

Fig. 4 Trajectory of the "demo" visitation along with snaps taken during it

Fig. 5 shows the full information provided by GO-CAD including all thus far described features. In the top left part there is the virtual building from avatar's viewpoint, below, the actual amount of consumed energy is charted. In the bottom part, there is important numerical information describing energy and financial intensiveness of the building under different lighting control types. Plan views of the building are on the left. They are used to draw in the actual location and trajectories of the user's movement in the house. In the bottom part, there is a console, in which all important events occurred in the virtual space are filled in.

Unlike existing systems, our GO-CAD includes a mechanism, enabling to process events that occurred in the virtual reality (Fig. 5). Based on signals from the environment it is possible to generate mathematical models in the programming language Java Script and/or Java. Mathematical model outputs can be represented in a 2D form using the Internet browser, however, it is also possible to change the virtual reality objects' properties. Both the 2D and the 3D representations are depicted in parallel, i.e. a part of data from the virtual environment are represented as a two-dimensional information, which allows a better 3D orientation.

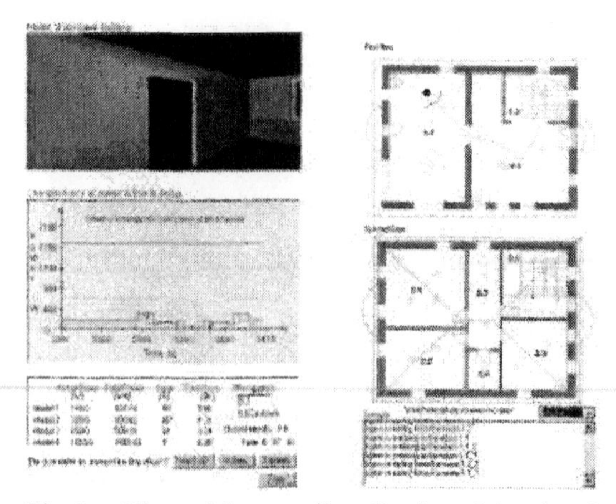

Fig. 5 View of the overall application of the virtual house

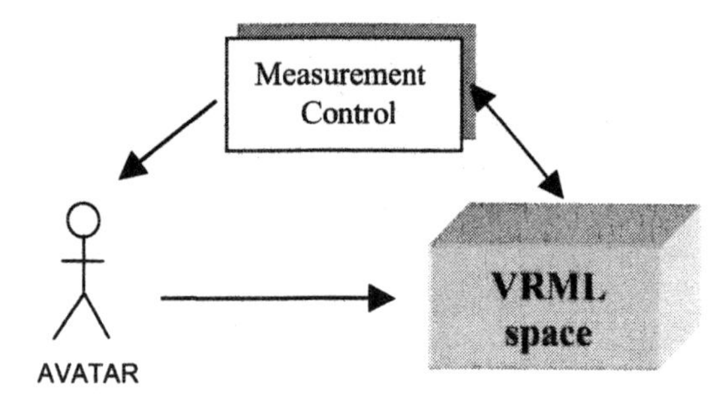

Fig. 6 Interactions between individual project parts

Since the beginning, GO-CAD has been developed for a multi-user environment. The whole interface is based upon and has been realized using the Internet explorer. To explore the projects, it is necessary to have installed a plug-in module for the VRML language. GO-CAD users are divided in three groups:

- administrator
- designer
- visitor

GO-CAD provides significant benefits mainly in project planning in the civil engineering field, therefore the case study and the import from CAD have been chosen from this field, too.

6 EXPERIMENT

To demonstrate possibilities of the GO-CAD based modelling in the virtual reality, a lighting model has been chosen. The experiment has been chosen such as to enable evaluation of the consumed energy and operation costs under variously defined model parameters. The model consists of the following basic blocks (Fig. 7):

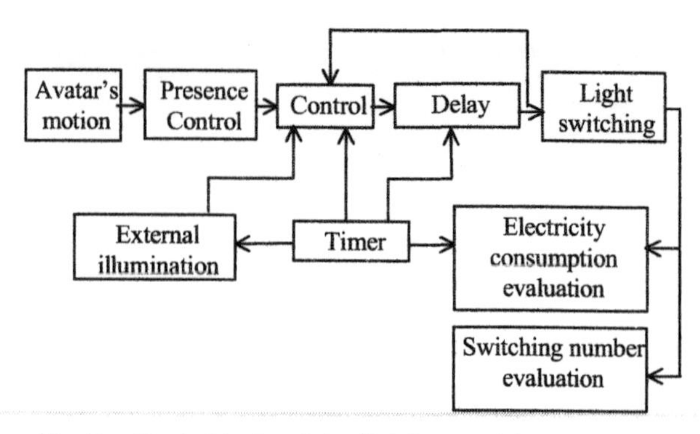

Fig. 7 Basic blocks of the lighting model

The experiment has been divided in two parts. The building was utilized as:

- business premises
- housing premises

A basic difference between these two approaches consists in the time during which visitors dwell in the building. In a company it is mostly from 8:00 to 18:00 (Fig. 8) and in a dwelling house, the time interval is split in two parts, the first one from 6:00 to 9:00 and the second one from 18:00 to 22:00 (Fig. 9).

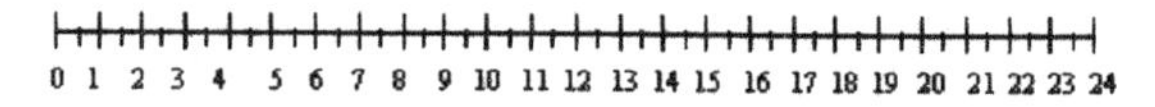

Fig. 8 Business premises utilization

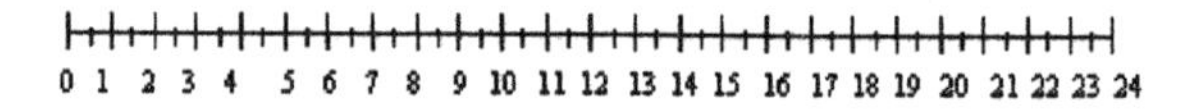

Fig. 9 Housing premises utilization

Whole-day trajectories have been generated through housing as well as business premises, along which moved the virtual space visitors. Resulting trajectories have been generated as follows: Time, during which the visitor stayed in the house, was split in half-hour intervals. Every half-hour, either one of the elementary trajectories or no trajectory were activated. After completing the elementary trajectory, the avatar stayed at the last location until a next half-hour passed, during which another elementary trajectory was completed. The following tables show the whole-day trajectories for individual person types (Tab. 1, Tab. 2).

	6^{00}	6^{30}	7^{00}	7^{30}	8^{00}	8^{30}	9^{00}	...	18^{00}	18^{30}	19^{00}	19^{30}	20^{00}	20^{30}	21^{00}	21^{30}	22^{00}	...
Aktivny	T_1	T_2	T_1	T_2	T_1	T_3	T_1		T_2	T_1	T_1	T_2	T_3		T_1	T_2	T_1	
Médium	T_2		T_2		T_2		T_3		T_2		T_3		T_2	T_3		T_2	T_1	
Pasivny	T_3						T_2		T_3								T_1	

Table 1 Housing premises: A whole-day trajectory composed of elementary trajectories

	8^{00}	8^{30}	9^{00}	9^{30}	10^{30}	11^{00}	11^{30}	12^{00}	13^{00}	13^{30}	14^{00}	15^{30}	16^{00}	16^{30}	17^{00}	17^{30}	18^{30}
Aktivny	T_1	T_2	T_1	T_2	T_1		T_2	T_3	T_1	T_2	T_3	T_1	T_1	T_2		T_2	T_1
Médium	T_2		T_1		T_3			T_2		T_1		T_2			T_2		T_1
Pasivny	T_3				T_2				T_3			T_2					T_1

Table 2 Business premises: A whole-day trajectory composed of elementary trajectories

Further parameters affecting the model have been changed in the block for simulating the external illumination during the year. Extreme values have been studied, i.e. without using the external illumination model; using the external illumination model – both during winter and summer solstices as well as during the equinox (Table 3). Despite a sufficient external illumination, corridors and stairs had to be lighted also during the day, as there were not a sufficient number of windows.

Date	Sun-rise	Sunset	Max. illuminance
-------------	0:0	0:0	0
21 March (spring equinox)	5:44	17:53	800
21 June (summer solstice)	4:42	20:43	1000
21 December (winter solstice)	7:33	15:49	600

Table 3 Parameters of the external lighting model

The last parameter, which has been changed during the experiment is called the delay. Due to the delay, duration of the output value in the state 1 is extended by a time constant, i.e. the light remains to be turned on also when avatar leaves the room. The time constants have been chosen as follows: 0s, 120s and 600s. This parameter was used to study the dependence between the number of switchings and the time constant of the delay.

7 EVALUATION OF RESULTS

As expected, the best results with respect to the amount of consumed energy have been obtained using the model, which provided lighting control based upon the presence at the lighting appliance. Conversely, as expected, the worst results have been obtained using a central lighting switching. The type of person affected the electricity consumption mainly when larger delays have been adjusted. In case of a zero delay the results have been comparable. In most cases, movement of an active dweller inside the building was more energy demanding. In some cases, even the passive dweller acquired worse results with respect to electricity consumption than the active one, especially when the latter often nipped out from the building during the day. The passive dweller remained at one place almost all the time, i.e. if there is no sufficient illumination from outside in the room where the house dweller is present, the light needs to be turned on all the time.

In all cases, the worst model with respect to switching the light fittings was the model No.2 (switching by rooms). Originally, the worst results were expected for the model No. 1 (switching by individual lighting appliances). This was due to the definition of avatar's trajectories within the building. During the automatic visitation the avatar did not pass under all light fittings in the given room whereby in case of the model No. 2, on entering the room all light fittings have been turned on. To avoid a too frequent switching of light fittings, it is necessary to decompose the room into smaller units with respect to presence scanners. This approach should be used mainly for larger rooms.

For illustration, results of an experiment are reported below. The measured values represent the model of business premises during the winter solstice.

	A1	M1	P1	A2	M2	P2	A3	M3	P3	A4	M4	P4
0 s	761	670	707	1129	1328	1406	2943	4389	4591	7528	12438	12832
120 s	1157	1273	858	2014	2008	1556	3800	4831	4664	8009	12412	12805
600 s	4316	2981	1410	4859	3594	2172	6618	6625	5362	9445	12702	13310

Table 4 Electricity consumption in business premises for the winter solstice

	A1	M1	P1	A2	M2	P2	A3	M3	P3	A4	M4	P4
0 s	350	213	73	409	214	85	314	183	70	236	60	38
120 s	200	121	41	217	121	45	224	128	50	142	20	20
600 s	195	120	37	204	116	42	217	140	49	135	19	19

Table 5 Number of lamp switchings in business premises for the winter solstice

In the charts and tables, an abbreviated notation is used consisting of two symbols: the first one is a letter representing the user type (Active – A, Passive - P), the second one is a figure indicating the number of the electricity consumption mathematical model (1 – switching by appliances, 2 – switching by rooms, 3 – switching by floors, 4 – switching the building as a whole). For example, the abbreviation P2 stands for a passive user and the lighting control model based upon switching the whole room.

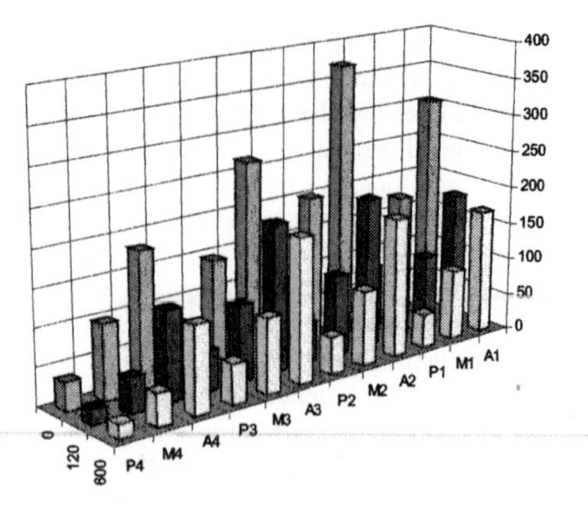

Fig. 10 Electricity consumption in business premises (21 December)

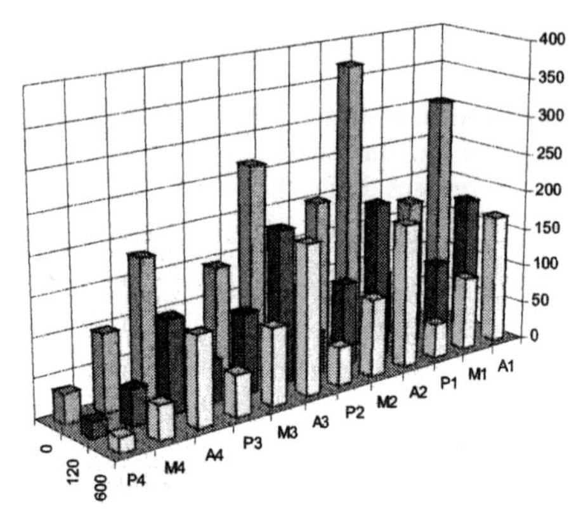

Fig. 11 Number of lamp switchings in business premises (21 December)

A delayed switch-out has significantly influenced the number of light appliance switchings. In our experiment, the average building visitation time was set to 150s. Maybe due to this reason the difference between the 120s and 600s switch-out time constants was small. The switching number intensiveness for a zero switch-out was about twofold than for the 120s and 600s constants. In a real control system implementation it will be necessary to consider how often the dwellers move about the particular room. For example, a presence-based lighting control in a busy corridor would not be well founded, as the light would remain to be turned on almost continuously. The control would be rather based exclusively on sensing the external illuminance, however only in such a case if there are a sufficient number of windows in the corridor. If this requirement is not fulfilled, just time-control can be considered.

When using the external illumination model, the electricity consumption in the housing premises was generally higher than in the business ones. Time intervals of movement for dwellers in a business building fell within the time of a higher external illuminance. During experimentation with the external illumination, some rooms (in particular corridors and stairs) were lighted up with no regard to external illuminance, i.e. also during the summer solstice, when the external illuminance is sufficient. During the whole time period of employees' moving about in business premises, the interior was lighted optimally. When persons moved in corridors the light was switched on.

The aim of the above experiment was to point at some novel possibilities of modelling under VRML using GO-CAD. The considered lighting example was simplified and its objective was not to solve a real situation.

8 CONCLUSION

A possibility to model and study product properties during the design is a novel quality. A further benefit brought about by the design in virtual reality placed at Internet is a possible communication among several designers of various professions, and their work on a common project, whereby they do not need to regularly meet in the real-world. They can communicate e.g. by short text messages and can exchange information during virtual sessions convened by the executive designer. A side effect is an extended possibility of home working.

An original contribution consists in a possible adding of scanners in the project under VRML, which provides a feedback from the environment with the avatar and the mathematical model. Information from such interactions can be used for simulation in various physical processes. To control the process several algorithms can be designed, which can be evaluated from a short-term as well as a long-term viewpoint.

GO-CAD consists of two parts: The first one represents a mechanism of adding completed projects on the server, and the second one enables modelling, e.g. modelling of lighting based upon the presence and external weather conditions. For the first part, it was necessary to generate access rights for three access levels: administrator, designer, visitor.

The second part is the lighting control model as such. In the experiment, four ways of lighting control have been considered:

- control by individual lighting appliances
- control by rooms
- control by groups, in our case by floors
- control of a whole, switching the whole building

The experiment results comply with the practice. Mainly as for electricity consumption they respond to assumptions. The experiment is secondary, primary was to prove that our SW environment enables the simulation/modelling of avatar-process/environment.

Unlike the existing systems, GO-CAD involves a bilateral communication between the VRML environment and the avatar. Besides visualization, we are able to evaluate and even respond to events arisen due to an event occurred in the virtual reality.

As original solutions are considered the following:

- utilization of VRML and other SW techniques for designing, modelling and simulation
- optimisation under the VRML environment
- corporate designing in the virtual reality

REFERENCES

Capkovic, M., Kozák, Š.: Neural networks for SISD real time control.1. vyd.Amsterdam, Netherlands : IOSPress, 2002. ISBN 1 58603 256 9.

Hantuch, I. *Automatic Synthesis of a Packaged Program for Real Time Technological Process Control.* IFAC Symposium on Computer Aided Design of Control Systems Preprints of IFAC-CAD/CS Symposium, p.285-298, Zuerich 1978

Hantuch, I. *CACSD/SW.* 1st IFAC Workshop on New Trends in Design of Control Systems, p. 84-86

Hantuch, I., Mozola M. Development of DEDS Control Methods using SW-Agent Approaches. *Preprints of IFAC Conference Control System Design, Bratislava, pp378-383, June 2000*

Hejdiš, J., Kozák, Š., Juráčková, L.: Self-tuning controllers based onorthonormal functions. In Journal Kybernetika, vol. 36, 2000, No. 4, s.477-491.

Hietikko, E.; Rajaniemil, E. Visualized data--tool to improve communication in distributed product development projects. Journal of Engineering Design, Mar2000, Vol. 11 Issue 1, p95, 7p, 2 diagrams

Hrúz, B., L. Mrafko, V. Bielko (2002). A comparison of the AGV controlsolution approaches using Petri nets. Proc. of the 2002 IEEEInternational Conference on Systems, Man and Cybernetics, Yasmine Hammamet,Tunisia, October 6-9, 2002.

Hubinský, P.: Virii and Internet. 5th Seminar of Slovak association for information security held during the computer exhibition COFAX'98, april 1998,pp.25-31.

Kozák, Š.: Development of control engineering methods and their applicationsin industry (Invited Lectures) 5th International Scientific-Technical Conference Process Control 2002. Kouty nad Desnou, Czech Republic:June 9-12,2002, s. R218. (in English)

Phillips, C. : Planning 'intelligent buildings.'. HD: Hospital Development, Jul 2001, Vol. 32 Issue 7, p31, 1p, 1bw

Vörös, J. : Low-cost implementation of distance maps for path planning using matrix quadtrees and octrees(Robotics and Computer-Integrated Manufacturing, Vol. 17,No. 6, pp 447-459, 2001)

Vörös, J.: Top-down generation of quadtree representation using color-change-code (Computers and Artificial Intelligence, Vol. 13, No.1, pp 91-103, 1994)

Blaxxun interactive, www.blaxxun.com

Browsers and plug-ins http://www.web3d.org/vrml/browpi.htm

Columnia university www.arch.columbia.edu/DDL/projects/vrml /vrml.html

Parallel Graphics www.parallelgraphics.com

WEB3D consortium www.vrml.org

VRML Applications in Construction http://www.new-technologies.org/ECT/Internet/vrml.htm

VRML Examples http://www.external.hrp.no/vr/vrml/

www.elsevier.com/locate/ifac

NONLINEAR SYSTEM MODELS BASED ON INTERVAL LINEARIZATION: A MEDICAL APPLICATION

Jaroslav Kultan

Abstract: The paper deals with an application of models based upon interval linearization in modelling human organism parameters, especially during long-term medical treatments. The generated model enables to predict patient's state as well as the impact of taken therapeutics on it. Application of such models brings about improvement of the treatment quality and helps physicians in the decision making process. *Copyright © 2003 IFAC*

Keywords: identification, interval linearization, parameters of human organism, state prediction, nonlinear systems

1. INTRODUCTION

In the medical practice there are often encountered diseases requiring a long-term treatment whereby it is almost impossible to continuously monitor the organism parameters. Very often, physicians are obliged to go repeatedly through a huge amount of data acquired from the blood, urine and other analyses, to be able to take decisions based upon the patient´s state. This process can be supported by creating a model of patient's state, based on which it is possible to take a correct decision. Therefore we focused on the identification of human organism parameters under long-term medical treatments. Results of blood tests, creation of their model and its application in the patient's state prediction have facilitated the decision making about additional interventions during the treatment, which helped to improve the whole process. For the identification and model generation we have used the interval linearization method.

2. NONLINEAR SYSTEM IDENTIFICATION

2.1. *Identification using the interval linearization method*

The interval linearization method is one of relatively new methods for nonlinear systems modelling. Basically, it differentiates from other identification and modelling approaches for such systems, one of which consists in substituting the real system by a set of nonlinear functions and relations which describe processes in the system (the human organism can be considered as a system); another possibility is substitution of the real system by a mathematical model acquired by its linearization in a chosen operating point, or it is possible to create abstract nonlinear models.

The interval linearization method offers a substantially different approach, generating a linear model created not only for one operating point but for a whole region called a linearization interval [1]. Such a model is able to include properties of the given system, whereby it requires neither to create so many models as in case of linearization in the operating point, nor to search intricate nonlinear functions which implementation is often questionable.

Identification of nonlinear dynamic systems by the interval linearization method has been invented in 1986. The method has been updated several times by improving some of their properties or the identification process itself [2] [3]. The presented method has been applied not only in nonlinear system modelling but also in designing controllers for nonlinear systems [4].

Identification of nonlinear dynamic systems consists in splitting the whole working range of the input and output variables by the limits uk_i and yk_i, respectively, into several intervals (from where the name of the method). Intersect of these intervals defines the linearization interval (Fig. 1).

Generating a system model using this method requires availability of the measured input and output data. Measurements are to be carried out during the system operation and organized in a table. Next, the limits of individual intervals for the input and output variables are to be specified. Intersect of these intervals defines the linearization section. For every nonlinear system there can be several such sections and the identification yields model of each linearization section in the following form

$$y(t) = q_0^k + \sum_{i=1}^{ny} q_i^k y(t - iT) + \sum_{j=1}^{nu} q_{ny+j}^k u(t - (j - d + 1)T) \quad (1)$$

or concisely

$$\mathbf{y}(t) = \mathbf{z}(t)\,\mathbf{q}^k \qquad k=1,2...,kk \quad (2)$$

where

$$\mathbf{z}(t)=[1,y(t-1),...,y(t-ny),u(t-d-1),...,u(t-d-nu+1)] \quad (3)$$

$\mathbf{q}^k$ is vector of model coefficients in the k-th linearization section, whereby

$$\mathbf{q}^k = (q_0^k, q_1^k ..., q_{nu+ny}^k) \quad (4)$$

nu – number of samples of the input variable

ny - number of samples of the output variable

t - time

T – sampling time

d – input variable shift

k - index of the linearization interval

A total number m of linearization intervals for nu =1 and ny= 1 is

m = mu.my

and for nu> 1 and ny > 1 the number of linearization intervals is given by

$$kk = mu^{nu}.my^{ny} \tag{5}$$

The procedure for determining the index "k" of the linearization interval is as follows. If the output variable y(t-i), i=1,2,..,ny is from the amplitude band (yk_{s-1}, yk_s), s =1,2,...,my, then this band is denoted by I_i=s. Similarly, if the variable u(t-j-d+1), j=1,2..,nu is from the band $(uk_{r-1}, uk)_r$, r =1,2,...,mu, the corresponding band is denoted I_{ny+j} = r. These bands notations are collected in a vector $I=(I_1, I_2,..I_{nu+ny})$. The linearization interval index is calculated as follows

$$k = I_1 + \sum_{l=2}^{nu+ny} (I_l - 1).\delta_l \tag{6}$$

where

$$\delta_l = my^{(l-1)} \quad \text{for } l=1,2,...,ny \tag{7}$$

$$\delta_{l+ny} = my^{ny}.mu^{(l-1)} \quad \text{for } l=1,2,...,nu \tag{8}$$

I – auxiliary index

For many systems, long-term operation measurements of u(t), y(t) proved, that in practice, the number of linearization intervals does not comply with their theoretical number and strongly depends on the appropriateness of chosen limits and the volume of measured data available.

Calculation of the linear model (1) or (2) for the k-th linearization interval consists in determining the vector of linearization coefficients (4) from measured samples y(t-i) and u(t-j-d+1) for i=1,...,ny, j=1,...,nu, within the k-th inearization interval.

The model itself is then specified from the relation $\mathbf{A}^k\mathbf{q}^k= \mathbf{b}^k$, from which we express the vector of the model $\mathbf{q}^k$ in the k-th linearization section using the least squares algorithm. To apply it, it is necessary to se tup the matrix $\mathbf{A}^k$ from measured values y(t-i) and u(t-j-d+1) and the vector b^k from measured values y(t). For the k-th row of the matrix $\mathbf{A}^k$ holds

$$a^k_{r,i+1} = y(t-i) \quad \text{for } i=1,2,..,ny \tag{9}$$

$$a^k_{r,j+1+ny} = u(t-j-d+1) \quad \text{for } j=1,2,...,nu$$

$$a^k_1 = 1$$

The r-th entry of the vector $\mathbf{b}^k$ satisfies

$$b^k_r = y(t-0) \tag{10}$$

where r = 1,2... n ; ny+nu+1 < n
The identification result is the matrix **Q**

$$\mathbf{Q} = \begin{pmatrix} q^1_0 & q^1_1 & \cdots & q^1_{nu+ny} \\ \vdots & & \cdots & \vdots \\ q^k_0 & q^k_1 & \cdots & q^k_{nu+ny} \\ \vdots & \vdots & \cdots & \vdots \\ q^m_0 & q^m_1 & \cdots & q^m_{nu+ny} \end{pmatrix} \tag{9}$$

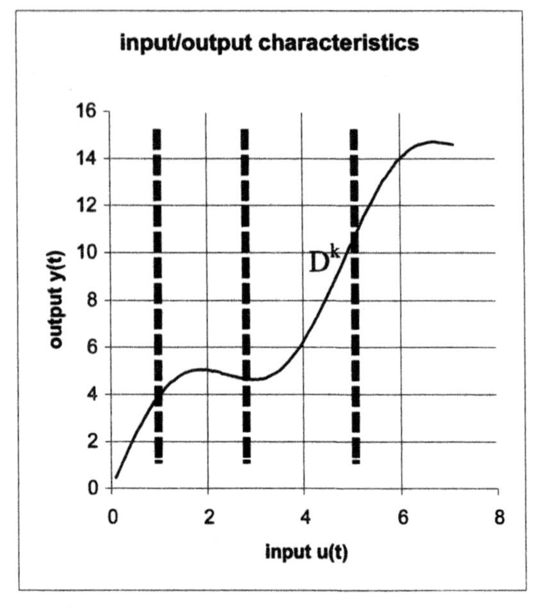

Fig.1. Input/output characterisics of the system

3. IDENTIFICATION OF BLOOD SYSTEM PARAMETERS

3.1. Data measurement

The considered method was verified in modelling leucocytes parameters for one year. During a one-year chemotherapeutic treatment blood samples were taken sporadically and individual parameters were measured. Values of individual parameters are organized in a table (Tab. 1). Input variables are values of used therapeutics (V - vincristin, CP – cysplatinum), or of certain supporting preparations (S), applied in cases when the organism parameters have decreased considerably. Output parameters are HG, Ery, HKG, LE, TR. Medicine intake is denoted by „1" in the day of application.

Tab.1. Measured values of blood parameters during the initial period

			INPUTS				OUTPUTS				
Date	N		Cyk	V	CP	S	HG	Ery	HKG	LE	TR
14.07.99	1	1	1	1	1		118	3.7	0.29	2.1	186
20.07.99	6	7					124	4.4	0.37	1.6	225
22.7.99	2	9		1			117	4.09	0.37	1.8	205
29.7.99	7	16		1			93	3.26	0.37	1.52	143
31.7.99	2	18					98	3.37	0.38	1.2	149
3.8.99	3	21					64	2.32	0.1	1	110
5.8.99	2	23					82	2.86	0.37	1.24	143
25.8.99	20	43	2	1	1		126	4.13	0.35	2.9	224
2.9.99	8	51		1			113	3.7	0.37	1.58	184
6.9.99	4	55					95	3.09	0.38	1.04	108
9.9.99	3	58		1			95	3.41	0.27	1.3	145
21.9.99	12	70					109	3.49	0.38	1.33	117
27.9.99	6	76					104	3.4	0.38	1.2	123
6.10.99	9	85					113	3.74	0.36	1.08	281
13.10.99	7	92					117	3.76	0.37	1.65	243
2.11.99	5	112					96	3.18	0.38	0.74	138
5.11.99	3	115					94	3.3	0.27	0.8	149
9.11.99	4	119					101	3.35	0.36	1.1	187

A graph of some monitored quantities is in Fig.2 whereby the leucocytes, which characterize patient's immunity are the most interesting parameter.

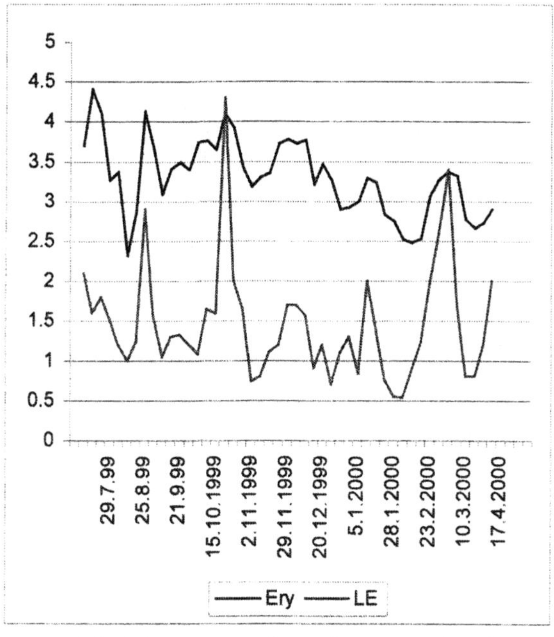

Fig.2 Measured values of some monitored quantities

3.2. Processing of measured data

As the input data were measured in various time intervals, they need to be pre-processed and appropriately adapted for next computations. Computed intervals between individual measurements and missing values have been interpolated so as to reflect in the best way the values expected in the considered time.

The simplest interpolation method is the linear interpolation. For values in individual days it substitutes the values obtained by a uniform distribution of the states between two measured values.

The quadratic interpolation method is a more advantageous one, using a quadratic approximation of values between two measurements. Though it is quite computationally demanding it follows better continuous changes of individual parameters.

Input variables have been substituted by exponentials reaching their maximum at application time of basic therapeutics when their concentration in organism decreases successively. The exponential forgetting factor changes with each considered variable.

	V	CP
A.	1	1
D	0.3	0.04

Table 2 summarizes values of measured, and interpolated (linearly and quadratically) values. Measured values of individual variables are shown in Fig. 3. Comparison of linearly and quadratically interpolated values is depicted in Fig. 4. In measurement instants values of individual functions coincide. Leucocytes and input variable values used for modelling are shown in Fig. 5

Tab. 2. Approximation of selected parameters of the organism

days	V	CP	S	HG	Ery	HKG	LE	HG	Ery	HKG	LE
1	1	1	0	118	3.7	0.29	2.1	119	3.73	0.29	2.08
2	0.7408	0.9608	0	119	3.82	0.3	2.02	119	3.73	0.29	2.08
3	0.5488	0.9231	0	120	3.93	0.32	1.93	122	4.01	0.31	1.89
4	0.4066	0.8869	0	121	4.05	0.33	1.85	125	4.22	0.33	1.74
5	0.3012	0.8521	0	122	4.17	0.34	1.77	126	4.37	0.35	1.64
6	0.2231	0.8187	0	123	4.28	0.36	1.68	127	4.44	0.36	1.58
7	0.1653	0.7866	0	124	4.4	0.37	1.6	126	4.46	0.37	1.57
8	0.1225	0.7558	0	121	4.25	0.37	1.7	124	4.4	0.37	1.61
9	1.0907	0.7261	0	117	4.09	0.37	1.8	120	4.24	0.37	1.72
10	0.808	0.6977	0	114	3.97	0.37	1.76	116	4.07	0.37	1.81
11	0.5986	0.6703	0	110	3.85	0.37	1.72	109	3.84	0.37	1.85
12	0.4435	0.644	0	107	3.73	0.37	1.68	103	3.65	0.37	1.86
13	0.3285	0.6188	0	103	3.62	0.37	1.64	98.6	3.5	0.37	1.85
14	0.2434	0.5945	0	99.9	3.5	0.37	1.6	95.2	3.38	0.37	1.8
15	0.1803	0.5712	0	96.4	3.38	0.37	1.56	93.2	3.3	0.37	1.74
16	1.1336	0.5488	0	93	3.26	0.37	1.52	92.4	3.26	0.37	1.64
17	0.8398	0.5273	0	95.5	3.32	0.37	1.36	95.5	3.33	0.39	1.5
18	0.6221	0.5066	0	98	3.37	0.38	1.2	99.3	3.43	0.4	1.33

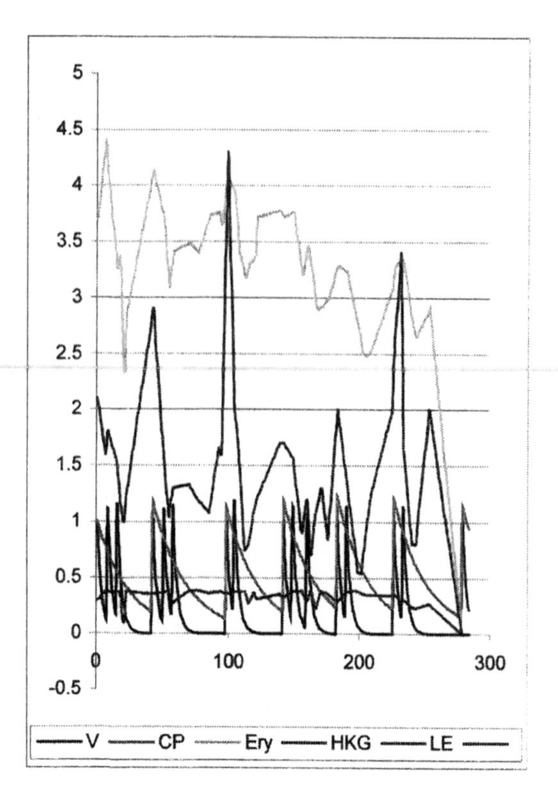

Fig. 3 Graphical representation of the linear approximation of selected organism parameters

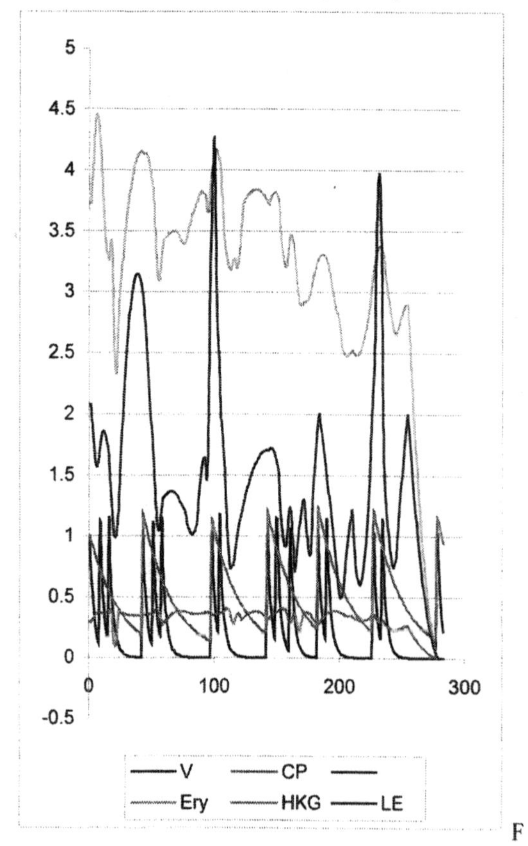

Fig. 4. Quadratic approximation of selected parameters of the organism

For modelling blood parameters we have chosen one of its most important representatives, namely the number of leucocytes indicating patient's state with respect to the possible application of chemotherapeutics. Even the state of leucocytes (along with other parameters, of course) determines whether it is possible to give the patient the therapeutics or not.

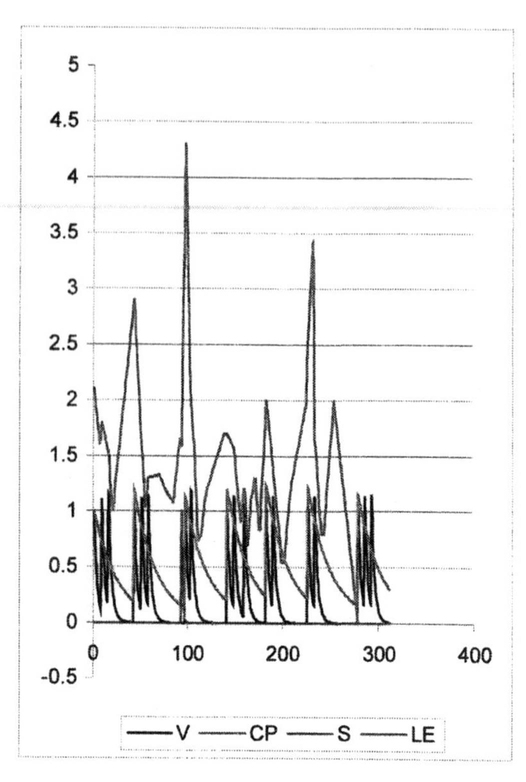

Fig. 5 Parameters of the input therapeutics and the response of the organism - leucocytes

3.3. Search of linearization intervals

A serious drawback of the described method is the search for linearization intervals, i.e. specification of limits for individual intervals. In practice this problem is being solved in various ways. As medical processes are relatively slow a stepwise changing the limits of domains and ranges of individual variables can solve this problem.

Next, a model is calculated for all obtained intervals, and a simulation is carried out with the already available samples. Individual models are compared using the classical method of summing squared differences between the measured and the calculated values. Using a graphical representation of individual input variables and modified values of squared error sums for each model we can find the intervals with the least difference between the model outputs and the real values. Table 3 shows values of some limits (hr1, hr2, hr3, hr4 and the "sqe" – sum of squared errors).

Tab.3 Sum of squared errors (sqe) as a function of limits

pc	hr1	hr2	hr3	hr4	sko
1	0.1	0.5	0.7	2.4	0.39
2	0.1	0.9	0.4	2.4	0.58
3	0.1	0.5	0.7	1.4	0.71
4	0.1	0.1	1	0.4	1.08
5	0.1	0.5	1	0.8	1.38
6	0.3	0.1	0.4	0.2	2.44
7	0.1	0.5	0.7	2	25
8	0.1	0.3	0.1	0.4	97.3
9	0.1	0.7	0.1	0.4	124
10	0.1	0.3	1	1.2	282
11	0.1	0.1	1	0.6	340
13	0.1	0.7	1	0.6	388
14	0.1	0.1	1	1.2	
15	0.1	0.9	0.7	2.6	0.41
16	0.1	0.7	0.7	0.4	0.61
17	0.1	0.3	0.7	1.2	0.83
18	0.1	0.7	1	1.2	1.12
19	0.1	0.5	0.1	2	1.47

For the next analysis a model with the limits (hr1, hr2, hr3,hr4) = (0,3;0,3;0,7;0,5) has been chosen. Mathematical model and simulated data are in Fig. 6.

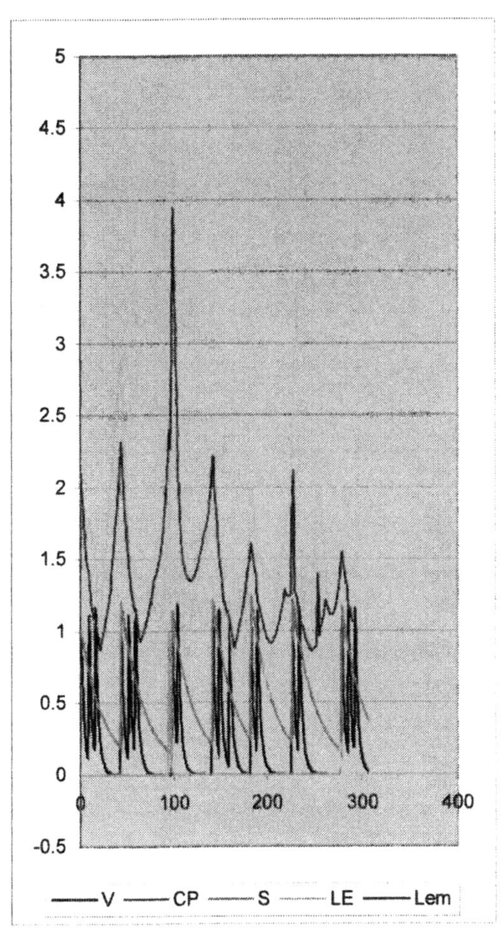

Fig. 6 Simulation of LE parameters

4. APPLICATION OF GENERATED MODELS

4.1. Problems of patient's treatment.

Application of some therapeutics may bring about a considerable worsening of some human organism parameters and thus their intake is possible only under the assumption that the patient tolerates such a treatment. One of such procedures is the application of cytostatics in curing oncogenous diseases. Application of therapeutics is possible only if the leucocytes parameters have reached some specified value (2000). To find out the patient's actual state it is necessary to take his blood, evaluate the parameters and to continue the treatment under the assumption that the state is sufficiently good. If there are low leucocytes, the blood taking has to be repeated after 1-3 days. Such frequent blood takings disturb the patient, and make him suffer. Taking into account his overall state, such a large number of blood withdrawals cannot contribute to an improved recovery. On the other hand frequent blood takings are important because it is necessary to apply the therapeutics in the specified time (necessity to keep the curative procedures) or with a least possible delay.

It use to happen that despite numerous tests, the patient's state does not improve and in order to keep certain curative procedures it is necessary to apply supporting and stimulating pharmaceuticals.

It is especially the decision making about whether to make the patient suffer or stimulate the organism all the same, which can be facilitated by the proposed parameter model.

From Fig. 6 it is evident that in the last 250 –300 days of curing, the leucocytes (LE) number drops and at the time of therapeutics application it even does not reach the lower limit 2.

4.2. Simulation of the medical treatment

Based on the analysis of measured data and simulation of the possible patient's state (Fig. 6) it is possible to assume that the patient's state will remain on a low level for a long-time and he cannot be given the relevant therapeutics due to it.

In a standard treatment the doctor takes the blood samples several times and after having gathered long-term results he decides to skip one cycle. Some of the following facts underlie such a decision:

- on checking parameters he supposed that the patient's state will improve, however this did not happen;
- there was a possibility of helping by other means, however relatively much time has passed since then and therefore such a stimulation is already inappropriate.

Of course, there are a lot of other facts to be considered in a physician's decision making. Simulation of patient's state before starting a new curing cycle could be one possible tool for helping his decision taking. The obtained model shows that the organism parameters are relatively low and the patient is not able to create a sufficient amount of necessary sub-

stances. Should these facts be confirmed by the results from the first blood test, the physician can consider application of stimulating therapeutics, supporting this decision by a simulation on the models.
In a real situation we have first simulated the impact of the blood transfusion on other parameters. Transfusion was carried out already at about the 100th day of the curing procedure. Simulation results under stimulator application are depicted in Fig. 7. Based on obtained results (growth of the LE value above 3) correctness of such a decision could have been anticipated.

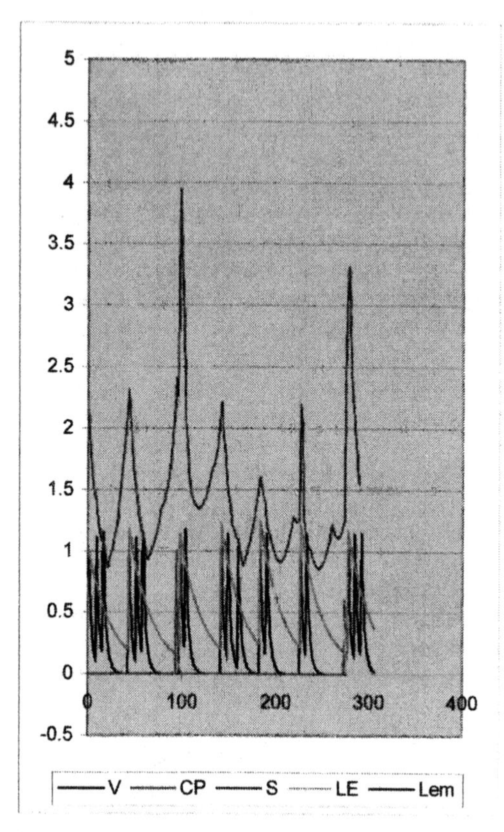

Fig.7 Simulation of LE parameters after application of a stimulator

Then the transfusion was completed and the whole curative cycle was completed successfully.
Of course, the experience of physicians played an important role in the whole treatment process as well, and the proposed model has served only as one supporting tool for the decision-making.

CONCLUSION

The aim of this paper was to present a practical application of one of the dynamic systems identification methods in medicine. At the same time we wanted to emphasize that implementation of identification, modelling, simulation and control methods in this field not only improves the treatment process but can remove lot of ache in this treatment as well, and enables a more exact decision making, often even saving the patient's life.
Certainly, the interval linearization method is not the only suitable one but hopefully it is not the last one applied in medicine, and its implementation will stimulate application of other methods as well.

REFERENCES

S.A. Bilings and W.S.F. Voon: *Piecewise linear nonlinear system identification. Internation Journal of Control, Vol 46,N.1,1987*

Harsányi L., Kultan J.: *Identifikácia nelineárnych systémov metódou intervalovej linearizácie. Journal of Electrical Engineering, Vol.41., No 11, 1990, str. 825-836.*

Harsányi L., Kultan J.: *Method of selective forgetting for nonlinear system identification Journal of Electrical Engineering,, Vol 43, No 7, 207-210, Bratislava,1992*

Veselý V., Kultan J.: *Regulator synthesis for nonlinear systems. Technical report EF SVŠT, Department of Automatic Control Systems, 1992*

VIRTUAL SENSORS BASED ON NEURAL NETWORKS WITH ORTHOGONAL ACTIVATION FUNCTIONS

J. Jovankovič * M. Žalman**

**Department of Automation and Control*
Faculty of Electrical Engineering
Slovak University of Technology
Slovakia
jovankovic@nov1.kar.elf.stuba.sk
***Department of Automation and Control*
Faculty of Electrical Engineering
Slovak University of Technology
Slovakia
zalman@nov1.kar.elf.stuba.sk

Abstract: The paper deals with an application of neural networks with orthogonal activation functions (OAFNN) in the sensorless field oriented control structure with induction motor (IM). The OAFNN has been trained to estimate rotor speed and has been used in the speed control structure as the speed virtual sensor. The Chebychev orthogonal polynomials of fourth order were applied as activation functions for each input (i_{s2}, u_{s2}^*). The speed structure of servo drive with IM has been realized on the dSPACE DS1102 development system with 1,1kW IM. *Copyright © 2003 IFAC*

Keywords: Artificial neural network (ANN), virtual sensor (VS), neural networks with orthogonal activation functions (OAFNN), field oriented control of IM, sensorless speed control.

1. INTRODUCTION

The dynamic control structure of IM brings high precision in controlling of electromechanic values (torque, speed or position). The dynamic control is classified according to physical principles on to:

- Field oriented control (FOC),
- Direct torque and flux control (DTFC).

In both cases the aim is to achieve linear relationship between the motor current and generated torque by means of independent controlling of the magnetic flux and torque.

The FOC structures (Fig. 1) using model of IM (Fig. 2) in reference frame of the rotor magnetic flux (*Fig. 3*). The rotor magnetic flux angle of swing out (υ_s) is computed by using the model of IM (*Fig. 4*). The dynamic control can be realized:

- In closed control loop, where the electrical and mechanical values are measured,
- In the open control loop, where only the electric values are measured.

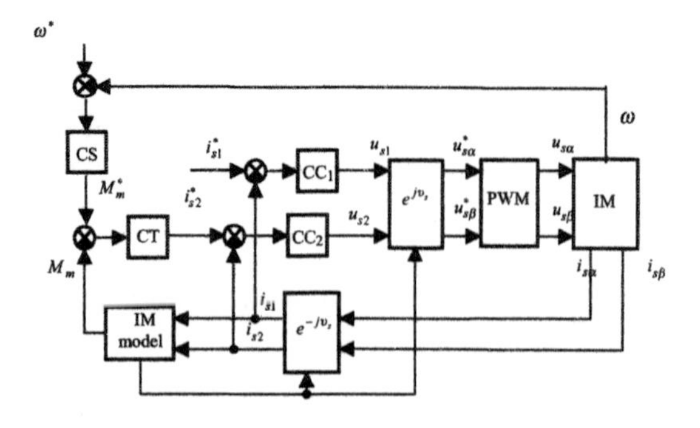

Fig. 1. Direct field oriented control scheme
CC1, CC2 -current controllers, CT – torque controller, CS – speed controller, $e^{-j\upsilon_s}$, $e^{+j\upsilon_s}$ -Park transformations, PWM – puls width modulation,

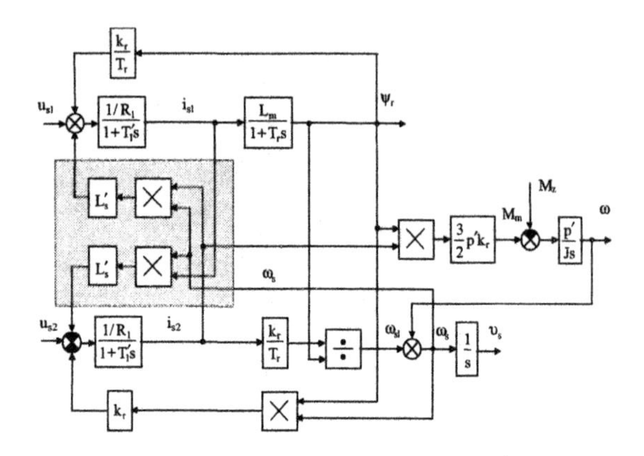

Fig. 2. Block diagram of IM in reference frame (1, 2)

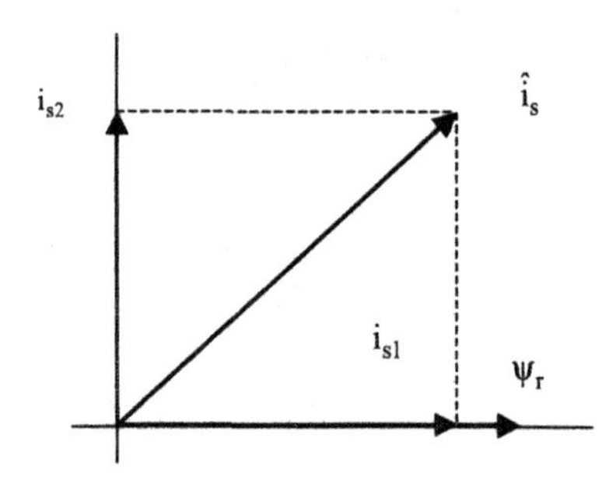

Fig. 3. Vector diagram of the stator current and rotor magnetic flux

Sensorless speed field oriented control structures of the servo drive with IM have been a subject of research on many working-place more than ten years. The speed control structures without direct measuring of the rotor speed (speed was estimated only from measured electric values) should have high dynamic and static accuracy for very low and high velocities even in case of the load torque changes. It was detail elaborated several systems based on MRAS (Tab. 1) structures [1], [2], [4]. Each reference model is very sensitive to precise parameters identification. Some parameter values (in this structure stator resistance R_s) are being markedly changed during running and their on-line identifications has to be ensured [3], [5]. Nevertheless most of these structures fail at very small or high velocities.

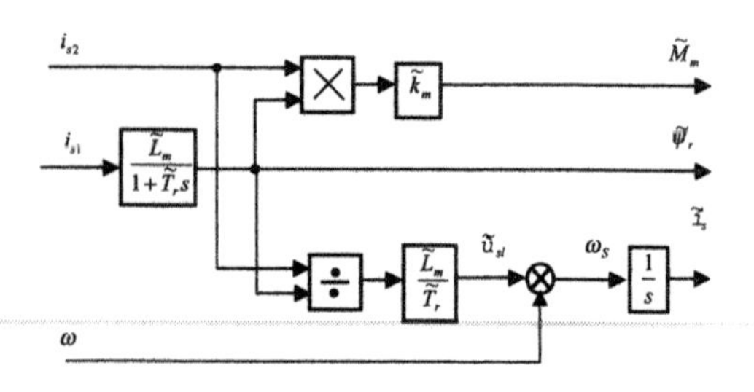

Fig. 4. Direct observer of torque magnetic flux and angle of swing out

Table 1 Flux MRAS system for speed estimation

The stator eq.- reference model
$\psi_{r\alpha} = \frac{L_r}{L_m}\left[\int (u_{s\alpha} - R_s i_{s\alpha})dt - \sigma L_s i_{s\alpha}\right]$ $\psi_{r\beta} = \frac{L_r}{L_m}\left[\int (u_{s\beta} - R_s i_{s\beta})dt - \sigma L_s i_{s\beta}\right]$
The rotor eq.- adjustable model
$s\psi_{r\alpha} = -\frac{1}{T_r}\psi_{r\alpha} - \omega\psi_{r\beta} + \frac{L_m}{T_r} i_{s\alpha}$ $s\psi_{r\beta} = -\frac{1}{T_r}\psi_{r\beta} + \omega\psi_{r\alpha} + \frac{L_m}{T_r} i_{s\beta}$

where,

$\psi_{r\alpha}, \psi_{r\beta}$ - α–axis and β–axis rotor fluxes
$i_{s\alpha}, i_{s\beta}$ - α–axis and β–axis stator currents
$u_{s\alpha}, u_{s\beta}$ - α–axis and β–axis stator voltages
R_s - stator resistance
L_r, L_s, L_m - rotor stator and mutual inductance
T_r - rotor time constant

Lately there have been some attempts to employ virtual speed sensors based on the artificial neural networks (ANN) [6], [7]. The structures with virtual sensor in the feedback are robust to state variable variations and have good performance at small velocities. Compared with some sensorless structures based on MRAS their advantage is the possibility to be trained off-line through the simulation models. The virtual sensors obtained in this way give also good experimental results.
Unlike feed forward ANN with sigmoidal (tansig) activation functions OAFNN have an easier representation. Moreover, the precise approximation of input-output relations realized with a fairly lower number of inputs and usually needs a few steps to be trained.

2. VIRTUAL SENSOR BASED ON OAFNN

Artificial neural network trained to estimate some state variables so called *virtual sensor* (VS), because after they had learned input-output pattern, they were taking over functions of the sensors. During creations of the virtual sensor it's very important to have good input-output relation for ANN. In many cases the relations are obtained by expert's knowledge of the process. For the speed VS based on OAFNN we have chosen the following relation:

$$\tilde{\omega}(k+1) = f(u_{s2}^*(k), i_{s2}(k)) \qquad (1)$$

where,

$\tilde{\omega}(k+1)$ is the rotor speed in $(k+1)^{th}$ discrete step,

$i_{s2}(k)$ is the torque making part of the stator currents in reference frame (1, 2),

$u_{s2}^*(k)$ is a second part of the reference stator voltages in the reference frame (1, 2).

Neural network structure

The Chebychev orthogonal polynomials (Tab. 2) were chosen as activation functions for each of the inputs (i_{s2}, u^*_{s2}). The current and voltage input variables have to be normalized to the range <-1,1>. With respect to the approximation accuracy it is desirable to chose the highest possible order. Concerning that with increasing the input order, the number of neurons in the hidden layer grow exponential, we decided for forth order for each inputs (*Fig. 5*).

Table 2 The Chebychev orthogonal polynomials

Explicit pattern	Recursive pattern
$\phi_i = \cos(i.\arccos(x))$	$\phi_0 = 1$, $\phi_1 = x$
	$\phi_i = 2x.\phi_{i-1} - \phi_{i-2}$

The output from the ANN can be described as

$$\tilde{\omega} = \ddot{\text{O}}^T(\mathbf{x}).\mathbf{w} \qquad (2)$$

where,

$\mathbf{w}$ is the vector of weights (dimension 1x16),

$\mathbf{x}$ is the vector of inputs (dimension 1x8)

$\mathbf{x} = \lfloor i_{s2}, u_{s2}^* \rfloor$,

$\ddot{\text{O}}^T(\mathbf{x}) = \prod_{i=1, j=1}^{i\le4, j\le4} [\phi_i(u_{s2}^*)\cdot\phi_j(i_{s2})]$ are the orthogonal functions, while $\phi_{j,i}$ are the one dimension orthogonal functions produced by each of the hidden neurons.

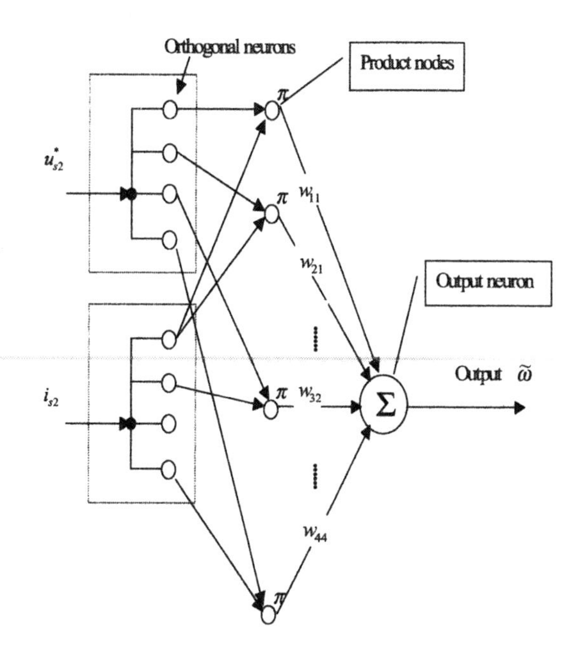

Fig. 5. Proposed orthogonal activation function-based neural network

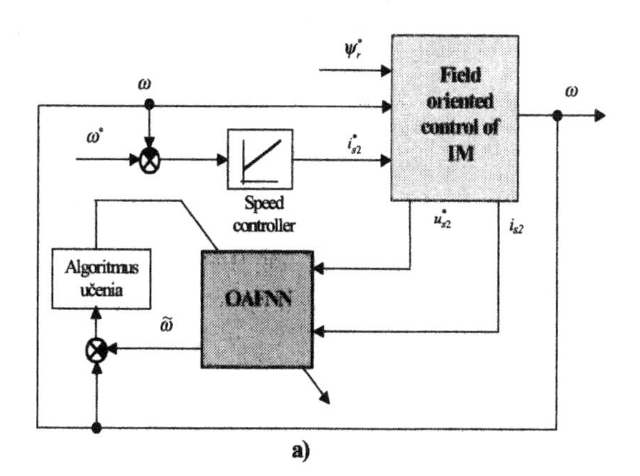

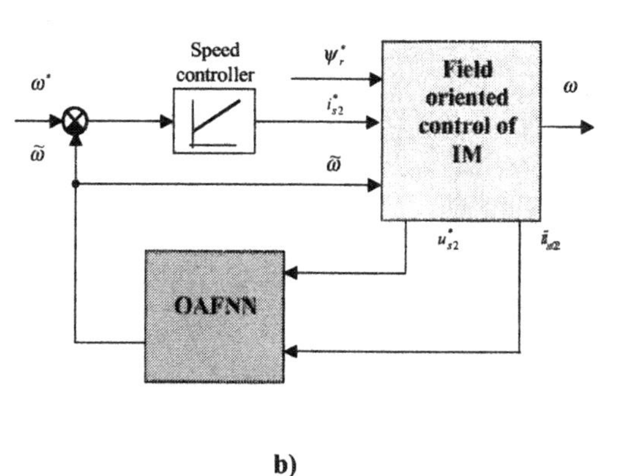

Fig.6. a) Learning algorithms, b) Sensorless field oriented control of IM with VS

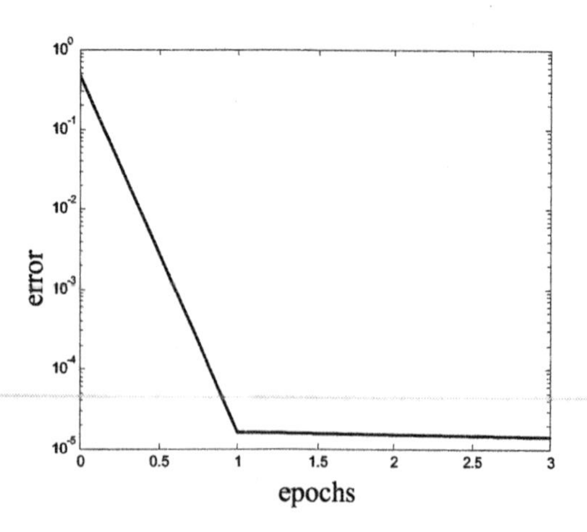

Fig. 7. Training process of OAFNN

Neural network training

The learning in neural networks is performed by adapting the network weights (**w**) such that the expected value of the mean squared error between network output and training output is minimized.

The gradient descent-based learning algorithm in supervised learning has been used. The cost function for learning performance evaluation is given by

$$J = E[(\omega - \widetilde{\omega})^2] = E[(\omega - \ddot{o}^T(\mathbf{x}).\widetilde{\mathbf{w}})^2] = \\ E(\omega^2) + \widetilde{\mathbf{w}}^T \mathbf{R} \widetilde{\mathbf{w}} - 2\mathbf{P}^T \widetilde{\mathbf{w}} \quad (3)$$

where ω is the training output, $\widetilde{\mathbf{w}}$ is the estimate of the network weight vector, $\mathbf{R} = e(\ddot{o}\ddot{o}^T)$ is the autocorrelation matrix of activation functions, and $\mathbf{P} = e(\omega\ddot{o})$ denotes the cross-correlation vector of training outputs and activation functions. For orthonormal activation functions, the autocorrelation matrix is a nth -order diagonal matrix with equal elements on the diagonal. The perfect symmetry of the error contour projection surface contributes to the fast convergence of the learning process [8].
The inputs-output pair set has been obtained from simulation of the speed control structure (Fig.6a)). As we can see from Fig. 7, one training epoch is sufficient for the OAFNN.

3. SIMULATIONS

Fig.6b) shows block diagram of the control structure with speed VS based on OAFNN. The VS was placed in the feedback loop of the rotor speed through simulations.
Fig. 8 displays behavior of speed estimation for different step changes in speed reference. Influence of load torque step amount 1Nm on speed estimation displays Fig. 9. From the simulation experiments we can see quality of the speed estimation with very good characteristics in dynamic as well as static mode. Static error of estimation is less than 2%.

4. EXPERIMENTAL RESULTS

The field oriented control structure with VS of the rotor speed was realized on the dSPACE DS1102 development system with 1,1 kW IM according to the block diagram in Fig.2b). The real-time control was realized by means of the internal PC ISA card based on the signal processor TMS320C31. The sampling time was set to 0.9ms. The measured rotor speed was used only to comparison and quality interpretation of the VS. Experimental behavior off speed estimation for different step changes in speed reference displays fig.6.

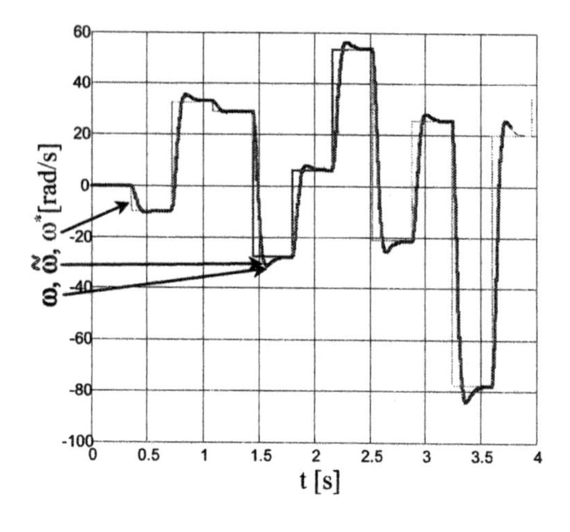

Fig. 8. Actual and estimated speed response for different step changes in speed reference

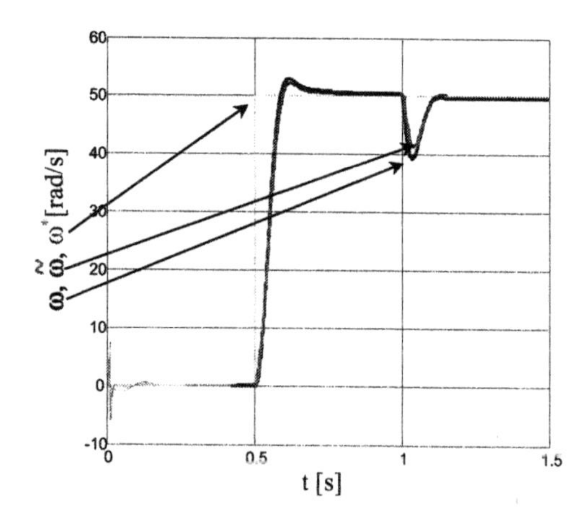

Fig. 9. Actual and estimated speed response for step load torque at 1s (1Nm)

The OAFNN has good model generalization and the sensorless structure with speed VS in feedback loop is stable for whole training range <-200, 200> rad/s. Estimated speed is rippled but oscillation wasn't transfer to mechanical speed.

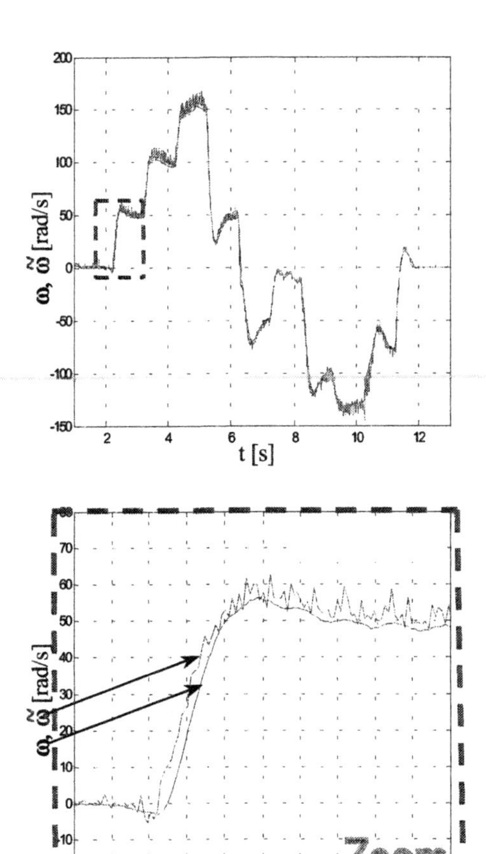

Fig. 10. Actual and estimated speed response for different step changes in speed reference

5. CONCLUSION

We were limited by the computational speed and capacity of the development system during designing of OAFNN. For the better functional approximation should be chosen higher order of orthogonal polynomials. Our real-time algorithm overrun when the Chebychev's polynomials have higher than 4th order. In general, the used OAFNN VS have a good dynamic as well as static accuracy. Also have a good generalization property and represent a powerful supply, robust on parametric variation. The OAFNN likewise the feed forward NN structures have a perspective future as the virtual sensors in a servodrives. Their capability learning from simulation examples is enabling to create database of virtual sensors for definite motors power and estimating state variables. It could be transferred to frequency transducers by means of external memory. This should bring higher quality and powerful features to control of servosystems.

This paper was partially supported by the Slovak Scientific Grant Agency grant no. 1/0153/03.

REFERENCES

Schauder, C.: *Adaptive Speed Identification for Vector Control of Induction Motors without Rotational Transducers*, In IEEE Trans. on Ind. Appl., Sept./Oct., 1992, Vol. 28, No. 5, pp.: 1054-1061.

Perng, S.-S., Lai, Y.-S., Liu, C.-H.: A Novel *Sensorless Controller for Induction Motor Drives*, In EPE'97 Trondheim, 1997, No. 0, pp.: 4.480-4.485.

Peng, F.-Z., Fukao, T.: *Robust speed indentification for speed-sensorless Vector control of induction motors*, In IEEE Trans. on Ind. Appl., Sept./Oct., 1994, Vol. 30, No. 5, pp.: 1234-124.

Oualha, A., Benmessaoud, M., Slama-Belkhodja, I., Sellami, F.: *Discrete Speed Sensorless Drive of IM: Structure and Stability*, In EPE'99 Lausanne, 1999.

Tungpimolrut, K., Peng, F.-Z., Fukao, T.: *Robust Vector Control of Induction Motor without Using Stator and Rotor Circuit Time Constants*, In IEEE Trans. on Ind. Appl., Sept./Oct., 1994, Vol. 30, No. 5, pp.: 1241-1246.

Vas P. : *Artificial-Inteligence-Based Electrical Machines and Drives*. Oxford University Press, Inc., New York 1999.

Jovankovič J., Žalman M., Abelovský M.: *Estimation of the rotor speed using feed-forward neural network*, Internacional Conference on EDPE, Podbaltské 2001.

Leondes C. T.: *Control and Dynamic Systems*. Section 2.- Zhu C., Shukla D., Paul F.-W.: Orthogonal Functions for System Identification and Control. Academic Press, San Diego 1998.

www.elsevier.com/locate/ifac

OPTIMAL CHOICE OF HORIZONS FOR PREDICTIVE CONTROL BY USING GENETIC ALGORITHMS

R. Haber*, U. Schmitz*, R. Bars**

**Department of Plant and Process Engineering, Laboratory of Process Control, University of Applied Science Cologne, D-50679 Köln, Betzdorfer Str. 2, Germany fax: +49-221-8275-2836 and e-mail: robert.haber@fh-koeln.de, ulrich.schmitz@fh-koeln.de*
***Department of Automation and Applied Informatics, Budapest University of Technology and Economics, H-1111, Budapest, Goldmann Gy. tér 3., Hungary; fax : +36-1-463-2871.and e-mail : bars@aut.bme.hu*

Abstract: For predictive control in industry often very long horizons for control error and manipulated signal are used because of the slow processes which take place in the petrochemical industry. In order to reduce the computational effort some commercial predictive control program packages offer the ability to reduce the number of points in both horizons but do not recommend how to select the points which have to be considered in the horizon of the control error and manipulated variable. In this work the authors introduce an optimal choice not only of the horizon lengths itself but also for the strategy of reducing the number of points in the horizons. A genetic optimization algorithm was used both for the search for the optimal length of the horizons and for the best allocation of the points in the horizons. The results of the optimization process where used to deduct a simple rule. *Copyright © 2003 IFAC*

Keywords: Predictive control, long horizons, blocking, coincidence points, compaction points, optimal choice of horizons

1. INTRODUCTION

Predictive control is often used with very long horizons, once because of the often deployed slow processes, but also to achieve a better robustness of the control loop. The better robustness can be achieved by longer horizons because the predictive control algorithm allocates more changes in the horizon which result in a more smooth, slower and more robust control of the plant.

On the other hand the problem with long horizons is, that the computational effort of the controller rises with the length of both horizons. The length of the control error horizon raises the computational effort for the necessary simulations while the length of the manipulated variable horizon raises the computational effort for the optimization of the cost function by increasing the optimization dimension.

In order to handle such long horizons a common technique is to consider a reduced number of points in the control error horizon as well as in the manipulated signal horizon. The technique which views only a reduced number of points in the manipulated signal horizon is named *blocking technique* while the technique which takes care only of a reduced number of points in the control error horizon is named *using coincidence points*.

Both, the length of horizons as well as the allocation of the reduced number of points in both horizons are optimized by a genetic algorithm in subject to the minimum overshoot and the shortest settling time. The optimization is done in a two-stage process:

1. the length of the control error and manipulated variable horizon were optimized without reducing the computation points in the horizons,

2. the allocation of the computation points in both horizons were optimized so that the controlled signal should approximate the controlled signal without reducing the computation points in the horizons.

2. PREDICTIVE CONTROL ALGORITHM

The objective of the algorithm is to minimize the cost function of the generalized predictive control

$$J = \sum_{n_e = n_{e_1}+1}^{n_{e_2}+1} \left[y_{ref}(k+n_e+d) - \hat{y}(k+n_e+d) \right]^2$$

$$+ \lambda_u \sum_{j=1}^{n_u} \Delta u(k+j-1)^2 \Rightarrow \min_{\Delta u}$$

with

n_{e1}: begin of the control error horizon over the dead time

n_{e2}: end of the control error horizon over the dead time

n_u: number of manipulated signal increments (length of the manipulated signal horizon)

y_{ref}: future reference signal

$\hat{y}$: predicted control signal

Δu: manipulated signal increments.

d: mathematical discrete dead time

λ_u: weighting factor for manipulated signal increments

The analytical solution is given by (Clarke, 1987)

$$\mathbf{\Delta u} = \left(\mathbf{H}_\Delta^T \mathbf{H}_\Delta + \lambda_u \right)^{-1} \mathbf{H}_\Delta^T \left(\mathbf{y}_{ref} - \hat{\mathbf{y}}_{free} \right)$$

with

$\mathbf{H}_\Delta$ impulse transfer function matrix

$\mathbf{\Delta u}$ manipulated signal increment vector

$\mathbf{y}_{ref}$ future reference signal vector

$\hat{\mathbf{y}}_{free}$ free response

As usually only the first manipulated variable of the hole computed manipulated variable horizon is applied to the process (receding horizon strategy).

3. OPTIMAL CHOICE OF THE NOT COMPACTED HORIZONS

The first task was the optimal choice of the horizons (both the control error horizon as well as the manipulated signal horizon). As most processes in industry can be approximated by a aperiodic second order process, such processes were used as basis for the optimization approach. No dead time was considered because with predictive control the behavior remains the same as without dead time – of course delayed by the dead time. The examined processes are shown in detail in Table 1 with their small (T_{small}) and big (T_{big}) time constants.

Table 1 Transfer functions and step responses of the processes

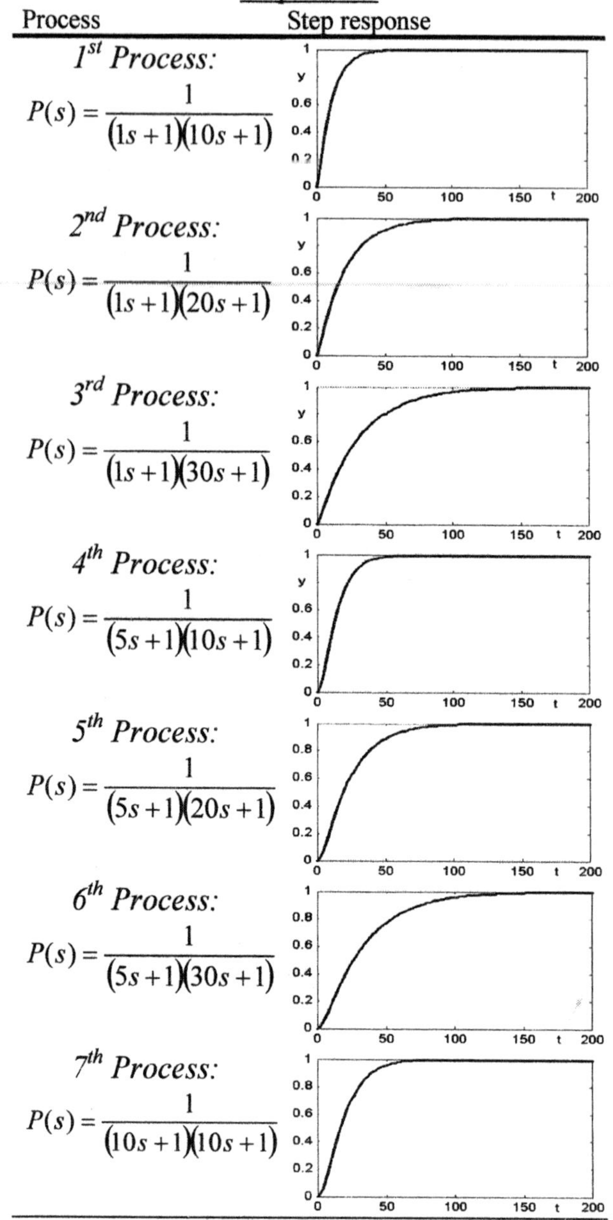

In all cases the sampling time was ΔT=0.2sec.

The length of the horizons were optimized according to three criteria in case of a reference step from 0 to 1:

- the settling time,
- the maximum overshoot,
- the length of the horizons.

The third criterion was taken into consideration to prevent the optimization algorithm to make the horizons longer than necessary (a horizon longer than necessary does not have a negative influence on the control result but on the computation time). Therefore, the lengths of the horizons were considered also, but only if they are longer than a default value:

- control error horizon length: four times the sum of the time constants ($n_{e2,max} > 4T_\Sigma$)
- manipulated signal horizon length: the sum of the time constants ($n_{u,max} > T_\Sigma$)

These default values have been proven empirically as satisfactory limits.

The optimization itself was performed by means of the genetic optimization algorithm written by (Sekaj, 2002). The cost function to be optimized was in all cases

$$
\begin{aligned}
J &= J_{set} + J_{ov} + J_h \\
&= \lambda_{98\%} t_{98\%} \\
&+ \lambda_{ov} \left| y_{\max} - y_{ref} \right| \\
&+ \lambda_{e,hor} \sum \left(n_e > 4 \frac{T_\Sigma}{\Delta T} \right) \\
&+ \lambda_{u,hor} \sum \left(n_u > \frac{T_\Sigma}{\Delta T} \right) \\
&\Rightarrow \min_{n_{e1}, n_{e2}, n_u}
\end{aligned}
$$

In the simulation all weighting factors were set to $\lambda_{98\%} = \lambda_{ov} = \lambda_{e,hor} = \lambda_{u,hor} = 1$. The weighting of the change in the manipulated variable was $\lambda_u = 0.1$.

The figures 1 to 7 show the results of the optimization for the seven processes. In all figures the maximum overshoot and the settling time are marked with thin lines.

The cost values for all optimization processes are shown in Table 2

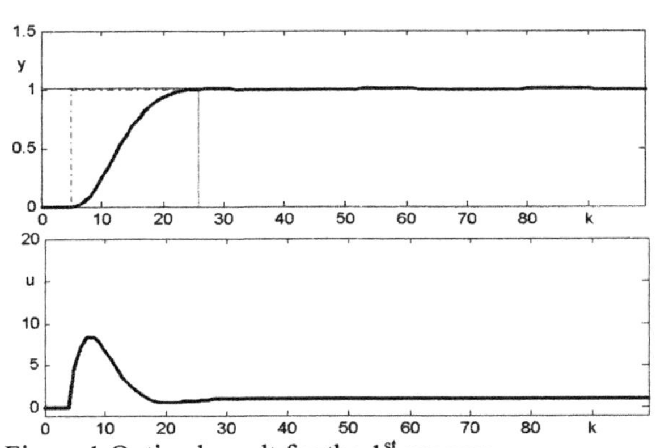

Figure 1 Optimal result for the 1st process

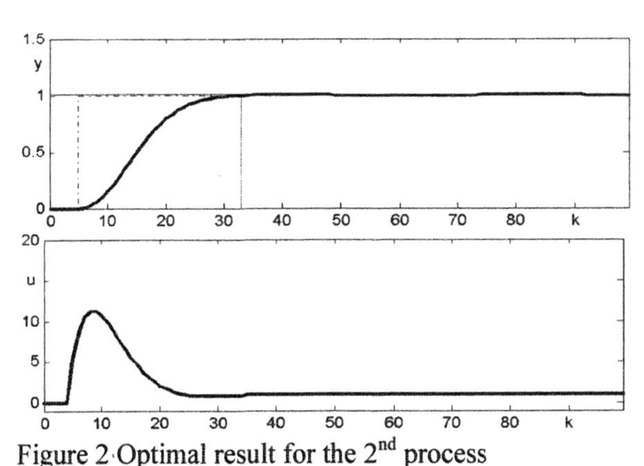

Figure 2 Optimal result for the 2nd process

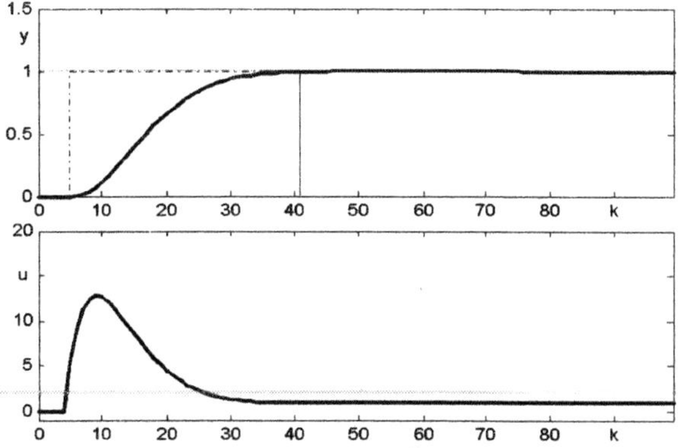

Figure 3 Optimal result for the 3rd process

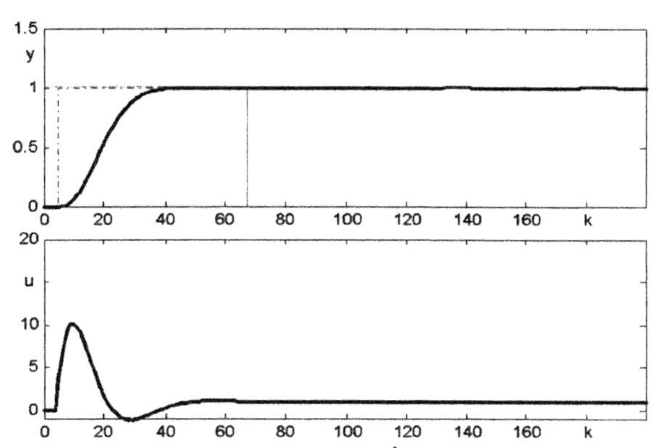

Figure 4 Optimal result for the 4th process

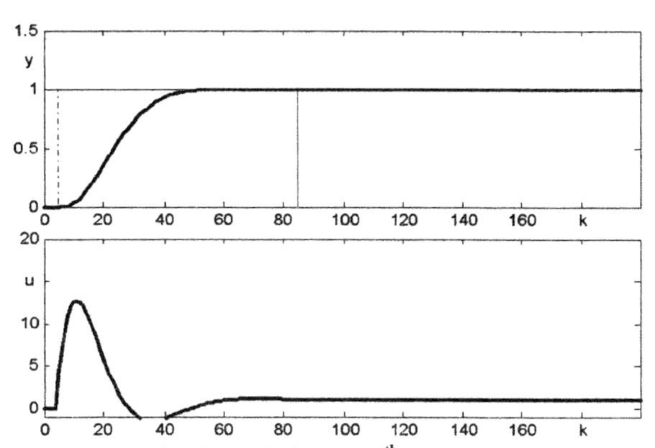

Figure 5 Optimal result for the 5th process

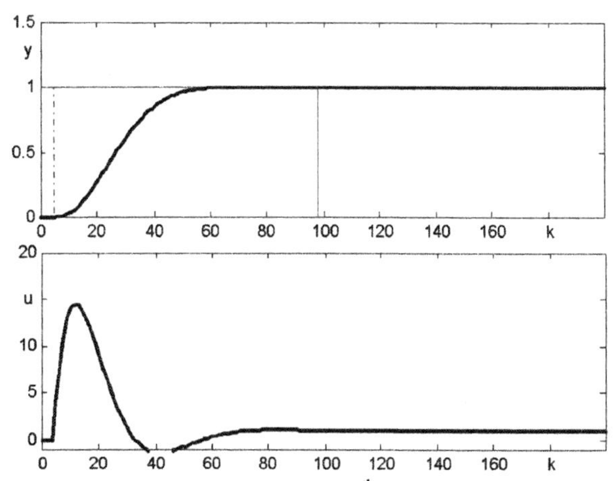

Figure 6 Optimal result for the 6th process

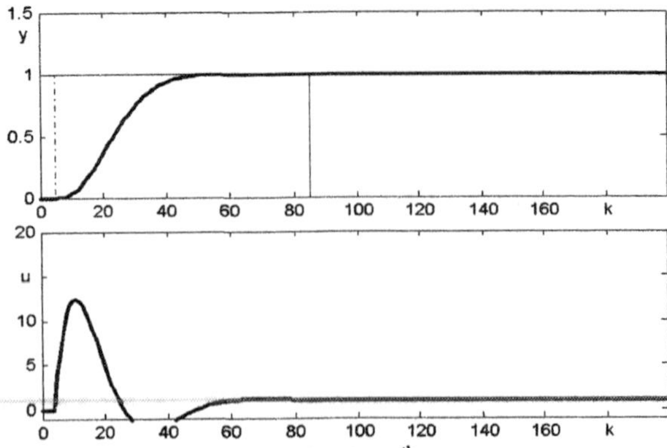

Figure 7 Optimal result for the 7th process

Table 2 Cost values of the optimization process

Process	Overall cost factor	Cost factor for overshoot	Cost factor for settling time	Cost factor for horizon length
1st	5.27	0.07	5.20	0.00
2nd	6.76	0.16	6.60	0.00
3rd	8.27	0.07	8.20	0.00
4th	13.40	0.00	13.40	0.00
5th	17.00	0.00	17.00	0.00
6th	19.60	0.00	19.60	0.00
7th	17.00	0.00	17.00	0.00

The results of the horizon lengths are shown in Table 3. For better comparison the sums of the time constants in discrete time are also shown in Table 3.

Table 3 Resulting horizon lengths of the optimization process.

Process	n_{e1}	n_{e2}	n_u	$T_{big}/\Delta T$	$T_{small}/\Delta T$
1st	2	181	4	50	5
2nd	4	419	4	100	5
3rd	6	543	4	150	5
4th	1	68	9	50	25
5th	3	91	9	100	25
6th	5	113	9	150	25
7th	0	81	10	50	50

As a rule of thumb the horizon lengths can be chosen as follows:

- The begin of the control error horizon should be 2% of the big time constant, thus: $n_{e_1} \approx 0.04\left(\frac{T_{big}}{\Delta T} - \frac{T_{small}}{\Delta T}\right)$
- The end of the control error horizon should be nearly

$$n_{e_2} \approx \begin{cases} 18\frac{T_{big}}{T_{small}}, & \text{for } \frac{T_{big}}{T_{small}} \geq 10 \\ \frac{1}{2}\left(\frac{T_{big}}{\Delta T} + \frac{T_{small}}{\Delta T} + 60\right), & \text{for } \frac{T_{big}}{T_{small}} < 10 \end{cases}$$

- The length of the manipulated variable horizon should be

$$n_u \approx \begin{cases} 0.8\frac{T_{small}}{\Delta T}, & \text{for } \frac{T_{big}}{T_{small}} \geq 10 \\ 0.4\frac{T_{small}}{\Delta T}, & \text{for } 1 < \frac{T_{big}}{T_{small}} < 10 \\ 0.2\frac{T_{small}}{\Delta T}, & \text{for } \frac{T_{big}}{T_{small}} \leq 1 \end{cases}$$

4. OPTIMAL CHOICE OF THE COMPACTED HORIZONS

The next task was to reduce the large amount of point to P_e=20 for the control error horizon and P_u=10 for the manipulated variable horizon. As it can be seen in Table 3, only the horizon lengths for the processes with a time constant ratio of 10 or more result in very long horizons. Thus the further examinations are only evaluated on these (three) processes.

The aim of the optimization process was to achieve a control with reduced number of points (a maximum of 20 points for control error horizon, thus 20 coincidence points, and a maximum of 10 for the manipulated variable horizon, thus 10 blocking points) that is nearest to the control with the optimized size of the horizon but without blocking or using coincidence points, computed in chapter 3 of this paper. The cost factor to be optimized is described by

$$\begin{aligned} J &= J_y + J_u + J_h \\ &= \lambda_{y,dev}\sum_{k=0}^{100}\left(y_{full}(k) - y_{red}(k)\right)^2 \\ &+ \lambda_{u,dev}\sum_{k=0}^{100}\left(u_{full}(k) - u_{red}(k)\right)^2 \\ &+ \lambda_{Pe}\sum\left(P_e > 4\frac{T_\Sigma}{\Delta T}\right) \\ &+ \lambda_{Pu}\sum\left(P_u > \frac{T_\Sigma}{\Delta T}\right) \\ &\Rightarrow \min_{P_e,P_u} \end{aligned}$$

with

- y_{full}: controlled variable, controlled with all points in the horizons
- y_{red}: controlled variable, controlled with a reduced number of points in the horizons
- u_{full}: manipulated variable with all points in the horizons
- u_{red}: manipulated variable with a reduced number of points in the horizons
- P_e: allocation of the points in the control error horizon
- λ_{Pe}: weighting factor for the allocation of the points in the control error horizon
- P_u: allocation of the points in the manipulated variable horizon
- λ_{Pu}: weighting factor for the allocation of the points in the manipulated variable horizon

$T_\Sigma/\Delta T$: sum of all time constants of the process relative to the sampling time

$\lambda_{y,dev}$: weighting factor for the error between the controlled variables with and without reduced horizons, here chosen to be $\lambda_{y,dev} = 1$

$\lambda_{u,dev}$: weighting factor for the error between the manipulated variable with and without reduced horizons, here chosen to be $\lambda_{u,dev} = 0.1$

The Figures 8 to 10 shows the control result of the genetic optimization process by solid lines. The dashed lines are the reference of the control without the compacted horizons and the thin dash-dotted lines are the reference signal for the control. The Figures 11 to 13 show the position of the optimal coincidence points both, in the step response and in the impulse response of the corresponding processes, marked as circles.

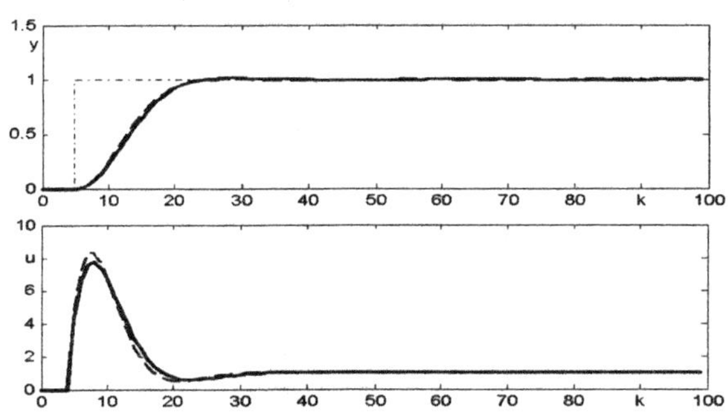

Figure 8 Control with the optimal allocation of the reduced number of points in the horizons for the 1st process

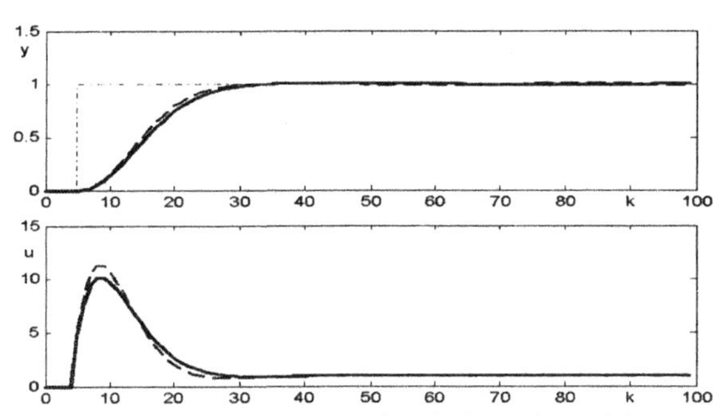

Figure 9 Control with the optimal allocation of the reduced number of points in the horizons for the 2nd process

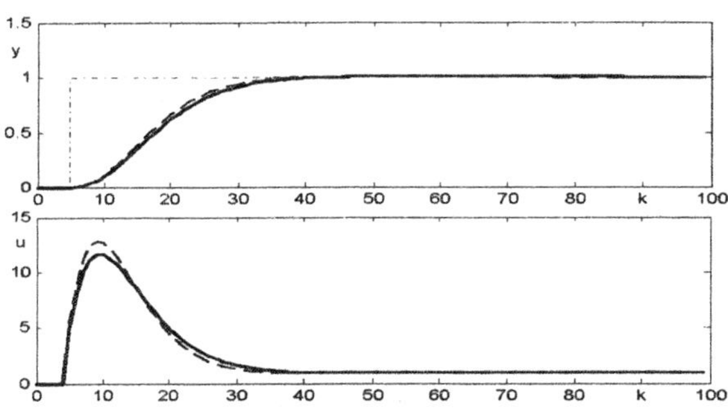

Figure 10 Control with the optimal allocation of the reduced number of points in the horizons for the 3rd process

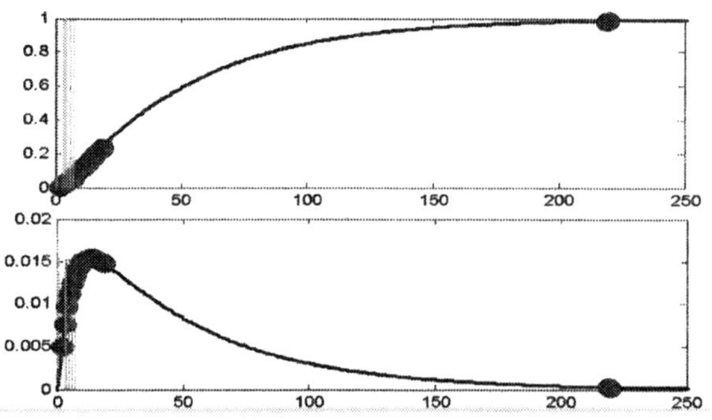
Figure 11 Optimal allocation of the coincidence and blocking points for the 1st process

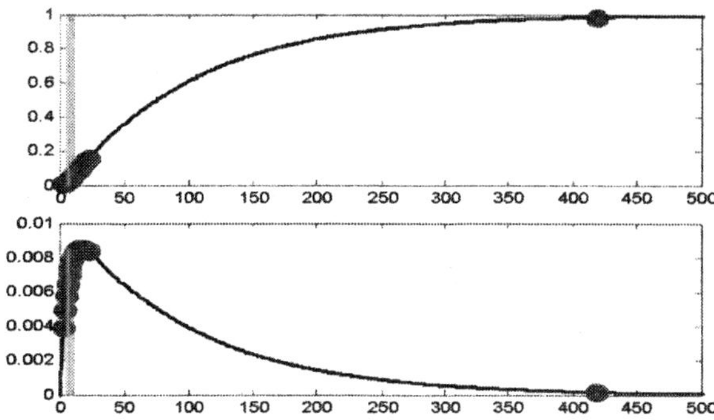
Figure 12 Optimal allocation of the coincidence and blocking points for the 2nd process

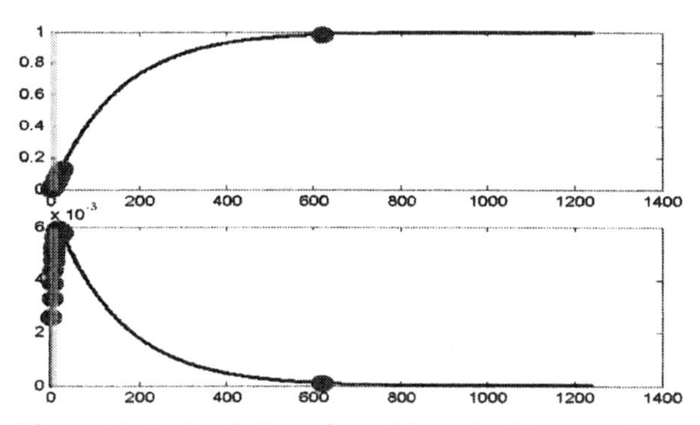

Figure 13 Optimal allocation of the coincidence and blocking points for the 3rd process

The optimal coincidence and blocking are shown in Table 4 for the investigated cases.

The cost factors with their components (the sub-cost-factors) are shown in Table 5.

As it can be seen in the Figures 10 to 13 and in Table 4, the best result can be achieved if the coincidence points are settled at the rising part of the impulse response of the process. Also it can be seen that the coincidence points should be more on the rising part of the impulse response if the ratio between the big and the small time constant becomes bigger - the 3rd process has the biggest ratio of 30 and all points are allocated on the rising part of the impulse response, see Figure 13. On the other hand, if the ratio between the time constants is lower, the points are located both, on the rising part of the impulse response and also at the maximum and on the early part of the falling part of the impulse response. This behavior may be founded in the fact that the bigger the ratio between the time constants of a second-order process is, the more the process can be approximated by a first-order process. In addition to this some few points are settled at the very end of the step response, in a part of the response, where the process acts mostly static to the change in the manipulated signal (see Table 4). Remark that the sum of the discrete time constants are 55 for the 1st, 105 for the 2nd and 155 for the 3rd process and therefore, the control error horizon length limits (to achieve J_h=0) were 220 for the 1st, 420 for the 2nd and 620 for the 3rd process, respectively.

Table 4 Optimal allocation of coincidence and blocking points

1st process		2nd process		3rd process	
P_e	P_u	P_e	P_u	P_e	P_u
2	1	3	1	3	1
3	3	4	5	4	7
4	4	5	6	5	8
5	5	6	7	6	9
6	6	7	8	7	10
7	7	8	9	8	11
8		9	10	9	12
9		10	11	10	13
10		11		11	14
11		12		12	15
12		13		13	
13		14		14	
14		15		15	
15		16		16	
16		17		17	
17		18		18	
18		19		20	
19		20		22	
219		22		26	
220		419		620	

Table 5 Resulting cost factors of the optimization processes for the reduced number of points in the horizons

Process	J	J_y	J_u	J_h
1st	0.332	0.00971	0.322	0.00
2nd	1.07	0.0329	1.04	0.00
3rd	1.17	0.0291	1.15	0.00

The allocation of the blocking points show an other behavior. The first point is fixed to the value 1 because if there is no change in the first manipulated signal allowed, the controller cannot work with a receding horizon strategy where only the first signal of the manipulated signal sequence is put out. Remark that the blocking points denote "allowed movements in the manipulated at the given discrete time".

The second to the last blocking-points spread nearly linearly over the time constants. The coincidence points also spread nearly linearly over the time constants but the data are to few to give an authentic equation for the begin and the end of the blocking and/or the coincidence points.

CONCLUSION

For predictive control often very long horizons for the control error and the manipulated signal have to be used. In order to reduce the computational effort the control error is considered in not all sampling points of the control error horizon and changes in the manipulated variable are not allowed in all sampling points of the manipulated variable horizon. Thus, there are two kinds of controller parameters to be chosen:

1. length of the control error and manipulated variable horizon and
2. allocation of coincidence and blocking points in the horizons.

A tow-stage optimization procedure was applied:

1. the length of the control error and manipulated variable horizon were optimized without reducing the computation points in the horizons,
2. the allocation of the computation points in both horizons were optimized so that the controlled signal should approximate the controlled signal without reducing the computation points in the horizons.

A genetic optimization algorithm was used both for the search for both the optimal length of the horizons and the best allocation of the points in the horizons. The results are interpreted for a second-order linear process which is a typical model in the process industry.

ACKNOWGLEDGMENT

The work has been supported earlier by the Ministry of Science and Research of NRW (FRG) in the program „Support of the European Contacts of the Universities / Förderung der Europafähigkeit der Hochschulen" and now is supported by the University of Applied Science Cologne in the program "Advanced Process Identification for Predictive Control" and by the program of EU-Socrates. The third author's (R.B.) work was also supported by the fund of the Hungarian Academy of Sciences for control research and partly by the OTKA fund TO29815. All supports are kindly acknowledged.

Authors are thankful to Dr. I. Sekaj (Department of Automatic Control Systems, Faculty of Electrical Engineering and Information Technology, Slovak University of Technology, Slovak Rep.) for his help in applying the genetic algorithm for the case presented.

LITERATURE

Clarke, D.W., C. Mohtadi, P.S. Tuffs (1987). Generalized predictive control – Part I. The basic algorithm. Automatica, 24, 2, 137-148.

Sekaj, I. (2002), Genetic Algorithms with Changing Criterion Functions, Intelligent Technologies - Theory and Applications, P.Sincak et al. (Eds), IOS Press, 2002, ISBN 1 58603 256 9, pp. 183-188

MPC AS A TOOL FOR SUSTAINABLE DEVELOPMENT INTEGRATED POLICY ASSESSMENT

B. Kouvaritakis *,** **M. Cannon** * **G. Huang** *

* *University of Oxford, Department of Engineering Science, Parks Rd, Oxford, OX1 3PJ, UK*
** *Corresponding author, basil.kouvaritakis@eng.ox.ac.uk*

Abstract: Recent work has laid the foundations of a quantitative approach to sustainable development policy assessment through a formulation which is both stochastic and multi-objective. The key to this development was a static open-loop optimization. The current paper explores the connections between such an approach and model predictive control (MPC), and proposes a reformulation which retains the stochastic and multi-objective nature of the problem, but introduces dynamics and gives consideration to closed-loop performance and stability. The proposed approach has wide applicability but it is hoped that it will provide sustainable development practitioners in particular with new insights.

Keywords: Sustainable development; model predictive control; stochastic systems

1. INTRODUCTION

The aim of this paper is to explore the connections that may exist between predictive control and policy assessment in problems of sustainable development. Recent consideration of the latter (Kouvaritakis, 2000*a*) proposed a receding horizon application of a static optimization. A landmark in its own right, this work pioneered a quantitative and rigorous approach to the problem of assessing policy. However the optimization allowed only for one policy adjustment (to be implement in the near future), although of course receding horizon application of this strategy could be used to account for model mismatch at future time instants. Furthermore, both the constraints and objective functions placed all the emphasis on cumulative effects at the end of the prediction horizon. More crucially, no explicit consideration was given to the closed-loop effects of the receding horizon application of what in essence is an open-loop optimization. Whereas it is possible that discounting and the effect of constraints dictated by the problem may avert the onset of instability, there is no guarantee that the results thus obtained would be optimal. It is nevertheless still the case that this approach is a major step forward in that it poses the problem in a strongly stochastic formulation and casts it in a multi-objective formulation.

The current paper retains the stochastic and multi-objective attributes of the earlier formulation, but also introduces dynamics and gives guarantees of closed-loop stability. This is achieved by adapting the methodology of model predictive control to the problem. After a review of the earlier work in Section 2, a discussion follows in Section 3 on a reformulation which permits the introduction of dynamics. The choice of suitable models is explored in Section 4 and is followed in Section 5 by the discussion of a decomposition of the predictive control strategy into a two-phase optimization (one performed offline with the view to defining a suitable setpoint, the other concerning online optimization of tracking). Both phases are cast in a stochastic setting and are applicable to a multitude of problems, other than those related specifically to sustainable development. The results of the paper are illustrated by a simple example in Section 6.

2. EARLIER WORK

Integrated policy assessment to be investigated under MINIMA-SUD (Kouvaritakis, 2002) involves identifying measurable sustainability indices and common instruments, modelling the impact of instruments on these indices, and the development of a tool that helps assess the efficacy of policy in minimizing risk and maximizing benefit. Key ingredients in this approach are the strongly stochastic nature of models and the formulation of the problem as a multi-objective optimization (Kouvaritakis, 2000*a*)

One area within this very wide framework concerns the assessment of public Research and Development budget allocations, u_i, within a constrained total allocation, between alternative power generating technologies: e.g. BGT, Biomass Gasification Gas Turbines (u_1); GTC, Gas Turbine in Combined Cycle (u_2); SCC, Super Critical Coal (u_3); or Wind Turbines (u_4). Pre-specified targets are to be met by indicators

y_j: e.g. total CO_2 emissions (y_1); cumulative total emissions (y_2); cumulative discounted energy costs (y_3); or energy cost per capita for particular regions (y_4). Discounting is commonly introduced to account for the fact that future benefit is not as important as benefit at current time, and conversely that near future cost is more critical than far future cost. Sustainability implies that development should occur now without compromising the potential for development in the future, but clearly the two are not given equal weighting.

The effect of u_i on y_j is dynamic and strongly stochastic. Earlier work (ISPA) (Kouvaritakis, 2000*a*) concentrated on the latter aspect and ignored the former in deriving a static optimization formulation of the problem. Unlike MPC, which optimizes the trajectories of u_i, y_j over a horizon, ISPA considered the optimization of output behaviour at a single future instant over a budget allocation being made at current time only. Given the strongly stochastic nature of the dependence of outputs on inputs the problem was formulated as:

$$\max_{u_i,\, i=1,\dots,n_u} \quad \Pr(y_1 > A_1) \tag{1a}$$

$$\text{subject to} \quad \Pr(y_j \le A_j) \ge p_j \quad j=2,\dots,n_y \tag{1b}$$

$$\sum_{i=1}^{n_u} u_i \le B, \quad u_i \ge 0 \quad i=1,\dots,n_u \tag{1c}$$

where (1c) ensures that R&D allocations are positive and do not exceed a total allocation bound B. The inequalities (1b) are conventionally imposed on the sum of predicted output values over a horizon. However given that the predicted outputs are assumed to depend linearly on the actions u_i, this does not affect the overall nature of the problem. The formulation is likewise unchanged by the consideration that all indicators y_j (apart from CO_2 emissions) are discounted.

Implicit in (1a) and (1b) is a stochastic dependence of y_j on u_i, $i=1,\dots,n_u$, which is modelled statically as

$$y_j = r_j^T u, \qquad u = [u_1 \ \cdots \ u_{n_u}]^T \tag{2}$$

where r_j is a vector of impact coefficients. Extensive simulations on a detailed model (Kouvaritakis, 2000*b*) were used to compute the probability distributions for r_j, $j=1,\dots,n_y$, and it was determined that a reasonable approximation was given by

$$r_j \sim \mathcal{N}(\bar{r}_j, R_j) \tag{3}$$

where the expectations $\bar{r}_j$ and covariance matrices R_j, $j=1,\dots,n_y$, are computable constants. It was further confirmed that the error in the above approximation made little difference to the optimal solution of (1a-c) and therefore the model (2-3) was used to formulate the equivalent optimization problem:

$$\max_{u} \quad \mathfrak{N}\left(\frac{\bar{r}_1^T u - A_1}{\|u\|_{R_1}}\right) \tag{4a}$$

$$\text{subject to} \quad \mathfrak{N}\left(\frac{A_j - \bar{r}_j^T u}{\|u\|_{R_j}}\right) \ge p_j \quad j=2,\dots,n_y \tag{4b}$$

$$\sum_{i=1}^{n_u} u_i \le B, \quad u_i \ge 0 \quad i=1,\dots,n_u \tag{4c}$$

where $\mathfrak{N}(z) = \Pr(Z \le z)$ for $Z \sim \mathcal{N}(0,1)$, and $\|u\|_{R_j}^2 = u^T R_j u$. To avoid the unacceptable position of gambling, the probabilities in (4a) and (4b) must not be less than 0.5 and thus (4a-c) can be expressed as

$$\min_{u} \quad \frac{A_1 - \bar{r}_1^T u}{\|u\|_{R_1}} \tag{5a}$$

$$\text{subject to} \quad \|u\|_{R_j} \le \frac{A_j - \bar{r}_j^T u}{\alpha_j} \quad j=2,\dots,n_y \tag{5b}$$

$$\sum_{i=1}^{n_u} u_i \le B, \quad u_i \ge 0 \quad i=1,\dots,n_u \tag{5c}$$

where $\mathfrak{N}(\alpha_j) = p_j$. It can be shown that (5a-c) defines an efficiently solvable convex optimization problem. The strong stochastic nature of the problem is reflected in the probabilistic objective and constraints (1a-b). This is distinct from formulations of the problem in terms of expected values, which results in a deterministic optimization, thus ignoring the highly stochastic nature of sustainable development indicators.

3. INTRODUCING DYNAMICS

The connections between the above formulation and MPC are obvious, although there are some distinctive differences. The most significant of these is that the dependence of y_j on the actions u_i is modelled through a static relationship of the form (2), which has no explicit dependence on time, whereas it is expected that the indicators y_j depend on the profile of budget allocations u_i over the prediction horizon. Features allied to the non-dynamic formulation are: (i) the use of a single future instant in the output prediction (which implies no guarantee of closed-loop stability); and (ii) the assumption that budget allocation adjustments are performed once at current time (which implies a suboptimal mean-level predicted input sequence).

This paper introduces dynamics into the problem and considers the effects of a receding horizon strategy on closed-loop behaviour. For simplicity we initially consider the SISO case, with inputs and outputs at instant k denoted by $u(k), y(k)$. Simulations of shocks (superimposed on baseline values) delivered to actions (Kouvaritakis, 2000*b*) associated with BGT, GTC, SCC, or Wind, show that the output responses (relative to baseline responses) can be modelled adequately by Hammerstein models. For shocks of up to 10% such models can be reduced to linear ARMA (autoregressive moving average) representations:

$$y(k+1) = \frac{b(z)}{a(z)} u(k), \quad \begin{matrix} b(z) = b_0 + \cdots + b_{m-1} z^{-m+1} \\ a(z) = 1 + \cdots + a_n z^{-n} \end{matrix} \tag{6}$$

or MA (moving average) representations:

$$y(k+1) = g(z)u(k), \quad g(z) = g_0 + \cdots + g_{q-1} z^{-q+1} \tag{7}$$

The coefficients of (6) or (7) can be identified via black-box identification procedures which introduce a random process ε to account for the presence of exogenous disturbances and/or modelling errors

$$y(k+1) = \frac{b^0(z)}{a^0(z)} u(k) + \frac{T(z)}{a^0(z)} \varepsilon(k) \tag{8}$$

and

$$y(k+1) = g^0(z)u(k) + \varepsilon(k) \quad (9)$$

where $T(z)$ is introduced in (8) to improve model accuracy given some prior knowledge of the system dynamics and the power spectral density function of ε. Under a Gaussian assumption on ε, the application of generalized least squares to (8) or least squares to (9) generates estimates for the vectors of model parameters, which have distribution

$$\theta \sim \mathcal{N}(\bar{\theta}, \Theta) \quad (10)$$

where $\theta = [b_{m-1} \cdots b_0 \; a_n \cdots a_1]^T$ for (6) or $\theta = [g_{q-1} \cdots g_0]^T$ for (7), and $\bar{\theta}, \Theta$ denote the expected value and covariance of θ. To achieve the desired trade-off between variability and bias, the model order n (and m) for (6), or q for (7), can be selected on the basis of well-known criteria (e.g. (Akaike, 1974)).

This modelling framework provides a direct bridge between the control problem and policy assessment. To see this, consider for example the i-step ahead prediction generated at time k by (7):

$$\begin{aligned} y(k+i|k) &= \theta^T \tilde{u}(k+i-1|k), \\ \tilde{u}(k+i-1|k) &= [u(k+i-q|k) \cdots u(k+i-1|k)]^T \end{aligned} \quad (11)$$

For $i \geq q$, (11) has the form of (2) since the the predicted values of y depend linearly on the decision variables u, and the distribution for θ in (10) is analogous to the distribution (3) for r. For $i < q$, only the last i elements of $\tilde{u}(k+i-1|k)$ are free variables, the remaining elements being past values of u which are known at time k. Thus partitioning $\tilde{u}(k+i-1|k)$:

$$\tilde{u}(k+i-1|k) = [u^{pT}(k) \; \hat{u}^T(k+i-1|k)]^T, \quad (12)$$

where $\hat{u}$ contains the free variables in $\tilde{u}$ and u^p contains past values of u, we have

$$y(k+i|k) \sim \mathcal{N}\left(\bar{\theta}^T \begin{bmatrix} u^p(k) \\ \hat{u}(k+i-1|k) \end{bmatrix}, \left\| \begin{matrix} u^p(k) \\ \hat{u}(k+i-1|k) \end{matrix} \right\|_{\Theta}^2 \right) \quad (13)$$

This leads to a formulation in which optimization problems based on probabilistic objective functions and stochastic constraints such as (1a-c) remain efficiently solvable convex programs.

4. MODEL SELECTION

Each of the alternative models of (6) and (7) has distinct advantages and disadvantages. In particular the dimension of the parameter space of (7) is in general much greater than that of (6). However, the identification of (6) through (8) is more difficult since regression on the past output values implies that the effective noise component is correlated.

More importantly however, the prediction equations generated by (6) are of the form:

$$\begin{aligned} y(k+i+1|k) &= -a^T \tilde{y}(k+i|k) + b^T \tilde{u}(k+i|k) \\ \tilde{y}(k+i|k) &= [y(k+i-n+1|k) \; \cdots \; y(k+i|k)]^T \\ \tilde{u}(k+i|k) &= [u(k+i-m|k) \; \cdots \; u(k+i|k)]^T \\ a = [a_n \cdots a_1]^T&, \; b = [b_{m-1} \cdots b_0]^T \end{aligned} \quad (14)$$

This form of prediction model has the major disadvantage that both a and $\tilde{y}(k+i|k)$ for $i \geq 1$ contain random variables, and the term $a^T \tilde{y}(k+i|k)$ appearing in (14) is therefore not normally distributed. For this reason, it is convenient to adopt (7) in preference to (6).

Remark 1. It may appear that MA models restrict attention to open-loop stable systems only, but in the context of MPC this difficulty can be overcome through the use of bi-causal models (Kouvaritakis and Rossiter, 1991). Given that the predicted input trajectories must be stabilizing to ensure closed-loop stability (implying a dependence of future inputs on the past values of inputs/outputs), it can be shown that truncation errors arising from use of bi-causal MA representations of unstable prediction models converge to zero as the order of the MA model increases.

Remark 2. Earlier investigations (e.g. (Rossiter and Kouvaritakis, 2001)) have shown that it is possible to improve prediction accuracy by using a different model for each different number of steps-ahead prediction. These investigations were carried out for ARMA models but clearly also apply to MA models.

An alternative treatment of (1a-c) to that of (5a-c) can be derived from output bounds based on ellipsoidal confidence regions for θ, as we now discuss.

Lemma 3. Let θ in (11) have the distribution (10). Then $\|\theta - \bar{\theta}\|_{\Theta^{-1}}^2$ is distributed as χ^2 with q degrees of freedom, and the prediction $y(k+i|k)$ lies with probability p in the interval

$$\begin{aligned} \bar{\theta}^T \tilde{u}(k+i-1|k) - r_p \|\tilde{u}(k+i-1|k)\|_{\Theta} &\leq y(k+i|k) \\ \leq \bar{\theta}^T \tilde{u}(k+i-1|k) + r_p &\|\tilde{u}(k+i-1|k)\|_{\Theta} \end{aligned} \quad (15)$$

where r_p is defined by $\Pr(\|\theta - \bar{\theta}\|_{\Theta^{-1}}^2 \leq r_p^2) = p$.

Proof: From (10) we have $\Theta^{-1/2}(\theta - \bar{\theta}) \sim \mathcal{N}(0, I)$, so that $\|\theta - \bar{\theta}\|_{\Theta^{-1}}^2$ is distributed according to a χ^2 distribution with q degrees of freedom. Therefore θ lies with probability p within the ellipsoid

$$\mathcal{E}_p = \{\theta : (\theta - \bar{\theta})^T \Theta^{-1} (\theta - \bar{\theta}) \leq r_p^2\}. \quad (16)$$

It follows that, with probability p:

$$\begin{aligned} |(\theta - \bar{\theta})^T \tilde{u}(k+i-1|k)| &\leq \|(\theta - \bar{\theta})\|_{\Theta^{-1}} \|\tilde{u}(k+i-1|k)\|_{\Theta} \\ &\leq r_p \|\tilde{u}(k+i-1|k)\|_{\Theta}, \end{aligned}$$

which implies the bounds (15) since (11) gives $y(k+i|k) - \bar{\theta}^T \tilde{u}(k+i-1|k) = (\theta - \bar{\theta})^T \tilde{u}(k+i-1|k)$. □

Remark 4. Confidence ellipsoids of the form (16) are conventionally used to bound the effects of stochastic disturbances or model coefficients (e.g. (van Hessem *et al.*, 2001; Cloud and Kouvaritakis, 1986)). However the output prediction bounds used in (5a-c) lead to a less conservative treatment than equivalent bounds based on confidence ellipsoids for two reasons. Firstly, (5a-b) are derived directly from the distributions (13) of output predictions rather than from bounds (15) inferred indirectly from confidence ellipsoids for θ. This leads to tighter bounds than (15) since (13) implies that $y(k+i|k)$ lies with probability p in the interval

$$\overline{\theta}^T \tilde{u}(k+i-1|k) - \alpha_p \|\tilde{u}(k+i-1|k)\|_\Theta \leq y(k+i|k)$$
$$\leq \overline{\theta}^T \tilde{u}(k+i-1|k) + \alpha_p \|\tilde{u}(k+i-1|k)\|_\Theta \quad (17)$$

where α_p is defined by $\mathfrak{N}(\alpha_p) = \frac{1}{2}(p+1)$, so that $\alpha_p < r_p$ for all $q > 1$. Secondly, confidence ellipsoids necessarily provide interval bounds for $y(k+i|k)$ that are symmetric about the expected value $\overline{\theta}^T \tilde{u}(k+i-1|k)$, whereas direct use of (13) allows the derivation of semi-infinite or asymmetric interval bounds on $y(k+i|k)$ for any given probability.

The MA model (11) involves past and future input values only, providing no mechanism for using measurements to correct for model mismatch. To overcome this a disturbance term d can be introduced into (7):

$$y(k+1) = g(z)u(k) + d(k), \; d(k) = y(k) - \sum_{i=0}^{q-1} g_i u(k-i-1).$$

In DMC (Cutler and Ramaker, 1980), $d(k)$ is treated deterministically and computed on the basis of the expected value of θ. Alternatively it can be treated as a random variable with distribution

$$d(k) \sim \mathcal{N}\big(y(k) - \overline{\theta}^T u^p(k), \|u^p(k)\|_\Theta^2\big).$$

Another approach with some advantages (namely robustness in that it uses many past output values) is to redefine the past q values of inputs to be consistent with the current and past $q-1$ values of the outputs. This approach is used in the simulations of Section 6.

5. THE PREDICTIVE CONTROL ALGORITHM

The proposal here is to split the MPC strategy into two phases, the first defining an optimal setpoint, the second concerning setpoint tracking. The setpoint definition provides the mechanism for imposing an equality stability constraint. Given the stochastic nature of the problem, this can be invoked only in respect of expected values. Future research will address the possibility of replacing this with an inequality constraint imposed at a given confidence level.

5.1 Definition of setpoint

To allow predicted inputs to assume different values at different prediction times, and to impose the stability constraint that the output expected values reach a preassigned steady-state value y_1^{ss} at the end of the prediction horizon of N steps, the predicted input sequence at time k is defined as

$$\hat{u}(k) = \{u(k|k), \ldots, u(k+N_u|k), u^{ss}, \ldots, u^{ss}\} \quad (18)$$

where $N \geq N_u + q - 1$.

The setpoint optimization amounts to the determination of u^{ss} and $r = y_1^{ss}$ such that in the steady-state the lower bound on the primary output y_1 for a given confidence level p_1 is maximized, subject to bounds imposed on y_j $j = 2, \ldots, n_y$ with probability p_j, and subject to the budget constraint on discounted inputs. This can be formulated as

$$\max_{u^{ss}, A_1^{ss}} \; A_1^{ss} \quad (19a)$$
$$\text{subject to} \;\; \Pr(y_1(k+N|k) \geq A_1^{ss}) \geq p_1 \quad (19b)$$
$$\Pr(y_j(k+N|k) \leq A_j) \geq p_j \;\; j > 1 \quad (19c)$$
$$\sum_{i=0}^{N-1} \rho^i u^{ss} \leq B, \;\; u^{ss} \geq 0 \quad (19d)$$

where ρ is the discounting factor. As in Section 3, we assume that θ_j is normally distributed:

$$\theta_j = \mathcal{N}(\overline{\theta}_j, \Theta_j). \quad (20)$$

Therefore (19a-d) is a convex problem (expressing constraints (19b-c) in the form of (5b) results in a second order cone program (SOCP) (Lobo *et al.*, 1998)). The solution for u^{ss} defines the optimal setpoint

$$r = \overline{\theta}_1^T \mathbf{1} u^{ss}, \quad \mathbf{1} = [1 \cdots 1]^T. \quad (21)$$

Lemma 5. Let $t^{ss} = r - A_1^{ss}$, then the solution of (19a-d) satisfies

$$\Pr(y_1(k+N|k) \geq r - t^{ss}) \geq p_1$$
$$\Pr(y_1(k+N|k) \leq r + t^{ss}) \geq p_1. \quad (22)$$

Proof: (19b) is equivalent to $\mathfrak{N}(t^{ss}/\|\mathbf{1}u^{ss}\|_{\Theta_1}) \geq p_1$, which implies (22). □

5.2 The tracking phase

The second phase of the MPC strategy minimizes the deviation of y_1 from the setpoint r while respecting constraints on y_j, $j = 2, \ldots, n_y$ and u. Given the stochastic nature of the problem, the minimization of tracking error is performed in terms of upper bounds on $r - y_1$ corresponding to the probability p_1 employed in (19a-d). This is done by defining variable error bounds $t(i|k)$ such that $r - y_1(k+i|k) \leq t(i|k)$ and $y_1(k+i|k) - r \leq t(i|k)$ for $i = 1, \ldots, N$, both bounds being invoked with probability p_1. To obtain a guarantee of closed-loop stability despite uncertainty in θ_1, we use the approach of including in the MPC objective only the bounds on $y_1(k+i|k)$ that lie outside the interval $[r - t^{ss}, r + t^{ss}]$ (see e.g. (Kerrigan and Maciejowski, 2003)). The objective is therefore defined as the norm of an error measure s defined by

$$s = [s(1|k) \cdots s(N|k)]^T, \; s(i|k) = \max\{t(i|k) - t^{ss}, 0\} \quad (23)$$

Intuitively, this form of objective avoids penalizing errors below the threshold of what is achievable in steady-state with confidence level p_1.

A further requirement for closed-loop stability is that, at each instant k, there exists a predicted input trajectory $\hat{u}(k)$ of the form (18) which is feasible with respect to the discounted budget constraint

$$\sum_{i=0}^{N-1} \rho^i u(k+i|k) \leq B. \quad (24)$$

Given feasibility at time k, feasibility at $k+j$ for all $j \geq 1$ can be ensured by constructing a set of predicted input sequences $\hat{u}^f(k+j|k)$ on the basis of $\hat{u}(k)$:

$$\hat{u}^f(k+j|k) = \{u(k+j|k), \ldots, u(k+N_u-1|k), u^{ss}, \ldots, u^{ss}\} \quad (25)$$

and then imposing the constraints that (24) is satisfied by $\hat{u}^f(k+j|k)$ for $j = 1, \ldots, N_u - 1$. These constraints together with (24) are equivalent to

$$\sum_{i=j}^{N-1} \rho^{i-j} u(k+i|k) + \rho^{N-j}\left(\frac{1-\rho^j}{1-\rho}\right) u^{ss} \leq B \quad (26)$$

for $j = 0, \ldots, N_u - 1$.

Defining the MPC objective as the l_2-norm of the error measure s defined in (23) and invoking probabilistic constraints on y_j, $j > 1$, of the form (19c) and the budget constraints (26), we obtain the following MPC optimization problem to be solved at times $k = 0, 1, \ldots$:

$$\min_{\substack{\hat{u}(k) \\ s(i|k), t(i|k), i=1,\ldots,N}} \quad \bar{s}^2(k) = \sum_{i=1}^{N} s^2(i|k) \quad (27a)$$

$$\text{subject to } s(i|k) \geq t(i|k) - t^{ss}, \; s(i|k) \geq 0 \quad (27b)$$

$$\Pr(y_1(k+i|k) \geq r - t(i|k)) \geq p_1 \quad (27c)$$

$$\Pr(y_1(k+i|k) \leq r + t(i|k)) \geq p_1 \quad (27d)$$

$$\Pr(y_j(k+i|k) \leq A_j) \geq p_j \;\; j > 1 \quad (27e)$$

$$\sum_{j=i}^{N-1} \rho^{j-i} u(k+i|k) + \rho^{N-i}\left(\frac{1-\rho^i}{1-\rho}\right) u^{ss} \leq B \quad (27f)$$

$$\text{for } i = 1, \ldots, N.$$

The optimal solution for $\hat{u}(k)$ defines the receding horizon control law $u(k) = u(k|k)$.

Remark 6. The optimization (27a-f) is convex and can be formulated as a SOCP by replacing the objective with min $\bar{s}(k)$ subject to the additional constraint: $\|[s(1|k) \cdots s(N|k)]\| \leq \bar{s}(k)$, and by using the distributions (20) to express (27c-e) in the form of (5b).

Theorem 7. Assuming no model mismatch (i.e. in the absence of errors in model order and parameter distributions (20)), and under the receding horizon control law $u(k) = u(k|k)$, feasibility of the optimization (27a-f) at $k = 0$ implies feasibility for all $k > 0$. Furthermore the optimal value of the objective $\bar{s}(k)$ decreases monotonically with k, and the output y_1 of the closed-loop system converges to the interval $[r - t^{ss}, r + t^{ss}]$ in the sense that

$$\Pr(y_1(k+1|k) \geq r - t^{ss}) \geq p_1$$
$$\Pr(y_1(k+1|k) \leq r + t^{ss}) \geq p_1 \quad (28)$$

in the limit as $k \to \infty$.

Proof: The recurrence of feasibility is ensured by the feasibility, with respect to (27e-f) at time $k+1$, of the predicted input sequence $\hat{u}^f(k+1|k)$ defined in (25). Given the discussion above, the feasibility of (27f) is obvious. Furthermore, the definition of $\hat{u}^f(k+1|k)$ implies that the open-loop predictions of y_j reach steady-state at time $k+N+1$ under $\hat{u}^f(k+1|k)$, and the feasibility of (27e) therefore follows from the setpoint optimization (19a-d). This argument also implies the monotonic decrease of the optimal objective $\bar{s}(k)$, since from Lemma 5 it follows that $s(N+1|k) = 0$ under $\hat{u}^f(k+1|k)$, so that the optimum $\bar{s}(k+1)$ necessarily satisfies

$$\bar{s}^2(k+1) \leq \bar{s}^2(k) - s^2(1|k). \quad (29)$$

Summing this inequality over $k = 0, 1, \ldots$ yields $\sum_{k=0}^{\infty} s^2(1|k) \leq \bar{s}^2(0)$, which implies $s(1|k) \to 0$, and hence the bounds (28), asymptotically as $k \to \infty$. □

Remark 8. To avoid penalizing desirable overshoots in the predicted response for y_1, the upper bound on tracking error imposed by (27d) can be omitted from the MPC optimization (27a-f). The resulting control law has the closed-loop feasibility and stability properties described in Theorem 7, but with (28) replaced by the bound $\Pr(y_1(k+1|k) \geq r - t^{ss}) \geq p_1$ as $k \to \infty$.

In order to compare closed-loop performance of a control law based on the static ISPA optimization (based on maximizing the sum of the primary output over the prediction horizon) with the MPC laws developed in this section, we define an alternative MPC law based on the minimization of an l_1-norm of the tracking error measure s. The corresponding receding horizon optimization can be stated as follows:

$$\min_{\substack{\hat{u}(k) \\ s(i|k), t(i|k), i=1,\ldots,N}} \quad \bar{s}(k) = \sum_{i=1}^{N} s(i|k) \quad (30a)$$

$$\text{subject to } (27b), (27c), (27e), (27f) \quad (30b)$$
$$\text{for } i = 1, \ldots, N.$$

Clearly (30a-b) defines a convex problem (which can be formulated as a SOCP), and the associated control law has the feasibility and stability properties of the control law described in Remark 8.

6. NUMERICAL EXAMPLE

For the purposes of illustration, we consider here a simple example with $n_u = 1$, $n_y = 2$, and dynamics:

$$g_1(z) = \frac{-2(1 - 1.2z^{-1})}{(1 - 0.6z^{-1})(1 - 0.2z^{-1})}$$
$$g_2(z) = \frac{0.2}{(1 - 0.4z^{-1})}. \quad (31)$$

The MA models identified for (31) were truncated after $q = 10$ terms. The measurement noise (during identification) was taken to be white with zero mean and variance of 0.3 for both y_1 and y_2. The bounds on y_2 were chosen as $A_2 = 1$, and the probabilities p_1, p_2 were taken to be equal. A prediction horizon of $N = 19$ was used, with $N_u = 10$, and the budget constraint was set at $B = 20$ with discounting factor $\rho = 0.9$.

The same random processes were used as output disturbances during the closed-loop simulations of

Figs. 1-2, for which $p_1 = p_2 = 0.8$ (Fig. 1) and 0.95 (Fig. 2). The setpoints r defined by (19) are indicated by the dashed lines on the plots of y_1; dotted lines show the values of $r \pm t^{ss}$. It can be seen that the online optimization (27) succeeds in meeting the constraint on y_2 while producing small closed-loop errors for y_1. Due to the probabilistic statement of the problem, constraints on y_2 are exceeded at some instants (Fig. 1). An increase in p_2 makes the constraints on y_2 more stringent and thus results in less control authority being available with which to improve the performance of y_1 (Fig. 2). However the concurrent increase in p_1 results in a lower incidence of time instants during which the error bounds $r \pm t^{ss}$ are exceeded.

As expected, a single horizon strategy (such as that considered in the earlier sustainable development work (Kouvaritakis, 2000*a*)) produces poorer results, and indeed for some realizations of the random output disturbances leads to closed-loop instability. For the purposes of comparison, the same realization of the output random disturbance was used in closed-loop simulations of the receding horizon control laws based on (27) and (30) with $p_1 = p_2 = 0.8$. These were compared with a single horizon strategy (ISPA) of the form (1), which was modified to provide a meaningful comparison by incorporating the pointwise constraints on y_2 and budget constraints of (27e-f). The bound A_1 in (1a) was determined from the setpoint optimization of (19), and $p_2 = 0.8$ was used. Table 1 shows the closed-loop values of the costs for a simulation length of $N_f = 60$ samples. Predictably the single horizon strategy is suboptimal in terms of both l_1 and l_2 costs, this is partly due to poorer performance during transients resulting from the use of a single degree of freedom in predicted input trajectories.

Table 1. Closed-loop simulation costs

	Optimization		
	(27)	(30)	ISPA
$\sum_{k=0}^{N_f} y_1(k)$	128	138	135
$\sum_{k=0}^{N_f} (r - y_1(k))^2$	74.6	101	104

Acknowledgments This work is supported by the European Commission under EVG1-CT-2002-00082.

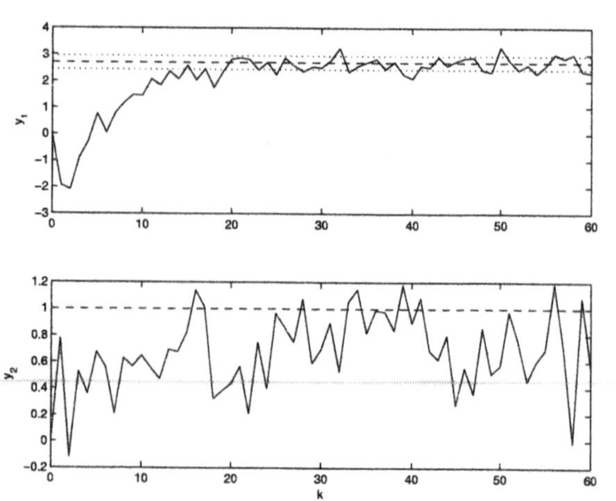

Fig. 1. Output responses for (27) with $p_1 = p_2 = 0.8$.

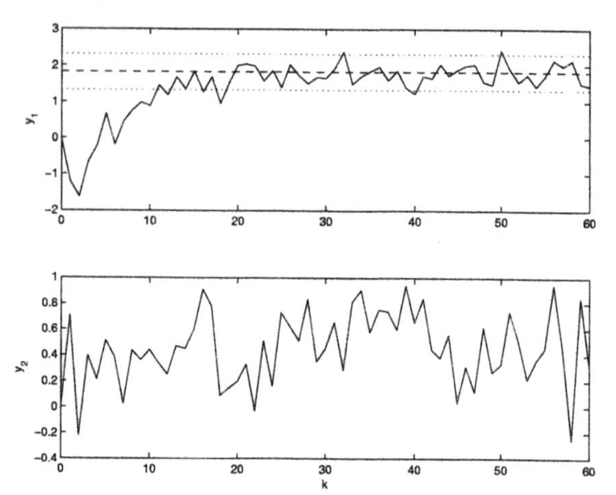

Fig. 2. Output responses for (27) with $p_1 = p_2 = 0.95$.

REFERENCES

Akaike, H. (1974). New look at the statistical model identification. *IEEE Trans. Aut. Control* **19**(6), 716–723.

Cloud, D.J. and B. Kouvaritakis (1986). Statistical bounds on multivariable frequency response: an extension of the generalized Nyquist criterion. *Proc. IEE pt. D* **133**(3), 97–110.

Cutler, C.R. and B.L. Ramaker (1980). Dynamic matrix control - a computer control algorithm. In: *Proc. JACC*. San Francisco.

Kerrigan, E. and J.M. Maciejowski (2003). Robustly stable feedback min-max model predictive control. In: *Proc. American Control Conf.*. pp. 3490–3495.

Kouvaritakis, B. and J.A. Rossiter (1991). Use of bi-causal weighting sequences in least squares identification of open-loop unstable dynamic systems. *Proc. IEE pt. D* **139**(3), 328–336.

Kouvaritakis, N. (2000*a*). The ISPA meta model. Technical report. European Commission project ENG2-CT-1999-00003.

Kouvaritakis, N. (2000*b*). Prometheus. Model developed under European Commission project ENG2-CT-1999-00003.

Kouvaritakis, N. (2002). Methodologies for integrating impact assessment in the field of sustainable development (MINIMA-SUD). Technical report. European Commission project EVG1-CT-2002-00082.

Lobo, M., L. Vandenberghe, S. Boyd and H. Lebret (1998). Applications of second-order cone programming. *Linear Algebra and its Applications* **284**, 193–228.

Rossiter, J.A. and B. Kouvaritakis (2001). Modelling and implicit modelling for predictive control. *Int. J. Control* **74**(11), 1085–1095.

van Hessem, D.H., C.W. Scherer and O.H. Bosgra (2001). LMI-based closed-loop economic optimization of stochastic processes under state and input constraints. In: *Proc. 40th IEEE Conf. Decision and Control*. pp. 4223–4228.

ONLINE NMPC OF A LOOPING KITE USING APPROXIMATE INFINITE HORIZON CLOSED LOOP COSTING

Moritz Diehl* Lalo Magni Giuseppe De Nicolao****

**Interdisciplinary Center for Scientific Computing (IWR), University of Heidelberg, Germany*
*** Dipartimento di Informatica e Sistemistica, University of Pavia, Italy*

Abstract: We consider a dual line kite that shall fly loops and address this periodic control problem with a state-of-the-art NMPC scheme. The kite is described by a nonlinear unstable ODE system, and the aim is to let the kite fly a periodic figure. Our approach is based on the 'infinite horizon closed loop costing" NMPC scheme to ensure nominal stability. To be able to apply this scheme, we first determine a periodic LQR controller to stabilize the kite locally in the periodic orbit. Then, we formulate a two-stage NMPC optimal control problem penalizing deviations of the system state from the periodic orbit, which also contains a state constraint that avoids that the kite collides with the ground. To solve the optimal control problems reliably and in real-time, we apply the newly developed "real-time iteration scheme" for fast online optimization in NMPC. The optimization based NMPC leads to significantly improved performance compared to the LQR controller, in particular as it respects state constraints. The NMPC closed loop also shows considerable robustness against changes in the wind direction. *Copyright © 2003 IFAC*
Keywords: periodic control, LQR, nonlinear control systems, predictive control, online optimization, stability, numerical methods, optimal control

1. INTRODUCTION

Nonlinear model predictive control (NMPC) is a feedback control technique that is based on the real-time optimization of a nonlinear dynamic process model on a moving horizon that has attracted increasing attention over the past decade (Qin and Badgwell, 2001). Important challenges that need to be addressed for any NMPC application are stability of the closed loop system and the numerical solution of the optimal control problems in real-time. In this paper we show how state-of-the-art NMPC techniques addressing these challenges can be applied to control a strongly unstable periodic system, namely a dual line kite that shall fly loops. The aim of our automatic control is to make the kite fly a figure that may be called a "lying eight". The corresponding orbit is not open loop stable, so that feedback has to be applied. We assume the state is fully accessible for control.

Since the natural setting of the problem is in continuous time, the NMPC implementation proposed here is developed for continuous-time systems. However, it basically differs from the continuous time NMPC algorithms for nonlinear systems previously published in the literature, see e.g. (Mayne and Michalska, 1990; Chen and Allgöwer, 1998). Continuous time methods usually assume that the NMPC law is continuously computed by solving at any time instant a difficult optimization problem. This is impossible in practice, as any implementation is performed in digital form and requires a non-negligible computational time. The NMPC setup proposed here is based on the method proposed in (Magni *et al.*, 2002), where a continuous

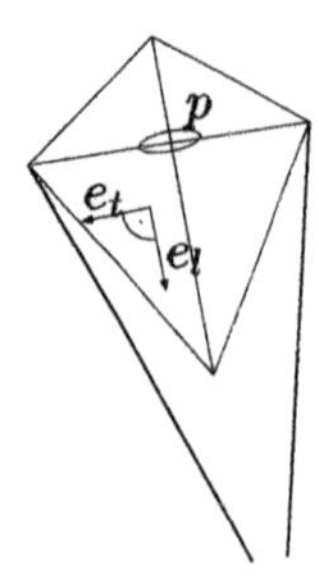

Fig. 1. A picture of the kite, and the unit vectors e_l and e_t along longitudinal and transversal kite axis.

time locally stabilizing control law is first designed. Then, a piecewise constant term computed via NMPC is added to the control signal provided by the stabilizing control law, in order to achieve some specific goals, such as the minimization of a prescribed cost or the enlargment of the output admissible set. In so doing, it is assumed that the signal computed by NMPC is piecewise constant and with a limited number of free moves in the future. Nominal stability of the overall system is preserved using the "infinite horizon closed loop costing" scheme proposed in (De Nicolao *et al.*, 1998). In the usual setting of this scheme, the optimization problems are solved up to a prespecified accuracy during each sampling time so that a feedback delay of one sampling time is introduced in the closed loop.

In this paper, however, we avoid this feedback delay by using the recently developed "real-time iteration" scheme (Diehl *et al.*, 2002*b*) for online optimization. The algorithm is based on the direct multiple shooting approach to optimal control problems (Bock and Plitt, 1984; Leineweber, 1999), but is characterized by the following features: first, the scheme efficiently initializes each new problem and performs only one optimization iteration per optimization problem. Thus it reduces sampling times to a minimum. Second, the computations of each "real-time iteration" are divided into a very short "feedback phase", and a much longer "preparation phase", which uses the sampling time to *prepare* the next feedback. Thus, each NMPC feedback is directly applied to the system, with a negligible delay that is orders of magnitude shorter than the sampling time. For details see e.g. (Diehl *et al.*, 2002*c*).

The paper is organized as follows. In Section 2 we derive the model equations for the kite model. The periodic reference orbit is analysed in Section 3 and we show how to design a stabilizing periodic linear controller based on LQR techniques. In Section 4 we finally describe the NMPC setup. Simulated closed loop experiments are presented and briefly discussed in Section 5.

2. KITE MODEL

The kite is held by two lines which allow to control the lateral angle of the kite, see Fig. 1. By pulling one line the kite will turn in the direction of the line being pulled. In this paper we employ a kite model that was originally developed in (Diehl, 2002; Diehl *et al.*, 2002*a*).

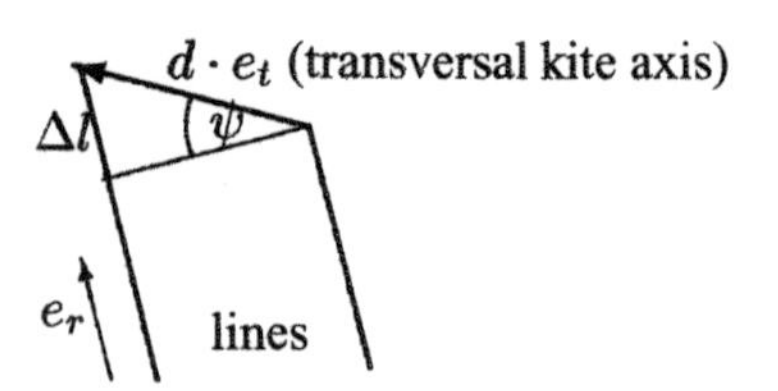

Fig. 2. The kite and its lines seen from the top, and visualization of the lateral angle ψ.

2.1 Kite Dynamics in Polar Coordinates

The movement of the kite at the sky can be modelled by Newton's law of motion and a suitable model for the aerodynamic force. Let us introduce polar coordinates θ, ϕ, r so that the position p of the kite relative to the kite pilot (in the origin) is given by: $p = (r\sin(\theta)\cos(\phi), r\sin(\theta)\sin(\phi), r\cos(\theta))$ with the last component being the height of the kite over the ground, and θ being the angle that the kite lines form with the vertical. We introduce a local right handed coordinate system with the basis vectors e_θ, e_ϕ, e_r, each pointing in the direction where the corresponding polar coordinate increases. Defining the corresponding components of the total force F acting on the kite, we can write Newton's law of motion for constant r in the form

$$\ddot{\theta} = \frac{F_\theta}{rm} + \sin(\theta)\cos(\theta)\,\dot{\phi}^2, \tag{1}$$

$$\ddot{\phi} = \frac{F_\phi}{rm\sin(\theta)} - 2\cot(\theta)\dot{\phi}\dot{\theta}, \tag{2}$$

with m denoting the mass of the kite. The force consists of two contributions, gravitational and aerodynamic force, so that we obtain $F_\theta = \sin(\theta)mg + F_\theta^{\text{aer}}$ and $F_\phi = F_\phi^{\text{aer}}$, where $g = 9.81$ m s^{-2} is the earth's gravitational acceleration. It remains to determine the aerodynamic forces F_ϕ^{aer} and F_θ^{aer}.

2.2 Kite Orientation

To model the aerodynamic force we first determine the kite's orientation. We assume that the kite's trailing edge is strongly pulled by the tail into the direction of the effective wind at the kite. Under this assumption the kite's longitudinal axis is always in line with the effective wind vector $w_e := w - \dot{p}$, where $w = (v_w, 0, 0)^T$ is the wind as seen from the earth system, and $\dot{p}$ the kite velocity. If we introduce a unit vector e_l pointing from the front towards the trailing edge of the kite (cf. Fig. 1), we therefore assume that $e_l = \frac{w_e}{\|w_e\|}$. The transversal axis of the kite can be described by a perpendicular unit vector e_t that is pointing from the right to the left wing tip (as seen from the kite pilot, in upright kite orientation). The orientation of e_t can be controlled, but it has to be orthogonal to e_l (cf. Fig. 1),

$$e_t \cdot e_l = 0. \tag{3}$$

However, the projection of e_t onto the lines' axis (which is given by the vector e_r) is determined from the length difference Δl of the two lines, see Fig. 2. If the distance between the two lines' fixing points on the kite is d, then the vector from the right to the left fixing point is $d \cdot e_t$, and the projection of this vector onto the lines' axis should equal Δl (being positive if the left wing tip is farther away from the pilot), i.e., $\Delta l = d\, e_t \cdot e_r$. Let us define the *lateral angle* ψ to be $\psi = \arcsin\left(\frac{\Delta l}{d}\right)$. For simplicity, we assume that we control this angle ψ directly. It determines the orientation of e_t which has to satisfy:

$$e_t \cdot e_r = \frac{\Delta l}{d} = \sin(\psi). \qquad (4)$$

A third requirement that e_t must satisfy and which determines its sign is

$$(e_l \times e_t) \cdot e_r > 0. \qquad (5)$$

This ensures that the kite is always in the same orientation with respect to the lines. We will now give an explicit form for the unit vector e_t that satisfies (3)–(5). Using the projection $\bar{w}_e$ of the effective wind vector w_e onto the tangent plane spanned by e_θ and e_ϕ, $\bar{w}_e := w_e - e_r(e_r \cdot w_e)$, we can define the orthogonal unit vectors $e_1 := \frac{\bar{w}_e}{\|\bar{w}_e\|}$ and $e_2 := e_r \times e_1$, so that (e_1, e_2, e_r) forms an orthogonal right-handed coordinate basis. In this basis the effective wind w_e has no component in the e_2 direction: $w_e = \|\bar{w}_e\| e_1 + (w_e \cdot e_r) e_r$. The definition

$$e_t := e_1(-\cos\psi \sin\eta) + e_2(\cos\psi \cos\eta) + e_r \sin\psi$$

with $\eta := \arcsin\left(\frac{w_e \cdot e_r}{\|\bar{w}_e\|} \tan(\psi)\right)$ indeed satisfies (3)–(5), as can be verified by direct substitution. Therefore we are now able to determine the orientation of the kite depending on the control ψ and the effective wind w_e only.

2.3 Aerodynamic Lift and Drag

The two vectors $e_l \times e_t$ and e_l are the directions of aerodynamic lift and drag, respectively. To compute the magnitudes F_L and F_D of lift and drag we assume that the lift and drag coefficients C_L and C_D are constant, so that we have

$$F_L = \frac{1}{2}\rho\|w_e\|^2 A C_L \quad \text{and} \quad F_D = \frac{1}{2}\rho\|w_e\|^2 A C_D,$$

with ρ being the density of air, and A being the characteristic area of the kite. Given the directions and magnitudes of lift and drag, we can compute F^{aer} as their sum, yielding $F^{\text{aer}} = F_L(e_l \times e_t) + F_D e_l$ or, in the local coordinate system

$$F_\theta^{\text{aer}} = F_L((e_l \times e_t) \cdot e_\theta) + F_D(e_l \cdot e_\theta)$$
$$F_\phi^{\text{aer}} = F_L((e_l \times e_t) \cdot e_\phi) + F_D(e_l \cdot e_\phi).$$

The system parameters that have been chosen for the simulation model are listed e.g. in (Diehl, 2002; Diehl *et al.*, 2002*a*). Defining the system state $x := (\theta, \phi, \dot{\theta}, \dot{\phi})^T$ and the control $u := \psi$ we can summarize the four system equations, i.e., (1)–(2) and the trivial equations $\frac{\partial \theta}{\partial t} = \dot{\theta}$, $\frac{\partial \phi}{\partial t} = \dot{\phi}$, in the short form $\dot{x} = f(x, u)$.

3. REFERENCE ORBIT AND LQR

Using the above system model, a periodic orbit was determined that can be characterized as a "lying eight" and which can be seen e.g. in Figure 3, as a (ϕ, θ)-plot. The wind is assumed to blow in the direction of the p_1-axis ($\theta = 90^o$ and $\phi = 0^o$). The periodic solution was computed using an off-line variant of the direct multiple shooting method, MUSCOD-II (Leineweber, 1999), imposing periodicity conditions with period $T := 8$ seconds and suitable state bounds and a suitable objective function in order to yield a solution that we considered to be a nice reference orbit. We denote the periodic reference solution by $x_r(t)$ and $u_r(t)$. This solution is defined for all $t \in (-\infty, \infty)$ and satisfies the periodicity condition $x_r(t+T) = x_r(t)$ and $u_r(t+T) = u_r(t)$.

3.1 Open Loop Stability Analysis

Numerical simulations of the kite using the open loop inputs $u_r(t)$ show that the kite crashes onto the ground very quickly after small disturbances, cf. Fig. 3. To analyse the asymptotic stability properties of the open loop system along the periodic reference orbit theoretically, let us consider the linearization of the system along the orbit. An infinitesimal deviation $\delta x(t) := x(t) - x_r(t)$ and $\delta u(t) := u(t) - u_r(t)$ would satisfy the periodically time-varying linear differential equation

$$\delta\dot{x}(t) = A(t)\delta x(t) + B(t)\delta u(t), \qquad (6)$$

with

$$A(t) := \frac{\partial f}{\partial x}(x_r(t), u_r(t)) \quad \text{and} \quad B(t) := \frac{\partial f}{\partial u}(\cdot).$$

Based on the linear time variant periodic system $\dot{x}(t) = A(t)x(t)$ we can compute its fundamental solution and the sensitivity of the final state of each period with respect to its initial value, which is called the "monodromy matrix". A numerical computation of the monodromy matrix for the kite orbit and eigenvalue decomposition yields two Eigenvalues ("Floquet multipliers") that are greater than one, confirming the observation that the system is unstable.

3.2 Design of a Periodic LQR Controller

In order to design a locally stabilizing controller (which is needed if we want to apply the infinite horizon closed loop costing NMPC scheme), we use the classical LQR design technique, applied to the periodic linear system (6). We introduce diagonal weighting matrices

$$Q := \text{diag}(0.4,\ 1,\ \text{s}^2,\ \text{s}^2)\frac{1}{\text{s}} \text{ and } R := 33\,\frac{1}{\text{s}}. \qquad (7)$$

To determine the optimal periodic LQR controller that minimizes the objective $\int \frac{1}{2}(x - x_r(t))^T Q(x - x_r(t)) + \frac{1}{2}(u - u_r(t))^T R(u - u_r(t))dt$, we find the symmetric periodic matrix solution $P(t)$ for the differential Riccati equation

$$-\dot{P} = Q + A(t)^T P(t) + P(t)A(t) - P(t)B(t)R^{-1}B(t)^T P(t)$$

by integrating the equation backwards for a sufficiently long time, starting with the unit matrix as final value. Once the periodic $P(t)$ is determined, the optimal LQR controller for (6) and (7) is given by $\delta u(t) = -K(t)\delta x(t)$ with $K(t) := R^{-1}B(t)^T P(t)$. We finally define the linear periodic feedback for the original system as

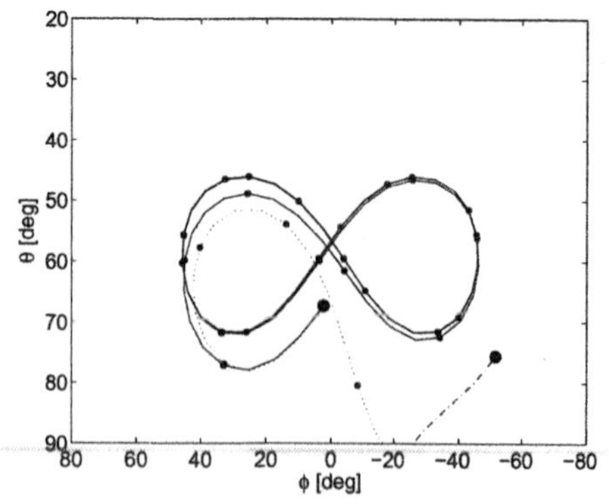

Fig. 3. The linear periodic controller is able to stabilize the system locally from slightly disturbed system states (solid line), in contrast to the open loop system (dotted), but lets the kite crash onto the ground for a larger deviation (dash dotted).

$$u_{\mathrm{LQR}}(x,t) := u_r(t) - K(t)(x - x_r(t)). \quad (8)$$

The linearly stabilized system $\dot{x}(t) = f(x(t), u_{\mathrm{LQR}}(x(t),t))$ is locally stable, as illustrated in Figure 3.

4. NMPC CONTROLLER SETUP

The aim of the NMPC controller is to stabilize the system in a larger region of attraction and to respect certain bounds that the linear controller may violate, and, furthermore, to lead to an improved performance with respect to a user defined criterion. In our case, the bounds arise because we want the closed loop kite to respect a security distance to the ground ($\theta = 90$ deg). We achieve this by requiring that $h(x,u,t) := 76.5\ \mathrm{deg} - \theta \geq 0$. The performance of the controller is measured by the integral of a function $L(x(t), u(t), t)$, which is in our case the squared deviation of the state x at time t from the reference orbit,

$$L(x,u,t) := \frac{1}{2}(x - x_r(t))^T Q (x - x_r(t)).$$

We introduce a sampling interval δ and give NMPC feedback to the system only at the times $t_k := k \cdot \delta$. At each sampling time t_k the NMPC shall deliver new controls $u(t), t \in [t_k, t_{k+1}]$ that depend on the current system value $x(t_k)$, where the optimization is based on a prediction of the future system behaviour. Many NMPC schemes exist that guarantee nominal stability, see e.g. (Allgöwer *et al.*, 1999; De Nicolao *et al.*, 2000); they mainly differ in the way the optimal control problems are formulated. Here, we work in the framework of the infinite horizon closed loop costing scheme (De Nicolao *et al.*, 1998).

4.1 Infinite Horizon Closed Loop Costing

In the infinite horizon closed loop costing scheme we express the control $u(t)$ that is actually applied to the plant at time $t \in [t_k, t_{k+1}]$ by the sum

$$u(t) = u_{\mathrm{LQR}}(x(t),t) + v_k \quad \forall t \in [t_k, t_{k+1}],$$

where the constant vector v_k is determined by the NMPC optimizer and implicitly depends on $x(t_k)$. Note that $v_k \equiv 0$ yields the linearly controlled closed loop. In the sequel we will use a bar to distinguish the predicted system state and controls $\bar{x}(t)$ and $\bar{u}(t)$ from the state and control vector of the real system.

Given the state $x(t_k)$ of the "real" kite at time t_k, we formulate the following optimal control problem, with control horizon $T_c = M\delta$ and prediction horizon $T_p > T_c$ (where T_p shall ideally be infinity).

$$\min_{\bar{v}_i, \bar{u}(\cdot), \bar{x}(\cdot)} \int_{t_k}^{t_k+T_p} L(\bar{x}(t), \bar{u}(t), t)\, dt \quad (9)$$

subject to

$$\begin{aligned}
&\dot{\bar{x}}(t) = f(\bar{x}(t), \bar{u}(t)), \quad \forall t \in [t_k, t_k + T_p],\\
&\bar{x}(t_k) = x(t_k),\\
&\bar{u}(t) = u_{\mathrm{LQR}}(\bar{x}(t),t) + \bar{v}_i, \quad \forall t \in [t_i, t_{i+1}],\\
&\qquad (i = k, \ldots, k+M-1),\\
&\bar{u}(t) = u_{\mathrm{LQR}}(\bar{x}(t),t), \ \forall t \in [t_k + T_c, t_k + T_p]\\
&0 \leq h(\bar{x}(t), \bar{u}(t), t), \ \forall t \in [t_k, t_k + T_p]. \quad (10)
\end{aligned}$$

In the case that $T_p = \infty$ and if the optimal control problem has a solution for $x(t_0)$, stability of the closed loop trajectory can be proved in a rigorous way (De Nicolao *et al.*, 1998; Magni *et al.*, 2002).

4.2 Real-Time Optimization Scheme

We choose a sampling interval $\delta = 1$ s and $M = 8$ sampling intervals as control horizon, $T_c = 8$ s. As the simulation of the periodic system over an infinite horizon is impossible, we employ here a finite $T_p = 24$ s, that we believe to be sufficiently long to deliver a fair approximation to the infinite horizon cost. For a theoretical discussion on how to truncate the series expressing the Infinite Horizon cost associated with the auxiliary linear control law without losing stability see (Magni *et al.*, 2001). Furthermore, to avoid a semi-infinite optimization problem, the problem is changed by imposing the inequality path constraints (10) only at prespecified points in time, here chosen to be the sampling times t_i on the control horizon, as well as start, center and end point of the prediction horizon.

The numerical solution of the optimization problems is achieved by the recently developed real-time iteration scheme (Diehl *et al.*, 2002*b*; Diehl *et al.*, 2002*c*) that is based on ideas developed in (Bock *et al.*, 2000). This scheme is based on the direct multiple shooting method (Bock and Plitt, 1984) that reformulates the optimization problem as a finite dimensional nonlinear programming problem with a special structure. We use an efficient implementation of the scheme within the optimal control package MUSCOD-II (Leineweber, 1999). One advantage of the scheme is that it nearly completely avoids the feedback delay of one sampling time present in most NMPC optimization schemes.

5. CLOSED LOOP EXPERIMENTS

In order to test the NMPC closed loop we have performed several numerical experiments. Here, the "real" kite is simulated by a model that coincides with the optimization model, but is subject to disturbances of different type.

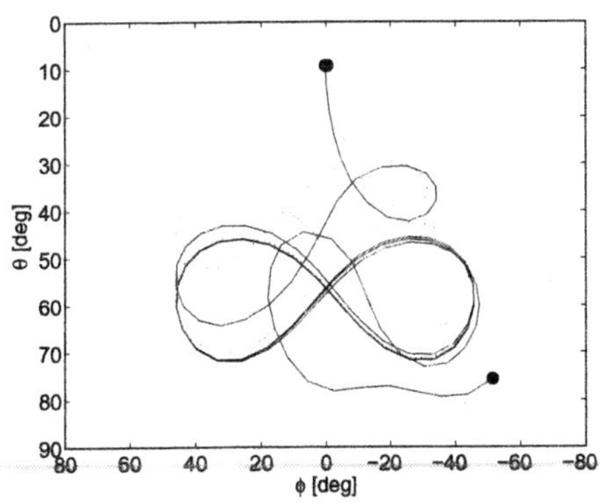

Fig. 4. The NMPC controller is able to control the kite even for the largely disturbed states at the bottom (solid line), in contrast to the LQR controller (dotted, cf. Fig. 3). For the disturbed state at the top, the performance for the NMPC (solid line, integrated costs 1.51) is better that that of the LQR (dotted, costs 1.75), as expected.

5.1 Comparison with LQR

First, let us in Figure 4 compare the NMPC with the periodic LQR. It can be seen that the NMPC is able to respect the state constraint $\theta \leq 76.5$ deg even for the scenario with the largely disturbed initial state at the bottom (cf. Fig. 3), in contrast to the LQR. For another scenario, where the system kite starts much too high at the sky, both controllers are able to stabilize the system. However, NMPC leads to a reduced objective with the cost integral $\int_0^\infty L(x(t), u(t), t)dt$ being 1.51 in contrast to 1.75 for the LQR. This difference can be seen in form of a considerably faster convergence towards the periodic orbit.

5.2 Strong Sidewind

In another scenario, the closed loop is tested against model uncertainty: we consider a continuing disturbance resulting from a change in the wind direction. The wind component in p_2-direction, that is assumed by the optimizer to be zero, is for the "real" kite set to a value 3 m/s that is 50% of the nominal wind v_w in p_1 direction. The NMPC closed loop results in a considerably disturbed but stable periodic orbit, as can be seen in Figure 5; the disturbed periodic orbit is reached after a very short transient. This contrasts sharply with the open loop and the LQR closed loop response which both result in a crash after a short time.

6. CONCLUSIONS

We have presented a method to design a nonlinear model predictive controller for periodic unstable systems, and have applied the method to a kite that shall fly loops. The method is based on the "infinite horizon closed loop costing" which requires a locally prestabilizing feedback. This prestabilization is achieved by a periodic LQR controller based on a system linearization along the periodic orbit. The NMPC controller uses an objective which only penalizes state deviations and a state constraint is formulated to ensure

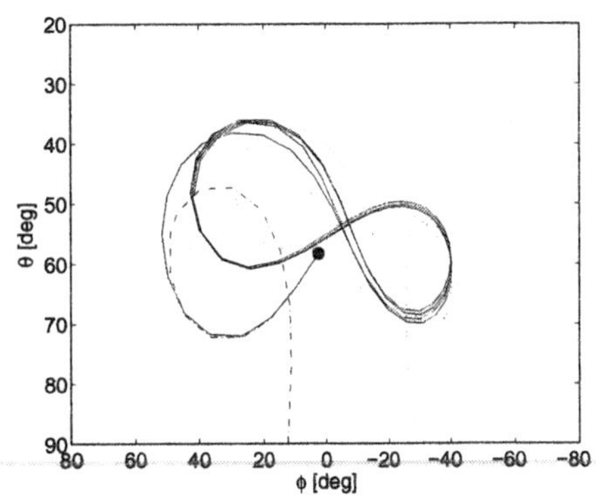

Fig. 5. Effect of model uncertainty in form of a strong side wind with 50 % the nominal wind speed. After a very short transient, the NMPC controlled kite loops in a considerably disturbed periodic orbit, but remains stable (solid line). The LQR closed loop response results in a crash after two periods (dotted), and the open loop crashes after 4 seconds (dash dotted).

that the kite does not crash onto the ground. The resulting optimal control problems are solved in real-time, once a second, by a state-of-the-art online optimization algorithm, the "real-time iteration scheme". This numerical scheme avoids the large feedback delay present in most optimization approaches to NMPC and allows to reduce sampling times to a minimum. The NMPC closed loop gives an excellent response to strong disturbances. Furthermore, it shows good robustness against model plant mismatch: in the presence of additional sidewind of 50% the nominal wind velocity the periodic orbit changes shape, but remains stable.

We want to mention here that the real-time iteration NMPC scheme used for the computations in this paper has also been successfully applied to a real pilot scale distillation column described by a stiff differential-algebraic equation model with 200 states, making a sampling time of 20 seconds possible (Diehl *et al.*, 2003).

Acknowledgements:
Financial support by the DFG within priority program 469 is gratefully acknowledged, as well as support by the "Institute of Mathematics and its Applications", University of Minnesota, which hosted the first author while parts of the paper have been developed. The second and the third authors acknowledge the partial financial support by MURST Project "New techniques for identification and adaptive control of industrial systems".

REFERENCES

Allgöwer, F., T. A. Badgwell, J. S. Qin, J. B. Rawlings and S. J. Wright (1999). Nonlinear predictive control and moving horizon estimation – An introductory overview. In: *Advances in Control, Highlights of ECC'99* (P. M. Frank, Ed.). pp. 391–449. Springer.

Bock, H. G. and K. J. Plitt (1984). A multiple shooting algorithm for direct solution of optimal control problems. In: *Proc. 9th IFAC World Congress Budapest*. Pergamon Press.

Bock, H. G., M. Diehl, D. B. Leineweber and J. P. Schlöder (2000). A direct multiple shooting method for real-time optimization of nonlinear DAE processes. In: *Nonlinear Predictive Control* (F. Allgöwer and A. Zheng, Eds.). Vol. 26 of *Progress in Systems Theory*. Birkhäuser. Basel. pp. 246–267.

Chen, H. and F. Allgöwer (1998). A quasi-infinite horizon nonlinear model predictive control scheme with guaranteed stability. *Automatica* **34**(10), 1205–1218.

De Nicolao, G., L. Magni and R. Scattolini (1998). Stabilizing receding-horizon control of nonlinear time varyng systems. *IEEE Trans. on Automatic Control* **AC-43**, 1030–1036.

De Nicolao, G., L. Magni and R. Scattolini (2000). Stability and robustness of nonlinear receding horizon control. In: *Nonlinear Predictive Control* (F. Allgöwer and A. Zheng, Eds.). Vol. 26 of *Progress in Systems Theory*. Birkhäuser. Basel. pp. 3–23.

Diehl, M. (2002). *Real-Time Optimization for Large Scale Nonlinear Processes*. Vol. 920 of *Fortschr.-Ber. VDI Reihe 8, Meß, Steuerungs- und Regelungstechnik*. VDI Verlag. Düsseldorf. Download also at: http://www.ub.uni-heidelberg.de/archiv/1659/.

Diehl, M., H. G. Bock and J. P. Schlöder (2002*a*). Newton-type methods for the approximate solution of nonlinear programming problems in real-time. In: *High Performance Algorithms and Software for Nonlinear Optimization* (G. Di Pillo and A. Murli, Eds.). Kluwer Academic Publishers B.V.

Diehl, M., H. G. Bock, J. P. Schlöder, R. Findeisen, Z. Nagy and F. Allgöwer (2002*b*). Real-time optimization and nonlinear model predictive control of processes governed by differential-algebraic equations. *J. Proc. Contr.* **12**(4), 577–585.

Diehl, M., R. Findeisen, S. Schwarzkopf, I. Uslu, F. Allgöwer, H. G. Bock, E. D. Gilles and J. P. Schlöder (2002*c*). An efficient algorithm for nonlinear model predictive control of large-scale systems. Part I: Description of the method. *Automatisierungstechnik*.

Diehl, M., R. Findeisen, S. Schwarzkopf, I. Uslu, F. Allgöwer, H. G. Bock, E. D. Gilles and J. P. Schlöder (2003). An efficient algorithm for nonlinear model predictive control of large-scale systems. Part II: Application to a distillation column. *Automatisierungstechnik*.

Leineweber, D. B. (1999). *Efficient reduced SQP methods for the optimization of chemical processes described by large sparse DAE models*. Vol. 613 of *Fortschr.-Ber. VDI Reihe 3, Verfahrenstechnik*. VDI Verlag. Düsseldorf.

Magni, L., G. De Nicolao, L. Magnani and R. Scattolini (2001). A stabilizing model-based predictive control for nonlinear systems. *Automatica* **37**(9), 1351–1362.

Magni, L., R. Scattolini and K. J. Astrom (2002). Global stabilization of the inverted pendulum using model predictive control. In: *15th IFAC World Congress*. Barcelona.

Mayne, D.Q. and H. Michalska (1990). Receding horizon control of nonlinear systems. *IEEE Trans. Automat. Contr.* **35**(7), 814–824.

Qin, S.J. and T.A. Badgwell (2001). Review of nonlinear model predictive control applications. In: *Nonlinear model predictive control: theory and application* (B. Kouvaritakis and M. Cannon, Eds.). The Institute of Electrical Engineers. London. pp. 3–32.

www.elsevier.com/locate/ifac

USING CONTROLLER KNOWLEDGE IN PREDICTIVE CONTROL

M. Fikar * **H. Unbehauen** ** **J. Mikleš** *

* *Department of Information Engineering and Process Control, FCHPT STU, Radlinského 9, SK-812 37 Bratislava, Slovakia, e-mail: fikar@cvt.stuba.sk, Fax: ++421/2 52.49.64.69*
** *Control Engineering Laboratory, Faculty of Electrical Engineering and Information Sciences, Ruhr-University Bochum, D44780 Universitätsstr. 150 Germany, Fax: ++49/234 7094-101*

Abstract: The paper discusses possible ways how to incorporate an existing controller into the predictive control framework. This idea is used in anti-windup framework, where the performance can degrade in some cases. The basic requirements are to meet the performance of the nominal controller in the unconstrained case and to guarantee stability in the constrained case. *Copyright © 2003 IFAC*

Keywords: predictive control, stability, controller design

1. INTRODUCTION

Constraints are omnipresent in real world control systems. When the knowledge about the constraints is neglected, degradation of performance and in some cases even instability may occur. To counteract this, several techniques for constraints handling have emerged recently. The main approaches include anti-windup and bumpless transfer (AWBT) design (Kothare *et al.* 1994, and references therein), reference governor (RG) (Bemporad 1998), and model based predictive control (MBPC) (Clarke *et al.* 1987, Zelinka *et al.* 1999).

In AWBT and RG approaches it is assumed that the controller is known whereas MBPC actually synthesises the controller and thus combines the two steps (nominal controller design + constraints handling) into the one. In some cases this may be considered as a drawback, especially, if a nominal controller was designed to guarantee some special requirements on the closed-loop system not easily attainable with predictive control.

In this paper a combined strategy to deal with constraints is proposed. It is assumed that the plant and the controller descriptions are known and that the closed-loop without constraints is stable (as in AWBT and RG). The model of the plant is used actively for predictions and thus the whole problem is posed in the MBPC framework Similar approaches have been published by (Rossiter and Kouvaritakis 1993, Scokaert *et al.* 1999, Chmielewski and Manousiouthakis 1996).

Two possible design methods are proposed. The first one manipulates the future setpoint sequence. In the second approach, knowledge about the desired closed loop poles is used. Both methods guarantee in the unconstrained case behaviour specified by the choice of the nominal controller. Moreover, in the constrained case stability of the closed-loop system is assured.

2. THE BASIC MATHEMATICAL SETUP

Let us consider a time-invariant, single input single output plant expressed in discrete-time form

$$Ay = Bu, \tag{1}$$

where y, u are the process output and manipulated input sequences, respectively. A and B are polynomials in z^{-1} that describe the input-output properties of the plant.

We assume that a class of references w is generated via

$$Fw = G, \tag{2}$$

where F, G are coprime. Here, F specifies a desired class of references (steps, ramps, harmonic signals,...) and G represents the initial conditions of the concrete reference.

In order to track the class of references given above (and to reject disturbances of the same class), an additional term $1/F$ is added before the controlled system (Kučera 1979, Chmúrny *et al.* 1988)

$$u = \frac{1}{F}\tilde{u}. \tag{3}$$

If we assume step changes in references, which is most often the case in predictive control, then $F = 1-z^{-1}$ and the signal $\tilde{u} = \Delta u$ is a sequence of control increments. However, other specifications for F can also be considered.

Hence, the overall plant is described by the transfer function B/AF and $y, \tilde{u}$ are its output and input sequences, respectively. It is assumed that this plant is free of hidden modes, thus AF, B are coprime.

As a controller, we consider the two-degree-of-freedom (2DoF) configuration described by the equation

$$P\tilde{u} = Rw - Qy, \tag{4}$$

where P, Q, R are controller polynomials that are coprime and $P(0)$ is nonzero. The 2DoF controller has been chosen due to its flexibility. However, any other controller structure could have been chosen.

2.1 *Alternate system description*

For the purposes of predictive control, the transfer function description of the controlled system will be transformed into vector-matrix notation. This is a standard procedure in predictive control. Considering the number of predicted outputs into the future being equal N (prediction horizon) this leads to the system description of the form

$$\boldsymbol{y} = \boldsymbol{G}\boldsymbol{u} + \boldsymbol{f}, \tag{5}$$

where

$$\boldsymbol{y} = [y_{t+1} \ \cdots \ y_{t+N}]^T, \tag{6}$$

$$\boldsymbol{u} = [\tilde{u}_t \ \cdots \ \tilde{u}_{t+N-1}]^T, \tag{7}$$

$$\boldsymbol{f} = [f_{t+1} \ \cdots \ f_{t+N}]^T, \tag{8}$$

$$\boldsymbol{G} = \begin{pmatrix} g_1 & \cdots 0 \\ \vdots & \ddots \vdots \\ g_{N-1} & \cdots g_1 \end{pmatrix}. \tag{9}$$

The matrix $\boldsymbol{G}$ and vector $\boldsymbol{f}$ can be calculated as usual from recursive Diophantine equations or by simulating the system recursively (Fikar 1998).

Finally, we consider constraints on the signals that may correspond to lower and upper hard constraints on the control signal, on the rate of change of the control signal, and to recommended lower and upper limits on the output signal. All these can be transformed into linear inequality constraints on the vector $\boldsymbol{u}$ and are generally written as

$$\boldsymbol{A}\boldsymbol{u} \geq \boldsymbol{b}, \tag{10}$$

where the matrix $\boldsymbol{A}$ and the vector $\boldsymbol{b}$ are of appropriate dimensions.

3. APPROACH #1

The first predictive control approach consists in a suitable generation of the future setpoint trajectory $\boldsymbol{y}^u = [y^u_{t+1} \ \cdots \ y^u_{t+N}]^T$ which is the trajectory of the unconstrained closed-loop plant output, i.e., future trajectory of the signal y based on equations (1)–(4). The cost function then minimises the surface between constrained and unconstrained output trajectories

$$J = \sum_{i=1}^{N}(y^u(t+i) - y(t+i))^2 = (\boldsymbol{y}^u - \boldsymbol{y})^T(\boldsymbol{y}^u - \boldsymbol{y}). \tag{11}$$

In the unconstrained case, minimisation of the cost function given by equation (11) leads to the optimum with $y^u(t+i) = y(t+i)$ and $J^* = 0$. This can easily be proved as the sequence y^u is calculated for the controlled plant and the given controller and hence is admissible.

Perhaps the simplest solution is to assume GPC settings taking no penalty on the control increments $\lambda = 0$ and the control horizon $N_u = N$. Substituting (5) into (11) leads to the quadratic programming problem

$$\min_{\boldsymbol{u}} J = -2\left(\boldsymbol{G}^T(\boldsymbol{y}^u - \boldsymbol{f})\right)^T\boldsymbol{u} + \boldsymbol{u}^T\boldsymbol{G}^T\boldsymbol{G}\boldsymbol{u}$$
$$\text{subject to } \boldsymbol{A}\boldsymbol{u} \geq \boldsymbol{b}. \tag{12}$$

The difficulty with this method is the lack of stability properties in the constrained case. However, even with this drawback, GPC is actively used in academia as well as in industry due to its easy implementation.

To assure stability also in the constrained case, the last part of the optimised trajectory will be generated by the linear controller (4). The minimum number m of sampling times follows from the stability requirements and should be the state dimension of the controlled system, in our case $m = \max(\deg(AF), \deg(B))$. This removes m

degrees of freedom from the optimisation problem, hence the prediction horizon N must be greater than m.

Therefore, we optimise only the first N_u control increments and constrain the last $N - N_u$ steps ($N - N_u \geq m$) to be generated by the controller (4).

The sequence of the control increments to be determined can be divided accordingly into two parts. The first part is optimised and the second one is the linear part, that is,

$$\boldsymbol{u} = \begin{pmatrix} \boldsymbol{u}_o \\ \boldsymbol{u}_l \end{pmatrix}, \tag{13}$$

$$\boldsymbol{u}_o = [\tilde{u}_t \ \dots \ \tilde{u}_{t+N_u-1}]^T, \tag{14}$$

$$\boldsymbol{u}_l = [\tilde{u}_{t+N_u} \ \dots \ \tilde{u}_{t+N-1}]^T. \tag{15}$$

The linear part $\boldsymbol{u}_l$ can be determined from the controller equation (4). Careful inspection of terms in (4) shows that $\boldsymbol{u}_l$ is a linear combination of $\boldsymbol{y}, \boldsymbol{u}_o$, and $\boldsymbol{w} = [w_{t+N_u-\deg(R)} \ \dots \ w_{t+N-1}]^T$. Hence, it can be written in matrix form as a sum of free and forced responses

$$\boldsymbol{u}_l = \boldsymbol{G}_{lw}\boldsymbol{w} + \boldsymbol{G}_{ly}\boldsymbol{y} + \boldsymbol{G}_{lo}\boldsymbol{u}_o + \boldsymbol{f}_{lu}. \tag{16}$$

This vector-matrix representation can be obtained in analogy to the system description (5) assuming the controller equation (4).

Combining (16) with (5) and eliminating intermediate variables yields

$$\begin{aligned} \boldsymbol{y} &= \boldsymbol{G}\boldsymbol{u} + \boldsymbol{f} = \boldsymbol{G}_1\boldsymbol{u}_o + \boldsymbol{G}_2\boldsymbol{u}_l + \boldsymbol{f} \\ &= \boldsymbol{G}_y\boldsymbol{u}_o + \boldsymbol{f}_y, \end{aligned} \tag{17}$$

where

$$\boldsymbol{G}_y = (\boldsymbol{I} - \boldsymbol{G}_2\boldsymbol{G}_{ly})^{-1}(\boldsymbol{G}_1 + \boldsymbol{G}_2\boldsymbol{G}_{lo}), \tag{18}$$

$$\boldsymbol{f}_y = (\boldsymbol{I} - \boldsymbol{G}_2\boldsymbol{G}_{ly})^{-1}(\boldsymbol{G}_2[\boldsymbol{G}_{lw}\boldsymbol{w} + \boldsymbol{f}_{lu}] + \boldsymbol{f}) \tag{19}$$

As $\boldsymbol{G}_2$ is zero on and above the main diagonal, the inverse matrix exists.

In the same manner for $\boldsymbol{u}_l$ we obtain,

$$\begin{aligned} \boldsymbol{u}_l &= (\boldsymbol{G}_{lo} + \boldsymbol{G}_{ly}\boldsymbol{G}_y)\boldsymbol{u}_o + (\boldsymbol{G}_{ly}\boldsymbol{f}_y + \boldsymbol{f}_{lu}) \\ &= \boldsymbol{G}_u\boldsymbol{u}_o + \boldsymbol{f}_u. \end{aligned} \tag{20}$$

The constraint description (10) holds for both components $\boldsymbol{u}_o, \boldsymbol{u}_l$. Substituting for $\boldsymbol{u}_l$ from (20) we get,

$$\begin{aligned} \boldsymbol{A}\boldsymbol{u} &\geq \boldsymbol{b} \\ (\boldsymbol{A}_1 \ \boldsymbol{A}_2)\begin{pmatrix} \boldsymbol{u}_o \\ \boldsymbol{u}_l \end{pmatrix} &\geq \boldsymbol{b} \\ (\boldsymbol{A}_1 + \boldsymbol{A}_2\boldsymbol{G}_u)\boldsymbol{u}_o &\geq \boldsymbol{b} - \boldsymbol{A}_2\boldsymbol{f}_u. \end{aligned} \tag{21}$$

The resulting quadratic programming problem will be obtained by substituting (17) into (11).

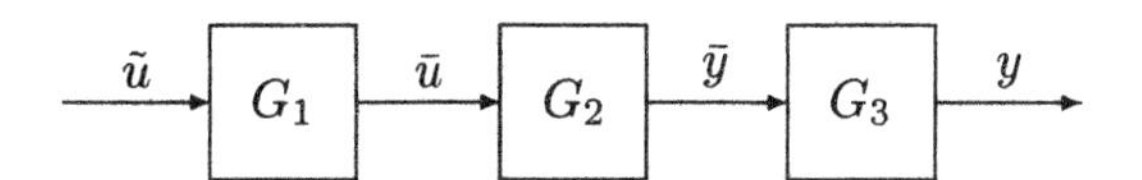

Fig. 1. System decomposition

Neglecting the constant term and using the inequality constraint (21) leads to the minimisation problem

$$\min_{\boldsymbol{u}_o} J = -2\left(\boldsymbol{y}^u - \boldsymbol{f}_y\right)^T \boldsymbol{G}_y\boldsymbol{u}_o + \boldsymbol{u}_o^T\boldsymbol{G}_y^T\boldsymbol{G}_y\boldsymbol{u}_o$$
$$\text{subject to } (\boldsymbol{A}_1 + \boldsymbol{A}_2\boldsymbol{G}_u)\boldsymbol{u}_o \geq \boldsymbol{b} - \boldsymbol{A}_2\boldsymbol{f}_u. \tag{22}$$

4. APPROACH #2

The second approach uses the fact that any predictive controller can be described as the 2DoF controller given by (4) that generates the closed-loop system characterised by the closed-loop pole polynomial M.

Let us decompose the plant B/AF into three subsystems G_1, G_2, G_3 in a series (see Fig. 1) where

$$G_1 = \frac{M}{A^+}, \quad G_2 = \frac{B^-}{FA^-}, \quad G_3 = \frac{B^+}{M}, \tag{23}$$

where the superscript $^+$ denotes the stable and the superscript $^-$ the anti-stable part of a polynomial. We assume that M is a stable polynomial. From the decomposition it follows that the sequences $\tilde{u}, y$ are generated from the internal signals $\bar{u}, \bar{y}$ filtered by the stable transfer functions

$$\tilde{u} = \frac{A^+}{M}\bar{u}, \quad y = \frac{B^+}{M}\bar{y}. \tag{24}$$

We will now apply the method CRHPC (Constrained Receding Horizon Predictive Control) (Clarke and Scattolini 1991) to the transfer function G_2 and the signals $\bar{u}, \bar{y}$.

To this end, let us define the number of degrees of freedom $n \geq 0$, the output horizon N, the control horizon N_u, and the system state dimension m as

$$N_u = \deg(FA^-) + n, \tag{25}$$

$$N = \deg(B^-) + n, \tag{26}$$

$$m = \max(\deg(FA^-), \deg(B^-)), \tag{27}$$

and the vectors

$$\bar{\boldsymbol{u}}^T = (\bar{u}_t \ \dots \ \bar{u}_{t+N_u-1}), \tag{28}$$

$$\bar{\boldsymbol{y}}^T = (\bar{y}_{t+1} \ \dots \ \bar{y}_{t+N-1}), \tag{29}$$

$$\bar{\boldsymbol{y}}_1^T = (\bar{y}_{t+N} \ \dots \ \bar{y}_{t+N+m-1}), \tag{30}$$

$$\bar{\boldsymbol{w}}^T = (w_{t+1} \ \dots \ w_{t+N-1})M(1)/B^+(1), \tag{31}$$

$$\bar{\boldsymbol{w}}_1^T = (\bar{w}_{t+N} \ \dots \ \bar{w}_{t+N+m-1})M(1)/B^+(1) \tag{32}$$

The polynomial formulation of the system G_2 is rewritten in the vector-matrix form similar as in (5)

$$\begin{pmatrix} \bar{\boldsymbol{y}} \\ \bar{\boldsymbol{y}}_1 \end{pmatrix} = \begin{pmatrix} \boldsymbol{G} \\ \boldsymbol{G}_1 \end{pmatrix} \bar{\boldsymbol{u}} + \begin{pmatrix} \bar{\boldsymbol{f}} \\ \bar{\boldsymbol{f}}_1 \end{pmatrix} \tag{33}$$

Finally, let us introduce the following cost function with the weighting matrices $\boldsymbol{W}_e > 0$ and/or $\boldsymbol{W}_u > 0$

$$J(\bar{\boldsymbol{u}}) = \bar{\boldsymbol{e}}^T \boldsymbol{W}_e \bar{\boldsymbol{e}} + \bar{\boldsymbol{u}}^T \boldsymbol{W}_u \bar{\boldsymbol{u}}, \tag{34}$$

where $\bar{\boldsymbol{e}} = \bar{\boldsymbol{w}} - \bar{\boldsymbol{y}}$.

The stable predictive controller is then defined as the following optimisation problem: Minimise (34) subject to the inequality constraints (10), the equality constraints

$$\bar{\boldsymbol{y}}_1 = \bar{\boldsymbol{w}}_1, \tag{35}$$

$$\bar{u}_{t+N_u+j} = 0, \quad j = 0, \ldots, N - N_u + m, \tag{36}$$

and the system equality constraint (33). The actual control increment $\tilde{u}(t)$ is calculated from (24).

It can easily be shown that the predictive controller without degrees of freedom ($n = 0$) is equivalent to a 2DoF controller with given closed-loop poles M. For the case of $n > 0$, the task is to construct a controller (and a cost function) such that in the unconstrained case the control actions are those of the nominal controller with $n = 0$.

This can be obtained by minimising the cost function

$$J = \sum_{i=1}^{\infty} \tilde{e}^2(t+i), \tag{37}$$

where

$$\tilde{e} = \frac{B^+}{M_1} \bar{e} \tag{38}$$

and the stable polynomial M_1 is given by the spectral factorisation equation (Kučera 1979)

$$B^* B = M_1^* M_1. \tag{39}$$

Although the infinite horizon cost function (37) is minimised, the optimisation problem has only a finite number of variables. For further details, see (Fikar and Unbehauen 1999).

5. SIMULATION RESULTS

In this section some of the properties of the proposed algorithms are studied by means of simulations. Let us consider the class of step-change references and the following unstable discrete-time system

$$G = \frac{z^{-2}}{(1 + 3z^{-1})^2}, \tag{40}$$

and control constraints $\Delta u \leq 5$, $u \geq -1$. For the controller consider 2DoF dead-beat controller with integral action of the form

$$u_t = 6u_{t-1} - 5u_{t-2} + w_t - 22y_t - 24y_{t-1} + 45y_{t-2} \tag{41}$$

The first simulation given in Fig. 2 shows the performance of this nominal controller in the unconstrained case (subscripts u) and in the constrained case (subscripts c). It is interesting to notice in both cases the same control action at the time $t = 1$ because it is within the constraints. However, this has the consequence that the system states are moved into an unstabilisable region with respect to the constraints.

Thus, the example shows a case where even the best anti-windup strategy would be unsuccessful because it lacks information about the future behaviour of the system.

The second simulation compares the proposed approaches and the results are shown in Fig. 3. One can observe that in both cases the closed loop is stabilised in the first part of the trajectory. During the second step change the control actions are within constraints and both predictive controllers generate the same actions as those of the nominal unconstrained controller.

The small differences in the first part are caused by different design aims while dealing with the constraints. Recall that the first controller tries to minimise the difference between unconstrained and constrained output trajectories whereas the second one gives more importance to the location of the desired closed-loop poles.

6. CONCLUSIONS

The article has discussed two different ways how to implement a given controller within the framework of predictive control. Of course, identical performance can only be achieved in the unconstrained case.

The advantage of the proposed methods has been shown by the simulation examples where for a given controller, conventional anti-windup methods cannot stabilise the constrained closed-loop system due to the lack of their predictive properties. More precisely, to guarantee stability for certain systems, a modification of the control trajectory has to be performed some steps before the constraints are reached.

Two possible methods how to incorporate the controller into the predictive control have been presented. In the first approach, the future set-point sequence has been manipulated and the cost function mininising the difference between constrained and unconstrained output trajectories has been chosen. In the second approach, terminal equality constraints leading to the desired closed-loop pole locations have been specified.

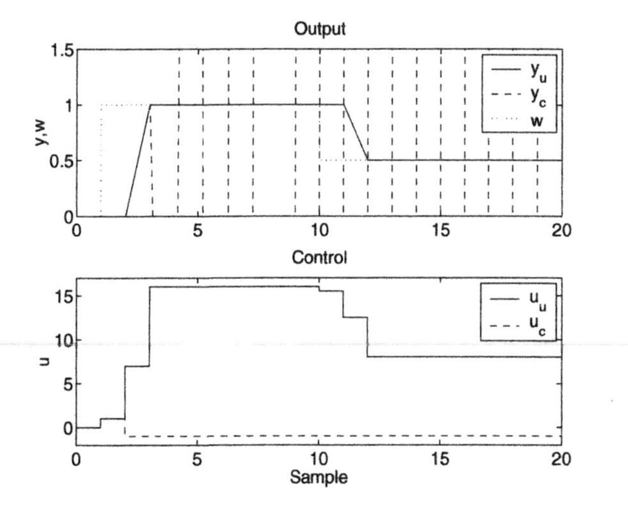

Fig. 2. Nominal controller performance

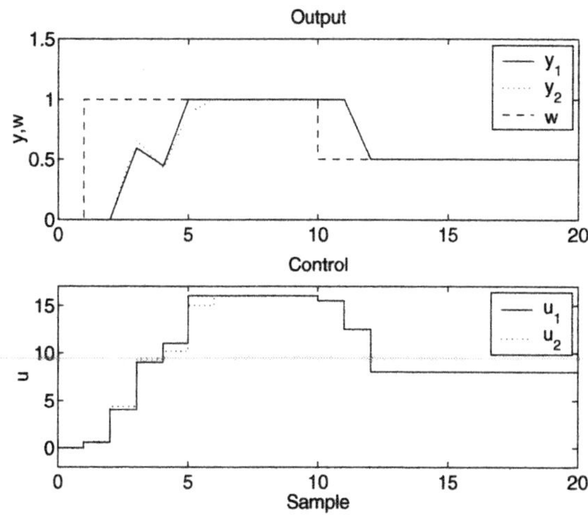

Fig. 3. Performance of the proposed controllers

In both approaches, stability of the constrained closed-loop system is guaranteed as well as a bumpless transfer between constrained and unconstrained modes. The behaviour of the controllers in the constrained mode is not the same due to their different design formulations.

Acknowledgements

The work has been supported by the Alexander von Humboldt Foundation and by the grant agency VEGA MŠ SR (1/135/03, 1/8108/01). This support is very gratefully acknowledged.

7. REFERENCES

Bemporad, A. (1998). Reference governor for constrained nonlinear systems. *IEEE Trans. Automatic Control* **43**(3), 415–419.

Chmielewski, D. and V. Manousiouthakis (1996). On constrained infinite-time linear quadratic optimal
control. *Systems Control Lett.* **29**(3), 121 - 129.

Chmúrny, D., R. Prokop and M. Bakošová (1988). *Automatic Control of Technological Processes.* Alfa. Bratislava. (in Slovak).

Clarke, D. W. and R. Scattolini (1991). Constrained receding-horizon predictive control. *IEE Proc. D* **138**(4), 347 - 354.

Clarke, D. W., C. Mohtadi and P. S. Tuffs (1987). Generalized predictive control - part I. The basic algorithm. *Automatica* **23**(2), 137 - 148.

Fikar, M. (1998). Predictive control - an introduction. Technical report KAMF9801. Department of Process Control, FCT STU, Bratislava, Slovakia.

Fikar, M. and H. Unbehauen (1999). Some new results in linear predictive control. Technical report MF9902. Control Engineering Laboratory, Faculty of Electrical Engineering, Ruhr-University Bochum, Germany.

Kothare, M. V., P. J. Campo, M. Morari and C. N. Nett (1994). A unified framework for the study of anti-windup designs. *Automatica* **30**(12), 1869–1883.

Kučera, V. (1979). *Discrete Linear Control: The Polynomial Equation Approach.* Wiley. Chichester.

Rossiter, J. A. and B. Kouvaritakis (1993). Constrained stable generalised predictive control. *IEE Proc. D* **140**(4), 243 - 254.

Scokaert, P. O. M., D. Q. Mayne and J. B. Rawlings (1999). Suboptimal model predictive control (feasibility implies stability). *IEEE Trans. Automatic Control* **44**(3), 648–654.

Zelinka, P., B. Rohal̆-Ilkiv and A.G. Kuznetsov (1999). Experimental verification of stabilising predictive control. *Control Engineering Practice* **7**(5), 601–610.

REDUCED DIMENSION APPROACH TO APPROXIMATE EXPLICIT MODEL PREDICTIVE CONTROL

Alexandra Grancharova[1, 2, 3] and Tor A. Johansen[4]

Department of Engineering Cybernetics, Norwegian University of Science and Technology, N-7491 Trondheim, Norway

Abstract: Exact or approximate solutions to constrained linear model predictive control (MPC) problems can be pre-computed off-line in an explicit form as a piecewise linear state feedback defined on a polyhedral partition of the state space. However, the complexity of the polyhedral partition often increases rapidly with the dimension of the state vector, and the number of constraints. This paper presents an approach for reducing the dimension of the approximate explicit solution to linear constraint MPC problems.

Keywords: predictive control, constraints, piecewise linear controllers.

1. INTRODUCTION

Model predictive control (MPC) is an efficient methodology to solve complex constrained multivariable control problems. The requirement to perform on-line optimization however limited the applicability of MPC mostly to slowly varying processes. Recently, several methods for explicit solution of MPC problems have been developed. The main motivation behind explicit MPC is that an explicit state feedback law avoids the need for real-time optimization, and is therefore potentially useful for applications with fast sampling where MPC has not traditionally been used. In (Bemporad, *et al.*, 2000) it was recognized that the constrained linear MPC problem is a multi-parametric quadratic program (mp-QP), when the state is viewed as a parameter to the problem. It was shown that the solution (the control input) has an explicit representation as a piecewise linear (PWL) state feedback on a polyhedral partition of the state space, see also (Bemporad, *et al.*, 2002; Seron, *et al.*, 2000; Tøndel, *et al.*, 2003), and they develop an mp-QP algorithm to compute a representation of this function. However, the complexity of the polyhedral partition often increases rapidly with the dimension of the state vector, and the number of constraints. This led to the investigation of efficient implementation of piecewise linear function evaluation (Borrelli, *et al.*, 2001; Tøndel and Johansen, 2002) as well as input trajectory parameterization (Tøndel and Johansen, 2002) and restrictions on the active constraint switching (Johansen, *et al.*, 2002) in order to reduce the complexity. In (Johansen, 2002), a method for reducing the dimension of the mp-QP solutions to the explicit constrained LQR problems has been investigated. It has been shown that for systems with certain structures such dimension reduction can be achieved by state space projections and that the performance degradation may be acceptable in some applications.

Recently, several approximate algorithms for solving mp-QP problems have been investigated, (Bemporad and Filippi, 2001; Johansen and Grancharova, 2002; Johansen and Grancharova, 2003; Grancharova and Johansen, 2002), with significant reduction in complexity. The algorithms developed in (Johansen

[1] This work was sponsored by the European Commission through the Research Training Network **MAC** (Multi-Agent Control, HPRN-CT-1999-00107)
[2] Present address: Institute of Control and System Research, Bulgarian Academy of Sciences, Acad. G. Bonchev str., Bl.2, P.O.Box 79, Sofia 1113, Bulgaria
[3] `alexandra@icsr.bas.bg`
[4] `Tor.Arne.Johansen@itk.ntnu.no`

and Grancharova, 2002; Johansen and Grancharova, 2003; Grancharova and Johansen, 2002) determine an approximate explicit PWL state feedback solution by imposing an orthogonal search tree structure on the partition. They lead to more efficient real-time computations and admit implementation at high sampling frequencies in embedded systems with inexpensive processors and low software complexity. This paper presents an approach for reducing the dimension of the approximate explicit solution to linear constraint MPC problems and can be considered as an extension of the approximate mp-QP algorithm suggested in (Johansen and Grancharova, 2002; Johansen and Grancharova, 2003).

2. EXPLICIT MPC AND EXACT MP-QP

Formulating a linear MPC problem as an mp-QP is briefly described below, see (Bemporad, *et al.*, 2000) for further details. Consider the discrete-time linear system:

$$x(t+1) = Ax(t) + Bu(t) \tag{1}$$

where $x(t) \in R^n$ is the state variable, $u(t) \in R^m$ is the input variable, $A \in R^{n \times n}$, $B \in R^{n \times m}$ and (A,B) is a controllable pair. For the current $x(t)$, MPC solves the optimization problem:

$$V^*(x(t)) = \min_{U \equiv \{u_t, \dots, u_{t+N-1}\}} J(U, x(t)) \tag{2}$$

subject to:

$$y_{\min} \le y_{t+k|t} \le y_{\max},\ k = 1, \dots, N \tag{3}$$

$$u_{\min} \le u_{t+k} \le u_{\max},\ k = 0, 1, \dots, N-1 \tag{4}$$

$$x_{t|t} = x(t) \tag{5}$$

$$x_{t+k+1|t} = Ax_{t+k|t} + Bu_{t+k},\ k \ge 0 \tag{6}$$

$$y_{t+k|t} = Cx_{t+k|t},\ k \ge 0 \tag{7}$$

with the cost function given by:

$$J(U, x(t)) = \sum_{k=0}^{N-1} \left[x^T{}_{t+k|t} Q x_{t+k|t} + u^T{}_{t+k} R u_{t+k} \right] + x^T{}_{t+N|t} P x_{t+N|t} \tag{8}$$

and symmetric $R > 0$, $Q \ge 0$, $P > 0$. The final cost matrix P may be taken as the solution of the algebraic Riccati equation. With the assumption that no constraints are active for $k \ge N$ this corresponds to an infinite horizon LQ criterion (Chmielewski and Manousiouthakis, 1996). This and related problems can by some algebraic manipulation be reformulated as:

$$V_z^*(x) = \min_z \frac{1}{2} z^T H z \tag{9}$$

subject to:

$$Gz \le W + Sx \tag{10}$$

where $z \equiv U + H^{-1}F^T x$. Note that $H > 0$ since $R > 0$. The vector x is the current state, which can be treated as a vector of parameters. For ease of notation we write x instead of $x(t)$. The number of inequalities is denoted q and the number of free variables is $n_z = m \cdot N$. Then $z \in R^{n_z}$, $H \in R^{n_z \times n_z}$, $G \in R^{q \times n_z}$, $W \in R^{q \times 1}$, $S \in R^{q \times n}$. The solution of the optimization problem (9)-(10) can be found in an explicit form $z^* = z^*(x)$, (Bemporad, *et al.*, 2000):

Theorem 1 *Consider the mp-QP (9)-(10) and suppose* $H > 0$. *The solution* $z^*(x)$ *(and* $U^*(x)$*) is a continuous PWL function of* x *defined over a polyhedral partition of the parameter space, and* $V_z^*(x)$ *is a convex (and therefore continuous) piecewise quadratic function.*

3. APPROXIMATE REDUCED DIMENSION MP-QP ALGORITHM

In (Johansen and Grancharova, 2002; Johansen and Grancharova, 2003; Grancharova and Johansen, 2002), algorithms that determine an approximate explicit PWL state feedback solution of possible lower complexity are developed. The idea is to require that the state space partition is represented as a binary search tree (quad-tree partition (Johansen and Grancharova, 2002; Johansen and Grancharova, 2003) or *k-d*-tree partition (Grancharova and Johansen, 2002), cf. Fig. 1), i.e. to consist of orthogonal hypercubes organized in a hierarchical data-structure. This allows extremely fast real-time search.

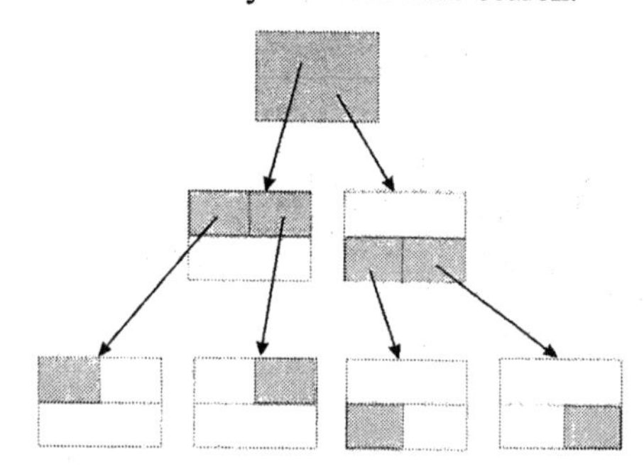

Fig. 1. *k-d*-tree partition of a rectangular region in a 2-dimensional state-space.

However, like the exact solution, the number of regions increases significantly with the order n of the system. Therefore, it is reasonable to search for modification of this algorithm that would reduce significantly the complexity of partition.

The present paper extends the authors work (Johansen and Grancharova, 2002; Johansen and Grancharova, 2003), where the approximate mp-QP algorithm is guaranteed to terminate with an approximate solution that satisfies a specified maximum allowed error in the cost function. Here, a modified approximate mp-QP algorithm is developed that is to be applied to high-order systems. The idea is to build partition in a sub-space of the state space, where each region will be characterized by a set of PWL feasible full-state feedback laws, since the unique optimal active set can not be determined from a reduced-dimension state, in general. In the real-time implementation of the explicit MPC, it will be necessary to check which of the feasible feedback laws associated to the current region is the optimal one (minimizes the performance index), see also (Johansen, *et al.*, 2002). However, keeping the number of these feedback laws small and searching in a reduced dimension partition, will guarantee an efficient real-time implementation.

Let the full state vector be $x \in R^n$, the reduced state vector on which the partition will be made be $\tilde{x} \in R^{\tilde{n}}$, where $\tilde{n} < n$, and the vector with the remaining part of the state variables be $\breve{x} \in R^{n-\tilde{n}}$. Denote the hypercube defining the whole region of interest in the full state space as X and the corresponding hypercube in the reduced state space as $\tilde{X}$, where $\tilde{X} \subset X$. Let also for convenience call the state variables $\tilde{x}$ "partitioned" and the state variables $\breve{x}$ "non-partitioned".

Lemma 1. *Consider the bounded hypercube* $\tilde{X}_0 \subset \tilde{X} \subset R^{\tilde{n}}$ *with vertices* $\{\tilde{v}_1, \tilde{v}_2, \dots, \tilde{v}_L\}$, *where* $L = 2^{\tilde{n}}$. *Let the lower and upper bounds of the non-partitioned state variables* $\breve{x} \in R^{n-\tilde{n}}$ *be* $\breve{x}^l$ *and* $\breve{x}^u$. *Consider the bounded hypercube* $X_0 \subset X \subset R^n$ *with vertices* $\{v_1, v_2, \dots, v_K\}$ *defined by:*

$$v_k = [\tilde{v}_i^T \ \breve{x}^{j^T}]^T, i = 1, 2, \dots, L, \ j = 1, 2, \dots, 2^{n-\tilde{n}} \quad (11)$$

where $\breve{x}_j = [\breve{x}_1^j \ \breve{x}_2^j \ \dots \ \breve{x}_{n-\tilde{n}}^j]^T, j = 1, 2, \dots, 2^{n-\tilde{n}}$ *represent all possible combinations of lower and upper values of the non-partitioned state variables and* $K = L \cdot 2^{n-\tilde{n}} = 2^n$. *Consider a set of L vertices:*

$$w_i^s = [\tilde{v}_i^T \ \breve{x}^{s^T}]^T \ , i = 1, 2, \dots, L \quad (12)$$

where $\breve{x}^s$ *satisfies* $\breve{x}^l \le \breve{x}^s \le \breve{x}^u$. *Let* $z_s^*(w_i^s)$ *be the optimal solution of the mp-QP (9) – (10) at the vertex* w_i^s *defined by (12). If* $\hat{K}_s$ *and* $\hat{g}_s$ *solve the QP:*

$$\min_{\hat{K}_s, \hat{g}_s} \sum_{i=1}^{L} \left(z_s^*(w_i^s) - \hat{K}_s w_i^s - \hat{g}_s\right)^T H \left(z_s^*(w_i^s) - \hat{K}_s w_i^s - \hat{g}_s\right) (13)$$

subject to:

$$G\left(\hat{K}_s v_k + \hat{g}_s\right) \le S v_k + W \ , k \in \{1, 2, \dots, K\} \ (14)$$

then the least squares approximation $\hat{z}_s(x) = \hat{K}_s x + \hat{g}_s$ *is feasible for the mp-QP (9) – (10) for all* $x \in X_0$.

Proof: It follows from convexity and linearity similar to Lemma 1 from (Bemporad and Filippi, 2001).

Lemma 1 is illustrated on Figure 2, where the full state vector is $x = [x_1 \ x_2 \ x_3]^T$, the partitioned state variables are $\tilde{x} = [x_1 \ x_2]^T$ and the non-partitioned state variable is $\breve{x} = x_3$.

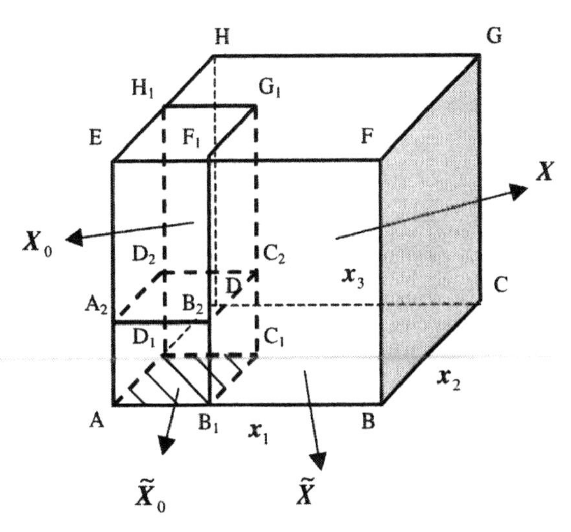

Fig. 2. Illustration of Lemma 1.

The hypercube X with vertices $\{A, B, C, D, E, F, G, H\}$ denotes the whole region of interest in the full state space and $\tilde{X}$ with vertices $\{A, B, C, D\}$ denotes the whole region in the reduced state space. Consider a region $\tilde{X}_0$ in the partition of $\tilde{X}$, with vertices $\{\tilde{v}_1, \tilde{v}_2, \tilde{v}_3, \tilde{v}_4\} = \{A, B_1, C_1, D_1\}$. The corresponding region X_0 in the full state space has vertices $\{v_1, \dots, v_8\} = \{A, B_1, C_1, D_1, E, F_1, G_1, H_1\}$. Let us consider the case where the set of vertices (12) correspond to the minimal allowed value of the non-partitioned state variable x_3, i.e. $\{w_1^1, w_2^1, w_3^1, w_4^1\} = \{A, B_1, C_1, D_1\}$. Conditions (13) and (14) of Lemma 1 mean that the candidate approximate solution $\hat{z}_1(x) = \hat{K}_1 x + \hat{g}_1$ is chosen as the best fit to the optimal solutions at the vertices $\{w_1^1, w_2^1, w_3^1, w_4^1\} = \{A, B_1, C_1, D_1\}$, while satisfying all constraints at the vertices $\{v_1, v_2, \dots, v_8\} = \{A, B_1, C_1, D_1, E, F_1, G_1, H_1\}$. By choosing another value for x_3 within the range $[x_3^l ; x_3^u]$, we will have another set of vertices $\{w_1^2, w_2^2, w_3^2, w_4^2\} = \{A_2, B_2, C_2, D_2\}$ generated according to (12) and therefore another candidate feedback law $\hat{z}_2(x) = \hat{K}_2 x + \hat{g}_2$ that will be the best fit to the optimal solutions at these vertices. Both candidate control laws $\hat{z}_1(x)$ and $\hat{z}_2(x)$ will be feasible in the region X_0 in the full state space. We therefore need to consider all candidate control laws corresponding to all possible values of the non-partitioned state variable $\breve{x} = x_3$.

This leads to the following algorithm for design of approximate explicit MPC controller:

Off-line algorithm

Algorithm 1 (approximate reduced dimension mp-QP)

Step 1. Initialize the partition in the reduced state space to the whole hypercube, i.e. $\tilde{P} = \{\tilde{X}\}$. Mark the hypercube $\tilde{X}$ as unexplored.

Step 2. Select any unexplored hypercube $\tilde{X}_0 \in \tilde{P}$. If no such hypercube exists, go to step 12.

Step 3. Determine several representative points $\breve{x}^s, s = 1,...,M$ for the non-partitioned state variables by gridding the space of these variables.

Step 4. Compute the solution to the QP (9) – (10) for $\tilde{x}$ fixed to each of the $2^{\tilde{n}}$ vertices of the hypercube $\tilde{X}_0$ and for $\breve{x}$ fixed to $\breve{x}^s$. Thus the QP will be solved for the following points $x^{i,s}$ in the full state space:

$$x^{i,s} = [\tilde{x}^{i^T} \; \breve{x}^{s^T}]^T, \; i = 1,...,2^{\tilde{n}} \tag{15}$$

If all QPs have a feasible solution, go to step 6. Otherwise, go to step 5.

Step 5. Compute the size of $\tilde{X}_0$ using some metric. If it is smaller than some given tolerance, mark $\tilde{X}_0$ infeasible and explored. Go to step 2. Otherwise, go to step 11.

Step 6. From the optimal solutions at the points $x^{i,s}$, compute s-th feasible linear state feedback law to be used as a candidate control law in the reduced dimension region $\tilde{X}_0$, by applying Lemma 1. This control law is denoted:

$$\hat{z}_s(x) = \hat{K}_s x + \hat{g}_s \tag{16}$$

Step 7. Repeat steps 4, 5 and 6 for all points $\breve{x}^s, s = 1,...,M$ of the non-partitioned state variables $\breve{x}$.

Step 8. Form the set $Z(\tilde{X}_0)$ of the candidate feedback laws to be used in the hypercube $\tilde{X}_0$:

$$Z(\tilde{X}_0) = \{\hat{z}_s(x), s = 1,...,M+1\}, \tag{17}$$

where $\hat{z}_{M+1}(x) = -H^{-1}F^T x$ is the unconstrained solution to the QP problem (9) – (10).

Step 9. For the hypercube $\tilde{X}_0$, compute the maximal approximation error in the cost function value among the candidate control laws in the subset $Z(\tilde{X}_0)$:

$$\varepsilon = \max_{x \in X_0} \{ \min_s \varepsilon_1(\hat{z}_s(x)) \} \tag{18}$$

where $\varepsilon_1(\hat{z}_s(x))$ is the upper bound of the approximation error obtained with the control law $\hat{z}_s(x) = \hat{K}_s x + \hat{g}_s$. The way to compute $\varepsilon_1(\hat{z}_s(x))$ is given in (Johansen and Grancharova, 2002).

Step 10. Determine if the hypercube $\tilde{X}_0$ needs to be split in order to reduce the cost function approximation error ε. If $\varepsilon > \bar{\varepsilon}$, go to step 11. Otherwise, mark $\tilde{X}_0$ explored and go to step 2.

Step 11. Split the hypercube $\tilde{X}_0$ into two hypercubes $\tilde{X}_1$ and $\tilde{X}_2$ by applying the heuristic rule from (Grancharova and Johansen, 2002). Mark them unexplored, remove $\tilde{X}_0$ from $\tilde{P}$, add $\tilde{X}_1$ and $\tilde{X}_2$ to $\tilde{P}$ and go to step 2.

Step 12. For every hypercube $\tilde{X}_i$ in the final partition $\tilde{P} = \{\tilde{X}_1, \tilde{X}_2, ..., \tilde{X}_R\}$, determine explicitly the sub-optimal cost functions $\hat{J}_s^i(x)$ that correspond to the candidate feedback laws $\hat{z}_s(x) = \hat{K}_s x + \hat{g}_s$, $\hat{z}_s(x) \in Z(\tilde{X}_i)$ associated to this region:

$$\hat{J}_s^i(x) = \frac{1}{2}\hat{z}_s^T(x) H \hat{z}_s(x) = x^T A_s x + b_s x + c_s, \tag{19}$$

where $A_s = \frac{1}{2}\hat{K}_s^T H \hat{K}_s$, $b_s = \hat{g}_s^T H \hat{K}_s$ and $c_s = \frac{1}{2}\hat{g}_s^T H \hat{g}_s$.

This algorithm will terminate with a set of PWL functions in each region in the reduced state space partition.

The real-time implementation of the explicit MPC controller is described with the following algorithm:

On-line algorithm

Algorithm 2 (selection of the best control law)

Step 1. Determine the hypercube $\tilde{X}_i$ in the reduced state space, to which the current state $\tilde{x}$ belongs. This region is characterized with M-number of candidate control laws $\hat{z}_s(x) = \hat{K}_s x + \hat{g}_s$ and their corresponding cost functions $\hat{J}_s^i(x) = x^T A_s x + b_s x + c_s$, $s = 1,...,M$.

Step 2. Compute the cost functions values $\hat{J}_s^i(x)$, $s = 1,...,M$ at the current state $x = [\tilde{x}^T \; \breve{x}^T]^T$ corresponding to the candidate control laws in the hypercube $\tilde{X}_i$. Denote the index of the cost function which has minimal value as r.

Step 3. Apply to the plant r-th state feedback law $\hat{z}_r(x) = \hat{K}_r x + \hat{g}_r$.

4. EXAMPLE

A laboratory model helicopter (Quanser 3-DOF Helicopter) is sampled with interval $T = 0.05s$ and the following state space representation is obtained:

$$A = \begin{pmatrix} 1 & 0 & 0.05 & 0 & 0 & 0 \\ 0 & 1 & 0 & 0.05 & 0 & 0 \\ 0 & 0 & 1 & 0 & 0 & 0 \\ 0 & 0 & 0 & 1 & 0 & 0 \\ 0.05 & 0 & 0.0013 & 0 & 1 & 0 \\ 0 & 0.05 & 0 & 0.0013 & 0 & 1 \end{pmatrix} \tag{20}$$

$$B = \begin{pmatrix} 0.0002 & 0.0002 \\ 0.0017 & -0.0017 \\ 0.0095 & 0.0095 \\ 0.0662 & -0.0662 \\ 0 & 0 \\ 0 & 0 \end{pmatrix} \qquad (21)$$

The states of the system are x_1 - elevation, x_2 - pitch angle, x_3 - elevation rate, x_4 - pitch angle rate, x_5 - integral of elevation error, and x_6 - integral of pitch angle error. The inputs to the system are u_1 - front rotor voltage and u_2 - rear rotor voltage. The system is to be regulated to the origin with the following constraints on the inputs and pitch and elevation rates $-1 \le u_1 \le 3$, $-1 \le u_2 \le 3$, $-0.44 \le x_3 \le 0.44$ and $-0.6 \le x_4 \le 0.6$. The cost function is given by $Q = \text{diag}\{100,100,10,10,400,200\}$, $R = I_{2\times 2}$ and P is given by the algebraic Riccati equation. We consider MPC problem with time horizon $N = 20$. The complexity of the optimization problem is reduced by implementing control input trajectory parameterization as it is described in (Tøndel and Johansen, 2002). The idea is to use an input trajectory parameterization with less degrees of freedom in order to reduce the dimension of the optimization problem. The most common approach is to pre-determine the time-instants at which the control input u_i is allowed to change (input blocking):

$$N_{change}^{u_i} = [1 \;\; N_1^{u_i} \;\; N_2^{u_i} \;\; \ldots \;\; N_l^{u_i}] \qquad (22)$$

Here, the time instants at which the control inputs can change are $N_{u_1} = N_{u_2} = [1\ 3\ 5\ 7\ 9\ 11\ 13\ 15]$, which makes totally 16 optimization variables. The full state space to be considered is 6-dimensional and is defined by X=[-0.5,0.5]×[-0.5,0.5]×[-0.5,0.5]×[-0.65,0.65]×[-0.7,0.7]×[-0.8,0.8]. The reduced state space $\tilde{X}$ (in which the partition is made) includes the state variables x_3 and x_4, and is defined by $\tilde{X}$ = [-0.5, 0.5]×[-0.65, 0.65]. The choice of these state variables is made according to the method described in (Johansen, 2002).

The partition in the reduced state space is shown in Fig. 3 and it has 72 regions with up to 15 candidate PWL control laws for each region. In Fig. 4 to 11 the time-domain trajectories for initial state $x(0) = [0.5\ 0.5\ 0\ 0\ 0\ 0]^T$ are given, where the solid line corresponds to the approximate MPC and the dotted line corresponds to the exact MPC. In Fig. 3, the two curves represent the approximate and the exact trajectories in the reduced state space (x_3, x_4) obtained for the same initial condition. It can be seen that the approximate MPC controller designed in the reduced state space gives feasible trajectories (all input and state constraints are satisfied) and the sub-optimality is acceptable.

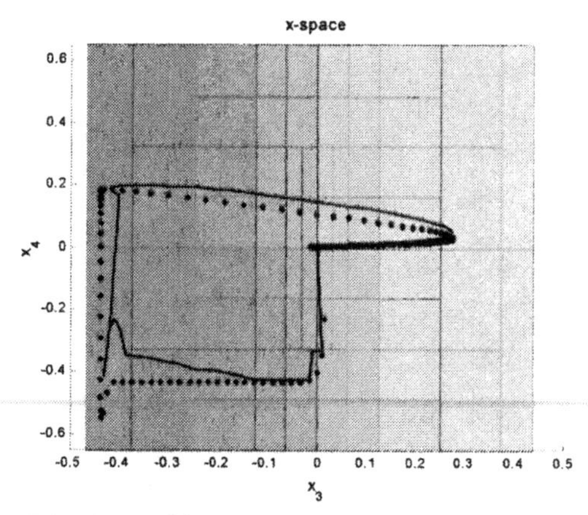

Fig. 3. Partition in (x_3, x_4) space.

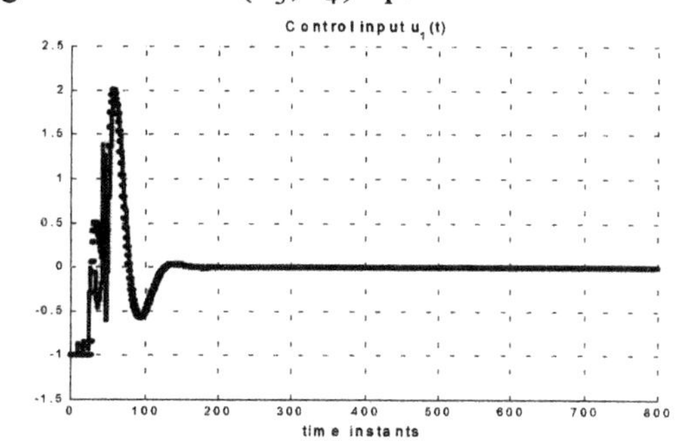

Fig. 4. Control input u_1.

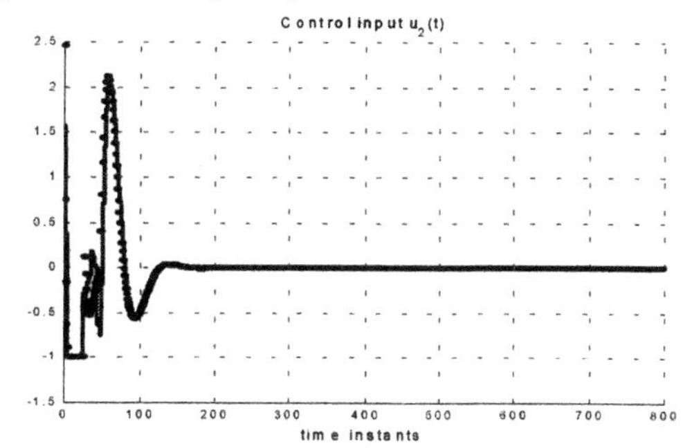

Fig. 5. Control input u_2.

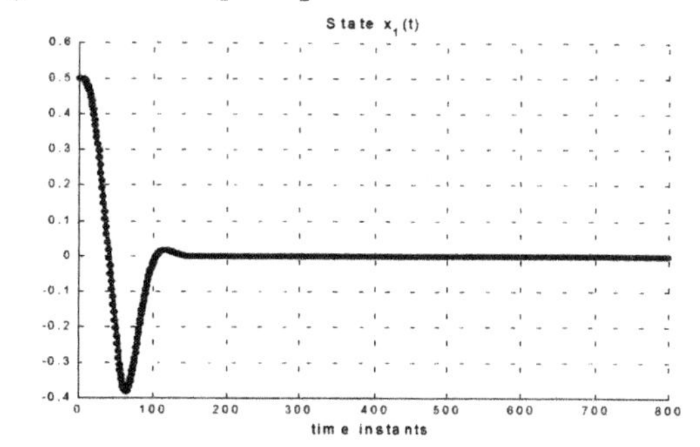

Fig. 6. State x_1 (elevation).

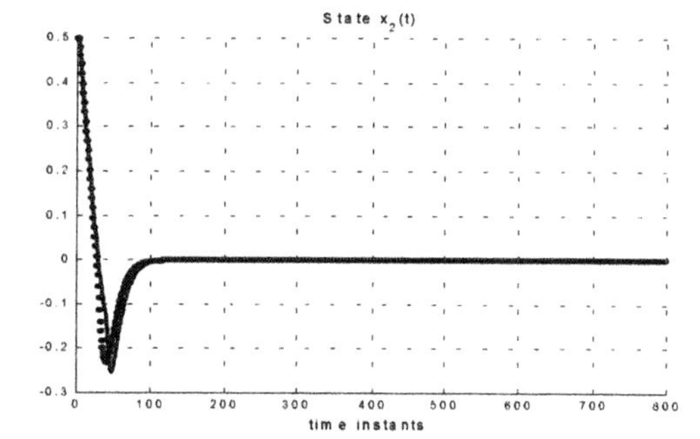

Fig. 7. State x_2 (pitch angle).

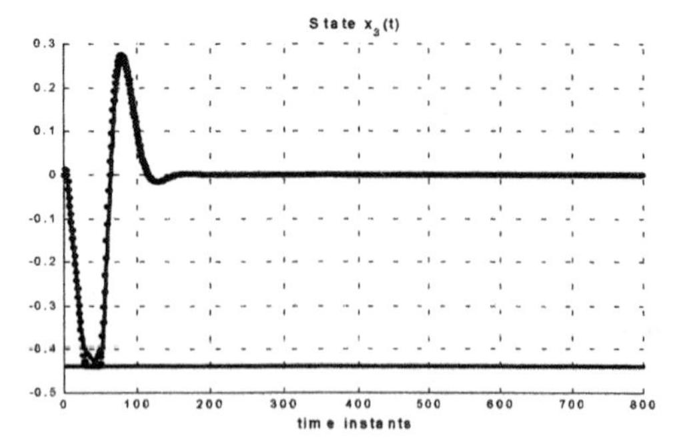

Fig. 8. State x_3 (elevation rate).

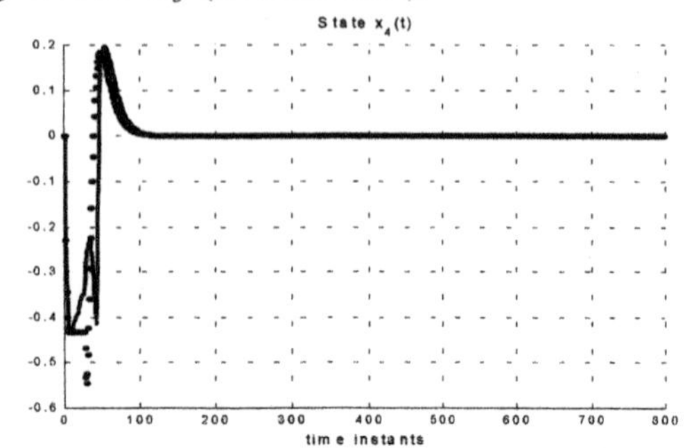

Fig. 9. State x_4 (pitch angle rate).

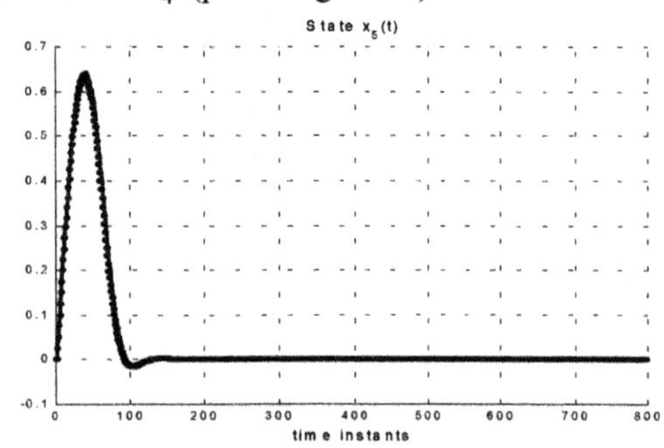

Fig. 10. State x_5 (integral error of elevation).

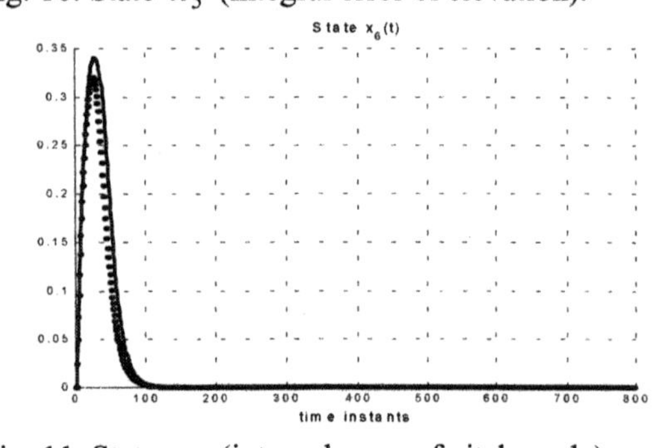

Fig. 11. State x_6 (integral error of pitch angle).

5. CONCLUSIONS

In this paper, an approach for reducing the dimension of the approximate explicit solution to linear constraint MPC problems is proposed. The algorithm builds a partition in a sub-space of the state space, where each region is characterized by a set of PWL feasible state feedback laws. Keeping the number of the candidate feedback laws small and imposing an orthogonal structure of the reduced dimension partition would guarantee an efficient real-time implementation. It is shown that the sub-optimality of the resulting approximate explicit MPC is acceptable.

REFERENCES

Bemporad, A., M. Morari, V. Dua and E. N. Pistikopoulos (2000). The explicit solution of model predictive control via multiparametric quadratic programming. In: *Proc. of American Control Conference, Chicago*, pp. 872-876.

Bemporad, A., M. Morari, V. Dua, and E. N. Pistikopoulos (2002). The explicit linear quadratic regulator for constrained systems. *Automatica*, **vol. 38**, pp. 3-20.

Bemporad, A. and C. Filippi (2001). Suboptimal explicit MPC via approximate quadratic programming. In: *Proc. of IEEE Conf. Decision and Control, Orlando*, pp.4851-4856.

Borrelli, F., M. Baotic, A. Bemporad, and M. Morari (2001). Efficient on-line computation of explicit model predictive control. In: *Proc. of IEEE Conf. Decision and Control, Orlando*, pp. TuP11-2.

Chmielewski, D. and V. Manousiouthakis (1996). On constrained infinite-time linear quadratic optimal control. *Systems & Control Letters*, **vol. 29**, pp. 121-130.

Grancharova, A. and T. A. Johansen (2002). Approximate explicit model predictive control incorporating heuristics. In: *Proc. of IEEE International Symposium on Computer Aided Control System Design, Glasgow, Scotland, U.K*, pp. 92-97.

Johansen, T. A. (2002). Structured and reduced dimension explicit linear quadratic regulators for systems with constraints. In: *Proc. of IEEE Conf. Decision and Control, Las Vegas*, pp. FrA12-3.

Johansen, T. A., I. Petersen and O. Slupphaug (2002). Explicit sub-optimal linear quadratic regulation with state and input constraints. *Automatica*, **vol. 38**, pp. 1099-1111.

Johansen, T. A. and A. Grancharova (2002). Approximate explicit model predictive control implemented via orthogonal search tree partitioning. In: *Proc. of 15-th IFAC World Congress, Barcelona, Spain*, session T-We-M17.

Johansen, T. A. and A. Grancharova (2003). Approximate explicit constrained linear model predictive control via orthogonal search tree. *IEEE Trans. Automatic Control*, **vol. 48**, pp. 810-815.

Tøndel, P. and T. A. Johansen (2002). Complexity reduction in explicit linear model predictive control. In: *Proc. of 15-th IFAC World Congress, Barcelona, Spain*, session T-We-M17.

Tøndel, P., T. A. Johansen and A. Bemporad (2003). An algorithm for multi-parametric quadratic programming and explicit MPC solutions. *Automatica*, **vol. 39**, pp. 489-497.

Seron, M., J. A. De Dona and G. C. Goodwin (2000). Global analytical model predictive control with input constraints. In: *Proc. of IEEE Conf. Decision and Control, Sydney*, pp. TuA05-2.

STABILIZATION OF SAMPLED-DATA NONLINEAR SYSTEMS BY RECEDING HORIZON CONTROL VIA DISCRETE-TIME APPROXIMATIONS

E. Gyurkovics[1] **and A. M. Elaiw**

Budapest University of Technology and Economics, School of Mathematics, Budapest, H-1521, Hungary, Fax:(36-1)463-1291, e-mail: gye@math.bme.hu, elaiw@math.bme.hu

Abstract: Results on stabilizing receding horizon control of sampled-data nonlinear systems via their approximate discrete-time models are presented. The proposed receding horizon control is based on the solution of Bolza-type optimal control problems for the parametrized family of approximate discrete-time models. This paper investigates the situation when the sampling period and the integration parameter used in obtaining approximate model coincide and can be chosen arbitrarily small. Sufficient conditions are established which guarantee that the controller that renders the origin to be asymptotically stable for the approximate model also stabilizes the exact discrete-time model for sufficiently small sampling parameters. *Copyright © 2003 IFAC*

Keywords: Controller design, Predictive control, Feedback stabilization, Sampled-data systems, Numerical methods

1. INTRODUCTION

The stabilization problem of nonlinear systems has received considerable attention in the last decades. The use of computers in the implementation of the controllers necessitated the investigation of sampled-data systems. One way to design a sampled-data control is to implement a continuous-time algorithm with sufficiently small sampling intervals. This approach is proposed in connection with the receding horizon method among the others in (Chen, *et al.*, 2000; Fontes, 2001; Jadababaie and Hauser, 2001). However, some difficulty may arise during the application of this method: 1) the exact solution of the nonlinear continuous-time model is typically unknown, therefore an approximation procedure is unavoidable; 2) it may be difficult to implement an arbitrarily time-varying control function. An overview and analysis of existing approaches for the stabilization of sampled data systems can be found in the recent papers (Nešić, *et al.*, 1999) and (Nešić and Teel) (see also the references therein). In these papers a systematic investigation of the connection between the exact and approximate models are carried out, numerous examples and counter-examples illustrating the effect of the sampling procedure are reported, and conditions are presented which guarantee that the same family of controllers that stabilizes the approximate discrete-time model also practically stabilizes the exact discrete-time model of the plant both for the cases of fixed sampling period and varying integration parameter (Nešić and Teel) and for the

[1] Partially supported by the Hungarian National Foundation for Scientific Research, grant no. T029893 and T037491

case when these two parameters coincide (Nešić, *et al.*, 1999) and (Nešić and Teel).

There are several ways to design controllers satisfying the conditions given in (Nešić, *et al.*, 1999) and (Nešić and Teel). In (Grüne and Nešić 2002), optimization based methods are studied; the design is carried out either via an infinite horizon optimization problem or via an optimization problem over a finite horizon with varying length. Another possibility is the application of the widespread receding horizon or model predictive control method. This method obtains the feedback control by solving a finite horizon optimal control problem at each time instant using the current state of the plant as the initial state for the optimization and applying "the first part" of the optimal control. The study of stabilizing property of such schemes has been the subject of intensive research in recent years. From the vast literature we mention here but a few (Mayne and Michalska, 1990; Gyurkovics, 1996; Chen and Allgöwer, 1998; De Nicolao, *et al.*, 1998a; Gyurkovics, 1998; Jadababaie and Hauser, 2001; Fontes, 2001; Hu and Linnemann, 2002) and we refer the reader to the excellent overview papers (De Nicolao, *et al.*, 1998b; Allgöwer, *et al.*, 1999; Mayne, *et al.*, 2000) and to the references therein. A great majority of works deals either with continuous-time systems with or without taking into account any sampling or with discrete-time systems considering the model given directly in discrete-time. To the best of our knowledge, the only exception is the very recent work of (Ito and Kunisch, 2002), where the effect of the sampling and zero-order hold is considered assuming the existence of a global control Lyapunov function. Relying on the results of (Nešić, *et al.*, 1999) and (Nešić and Teel), the present work studies the conditions under which the stabilizing receding horizon control computed for the approximate discrete-time model also stabilizes the exact discrete-time system in the case when sampling period and the integration parameter coincide and can be chosen arbitrarily small. It should emphasized that these conditions concern directly the data of the problem and the design parameters of the method, but not the result of the design procedure. From practical point of view, it is important to know whether the basin of attraction is sufficiently large when some stabilizing controller is applied. This set is frequently compared with that of the infinite horizon regulator. Here we shall show that the basin of attraction contains any compact subset of the set of initial point which are practically asymptotically controllable to the origin with piecewise constant sampled controllers. In what follows, the notation $\mathcal{B}_\Delta = \{z \in \mathbb{R}^p : \|z\| \leq \Delta\}$ will be used both in $\mathbb{R}^n$ and $\mathbb{R}^m$ and $\mathcal{K}$, $\mathcal{K}_\infty$ and $\mathcal{KL}$ denote the usual class-$\mathcal{K}$, class-$\mathcal{K}_\infty$ and class-$\mathcal{KL}$ functions (see e.g. Nešić, *et al.*, 1999).

2. PRELIMINARIES AND PROBLEM STATEMENT

2.1 The model

Consider the nonlinear control system described by

$$\dot{x}(t) = f\left(x(t), u(t)\right), \tag{1}$$

where $x(t) \in \mathbb{R}^n$, $u(t) \in U \subset \mathbb{R}^m$, $f : \mathbb{R}^n \times U \to \mathbb{R}^n$, with $f(0,0) = 0$, U is closed and $0 \in U$. The system is to be controlled digitally using piecewise constant control functions $u(t) = u(kT) =: u_k$, if $t \in [kT, (k+1)T)$, $k \in \mathbb{N}$, where $T > 0$ is the sampling period.

Assumption A1 (i) Function f is continuous, two times continuously differentiable at least in a neighborhood of the origin in both variables, and for any pair of positive numbers (Δ_1, Δ_2) there exist a $T^* > 0$ such that for any $x_0 \in \mathcal{B}_{\Delta_1}$, $\overline{u} \in U \cap \mathcal{B}_{\Delta_2}$ and $T \in (0, T^*]$ equation (1) with $\overline{u} \equiv u(t)$,and $x(0) = x_0$ has a unique solution on $[0,T]$ denoted by $\phi(., x_0, \overline{u})$.

(ii) For any $\Delta > 0$ there exists an $L_f = L_f(\Delta)$ such that

$$\|f(x,u) - f(y,u)\| \leq L_f\|x - y\|,$$

for all $x, y \in \mathcal{B}_\Delta$ and $u \in \mathcal{B}_\Delta$.

(iii) f is sufficiently smooth. □

Then, the exact discrete-time model of the system can be defined as

$$x_{k+1} = F_T^e(x_k, u_k), \tag{2}$$

where $F_T^e(x,u) := \phi(T, x, u)$. (A discussion about the case of finite escapes can be found e.g. in (Nešić, *et al.*, 1999). It has to be emphasized that F_T^e in (2) is generally not known.

Remark 1 If Assumption A1 is valid, then F_T^e is continuous in x and u and it satisfies a local Lipschitz condition of the following type: for each $\Delta > 0$ there exist $T^* > 0$, and $L > 0$ such that

$$\|F_T^e(x,u) - F_T^e(y,u)\| \leq e^{LT}\|x - y\|, \tag{3}$$

holds for all $u \in \mathcal{B}_\Delta$, all $T \in (0, T^*]$, and all $x, y \in \mathcal{B}_\Delta$.

Suppose that a parametrized family of approximate discrete-time plant models is given

$$x_{k+1} = F_T^a\left(x_k, u_k\right), \tag{4}$$

where T is the sampling parameter. We shall assume that the numerical approximation scheme preserves the properties of f in the following sense.

Assumption A2 (i) $F_T^a(0,0) = 0$, F_T^a is continuous in both variables and it satisfies the same kind of local Lipschitz condition as F_T^e (see Remark 1);

(ii) there exist a $r_0 > 0$, a $T^* > 0$ and a $L_{r_0}^a > 0$ such that for any $T \in (0, T^*]$ we have

$$\|F_T^a(x,u) - x\| \le TL_{r_0}^a(\|x\| + \|u\|) \qquad (5)$$

if $\|x\| + \|u\| \le r_0$.□

Remark 2 Observe that, if (i) Assumption A1 hold true then for many one-step numerical methods, the assertions of Assumption A2 can be proven.

2.2 Practical asymptotic controllability and stabilizability

Let $\Gamma \subset \mathbb{R}^n$ be a given compact set with a bound Δ_0, containing a neighborhood of the origin.

Definition 1. System (2) is *practically asymptotically controllable* (PAC) *from* Γ *to the origin*, if there exist a $\beta(.,.) \in \mathcal{KL}$ and a continuous positive and increasing function $\sigma(.)$ such that for all $x \in \Gamma$ and for all $r > 0$ there exists a control sequence $\mathbf{u}^r(x) = \{u_0^r(x), u_1^r(x), ...\}$, $u_t^r(x) \in U$, $\|u_t^r(x)\| \le \sigma(\|x\|)$ such that the corresponding solution ϕ of (2) satisfies

$$\|\phi_t(x, \mathbf{u}^r(x))\| \le \max\{\beta(\|x\|, tT), r\}, \qquad (6)$$

for all $t \ge 0$. Moreover, system (2) is *semiglobally practically asymptotically controllable to the origin* if it is PAC from any compact set $\Gamma \subset \mathbb{R}^n$.

Definition 2. System (2) is *practically asymptotically stabilizable* (PAS) *in* Γ *about the origin*, if there exist a $\beta(.,.) \in \mathcal{KL}$ and a continuous positive and increasing function $\sigma(.)$ such that for any $r > 0$ there exists a feedback $k^r : \Gamma \to U$, $\|k^r(x)\| \le \sigma(\|x\|)$ such that for any $x \in \Gamma$ the solution ϕ^c of $x_{t+1} = F_T^e(x_t, k^r(x_t))$, $x_0 = x$ inequality

$$\|\phi_t^c(x)\| \le \max\{\beta(\|x\|, tT), r\} \qquad (7)$$

holds true for all $t \ge 0$. Moreover, system (2) is *semiglobally practically asymptotically stabilizable about the origin* if it is PAS in Γ for any compact set $\Gamma \subset \mathbb{R}^n$.

Theorem 1 System (2) is practically asymptotically stabilizable in Γ about the origin if and only if it is practically asymptotically controllable from Γ to the origin.

Proof: Because of lack of space, the proof is omitted.

Remark 3 In view of Theorem 1, it is reasonable to consider the problem of PAS for system (2) over a compact set Γ from which it is PAC. If this latter property is semiglobal, then the same will also be true with respect to stabilization, otherwise semiglobal stabilization is impossible.

Remark 4 Theorem 1 remains valid if, similarly to (Kreisselmeier and Birkhölzer, 1994), the properties PAC and PAS are required with vanishing controllers, i.e. the left hand sides of (6) and (7) are substituted by $\|\phi_t(x, \mathbf{u}^r(x))\| + \|u_t^r(x)\|$ and $\|\phi_t^c(x)\| + \|k^r(\phi_t^c(x))\|$, respectively.

2.3 Basic definitions and assumptions

In what follows, a stabilizing feedback will be constructed for the *approximate model* and conclusion about the stability of the closed-loop exact model is drawn on the basis of the closeness of solutions of the two models. This closeness is characterized by the following definition:

Definition 3 Let a pair of strictly positive numbers (Δ_1, Δ_2) be given and suppose that there exist $\gamma \in \mathcal{K}$ and $T^* > 0$ such that

$$(x,u) \in \mathcal{B}_{\Delta_1} \times \mathcal{B}_{\Delta_2}, \quad T \in (0, T^*] \Longrightarrow \\ \| F_T^a(x,u) - F_T^e(x,u) \| \le T\gamma(T), \quad (8)$$

then the family F_T^a is said to be (Δ_1, Δ_2)-*consistent* with F_T^e. Moreover, if for any pair of strictly positive numbers (Δ_1, Δ_2) there exist $\gamma \in \mathcal{K}$ and $T^* > 0$ such that (8) holds true, then F_T^a is said to be *semiglobally consistent* with F_T^e.

Sufficient checkable conditions for consistency properties can be found in (Nešić, *et al.*, 1999) and (Nešić and Teel).

In order to define a receding horizon feedback controller, let (4) be subject to the cost function

$$J_T(N, x, \mathbf{u}) = \sum_{k=0}^{N-1} Tl_T(x_k^a, u_k) + g(x_N^a),$$

where $x_k^a = \phi_k^a(x, \mathbf{u})$, $k = 0, 1, ..., N$, l_T and g are given functions.

Assumption A3 (i) g is continuous, there exists a class-$\mathcal{K}_\infty$ function γ_1 such that $\gamma_1(\|x\|) \le g(x)$ and for any $\Delta > 0$ there exists a $L_g = L_g(\Delta) > 0$ constant such that $|g(x) - g(y)| \le L_g\|x - y\|$ for all $x \in \mathcal{B}_\Delta$.

(ii) l_T is continuous with respect to x and u, uniformly in small T, and for any $\Delta_1 > 0$, $\Delta_2 > 0$

there exist $T^* > 0$ and $L_l = L_l(\Delta_1, \Delta_2) > 0$ such that

$$|l_T(x,u) - l_T(y,u)| \leq L_l \|x - y\|$$

for all $T \in (0, T^*]$, $x, y \in \mathcal{B}_{\Delta_1}$ and $u \in \mathcal{B}_{\Delta_2}$

(iii) There exist a $T^* > 0$ and two class-$\mathcal{K}_\infty$ functions φ_1 and φ_2 such that the inequality

$$\varphi_1(\|x\|) + \varphi_1(\|u\|) \leq l_T(x,u) \leq \varphi_2(\|x\|) + \varphi_2(\|u\|),$$

holds for all $x \in \mathbb{R}^n$, $u \in U$ and $T \in (0, T^*]$.

Assumption A4 There exists a $T^* > 0$ such that the exact discrete-time model (2) is PAC from Γ to the origin for all $T \in (0, T^*]$.

Let $\beta(.,.)$ and $\sigma(.)$ be functions generated by Assumption A4 and let Δ_1 be such that $\Delta_1 \geq 1 + \max_{x \in \Gamma} \beta(\|x\|, 0)$. Moreover, for $0 < r \leq \Delta_1$, we introduce the notation

$$\mathcal{U}_r = U \cap \mathcal{B}_{\sigma(r)} \quad (9)$$

Assumption A5 The family F_T^a is semiglobally consistent with F_T^e.

Assumption A6 There exist $T^* > 0$ and $\eta > 0$ such that for all $x \in \mathcal{G}_\eta = \{x : g(x) \leq \eta\}$ there exists a $k_T(x) \in \mathcal{U}_{\rho_0}$ such that inequality

$$T l_T(x, k_T(x)) + g\left(F_T^a(x, k_T(x))\right) \leq g(x) \quad (10)$$

holds true for all $T \in (0, T^*]$, where ρ_0 is such that $\mathcal{G}_\eta \subset \mathcal{B}_{\rho_0}$.

In what follows, T_0^* denotes the minimum of T^* generated by Assumptions A1-A6 and Remark 1.

Proposition 1. If Assumptions A4-A6 hold true, then there exists $T^* > 0$ such that, for any $T \in (0, T^*]$ system (4) is asymptotically controllable from Γ to the origin.

Proof. The proof is an immediate consequence of Theorem 1 and A6.

Suppose that Assumption A4 holds true. Consider the optimization problem

$$P_T^a(N, x) : \min\{J_T(N, x, \mathbf{u}) : u_k \in \mathcal{U}_{\Delta_1}\}$$

where $\mathbf{u} = \{u_0, u_1, ..., u_{N-1}\}$. (We note that in problem $P_T^a(N, x)$ no terminal constraint is given.)

If this optimization problem has a solution denoted by $\mathbf{u}^*$ then its first element, i.e. u_0^* is applied at the state x. Since the optimal solution of $P_T^a(N, x)$ naturally depends on x, in this way a feedback has been defined on the basis of the approximate discrete-time model, i.e. $v_T^a(x) := u_0^*$.

Conditions, under which v_T^a asymptotically stabilizes the origin for the approximate model (4) with a fixed $T > 0$, are well-established (see e.g. (Gyurkovics, 1998, Chen and Allgöwer, 1998; De Nicolao, *et al.*, 1998a,b; Jadababaie and Hauser, 2001; Hu and Linnemann, 2002), the review papers (Allgöwer, *et al.*, 1999; Mayne, *et al.*, 2000) and the references therein).

In order to ensure the stabilizing property of v_T^a for the exact model (2) in the given set Γ, one needs somewhat stronger conditions than it is usual in receding horizon investigations: in fact, one has to derive certain estimations for the value of $P_T^a(N, x)$ which are *uniform* in small T.

3. MAIN RESULT

For any $x \in \mathbb{R}^n$, let

$$V_{N,T}^a(x) = \inf\{J_T(N, x, \mathbf{u}) : u_k \in \mathcal{U}_{\Delta_1}\}$$

where $\mathbf{u} = \{u_0, u_1, ..., u_{N-1}\}$, if the right hand side is finite and let $V_{N,T}^a(x) = \infty$ otherwise.

From assumptions A1-A4, it follows immediately that for any $T \in (0, T_0^*]$, any $N > 0$ and any $x \in \mathcal{B}_{\Delta_1}$, ($\Gamma \subset \mathcal{B}_{\Delta_1}$), $P_T^a(N, x)$ has a solution $\mathbf{u}^*(x)$, $V_{N,T}^a(.)$ is positive definite and continuous. With argumentations standard in receding horizon literature one can prove the following lemma.

Lemma 1 Suppose that Assumptions A1-A3 and A6 hold true. Then for any $T \in (0, T_0^*]$ and $N \geq 1$, the following statements are valid:

1.) For any $x \in \mathcal{G}_\eta$, $\phi_N^a(x, \mathbf{u}^*(x)) \in \mathcal{G}_\eta$ and $V_{N,T}^a(x) \leq g(x)$.

2.) If $\phi_N^a(x, \mathbf{u}^*(x)) \in \mathcal{G}_\eta$ for some $x \in \Gamma$, then $V_{\overline{N},T}^a(x) \leq V_{N,T}^a(x)$ for all $\overline{N} \geq N$, and

$$V_{N,T}^a(F_T^a(x, v_T^a(x))) - V_{N,T}^a(x) \leq -T l_T(x, v_T^a(x))$$

3.) If for $x \in \Gamma$ and for some t, $0 \leq t < N$, $\phi_t^a(x, \mathbf{u}^*(x)) \in \mathcal{G}_\eta$, then $\phi_N^a(x, \mathbf{u}^*(x)) \in \mathcal{G}_\eta$.□

Lemma 2 If Assumptions A1-A6 hold true, then there exist a T_1^*, $0 < T_1^* \leq T_0^*$ and a constant $V_{\max}^a$ such that $V_{N,T}^a(x) \leq V_{\max}^a$ for all $x \in \Gamma$ and $T \in (0, T_1^*]$

Proof. Because of lack of space, the proof is omitted.

Lemma 3 Suppose that Assumptions A1-A6 hold true. If $T \in (0, T_1^*]$ and $N \in \mathbb{N}$ are chosen so that

$$TN > \frac{V_{\max}^a - \eta}{c} =: \tau_2^*$$

where $c > 0$ is such a positive constant that $\varphi_1(\|x\|) \geq c$ for all $x \notin \mathcal{G}_\eta$, then

$$\phi_N^a(x, \mathbf{u}^*(x)) \in \mathcal{G}_\eta \qquad \text{for all } x \in \Gamma. \quad (11)$$

Proof. Because of lack of space, the proof is omitted.

In what follows, we shall assume that

$$\tau_2^* \leq NT \leq \tau_2^* + 1 \tag{12}$$

Assumption A7 There exist two positive constants c_1, c_2 such that

$\|f(x,u)\| \leq c_1 + c_2 l_T(x,u)$, for all $x \in \mathbb{R}^n$ and $u \in U$.

Let

$$\Gamma_{\max} = \left\{x \in \mathbb{R}^n : V^a_{N,T}(x) \leq V^a_{\max}\right\}. \tag{13}$$

Clearly, $\Gamma \subset \Gamma_{\max}$.

Lemma 4 Suppose that Assumptions A1-A7 hold true. For any $r > 0$ let

$$K_r = 2\left(c_1/\varphi_1(r) + c_2\right) + 1$$

and let $R_r : \mathbb{R}_{\geq 0} \to \mathbb{R}_{\geq 0}$ be defined as

$$R_r(s) = \min\left\{\gamma_1\left(s/2\right), s/(4K_r)\right\}. \tag{14}$$

Then there exists a $T_2^* > 0$ such that

$$\|F^a_T(x,u) - x\| \leq K_r T l_T(x,u)$$

for all $x \in \mathcal{B}_{\Delta_r}$, $u \in \mathcal{U}_{\Delta_1}$, where $\Delta_r = R_r^{-1}\left(V^a_{\max}\right)$.

Proof. Because of lack of space, the proof is omitted.

Lemma 5. Suppose that Assumptions A1-A7 hold true. Then there exist a $T_3^* > 0$ and a class-$\mathcal{K}_\infty$ function ψ_2 such that

$$V^a_{N,T}(x) \leq \psi_2(\|x\|) \tag{15}$$

for all $x \in \Gamma_{\max}$ and all $T \in (0, T_3^*]$. Moreover, for any $r_0 > 0$ there exist a class-$\mathcal{K}_\infty$ function $\psi_1^{r_0}$ such that

$$V^a_{N,T}(x) \geq \psi_1^{r_0}(\|x\|) \tag{16}$$

for all $x \in \Gamma_{\max} \backslash \mathcal{B}_{r_0}$ and all $T \in (0, T_3^*]$.

Proof. Let $(r_0, \overline{T} = T^*$ and $L^a_{r_0}$ be generated by Assumption A2) and let NT satisfy (12). Let us choose a positive r so that

$$0 < r \leq \min\left\{r_0, r_0/\left(4L^a_{r_0}(\tau_2^* + 1)\right)\right\}$$

and let $T_3^* = \min\left\{\overline{T}, T_1^*, T_2^*\right\}$ where T_2^* is generated by Lemma 4. Let $x_0 \in \Gamma_{\max}$ be arbitrary with $\|x_0\| \geq r_0$ and let the corresponding optimal trajectory be $\xi_k^* = \phi_k^a(x_0, \mathbf{u}^*(x_0))$. Then there are two possibilities:

a.) If $\|\xi_N^* - x_0\| \leq \frac{1}{2}\|x_0\|$, then $\|\xi_N^*\| \geq \frac{1}{2}\|x_0\|$, therefore

$$V^a_{N,T}(x_0) \geq g(\xi_N^*) \geq \gamma_1(\frac{1}{2}\|x_0\|)$$

b.) If $\|\xi_N^* - x_0\| > \frac{1}{2}\|x_0\|$, then we shall introduce the set of integers

$$\begin{aligned} \iota_1 &:= \{k : 0 \leq k \leq N-1, \\ &\qquad \|\xi_k^*\| + \|u_k^*(x_0)\| > r\} \\ \iota_2 &= \{0, 1, ..., N-1\} \backslash \iota_1. \end{aligned}$$

With this definition we have that

$$\begin{aligned} \frac{1}{2}\|x_0\| &\leq \|\xi_N^* - x_0\| \\ &\leq \sum_{k=0}^{N-1} \|F^a_T(\xi_k^*, u_k^*(x_0)) - \xi_k^*\| \\ &\leq \sum_{k\in\iota_1} TK_r l_T(\xi_k^*, u_k^*(x_0)) \\ &\quad + \sum_{k\in\iota_2} TL^a_{r_0}(\|\xi_k^*\| + \|u_k^*(x_0)\|) \\ &\leq K_r V^a_{N,T}(x_0) + rNTL^a_{r_0} \\ &\leq K_r V^a_{N,T}(x_0) + r(\tau_2^* + 1)L^a_{r_0} \end{aligned}$$

By the choice of r , $r(\tau_2^* + 1)L^a_{r_0} \leq \frac{1}{4}\|x_0\|$, thus $V^a_{N,T}(x_0) \geq \frac{1}{4K_r}\|x_0\|$.

Let $\psi_1^{r_0} : [0,\infty) \to \mathbb{R}_{\geq 0}$ be defined as $\psi_1^{r_0}(s) = R_r(s)$, then $\psi_1^{r_0} \in \mathcal{K}_\infty$ and (16) is valid.

On the other hand, we define

$$\mu(s) = \frac{s^2}{2} + \max_{\|x\|\leq s} g(x)$$

$$\nu(s) = \mu(\frac{\rho_1}{2}) + \frac{2}{\rho_1}(s - \frac{\rho_1}{2})V^a_{\max}$$

and $\psi_2 : [0,\infty) \to \mathbb{R}_{\geq 0}$ by

$$\psi_2(s) = \begin{cases} \mu(s), & \text{if } 0 \leq s \leq \rho_1/2, \\ \max\{\mu(s), \nu(s)\}, & \text{if } s > \rho_1/2, \end{cases}$$

then $\psi_2 \in \mathcal{K}_\infty$ and (15) holds true.□

Corollary 1 Under the assumptions of Lemma 5 $\Gamma_{\max} \subset \mathcal{B}_\Delta$, where $\Delta = R_{\rho_1}^{-1}(V^a_{\max})$ and $\rho_1 > 0$ is such that $B_{\rho_1} \subset \mathcal{G}_\eta$ and R_r given by (14)

Lemma 6 Suppose that Assumptions A1-A6 hold true. Then $V^a_{N,T}(.)$ is locally Lipschitz continuous in $\Gamma_{\max}$, uniformly in small T, i.e. there exist $L_v > 0$ and $\delta_v > 0$ such that for all $T \in (0, T_3^*]$ and $N \in \mathbb{N}$ with $TN \leq (\tau_2^* + 1)$, inequality

$$\left|V^a_{N,T}(x) - V^a_{N,T}(y)\right| \leq L_v\|x - y\| \tag{17}$$

holds true for all $x, y \in \Gamma_{\max}$ with $\|x - y\| \leq \delta_v$.

Proof. Because of lack of space, the proof is omitted.

We summarize the basic properties of $V^a_{N,T}$ in the following theorem.

Theorem 2 Suppose that Assumptions A1-A7 are valid. Then there exist such positive numbers

τ^* and T^* that for any $T \in (0, T^*]$ and $N \in \mathbb{N}$ with $\tau^* \leq NT \leq \tau^* + 1$

1.) there exists a function $\psi_2 \in \mathcal{K}_\infty$ such that $V^a_{N,T}(x) \leq \psi_2(\|x\|)$ for any $x \in \Gamma_{\max}$;

2.) for any $r_0 > 0$ there exists a function $\psi_1^{r_0} \in \mathcal{K}_\infty$ such that

$$\psi_1^{r_0}(\|x\|) \leq V^a_{N,T}(x), \tag{18}$$

for any $x \in \Gamma_{\max} \backslash \mathcal{B}_{r_0}$. Moreover $V^a_{N,T}(0) = 0$ and $V^a_{N,T}(x) > 0$ for any $x \in \mathcal{B}_{r_0} \backslash \{0\}$;

3.) for any $x \in \Gamma_{\max}$,

$$V^a_{N,T}(F^a_T(x, v^a_T(x))) - V^a_{N,T}(x) \leq -T\varphi_1(\|x\|);$$

4.) $V^a_{N,T}(.)$ is locally Lipschitz continuous in $\Gamma_{\max}$ uniformly in small T.

Proof. The assertions of the theorem follow from Lemmas 1-6 by taking $T^* = T_3^*$ and $\tau^* = \tau_2^*$.□

Theorem 3 Suppose that Assumptions A1-A7 hold true. Then there exist such positive numbers τ^* and T^* that for any $T \in (0, T^*]$ and $N \in \mathbb{N}$ with $\tau^* \leq NT \leq \tau^* + 1$, the exact discrete-time model with the receding horizon controller

$$x_{t+1} = F^e_T(x_t, v^a_T(x_t)), \quad x_0 \in \Gamma \tag{19}$$

is practically asymptotically stable about the origin.

Proof. The proof can follow the same line as that of Theorem 2 in (Nešić, *et al.*, 1999) with slight modifications. In fact, one has only to take care of the domain of validity of (18) and the local character of the Lipschitz continuity of $V^a_{N,T}$. Because of lack of space, the details are omitted.

REFERENCES

Allgöwer, F., Badgwell, T.A., Qin, J.S., Rawlings, J.B. and Wright, S.J. (1999). Nonlinear predictive control and moving horizon estimation-an introductory overview, in , *Advances in Control* (Frank, P.M. (Ed.)), 391-449, Springer, Berlin.

Birkhölzer, T. and Kreisselmeier, G. (1994). Numerical nonlinear regulator design, *IEEE Trans. Automatic Control*, 9, 33-46.

Chen, H., and Allgöwer, F. (1998). A quasi-infinite horizon nonlinear model predictive control scheme with guaranteed stability, *Automatica*, 34, 1205-1217.

Chen, W.H., Ballance, D. J. and O'Reilly, J. (2000). Model predictive control of nonlinear systems: Computational burden and stability, *IEE Proc.-Control Theory Appl.*, 147, 387-394.

De Nicolao, G., Magni, L. and Scattolini, R. (1998a). Stabilizing receding horizon control of nonlinear time-varying systems, *IEEE Trans. Automatic Control*, 43, 1030-1036.

De Nicolao, G., Magni, L. and Scattolini, R. (1998b). Stability and robustness of nonlinear receding horizon control , in *Nonlinear model predictive control. Papers from the workshop held in Ascona, June 2–6, 1998.* Ed. F. Allgöwer and A. Zheng. Progress in Systems and Control Theory, 26. Birkhäuser Verlag, Basel, 2000.

Fontes, F.A.C.C. (2001). A framework to design stabilizing nonlinear model predictive controllers, *System and Control Letters*, 42, 127-143.

Grüne, L. and Nešić, D. (2003). Optimization based stabilization of sampled-data nonlinear systems via their approximate discrete-time models, *SIAM J. of Control Optim.* 42, 98-122.

Gyurkovics, E. (1996). Receding horizon control of nonlinear uncertain systems described by differential inclusions, *J. Math. Systems, Estimation Control*, 6, 1-16.

Gyurkovics, E. (1998). Receding horizon control via Bolza-type optimization, *System and Control Letters*, 35, 195-200.

Hu, B. and Linnemann, A. (2002). Toward infinite-horizon optimality in nonlinear model predictive control, *IEEE Trans. Automatic Control*, 47, 679-682.

Ito, K. and Kunisch, K. (2002). Asymptotic properties of receding horizon optimal control problems, *SIAM J. of Control optim.*, 40, 1585-1610.

Jadababaie, A. and Hauser, J. (2001). Unconstrained receding horizon control of nonlinear systems, *IEEE Trans. Automatic Control*, 46, 776-783.

Mayne, D.Q., Rawling, J. B., Rao, C.V., and Scokaert, P.O.M. (2000). Constrained model predictive control: Stability and optimality, *Automatica*, 36, 789-814.

Mayne, D.Q., and Michalska, H. (1990). Receding horizon control of nonlinear systems, *IEEE Trans. Automatic Control*, 35, 814-824.

Nešić, D., Teel, A.R. and Kokotović. (1999). Sufficient conditions for stabilization of sampled-data nonlinear systems via discrete-time approximation, *System and Control Letters*, 38, 259-270.

Nešić, D. and Teel, A.R. A framework for stabilization of nonlinear sampled-data systems based on their approximate discrete-time models, *IEEE Trans. Automatic Control*, Submitted for publication.

Parisini, T., M. Sanguineti R. Zoppoli (1998). Nonlinear stabilization by receding-horizon neural regulators, *Int. J. Control*, 70, 341-362.

A MULTI-MODEL STRUCTURE FOR MODEL PREDICTIVE CONTROL

F. Di Palma and L. Magni [1]

Dipartimento di Informatica e Sistemistica, Università di Pavia, via Ferrata 1, 27100 Pavia, Italy
e-mail: {federico.dipalma, lalo.magni}@unipv.it
WEB: http://sisdin.unipv.it/lab/

Abstract: Model Predictive Control (MPC) is a wide popular control technique that can be applied starting from several model structures. In this paper black box models are considered. In particular it is analysed the sets of regressors that it is better to use in order to obtain the best model for multi step prediction. It is observed that for each prediction a different set of real data output and predicted output are available. Based on this observation a multi-model structure is proposed in order to improve the predictions needed in the computation of the MPC control law. A comparison with a classical one-model structure is discussed. A simulation experiment is presented. *Copyright © 2003 IFAC*

Keywords: Model Predictive Control, Black-box identification, Multiple models

1. INTRODUCTION

Model Predictive Control (MPC) has gained wide popularity in industrial process control due to the possibility of reformulating the control problem as an optimization problem in which many physical constraints and nonlinearities can be allowed for (Clarke, 1994), (Camacho and Bordons, 1995), (Kouvaritakis and Cannon, 2001). Another reason for the success is its ability to obtain good performances starting from rather intuitive design principles and simple models, such as truncated impulse responses or step responses, see e.g. (Richalet *et al.*, 1978), (Cutler and Ramaker, 1980) algorithms. In order to reduce the modeling and identification phase black-box model are often used in MPC packages. Black-box identification techniques, see e.g. (Ljung, 1987), (Sjoberg *et al.*, 1995), (Nelles, 2001), can in fact used to quickly derive models from experimental data by means of an estimation parameters procedure. The general problem is to find a relationship between past observation and future outputs. The choice of this map can be decomposed into two partial problems: how to choose the regressor vector from past inputs and outputs and how to choose the mapping from the regressor space to the output space. The regressors are generally given by past observations of the input signal, past observations of the output, simulated outputs from past input only, prediction errors, simulation errors. Each set of regressors defines a particular class of models that can be suitable for different issues. One of the most common application of a model is the forecast of process behavior. Two cases have to be distinguished: simulation and prediction. If the response of the model to an input sequence has to be calculated while the process outputs are unknown, this is called simulation. If, however, the process outputs are known up to the current time instant and it is asked for the model output at the next step, this is called one-step-ahead prediction. It is well known that a good model for prediction can be not suitable for simulation and vice versa. The aim of this paper is to characterize the most suitable class of models for model predictive control. In particular because, in order to compute the cost function to be minimized, the output must be predicted l step in the

[1] The authors acknowledge the partial financial support by $MURST$ Project "New techniques for the identification and adaptive control of industrial systems"

future with $l = 1, ..., N_p$ where N_p is the prediction horizon, the use of different models to make each prediction is proposed. Each model is obtained with a proper set of regressors that contain the different composition of the past observations of the output and of the simulated outputs. The organization of the paper is as follows. In Section 2 we introduce the main model structure used in system identification. In Section 3 the MPC control strategy is proposed and the use of an internal model is discussed. Section 4 describes the new multi-model structure. The identification and prediction algorithms for this structure are discussed. In Section 5 a simulation example is given.

2. NONLINEAR BLACK-BOX STRUCTURES

In this section a general black-box identification problem is introduced for nonlinear systems. Given two vectors

$$u^{t,n_u} = [u(t) \quad u(t-1) \quad \dots \quad u(t+1-n_u)] \quad (1)$$

$$y^{t,n_y} = [y(t) \quad y(t-1) \quad \dots \quad y(t+1-n_y)] \quad (2)$$

where t is the discrete time index, $u \in R^m$ is the input, $y \in R^m$ is the output while n_u and n_y are two real positive constants, we are looking for a relationship between present and past observations $[u^{t,n_u}, y^{t,n_y}]$ and future output $y(t+1)$

$$y(t+1) = g(u^{t,n_u}, y^{t,n_y}) + v(t+1)$$

The term $v(t+1)$ accounts for the fact that the next output $y(t+1)$ will not an exact function of past data. However, a goal must be that $v(t+1)$ is small, so that we may think of $g(u^{t,n_u}, y^{t,n_y})$ as a good prediction of $y(t+1)$ given past data. In order to find the function g let us parametrize a family of function within we have to search for it:

$$g(u^{t,n_u}, y^{t,n_y}, \theta) \quad (3)$$

The target is to find a suitable parametrization in order to obtain a good model. Because a parametrization of the function g with a finite- dimension vector θ is an approximation, in order to decide the structure it is crucial to know what the model will be used for. Once we have decided the structure and we have collected N data the parameter θ can be obtained by means of the fit between the model and the data record:

$$\sum_{t=\max(n_u,n_y)}^{N} \left\| y(t) - g(u^{t-1,n_u}, y^{t-1,n_y}, \theta) \right\|$$

The model structure family (3) is quite general, and it turns out to be useful to write g as a concatenation of two mappings: one that takes the increasing number of past observations and maps them into a finite-dimensional vector $\varphi(t)$ of fixed dimension and one that takes vector to the space of outputs:

$$g(u^{t,n_u}, y^{t,n_y}, \theta) = g(\varphi(t), \theta) \quad (4)$$

According to a rather classical terminology, we shall call the vector $\varphi(t)$ the regression vector, and its components will be referred to as regressors. The choice of the nonlinear mapping (3) has thus been decomposed into two partial problems for dynamic systems: how to choose the regression vector $\varphi(t)$ from past inputs and outputs and how to choose the nonlinear mapping $g(\varphi)$ from the regressor space to the output space. We thus work with prediction models of the kind

$$\hat{y}(t+1, \theta) = g(\varphi(t), \theta) \quad (5)$$

The regressors, in the general case, are given by:

(i) $u(t-k)$, past inputs;
(ii) $y(t-k)$, past outputs;
(iii) $\hat{y}_u(t-k)$, simulated outputs from past u only;
(iv) $\varepsilon(t-k) := y(t-k) - \hat{y}(t-k)$, prediction errors;
(v) $\varepsilon_u(t-k) := y(t-k) - \hat{y}_u(t-k)$, simulation errors;

The simulated output $\hat{y}_u(t-k, \theta)$ is the output from the model (4) if all measured outputs $y(t-k)$ in the regressors are replaced by the last computed $\hat{y}_u(t-k, \theta)$. The regressors $(iii) - (v)$ depend on the black-box model (3), so we should write $\varphi(t, \theta)$ instead of $\varphi(t)$ in (4).

Based on different combinations of the regressors we could thus distinguish between the following model classes:

- $NFIR$ (Nonlinear Finite Impulse Response) models, which use only $u(t-k)$ as regressors;
- $NARX$ (Nonlinear AutoRegressive with eXogenous input) models, which use $u(t-k)$ and $y(t-k)$ as regressors;
- $NARXAR$ (Nonlinear AutoRegressive with eXogenous AutoRegressive input) models, which use $u(t-k)$, $y(t-k)$ and $\hat{y}_u(t-k, \theta)$ as regressors;
- NOE (Nonlinear Output Error) models, which use $u(t-k)$ and $\hat{y}_u(t-k, \theta)$ as regressors;
- $NARMAX$ (Nonlinear AutoRegressive Moving Average with eXogenous input) models, which use $u(t-k)$, $y(t-k)$ and $\varepsilon(t-k, \theta)$ as regressors;
- NBJ (Nonlinear Box-Jenkins) models, which use $u(t-k)$, $\hat{y}(t-k, \theta)$, $\varepsilon(t-k, \theta)$ and $\varepsilon_u(t-k, \theta)$ as regressors; in this case the simulated output $\hat{y}_u$ is obtained as the output from (3), by using the same structure, replacing ε and ε_u by zeros in the regression vectors $\varphi(t, \theta)$.

The aim of this paper is to find the better model structure in order to make prediction for nonlinear model predictive control. In the next section the control problem and its main peculiarity are discussed.

3. NONLINEAR MODEL PREDICTIVE CONTROL

The system under control is assumed to be described by an unknown state equation of the form

$$x(t+1) = f(x(t), u(t), e(t)), \quad x(t_0) = \bar{x} \quad (6)$$
$$y(k) = h(x(t)) \quad (7)$$

where t_0 is the initial state, $x \in R^n$ represents the system state, $u \in R^m$ is the input vector, $y \in R^m$ is the output and $e(t) \in R^p$ is the noise.

The problem here considered is to design a control algorithm such that the output must track the reference signal $y_{ref}(\cdot)$.

If the model of the system (6)-(7) is known, $e(t) = 0$ and any equilibrium associated with a constant input is asymptotically stable, the MPC control law (Mayne *et al.*, 2000) can be solved by the following *Finite Horizon Optimal Control Problem* ($FHOCP$): given the state system $\bar{x}$, the last control value $u(t-1)$, the positive integers N_c (*control horizon*) and N_p (*prediction horizon*), $N_c \leq N_p$, the positive definite matrices Q, R, minimize, with respect to u^{t+N_c-1,N_c}, the performance index

$$J(\bar{x}, u(t-1), u^{t+N_c-1,N_c}, N_c, N_p) \quad (8)$$
$$= \sum_{k=t}^{t+N_p-1} \{(y(k) - y_{ref}(k))'Q(y(k) - y_{ref}(k))$$
$$+ (u(k) - u(k-1))' R (u(k) - u(k-1))\}$$

subject to

(i) the model system dynamics (6)-(7);
(ii) the control signal

$$u(t+k) = \begin{cases} u^{t+N_c-1,N_c} & k \in [0, N_c - 1] \\ u(t+N_c-1) & k \in [N_c, N_p - 1] \end{cases} \quad (9)$$

■

According to the *Receding Horizon* approach, the state-feedback MPC control law is derived by solving the $FHOCP$ at every time instant t, and applying the control signal $u = u^{t,t}_{OPT}$, where u^{t+N_c-1,N_c}_{OPT} is the solution of the $FHOCP$. In so doing, one implicitly defines the state-feedback control law

$$u(t) = \kappa^{RH}(x(t)) \quad (10)$$

In order to compute the cost function (8) the future output values y^{t+N_p-1,N_p-1} must be predicted according to an internal model of the system. In this paper we assume that the state-space model (6)-(7) is unknown and that a black-box input-output models in the form (5) described in Section 2 has been identified. Then the cost function (8) must be re-written as

$$J(\varphi(t), u(t-1), u^{t+N_c-1,N_c}, N_c, N_p) \quad (11)$$
$$= \sum_{k=t}^{t+N_p-1} \{(\hat{y}(k \mid t) - y_{ref}(k))'Q(\hat{y}(k \mid t) - y_{ref}(k))$$
$$+ (u(k) - u(k-1))'R(u(k) - u(k-1))\}$$

where $\hat{y}(k \mid t)$, indicates the prediction output values at time k based on the input values $u(i)$, $i \leq k$, on the measured output values up to time t (on other word $t+i, i \leq 0$) and on a model in the form (5). Note that in this case the role of the initial state $\bar{x}$ is up to the regression vector at time t, $\varphi(t)$.

Usually, (Nelles, 2001) a single model is used to evaluate the whole future output's sequence $\hat{y}^{t+N_p-1,N_p-1}$ associated with the input sequence (9). Employing only one model, the requested sequence is obtained iterating the internal model N_p times with the available data. To generate the first prediction $\hat{y}(t+1 \mid t)$ according with (5) the internal model must be feed with the past regressors values $\varphi(t \mid t)$. The notation $\mid t$, as previously used, indicate that the vector $\varphi(t)$ may be built bounding the old system output only up to time t.

Once $\hat{y}(t+1 \mid t)$ has been computed to obtain the next prediction $\hat{y}(t+2 \mid t)$ a new vector $\varphi(t+1 \mid t)$ must be provided. The vector $\varphi(t+1 \mid t)$ may be built from $\varphi(t \mid t)$ using a *Shift Register* (SR) for each kind of regressors. For each SR the update is made losing the oldest value and putting the new element. Obviously the SR number is due to the kind of the chosen model. The process is repeated N_p time until the prediction $\hat{y}(t+N_p \mid t)$ is reached.

It is important to notice that the whole procedure must be executed at any time t and for every input sequences (9) used to solve the $FHOCP$. In fact at system time $t+1$ each SR must be re-initialized with the proper component of regression vector $\varphi(t+1 \mid t+1)$.

The prediction algorithm can be summarized in the following algorithm:

(1) Initialize the SR to the proper component of regression vector $\varphi(t \mid t)$, set $i = 1$.
(2) Compute $\hat{y}(t+i \mid t)$ according with (5).
(3) Set $i = i + 1$ and update the SR:
    (a) for each SR, remove the oldest regressor values;
    (b) for each SR, insert the new regressor values.
(4) If $i \leq N_p$ then go to point 2.

At point 3.b the new regressor values are asked in order to compute the next prediction. The problem is that the past real output values are available only up to current time t. Then, if a model which has the past outputs among its regressors is used, the new regressors are not available in order to implement point 3.b. This happens, for example, when a general $NARX$, $NARMAX$, $NARXAR$ and NBJ models are considered (see Section 2). In practice the only thing that one can do in order to avoid this problem without modifying the model structure is to use the prediction at the previous steps instead of the unknown output real values. It is clear that following this procedure the model is identified considering a particular set of regressors and used for prediction with a different set of regressors (i.e. predictions are used instead of real outputs).

It is well know that a model identified in order to minimize the one-step-ahead prediction error is not in general a good model for simulation. For linear systems, for example, it is possible that the model of a stable system that minimize the one step-ahead prediction error is even unstable (Nelles, 2001).

Another possibility is the use of NOE or $NFIR$ models that can be correctly used because they do not count among its regressors past outputs as remarked in Section 2. Otherwise, using a NOE or FIR models the information of the real output value up to time t is not used.

In the next section we propose a Multi-Model (MM) structure that fully used the available information with a correct chosen of the regressors.

4. MULTI-MODEL STRUCTURE

To avoid the difficulty explained in the precedent section, we propose the use of different models for each requested prediction. The basic observation that motivates the proposed structure is that the set of the regressors that is possible to use in order to make the prediction is different for each k. In fact increasing k the number of past known outputs decreases, while one can use only the output simulated with the identified model. This means that for $k = t + 1$, the regressors we can use are $u(i), i < k$ and $y(i), i < k$ so that we are considering a $NARX$ model. On the contrary for $k > t + 1$ the regressors we can use are $u(i), i < k$, $y(i), i \leq t$ and $\hat{y}_u(i \mid t), t < i < k$ so that we are considering a $NARXAR$ model with a particular set of regressors. In view of this consideration, in the MM scheme N_p models are identified and each model is characterized by a different regressor vector φ_j composed by a different composition of real outputs and simulated outputs.

If only a finite number of regressors is considered, when k is strictly greater than $t + n_y$, where n_y is the maximum delay in the output regressors (2), only input and simulated output can be considered in the regressors so that a NOE model must be considered. In this case, if $n_y < N_P$ only $n_y + 1$ models can be used to compute the performance index (11).

Remarkably each model must be initialized with its own regression vector $\varphi_j (t \mid t - j + 1)$, and then run up for j steps in order to obtain the predicted value at time $t + j$. This means that the first model in one simulation step reaches the requested predicted output $\hat{y}(t + 1 \mid t)$ while $\hat{y}(t + 2 \mid t)$ is obtained in two simulation steps of the second model, and so on. Moreover you can note that, in order to obtain a correct use of the model, the vector $\varphi_j (t \mid t - j + 1)$ must be formed by the old simulation even if some real output is known.

Each model is then defined by the regression vector $\varphi_j (t \mid t - j + 1)$, the function $g_j(\cdot, \cdot)$ (5) and the parameter vector $\theta(j)$. The predicted output will be called $\hat{y}_j$ and its Shift Register SR_j.

4.1 *Multi-model prediction algorithms*

Once understood the MM scheme aim, the prediction algorithm is naturally obtained extending the one reported in Section 3.1. In fact in the MM scheme $n_y + 1$ models are present, and the prediction algorithm is the parallel of $n_y + 1$ prediction algorithm with progressive arrest time.

(1) Set $i = 1$ and for each model j with $1 \leq j \leq n_y + 1$, initialize the SR_j to the proper component of regression vector $\varphi_j (t \mid t - j + 1)$.
(2) For $j \geq i$ calculate $\hat{y}_j(t+i \mid t-j+i)$ according with (5).
(3) Set $\hat{y}(t + i \mid t) = \hat{y}_i(t + i \mid t)$.
(4) Set $i = i + 1$ and update the SR_j, with $j \geq i$:
    (a) for each SR_j, remove the oldest regressor value;
    (b) for each SR_j, insert the new regressor value.
(5) If $i \leq n_y$ then go to Point 2.
(6) Calculate $\hat{y}_{n_y+1}(t + i)$ according with (5).
(7) Set $\hat{y}(t + i \mid t) = \hat{y}_{n_y+1}(t + i)$.
(8) Set $i = i + 1$ and update the SR_{n_y+1}:
    (a) for each SR_{n_y+1}, remove the oldest regressor value;
    (b) for each SR_{n_y+1}, insert the new regressor value.
(9) if $i \leq N_p$ then go to Point 6.

4.2 *Multi-model identification algorithm*

In order to reduce the computational burden while maintain the main properties of the multi model structure (i.e. the use of a different mix of output data and simulated data among the regressors) we assume that all the functions $g_j(\cdot, \cdot)$ are equal to $g(\cdot, \cdot)$. On the other hand, each model will be characterized by a possible different set of parameter $\hat{\theta}(j)$ for each j. In this way, the problem to find a proper structure is solved only one time.

Then, given the non linear mapping $g(\cdot,\cdot)$ and the integer constants n_u end n_y, the MM identification algorithm is based on the following steps.

(1) Given the regression vector $\varphi_1(t \mid t)$, function of $y(t+i)$, $-n_y \le i < 0$, $u(t+i)$, $-n_u \le i < 0$, find the optimal value of $\hat{\theta}(1)$ for the $NARX$ model with a prediction criterion.
(2) For each $1 < j \le n_y$, find the optimal value of $\hat{\theta}(j)$ using the regression vector $\varphi_j(t \mid t-j+1)$ given by $\varphi_{j-1}(t \mid t-j+2)$ with $y(t-j)$ substituted by $\hat{y}_{u_j}(t-j \mid t-j+1)$ for the $NARXAR$ model with a prediction criterion.
(3) Find the optimal value of $\hat{\theta}(n_y+1)$ using the regression vector $\varphi_{n_y+1}(t)$ given by $\varphi_1(t \mid t)$ with $y(t-i)$ substitute by $\hat{y}_{u_{n_y+1}}(t-i)$ $\forall i$ for the NOE model with a prediction criterion.

Note that in step 3 $\varphi_{n_y+1}(t)$ does not contain past output but only simulated output from past input.

The real importance of these two algorithms is that the different models are identified and used with the same sets of regressors differently from the standard approach described in Section 3. It is obvious that in this way the obtained cost function (11) will be more carefully computed because each prediction is based on the "best" model within the considered class of function determined by $g(\cdot,\cdot)$.

5. SIMULATION EXAMPLE

In this section a simulation example based on the single-input, single-output linear time invariant system

$$A(d)\,y(t) = C(d)\,e(t) + \frac{B(d)}{F(d)}u(t) \qquad (12)$$

where d is the backward shift operator (Ljung, 1987), $A(d) = 0.3d^2 - 1.1d + 1$, $B(z) = -d^2 + d$, $C(d) = -1.5d + 1$, $F(d) = 0.08d^2 - 0.4d + 1$ is given.

It is shown that the proposed MM structure guarantees significant advantages even for a very simple model. In particular the MM structure makes the prediction at each step with the best model so that is gained more accuracy than with a single model structure such as $NARX$ or NOE that are usually used in literature.

5.1 *MM structure identification*

First of all the identification data (1), (2) was obtain feeding the system (12) with a *Multi-level Pseudo-Random Signal* (*MPRS*) for the input $u(t)$ (Braun *et al.*, 1999) and a *Random Gaussian Signal* (*RGS*) for the error $e(t)$ (Ljung, 1987). The signals are reported in Figure 1 with solid line.

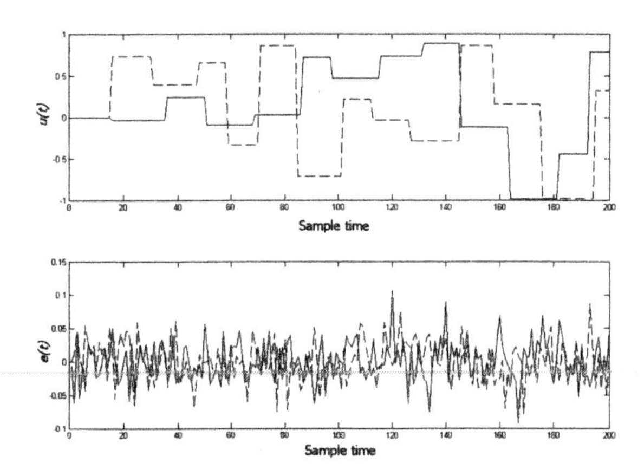

Fig. 1. Identification (solid line) and validation (dashed line) input signals.

Then the mapping $g_j(\cdot,\cdot)$ has been chosen as a linear map and the constants n_u and n_y has been respectively fixed equal to 2 and 4. Then the identification algorithm of Section 4.2 was applied and the next five models are obtained:

$$\begin{aligned}\hat{y}_1(t+1 \mid t) = {} & +0.6766u(t) - 0.6739u(t-1) + \\ & -0.8269y(t) - 0.1674y(t-1) + \\ & +0.296y(t-2) - 0.08282y(t-3)\end{aligned}$$

$$\begin{aligned}\hat{y}_2(t+2 \mid t) = {} & +1.957u(t+1) - 1.951u(t) + \\ & +0.2193\hat{y}_2(t+1) - 1.134y(t) + \\ & +0.62y(t-1) - 0.09467y(t-2)\end{aligned}$$

$$\begin{aligned}\hat{y}_3(t+3 \mid t) = {} & +1.523u(t+2) - 1.522u(t+1) + \\ & -0.04702\hat{y}_3(t+2) - 0.4564\hat{y}_3(t+1) + \\ & -0.5582y(t) + 0.2463y(t-1)\end{aligned}$$

$$\begin{aligned}\hat{y}_4(t+4 \mid t) = {} & +0.7386u(t+3) - 0.7347u(t+2) \\ & -1.122\hat{y}_4(t+3) + 0.1193\hat{y}_4(t+2) + \\ & +0.1717\hat{y}_4(t+1) - 0.03102y(t)\end{aligned}$$

$$\begin{aligned}\hat{y}_5(t+5) = {} & 0.7168u(t+4) - 0.7141u(t+3) + \\ & -1.298\hat{y}_5(t+4) + 0.2712\hat{y}_5(t+3) + \\ & +0.2308\hat{y}_5(t+2) - 0.068\hat{y}_5(t+1)\end{aligned}$$

5.2 *MM prediction analysis*

Once identified the MM structure, we test them on the validation data set reported in Figure 1 with dashed line. As quality index we consider the *Sum of Prediction Error* (*SPE*) that, given the prediction step i and the model j, is defined as

$$SPE(i,j) = \sum_{t=n_m}^{N} |\hat{y}_j(t+i \mid t) - y(t+i)|$$

where $n_m = \max(n_u, n_y)$. In order to appraise the different model's quality the values of $SPE\,(i,j)$ are reported in Figure 2.

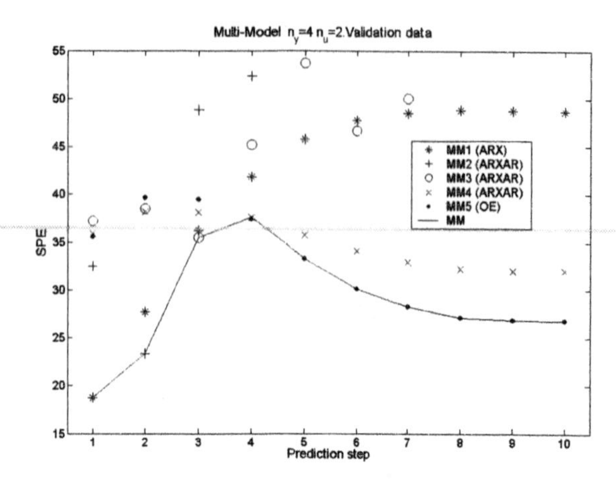

Fig. 2. $SPE\,(i,j)$ for singol models and MM scheme.

To globally evaluate the model effectiveness in computing the MPC cost function (11), where we fixed $N_p = 10$, the SPE must be extended along the prediction horizon. Then for a single model based structure the total cost is given by

$$SPE_j = \sum_{i=1}^{N_p} (SPE\,(i,j))$$

while for the MM structure is given by

$$SPE_{MM} = \sum_{j=1}^{N_p} (SPE\,(j,j))$$

In Table 1 the SPE variation of the different model related to the SPE of the ARX model are reported. It is apparent that the MM structure obtains better performance that single models, in view of the possibility to use different model at each step.

$\frac{SPE_2}{SPE_1}$	$\frac{SPE_3}{SPE_1}$	$\frac{SPE_4}{SPE_1}$	$\frac{SPE_5}{SPE_1}$	$\frac{SPE_{MM}}{SPE_1}$
262.5%	116.3%	84.5%	78.6%	69.7%

Table 1 - Confront single-model/MM structure.

6. CONCLUSION

In this paper a new structure for system identification of models for MPC is presented. This structure, called MM structure, is based on the identification of different model in order to optimize the prediction on each step involved in the computation of the cost function. An example is presented to show the potentiality of this approach.

7. REFERENCES

Braun, M. W., D. E. Rivera, A. Stenman, W. Foslien and C. Hrenya (1999). Multi-level pseudo-random signal design and "model-on-demand" estimation applied to nonlinear identification of a rtp wafer reactor. In: *American Control Conference*. San Diego, California.

Camacho, E. and C. Bordons (1995). *Model Predictive Control in The Process Industry*. Springer.

Clarke, D. W. (1994). *Advances in Model-Based predictive Control*. Oxford University Press.

Cutler, C. R. and B. C. Ramaker (1980). Dynamic matrix control-a computer control algorithm. In: *Automatic Control Conference*.

Kouvaritakis, B. and M. Cannon (2001). *Nonlinear Predictive Control: Theory and Practice*. IEE.

Ljung, L. (1987). *System Identification, Theory for the user*. Prentice-Hall, Englewood Cliffs, NJ.

Mayne, D. Q., J. B. Rawlings, C. V. Rao and P. O. M. Scokaert (2000). Constrained model predictive control: Stability and optimality. *Automatica* **36**, 789–814.

Nelles, O. (2001). *Nonlinear System Identification: From Classical Approaches to Neural Networks and Fuzzy Models*. Springer.

Richalet, J., A. Rault, J. L. Testud and J. Papon (1978). Model predictive heuristic control: applications to industrial processes. *Automatica* **14**, 413–428.

Sjoberg, J., Q. Zhang, L. Ljung, A. Benveniste, B. Delyon, P. Glorennec, H. Hjalmarsson and A. Juditsky (1995). Nonlinear black-box modeling in system identification: a unified overview. *Automatica* **31**, 1691–1724.

MULTIOBJECTIVE EVOLUTIONARY DESIGN OF PREDICTIVE CONTROLLERS

Lavinia Ferariu

"Gh. Asachi" Technical University of Iaşi
Department of Automatic Control and Industrial Informatics
Bd. D. Mangeron 53A, RO-6600 Iaşi, Romania
Fax: +40-232-230751, E-mail: lferaru@ac.tuiasi.ro

Abstract: The paper presents an evolutionary predictive control algorithm. The model of the process (linear or nonlinear) is used in order to forecast the process outputs at future time instants. At each sample time, an evolutionary algorithm calculates the optimal control action, subject to the imposed cost functions and the specified constraints. The multiobjective optimisation is efficiently solved using special rules for fitness values' computation. The proposed method is able to support a flexible formulation of the design specifications. Assuming that the controlled system is slow and robust, the approach offers satisfactory results. The algorithm can efficiently cope with the environmental changes and the known variations of the process model. *Copyright © 2003 IFAC*

Keywords: multiobjective optimisation, genetic algorithms, model based predictive control.

1. INTRODUCTION

Model based predictive control is increasingly gaining acceptance in industry, due the fact that it allows a good trade-off between performance and complexity and it represents an efficient and flexible tool in solving complex control problems. The power of this methodology is related to the fact that it uses an accurate dynamic model of the controlled system in order to predict the consequences of the control actions, before they happen (Clarke, 1994; De Keyser and Donald, 1999). The strategy can be applied for linear or nonlinear plants.

In a realistic problem formulation, the desired control performances have to be specified by means of a set of objective functions and constraints. In that case, the algorithm has to solve complex optimisation problems in order to find an adequate control policy.

Many authors suggested the evolutionary techniques as a promising alternative for solving difficult real-world problems in control system engineering (Dasgupta and Michalewicz, 1997; Miettinen, *et al.*, 1999; Fleming and Purshouse, 2002). Based on a stochastic search and on mechanisms similar to biological evolution, evolutionary algorithms represent efficient optimisation methods with satisfactory results in nonlinear, constraint and multiobjective optimisations (Bäck, *et al.*, 1997). The evolutionary algorithms can perform a robust search and are able to cope with multimodality, discontinuity, time-variance, randomness and noise. Unfortunately, these algorithms require large computational resources, so their applicability to real-time systems is quite limited.

This approach suggests an evolutionary predictive control algorithm that is able to support flexible design specifications. A set of objective functions and constraints can be considered, assigned with different levels of priorities. The multiobjective optimisation problem is addressed by means of non-dominated sorting procedures (Deb, 2001). A set of special rules is proposed for the computation of adequate fitness values.

The paper is organised as follows. Section 2 presents the mechanisms used for the multiobjective optimisation. Details regarding the evolutionary design of the predictive control scenario are given in Section 3. The applicability of the approach is studied in Section 4, with respect to a simulated thermal process. Finally, in section 5, several conclusions are outlined.

2. THE MULTIOBJECTIVE OPTIMISATION

The design of appropriate control actions is formulated as a multiobjective optimisation problem.

One considers a dynamic process (possibly nonlinear), described by the discrete-time model (1), corresponding to the sample period T :

$$y(k) = f(y(k-1),...y(k-n_A), \\ u(k-1),....,u(k-n_B), \\ e(k-1),......e(k-n_E)), \quad (1)$$

where $y \in \Re^n$, $y = [y_j]_{j=1,..,n}$ denotes the process output, $u \in \Re^m$, $u = [u_j]_{j=1,..,m}$ represents the process input, $e \in \Re^{m_e}$ indicates the disturbances that affect the system behaviour and $k > 0$ specifies the current time instant.

The prediction is done by means of model (1), considering the past process inputs, the past process outputs and the future control scenario:

$$\{u(k+i/k), i = \overline{0, N-1}\}. \quad (2)$$

Here, $u(k+i/k) \in \Re^m$ specifies the future postulated values of the input at time instant k. The predicted output values are denoted by

$$y(k+1/k) \in \Re^n, \; i = \overline{1+d, N+d}, \quad (3)$$

where $N+d$ represents the prediction horizon and d indicates the dead time of the controlled process.

The reference trajectory defined over the prediction horizon, $\{r(k+i/k) \in \Re^n\}$, $i = \overline{d+1, N+d}$, describes the desired trajectory from the current output value to the imposed setpoint. The process output must follow, as close as possible, the specified reference trajectory.

At each sample time, the control vector (2) is built with respect to the objective O_1, namely the minimisation of the squared predicted output errors:

$$f_1 = \sum_{i=d+1}^{N+d} \{\lambda_i[y(k+i/k) - r(k+i/k)]^T \cdot \\ \cdot [y(k+i/k) - r(k+i/k)]\} \quad (4) \\ \lambda_i \geq 0$$

Here, $[\cdot]^T$ denotes the transposed matrix.

The optimisation problem also demands the minimisation of the required control energy:

$$f_2 = \sum_{i=0}^{N-1} \alpha_i[u(k+i/k)]^T \cdot [u(k+i/k)] \quad (5) \\ \alpha_i > 0$$

and this objective, denoted with O_2, is considered to be less important than the objective O_1.

The designed control scenario must satisfy some constraints. It is specified the maximum allowed variation between two successive components of the control action:

$$\left|\Delta u_j(k+i/k)\right| \leq LU_j, \\ i = 0,.., N-1, \; j = 1,...,m, \; LU_j > 0, \quad (6)$$

where $\Delta u_j(k+i/k) = u_j(k+i/k) - u_j(k+i-1/k)$ represents the sequence of the postulated future control input variation.

Also, the process inputs have to be set between predefined lower and upper boundaries:

$$u_L \leq u_j(k+i/k) \leq u_H \\ i = 1,..., N, \; j = 1,...,m. \quad (7)$$

The evolutionary procedure can easily handle the constraint (7), based on the considered chromosome encoding and genetic operators, as it is described in Section 3.

In order to find feasible solutions subject to the inequality constraint (6), the proposed algorithm introduce a new objective, O_3, which demands the minimisation of:

$$f_3 = \sum_{i=0}^{N-1} \beta_i \min\{0,[LU - |\Delta u(k+i/k)|]^T\} \cdot \\ \cdot \min\{0,[LU - |\Delta u(k+i/k)|]\}, \quad (8) \\ \beta_i > 0.$$

The global minimum value of f_3 is achieved only for the solutions that satisfy (6). This objective is assigned with the highest priority during the multiobjective optimisation.

Concluding, the design procedure considers a set of three objectives, each one assigned with a different priority. The highest priority objective, namely O_3, forces the algorithm to search for feasible solutions subject to the inequality constraint (6). The intermediary priority objective, O_1, guarantees an adequate accuracy for the automatic system. The lowest priority objective, O_2, demands the minimisation of the control energy.

The approach can be easily extended in order to consider supplementary objectives and inequalities constraints. The ineqalities constraints can be solved by including new objectives, assigned with the highest level of priority, as illustrated for O_3.

If all objectives would have with the same priority, the multiobjective optimisation problem could be solved in the Pareto sense (Deb, 2001; Fonseca and Fleming, 1998). In that case, the solution is represented by a family of points, named the Pareto - optimal set. Each point of this surface is optimal in the sense that no improvement can be achieved in one component without degradation in at least one of the remaining components.

If the objectives are assigned with different priorities, one problem is to assure an appropriate trade-off between them. The present approach adapts the distances existing between the considered levels of priorities using the information encoded by the population. The decision mechanism and the search procedure are combined according to a progressive articulation of preference. For each objective function f_i, $i=1,2,3$ a goal g_i is associated (Deb, 2001; Fonseca and Fleming, 1998; Rodriguez-Vazquez and Fleming, 1997). The goals define the desired area for the objective values. The individuals placed beyond the specified area are not encouraged to produce offspring and to survive. During the evolutionary loop, the goals are adapted according to the mean performances of the current population (Marcu, 1997). Considering f_i^{mean}, $f_i^{\min}$ and $f_i^{\max}$ the mean value, the minimum value and the maximum value achieved by the objective function f_i, $i=1,2,3$ over the current population of individuals, the goals' adaptation rule can be written as:

$$g_i^{new} = \min\{g_i^{old}, \min\{f_i^{mean}, (f_i^{\max} + f_i^{\min})/2\}\}, \quad i = 1,\dots,Q, \tag{9}$$

where g_i^{new}, g_i^{old} represent the new and the old value computed for the goal g_i associated to the objective function f_i.

The approach uses a ranking selection. In order to assure a convenient trade-off between the imposed levels of objectives' priorities, the individuals are classified in four different groups (C_1, C_2, C_3, C_4), based on the values of their objective functions,. The ranks are computed with respect to the rules described in the following.

The individuals included in $C_i, i=1,2,3$ dominate all individuals of groups $C_j, j=i+1,..,4$.

The group C_1 contains the best individuals x of the current population, for which all objectives satisfy the imposed goals and the constraint (6) is not violated ($f_i(x) < g_i$, $i=1,2$ and $f_3(x) = 0 \le g_3$). Inside this group, the individuals are firstly sorted according to f_1. The chromosomes having the same f_1 values are then sorted subject to f_2.

The second group, C_2, includes the individual x for which all objectives satisfy the imposed goals, but the constraint (6) is violated ($f_i(x) < g_i$, $i=1,3$ and $f_3(x) \ne 0$). Because these individuals have satisfactory performances with respect to all imposed objectives, the algorithm reconfigures the levels of priorities inside this group. An individual included in the second group dominates another individual of the second group, if it has better values for the objective functions f_1 and f_3. As consequence, the selection pressure imposed by O_2 is decreased, in order to simplify the search procedure. Also, the algorithm encourages the search process toward the achievement of the feasible solutions which guarantee a good accuracy of the automatic system.

The third group, C_3, includes the individual x for which the highest priority objective satisfies the imposed goal ($f_3(x) < g_3$), but at least one of the other objective function values is bigger than the associated goal ($\exists i \in \{1,2\}$ with $f_i(x) > g_i$). An individual included in the second group dominates another individual of the second group, if it has better values for the objective functions f_1 and f_3.

The forth group, C_4, includes individuals x for which the highest priority objective does not satisfy the imposed goals ($f_1(x) > g_1$). An individual included in the forth group dominates another individual of the forth group, if it has better values for all the considered objective functions.

This strategy adapts the selection pressures inside each group, based on the phenotipic properties of the individuals. It encourages the exploration toward the achievement of feasible solutions, able to guarantee a convenient accuracy of the automatic system. The mechanism can improve the convergence rate, with benefits for the on-line design of the control law.

3. THE EVOLUTIONARY DESIGN OF THE PREDICTIVE CONTROL POLICY

The evolutionary predictive control strategy can be described as follows.

At each sample time, the future optimum control actions are searched by means of an evolutionary process. The evolutionary algorithm acts for MAX_GEN generations on a population of *Nind* possible solutions, named individuals or chromosomes. Each individual included into the population directly encodes a possible control scenario:

$\mathbf{u}^j(k/k)$	..	$\mathbf{u}^j(k+i/k)$	..	$\mathbf{u}^j(k+N-1/k)$

$$i=\overline{0,N-1},\ j=\overline{1,N_{ind}} \tag{10}$$

The algorithm considers the float encoding.

At $k=1$, the initial population is randomly generated. Its chromosomes are uniformly distributed within the search space allowed by the inequality constraint (7).

For each individual, all objective functions $f_i, i=1,2,3$ are evaluated and, then, the corresponding fitness values are specified, as indicated in Section 2. The predicted plant outputs values are computed according to the available model, the process measurements and the encoded future postulated inputs. The best chromosomes are encouraged to be selected into the reproduction pool and to become parents, according to the stochastic universal sampling procedure (Bäck, *et al*., 1997). Their genetic material is combined in order to proceduce new solutions, named offspring. Adequate internal genetic operators are used, namely the arithmetic crossover and the uniform mutation (Michalewicz, 1996). They can maintain an appropriate diversity for the genetic material and guarantee the achievement of feasible solutions subject to constraint (7).

The offspring are evaluated and some of them replace the worst parents, in order to form the population of the next generation. The insertion is solved by means of *Pareto reservation strategy* (Tamaki, *et. al*, 1996). S_{off} offspring and $Nind - S_{off}$ individuals contained by the current population are selected to survive to the next generation. The method encourages the survival of non – dominated chromosomes. If the number of non - dominated individuals exceeds, a parallel selection is considered upon the set of non – dominated individuals, meaning that the selection is applied separately for each objective. If the number of the non - dominated individuals is less than the number of individuals that must survive, the rest of the population is filled in, using a parallel selection, applied to the set of dominated chromosomes.

The evolutionary procedure is implemented according to the island model. That means, the population is separated into several isolated subpopulations. Once at *No_migr* generation, an exchange of information is permitted between them. The mechanism improves the performances of the algorithm, especially in the case of multimodal objective functions.

One of the best chromosomes contained by the population of the last generation is selected and its component $u(k/k)$ represents the postulated control action.

At the next sampling instant, all time-sequences are shifted, a new output measurement is obtained and the whole procedure is repeated, because the algorithm is based on the "receding horizon" strategy. This means that the new control input $u(k+1/k+1)$ is recomputed, considering a new evolutionary process with *MAX_GEN* generations. Usually, the result indicates that $u(k+1/k+1) \neq u(k+1/k)$.

The genetic material is efficiently passed between two successive evolutionary processes. At sample time $k+1$, $k>1$, the initial population is generated as follows: firstly, the genetic material achieved at the previous sample time k is shifted and, then, the mutation operator is applied. The mutation rate P_m is adapted, taking into consideration the variations of the reference trajectory over the prediction horizon. The sequence $\{r(k+i/k) \in \Re^n\}$, $i=\overline{d+1,N+d}$ is compared with sequence $\{r(k+i+1/k+1) \in \Re^n\}$, $i=\overline{d+1,N+d}$. The mutation rate P_m is higher if the reference values are strongly modified from a sample time to another. Based on this strategy, the search procedure can benefit from the experience achieved by the algorithm at the previous time instants.

A schematic description of the proposed evolutionary predictive control algorithm is presented in the following:

1. Initialise the set S_u of previous control inputs and the set S_y, containing available past measurements; initialise the reference trajectory over the prediction horizon.
2. For all objective functions, initialise g^{old} with maximum allowed float value.
3. Create the initial the population of chromosomes subject to (7).

4. Evaluate the individuals, adapt the goals and compute the fitness values.
5. At each sample time
   5.1. For *MAX_GEN* generations
      5.1.1. For each subpopulation
         i. Select parents for recombination.
         ii. Apply crossover and mutation operators.
         iii. Evaluate the offspring.
         iv. Create the population of the next generation.
         v. Update the goals and the fitness values.
         vi. Once at *No_migr* generations, exchange individuals with the other subpopulations (migration stage). Then, update the goals and the fitness values.
   5.2. Select one of the best chromosomes.
   5.3 Shift the genetic material.
   5.4. Update the sets S_y, S_u and the reference values for the considered prediction horizon.
   5.5. Update P_m and create the initial population of next evolutionary process.
   5.6. Evaluate the individuals, adapt the goals and compute the fitness values.
6. End of the algorithm.

Supplementary, the convergence rate of the algorithm can be increased, by applying a local optimisation procedure, in Lamarckian sense. The designer can consider, for instance, Hook – Jeeves or hill climbing methods.

4. STUDY CASE

The experiments were carried out on a simulated. multi input multi output furnace system (Lazăr, *et al.*, 1998). The heating process is based on radiation and convection, produced by means of two heating resistors. The system is equipped with two sensors, measuring the temperature achieved inside the furnace, in two different temperature zones. The plant has two inputs, namely the voltages applied on the heating resistors and two outputs (the temperatures resulted in the considered heating zones). A coupling controller is included. Thus, the complex optimisation problem is split into two simpler optimisation problems. As consequence, the design algorithm has greater chances to find an adequate control sequence during the on-line configuration.

The process model is indicated in (11), (12), (13), for the sampling period $T = 40\,\text{sec}$:

$$G(q^{-1}) = \begin{bmatrix} G_{11} & 0 \\ 0 & G_{22} \end{bmatrix}, \tag{11}$$

$$G_{11}(q^{-1}) = q^{-2}\frac{0.0127q^{-1}}{1-0.9941q^{-1}-0.0064q^{-2}}, \tag{12}$$

$$G_{22}(q^{-1}) = q^{-4}\frac{0.0067q^{-1}}{1-0.7631q^{-1}-0.2329q^{-2}}. \tag{13}$$

The optimal control scenario was computed subject to

$$\begin{gathered} 0 \le u_j(k+i/k) \le 5 \\ i = 0,\dots,N-1,\ j = 1,2 \end{gathered}. \tag{14}$$

The performances of the algorithm were investigated with respect to different sets of parameters.

As expected, better results were achieved if the weights λ_i, with $i = d+1,\dots,N+d$, indicated in equation (4), respectively the weights α_i and β_i with $j = 0,\dots,N-1$, indicated in equations (5) and (8), had not the same values for the whole prediction horizon. The optimisation procedure can be forced to improve the components $u(k/k)$, by setting larger values for λ_{d+1}, α_0 and β_0. Even the other components encoded into the chromosomes are considered to be less important for the current evolutionary process, they could be modified at the next sample instants.

If a larger prediction horizon is considered, better values for the objective function O_1 can be obtained. Unfornately, this also implies the increasing of the chromosome length. As consequence, the evolutionary process has to work on larger populations and/or for a larger number of generations in order to produce adequate results, so it would requires larger computational resources.

The performances of the algorithm are strongly dependent on the maximum allowed variation between two successive components of the control action, denoted as $[LU_j]_{j=1,\dots,m}$ in (6). If too small values of $[LU_j]_{j=1,\dots,m}$ are considered, then the objective O_3 is very difficult to be minimised at the sample instant $k=1$, due the fact that larger variation are demanded by the objective O_1. In that situation two supplementary strategies can be used, in order to guarantee the achievement of feasible solution: one possibility consists in applying a local optimisation to the individuals of the generated initial population, with respect to O_3; another possibility is to generate the initial population considering smaller values for the upper boundaries of the control actions.

In figure 1, a set of convenient results are illustrated for the first and the second heating zhone. They corresponds to the following algorithm parameters: the number of individuals contained into the population $N_{ind} = 120$; the number of generations for an evolutionary loop $MAX\_GEN = 35$; the number of subpopulations $SUBPOP = 2$; the length of the prediction horizon $N=15$; 50% individuals selected for reproduction; 50% offspring inserted into the population; the maximum allowed variation between two successive components of the control action $LU_{1,2}=1$; the weights considered by the objective functions: $\alpha_o=1, \alpha_1=...=\alpha_{n-1}=0.01$, $\lambda_{d+1}=50, \lambda_{d+2}=5, \lambda_{d+3}=5, \lambda_{d+4}=...=\lambda_{d+N}=1$, $\beta_o=10, \beta_1=...=\beta_{n-1}=1$.

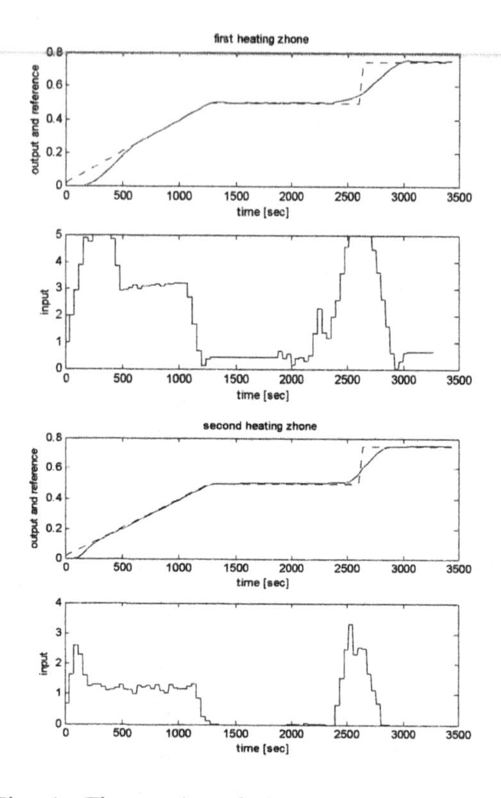

Fig. 1. The results of the evolutionary control algorithm for $N_{ind}=120$, $SUBPOP=2$, $N=15$, $MAX\_GEN=35$ (process input and output – solid line; reference – dotted line).

5. CONCLUSIONS

The paper suggests a design algorithm dedicated to assist the on-line configuration of a model based predictive control law. It can be applied for linear or nonlinear process models and needs no information regarding the derivatives of the objective functions.

The optimum control action is computed by an evolutionary procedure. This combination permits a flexible formulation of the design specifications and guarantees satisfactory results even in the case of noisy or time-variant environment. The multiobjective optimisation problem is solved by means of a non-dominated sorting procedure. The set of rules proposed for the computation of the fitness values establishes a good trade-off between the priorities of the imposed objectives. It encourages the achievement of feasible control policies, with good results subject to the accuracy of the automatic system. It also assures a convenient convergence rate for the algorithm.

Although, the approach can be applied only for robust plants, characterised by large time-constants.

REFERENCES

Bäck, T., D. Fogel and Z. Michalewicz (1997). *Handbook of Evolutionary Computation*. Oxford University Press, UK.

Clarke, D., Ed. (1994). *Advances in Model Based predictive Control*, Oxford University Press.

Dasgupta, D. and Z. Michalewicz (1997). *Evolutionary Algorithms in Engineering Applications*, Springer.

Deb, K. (2001). *Multi-Objective Optimization using Evolutionary Algorithms*, Wiley.

De Keyser R. and J. Donald (1999). Model Based Predictive Control in RTP Semiconductor Manufacturing. Proceedings of *IEEE International Conference on Control Applications*, USA.

Fleming, P. J. and R.C. Purshouse (2002). Evolutionary algorithms in control systems engineering: a survey. *Control Engineering Practice*, **10**, 1223-1241.

Fonseca, C.M. and P.J. Fleming (1998). Multiobjective Optimisation and Multiple Constraint Handling with Evolutionary Algorithms – Part I: A Unified Formulation. *IEEE Transactions on Systems, Man, and Cybernetics – Part A*, **28** (1), 26-37.

Lazăr, C., E. Poli and B. Mustaţă (1998). Implementation of predictive controller for thermal treatment processes. Proceedings of *IFAC Symposium on Automation in Mining, Mineral and Metal Processing*, Koln, Germany, 379-384.

Marcu, T. (1997). A Multiobjective Evolutionary Approach to Pattern Recognition for Robust Diagnosis of Process Faults. Preprints of *IFAC Symposium on Fault Detection, Supervision and Safety for Technical Processes*, Hull, UK, **2**, 1193-1198.

Michalewicz, Z. (1996). *Genetic Algorithms + Data structures = Evolution Programs* (3rd Edition), Springer Verlag, USA.

Miettinen, K., M. Markela, Pekka Neittaanmaki, J. Periaux, Eds. (1999). *Evolutionary Algorithms in Engineering and Computer Science*, Wiley.

Rodriguez-Vazquez, K. and P.J. Fleming (1997). A Genetic Programming NARMAX Approach to Nonlinear System Identification. Proceedings of *GALESIA '97*, Sheffield, UK, 409-413.

Tamaki, H., H. Kita and S. Kobayashi (1996). Multiobjective Optimisation by Genetic Algorithms: A Review. Proceedings of *Conference on Evolutionary Computation*, Nagoya, Japan, 517-522.

A NOTE TO CALCULATION OF POLYTOPIC INVARIANT AND FEASIBLE SETS FOR LINEAR CONTINUOUS-TIME SYSTEMS

B. Rohaľ Ilkiv *

* *Department of Automation & Measurement*
Faculty of Mechanical Engineering, Slovak University of Technology
Nám. Slobody 17, 81231 Bratislava, Slovakia
fax : +421 2 52495315 and e-mail : rohal@kam.vm.stuba.sk

Abstract: A common way of guaranteeing stability in model predictive control (MPC) is to add into the open-loop optimization problem a terminal constraint set which is invariant and feasible. This note presents an algorithm for calculation of low-complexity polytopic invariant sets for linear continuous-time systems with complex conjugate closed-loop eigenvalues. Non-symmetrical amplitude and rate input constraints are considered. *Copyright © 2003 IFAC*

Keywords: Constrained control, Stability, invariant sets

1. INTRODUCTION

Any locally stable time-invariant dynamical system admits some domains in its state-space from which any state-vector trajectory cannot escape. These domains are called positively invariant sets of the system and are presently widely used in MPC for designing of terminal constraint sets - *target sets* - as a one tool imperative for guarantee of the system closed-loop stability (Kouvaritakis *et al.*, 2000), (Mayne *et al.*, 2000), (De Dona *et al.*, 2002), (Marruedo *et al.*, 2002). The existence, characterization and practical calculation of positively invariant sets of dynamical systems is therefore a basic issue for many constrained MPC schemes utilizing dual mode prediction strategies. The main idea is to determine an invariant set in the state space having the property that no constraints violation occurs as long as the state remains in such a set. More recent contributions (Blanchini, 1994),(Blanchini, 1999) have proposed constructive methods to deal with the problem.

Among the candidate invariant sets in literature there are two kind of families which have been essentially considered, ellipsoidal and polyhedral sets. Ellipsoids are classical invariant sets in control theory closely related to quadratic Lyapunov functions. Polyhedral sets have been involved in the solution of control problem starting from the 70's. Their importance is due to the fact that they are often natural expression of physical constraints on control, state or output variables. As invariant sets, they can have significantly larger volumes than the invariant ellipsoidal ones, also. Their shape is in some sense *more flexible* than that of the ellipsoids, this fact leads to their better ability in the approximation of reachability sets and domains of attraction of dynamical systems. The trade off of this flexibility is that they have in general a more complex representation.

This note deals with a problem of calculation of *low-complexity* polytopic invariant sets for linear

[1] Partially supported by the Slovak Scientific Grant Agency VEGA - project No. 1/9278/02

continuous-time systems with complex conjugate closed-loop eigenvalues. It is assumed that the system is subject to in general non-symmetric amplitude and rate input constraints, what is the most frequent case in practice. The extension to state or output constraints is straightforward and will not be considered here. Compare to maximal admissible sets (Kouvaritakis *et al.*, 2000), the low complexity invariant sets are more numerically tractable and their definition does not depend on higher powers of the system closed-loop matrix.

2. DEFINITIONS AND NOTATIONS

Generally a polyhedral set (polyhedron) can be introduced in the form of plane or vertex representation. In the plane representation the polyhedral set can be defined as an intersection of a finite number of halfspaces as follows

$$\mathscr{S}(\mathbf{G},\mathbf{w}_1,\mathbf{w}_2) = \{\mathbf{x} \in \mathbb{R}^n; -\mathbf{w}_1 \leq \mathbf{G}\mathbf{x} \leq \mathbf{w}_2\} \quad (1)$$

where the matrix $\mathbf{G} \in \mathbb{R}^{gxn}$, and vectors $\mathbf{w}_1, \mathbf{w}_2 \in \mathbb{R}^g$. According to the selected triple $[\mathbf{G},\mathbf{w}_1,\mathbf{w}_2]$, $\mathscr{S}(\mathbf{G},\mathbf{w}_1,\mathbf{w}_2)$ can be any type of the polyhedral set - bounded or unbounded, including or not including the origin point, $\mathbf{x} = 0$, in the state space. In particular, if $\mathbf{w}_1, \mathbf{w}_2 > 0$, the origin point belongs to the interior of the polyhedral set, i.e. $0 \in int(\mathscr{S}(\mathbf{G},\mathbf{w}_1,\mathbf{w}_2))$. Observe that any polyhedron $\mathscr{S}(\mathbf{G},\mathbf{w}_1,\mathbf{w}_2)$ can be characterized by matrix $\mathbf{G}^*$ and vector $\mathbf{w}^*$, $\mathbf{G}^* = [\mathbf{G}^T, -\mathbf{G}^T]^T$, $\mathbf{w}^* = [\mathbf{w}_2^T, \mathbf{w}_1^T]^T$ and defined in equivalent, frequently used form

$$\mathscr{S}(\mathbf{G}^*,\mathbf{w}^*) = \{\mathbf{x} \in \mathbb{R}^n; \mathbf{G}^*\mathbf{x} \leq \mathbf{w}^*\} \quad (2)$$

The linear inequality constraints applied to input, state or output variables define a polyhedral set in the state space of given dynamical system. If the polyhedral set can by made positively invariant by a state feedback, then these constraints are satisfied all along the state trajectory for all initial point in the set.

Consider that the continuous-time linear system to be controlled (under a dual-mode prediction scheme) is represented by the following state equation

$$\dot{\mathbf{x}}(t) = \mathbf{A}\mathbf{x}(t) + \mathbf{B}\mathbf{u}(t) \quad (3)$$

with $(\mathbf{A},\mathbf{B})$ stabilizable, $\mathbf{A} \in \mathbb{R}^{nxn}$, $\mathbf{B} \in \mathbb{R}^{nxm}$, rank$(\mathbf{B}) = m \leq n$, $\mathbf{x}(t) \in \mathbb{R}^n$, $\mathbf{u}(t) \in \mathbb{R}^m$, and $t \in [0,\infty]$. Assume that only the control vector $\mathbf{u}(t)$ is constrained, hence an admissible set of control signals is defined by the following system of linear inequalities

$$-\mathbf{u}_{min} \leq \mathbf{u}(t) \leq \mathbf{u}_{max} \quad (4)$$

$$-\dot{\mathbf{u}}_{min} \leq \dot{\mathbf{u}}(t) \leq \dot{\mathbf{u}}_{max} \quad (5)$$

The vectors $\mathbf{u}_{min}, \mathbf{u}_{max}, \dot{\mathbf{u}}_{min}, \dot{\mathbf{u}}_{max}$ are considered real with positive components. Under a closed-loop linear state control law

$$\mathbf{u}(t) = \mathbf{F}\ \mathbf{x}(t) \quad (6)$$

with gain $\mathbf{F} \in \mathbb{R}^{mxn}$, the evolution of the system is described by

$$\dot{\mathbf{x}}(t) = \mathbf{A}_0\ \mathbf{x}(t) \qquad \mathbf{A}_0 = \mathbf{A} + \mathbf{BF} \quad (7)$$

In the state space of the autonomous system (7), the inequalities (4) and (5) define nonsymmetrical polyhedral domains

$$\mathscr{D}_0(\mathbf{F},\mathbf{u}_{min},\mathbf{u}_{max}) = \\ = \{\mathbf{x} \in \mathbb{R}^n; -\mathbf{u}_{min} \leq \mathbf{F}\mathbf{x} \leq \mathbf{u}_{max}\} \quad (8)$$

$$\mathscr{D}_1(\mathbf{F}\mathbf{A}_0,\dot{\mathbf{u}}_{min},\dot{\mathbf{u}}_{max}) = \\ = \{\mathbf{x} \in \mathbb{R}^n; -\dot{\mathbf{u}}_{min} \leq \mathbf{F}\mathbf{A}_0\mathbf{x} \leq \dot{\mathbf{u}}_{max}\} \quad (9)$$

Note that these polyhedrons can be bounded or unbounded. The latter case is more obvious in real applications, due to fact that $m \leq n$.

For further explanation the following notation will be employed. $\mathbf{A}^+$ is a matrix of components (a_{ij}^+) defined as $(a_{ij}^+) = max(a_{ij},0)$, where a_{ij} are components of a real matrix $\mathbf{A}$. $\mathbf{A}^-$ is a matrix of components (a_{ij}^-) defined as $(a_{ij}^-) = -min(a_{ij},0)$. $\mathbf{A}^\oplus$ is a matrix of components $(a_{ij}^\oplus)$ defined as $a_{ii}^\oplus = a_{ii}$ and $(a_{ij}^\oplus) = max(a_{ij},0)$. $\mathbf{A}^\ominus$ is a matrix of components $(a_{ij}^\ominus)$ defined as $a_{ii}^\ominus = 0$ and $(a_{ij}^\ominus) = -min(a_{ij},0)$.

For the polyhedral set of type (1) the positive invariance conditions are given by following theorem, see e.g. (Bitsoris, 1991), (Blanchini, 1999):

Theorem 2.1. The polyhedral set $\mathscr{S}(\mathbf{G},\mathbf{w}_1,\mathbf{w}_2) = \{\mathbf{x} \in \mathbb{R}^n; -\mathbf{w}_1 \leq \mathbf{G}\mathbf{x} \leq \mathbf{w}_2\}$ where $\mathbf{G} \in \mathbb{R}^{gxn}$, rank$\mathbf{G} = g$, $\mathbf{w}_1 \in \mathbb{R}^g$, $\mathbf{w}_2 \in \mathbb{R}^g$, $\mathbf{w}_1 > 0$ and $\mathbf{w}_2 > 0$ is a positively invariant set of system (7), if and only if there exists a matrix $\mathbf{H} \in \mathbb{R}^{gxg}$, such that

$$\mathbf{G}\mathbf{A}_0 - \mathbf{H}\mathbf{G} = 0 \quad (10)$$

$$\begin{bmatrix} \mathbf{H}^\oplus & \mathbf{H}^\ominus \\ \mathbf{H}^\ominus & \mathbf{H}^\oplus \end{bmatrix} \begin{bmatrix} \mathbf{w}_2 \\ \mathbf{w}_1 \end{bmatrix} \leq 0 \quad (11)$$

These invariance relations can by also obtained from a repeated use of the following Farkas lemma, (Castelan and Hennet, 1992):

Lemma 2.1. Extended Farkas Lemma. Given two polyhedra $\mathscr{S}_1(\mathbf{G}_1,\mathbf{g}_1) = \{\mathbf{x} \in \mathbb{R}^n; \mathbf{G}_1\mathbf{x} \leq \mathbf{g}_1\}$ and $\mathscr{S}_2(\mathbf{G}_2,\mathbf{g}_2) = \{\mathbf{x} \in \mathbb{R}^n; \mathbf{G}_2\mathbf{x} \leq \mathbf{g}_2\}$ then $\mathscr{S}_1(\mathbf{G}_1,\mathbf{g}_1) \subseteq \mathscr{S}_2(\mathbf{G}_2,\mathbf{g}_2)$ if and only if there exists a non-negative matrix $\mathbf{M}$ such that $\mathbf{M}\mathbf{G}_1 = \mathbf{G}_2$ and $\mathbf{M}\mathbf{g}_1 \leq \mathbf{g}_2$.

3. INVARIANT AND FEASIBLE TERMINAL SETS

An invariant and feasible set of states is such a set that, given the state feedback control law $\mathbf{u}(t) = \mathbf{F}\mathbf{x}(t)$ operating under the second mode of a dual-mode MPC scheme, satisfies input constraints (4), (5) and makes the state trajectory of the closed-loop system remain in the set continuously. The set can then be used as the terminal or target set for the first mode of predictive control.

A polyhedral invariant set of a dynamical system can be described as an intersection of a number of invariant halfspaces. This technique leads to the *low-complexity* polytopic invariant sets. For linear systems $\dot{\mathbf{x}}(t) = \mathbf{A}_0\mathbf{x}(t)$ these halfspaces can be supplied by real eigenvalues and left eigenvectors of the system matrix $\mathbf{A}_0$. The following lemma (Jirstrand, 1998) characterizes the connection between invariant halfspaces and left eigenvectors of the system matrix $\mathbf{A}_0$.

Lemma 3.1. Let $\varphi \in \mathbb{R}^n$ be a left eigenvector of $\mathbf{A}_0 \in \mathbb{R}^{nxn}$ with a real eigenvalue $\lambda \in \mathbb{R}$, i.e. $\varphi^T\mathbf{A}_0 = \lambda\varphi^T$. Then the state space can be decomposed into three invariant sets given by $\varphi^T\mathbf{x}(t) < 0$, $\varphi^T\mathbf{x}(t) = 0$ and $\varphi^T\mathbf{x}(t) > 0$.

The invariance of these sets in the lemma follows directly from properties of exponential functions. Computing left eigenvectors corresponding to real eigenvalues gives a number of invariant slabs for negative eigenvalues, half spaces for positive eigenvalues and hyperplanes for zero eigenvalues. Since intersections of invariant sets are invariant, the real left eigenvectors define normal vectors of an invariant polyhedra. If the closed-loop system matrix $\mathbf{A}_0$ is diagonalizable (i.e. there exist n independent left eigenvectors $\phi_i, i = 1 \dots n$) and all eigenvalues $\lambda_i, i = 1 \dots n$, are real and negative, the resulting invariant polyhedra is *polytope*, let us denote that as

$$\mathscr{P}(\mathbf{W}, \mathbf{w}_1, \mathbf{w}_2) = \{\mathbf{x} \in \mathbb{R}^n; -\mathbf{w}_1 \le \mathbf{W}\mathbf{x} \le \mathbf{w}_2\} \tag{12}$$

with

$$\mathbf{W} = \begin{bmatrix} \varphi_1^T \\ \vdots \\ \varphi_n^T \end{bmatrix}, \qquad \mathbf{W} \in \mathbb{R}^{nxn} \tag{13}$$

For the polytope the *definition* matrix $\mathbf{H}$, defined by the invariance condition (10) always exists

$$\mathbf{H} = \mathbf{W}\mathbf{A}_0\mathbf{V}, \qquad \mathbf{V} = \mathbf{W}^{-1} \tag{14}$$

The polytope $\mathscr{P}(\mathbf{W}, \mathbf{w}_1, \mathbf{w}_2)$ is positively invariant if for the calculated matrix $\mathbf{H}$ the relation (11) holds for any positive $\mathbf{w}_1, \mathbf{w}_2$. This is always true under the assumption that all eigenvalues are real and negative.

Considering that the definition matrix $\mathbf{H}$ is defined in advance, the task is then to determine $\mathbf{w}_1, \mathbf{w}_2$ in such manner that the resulting polytope may obtain desired properties. Our aim is then to find conditions under which the invariant polytope (12) can be inscribed into the possibly unbounded polyhedrons (8) and (9), induced in the state space with the used constraints, and to find such positive parameters $\mathbf{w}_1, \mathbf{w}_2$ maximizing the polytope volume.

A condition for inscribing the invariant polytope (12) to the amplitude constraints polyhedron (8) provides the following lemma:

Lemma 3.2. Given a definition matrix $\mathbf{W}$ of a polytope $\mathscr{P}(\mathbf{W}, \mathbf{w}_1, \mathbf{w}_2) = \{\mathbf{x} \in \mathbb{R}^n; -\mathbf{w}_1 \le \mathbf{W}\mathbf{x} \le \mathbf{w}_2\}$ and a polyhedron $\mathscr{D}_0(\mathbf{F}, \mathbf{u}_{min}, \mathbf{u}_{max}) = \{\mathbf{x} \in \mathbb{R}^n; -\mathbf{u}_{min} \le \mathbf{F}\mathbf{x} \le \mathbf{u}_{max}\}$ then $\mathscr{P}(\mathbf{W}, \mathbf{w}_1, \mathbf{w}_2) \subseteq \mathscr{D}_0(\mathbf{F}, \mathbf{u}_{min}, \mathbf{u}_{max})$ if and only if there exist non-negative vectors $\mathbf{w}_1, \mathbf{w}_2$ such that

$$\begin{bmatrix} \mathbf{M}_0^+ & \mathbf{M}_0^- \\ \mathbf{M}_0^- & \mathbf{M}_0^+ \end{bmatrix} \begin{bmatrix} \mathbf{w}_2 \\ \mathbf{w}_1 \end{bmatrix} \le \begin{bmatrix} \mathbf{u}_{max} \\ \mathbf{u}_{min} \end{bmatrix}, \quad \text{and } \mathbf{M}_0 = \mathbf{F}\mathbf{V}$$

Proof. To prove the lemma a slight modification of the technique introduced in (Bitsoris, 1991) can be used. Observe that the polyhedral sets $\mathscr{P}(\mathbf{W}, \mathbf{w}_1, \mathbf{w}_2)$, $\mathscr{D}_0(\mathbf{F}, \mathbf{u}_{min}, \mathbf{u}_{max})$ can be written in the equivalent form (2), as $\mathscr{P}(\mathbf{W}^*, \mathbf{w}^*)$, $\mathscr{D}_0(\mathbf{F}^*, \mathbf{u}^*)$. For $\mathscr{P}(\mathbf{W}^*, \mathbf{w}^*) \subseteq \mathscr{D}_0(\mathbf{F}^*, \mathbf{u}^*)$ the Farkas lemma implies the existence of a non-negative matrix $\mathbf{M}^* \in \mathbb{R}^{2mx2n}$, such that

$$\mathbf{M}^*\mathbf{W}^* = \mathbf{F}^* \qquad \mathbf{M}^*\mathbf{w}^* \le \mathbf{u}^* \tag{15}$$

If $\mathbf{M}^*$ is decomposed as

$$\mathbf{M}^* = \begin{bmatrix} \mathbf{M}_{11} & \mathbf{M}_{12} \\ \mathbf{M}_{21} & \mathbf{M}_{22} \end{bmatrix}, \qquad \mathbf{M}_{ij} \in \mathbb{R}^{mxn}$$

then from (15), it follows that

$$\mathbf{F} - (\mathbf{M}_{11} - \mathbf{M}_{12})\mathbf{W} = 0$$

$$\mathbf{F} - (\mathbf{M}_{22} - \mathbf{M}_{21})\mathbf{W} = 0$$

Since $\mathbf{F} \in \mathbb{R}^{mxn}$ and rank$\mathbf{F} = m$, from the above relations it follows that

$$(\mathbf{M}_{11} - \mathbf{M}_{12}) = (\mathbf{M}_{22} - \mathbf{M}_{21})$$

Consequently, condition (15) is satisfied by

$$\mathbf{M}_0 = (\mathbf{M}_{11} - \mathbf{M}_{12}) = (\mathbf{M}_{22} - \mathbf{M}_{21})$$

and for $\mathbf{M}_0$ holds

$$\mathbf{M}_0 = \mathbf{F}\mathbf{W}^{-1} = \mathbf{F}\mathbf{V}$$

Selecting $\mathbf{M}_{11} = \mathbf{M}_{22} = \mathbf{M}_0^+$ and $\mathbf{M}_{12} = \mathbf{M}_{21} = \mathbf{M}_0^-$ it is possible to conclude, that

$$\mathbf{M}^* = \begin{bmatrix} \mathbf{M}_0^+ & \mathbf{M}_0^- \\ \mathbf{M}_0^- & \mathbf{M}_0^+ \end{bmatrix} \tag{16}$$

is always a non-negative matrix and for the defined non-negative vector $\mathbf{u}^*$ this implies the statement of the lemma. □

Analogically, a similar condition for inscribing the invariant polytope (12) to the rate constraints polyhedron (9) provides the lemma:

Lemma 3.3. Given a definition matrix $\mathbf{W}$ of a polytope $\mathscr{P}(\mathbf{W}, \mathbf{w}_1, \mathbf{w}_2) = \{\mathbf{x} \in \mathbb{R}^n; -\mathbf{w}_1 \leq \mathbf{W}\mathbf{x} \leq \mathbf{w}_2\}$ and a polyhedron $\mathscr{D}_1(\mathbf{FA}_0, \dot{\mathbf{u}}_{min}, \dot{\mathbf{u}}_{max}) = \{\mathbf{x} \in \mathbb{R}^n; -\dot{\mathbf{u}}_{min} \leq \mathbf{FA}_0\mathbf{x} \leq \dot{\mathbf{u}}_{max}\}$ then $\mathscr{P}(\mathbf{W}, \mathbf{w}_1, \mathbf{w}_2) \subseteq \mathscr{D}_1(\mathbf{FA}_0, \dot{\mathbf{u}}_{min}, \dot{\mathbf{u}}_{max})$ if and only if there exist non-negative vectors $\mathbf{w}_1, \mathbf{w}_2$ such that

$$\begin{bmatrix} \mathbf{M}_1^+ & \mathbf{M}_1^- \\ \mathbf{M}_1^- & \mathbf{M}_1^+ \end{bmatrix} \begin{bmatrix} \mathbf{w}_2 \\ \mathbf{w}_1 \end{bmatrix} \leq \begin{bmatrix} \dot{\mathbf{u}}_{max} \\ \dot{\mathbf{u}}_{min} \end{bmatrix}, \text{ and } \mathbf{M}_1 = \mathbf{FA}_0\mathbf{V}$$

The following lemma summarizes obtained results and provides an inclusion of amplitude and rate constraints with positive invariance conditions of polytope (12):

Lemma 3.4. The polytope $\mathscr{P}(\mathbf{W}, \mathbf{w}_1, \mathbf{w}_2)$ is invariant and feasible for a given state feedback $\mathbf{F}$ if and only if there exist non-negative vectors $\mathbf{w}_1, \mathbf{w}_2$ such that

$$\begin{bmatrix} \mathbf{M}_1^+ & \mathbf{M}_1^- \\ \mathbf{M}_1^- & \mathbf{M}_1^+ \end{bmatrix} \begin{bmatrix} \mathbf{w}_2 \\ \mathbf{w}_1 \end{bmatrix} \leq \begin{bmatrix} \dot{\mathbf{u}}_{max} \\ \dot{\mathbf{u}}_{min} \end{bmatrix} \quad \mathbf{M}_1 = \mathbf{FA}_0\mathbf{V}$$

$$\begin{bmatrix} \mathbf{M}_0^+ & \mathbf{M}_0^- \\ \mathbf{M}_0^- & \mathbf{M}_0^+ \end{bmatrix} \begin{bmatrix} \mathbf{w}_2 \\ \mathbf{w}_1 \end{bmatrix} \leq \begin{bmatrix} \mathbf{u}_{max} \\ \mathbf{u}_{min} \end{bmatrix} \quad \mathbf{M}_0 = \mathbf{FV}$$

$$\begin{bmatrix} \mathbf{H}^\oplus & \mathbf{H}^\ominus \\ \mathbf{H}^\ominus & \mathbf{H}^\oplus \end{bmatrix} \begin{bmatrix} \mathbf{w}_2 \\ \mathbf{w}_1 \end{bmatrix} \leq \begin{bmatrix} 0 \\ 0 \end{bmatrix} \quad \mathbf{H} = \mathbf{WA}_0\mathbf{V}$$

The proof of the lemmas is straightforward and follows from previous Lemma 3.2 and the Theorem 2.1. Notice that for real and stable eigenvalues of $\mathbf{A}_0$ the matrix $\mathbf{H} = \mathbf{WA}_0\mathbf{V}$ is a diagonal matrix with non-positive entries. Consequently, isolation in, the invariance condition (11) for the polytope $\mathscr{P}(\mathbf{W}, \mathbf{w}_1, \mathbf{w}_2)$ is always true for any non-negative choice of $\mathbf{w}_1, \mathbf{w}_2$.

3.1 *Complex conjugate eigenvalues*

A critical point of presented considerations is the case when a pair (or several pairs) of stable complex conjugate eigenvalues $a_i \pm jb_i$, $Re(a_i \pm jb_i) < 0$, appears among the other stable real eigenvalues λ_i of the closed-loop system matrix $\mathbf{A}_0$. Then in the definition matrix $\mathbf{W}$ of plane representation of the polytope $\mathscr{P}(\mathbf{W}, \mathbf{w}_1, \mathbf{w}_2)$, there appear left complex conjugate eigenvectors $\varrho_i \pm j\sigma_i$ and the polyhedron is not further bounded in that directions. In the similar case it is necessary to look for a reasonable technique that maintains the polyhedron $\mathscr{P}(\mathbf{W}, \mathbf{w}_1, \mathbf{w}_2)$ bounded in $\mathbb{R}^n$. One way is to apply well-known elementary rotation matrices (Kouvaritakis *et al.*, 2002)

$$\mathbf{E}^* = \frac{1}{\sqrt{2}} \begin{bmatrix} 1 & -j \\ 1 & j \end{bmatrix}, \quad \mathbf{E} = \frac{1}{\sqrt{2}} \begin{bmatrix} 1 & 1 \\ j & -j \end{bmatrix}$$

to the *complex conjugate* rows and columns of matrices $\mathbf{W}$ and $\mathbf{V}$. This decomposes these complex conjugate pairs of eigenvectors to their real and imaginary parts (multiplied by a constant). Define following diagonal unity *transformation* matrices $\mathcal{E}^*$ and $\mathcal{E}$

$$\mathcal{E}^* = \begin{bmatrix} \mathbf{I} & 0 \\ 0 & diag(\mathbf{E}^*) \end{bmatrix}, \quad \mathcal{E} = \begin{bmatrix} \mathbf{I} & 0 \\ 0 & diag(\mathbf{E}) \end{bmatrix}$$

with elementary rotation matrices $\mathbf{E}^*$ and $\mathbf{E}$, placed on that positions where these complex conjugate eigenvalues are arranged. Computing firstly a transformed matrix $\tilde{\mathbf{V}} = \mathbf{V}\mathcal{E}^*$ with the help of matrix $\mathbf{V}$, which columns are right eigenvectors, primarily calculated by a usual software support, a new definition matrix $\tilde{\mathbf{W}}$ of polytope (12) can be written as

$$\tilde{\mathbf{W}} = [\tilde{\mathbf{V}}]^{-1} = [\mathbf{V}\mathcal{E}^*]^{-1} = \mathcal{E}\mathbf{W} \qquad (17)$$

Each pair of these complex conjugate eigenvectors in the matrix $\mathbf{W}$ is then transformed to a new pair of real normal vectors of matrix $\tilde{\mathbf{W}}$, consisting of real and imaginary parts of original complex eigenvectors (multiplied by a constant). The solution of the invariance condition (10) for polytope (12) then looks like

$$\tilde{\mathbf{H}} = \tilde{\mathbf{W}}\mathbf{A}_0\tilde{\mathbf{V}} = \mathcal{E}\mathbf{W}\mathbf{A}_0\mathbf{V}\mathcal{E}^* = \mathcal{E}\mathbf{H}\mathcal{E}^* \qquad (18)$$

It is easy to see that this matrix is not further diagonal and has the following diagonal *block* structure

$$\tilde{\mathbf{H}} = \begin{bmatrix} \text{diag}(\lambda_i) & 0 \\ 0 & \text{diag}(\Lambda_i) \end{bmatrix}, \quad \Lambda_i = \begin{bmatrix} a_i & b_i \\ -b_i & a_i \end{bmatrix} \qquad (19)$$

This structure causes that the positive invariance condition (11) applied to $\tilde{\mathbf{H}}$ need not be met for all complex eigenvalues $a_i \pm jb_i$, $Re(a_i \pm jb_i) < 0$, i.e. not for any stabilizing feedback gain $\mathbf{F}$.

Theorem 3.1. Let $\mathbf{F}$ be such that $\mathbf{A}_0$ has distinct or simple eigenvalues and let (λ_i, φ_i) and $(a_i \pm jb_i, \varrho_i \pm j\sigma_i)$ denote the real and complex left (eigenvalues, eigenvector) pairs. Then for $\tilde{\mathbf{W}} = [\tilde{\mathbf{V}}]^{-1}$, where $\tilde{\mathbf{W}}$ is the matrix comprising as columns the vectors φ_i and ϱ_i, σ_i (each of the ϱ_i, σ_i taken only once), there exist non-negative vectors $\mathbf{w}_1, \mathbf{w}_2$ such that $\mathscr{P}(\tilde{\mathbf{W}}, \mathbf{w}_1, \mathbf{w}_2)$ is positively invariant under (7) for the feedback gain $\mathbf{F}$ if and only if $\mathbf{A}_0$ is Hurwitz and for all its complex eigenvalues holds

$$\frac{Im(a_i \pm jb_i)}{Re(a_i \pm jb_i)} \in [-1, 1] \qquad (20)$$

The condition (20) defines for complex conjugate eigenvalues an admissible sector of the complex plane, outlined in Figure 1, on which the theorem is valid all the time.

Proof. The proof directly follows from (11), using the diagonal block structure of $\tilde{\mathbf{H}}$ and the non-negative property of vectors $\mathbf{w}_1, \mathbf{w}_2$.

In order to $\mathbf{A}_0$ be Hurwitz all eigenvalues must lie in the left-half of the complex plane. It means that all λ_i in the block diagonal matrix $\tilde{\mathbf{H}}$, relation (19), are always negative together with all a_i belonging to square blocks Λ_i. The negativeness of the diagonal segment diag(λ_i) of matrix $\tilde{\mathbf{H}}$, which consists of λ_i, clearly implies the existence of corresponding non-negative entries at vector $[\mathbf{w}_2^T, \mathbf{w}_1^T]^T$ in the positive invariance condition (11). The existence of the rest of the non-negative entries in the vector corresponding to diagonal segments diag(Λ_i) of matrix $\tilde{\mathbf{H}}$, must be analyzed separately.

Define the decomposition of Λ_i to $\Lambda_i^{\oplus}$, $\Lambda_i^{\ominus}$ and determine under what conditions there exists a non-negative tetrad $[w_{i+1}, w_{i+2}, w_{i+3}, w_{i+4}]^T$ belonging to the decomposition and fulfilling the invariance condition (11), introduced as

$$\begin{bmatrix} \Lambda_i^{\oplus} & \Lambda_i^{\ominus} \\ \Lambda_i^{\ominus} & \Lambda_i^{\oplus} \end{bmatrix} \begin{bmatrix} w_{i+1} \\ w_{i+2} \\ w_{i+3} \\ w_{i+4} \end{bmatrix} \leq 0$$

Substitute for the $\Lambda_i^{\oplus}$, $\Lambda_i^{\ominus}$ and write the real part of the complex conjugate eigenvalue with the negative sign. Then the above inequality takes the form

$$\begin{bmatrix} -ai & b_i & 0 & 0 \\ 0 & -a_i & b_i & 0 \\ 0 & 0 & -a_i & b_i \\ b_i & 0 & 0 & -a_i \end{bmatrix} \begin{bmatrix} w_{i+1} \\ w_{i+2} \\ w_{i+3} \\ w_{i+4} \end{bmatrix} \leq 0$$

which can be further expressed as

$$w_{i+1} \geq \frac{b_i}{a_i} w_{i+2}, \quad w_{i+2} \geq \frac{b_i}{a_i} w_{i+3}$$
$$w_{i+3} \geq \frac{b_i}{a_i} w_{i+4}, \quad w_{i+4} \geq \frac{b_i}{a_i} w_{i+1}$$

A simple inspection of resulting inequalities directly implies the validity of condition (20) for the ith complex conjugate eigenvalues pair in the case we want to obtain any non-negative solution $[w_{i+1}, w_{i+2}, w_{i+3}, w_{i+4}]^T$. This closes the proof.□

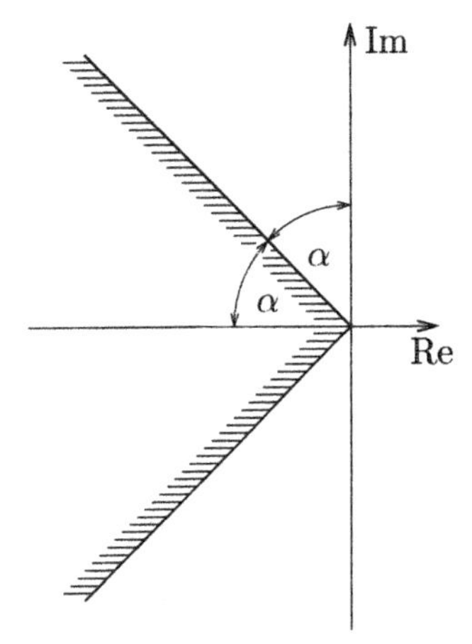

Fig. 1. The admissible sector for closed-loop eigenvalues

The above theorem is a generalization to continuous-time systems of a similar theorem formulated for discrete-time systems in (Kouvaritakis *et al.*, 2002).

4. ALGORITHMS FOR INVARIANT AND FEASIBLE SETS

The task now is to construct a possibly largest polytope $\mathscr{P}(\tilde{\mathbf{W}}, \mathbf{w}_1, \mathbf{w}_2)$ satisfying positive invariance and feasibility conditions of Lemma 3.4 that means to design algorithms for solving the following problem:

Problem 4.1. Find for the given state feedback gain $\mathbf{F}$ optimal values $\hat{\mathbf{w}}_1$, $\hat{\mathbf{w}}_2$ of $\mathbf{w}_1$, $\mathbf{w}_2$, *maximizing in a defined sense* the polytope $\mathscr{P}(\tilde{\mathbf{W}}, \mathbf{w}_1, \mathbf{w}_2)$ fulfilling simultaneously conditions of Lemma 3.4.

Usually the aim is to design a procedure intended to enlarge the polytope *volume*. Based on linear and convex programming in (Kouvaritakis *et al.*, 2002) and (Karas, 2002) there were suggested for discrete-time systems efficient algorithms, aimed at finding the largest values of individual entries in $\mathbf{w}_1$, $\mathbf{w}_2$ and consequently to enlarge the polytope volume. These algorithms, after a slight modification, can be augmented also to the continuous-time case. As an example an algorithm based on linear programming is introduced.

Algorithm 4.1

(1) set $\mathbf{x}_0$, choose ϵ to be close to zero ($\epsilon \to 0$), define $\mathbf{P}, \mathbf{p}$ and $\mathbf{x}$ in following way:

$$\mathbf{P} = \begin{bmatrix} \mathbf{M}_1^+ & \mathbf{M}_1^- \\ \mathbf{M}_1^- & \mathbf{M}_1^+ \\ \mathbf{M}_0^+ & \mathbf{M}_0^- \\ \mathbf{M}_0^- & \mathbf{M}_0^+ \\ \tilde{\mathbf{H}}^{\oplus} & \tilde{\mathbf{H}}^{\ominus} \\ \tilde{\mathbf{H}}^{\ominus} & \tilde{\mathbf{H}}^{\oplus} \end{bmatrix} \quad \mathbf{p} = \begin{bmatrix} \dot{\mathbf{u}}_{max} \\ \dot{\mathbf{u}}_{min} \\ \mathbf{u}_{max} \\ \mathbf{u}_{min} \\ 0 \\ 0 \end{bmatrix} \quad \mathbf{x} = \begin{bmatrix} \mathbf{w}_2 \\ \mathbf{w}_1 \end{bmatrix}$$

(2) minimize:

$$t_{min} = \min t$$

subject to:

$$\mathbf{P}\,\mathbf{x} \leq \mathbf{p} + t\,\mathbf{I}$$
$$\mathbf{x} \geq \mathbf{x}_0 - t\,\mathbf{I}$$

where $\mathbf{x}$ is assumed as a free variable

(3) set $\mathbf{x}_0 = \mathbf{x}$
(4) repeat steps 2., 3., until $| t_{min} | < \epsilon$
(5) finally $\tilde{\mathbf{x}} = \mathbf{x}$.

The algorithm tries to maximize a minimal *entry* in $\hat{\mathbf{w}}_2, \hat{\mathbf{w}}_1$. For alternative procedures, see the cited literature. Using the algorithms, the resulting polytope $\mathscr{P}(\tilde{\mathbf{W}}, \hat{\mathbf{w}}_1, \hat{\mathbf{w}}_2)$ can be used as the invariant and feasible terminal - target set for a predicted state under the first mode of predictive control and with this to provide means for ensuring the closed-loop stability.

5. CONCLUSIONS

In this note the construction of low-complexity invariant and feasible polytopic sets as terminal constraint sets for stable predictive control of continuous-time linear systems with amplitude and rate input constraints is discussed. An algorithm has been suggested trying to maximize volumes of these sets. The Theorem 3.1 defines a condition which is necessary for its solvability in the case of closed-loop complex conjugate eigenvalues. The theorem requires to select the state feedback gain **F**, fictitiously acting in the tail of the prediction horizon, in such a way that those eigenvalues must be placed within the sector region prescribed in the left-half complex plane by the condition (20). The standard LQ design may not be able to fulfill the condition, through an appropriate weighting in the quadratic cost, especially at a multivariate design. In such a case techniques of LQ controllers design with regional pole constraints are well applicable, see e.g. (Haddad and Bernstein, 1992).

In spite of the fact that the presented solution has been concentrated on control input constraints only, potential state or output constraints can be handled in a similar way.

REFERENCES

Bitsoris, G. (1991). Existence of positively invariant polyhedral sets for continuous-time linear systems. *Control theory and advanced technology* **7**(3), 407–427.

Blanchini, F. (1994). Ultimate boundedness control for uncertain discrete-time systems via set-induced lyapunov functions. *IEEE Transactions on Automatic Control* **39**(2), 428–433.

Blanchini, F. (1999). Set invariance in control. *Automatica* **35**(11), 1747–1767.

Castelan, E. B. and J. C. Hennet (1992). Eigenstructure assignment for state constrained linear continuous time systems. *Automatica* **28**(3), 605–611.

De Dona, J. A., M. M. Seron, D. Q. Mayne and G. C. Goodwin (2002). Enlarged terminal sets guaranteeing stability of receding horizon control. *Systems and Control Letters* **47**, 57–63.

Haddad, W. M. and D. S. Bernstein (1992). Controller design with regional pole constraints. *IEEE Transactions on Automatic Control* **37**(1), 54–69.

Jirstrand, M. (1998). Constructive Methods for Inequality Constraints in Control. PhD thesis. Department of Electrical Engineering, Linköping University. Linköping, Sweden.

Karas, A. (2002). Stabilizing predictive control with constraints. PhD thesis. Slovak technical University. Bratislava, Slovak republic. (In Slovak).

Kouvaritakis, B., M. Cannon, A. Karas, B. Rohaľ-Ilkiv and C. Belavý (2002). Asymmetric constraints with polyhedral sets in MPC with application to coupled tanks system. In: *Submitted to IEEE 2002 Conference on Decision and Control.* Las Vegas, Nevada, USA. Also submitted to International Journal of Robust and Nonlinear Control.

Kouvaritakis, B., Y. I. Lee, G. Tortora and M. Cannon (2000). MPC stability constraints and their implementations. In: *Preprints from the IFAC Conference Control System Design* (Š. Kozák and M. Huba, Eds.). Slovak University of Technology. Bratislava, Slovak republic. pp. 205–212.

Marruedo, D. L., T. Álamo and E. F. Camacho (2002). Enlarging the domain of attraction of mpc controller using invariant sets. In: *15th IFAC Triennial World Congress.* Barcelona, Spain.

Mayne, D. Q., J. B. Rawlings, C. V. Rao and P. O. M. Scokaert (2000). Constrained model predictive control: Stability and optimality. *Automatica* **36**(6), 789–814.

AUTHOR INDEX

Continued from outside back cover

Title/Year of publication	Editor(s)	ISBN
2002 continued		
Periodic Control Systems (W)	Bittanti & Colaneri	0 08 043682 X
Modeling and Control in Environmental Issues (W)	Sano, Nishioka & Tamura	0 08 043909 8
Computer Applications in Biotechnology (C)	Dochain & Perrier	0 08 043681 1
Time Delay Systems (W)	Gu, Abdallah & Niculescu	0 08 044004 5
Control Applications in Post-Harvest and Processing Technology (W)	Seo & Oshita	0 08 043557 2
Intelligent Assembly and Disassembly (W)	Kopacek, Pereira & Noe	0 08 043908 X
Adaptation and Learning in Control and Signal Processing (W)	Bittanti	0 08 043683 8
New Technologies for Computer Control (C)	Verbruggen, Chan & Vingerhoeds	0 08 043700 1
Internet Based Control Education (W)	Dormido & Morilla	0 08 043984 5
Intelligent Autonomous Vehicles (S)	Asama & Inoue	0 08 043899 7
2003		
Proceedings of the 15th IFAC World Congress 2002 (CD + 21 vols)	Camacho, Basanez & de la Puente	008 044184 X
Modeling and Control of Economic Systems (S)	Neck	0 08 043858 X
Mechatronic Systems (C)	Tomizuka	0 08 044197 1
Programmable Devices and Systems (W)	Srovnal & Vlcek	0 08 044130 0
Real Time Programming (W)	Colnaric, Adamski & Wegrzyn	0 08 044203 X
Lagrangian and Hamiltonian Methods in Nonlinear Control (W)	Astolfi, Gordillo & van der Schaft	0 08 044278 1
Intelligent Control Systems and Signal Processing (C)	Ruano, Ruano & Fleming	0 08 044088 6
Guidance and Control of Underwater Vehicles (W)	Roberts, Sutton & Allen	0 08 044202 1
Analysis and Design of Hybrid Systems (C)	Engell, Gueguen & Zaytoon	0 08 044094 0
Intelligent Manufacturing Systems (W)	Kadar, Monostori & Morel	0 08 044289 7
Control Applications of Optimization (W)	Gyurkovics & Bars	0 08 044074 6
Fieldbus Systems and Their Applications (C)	Dietrich, Neumann & Thomesse	0 08 044247 1
Intelligent Components and Instruments for Control Applications (S)	Almeida	0 08 044010 X
Modelling and Control in Biomedical Systems (S)	Feng & Carson	0 08 044159 9
2004		
Advances in Control Education (S)	Lindfors	0 08 043559 9
Robust Control Design (S)	Bittanti & Colaneri	0 08 044012 6
Fault Detection, Supervision and Safety of Technical Processes (S)	Staroswiecki & Wu	0 08 044011 8
Technology and International Stability (W)	Kopacek & Stapleton	0 08 044290 0
System Identification (SYSID 2003) (S)	van den Hof, Wahlberg & Weiland	0 08 044249 8
Control Systems Design (C)	Kozak & Huba	0 08 044175 0
Robot Control (S)	Duleba & Sasiadek	0 08 044009 6
Time Delay Systems (W)	Garcia	0 08 044238 2
Transportation Systems (S)	Aoki & Tsugawa	0 08 0440592
Manoeuvring and Control of Marine Craft (C)	Batlle & Blanke	0 08 044033 9
Power Plants and Power Systems Control (S)	Lee & Shin	0 08 044210 2
Automated Systems Based on Human Skill (S)	Stahre & Martensson	0 08 044291 9
Automatic Systems for Building the Infrastructure in Developing Countries (Knowledge and Technology Transfer) (W)	Istefanopulos	0 08 044204 8
Intelligent Assembly and Disassembly (W)	Borangiu & Kopacek	0 08 044065 7
New Technologies for Automation of the Metallurgical Industry (W)	Wei Wang	0 08 044170 X
Advanced Control of Chemical Processes (S)	Allgöwer & Gao	008 044144 0
Computer Applications in Biotechnology – CAB 8 (S)	Pons	0 08 044251 X
Information Control Problems in Manufacturing (S)	Kopacek, Pereira & Morel	0 08 044249 8
Advances in Automotive Control (S)	Rizzo, Glielmo, Gianese & Vasca	0 08 044250 1
Cost Oriented Automation (S)	Zaremba, Sasiadek & Erbe	0 08 044309 5

Printed and bound by CPI Group (UK) Ltd, Croydon, CR0 4YY
08/05/2025
01864925-0002